Hybrids & Haecceities

EDITORS
Dr. Masoud Akbarzadeh, Dr. Dorit Aviv, Hina Jamelle, Robert Stuart-Smith

COPYEDITOR & LAYOUT EDITOR
Gabi Sarhos

GRAPHIC IDENTITY
Madison Green, Paul Germaine McCoy, Peik Shelton

EDITORIAL ASSISTANTS
Anna Ji-Eun Lim, Minyang Yuan

PRINTER
IngramSpark

© Copyright 2023
Association for Computer Aided Design in Architecture (ACADIA)

All rights reserved by individual project authors who are solely responsible for their content.

No part of this work is covered by copyright may be reproduced or used in any form, or by any means graphic, electronic, or mechanical, including recording, taping or information storage and retrieval systems without prior permission from the copyright owner.

Conference hosted by the University of Pennsylvania Stuart Weitzman School of Design in Philadelphia, Pennsylvania.

ISBN 979-8-9860805-8-1

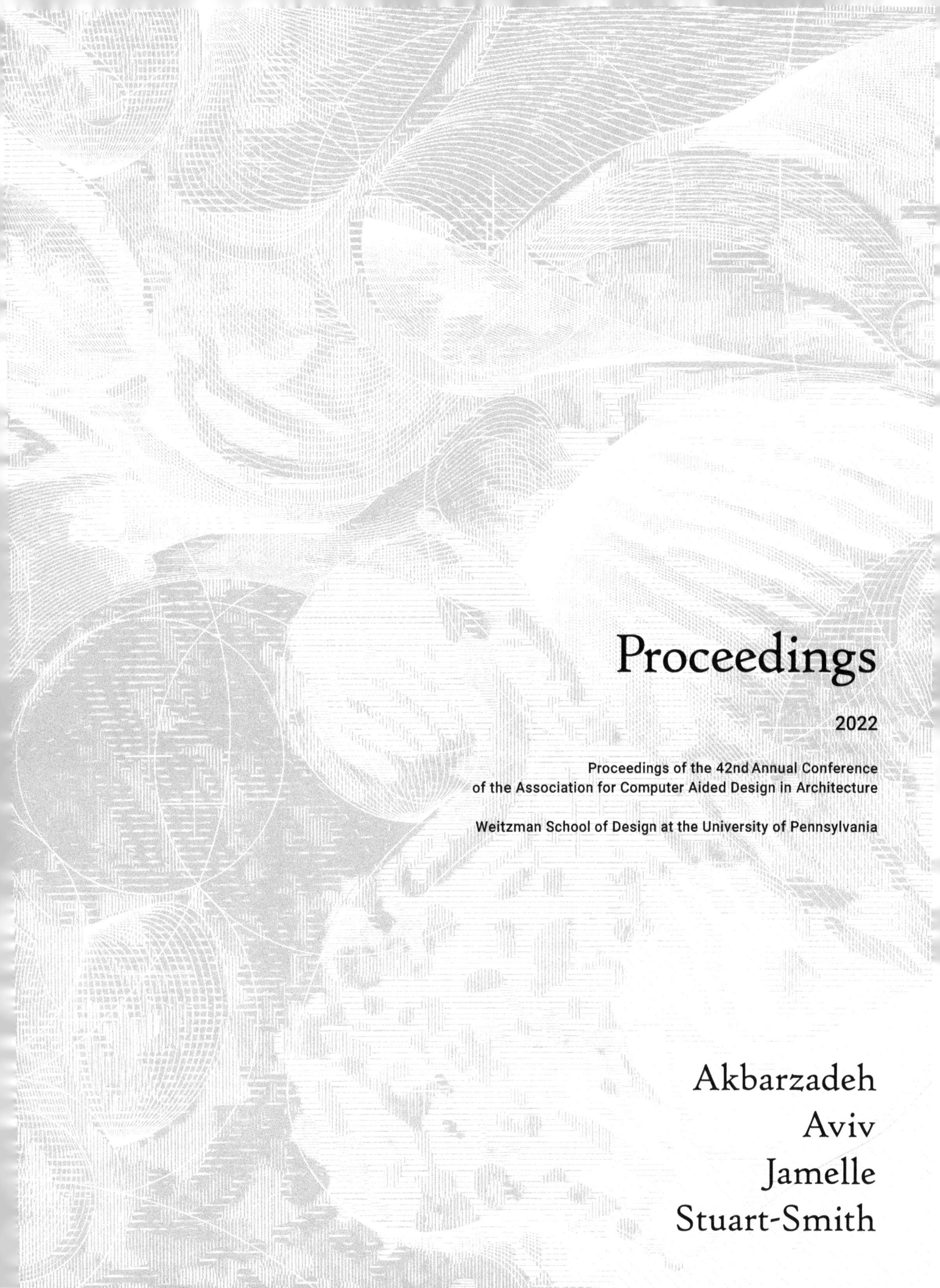

Proceedings

2022

Proceedings of the 42nd Annual Conference
of the Association for Computer Aided Design in Architecture

Weitzman School of Design at the University of Pennsylvania

Akbarzadeh
Aviv
Jamelle
Stuart-Smith

Table of Contents

INTRODUCTION

xiii   Foreword: ACADIA Foundations and Critical Innovation in a New Technological Paradigm
Jenny Sabin

xxi   Introduction: *Hybrids & Haecceities*: Towards More Diverse and Considered Forms of Embodiment and Participation in Design
Masoud Akbarzadeh, Dorit Aviv, Hina Jamelle, Robert Stuart-Smith

ROBOTICS AND ARCHITECTURAL DESIGN: CONTROL STRATEGIES

3   Session Chair's Introduction
Arash Adel

4   Tolerance-Aware Design of Robotically Assembled Spatial Structures
Augusto Gandia, Fabio Gramazio, Matthias Kohler

24   Approaching Architectonic Interfaces
Nicolas Stephan, Marine Lemarié, Alexandra Moisi, Stefan Rutzinger, Kristina Schinegger

34   Robotic 3D Printing Multilayer Building Envelope
Mania Aghaei Meibodi, Wes McGee, Alireza Bayramvand

44   Collective Aerial Additive Manufacturing
Robert Stuart-Smith, Durgesh Darekar, Patrick Danahy, Basaran Bahadir Kocer, Vijay Pawar, Mirko Kovac

ADVANCED PRODUCTION METHODS I: ROBOTIC FABRICATION AND MATERIAL FORMATION

57   Session Chair's Introduction
Jose Luis Garcia del Castillo López

58   Column-Slab Interfaces for 3D Concrete Printing
Ana Anton, Eleni Skevaki, Patrick Bischof, Lex Reiter, Benjamin Dillenburger

68   Non-Linear Fabrication
Sulaiman Al Othman, Martin Bechthold

76   3D Printing and Shape Memory Alloys
Hyunchul Kwon, Priyank Soni, Ali Saeedi, Moslem Shahverdi, Benjamin Dillenburger

90   Smart Branching
Chenxiao Li, Mingyang Yuan, Zilong Han, Billie Faircloth, Jeffrey S. Anderson, Nathan King, Robert Stuart-Smith

98   Intentional Folds
Gosia (Malgorzata) Pawlowska

DISRUPTIVE MODES OF PRACTICE AND PEDAGOGY

107 Session Chair's Introduction
Vernelle A. A. Noel

108 Implementation of Interactive Virtual Sites for Architectural Education
Anastasia Globa, Kun Lyu, Ozgur Gocer, Muhammed Yildirim

120 Entangled Simulations
Jose Sanchez

128 Closing the Gap
Nicolas Azel, Brandon Pachuca, Lucien Wilson

138 Learning from Players
Guzden Varinlioglu, Ozgun Balaban, Daniel Tsai, Charles Ngai-Hang Wu, Takehiko Nagakura

ADVANCED PRODUCTION METHODS II: KNITTED ARCHITECTURE

149 Session Chair's Introduction
Tsz Yan Ng

150 From Garment to Building
Leanne Zilka, Jenny Underwood

156 BioKnit
Armand Agraviador, Jane Scott, Romy Kaiser, Elise Elsacker, Aileen Hoenerloh, Ahmet Topcu, Ben Bridgens

168 Design-to-production Workflows for CNC-knitted Membranes
Yuliya Sinke, Mette Ramsgaard Thomsen, Martin Tamke

182 3D Knit Spacesuit Sleeve
Lavender Tessmer, Ganit Goldstein, Guillermo Herrera-Arcos, Volodymyr Korolovych, Rachel Bellisle, Cody Paige, Christopher Shallal, Atharva Sahasrabudhe, Hugh Herr, Svetlana V. Boriskina, Dava Newman, Skylar Tibbits

ADVANCED MATERIALS

195 Session Chair's Introduction
Ezio Blasetti

196 Inventory
Gil Sunshine

208 Parametric Matter
Adam Blaney, Dilan Ozkan, Emel Pelit, Mariana Fonseca Braga, John G. Hardy, Mark Ashton

224 Nanotectonica SEM-GAN
Jonas Coersmeier, James Nanasca, Ivan Yan Man Hin, Ezio Blasetti

244 Introducing Bespoke Properties to Slip-Cast Elements
Davis Dunaway, Dan Rothbart, Layton Gwinn, Nathan King, Robert Stuart-Smith

256 Designer Agency in 3D Packing of Irregular Material Stock
Patricia Dueñas Gerritsen, Emily Wissemann, Jose Luis García del Castillo y López

FIELD NOTES

267 Session Chair's Introduction
Melissa Goldman, Chair

268 Alternative Typographic Histories
Levi Hammett, Fatima Abbass, Hind Al Saad, Mohammad Suleiman

272 Straddling the Boundary
Julianna Cano

276 Collateral Computation
Misri Patel

280 Nemagari-no-Takumi Workshop
Nicholas Bruscia, Daiki Kanaoka, Hideaki Asaoka, Kotaro Iwaoka

284 Composite Mies
Nick Safley

290 Experimentations in Neuroscience for Architecture
Kristine Mun, Biayna Bogosian

294 Field Guide to Meta-Architecture
Cameron Nelson

COMPUTATIONAL METHODS FOR STRUCTURAL DESIGN

299 Session Chair's Introduction
Jonas Coersmeier

300 A Web-based Interactive Structural Pattern Generation Tool with Graphic Statics and Machine Learning of Dragonfly Wings
Hao Zheng, Masoud Akbarzadeh

310 Building Synthetic Data Sets or How to Learn from Future Architectures?
Daniel Koehler

318 Methods for Integrating Architectural Design Intent into the Agent-based Design of (Adaptive) Truss Structures
Mathias Maierhofer, Achim Menges

326 Form-finding of Architectural Knitted Tensioned Structures
Farzaneh Oghazian, Sam Moradzadeh, Felecia Davis

NEW ECOLOGIES I: BIOMATERIALS / BIOTECH

335 Session Chair's Introduction
Laia Mogas-Soldevila

336 RePrint
Rebecka Rudin, Malgorzata A. Zboinska, Sanna Sämfors, Paul Gatenholm

346 Integrated Design Strategies for Multi-scalar Biopolymer Robotic 3D Printing
Gabriella Rossi, Ruxandra-Stefania Chiujdea, Laura Hochegger, Ayoub Lharchi, Paul Nicholas, Martin Tamke, Mette Ramsgaard Thomsen

356 3D Printed Formwork for Mycelium Bound Composites
Matthias Leschok, Benjamin Dillenburger

366 Developing a Digital Design Workflow for Nexorade Bamboo Structure
Jonas Hauptman, Ramtin Haghnazar, Sara Saghafi Moghaddam

378 Tactical Sedimentation of Architectural Reef System
Colleen Duong, Dana Cupkova, Azadeh Sawyer, Marantha Dawkins

PERFORMATIVE DESIGN: STRUCTURAL AND MATERIAL SYNTHESIS

391 Session Chair's Introduction
Shelby Doyle

392 LOOPS; A Mobile, Shape-Changing Architectural System: Robotically-Actuated Bending-Active Tensile Hybrid Modules
Valentina Soana, Yichao Shi, Tongyao Lin, Yiting Ma, Ling Dai

406 Digital Bamboo
Marirena Kladeftira, Matthias Leschok, Eleni Skevaki, Davide Tanadini, Patrick Ole Ohlbrock, Pierluigi D'Acunto, Benjamin Dillenburger

418 Curved-crease Folding of Bending-active Plates as Formwork
Lotte Scheder-Bieschin, Tom Van Mele, Philippe Block

432 HoloWall
Leslie Lok, Jiyoon Bae

444 Tailoring Bending Behavior
Lei Gong, Xinjie Zhou, Hua Chai, Junguang Liu, Philip F. Yuan

NEW ECOLOGIES II: RESPONSIVE/ ADAPTIVE DESIGN METHODS

455 Session Chair's Introduction
William Braham

456 Discretizing Low-tech Adaptive Rammed Earth Formwork
Pragya Gupta, Dana Cupkova

468 Robotic Fabrication of 3D Printed Clay Opening as a Passive Cooling System
Deena El-Mahdy, Marwa Abd ElRahim, Adel AlAtassi

474 Integrative Green Building Envelope
Juliette Zidek, Laurin Aman, Xinran Li, Jumaanah Alhashemi, Mania Aghaei Meibodi

486 Lines of Flight; Facade Design for Multispecies Migrations
John Kim, Adam Marcus, Molly Reichert

498 Computer-aided Ecological Connectivity
Mathilde Marengo, Iacopo Neri

ENVIRONMENTAL PERFORMANCE AND SIMULATION

505   Session Chair's Introduction
Billie Faircloth

506   Passive Cooling Strategies for Thriving in a Changing Climate
Bertug Ozarisoy, Hasim Altan

524   High-Density Building Form Generation Considering Daylight Performance
Jun Xiao, Yubo Liu, Qiaoming Deng

536   The Potential of Mitigating Urban Heat Island with Vacant Lands in Philadelphia
Hui Tian, Jiali Yao, Shimin Tu

546   Resonant Hexagon Diffuser
John Nguyen, Philipp Cop, Nicholas Hoban, Brady Peters, Ted Kesik

558   The Sound of Kerfing
Alireza Borhani, Negar Kalantar, Erfan Rezaei Azari, Anastasia Muliana, Zaryab Shahid, Ed Green

CIRCULARITY, CARBON-NEGATIVE DESIGN, AND FABRICATION

573   Session Chair's Introduction
Franca Trubiano

574   Lithopic House
Dana Cupkova, Han Meng, Jinmo Rhee

586   ZeroWaste
Edvard P. G. Bruun, Erin Besler, Sigrid Adriaenssens, Stefana Parascho

598   The Reproduction of Chinese Traditional Timber Structure
Yang-Ting Shen, Mi-Chi Wang, Lien-Kai Huang, You-Min Gao, Chia-Chin Yen

604   Co-Robotic Assembly of Nonstandard Timber Structures
Arash Adel

614   Tangential Timber
Kyle Schumann, Katie MacDonald, Abigail Hassell

MACHINE LEARNING AND ARTIFICIAL INTELLIGENCE

627   Session Chair's Introduction
Matias del Campo

628   Design Contextualism by AI
Woongki Sung, Takehiko Nagakura, Daniel Tsai

638   Integrated Reconfigurable Autonomous Architecture System
Tyson Hosmer, Jiaqi Wang, Wanzhu Jiang, Ziming He

652   Deep Relief
Andrew Saunders, Riley Studebaker, Claire Eileen Moriarty

662   Towards an Adversarial Architecture
Antonio Furgiuele, Mehmet Ergezer, Cagri Hakan Zaman

672   Latent Isovist
Mikhael Johanes, Jeffrey Huang

BIG DATA AND AUGMENTED ENVIRONMENTS

683   Session 15 Chair's Introduction
Biayna Bogosian

684   Depth Camera Feedback for Guided Fabrication in Augmented Reality
Gwyllim Jahn, Cameron Newnham, Nick van den Berg

694   BIMxAR: Building Information Modeling-Powered Augmented Reality
Ziad Ashour, Wei Yan

704   Visualization Methods for Big and High-Dimensional Acoustic Data
Achilleas Xydis, Chaoyu Du, Romana Rust, Fabio Gramazio, Matthias Kohler

714   Measuring Street Vitality Based on Video-image Using Deep Learning
Yunqin Li, Jiaxin Zhang, Xueqiang Wang, Kai Ma

CRITICAL ANALYSIS OF ARCHITECTURAL DESIGN AND PRODUCTION

725   Session 16 Chair's Introduction
Kathy Velikov

726   Setting Historic Computer Systems in Motion
Theodora Vardouli

736   Parsed Precedent
Paul Howard Harrison

742   What is Creativity?
Neil Leach

752   Returning the Gaze
Behnaz Farahi

KEYNOTES

764   H&H Prologue Panel
Marcos Cruz, Winka Dubbledam, Rashida Ng,
Mette Ramsgaard Thomsen, Moderator: Robert Stuart-Smith

768   Origins and Destinations Beyond Midjourney
Chigozie Nri, Joel Simon, Kyle Steinfeld,
Moderator: Masoud Akbarzadeh

772   Artificial Intelligence and the Future of Architectural Design
Antoine Picon

782   H&H Epilogue Panel: New Technologies and their Effect on Architectural Design and Culture
Ferda Kolatan, Andrew Kudless, Antoine Picon, Jenny Sabin,
Moderator: Hina Jamelle

AWARDS

786   Design Excellence Award
Höweler + Yoon Architecture

790   Society Award for Leadership
Jason Kelly Johnson

794   Teaching Award of Excellence
Rajaa Issa

800   Innovative Research Award of Excellence
Felecia Ann Davis

WORKSHOPS

807   Workshop Chair's Introduction

808   Robotic Fabrication for Building Components
ZHACODE - Cesar Fragachan and Tim Fu

809   Augmented Vision: Realtime Feedback for Guided Fabrication in Augmented Reality
FOLOGRAM - Cameron Newnham and Nick van den Berg

810   Bio-polymer Printing: Strategies for Material Grading of Bio-based Material
Mette Ramsgaard Thomsen, Paul Nicholas,
Gabriella Rossi, Carl Eppinger

811   A Molten Gesture: Expanding 'Hand-Craft' Through Body Tracking and Robotic Arms
Claire Moriarty and Riley Studebaker

812   Robotic Mark Making
Sara Codarin and Karl Daubmann

813   Form-finding Explorations in CLT
Amin Adelzadeh and Hamed Karimian A.,
and Christopher Robeller

814   Data-driven Urban Design
Grimshaw - Jorge Sainz de Aja and Esther Rubio Madronal

815   Marginalized Craft Traditions and Advanced Fabrication
Duane McLemore

816 Power Automate Data Workflows in AEC
Cesar Escalante, Philippe Videau, Geng Wang, and Sagar Bave

817 Diffusion: Architecture, AI
Matias del Campo and Sandra Manninger

818 AI, Literature, and the Mind's Eye
Karel Klein

819 Generating Spatial Hybrids: 3DGANS
Benjamin Ennemoser and Ingrid Mayrhofer-Hufnagl

820 Digital Sculpting
Patrick Danahy and Caleb Ehly

821 Vibrant Artefacts
Barry Wark

ACADIA CREDITS

824 Conference Chairs

826 Workshop, Exhibition, and Media Chairs

828 Departmental Chair

830 ACADIA Organization

831 Conference Management

832 Peer Review Committee

837 ACADIA 2022 Sponsors

839 The Graphic Identity of *Hybrids and Haecetties*

Foreword

ACADIA Foundations and Critical Innovation in a New Technological Paradigm

Jenny Sabin

ACADIA President January 1, 2021- December 31, 2022
Arthur L. and Isabel B. Wiesenberger Professor in Architecture, Cornell University
Chair, multi-college Department of Design Tech, Cornell University
Director, Sabin Lab
Principal, Jenny Sabin Studio

Hybrids & Haecceities, ACADIA's 2022 and 42nd annual conference, hosted by the Weitzman School of Design at the University of Pennsylvania, marked our first in-person conference since 2019. After life-changing events experienced during the pandemic and two years of online ACADIA conferences, the excitement of our community gathering again was palpable on the first day of the conference. The Weitzman School of Design was slated to host the ACADIA conference in 2020, but due to the complexities of the pandemic and a commitment to hosting an in-person conference, this was pushed to 2022. As ACADIA's primary activity is the exchange of knowledge through an annual conference (mandated in the organization's by-laws), in December 2021 the ACADIA Board of Directors resumed planning discussions with the Penn conference team. Together, we aimed to build upon and make connections with the critical discourse and discussions generated over the last two years of online ACADIA conferences and to recharge the discussions we have at ACADIA considering paradigm shifting innovations in technology such as generative AI, and to amplify our efforts to expand and diversify the ACADIA community.

During a complex and uncertain time, the original Penn team persevered from 2020 to organize and host an extraordinary conference in the fall of 2022. On behalf of the ACADIA community, I would like to sincerely thank Dorit Aviv, Hina Jamelle, Rob Stuart-Smith, Masoud Akbarzadeh, workshop chair, Andrew Saunders, exhibition chair, Ferda Kolatan, and media chair, Nate Hume for volunteering their time to take on the incredible amount of work to critically and creatively conceptualize, organize, support, and run ACADIA's 42nd conference event, the 2022 Hybrids & Haecceities conference. We would also like to thank the Weitzman School of Design leadership, Chair of the department of Architecture, Winka Dubbeldam, and Dean Fritz Steiner, for their support, enthusiasm, and commitment to hosting and taking on the tremendous collective effort that goes into organizing an international peer reviewed conference and to navigating the complexity of our first post pandemic in-person conference. In addition to the team, I would like to thank the ACADIA Steering Committee, including Shelby Doyle (Chair), Biayna Bogosian, Vernelle A. A. Noel, and Behnaz Farahi for their service and support to the conference team in the organization of the conference. I would also like to sincerely thank ACADIA's former President and outgoing Vice President, Kathy Velikov, for her dedication and guidance to the organization of this year's in-person conference and support of the ACADIA community. Kathy's contribution to and institutional knowledge of ACADIA is extraordinary. I also want to recognize and thank Cameron Nelson for their dedication and tireless efforts as conference production assistant. Thank you also to the entire Board for stepping up and leading new initiatives and taking on the workload of committee work that this conference brought – from expanding our diversity efforts, innovating new development strategies, jurying the submission formats, to reviewing and awarding scholarship grants and paper awards programs. I would also like to extend a huge thanks to the Weitzman School of Design staff and students for their tireless efforts and commitment over the last year plus in planning and organizing an excellent conference. The conference chairs, Weitzman leadership, ACADIA steering committee, staff, and students have given countless hours of time to co-envision the conference and workshops, run the peer review process, develop the proceedings, mount an exhibition, and plan, organize, and manage everything that we experienced during the conference week.

Starting with the successful in-person and online workshops at the start of the conference week, followed by 3 days of presentations and engaging discussions, the conference is a testament to the positive energy, collaborative spirit, optimism, leadership, and commitment of these amazing individuals. The conference co-chairs hosted an incredible array of 14 online and in-person workshops which took place earlier in the week preceding the conference and concluding with a colloquium with short presentations by the workshop leaders and participants. Through the summary presentations of the workshop outcomes, we saw the cutting edge of computational design research and practice, including advancements in additive and robotic manufacturing of biopolymers and timber structures, guided fabrication through augmented reality, the cutting edge of generative AI, data-driven urban design, and design-to-assembly of lightweight shell structures. 151 people attended the workshops. In this conference, the co-chairs raised our ambitions not only in terms of content, but also the conference organization. In many ways, we had to rethink the in-person conference experience considering the positive outreach and accessibility that our online conferences afforded. We also grappled with how to model and estimate how many people from our community would attend the conference in a post-pandemic environment. This was an ambitious conference in terms of programming with our first online streaming option. It was an immense organizational effort on the part of the conference team, and each session and keynote panel were excellent, engaging, thought-provoking, and impactful. Gathering again in person with our community was indeed special and marked by many things to celebrate, reflect upon, and look forward to.

The critical discourse that the conference theme framed, and the content presented in the papers, projects, and field notes sessions underscore the design innovation, rapid technological change, and complexity of our time. A beleaguered planet recovering from the pandemic and an urgent climate emergency demands that computational design as a field and practice cultivate new collaborative models for research and pedagogy to comprehend key social, environmental, and technological issues. As I wrote in my 2021 foreword to Realignments: Toward Critical Computation, we need to continue to strive for stronger integration and collaboration between research and practice, integration and new methods and models that are reflected in the ACADIA 2022 conference proceedings. For example, can distributed design networks and micro-factories contribute to next generation methods for low-cost, efficient, sustainable, and innovative fabrication and construction strategies? How might we rethink building materials through 3D bioprinting and biologically informed design? Can buildings be made more responsive through the integration of artificial intelligence and big data? 3D printing, advanced manufacturing, and robotics are transforming how we live, work, do business, and engage with our communities. Recent advances in computation, visualization, material intelligence, and fabrication technologies have begun to alter fundamentally how we design, construct, and make from the nano to macro scales. We are at the beginning of the 4th Industrial Revolution; a paradigm shift that some historians argue is the biggest to impact design since the medieval period. And as technology has changed the design paradigm, we are facing a global climate crisis. According to the World Green Building Council (WGBC 2017), building and construction account for 40 percent of annual global carbon emissions, which are significantly contributing to global climate change. At the same time, the COVID-19 crisis forced a creative reconsideration of workspaces and how collaborative design work is done across informal and lateral networks and online platforms, while also revealing and amplifying ongoing entanglements of systemic racism and persistent issues impacting diversity, equity, and inclusion. The diverse, experimental, and innovative array of ACADIA proceedings this year through paper and project presentations, keynote panels, field notes, and beyond reflect the complexity of our time and the urgency of now, and that we must come together to address these crises with radical new models for design research and collaboration across disciplines, alternative forms of practice, and communities.

The last two years of online conferences marked an important shift that this year's conference continues. Last year also marked the 40th anniversary of the ACADIA organization and we were able to continue to celebrate this at the 2022 conference through a table dedicated to the ACADIA Cultural History project featuring past proceedings and printed textiles reflecting upon the 40 plus years of innovation and work by the ACADIA community. Tremendous thanks go to Board members and President-elect, Shelby Doyle, and Melissa Goldman for all their efforts towards this initiative, which included completing the daunting task of amassing the entire archive of ACADIA proceedings and making them free and accessible to all on the ACADIA website.

Hybrids & Haecceities continues to build upon and add to the important critical topics and questions that the 2020 and 2021 conferences raised, introducing conversations, people, and perspectives in the keynotes through prologue and epilogue panels as a new and important trajectory to the ACADIA conference format. The prologue and epilogue panels allowed for extended dialogue on many topics presented during the paper, project, and field notes sessions. Topics range from robotics and architectural design, robotic fabrication, knitted architecture, advanced materials, biomaterials and ecological design, performative design and adaptive design methods, environmental performance, circularity, machine learning and artificial intelligence, generative AI, big data and augmented environments, and critical computation. In the prologue panel, Keynote Antoine Picon kicked off the discussion with his provocative lecture, Artificial Intelligence and the Future of Architectural Design, which touched upon issues and topics raised by industry experts in the earlier Special Topic Panel: Origins and Destinations Beyond Midjourney.

Over the last 3 years, these important questions and topics have opened new conversations that we hope will persist, mature, and continue to contribute to the diversity of the ACADIA community. The conference theme inspired high quality and exciting peer reviewed papers and projects and positive response to relatively new submission initiatives such as the Field Notes. The conference hosted 14 workshops, received 223 paper submissions, 91 projects, and 21 field notes. Thank you to this year's scientific committee composed of over 170 peer reviewers. Our scientific committee ensures that the quality and caliber of the work at ACADIA is excellent and rigorous. This year our acceptance rate for the papers was 29 percent and 32 percent for the projects. Thank you to the conference chairs, Dorit Aviv, Hina Jamelle, Rob Stuart-Smith, and Masoud Akbarzadeh for running a rigorous peer review process. The technical submissions resulted in 16 sessions. We would also like to thank the 16 session chairs who built on the conference theme and facilitated rich conversations with the authors.

As a non-profit, ACADIA's activities are only supported by membership, conference registration, and sponsorship donation. For a third year, we wanted to radically reduce the cost of conference registration, and so we have depended on our sponsors for their generous financial support. Over the last two years, the virtual format of this conference enabled us to do things that we could not have done in person. We have been able to welcome a much more global and diverse audience who in previous years may not have had the capacity or means to travel to the annual conference. Now that we are back in person, we have built upon this global outreach and opened conference registration for free via live streaming to 612 global students thanks to generous support from Autodesk. We could not have done all of this without the powerful financial support of our sponsors. Their generosity is particularly appreciated during this period. Without the generous support of our sponsors, it would not be possible to stage a conference like ACADIA. I would like to thank our platinum sponsors: Autodesk, NVIDIA, Dell, intel, and Zaha Hadid Architects; our silver sponsor Grimshaw; our bronze sponsor EVENTSCAPE; and chaos, ORO Editions, and Coop Himmelb(l)au at the sponsor level. This year, we saw the highest amount of support from our sponsors! Thank you to our Development Officer and committee members, Matias del Campo, and Sina Mostafavi. We would like to thank our media partner, Architects Newspaper. And special thanks go to our communications officer Melissa Goldman and vice communications officer Shelby Doyle for extensive media outreach and management.

612 students from all over the world have been able to register for the conference for free via live stream, which demonstrates ACADIA's commitment to democratizing knowledge and access. The sponsorship from Autodesk also enabled us to initiate a series of more targeted grants and awards. We continued to award special grants for students from Mexico and for international students to attend the workshops and conference. With this support, we were able to award 19 workshop scholarships and 18 scholarships to attend the conference. These scholarships went to students from Mexico to India to Iran and beyond. I also want to thank our members who have contributed to the scholarship fund

this year. I'd like to thank the scholarships committee chaired by Kathrin Dörfler, Daniel Bolojan, Vernelle A. A. Noel, and Melissa Goldman.

In 2019 and 2020, we initiated a partnership with NOMA, the National Organization of Minority Architects. And thanks to the work of Jason Johnson, Kathy Velikov, myself, June Grant, and our Diversity Committee, ACADIA and Autodesk were able to award grants to attend the workshops for NOMA students and professionals, as well as conference registration for NOMA professionals and academics. This year we launched the inaugural ACADIA x NOMAS Workshop, an introduction to computational design. I'd like to thank the workshop chairs from our Board of Directors Shelby Doyle, Biayna Bogosian, Sina Mostafavi, and conference coordinator and assistant, Cameron Nelson. In total 42 students have been able to attend the workshops for free.

This year, ACADIA awarded 4 awards to 4 exceptional individuals. The awards are Design Excellence Award to Howeler + Yoon, Innovative Research Award of Excellence to Felecia Ann Davis, Society Award for Leadership to Jason Kelly Johnson, and Teaching Award of Excellence to Rajaa Issa. I want to thank the awards committee chaired by Maria Yablonina along with Biayna Bogosian, Leslie Lok, and June Grant for their leadership and effort towards the ACADIA awards process. This year, we continued to award best paper and project awards. The ACADIA Annual Conference Awards for peer-review submissions have three categories: Best Paper, Best Project, and the Vanguard Award (for paper submissions). A selected jury, composed of individuals from the conference co-chairs, the scientific committee, and members of the ACADIA community, deliberated on the award selections. This year, the jury decided to award a third project award, the Conference Chair Award. Thank you to the jury and the scientific committee chaired by Tsz Yan Ng for their hard work in coordinating the awards process.

ACADIA continued year three of a new initiative and collaboration with our sibling CAAD organizations that we launched in 2020. Working with the Presidents and colleagues of our sibling CAAD organizations we organized the third World CAAD PhD workshop. The aim of the World CAAD PhD Workshop is to introduce junior researchers at the PhD stage to different schools and different research cultures within the global CAAD community. The workshop offers students an opportunity to receive constructive feedback from prominent researchers and academics of the CAAD community and provides students with an occasion to position their research within the global CAAD research arena. Each sibling organization is represented at the workshop by three PhD student delegates, who were chosen through a competitive submission process, as well as experts from each community. Thank you to the 2022 ACADIA delegates and to Board member Vernelle A. A. Noel for spearheading and coordinating these important efforts.

The 2022 in-person conference built upon the last two years of successful online conferences in unprecedented ways, including increasing attendance and international reach, which has positively impacted gender parity. This included in-person and live-stream options for registration, including free live-streaming access to students globally, and the quantitative difference can be seen when we look at some of the attendance statistics. Our typical in-person conferences attract 300-350 participants. This year, we had 978 registered attendees with 612 online streaming registrations. 755 of these registrations were students, and we are so excited to see such overwhelming interest in the student community! Incredibly, the open and accessible format of the conference has shifted the gender balance of our attendees. Whereas typically female attendees have remained at about 29 to 30% in previous in-person conferences over the past three years, this year, of attendees who chose to report their gender, we see a significant increase to 49% female attendance! This represents the highest percentage of female attendance to the annual conference in ACADIA's history, which is very encouraging for the future of this community in terms of gender parity. This is the first year that ACADIA collected information on race and demographics through the Eventbrite ticket platform. Conference attendees reporting information on race, including Asian, Black or African American, Native Hawaiian or Pacific Islander, American Indian or Alaska Native account to 33.7% of the total conference attendance. BIPOC conference attendance is a small percentage of this, pointing to the necessity for long-term multi-year investment in outreach and increasing access to reach these communities through online platforms and important initiatives such as the ACADIA X NOMAS workshop and ACADIA Autodesk workshop.

As we celebrate our first in-person ACADIA conference since 2019, it is also an important time to reflect, celebrate, critically assess, and generate new approaches to our conference format and where we are headed as a community and organization in the future. Over 40 years ago and during the first decade of the ACADIA organization, there was a strong focus on pedagogy, CAAD tools, and design computing where topics such as "Introducing Computer-Aided Design into the Architecture Curriculum", "The Interactive Effect in Technical Education", "Integrating Computers into the Architectural Curriculum", and "Computing in Design Education" framed the conference proceedings and discourse in the field.[1] During the early 90s, a shift was made towards understanding the relationships between research and practice through the lens of "Design and Representation", "Visualization Techniques", "Reality and Virtual Reality", "Intelligent Models", "Architectural Graphics", and alternative models of reality.[2] In the preface of the 1991 proceedings, co-editors and chairs, Glenn Goldman and Michael Stephen Zdepski write, "During the past ten years computers in architecture have evolved from machines used for analytic and numeric calculation, to machines used for generating dynamic images, permitting the creation of photo-realistic renderings, and now in a preliminary way, permitting the simulation of virtual environments. Digital systems have evolved from increasing the speed of human operations to providing entirely new means for creating, viewing, and analyzing data."[3] We also saw new developments in generative systems and component-based spatial reasoning.

In 1993, the title of the ACADIA conference was, "Education and Practice: The Critical Interface", again with a focus on education, methods and meaning, computers in the urban landscape and a call for greater integration between research and practice and new "ways of seeing" considering quickly changing technologies and methods.[4] During this conference it was argued that with more information comes more responsibility in the design and construction of spaces and buildings that are "successful aesthetically, economically, energetically, and socially" and that perhaps the computer could help in this context.[5] The proceedings explored the "tight, complex, and critical relationship between computers, education, and practice."[6] In the second half of the 90s, topics explored included "Computing in Design: Enabling, Capturing, and Sharing Ideas" and the relationship between collaboration, reasoning, and pedagogy in design computation, representation, and design.[7,8]

At the threshold of the millennium and moving into the early 2000s, new developments in digital fabrication, parametric design, digital data and simulations, and design computation processes allowed for emergent and bottom-up design processes. These new methods framed conversations around controlled indeterminacy, responsive architecture, technologies that hybridize physical and cyberspaces, as well as questions around access and the availability of computers in the design studio. The 2004 conference, co-chaired by Philip Beesley, Nancy Cheng, and Shane Williamson, examined the digital practice of architecture through fabrication where cutting-edge built projects such as the SmartWrap Pavilion by Kieran Timberlake, the Bahai Temple / Temple of light by Hariri Hariri Architects, and the innovations in digital fabrication in construction processes on the Sagrada Familia set the stage.[9] In 2008, Silicon+Skin: Biological Processes and Computation, co-chaired by Andrew Kudless, Neri Oxman, and Marc Swackhamer, brought biology to design computation to explore new forms of making in collaborative contexts.[10] In their opening introduction, the editors write, "It is curious that in order to define a discipline one must describe its boundaries. This delineation is especially problematic in the field of design computation as it encounters an expanding constellation of related domains. It has become rhizomatic, spreading horizontally across the various disciplinary landscapes, creating a multiplicity of creative nodes." [11]

Over the last decade, with the integration of materials and making, and more recently robotic fabrication, we see innovative developments that embed ecological design thinking, concepts of responsiveness, smart geometry, digital craft, embodied fabrication, responsive and social expression in architecture, synthetic landscapes, pervasive and ubiquitous computing, feedback, interactive and sensing technologies, and adaptation to unearth new reciprocal models for engagement and entanglements between architecture, humans, machines, and the environment.[12] The 2022 conference not only built upon this robust foundation and expanded the critical discourse framed over the last two years of online conferences, but also importantly critically dialogued the rapidly expanding and changing frontier of generative AI and AI in architecture research and practice through presentations, panel discussions, and keynotes. From the first decade of the organization, ACADIA has maintained a commitment to evolve and at the same time maintain a foundation of rigor and excellence. The diverse, experimental, and innovative array of ACADIA proceedings this year through papers and projects, keynote panels, field notes, and beyond reflect the complexity and urgency of our time. At a time of ecological catastrophe and sociopolitical challenges, the result of our deeply entrenched misconceptions of human dominance over nature, I am encouraged by the optimism, critical discourse, technological innovation, commitment to diversity, equity, and inclusion, and creative ingenuity of our community as expressed through the proceedings of Hybrids & Haecceities. Dialogue and novel approaches to design and research were presented and discussed with the aim of dissolving binary conditions and inherent hierarchies to embrace new modes of practice and research built upon a robust foundation and response to a rapidly changing context and new technological paradigm for computational design and our community.

NOTES

1. See McIntosh, G. Patricia, ed. ACADIA Workshop 85' Proceedings. Tempe: The Association for Computer-Aided Design in Architecture, 1985.

2. See Goldman, Glenn, and Michael Stephen Zdepski, eds. Reality and Virtual Reality. Newark: The Association for Computer-Aided Design in Architecture, 1991.

3. Ibid., 6.

4. See Morgan, Fred, and Richard W. Pohlman, eds. Education and Practice: The Critical Interface. College Station: The Association for Computer-Aided Design in Architecture, 1993.

5. Ibid., 7.

6. Ibid.

7. See Kalisperis, N. Loukas, and Branko Kolarevic, eds. Computing in Design: Enabling, Capturing and Sharing Ideas. Seattle: The Association for Computer-Aided Design in Architecture, 1995.

8. See McIntosh, Patricia, and Filiz Ozel, eds. Design Computation: Collaboration, Reasoning, Pedagogy. Tempe: The Association for Computer-Aided Design in Architecture, 1996.

9. See Beesley, Philip, Nacy Yen-Wen Cheng, and R. Shane Williamson, eds. Fabrication: Examining the Digital Practice of Architecture. Toronto: University of Waterloo School of Architecture Press, 2004.

10. See Kudless, Andrew, Neri Oxman, and Marc Swackhamer, eds. Silicon+Skin: Biological Processes and Computation. Minneapolis: The Association for Computer-Aided Design in Architecture, 2008.

11. Ibid., 15.

12. See ACADIA Proceedings 2012 – 2019.

Hybrids & Haecceities

Towards More Diverse and Considered Forms of Embodiment and Participation in Design

Dr. Masoud Akbarzadeh, Dr. Dorit Aviv, Hina Jamelle, Robert Stuart-Smith

Hybrids & Haecceities provides a unique opportunity to reflect on the challenges and opportunities in undertaking computation-led research, practice or theory in design today. During the last few years alone, the US and global communities have experienced considerable social, economic, environmental and political upheaval, from the COVID-19 pandemic, natural disasters, war, famine, to racial oppression and an increasingly polarized society that struggles to achieve equity, means of community, or a common story. Many of these issues are further exasperated by the ever more apparent effects of climate change.

In this context, the design and construction industries continue to have profound global effects with significant political, economic, and environmental consequences. Today's social and ecological crises are vast and complex, however, and cannot be addressed through design alone. To confront these issues, designers must forge new collaborations, and extend their means of analysis, evaluation and action, furthering design agency. Perhaps one of the greatest challenges we must face is the fact that design problems and opportunities are not easily integrated within computational paradigms and when generalized—can have profound impacts. Joy Buolamwini's research into algorithmic bias highlights how even simple oversights in the creation of narrow data sets can easily lead to biases that can have broad and unintended impacts such as unfair racial profiling in policing and law enforcement (Buolamwini and Gebru 2018).

Cathy O'Neil describes such technologies as "weapons of math destruction" (WMDs), where overly simplified heuristics within computational models prevalent in almost every aspect of society have led to exclusionary or inequitable treatment. O'Neil suggests that most of today's computational models fail to sufficiently represent, describe or include the diversity of today's world and communities (O'Neil 2016).

Since the middle of the twentieth century, pioneers in the architecture and design professions have sought to leverage computing to integrate diverse performances and effects in the design of the built environment. Nicholas Negroponte's Architecture Machine Group (Negroponte 1972) and others developed computational and robotics research in the early 1950s, while Cedric Price together with John and Julia Frazer, explored the integration of computing in speculative building proposals to support user flexibility and play (Frazer 1995). Several Viennese architects in the 60s and 70s utilized digital media, microprocessors and sensor-feedback to support a wearable, event-based architecture (Haraway and Wolfe 2016). Together these explorations might support Donna Harraway's "Cyborg Manifesto," if they were directed more towards empowerment and eclecticism. The digital design movement of the 1990s and early 2000s, extended this endeavor to address complex and diverse architectural considerations by leveraging computing for both performance modelling, and its inherent ability to produce geometric variability that has been furthered this century into material concerns. Beyond the automation of tasks, many designers explored the encoding of degrees of difference in families of buildings, building elements, and in material composition through approaches to simulation, design or fabrication. Computation offered a means to address variable conditions and produce variable outcomes or outcomes that embody internal variation.

Variability, however, does not necessarily provide considered responses to different user groups, environments or site scenarios. Computation and design methods can easily generalize a design problem, as can be seen when a suite of projects follow shortly after a new software plugin or algorithm is made publicly available. For variability to be meaningful, it could perhaps engage in the specifics of each project and its stakeholders, and offer bespoke, tailored, site-specific, personalized of user-customized solutions. Until recently, such one-off design outcomes would have been cost-prohibitive to produce, and the computational models arguably not sufficiently robust for deep levels of customization to avert potential pitfalls of generalization.

However, with the emerging Fourth Industrial Revolution (Industry 4.0), a fundamental shift away from abstract generalized models of mass production is presently taking place, toward greater degrees of customization at unprecedented scales. Its key enabling technologies include autonomous robotics, autonomous manufacturing, and 3D printing, amongst others. These are supported by cyber-physical systems, internet of things (IoT) technologies, decentralized infrastructure (Schwab 2017), and are also increasingly being augmented by deep learning models . The vast interconnectedness of these technologies will be disruptive and potentially risks an exponential rise in production, consumption and waste that poses an existential threat to our climate and its delicately balanced ecosystems, and to social equity. There is also an urgent need to decarbonize buildings, to reduce cost and increase production of housing to meet demand, and to provide equitable infrastructure to communities at risk. Concurrently, however, Industry 4.0 technologies also support a shift to more diverse and considered forms of embodiment and participation in the built environment.

Can alternative models of design that insufficiently describe or include so much of today's diverse world; its genders, races, ethnicities, species, landscapes, augmented worlds, cyborgs, bio-synthetic or genetically engineered conditions?

How could one cater for, or engage meaningfully with such diversity? This year's theme, *Hybrids & Haecceities* seeks to contribute ideas to this question. The theme's use of these two terms comes with a call to action to critically and creatively re-assess the means in which we conceive, undertake and implement design. Both "hybrids" and "haecceities" carry several associations that operate together to provide fertile ground for framing future design work and discourse.

Hybrids are entities with characteristics enhanced by the process of combining two or more elements with different properties. Bred hybrids such as mules or ligers (horse-donkey and lion-tiger hybrids) embody different physical and behavioral characteristics from their parents. A liger might not only appear to be something between a lion and a tiger, but also inherit a tiger's swimming capabilities and a lion's social skills. More surprisingly though, the mix of the two species genes can produce characteristics not apparent in either parent species – Liger's are typically larger than lions or tigers (Bonnicksen 2009). Hybrids can exist in varying ratios, and be more than their constituent parts. Hybrids, therefore, dispel binary thinking, and offer opportunities for the combining of materials (such as novel composites), technologies or objects, for a valuable outcome. An interest in hybrids is a rejection in dualistic binary conditions, and asserts new forms of unity that might address ethical, social, theoretical, or environmental concerns, or be a means to approach engineering, design or aesthetics. Hybridization can lead to untold possibilities.

The first known use of word *Haecceity* is attributed to Philosopher-theologian-priest John Duns Scotus in the thirteen century CE, to describe a non-qualitative property that defines something as unique and indivisible. For Scotus, haecceity was essentially the "thisness" (derived from the Latin "haec", meaning "this") as opposed to a "whatness" or essence of an object or substance (Cross 2022). In *The Rise of Realism*, philosophers Manuel Delanda and Graham Harman agree that haecceities describe uniquely identifying features, or "thisness" of a specific cat, as opposed to its breed or species that equally describe several different cats. They differ on whether haecceities also involve history, generative processes, artefact or agency (DeLanda and Harman 2018). Haecceities as a term, challenges a generalization of things by focusing on what is unique.

In concert, *Hybrids & Haecceities* offers a provocation towards more inclusive and specific forms of computational design. *Hybrids & Haecceities* rejects binary thinking at all levels, is critically optimistic towards technology, and seeks more open and diverse engagement with the world-at-large, where each engagement is specific. *Hybrids & Haecceities* can and should also be interpreted as an agenda that seeks more considered design outcomes, that lead us towards new impactful, aesthetic, and ethical solutions leveraged by new technologies.

ACADIA 2022 *HYBRIDS & HAECCEITIES* PROCEEDINGS: PAPERS & FIELD NOTES

The *ACADIA 2022 Hybrids & Haecceities Proceedings: Papers & Field Notes* is the second of two volumes that include all of ACADIA 2022's academic endeavors. It documents double-blind peer-reviewed research paper and field note submissions together with documentation on the conference workshops, keynote lecture and panel discussion transcripts. As an AIA Continuing Education Provider, these ACADIA 2022 events also offered attendees AIA Learning Units (LUs) across the conference sessions and in workshop sessions. Due to a particularly strong year of submissions, fewer than 30% of submitted papers were competitively selected for inclusion in the conference proceedings and presentations. Papers comprise of full-paper and work-in-progress submissions covering a diverse range of research interests. The accepted sixty-six papers were grouped into sixteen sessions and distributed across the three days of the conference. Papers are organized in the *Proceedings* book according to these conference sessions. Each session focused on a different H&H topic that included New Ecologies, Advanced Production Methods, Robotics and Architectural Design, Advanced Materials, Machine Learning and Artificial Intelligence and others. Accompanying the papers, the proceedings includes Field Notes submissions; these shorter pieces are focused on provocative writing and works that seek to provide a discursive contribution to computational design through critical ideas, observations, narratives, manifestoes, or other forms of publication such as image-based essays or annotated images. *ACADIA 2022 Hybrids & Haecceities Proceedings: Papers & Field Notes* embodies a diverse and ambitious set of research challenges, methods, and design outcomes that seek to contribute to a shared knowledge and exploration into computational design and its diverse and real-world impacts.

As a whole, it is an awe-inspiring body of research and speaks volumes to the caliber of the authors and the design communities it represents.

REFERENCES

Bonnicksen, A. L. 2009. *Chimeras, Hybrids, and Interspecies Research: Politics and Policymaking*. Washington, DC: Georgetown University Press.

Buolamwini, J., and T. Gebru. 2018. "Gender Shades: Intersectional Accuracy Disparities in Commercial Gender Classification." In *Proceedings of the 1st Conference on Fairness, Accountability and Transparency*, vol. 81, edited by S. A. Friedler and C. Wilson. New York: PMLR. 77–91. https://proceedings.mlr.press/v81/buolamwini18a.html.

Cross, R. 2022. "Medieval Theories of Haecceity." In *The Stanford Encyclopedia of Philosophy*, Spring 2022, edited by E. N. Zalta. Stanford, CA: Metaphysics Research Lab, Stanford University.

DeLanda, M., and G. Harman. 2018. *The Rise of Realism*. New York: John Wiley & Sons.

Frazer, J. 1995. "An Evolutionary Architecture." In *An Evolutionary Architecture*. London: Architectural Association Publications.

Haraway, D. J., and C. Wolfe. 2016. *Manifestly Haraway*. Minneapolis: University of Minnesota Press.

Negroponte, N. 1972. *The Architecture Machine: Toward a More Human Environment*. Cambridge, Mass.: The MIT Press.

O'Neil, C. 2016. *Weapons of Math Destruction: How Big Data Increases Inequality and Threatens Democracy*. New York: Crown.

Schwab, K. 2017. *The Fourth Industrial Revolution*. New York: Crown Business.

Robotics and Architectural Design: Control Strategies

Arash Adel, Chair

The building industry contributes to roughly one-third of global CO_2 emissions. Investigating and developing more efficient design and construction methods is critical in reducing the building industry's carbon footprint. When coupled with integrative computational design methods and optimization techniques (e.g., material minimization), robotic construction and additive manufacturing techniques could unlock the realization of efficient structures and low-carbon construction processes. Moreover, robotic technologies provide immense freedom for constructing bespoke structures and buildings; however, they also impose constraints (e.g., fabrication constraints) on the design process that need to be formalized and satisfied. For instance, working with robotic assembly technologies requires investigating methods to account for material imperfections and fabrication tolerances automatically (or semi-automatically). The paper by Gandia et al. presents "a computational design method that integrates capabilities to manage material and fabrication tolerances occurring during the robotic assembly of spatial timber structures with tight-fit connections." Furthermore, working with robotic additive manufacturing techniques requires developing adequate control strategies coupled with computational design processes to facilitate the design and manufacturing processes, exemplified in this session with two different robotic approaches.

Within the realm of aerial additive manufacturing, the paper by Smith et al. presents a "multi-agent mission-planning robot control framework and simulation environment for Collective Aerial Additive Manufacturing (Collective AAM) together with an approach to the design of scaffold-free 3D shell geometries." And with regards to additive manufacturing using industrial robots, the paper by Meibodi et al. presents "the use of robotic pellet extrusion 3d printing (3DP) in the production of a multifunctional thermoplastic building envelope." When coupled with optimization techniques (e.g., material minimization), additive manufacturing could unlock the realization of efficient structures and low-carbon construction processes. The paper by Nasiri and Rakha presents "a theoretical and practical method of optimization and fabrication of multi-agent fabricated clay brick to be used in a bottom-up approach to manufacturing the environment." And lastly, these fabrication technologies require investigating new interfaces and design methods. To address this, the paper by Stephen et al. "describes the notion of the interface and its design implementations and concludes with a corresponding design methodology."

Tolerance-Aware Design of Robotically Assembled Spatial Structures

Augusto Gandia
Massachusetts Institute of Technology

Fabio Gramazio
ETH Zurich

Matthias Kohler
ETH Zurich

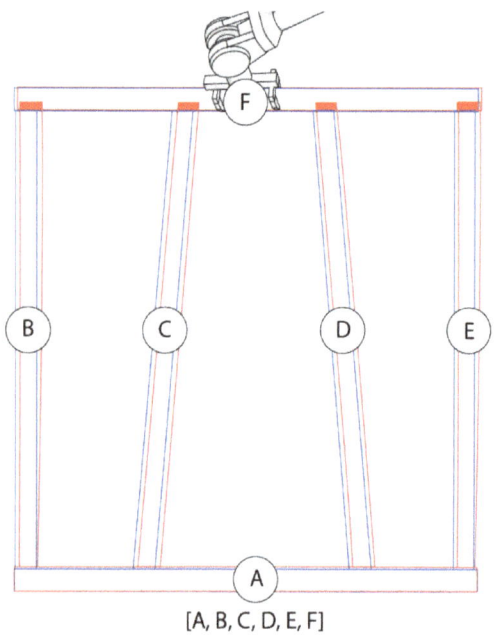 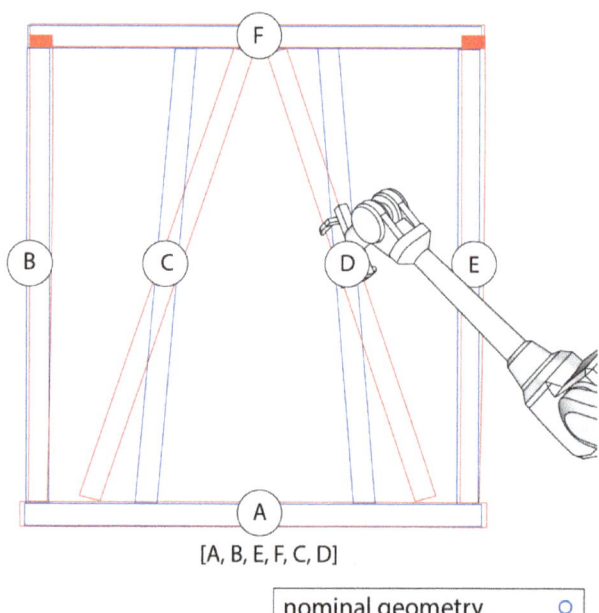

1 Illustration of two different assembly sequences for the construction of a timber frame. The first sequence (left) leads to a misfit when placing beam F, and therefore, it requires cutting beams B-E, since they are longer than expected (red). The second sequence (right) presents a more efficient approach to overcome the previously introduced misfit by cutting only beams B and E. Beam F is later placed and beams C-D are rotated in.

ABSTRACT

The robotic assembly of spatial timber structures with tight-fit connections, such as T-butt joints with screws lead to tolerances that add up after each placed building member. As such the built structure incrementally deviates from its digital counterpart throughout the assembly-sequence, which results in misfits or even deadlocks during the assembly. This problem has been often addressed through adaptive feedback strategies, which consist of measuring each placed beam and adjusting the fabrication model accordingly. Their main limitation of feedback strategies is that they only react to each previous step of the assembly sequence, and therefore, do not allow strategic decisions to propagate tolerances along several steps of the assembly.

This paper presents a computational design method that integrates capabilities to manage material and fabrication tolerances occurring during the robotic assembly of spatial timber structures with tight-fit connections. This is achieved by building a data-base of tolerances measured during the robotic assembly process, which then allow for tolerance simulation as part of an assembly sequence planning method based on the Kruskal algorithm. Through a combination of optimization and linear regression techniques, the developed method enables designers to minimize deviations of their designs and diminish the risks of misfits during fabrication. In consequence, it allows for tolerance-aware designs.

INTRODUCTION

During the last two decades, the accessibility to industrial robots has enabled the investigation of novel assembly processes, such as for the assembly of space-frames with freely oriented building members (Helm et al. 2016; Piskorec et al. 2012; Eversmann et al. 2016; Parascho et al. 2017; Thoma et al. 2018). In contrast to manually assembled prefabricated systems, these robotic processes could potentially enable the production of the custom parts through less wasteful/invasive approaches (e.g. CNC cutting) and automatic assembly without the need for scaffoldings or labeling systems.

These processes, however, introduce several new challenges, including the robotic manipulation of building members while avoiding obstacles, the handling of tolerances that build up after the placement of each building member, the counteraction of gravitational sagging and the avoidance of unstable configurations during assemblage. This paper builds upon a previous publication that tackles the planning of collision-free robot paths, therefore it focuses on the second challenge, which is the handling of tolerance build-up (Gandia et al. 2019). In contrast to existing feedback-based approaches typically used in robotic fabrication, this investigation aims to tackle tolerance build-up at early design stages, when it is possible to avoid unwanted design changes that may be caused later by fabrication misfits or deadlocks.

The investigation introduces a computational method, based on the matrix method (Marziale and Polini 2009) and the Kruskal algorithm, that allows planning assembly sequences that consider the build-up of tolerances. The approach is investigated and developed through several experiments and evolves from a tool for refining designs according to visualized tolerances towards a method for automatically searching assembly sequences while minimizing displacement and misfits during assemblage.

The research is conducted in collaboration with the Spatial Timber Assemblies (Thoma et al 2018), as part of the *DFAB HOUSE* (Grasser et. al 2020). and the work *Complex Timber Structures from Simple Elements* (Apolinarska 2018). They identify fabrication constraints, provide geometry-generation custom setups, and enable the collection of fabrication-data. This is conducted in the large-scale multi Robotic Fabrication Laboratory (RFL) (Bonwetsch and Lyrenmann 2010) during the construction of the timber modules of the *DFAB HOUSE*.

2 Compensation of tolerance build-up through manual adjustments for the assembly of timber frames

3 Tolerance measurement procedure for the robotic assembly of the Complex Timber Structures from Simple Elements

STATE OF THE ART

In conventional approaches used in industry for the manual assembly of timber frames, the tolerances are tackled by precisely cutting building members and by adjusting them through cuts, extra timber chips, or reorientation during assembly (Figure 2). In robotic assembly, projects with tight-fit connections, such as the *Complex Timber Structures from Simple Elements* (Figure 3) (Apolinarska 2018) relied on the previously introduced feedback based methods.

While such strategies are highly effective for handling tolerances during fabrication, as they operate directly on the physical object, they are not suitable to manage tolerances throughout several steps of the assembly process. In turn, unforeseen misfits and deadlocks can occur, which would often need unwanted design modifications for their resolution. This paper presents a method for handling tolerances while designing the structure, which is when design adjustments can be planned (Chase 1999). In contrast to existing tolerance simulation solutions, such as Sigmetrix CETOL (Sigmetrix 2020), Inventor Tolerance Analysis (Inventor n.d.), CATIA V5 (Dassault Systèmes. n.d), the developed method considers the robotic processes, but more importantly, it is integrated in computational design. Additionally, the method builds upon other solutions developed in COMPAS (Van Mele et al. 2017-2021), such as Data Management and Modeling of Complex Interfaces in Imperfect Discrete Element Assemblies (Frick et al. 2017).

METHOD

Robotically assembled spatial timber structures with tight-fit connections present tolerances from different sources. The tolerance-scope of this paper is formed by those that directly affect the robotic assembly of spatial structures and can be classified into two main categories: (a) Tolerances of the Building Member, which are caused by deviations intrinsic to timber, through inaccuracies of the machining of raw material into standard beams and calibration errors of the cutting procedure; and (b) Tolerances of the Assembly Process, which are caused by inaccuracies of the robotic placement process.

TOLERANCES OF THE BUILDING MEMBERS

- Axial Deformation caused by the natural deviations of timber and are identified on the longitudinal axis of a beam by comparing the center point of its ends with the one of their digital counterpart (Figure 4a)
- Cross Section Deviations caused by manufacturing deviations and occur on the transverse axes of the beam (Figure 4b)
- Cut Angle Error caused by decalibration of the saw and the robot positioning of the beam; this can be quantified by measuring the plane at the ends of the beam (Figure 4c)

TOLERANCES OF THE ASSEMBLY PROCESS

- Beam Positioning Deviations caused by positioning and rotational inaccuracies of the robotic setup (Figure 4d)

Tolerances are measured through a measurement procedure, which entails a proper understanding of their magnitude and frequency. It was implemented into the robotic fabrication of the Spatial Timbers Assemblies to measure deviations of each machined beam by using a gantry robot of the RFL (Figure 5). It integrated an edge detection optic sensor, Baumer PosCon 3D Edge (Baumer n.d.) (Figure 5a) that allowed measuring beam ends after being cut. Additionally, the position and orientation of the robot Tool Center Point (TCP) was minimized within submillimeter accuracy by using an external tracking system iGPS (Nikon Metrology NV n.d.) (Figure 5d).

The measurement procedure begins by robotically placing a recently cut beam at one of the eight pre-calculated positions (Figure 6, Steps 1-8). Each position-pair (e.g. 1-2) is measured within the sensor distance range, which in this case is 15-25 cm. This is performed by positioning the beam relative to the sensor (Figure 6, red), which allows reconstructing the geometry of its four edges for both beam ends and for the beam center, which is derived from the robot gripping pose (Figure 6c).

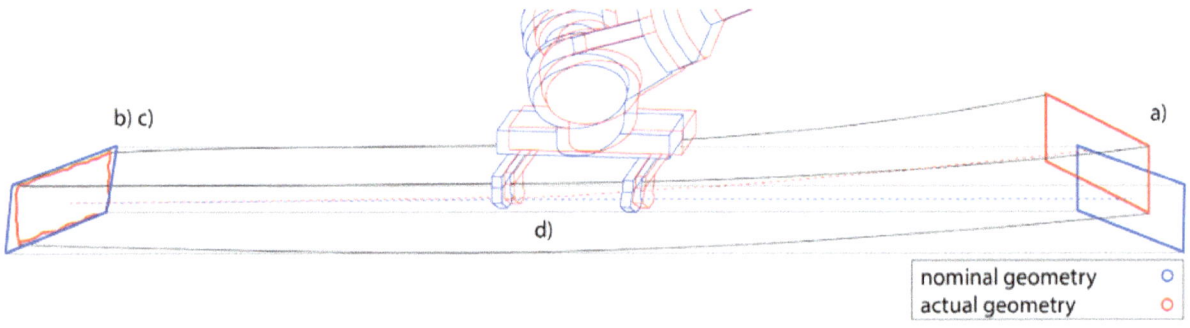

4 Visualization of (a) the Axial Deformation; (b) the Cross Section Deviations; (c) the Cut Angle Error; and (d) the Beam Positioning Deviations registered during the robotic fabrication of the DFAB House

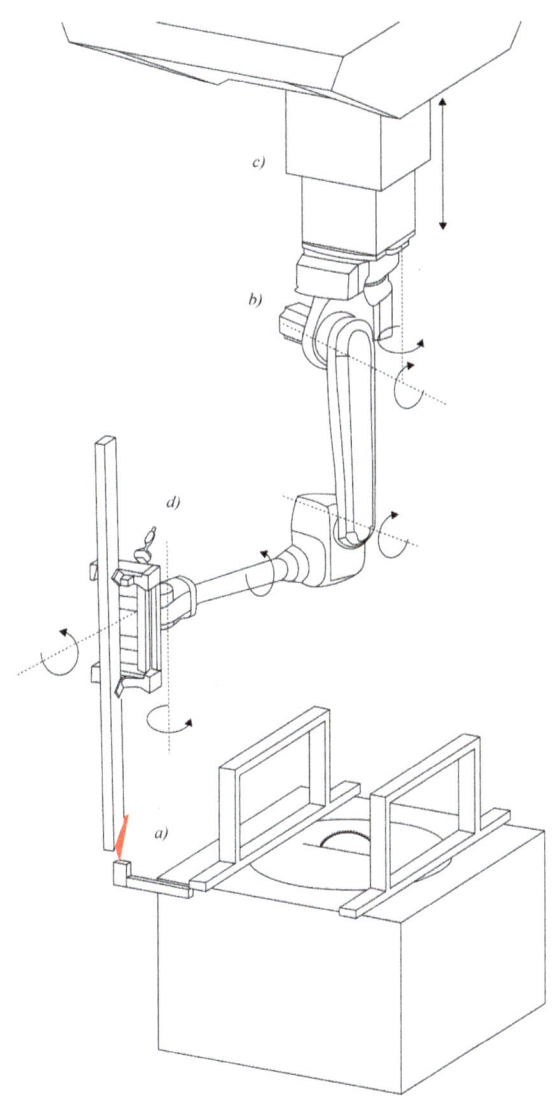
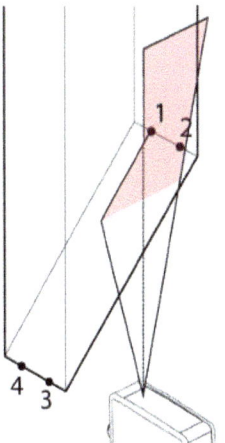
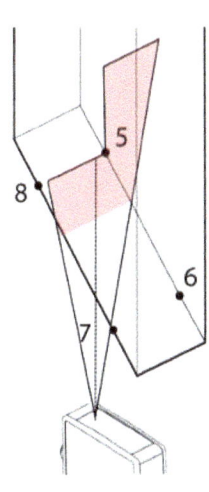

This procedure was performed on twenty-two beams and compared to their corresponding nominal digital counterpart in order to calculate tolerance ranges. While this amount of data may not be statistically significant for a method that aims at visualizing actual tolerances, it was considered sufficient data to validate the experiment on *The Complex Timber Structures from Simple Elements*, since the method pursues visualizing a tendency of the three previously introduced tolerance sources during early design phases.

The developed computational method requires inputting the design model of a spatial structure, which then is used to generate assembly sequences. These sequences are used to propagate tolerances along the building members and in turn, it allows researchers to estimate the tolerance effects on the structure, which serves to identify the most suitable assembly sequences (Figure 7).

5   Visualization of the measurement setup with (a) optic sensor; (b) six-axis robotic arm; (c) gantry; and (d) tracking system used during the data-collection procedure

6   Visualization of the eight points measured as strategy to quantify tolerances of a beam end

Specifically, the input structure is topologically represented through a COMPAS network, which defines each beam by a node and each connection by a directed edge pointing in the direction of the assembly sequence (Skiena 1998). The first parameter that needs to be set is the hierarchy of the beams in the structure, which allows the prioritization of the assembly of certain beams among others and is accomplished by assigning an ascendant ordered weight expressed by an integer (lower integers assign high priority in the assembly sequence, higher numbers assign low priority).

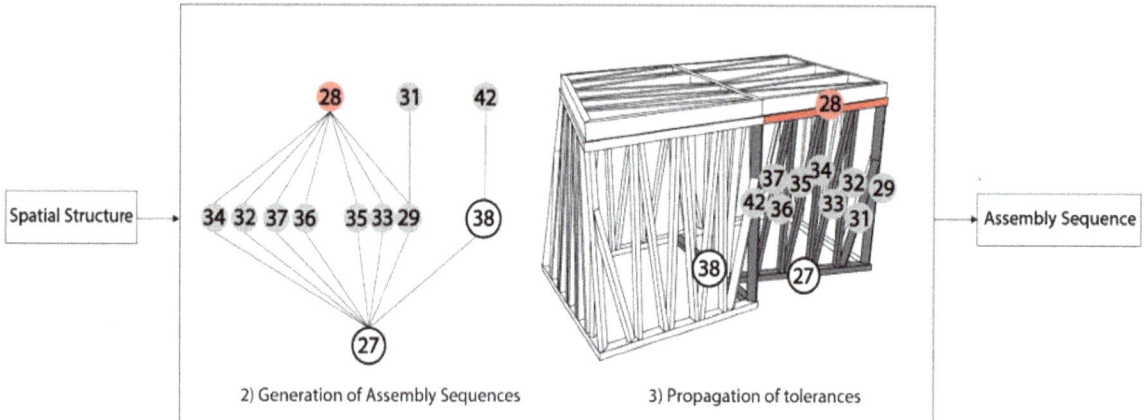

For example, setting the weights for the assembly of the beams of a timber frame wall (Figure 8a) can lead to an assembly sequence that prioritizes the placement of a top beam over the infill beams (Figure 8b) or vice versa (Figure 8c). The diagram represents the beam type by the color of the node and the resulting assembly sequence by the array with the beam names.

For example, a rule for a top cord beam (black) of a timber frame (Figure 9a) would prevent its placement if only one vertical support (light gray) has been placed, in case this is considered structurally unstable (Figure 9b). Thus, during the assembly sequence search, it is checked if two supports have been placed before the assembly of a top cord (Figure 9c).

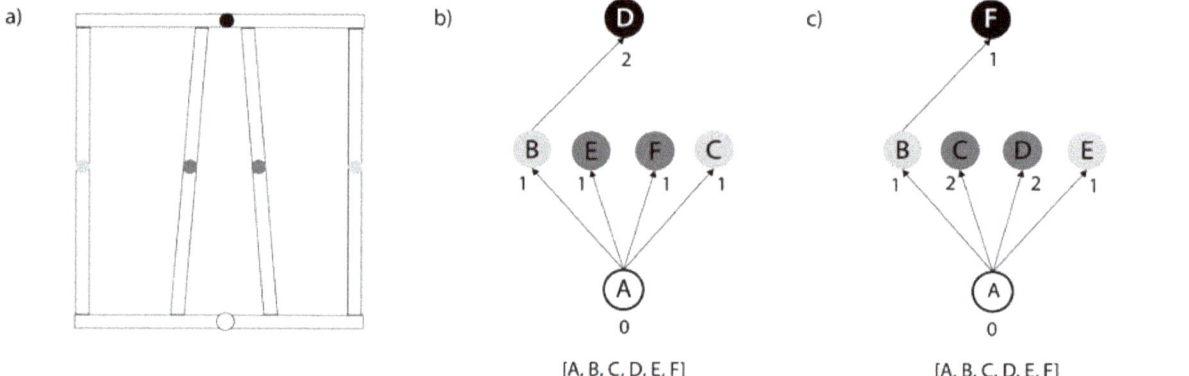

In a subsequent step, dependency rules between beams are specified to assure the stable assembly of each building member along the assembly sequence. These rules are hard-coded for each beam type of the structure for the sake of simplifying the method, which future research could be solved through the implementation of a structural analysis method.

The Kruskal algorithm is used to generate the assembly sequence while evaluating the previously defined conditions for hierarchy and dependencies between nodes. This procedure begins by seeking a connection (directed edge) between the start (node with weight = 0) and an adjacent building member (node). If both introduced conditions are fulfilled, a connection is established, which serves to set their order of assembly.

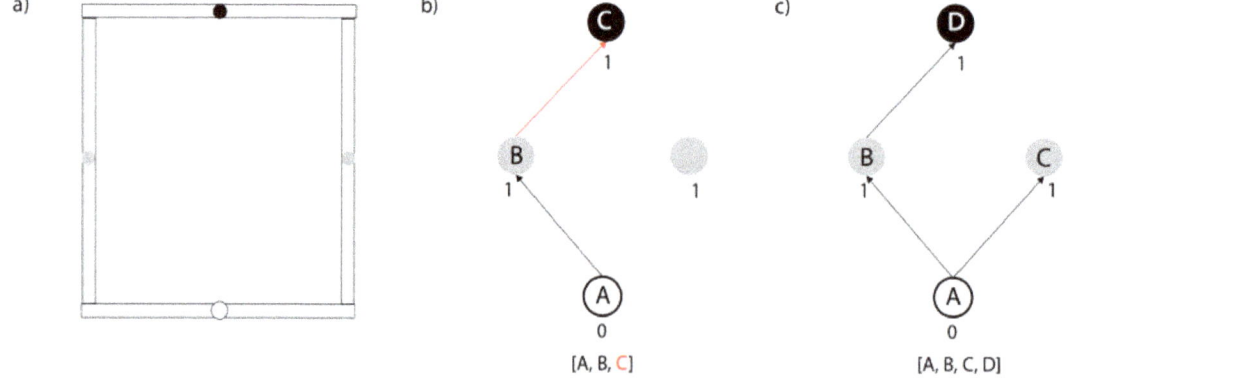

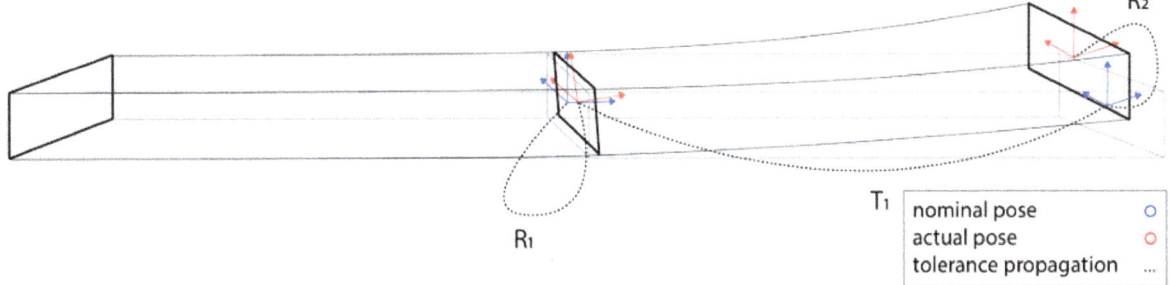

The procedure is repeated with all nodes, and finalizes if all building members are connected. The outcome is a tree of connected nodes with edges that are directed from the first to the second visited node. The method propagates tolerances through translation (See Fig. 10, T) and rotation (See Fig. 10, R) matrices from the middle towards the ends of each beam. While beam positioning deviations are assigned through R1, axial deformation, cross section deviation, cut angle error, and sensor linearity error are implemented through T1 and R2.

After a first building member is placed (See Fig. 11, Beam 1), its tolerances (See Fig. 11, R1, T1 and R2) propagate towards the adjacent member (See Fig. 11, Beam 2). Therefore, its initial position needs to be readjusted (See Fig. 11, F1) to be placed. This shift is measured as the displacement from the initial midpoint of the beam towards the shifted midpoint and is measured as the magnitude of its vector. The sum from all building members is the cumulative displacement of the structure, which can be used to compare assembly sequences.

7 Diagram of the process for planning a valid assembly sequence; it starts by inputting a design model then generating assembly sequences (left) on which estimated tolerances are propagated (right) until finding a valid sequence / output.

8 Example of setting the hierarchies to assemble a timber frame (a). It consist of setting a weight (0-2) for each beam (bottom beam = white, vertical support = light gray, infill beam = dark gray, and top cord beam = black), which can result into two assembly sequences (a and b) that are represented by the nodes size (descendant order).

9 Example of how the dependency rules assigned for a timber frame (a) steer a first assembly sequence (b) from unsuccessfully pursuing to place a top cord (black) on a single vertical support (light gray) to a second sequence (c) in which both supports (light gray) are placed and then the top cord.

10 Visualization of tolerance propagation on a timber beam.

11 Visualization of tolerance propagation from a first placed timber beam (Beam 1) to the next one (Beam 2). This consists of tolerances that propagate from the center of the first beam to its end (R1, T1, and R2) and from there to the center of the second beam while considering tolerances of the shared end (T2, R3, T3, and R4), which leads to a shift on the position (from F1 to F2).

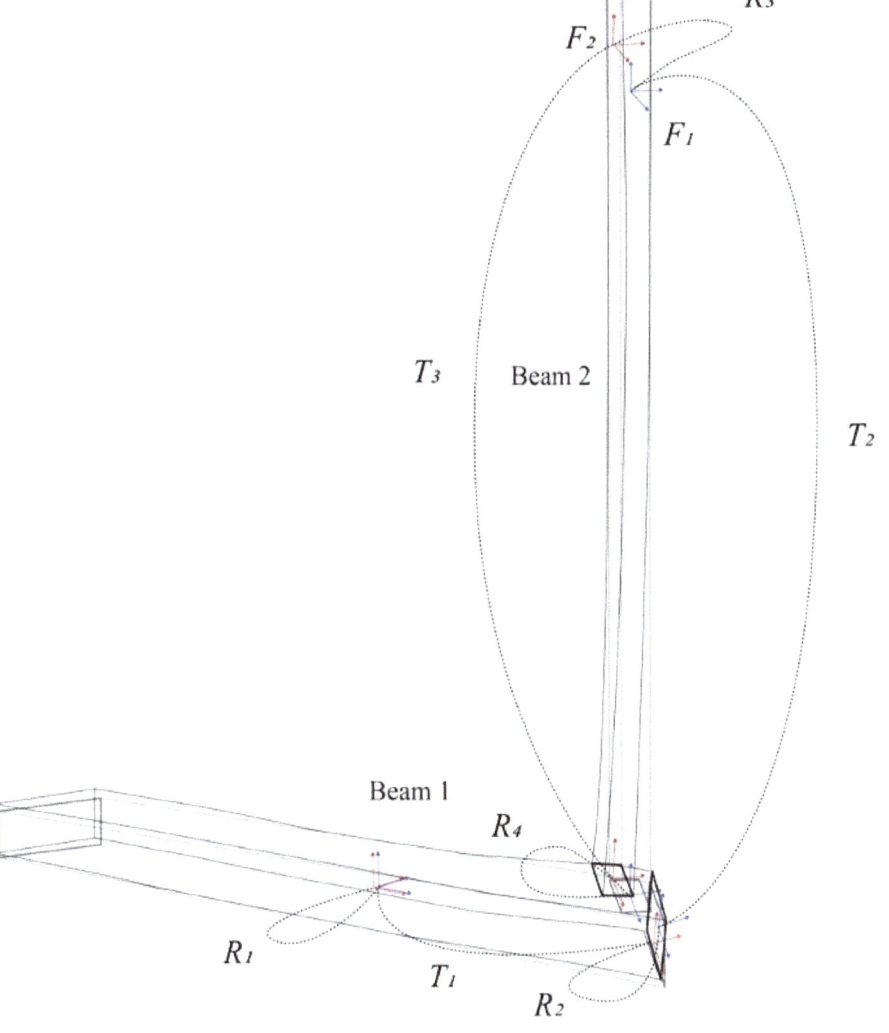

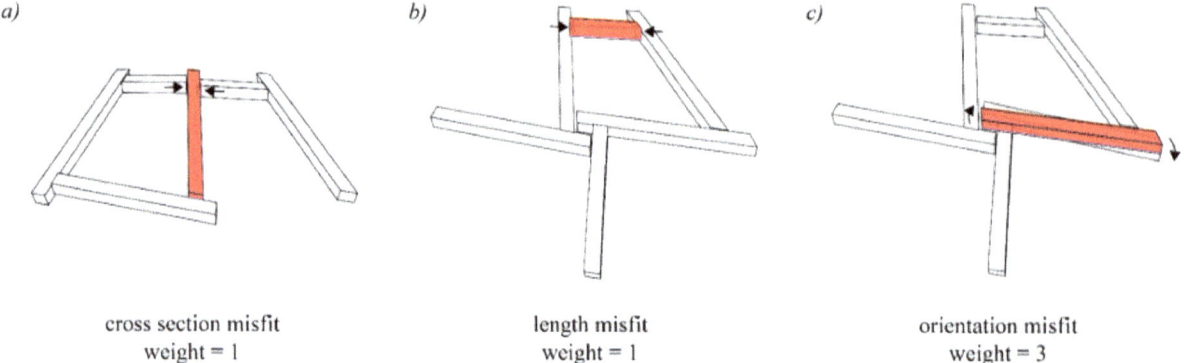

The misfits, which result from tolerances propagating along several timber building members, are classified into three types. Cross section misfits (Figure 12a), length misfits (Figure 12b), and orientation misfits (Figure 12c). Each misfit type has a different weight according to the difficulty to overcome it later in the actual robotic assembly of the part. For example, correcting the cross-section misfit would require the selection of a piece of wood from a different batch with a different cross section, and correcting the length misfit would need cutting the part shorter or longer; alternatively, correcting the orientation misfit would require calculating a new orientation, which is more complex. As indicated in Figure 12, the weight for the cross-section misfit and the length misfit is 1 and for the orientation misfit is 3. The weighted misfits occurring during the propagation of tolerances on a structure are summed up as the misfit factor, which serves together with the introduced accumulative displacement to compare assembly sequences.

The developed method relies on the following three assumptions: (1) the position and orientation of beams is within the sub millimeter accuracy of the external tracking system; (2) each placed beam is clamped to the structure to assure that tolerances build-up unidirectional from the structure towards the placed beam; and finally (3) deviations caused by the weight of the structure are not considered.

EXPERIMENT I:
Assembly Sequence Planning Using Visualized Tolerances

This experiment validated the developed method on a module designed by the Spatial Timber Assemblies (Adel 2020, 117) (Figure 13). It consisted of finding a strategy to assemble a module with 85 freely oriented beams of different lengths (2000 - 4600 mm) and cross-sections (200 x 80 mm, 120 x 100 mm, 120 x 60 mm) manually connected with screws.

12 Visualization of the three misfit types, including the cross-section misfit, length misfits, and orientation misfit with their corresponding weights

13 Visualization of the investigated module (thickened silhouette) from the Spatial Timber Assemblies, *DFAB HOUSE*

14 Visualization of tolerances that build up along the first strategy on a module of the Spatial Timber Assemblies

15 Digraph of the first strategy, which shows the connectivity (edges) between beams (nodes), their hierarchy and the assembly sequence (array of beam indexes)

16 Visualization of tolerances that build up along the second strategy on a module of the Spatial Timber Assemblies

17 Digraph of the second strategy showing the connectivity (edges) between beams (nodes), their hierarchy, and the assembly sequence (array of beam indexes)

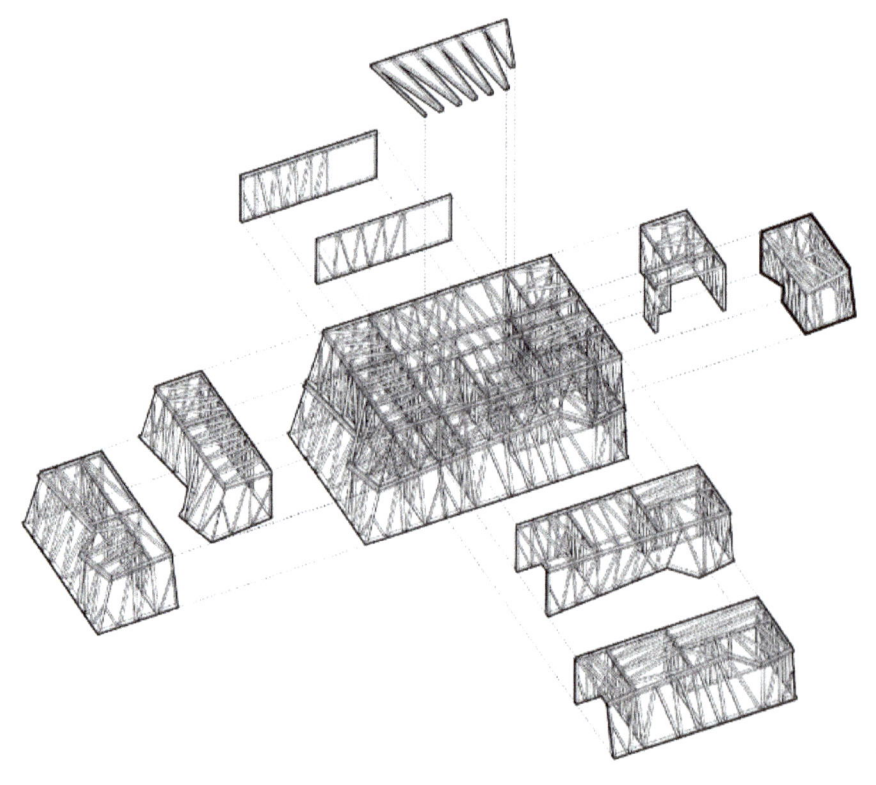

ROBOTICS AND ARCHITECTURAL DESIGN: CONTROL STRATEGIES

Hybrids & Haecceities

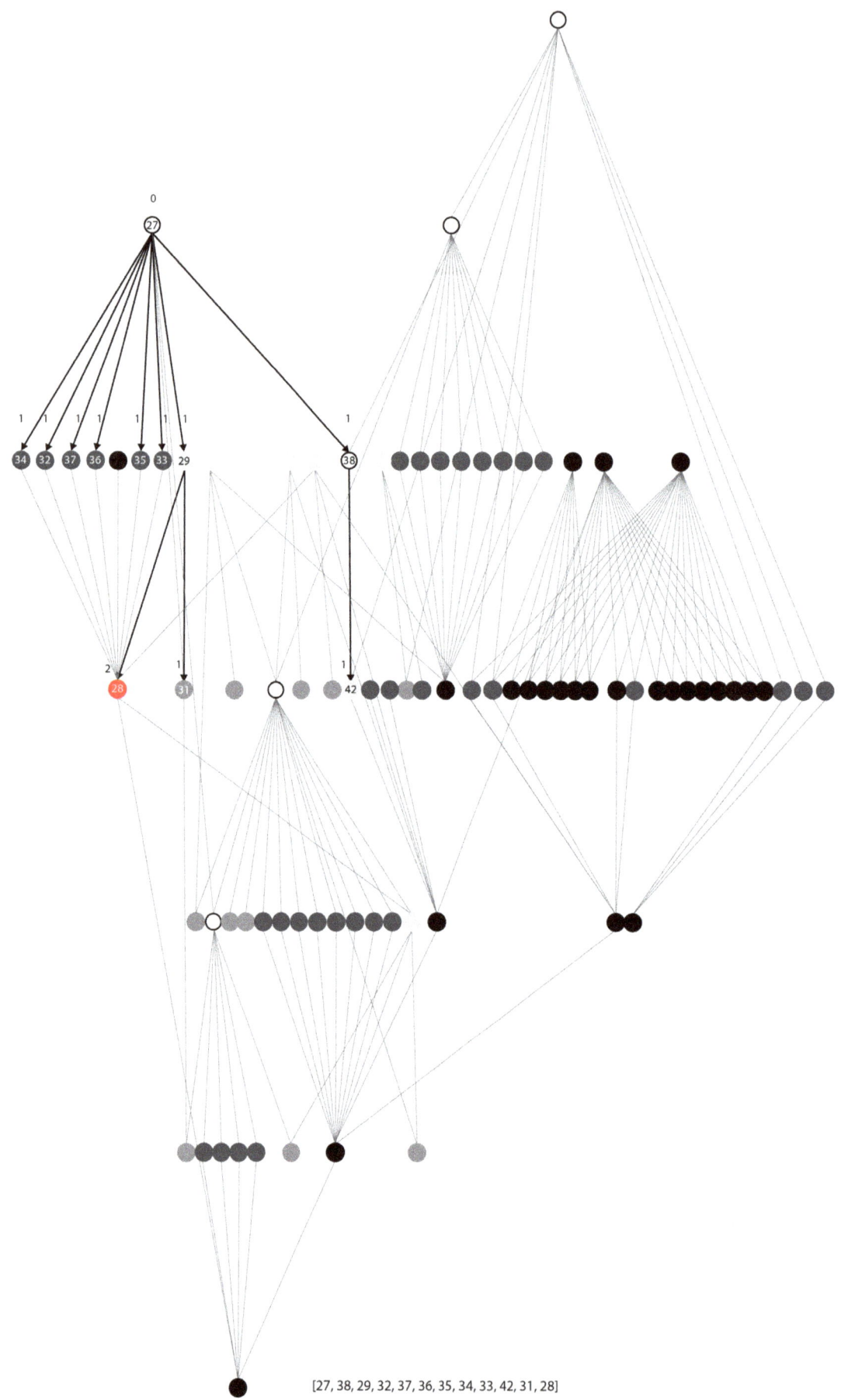

[27, 38, 29, 32, 37, 36, 35, 34, 33, 42, 31, 28]

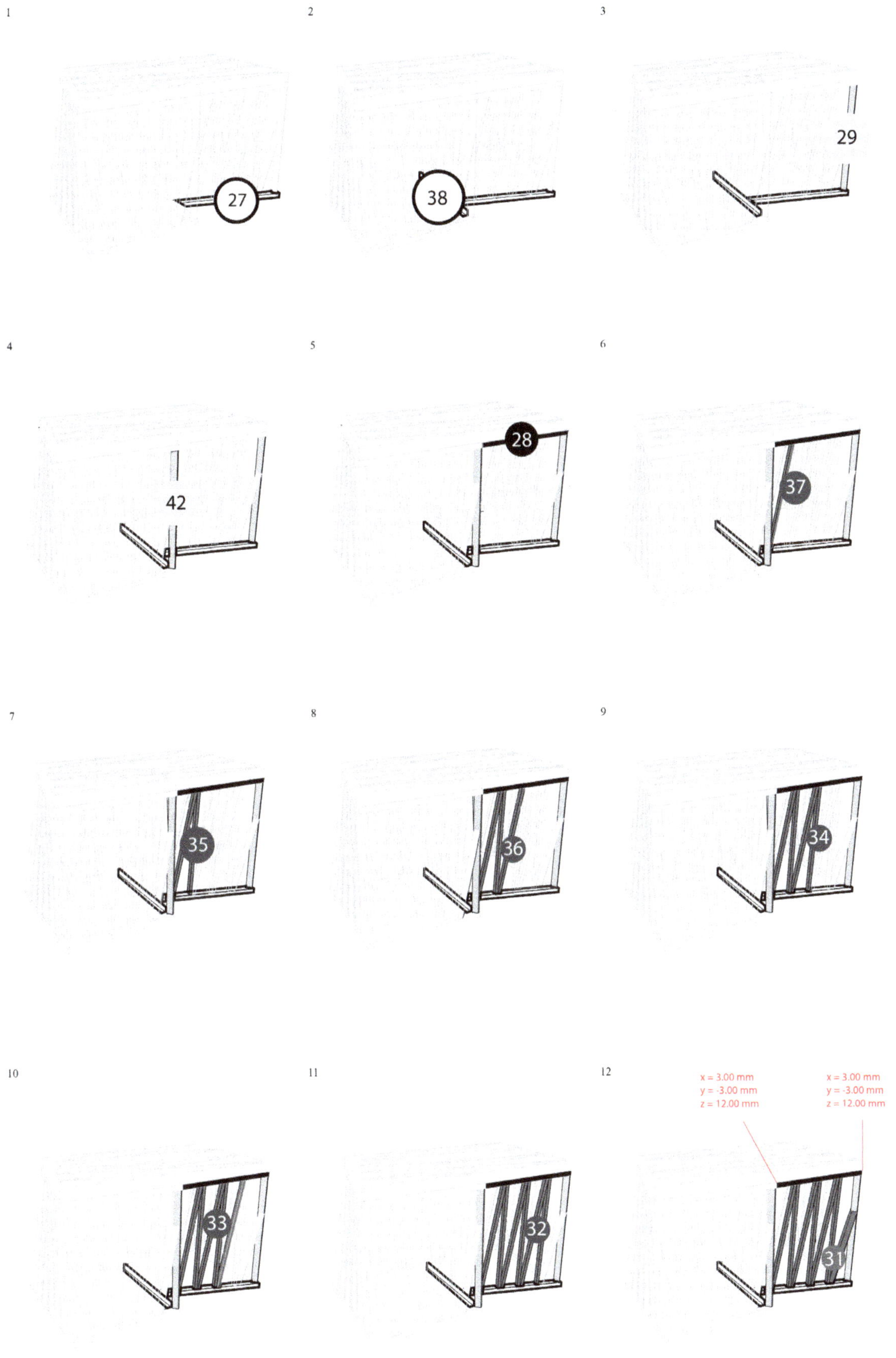

1	2	3
27	38	29

4	5	6
42	28	37

7	8	9
35	36	34

10	11	12
33	32	31

x = 3.00 mm
y = -3.00 mm
z = 12.00 mm

x = 3.00 mm
y = -3.00 mm
z = 12.00 mm

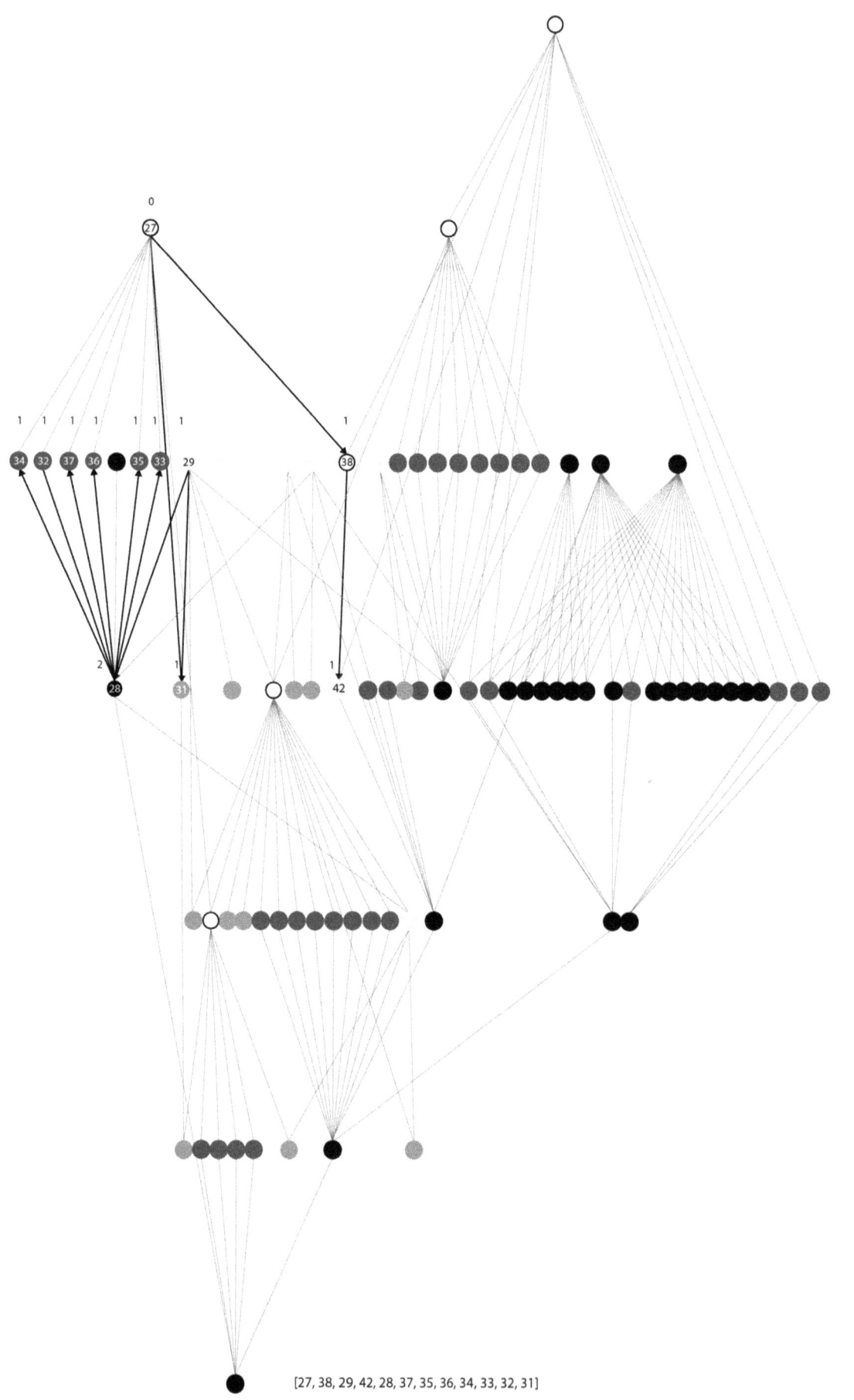

17

[27, 38, 29, 42, 28, 37, 35, 36, 34, 33, 32, 31]

We first tested an assembly strategy often used by woodworkers in industry, which consists of placing the beams from the bottom of the structure towards its top (Figure 14). Thus, we placed first the bottom beams, then the vertical and infill beams, and finally the top-cords. Setting up the parameters for generating such a sequence required the definition of three actions, the start, the hierarchy for prioritizing vertical and infill beams over the top-cord and the dependency rules. The start is the bottom beam (number 27). The dependency rules are defined by taking into account the fabrication requirements of the project, which consist of allowing the placement of vertical beams only on top of a previously placed bottom beam (Figure 14, step 3), infill beams on a bottom beam (Figure 14, step 4), diagonal beams after its adjacent, bottom and vertical beams (Figure 14, step 11), and finally, top-cords after both adjacent vertical beams are in place (Figure 14, step 12).

The tolerance build-up visualized shifts that were transferred from the bottom beam to the verticals and infill beams, resulting in a accumulative displacement of 3.0 mm in x, -3.0 mm in y and 12.00 mm in z. To be able to assemble the structure, then we needed to either update the final position of the top-cord, which would transfer significant tolerances to adjacent modules of the house. Thus, this option was considered as a misfit and therefore, it was displayed in red (Figure 14, step 12). To react to this, we proposed a second strategy that started by placing the beams of the frame and then the infill beams (Figure 16, steps 1-12). Thus, we set the start and hierarchy for placing the bottom beams first (Figure 16, steps 1-2), then the verticals (Figure 16, steps 3-4), top cords (Figure 16, step 5), and finally the infill beams (Figure 16, steps 6-11). It relied on the same dependency rules as in the previous strategy and the cumulative displacement value remained the same. The tolerance build up visualized deviations that were transferred from the bottom beams to the verticals, as in the previous strategy, which resulted in a cumulative displacement of 3.0 mm in x, -3.0 mm in y and 12.00 mm in z (Figure 16, step 12). This necessitated updating the final position of the top cord as in the previous experiment. However, in this case the top-cord could be placed as we assumed that the beam would admit certain bending (Figure 16, step 5). An alternative solution was to cut the two verticals before. In a last step, the infill beams were inserted through a robotic rotational movement (Figure 16, steps 8-12).

EXPERIMENT II:
Automatic Assembly Sequence Planning using Optimization and Machine Learning Techniques

This experiment validated the method on a module designed by the Complex Timber Structures. It consisted of a freestanding double layer canopy of approximately 10 by 5.5 m span and 3.0 m height (Figure 18). It comprised of 72 building members arranged in three layers (1st layer = bottom beams, 2nd layer = diagonal beams, and 3rd layer = top beams), which are connected through glued T-joints. Two assumptions regarding the fabrication process allowed the experiment. First, each robotically placed beam was clamped to the built structure after being placed to assure that tolerances are transferred unidirectionally from the built structure to the placed beam and not inversely. Second, gravitational sagging was not considered, since the cross section and length of building members assured minimum deviations.

Counteracting tolerances for such design was considerably more challenging than in the previous one due to the reciprocal

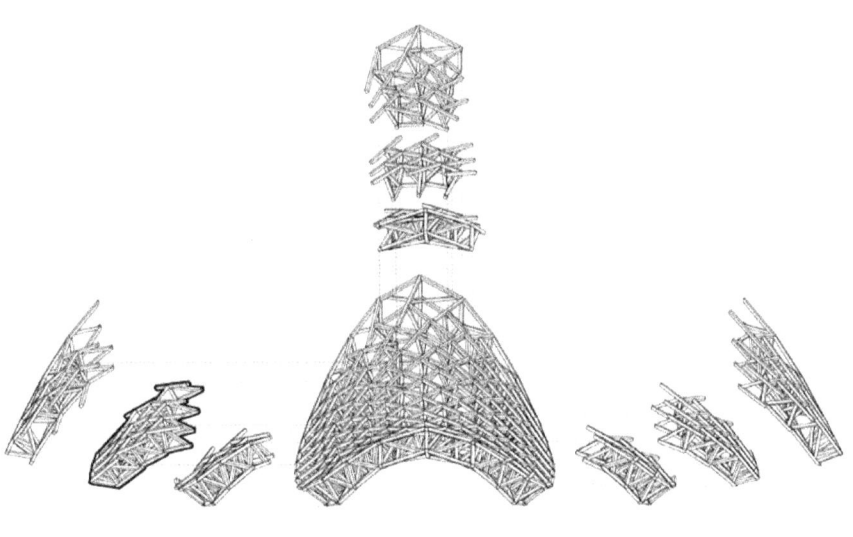

18 Visualization of the investigated modules (thickened silhouette) as part of the module of the Complex Timber Structures

19 Visualization of tolerances that build up along the first strategy along a module of The Pringle Structure

20 Digraph of the first strategy showing the connectivity (edges) between beams (circles), their hierarchy, and the assembly sequence (array of beam indexes)

21 Visualization of tolerances that build up along the second strategy on a module of the Spatial Timber Assemblies

22 Digraph of the second strategy showing the connectivity (edges) between beams (nodes), their hierarchy as in the previous image, and the assembly sequence (array of beam indexes)

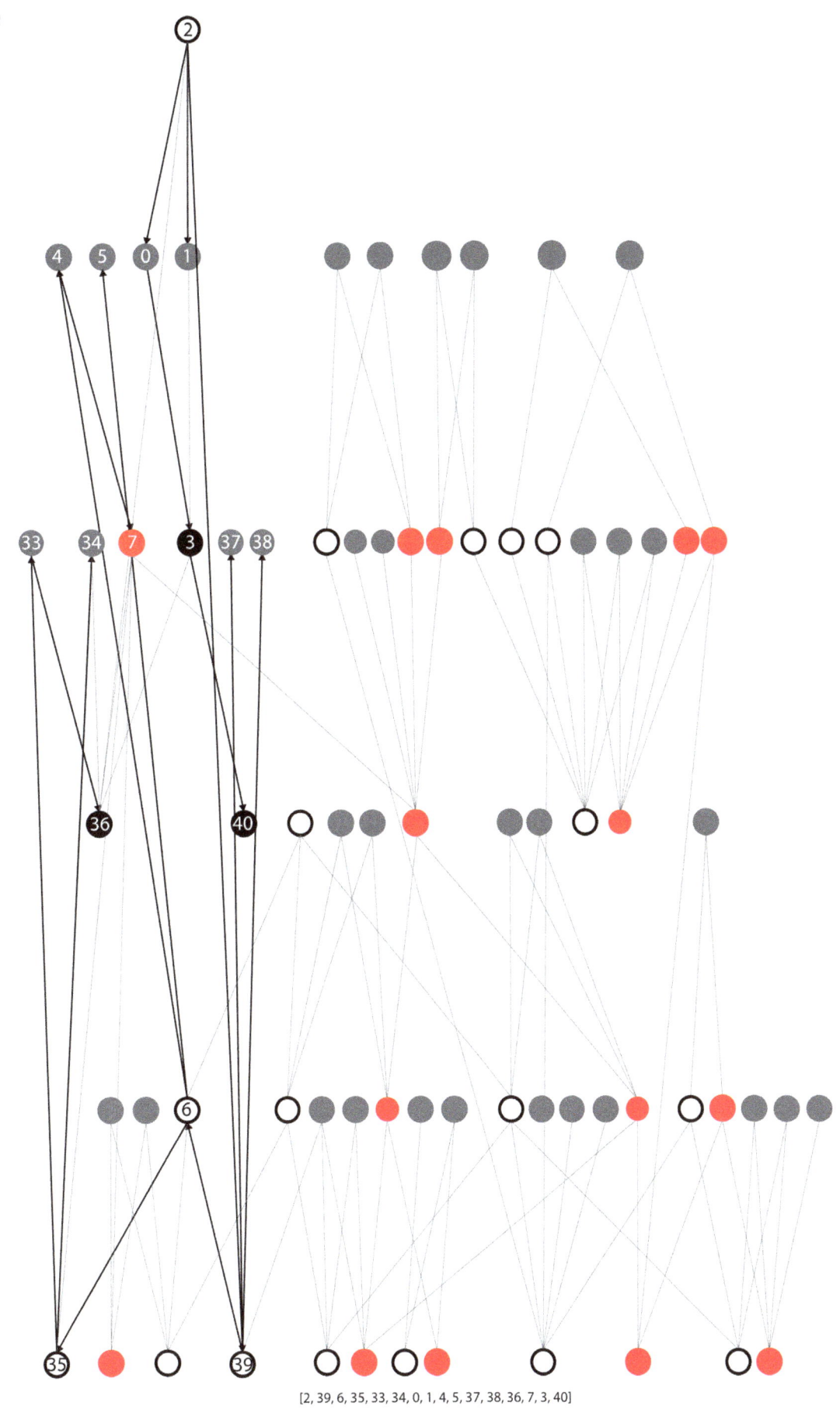

[2, 39, 6, 35, 33, 34, 0, 1, 4, 5, 37, 38, 36, 7, 3, 40]

ROBOTICS AND ARCHITECTURAL DESIGN: CONTROL STRATEGIES

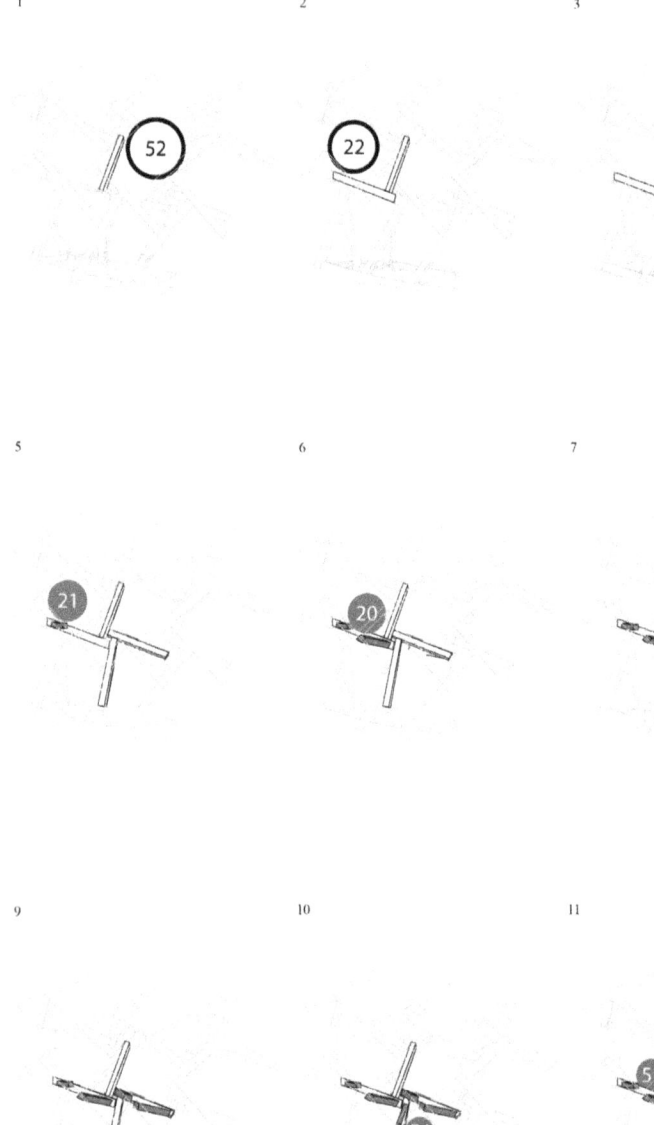
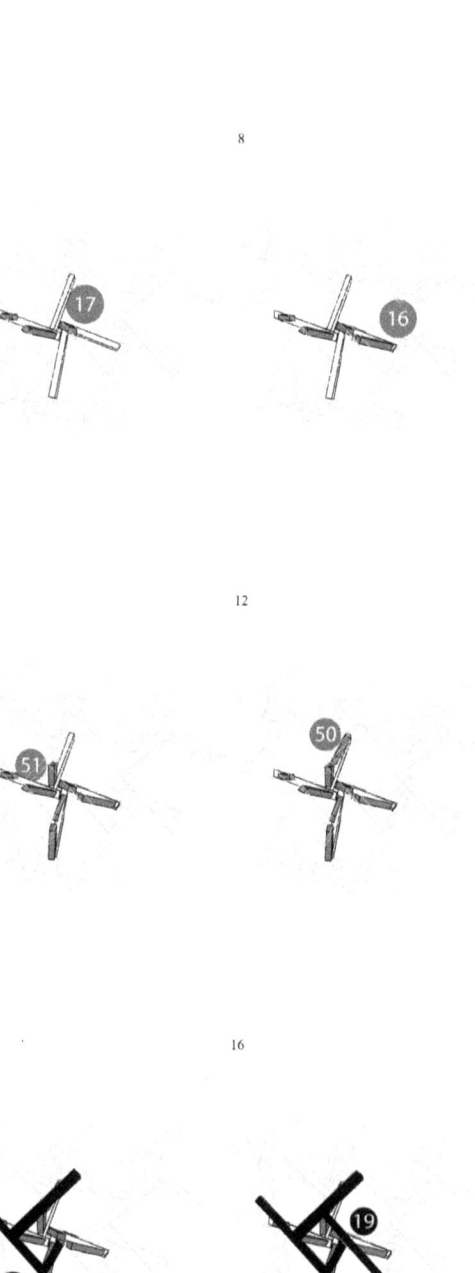

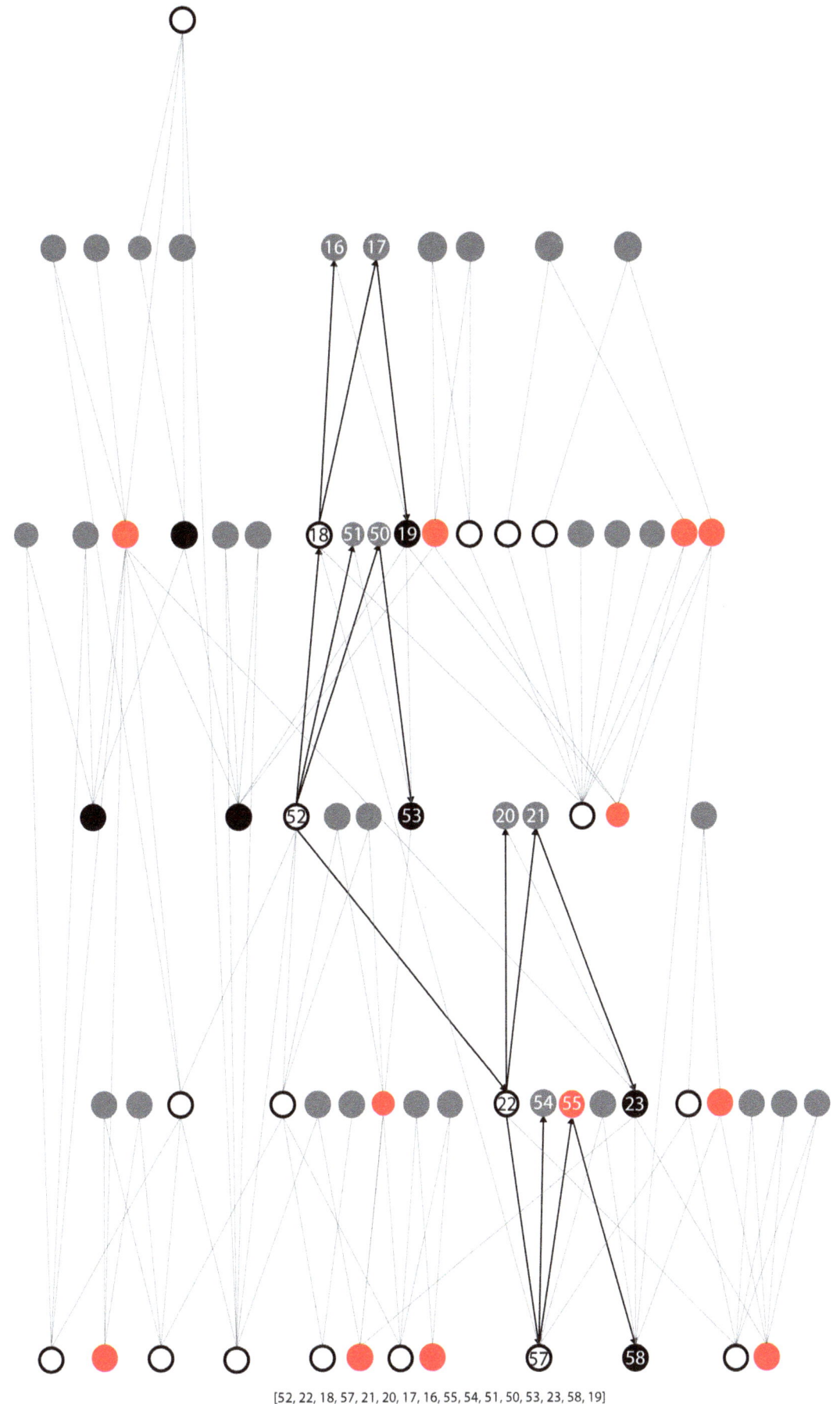

[52, 22, 18, 57, 21, 20, 17, 16, 55, 54, 51, 50, 53, 23, 58, 19]

ROBOTICS AND ARCHITECTURAL DESIGN: CONTROL STRATEGIES

connection that often led to a large accumulative displacement and misfit factor. We first tested a strategy that aimed to minimize tolerances transferred to adjacent modules. We started by placing all building members of the boundary. The start and hierarchy were set accordingly (Figure 16, step 1), as well as the dependency rules, which consisted of placing the bottom-layer beams only next to an adjacent bottom-layer beam (Figure 16, step 1-4), the middle-layer beams only on top of a bottom-layer (Figure 16, step 5-12), and the top-layer only on two supporting middle-layer beams (Figure 16, step 13). This resulted in an accumulative displacement of 200 mm and a misfit factor of 10.

In contrast to the previous strategy, we decided to test a sequence that built the module from its center towards the boundary (Figure 18, step 1-16). In turn, the accumulative displacement was 300 mm and the misfit factor 5, which proved the significant reduction of accumulative displacement through changes on the assembly sequence. However, this sequence also presented a higher misfit factor, which proved that for certain complexity of structures, larger numbers of valid assembly sequences need to be found and evaluated. Unfortunately, this could not be efficiently achieved through manual approaches and required the investigation of automated techniques.

To address this, we developed and tested a method for automatically searching all possible assembly sequences, propagating tolerances and evaluating their accumulative displacement as well as misfit factor. The procedure was implemented through the custom implementation of an optimization algorithm that generated sequences and propagated tolerances while automatically changing the start within the beams of the module's bottom layer, as they need to be placed first.

This resulted in 323 assembly sequences with their corresponding propagated tolerances (Figure 23). The range of accumulative displacement was 150 mm to 600 mm and the misfit factor was from 1 to 17 units. From this data an acceptable sequence could be chosen; however, after this process the best solution was still unknown. Thus, we required the implementation of a multivariate linear regression method within Octave (https://octave.org), which allowed us to find the best fit line for the input data (Figure 24). To improve the accuracy of our line we used a gradient descent method that allowed us to predict an optimal assembly sequence with 167 mm of accumulative displacement and 0 misfits.

Firstly, four assembly sequences with the worst case tolerances were selected, which were the ones at the boundaries of possible accumulative displacement and misfit factor. Specifically, this consisted of a sequence with high accumulative displacement and low misfit factor (Figure 25, red), a sequence with high cumulative displacement and high misfit factor (Figure 25, green), a sequence with low accumulative displacement and low misfit factor (Figure 25, blue) and a sequence with low accumulative displacement and high misfit factor (Figure 25, yellow).

Subsequently, randomized tolerances were propagated 30 times along each of the four selected assembly sequences (Figure 25, red, green, blue and yellow). These new 120 assembly sequences with randomized tolerances tended to

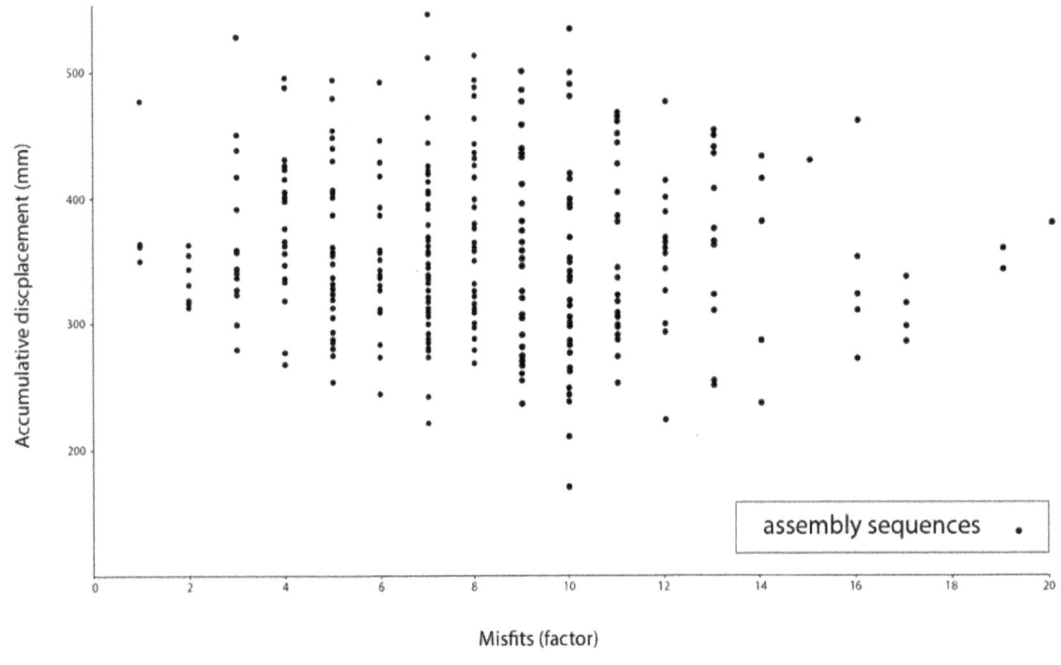

23

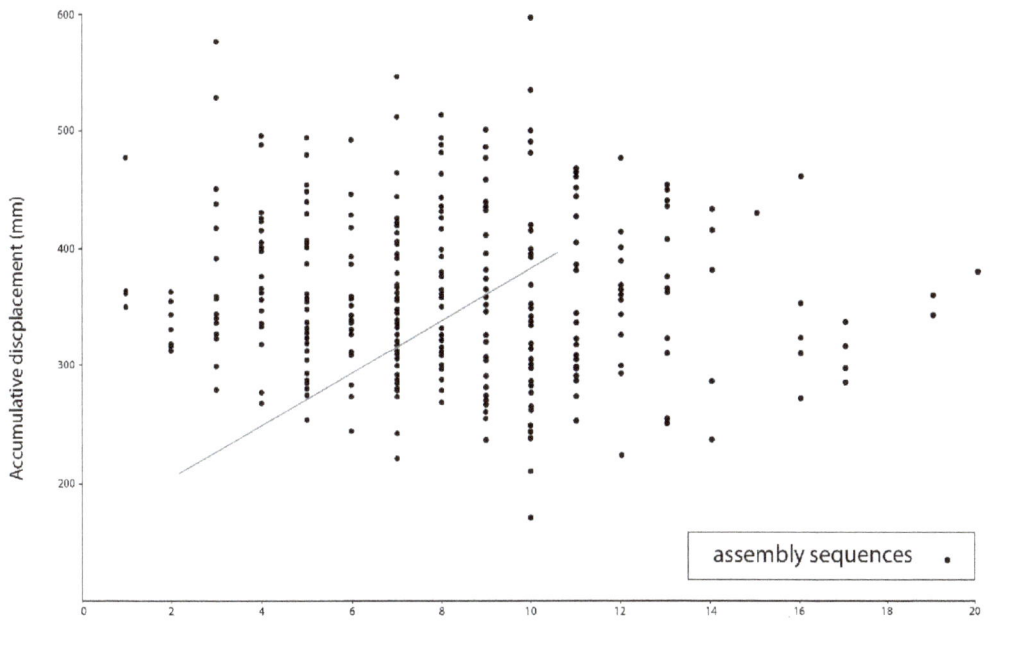

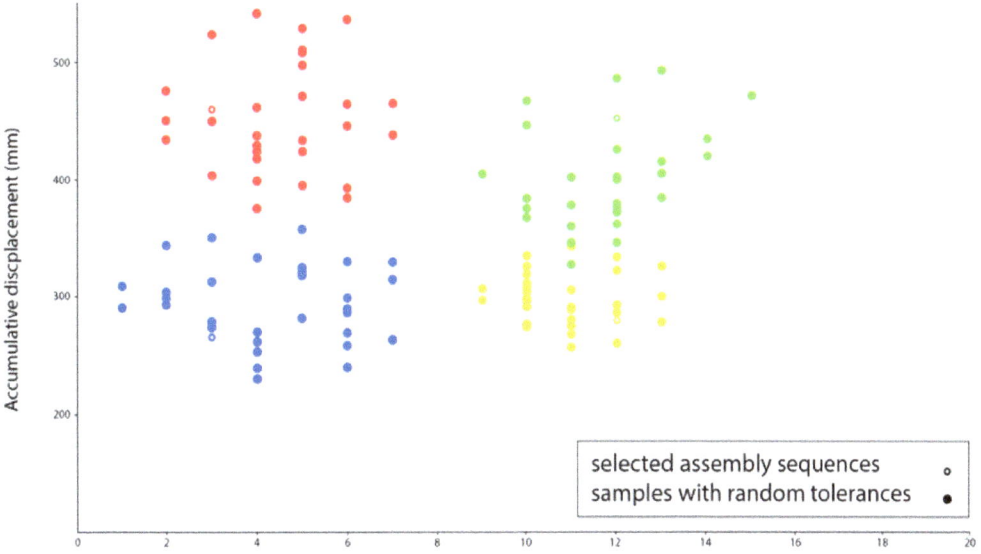

vary about 135 mm accumulative displacement and 4 misfit factor units with respect to the four selected samples of assembly sequences with worst case tolerances, which we considered an admissible range for fabrication and allowed for clustering the samples.

RESULTS AND DISCUSSION

An important goal of the investigation was to enable designing robotically-assembled spatial timber structures while visualizing tolerances. This could lead to the development of strategies that minimize displacement and misfits during fabrication and therefore, to counteracting unwanted design adjustments. That goal was achieved and validated during the

23 Scatter graph of the 323 generated assembly sequences

24 Best fit line representing the predicted relationship between cumulative displacement and misfit factor calculated through multi-variable linear regression

25 Visualization of our validation method presenting four assembly sequences with extreme accumulative displacement and misfit factor and with their corresponding 30 tolerance propagation (red, green, blue, yellow) that form clusters

assembly of the spatial timber units. This also resulted in design decisions, such as the vertical beams supported by diagonals, which allowed to collaborative build corners while avoiding tolerance transfer to adjacent modules (see Experiment I).

Additionally, the Experiment II aimed at planning convenient assembly sequences for robotically assembling complex designs, such as a reciprocal frame. This required an automatic optimization approach combined with a multi-variable linear regression method that allowed to find all solutions and filter the ones with minimal displacement and misfits.

CONCLUSION

The conducted experiments validated the developed method for planning robotically assembled spatial structures. It proved that the method can be used to estimate the assembly of the investigated structures and in turn, assist early design decisions. However, further investigation is required to fully enable fabrication-aware design of robotically assembled timber frames. This would require solving the other challenges introduced by the robotic assembly of such structures, which include gravitational sagging and the avoidance of cantilevered structures. Additionally, these methods, together with methods for automatically planning collision-free robot paths would need to be integrated into a coherent workflow, which through intuitive data visualizations could enable designers to navigate such complex design space.

The experiments also validated the approach to estimate the planning of assembly sequences that minimize deviations and misfits, thus, facilitating the robotic assembly of structures. However, assembly sequences that truly map the reality of the fabrication process should include larger amounts of tolerance data. The measurements could be constantly retrofitted in the tolerance database throughout the production of the entire building, resulting in a model that minimizes discrepancies between the digital design model and the fabricated structure along the process. This could lead to the development of a tolerance database, which could be available to other projects. Thus, leading to data rich computational methods for prefabrication.

ACKNOWLEDGMENTS

Special thanks goes to Prof. Dr. Ena Lloret Fritschi, who contributed by providing feedback during the editing of this text; Dr. Alexandra Anna Apolinarska, who provided the Pringle Structure for Experiment II; Dr. Romana Rust, who provided feedback during the conduction of experiments and regarding the graphical representation; Prof. Dr. Arash Adel, who provided the design of the Timber Modules for Experiment I; Gonzalo Casas, who suggested the Kruskal algorithm and helped in the data-collection; and Andreas Thoma, Matthias Helmreich, and Matteo Pacher, who shared their fabrication knowledge.

REFERENCES

Adel Ahmadian, Arash. 2020. "Computational Design for Cooperative Robotic Assembly of Nonstandard Timber Frame Buildings." PhD thesis, ETH Zurich. https://doi.org/10.3929/ethz-b-000439443. url: https://doi.org/10.3929/ethz-b-000439443.

Apolinarska, Aleksandra Anna. 2018. "Complex Timber Structures from Simple Elements Computational Design of Novel Bar Structures for Robotic Fabrication and Assembly." PhD thesis, ETH Zurich. https://doi.org/10.3929/ethz-b-000266723. url: https://doi.org/10.3929/ethz-b-000266723.

Autodesk. n.d. "Inventor Tolerances Analysis: CAD-embedded tolerance stackup analysis software." Accessed May 9, 2022. https://www.autodesk.com/products/inventor-tolerance-analysis/overview.

Baumer. n.d. *Baumer POSCON 3D*. Accessed May 9, 2022. http://vt.baumer.com/in-en/products/distance - measurement / light - section - sensors / poscon - 3d - edge - measurement.

Bonwetsch, Tobias, and Michael Lyrenmann. 2010. "Robotic Fabrication Laboratory, 2010-2016." Gramazio Kohler Research, ETH Zürich. Accessed March 3, 2023. https://gramaziokohler.arch.ethz.ch/web/d/projekte/186.html.

Chase, K. W., S. Magleby, and C. Glancy. 2004. "Tolerance analysis of 2-D and 3-D mechanical assemblies with small kinematic adjustments." *Advanced Tolerancing Techniques* 218: 1869–1873.

Dassault Systèmes. n.d. *CATIA V5*. Accessed May 9, 2022. https://www.3ds.com/products-services/catia.

Gandia, A., S. Parascho, R. Rust, G. Casas, F. Gramazio, and M. Kohler. 2019. Towards Automatic Path Planning for Robotically Assembled Spatial Structures. In *ROBARCH 2018; Robotic Fabrication in Architecture, Art and Design 2018*, edited by J. Willmann, P. Block, M. Hutter, K. Byrne, and T. Schork. Cham: Springer. https://doi.org/10.1007/978-3-319-92294-2_5

Gramazio Kohler Research. n.d. "Spatial Timber Assemblies, Zurich, 2016-2018." Accessed January 11, 2020. http://dfab.arch.ethz.ch/web/e/forschung/311.html.

Gramazio Kohler Research. n.d. "Robotic Fabrication Laboratory, 2010-2016." Accessed May 9, 2022. http://dfab.arch.ethz.ch/web/e/forschung/186.html.

Graser, K., M. Baur, A. A. Apolinarska, K. Dorfler, N. Hack, A. Jipa, E. Lloret-Fritschi, et a. 2020. "DFAB HOUSE: A Comprehensive Demonstrator of Digital Fabrication in Architecture." In *Fabricate 2020: Making Resilient Architecture*. London: UCL Press. 130–39. https://doi.org/10.2307/j.ctv13xpsvw.21.

DFAB HOUSE. n.d. "DFAB HOUSE Digital Fabrication and Living." Home Page. Accessed May 9, 2022. http://dfabhouse.ch/.

Eversmann, P., F. Gramazio, and M. Kohler. 2017. "Robotic prefabrication of timber structures: Towards automated large-scale spatial assembly." *Construction Robotics* 1: 49–60. https://doi.org/10.1007/s41693-017-0006-2.

Frick, Ursula, Tom Van Mele, and Philippe Block. 2016. "Data management and modeling of complex interfaces in imperfect discrete-element assemblies." In *Proceedings of the IASS Annual Symposium 2016: Spatial Structures in the 21st Century*. Tokyo: International Association for Shell and Spatial Structures (IASS). 26–30.

Gramazio Kohler Architects. n.d. "Spatial Aggregations 1, ETH Zürich, 2012; Elective course (4KP)." Accessed May 9, 2022. https://gramazio-kohler.arch.ethz.ch/web/e/lehre/228.html.

Gramazio Kohler Architects. n.d. "Robotic Pavilion, ETH Zurich, 2016; MAS Programme in Architecture and Digital Fabrication 2015-2016." Accessed May 9, 2022. http://gramaziokohler.arch.ethz.ch/web/e/lehre/309.html.

Gramazio Kohler Architects. n.d. "Mesh Mould and In Situ Fabricator, 2016-2017; in the DFAB HOUSE." Accessed May 9, 2022. https://gramaziokohler.arch.ethz.ch/web/e/forschung/324.html.

Helm, Volker et al. 2016. "Additive Robotic Fabrication of Complex Timber Structures." In *Advancing Wood Architecture: A Computational Approach*, edited by A. Menges, T. Schwinn, and O. D. Krieg. London: Routledge. 29–42. https://doi.org/10.4324/9781315678825.

Lussi, Manuel et al. 2018. "Accurate and Adaptive in Situ Fabrication of an Undulated Wall Using an on-Board Visual Sensing System." In *Proceedings of the 2018 IEEE International Conference on Robotics and Automation*. 3532–3539. https://doi.org/10.1109/ICRA.2018.8460480.

Marziale, Massimiliano, and Wilma Polini. 2009. "Review of Two Models for Tolerance Analysis: Vector Loop and Matrix." *International Journal of Advanced Manufacturing Technology* 43: 1106–1123. https://doi.org/10.1007/s00170-008-1790-0.

Nikon Metrology NV. n.d. *iGPS*. Accessed May 9, 2022. https://www.nikonmetrology.com/en-gb/product/igps.

Parascho, Stefana, Augusto Gandia, Ammar Mirjan, Fabio Gramazio, and Matthias Kohler. 2017. "Cooperative Fabrication of Spatial Metal Structures." In *Fabricate 2017*, edited by A. Menges, B. Sheil, R. Glynn, and M. Skavara. London: UCL Press. 24–29.

Piskorec, Luka, Thomas Cadalbert, Ralph Bärtschi. 2012. "Spatial Aggregations 1." Gramazio Kohler Research, ETH Zürich. Accessed March 4, 2023. https://gramaziokohler.arch.ethz.ch/web/d/lehre/228.html.

Simetrix. n.d. "CETOL 6σ Tolerance Analysis Software." Accessed May 9, 2022. https://www.sigmetrix.com/products/cetol-tolerance-analysis-software.

Skiena, Steven. 1998. *The Algorithm Design Manual*. Santa Clara, CA: TELOS, The Electronic Library of Science.

Thoma, Andreas, Arash Adel, Matthias Helmreich, Thomas Wehrle, Fabio Gramazio, and Matthias Kohler. 2019. "Robotic Fabrication of Bespoke Timber Frame Modules. In *ROBARCH2018: Robotic Fabrication in Architecture, Art and Design 2018*, edited by J. Willmann et al. Zurich, Switzerland: Springer. 447–458. https://doi.org/10.1007/978-3-319-92294-2_34.

Van Mele, Tom, et al. 2017-2021. "COMPAS: A framework for computational research in architecture and structures." https://doi.org/10.5281/zenodo.2594510.

Wu, Kaicong, and Axel Kilian. 2018. "Designing Natural Wood Log Structures with Stochastic Assembly and Deep Learning." In *ROBARCH2018: Robotic Fabrication in Architecture, Art and Design 2018*, edited by J. Willmann et al. Zurich, Switzerland: Springer. 16–30. https://doi.org/10.1007/978-3-319-92294-2_2.

IMAGE CREDITS

Figure 2: ©Timbeco Woodhouse
Figure 3: ©Alexandra Apolinarska
Figures 13: ©Arash Adel

All other drawings and images by the authors.

Augusto Gandia is an architect and researcher with interests in computational design of fabrication-aware architecture. He conducted studies at the Mendoza University and the Bauhaus University Weimar, and complete doctoral studies with NCCR Digital Fabrication and Gramazio Kohler Research, ETH Zurich. He worked at pioneering offices such as designtoproduction, and he is currently a Postdoctoral Associate in the Department of Architecture, Massachusetts Institute of Technology (MIT).

Fabio Gramazio is an architect with multidisciplinary interests ranging from computational design and robotic fabrication to material innovation. The recent research is outlined and theoretically framed in the book *The Robotic Touch: How Robots Change Architecture* (Park Books, 2014). Fabio Gramazio joined the NCCR Digital Fabrication in 2014 as a principal investigator.

Matthias Kohler is an architect with multidisciplinary interests ranging from computational design and robotic fabrication to material innovation. The recent research is outlined and theoretically framed in the book *The Robotic Touch: How Robots Change Architecture* (Park Books, 2014). Matthias Kohler initiated the NCCR Digital Fabrication and was its director from its official founding in the summer of 2014 until the summer of 2017. He now works as one of the NCCR's principal investigators.

Approaching Architectonic Interfaces

Considerations for an Interface-centered Design Methodology

Nicolas Stephan
i.sd, University of Innsbruck

Marine Lemarié
i.sd, University of Innsbruck

Alexandra Moisi
i.sd, University of Innsbruck

Stefan Rutzinger
i.sd, University of Innsbruck

Kristina Schinegger
i.sd, University of Innsbruck

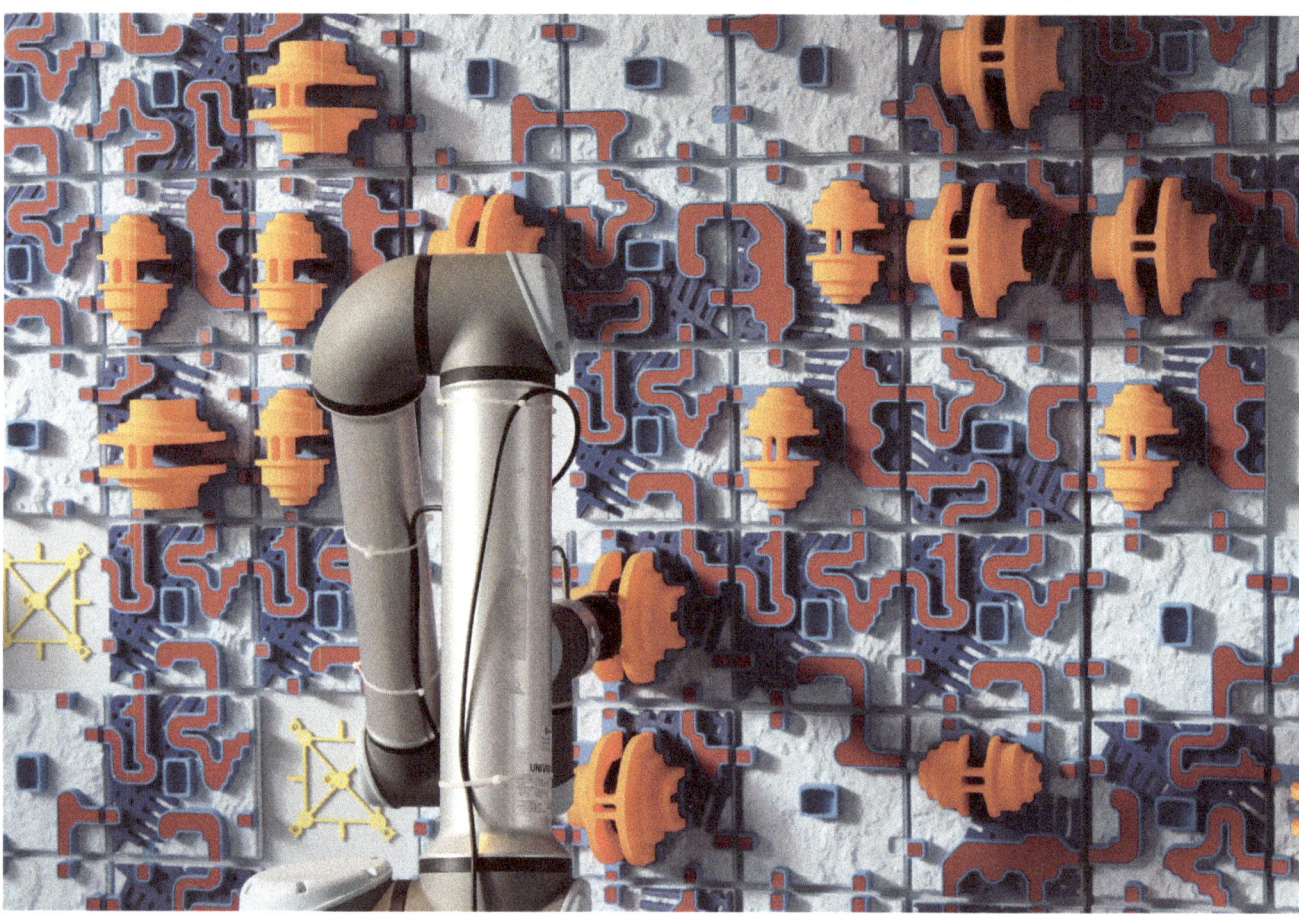

1 The robotic arm manipulates magnetic tiles that form a relief wall

ABSTRACT

The paper describes the notion of the "architectonic interface" and its design implementations and concludes with a corresponding design methodology. It summarizes key moments in interdisciplinary discourse in physics, ergonomics, philosophy, and cultural theory extended into the field of architecture, and further as related to a prototypical definition of an architectonic interface. This field-related definition guides the design process and informs it with four aspects: surface, relation, territorial space, and embodied interaction. These four aspects describe the associations between the elements of the interface, both animate and inanimate entities. To put to the test the architectonic interface, a robotic installation was developed as part of an architecture museum exhibition, which demonstrates the interface's ability to create a space of interaction between users, robots, objects, and movements in both the physical and the virtual reality. This effectively generates a territory of dependencies between these actors through the medium of the architectural object as a data carrier emblematic of the architectonic interface.

INTRODUCTION
This paper discusses the notion of the "architectonic interface" by tracing relevant historic roots and philosophical concepts and showcasing its potential for an interactive application in a first use case as an exhibition installation (Figure 1). The research aims to suggest an alternative definition of the architectonic interface and its constitution as a formal/aesthetic design task and to develop a corresponding design methodology.

In 1984 at MIT Media Lab, Nicholas Negroponte discussed qualities of interfaces and ultimately sought more pleasurable ways of interacting with them. Negroponte argues that "High-tech is high-touch" and that hands and fingers do not only have the advantages of direct control through the user's hands, but they also give high-resolution input (Negroponte, 1984). In the search for a more immediate, enjoyable, and high-res/high-touch interface, the discourse of the architectonic interface seems to provide a lot of potentials: as an intrinsically spatial expressive system (inter-face) it already understands human users as moving, and multimodally perceiving. The architectonic interface thereby widens the concept of the user interface, approaching interaction in a spatially complex, multisensory, and multilayered manner.

The use case installation described in this paper is the first development in a larger research framework that aims to establish a design methodology for architectonic interfaces, to trace the term's historical and theoretical underpinnings, and to discuss consequences and potential for design and aesthetics. The installation was exhibited in a museum space and focused not only on the tactile-optical (Riegl 1901) but also on the kinesthetic potential of the interface. The users of the installation could touch the exhibited artifacts and screens with their hands and fingers, therefore acting as a medium of direct input for the machine. The users could take several positions and roles in a play between human and machinic agents. Instead of understanding the interface as a GUI (Graphical User Interface), or an invisible device such as the brain-computer-interface (BMI) (Musk and Neuralink 2019), this installation constructs the interface as a social event connecting several human and machine agents.

STATE OF THE ART
Today the interface is often associated with a two-dimensional (G)UI (Butterfield et al. 2016, 1135). Historically, however, it was understood as an environment encompassing volumes, systems, and movements in space (Murrell 1965, xiii), making it an essentially architectural concept. In that sense, the interface would not be of predetermined structure or size, leading to a broad variety of "interfaces" within the design research discourse. One approach for an object-scale human-machine interface can be seen in the 1961 exhibition by Ray and Charles Eames (Harwood 2011, 181). The exhibition featured binoculars that combined physical mechanics with digital imagery (Eames Office, n.d.). In the same decade, Nicholas Negroponte developed interfaces for human-machine collaboration in architectural design (Negroponte 1970). In 2005, Georg Flachbart of mind(21)factory and Peter Weibel of the ZKM curated a discussion on architecture as an interface between the physical and the digital (Flachbart and Weibel 2005). Today, design theorist Benjamin Bratton of the University of California San Diego uses the term to conceptualize the sovereign power of planetary infrastructure (Bratton 2016), while Shannon Mattern of The New School addresses contemporary issues of mobility and urbanism through the notion of the interface (Mattern and Adams 2019). The book *Interface* by MIT's Branden Hookway (Hookway 2014) aims to deliver a comprehensive theory of the interface and constitutes the most extensive conceptualization of the topic to date.

For the purpose of this paper, the authors intend to extend Hookway's theory into a prototypical design theory and methodology, whereby the interface produces unconventional relationships between spaces, objects, technologies, humans, and non-humans while creating a complimentary formal design language.

THEORETICAL FRAMEWORK
The following sections outline the theoretical framework of the research, the foundation for the design experiments.

Etymology
The Latin term "inter-" means between, among, amid, in between, in the midst (Oxford English Dictionary 2022). It suggests that it relates to a condition inside a predefined area surrounded by a bounding condition (Hookway 2014, 8). The word face derives from the Latin "facies" which means outward appearance, form, or shape (Oxford English Dictionary 2022)—a hint to something representational, pointing towards an exterior condition. "Facies" in turn derives from the Latin verb "facere" which means to act. The word "face" can be read as an outward-oriented boundary that actively turns toward the exterior (Hookway 2014, 8). Therefore, the word interface describes an inward- and outward- orientation at the same time; a contained interior condition and an active agent towards the outside. Hookway describes it as follows: "In encompassing interiority and exteriority, passivity and activity, the interface governs transformations from interior state to exterior relation, from inward to outward expression." (Hookway 2014, 9)

Physics and Ergonomics

The term interface was initially used in the nineteenth century by physician James Thomson to characterize processes in fluid dynamics. "Interface" described the colliding condition of adjacent fluids (Thomson 1912, 327). The notion was later introduced to the field of ergonomics—the study of the relation between humans and their working environment (Murrell 1965, xiii)—in which it described a medium, connecting humans to their technological environment (Harwood 2012). When the advent of computers and logistics made efficient communication between humans and electronic databases increasingly relevant, the concept was tied to digital technology resulting in today's user interface (Harwood 2011, 9).

Even though the contemporary discussion is dominated by the (graphic) user interface, the term "interface" still contains aspects and qualities from physics and ergonomics (Hookway 2014, 59). A definition for the architectonic interface can profit from rediscovering and emphasizing aspects such as temporality, fluidity, and spatiality.

Philosophy and Cultural Theory

According to Malcolm McCullough, who connected interaction design with cultural anthropology and technology, the success of the computer as a design tool would be determined by its ability to achieve more continuity with the physical world, citing that a craftsperson has to touch their work and that virtual craft seems like an oxymoron (McCullough 1996). The cultural theorist Branden Hookway defines the interface as "[...] the processes by which it draws together two or more otherwise incompatible entities into a compatibility, within which they become available to one another to the extent allowable within the operation of the interface, and from this compatibility produces an overall governance or control." (Hookway 2014, 17). He further points out that the interface is not a physical object but the temporary relation between objects, actions, and ideas resulting in a process of communication or transmission (Hookway 2014, 4). This process in turn creates spatiality. The interface has at least three spatial implications: the spatiality of its constituting components, the spatiality of the interfaced entities, or its influence on the surrounding space. This paper focuses on the latter, which will be referred to as "territorial space."

Proposal for an Architectural Definition

Based on earlier references in the physics, ergonomics, philosophy, and cultural theory, this paper defines the architectonic interface as follows:

- The interface is not a physical object, but the translation process of how otherwise incompatible entities are set into relation with each other and, therefore, has the power to control the conditions of this relation (Figure 2).
- It provides a "face" and is, therefore, expressive and affective; consequently, it constitutes an aesthetic. It has at least two appearances/modi since it connects two incompatible entities.
- The interface is per se transformative and changeable; it can be in a stable state or in an altering one.
- The interface is already architectonic since it is based on a spatial definition; it is a meeting point between an inner and an outer condition.
- The architectonic interface fully exploits the experiential, multimodal, kinesthetic, and high-res/high-touch potential of a human-centered interaction.
- The architectonic interface has at least three spatial implications: the spatiality of its constituting components, the spatiality of the interfaced entities, or its influence on the surrounding space.

An Indirect Approach to Interfaces

The interface can either be described by its elements or its potential effects and affects. Hookway points out that because of its fluid and temporal being the interface remains hidden and can only be perceived indirectly (Hookway 2012, 16). Thus, a description of its elements does not reveal its potential and might lead to oversimplification; as Graham Harman points out, a magic trick is destroyed by its explanation. To avoid the trap of oversimplification, Harman proposes to describe objects indirectly: "[...] just as jokes or magic tricks are easily ruined when each of their steps is explained, thinking is not thinking unless it realizes that its approach to object can only be oblique" (Harman 2012, 12). The interface as a concept behaves very much like Harman's joke or magic trick. Talking about its technical setup and specificities reduces it to a merely technical problem. Speaking about its effects on its human users evaluates it only in terms of user-friendliness. Even though both aspects are legitimate, they remain only aspects and do not adequately describe the

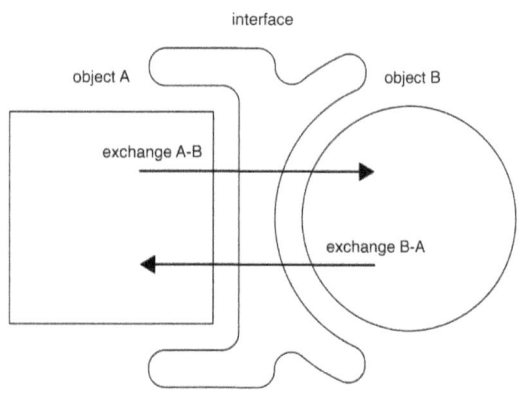

2   The interface makes incompatible entities compatible with each other.

interface itself. To maintain the puzzling immediacy of the interface, not every aspect should be pinpointed and scrutinized. Instead, it is productive to develop the project through a set of design considerations, while keeping certain elements deliberately undefined, allowing for unforeseen qualities to complement the interface. The paper outlines four of these considerations—surface, relation, territorial space, and embodied interaction—and discusses their implementation in a prototypical installation.

DESIGN CONSIDERATIONS

Following Herbert Simon's definition of the interface as a meeting point between inner and outer conditions (Simon

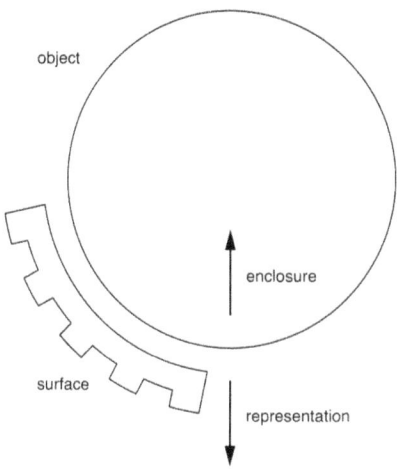

3   The surface represents an enclosed, inwards oriented object towards the outside.

1996, 6) and its constituent parts "inter" and "face," the authors start the design exploration with the aspects of surface, relation, territorial space, and embodied interaction. A formal strategy can negotiate between these four aspects by using the physical (tactile) and the imaginary (virtual) as translators.

Surface

Hookway introduces the concept of the surface (Figure 3), which takes the function of communicating towards the exterior through its appearance. It uses its own properties and characteristics to serve as a reference to the enclosed, while at the same time concealing it (Hookway 2014, 13). The design of the surface of an interface addresses representation, readability, and haptics, which control how the interface is perceived. Rudolf Arnheim notes that humans perceive the shapes they see via visual concepts. The assignment of shapes to concepts is done through perceptual patterns, meaning the seen object is not the object itself but an approximation to the concept of this object (Arnheim 1969, 27). The more these patterns contrast from their surroundings the easier they are identified as something known or expected (Arnheim 1969, 28). Arnheim further points out that human and machine perception is based on the same concept (Arnheim 1969, 31). Arnheim's observations on perception can be applied to the design of the interface's surface. The surface can be addressed in the design process through a set of dialectic parameters: abstract and figurative shapes, high and low contrasts, flatness and depth, uniqueness and repetition, 2D and 3D, continuity and discreteness. By negotiation between the parameters and their corresponding aesthetics, designers can control how the interface is perceived. This representation determines how users engage with the interface, which can be described as relation.

Relation

Hookway states that the interface exists by establishing relations between entities through their individual qualities (Hookway 2014, 39). It implies that interfaces need a logical underlying medium or structure to host, systematize, and describe these relations and interactions. Examples of such structures are the laws of physics, mechanisms for social interaction, mathematical calculations, and geometrical systems. A design-related example is the grid. It can be used to describe and manipulate the position and orientation of objects, their economic implications, and their relationship to surrounding objects. Computation allows informing grids with an underlying logic, effectively storing information in geometry. This information can be used to map and update the interface and further embed conditions on its functionality enabling a dynamic spatial approach to interconnection. The grid enables comprehension on a mathematical and visual

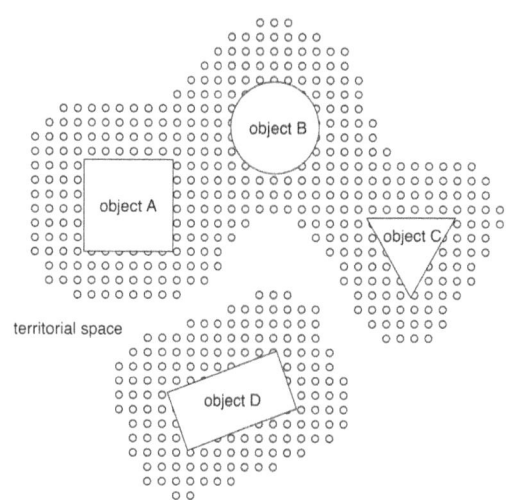

4   The interface's territorial space cannot be demarcated through a closed curve. It should be seen as a diffuse, perforated, and discontinuous zone, shifting in shape and intensity.

5   The main feature of the case study project is the modular relief wall, which is constantly rearranged by the robot.

level, making it equally suitable for machines and humans. Finally, it can stay hidden from the onlooker or be integrated into the formal approach of the surface. The integration of the system of relations into the formal approach visualizes the governing conditions of the interface, making it graspable for the onlooker.

Territorial Space

As interfaces govern the conditions by which the entities they connect can exchange information, they inherit power over the connected entities. Hookway points out that by doing so, interfaces open an immaterial space related to the dimension of time or culture (Hookway 2014, 17). This immaterial space, or small interventions, influences the physical one, effectively establishing a territory around the interface (Figure 4). By designing the interface's constituent components and their relation to each other, designers can indirectly control its governing conditions and, therefore, participate in the transformation of the surrounding space and can affect perception, circulation, and materiality. An example is the transformation of the Piazza San Pietro in Rome: it can be argued that placing a single obelisk in front of the Saint Peter's Basilica led to the complete restructuring of the piazza by serving as a central reference point (Bacon 1976, 131-135). Relatively simple operations like this can be used in the design of an interface to change the structuring of a surrounding space.

Embodied Interaction

Since the interface encompasses spaces, interactions, movements, and systems, in terms of human-computer interaction (HCI), it also includes body gestures, or "embodied interaction." As labeled by Malcolm McCullough (McCullough 2004) and Paul Dourish, embodied interaction refers to the integration of the user's body, senses, and natural gestures to achieve results in the interaction with technology (Dourish 2004). According to Dourish, embodiment is more than a physical feature; it is about our understanding of the world, our location, and participation in the environment. Human and non-human actors can be considered part of the interface since they, through their interaction with other constituting entities, activate or complete the interface. Therefore, the body is a suitable medium for this interaction and including its qualities in the design consideration can enrich the possibilities of interaction.

CASE STUDY PROJECT

To complement the theoretical research, the notion of the architectonic interface was tested with an installation piece.

The four guiding aspects "surface," "relation," "territorial space," and "embodied interaction" were implemented in the design of this installation.

GENERAL DESCRIPTION

The experimental prototype, "Phygital Space: An Interfac-Enabled Augmentation," was planned for a group exhibition at the architecture museum aut. architektur und tirol and was accessible to the public for four months. The installation consists of a constantly changing modular wall relief of 125 x 250 cm (Figure 5) and a virtual space available through a customary tablet (Figure 8). The seventy-two 3D printed modules of the relief are rearranged by a six-axis industrial robot mounted on an additional horizontal linear axis. The information about the location and type of every individual relief tile is linked to a Mixed Reality (MR) space, which reacts formally to changes within the physical matrix and enables visitors to interact with the relief through a set of commands, actively manipulating physical as well as digital components of the architectural space.

INDIVIDUAL TILES

There are four different component types, varying in depth, density, and placement of their constituent layers (Figure 6). Each component type consists of five layers, varying in formal language and color: A light blue "terrain," a dark blue "grid," orange "arcs," and red "figures" framed by a blue "envelope." These colors and formal languages create a higher contrast for easier visual and tactile differentiation.

WALL ARRANGEMENT

The tiles are arranged on a square grid with seventy-two cells. Every cell of the grid is 20 x 20 cm in size and fits one tile. They get snapped to the correct location through a yellow wall connector fitted with magnets on the back of each tile. The square shape of the cell as well as the placement of the magnets allows for four different rotations (0°, 90°, 180°, 270°) of the tile. These rotations in combination with the four different tile types allow for sixteen unique states per slot (Figure 7). These states can be used to "encode" information and apply rule-based transformation processes (e.g. "Cellular Automata" or "Shape Grammars") to the relief.

MIXED REALITY SPACE

Every 3D tile in the MR space (Figure 8) corresponds to a physical counterpart of the relief and is encoded with the physical data. The decoding process takes place through interaction and relations between humans and machines. The robot, tablet, users, and their relations form an interface where users become active space-makers. Changes in the virtual wall update the physical relief, which in turn again changes the MR spatial experience. In the MR environment, the selected tile unfolds above the user, while four additional modules snap into an "enclosed" space surrounding the user. This new MR room "overwrites" the initial exhibition space with information and qualities decoded from the relief wall. While machines read the relief wall directly, the generated 3D spaces help human users understand the meaning and layers of the tiles in a spatial and architectural scale.

VISITOR INTERACTION

Users can choose a tile from the available components and place it on a pedestal in front of the robot. The robot then picks the tile up and places it in an empty slot on the relief

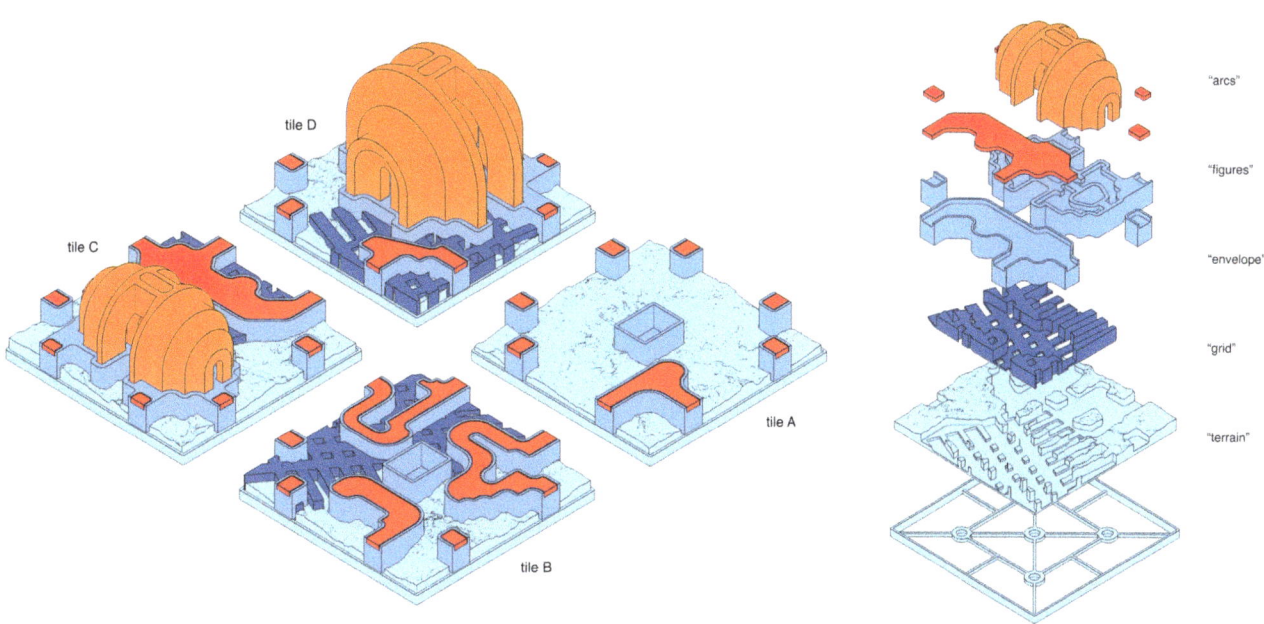

6   The elements of the tiles are fabricated separately and can be stacked.

wall. The resulting new configuration changes the computational relationship of the parts and triggers a change in the corresponding digital space. The same process works vice versa: users can press a tablet button that represents the position of a physical tile on the relief wall. This will trigger an instant change in the virtual space and send commands to the robot to replace the selected tile and, therefore, synchronize all physical and digital components of the installation.

Implementation of Guiding Principles

SURFACE

The physical relief was considered the representational medium or "surface" of the interface. Its contrast-heavy design was intended to be easily identified by both human and machine vision software and had a range of visually distinct features, allowing the public to follow the dynamics of the relief. It aimed to make them intuitively understand the relations between the relief, the robot, and the digital space and their own role in the translation process. The MR space constituted a virtual extension of this surface, offering visitors further interaction and spatial qualities.

RELATION

A digital version of the grid was used to store various states, describing the relationship between the tiles and their virtual counterparts. These were coupled with the protocols of the robot arm as well as triggers for digital elements on the tablet. Visitors were encouraged to experiment with the physical as well as digital components of the installation, which in turn had an effect on the protocols and the resulting formal arrangements. By participating in the translation process, visitors could become active parts of the interface.

TERRITORIAL SPACE

The installation defines a territory within the boundaries of the exhibition space. This territory is diffuse and changes according to the current configuration of tiles and virtual space, movement of the robot, and position of the visitor. Nevertheless, it influences the perception and behavior of the visitor. It draws attention to specific areas, the movement of the robot keeps visitors at a distance, while the pedestals attract them closer. The MR installation further extends the space away from the relief and defines virtual boundaries that can be respected or disregarded by the visitors.

EMBODIED INTERACTION

The installation is an invitation for users to actively engage with the pieces by picking and placing tiles on the surface that the robot "sees" and by moving around in space with the tablet to see the virtual objects surrounding them. The MR and the robot enable embodied interaction with the user, as a medium of physical experience away from the screen.

VISITOR FEEDBACK

Interviews were conducted with several exhibition visitors. The findings showed that the concept was intuitively graspable. Visitors noted that they could understand the concept by testing the effects of a button or by placing a physical tile on the pedestal. As a possible improvement, it was mentioned that it would be interesting to embed several layers of complexity, to make more graspable layers of understanding that require philosophical or architectural foreknowledge.

CONCLUSION

The work focuses on interface-centered design methodology and discusses four guiding factors, or aspects of this methodology: "surface," "relation," "territorial space," and "embodied interaction." The described robotic installation was the first prototype for this interface-centered design methodology. The installation involved high levels of dependencies between its actors: for the installation to function properly, the user has to put a relief cell on the robot's visible surface which the robot then recognizes, picks, and moves to an empty slot on the relief wall. This physical transformation of the relief wall was then encoded in a mixed reality space that included virtual counterparts of the physical cells. Users could interact with the virtual objects through a tablet, which activates feedback for the robot to add, move, or remove cells on the wall to other free slots.

Testing the ideas in the context of an exhibition project proved suitable since it raised many questions which could then be integrated into the research and used to further refine it. Exhibiting the project in a museum space enabled the testing of the embodied interaction with non-project-related users. The installation proved that using the mentioned aspects—"surface", "relation", "territorial space", and "embodied interaction"—as a basis for a design can create very distinct results, which creates a positive outlook on the future of an interface-centered design methodology. However, the implementation of these aspects proved problematic since they lead to very complicated technical setups, often drawing the designers' as well as onlookers' attention away from the concept toward the technicalities. Future case studies should address this problem starting at the conceptual level to avoid overcomplication. Further research will focus on conceptualizing other aspects of the interface, especially its various spatial implications.

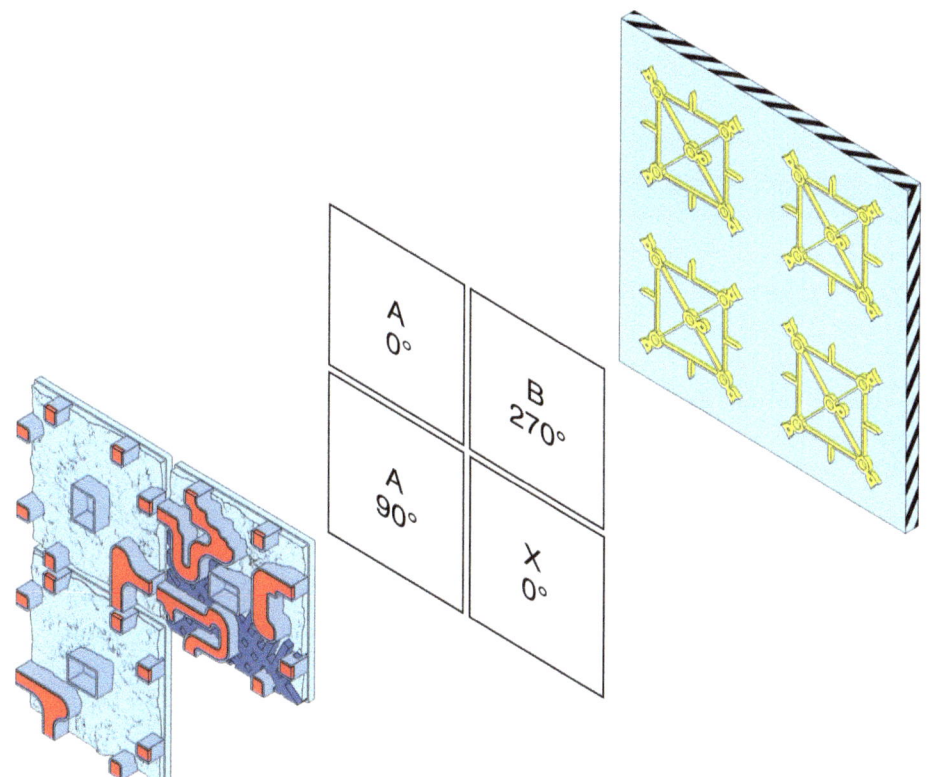

7 The wall is made of seventy-two tiles arranged in a grid: information about tile type and rotation is stored digitally

8 View displayed on the tablet: users manipulate the space by interacting with both its physical and its Mixed Reality components

ACKNOWLEDGMENTS

The code used to control the robot and the MR application was written by Adam Geraia and Robby Kraft. The project was developed at i.sd, University of Innsbruck.

REFERENCES

Arnheim, Rudolf. 1969. *Visual Thinking*. Berkeley: University of California Press.

Bacon, Edmund N. 1976. *Design of Cities*. Rev. ed. New York: Penguin Books.

Banham, Reyner. 1969. *The Architecture of the Well-Tempered Environment*. London: The Architectural Press.

Bratton, Benjamin. 2016. The Stack: On Software and Sovereignty. Cambridge, MA: The MIT Press. Kindle.

Butterfield, Andrew, Gerard Ekembe Ngondi, and Anne Kerr. 2016. *A Dictionary of Computer Science*. 7th ed. Oxford: Oxford University Press.

Dourish, Paul. 2004. *Where the Action Is: The Foundations of Embodied Interaction*. Cambridge, MA: MIT Press.

Eames Office. n.d. "Mathematica: A World of Numbers … and Beyond." Accessed May 17, 2022. https://www.eamesoffice.com/the-work/mathematica/.

Eddington, A.S. 1929. *The Nature of the Physical World*. New York: MacMillan.

Flachbart, Georg, and Peter Weibel, ed. 2005. *Disappearing Architecture: From Real to Virtual to Quantum*. Basel: Birkhäuser.

Harman, Graham 2012. *The Third Table / Der dritte Tisch*. Ostfildern: Hatje Cantz Verlag.

Harwood, John. 2011. *The Interface: IBM and the Transformation of Corporate Design, 1945-1976*. Minneapolis: University of Minnesota Press.

Harwood, John. 2012. "The Interface: Ergonomics and the Aesthetics of Survival." In *Governing by Design: Architecture, Economy, and Politics in the Twentieth Century*, edited by Dianne Harris. Pittsburgh: University of Pittsburgh Press. 70-92.

Hookway, Branden. 2014. *Interface*. Cambridge, MA: The MIT Press.

Mattern, Shannon, and Zed Adams. 2019. Thinking Through Interfaces 2019. Accessed March 25, 2022. https://interfaces.wordsinspace.net/2019/.

McCullough, Malcolm. 1996. *Abstracting Craft: The Practiced Digital Hand.* Cambridge, MA: MIT Press.

McCullough, Malcolm. 2004. *Digital Ground: Architecture, Pervasive Computing, and Environmental Knowing*. Cambridge, MA: The MIT Press.

Murrell, Hywel. 1965. *Human Performance in Industry*. New York: Reinhold Publishing Corporation.

Musk, Elon, and Neuralink. 2019. "An Integrated Brain-Machine Interface Platform with Thousands of Channels." *bioRxiv* 703801. https://doi.org/10.1101/703801.

Negroponte, Nicholas. 1970. *The Architecture Machine*. Cambridge, MA: The MIT Press.

Negroponte, Nicholas. 1984. "5 predictions, from 1984." TED Talks video, 25:10. https://www.ted.com/talks/nicholas_negroponte_5_predictions_from_1984. Accessed May 13, 2022.

Oxford English Dictionary. n.d. "'Inter-.'" Accessed March 29, 2022. https://www.oed.com/view/Entry/97516?rskey=XQ1TMC&result=5#eid.

Riegl, Alois. 1901. *Die spätrömische Kunst-Industrie nach den Funden in Österreich-Ungarn im Zusammenhange mit der Gesamtentwicklung der Bildenden Künste bei den Mittelmeervölkern*. Vienna: Hof- und Staatsdruckerei.

Simon, Herbert A. 1996. *The Sciences of the Artificial*. 3rd ed. Cambridge: The MIT Press.

Thomson, James. 1912. "Notes and Queries—On Gases, Liquids, Fluids." In *Collected Papers in Physics and Engineering*, edited by Joseph Larmor and James Thomson. 327-333. Cambridge, UK: Cambridge University Press.

IMAGE CREDITS

All drawings and images by the authors.

Nicolas Stephan is a PhD researcher at i.sd, University of Innsbruck. His work focuses on the potential of architecture to serve as an interface between users, machines, and the built environment. Further, he is a co-founder of DISTANT REALITIES, a research-based architecture and design studio.

Marine Lemarié is a senior lecturer at i.sd, University of Innsbruck, where she teaches design studios and focuses on the application of machine learning in design strategies. She is the co-founder of DISTANT REALITIES, a research-based architecture and media arts studio and a member of the General Context Network.

Alexandra Moisi is a PhD researcher at i.sd, University of Innsbruck. She is involved in a series of research projects on the experimental use of augmented reality in design and fabrication. Her main interests lie at the intersection between humans, architecture and technology, the latter being integrated as a creative medium for expression in the process of conceptual design.

Stefan Rutzinger is professor for structure and design and head of i.sd at the Faculty of Architecture, University of Innsbruck. He is a principal investigator in the Special Research Project "Advanced Computational Design" and co-project leader of the SFB subproject "Computational Immediacy."

Kristina Schinegger is a professor for structure and design and co-head of the research group i.sd. She is a principal investigator in the Special Research Project "Advanced Computational Design" and co-project leader of the SFB subproject "Computational Immediacy."

Robotic 3D Printing Multilayer Building Envelope

Geometric Design and Robotic Control Strategies

Mania Aghaei Meibodi
DART, University of Michigan

Wes McGee
University of Michigan

Alireza Bayramvand
DART, University of Michigan

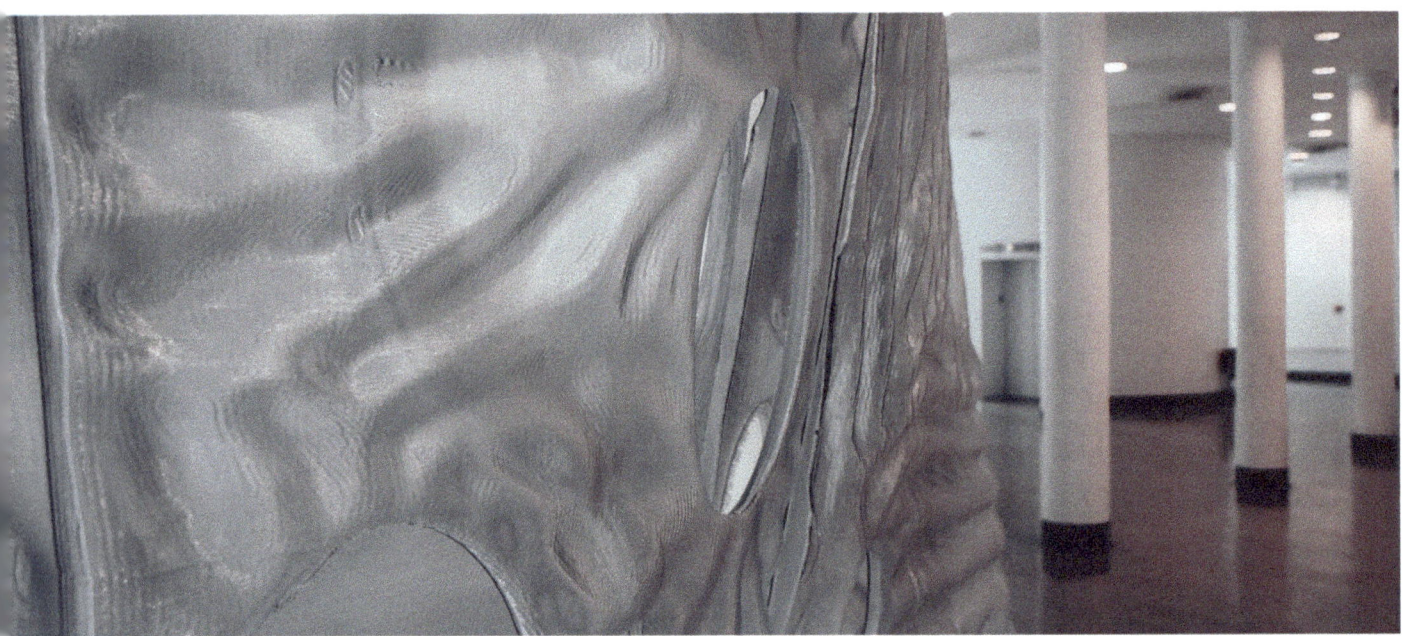

1 Multilayer thermoplastic building envelope, *Plastic Architecture* exhibition, Cooper Union, New York, 2021

ABSTRACT

This research explores the use of robotic pellet extrusion 3D printing (3DP) in the production of a multifunctional thermoplastic building envelope. A computational design method was developed to generate multi-layered systems of interconnected volumes using minimal surfaces, combined with a rib-stiffening approach that accommodates fabrication constraints. The investigation highlights the development of an integrated system that includes robotic end-effector tooling, programming/control methods to allow "endless" prints, as well as specific toolpath strategies to improve print speed and quality. These developments are then demonstrated through the design and fabrication of two 2.2 m by 1.1 m building envelope panels, and the results are discussed along with specific details of the printing process. The innovations of this research are: (1) a computational design tool that allows intuitive generation and adaptation of multilayer building envelopes to site criteria; and (2) a robust robotic control system allowing continuous, uninterrupted printing at architectural scale with minimal supervision and high-quality surface finish.

INTRODUCTION

Energy-efficient building design is one solution combating global warming that reduces energy consumption while utilizing renewable energy for building operations (Torcellini et al. 2006; Li et al. 2013). Building envelope design significantly affects a building's energy performance as the mediator between interior and exterior climatic control. Building envelopes account for 70% of a building's energy performance (Yang and Choi 2015). They are site-specific, and their geometry depends on various climate conditions and the surrounding natural or urban context. Envelope geometry influences a building's operational energy, and affects structural performance and material efficiency (Fang and Cho 2019). Designing envelopes for improved energy efficiency often results in geometrically complex forms, with complicated exterior surfaces and internal topologies. However, due to higher construction costs, the industry has adopted elementary shapes (Hemsath and Bandhosseini 2015). Innovations in the design, material, and construction process of building envelopes are critical to reduce embodied and operational carbon and enhance building life spans.

3DP technologies allow for the fabrication of topologically-complex building envelopes with intricate geometric features at multiple scales. This is demonstrated in binder jet 3DP of sand mold for casting lightweight freeform aluminum facades (Aghaei Meibodi et al. 2018; Aghaei Meibodi et al. 2019), binder jet 3DP for doubly curved glass fiber reinforced concrete facades (Aghaei Meibodi et al. 2021), robotic 3DP of mineral foam for lightweight facades (Bedarf et al. 2021), robotic extrusion 3DP of fiber reinforced concrete facade panels (McGee et al. 2022), and 3DP with different kinds of clay and ceramics (Rael and San Fratello 2017, 2018; Cruz et al. 2020; Peters 2013; Lekka Angelopoulou 2022).

3DP thermoplastic is an emerging field of research with radical innovations in facades and building envelopes. Thermoplastics are lightweight, recyclable, durable, versatile, watertight, can withstand severe weather conditions, and take nearly any shape. 3DP thermoplastics for envelopes often have been investigated using one of the two extrusion-based 3DP process: Fused Filament Fabrication (FFF), where thermoplastic filament is melted through a heated nozzle; and Fused Granular Fabrication (FGF), where thermoplastic granules are fed from a hopper into a barrel containing a motor-driven screw along multiple heat zones. This research focuses on FGF as a more viable method for scaling to real-world construction.

Robotic 3DP further offers higher geometric degrees of freedom and scalability than traditional three-axis gantry 3DP. Research in Robotic FGF, herein referred to as Robotic Pellet Extrusion (RPE), is increasingly being used in construction scale 3DP and has already been applied in projects related to building envelopes such as SensiLab and B515 Studios projects (Burry et al. 2020), thermoplastic concrete formwork from carbon fiber PETG and translucent PETG (Aghaei Meibodi et al. 2021), and well known projects by commercial fabricators like Nagami and Branch Technology. However, several key challenges such as low print speed, poor surface finish, and difficulty in starting and stopping the extrusion bead currently limit achieving the full potential of RPE for large scale building envelopes.

3DP thermoplastic has immense potential for integrating multiple functions within building envelope designs, particularly with respect to thermal energy efficiency (Mungenast 2017). Integration of cavities for air or phase-changing materials (PCM) have been shown to retard thermal conduction for better performance and energy efficiency of buildings. Notable precedents are "The Double Face 2.0" project integrating PCM in facades to store and delay thermal transmission (Tenpierik et al. 2018), and "The SPONG3D" integrating air cavities for thermal insulation along with a motile liquid for thermal storage and exchange (Sarakinioti et al. 2017).

However, what is most notably missing from most studies on 3DP of thermoplastic building envelopes is a comprehensive design tool developed in the context of fabrication methods to enable innovative multifunctional design solutions. This research contributes to the emerging field 3DP Thermoplastic by focusing on novel computational design strategies and large scale RPE of multifunctional integrative thermoplastic building envelopes.

APPROACH

This research advances existing research on 3DP thermoplastic envelopes by (1) developing a non-interrupted RPE system compatible with the scale of construction and geometrically-complex envelope components, (2) developing toolpathing approaches and control systems that improve the speed and efficiency of RPE, and (3) developing a novel computational design method for intuitive and automatic generation of multilayer building envelope systems with partitioned interiors based on minimal surfaces that is also adaptable to site-specific criteria. Minimal surfaces are a useful design medium for generating partitioned volumes and the ideal geometry for achieving complex topology while eliminating any need for building supports during 3DP. This research utilizes minimal surfaces to conceptualize a speculative multilayer building envelope system made of only thermoplastic for improved energy and thermal performance. We present and discuss a 1:1 scale prototype and case study for understanding and exploring the geometric complexities achievable and their fabrication.

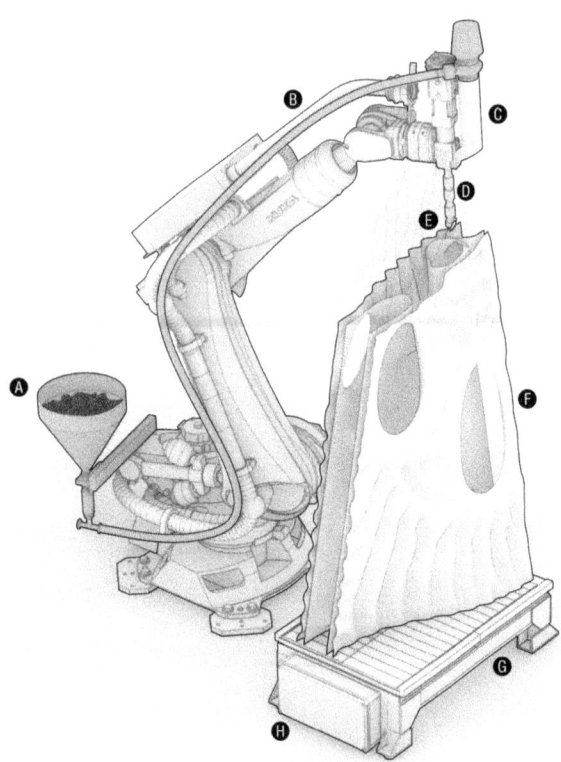 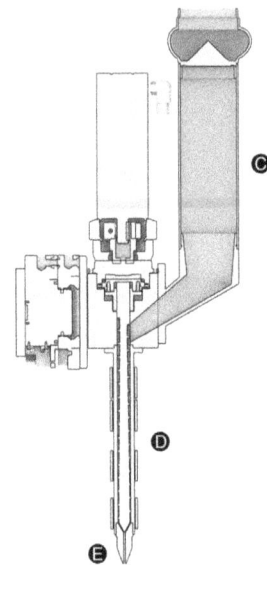

2   Diagram of the robotic pellet extrusion system with a continuous feeding system. The automated pellet feeding system transfers pellets from a 30L hopper (A) via a pipe (B) to a smaller hopper (C). The end effector consists of a servo-driven extrusion screw and three-zone heated barrel (D) and nozzle (E). The heated bed (H) is constructed of aluminum slats attached to a thick aluminum plate needed to ease the dismounting of the 3D printed part (F) after 3D printing (G).

Fabrication System: Non-Interrupted Robotic Pellet Extrusion

PETG is used as the 3D printing feedstock as it is a lightweight yet semi-rigid material with good impact resistance, enabling the production of durable and hard-wearing components that are easy to handle in the assembly process. It is highly water-resistant and enables 3D printing of a waterproof enclosure system with cladding and an integral vapor barrier made of the same material.

This research develops a customized robotic pellet extrusion system entailing a continuous feeding system to enable the large-scale production of geometrically complex plastic parts (Figure 2). The current version of the system is based on a 22 mm diameter extrusion screw with a 2:1 compression ratio and a 15:1 length to diameter ratio (Figure 2). It is driven by a synchronous servo motor, which is positionally controlled as an external robot axis. To enable continuous 3D printing on a construction scale, the system has been upgraded to include an automated pellet feeding system that transfers pellets from a 30L hopper on the robot's base to a smaller hopper mounted directly to the end effector (Figure 2). The system also utilizes an annular cooling nozzle fed from a passive refrigeration unit to solidify the deposited material rapidly. This improves production speed and the ability to print overhanging geometries without the need for 3D printing support (McGee and Peller 2020). The printing is performed on a heated bed constructed of precision aluminum slats attached to a thick aluminum platen. These slats allow for the removal of parts which include large "raft" areas while restraining the part during cooling (Figure 2).

Extrusion Control And Toolpathing Strategies

Extrusion-based 3DP requires controllable, steady material flow rates to produce consistent thickness walls at a given layer height. In the case where the travel velocity of the end effector is constant, this behavior is well-characterized and relatively simple to control. However, in this project where there are multiple closed contours per layer, as well as combinations of both large and small toolpath radii, the end effector travel speed and the extrusion rate need to vary significantly and synchronously. In the case of FFF processes, the dynamics of the extruder are generally fast and well modeled. It is significantly more complex to model and predict the behavior of pellet extruders, and most literature focuses on the steady-state as opposed to dynamic behavior. The high output rate (corresponding to higher end-effector travel speeds), coupled with the relatively small orifice diameter, leads to higher back pressure and a significant time lag between the motion of the extruder screw and the response of the output. Several techniques have been tested to compensate for this time lag in the system, discussed below.

Lead-In and Lead-Out Geometry
For the extrusion setup and travel speeds used in this research, an iterative approach was used to identify the "excess flow" that occurs at the end of an extrusion path. A decision was made to utilize a "wipe" motion on the inside of each contour, depositing the excess material within the 3DP cavity. During the toolpath generation stage, a "winding number" technique is used to determine inside from outside, which allows for multilayer systems, and avoids the material being deposited on the outside surface. The toolpath generation step allows for a parametric "lead-in" and "lead-out" behavior, providing control over the shape and speed of the path. The ideal path, in this case, was a balance between the extra time required to "bleed" the extrusion pressure and the desire to maintain the speed of the process. The target travel speed of the extruder along the path was 50 mm/s, to account for the necessary layer-to-layer cooling time and the speed that the end effector could maintain at this level of geometric complexity. During the "lead-out" motion, the speed was reduced to 12 mm/s resulting in a lead-out path which traversed a 15mm diameter circle in approximately 4 seconds before "wiping" the excess material on the completed layer and ramping away from the layer tangentially followed by a rapid traverse to the next contour or layer.

Online Control Methods and Innovations to Improve Extrusion Synchronization
While CNC machine controllers have developed significantly in their processing power and memory over the last few decades, industrial robots have not experienced the same demands, with primary use in repetitive motion applications with limited geometric complexity. With the overall accuracy of a large 6-axis robot typically ranging between 0.5 to 1.0 mm spherical error, it is important to use geometric filtering methods to reduce the size of toolpaths, particularly in the case of large-scale AM toolpaths. A geometrically complex, multi-layered facade panel sliced at heights of 1 to 2 mm might require over 1 million motion commands. This is well beyond what is supported by a robot controller such as the Kuka KRC4. Some manufacturers provide the ability to receive streamed robot instructions in a buffer (ABB, UR), and the research community has provided plugins to support this (HAL, ExMachina, COMPAS, et al). While the KRC4 controller does not support this functionality, it does, however, support data streaming at a lower level through the Kuka Robot Sensor Interface (RSI). RSI allows an external motion server to stream Cartesian or joint (servo) positions at either a 4 or 12 ms update rate. This technique has been used in prior research to enable adaptive fabrication methods (Vesey et al. 2015).

In this research, an industrial CNC motion controller is used to decode and plan the synchronous Cartesian motion trajectory (six degrees of freedom and extruder feed) while accommodating joint velocity limits. To accomplish this, the motion controller must perform the kinematic transformation to joint space in real-time (~1 ms) during the lookahead phase of planning, then adapt the overall motion profile of the Cartesian motion; this is particularly important in the vicinity of robot singularities. In this case, the motion controller is modeled as a "virtual" robot. The joint positions of the planned trajectory are then streamed to the KRC4 controller via the EtherCAT fieldbus where they are interpreted by the RSI interface. One advantage of using a "virtual" robot is the ability to time delay the robot servo positions relative to the extruder positions. This allows for partial compensation of the substantial physical lag between extruder motion and actual material extrusion. This research utilizes Beckhoff's TwinCAT, developed as a "digital twin" allowing bidirectional communication between external software such as Rhino or ROS (Liang et al. 2022). The TwinCAT CNC controller additionally provides an external streaming interface that buffers g-code instructions, allowing near real-time streaming of motion commands and eliminating memory limitations on the robot program while also opening up extensive possibilities for adaptive and responsive fabrication processes.

Computational Design
3DP enables production of the entire building envelope as one single component for integrating multiple functions such as insulation voids, vapor barriers, and exterior cladding. Design tools are needed that allow design exploration of envelope elements that can intricately separate interior and exterior zones with multiple functional internal layers. A computational model is developed that enables designers to generate: (a) envelope architecture with interior layers designed for the potential integration of different fluids such as air and phase changing material (PCM) that increase thermal insulation (Figure 3), and (b) an exterior shell with undulated detailing that can vary in size for structural stability. This computational design model is composed of the following algorithms:

- An algorithm that generates the topology of the envelope through a Minimal Surface with intertwined networks of channels and interconnected regions, then parametrically offsets the form inward to provide for separate zones or volumes.
- An algorithm that adapts the generated geometry to required functions then translates the resulted topology to rough mesh, then to high-resolution mesh
- An algorithm that generates a Rib-Stiffened Surface through a Reaction-Diffusion model

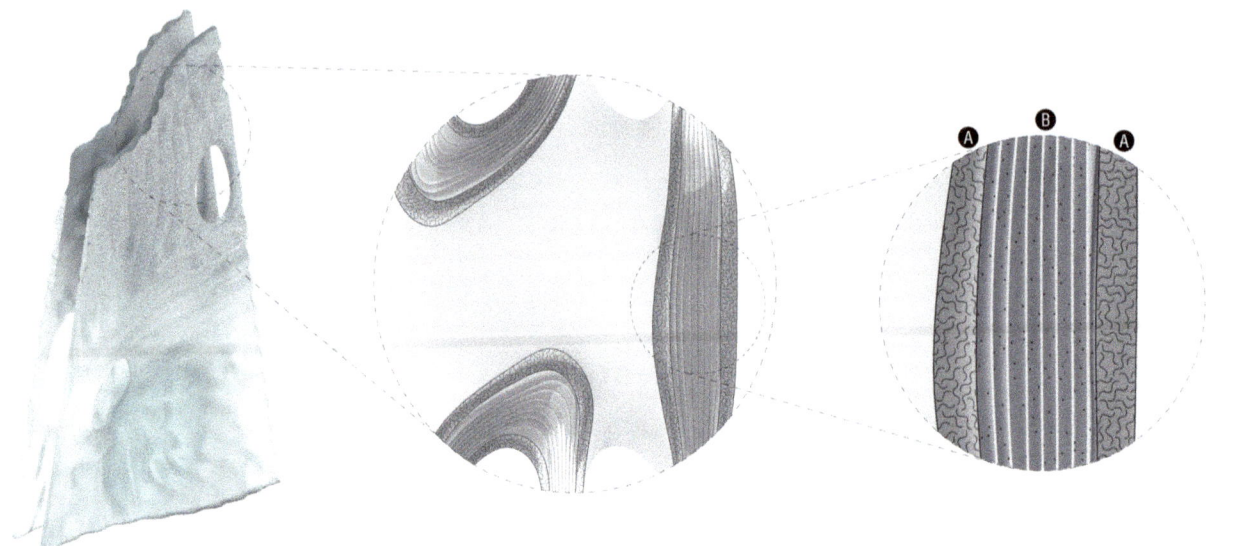

Minimal Surface to Generate Multifunctional Topologies

Minimal surfaces are a useful design medium for creating multi-layered building envelopes because they bisect space into two separate yet interwoven volumes (Figure 4). This geometric property is particularly useful for applications where two volumes interface across a plane, such as that with PCM and air space for thermal insulation. Instead of only containing one fluid, the inner volume (V1) can be offset inward to create multiple separate interior air layers providing designers flexibility in adjusting thermal resistance. The result is sandwiched layers where PCM may occupy the outer peripheral volume, while air pockets occupy interior volumes. This putative model would result in greater thermal resistance as the number of air filled interior layers is increased (Figure 3).

In contrast to FFF 3DP of complex topology that requires printing of support structures for cantilever areas, minimal surfaces do not require support structures as they have zero mean curvature, meaning every point in the print is supported by a doubly curved surrounding (Figure 6). This significantly improves production speed, reduces material waste, and ensures printability without deformation. Finally, the interconnected channels of the minimal surface provide structural stability, and furthermore, maintain stability as the height of the print increases during fabrication.

The integration of multifunctional behavior in an envelope system results in working with complex topologies. In this research, minimal surfaces are utilized as a geometric medium that generates a topology with two separate volumes at the interface of inside and outside climate zones. There are different ways to generate and visualize minimal surfaces digitally. Most methods use mathematical expressions or computational physics simulation with mesh relaxation to generate minimal surfaces in 3D design space, which can be difficult to use when a high degree of

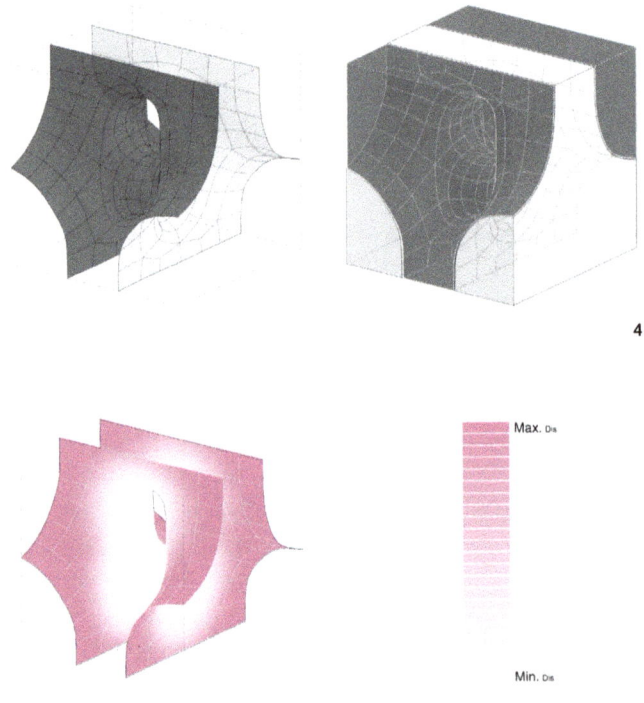

3   The left image shows how a minimal surface divides the inside and outside climate and provides for two interconnected regions and two separate volumes in the envelope; the middle image is a zoom into a region with sandwiched layers; the right image is a detail drawing of sandwiched PCM layers occupying the outer periphery volume (A) and air pockets occupying the interior (B).

4   A minimal surface (left) bisects the space into two separate yet interwoven volumes (right).

5   A deformation diagram demonstrating structural stability and stiffness as higher in areas closer to channels that create interconnected regions created in a minimal surface.

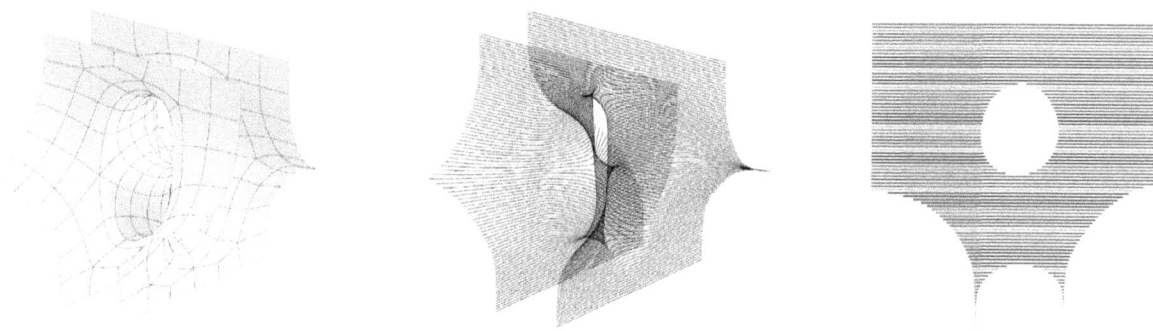

6   The perspective (middle) and front view (right) of the toolpath corresponding to the minimal surface (left) show that minimal surfaces do not require support structures.

7   Diagram showing MSSG method for generating a minimal surface using two sets of curves in red and blue. The left two diagrams show MSSG generating a simple plane surface when inputs are two sets of parallel curves. The two diagrams on the right show the creation of a tube and interconnecting separate layers to create a connection between zones by just introducing a curve.

formal control and articulation is needed. In this research, we use an approach—newly developed at Digital Architecture Research Technologies (DART)—named "Minimal Surfaces through Skeletal Graphs" (MSSG) (Aghaei Meibodi et al. 2022), which uses skeletal graphs to generate, alter, and manipulate minimal surfaces intuitively, enabling their geometric adaptation to functional and program requirements. A skeletal graph of a minimal surface can be viewed as a curve resulting from the shrinking of a minimal surface along the direction of its normal vectors while avoiding any pinching off that would change the topology of the surface (Hoffman n.d.). The MSSG algorithm takes two sets of curves—representing skeletal graphs of a minimal surface—as inputs and then generates a surface on which any point is equidistant from the closest curve in the two sets. A Marching Cube algorithm (Lorensen and Cline 1987) is employed to create triangle meshes by iterating, or marching, over a uniform grid of cubes. Generating minimal surfaces through MSSG allows for intuitive interaction and a higher degree of control in the design process as the designer can easily manipulate sets of curves with a live update in the topology of the minimal surfaces rather than modifying mathematical expressions (Figure 7).

GEOMETRIC ADAPTATION TO FUNCTION AND TRANSLATION FROM ROUGH TO HIGH-RESOLUTION MESH

Once the desired topology with different regions and their interconnecting channels is generated, it is input to an algorithm that regenerates the same topology from rough mesh faces with parametric variables; this allows adaptation of the topology to the functions and programs by enabling parametric changes in overall depth, spacing between different surfaces, dimensions of the interconnecting channel, and enabling an increase in the number of zones through offset surface (Figure 8). Once the variables are fixed, the rough mesh is translated to high-resolution mesh using Catmull–Clark technique (Catmull and Clark 1978).

A RIB-STIFFENED SURFACE THROUGH REACTION-DIFFUSION ALGORITHM (RDA)

A rib-stiffened surface geometry can be generated to provide sufficient structural performance while minimizing material usage. Combining this surface with the multi-layered, interconnecting topology is geometrically challenging due to the continuously variable length of the individual layer paths, resulting in a variable number of ribs along the length of the layer. To solve this globally for the surface, a reaction-diffusion (RD) algorithm can be used (Figure 9). Reaction-diffusion models are often used to simulate chemical reactions or the diffusion of substances over a surface. By controlling the basic parameters of the equations, they can be tuned to generate a wide range of surface patterns. In this case, a vector field was used to weigh the diffusion parameters and produce a branching pattern across the surface. By using a variation of the "heat method" (Crane et al. 2013), the vector field can be generated to follow the shortest path between

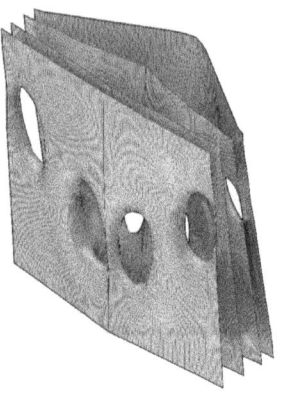 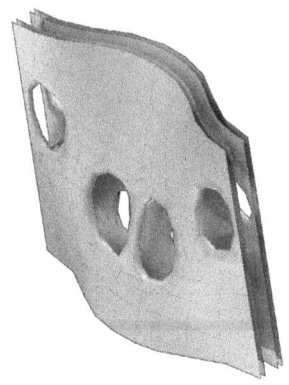 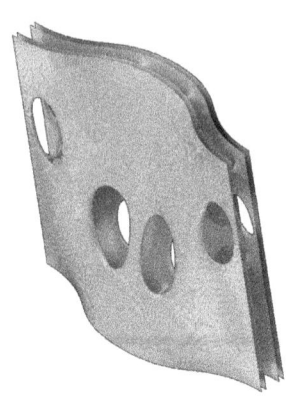

8   (Left) The resulting minimal surface with two topology regions and their interconnecting channels derived from the SGAMS algorithm; (center) a parametric adaptation of the topology to the functions and reconstruction of the minimal surface with rough mesh faces; and (right) a crease in the mesh faces to increase surface resolution.

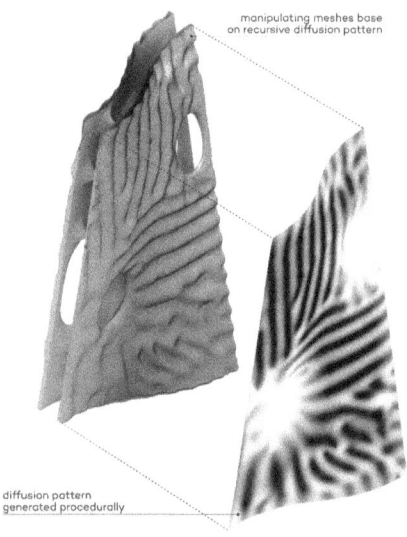

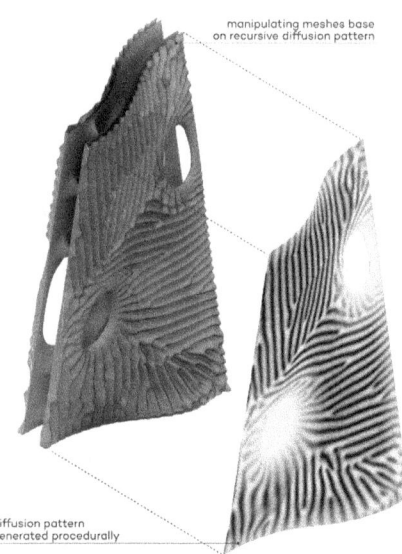

9   Manipulation of mesh based on the generated recursive diffusion pattern; a more detailed diffusion pattern would require higher mesh resolution (bottom) in comparison to the less branched and detailed pattern (top).

mesh boundaries, designated as "cold and hot," resulting in radial ribs advancing from the openings to the edge of the components. However, while radial ribs were the optimal solution for stiffening the elements with respect to the element's topology, printing radial ribs is significantly more challenging than vertical ones. To ensure successful RPE of the ribs, parameters such as depth, width, and overhang were refined and optimized based on feedback from physical prototyping.

CASE STUDY: MULTILAYER BUILDING ENVELOPE (MBE)

The proposed approach was further investigated by designing and constructing a lightweight building envelope prototype for the *Plastic Architecture* exhibition at Cooper Union in New York. The prototype was made from two thermoplastic components with dimensions of 1,170 mm by 2,268 mm by 178 mm (height, width, depth), 3D printed using RPE, and designed around the concept of a translucent multi-layer envelope consisting of separated volumes for natural ventilation and thermal insulation. The envelope system is composed of (a) an exterior shell with undulated detailing of varied size, and (b) multiple interior air and liquid PCM layers for thermal insulation.

The outermost volume is designed to contain PCM that stores latent heat HVAC equipment run times. Interior volumes are designed to contain a vacuum, air, or aerogel for thermal insulation (Figure 10). The air gap between the two main zones is an air-moisture barrier that removes water vapor by evaporation, producing a breathable envelope system. Interconnecting channels between the two main zones create passages between their volumes, and the undulating geometry generated by a reaction-diffusion algorithm growing outwards from the channels to the edges provides structural stability.

RESULTS AND DISCUSSION

The resulting 1:1 scale prototype consists of two 3D printed thermoplastic elements (2270 mm by 1170 mm by 200 mm;

Table 1: Geometric and production details of the low and high-resolution facade panels

Dimensions of each panel	2270 X 1170 X 200 to 230 mm (height X width X depth) with 8 to 50 mm thickness for the outset layers
Material	thermoplastic PETG
Volume of material for one element	41.3 liters of polymer
Weight of one element	44.5 kilograms
Printing time of each element	44 hours
Number of mesh faces in the low-resolution panel	70,000 triangles
Number of mesh faces in the high-resolution panel	840,000 quads
Toolpath kilometer in low-resolution panel	10,323 meters
Toolpath kilometer in high-resolution panel	10,333 meters

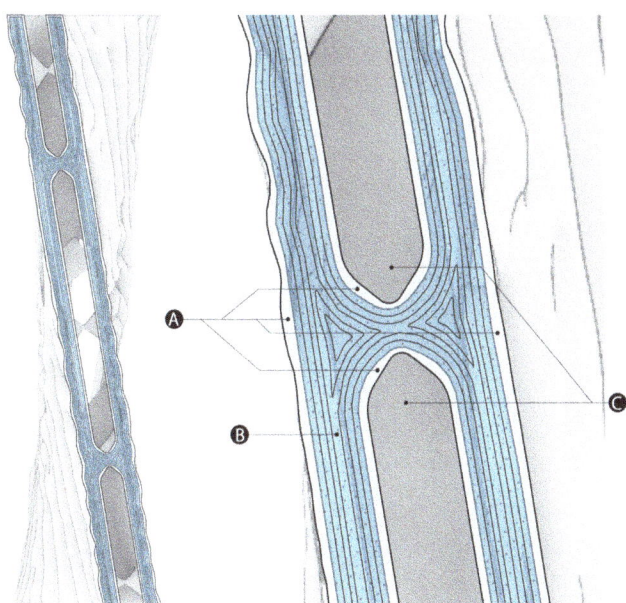

10 Section through the MBE system shows the outermost volume is designed to contain PCM (A), inner volumes (B) designed to contain a vacuum, air, or aerogel and can increase the number of layers by 3DP increases, and the air gap between the two main zones is an air moisture barrier (C).

height x width x depth) (Figure 11). It showcases the advantages of FGF 3DP for construction-scale components and functional integration in building envelopes. Each element entails a complex topology with multiple interior layers that separate volumes as well as connections that link related zones. Each panel weighs 44.5 kg and was 3D printed over the course of 44 hours (Table 1). To evaluate the impact of the mesh resolution on the production speed, one of the two elements entailed ten times as many mesh faces.

The design and development of a robust "raft" was necessary to ensure the stability of the prototype throughout the 3DP process. The six-layer thick raft successfully secured the print to the heated bed throughout the printing time. The minimal surface geometry with its multiple interconnecting layers provides adequate stability throughout the 3DP process, eliminating the need for additional internal infill. The minimal surface's double curvature effectively provides support for subsequent layers and eliminates the need for additional support. While minimal surfaces provide for supportless printing, successful 3D printing of overhanging geometries significantly relies on the layer thickness versus bead width of material extrusion.

The lead-in and lead-out design innovations allow excess extrusion material to collect within the printed geometry resulting in clean seams. Consequently, manual care during the fabrication process was no longer required.

CONCLUSION

This paper explores the fabrication of topologically complex geometries that have the potential to contribute to improved energy efficiency of buildings. It showcases that the Robotic Pellet Extrusion (RPE) of thermoplastic is a viable method to produce such topologies and can profoundly change the way we conceptualize and design the envelope system by allowing for the integration of multiple functions in a singular component. The fabrication and computational methods developed through this research resulted in the successful 3D print of a geometrically complex full-scale prototype with multiple layers that can be translated to other building envelope systems.

Future research will continue to advance research on functionally integrated sustainable building envelopes by (a) developing the fabrication system to allow for the high-speed production of thinner plastic layers, (b) performance testing for environmental factors like thermal resistance, natural ventilation, daylighting, and durability, and (c) explorations of circular economic models using recyclable thermoplastic and bio-plastic building envelopes.

11 The 1:1 scale prototype is made of two thermoplastic elements.

REFERENCES

Aghaei Meibodi, Mania, Alireza Bayramvand, and Mehrad Mahnia. 2022. "Minimal Surfaces through Skeletal Graphs (MSSG)." Collaboration between Digital Architecture Research Technologies (DART) at University of Michigan and Fabtory, *On Air* exhibition, curated by Kathy Velikov.

Aghaei Meibodi, Mania, Rena Giesecke, and Benjamin Dillenburger. 2018. "Digital metal: Additive manufacturing for casting metal parts in architecture." In *1st International Conference on 3D Construction Printing (3DcP)*. Melbourne, Australia.

Aghaei Meibodi, Mania, Marirena Kladeftira, Thodoris Kyttas, and Benjamin Dillenburger. 2019. "Bespoke Cast Facade: Design and Additive Manufacturing for Aluminum Facade Elements." In *ACADIA 19: Ubiquity and Autonomy; Proceedings of the 39th Annual Conference of the Association for Computer Aided Design in Architecture.* 100-109.

Aghaei Meibodi, Mania, Pietro Odaglia, and Benjamin Dillenburger. 2021. "Min-Max: Reusable 3D printed formwork for thin-shell concrete structures-Reusable 3D printed formwork for thin-shell concrete structures." In *ACADIA 21: Toward Critical Computation; Proceedings of the 41st Annual Conference of the Association for Computer Aided Design in Architecture.*

Aghaei Meibodi, M., R. Craney, and W. McGee. 2021. "Robotic Pellet Extrusion 3D Printing and Integral Computational Design of Reinforced Thin Shell Formwork System for Sandwich Concrete Walls." In *ACADIA 21: Toward Critical Computation; Proceedings of the 41st Annual Conference of the Association for Computer Aided Design in Architecture.*

Aghaei Meibodi, Mania, Andrei Jipa, Rena Giesecke, Demetris Shammas, Mathias Bernhard, Matthias Leschok, Konrad Graser, and Benjamin Dillenburger. 2018. "Smart Slab. Computational design and digital fabrication of a lightweight concrete slab." In *ACADIA 18: Recalibration, On imprecisionand infidelity; Proceedings of the 38th Annual Conference of the Association for Computer Aided Design in Architecture.* Mexico City: ACADIA. 434–443.

Akilo, M. A. 2018. "Design and analysis of a composite panel with ultra-thin glass faces and a 3D-printed polymeric core." PhD diss., MSc thesis, University of Bologna. https://amslaurea.unibo.it/15351/.

Bedarf, Patrick, Dinorah Martinez Schulte, Ayça Şenol, Etienne Jeoffroy, and Benjamin Dillenburger. 2021. "Robotic 3D Printing of Mineral Foam for a Lightweight Composite Facade Shading Panel." In *Proceedings of the 26th CAADRIA Conference*, vol. 1. Hong Kong. 603–612.

Catmull, E. and J. Clark. 1978. "Recursively generated B-spline surfaces on arbitrary topological meshes." Computer-Aided Design 10 (Sept): 350–355.

Crane, Keenan, Clarisse Weischedel, and Max Wardetzky. 2013. "Geodesics in heat: A new approach to computing distance based on heat flow." *ACM Transactions on Graphics (TOG)* 32 (5): 1–11.

Cruz, Paulo J. S., Bruno Figueiredo, João Carvalho, and Tatiana Campos. 2020. "Additive manufacturing of ceramic components for façade construction." *Journal of Facade Design and Engineering* 8 (1): 1–20.

Fang, Yuan, and Soolyeon Cho. 2019. "Design optimization of building geometry and fenestration for daylighting and energy performance." *Solar Energy* 191: 7–18.

Hemsath, T. L., and K. A. Bandhosseini. 2015. "Building design with energy performance as primary agent." *Energy Procedia* 78: 3049–3054.

Hoffman, James T. et al. n.d. "Skeletal Graphs of Triply Periodic Surfaces." *The Scientific Graphics Project* (online database). Accessed February 14, 2021. https://www.msri.org/publications/sgp/jim/geom/surface/global/skeletal/index.html.

Lekka Angelopoulou, Sofia. 2022. "Studio RAP 3D prints ceramic tiles and red bricks for amsterdam boutique facade." *designboom* (blog). January 28, 2022. Accessed February 6, 2022. https://www.designboom.com/architecture/studio-rap-3d-print-ceramic-tiles-red-bricks-amsterdam-boutique-facade-01-28-2022/.

Li, Danny H. W., Liu Yang, and Joseph C. Lam. 2013. "Zero energy buildings and sustainable development implications—A review." *Energy* 54 (June): 1–10.

Liang, Ci-Jyun, Wes McGee, Carol C. Menassa, and Vineet R. Kamat. 2022. "Real-time state synchronization between physical construction robots and process-level digital twins." *Construction Robotics* 6: 1–17.

Lorensen, William E., and Harvey E. Cline. 1987. "Marching cubes: A high resolution 3D surface construction algorithm." *ACM SIGGRAPH Computer Graphics* 21 (4): 163–169.

McGee, Wes, Tsz Yan Ng, Kequan Yu, and Victor C. Li. 2020. "Extrusion nozzle shaping for improved 3DP of engineered cementitious composites (ECC/SHCC)." In *RILEM International Conference on Concrete and Digital Fabrication*. Cham: Springer. 916–925.

McGee, W., K. Velikov, G. Thun, and D. Tish. 2017. "Infundibuliforms: Kinetic systems, additive manufacturing for cable nets and tensile surface control." In *Fabricate 2017: Rethinking Design and Construction*. London: UCL Press. 4–6.

McGee, Jonathan Wesley, and Asa Leland Peller. "Extrusion die and nozzle cooling system for large scale 3D additive manufacturing." U.S. Patent US10807292B2, filed 24 October 2017, and issued 20 October 2020.

Mungenast, M., O. Tessin, V. Blum, O. Khuraskina, L. Morroni, and T. Gutheil. 2017. "Fluid Morphology, 3D Printed Functional Integrated Building Envelope." Associate Professorship of Architectural Design and Building Envelope, TUM, support of Rodeca, Picco's 3D World, Delta Tower.

Peters, Brian. 2013. "Building bytes: 3D-printed bricks." In *ACADIA 13: Adaptive Architecture; Proceedings of the 33rd Annual Conference of the Association for Computer Aided Design in Architecture*. 433–434.

Rael, Ronald, and Virginia San Fratello. 2017. "Clay bodies: crafting the future with 3D printing." *Architectural Design* 87 (6): 92–97.

Sarakinioti, Maria Valentini, Michela Turrin, M. Teeling, Paul de Ruiter, Mark van Erk, Martin Tenpierik, Thaleia Konstantinou et al. 2017. "Spong3d: 3D printed facade system enabling movable fluid heat storage." *Spool* 4 (2): 57–60.

Snooks, Roland, and Laura Harper. 2020. "Printed Assemblages: A Co-Evolution of Composite Tectonics and Additive Manufacturing Techniques." In *Fabricate 2020*, edited by J. Burry, J. E. Sabin, B. Sheil, and M. Skavara. London: UCL Press.

Torcellini, Paul, Shanti Pless, Michael Deru, and Drury Crawley. 2006. "Zero energy buildings: a critical look at the definition." National Renewable Energy Lab (NREL) Conference Paper No. NREL/CP-550-39833. ACEEE Summer Study, Pacific Grove, California.

Torcellini, Paul A., and Drury B. Crawley. 2006. "Understanding zero-energy buildings." *ASHRAE Journal* 48 (9): 62–69.

Tenpierik, Martin, Michela Turrin, Yvonne Wattez, Tudor Cosmatu, and Stavroula Tsafou. 2018. "Double Face 2.0: A lightweight translucent adaptable Trombe wall." *Spool* 5 (2).

Vasey, L., E. Baharlou, M. Dörstelmann, V. Koslowski, M. Prado, G. Schieber, and J. Knippers. 2015. Behavioral design and adaptive robotic fabrication of a fiber composite compression shell with pneumatic formwork. In *ACADIA 15: Computational Ecologies: Design in the Anthropocene*. Cincinnati, OH. 296–309.

Yang, Chao, and Joon-Ho Choi. 2015. "Energy use intensity estimation method based on façade features." *Procedia Engineering* 118: 842–852.

IMAGE CREDITS

Photographer: Christopher Voltl, ©DART, University of Michigan.

Dr. Mania Aghaei Meibodi is Assistant Professor of Architecture and Chair of Digital Architecture Research and Technologies (DART) Laboratory at Taubman College of Architecture and Urban Planning at the University of Michigan.

Wes McGee is an Associate Professor in Architecture and the Director of the Fabrication and Robotics Lab at Taubman College of Architecture and Urban Planning at the University of Michigan.

Alireza Bayramvand is a research assistant at the Digital Architecture Research and Technologies (DART) Laboratory at Taubman College of Architecture and Urban Planning, University of Michigan, with a specialty in computational design and robotic construction.

Collective Aerial Additive Manufacturing

Incrementally Built Shell Structure Design

Robert Stuart-Smith
University of Pennsylvania
University College London

Durgesh Darekar
University College London

Patrick Danahy
University of Pennsylvania

Basaran Bahadir Kocer
Imperial College London

Vijay Pawar
University College London

Mirko Kovac
Imperial College London
Swiss Federal Laboratories of Material Science and Technology (Empa)

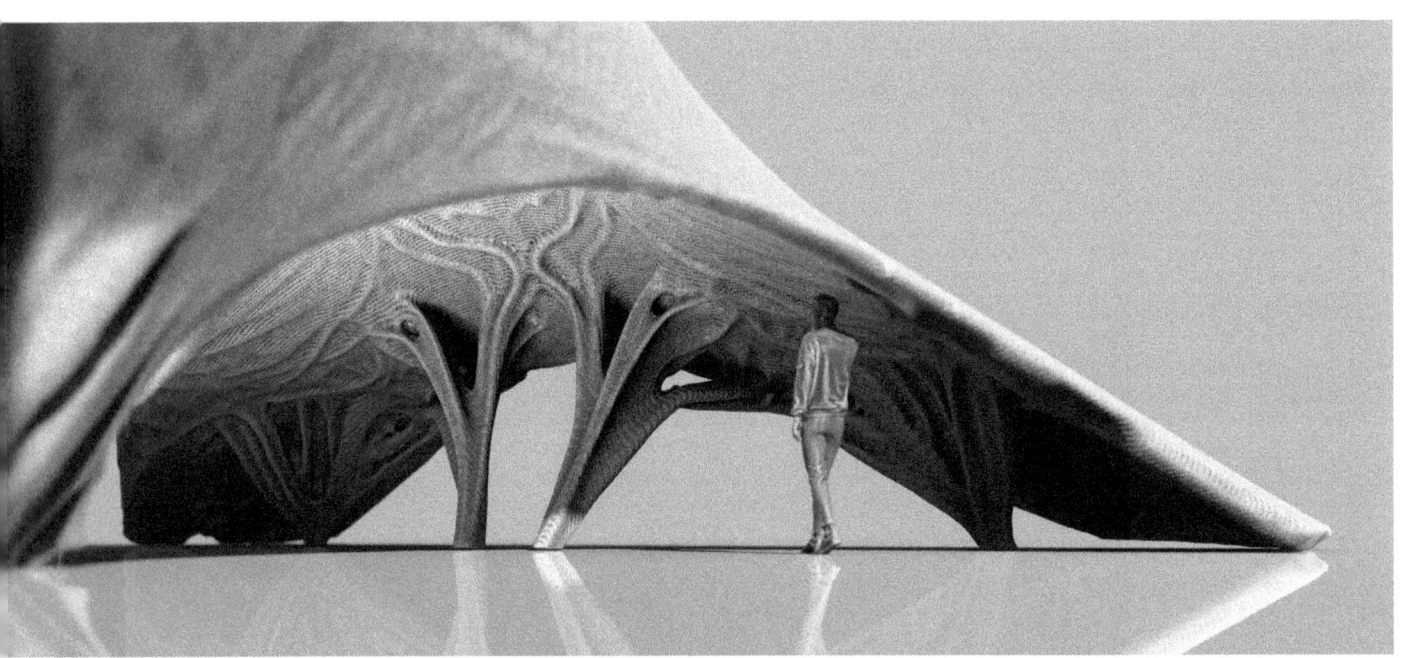

1 3D shell design for incremental construction by Collective AAM. The shell design method provides proof of concept for a design approach that mitigates instability throughout construction without the use of temporary supports or scaffolds.

ABSTRACT

Aerial robot additive manufacturing (AM) offers a means to supplement existing on-site AM approaches that are limited in build envelope and accessibility, to support automated construction in more diverse locations. Aerial AM has been recently demonstrated by the authors using quadcopters to extrude cementitious and foam materials within tolerances sufficient for building construction (Zhang et al. 2022). However, approaches to the distributed control and mission planning of multi-robot manufacturing and corresponding building design solutions that engage with this novel technology still need to be developed. Although AM supports substantial geometric design freedom, there are practical challenges that currently limit on-site AM to vertically orientated geometries that remain stable during incremental manufacture. The installation of roofing or other spanning structures is difficult to accomplish without support scaffolding or supplementary building methods, rendering complete automation of AM buildings impractical, whilst limiting the ability of designs to offer efficient monolithic structural solutions. Funicular shell geometries (Rippmann and Block 2013) offer a materially efficient solution to a monolithic spanning enclosure, however, they typically require scaffolding during construction, rendering them unsuitable for incremental manufacture by a team of robots, unless mid-construction stability challenges are addressed. This paper proposes a multi-agent mission-planning robot control framework and simulation environment for Collective Aerial Additive Manufacturing (Collective AAM) together with an approach to the design of scaffold-free 3D shell geometries suited to Collective AAM's adaptive and incremental approach to concurrent building. Individual agent and swarm-based vector steering behaviors are evaluated to support future integration with existing AAM manufacturing capabilities recently published in the journal *Nature* (Zhang et al. 2022).

INTRODUCTION

Additive Manufacturing (AM) is rapidly gaining traction in the construction industry, demonstrating a reduction in the time and cost of building whilst offering substantial geometric design freedom at practically no additional cost. On-site AM primarily utilizes Material Extrusion (MEX) AM methods, such as the continuous extrusion of concrete or clay-based materials (Gibson et al. 2021). Current methods typically require the use of a gantry system sufficiently large for the extrusion end-effector to reach all regions inside a desired build envelope. Although faster than historical construction methods, current AM methods could be improved by distributing the building task amongst several smaller AM machines that would enable parallel manufacturing and support more flexible methods of construction. Mobile robots can support concurrent operation and adapt building to different terrains, yet most are limited to circulate inefficiently around previously built material on grade and constrained in vertical reach. Aerial robot AM enables building in more remote and hard-to-access regions where existing machinery or ground robots could not operate. Aerial AM (AAM) has been demonstrated in cementitious and foam materials up to two meters in height within acceptable tolerances for building (Zhang et al. 2022), yet an approach to collective AAM and building designs realizable with the technology have not been developed or evaluated together.

Although AM supports diverse geometrical designs, not all designs are stable throughout an incremental MEX on-site building process. Where off-site AM permits the manufacture of parts at different orientations to their assembled state, on-site MEX provides no such opportunities. Temporary scaffolding employed in off-site AM is not feasible on-site due to increases in material, transportation, manufacturing, and disassembly time necessitated at the scale of a building's volume. To avoid the need for temporary scaffolding, on-site AM is typically limited to vertically orientated geometries that cannot span interior spaces to provide full enclosure (such as roofing), requiring additional building methods to complete a building. This prohibits AM design-engineering solutions that optimize material across an entire monolithic structure, constraining designs to less materially efficient solutions. It also limits full automation of AM buildings constructed by mobile robots, limiting their impact on site accessibility and cost. This paper proposes a mission planning aerial robot control framework and simulation environment for Collective Aerial Additive Manufacturing (Collective AAM) (Figure 2) and an approach to building shell design that alleviates the need for temporary support scaffolds during Collective AAM's distributed and incremental approach to building. No manufacturing is undertaken within the research; rather a novel approach to both construction and design is demonstrated through simulation modeling, design, and initial UAV flight tests to evaluate the feasibility of the approach supporting already demonstrated AAM (Zhang et al. 2022).

BACKGROUND

Collective Robotic Construction (CRC) enables a team of robots to build in parallel to construct a structure greater in size than individual robot builders (Petersen et al. 2019). To date, CRC has primarily focused on the assembly of discrete volumetric elements (Petersen, Nagpal, and Werfel 2011; Seo, Yim, and Kumar 2013) and filament winding/weaving (Augugliaro et al. 2013; Duque Estrada et al. 2020; Stuart-Smith 2016), with little research undertaken into CRC mission planning for additive manufacturing (AM) or its relation to building designs. Limited AM CRC research has been developed with mobile ground vehicles that remain on grade (Sustarevas et al. 2019; Napp and Nagpal 2014; Zhang et al. 2018), or constrained to move on top of previously built material (Jokic et al. n.d.). In both cases, circulation is restricted by building operations that increase congestion. As an alternative approach, aerial robots can circulate overhead to and from building tasks without being constrained to sharing the congested space of the building site.

Assembly-based aerial CRC (Augugliaro et al. 2014; Lindsey, Mellinger, and Kumar 2011) has been demonstrated using specific flight corridors, while Aerial AM (AAM) has only recently been demonstrated by the authors, manufacturing in cementitious and foam materials up to two meters in height (Zhang et al. 2022). Approaches to CRC by AAM and corresponding building design solutions are yet to be developed and would involve flight planning solutions capable of responding to diverse geometric designs and site conditions, requiring variations in flight routes between recharging stations and a manufacturing site. Given the volatile nature of manufacturing in-flight where the robustness of robot platforms, adverse weather, and other dynamic events could undermine a predetermined construction sequence, adaptive solutions are also needed that build in redundancy to individual robots and incremental building operations and offer a distributed approach to Collective AAM mission-planning.

Many building designs only achieve structural stability upon completion and would therefore be unsuitable for Collective AAM's incremental approach to building. Most on-site AM buildings to date are constrained to more vertically orientated geometries (Chen and Yossef 2015), with only off-site MEX AM offering a limited demonstration of partially inclined geometries (XtreeE n.d.). Structurally stable incremental building strategies have been demonstrated in CRC research using modular blocks (Andreen et al. 2016; Werfel, Petersen, and Nagpal 2011), yet these approaches utilize more material

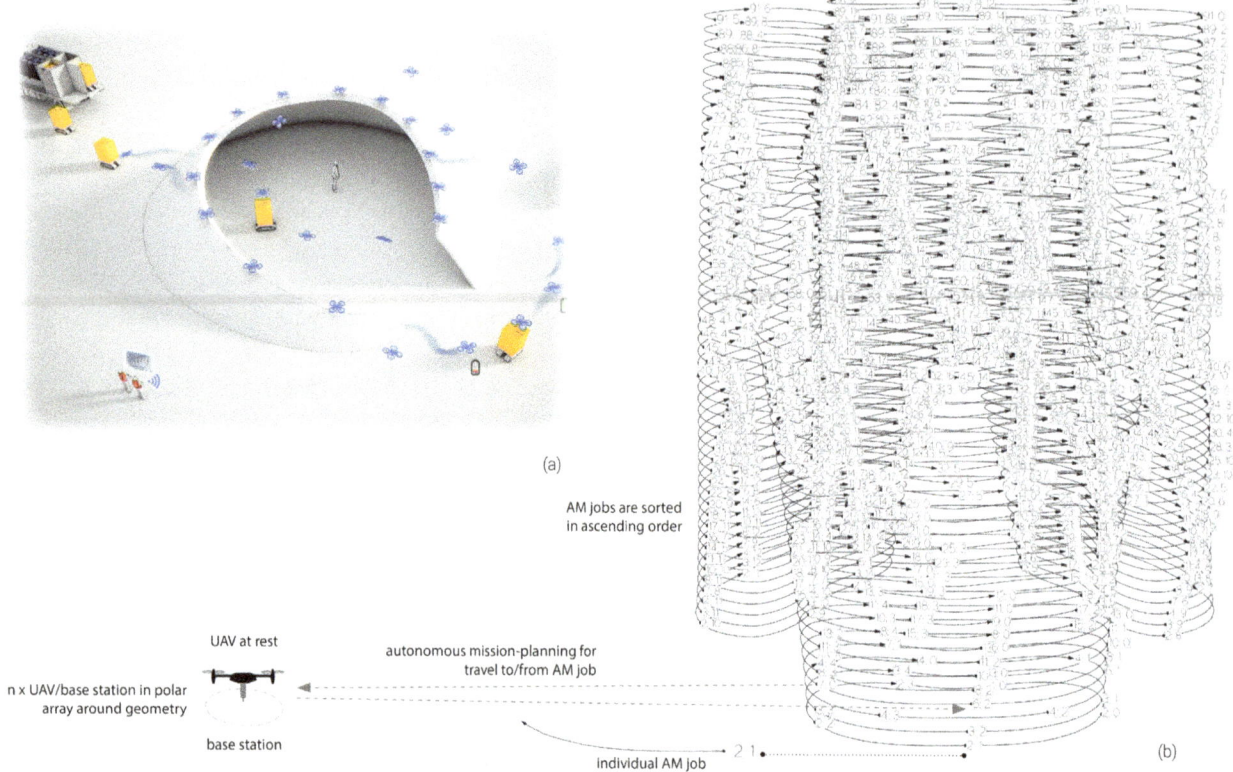

than non-incremental building designs such as 3D shell structures, that require only a thin layer of material oriented relative to structural forces. Funicular shell surface geometries (Cuito and Montes 2003) and related 3D graphics statics approaches to shells (Veenendaal and Block 2012; Rippmann and Block 2013; Nejur and Akbarzadeh 2021) are materially efficient, yet their inclined geometries would be unstable during incremental manufacturing without the employment of temporary scaffolding. However, scaffolding approaches add complexity to autonomous robot building operations and scale poorly as increases in building surface area or structural span require significant increases in scaffolding volume. To support greater geometrical freedom, design approaches that can mitigate mid-build instability are needed that can approach levels of material efficiency closer to surface-based designs that don't consider incremental manufacture.

METHODS

The research involved the development of a Collective AAM software framework (Figure 3) that supports both actual Unmanned Aerial Vehicles (UAVs) flight operations and digital simulation modeling together with an associated design approach to building shells suited to incremental building.

Collective AAM Software Framework

A multi-agent Collective AAM software framework with adaptive task determination was developed to support high-level swarm flight mission planning and building operations in both simulated and real UAVs. The flight behavior of individual UAVs was calibrated to align simulation modeling with physical flight data through a series of flight experiments using a single custom-built quadcopter, extending the authors' prior multi-robot demonstrations (Zhang et al. 2022) to support future manufacturing-orientated research. The software directs continuous flight via small incremental vector-based steering behaviors in lieu of a sequence of predetermined waypoints, enabling a UAV to make autonomous decisions to support adaptive flight and building operations.

Distributed AM Setup

Similar to desktop 3D printer software, the Collective AAM approach contours a surface geometry; however, to enable contours to be manufactured by several robots concurrently, each contour was broken up into several discrete curve trajectories. A 3D design can be input into a custom Rhino3D Grasshopper file to generate a set of AM tasks and a multi-robot set-up configuration. A sequence of horizontal contour curves are then created at vertical intervals related to a specified AM extrusion diameter (a larger layer height of 0.3 m is used for expediency in this research). Each contour is split into a series of smaller curves of length less than or equal to a maximum manufacturing length of 3 m that corresponds to a full material payload used in Aerial AM demonstrations (Zhang et al. 2022). Each curve is then divided into a series of equidistant waypoints, and stored in a nested Python list as an individual AM job. A circular boundary offset from the 3D

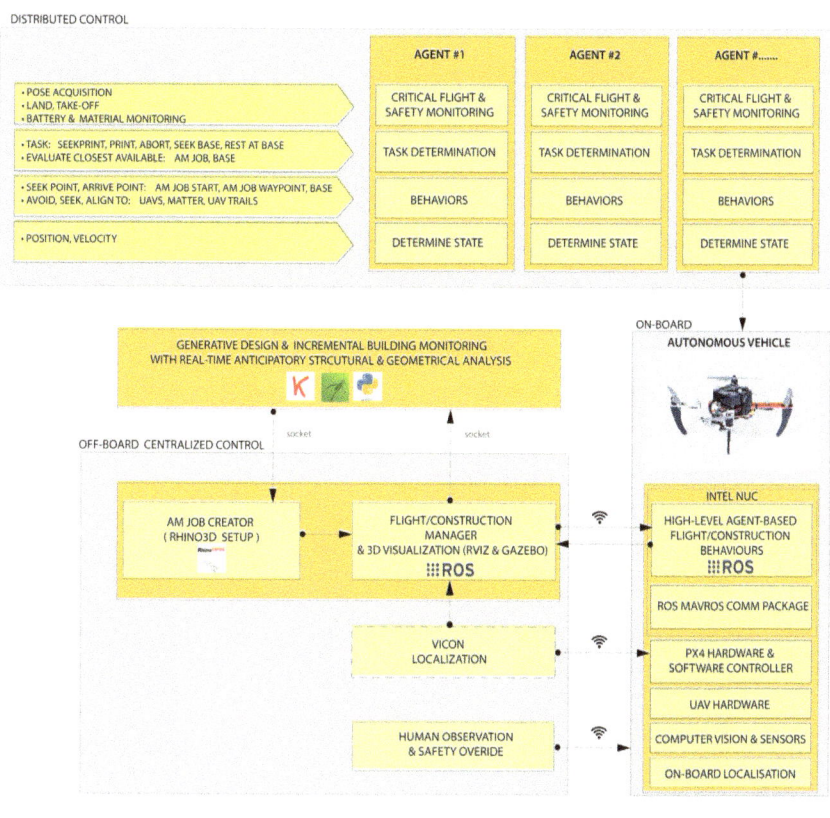

2 Collective AAM concept:
(a) Multiple aerial robots collectively additively manufacture a geometry in-flight
(b) Manufacturing is broken into small print trajectories to support distributed, parallel manufacturing

3 Collective AAM software framework involves a custom ROS package interfacing with other ROS packages and Rhino3D

4 UAV vector-based steering behaviors

5 Selection of event-based rules governing agent-based task determination to support adaptive flight and manufacturing

geometry was used to distribute a specified number of UAV home positions, equally distanced from each other. A custom Python script exports the list of AM jobs and UAV start positions to a text file.

ROS Collective AAM Package

A custom Robot Operating System (ROS) package (Open Robotics 2021) was created in Python using the rospy library to control actual UAV flight behavior for Collective AAM and for simulation modeling of Collective AAM (Figure 3). The framework imports data from a Rhino3D exported text file and creates a series of instances of Python classes, one for each UAV, base station, and AM job. The framework launches a ROS programming Node that determines each UAV's (virtual agent or real quadcopter) flight trajectories over time and visualizes their motion in ROS's RViz package. UAVs

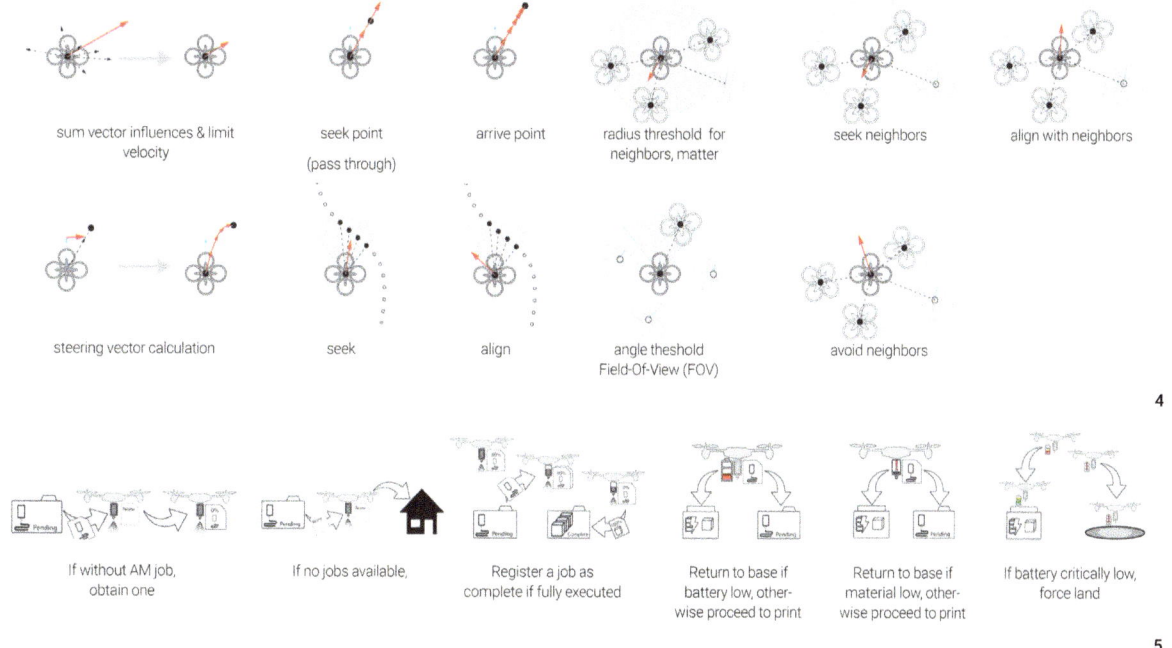

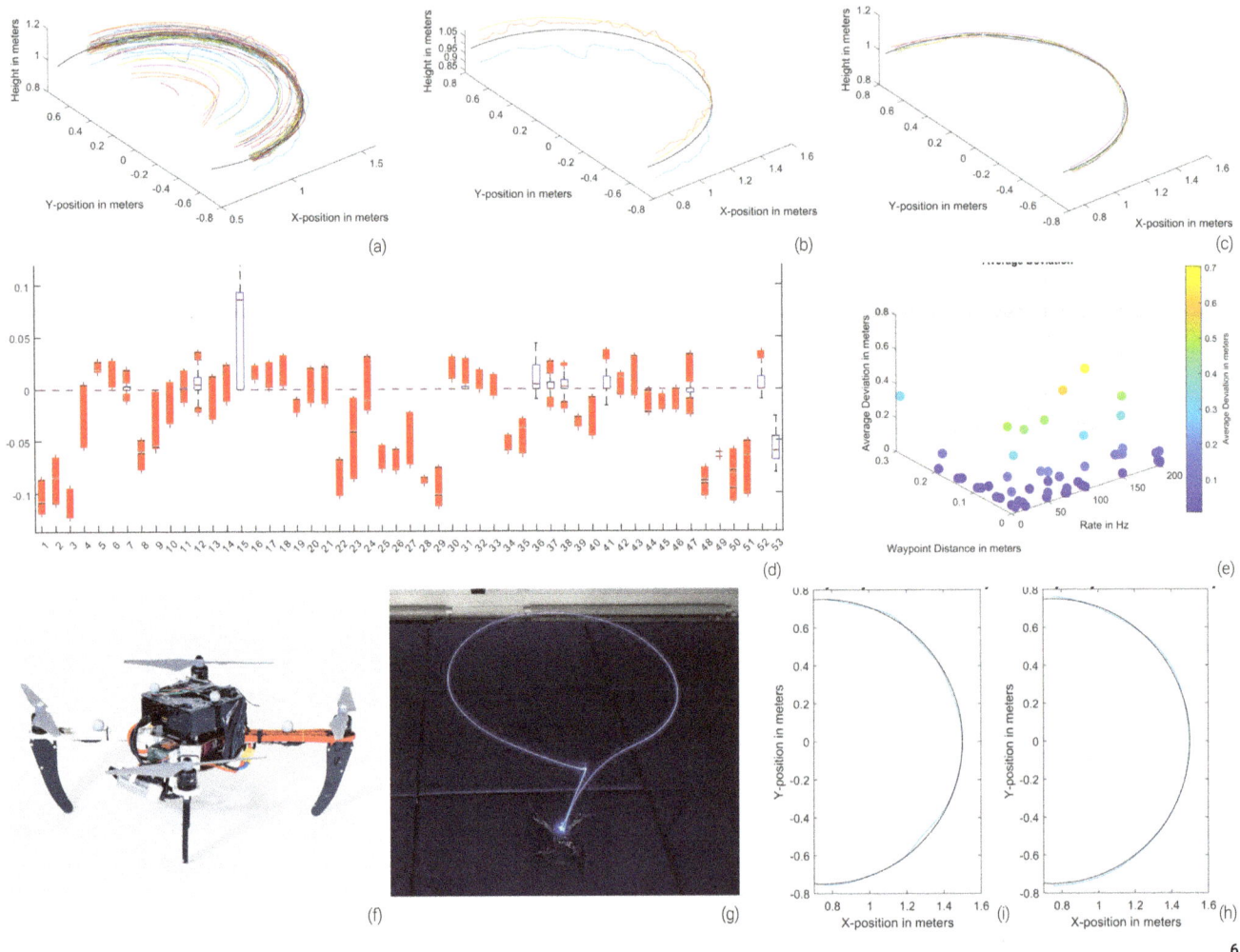

autonomously determine their tasks (seek print, print, abort, hover, seek home, land, take off, etc.) and select the closest, lowest available AM job to print before returning to the closest available home base for re-charging and material re-supply. UAVs can dynamically swap base locations and job locations throughout the manufacturing process and can incrementally build a geometry from the ground up. Manufacturing is not limited to one strata at a time but can proceed across several strata in parallel provided AM jobs are not skipped below and lower jobs are prioritized.

Mission-planning was achieved via a series of event-based goals that govern autonomous UAV decision making (Figure 5). A UAV selects the closest available job and flies to its start location. Upon finishing, it returns to the closest unoccupied base station before continuing with another AM job. During flight to and from manufacturing, UAVs seek a direct route to their destination while deviating locally to dynamically avoid each other and already manufactured geometry. A UAV currently manufacturing has right-of-way and does not undertake avoidance measures. UAVs are constrained to not pass over each other as such actions can create downwash turbulence and cause a UAV underneath to momentarily drift away from its desired trajectory. To avoid congestion where several UAV AM jobs are in close proximity, a UAV approaching an AM job opts to hover, and abort the job to seek a different job if more than a specified amount of time passes. A UAV will also abort a flight and cancel an AM job before or during its execution if experiencing low battery or material supply levels. In such events, the AM job will be returned to a global list and made available for another UAV to complete. UAVs will not take off from a base station unless AM jobs remain available, thus all UAV flights cease when manufacturing is complete.

While UAVs engage in dynamic avoidance of other UAVs and manufactured material, there are scenarios where a UAV could become trapped around built geometry as it seeks the shortest route to a base station. To avoid such conditions, a path planning algorithm such as A* could be implemented to ensure a viable route is followed; however, this would significantly increase computational complexity and latency. A more expedient alternative that supports on-board path-planning and comparatively less system-wide knowledge was developed that constrains UAVs to approach and depart print

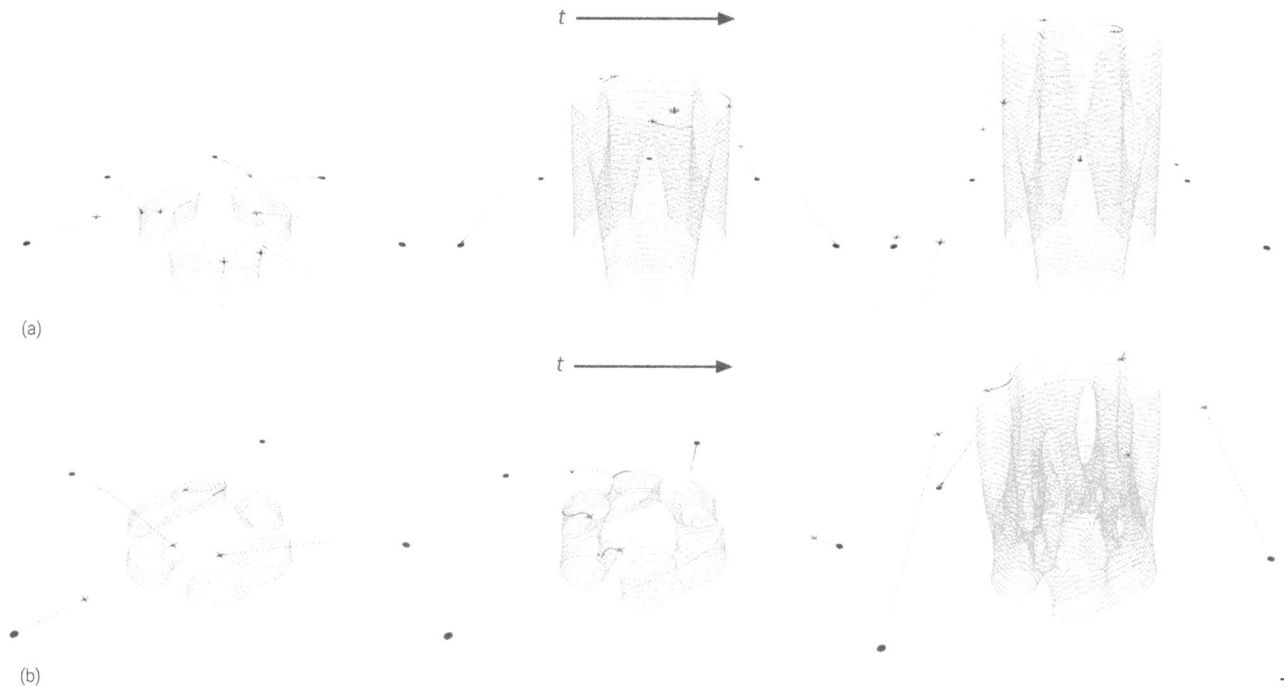

6 Physical flight tests on a single UAV using incremental vector motion adaptive flight: (a) Trajectories from all 53 flight tests, (b) Oscillating trajectories (low Hz rate), (c) High accuracy trajectories, (d) Y & Z deviation box plot from all 53 flight tests, (e) Average deviation in all 53 flight tests, (f) Custom built UAV testing platform, (g) Light trace time-lapse of UAV flight, (i-j) Top view of two most accurate trajectories (in blue)

7 Collective AAM simulations of two different multi-manifold geometries (a, b)

jobs at a small height above their AM job, within a circle radius surrounding the building geometry. UAVs approaching a job calculate a point of intersection between their destination and the circle and ascend to this position before passing over built matter to arrive at an AM job start position. On completing a job, a UAV will pass at the same altitude, a point of intersection with the circle prior to descending to the closest available base station. As the AM job position sits inside of the circle, intersections were calculated by finding the secant line that intersects the circle at two positions (Rhoad et al. 1991).

Each UAV incrementally adjusts its position over time to control flight using vector-based steering behaviors inspired by Craig Reynolds' Boids simulations (Reynolds 1999). Vector-based flight behaviors include: seek point (moving at full velocity), arrive point (reducing velocity to zero on arrival), seek, avoid and align to other UAVs, UAV trails (past locations), and previously manufactured geometry (Figure 4). Each behavior is calculated as a vector relative to a UAV's current position and converted to a steering vector by creating a vector relative to a UAV's current velocity vector. All behaviors are weighted and summed to create an acceleration vector that is added to a UAV's position each time frame as a nearby waypoint. In simulation, the frequency and magnitude of each waypoint solely determines UAV velocity and simulation time. In physical UAV flights, however, these impact overall flight stability and accuracy.

UAV Flight Controller Calibration

A custom UAV platform was developed that comprised of a DJI-F450™ quadcopter frame and components, Pixhawk PX4™ controller, Intel Realsense™ RGBD camera and an Intel NUC™ computer. For low-level flight control, each UAV instance in the ROS COLLECTIVE AAM package launches an instance of the PX4 controller ROS library. The current pose of the UAV platform is provided by a Vicon motion-capture camera system that tracks reflective markers distributed across the UAV frame. This pose was fed back to the PX4 controller at 150 Hz. The UAV's velocity vector was then added to its current position to determine its next waypoint. As the magnitude of this waypoint vector and the frequency it is sent to the PX4 controller impacts the stability and accuracy of the UAV's flight trajectory, a series of 53 experiments were conducted on a semi-circular trajectory to ascertain a suitable control frequency, waypoint magnitude and flight velocity (Figure 6). To ensure landing and take-off did not impact results, only the middle semi-circle arc of a circular flight was logged in ROSbags and Vicon motion tracking for comparative analysis. Trajectory deviations were evaluated relative to proof-of-concept aerial AM demonstrations that included pose-correction (Zhang et al. 2022) to determine whether deviation was within suitable ranges for an actual Collective AAM demonstration in future research. Optimal settings were incorporated into

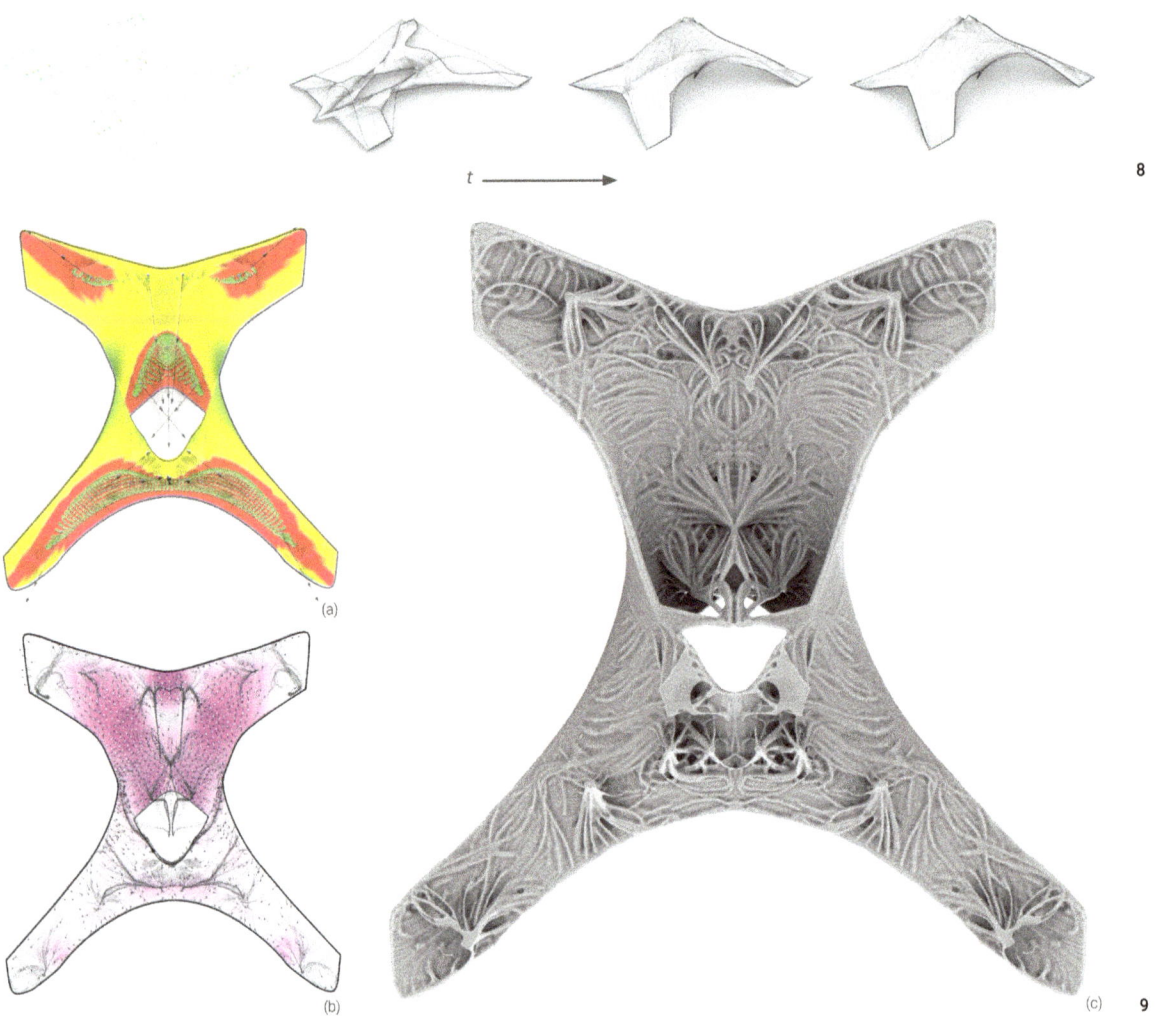

the Collective AAM software package and used in Collective AAM simulations of two different multi-manifold geometries to evaluate the adaptability of the framework (Figure 7).

3D Shell Design for Incremental Construction
Initial Shell Design Method

As AM MEX built designs are typically built in horizontal layers without re-enforcement, MEX is better suited to geometries that primarily operate in compression. As such, a funicular shell design method similar to Gaudi, Block and others (Rippmann and Block 2013) that reduces horizontal thrust and generates structurally efficient thin-shell designs was developed as a point of departure. The method involved the creation of a 2D particle-spring mesh tailored surface pattern and designated anchor positions and "stitching" springs that force edges of a mesh profile to be seamed together. During a physics simulation with inverse gravity, a catenary-based surface geometry is generated with a topology modified by seaming operations. The method is formally and topologically flexible and could be adapted to different design and site conditions. For the purposes of this study, one design outcome was utilized to demonstrate the overall research approach (Figure 8). As outcomes from this method are only efficient after construction is completed, additional methods had to be incorporated to address incremental manufacture.

Design for Incremental Manufacture

For the purposes of this study, incremental building viability was assessed solely through an evaluation of surface inclination and structural analysis data (deflection and principle stress) obtained in Rhino3D and Karamba3D. A design geometry containing no surface area below a 45° slope and no failure apparent in structural analysis (using a shell analysis with concrete of 50 mm thickness for proof of concept) was considered potentially viable. Future research would subject designs to further analysis throughout incremental build states and expand analysis criteria. Initial analysis of the selected design confirmed adjustments to support incremental building were required. A surface morphing method was considered that could adjust the form of the shell to reduce areas of shallow incline. However, this would result in a taller geometry of greater surface area and interior volume. A second approach considered identifying regions of shallow inclination or high structural deflection and extruding these

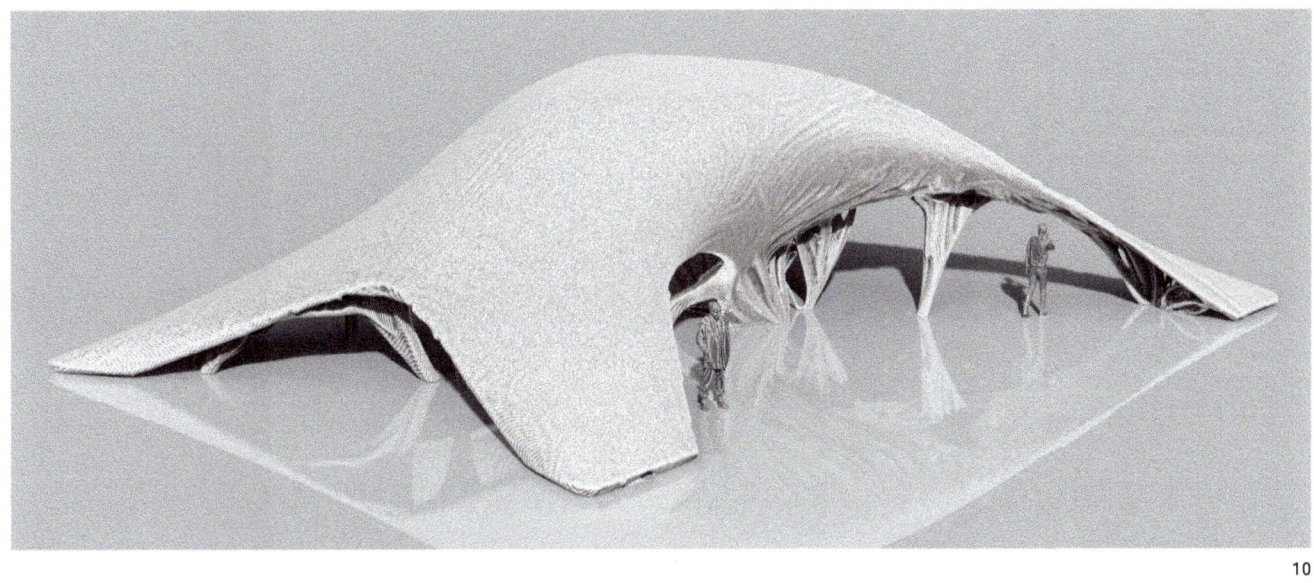

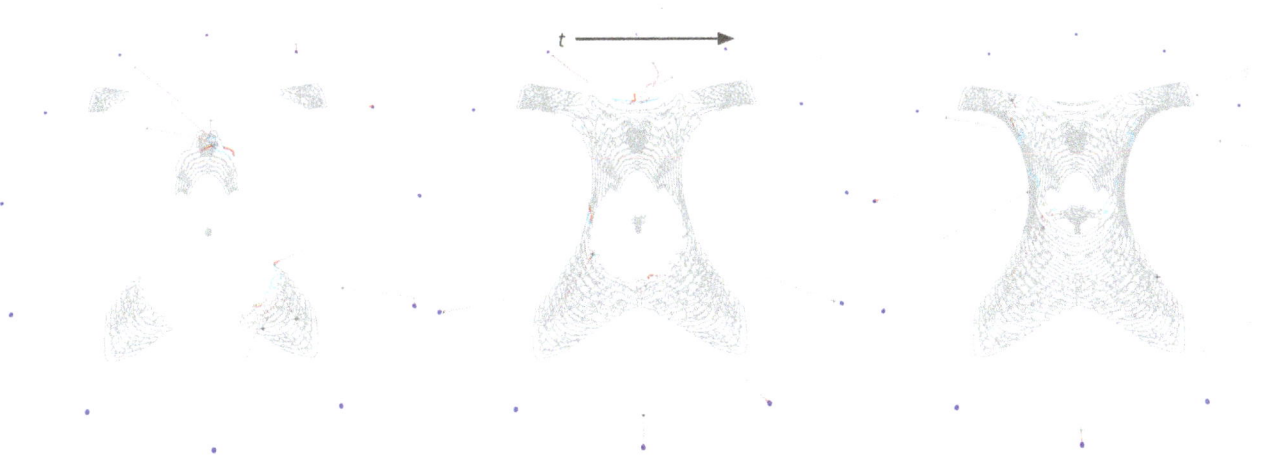

8 Funicular shell design: tailoring and seaming approach within a particle-spring physics simulation

9 Multi-Agent Local Reinforcement Method: (a) additional support vectors defined in shallow inclination regions; (b) agents are seeded in high deflection areas and move downwards towards support vectors; (c) Worm's-eye view of shell interior. Additional props and stiffening ribs primarily concentrated in shallow inclination areas that would collapse without the additional support during incremental construction.

10 Exterior of locally reinforced shell design with shallow ribbed reinforcement, showing reinforcement with greater depth to the interior

11 Collective AAM simulation of locally reinforced shell design

to the ground to create additional supports. Although structurally viable, this approach would result in increased surface area and reduced interior space. Beyond their shortcomings, both approaches also relied on global analysis, and were therefore not conducive to future incremental adjustment during building.

As an alternative, a local reinforcement method was developed (Figure 9)—which can operate prior to, and during construction—that tasked a virtual multi-agent system with local, additional material placement. The method extends prior research into multi-agent generative design (Stuart-Smith, Danahy, and Rotta 2020; Stuart-Smith and Danahy 2022), utilizing similar behaviors described in Figure 4 to seed new AM trajectories in areas of high structural stress or low inclination angle in a custom Python code. These trajectories, created by autonomous agent particles' motion over simulated time, were developed to interconnect as a networked series of 3D curves to form truss-like "prop" conditions while connecting back to the geometry in low-deflection and more vertically orientated surface conditions. Agent trajectories orientate to principle stress vectors across the geometry, seek to move downwards influenced by gravity, seek out vertical supports, and separate from the shell surface by a distance relative to local structural deflection and utilization values (increasing structural depth where it is needed). These behaviors are each described through steering vectors that are summed and weighted to an agent's velocity so that agent motion is influenced by multiple criteria. Additional vertical support locations were also created by identifying regions of

12 Worm's-eye Comparative Analysis: Initial Shell (IS) vs. Locally Reinforced Shell (RS)

Inclination analysis with red ≤ 55 deg inclination: (a) IS, (d) RS

Structural deflection
(b) IS max = 0.23 m
(e) RS max = 0.23 m

Principal stress
(c) IS
(f) RS

shallow inclination angle (below 45° slope) and high structural deflection. Support vector orientation is determined by weighting a local surface normal vector, the z-axis vector, and vectors towards adjacent support locations to ensure supports respond to structural gravitational and geometrical considerations.

Agent trajectories resulted in locally configured strut curve formations that reinforce the shell surface. These additional 3D curves were interpolated as a volumetric surface together with the shell surface by creating an isosurface in Houdini. Houdini was selected over alternative approaches due to its clustering and threshold controls over regions in the isosurface that ensured filigree features are safeguarded while overall smoothness and thickness are achieved.

Simulation Modeling and Testing of the Design

The multi-agent local reinforcement method was applied to a shell surface design outcome and analyzed for structural stress and deflection with results compared to the initial shell design (Figures 9, 12). A Collective AAM simulation was also undertaken to test the Collective AAM method's ability to adapt to the shell geometry (Figure 7). To estimate an incremental build process, structural analysis was performed on three mid-construction states to evaluate how well the initial reinforcement method faired. Future research aims to repeat the analysis and local reinforcement method frequently throughout simulated building operations. While only used prior to a Collective AAM simulation in this paper, the method was configured to support incremental use in future research, with a socket connection made between the ROS simulation environment and Grasshopper using a custom Python script. UAV positions and the completion of AM jobs are streamed to Rhino3D where building progress is able to be structurally and geometrically analysed at periodic intervals. In this scenario, analysis would evaluate already built material and several layers of future projected building activity. Areas identified as problematic in still-to-be-built regions could then be adjusted and a new Collective AAM start-up file exported. The Collective AAM ROS package can be notified of the update via the socket and update remaining AM jobs to those in the most recent Rhino3D export to ensure building operations can adapt to incremental design changes.

RESULTS AND DISCUSSION

The Collective AAM software framework effectively coordinated the autonomous local decision-making of several UAVs in simulation to concurrently undertake Collective AAM of several different geometries, including the locally reinforced shell design and effectively supported locally adaptive behavior in each UAV. Flight tests with a single DJI F450 quadcopter confirmed that an incremental vector-based approach to flight is feasible. The accuracy of the F450's flight trajectory was determined by the confluence of the waypoint vector magnitude, the frequency of waypoint instructions, and UAV velocity. Results in Figure 7 show extreme variations across 53 different flight tests. Trajectories significantly smaller than intended resulted from the UAV receiving a new waypoint prior to arriving at a previous one (illustrating a Hz rate too great for the magnitude of the waypoint). A more jagged flight that oscillates around the intended trajectory arose primarily from Hz rates too infrequent relative to waypoint vector magnitudes. Successful flights obtained an average deviation of 0.0096 m (Table 1), operating within a range that can be mitigated by pose correction on an already working Aerial AM platform (Zhang et al. 2022), validating the approach to flight and multi-agent adaptive mission-planning.

This distributed, incremental, and parallel approach to AM poses significant challenges to building design. As a first foray into this relatively unchartered area, funicular shell surface structures were identified as a reasonable departure point yet did not sufficiently meet stability requirements for

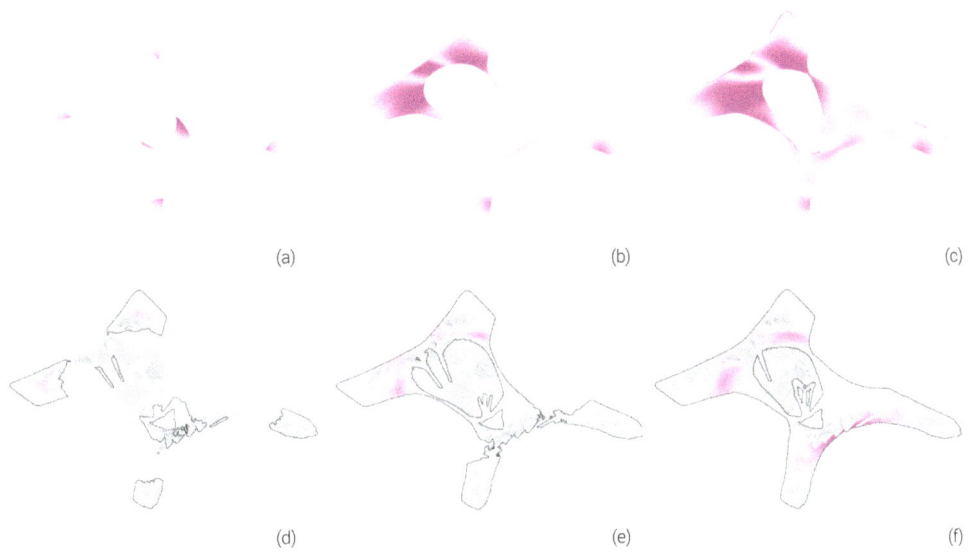

13 Worm's-eye comparative structural analysis throughout incremental building: Initial Shell (IS) vs. Reinforced Shell (RS)

Step 1: (a) IS Deflection max: 0.0057m, 30.92% within worst 25% range (d) RS Deflection max: 0.0057m, 5.26% within worst 25% range

Step 2: (b) IS Deflection max: 0.0279m, 37.56% within worst 25% range (e) RS Deflection max: 0.014m, 11.37% within worst 25% range

Step 3: (c) IS Deflection max: 0.0746m, 67.09% within worst 25% range (f) RS Deflection max: 0.0324m, 16.73% within worst 25% range

TABLE 1 UAV flight tests: best 5 of 53 flights

Trajectory Number	Hz	Waypoint Magnitude	Velocity	Average Deviation
38	100	0.001	0.1	0.0096
11	10	0.02	0.2	0.011
23	50	0.002	0.1	0.0116
12	10	0.05	0.5	0.0123
40	150	0.0005	0.075	0.0124

incremental manufacture. A local reinforcement method was developed and tested on a shell design and in a Collective AAM simulation. Structural deflection in all results was insufficient for building performance yet provided a clear demonstration in proof of concept that could easily be improved on. Results demonstrate that the initial reinforcement prior to manufacture improved the final condition's structural efficiency, reducing maximum deflection from 0.23 m to 0.06 m, while principal stress and shallow inclinations below 55° were significantly reduced. However, the method increased material volume by almost 50%. Structural analysis at three intervals during manufacture (Figure 10) highlights lower deflection levels up to 0.0324 m in the locally reinforced design, less than half that of the initial shell design which exhibited up to 0.0746 m maximum deflection. Also of note, up to 67.09% of the initial shell design geometry had deflection in its highest 25% range, while the locally reinforced design only contained 16.73% within its worst 25% range (Figure 13). Thus, the locally reinforced design's structural performance was considerably better in mid-construction stages compared to the initial shell design, indicating the design approach made significant improvements relative to incremental building considerations. Future work should implement the method in a feedback loop throughout Collective AAM to explore real-time design optimization throughout interim building steps. Further optimization is also possible of the analysis thresholds and local reinforcement variable settings that could reduce additional material volume.

The method embodied a unique aesthetic integral to the manufacturing approach (Figure 14), and while facilitating incremental construction, added considerably more material to the design, raising questions on how best to evaluate the method's utility relative to existing on-site AM and design approaches. At a fundamental level, the cost of an AM building could be considered to be the product of material volume (and type), the energy and time involved in construction, together with transportation to and from construction operations (transportation however is related to material volume and robot population). These same metrics also determine a design's affordability (social) and environmental impact. A fitness metric could therefore be described as:

$$\text{Collective AAM Efficiency} = a\left(\frac{\text{Floor Area}}{\text{Material Volume}}\right) + b\left(\frac{\text{Construction Time}}{\text{Robot Population} \times \text{Energy}}\right)$$

Where 'a' and 'b' are constants that weight the building outcome relative to the construction process. As an initial exploration of the method, the construction time and UAV energy expense are not yet sufficiently accurate. Much work can be done to improve the efficiency of the Collective AAM approach, such as by optimizing the number and distribution of UAVs and base stations. While flight seems to impose additional energy requirements, the small, distributed hardware of UAVs could support passive solar charging if total build time is not critical. Alternatively, hot-swapping batteries should be evaluated for energy draw relative to larger AM gantry or

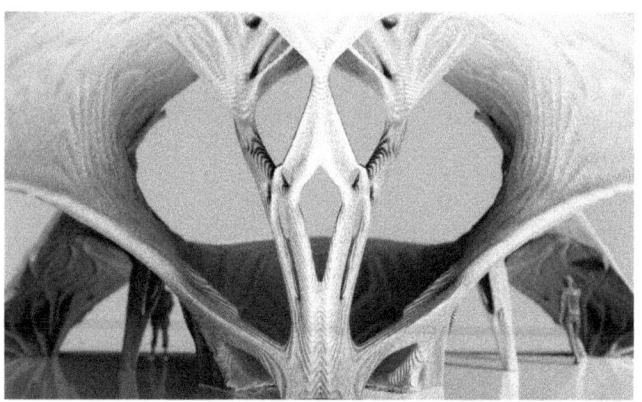

14 View from inside shell; additional supports prop up parts of the shell with a shallow inclination angle

crane systems that typically run on 480 V power. Although the incremental design approach could theoretically also be applied to a single end-effector AM gantry system, the advantages of the Collective AAM approach would come to the fore, where parallel manufacturing supports a building sequence that reduces the quantity of mid-build supports compared to a gantry system. Such studies should be explored across several different design geometries to evaluate Collective AAM relative to gantry AM approaches.

CONCLUSION

The Collective AAM mission-planning software extended CRC into aerial AM capabilities in both simulation modeling and actual flight tests. Collective AAM simulations illustrate an approach to distributed, adaptive, and concurrent building by a swarm of aerial robots, while the design approach to 3D shell geometries highlight the gap in capabilities between current on-site AM methods and those needed to support incremental distributed on-site AM. The proposed method provides an initial proof of concept yet leaves several issues on stability dynamics unexplored that could be pursued in ongoing research. Further optimization of the Collective AAM framework is also required, and implementation on a team of UAVs for a multi-robot flight demonstration where additional challenges in networking infrastructure and latency will need to be addressed. Following this, development and testing on several AAM platforms with manufacturing capabilities will be undertaken, to evaluate physical MEX AM by multiple robots in flight.

The proposed design and incremental local reinforcement method hybridizes permanent supports within a 3D shell design in lieu of temporary scaffolding. While this reduces material efficiency relative to traditional shell structures, it offers a means to build a complete structure using autonomous, distributed aerial robots without necessitating scaffolding or additional building systems. Further development should be undertaken in optimizing this approach and ensuring it operates in-the-loop throughout building. The research demonstrates an alternative approach to construction and design that is in early stages of development and does not address broader questions on the scenarios in which this technology is best placed to supplement existing building methods. Considerable work remains in establishing best-fit applications for the approach, and in evaluating the environmental and socio-economic impact of Collective AAM. While future work will explore these issues, it is hoped that this research offers some encouragement towards more materially efficient building designs that could support a site-specific approach to bespoke AM Design for Manufacture (DFMa).

ACKNOWLEDGMENTS

This research was undertaken in the Autonomous Manufacturing Lab at both the University of Pennsylvania (Architecture) and University College London (Computer Science), and in the Aerial Robotics Lab at Imperial College. The research is also supported by EPSRC funding EP/N018494/1 & EP/S031464/1. Mirko Kovac was also supported by a Royal Society Wolfson Fellowship.

REFERENCES

Andreen, David, Petra Jenning, Nils Napp, and Kirstin Petersen. 2016. "Emergent Structures Assembled by Large Swarms of Simple Robots." In *ACADIA 16: Proceedings of the 36th Annual Conference of the Association for Computer Aided Design in Architecture (ACADIA)*.

Augugliaro, Federico, Ammar Mirjan, Fabio Gramazio, Matthias Kohler, and Raffaello D'Andrea. 2013. "Building Tensile Structures with Flying Machines." In *2013 IEEE/RSJ International Conference on Intelligent Robots and Systems*. 3487-92.

Augugliaro, Frederico, Sergei Lupashin, Michael Hamer, Cason Male, Markus Hehn, Mark W. Mueller, Jan Sebastian Willmann, Fabio Gramazio, Matthias Kohler, and Raffaello D'Andrea. 2014. "The Flight Assembled Architecture Installation: Cooperative Contruction with Flying Machines." *IEEE Control Systems* 34 (4): 46–64. https://doi.org/10.1109/MCS.2014.2320359.

Chen, A., and M. Yossef. 2015. "Applicability and Limitations of 3D Printing for Civil Structures Applicability and Limitations of 3D Printing for Civil Structures." In *Proceedings of the 2015 Conference on Autonomous and Robotic Construction of Infrastructure*. 237-246.

Cuito, Aurora, and Cristina Montes. 2003. *Antoni Gaudi*. Cologne, Germany: DuMont.

Duque Estrada, Rebeca, Fabian Kannenberg, Hans Jakob Wagner, Maria Yablonina, and Achim Menges. 2020. "Spatial Winding: Cooperative Heterogeneous Multi-Robot System for Fibrous Structures." *Construction Robotics* 4 (3–4). https://doi.org/10.1007/s41693-020-00036-7.

Gibson, Ian, David Rosen, Brent Stucker, and Mahyar Khorasani. 2021. *Additive Manufacturing Technologies*. Cham: Springer International

Publishing. https://doi.org/10.1007/978-3-030-56127-7.

Jokic, Sasha, Petr Novikov, Shihui Jin, Stuart Maggs, Cristina Nan, and Dori Sadan. 2022. "Minibuilders." *IAAC*. Accessed April 14, 2022. http://robots.iaac.net/.

Lindsey, Quentin, Daniel Mellinger, and Vijay Kumar. 2011. "Construction of Cubic Structures with Quadrotor Teams." In *Robotics: Science and Systems Conference Proceedings VII*. https://doi.org/10.15607/RSS.2011.VII.025.

Napp, Nils, and Radhika Nagpal. 2014. "Distributed Amorphous Ramp Construction in Unstructured Environments." *Robotica* 32 (2): 279–90.

Nejur, Andrei, and Masoud Akbarzadeh. 2021. "PolyFrame, Efficient Computation for 3D Graphic Statics." *CAD Computer Aided Design* 134. https://doi.org/10.1016/j.cad.2021.103003.

Petersen, Kirstin H., Nils Napp, Robert Stuart-Smith, Daniela Rus, and Mirko Kovac. 2019. "A Review of Collective Robotic Construction." *Science Robotics* 4 (28). https://doi.org/10.1126/scirobotics.aau8479.

Petersen, Kirstin, Radhika Nagpal, and Justin Werfel. 2011. "TERMES: An Autonomous Robotic System for Three-Dimensional Collective Construction." In *Robotics: Science and Systems Conference Proceedings VII*.

Reynolds, C. W. 1999. "Steering Behaviors for Autonomous Characters." In *Proceedings of Game Developers Conference 1999*. San Jose, California. 763–82.

Rhoad, R., G. Milauskas, and R. Whipple. 1991. *Geometry for Enjoyment and Challenge*. Evanston: McDougal, Littell.

Rippmann, Matthias, and Philippe Block. 2013. "Funicular Shell Design Exploration." In *ACADIA 2013: Adaptive Architecture; Proceedings of the 33rd Annual Conference of the Association for Computer Aided Design in Architecture*. 337–346.

Seo, Jungwon, Mark Yim, and Vijay Kumar. 2013. "Assembly Planning for Planar Structures of a Brick Wall Pattern with Rectangular Modular Robots." In *2013 IEEE International Conference on Automation Science and Engineering (CASE)*. 1016–21. https://doi.org/10.1109/CoASE.2013.6653996.

Stuart-Smith, Robert, and Patrick Danahy. 2022. "Visual Character Analysis Within Algorithmic Design, Quantifying Aesthetics Relative To Structural And Geometric Design Criteria." In *CAADRIA 2022, Post-Carbon; Proceedings of the 27th CAADRIA Conference*, vol. 1, edited by Jeroen van Ameijde, Nicole Gardner, Kyung Hoon Hyun, Dan Luo, and Urvi Sheth. Sydney, Australia. 131-40.

Stuart-Smith, Robert, Patrick Danahy, and Natalia Revelo la Rotta. 2020. "Topological and Material Formation." In *ACADIA 2020: Distributed Proximities; Proceedings of the 40th Annual Conference of the Association for Computer Aided Design in Architecture*, vol. 1, edited by B. Slocum, V. Ago, S. Doyle, A. Marcus, M. Yablonina, and M. del Campo. 290–299.

Sustarevas, Julius, K. X. Benjamin Tan, David Gerber, Robert Stuart-Smith, and Vijay M. Pawar. 2019. "YouWasps: Towards Autonomous Multi-Robot Mobile Deposition for Construction." In *2019 IEEE/RSJ International Conference on Intelligent Robots and Systems (IROS)*, Macau, China. 2320–2327. https://doi.org/10.1109/IROS40897.2019.8967766.

Veenendaal, D., and P. Block. 2012. "An Overview and Comparison of Structural Form Finding Methods for General Networks." *International Journal of Solids and Structures* 49 (26): 3741–3753. https://doi.org/10.1016/j.ijsolstr.2012.08.008.

Werfel, Justin, Kirstin Petersen, and Radhika Nagpal. 2011. "Distributed Multi-Robot Algorithms for the TERMES 3D Collective Construction System." In *Proceedings of the IEEE/RSJ International Conference on Intelligent Robots and Systems (IROS 2011)*. 1–6.

XtreeE. n.d. "XtreeE | The Large-Scale 3D." Accessed February 10, 2022. https://xtreee.com/en/.

Zhang, K., P. Chermprayong, F. Xiao, D. Tzoumanikas, B. Dams, S. Kay, B. B. Kocer, A. Burns, L. Orr, C. Choi, D. D. Darekar, W. Li, S. Hirschmann, V. Soana, S. A. Ngah, S. Sareh, A. Choubey, L. Margheri, V. Pawar, R. Ball, C. Williams, P. Shepherd, S. Leutenegger, R. Stuart-Smith, and M. Kovac. 2022. "Aerial Additive Manufacturing with Multiple Autonomous Robots." *Nature* 609 (7928): 709–17. https://doi.org/10.1038/S41586-022-04988-4.

IMAGE CREDITS
All drawings and images by the authors.

Robert Stuart-Smith is Director of the MSD-RAS program at the University of Pennsylvania, Director of the Autonomous Manufacturing Lab (AML) at Penn (Architecture), and Co-Director of the AML at Univesity College London (Computer Science).

Durgesh Darekar is a researcher in the Autonomous Manufacturing Lab (AML) at Univesity College London (Computer Science).

Patrick Danahy is a researcher in the Autonomous Manufacturing Lab (AML) at University of Pennsylvania (Architecture) and Assistant Research Professor of Architecture and Design Innovation Fellow at Ball State University.

Basaran Bahadir Kocer is a researcher in the Aerial Robotics Lab at Imperial College London, UK.

Vijay Pawar is Co-director of the Autonomous Manufacturing Lab (AML), and co-director of the Touch Lab at Univesity College London (Computer Science).

Mirko Kovac is Director of the Aerial Robotics Lab at Imperial College London, UK, and Head of the Materials and Technologies Center of Robotics at the Swiss Federal Laboratories for Material Science and Technology (EMPA), and is a Royal Society Wolfson Fellow.

Session Introduction

Advanced Production Methods I: Robotic Fabrication & Material Formation

Jose Luis Garcia del Castillo López, Chair

The session "Advanced Production Methods I: Robotic Fabrication and Material Formation" explores the cutting-edge realm of computational fabrication methods, pushing the boundaries of our understanding of material making, and uncovering novel opportunities to create frameworks that encourage matter to form, shape, and even express itself. The research presented in this chapter not only delves into the intricacies of robotic fabrication and material formation, but also challenges standard assumptions of how we make things with machines. By reimagining traditional techniques and embracing unconventional material behavior, the authors present groundbreaking examples of interwoven understanding of material and fabrication processes.

One of the key topics investigated in this session is the innovative use of 3D concrete printing in architectural design, fabrication, and assembly strategies, particularly focusing on column-slab interfaces. By developing new approaches to optimize the structural and aesthetic properties of these integral elements, the authors demonstrate the significance of adopting unconventional techniques to enhance the potential of concrete as a material in architectural design.

Another area of exploration is the concept of non-linear fabrication, which aims to streamline the process of rapid data collection and toolpath recalibration in 3D printing of clay lattices. This approach challenges conventional linear methods and demonstrates the potential for employing real-time feedback to create more efficient, adaptable, and responsive fabrication processes. The authors showcase how these advances can lead to more intricate and complex structures, ultimately expanding the creative possibilities in the field of architecture.

The chapter also delves into the exciting intersection of 3D printing and shape memory alloys, specifically focusing on the development of 3D printed bi-stable geometries and shape memory alloys for energy-efficient kinetic elements. By exploiting the unique properties of these materials, the authors present innovative solutions for the creation of responsive and adaptable architectural components that can contribute to a more sustainable and dynamic built environment.

Lastly, the concept of smart branching is examined, offering a fresh perspective on material behavior and fabrication processes. By embracing the inherent characteristics of materials, such as sagging, squeezing, or slumping, and reconsidering what constitutes undesirable material behavior, the authors reveal how these natural properties can be harnessed and integrated into the design process to produce sophisticated, organic forms.

The authors' collective research and insights offer a glimpse into the future of architectural design and material innovation. By challenging conventional assumptions and embracing the unexplored potential of materials and fabrication processes, these studies lay the foundation for a new era in architectural design, one that is more adaptable, sustainable, and expressive than ever before.

Column-Slab Interfaces for 3D Concrete Printing

Design, Fabrication and Assembly Strategies

Ana Anton
D-ARCH/ETH Zurich

Eleni Skevaki
D-ARCH/ETH Zurich

Patrick Bischof
D-BAUG/ETH Zurich

Lex Reiter
D-BAUG/ETH Zurich

Benjamin Dillenburger
D-ARCH/ETH Zurich

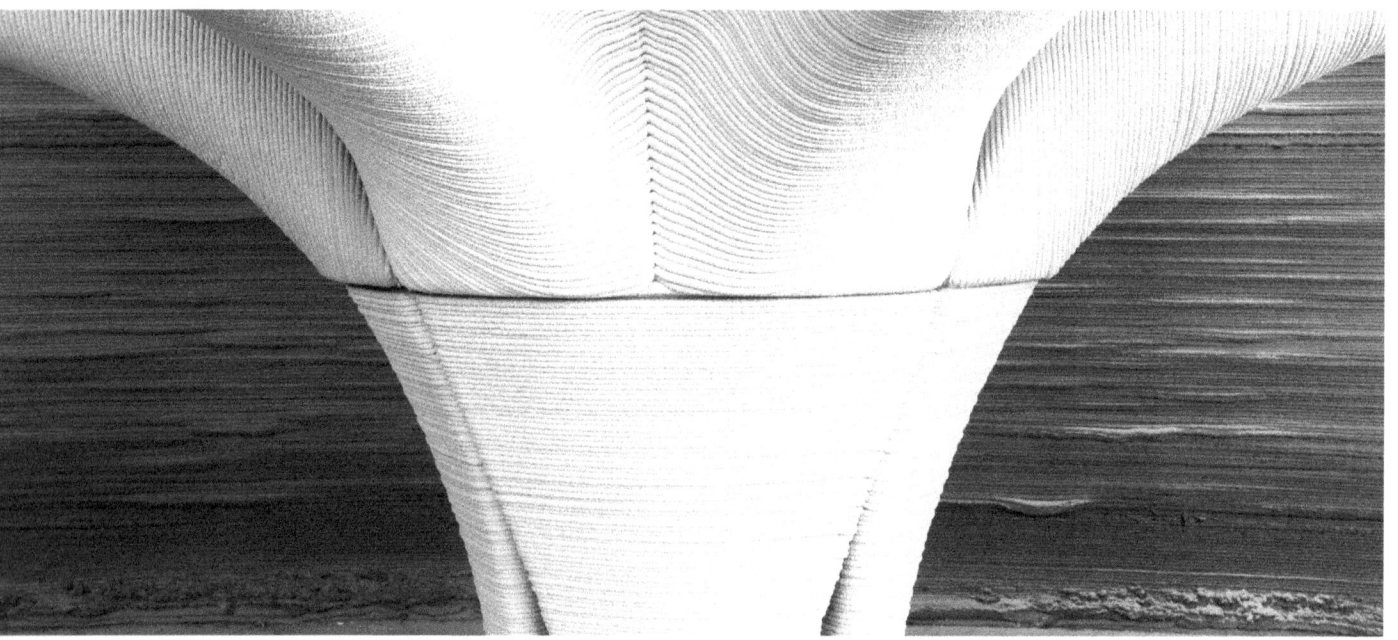

1 Detail of the column-slab connection

ABSTRACT

3D Concrete Printing (3DCP) currently dominates the scene of digital fabrication with concrete. 3DCP can be utilized on-site or in prefabrication setups. While prefabrication with 3DCP allows for more complex construction elements, it also requires the design for connections and assembly.

In the context of prefabrication using 3DCP, this paper illustrates the state of research in the design, construction, and assembly of 3D printed components. It proposes segmentation and fabrication strategies to produce horizontal and vertical structural members of a column-slab building system following the typology of mushroom slabs. The design proposal was experimentally investigated with two prototypes. Each prototype consists of five elements assembled into 1:1 scale, column-slab connection demonstrators. These prototypes were robotically fabricated using a three-component 3DCP material system and a hardware setup enhanced by a sensor system that supplies online fabrication data for process monitoring and quality control. Based on 3D scans, the recorded fabrication data set is benchmarked against the initial design models to assess the accuracy of the printed elements.

Conclusions about process robustness and prefabrication quality of the components are drawn, focusing on the interfaces between components. Finally, mitigation strategies are suggested to enhance the accuracy of the proposed fabrication method.

INTRODUCTION

Despite continuous innovation, recent studies show that concrete technology accounts for 8% of the global CO_2 emissions (Scrivener, John, and Gartner 2018). Nevertheless, the vast availability of raw materials, high structural performance, good fire resistance, and formability make reinforced concrete irreplaceable in many current applications at scale (Van Damme 2018).

Structural shape optimization is a promising direction toward lowering concrete consumption, ideally suited to digital fabrication. Geometric variation and mass customization at no economic disadvantage are just two of the predicted benefits of digital fabrication. As part of the growing repertoire of digital fabrication technologies in the construction sector (Wangler et al. 2019), 3D Concrete Printing (3DCP) can produce freeform concrete elements, without formwork, directly from CAD data (Khoshnevis 2004).

Designs resulting from an optimization process tend to have complex shapes. 3D printing them in the same orientation as their final use is rarely possible, a statement specifically relevant for shape optimized ribbed slabs. This is where segmentation and reorientation of the components on the print bed becomes an important tool enabling the fabrication of geometrically-complex assemblies. Furthermore, the segmentation entails the need for designing connections to (i) transfer loads and (ii) rationalize the assembly. Structural connections can be realized as a reinforced or unreinforced construction joint (wet joint) or dry joint. A construction joint may be understood as an interface with new concrete, mortar, or adhesive applied against existing concrete. Dry joints are utilized when assembling two precast elements without adding on-site concrete or adhesives. Dry joints often incorporate fitting keys to (i) transfer shear stresses across the joint and (ii) facilitate the assembly of precast elements (cf. keyed connections in precast segmental bridges).

Several available examples of columns (Gaudillière et al. 2019; Anton et al. 2021) and slabs (Sliskovic et al. 2020; Anton et al. 2020) demonstrate the scalability of 3DCP technology for architectural applications. Nevertheless, despite numerous attempts to realize individual construction elements, complete structurally valid column-slab building systems fully integrating architectural design, structural design, and fabrication are in the early stages of development. Such a column-slab generally requires reinforcement integration because the structural requirements exceed the structural capacity of the 3DCP material. However, suitable reinforcement applications for 3DCP are still under investigation (Mechtcherine et al. 2021). However, knowledge on (i) tectonics, (ii) segmentation, printability, assembly strategies, all informed by structural considerations, (iii) fabrication constraints, (iv) handling, transportation, building sequence, and logistics, is available and could foster the construction of bespoke concrete structures if adequately adapted to the novel 3DCP technology. Given this insight, we hypothesize that applying tectonic principles is a practical approach to developing a column-slab building system. We start this approach by addressing the column-slab connection, a crucial structural detail given the high bending moments and normal forces often prevailing in this region. This design-driven investigation prioritizes fabrication strategies, leaving the dimensioning of the structural elements as an open design parameter of the system.

AN OVERVIEW OF 3DCP ASSEMBLIES

Production constraints for joints define the design of the actual structure, which is especially relevant for prefabricated construction (Bischof, Mata-Falcón, and Kaufmann 2022). Especially when using dry joints for prefabricated construction, precision is critical in prefabrication because components are expected to fit without on-site modifications, in order to accelerate their assembly.

The typical tolerance for construction joints is +/- 10 mm (Ganz and Fischer 2014), while the tolerances for fitting dry joints are far lower. Nevertheless, small tolerances are very demanding for 3DCP: besides minor challenge of positioning the robot, the deformations usually occurring after the material is deposited are difficult to quantify. 3DCP utilizes freely-deposited fluids

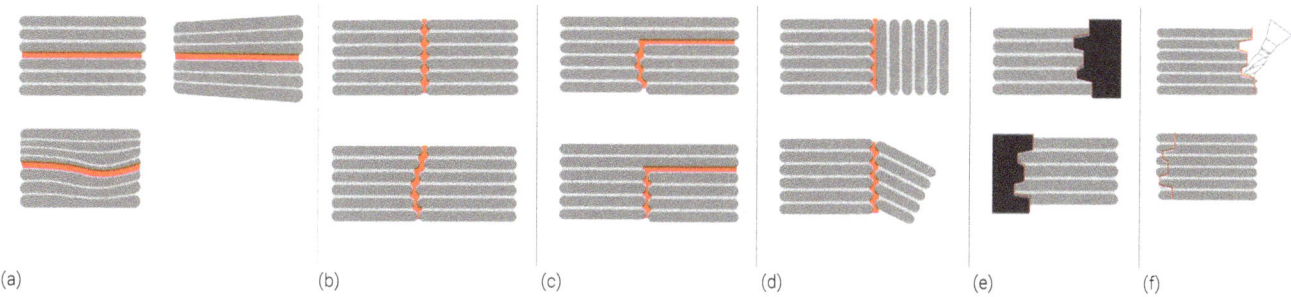

2   Fabrication strategies for 3DCP interfaces: (a) along the layers; (b) transversal to the layers; (c) combining along and transversal layers; (d) independent of layer orientation; (e) high precision formwork or match casting; (f) post-processing with milling or cutting

that tend to warp, shrink, or crack. The most important parameters driving these deformations are the initial yield stress of the filament, strength buildup, temperature and humidity of the environment, surface water evaporation, shrinkage during hardening, and resulting internal stresses. Furthermore, most 3DCP configurations entail a stepped surface texture characteristic of the layered deposition.

The challenge of complying with small tolerances becomes essential when working with non-standard forms. Building on the competitive advantage of 3DCP in the realm of bespoke and freeform geometries (De Schutter et al. 2018) implies the necessity to identify practical solutions for the fabrication of connections. These solutions may be specific to loading considerations: this paper considers solutions for the transfer of compressive stresses and shear stresses but does not discuss solutions for the transfer of tensile stresses.

Fabrication Strategies
Figure 2 shows various fabrication strategies that consider positioning on the print bed and layer orientation when deciding the segmentation strategy for creating unreinforced construction and dry joints. Unreinforced construction joints require surfaces with high roughness. Such a high roughness may be generated by exploiting the characteristic stepped surface texture of 3DCP elements. A precise dry joint interface (Figure 2e) may be generated by printing on top or against (i) a high-definition formwork (Anton et al. 2020) or (ii) a previously produced 3DCP element, simulating the match casting process (Gebhard et al. 2022). Furthermore, strategies to fabricate dry joints include post-processing, e.g. by milling, as shown in Figure 2f (Battaglia, Miller, and Zivkovic 2019; Muñiz et al. 2021).

When segmenting structural members or components following the layer orientation and aiming for an interface defined by the slicing strategy (Figure 2a), the layers can be planar and parallel (Bos et al. 2019), planar with an inclination (Grasser et al. 2020; Bhooshan et al. 2022) or curved (Anton et al. 2019). When segmenting structural members or components in the direction perpendicular to the layers (Figure 2b), the resulting elements meet on interfaces transversal to the layer orientation (Figure 2c). This is the case of a design investigation for a furniture piece (Shaker et al. 2021). Another consideration, especially relevant when printing customized elements with high geometric complexity, is the segmentation incorporating an angle between the layering of adjacent elements (Figure 2d).

Joining Strategies
Figure 3 summarizes the joining strategies for connections. Unreinforced construction joints and dry joints serve to transfer compressive and shear stresses. While the capacity to transfer compressive stresses mainly relies on the accuracy of joining surfaces, the capacity to transfer shear stresses depends on the roughness to enhance frictional resistance and provide mechanical interlock, which is generally achieved by using shear keys as shown in Figure 3c. Tensile stresses can only be transferred with reinforcement continuity across the joint, a strategy which is not discussed in this paper.

The methods to provide high accuracy of joining surfaces include the provision of (i) means to fabricate matching surfaces (see Fabrication Strategies section), (ii) mortar or adhesive between the two joining surfaces, or (iii) neoprene layer to ensure uniformly distributed compressive stresses, as applied by Bos et al. (2019). Such accurate joining surface may be post-tensioned, as shown by Bos et al. (2019), or shaped to generate compressive stresses (Bhooshan et al. 2022).

3DCP components are generally not precise enough for direct assembly after printing, given their layered surface texture. With construction joints, these imprecisions are usually mitigated by the use of mortar or adhesive (Figure 3a). Note that additional formwork efforts may be necessary on-site when applying mortar or adhesive to connect adjacent surfaces.

DESIGN AND FABRICATION METHODS FOR THE COLUMN-SLAB CONNECTION DETAIL

This paper presents a segmentation and assembly strategy using mortar to connect the 3DCP components forming a mushroom column-slab connection detail. With the application of mortar, the assembled components are not dismountable. The structural continuity is achieved through casting regular reinforced concrete inside the 3DCP shell.

3 Joining strategies for 3DCP components: (a) construction joints using mortar or other adhesives; (b) dry joints using neoprene membranes; (c) dry joints using shear keys

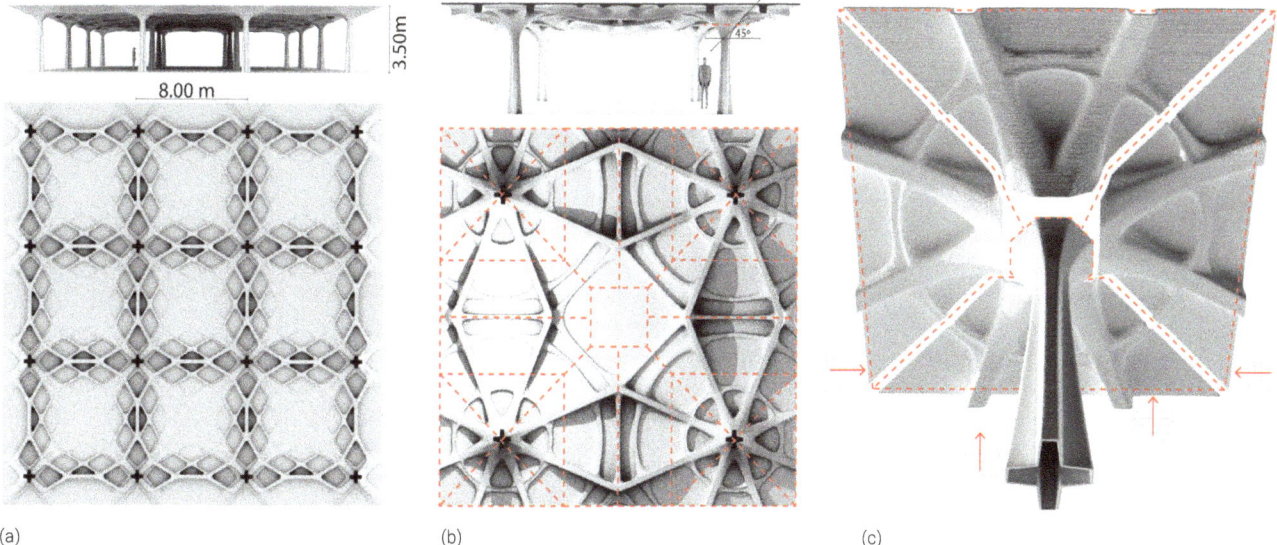

4 Design of the column-slab system: (a) design variation of the column-slab building system; (b) segmentation strategy for the space between 4 columns; and (c) connection detail between column and slab, including printing orientation for each component. The dotted lines represent the segmentation trajectories; arrows represent layers' direction.

The geometric freedom enabled by 3DCP offers the opportunity to rethink a column-slab connection and produce seamless transitions between vertical and horizontal building members. The proposed methodology illustrates the segmentation, fabrications, and assembly steps necessary to create such a column-slab connection detail. The workflow consists of the following steps: design, segmentation, print-path generation, fabrication, assembly, rebar placement, and casting.

Design and Segmentation

The presented case study consists of a 3DCP column-slab building system with a mushroom column capital and a two-way ribbed slab (Figure 4) (Mata-Falcón et al. 2022). The structural system, consisting of a ribbed slab and punctual supports, follows the direction of principle moments, as introduced by Pier Luigi Nervi and Aldo Arcangeli for The Gatti Wool Factory in Rome (Bologna and Gargiani 2016). This system consists of radially distributed main ribs starting from the column support connected by secondary rings.

The design employs mesh subdivision as a modeling strategy to generate the overall form of the ensemble, which can be scaled up and adapted for different spans and spatial configurations. The segmentation process unites printability constraints and

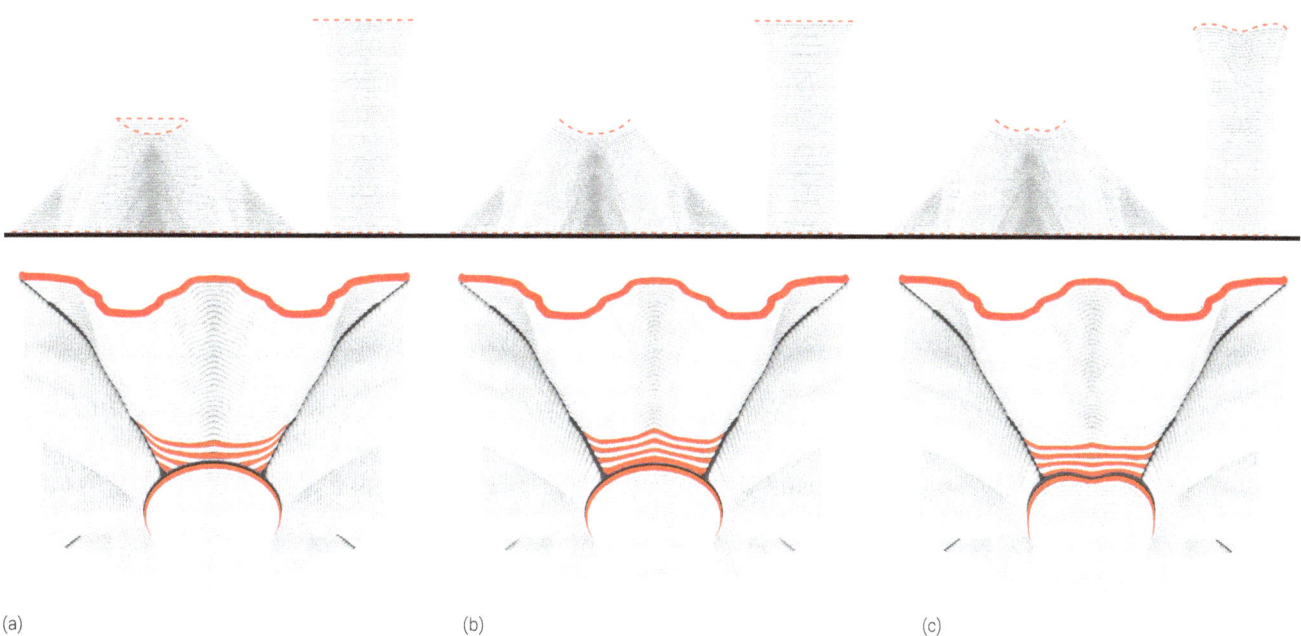

5 Fabrication of the column-slab detail: (a) with horizontal layers; (b) column with horizontal layers and slab with curved layers; (c) with curved layers.

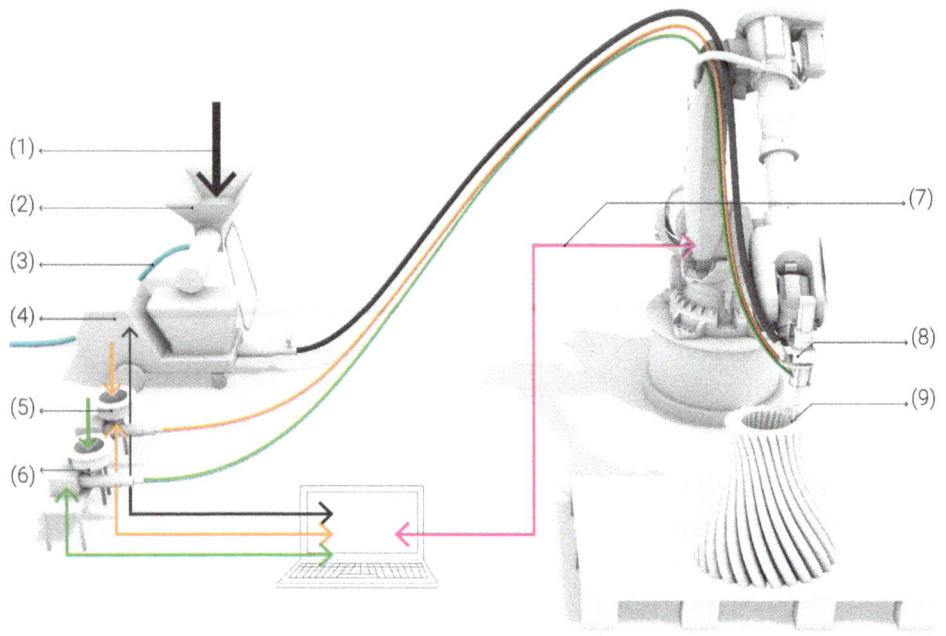

6 3DCP Fabrication setup: (1) dry mix inlet; (2) concrete mixer; (3) water connection; (4) concrete pump; (5) accelerator pump; (6) thickener pump; (7) motion control and feedback using the COMPAS FAB (Rust et al. 2018) and the COMPAS RRC (Fleischmann and Casas 2020) framework; (8) extruder tool; (9) TCP of the extruder tool

7 Prototyping the slab segments: (a) two by two; (b) as individual segments; (c) using curved layers

8 Prototypes: (a) using print-path of type 2; (b) assembly of components using silicone (top) and mortar (bottom); (c) 3DCP formwork assembly, rebar placement (top), and concrete casting (bottom); (d) using curved layers as connection detail

aesthetic considerations. The segments were selected such that their maximum overhang remains under 45° as a fabrication requirement. The 3DCP formwork of the column capital is detached from the slab by a horizontal segmentation plane. The formwork of the remaining slab segment is split into four pieces, following the diagonals of the square perimeter.

This approach aims to utilize principles of stereotomy and segmentation to inform the subdivision operation. Consequently, the resulting four slab segments (Figure 4c shows position within the structure; Figure 7 shows position during printing) are rotated 90° after printing while the column capital maintains its original orientation.

Print Path Design
Virtual prototyping was utilized first as a method to develop the column-slab connection detail. This means that possible variations of how the five components come together were modeled up to the layer's scale (Figure 5). The three variations depend on how layers are deposited to create a tectonic continuity from component to component. The three resulting variations, illustrated in Figure 6 synthesize the approach of layer design, which ranges from fully planar layers (Figure 5a) to only curved layers for the connection between slab and column (Figure 5c). Physical prototyping was tested for solutions shown in Figures 5a and 5b.

Fabrication Setup
The printing system is schematically illustrated in Figure 6. It utilizes a three components (3K) material processing strategy, where each of the components is individually dosed to a custom extruder tool mounted on the sixth axis of an ABB 6700, with an arm reach of 3.05 m and a payload of 175 kg. A continuous horizontal mixer PFT HM 24 with a hopper is used to prepare the concrete by mixing a proprietary dry mix with water. Three progressive cavity pumps are used for pumping the material to the extruder tool: a PFT Swing L FU 400V concrete pump and two Viscotec ViscoPro-C for accelerator and thickener.

The dry mix consists of quartz aggregates with a particle size between 0 to 1.2 mm, fine limestone, white CEM I binder, superplasticizer, and stabilizer. The accelerator is based on a white calcium aluminate cement paste, and the thickener is a sodium metasilicate water suspension (Reiter et al. 2022).

The printing parameters were defined by a 0.5 L/min base mix, 0.03 to 0.04 L/min accelerator, and 0.007 L/min thickener, printed with a robot Tool Central Point (TCP) speed of 80 mm/s. These parameters generate a 20 mm layer width and 5 mm height filament cross-section. This technique is suitable for a free flow concrete filament that adjusts its size based on the parameters. For example, variable layer height is achieved using variable print speed while keeping the flow rates constant (Anton et al. 2019).

The robot motion and hardware control parameters, such as flow rate and start/stop operations, are integrated into a centralized control that operates using the COMPAS FAB (Rust et al. 2018) and the COMPAS RRC (Fleischmann and Casas 2020) open-source computational framework for robotic fabrication.

3D Printing for the Column-Slab Connection Detail
The prototyping phase includes two fabrication strategies for creating the print path of the slab segments. The first strategy

(a) (b) (c) 7

entails printing the slab elements two by two into a continuous print path (Figure 7a). Here, the components have more stability during printing, owing to the large contact surface with the print bed and a more extended contour. Moreover, the connection between the two elements is able to direct some tension efforts through the connecting concrete, resulting in a lower bending torque on the supporting base. However, a drawback of this strategy is that it requires an additional post-processing step: cutting the elements into their final shape, thus generating extra work and waste.

The second strategy involves printing every slab segment individually (Figure 7b). In this case, the contour of one layer does not form a closed polygon. Hence, once the print path reaches the end of one outline, the tool moves in the vertical direction with a distance corresponding to the layer height. The subsequent layer is printed in a reversed direction with respect to the former. Such a print path requires more stringent fabrication parameters: one contour is shorter; therefore, a faster buildup rate is necessary, and when the extruder returns on a freshly deposited concrete, the initial filament yield stress needs to be higher. This print-path strategy requires no additional post-processing of the finished print, which may constitute a distinct advantage over the previously presented approach. However, the challenging fabrication highly relied on the applied 3K fabrication setup.

These two print-path strategies were tested for both horizontal and curved layers. Nevertheless, the success rate of printing with curved layers remained relatively low, as the print quality of objects in Figures 7c and 8d show. Ultimately, the fabrication inaccuracies resulting from printing with curved layers can be mitigated through an independent and extensive study of the printing parameters combined with tool orientations following the normal vector to the print path.

Data Acquisition

The acquisition of fabrication data is essential for assessing the quality of what was 3D printed. Feedback during printing and scanning after printing contribute to evaluating the fabrication accuracy of our system.

A fabrication data set was recorded during the printing process with the help of the COMPAS FAB open-source software (Rust et al. 2018) and COMPAS RRC (Fleischmann and Casas 2020), the ABB communication library. One data point consists of

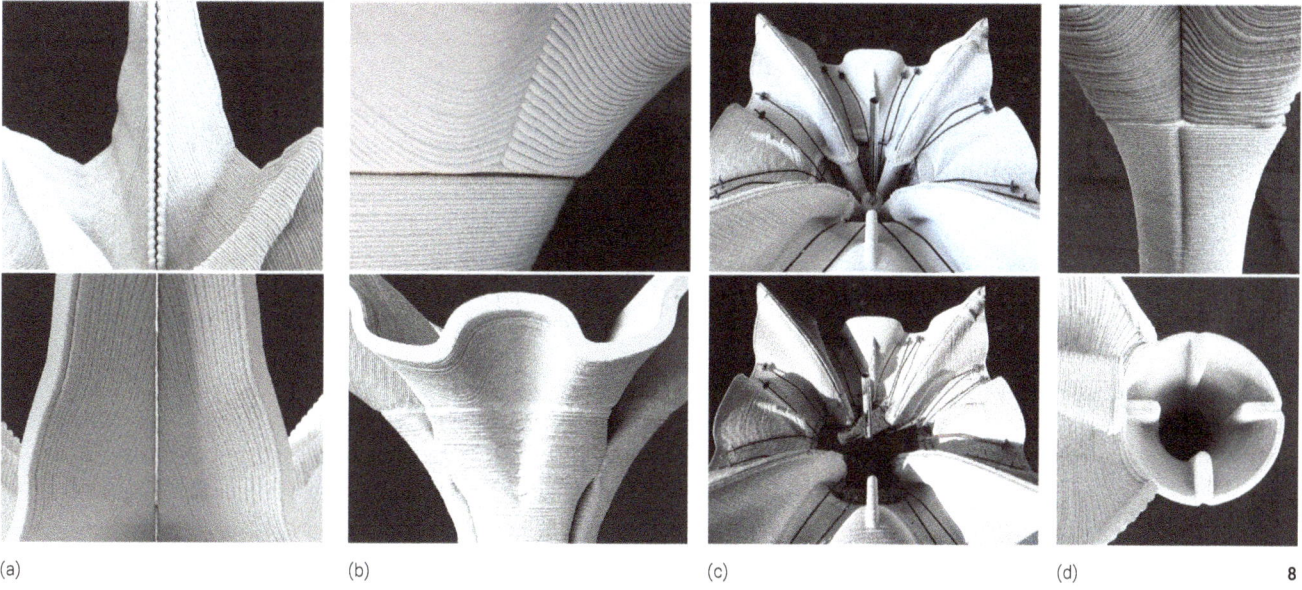

(a) (b) (c) (d) 8

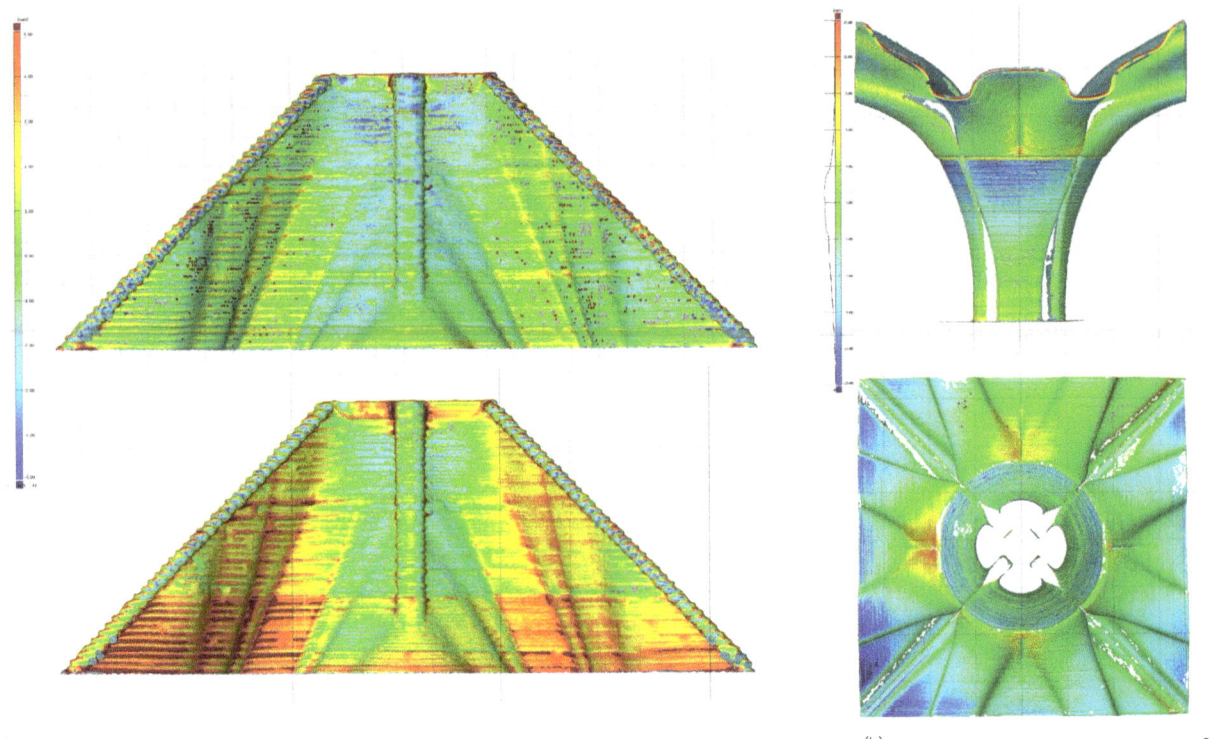

(a) (b) 9

information on the x-, y-, and z- coordinates of the TCP, the speed at the TCP of the robot, the concrete flow rate, and the accelerator flow rate. This data set is sufficient to create a 3D model of the printed element using the recorded printing parameters. Here, the thickness of the filament is not geometrically pre-defined but calculated as a function of the target layer height, recorded TCP speed, and recorded flow rates.

After printing, the samples were scanned with a structured light 3D scanner, Creaform GoScan 50. The collected data was further processed and compared with the designed meshes inside GOM Inspect software (GOM Metrology 2022) (Figure 9).

This study includes the following comparisons: geometrical deviations resulting from (i) recorded print data and the scanned geometry and (ii) the designed geometry and the scanned geometry. The first comparison evaluates if the feedback data sufficiently describes the physical object, potentially eliminating the need to scan altogether. The second comparison serves to identify what are the deviations of the printed element from the designed element. This process was applied both to each 3DCP element and to the assembly.

RESULTS & DISCUSSION

The assembly process followed the successful fabrication of one column capital and four identical slab segments. After printing, the segments are left to harden for seven days, when the concrete acquires sufficient strength to allow moving and post-processing the components. The excess material from the slab elements (Figure 7a) was manually cut with a circular saw. The cutting lines were positioned along a purposely-designed geometric fold to make the manual cut invisible in the final assembly; thus, the high precision of this post-processing step is not a strict fabrication demand.

The four slab elements were placed on top of the column capital and held falsework, including punctual props with adjustable heights. The joints were sealed with a 20 mm layer of mortar (Figure 8b bottom) or with a 5 mm layer of silicone (Figure 8b top). The reinforcement was manually inserted into the assembled formwork before standard concrete was used to fill the 3DCP formwork (Figure 8c).

The proposed design of the column-slab connection was 3D printed and assembled as a demonstrator for the column-slab connection detail. The final prototype occupies a 1 m³ volume. The analysis of the recorded data set allowed to benchmark the assembled 3DCP elements against the initial 3D model.

When looking at the comparison between designed and scanned meshes, we obtain a mean distance of +2.00 mm with a standard deviation of +1.95 mm (Figure 8a bottom). When comparing the 3D model obtained from recording the feedback with the scanned mesh of the 3D printed element, we obtain a mean distance of +0.06 mm with a standard deviation of + 1.83 mm (Figure 9a top). The standard deviation of 1.83 mm results from high deviation peaks located at the end points of each contour, exactly where the path goes back in the opposite direction, as visible in Figure 9a (top). Given the sub-millimeter mean distance, we can conclude that recording the feedback is an excellent tool for gathering a fitting digital representation of what was 3D printed. Nevertheless, it is important to mention that absolute accuracy is hard to achieve, knowing that the flow rates of the progressive cavity pumps are theoretical and not measured online.

The assembled mushroom column-slab connection is within +/- 15 mm accuracy to the 3D model surface (Figure 9b), with a mean distance of -1.4 mm and a standard deviation of +5.24 mm. The precision of the proposed method is acceptable for current building standards. In the case of the assembly, the source of this deviation comes from a tilt of the slab on top of the column. Therefore, measures to correct these deviations can be mitigated by integrating positioning details in the 3DCP parts that can be easily aligned with the necessary temporary supports of the falsework.

CONCLUSION AND OUTLOOK

The ambition to develop a column-slab connection is rooted in the current need to create scalable structural building systems that answer the need for reducing material consumption. The presented strategy, including segmentation, printing, and assembly, focuses on a tectonic approach for designing prefabricated structural assemblies using 3DCP to meet this need. The proposed column-slab connection is an essential design detail of a structural system based on ribbed slabs. A strong focus of this design-driven research is to create continuity in the layers' disposition from the column to the slab and the translation from a vertical to a horizontal layer orientation.

On the one hand, this paper presents a novel 3DCP design-to-fabrication strategy for a column-slab connection detail. The first specific contribution of the research is the correlation between the segmentation strategy and the design of the print path. A tectonic approach to segmentation combined with adaptable print paths brings us closer to developing a genuine 3DCP column-slab system. The fabricated column-slab connection detail displays a geometric complexity exceeding the current state of the art in conventional concreting technologies, both in customization and design for the specific layered aesthetics of 3DCP (Figure 10).

On the other hand, this paper introduces a 3D concrete printing process based on three ingredients simultaneously intermixed inside the printing tool. Moreover, the printing process is equipped with sensor monitoring and feedback, enabling quality control of the 3DCP components. Furthermore, the used material processing platform for 3D printing geometries was explored with (i) a higher degree of overhang and (ii) monitoring the produced results.

9 Feedback and scanning: (a) one slab segment: (top) comparison between feedback and scanning; (bottom) comparison between designed CAD and scanning; (b) scanning of the assembled prototype

10 Assembled prototype of the mushroom column-slab connection detail

An outlook for this research includes upscaling the method to the entire column-slab building system, a proper dimensioning and evaluation of structural performance, an assessment of recourse consumption, and a comparison to other column-slab building systems. Furthermore, the integration of minimal reinforcement into the 3DCP components is essential to ensure safe handling, transportation, and assembly of the 3DCP components.

To conclude, the long-term goal of this research is to improve the efficiency and sustainability of concrete technology by enabling structures that utilize less material and outperform traditional concrete elements through optimized geometries and embedded functional and aesthetic features.

ACKNOWLEDGMENTS

This research was supported by the NCCR Digital Fabrication, funded by the Swiss National Science Foundation (NCCR Digital Fabrication Agreement #51NF40-141853). Additional funding was received from Knauf AG and BASF.

The authors acknowledge the essential contribution of Dr. Timothy Wangler and Prof. Robert Flatt to the fabrication setup and material system development. We are equally grateful to Lukas Gebhard, Dr. Jaime Mata-Falcón, and Prof. Walter Kaufmann for their advice in structural design. We want to thank Michael Lyrenmann, Philippe Fleischmann, Tobias Hartmann, Heinz Richner, Andreas Reusser, and Robert Presl for their support in the technical implementation, robotic setup, and scanning.

REFERENCES

Anton, A., A. Jipa, L. Reiter, and B. Dillenburger. 2020. "Fast Complexity: Additive Manufacturing for Prefabricated Concrete Slabs." In *Digital Concrete 2020; Paper Proceedings of the Second RILEM International Conference on Concrete and Digital Fabrication*. Cham: Springer. 1067–77. https://doi.org/10.1007/978-3-030-49916-7_102.

Anton, A., A. Yoo, P. Bedarf, L. Reiter, T. Wangler, and B. Dillenburger. 2019. "Vertical Modulations:Computational Design for Concrete 3D Printed Columns." In *ACADIA 2019: Ubiquity and Autonomy; Paper Proceedings of the 39th Annual Conference of the Association for Computer Aided Design in Architecture (ACADIA)*. 596–605.

Anton, A., L. Reiter, T. Wangler, V. Frangez, R. J. Flatt, and B. Dillenburger. 2021. "A 3D Concrete Printing Prefabrication Platform for Bespoke Columns." *Automation in Construction* 122: 103467. https://doi.org/10.1016/j.autcon.2020.103467.

Battaglia, Christopher A., Martin Fields Miller, and Sasa Zivkovic. 2019. "Sub-Additive 3D Printing of Optimized Double Curved Concrete Lattice Structures." In *Robotic Fabrication in Architecture, Art and Design 2018*. Cham: Springer. https://doi.org/10.1007/978-3-319-92294-2_19.

Bhooshan, S., V. Bhooshan, A. Dell'Endice, J. Chu, P. Singer, J. Megens, T. Van Mele, and P. Block. 2022. "The Striatus bridge." *Architecture, Structures and Construction* 2: 521–543. https://doi.org/10.1007/s44150-022-00051-y.

Bischof, Patrick, Jaime Mata-Falcón, and Walter Kaufmann. 2022. "Fostering Innovative and Sustainable Mass-Market Construction Using Digital Fabrication with Concrete." *Cement and Concrete Research* 161 (Feb): 106948. https://doi.org/10.1016/j.cemconres.2022.106948.

Bologna, Alberto, and Roberto Gargiani. 2016. *The Rhetoric of Pier Luigi Nervi Forms in Reinforced Concrete and Ferro-Cement*. Lausanne: EPFL Press.

Bos, Freek, Rob Wolfs, Zeeshan Ahmed, and Theo Salet. 2019. "Large Scale Testing of Digitally Fabricated Concrete (DFC) Elements." In *RILEM Bookseries*. Cham: Springer. 129–47. https://doi.org/10.1007/978-3-319-99519-9_12.

Damme, Henri Van. 2018. "Concrete Material Science: Past, Present, and Future Innovations." *Cement and Concrete Research* 112. https://doi.org/10.1016/j.cemconres.2018.05.002.

De Schutter, Geert, Karel Lesage, Viktor Mechtcherine, Venkatesh Naidu Nerella, Guillaume Habert, and Isolda Agusti-Juan. 2018. "Vision of 3D Printing with Concrete — Technical, Economic and Environmental Potentials." *Cement and Concrete Research* 112: 25–36. https://doi.org/10.1016/j.cemconres.2018.06.001.

Fleischmann, Philippe, and Gonzalo Casas. 2021. *COMPAS RRC*. v.1.1.0 Zenodo. https://doi.org/10.5281/zenodo.4639419.

Gaudillière, Nadja, Romain Duballet, Charles Bouyssou, Alban Mallet, Philippe Roux, Mahriz Zakeri, and Justin Dirrenberger. 2019. "Large-Scale Additive Manufacturing of Ultra-High-Performance Concrete of Integrated Formwork for Truss-Shaped Pillars." In *Robotic Fabrication in Architecture, Art and Design 2018*. Cham: Springer. 459–72. https://doi.org/10.1007/978-3-319-92294-2_35.

Ganz, Hans-Rudolf, and Jürg Fischer. 2014. "SIA 262: Concrete Structures." In *Concrete Structures Norms*. Zurich: TEC 21.

Gebhard, L., P. Bischof, A. Anton, J. Mata-Falcón, B. Dillenburger, and W. Kaufmann. 2022. "Pre-installed Reinforcement for 3D Concrete Printing." In *Third RILEM International Conference on Concrete and Digital Fabrication*. RILEM Bookseries, vol. 37, edited by R. Buswell, A. Blanco, S. Cavalaro, and P. Kinnell. Cham: Springer. https://doi.org/10.1007/978-3-031-06116-5_64.

GOM Metrology, Carl Zeiss GmbH. 2022. "GOM Inspect Pro." Zeiss. https://www.gom.com/en/products/zeiss-quality-suite/gom-inspectpro.

Grasser, G., L. Pammer, H. Köll, E. Werner, and F. P. Bos. 2020. "Complex Architecture in Printed Concrete: The Case of the Innsbruck University 350th Anniversary Pavilion COHESION." In Second RILEM International Conference on Concrete and Digital Fabrication. RILEM Bookseries, vol. 28, edited by F. Bos, S. Lucas, R. Wolfs, and T. Salet. Cham: Springer. Cham. https://doi.org/10.1007/978-3-030-49916-7_106.

Khoshnevis, Behrokh. 2004. "Automated Construction By Contour Crafting; Related Robotics and Information Technologies." *Automation in Construction* 13 (1): 5–19. https://doi.org/10.1016/j.autcon.2003.08.012.

Mata-Falcón, J., P. Bischof, T. Huber, A. Anton, J. Burger, F. Ranaudo, A. Jipa, L. Gebhard, L. Reiter, E. Lloret-Fritschi, T. Van Mele, P. Block, F. Gramazio, M. Kohler, B. Dillenburger, T. Wangler, and W. Kaufmann. 2022. "Digitally Fabricated Ribbed Concrete Floor Slabs: A Sustainable Solution for Construction." *RILEM Technical Letters* 7: 68-78.

Reiter, L., A. Anton, T. Wangler, B. Dillenburger, and R. J. Flatt. 2022. "A 3D Printing Platform for Reinforced Printed-Sprayed Concrete Composites." In *Third RILEM International Conference on Concrete and Digital Fabrication*. RILEM Bookseries, vol 37, edited by R. Buswell, A. Blanco, S. Cavalaro, and P. Kinnell. Cham: Springer. https://doi.org/10.1007/978-3-031-06116-5_37.

Rust, R., G. Casas, S. Parascho, D. Jenny, K. Dörfler, M. Helmreich, and A. Gandia. 2018. COMPAS FAB. V.0.27.0. Zenodo. https://doi.org/10.5281/zenodo.7178886.

Scrivener, Karen L., Vanderley M. John, and Ellis M. Gartner. 2018. "Eco-Efficient Cements: Potential Economically Viable Solutions for a Low-CO2 Cement-Based Materials Industry." *Cement and Concrete Research* 114: 2–26. https://doi.org/10.1016/j.cemconres.2018.03.015.

Shaker, Aya, Noor Khader, Lex Reiter, and Ana Anton. 2021. "3D Printed Concrete Tectonics: Assembly Typologies for Dry Joints." In *ACADIA 21: Realignments: Toward Critical Computation; Paper Proceedings of the 40th Annual Conference of the Association for Computer Aided Design in Architecture (ACADIA)*. 1–9.

Sliskovic, Valentino, Bernhard Freytag, Andreas Trummer, and Stefan Peters. 2020. "Additive Fabrication of Concrete Elements by Robots: Lightweight Concrete Ceiling." In *Fabricate: Making Resilient Architecture*, edited by J. Burry, J. Sabin, B. Sheil, M. Skavara. Antwerp: UCL Press. 124–29.

Van Damme, H. 2018. "Concrete Material Science: Past, Present, and Future Innovations." *Cement and Concrete Research* 112 (Oct): 5–24. https://doi.org/10.1016/j.cemconres.2018.05.002.

Wangler, Timothy, Nicolas Roussel, Freek P. Bos, Theo A.M. Salet, and Robert J. Flatt. 2019. "Digital Concrete: A Review." *Cement and Concrete Research* 123: 105780. https://doi.org/10.1016/j.cemconres.2019.105780.

IMAGE CREDITS

Figures 1, 10: Andrei Jipa, Digital Building Technologies, January 2022. All other drawings and images by the authors.

_____

Ana Anton is a PhD Candidate at the chair for Digital Building Technologies, Institute of Technology in Architecture, ETH Zurich, and is associated with the National Centre for Competence in Research – Digital Fabrication, where she leads the research in 3D concrete printing. She received her architectural degree, Cum Laude, at TU Delft in 2014. Her current research, Tectonics of Concrete Printed Architecture, focuses on robotic concrete extrusion processes for large scale building components.

_____

Eleni Skevaki is a Research Assistant at the chair of Digital Building Technologies, Institute of Technology In Architecture (ITA) of ETH Zurich. She received her Diploma in Architecture from the National Technical University of Athens. She has also studied in ENSA Paris Malaquais before joining the MAS in Architecture and Digital Fabrication in ETH. Through her work, academic research, and teaching activities, she aims to rethink construction through new technologies, computational geometry, robotic fabrication, and 3D printing. She has worked on a range of large-scale research projects that employ digital fabrication for concrete and other materials.

_____

Patrick Bischof is a PhD Candidate at the chair of Concrete Structures and Bridge Design, Institute of Structural Engineering, ETH Zurich, and is associated with the National Centre for Competence in Research – Digital Fabrication. He graduated in 2011 as a civil engineer at ETH Zurich before working in practice for four years. His current research focuses on rethinking classical concepts of structural concrete and on identifying new opportunities for sustainable design and production offered by digital fabrication.

_____

Lex Reiter, PhD is a post-doctoral researcher at ETH Zürich working on early age strength build-up and its control for digital fabrication processes with concrete, among which is layered extrusion. His research interest is in the physical and chemical processes that allow building without formwork and at high vertical rate as well as associated processing challenges.

_____

Benjamin Dillenburger is Professor for Digital Building Technologies at the Institute of Technology In Architecture at the Department of Architecture, ETH Zurich. His research focuses on the development of building technologies based on the close interplay of computational design methods, digital fabrication, and new materials. In this context, he searches for ways to exploit the potential of additive manufacturing for building construction.

Non-Linear Fabrication

Sulaiman Al Othman
Harvard University

Martin Bechthold
Harvard University

Rapid Data Collection and Toolpath Recalibration
in 3D Printing of Clay Lattices

1   An extreme case of clay lattice geometry is printed using the proposed scanning method, promising an important fast and accurate calibration step in 3D printing.

ABSTRACT

Non-Linear Fabrication presents a novel method for the calibration of complex toolpath geometries based on rapid data collection in additive manufacturing of clay material. This fabrication method addresses the inherent variations of the clay-extrusion process imposed by the material's wet state, which often leads to large tolerances, failures, and undesirable material outcomes. Clay, just like other natural, paste-like materials, offers a potential reduction in the embodied CO_2 emitted by the production of buildings using conventional materials, yet its large tolerances during printing remain an obstacle. The ability to anticipate and correct the complex material behavior during the extrusion process is important in the effort to achieve accurate building components. This paper describes an improved data collection methodology in the context of clay 3D printing that integrates structured light scanning technology. The ultimate goal is to use this data for toolpath calibration during the next step of the research. The integrated process measures and then addresses the deflections caused by the successive build-up of clay layers that cause changes in stiffness across the lower printed layers, distortions and shifting of clay beads caused by extrusion pressure and nozzle maneuvering, and air gaps in the clay mix that affect the material flow rate. Two cases of lattice printing demonstrate the viability of this method, presenting an important step towards next-generation solutions for fast toolpath calibration in 3D printing.

INTRODUCTION

Recent advances in additive manufacturing of earth-based materials show promising opportunities for sustainable building construction. However, the inherent property variations of these materials and their complex behavior during the extrusion and drying process pose significant challenges when attempting to produce accurate building components. Traditionally designers seek to understand these fabrication uncertainties through trial-and-error prototyping; they address excessive tolerances through the allocation of safety margins or appropriate detailing strategies. But these common approaches are time-consuming and costly. Advanced simulation methods such as finite-element analysis (FEA) cannot presently capture the highly non-linear process of 3D printing a variable paste-based material: the clay dries and shrinks at unpredictable rates while new layers continue to be deposited, thereby imposing increasing self-weight onto the lower layers in a highly non-linear process.

This paper addresses the non-linear behavior of paste-extrusion materials through the microcosm of clay 3D printing, which was initially used to develop the early contour crafting methods for concrete (Khoshnevis 2004). With the advent of robotic and digital fabrication technologies, designers can control many fabrication parameters of the clay extrusion process and its computer-generated geometry to produce functional architectural objects with controlled textures (Rael et al. 2016), custom façade components with voids (Bechthold et al. 2015 ; Rosenwasser et al. 2017), and 3D modularized systems for architectural wall constructions (Seibold et al. 2018). Yet, attempts to compensate for the complex deformation behavior of clay during the extrusion are rarely documented, and the goal of producing more accurate architecture building components remains elusive. This research builds on previous work on calibrating toolpath geometry for 3D printing of clay lattices (Im et al. 2019). Lattices represent an extremely challenging morphology to generate in clay, thus presenting a suitable microcosm that serves as a vehicle to address the more general goal of shape accuracy for clay 3D printed shapes. The work of Im et al. (2019) integrated an industrial displacement laser sensor within the clay-extrusion process to measure the material deflection of the printed lattice layer in order to correct the toolpath geometry of subsequent layers. This created a feedback loop between the fabrication and design processes that allowed large vertical structures to be printed successfully. However, several drawbacks and limitations remained. A toolpath design was required to robotically guide the distance sensor to the exact locations where measurements were needed, yet these locations remained uncertain as the printed shape deviated significantly from the designed shape. The force of the extrusion caused by the pneumatic system and

2 The deflection of lower vertical pylons often causes the nozzle to deposit clay in the air, which leads to improper anchoring of the next layer and further deflection and undesired material outcomes

3 The nozzle maneuvering and material extrusion pressure cause distortion and shifting along the pylon's horizontal and vertical axes.

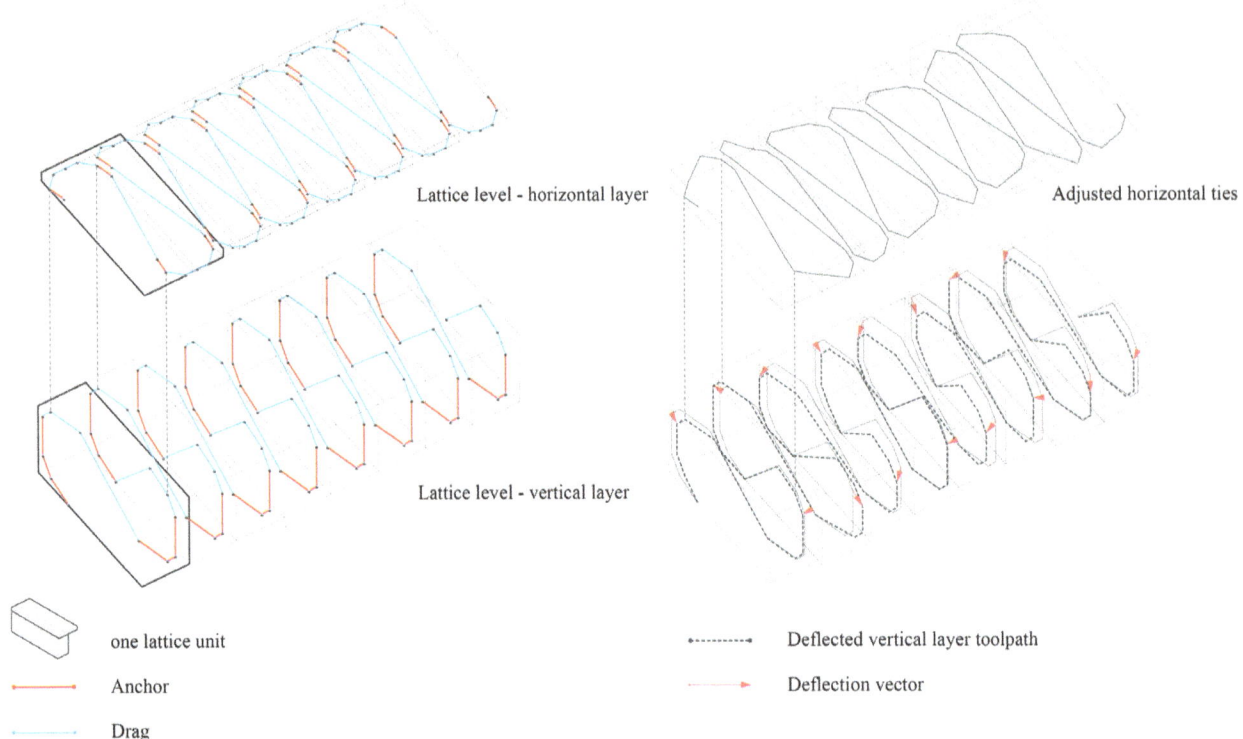

4 (left) Lattice toolpath involves two layers of alternating speed and material deposition behavior of anchoring and dragging;
(right) Horizontal ties toolpath recalibration based on the vertical layer's deflection vectors

the pulling force on the extruded clay bead caused by the robotic nozzle maneuvering along the 3D toolpath contributed to distortions and shifts: not only along the vertical dimension, but also on the other two horizontal dimensions—the X and Y dimensions— that caused the sensor to frequently miss those locations. In addition, the sensor needed to linger for several seconds to collect many measurements that would be averaged, slowing down the printing process. Finally, the sensor was limited to record Z displacement values only, thereby missing the other, equally important coordinates needed for an accurate calibration process.

This paper presents the first component of ongoing research geared towards using machine learning methods to predict material uncertainties in 3D printing of clay. The authors present a method involving a 3D structured light scanner that significantly improves the calibrated toolpath's precision while reducing the time needed for data collection. This approach can offer an efficient method for creating training and testing sets for machine learning applications in contemporary fabrication technologies, potentially allowing for precise fabrication with highly variable, natural or even living materials.

BACKGROUND

Methods for monitoring the material behavior and correcting for them in extrusion-based applications have been identified. Kazemian et al. (2019) integrated a machine-vision system to control and modulate the flowrate of concrete for robotic construction. The vision system has a video camera that sends still images in real-time to a microcontroller and processing algorithm, which calculates the changes in the volumetric flow rate of material deposition through edge-detection techniques and corrects the material extrusion rate to compensate for any discrepancies. However, this system is limited to adjusting the material extrusion parameters and does not consider the correction of toolpath geometry, which poses a challenge when building up many layers. Dritsas et al. (2018) used 3D scanning to measure the displacement of printed cylindrical objects out of natural composites, creating training and validation sets for machine learning models. However, scanning was performed after printing the entire object, and the intention was not for real-time calibration. To evaluate shape distortions and surface texture, Seibold et al. (2018) and Wi et al. (2020) used a 3D structured light scanner to reconstruct high-resolution 3D models of printed clay objects. This technology can potentially be integrated into paste-based extrusion workflows to calibrate toolpath geometry in real-time, reducing surveying time and increasing accuracy. This creates a novel method to rapidly generate more geometric data rapidly for machine learning applications.

METHOD

Lattice Toolpath Geometry and Required Calibration

Clay lattices are an extreme case of clay 3D printing, that uses

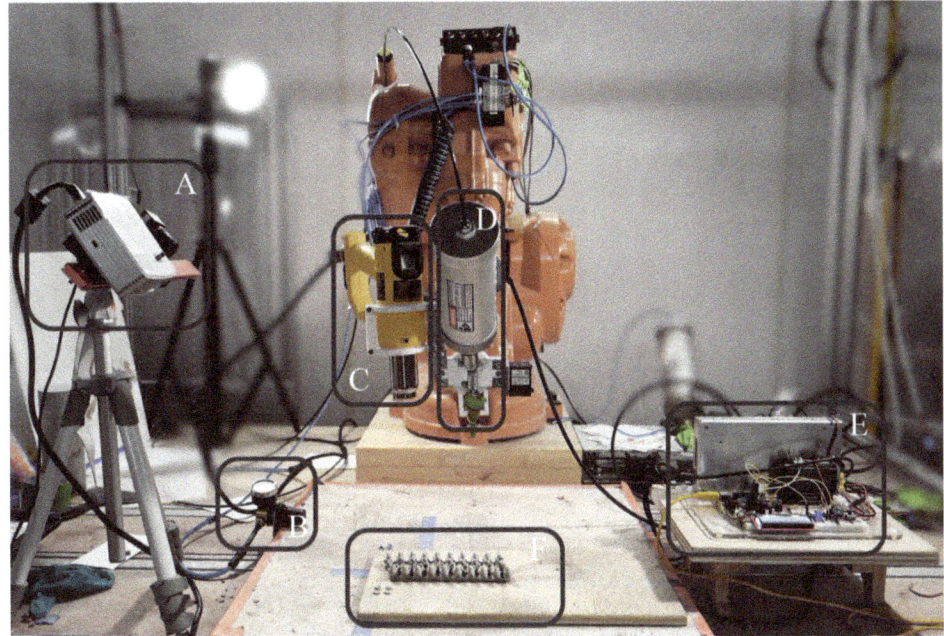

5 Clay lattice printing and scanning setup: A) 3D structured light scanner, B) Air pressure regulator, C) Detachable heat gun, D) Clay extrusion system, E) Microcontroller unit for extrusion, scanning, and heat gun control, and F) Printing area

a special process called "spatial print trajectories," developed by the authors (Al Othman et al. 2018). The lattice geometry best represents extreme fabrication uncertainty due to its overall delicate form that is subject to large deformations during the printing and drying process. We expect processes and methods developed for this extreme material geometry to be generalizable to other, more conventional forms. Printing clay lattices involves two sets of alternating material-controlled behavior: anchoring and dragging to produce a lattice unit (Figure 4 left). These are achieved by controlling fabrication parameters, such as the printing head speed and the material extrusion rate. The robotically guided extrusion system prints a lattice unit by first depositing (anchoring) material at a low speed and a high extrusion rate along the vertical direction to create a pylon. Then, the material is dragged in a complex movement to the opposite side at a high head speed and a low extrusion rate to start printing another vertical pylon. These pylons are braced with horizontal ties following a similar printing behavior. The toolpath calibration includes adjusting the horizontal ties' start position for each lattice unit according to the anticipated deflection of the vertical pylons to ensure that each subsequent lattice level is accurately built up (Figure 4 right).

3D Structured Light Scanning Process

A David Structured Light Scanner (SLS-3) is used to scan and reconstruct a high-resolution mesh of the physical state of each lattice level after it is printed. It works by projecting different fringe patterns from a projector—generated by the scanner software—onto the object. A calibrated camera then captures the distorted patterns and sends the data to the software to calculate the 3D information using triangulation (Zhang 2010). This entire process—from scanning to 3D shape reconstruction—takes a maximum of two seconds and generates more than a million data points (vertices) for scans at 50 μm resolution, making it a very efficient and highly accurate data collection method for calibrating real-time printing in the next step. Multiple images of objects from various angles are required to build the entire 3D model of the scanned object. However, in the context of clay lattice printing—also applicable to other 3D printing methods—only the print's top surface features are important for the calibration process, and thus one scan is sufficient.

Lattice Printing and Scanning Setup

The clay lattices are printed using a pneumatic paste-extrusion system from VormVrij 3D that is mounted to an ABB-IRB140 industrial robot. Standard wet clay is loaded into a pressure tank attached to the extruder assembly under fixed supply of air pressure at 2.7 bar. The clay is pushed to the extruder assembly, where a motorized augur with controlled RPM regulates and sets the flow of clay. The clay body is prepared using cone 10 stoneware clay and water at 1 kg of clay to 70 g of water. The toolpath for the lattice geometry is generated in the Grasshopper3D environment and the Machina Bridge application (García del Castillo y López 2016) is used to program the toolpath's positions and speed data and send them in real-time to the robot controller for execution. After the first lattice level is printed, a heat gun is robotically guided to dry the anchor locations where the next lattice level is deposited, providing immediate stabilization to reduce deflections and, in the worst case scenario, avoid major collapses. The scanning then starts on the David 5 scanning software (Hewlett Packard 2016) to obtain a 3D mesh of the printed lattice, which takes less than

6 Toolpath real-time calibration and printing of three lattice levels: (left) Scanning; (center) Printing the calibrated horizontal ties; (right) Printing the calibrated vertical pylons

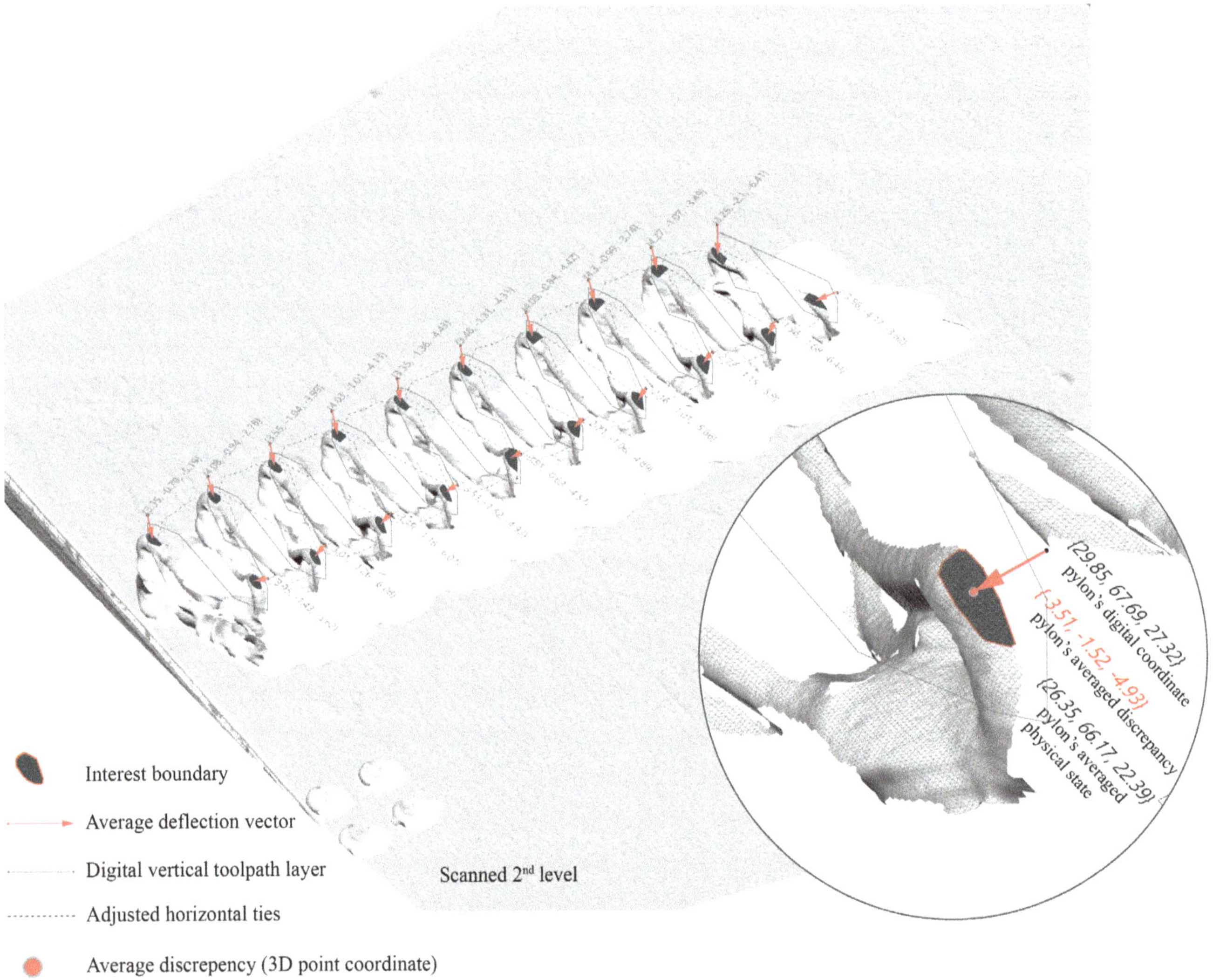

7 Interest boundaries around the scanned vertical pylons are selected with a customized processing algorithm; an average discrepancy vector translates the next layer and adjusts the next toolpath

LUW (mm)	LUL (mm)	LUH (mm)	AHS (mm/s)	DHS (mm/s)	RPM (rpm)	ST (mm)	FL (index)	ND (mm)
14	28	15	8	30	56	0	2	4

# data point	Discrepancy – X (mm)	Discrepancy – Y (mm)	Discrepancy – Z (mm)
1	3.35	-0.79	-5.19
2	-3.95	-1.43	-3.62
3	4.08	-0.94	-2.78
⋮	⋮	⋮	⋮
20	-3.56	-4.17	-3.82

8 Lattice level of ten lattice units (each unit has two vertical pylons) has 20 data points; collecting data from physical lattices is easily achievable with the scanning workflow.

2 seconds. The scanner is stationary and positioned to point at the print's top surface (Figure 5).

Toolpath Recalibration

The scanned mesh is exported to Grasshopper3D as an .obj file, where a customized Python script reads and orients it to match and align with the coordinate system of the printing area. The aligning process involved printing a rectangle of a known dimension and distance to the print's digital origin. The scanned mesh is then translated and rotated so that the rectangle it contains aligns with its digital location. This process is only carried out one time after the scanner is shifted from its original position. The mesh vertices of the aligned scan are extracted, but only those around the vertical pylons are selected. Afterwards, a script selects the highest two hundred 3D points from each vertical pylon and averages them (Figure 7). The average points present the anchor location for the horizontal ties of the same lattice level toolpath. The difference between the average point and the top of the vertical layer, or pylon, of the digital toolpath presents the discrepancy value that the horizontal ties are adjusted to accordingly. The base of the next lattice level will adjust and follow the described scanning and printing procedure until the entire lattice geometry is printed (Figure 6).

Data Parsing for Training Set

Discrepancy values (output features or response variables) are recorded for each lattice unit after the printed lattice level is scanned. Essentially, these are exported with the input features of that level to a .csv file. Here we define our input features with their ranges to include the lattice unit dimension: width (LUW) (10 – 50 mm), length (LUL) (10 – 100 mm), and height (LUH) (10 – 40 mm), anchor head speed (AHD) (3 – 20 mm/s), and drag head speed (DHS) (10 – 80 mm/s), extrusion rate controlled by the stepper motor (RPM) (20 – 120 rpm), stepping distance of each level (ST) (-5 – 5 mm), lattice unit level index (FL) (1 – 8), and nozzle diameter (ND) (2 – 10 mm) (Figure 8). The parameters' ranges were defined based on expert knowledge from previous experimental studies about lattice clay printing (Im et al. 2019). In the next stage of this research, we will print several lattice prototypes with different characteristics and forms and collect data from them following the scanning pipeline described. We will split the collected data points into training and testing sets to train a neural network model to predict the discrepancies for any input lattice features, eliminating the need for the scanning step.

RESULTS

We tested the improved calibration pipeline with the structured light scanner on three lattice forms with different parameters, two of which were extreme cases that involve setting the value of a few lattice parameters to their high or low ranges. The first one presents an approximate center point, which means it is generated by setting the lattice parameters' (input features) ranges at their mid-point. The pipeline successfully printed and calibrated the lattice toolpath in real-time. We observed that the calculated average point tends to shift inward at both of the lattice's sides (longer direction), reflecting a high x-axis discrepancy (Figure 9 top). This is because the vertical pylons fold onto themselves upon dragging the extruded material to the opposite side, creating longer surface area on that axis. Thus, the calculated average point inside the interest boundary tends to be in the center of that surface. Yet, it was still possible to correctly anchor the horizontal ties. Other strategies of obtaining a point from the vertical pylon's interest boundary could be tested to improve the quality of the x-axis discrepancy measurements. Since this lattice prototype presents the center point of the nine parameters' ranges, the mean deflection around the other axes were expected to become low: under 1 mm (Figure 9 top). The second lattice prototype has unit width (LW), head speeds (DHS and AHS) and RPM on the low range. Although the scanned data showed a maximum z-discrepancy of 8.2 mm—the vertical pylons deflected because of insufficient material volume due to the low RPM—each lattice level was successfully calibrated and printed to account for the high discrepancy

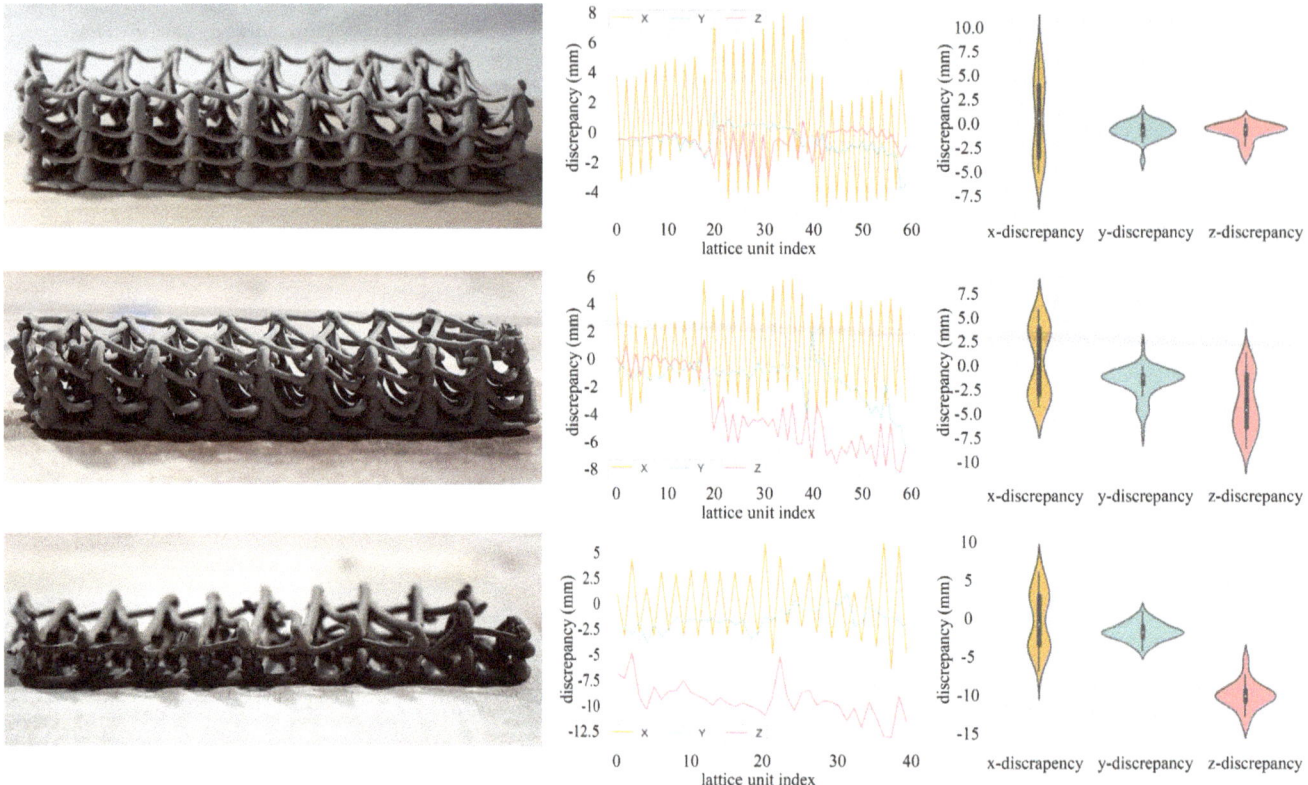

9 (top) Printed lattice of mid-point parameters' ranges, showing high x-discrepancy values due to the average point's location inside the interest boundary; (center and bottom) Printed lattice of extreme parameter ranges showing high z-discrepancy values, yet it was successfully calibrated and printed

values (Figure 9 center). The third lattice prototype has unit height (LH) and head speeds (DHS and AHS) on the high range. The lattice level average z-deflection was more than half of the original height with a maximum deflection of 13.1 mm, yet the second level toolpath geometry was successfully calibrated and anchored correctly onto the first level (Figure 9 bottom). Altogether, the steps of scanning, reading, and calibrating took less than 30 seconds for each lattice level, which was a significant improvement compared to the previous spot laser sensor. The new process also allowed the toolpath geometry to adjust in all three dimensions instead of only in the vertical direction.

CONCLUSION

This paper presents an improved real-time toolpath recalibration workflow for an extreme case of paste-based additive manufacturing processes: 3D printing of clay lattices. The deflection caused by the material build-up, the distortion and shifting from the extrusion pressure and robotic movements, and the dimensional changes after printing each layer caused by irregular material drying and shrinkage rates motivated the need to enhance the material surveying component with a 3D structured light scanner for accurate calibration of toolpath geometry. The results suggest that the enhanced calibration systems can (1) rapidly provide high resolution 3D reconstructions of each printed lattice layer in real-time with high-level calibration fidelity, and (2) collect high-quality geometrical information that could potentially be used for training machine learning models. This enhanced calibration workflow can be generalized and transferred to similar applications that exhibit high material and fabrication uncertainties, such as 3D printing of concrete or living materials.

REFERENCES

Al Othman, Sulaiman, Hyeonji Claire Im, Francisco Jung, and Martin Bechthold. 2018. "Spatial Print Trajectory." In *Robotic Fabrication in Architecture, Art and Design 2018*, edited by J. Willmann, P. Block, M. Hutter, K. Byrne, and T. Schork. Cham: Springer. 167–180.

Bechthold, Martin, Anthony Kane, and Nathan King 2015. "Ceramic Material Systems." In *Architecture and Interior Design*. Basel: Birkhäuser.

Dritsas, S., S. E. P. Halim, Y. Vijay, N. G. Sanandiya, and J. G. Fernandez. 2018. "Digital fabrication with natural composites." *Construction Robotics* 2 (1-4): 41–51.

García del Castillo y López, Jose Luis. 2019. "Robot Ex Machina: A Framework for Real-Time Robot Programming and Control." In *ACADIA 19: Ubiquity and Autonomy; Proceedings of the 39th Annual Conference of the Association for Computer-Aided Design in Architecture*. University of Texas at Austin.

Hewlett Packard. *David 5 3D Scanner*. V.5.0.6. 2016.

Im, Hyeonji Claire, Sulaiman Al Othman, and Jose Luis García del

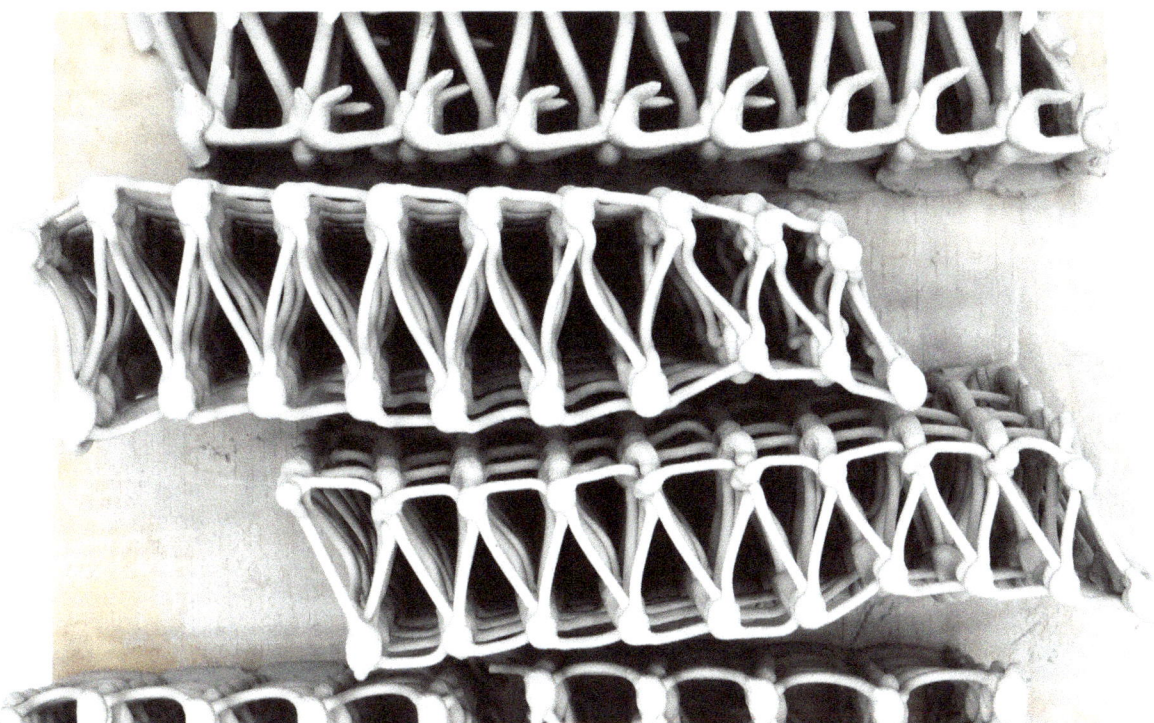

10 Calibrated clay lattices

Castillo. 2018. "Responsive Spatial Print; Clay 3D Printing of Spatial Lattices Using Real-Time Model Recalibration." In *ACADIA 18: Recalibration, On Imprecision and Infidelity; Proceedings of the 38th Annual Conference of the Association for Computer Aided Design in Architecture*, 286–293. Mexico City, Mexico.

Kazemian, A., X. Yuan, O. Davtalab, and B. Khoshnevis. 2019. "Computer vision for real-time extrusion quality monitoring and control in robotic construction." *Automation in Construction* 101: 92–98.

Khoshnevis, B. 2004. "Automated construction by contour crafting; Related robotics and information technologies." *Automation in Construction* 13 (1): 5–19.

Rael, R., and V. San Fratello. 2017. "Clay Bodies: Crafting the Future with 3D Printing." *Architectural Design* 87 (6) 92–97.

Rosenwasser, David, Sonya Mantell, and Jenny E. Sabin. 2017. "Clay Non-Wovens: Robotic Fabrication and Digital Ceramics." In *ACADIA 2017: Disciplines & Disruption; Proceedings of the 37th Annual Conference of the Association for Computer Aided Design in Architecture*. Cambridge, MA.

Seibold, Z., K. Hinz, J. López, N. Alonso, S. Mhatre, and M. Bechthold. 2018. "Ceramic Morphologies: Precision and Control in Paste-Based Additive Manufacturing." In *ACADIA 18; Proceedings of the 28th Annual Conference of the Association for Computer Aided Design in Architecture*. Mexico: ACADIA. 351–358.

Wi, K., V. Suresh, K. Wang, B. Li, and H. Qin. 2020. "Quantifying quality of 3D printed clay objects using a 3D structured light scanning system." *Additive Manufacturing* 32: 100987.

Zhang, S. 2010. "Recent progresses on real-time 3D shape measurement using digital fringe projection techniques." *Optics and Lasers in Engineering* 48 (2): 149–158.

IMAGE CREDITS

Figure 1 - 10: © MaP+S Group

Sulaiman Al Othman is an architect and computational designer. He is a Doctor of Design candidate at the Harvard Graduate School of Design. His research focuses on developing predictive models to address material uncertainty in additive manufacturing of earth-based materials for various design applications.

Martin Bechthold is the Kumagai Professor of Architectural Technology at the Harvard Graduate School of Design. He co-directs the Master in Design Engineering program at Harvard, and is the founding director of the Materials Processes and Systems (MaP+S) Group. MaP+S advances the understanding of materiality in the built environment, leveraging design computation, robotics, and pursuing collaborative work with material scientists. Bechthold also founded Harvard's Laboratory for Design Technologies as a platform for industry and academy to connect in the effort to expand knowledge in design and construction. His work has been widely published nationally and internationally.

3D Printing and Shape Memory Alloys

3D-printed Bi-stable Geometries and Shape Memory Alloys for Energy Efficient Kinetic Elements

Hyunchul Kwon
D-ARCH/ETH Zürich

Priyank Soni
D-ARCH/ETH Zürich

Ali Saeedi
Structural Engineering Lab/Empa

Moslem Shahverdi
Structural Engineering Lab/Empa

Benjamin Dillenburger
D-ARCH/ETH Zürich

1   SMA-embedded 3D prints for kinetic elements

ABSTRACT

This paper presents a novel method combining the use of 3D printing (3DP) and shape memory alloys (SMAs) to compose kinetic architectural elements that are energy- and material-efficient within compact-integrated composites. Kinetic systems for architectural use have been explored since the late twentieth century using motor mechanics. However, the primary challenges of this method include maintenance of mechanical units, their high energy demand, and noise during actuation. To address these shortcomings, this research explores a hybrid of 3DP motion-optimized parts with embedded SMAs as a muscle that changes shape with temperature stimulus. This combination leverages 3DP to geometrically control shape-morphing behavior for material-efficient, compact-integrated parts, and SMA to allow for low maintenance and soundless actuation. However, current SMA applications permanently require energy to stabilize one geometric state. To reduce required energy to a minimum, we present a novel method that combines embedded SMA with 3DP bi-stable mechanism. This approach only requires energy for switching between states, dramatically reducing energy consumption. This could be the key to efficient architectural applications. As part of the evaluation, factors such as controllability of shape morphing behavior, repeatability, materials, and energy efficiency are investigated. An experimental program is developed with different SMA-embedded, 3D-printed specimens. The program then explores a possible approach to scaling up with two prototypes. The presented synthesis of smart materials with additive manufacturing of bi-stable geometries could contribute to the field of composites in kinetic architecture by reducing the operational energy, thus opening a path towards more sustainable real-world applications.

INTRODUCTION

Kinetic architecture describes spaces and building elements that can physically reconfigure or adapt themselves to meet the changing needs of their users and surroundings. The kinetic design depends on motion for its effects (Razaz 2010). The last decades show an ever-growing interest in kinetic architecture due to its ability to control environmental factors such as solar exposure, ventilation, and acoustics for spatial comfort (Velikov and Thün 2013).

However, the predominant reliance on mechanical systems such as motors and engines for movement is still a hurdle to the widespread adoption of this concept in architecture (Maragkoudaki 2013). These traditional systems are highly customized and pose various drawbacks. These include high installation cost, low energy efficiency, high maintenance requirements, bulkiness, and noise during activation (İlerisoy and Başeğmez 2018). Therefore, looking toward new fabrication techniques and non-mechanical alternatives can bring greater possibilities to kinetic architecture.

Although there are various novel methods of producing motion without mechanical parts, material systems are simpler and more energy-efficient. These include shape memory alloys (SMAs), electroactive polymers, thermobimetals, and hygroscopes that can bring about motion with an electrical, thermal, or chemical stimulus (Maragkoudaki 2013; Correa et al. 2015). With the 4D printing method, a recently developed field originating from 3D printing (3DP), some research utilizes shape memory polymers or hygroscopic materials to directly and additively manufacture kinetic parts (Ge, Qi and Dunn 2013; Tibbits 2014; Correa et al. 2015; Momeni, Liu and Ni 2017). Despite significant innovations in direct 3DP shape-changing materials, the studies conducted on this technology are focused on environmental shape-morphing controls such as heat, light, or water, which are limited in certain environmental conditions or passive in nature (Tibbits 2014; Correa et al. 2015). In this context, this study aims to investigate the active control of kinetic systems with electroactive SMA materials. SMAs further have the advantage of higher accessibility and durability for architectural use (Jani et al. 2014). SMAs are categorized as a type of smart material widely utilized in the form of sheets and wires to fabricate intelligent and advanced structures. Due to transformations between the austenite and martensite phases, SMAs can return to their original shape through electric resistance heating. It has been demonstrated that SMAs can withstand more than 500,000 working cycles without a significant reduction in strain recovery behavior (Mammano and Dragoni 2014).

However, there are shortcomings, such as high initial cost and low material abundance, which makes it difficult to scale up for architectural use. Thus, the requirement to use SMA efficiently—such as in the form of wires and by embedding it in another shape-morphing enabler—becomes crucial. This research tries to tackle this disadvantage by hybridizing two-way SMA wires as a tendon, or the actuated deformer embedded into 3D-printed parts as its body or shell, creating resource-efficient kinetic elements. Unlike one-way SMAs, two-way SMAs change their shape after activation upon cooling. One-way SMAs do not change shape once they have cooled after activation. Moreover, 3DP further enables the customizability and geometrical complexity of kinetic systems, acting as a fast interface between design-fabrication processes. From various 3DP technologies, this research focuses on fused deposition modeling (FDM) which is a widely available 3DP method in which molten plastic material is extruded and hardens instantly after deposition. FDM can utilize a wide variety of plastics, and its hardware setup is simple (Kwon et al. 2018). Specifically, FDM is suited to shape-morphing behavior due to the printability of flexible and elastic polymers.

STATE OF THE ART

The idea of employing SMAs as actuators for controlling the shape of flexible polymers was first introduced more than thirty years ago (Rogers, Liang and Jia 1991). Since then, much research has been conducted to achieve more efficient shape-morphing behavior (Ahn et al. 2012; Jung et al. 2013; Naghashian, Fox and Barnett 2014; Lacasse et al. 2015; Bodkhe et al. 2020). In a study developing a theoretical model for activating curved composite panels using embedded SMA wires, the researchers verified their model by setting up experiments and reporting the high efficiency of SMA actuators for the fabrication of kinetic structures (Jung et al. 2013). Another study presented SMA-based composite structures with coupled bending-twisting shape-morphing behavior, where the researchers combined the anisotropic behavior of laminated composites with SMA actuators to achieve complex structural deformation (Ahn et al. 2012). Recently, an academic group proposed a 3DP method to fabricate SMA-embedded composites by embedding the polymer-encapsulated actuator elements within the polymer-based structures (Bodkhe et al. 2020).

In an architectural context, research using SMA actuation for façades shows the potential to reduce the amount of labor required for maintenance and eliminate the need for mechanical part replacement with low energy consumption (Yi et al. 2020). Despite this potential, only a few studies have used SMAs for building environments. The elements studied are limited in scale (< 0.5 m × 0.5 m) or some studies have focused only on conceptual investigations (Lignarolo, Lelieveld and Teuffel 2011; Chamilothori, Kampitaki and Oungrinis 2013; Yi

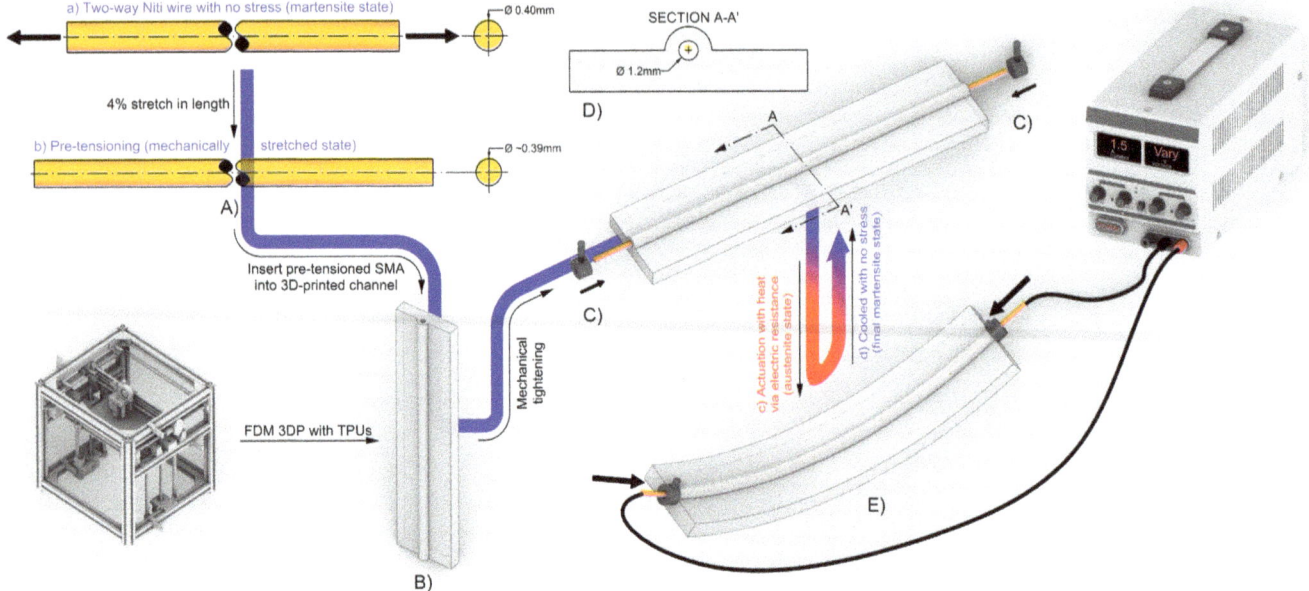

2   Schematics of the SMA-embedded composite system: (A-C) Preparation of parts for an SMA-embedded composite structure; (D) an exemplary cross-section of the structure; and (E) its expected motion

et al. 2020). In one of the few studies on large-scale elements, costly SMA sheets, with higher energy consumption than SMA wires, were used (Ghomshei et al. 2005).

A review of published investigations in the field of SMA-embedded composites has revealed that there are some challenges in the design, fabrication, and actuation of such structures:

- The majority of studies have been performed on simple geometries, such as flat panels or straight beams.
- Only small-scale individual specimens were considered for the experiments.
- The introduced SMA-embedded structures cannot fully recover their initial shape upon cooling. They needed an additional system to complete the shape-morphing cycle.
- The introduced SMA-embedded structures require a continuous flow of energy; they needed constant heat to maintain the actuated shapes.

In this context, further investigations should focus on more complex SMA-embedded structures. For the design and fabrication of such structures, the spectacular physical properties of SMAs and the ability of 3DP can be employed to overcome the weight, noise, and complications of previously available actuators. Besides, this research includes case studies exploring a possible approach to applying the system on a larger scale. Furthermore, new energy-efficient mechanisms are needed that allow for switching between states in order to dramatically reduce energy consumption, as this could be the key to efficient architectural applications.

METHODS

The aim of the research is to investigate the effective and efficient shape-morphing control of kinetic elements with the advantages offered by 3DP and SMA. To address the limited geometric shapes found in the precedent works, some case studies are performed on complex, 3-dimensional geometries. The studies include the exploration of an assembly approach of multiple, complex SMA-embedded composites to form larger-scale prototypes.

Elemental Principle

Figure 2 shows the basic composition of an SMA-embedded composite, the utilized components of the actuation setup, and a complete actuation cycle.

The following experiments are conducted with a thin Ø 0.4 mm SMA wire. Although iron- and copper-based SMAs also exist, the work presented emphasizes nickel-titanium alloy (NiTi) SMAs due to their high thermo-mechanic deformation under heating. NiTi has various other advantages for use in architecture including high weather resistance, low maintenance, high energy efficiency, low transformation temperature requirement, and long repeatability lifespan (Jani et al. 2014). Because the maximum recoverable pre-strain for this type of SMA is 6.7%, the wires are stretched by 4% to enhance the amount of actuation (Figure 2-A). This pre-strained wire contracts in length by 4% when heat is applied; the heating is done directly via electric resistance. The resistance of the SMA wire is 1 Ω per 150 mm. A power supply is used with a 1.5 ampere flow of an electric current.

In regards to FDM 3DP material, thermoplastic polyurethanes (TPUs) are used (Figure 2-B). TPU, with properties that range from a high-performance elastomer to tough thermoplastic, has been extensively used due to its superior physical properties, e.g. high tensile strength, abrasion and tear resistance, and low-temperature flexibility (Lu and Macosko 2004). Four different types of TPUs are 3D-printed and compared to identify the characteristics of each and their suitability for kinetic applications. The SMA wires are then inserted into the 3D-printed channels (Ø 1.2 mm) and mechanically anchored at both ends of the channels with adjusting rings and grub screws (Figure 2-C).

To allow SMA-embedded 3D-printed polymers to be actuated to the desired deflection, it is crucial to define the position of SMA wires. Since the specimen with a centrally located SMA wire just remains in the position without deflection, as a premise the wires need to lean to one side of the shell of 3D prints (Figure 2-D and 2-E).

The following sections describe the specific methods investigated, utilizing 3D-printed complex geometries for high controllability of shape-morphing and developing a bi-stable geometric system for radical energy-efficient actuation.

3D-printed Complex Geometries for High Shape-morphing Controllability

Unlike conventional manufacturing, which requires large, complex machinery, 3DP enables the integrated fabrication of geometrically complex parts at low cost, facilitating rapid prototyping for mass customization (Yi et al. 2020). Another significant advantage of 3DP is that one can precisely control geometric structure, resulting in control of stiffness, and therefore gaining high control over the shape-morphing behavior (Figure 3).

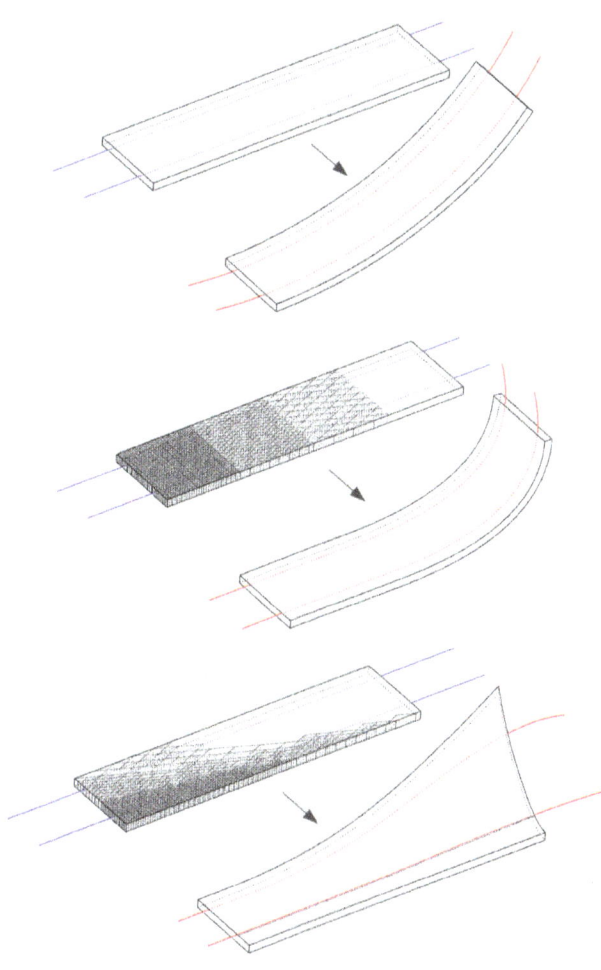

3 Possible control of shape-morphing behaviors from monoclastic curvatures to double curved surfaces through differentiated and graded geometric structures

While the precedent research of such 3D-printed graded structures has focused on their potential for structural reinforcement (Kwon and Dillenburger 2019; Taseva et al. 2020; Li et al 2020), material gradients are utilized in this research to achieve higher shape-morphing control over existing mechanical methods. The material gradient further enables the efficient use of resources, applying polymer materials only where needed. This method allows for more efficient shape-changing behavior in terms of material and energy use, even compared to commonly non-graded polymer matrices (Hazlehurst, Wang and Stanford 2014).

Bi-stable Geometries for Energy and Material Efficiency

The actuation of two-way SMAs need a relatively small but continuous flow of electric energy (1.5 watts per 100 mm SMA wire), which is crucial for applying the system to building environments. Figure 2 further describes the states of two-way SMA wires and their energy requirements during the actuation of an SMA-embedded composite structure (Figures 2-A to 2-D).

To address this challenge, the bi-stable system is studied, which signifies the system has two stable equilibrium states (Morris 1992). Something that is bi-stable can result in either of two states, i.e., a light switch. Bi-stable behavior can occur in mechanical linkages with specific geometric shapes. An efficient shape-morphing response, in this context, is obtained in the proposed adaptive system by employing a set of two 3D-printed, bi-stable geometries with two sets of embedded SMA wires. The concept, namely the SMA-embedded bi-stable geometry (SMABG) system, is illustrated in Figure 4.

A complete shape-morphing cycle consists of (A) forward activation for generating the desired shape in the structure and (B) reverse activation for returning the structure to its initial shape. A minimum amount of energy is required for the morphing cycles because there is no need for permanent

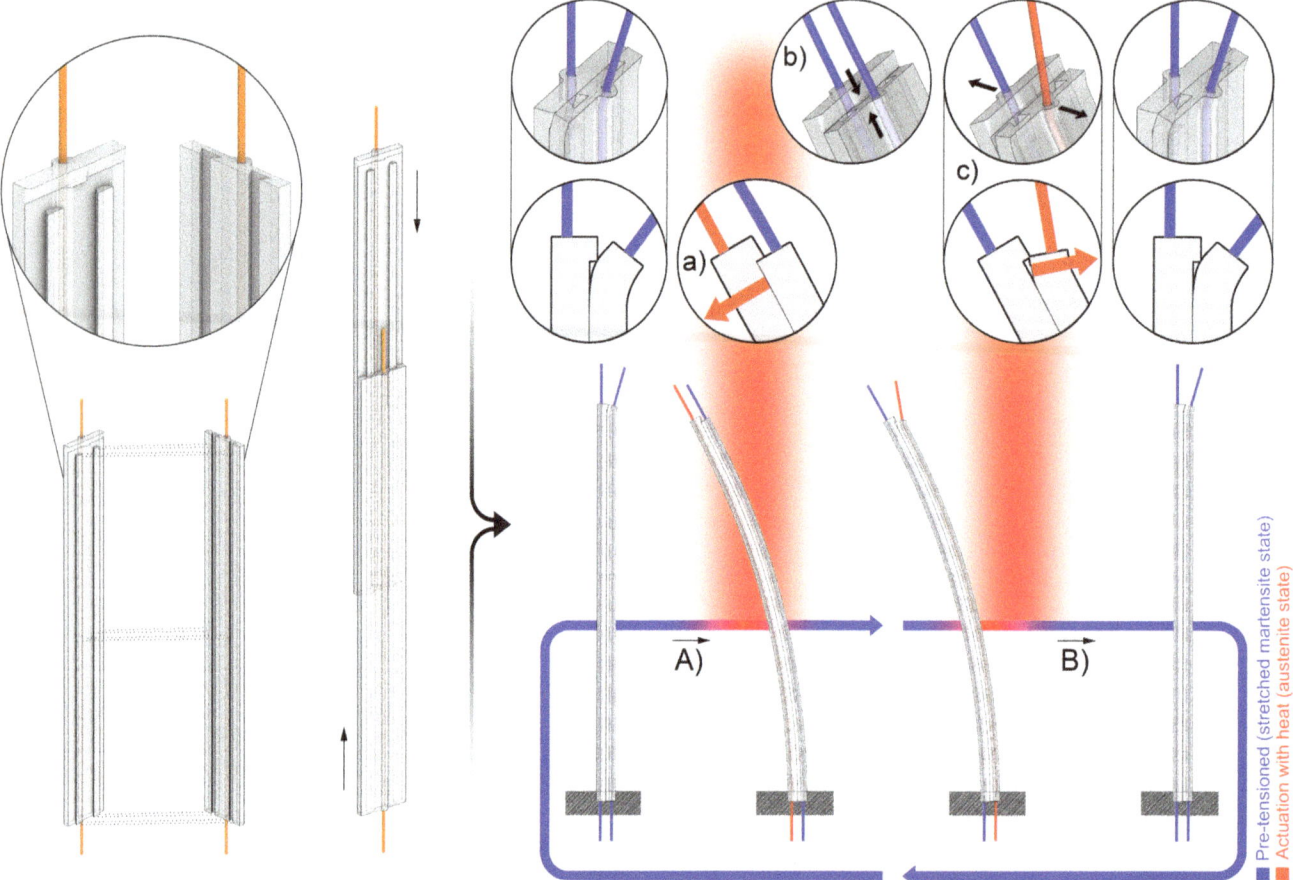

4 Schematics of bi-stable geometries for SMA embedding: A) forward actuation; B) reverse actuation; a-b) interlock mechanism; c) release mechanism

activation power (electrical current) to hold the structure in either the initial or secondary shape, as follows, and illustrated in Figure 4: (a) the forward set of SMA wire is initially activated by applying the electric current to generate the structure's required deflection; (b) once the deflection reaches its desired value, the electric current is turned off. In this step, the structure remains in its deflected shape, and the two 3D prints interlock each other stably due to a linkage feature. To release the linkage (c), the reverse set of SMA wire is activated with reverse shape-morphing. During a complete cycle, an intricate 3D-printed male-to-female railing enables a smooth motion transition, allowing the two 3D prints to operate as one part. Such switch-like mechanical bistability can be repeated continuously for many cycles. This further enhances the energy sustainability of the proposed system.

Based on these methods investigated, an experimental program is developed. Three interrelated sets of experiments are identified as follows:

- *Fundamental experiments*. With a series of small-scale specimens, the studies focus on identifying the characteristics of FDM 3DP applicable TPU materials, proving the concept of shape-morphing effectivity through the material gradient, and diversification of motions through wire-location and geometric differentiations.
- *Case studies*. Two prototypical applications are materialized, exploring 3-dimensional geometries and complex shapes, respectively. The studies are then focused on an assembly method of such components to compose an adjustable lighting instrument and a facade element, investigating an integrative strategy for mechanical, electrical, and architectural design.
- *Bi-stable geometric systems*. This experiment aims at proving the concept of adapting bistability in SMA-actuated kinetic elements and its ability to radically reduce energy consumption.

RESULTS AND DISCUSSION

The following experiments incorporate optimal parameters derived from early-stage 3DP tests with multiple TPU 3DP materials with Raise3D V2 Plus FDM 3D printers.

Fundamental Experiments

Initially, three sets of experiments are conducted to validate the feasibility of combining 3D prints with SMA wires for kinetic elements. Small-scale specimens are 3D-printed and subsequently, NiTi SMA wires are embedded.

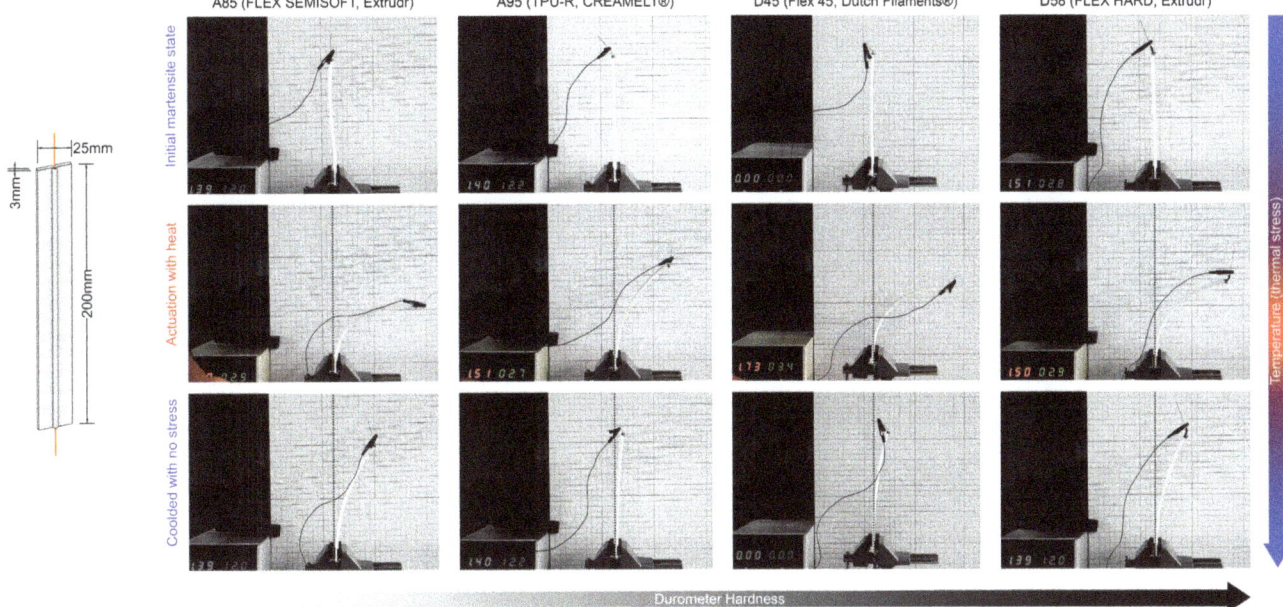

5 FDM 3DP TPU material investigations

The objective of the first experiment is to identify the characteristics of TPU materials that are commercially available for 3DP and consequently verify their suitability for kinetic elements (Figure 5).

The evaluation criteria include range of motion (whether the specimens actuate as expected), recoverability (whether the specimens fully recover their initial shape upon cooling), and elasticity. Four TPU materials are selectively chosen according to the given results from the manufacturers' durometer hardness tests (ASTM D2240). The A scale is softer, while the D scale is harder, and higher numbers indicate a greater resistance to indentation. Table 1 shows the specifications of test materials and their results which are empirically analyzed.

Overall, the four commercially available FDM TPU materials show a good range of motion during SMA actuation. However, the lower the durometer hardness value is, the tougher it is for the specimen to recover to its initial shape (i.e., A85).

Low recoverability is also observed when the material is less elastic (FLEX HARD, Extrudr compared with Flex 45, Dutch Filaments®). These results indicate that TPUs with durometer hardness values under A95 are unsuitable for the given composition of kinetic specimens, while the elasticity of the material is a crucial factor in achieving highly repeatable SMA-embedded 3D-printed composites. Based on the results, further experiments are conducted except for A85 (FLEX SEMISOFT, Extrudr).

The second test investigates the controllability of shape-morphing behavior through 3D-printed complex geometries. The concept of controlling the behavior by differentiating the topology of a structure (graded-material) was verified by fabricating a set of small-scale prototypes (Figure 6). The set consists of specimens with a gradual increase in structural density and one reference specimen that has a consistent structural hierarchy. The schematic of the graded-material specimens is illustrated in Figure 6-A. Subsequently, the

TABLE 1 Specifications of FDM 3DP TPU materials and their relative comparisons

Durometer Hardness (Name, Manufacturer)	Range of Motion	Recoverability	Elasticity
A85 (FLEX SEMISOFT, Extrudr)	Very High	Low	Low
A95 (TPU-R, CREAMELT®)	Middle	High	Very High
D45 (Flex 45, Dutch Filaments®)	High	High	High
D58 (FLEX HARD, Extrudr)	High	Middle	Low

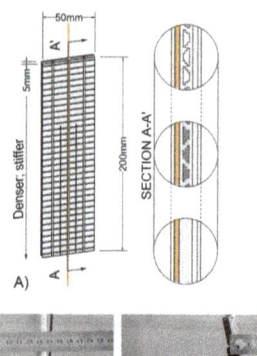
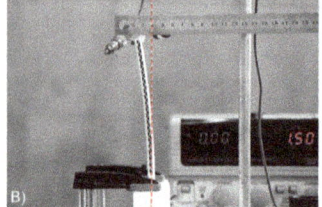
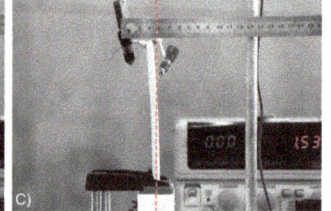

behaviors of the 3D-printed specimens with embedded SMA wires are observed and compared. The comparison of behavior is illustrated in Figures 6-B and 6-C. As expected, the specimens with differentiated density show relatively asymmetrical shape-morphing behavior, whereas the comparison group shows gradual morphing over the shape. Consequently, this experiment verifies the potential of 3DP for effective shape-morphing control with enabled material gradation.

Furthermore, an additional geometric feature is tested. The specimens contain a feature where tensional cables are on one side of the shell, intentionally opposite from the desired direction of motion (Figure 7). The actuation tests show motion only in the area where the cables are not placed, as predicted (Figure 7-A). This feature allows for the more intuitive motion control from the early design stage. Particularly, a specimen with predefined tension cables showcases a dynamic movement within a compact specimen (Figure 7-A-d). A test was also performed with differentiated SMA wire location to enable 3-dimensional motion such as twisting (Figure 7-B). The freeform shape of 3D-printed channels for the wire allows the 3D print to contract asymmetrically.

Nevertheless, a critical issue that should be considered regarding the behavior of SMA-embedded composite structures is functional fatigue during working cycles. The ability of the TPU elements to recover from strain may decrease due to the actuation heat and repeated cycles. Each of the previously discussed specimens is tested, repeating the actuation cycle more than 10 times on average, while a few samples are tested more than 100 times, observing good repeatability.

6   Tests of shape-morphing controllability through differentiated geometries: A) To control shape-morphing behavior, three different densities of infill structures are distributed; to compare the controlled behavior, a specimen consisting of a homogeneous infill is additionally prepared; B) controlled shape-morphing behavior with differentiated geometric structure; C) homogeneous behaviors of shape-changing

7   Tests of shape-morphing controllability through differentiated geometries: A-a to d) Diverse motion behaviors through 3D-printed tensional cable features; B) unsymmetrical double curved motion control through differentiating SMA wire location

8   Actuation of the light regulator; four sets of PTFE insulated SMA wires (1.2 m) are prepared, and each of them is inserted into a set of four 3D-printed components (series connection); these four sets are then connected in parallel with electric jump wires, resulting in the use of 72 watts

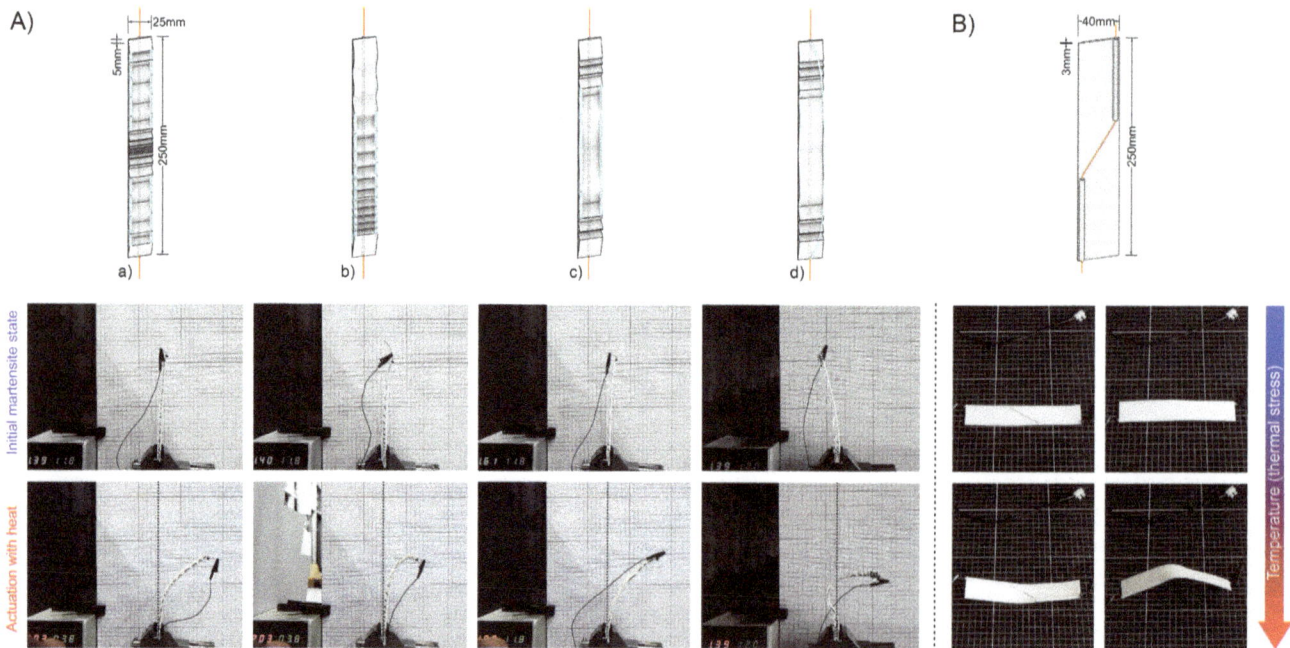

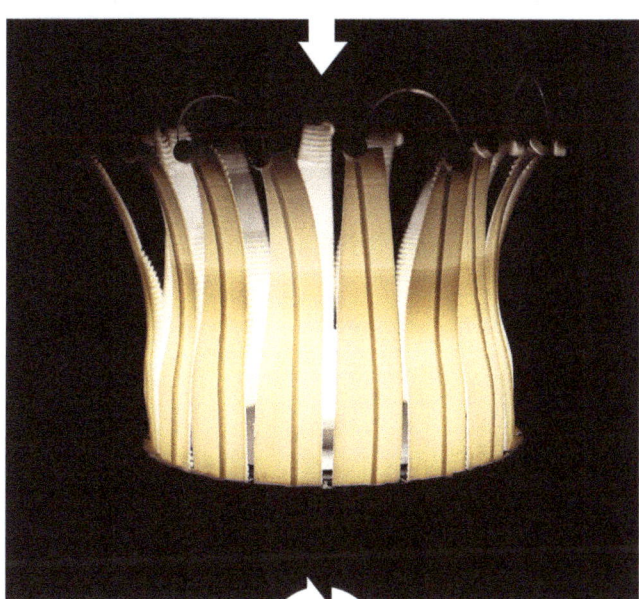
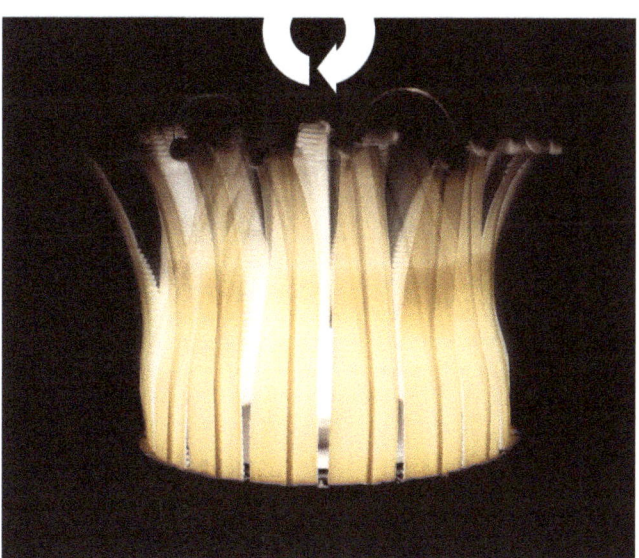

However, the issue of permanent thermal deformation of TPU arises. Several short-term activations (~3 minutes) could be repeated; however, showing the issue when the specimen is actuated and remains in the desired position for longer than 10 minutes. Even upon cooling, the specimen maintains the actuated shape without recovering back to its original state. This problem of thermal deformation necessitates the investigation of thermal insulation of SMA wires, as well as the bistability. Moreover, further tests could be conducted to verify such asymmetrical shape-morphing behavior (Figure 7-B) in various geometries, aiming at achieving synclastic and anticlastic curvatures.

Case Studies

In an exploratory manner, the further study aims at proposing a possible scale-up strategy by integrating multiple SMA-embedded components into two functional applications: An adjustable lighting regulator and a facade panel prototype. Overall, to avoid the thermal deformation of the 3D prints, a polytetrafluoroethylene (PTFE) tube (Øi 0.5 mm, Øa 0.96 mm) is utilized. The SMA wires are first inserted into the PTFE tubes, which are then embedded into 3D-printed channels. The thermosetting properties of PTFE insulate the TPU from the electric resistance heat of the wires. Because the stress is concentrated at both ends of the channels, bootlace ferrules are additionally inserted. In contrast to the small-scale specimens that are independently actuated with an electrical source, the assembly of multiple kinetic components is crucial to implementing an intricate electrical design procedure in the early stage of planning. The idea is to electrically connect some components in series as a set, then the sets of series components are connected in parallel.

For the lighting regulator, a holder is mounted onto a lamp with sixteen tentacle or petal-like 3D-printed TPU components, in which the design of each is a successor to previously discussed specimens (Figures 6, 7). The 3D-printed components further reflect the exploration of 3-dimensional geometric freedom. These components are specifically designed with translucent D45 (Flex 45, Dutch Filaments®) to disperse light when closed and create shadowed patterns when opened (actuated). To fully bloom (actuation) during the on-state, variable material density is programmed on one side of the component. To achieve this, multiple samples are tested with different densities and thicknesses, calibrating the opening angles and closed positions of the petals. After embedding PTFE insulated SMA wire into 3D prints, the entire components are mounted onto the 3D-printed polylactic acid holder and tightened with adjusting rings and grub screws. Consequently, the functional, adjustable lighting regulator (50 cm × 25 cm × 25 cm) is materialized and the actuation demonstration is performed (Figure 8). Most of the components behave as expected. However, the uneven tightness of manually assembled anchors leads the overall structure to be observed as disorganized. Besides, a few components deflect relatively more or less than desired for the same reason.

Following the development of the light regulator, an active façade panel is designed, aimed at adding functionality to control solar exposure. Inspired by the visual patterns of fluid flow, simulations are run on a panel size of 900 mm × 600 mm. The design process follows a generative approach and the panel with the most suitable opening patterns is selected. Subsequently, forty-seven unique kinetic components are defined. The design of each is a successor to the previously discussed specimen (Figure 7-B) with more various wire locations as well as complex shapes. This allows for every component to behave differently, e.g., motion directions, curvatures, and ranges, enabling the generation of diversified opening patterns and hence the solar exposure. Two different TPU materials are used, the translucent D45 (Flex 45, Dutch Filaments®) and transparent D58 (FLEX HARD, Extrudr), for different transparencies to diversify the light and shadow conditions whilst allowing a diverse range in motion. The 3D-printed components are first assembled and bolted onto a laser-cut fiberboard. Insertion and wiring of PTFE insulated SMA wire are performed, with the intention of enclosing complete circuitry inside the 3D prints and insulating electrically. This allows the facade element to avoid short circuits in environmental conditions. Consequently, the total parts are consolidated into an active facade panel and the actuation demonstration is performed, showcasing the desired shape-morphing behavior (Figures 9, 10).

The method of complex assembly could be the first step towards up-scaling. Nonetheless, the current process of assembly, including SMA wire insulation and insertion into 3D prints not only causes manual errors of anchor tightness but also is highly labor-intensive, revealing the necessity of automation.

SMA-embedded Bi-stable Geometry (SMABG)

Another key aspect of adapting the proposed method for architectural applications is energy efficiency, concerning the constant electrical energy use during actuation. Preliminary tests are made based on the existing studies of bi-stable geometries (Bende et al. 2015; Jeong et al. 2019), but without successful state-change when it comes to combining with SMA wires. Therefore, the proposed method of the state-transferring mechanism is studied (see section 3D-printed Complex Geometries for High Shape-morphing Controllability), which is similar to the pawl release mechanism. The specimen (200mm x 25mm x 6mm) is fabricated when the two 3D prints for forwarding and reversing are assembled.

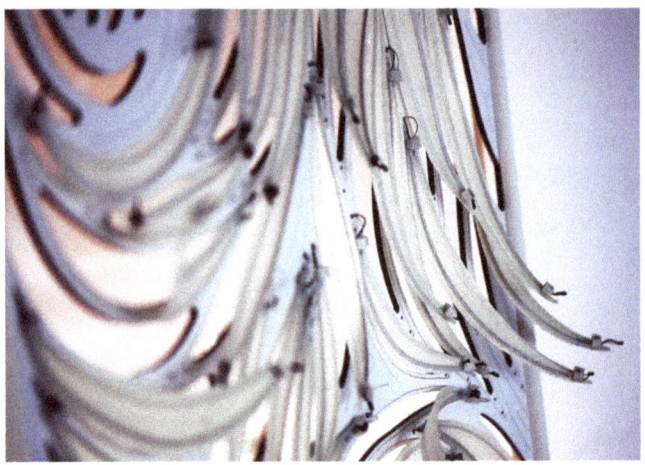

9

9 Actuation of the facade panel

10 Forty-seven PTFE insulated SMA wires are prepared with the matching lengths for each 3D print's channel; each of them is then inserted into 3D-printed components; another channel of each 3D print is used for enclosing the electric jump wire; according to predefined sets of 3D prints, each set is connected in series, which the sets are then connected in parallel, resulting in the use of 180 watts.

TABLE 2 Energy use of SMA-embedded bi-stable structure and comparison

Energy of a SMABG Specimen (kilowatt-hour)		Energy of a 100 watt Light Bulb (kilowatt-hour)		Equivalence	
				Energy-compatible Numbers of SMABG Specimens to a 100 watt Bulb	Compatible Actuation Cycles of SMABG to an Hour of Using a 100 watt Bulb
9 (watts) × 3600 (seconds) = 32.4	=	100 (watts) × 3600 (seconds) = 360		360 (kilowatt-hour) ÷ 32.4 (kilowatt-hour) ≈ 11 specimens	3600 (seconds) ÷ 15 (seconds) = 240 cycles

Relatively smooth-faced TPU A95 (TPU-R, CREAMELT®) is chosen because the SMABG system requires a material with not only high elasticity, but also a low coefficient of friction between contact areas. Since the expected actuation time to reach the interlock state (desired deflection) is short, two SMA wires (each 300 mm) are embedded without the PTFE insulation into each 3D print (9 watts per specimen). The experiment is conducted with a good agreement between expectation and the physical result (Figure 11).

When the forward set of SMA wire is heated (Figure 11-B), the linkage feature of 3D prints successfully holds the structure in its deflected shape (observed > 2 hours) (Figure 11-C). The releasing mechanism performs smoothly by activating the reverse SMA wire (Figure 11-D). The actuation cycles are performed ~100 times without failure. The actuation time for the interlock state requires 10 seconds, while release actuation takes 5 seconds. From the energy consumption perspective, the required energy of repeatedly actuating 11 SMABG specimens 240 times (cycles of on and off) is equivalent to constantly turning a conventional 100 watt light bulb on. Table 2 specifies the comparison.

Consequently, high energy efficiency is enabled by combining the proposed bi-stable geometry with SMA wire actuation. In this case, bi-stable 3D-printed geometries not only reduce energy consumption but also minimize the time of heat exposure, eliminating the issue of permanent thermal deformation. Moreover, the proposed SMABG structures have the ability to be designed within thin, compact-integrated kinetic elements.

CONCLUSIONS

This paper presents a novel, energy-efficient method for kinetic architectural elements by combining the use of 3DP to fabricate the body of motion and SMA to act as a muscle-like actuated deformer. The novelty of the method includes the investigation of material gradients for higher deflection control, the exploration of an up-scaling approach with assembly, and the development of the bi-stable geometric system for energy reduction. With a number of specimens, the benefits of 3D-printed graded materials are proven to control the shape-morphing behaviors of SMA-embedded composite structures. Further prototypes assembled with multiple kinetic components showcase the possibility of scaling the system up. The developed bi-stable geometric system and its experiment demonstrate the low energy-consuming motion, which could be fundamental to efficient architectural adaptations.

This research is a first approach to developing design, fabrication, and assembly methods for energy-efficient kinetic elements. One could further investigate relevant aspects, such as automation of SMA wire embedding and insulation coating to tackle currently labor-intensive assembly procedures. Furthermore, the developed bi-stable geometric system should be tested on a larger scale architectural application, i.e., an adaptive façade element, with validation of short- and long-term mechanical, thermal, and functional behaviors. This could involve testing weather resistance. Last but not least, the integration of a photovoltaic system could be investigated. This may accelerate the adaptation of the system to the real world, with its possibility of zero-energy actuation for kinetic building structures.

ACKNOWLEDGMENTS

The authors would like to thank a number of partners and collaborators whose dedication helped us fulfill the research described in this paper, including Dr. Zafeirios Triantafyllidis, Dr. Rudolf Hufenus (Empa), Hyuk Sung Kwon, and Angela Yoo (Digital Building Technologies, ETH Zurich). Partially, the research was developed for a thesis of the 2020/21 MAS ETH DFAB. The MAS ETH DFAB is a one-year full-time educational program of the NCCR Digital Fabrication, jointly organized by Digital Building Technologies (DBT) and Gramazio Kohler Research.

REFERENCES

Ahn, Sung-Hoon, Kyung-Tae Lee, Hyung-Jung Kim, Renzhe Wu, Ji-Soo Kim, and Sung-Hyuk Song. 2012. "Smart soft composite: An integrated 3D soft morphing structure using bend-twist coupling of anisotropic materials." *International Journal of Precision Engineering and Manufacturing* 13 (4): 631–634.

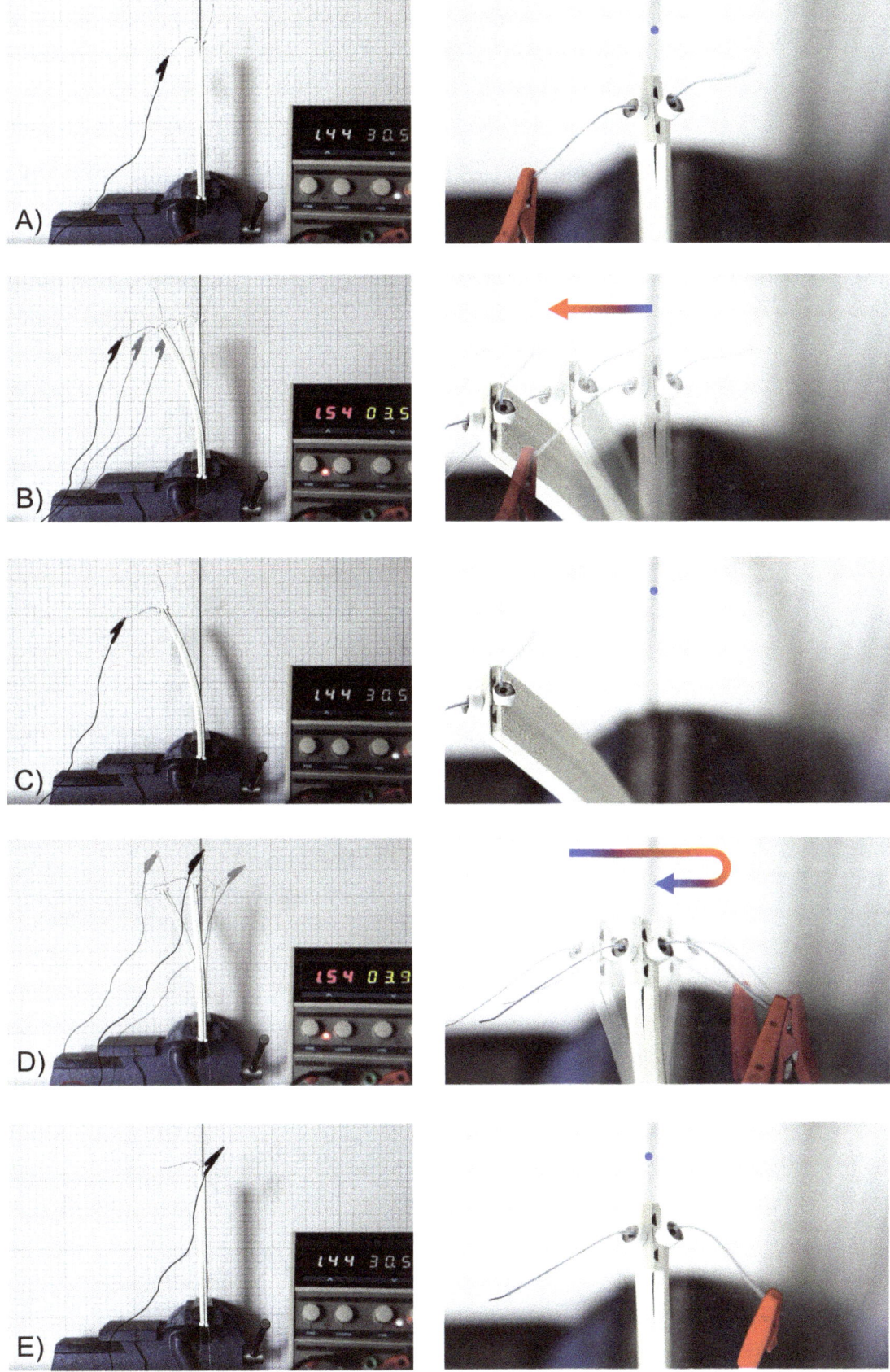

11 Demonstration of bistability employing a set of two 3D-printed bi-stable geometries with two sets of embedded SMA wires: A) Initial martensite state; B) Actuation with heat for the desired deflection (austenite state); C) Maintaining the desired deflection with no heat (interlock state); D) Reverse actuation to return to the initial state (release state); and E) initial martensite state

Bende, Nakul Prabhakar, Arthur A. Evans, Sarah Innes-Gold, Luis A. Marin, Itai Cohen, Ryan C. Hayward, and Christian D. Santangelo. 2015. "Geometrically controlled snapping transitions in shells with curved creases." *Proceedings of the National Academy of Sciences* 112 (36): 11175–11180.

Bodkhe, Sampada, Lorenzo Vigo, Shengyun Zhu, Oleg Testoni, Nicole Aegerter, and Paolo Ermanni. 2020. "3D printing to integrate actuators into composites." *Additive Manufacturing* 35: 101290.

Chamilothori, K., A. M. Kampitaki, and K. A. Oungrinis. 2013. "Climate-Responsive Shading Systems with Integrated Shape Memory Alloys (SMA)." In *Proc., 8th Energy Forum on Solar Building Skins Conf. Munich, Germany: Economic Forum*. Munich.

Correa, David, Athina Papadopoulou, Christophe Guberan, Nynika Jhaveri, Steffen Reichert, Achim Menges and Skylar Tibbits. 2015. "3D-printed wood: programming hygroscopic material transformations." 3D Printing and Additive Manufacturing 2 (3): 106–116.

Ge, Qi, H. Jerry Qi, and Martin L. Dunn. 2013 "Active materials by four-dimension printing." *Applied Physics Letters* 103 (13): 131901.

Ghomshei, M. M., N. Tabandeh, A. Ghazavi, and F. Gordaninejad. 2005. "Nonlinear transient response of a thick composite beam with shape memory alloy layers." *Composites Part B: Engineering* 36 (1): 9–24.

Hazlehurst, Kevin Brian, Chang Jiang Wang, and Mark Stanford. 2014. "An investigation into the flexural characteristics of functionally graded cobalt chrome femoral stems manufactured using selective laser melting." *Materials & Design* 60: 177–183.

El Razaz, Zeinab. 2010. "Sustainable Vision of Kinetic Architecture." *Journal of Building Appraisal* 5: 341–356.

İlerisoy, Zeynep Yeşim, and Merve Pekdemir Başeğmez. 2018 "Conceptual Research of Movement in Kinetic Architecture." *Gazi University Journal of Science* 31 (2): 342–352.

Jani, Jaronie Mohd, Martin Leary, Aleksandar Subic, and Mark A. Gibson. 2014. "A review of shape memory alloy research, applications and opportunities." *Materials & Design (1980-2015)* 56: 1078–1113.

Jeong, Hoon Yeub, Soo-Chan An, In Cheol Seo, Eunseo Lee, Sangho Ha, Namhun Kim, and Young Chul Jun. 2019. "3D printing of twisting and rotational bistable structures with tuning elements." *Scientific Reports* 9 (1): 1–9.

Jung, Beom-Seok, Jung-Pyo Kong, NingXue Li, Yoon-Mi Kim, Min-Saeng Kim, Sung-Hoon Ahn, and Maenghyo Cho. 2013. "Numerical simulation and verification of a curved morphing composite structure with embedded shape memory alloy wire actuators." *Journal of Intelligent Material Systems and Structures* 24 (1): 89–98.

Kwon, Hyunchul, and Benjamin Dillenburger. 2019. "Optimized Internal Structures for 3D-Printed Sandwich Elements." In *Proceedings of the IASS Annual Symposium 2019 – Structural Membranes 2019. 1278–1285*. Barcelona: International Association for Shell and Spatial Structures (IASS).

Kwon, Hyunchul, Martin Eichenhofer, Thodoris Kyttas, and Benjamin Dillenburger. 2018. "Digital Composites: Robotic 3D Printing of Continuous Carbon Fiber-Reinforced Plastics for Functionally-Graded Building Components," In *Robotic Fabrication in Architecture, Art and Design 2018*, edited by J. Willmann, P. Block, M. Hutter, K. Byrne, and T. Schork. 363–376. Cham: Springer.

Lacasse, Simon, Patrick Terriault, Charles Simoneau, and Vladimir Brailovski. 2015. "Design, manufacturing, and testing of an adaptive composite panel with embedded shape memory alloy actuators." *Journal of Intelligent Material Systems and Structures* 26 (15): 2055–2072.

Li, Yan, Zuying Feng, Liang Hao, Lijing Huang, Chenxing Xin, Yushen Wang, Emiliano Bilotti et al. 2020. "A review on functionally graded materials and structures via additive manufacturing: from multi-scale design to versatile functional properties." *Advanced Materials Technologies* 5 (6): 1900981.

Lignarolo, Lorenzo, Charlotte Lelieveld, and Patrick Teuffel. 2011. "Shape morphing wind-responsive facade systems realized with smart materials." In *Adaptive Architecture: An International Conference*. London.

Lu, Qi-Wei, and Christopher W. Macosko. 2004. "Comparing the compatibility of various functionalized polypropylenes with thermoplastic polyurethane (TPU)." *Polymer* 45 (6): 1981–1991.

Mammano, G. Scirè, and Eugenio Dragoni. 2014. "Functional fatigue of Ni–Ti shape memory wires under various loading conditions." *International Journal of Fatigue* 69: 71–83.

Maragkoudaki, Anna. 2013. "No-mech kinetic responsive architecture: Kinetic responsive architecture with no mechanical parts." In *2013 9th International Conference on Intelligent Environments*. 145–150. Athens: IEEE.

Momeni, Farhang, Xun Liu, and Jun Ni. 2017. "A review of 4D printing." *Materials & Design* 122: 42–79.

Morris, Christopher W. 1992. *Academic press dictionary of science and technology*, vol. 10. Oxford: Gulf Professional Publishing. 267.

Naghashian, S., B. L. Fox, and M. R. Barnett. 2014. "Actuation curvature limits for a composite beam with embedded shape memory alloy wires." *Smart Materials and Structures* 23 (6): 065002.

Rogers, C. A., C. Liang, and J. Jia. 1991. "Structural modification of simply-supported laminated plates using embedded shape memory alloy fibers." *Computers & Structures* 38 (5-6). 569–580.

Taseva, Yoana, Nik Eftekhar, Hyunchul Kwon, Matthias Leschok, and Benjamin Dillenburger. 2020. "Large-Scale 3D Printing for Functionally-Graded Façade." In *Proceedings of the 25th International Conference of the Association for Computer-Aided Architectural Design Research in Asia (CAADRIA)*, edited by D. Holzer, W. Nakapan, A. Globa, and I. Koh. 183–102. Bangkok: CAADRIA.

Tibbits, Skylar. 2014. "4D printing: multi-material shape change." *Architectural Design* 84 (1): 116–121.

12 Details of the kinetic facade panel during actuation

Velikov, Kathy, and Geoffrey Thün. 2013. "Responsive building envelopes: characteristics and evolving paradigms." In *Design and Construction of High Performance Homes*, edited by F. Trubiano. Oxfordshire: Routledge. 75–92.

Yi, Hwang, Dongyun Kim, Yuri Kim, Dongjin Kim, Je-sung Koh, and Mi-Jin Kim. 2020. "3D-printed attachable kinetic shading device with alternate actuation: Use of shape-memory alloy (SMA) for climate-adaptive responsive architecture." *Automation in Construction* 114: 103151.

IMAGE CREDITS

Figure 1, 3-4, 6, 8-12: ©Hyunchul Kwon, 2021-2022
Figure 2: ©Mohammad Muzeem, 2020; Wojciech Gil, 2021; Hyunchul Kwon, 2022
Figure 5, 7: ©Priyank Soni, 2021
All other drawings and images by the authors.

Hyunchul Kwon is a doctoral researcher at the chair of Digital Building Technologies, ETH Zürich. He received a Master of Architecture (Distinction) from The Bartlett School of Architecture UCL, where he taught and lectured thereafter. During his PhD study, he has developed a novel approach to design and fabricate large-scale composite building components utilizing seamless integration of robotic multi-material add-on 3D printing technology and computational design system.

Priyank Soni is an architect-scenographer with a specialization in digital fabrication technologies. He became an architect in India at School of Planning and Architecture, Bhopal, and completed his Masters of Advance Studies in Digital Fabrication at ETH Zürich. He was associated with Atelier Brueckner in Stuttgart to explore scenography and exhibition design where he has been a part of many renowned museum projects worldwide. He is currently associated with Uniplan, Cologne for conceptualizing digital design strategies in branded spaces.

Ali Saeedi, PhD is a scientist at the laboratory of structural engineering, Empa. He is an experienced researcher with a demonstrated history of working at both academic and industrial sections. He is skilled in structural design, finite element analysis, and scientific research. He has several years of experience working on various types and mechanical properties of smart and advanced composite structures. His PhD study focused on mechanical engineering.

Moslem Shahverdi, PhD is the leader of the Advanced Structural Materials group at Empa structural laboratory. He is an expert in FRP composites, SMAs, fatigue-fracture mechanics, material characterization, large-scale testing, and numerical modeling. He is involved in teaching at ETH Zürich and at the University of Tehran.

Benjamin Dillenburger, PhD is Professor at the Institute of Technology in Architecture at the Department of Architecture, ETH Zürich. He is leading the research group Digital Building Technologies, which investigates computational design and digital fabrication with a focus on large-scale additive manufacturing in architecture.

Smart Branching

An Experimental Method for Heterogeneous Branching Networks using Non-planar 3D Printed Clay Deposition

Chenxiao Li
University of Pennsylvania

Mingyang Yuan
University of Pennsylvania

Zilong Han
University of Pennsylvania

Billie Faircloth
KieranTimberlake
& University of Pennsylvania

Jeffrey S. Anderson
University of Pennsylvania
& Pratt Institute

Nathan King
University of Pennsylvania
& Virginia Tech

Robert Stuart-Smith
University of Pennsylvania
& University College London

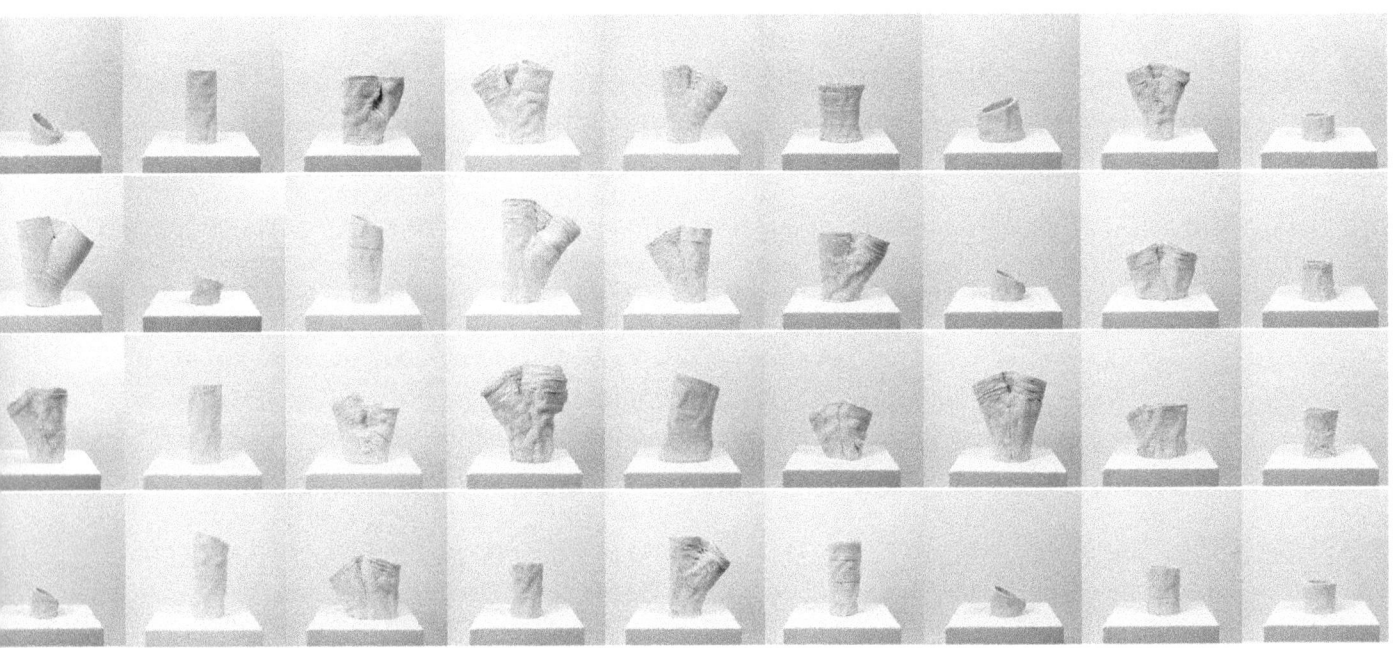

1 Photograph collection of 36 physical models produced using non-planar 3D printed clay deposition

ABSTRACT

Clay extrusion 3D printing with 6-axis industrial robots and ceramic firing pipelines has inspired designers' reflection on material properties, design methodologies, automatic manufacturing, and logistics of fabrication. It brings the potential for innovative industrial bespoke production of architectural components. Plasticity and malleability are merits to the creative freedom of the form, yet they also pose technical challenges. The goals of the research are designing specifically printable geometries at a durable scale and optimizing methodologies for fabrication to ensure both quality and efficiency.

Through the design and fabrication of a 1.3m-high physical prototype sampled from our facade proposal, we developed a relatively automated project pipeline. It aims to achieve the generative and evolutionary design and a non-planar clay deposition method for tubular branching components.

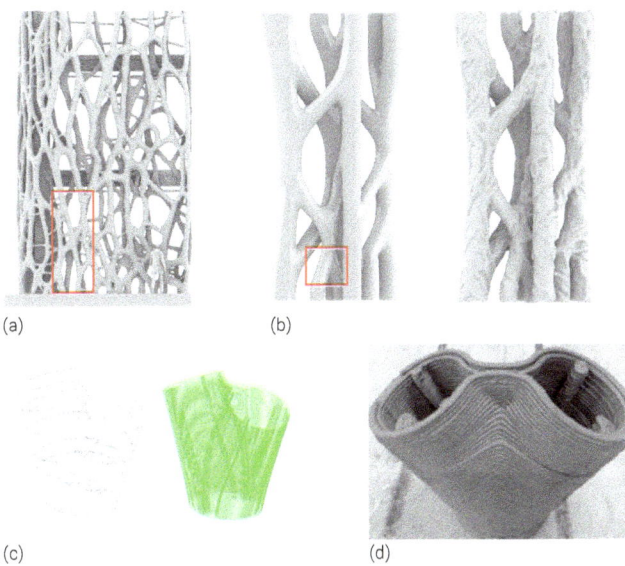

2   Fabricating branching structures using clay extrusion 3D printing presents technical challenges at bifurcations: (a) Digital model of full scale architectural envelope proposal based on a customized method; (b) Digital model of the 1.3 m high prototype; (c) Double-layer print path for a 45° bifurcated branch component; and (d) Outcome of 45° bifurcated branch component produced using non-planar 3d printed deposition

INTRODUCTION

Branch-like geometries provide diverse structural, functional, and aesthetic potential for trussed, diagrid, or funicular structures. Clay extrusion 3D printing allows for variable manufacturing and can support heterogeneous arrangements of branch-like geometries (Bechthold 2016); however, several technical challenges exist in the fabrication of these geometries with clay extrusion 3D printing due to clay's plasticity and drying time.

Background and State of the Art

3D printed concrete and clay have similar types of plasticity when extruded; therefore, it is helpful to review advances in concrete 3D extrusion printing. 3D printing methods for concrete and ceramics both involve continuous extrusion. 3D printed concrete, often referred to as Chemical Reaction Bonding Concrete (CRB/C) and ceramic Liquid Deposition Modeling (LDM) both involve the use of an industrial extrusion screw and a liquid material reservoir that feeds the extruder. While concrete incorporates a mixer, clay LDM utilizes a cylindrical tank located on the robot adjacent to the extruder. Both can be performed on an industrial robotic arm to print on a flat floor, where the extrusion can be switched on or off by a digital signal (Kontovourkis and Tryfonos 2018). With a 6-axis industrial robotic arm, printing has fewer limitations: each extrusion layer is not necessarily limited to horizontal deposition. Therefore, it is possible to print ceramics and concrete on complex surfaces such as freeform foam shaped by a hotwire cutter on a robotic arm (Ko et al. 2018).

Research on the non-planar deposition of 3D printed concrete is far ahead of similar research in clay. In some projects, the more complex printing method of rotating the tool center point (TCP) plane is applied to bifurcated cylinders using sidestepping (Cruz et al. 2022). Helpful techniques have been developed in the past five years including changing the toolpath to optimize the surface (Zhong et al. 2020; Nisja et al. 2021) and exploring the printing of smaller bifurcated parts using planar 3D ceramics extrusion printing (Xing et al. 2021). Apart from non-planar 3D printing, printing on a movable build plate is also a notable FDM method (Nayyeri et al. 2022). However, remaining limitations are managing the plasticity and semi-liquid properties of clay to maintain balance during the printing process; the allowable inclination angle of geometries to avoid collapse; the need to prevent slumping of inclined parts due to gravity; and increasing the surface quality of the prints.

Experimental Overview

Our experiments prioritize the following questions:

- What are the precautions to prevent excessive deposition of clay during the printing to ensure the surface quality of each print?
- How can parts be balanced during the printing process? Through an automatically updated feedback loop with real sense scanning, is it possible to update the tool path according to design intent?
- What aspects of this fabrication pipeline could contribute to future industrial bespoke production?

Firstly, we conducted experiments to quantify the printing parameters, including changing TCP plane angles. Secondly, we utilized a sloped bed and differentiated the deposition thickness of the clay. We also used depth tracking (Intel RealSenseTM depth camera D435) to detect deformations in inclined prints and adjust the printing path as needed. Finally, we completed the design and robotic fabrication of a prototype branching structure, incorporating findings from former experiments on the branching shape. The 1.3 m tall prototype is comprised of 37 components that demonstrate our approach to non-planar 3D clay extrusion printing (Figure 2).

MATERIALS AND METHODS

Clay Body and Firing Process

All tests are based on MC10G earthenware clay, and the shrinkage rate of 4.5% and a water absorption of ±1%. The ratio we used for the mixture is 25 lbs clay to 16 oz water. They are inserted in a pug mill, Peter Pugger VPM-60 Power Wedger, for mixing with water for at least 2 hours. It is also worth noting that, from our tests, it is better to let the clay sit for more than 2 hours after being pugged out of the mill. The whole prototype can take up to 3 days for printing, and the printing time for each component is approximately 5-15 minutes. Each component took 12-24 hours to be fully bone-dried depending on the amount of clay accumulated. After being bone dried, components are fired under cone 04 with temperature 1945 °F and then cooled down. The whole process of firing and cooling takes 24-36 hours.

Robotic Clay Extrusion

All robotic clay extrusion was done using 6-axis ABB IRB4600 industrial robot (ABB robot), with LDM WASP Extruder XL (WASP extruder) and a 5 liter Clay Tank (Figure 3). The cooperation with robots is through RobotStudio Online and FlexPendant.

Early Experiments

We conducted a series of tests on 3D printing parameters specific to ABB robot and WASP extruder, and observed that nozzle diameter (ND) of 4 mm, moving speed (MS) of 80%, layer height (LH) of 5.5 mm, and extruding speed (ES) of 9 worked best in all results for vertical cylinders of 10 cm diameter.

Subsequently, we conducted two experiments to test the limits of non-planar 3D deposition of a bifurcated tubular geometry: the first one testing the limitation of height, and the second testing the slope as the printing bed, as well as the branching angle of the bifurcation.

Experiments of Printing Cylinders

We printed cylinders of different heights with the same diameter and printing parameters, and concluded that the maximum viable height of printing a basic cylinder without any collapse is around 18 - 20 cm (Table 1).

Experiments of Printing Bifurcation on Slope

We printed cylinders and bifurcated tubes of 30°, 45°, and 60°

TABLE 1 Experiments on Scale Limitation

Test Number	Invariants	Height (centermeter)	Result	Detail
1	NZ = 4mm, MS = 80%, LH = 5.5mm, ES = 9	10	√	no obvious deformation
2		15	√	bottom is about 5% thicker
3		17.5	√	bottom is about 15% thicker
4		20	×	unacceptable collapsing

on horizontal surfaces and slopes (Figure 4). To exclude interference of other uncertainties, such as different humidity of the clay body, we tested each print twice, and the result was recorded as a "√" only if both prints were successful. Thus we found out that the relationship between and angle and the slope is that the best choice for 30°, 45°, 60° bifurcation is a 10° slope.

FINAL PROTOTYPE

Design to Fabrication Methodology

We describe a muti-step design to fabrication workflow that incorporates: 1) The generation of the branching pattern using cellular automata (CA), solar radiation analysis, and a self-organization algorithm; 2) The modelling of components uses SubD Multipipe in Rhinoceros 3D and an evolutionary algorithm to generate a structural augmented lattice network on top of the geometry; 3) A toolpath using non-planar slicing methods; 4) A bespoke 3D printing process using non-planar 3D printing with a sloped bed.

Branching Pattern Generation

To generate the branching pattern, we created a custom combination of cellular automata CA rules (Figure 5) based on Stephen

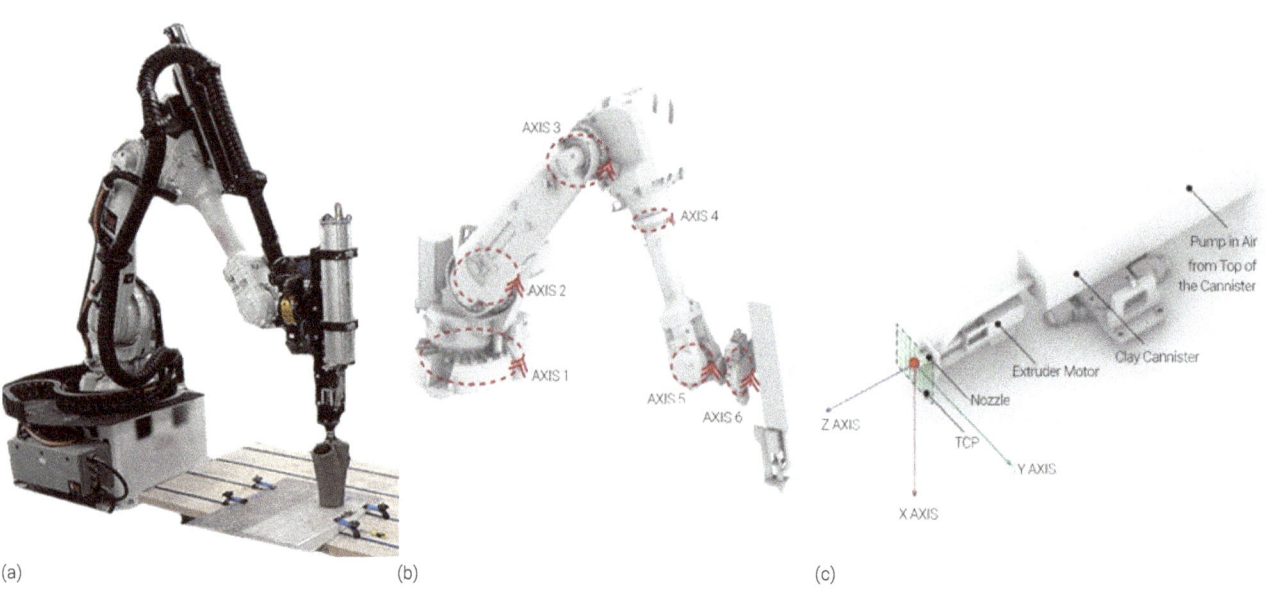

3 Robotics elements of the experiments: (a) 6-axis ABB IRB4600 industrial robot, (b) Digital model of 6-axis ABB IRB4600 noting each axis of rotation with clay extruder, (c) Details of the clay extruder with motor, nozzle, and canister.

Wolfram's 1D CAs with three possible CA cell states (0, 1 or 2) (Wolfram 2002 within a two-dimensional array of cells to produce a branching organization of cells of the same state. This first branching pattern was used to generate center-curves of the proposed networked branching ceramic assemblage.

To generate site specific bespoke design that respond to the related climate condition, we performed a solar radiation analysis with Ladybug Tools 1.4.0 (Ladybug Tools LLC 2022) on the branching pattern using the values to either attract or repel branches. The final branching pattern was used to generate center-curves of the proposed networked branching ceramic assemblage (Figure 6).

Modeling the Branching Prototype Components

All components were printed with a double-wall thickness: one 8 cm diameter tube for structural reinforcement, while the outer 10 cm diameter tube also embodied a variable ornamental pattern (Figure 2). The prototype is composed of 37 components that range from 5 cm to 30 cm tall, generated by splitting the interconnected branches based on two rules (Figure 10). Overall, "X" splits were horizontal and "Y" splits were vertical, and some components were generated by using both rules (Figure 9).

Generating Component Tool Paths

For branches whose centerline symmetry axis deviates from the absolute vertical line by an angle greater than 15° and whose height is greater than 10 cm, the digital model is first rotated by 10° and then the tool path is generated; if the height

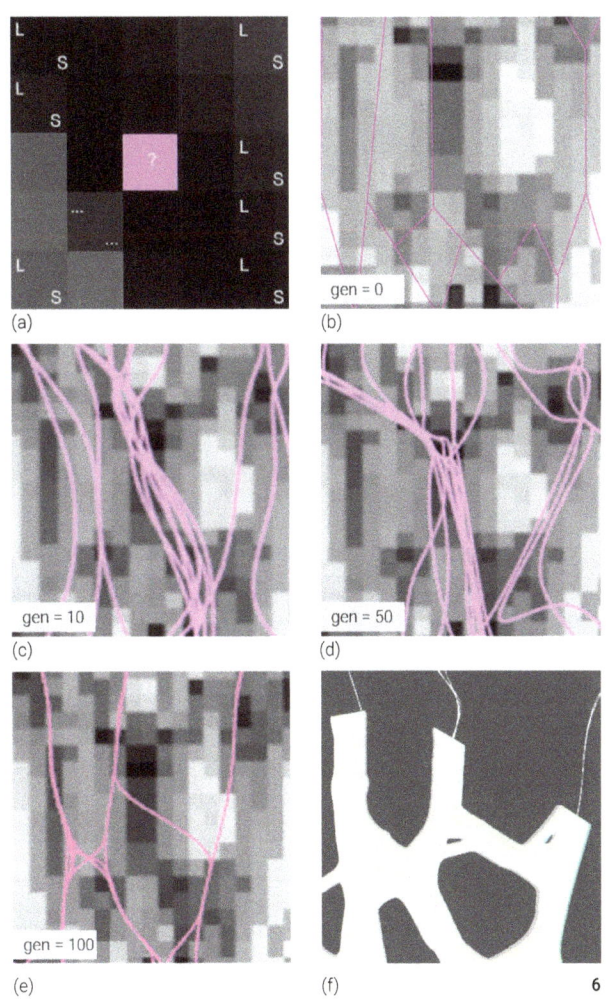

4 Basic bifurcations were printed on different slopes of 0°, 5°, 10°, 15°, and 20°. Only successed experiments were listed in this diagram, and we can see that, firstly, slope can help balance the print; secondly, each slope of same degree has certain tolerance of branching degree of the prints.

5 Generation of the branching pattern: (a) Each CA cell has three possible states (0, 1 or 2), and adjusts its own state based on the states of neighboring cells; (b) This two-dimensional array of cells can produce a branching organization of cells of the same state; and (c) Branching pattern from cells: the state 0 cell acts as a void, state 1 influencing a displacement of 5 cm, and state 2 influencing displacement of 10 cm

6 Generation of branching pattern: (a) Firstly, translate solar radiation analysis results into hue, saturation, lightness (HSL) images, so that the CA point can find the closest point on mesh (HSL image) and inherit its L-value, and then the CA points will attract or repel the control points on original curves (pink square); (b) Original curves based on the first branching pattern; (c)(d)(e) Iterations of self-organization with attractors from solar radiation analysis; and (f) Use the curves produced as center-curves of the basic pipes

is greater than 15cm, additional reinforcement structures need to be added internally for support. If the height is greater than 15cm, it is necessary to add an additional reinforcement structure for support inside. All others can generate toolpath routinely (Figure 9).

Printing Process

For truncated cylinders, non-planar 3D printing was implemented (Figure 10); bifurcated tubes were printed on a 10° slope with non-planar 3D printing (Figure 11). For truncated cylinders that are taller than 15 cm, a real-sense camera was used to 3D scan already printed layers so that the print-path could be automatically adjusted after finishing every 10 layers to ensure the desired geometry for each component could be successfully fabricated irrespective of individual layer settlement (Figure 12).

For assembly of the prototype, we adopted a post-tension system (Figure 13). For the connection, we use laser-cut acrylic sheets and 3D printed spherical nuts (Figure 14). The spherical nuts can rotate to embed in the acrylic holes (Figure 7).

RESULTS

We printed more than 40 components for the final prototype. Some parts failed due to deformations caused by incorrect storage or due to accidental breakage. Finally we printed 37 piece of clay extruded components within three days of duration (Figure 1)

Clay Extrusion 3D Printing Parameter

For truncated and bifurcated branches, we found general printing parameters that allowed for successful production of components: nozzle diameter (ND) = 4 mm, moving speed (MS) = 80%, layer height (LH) = 5.5 mm, extruding speed (ES) = 9 (see Table 2). The most important note is that if LH = 70%-90%*ND, the clay deposition performed best. For the toolpath design, we found that cracks easily appear at the corners of our tests. The comparison results show that a more rounded curve/chamfer tool path can avoid cracking at sharp corners (fillet radius > nozzle diameter).

For the printing bed tests, we found that for our bespoke bifurcation—one vertical branch, another one bent (Figure 16)—when we ignore the potential uneven deposition or slight deformation in the printing process, the 10° slope works for

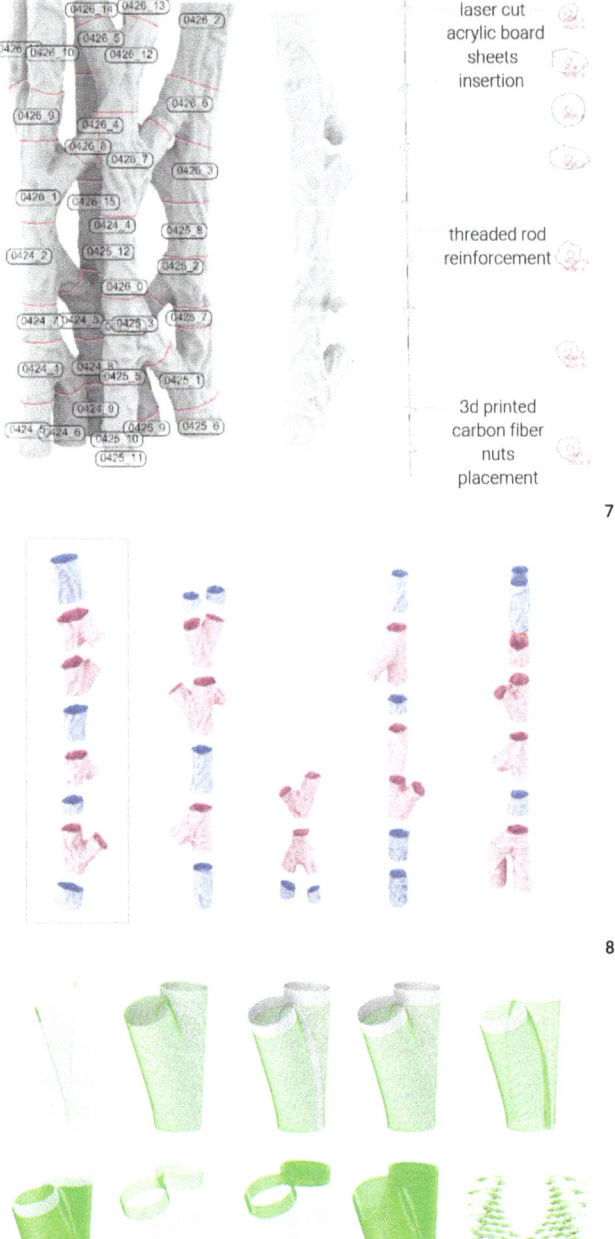

7   (left) The slicing follows two rules: split branches horizontally if they incorporate a post-tensioning rod, if not, then split branches perpendicular to their normal; (right) Example of one column: Post-tensioned system inside one vertical branch and customized partitions, sphere nuts and acrylic sheets, for these components

8   By following the slicing rules, all sliced segments for printing only have two different basic shapes: "I" (cylinder or trucated cylinder) or "Y" (bifurcated cylinder)

9   The generation routine of non-planar print-path for "Y"

TABLE 2 Experiments on printing variables

Test Number	Invariants	Value	Detail
1-1	Layer Height	[2.0mm, 3.5mm, 5.0mm]	
1-2	Extruding Speed	[3, 6, 9]	
1-3	Moving Speed	[30%, 60%, 90%]	

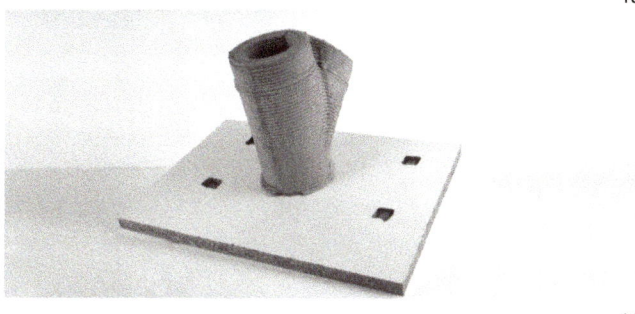

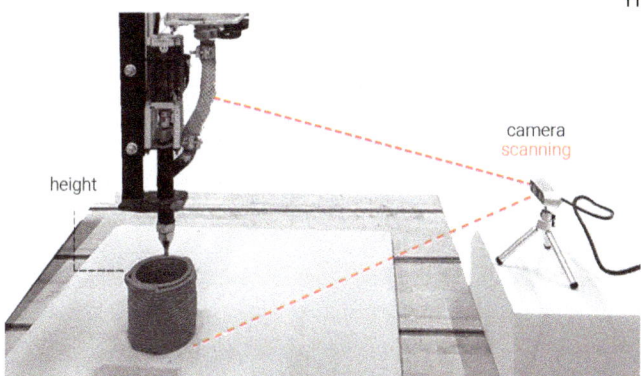

10 Non-planar 3D print deposition Method 1: Non-planar 3D print
11 Non-planar 3D print deposition Method 2: Printing on slope
12 Non-planar 3D print deposition Method 3: Using Intel RealSense RGBD camera to recalibration
13 Assembly of clay segments: First, we used customized sphere nuts to fix the four post-tensioned rods on the base, and then place each layer of clay segments, acrylic sheets and nuts from bottom to top
14 Physical partitions samples using 1/8 inch acrylic sheets and 3D printed 1 inch diameter and 1/2 inch thick spherical nuts

all branching prints (Table 3). It significantly improves the balance in the printing process (Figure 11).

Workflow

The non-planar deposition methods have been beneficial in our bespoke process. Firstly, non-planar printing effectively avoids over-squeezing and collapsing and improves the performance of truncated and bifurcated segments (Figure 16). Also, a continuous non-planar print path ensured the high efficiency of production. The addition of the slope helped some bifurcations that were difficult to balance to be produced successfully (Figure 2).

However, some identified issues necessitate further research. The real-sense camera used for recalibration was not accurate enough. It proved difficult to check the scan height in the camera's real-time feedback image during continuous clay extrusion. Therefore, we tried to introduce a pause in printing after each 10 completed layers to conduct a periodic scan and update the height of the robot's tool path for subsequent layers. The algorithm at this stage is not fully automated, with many parameters still requiring manual inputs.

CONCLUSION

In the three non-planar deposition methods we used for our bifurcated tubes, non-planar printing has been successfully achieved without substantial collapse or material settlement (Figure 15); however, further experiments should be carried out to as certain a universal solution for more complex branching forms.

In the printing-on-slope method, the slope setting clearly brings a better balance and improves the success rate. However, there are also problems of complex calibration, increased resources, and wasted time. The initial position of the print is prone to errors, and even a few millimeters of error can result in insufficient adhesion or over-squeezing of the initial layers, thus affecting the stability and quality of subsequent prints. Following every printed component on a slope, one needs to wait for the clay body to dry out before transferring it to the flat floor. Therefore, we needed to prepare a lot of additional slopes.

Due to the low-resolution real-sense camera, we have only conducted the digital part of the experiment so far. The actual simulation requires a very clean background, and the presence of stray colors can lead to significant scanning errors, which left us with a question. If this technical problem can be solved, we can further combine non-planar printing, printing on the slope, and real-time recalibration to create a truly autonomous non-planar deposition printing system. This helps enrich the morphological diversity of ceramic branching structures, increasing the applicability of clay extrusion 3D printing.

15 Comparison between different approaches to non-planar 3D printing shows that we are able to control non-planar 3D printing for the desired results through updates in TCP and toolpath

16 Two representative comparison of before and after adopting our methods of improving the printing equality: (a) 1-1 test slumping result; (b) 1-3 test result, successfully maintaining the balance; (c) Horizontal 3D printing; and (d) non-planar printing results without over-squeezing

TABLE 3 Experiments on different degrees of branching and printing beds.

Test Number	Degree of Slope	Degree of Branching	Result	Observation
1-1	0°	30°	×	slumped at very last moment
1-2	5°	30°	√	no slumping observed
1-3	10°	30°	√	no slumping observed
1-4	15°	30°	×	collapsed from the bending side
1-5	20°	30°	×	collapsed from the bending side
2-1	0°	45°	×	slumped at 60% printing
2-2	5°	45°	√	no slumping observed
2-3	10°	45°	√	no slumping observed
2-4	15°	45°	√	sinking at bending branch slightly
2-5	20°	45°	×	collapsed from the bending side
3-1	0°	60°	×	slumped
3-2	5°	60°	×	slumped
3-3	10°	60°	√	sinking at bending branch slightly
3-4	15°	60°	√	no slumping observed
3-5	20°	60°	√	no slumping observed

ACKNOWLEDGMENTS

This research was undertaken in the Masters of Science in Design: Robotics and Autonomous Systems (MSD-RAS) program, at the University of Pennsylvania. Authors Li, Yuan, and Han were students in the program. Faircloth, Anderson, King, and Stuart-Smith were instructors who contributed to the research project and paper during and post studies. The authors also wish to thank teaching assistant David Forero and former student Yuxuan Wang for their support and input into the research.

REFERENCES

Bechthold, Martin, Anthony Kane, and Nathan King. 2015. *Ceramic Material Systems in Architecture and Interior Design.* Basel: Birkhauser.

Cruz, Paulo J. S., Bruno Figueiredo, João Moreira, and Samuel Ribeiro. 2022. "Ficus Columns." Accessed June 1, 2022. https://www.aclab-idegui.org/ficus-columns.

Ko, Minjae, Donghan Shin, Hyunguk Ahn, and Hyungwoo Park. 2018. "InFormed Ceramics: Multi-Axis Clay 3D Printing on Freeform Molds." In *Robotic Fabrication in Architecture, Art and Design 2018*, edited by J. Willman, P. Block, M. Hutter, K. Byrne, and T. Schork. Cham: Springer Nature. 297–308.

Kontovourkis, Odysseas, and George Tryfonos. 2018. "Integrating Parametric Design with Robotic Additive Manufacturing for 3D Clay Printing: An Experimental Study." In *2018 Proceedings of the 35th International Symposium on Automation and Robotics in Construction (ISARC).* Berlin, Germany. 918–925.

Ladybug Tools LLC. 2022. Ladybug Tools V. 1.4.0. http://www.ladybug.tools/.

Nayyeri, Pooyan, Kourosh Zareinia, and Habiba Bougherara. 2022. "Planar and Nonplanar Slicing Algorithms for Fused Deposition Modeling Technology: A Critical Review." *The International Journal of Advanced Manufacturing Technology* 119 (5–6): 2785–2810.

Nisja, Georg Aarnes, Anni Cao, and Chao Gao. 2021. "Short Review of Nonplanar Fused Deposition Modeling Printing." *Material Design & Processing Communications* 3 (4): e221.

Wolfram, Stephen. 2002. *A New Kind of Science.* Champaign, IL: Wolfram Media.

Xing, Yu, Yu Zhou, Xin Yan, Haisen Zhao, Wenqiang Liu, Jingbo Jiang, and Lin Lu. 2021. "Shell Thickening for Extrusion-Based Ceramics Printing." *Computers & Graphics* 97: 160–69. https://doi.org/10.1016/j.cag.2021.04.031.

Zhong, Fanchao, Wenqiang Liu, Yu Zhou, Xin Yan, Yi Wan, and Lin Lu. 2020. "Ceramic 3D Printed Sweeping Surfaces." *Computers & Graphics* 90: 108–15. https://doi.org/10.1016/j.cag.2020.05.007.

IMAGE CREDITS

All drawings and images by the authors.

(a)
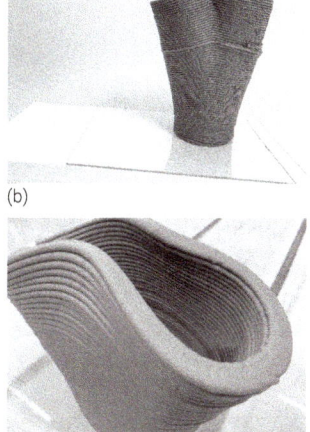
(b)

(c)

(d)

16

Chenxiao Li is a recent graduate of Master of Science in Design: Robotics and Autonomous Systems program from University of Pennsylvania. She is a Research Assistant focusing on computational design and robotics fabrication.

Mingyang Yuan is a Research Assistant in the Autonomous Manufacturing Lab at the University of Pennsylvania. She graduated with a Master of Architecture, and continued studying at the Master of Science in Design: Robotics and Autonomous Systems program.

Zilong Han is an architecture graduate student from University of Pennsylvania who believes technology will change the world.

Billie Faircloth is a practicing architect, and partner at KieranTimberlake, where she leads transdisciplinary research, design, and problem-solving processes across fields, including environmental management, chemical physics, materials science, and architecture. She is also a Adjunct Professor at University of Pennsylvania Weitzman School of Design.

Jeffrey S. Anderson currently teaches design studios and advanced media seminars in the Graduate Architecture and Urban Design program at Pratt Institute and the Graduate Architecture Program at the University of Pennsylvania.

Nathan King is the Co-Director of the Center for Design Research at Virginia Tech, an Instructor at the University of Pennsylvania and Harvard University, and leads the Autodesk Research organization focusing on the Industrialization of Construction.

Robert Stuart-Smith is Director of the MSD-RAS program at the University of Pennsylvania, Director of the Autonomous Manufacturing Lab (AML) at Penn (Architecture), and Co-Director of the AML at Univesity College London (Computer Science).

17

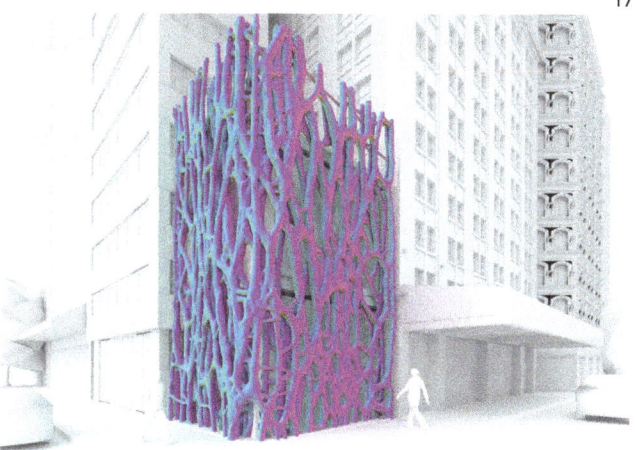

18

19

17 Final 1.3 m high ceramic assembly prototype
18 Facade proposal: exterior
19 Facade proposal: interior

Intentional Folds

Gosia (Malgorzata) Pawlowska
Consortium for Research and Robotics, Pratt Institute

Robotic Incremental Sheet Forming as a Mold for Slumped Glass

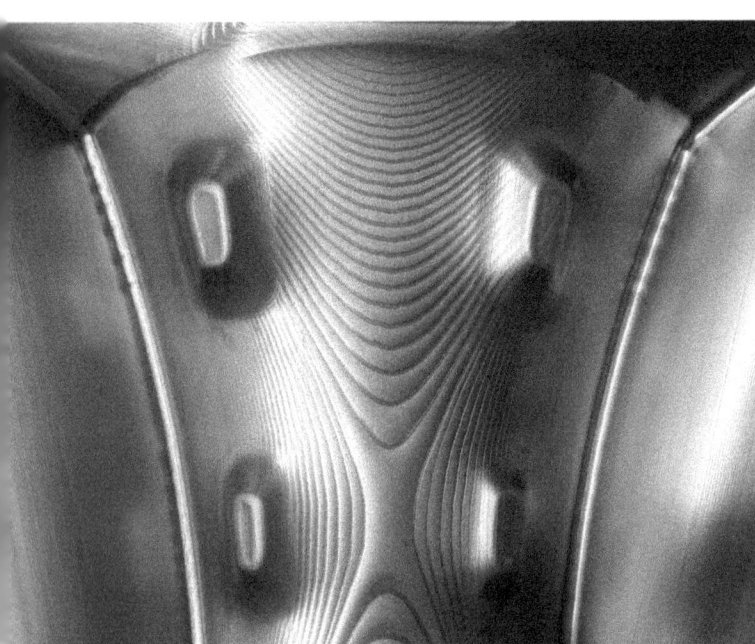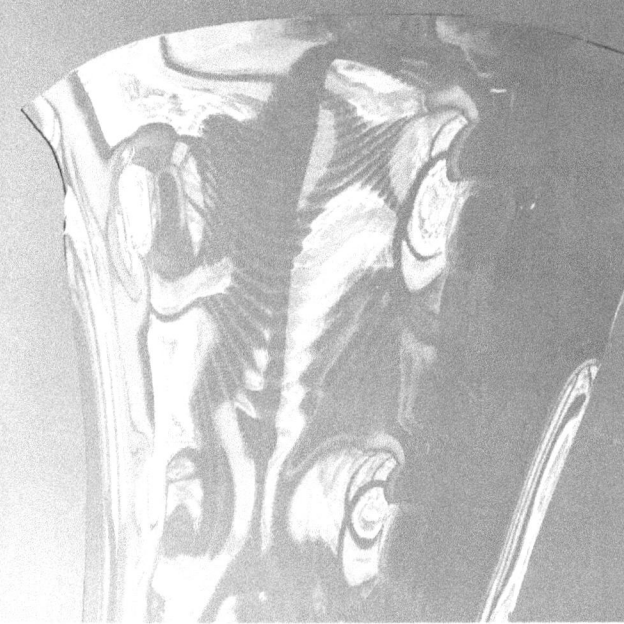

1 Steel mold formed by robotic ISF and corresponding slumped glass panel

ABSTRACT

This paper investigates new techniques for the production of creative forms in glass by the novel application of robotic incremental sheet forming (ISF) to make steel molds for slumped glass. Also known as single point incremental forming (SPIF), ISF is an industrial fabrication process that uses a robotic arm and end-of-arm tool to press a three-dimensional shape into a sheet material by applying concentrated force along a given toolpath. While precedent exists for architectural assemblies of metal panels formed by ISF, this work proposes an original prototype whereby three-dimensional steel panels are used as formwork for architectural glass. Glass slumping is a kiln-forming technique wherein a sheet of glass is heated to temperatures between 1250-1400 degrees Fahrenheit, softening it enough to slump or drape over a mold. It is common to make formwork for glass slumping out of refractory, ceramic, or steel that is rolled or welded. Robotic ISF offers an approach to innovate traditional methods of mold-making for glass. The enhanced capabilities of digital tooling enable the glass to achieve complex geometries with a greater degree of precision. Understanding degrees of tolerance and typical deviations in both the steel and glass at each stage of the process is essential towards design for assembly of multiple glass panels. The versatile forming process of ISF can establish a set of precise features in the formed panel while retaining some material characteristics like organic curvature, folds, and textures beyond the connection nodes. Departing from the typical level of standardization and constraint in architectural glass structures, this work aims to advance conversations around technology and craft and maintain a materially-driven approach to glass fabrication via digital tooling.

INTRODUCTION

The work presented in this paper explores a novel fabrication method: the combination of steel formwork made by 6-axis robotic incremental sheet metal forming and slumped glass. Digital design and advanced tooling bring new levels of precision to the traditional method of kiln-forming glass, resulting in a hybrid, materially-driven workflow for novel architectural glass assemblies—the design of which is informed by inherent properties of both steel and glass alongside the strengths and limitations of the technology and craft techniques.

CONTEXT: Incremental Sheet Forming

Incremental sheet forming (ISF) is an industrial fabrication process that uses a robotic arm and end-of-arm tool to press a three-dimensional shape into a sheet of metal by applying concentrated force along a given toolpath. Also known as single point incremental forming (SPIF), the robotic toolhead is comprised of a round-tipped forming tool, and presses the metal into a given shape by incrementally stepping down along contour lines (Figure 1). It is a cost-effective method for prototyping and short-run production of versatile forms in sheet metal compared to traditional stamping with a die and punch.

Although ISF has not been previously explored in the context of glass fabrication, it has been successfully applied in the form of metal architectural skins (Kaalo et al. 2014) and shell structures. "A Bridge Too Far" demonstrated the structural capacity of three-dimensional aluminum panels formed by ISF (Nicholas et al. 2018). The double-skin bridge was assembled from individual panels formed according to simulated rigidization patterns and connection points. Research by Kaalo and Newsom at the University of Michigan applied a ribbed texture to doubly-curved steel panels to generate an aggregate cladding system, implying viable applications of ISF in building envelope design.

Efforts have been made to improve the geometrical accuracy of ISF, such as measuring deviations in the formed geometry by 3D scanning and adaptive toolpath design (Yu et al. 2022).

CONTEXT: Slumped Glass

Developments in CAD technology and robotic fabrication have expanded the breadth of bespoke and versatile material practice in architecture, but examples of digital tooling are less common in glass forming applications. The extreme temperatures required to manipulate glass are a challenge to experimentation and understanding of the material. A unique attempt to digitize glass manufacturing was made by "Glass Cast" at the University of Michigan (Newell et al. 2012). Their reconfigurable tooling system for freeform slumped glass required a custom kiln with integrated pin-mold: supports

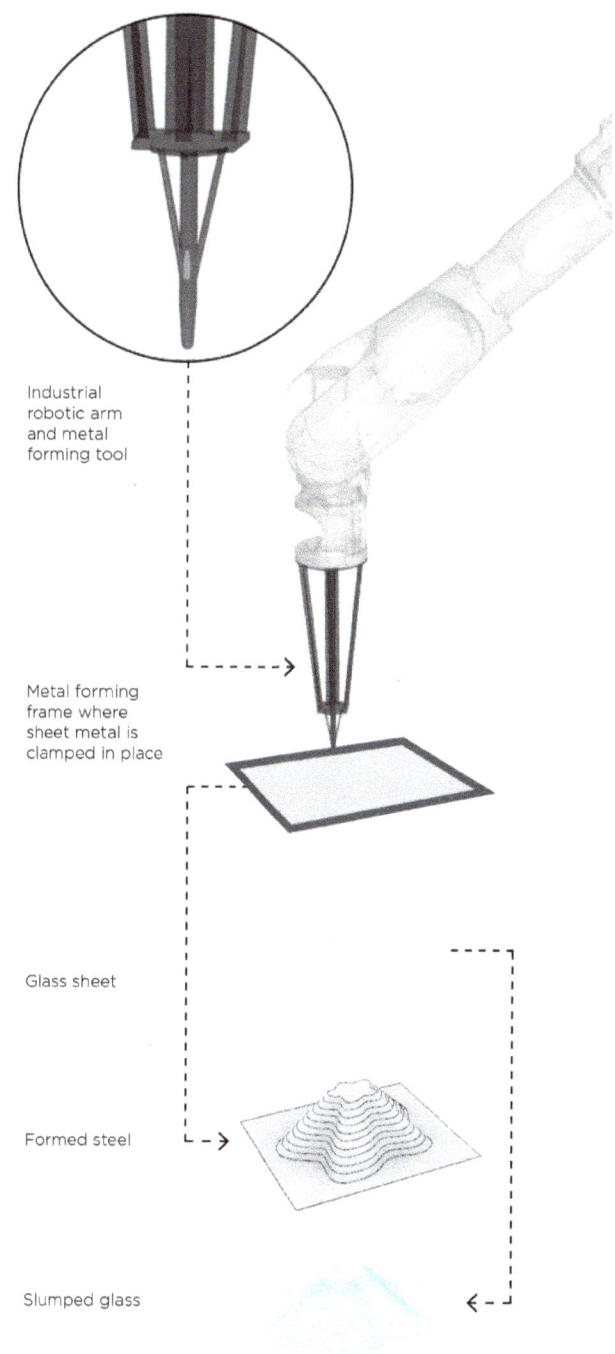

2   Workflow diagram

could be raised and lowered using servo motors, which would then modify the points over which a sheet of float glass in the kiln would drape.

A more economical approach was explored in the author's own work "Viscous Catenary," where a repositionable set of waterjet steel fins acted as a mold for slumped glass panels formed in a typical kiln, applying digital design tools to indirectly manipulate the curvature of the glass and establish points of connection for a freeform laminated channel glass system (Pawlowska 2020). Steel proved to be an effective mold material for shaping glass due to its resilience at peak kiln temperatures.

OBJECTIVES

How can ISF enable a novel yet accessible approach to molding the fabrication and assembly of complex geometries in glass? Building on previous research in digitally designed architectural glass, "Intentional Folds" introduces robotic incremental sheet forming as the production method of steel molds for glass.

Without involving any modifications to a kiln or tooling at extreme temperatures, the robotically-formed steel can translate the complexity and relative precision of new technologies to the glass indirectly. Feedback from the material at each stage of the process determines adjustments to improve geometric accuracy in both the steel molds and glass panels, while a measured relationship between controlled and freeform regions in the slumped glass addresses the material's unpredictable nature.

While conventional techniques for manufacturing architectural glass necessitate high levels of standardization and redundancy, robotic fabrication can establish a set of control points in the desired geometry, absorbing tolerances and accounting for inherent deviations in the material while designing for assembly. The hybrid approach of incremental sheet metal forming and glass slumping relies on an ability to discern patterns of behavior in the steel and glass and adjusting variables in the fabrication process accordingly.

METHOD A: Steel Formwork

Design of the steel formwork and toolpaths for incremental sheet forming was accomplished in Rhinoceros and Grasshopper (McNeel et al. 2010). A contouring script was used to generate a single toolpath curve that was loaded into the Robots plugin for Grasshopper (McNeel et al. 2017) for translation into a PGF file format compatible with the industrial robotic arm. At the design stage, parameters such as the density of features, depth, distance to the edge, and sharpness of angles would be set within the limits observed and established by initial testing of forms in 22 ga steel (Figure 3).

The ductile nature of steel enabled localized deformation while inevitably resulting in some deflection and springback. The findings from early experiments established a maximum achievable angle of approximately 120 degrees before excessively stretching and tearing the steel. Deflections and pinch points suggested that the appropriate stepdown, or distance between pressed layers, correlates to the complexity of the geometry and its proximity to the edges where the metal sheet

3   Test geometries formed in 22 ga steel

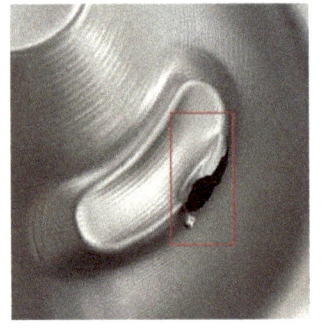 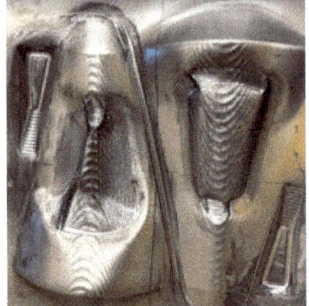

4A   4B

4   Examples of failure conditions: A) Tear due to steep angle; and B) Buckling from lack of resistance

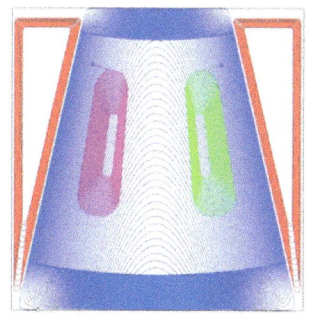
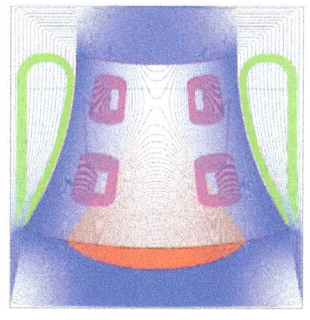

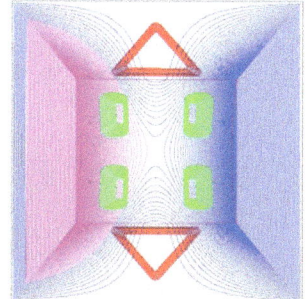
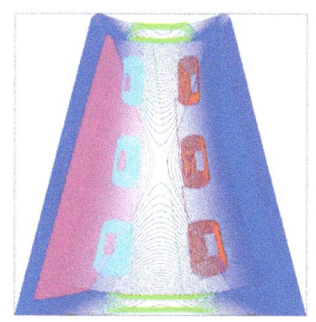
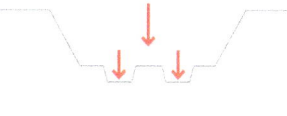
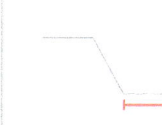

a) Maximum forming angle approaches 120° increased risk of tearing

b) Overlayed toolpaths of varying slope to improve formability of steep angles

c) Overlayed toolpaths to create geometric features, same forming direction; steeper forming angles are tolerated at a greater distance from panel edge

d) Overlayed toolpaths in opposite direction, risk of buckling if proportions exceed limit of original shape

6

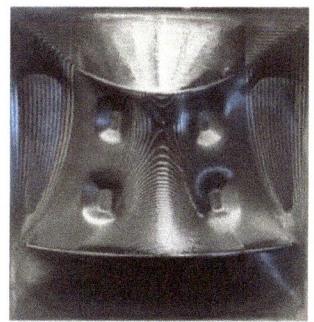

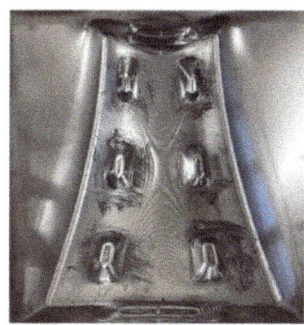
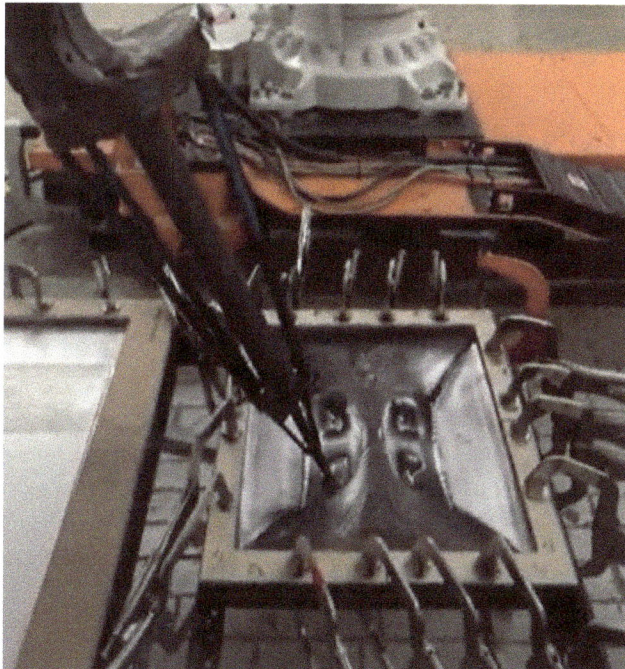

5

7

is clamped into a frame. The stepdown distance ranged from 1.3 mm to 0.4 mm. Where the desired form was more steep or extreme relative to sheet size, a finer stepdown would reduce strain in the steel or excess torsion on the joints of the robotic arm, but increase the fabrication time.

The design process evolved to enable more complex geometries by layering toolpaths (Figure 5). Gradually increasing the depth and slope could be achieved by forming in multiple passes. To compensate for deflection, certain areas in a panel could be formed from both sides by flipping the piece.

5 Layered toolpaths and formed steel molds with complex features (22 ga, 16-inch)

6 Typical parameters for incremental sheet forming steel: adaptive toolpath strategies

7 Incremental sheet forming setup

ADVANCED PRODUCTION METHODS I: ROBOTIC FABRICATION AND MATERIAL FORMATION

8 Sheet of glass placed on mold before slumping

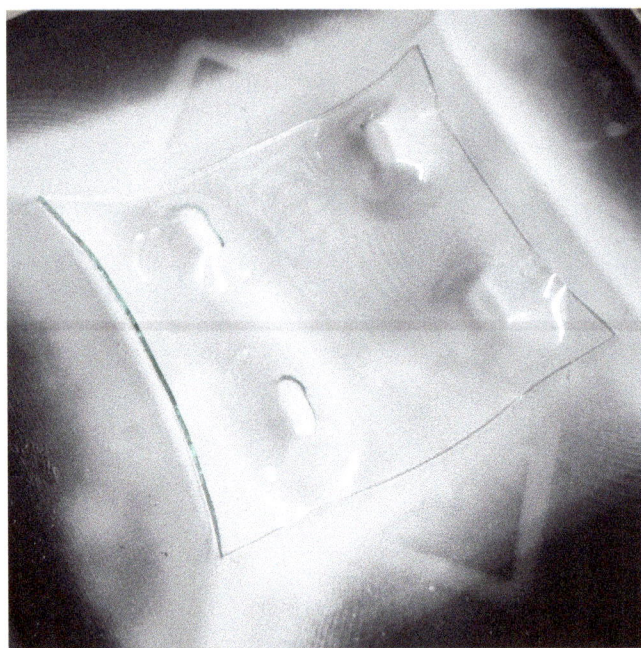

9 Glass panel after slumping

However, this introduced another potential for misalignments in calibration of the toolpath. When layering toolpaths in opposing directions, there was a limit to the relative size of additional geometry before the initial form would buckle under excessive strain (Figure 6).

METHOD B: Slumped Glass

Venturing a step further than previous architectural work in robotic incremental sheet forming (ISF/SPIF), the intention of this project was to apply the formed steel panels as molds for slumped glass. The combined processes would result in uniquely designed glass panels that can incorporate areas of precise folding and connection points. A hybrid, materially-informed workflow emerged that was informed by both characteristics of the steel and glass, as well as parameters in the tooling. To function successfully as a mold for glass, the steel had be coated with refractory mold-release to prevent fusing of glass to the mold at high temperatures. When fired over the mold, a flat sheet of glass would soften, drape, and assume the approximate shape of the steel (Figures 8, 9).

The length of each temperature segment in a glass firing schedule is crucial for obtaining desired results in the final form, as well as cooling the glass slowly enough to prevent cracks from stress embedded in the material (Figure 10).

The prototypes in this research used 1/8 inch float glass, manually pre-cut when necessary using a typical diamond-wheeled glass cutter (Figure 10). As the slumped glass panels conform to the shape of the steel mold, unpredictable features can occur

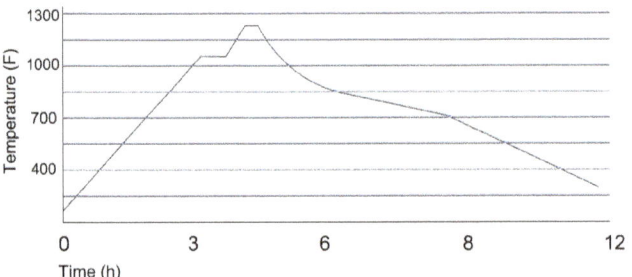

10 Tyical glass slumping cycle

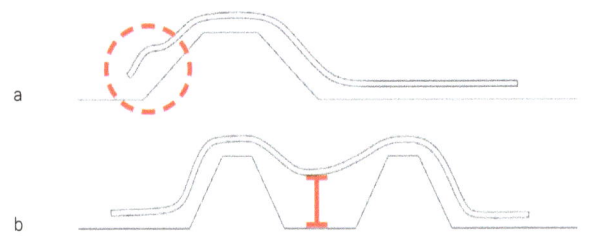

11 Deviations in slumped glass at edge (a) and between dense supports (b)

such as folds and textures in the glass. Some of these constraints can be eliminated by carefully tracking the temperature levels and hold times. If the glass is fired too hot, it will pick up more surface imperfections and deeper textures. However, firing to lower temperatures could prevent the glass from draping accurately over given features in the mold.

Other factors that would impact the accuracy between geometries formed in the steel and glass included distance of raised features to the edge of the glass, the distance between

12 Early prototypes of slumped glass assemblies joined by adhesive

adjacent features, and their depth. When an elevated feature occurred close to the edge of the glass, it was more likely for the material to fold, since it lacked the self-weight to fall flat beyond the peak of the geometry. Supports placed too close to each other would prevent the glass from conforming to the lower points of the mold between them (Figure 11).

ASSEMBLY

Having established a hybrid, materially-driven workflow between forming steel panels by ISF and slumping glass-the challenge became to design an intelligent and reliable assembly system to generate more complex curved glass structures. A series of early prototypes were joined using UV-curing glue; however, this involved a messy application process and had the tendency to yellow over time. A cleaner alternative was found to be VHB (very high bonding) tape—a double-sided transparent silicone adhesive made from an acrylic polymer that cures to be virtually indestructible, and has even been tested in curtain wall installations on building exteriors (Townsend et al. 2011).

Since it is preferable to join glass over a larger area to avoid the concentration of point loads (Drass 2018), the approach was taken to design connections at a series of nodes. These raised geometric features would also capture the most precise imprint from the steel mold in the glass, while areas of lower elevation or edges tended to have imperfections. Thus, the assembly system was driven by material properties. To form an aggregate structure, the connection nodes were mapped onto a pair of offset surfaces that were sub-divided to enable splice-lamination of the glass panels. A self-supporting cone structure was designed and built as final proof of concept (Figures 13-15).

OUTLOOK

This body of research demonstrates novel methods for the fabrication of bespoke slumped glass panes, using steel formwork made by robotic incremental sheet forming (ISF). To make informed adjustments to the process, material tendencies were observed. The assembled prototypes suggest new opportunities for the design of complex glass skins or structures, pending further investigation into geometric accuracy and assembly.

Future work could benefit from introducing digital methods for measurement and feedback, such as 3D scanning, to quantify and track deviations between the digital model, steel, and glass. Information gained from a precise scan might

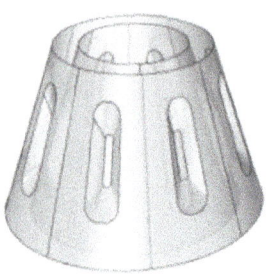

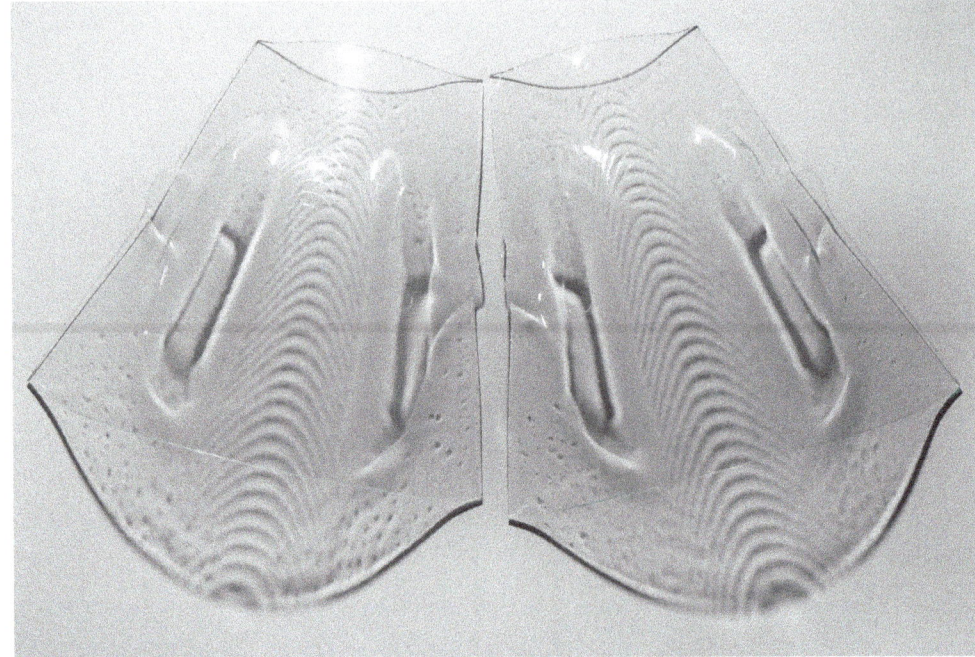

13 Offset surface glass cone diagram **14** Slumped glass panel with traces of toolpath texture from ISF steel mold

identify areas where the geometry needs to compensate for deflections, or improve the assembly by providing a model for bespoke connections that conform to emergent features in the material more accurately (3D printed or cast silicone gaskets, sub-structures).

Further research is currently in progress exploring the same method of metal forming (ISF) applied to blow molds for glass, another traditional craft technique where a glass bubble is inflated into a multi-sided mold. Initial experiments in this process are revealing how the glass is better able to pick up surface features with the added pressure of air inflating and pressing the walls of the glass vessel against the mold.

Novel glass structures provide an opportunity to explore the challenges presented by designing with this highly versatile and adaptable material, and how these can be overcome with appropriate material understanding. The challenges of making a complex geometric glass assembly can be overcome with appropriate material understanding and the power of new technologies such as robotic ISF for the fabrication of molds to effectively shape the glass.

ACKNOWLEDGMENTS

Thanks to the Consortium for Research and Robotics without whom this work would not have been possible, including: M. Parsons. A. Beebe, P. Degroot, J. Nanasca, and D. Basireddi.

REFERENCES

Cui, Q., S. Pawar, M. He, and C. Yu. 2022. "Forming Strategies for Robotic Incremental Sheet Forming." In *POST-CARBON; Proceedings of the 27th CAADRIA Conference*, edited by J. van Ameijde, N. Gardner, K.H. Hyun, D. Luo, and U. Sheth. 171–180.

Drass, M., G. Schwind, and J. Schneider. 2018. "Adhesive connections in glass structures—Part I: Experiments and analytics on thin structural silicone." *Glass Struct Eng* 3: 39–54.

Townsend, Benjamin, et al. 2011. "Characterizing acrylic foam pressure sensitive adhesive tapes for structural glazing applications—Part I: DMA and ramp-to-fail results." *International Journal of Adhesion and Adhesives* 31 (7): 639–649.

Kalo, A., and M. J. Newsum. 2014. "Robotic Incremental Sheet Metal Fabrication." In *ACADIA 14: Design Agency; Projects of the 34th Annual Conference of the Association for Computer Aided Design in Architecture (ACADIA)*. 71–74.

McGee, W., C. Newell, and A. Willette. 2012. "Glass Cast: A Reconfigurable Tooling System for Free-form Glass Manufacturing." In *ACADIA 12: Synthetic Digital Ecologies; Proceedings of the 32nd Annual Conference of the Association for Computer Aided Design in Architecture*, edited by J. K. Johnson, M. Cabrinha, and K. Steinfeld. 47–49.

McNeel, Robert, et al. 2010. Rhinoceros 3D. V. 6.0. Robert McNeel & Associates, Seattle, WA. Windows.

McNeel, Robert, et al. 2017. Grasshopper Build 1.0.007. Robert McNeel & Associates, Seattle, WA. Windows.

Nicholas, P., D. Stasiuk, E. C. Nørgaard, C. Hutchinson, M. Ramsgaard

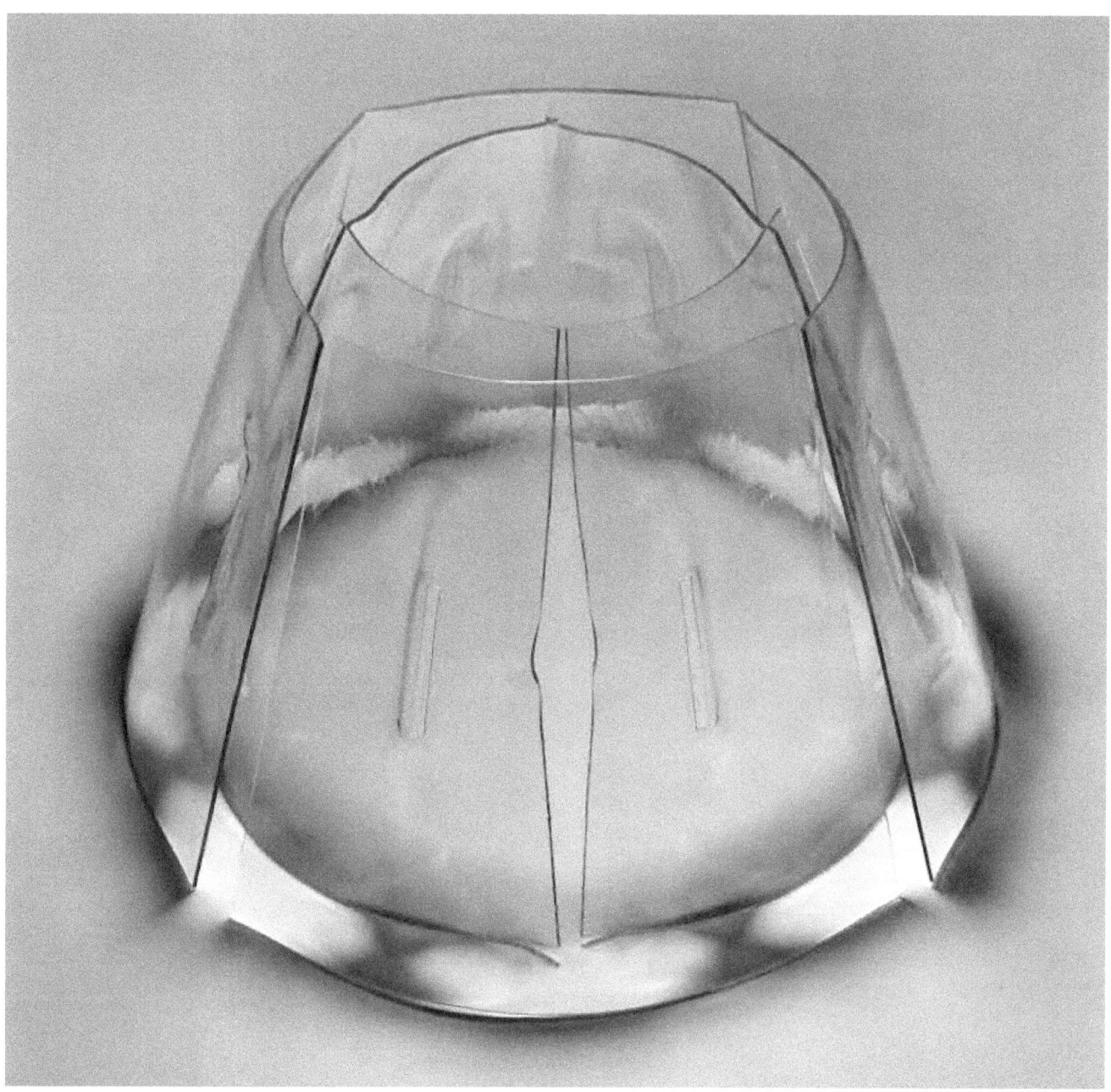

15 Assembled conical structure of 8 slumped glass panels joined by VHB tape

Thomsen. 2016. "An Integrated Modelling and Toolpathing Approach for a Frameless Stressed Skin Structure, Fabricated Using Robotic Incremental Sheet Forming," In *RobArch: Robotic Fabrication in Architecture, Art and Design*, edited by D. Reinhardt, R. Saunders, and J. Burry. Cham: Springer. 62–67.

Pawlowska, Gosia. 2020. "Viscous Catenary." In *ACADIA 20: Distributed Proximities; Proceedings of the 40th Annual Conference of the Association of Computer Aided Design in Architecture*, vol. 2, edited by M. Yablonina, A. Marcus, S. Doyle, M. del Campo, V. Ago, B. Slocum. 170–175.

IMAGE CREDITS

All drawings and images by the author.

Gosia (Malgorzata) Pawlowska holds a B.Arch from Cornell University (2016) and an M.Arch from The Bartlett School of Architecture (2019). She is an architectural designer and glass artist based in New York, with an experimental practice in glass craft and digital fabrication techologies as applied to mold-making.

Session Introduction

Disruptive Modes of Practice and Pedagogy

Vernelle A. A. Noel, Chair

The session "Disruptive Modes of Practice and Pedagogy" brought together architecture, computation, and narrative for design education, practice, and research. Innovations in this paper session lay in the researchers' creation and use of computational design tools and methods to address accessibility and experience in and about real-world conditions.

The first paper, "Implementation of Interactive Virtual Sites for Architectural Education," addresses the problem of engagement and learning associated with limited access to physical sites. Authors used VR tools and methods to facilitate digital site visits for publics engaged in online learning. The authors demonstrate that when hybrid approach to architectural education considers students' skills, time, and financial resources, online engagement in architectural education can improve. The second paper, "Entangled Simulations," speaks to the fallacy of digital twins that erase social, economic, and political realities of design and the built environment. By employing reenactment and narratives, the author creates a gaming environment in which players must grapple with social questions and narratives in the creation of architectural projects that include computational design and fabrication. Drawing on the work of Jeremy Deller, and action as cognition, the author presents reenactment as a way of understanding and working with historical and social events for new frames, descriptions, and entanglements in computational design. How do we shift and center those not considered when we create? The work demonstrates that we can problematize and develop understandings of the entanglements in technological practices instead of framing them as abstract within narrative video game spaces. The third paper, "Closing the Gap," focuses on the problem of specialized expertise and subject matter knowledge requirements for users of computational design platforms and environments in architectural firms. While the authors' platform is currently internal, they argue that creating cloud-based simulation and tool creation platforms outside proprietary software environments broadens participation in the computational design space by non-specialists. The final paper, "Learning from Players," employs collective intelligence through serious gaming to engage with architectural history. The authors conclude that by reconstructing networks and architectural histories through serious board games, we can uncover lost narratives and use the behaviors of players as a source of data.

Themes of narratives, reenactment, and experience are evident in these works. For deeper learning, engagement, and impact in our field, we must move beyond the visual. We cannot choose to turn a blind eye to the social, historical, and political structures that influence and impact our work and societies.

Implementation of Interactive Virtual Sites for Architectural Education

Anastasia Globa
The University of Sydney

Kun Lyu
The University of Sydney

Ozgur Gocer
The University of Sydney

Muhammed Yildirim
The University of Sydney

1 Face-to face site visit vs. virtual site visit in the context of architectural education; site: Cadigal Green Park, Sydney, NSW, Australia

ABSTRACT

Recently, interactive technology and delivery through online platforms have increasingly become key components of teaching and learning practices worldwide. Unfortunately, students who study online often demonstrate significantly reduced levels of engagement, struggling to achieve learning outcomes. The affordances of Virtual Reality (VR) technologies have a clear potential to improve engagement levels and deliver a sense of physical experience to online students. However, despite the potential benefits of VR applications for e-learning and new modes of practice, there are significant barriers to using it. The implementation of VR often requires significant time and financial investment. The use of gaming engines typically adopted for VR development also requires considerable levels of skills/knowledge. As a result, VR systems have very limited uptake in the education sector. This study proposes that an alternative development workflow for hybrid learning can be implemented using a commercially available VR tour software application, 3DVista, that requires much less time and effort to produce virtual environments. The proposed alternative was compared with a popular gaming engine Unity. This paper presents a comparative experimental study that tests the implementation of these two workflow approaches used for site visits and design explorations for architectural education. This investigation focuses on four key aspects of development and integration: (1) functionality, affordances, and limitations, (2) skills and knowledge requirements, (3) time, and (4) cost. This study aims to provide a better understanding of the scope of resources that would be required for future potential VR integrations in the field.

INTRODUCTION

As a result of the ongoing COVID-19 pandemic and consequent rapid shift to delivery through online platforms, interactive technology has increasingly become one of the key components of teaching and learning practices worldwide (Peimani and Kamalipour 2021). The University of Sydney student surveys for Bachelor's and Master's levels within the School of Architecture, Design, and Planning (2020-2022) clearly show that students who study online, demonstrate significantly reduced levels of engagement, struggle to achieve learning outcomes, and often receive lower marks compared to their on-campus peers. Based on the students' feedback, the fundamental challenge identified for these issues was the ability to deliver a sense of physical experience to architectural students who study online. The question arises: how can we deliver better engagement and improved physical experiences for online learning cohorts?

Recently, due to advances and exponential ubiquity of Virtual Reality (VR) technologies, particularly in terms of interactions and quality of visualizations, immersive virtual applications became especially attractive for the educational sector. VR has a high potential for teaching applications because it allows students to engage in the task more deeply by providing higher levels of immersion (Radianti et al. 2020). It has been proven time and again that the levels of engagement and learning outcomes have a strong positive correlation (Walker et al. 2021; Sweetman 2021). In the context of architectural e-learning, VR systems can provide higher levels of interactivity and immersion, enabling online student cohorts to experience some aspects of physical presence while being in the virtual world. However, there is still an apparent lack of rigor and understanding of the relationship between learning experiences and VR applications, and a lack of framework for the design, development, and implementation of VR systems for education.

Despite the potential benefits of VR for e-learning, there are significant barriers to using this technology. The implementation of VR often requires considerable time and financial investments including the labor cost (development), purchasing of software and equipment, as well as the upskilling of teaching staff. As a result, the implementation of VR is not viable en masse, having very limited current uptake in education.

This experimental research project aims to contribute to new knowledge and a better understanding of VR implementation strategies for architectural education. The paper discusses how VR applications can be used to support online tutorials that have been traditionally taught face-to-face. The manuscript focuses on two aspects in particular: a) integration of VR experiences into teaching activities, and 2) their development and integration from the aspects of functionality, affordances, limitations, skills and knowledge requirements, time, and cost.

BACKGROUND / RECENT STUDIES

Increasingly educators are motivated to "add an extra dimension to the classroom" by creating engaging virtual experiences for their students (Radianti et al. 2020). Digital technology and devices are currently being adopted for education, teaching, and training purposes across various fields and application domains. Users often describe their VR experiences as getting the sense of "being there," immersed in the virtual scene, and disconnected from the real physical world. The perceived affordances of VR environments include such elements as presence, immersion, usability, empathy, and embodiment (Norman 1999). Gibson (1979) coined the concept of "affordance" in his ecological approach to visual perception to describe the capacities of the environment to support human actions. The term was then adapted and became widely used in the field of Human-Computer Interaction as a technological affordance, referring to the potential and limitations of the technological tools for user interaction (Gaver 1991).

Virtual Reality provides an immersive experience for its users to perceive and interact with the virtual world with the various display devices (Lowood 2021). Combining with the recent development of sensory apparatus, more senses (olfactory, tactile, thermal, kinesthetic, etc.) can be implemented in the multisensory VR experience to promote the sense of presence, feeling, and behaving in VR as if in the real world (Cooper et al. 2018; Globa et al. 2022). For example, wind and radiative heating has been incorporated into Virtual Reality systems to provide thermal feedback to the users (Hülsmann et al. 2014; Verlinden et al. 2013). There have also been attempts to integrate olfactory sense into multisensory VR experience by means ranging from low-tech air diffusers with essential oil (Jiang et al. 2018) to precisely controlled olfactometers (Hedblom et al. 2019). These technologies have the potential to deliver a more diverse range of physical experiences for online audiences, addressing research problems identified by this study.

Researchers and practitioners from a wide range of fields including medicine, psychology, sociology, entertainment, arts, architecture, construction, and education have adopted VR for various purposes. Although VR has been widely used as a training, visualization, and communication tool in education, limited attention has been given to the application of VR in architecture education. Hossain Maghool et al. (2018) have developed a VR-based educational tool to enhance architecture students' learning experience of building processes

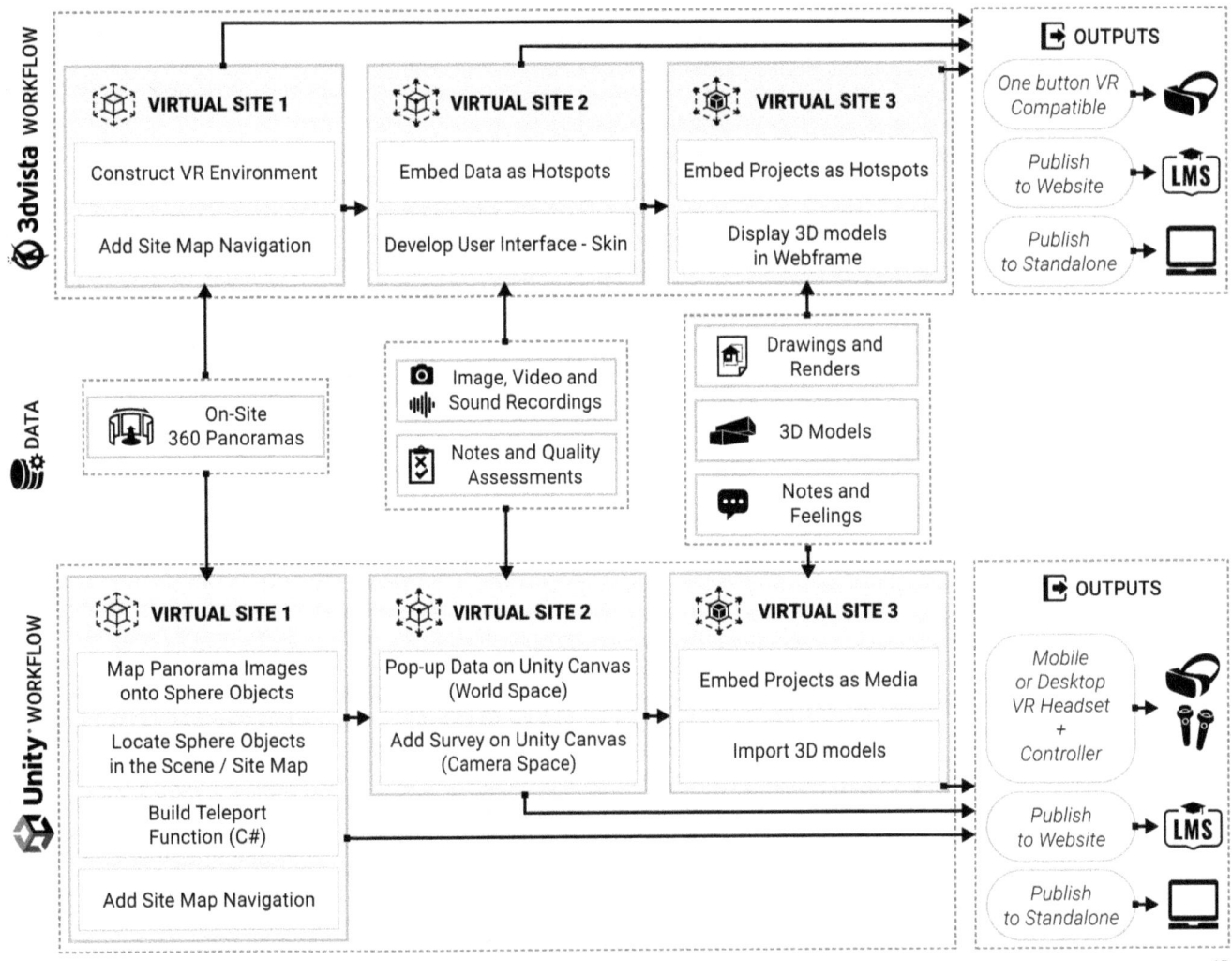

and architectural details. In another study, immersive virtual reality was utilized as a part of an architectural history course taught remotely to provide students with the opportunity to virtually visit historical buildings (Chan et al. 2022). Abu Alatta & Freewan (2017) explored how immersive virtual reality can assist students with the spatial understanding, exploration, and experience of design concepts, form-finding, imagination, and creative thinking in the design process. Özgen et al. (2021) argued that the VR-based design education tool might be a surrogate for the paper-based basic design education approach to develop students' problem-solving skills.

In industry sectors, VR is used for entertainment as well as virtual tours that offer an interactive, engaging, and realistic experience of visiting a place without the limit of time and space (El-Said and Aziz 2022). Previous research has demonstrated that virtual tours of museums (Resta et al. 2021), heritage sites (Mah et al. 2019), touristic places (Wu and Lai 2022), retail stores (Baek et al. 2020), and properties (Yan et al. 2020) have been an effective way to stay connected with the visitors and engage with more customers. Several software applications were developed to accommodate virtual tour functionality for the industry, such as 3DVista (2022), Kuula (2022), and Klapty (2022). These are rarely adopted for VR development outside of their target domains. Most of the VR environments in both industry and education sectors utilize gaming engines to develop their applications.

In terms of hardware, VR can be displayed using standalone or tethered head-mounted displays (HMDs), such as Oculus Quest 2, HTC Vive Pro 2, etc.; mobile VR, such as Google Cardboard (2022); desktop VR; and CAVE systems (Radianti et al. 2020). HMDs have widespread use among researchers, practitioners, and the public because they are powerful and convenient and ensure that users have a high level of immersion, presence, and realism (Di Natale et al. 2020). For remote education purposes, VR environments can be output to personal mobile devices using a range of affordable VR headsets for smartphones, such as Google Cardboard and VR Shinecon (2022), making this technology available to online students. Recently more interest is given to mobile application development for VR in education (Virmani et al. 2022).

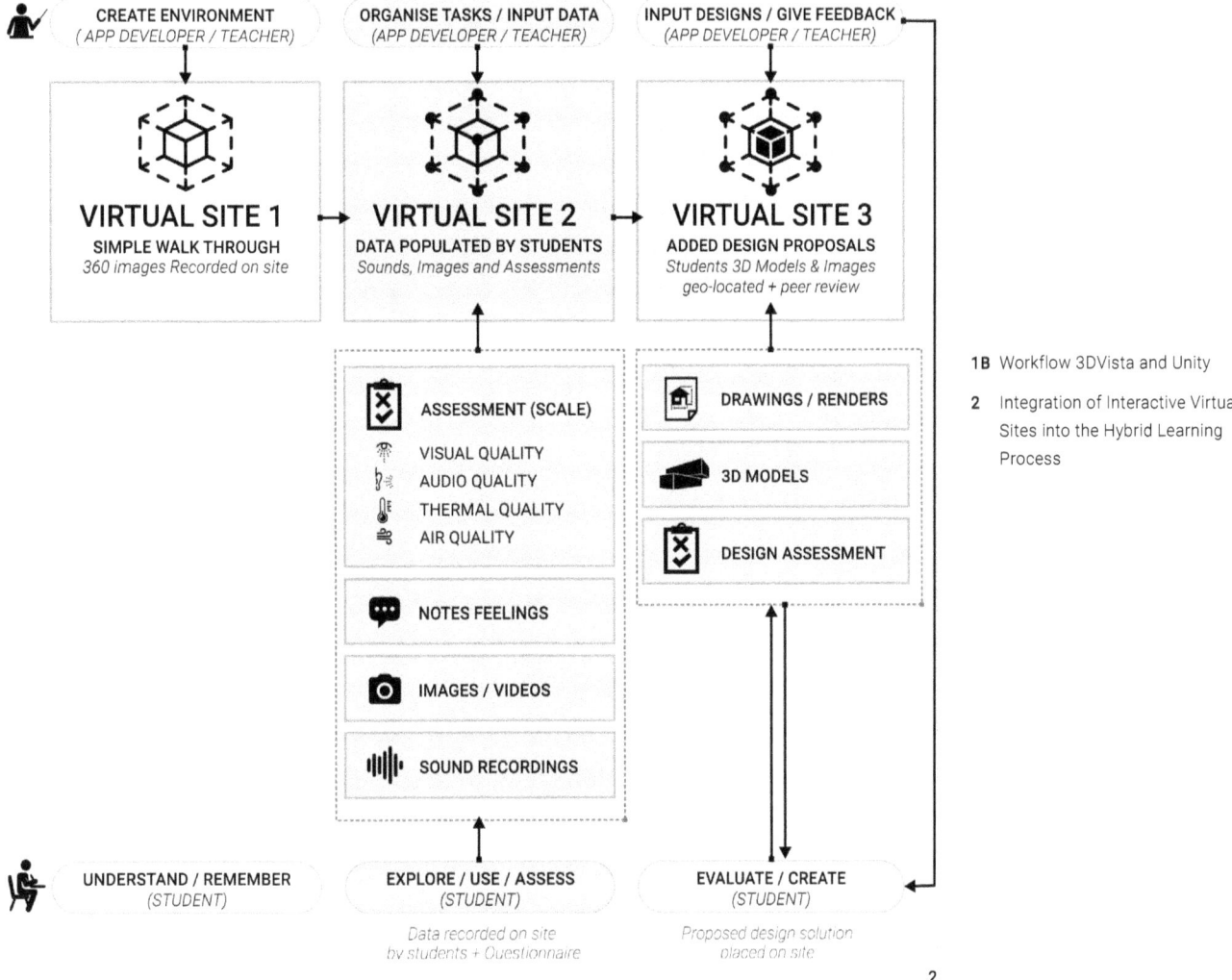

1B Workflow 3DVista and Unity

2 Integration of Interactive Virtual Sites into the Hybrid Learning Process

METHODOLOGY

Major barriers to VR implementation for education relate to significant investments that they require in terms of time, cost, and understanding of the relationship between VR experiences integrated into teaching activities and learning outcomes. In this regard, the main research questions explored in this paper are:

- What are the technological benefits and limitations of two alternative VR site development and implementation workflows, comparing virtual tour software to using a gaming engine?
- What investments and resources—such as skilled labor, hardware, and software—are required for the two explored VR integration/workflow scenarios?

To answer these research questions, our hypothesis states that alongside gaming engines that are often used to develop VR applications, but require considerable levels of development skills and time commitment, an alternative development workflow can be implemented using a commercially available VR tour software application that requires much less time and effort to produce virtual environments.

The comparative experimental study presented in this paper tests this hypothesis using a series of case studies that implemented VR environments for site visits and design explorations in the context of hybrid learning in architectural education varied between two workflow approaches: (1) using a typical commercially available virtual tour software, 3DVista and (2) using a popular gaming engine, Unity (2022). The study investigates the proposition of using iterative VR environments to provide interactive experiences for online student cohorts.

Figure 1B presents two alternative workflows for VR implementation for virtual site activities integrated into two architectural design studio study units. VR activities and subsequent VR environments are split into three steps that follow the learning objectives of the unit (Figure 2). The first workflow uses 3DVista software, and the second uses Unity (Figure 1B). Both approaches/workflows use the same source

3

data (Figure 1B middle) that include 360 panoramas that are taken on-site and used to create Virtual Site 1 (VS1); images, sound and video recordings, qualitative assessments, and notes used to populate Virtual Site 2 (VS2); and students' design projects placed on VR site used for Virtual Site 3 (VS3). As outputs VR environments can be published to smartphones or integrated VR headsets, published to websites or standalone computers.

3DVista Workflow

After taking 360 panoramas on-site (using a Ricoh Theta Z1 camera), the workflow for VS1 consists of two steps: creating a VR environment using panoramas, and adding site map navigation. VS2 development includes embedding data as hotspots and developing User Interface (UI). VS3 development requires embedding students' projects as Hotspots and displaying 3D models as Webframes.

Unity Workflow

The workflow for creating the initial scene, VS1, involves four steps: mapping panoramas onto sphere objects, locating speeches in the scene, building a teleport function (that requires C# programming), and adding site map navigation. VS2 development includes adding pop-up data on Unity Canvas (world space) and adding surveys (camera space).

VS3 development consists of importing 3D models as mesh objects/assets and embedding students' projects as media.

CASE STUDIES / CONTEXT / EXPERIMENTAL SET-UP

Since the start of the COVID-19 pandemic, educational institutions, including the University of Sydney, started to adopt the practice of offering unit content for both online and on-campus student cohorts. Classes for these units often run in parallel: one teaching stream meeting in-person sharing the same physical space, and the other happening online—in a digital space that offers a hybrid learning environment. This research study investigates an opportunity of using semi- and fully immersive interactive virtual environments to provide interactive experiences for online student cohorts, using site visit activity as an example.

To achieve this, it was proposed to develop and integrate three iterations of virtual sites into the learning process of two study units in the School of Architecture, The University of Sydney. Figure 2 illustrates the teaching and learning objectives of each virtual site in the context of an architectural design studio, where the top row shows the app developer/teacher's perspective, and the bottom row shows the students' perspective. VS1 (Virtual Site 1) is created as a simple walkthrough

VIRTUAL SITE 2

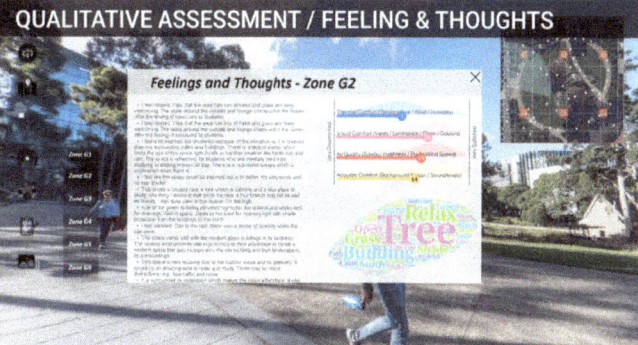

VIRTUAL SITE 3

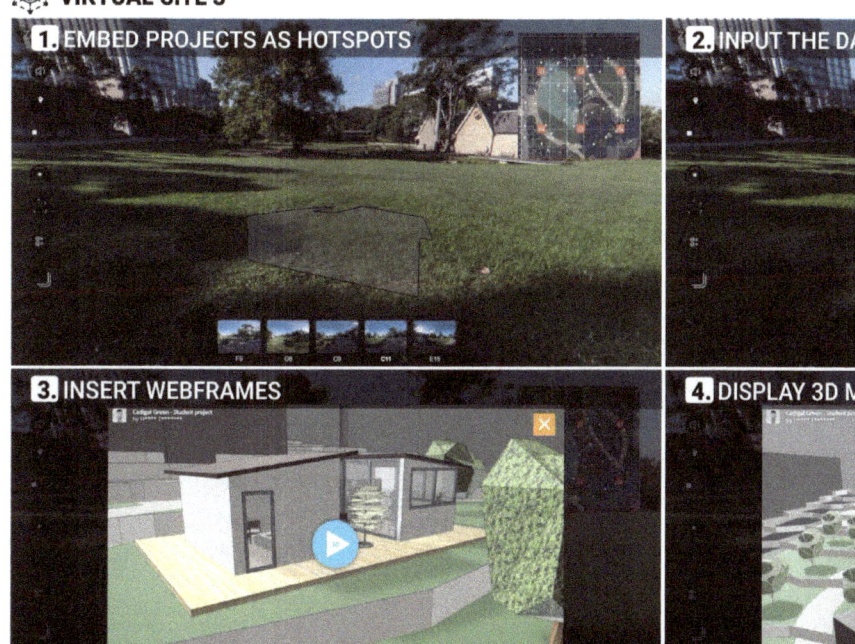
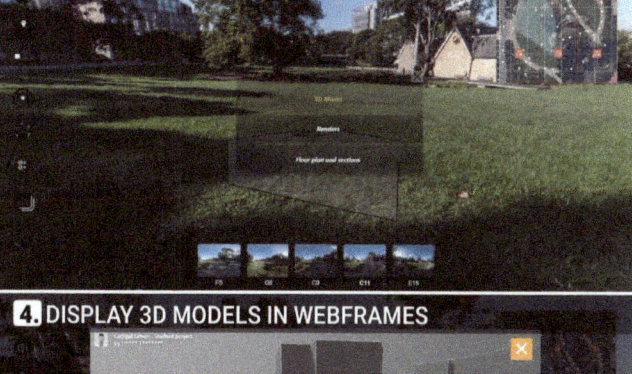
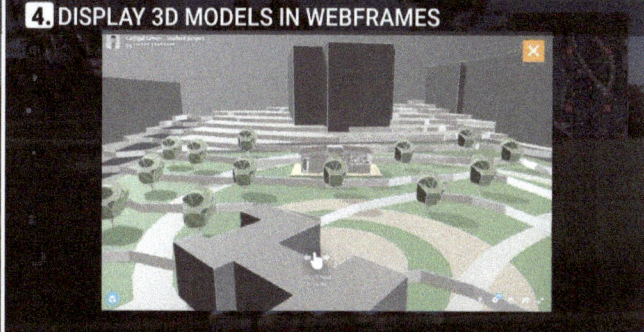

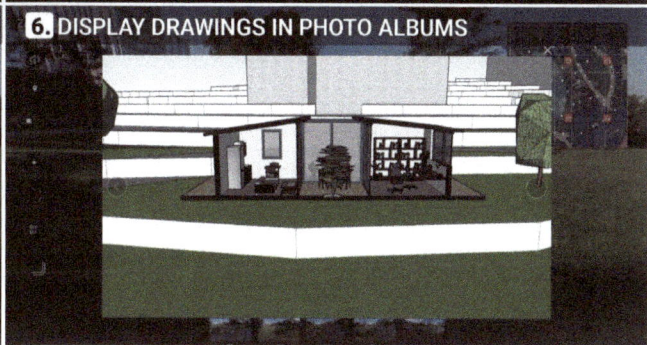

3   Case Study 1: Cadigal Green site; physical site visit, desktop, and VR applications

4   Cadigal Green site; VS2 hotspots and qualitative data communication

5   Cadigal Green site; VS3 functionality, 3D Vista application example

environment aiming to familiarize online students with the existing site context where their design / proposed building will be located. With Virtual Site 2 (VS2) we explored how VR systems could be used to enable the communication of experiences of on-campus students who were able to visit sites in person with those who could only visit remotely. For each zone on the site map on-campus students have recorded and

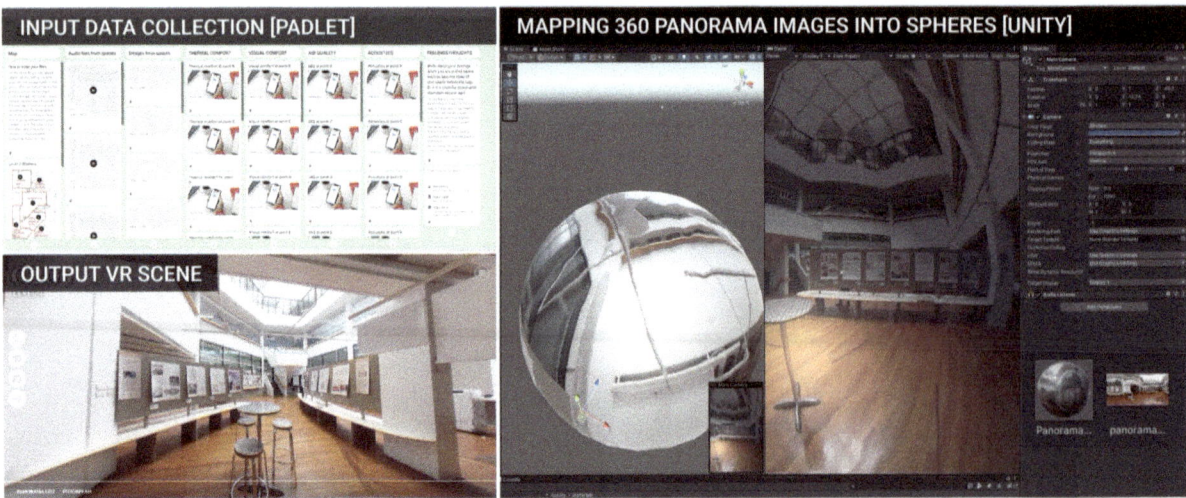

6

submitted their: (1) qualitative assessments of visual, audio, thermal, and air quality; (2) notes and feelings; (3) 1-3 key images or videos and; (4) sound recording of ambient noise on site. Finally, VS3 was aimed to close a physical-to-digital loop by offering both on-campus and online students the ability to explore their proposed design projects placed within the existing site context in VR. For this student's projects, drawings renders, and 3D models were mapped into the virtual site. VS3 also allowed students to actively engage with the projects in context by leaving feedback via comments (Figure 2).

This experimental set-up was implemented for two case studies: Case Study 1, Cadigal Green site (exterior, urban park context) that was used for the 'Architectural Communication 1' unit (Figure 3); and Case Study 2, Wilkinson Building site (interior spaces) used for the 'Designing for Environmental Quality' unit.

CASE STUDY 1: CADIGAL GREEN SITE
Site Context
Cadigal Green site represents an outside urban park context located at the Camperdown Campus, University of Sydney, Australia. It was originally recorded (360 panorama photography with Ricoh Theta Z1) in November 2021. Conditions recorded included: sunny afternoon, evergreen vegetation, surrounded by public buildings, and sparsely populated with students.

Participants / Educational Context
A cohort of a total of 150 students—25% online students and 75% on-campus students—participated in the Architectural Communication 1 unit of the Bachelor of Architecture Environments year one curriculum. The design studio project task was to choose any location within the Cadigal Green site and design a study space for two students. VS1 was developed for the beginning of the semester, VS2 for mid-semester, and VS3 for the very end of the semester.

Design / Integration
VS1 was used solely for the online students to get them an initial idea of space and an understanding of the existing site context. VS2 was created to integrate site-related data for online students to have a more comprehensive understanding of the site conditions. Data about the Cadigal Green site was collected during on-campus students' site visit activity, including photographs, audio/video recordings, subjective feelings, and multisensory environmental assessment. Photographs and audio/video recordings of specific site features were embedded in their corresponding locations in the virtual environment using hotspots. Subjective feeling notes and environmental assessments for sub-areas of the site were integrated into the global interface (Figures 2-4).

The development of VS3 is based on the design proposals produced by all student cohorts and feedback given by the teacher and student peers. The data collected in the form of drawings, renders, and 3D models were embedded in the VR scene and placed within the chosen location on site. Figure 5 illustrates the 3D Vista workflow integration that is achieved firstly through the use of polygon hotspots. External webframes and built-in photo albums were then utilized to display the projects. Subsequently, interactive design assessments were performed using HMDs or desktops to enhance the designs.

CASE STUDY 2: WILKINSON BUILDING SITE
Site Context
Wilkinson Building is located at the Camperdown Campus of the University of Sydney, Australia. It is the faculty building for Architecture, Design, and Planning students. It was recorded (with Ricoh Theta Z1 360 camera) in February 2022. Conditions included: indoor, semi-outdoor conditions on a sunny morning, a few students were in the courtyard and indoors.

FUNCTIONALITY		Unity Workflow	3dvista Workflow
	Construct VR Scene (VS1)	(1) 360 Panoramas/Videos; (2) Digital model of the virtual scene at different LODs and styles; (3) Ability to integrate another sensory component (thermal, olfactory, haptic);	(1) 360 Panoramas/Videos;
	Navigation (VS2)	(1) Teleportation; (2) Continuous movement; (3) Integrate kinesthetic interaction (sit, run, walk, hand action, etc)	(1) Teleportation (newest version click&go navigation simulate the feeling of continuous movement)
	Data Integration (VS2 & VS3)	(1) Embed various forms of data (text, picture, videos, audio recordings); (2) Spatial representation of audio recording; (3) Pop-up data panel based on flexible conditions (click, user's location, timer, etc.); (4) Directly import 3D model with the ability to add material texture, lighting, etc., allow walk-in for design outputs; (5) In-game survey; (6) Record user behavior.	(1) Embed various forms of data (text, picture, videos, audio recordings); (2) Pop-up data panel based on pre-defined conditions (click, rollover, rollout, on start, etc.); (3) Present 3D model only in external Webframe, only allow rotate around 3D model; (4) Survey via external tools.
	Output Hardware	Output to desktop applications (PC, Mac, & Linux) mobile devices (iOS, Android), websites (WebGL), VR devices; The ability to configure VR headset controller and other related apparatus (eye tracker, sensory grove, etc.)	Output to desktop applications (PC, Mac) mobile devices (iOS, Android), websites, VR devices, google streetview; Easy to integrate to Learning Management System (LMS) via SCORM.

6   Case Study 2: Wilkinson Building site; example panorama mapping in Unity

7   Table 1. VR Site Application / Functionality: comparison of Unity vs. 3DVista

Participants / Educational Context

'Designing for Environmental Quality' unit is offered to the students of the Bachelor of Architecture and Environments degree. 120 students were enrolled in this unit and 46 of them undertook their studies online. This unit of study focuses on indoor environmental quality, and how it may affect people's experience and perception of their surroundings. This is a design studio where students need to propose a detailed design for retrofitting two floors of an existing office building into a contemporary open-plan office space.

Design / Integration

VS1 was used solely for the online students to get them an initial idea of spatial characteristics and building systems and appliances. VS2 was created to integrate site-related data for campus and online students to have more comprehensive information about user experience to perform further analysis. Data about the Wilkinson Building site was collected by on-campus students during their visits, including photographs, audio/video recordings, subjective feelings, and multisensory environmental assessment collected via Padlet within Canvas (Figure 6). The objective of VS3 was to represent students' project proposals within the Wilkinson Building context.

RESULTS / COMPARISONS UNITY VS 3DVISTA WORKFLOWS

The Results section presents a holistic comparison of VR site integration within an architectural education setting between two development workflows: first using Unity software and second using 3DVista. This comparative investigation focuses on four key aspects of development and integration: (1) functionality, affordances, and limitations, (2) skills and knowledge requirements, (3) time, and (4) cost.

Table 1 (Figure 7) illustrates the scope of the available functionality of each software application in terms of initial construction of the VR scene, navigation, data integration, and hardware output options. In terms of VR scene functionality, Unity has a much wider range of options compared to 3DVista. 3DVista only uses 360° panoramas and videos as scene construction inputs, while Unity is designed to accommodate various data inputs including digital models (3D objects imported as mesh), images and video media, tables, and live data inputs, linking Grasshopper3D definitions, and importing existing assets/packages, etc. Unity also allows the integration of a variety of sensory components (haptic, thermal, and even olfactory) as well as connecting with physical interactive systems (Arduino).

In terms of navigation, 3DVista only allows teleportation function, while Unity also offers continuous movement (using 1st or 3rd person controllers) and such kinesthetic interactions as sit, run, jump, fly, walk, use hand actions, etc. Unity is also much superior in terms of data integration. 3DVista can only embed data (text, images, video, and audio) using pop-up data panels and webframes for 3D models. Embedding surveys into VR scenes in Unity can happen internally, while 3DVista can only use links to external tools such as a Mentimeter. For VR outputs both platforms can offer a wide range of options (mobile, desktop, website, headset), although the Unity application enables to use of different controller options and eye-tracking.

SKILL & KNOWLEDGE REQUIREMENT		Unity Workflow	3dvista Workflow
	Construct VR Scene (VS1)	(1) Panoramic picture&video editing skills; (2) 3D modelling skills; Rendering knowledge; (3) Programming skills (C#, arduino), knowledge about circuit design, sensory output device.	No prerequisite knowledge required.
	Navigation (VS2)	Programming Skills (C#)	Knowledge about 3DVista hotspot integration
	Data Integration (VS2 &3)	(1) Knowledge about Unity UI; (2) Knowledge about Unity audio system and audio effect; (3) Programming Skills (C#); (4) 3D modelling and rendering skills; (5) Programming Skills (C#); (6) Programming Skills (C#)."	(1) (2) Knowledge about 3DVista hotspot and Skin; (3) (4) Knowledge about embed webframe in 3DVista
	Output	Knowledge about output procedure and related device.	Knowledge about output procedure and related devices.

8 Table 2. VR Site Application / Skills and Knowledge: comparison of Unity vs. 3DVista

TIME COMMITMENT		Unity Workflow	3dvista Workflow
	Construct VR Scene (VS1)	Expert: (1) 0.5 day; (2) 5-10 days; (3) 10 days; Starter: (1) 2 day; (2) 20 - 25 days; (3) 30 - 40 days	Expert: 0.5 day Starter: 1 day
	Navigation (VS1)	Expert: (1) (2) 1 day; (3) 2 days; Starter: (1) (2) 4 days; (3) 5 days.	Starter / Expert: 0.5-1 day
	Data Integration (VS2 &3)	Expert: 7 days Starter: 15 days	Expert: 6 days Starter: 7 Days
	Output	Starter / Expert: 0.5 day	Starter / Expert: 0.5 day
	Total (For integrating panoramas into VR workflow)	**Expert: 9 days** Breakdown: VS1: 1.5 days; VS2: 4 days; VS3: 3 days; Output: 0.5 day **Starter: 21.5 days** Breakdown: VS1: 6 days; VS2: 10 days; VS3: 5 days; Output: 0.5 day *(Note: for Unity it would take much more time for starters to implement the workflow compared to experts)*	**Expert: 7.5 days** Breakdown: VS1: 1 day; VS2: 4 days; VS3: 2 days; Output: 0.5 day **Starter: 9 days** Breakdown: VS1: 1.5 days; VS2: 5 days; VS3:2 days; Output: 0.5 day *(Note: for 3DVista it is easy to learn, so the time difference is small)*

9 Table 3. VR Site Application / Time: comparison of Unity vs. 3DVista

COST		Unity Workflow	3dvista Workflow
	Software	Free	Virtual Tour Pro licence $499
	Work (expert developer)	**Total $3088** VS1 $514 VS2 $1372 VS3 $1029	**Total $2574** VS1 $343 VS2 $1372 VS3 $686
	Work (academic level A-B)	**Total $8149** VS1 $2273 VS2 $3789 VS3 $1894	**Total $3410** VS1 $568 VS2 $1895 VS3 $758
	Hardwear	Publish to Web - free, Publish to standalone - free, smartphone VR headset $3 - $100, integrated VR headset - $350 +	

10 Table 4. VR Site Application / Cost: comparison of Unity vs. 3D Vista workflows, values shown in U.S. dollars

The benefits of using 3DVista workflow become apparent when we examine the skills and knowledge that are required for VR scene development and application. Table 2 (Figure 8) shows that the construction of a simple VR in the 3DVista scene is very simple and does not require any preexisting knowledge and can be used intuitively. The use of Unity on the other hand would require an extensive set of skills and knowledge associated with scene construction, navigation, and data integration, including the ability to use C# programming.

Tables 3 and 4 (Figures 9, 10) illustrate that the use of the Unity workflow will require a bigger time and cost commitment compared to the 3DVista workflow. Table 3 shows that on average it would take an expert 9 days (calculated for case studies 1 and 2) to deliver all three VR scenes (VS1, VS2, and VS3), while 3D Vista workflow would require 7.5 days. The dramatic difference in time investment becomes apparent when we compare the non-expert labor involved in VR scene development, 21 days for Unity compared to 9 days for 3DVista. It is apparent that for Unity it would take much more time for novices to implement the workflow compared to experts. 3DVista on the other hand is easy to learn and implement.

Table 4 (Figure 10) presents the cost calculation for VR site visit development and integration for both workflows. Note that for our study the cost was calculated based on the development time (Table 3) and salary scales provided by the University of Sydney for 2022 (including +30% on cost). Expert developer-level was set as 'Graduate with experience Technical Specialist (HEO Level 6)' and novice/starter developer was set as 'Research Associate, Level A', representing a beginner level academic (university tutor or lecturer). All costs were sources and calculated as Australian dollars (AUD) and then converted and presented in the table as U.S. dollars (USD) based on exchange rate values from Bloomberg Markets for May 19, 2022.

DISCUSSION AND FUTURE STUDIES

The implication of the findings discussed in the results sections is that the use of alternative VR development workflows could result in significant differences in time and cost investments. Undoubtedly, in terms of functionality and technological benefits, gaming authoring software such as Unity allow a much wider range of options for interactions and input/output data compared to virtual tour creation software such as 3DVista (Table 1). Gaming engines also allow the construction of VR scenes using 3D models/mesh inputs or programming rather than 360° panoramas, which could be required for certain educational applications (Globa et al. 2022). However, for architectural education in the context of online site visit activities, sharing of occupancy data, and on-site design exploration (Case Studies 1 and 2), the 3DVista workflow serves as a viable alternative. Although virtual tour software does not allow continuous navigation, supporting only the teleportation or map/panorama navigation within the scene, it somewhat negates the potential of motion sickness occurrence that is often associated with the use of VR.

One of the key advantages of using the 3DVista development and implementation workflow was the fact that it did not require any preexisting skills or knowledge about 3D modeling, VR integration, or game development and could be easily used by members of the teaching staff. According to the time and cost investment options, hiring an expert Unity developer could be more feasible than paying a teaching staff member to work with 3DVista (novice level); and the use of Unity is free, whereas a 3DVista license requires payment (Tables 2-4).

However, our experience shows that it could be very challenging to find experienced Unity developers for such relatively small-scale projects locally. In the longer run, it also could be much harder to streamline updating the scenes and supporting the use of VR systems created with gaming engines. More research is needed to investigate the feasibility of creating reusable VR application templates for education and testing those for both Unity and 3DVista workflows.

This research project was driven by the proposition that the systematic integration of VR applications into teaching activities could provide improved levels of engagement for online architectural students, dissolving binary conditions of online and physical education spaces and processes. The VR applications described in this paper provided the opportunity for the online students to have an interactive, engaging, and realistic experience of visiting indoor and outdoor sites whenever they wanted. From the user/student perspective, VR hardware for education can cost as little as US$3 (Google Cardboard) or completely free if we consider web and desktop applications, making it both feasible and accessible for most online students (Table 4). Future research and user studies are required to test whether virtual environments could deliver a higher degree of physical experiences for online tutorials that have been traditionally taught face-to-face, such as site visits in architecture.

This study presented in this manuscript intends to provide a better understanding of the scope of resources that would be required for future potential integrations in the field. It should be acknowledged that different educational objectives and contexts could require different functionality for VR development. However, the case studies presented in this manuscript can serve as a valuable initial example of alternative integration workflows and associated investments and resources, generalizable and applicable to a wider field of knowledge. The next stage of this experimental project (in progress) will investigate whether multi-sensory experiences in VR can provide an opportunity for experiential hybrid learning. Bridging the gap between theoretical knowledge and its real-world applications, engaging with learners in such a way that their interaction with content becomes personalized and meaningful. In-depth user studies are currently being conducted at the University of Sydney targeting a sample size of 30 participants/architecture students, providing empirical evidence on the degree of effectiveness of this approach to the students. The findings of these user studies are planned to be converted into a journal paper, ensuring further dissemination of this work.

ACKNOWLEDGEMENTS

The authors would like to thank the Association of Architecture Schools of Australia for providing leadership and advocacy for architectural education in Australia. We also would like to thank the School of Architecture, Design, and Planning at the University of Sydney for their support and encouragement with the development of our research project.

REFERENCES

3DVista. 2022. Accessed May 22, 2022. https://www.3dvista.com/.

Abu Alatta, Rawan, and Ahmed Freewan. 2017. "Investigating the Effect of Employing Immersive Virtual Environment on Enhancing Spatial Perception within Design Process." *Archnet-IJAR* 11 (2): 219–38. https://doi.org/10.26687/archnet-ijar.v11i2.1258

Baek, Eunsoo, Ho Jung Choo, Xiaoyong Wei, and So Yeon Yoon. 2020. "Understanding the Virtual Tours of Retail Stores: How Can Store Brand Experience Promote Visit Intentions?" *International Journal of Retail and Distribution Management* 48 (7): 649–66. https://doi.org/10.1108/IJRDM-09-2019-0294.

Chan, Chiu Shui, Jelena Bogdanovic, and Vijay Kalivarapu. 2022. "Applying Immersive Virtual Reality for Remote Teaching Architectural History." *Education and Information Technologies* 27: 4365–4397. https://doi.org/10.1007/s10639-021-10786-8.

Cooper, Natalia, Ferdinando Milella, Carlo Pinto, Iain Cant, Mark White, and Georg Meyer. 2018. "The effects of substitute multisensory feedback on task performance and the sense of presence in a virtual reality environment." *PLoS ONE* 13 (2): 1–25. https://journals.plos.org/plosone/article?id=10.1371/journal.pone.0191846.

Di Natale, Anna Flavia, Claudia Repetto, Giuseppe Riva, and Daniela Villani. 2020. "Immersive Virtual Reality in K-12 and Higher Education: A 10-Year Systematic Review of Empirical Research." *British Journal of Educational Technology* 51 (6): 2006–33. https://doi.org/10.1111/bjet.13030.

El-Said, Osman, and Heba Aziz. 2022. "Virtual Tours a Means to an End: An Analysis of Virtual Tours' Role in Tourism Recovery Post COVID-19." *Journal of Travel Research* 61 (3): 528–48. https://doi.org/10.1177/0047287521997567.

Gaver, William W. 1991. "Technology Affordances." *CHI '91: Proceedings of the SIGCHI Conference on Human Factors in Computing Systems*. 79–84. https://doi.org/10.1145/108844.108856.

Globa, Anastasia, Beau B. Beza, and Rui Wang. 2022. "Towards Multi-Sensory Design: Placemaking through Immersive Environments–Evaluation of the Approach." *Expert Systems with Applications* 204 (January): 117614. https://doi.org/10.1016/j.eswa.2022.117614.

Google Cardboard. 2022 "Google VR." Accessed May 22, 2022. https://arvr.google.com/cardboard/.

Hedblom, Marcus, Bengt Gunnarsson, Behzad Iravani, Igor Knez, Martin Schaefer, Pontus Thorsson, and Johan N. Lundström. 2019. "Reduction of Physiological Stress by Urban Green Space in a Multisensory Virtual Experiment." *Scientific Reports* 9 (1): 1–11. https://doi.org/10.1038/s41598-019-46099-7.

Hossain Maghool, Sayyed Amir, Seyed Hossein (Iradj) Moeini, and Yasaman Arefazar. 2018. "An Educational Application Based on Virtual Reality Technology for Learning Architectural Details: Challenges and Benefits." *Archnet-IJAR* 12 (3): 246–72. https://doi.org/10.26687/archnet-ijar.v12i3.1719.

Hülsmann, Felix, Julia Fröhlich, Nikita Mattar, and Ipke Wachsmuth. 2014. "Wind and Warmth in Virtual Reality: Implementation and Evaluation." *VRIC '14: Proceedings of the 2014 Virtual Reality International Conference*. https://doi.org/10.1145/2617841.2620712.

Jiang, Like, Massimiliano Masullo, Luigi Maffei, Fanyu Meng, and Michael Vorländer. 2018. "A Demonstrator Tool of Web-Based Virtual Reality for Participatory Evaluation of Urban Sound Environment." *Landscape and Urban Planning* 170 (Feb): 276–82. https://doi.org/10.1016/j.landurbplan.2017.09.007.

Johnson, Timothy D. 1983. "The Ecological Approach Revisited." *Behavioral and Brain Sciences* 6 (1): 184–187. https://doi.org/10.1017/S0140525X00015466.

Klapty. 2022. Virtual Tour. Accessed October 26, 2022. https://www.klapty.com/.

Kuula. 2022. Virtual Tours made easy. Accessed October 26, 2022. https://kuula.co/.

Lowood, H. E. 2022. "Virtual reality." *Encyclopedia Britannica*. Accessed May 3, 2022. Last updated Sep 8, 2022. https://www.britannica.com/technology/virtual-reality.

Mah, Osten Bang Ping, Yingwei Yan, Jonathan Song Yi Tan, Yi Xuan Tan, Geralyn Qi Ying Tay, Da Jian Chiam, Yi Chen Wang, Kenneth Dean, and Chen Chieh Feng. 2019. "Generating a Virtual Tour for the Preservation of the (in)Tangible Cultural Heritage of Tampines Chinese Temple in Singapore." *Journal of Cultural Heritage* 39: 202–11. https://doi.org/10.1016/j.culher.2019.04.004.

Norman, Donald A. 1999. "Affordance, Conventions, and Design." Interactions 3 (6): 38–43. https://doi.org/10.1145/301153.301168.

Özgen, Dilay Seda, Yasemin Afacan, and Elif Sürer. 2021. "Usability of Virtual Reality for Basic Design Education: A Comparative Study with Paper-Based Design." *International Journal of Technology and Design Education* 31 (2): 357–77. https://doi.org/10.1007/s10798-019-09554-0.

Peimani, Nastaran, and Hesam Kamalipour. 2021. "Online education and the COVID-19 outbreak: A case study of online teaching during lockdown." *Education Sciences* 11 (2): 72.

Radianti, Jaziar, Tim A. Majchrzak, Jennifer Fromm, and Isabell Wohlgenannt. 2020. "A Systematic Review of Immersive Virtual Reality Applications for Higher Education: Design Elements, Lessons Learned, and Research Agenda." *Computers and Education* 147 (April): 103778. https://doi.org/10.1016/j.compedu.2019.103778.

Resta, Giuseppe, Fabiana Dicuonzo, Evrim Karacan, and Domenico Pastore. 2021. "The impact of virtual tours on museum exhibitions after the onset of Covid-19 Restrictions: Visitor Engagement and Long Term Perspectives." *Scires-It* 11 (1): 151–66. https://doi.org/10.2423/i22394303v11n1p151.

Sweetman, David S. 2021. "Making Virtual Learning Engaging and Interactive." *FASEB BioAdvances* 3 (1): 11–19. https://doi.org/10.1096/fba.2020-00084.

Unity3D. 2022. "The world's leading platform for real-time content creation." Unity V.2018.1.4f1. Accessed May 22, 2022. https://unity.com/.

Verlinden, Jouke C., Fabian A. Mulder, Joris S. Vergeest, Anna De Jonge, Darina Krutiy, Zsuzsa Nagy, Bob J. Logeman, and Paul Schouten. 2013. "Enhancement of Presence in a Virtual Sailing Environment through Localized Wind Simulation." *Procedia Engineering* 60: 435–41. https://doi.org/10.1016/j.proeng.2013.07.050.

Virmani, Niyati, Sapna Sampath, Yash Vasudeo, Sankalp Shinde, Swati Sharma, and Arvind Mathur. 2022. "Mobile Application Development for VR in Education." In *Proceedings of the 2nd International Conference on Recent Trends in Machine Learning, IoT, Smart Cities and Applications*. Singapore: Springer. 431–441.

VR Shinecon. 2022. "VR Glasses for Mobile Phone." Shinecon. Accessed October 26, 2022. https://www.shinecon.com/vr-glasses/vr-glasses-for-mobile-phone/vr-shinecon-hot-seller-vr-headsets-vr-glasses.html.

Walker, Kristen A., and Katherine E. Koralesky. 2021. "Student and Instructor Perceptions of Engagement after the Rapid Online Transition of Teaching Due to COVID-19." *Natural Sciences Education* 50 (1): 1–10. https://doi.org/10.1002/nse2.20038.

Wu, Xiaohong, and Ivan Ka Wai Lai. 2022. "The Use of 360-Degree Virtual Tours to Promote Mountain Walking Tourism: Stimulus–Organism–Response Model." *Information Technology and Tourism* 24 (1): 85–107. https://doi.org/10.1007/s40558-021-00218-1.

Yan, Zhenbin, Zixuan Meng, and Yong Tan. 2020. "How Does Virtual Reality Matter? Evidence from an Online Real Estate Platform." In *International Conference on Information Systems, ICIS 2020 - Making Digital Inclusive: Blending the Local and the Global*. https://doi.org/10.2139/ssrn.3802243.

Dr. Anastasia Globa is a researcher, academic, and designer working in the field of architecture, with strong research interests in algorithmic design, interactive systems, and simulations. Dr. Globa currently holds the position of Lecturer in Computational Design and Advanced Manufacturing at the University of Sydney. She is a member of the CoCoA research lab and SydneyNano research center, and she closely collaborates with the Computer Aided Architectural Design in Asia community, having been elected CAADRIA President in 2022. Dr. Globa specializes in the development and testing of digital and physical prototypes, interdisciplinary design, and academic / industry collaborations.

Kun Lyu is a PhD research student at the School of Architecture, University of Sydney. Kun obtained his Master of Architecture degrees both in China and Australia, and has worked as an architect in east China architectural design and research institute, Shanghai and has been involved in some major public architectural projects including Hangzhou T4 Airport Terminal. His current research focuses on biophilic design and human multisensory experience of architecture, and employing Virtual Reality technology as a research approach. He is working on his thesis, titled "Restorative Properties of Semi-Outdoor Space in the Workplace: Towards an Atmospheric Quality of Architecture."

Dr. Ozgur Gocer is a lecturer at the School of Architecture Design and Planning at The University of Sydney. She holds a PhD in Building Science from Istanbul Technical University. Her research interest and expertise focus on investigating the built environment through the lenses of occupant experience with a view of informing design and performance improvements.

Muhammed Yildirim is a PhD Candidate at University of Sydney, School of Architecture Design and Planning. His PhD thesis explores the potential of multisensory VR as an experimental tool for investigating restorative effects of biophilic workplaces with multimodal cues. Muhammed master's dissertation (Firat University) investigated Sustainable and Energy-Efficient Housing Design through Building Information Modeling.

Entangled Simulations

Jose Sanchez
University of Michigan

The Affordances of Narrative within Computational Systems

1 *Common'hood* video game developed by Plethora Project, studio led by Jose Sanchez

ABSTRACT

This paper discusses the tensions and affordances of compounding narrative structures to simulation software. Operating within a video game environment, real-time simulations of resources management are confronted with narrative distortions that invite the player, an entangled actor in the simulation, to consider the payoff matrix that involves human and local values. This paper aims to establish a critique of models of simulation that argue for an objective reality and, in contrast, proposes the possibility of an entangled reading of simulation drawing from second-order cybernetics and Arturo Escobar's definition of "Political Ontology." This framework is demonstrated in the development of *Common'hood*, a video game software simulation that combines narrative structures with system dynamics questioning the capacity of cognitive and aspirational imaginaries to distort the mathematical foundations of the software.

BUTLER'S ENTANGLED REALISM

In her novel *The Parable of the Sower*, Octavia Butler (Butler 1993) introduces us to a bleak dystopian version of our world. In this setting, Butler exacerbates the instrumental logic of survival, taking us through the calculations for rationing and personal security. The setting forces the reader to consider precarity as an inescapable background, one that is capable, at any given moment, of permanently damaging the personhood of each one of the inhabitants of this world.

The rationality of survival constitutes a cybernetic exercise of self-provision, a model of a network with inputs and outputs that requires a baseline degree of satisficing in order to ensure self-preservation. Butler's characters are presented as constantly aware of the payoff matrix and the risks/rewards available in each transaction. In short, Butler's setting amplifies the transactional economic imperative of reality.

Nevertheless, Butler's characters do not operate as automatons in this backdrop. Butler juxtaposes this paradigm with vectors that escape the 'objective' logic of this world. Each character in the narrative is capable, in its own way, of distorting or even dismantling the rationality of survival. This is mainly demonstrated by the character of Lauren Oya Olamina, who understands that a community can mathematically survive through the administration of resources and following economic principles, but such a paradigm remains silent on what are the driving forces that would keep a community persisting or developing a sense of purpose for such a challenging form of living.

Butler develops the framework of 'Earth Seed,' as a spiritual paradigm that can supersede existing religions, and in turn is capable of re-orienting human subjects to one another to become aware of their own entangled realism. This is the foundation for any collective enterprise, a foundation for common ground and for ambitious projects of prosperity, to defeat scarcity.

In the development of software simulations, it is possible to simulate the initial framework proposed by Butler as a model of system dynamics, one in which the availability of resources is confronted with metabolic rates of consumption. This model should also include layers of game theory, where actors might want to support or betray one another based on calculations of self-preservation. The framework of survival dynamics leans into an abstract computational model that is scalable and potentially independent of locality. Butler's narrative, on the other hand, eludes the possibility of an abstract scalable model. By deeply entangling characters within the scarcity setting she developed, she demonstrates how the locality of the subjects is capable of warping and distorting the payoff matrix of rationality.

Computational modeling using System Dynamics

The framework of system dynamics is one of the possible methods that could allow us to computationally model the dynamics of scarcity; such a model understands the world as sources, sinks, flow, and rates (Page 2018). The availability of food, for example, could be modeled by the sources of food available and the growth rate or production rate of the produce. In this framework, human survival would be considered a sink with a specific caloric consumption. Systems dynamics is a well-established model for simulation. In the words of Scott Page (Page 2018, 212):

> "The great value of systems dynamics models resides in part in their ability to help us reason through the effects of our actions.... Though we cannot always anticipate every indirect effect—the positive and negative feedbacks, with models we can think more clearly and deeply through the implications of the feedbacks we do identify."

Nevertheless, at the core of abstract models such as system dynamics lies what Arturo Escobar considers an ontological problem; in other words, a problem of defining who gets a voice in such a model—what are the entities that exist within the model and what are their capacity to establish relationships with other units in the model (Escobar 2018, 112):

> "Political Ontology refers to the power-laden practices involved in bringing into being a particular world or ontology."

Escobar has actively developed the concept of political ontology to signal a problematization of the universal ontology of the dominant forms of modernity, resisting the notion of a "One-World World" (Escobar 2018).

This issue becomes particularly acute when considering simulations for what industry has denominated 'Digital Twins.' Digital twins are simulations that operate as digital counterparts of physical systems. Software packages such as AnyLogic, MatLab, or Nvidia's Omniverse have been supporting the construction of such systems. Digital twins are often valuable tools when replicating a manufacturing facility. The possibility to run simulations and configurations of a space offers a tremendous performative advantage in the possibility of optimizing and detect potential problems.

But as mentioned before, a simulation model needs a boundary; therefore, it is faced with an ontological problem. What is included in the simulation is at the absolute discretion of the researcher modeling the system.

While the popularization of digital twins has been recent, enabled by the computational capacities of simulation, the

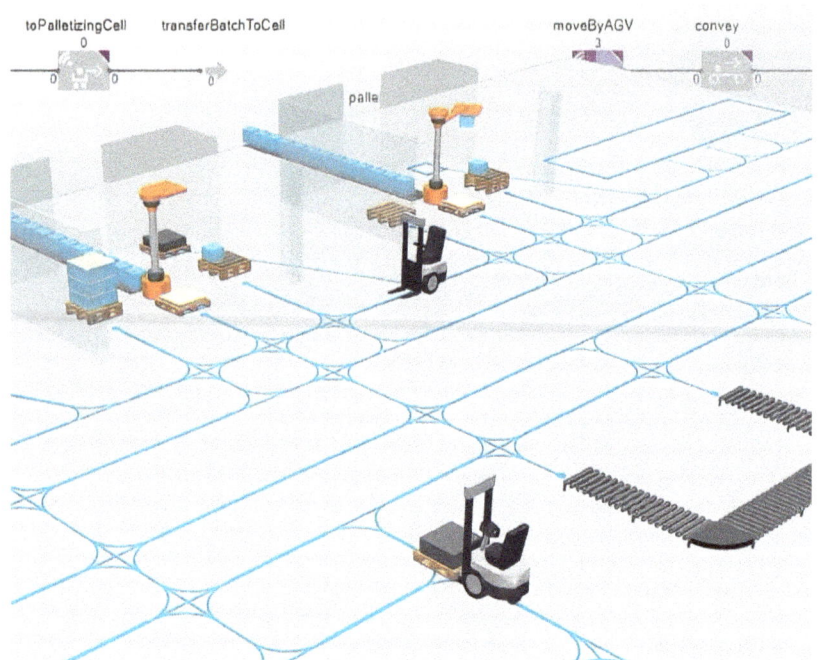
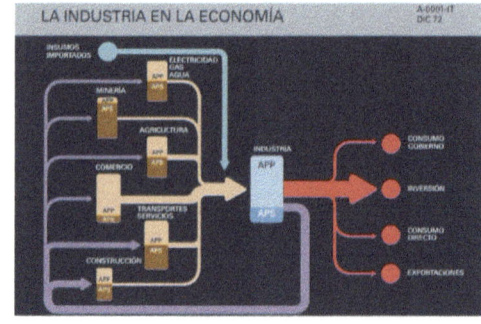

imaginaries of systems as counterparts for reality have been present since the inception of Cybernetics. In her exhaustive documentation, Eden Medina depicts the inception, motivations, and context that gave rise to project *Cybersyn* in the democratic socialist regime of Salvador Allende in Chile (Medina 2011). *Cybersyn* was an attempt to implement a computational model based on principles of cybernetics developed by Stafford Beer to regulate and control the Chilean economy between 1971 and 1973; Medina recounts how Cybersyn was based upon Beer's *Liberty Machine*, as she explains (Medina 2011, 87):

> "The Liberty Machine modeled a sociotechnical system that functioned as a disseminated network, not a hierarchy; it treated information, not authority, as a basis for action, and operated in close to real-time to facilitate instant decision making and eschew bureaucratic protocols."

Medina explains how Beer's idea of control is not founded on the notion of domination but rather on self-regulation, or the ability of a system to adapt to internal and external changes and survive (Medina 2011, 72):

> "While Beer's work was repeatedly criticized for using computers to create top-down control systems that his detractors equated with authoritarianism and the loss of individual freedom. Such criticism extended to the design of Project Cybersyn, but, as this books illustrates, they were to some extent ill-informed."

Cybersyn represents an early yet ambitious attempt to implement a digital twin paradigm for the regulation and coordination of a country's economy. This example, being a government enterprise, makes evident the ontological struggles and challenges of such an ideology.

Escobar establishes a clear divide between what he denominates modern thinking—an attitude toward 'objective models' that can be observed and to our interest simulated—and non-modern thinking, where the observer collapses and becomes part of the model being observed. This leans into ideas of second-order cybernetics, as framed by Heinz von Foerster (von Foerster 2022), who explains:

> "What appears to us today being most natural to see and think, was then not only difficult to see, but wasn't even allowed to be thought. Why? Because it would violate the basic principle of scientific discourse, which demands the separation of the observer and the observed. It is the principle of objectivity. The properties of the observer shall not enter the description of these observations."

The return to the consideration of principles of second-order cybernetics is aligned with the ontological struggles that attempt to challenge the 'One-World World' from modern imaginaries and enable what Escobar calls Pluriversal thinking (Escobar 2018).

For Escobar, the entanglement between the model and the observer radically changes what we constitute as Real, shifting from an objective reality that can be found independent from our interaction to a reality that is, in fact, enacted by our participation in it. Escobar draws from the studies

2 AnyLogic Digital Twin simulation as presented by Rainer Dronzek

3 Project *Cybersyn*, cybernetic simulation project developed by the government of Salvador Allende, 1971

4 Project Cybersyn, cybernetic simulation project developed by the government of Salvador Allende, 1971

5 Jeremy Deller, Re-enactment of The Battle of Orgreave, 2001

of Kriti Sharma (Sharma 2015), who sees participation as a form of enactment, one that allows reality to come into being (Escobar 2018).

The appeal for abstract models can be directly linked with issues of scalability. Contemporary technology and even entire computer programming paradigms such as Object-Oriented Programming have been founded upon notions of abstraction, reusability, and scalability, allowing systems to tackle problems of larger complexity and scale while maintaining human comprehension. Anna Lowenhaupt Tsing has problematized scalability, connecting its origins to colonial plantations and the economic attempt to reduce the messiness of the nonscalable; as she presents it (Lowenhaupt Tsing 2021, 46):

"Scalability, in contrast, is the ability of a project to change scales smoothly without any change in project frames. A scalable business, for example, does not change its organization as it expands. This is possible only if business relations are not transformative, changing the business as new relations are added. Similarly, a scalable research project admits only data that already fit the research frame. Scalability requires that project elements be oblivious to the indeterminacies of encounter; that's how they allow smooth expansion. Thus, too, scalability banishes meaningful diversity, that is, diversity that might change things."

Lowenhaupt Tsing recognizes the hard work necessary to make a project scalable, and how contemporary modern culture has reached a form of obsession with models that scale. By contrast, Lowenhaupt Tsing invites us to consider an ethnographic perspective, one that could become aware of the affordances of the nonscalable without demonizing any of these dualities (Lowenhaupt Tsing 2021):

"The challenge for thinking with precarity is to understand the ways projects for making scalability have transformed landscape and society, while also seeing where scalability fails—and where nonscalable ecological and economic relations erupt. It is key to take note of the careers of both scalability and nonscalability."

The pluriversal framework developed by Escobar presents us with critical challenges for researchers in the area of simulation, inviting us to question our role and entanglement in the systems we model. Moreover, instead of considering a digital twin a digital counterpart, perhaps we ought to think of digital infrastructure as extension of reality itself, one that further enmeshes the relationships between designers and socio-technical systems. Adding to this, based on Lowenhaupt Tsing's observation of scalable/nonscalable systems, could we conceive of simulations as software entanglements that invite a user to understand scalable models without having a blind eye for the nonscalable?

What follows is a consideration of a drastically different account for replicating reality: the practice of re-enactment. Such a study sheds light over the blindspots of computational modeling and reveals the utterly incomplete depiction of reality attempted by digital twins.

From Twins to Re-Enactment
In 2001, Jeremy Deller directed the re-enactment of the Battle

6　Jeremy Deller, re-enactment of The Battle of Orgreave, 2001

7　*Papers, Please* video game by Lucas Pop

of Orgreave. The Battle of Orgrave was a violent confrontation between protesters and the police as part of the 1984 UK miners' strike. In her recount of the event, Alice Correia explains (Correia 2006):

"The eventual confrontation at Orgreave was the first-time riot police had been used to contain an industrial strike in the UK. (...) Margaret Thatcher's classification of the strike as 'mob violence' and her branding of the miners as 'the enemy within' added a sense of menace to the dispute and acted as a way of 'othering' the miners from the majority of the country."

Deller's re-enactment of the event in 2001 required over 1,000 people, of whom about 200 were ex-miners, together with a handful of ex-policemen (Correia 2006). The re-enactment presents us with a different kind of simulated reality, one that is not trying to 'compute' causes, resources, or outcomes but rather get closer to the ground to the narratives, emotions, and aspirations of a community. Correia relates that Deller's performance has been interpreted within an ethnographic turn within art practices, and that the performance, while instigated by Deller, becomes its own form of reality in the enactment of the piece by ex-miners. As Correia recounts (Correia 2006):

"For many of the miners taking part in the performance, The Battle of Orgreave offered an opportunity to reassert their truth of what happened on that day in 1984."

Deller's ethnographic approach brings forward the territory in the practice of embodied participation. There is no abstract model or choreography to be performed but rather a recount of forces, emotions, and aspirations, all of which elude computational modeling. Perhaps we could consider Deller's performance as an altogether different form of cognition. Escobar has used the writings of Francisco Varela to understand a perspective of cognition as enaction (embodied action), as he explains (Escobar 2018):

"By linking cognition to experience, our authors lead us into an altogether different tradition. In this tradition we recognize in a profound way that 'the world is not something that is given to us but something we engage in by moving, touching, breathing, eating.'"

Deller's performance operates within the nonscalable domains of modeling, revealing layers that are in the blindspot of other computational methods. The Battle for Orgreave establishes a clear political ontology, giving voice to subaltern subjects, "people who might otherwise have been 'hidden from history'" (Correia 2006).

The question that follows regards the possibility of restructuring the well-established methods for computational design within the space of simulation and computational modeling, trading an ambition for perfectly scalable systems for

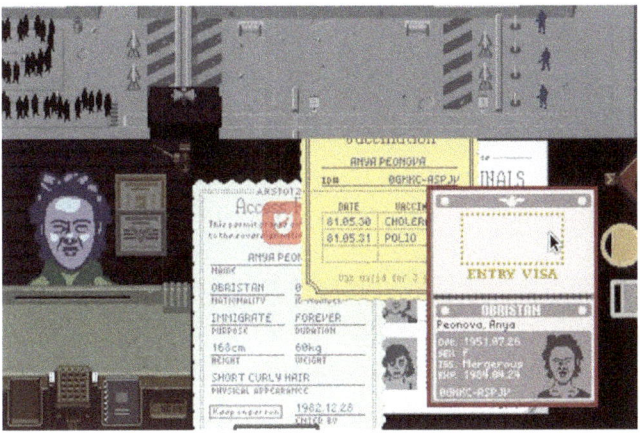

7

Papers Please successfully demonstrates how a narrative setting, meaning, and context distorts the otherwise mechanistic payoff matrix of a task. Human players, entangled entities within this system, are capable of layers of cognition that are not explicitly stated within the simulation's expectation. Setting and context can create ambiguity and difficulty in ethical choices for the player, to the degree that out of a population of players; we would end up with a large range of attitudes towards the simulation, each with a personal argumentation for the paths taken.

The encounter of two ontologies in *Papers Please* brings to the foreground the political nature of simulations and makes us question models that present fictional worlds as objective and neatly structured to be optimized.

systems that could be aware of layers of cognition that often escape non-territorial models. Below is an argument of how video games as a medium are capable of hybridizing forms of computational simulation with narrative and context. Within this medium, a player could be conceived not as a hierarchical user that is all-powerful but rather as an entangled participant within a narrative world.

The Entangled Player
Video games have a long-established tradition of role-playing; role-playing is the practice of embodying an avatar, often a character that belongs to and operates within an established world. While this is a common practice within narrative games, it is not often the case where the narrative is presented in opposition to the mechanical economy of a game. The example of *Papers Please* (Pope 2013), a video game designed by Lucas Pope, demonstrates how this is possible. In *Papers Please*, you (the player) play as an immigration officer with the power to grant or deny permit of entry to the people attempting to cross the border of a fictional country in conflict. The game asks you to revise the documentation of every citizen and determine if there are any missing documents or conflicting information. The game plays initially very much like a puzzle, where there is a right and wrong answer. This setup changes when you start realizing that the officer you are role-playing has a family and that you need to make ends meet. The officer's payroll is tied to how many people you can process in a day, so it is in your best interest to process citizens quickly. Errors are penalized, so you have to be careful as well. These elements define the mechanical framework of the game, one with a clear payoff matrix and rules for success. This is contrasted with the heart-breaking stories of the citizens that are trying to cross the border. The narratives of subaltern subjects and their aspirations to be reunited with family members within countries in conflict resonate as particularly attuned to current geopolitics.

What follows is a depiction of how the frameworks of what has been coined here as 'entangled simulations' has been instrumental in designing the narrative of *Common'hood*, offering an ontological conflict to economic growth.

Simulation and Narrative in *Common'hood*
The challenges presented above are the framework used for the design and development of the video game research project *Common'hood*. *Common'hood* is an economic simulation game that invites a player to construct a productive workshop, one that is able to allow for the means of economic sustainability. The game uses a system dynamics model of simulation, understanding the material inputs, labor requirements, and means of economic outputs to recreate a digital economy with various possible productive paths. The economic simulation behind *Common'hood* follows the principle of a digital twin, understanding how machines, labor, and knowledge can allow production chains that produce a surplus. The game digitally replicates the tools and practices already established with a digital fabrication community, establishing a meta-commentary on the ontological imaginaries that are afforded within this community.

While the economic model could be understood as a scalable and context-agnostic simulation, the game places the player within a narrative backdrop, one that, much like Butler's work, exacerbates the human struggles at the edge of precarity. The narrative backdrop entangles the player with the events of economic struggles, such as foreclosures and homelessness. The player is invited to role-play operating as a squatter occupying an abandoned factory. It is within this setting that the economic model is put into action, forcing the player to design forms of self-provision many of which include architecture. This is achieved initially through direct labor, performing material scavenging and fabrication, and operating under the imaginaries of the DIY (do-it-yourself) community. As

8   *Common'hood* video game developed by Plethora Project, studio led by Jose Sanchez
9   *Common'hood* screenshot
10 *Common'hood* screenshot

the game progresses and the player is introduced to many other simulated characters operating as fellow squatters, the game pivots from self-preservation toward community development. In this context, the same self-sufficient imperatives apply, but within each new interaction, social narratives emerge, many of which challenge or question economic imperatives of growth.

The narrative layer developed for *Common'hood* consists of over 40,000 lines of dialog that have been constructed as a nonscalable depiction of the social reasoning behind the economic affordances of the simulation. The narrative layer offers a second ontology, an alternative to economic growth and financial gains. This second ontology is that of mutual aid and conviviality. The game does not enforce for the player to engage with this second ontology but does make evident the impact of actions taken by the player. The game allows the player to seek fast economic growth at the cost of overworking community members, but not without being confronted with the physical and psychological impact this could have on them. In addition, the larger sense of purpose discussed within the narrative, that of constructing a 'Commons' or practice of stewardship and mutual aid, cannot be formally achieved by means of economic gains or technological advancement; it can only be achieved by deepening the relationships with each one of the eighteen characters that are narratively encountered within the game. Such a narrative confronts the player with stories of addiction and abuse and many aspirational narratives of what a community could have been or could become.

Common'hood presents the design of architecture within this setting. A setting that is not an agnostic canvas but rather a conscious attempt to offer an ontological conflict, asking a designer what are the world and imaginaries that we occupy at the moment of designing. The two ontologies presented bring to the foreground the political nature of simulation and an urgent call for questioning models of objective realism predicated under the banner of technological progress.

CONCLUSION – LOOKING AHEAD

While the framework presented above describes the affordances of hybridizing computational models with a narrative tradition, the development of such a framework in *Common'hood* is a first attempt to connect these two domains, which has not been clear since its inception in 2017. This framework has evolved out of the inability of computational models to engage with the non-scalar narratives pervasive under economic inequality. The maturation of this trajectory points to a tighter relation with territorial narratives where simulated characters could amplify the voices of subaltern subjects in a way that could engage in the question: Who has a voice in the simulations we produce? It is possible to envision a shift in the practice of computational design, one that becomes aware of the shortcomings of past decades of practice. Such a shift will only occur by challenging the ontological underpinnings of established models, opening the field to unexplored territories.

ACKNOWLEDGEMENTS

The *Common'hood* project has spanned five years and could not have been possible without the work and dedication of a large team which includes: Zach Day Scott (Lead Developer), Max Sabido (Marketing / Community Manager), Shuruq Tramontini (Lead Artist), Satrio Dewantono (Developer), Kellan Cartledege (Environmental Artist / Developer), Selma Mutal (Composer / Composer OST), Yucong Wang (3D Artist), Dane Matthew Carstens (Developer).

10

REFERENCES

Butler, Octavia E. 1993. *The Parable of the Sower*, 1st ed. New York: Four Walls Eight Windows.

Correia, Alice. 2006. "Interpreting Jeremy Deller's The Battle of Orgreave." *Visual Culture in Britain* 7 (2): 93–112.

Escobar, Arturo. 2018. *Designs for the Pluriverse Radical Interdependence, Autonomy, and the Making of Worlds*. Durham: Duke University Press Books.

Lowenhaupt Tsing, Anna. 2021. *The Mushroom at the End of the World: On the Possibility of Life in Capitalist Ruins*. Princeton, NJ: Princeton University Press.

Medina, Eden. 2011. *Cybernetic Revolutionaries: Technology and Politics in Allende's Chile*. Cambridge, Mass: The MIT Press.

Page, Scott E. 2018. *The Model Thinker: What You Need to Know to Make Data Work for You*. New York: Basic Books.

Pope, Lucas. 2013. *Papers, Please*. 3909 LLC.

Sharma, Kriti. 2015. *Interdependence: Biology and Beyond*. New York: Fordham University Press.

von Foerster, Heinz. 2002. "Ethics and Second-Order Cybernetics." In *Understanding Understanding: Essays on Cybernetics and Cognition*. New York: Springer.

IMAGE CREDITS

Figure 2: ©AnyLogic
Figure 3, 4: Public domain, Chilean Government
Figures 5, 6: ©Jeremy Deller
Figures 7: ©Lucas Pope

All other drawings and images by the author.

Jose Sanchez is an architect, game designer, and theorist based in Detroit, Michigan. He is the director of the Plethora Project, www.plethora-project.com, a research studio behind the video games *Block'hood* and *Common'hood*, digital social platforms that aid the authoring of architectural and ecological thinking to non-expert audiences. He is the author of the book *Architecture for the Commons: Participatory Systems in the Age of Platforms* published by Routledge in 2020 and the co-creator of Bloom Games. He is currently at the University of Michigan, where he serves as Associate Professor at the Taubman College School of Architecture.

Closing the Gap

Nicolas Azel
KPF, New York

Brandon Pachuca
KPF, New York

Lucien Wilson
KPF, New York

Leveraging Rhino Compute and Cloud Computing
to Distribute Custom Computational Tools

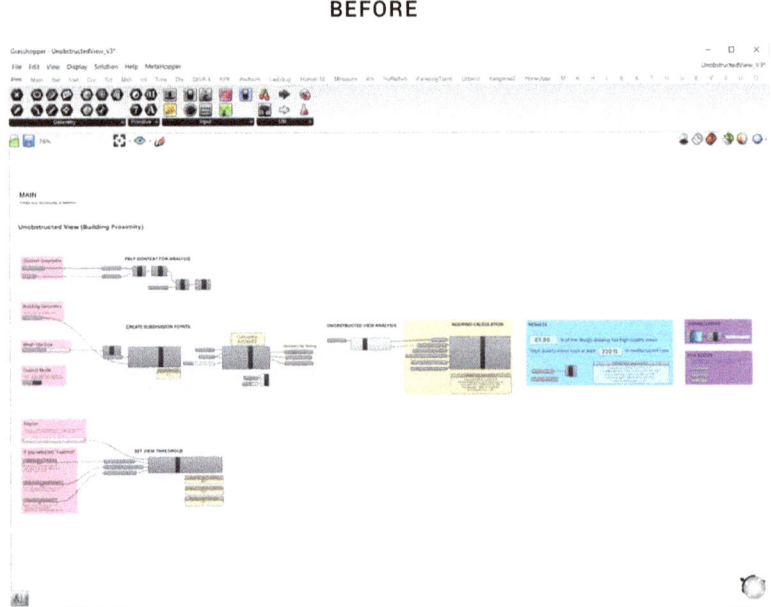
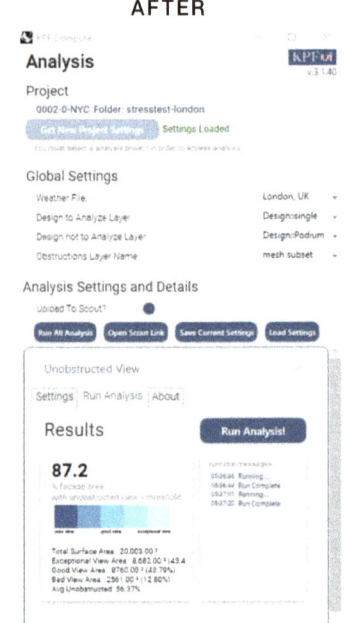

1 Side by side comparison of a computational tool in the Grasshopper interface vs. the same tool inside KPF Cloud Tools' Rhino plugin interface

ABSTRACT

In contemporary practice, there is a plethora of computational tools available to designers through software extensions customization tools, and software as a service platform. Examples of these tools include Climate Studio, Grasshopper, and Cove.Tool respectively. These computational tools apply to a wide range of tasks, from automating the repetitive to in-depth analysis of design behavior. These tools benefit designers by increasing efficiency through automation and drive high-performing design outcomes through simulation and analysis. While there is a wide range of toolsets available to designers, they frequently require specialized programming expertise or subject matter knowledge to apply effectively these tools in the design process, or they require designers to learn how to use new software.

This paper shares KPF Cloud Tools, a platform for using Rhino Compute (McNeel's REST API for RhinoCommon and Grasshopper) to run a library of Grasshopper tools through a cloud server via a Rhino plugin with a procedurally generated user interface, making it quick to deploy new tools (Robert McNeel & Associates 2010). We describe the professional challenges that the KPF Cloud Tools platform solves, document the technical implementation of the platform, and illustrate its benefit through the impact on a large architectural practice.

INTRODUCTION

Across architectural offices today, specialists are building custom computational tools in Grasshopper, a visual programming plugin for Rhinoceros 3D, one of the most common 3D modeling tools (American Institute of Architects 2021). Since being introduced in 2007, Grasshopper has grown into a robust platform to rdly develop custom analysis and modeling routines uniquely suited to a given project, work style, or firm standard. While the presence of advanced Grasshopper workflows has permeated across the industry and is readily available in most contexts where Rhino3D is professionally used, applicable computation tools are frequently underutilized due to challenges around software expertise and software deployment.

A common approach to using custom computational tools within a firm's design process is to establish specialist computational teams to support the creation and implementation of custom computational tools specific to the nuances of a given project. This has helped to increase the sophistication, and awareness of, computational tools but underdelivers on computation's potential by confining its use to experts and limiting access for general practitioners (Peters 2018).

Since there will never be enough specialists in a design firm to match the demand for these types of tools, it is critical that firms develop methods that bridge the gap to facilitate non-specialists using these tools. Additionally, most firms have spent years or even decades developing libraries of computational tools that are not being used broadly because of the inaccessibility of their interface. To address these challenges the KPF Urban Interface team (ui.kpf.com) has developed KPF Cloud Tools. KPF Cloud Tools is a platform and methodology for transitioning computational tools onto the cloud, streamlining tool access within designers' native modeling environment (Rhino3D), improve accessibility of tools for non-specialists, reduce administrative overhead, and cultivate more robust collaboration between specialists and designers.

Proposed Platform-based Solution

KPF Cloud Tools enables specialist teams to solve the last mile of delivering computational tools to users by removing barriers to entry, such as hardware requirements and Grasshopper knowledge. We do this through three core functionalities: (1) Removing complex interfaces from the equation and replacing them with an intuitive, simpler version within the software designers are already working in (Figure 1); (2) Establishing a framework to generate procedurally the tool's user interface, so that specialists can focus on building new tools without the burden of building custom user interfaces for each tool; and (3) Reducing administrative challenges of custom software setup, by hosting and running tools on the cloud, minimizing requirements for user's local machine. Simply put, we aim to meet the users where they are already working, and to manage and run the tools somewhere else.

Related Work

Other industries have started exploring the use of cloud computing to make complex tools and simulations available to non-experts, facilitate collaboration, and deliver the tools through intuitive, easy to use interfaces (Askary 2020). Walker et al. recognize the limitations of simulation tools only being available to experts. "Without direct access to models, decision makers and stakeholders often rely on modeling experts to develop, run, and interpret simulation models on their behalf. This process can create a bottleneck in the flow of knowledge attained through simulation modeling and hinder stakeholder creativity" (Walker 2022).

Buytaert et al. created a web interface for non-experts to run cloud-based environmental simulations to "turn the typical top-down flow of information from scientists to users into a much more direct, interactive approach." However, they also identify an important limitation of this approach, without an expert directly involved there is a challenge "to the communication of model results and their assumptions, shortcomings, and errors" (Buytaert 2009).

We find there are three major challenges for broader accessibility of Grasshopper tools: (1) There are often not enough specialists to support design teams, resulting in projects not having access to design or performance analytics on hand; (2) While the Grasshopper interface makes crafting complex computational workflows relatively quick and easy, it is intimidating to newcomers, and asking designers to learn a new software or interface is often a non-starter in the fast-paced context of architectural practice; and (3) There is a non-trivial amount of administrative overhead associated with managing Grasshopper plugins, tool versioning, and library dependencies needed to run every tool across a firm.

In the architectural industry cloud computing has been primarily used for three purposes—rendering, for hosting Building Information Models (BIM,) and for Virtual Reality (VR)—and except for the case with BIM, these uses were first developed in other industries. Cloud computing and more specifically render farms have long been used in the architectural industry to speed up the production of renderings using methods first established for animation in the film industry (Yao 2009). Since there will never be enough specialists in a design firm to match the demand for these types of tools, it is critical firms develop methods that bridge the gap to facilitate non-specialists using these tools.

Comparable Platform Solutions

The KPF Cloud Tools platform fits into a broader field of computational extensions for Rhino3D that help make Grasshopper analysis more accessible to non-specialists; as such, this section compares similar platforms to illustrate the novel contributions of KPF Cloud Tools.

HUMAN UI

Human UI is a Grasshopper plugin developed by Andrew Heumann and NBBJ's Design Computation Leadership Team and released in 2016 (Heumann 2022). Human UI allows users to build custom user interfaces (UI) from within Grasshopper and allows accessible UI to connect to, but graphically separate from, Grasshopper script. Both Human UI and our platform are built on Windows Presentation Foundation (WPF) and work as standalone window panels while modeling in Rhino software. Both methods provide more accessible interfaces to enter tool input parameters into a Grasshopper script and exist as child windows to Rhino built on WPF.

Limitations
- The Human UI interface is managed by Grasshopper, and this can lead to unexpected behaviors and frozen interfaces when Grasshopper is processing a solution.
- Human UI requires specialists to develop the user interface specifically for each Grasshopper script catering to the given inputs and outputs statically.
- Human UI executes within the Grasshopper script on a client computer and requires anyone operating the script to have all necessary plugins installed.

Platform Solutions
- Our platform is written in C# and runs independently of the client analysis which minimizes unexpected behavior or frozen interfaces.
- Our platform procedurally generates the user interface based on standard inputs specified by the Grasshopper script. This abstraction of the user interface allows it to adapt to various Grasshopper scripts to generate a user interface without additional programming or manipulation.
- Our platform abstracts the running of tools to a virtual cloud Windows instance, so a client device only needs the Rhino plugin to run Grasshopper tools.

GRASSHOPPER PLAYER

Grasshopper Player is a command-line interface for Grasshopper scripts released by McNeel in Rhino 7 (Robert McNeel & Associates 2010) that allows computational designers to package a Grasshopper script to run without a user launching Grasshopper. Both use Rhino Compute to process Grasshopper scripts, and have the same standard Hops input-output parameter formatting so that no custom interface needs to be written from script to script (Payne 2022).

Limitations
- Grasshopper Player requires users to install the plugin dependencies on their computer for any scripts they want to execute.
- Grasshopper Player uses a simple command-line interface to enter script input parameters which can be cumbersome for complex scripts and confusing for some users.

Platform Solutions
- Our platform procedurally generates a 'form style' user interface for entering script inputs, providing a more approachable format for entering parameters.

SWARM

Swarm is an online web platform developed by the Thornton Tomasetti core studio and recently merged with Shapedriver (Thornton Tomasetti 2022). Swarm provides mechanisms for users to upload personal Grasshopper scripts that are then hosted to the public as a community contributed library of functionality. Both Swarm and our Platform leverage Rhino Compute to run analysis headlessly in the cloud; both allow for easy upload and contribution of scripted tools to a central repository; and both provide an easy to navigate user interface for the entry of script parameters.

Limitations
- Swarm scripts are run from within an online 3D viewer, which limits its applications in existing architectural design workflows, such as within Rhino.

Platform Solutions
- Our platform works directly in the designers' native modeling environment of Rhino. This allows computational tools to be accessed more directly during the design process and provides more flexibility in creating input geometry.

KPF CLOUD TOOLS

The KPF Cloud Tools platform is built on an ecosystem of custom coded software and existing services implemented in novel ways. These have been brought together to form a continuous workflow that presents itself as a singular product to end users (Figure 2). There are four critical infrastructure components to our platform:

- The client-side Rhino plugin, which provides an interface to access and run the computational tools;
- A windows virtual server, which uses McNeel's Rhino Compute to run each analysis on the cloud;

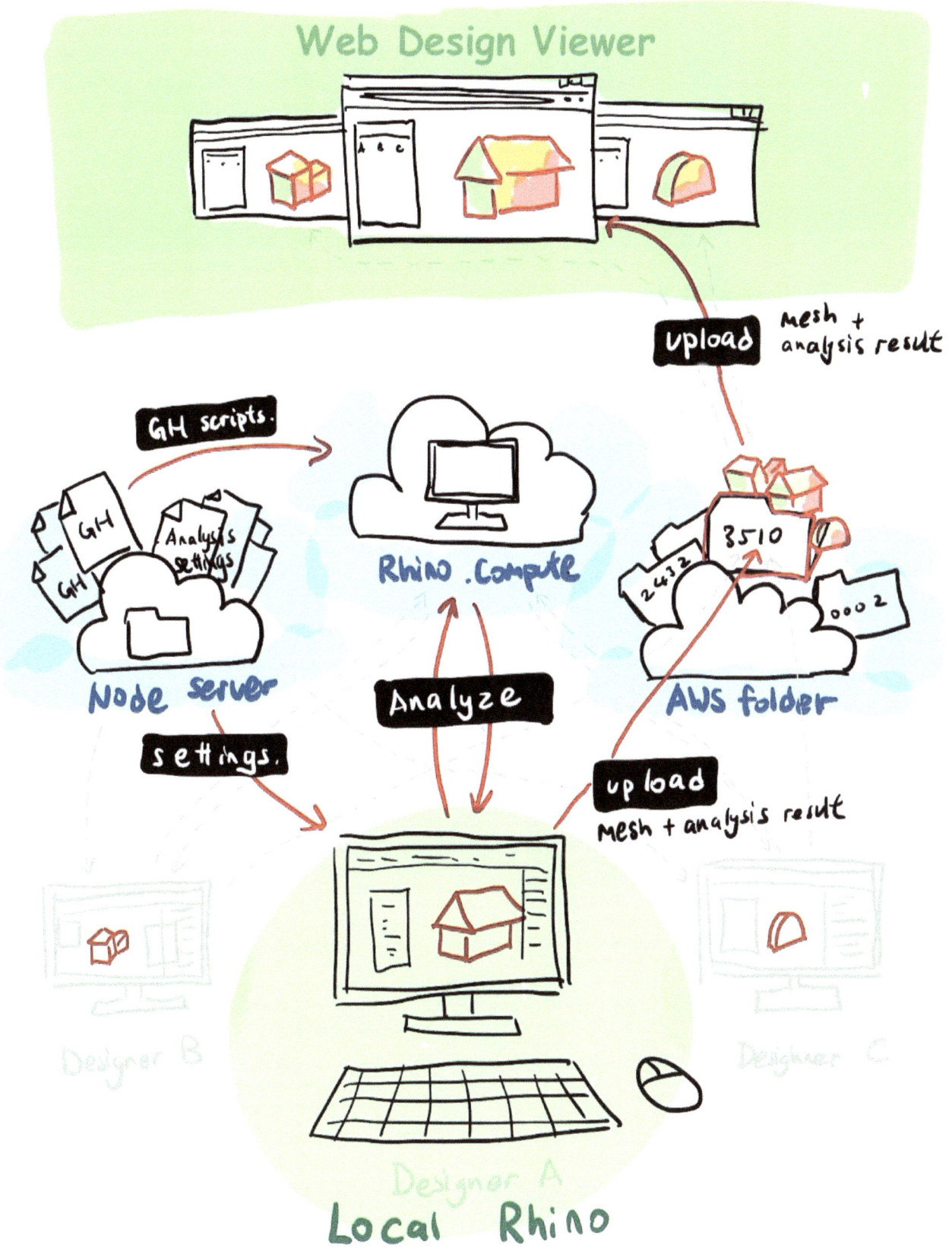

2 Interaction of platform components

- An Ubuntu node server that acts as authentication and a proxy between clients and the Windows server, as well as hosts a web interface for selecting desired tools; and
- Cloud storage buckets used to save tool results and serve design analysis to an in-house design explorer website for review and comparison.

Rhino Plugin

The Rhino plugin is users' primary interface for working with the platform and running tools. The plugin interface provides three types of interactions: (1) provides a gateway for tool selection (Figure 3); (2) enables the use of tools through setting tool inputs, running the tool, and viewing tool results (Figure 4); and (3) facilitates the uploading and storage of design options and tool results to a cloud storage system for design review in external web platforms.

The Rhino plugin launches from a simple Rhino toolbar or the command line and is managed as a standalone child window to Rhino. It is developed in C# and is for Windows operating system only. The interface is written with Windows Presentation Format (WPF) and procedurally generates the UI for the analysis inputs based on a generic tool schema. This generic tool schema describes the tool inputs, outputs, information about typical use cases, plus links to additional resources and is used to instruct the plugin on how to produce the UI for each tool. The schema is passed into the plugin from a local file or the tool selection web interface and is maintained along with the Grasshopper script for each tool in a centralized repository. The procedural generation of the tool UI is an important aspect of the platform, as it enables the development of diverse tools without the overhead of designing the user interface. It also means that new tools can be rdly delivered to designers without the need to update the plugin or otherwise put additional requirements on the end-user to access new tools.

Windows Rhino Compute Server

The role of the Windows Rhino Compute Server is to host McNeel's Rhino Compute Server and run all analysis requests sent from the Rhino plugin on the client's computer.

Amazon Web Services (AWS) (Amazon 2022a) hosts the Windows virtual server in production, running on a compute-optimized C5 EC2 (Amazon 2022b) virtual machine running Windows 2019 Server. CPU and memory resource allocations are static, based on predicted historical traffic throughout the Windows service due to long startup load times for Windows machines. Production setup of the Rhino Compute Windows service is documented by McNeel in the article "Deployment to Product Servers" (Payne 2022b). Internet Information Services (ISS) manages parent and child process threads, library updates, and starting a child process to handle incoming web requests. The startup command sets the number of child processes to handle web requests per parent Rhino Compute and is not limited to the number of available logical cores. The child process will timeout after a set idle span specified by the startup service command, currently defaulting to one hour. The ISS will start a child process if an incoming web request is received and no child process is currently active due to an idle span timeout. The ability to start a timed-out child process is helpful in effectively managing per core hour billing cost, which charges based upon the uptime of the Rhino Compute process.

Node Server and System Interaction

The role of the Node Server is to provide a clean interface for selecting desired tools and acting as an authentication proxy for passing tool requests and responses between the Rhino plugin and the Windows Rhino Compute Server.

The technology stack moves data from a front-end web browser or Rhino plugin through a middle-end Node.js authentication layer and then to the Rhino Compute Windows server back-end. After the front-end Rhino plugin collects user form inputs and design geometry, the data is encoded and sent via an HTTP request to the middle-end Node authentication server. The middle-end is a proxy server routing data requests to the back-end Windows Rhino Compute server, ensuring no unauthorized requests to back-end resources are permitted (Figure 5).

Once the request is received by the Rhino Compute Windows server, it handles the request via child process. The process thread calls the Grasshopper script and loads the script into memory. Inputs are assigned based on the script's corresponding 'Get Components' and once the Grasshopper script runtime is complete, the outputs specified within the script by 'RH_OUT' are returned to the middle-end Node server (Baer et al. 2022). The middle-end service then returns the analysis results back to the front-end Rhino plugin for visualization and user evaluation.

Cloud Storage Buckets

The platform's cloud storage buckets are used as a central location to store design options and tool results. This provides several benefits and extends the platform's use to include archiving, design communication, and design presentation:

- Systematized and consistent design archiving/storage for the increased quantity of design data generated by tools on the platform,
- Hosting project data for use in external web interfaces and presentation tools, and
- Multiple users can contribute to the same project with consistent presentation.

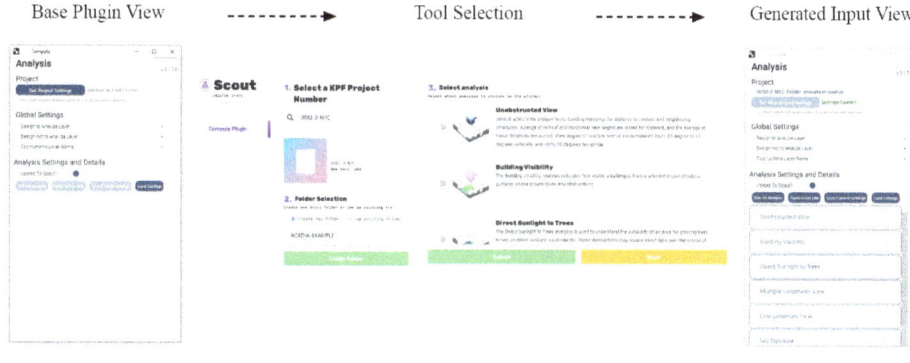

3   Tool selection process on both the web and the Rhino plugin

4   Example interfaces generated for a single tool

5   Interaction between platform components

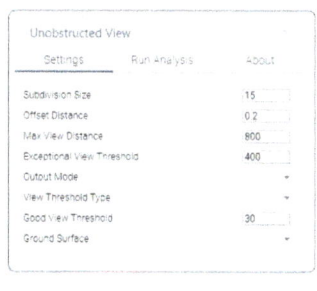

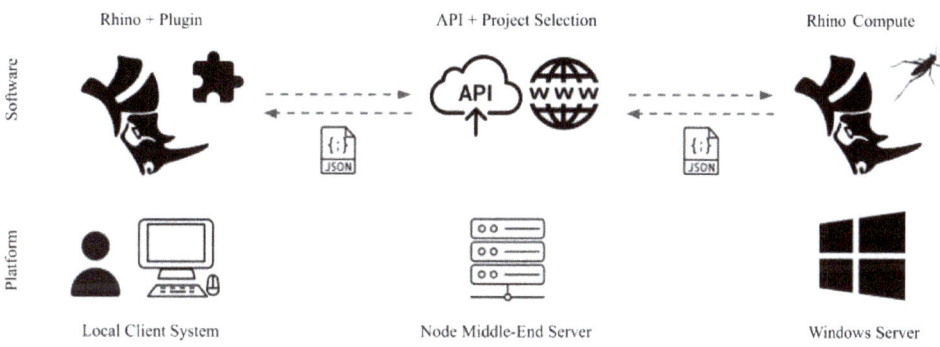

The first benefit is systemized archiving of the design process for easy and consistent access beyond the life of the project. This is particularly important as the platform allows for a greater variety of design options to be rapidly generated and analyzed across a wide range of tools. This greater quantity of information can quickly lead to a breakdown within human archiving in the absence of this cloud storage integration. The second benefit is the hosting of design options for access by external presentation platforms or design explorers that allow users to present many design options and their respective performance metrics or analysis results. Often these design explorers are critical to the evaluation of the data rich, option-based process that our platform enables.

Our cloud storage system is hosted by Amazon Web Services (AWS) and design materials are uploaded and accessed through the following process (Amazon 2022a). First, the client Rhino plugin submits data to the middle-end Node server, which acts as a proxy and authentication layer to the storage bucket; second, the Node server creates the appropriate folder hierarchy in the storage bucket and uploads the client's data as flat files; third, the Node server provides an access to retrieve data from the storage bucket and deliver them to any valid and authenticated requests.

IMPACT IN PRACTICE

KPF (Kohn Pedersen Fox) is a large (700+ employee) architectural and urban design firm. In this section we will demonstrate the benefits of KPF Cloud Tools through the impact on a large firm. The benefits were identified through dozens of user interviews, tracking of usage, and comparison with previous methods and can be grouped into three categories: 1) accessibility and adoption of tools by design teams, 2) administrative benefits, and 3) leveraging existing assets.

Accessibility and Adoption of Tools by Design Teams

The primary motivation of developing this KPF Cloud Tools was to make computational tools accessible to non-specialists and increase firm-wide computational adoption.

Prior to development of the KPF Cloud Tools, internal firm efforts to make computational tools more widely available focused on standardization and education for simulation tools such as views, daylight, walkability, and outdoor thermal comfort. This took the form of carefully packaged and documented Grasshopper files hosted from a central website with step-by-step instructions and example files. They were designed to be plug-and-play and usable with minimal Grasshopper expertise; however, the adoption rate remained extremely limited. The most frequent feedback from designers was a polite decline and requests for specialists to manage all use of Grasshopper-based tools.

While we are still in the early release phase of the KPF Cloud Tools platform, the initial adoption rate of computational tools has been much higher. Very few designers have declined to use the platform when offered, and many designers have proactively reached out to get set up with the Rhino plugin. This allows any designer access to a library of computational tools independent of the availability of specialists to run them. This both saves time and allows designers to quickly evaluate the performance of their design relative to key metrics making for better performing designs firm-wide (Figure 6).

Additionally, in the previous method of tool distribution, communicating how to use the tools, interpret results of simulations, and use them to make design decisions was difficult to do. With KPF Cloud Tools we can communicate this type information at various points along the user flow, from tool selection to tool tips explaining inputs and options, to pop-ups to help explain results.

Administrative Benefits

Significant time is spent maintaining complex software dependencies across a large firm like KPF. Computational tools developed by specialists often use Grasshopper plugins that must be deployed firm-wide to allow for any designer to use those tools. This process involves setting up and testing plugins on sample computer images, communicating any issues back to IT, and pushing a new computer image firm-wide. It typically takes two weeks to a month to get new Grasshopper plugins or software tested and distributed across the firm. The library of Grasshopper scripts, centrally hosted for the Rhino Compute solution, is easier to manage, version, and update when centralized. Comparing the management of a single centralized Windows Compute Server to that of deploying a consistent Windows environment across the entire firm, the centralized approach is far less time-consuming for multiple parties and easier to rapidly update.

Leveraging Existing Assets

Many architectural firms have spent years investing in the creation of computational design tools. At KPF, the firm has a library of 42 regularly used Grasshopper tools that have been developed for over ten years. Until the development of KPF Cloud Tools, many of these tools have been siloed within the specialty teams that created them. This platform opens access to these tools from a couple dozen specialists to an office of 700 designers. Additionally, custom tools can now have a longer operating life because scripts that may have been used once for a specific project, can now be centrally stored, easily re-accessed, and applied in other contexts when suitable, without technical setup or modification.

The KPF Cloud Tools platform also makes the deployment of new tools very fast. In the case of a tool that is formatted to match the inputs need for procedural Rhino UI generation, it takes just one day to add a new tool. For tools that have a structure that doesn't allow for procedural UI generation, it can take longer to deploy, from five to ten days; however, this is often much shorter than it would take to recreate the Grasshopper tool as a standalone Rhino plugin, the alternative method of accessible deployment without our platform. For example, we recently developed a Grasshopper tool for creating 3D context models, which leverages a Grasshopper plugin for querying the Open Street Maps API to get geometry and data. We put the tool on our platform and built a web map interface for using the tool. The time to build the Grasshopper definition, put it on the platform, and build the web interface was a little over a month. To get the same level of functionality in a standalone tool, either as a Rhino plugin or web map, would take upwards of four months of development time.

CONCLUSION

While the implementation of this platform within a large architectural practice has had a large beneficial impact, there are considerations and limitations worth discussing.

First, debugging Grasshopper definitions that fail to run on Rhino Compute is challenging. The error messages record the name of the component and GUID where the error occurred but not any details about its inputs or outputs. In the case of

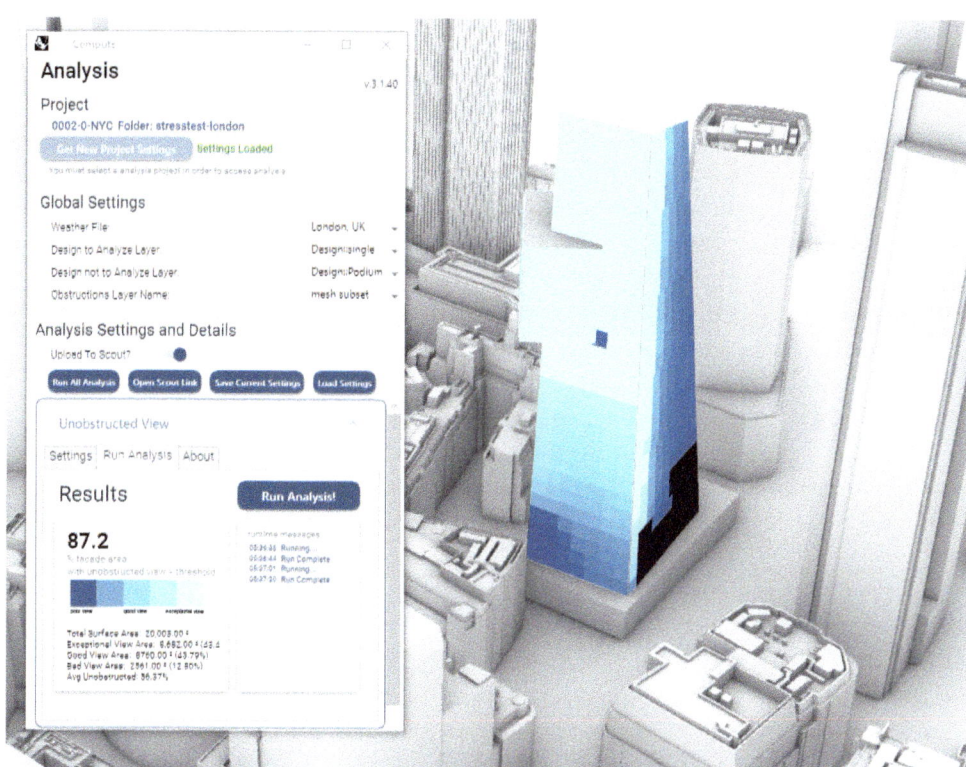

6 Examples of designers using the Rhino plugin to run solar radiation and view analysis through the Rhino Compute server

clusters, if the error occurs within a cluster, the error logged is only the name of the cluster and not the name of the component throwing the error. In both cases, if a script throws an error, a user will need to open the Grasshopper document, set up the scripts inputs, and search for that GUID to check why it failed. Additionally, it is not always possible to recreate an error when running a script in the typical Grasshopper interface because of differences between how Rhino operates in windowed versus headless mode. This means script debugging is more time intensive and requires extra effort from a script author to build in checks and custom error messaging if any details are to be passed to a platform user.

Next, the per core hour costs for the Rhino license and cloud computing might be prohibitively expensive for some companies. Additionally, web development expertise is needed to deploy this platform. While most architectural firms have computational specialists who could develop the Rhino plugin necessary for the platform, many are unlikely to have someone who can do the web development work required as it is uncommon expertise in the architectural industry.

Finally, there is a danger in making complex simulation tools easy to use without pairing it with the proper domain knowledge to calibrate the settings and interpret the results. For example, environmental simulation such as solar radiation and outdoor thermal comfort all require expertise to properly set up and interpret the results. The Urban Interface team is addressing this limitation by setting up the platform, calibrating it for their specific project, and then having a session to explain how to interpret the results. While the goal for KPF Cloud Tools is to be used firm-wide, rollout has been incremental so that plugin use is paired with the required education.

Despite some deployment challenges, the platform has achieved key goals for KPF. It has exposed a growing library of computational tools to a broader audience, has opened the contribution of tools to non-software developers, and gives everyone more time to focus on tool use or tool creation over learning and managing software. Beyond the benefits demonstrated in the previous section, we believe that this type of approach to deploying tools can disrupt the current mode of architectural practices by democratizing access to computation and tools traditionally the domain of experts. By moving easy-to-build computational processes—relative to software development—out of the Rhino desktop environment and onto the cloud, we decouple the relationship between the client software and the tools logic. This decoupling, and the programming of the plugin interface in WPF, allows the delivery of custom Grasshopper tools to be introduced in a wide range of Microsoft Windows softwares like Revit, ArcGIS, or Unity. As a software architecture, we believe this type of solution will grow across the industry in the coming years. Particularly with the introduction of Rhino Compute as an excellent platform for the delivery of custom workflows alongside the growing list of commercial design services like SimScale, InFraReD, and Autodesk Forge (City Intelligence Lab 2022). Finally, broad adoption of computational tools by non-experts has the potential to increase analysis and data literacy across the industry and drive better performing outcomes for built projects.

ACKNOWLEDGMENTS

Thank you to the Urban Interface team and designers across KPF who have helped us to test and evolve this platform. We would like to give special acknowledgment and thanks to Demi Chang who contributed significantly to the initial development of the platform and helped in the production of diagrams within this paper. We also would like to thank the team at McNeal for their encouragement and help throughout the development process including Steve Baer, Will Pearson, Brian Gillespie, Luis Fraguada, and Andy Payne. Finally, we would like to thank Thomas Mahon of BiMorph Digital Engineering for his guidance on software architecture and his contributions to the codebase.

REFERENCES

American Institute of Architects (AIA). 2021. "Technology, Culture & the Future of the Architectural Firm." AIA.org. Accessed May 11, 2022. https://content.aia.org/sites/default/files/2021-11/21002_AIA_Tech_Culture_Report_v4_10-10-21.pdf.

Amazon. 2022a. "AWS Home Page." Accessed May 16, 2022. https://aws.amazon.com/?nc2=h_lg.

Amazon. 2022b. "Amazon EC2 Instance Types." Accessed May 16, 2022. https://aws.amazon.com/ec2/instance-types/

Askary, Zareef, and Ravinder Kumar. 2020. "Cloud computing in industries: A Review." *Recent Advances in Mechanical Engineering. Lecture Notes in Mechanical Engineering*, edited by H. Kumar and P. Jain. Singapore: Springer. 107–116. https://doi.org/10.1007/978-981-15-1071-7_10.

Baer, Steve, Scott Davidson, Andy Payne. 2022. "The Hops Component." Rhino Developer. Last updated February 2, 2022. Accessed May 11, 2022. https://developer.rhino3d.com/guides/compute/hops-component/.

Buytaert, Wouter, Selene Baez, Macarena Bustamante, and Art Dewulf. 2012. "Web-Based Environmental Simulation: Bridging the Gap between Scientific Modeling and Decision-Making." *Environmental Science and Technology* 46 (4): 1971–76. https://doi.org/10.1021/es2031278.

City Intelligence Lab. 2022. "InFraReD 0.1 Launch." Accessed May 13, 2022. https://cities.ait.ac.at/site/index.php/2021/07/24/infrared/.

Heumann, Andrew. 2022. "Computation: Human UI." www.andrewheumann.com. Accessed May 11, 2022. http://andrewheumann.com/#computation.

Payne, Andy. 2022a. "What is Hops." Rhino Developer. Accessed March 21, 2022. https://developer.rhino3d.com/wip/guides/compute/what-is-hops/.

Payne, Andy. 2022b. "Deployment to Production Servers." Rhino Developer. Accessed March 21, 2022. https://developer.rhino3d.com/guides/compute/deploy-to-iis/.

Peters, Brady. 2018. "Defining Environments: Understanding Architectural Performance through Modelling, Simulation and Visualisation." *Architectural Design* 88 (1): 82–91. https://doi.org/10.1002/ad.2262.

Rhino Developer. n.d. "Compute Guides." Accessed May 16, 2022. https://developer.rhino3d.com/wip/guides/compute/.

Robert McNeel & Associates. *Rhinoceros 3D*, Version 7.14. Robert McNeel & Associates, Seattle, WA. 2010.

Thornton Tomasetti. 2022. "Swarm Alpha." Swarm Website. Accessed May 13, 2022. https://swarm.thorntontomasetti.com/.

Walker, Jeffrey D., and Steven C. Chapra. 2014. "A Client-Side Web Application for Interactive Environmental Simulation Modeling." *Environmental Modelling and Software* 55 (C): 49–60. https://doi.org/10.1016/j.envsoft.2014.01.023.

Yao, J., Pan, Z., Zhang, H. 2009. A Distributed Render Farm System for Animation Production. In *Entertainment Computing – ICEC 2009. Lecture Notes in Computer Science*, vol. 5709, edited by S. Natkin and J. Dupire. Berlin, Heidelberg: Springer. https://doi.org/10.1007/978-3-642-04052-8_31.

Nicolas Azel is an urban designer and educator whose work seeks to engage landscape systems for the promotion of egalitarian communities and increased ecological health. He specializes in design computation, with a focus on complex urban systems and landscape ecology. Some of Nico's past work includes community planning, infrastructural systems planning, and public parks design, as well as coding custom design software, visualizing complex datasets, and developing interactive simulations of stormwater runoff. He is currently based out of New York City, where he works as a Computational Designer.

Brandon Pachuca is an Urban Data Scientist and Web Developer with a background in architecture and urban planning. He creates cloud-enabled web-based data visualizations to help facilitate collaborative learning and insight. Brandon holds a Master of Science from the NYU Center for Urban Science and Progress, focusing on how technology, AI, and policy fit together to overcome our community's most complex challenges. Brandon previously worked at Kohn Pedersen Fox Associates on the KPFui team as an Urban Data Analyst and Web Developer. He created award-winning data visualization software for architects to better understand performance-based trade-off decisions throughout the design process.

Lucien Wilson is a Director at Kohn Pedersen Fox and leads KPF Urban Interface (ui.kpf.com), an urban technology team focused on computational design, software development, and data analytics. The UI team has worked on over 400 projects globally ranging from master plans to rezonings to supertall towers. His work and research has been published in both academic and popular publications from *Architectural Science Review* and the *Economist* to *Wired* and the *New York Times*.

Learning from Players

Exploring Collective Intelligence in a Cultural Heritage Board Game through Geo-Spatial and Behavior Analysis

Guzden Varinlioglu
Massachusetts Institute of Technology (MIT) and Izmir University of Economics (IUE)

Ozgun Balaban
Columbia University

Daniel Tsai
Massachusetts Institute of Technology (MIT)

Charles Ngai-Hang Wu
Massachusetts Institute of Technology (MIT)

Takehiko Nagakura
Massachusetts Institute of Technology (MIT)

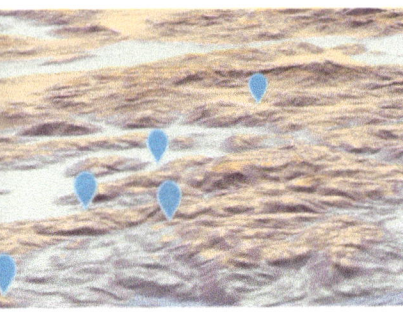

1 The real-world topography in: (a) GIS; (b) the board game; (c) players during the sessions and their recording process; and (d) game elements

2 Data collection for creating: (a) distribution map; (b) abstraction of the data in GIS to create a game board; (c) gamification/design of game mechanics; (d) test plays and recording of the collective intelligence/decision making; (e) the spatial analysis, game analytics, and behavior analysis of gamers

ABSTRACT

The time spent playing games continues to increase, and educators and researchers have realized the potential of using gamification as a learning tool, and less often, using the collaborative environment of games to solve design problems. The unique contribution of this research is to employ gamers' collective intelligence to find the answers to questions of architectural history. This project is about mapping ancient trade routes, creating a credible geospatial narrative of travel and places of transported materials in the past. To uncover the intercity trade networks of the Middle Ages in Anatolia, we documented locations of an-iconic heritage building type, the caravanserai, a roadside inn used by travelers to rest during long journeys. This paper explores the unconventional decision-making environment of gamification to uncover past urban networks. To test the validity of decision making behavior of humans in the simulated topography, we designed, implemented, and tested a serious board game simulating the urban networks of the past. Using the topography, we abstracted, simplified, and represented several layers of GIS data into hexagon tiles to design a board game. The game also employs playing cards, divided into chance and trade cards, which are used to determine players' movement on the board and their scores. In the game environment, we simulate the movement of trade, while the players, as agents, explore and reveal possible intercity networks. We monitor and document the gamers' pathfinding/pathmaking decisions and use these to make comparisons with computational simulations. By tracking users' movements and behaviors, we were able to create data for spatial analysis, game statistics, and user behaviors. Based on experiments and employing gamers' stigmergy, the research provides predictions for lost urban networks of Anatolia.

INTRODUCTION

Increasing interest in games has led educators and academicians to see their potential as a learning tool. Especially during the pandemic period, games became a tool to complement traditional teaching methods to improve the learning experience, while also teaching other skills, such as following rules, problem solving, interaction, critical thinking skills, creativity, teamwork, and decision-making. Using games as an educational tool provides opportunities for deeper learning. Games, along with certain innovations, such as VR/AR technologies, can also bring a clearer understanding of human perceptions of problems and decision-making processes in close-to-real world conditions.

Using games as participatory design tools is an approach to design that invites all stakeholders into the design process, bringing a better understanding of needs, and allowing them to be better met, and even preempted. Similarly, creating a collective intelligence is a utopian vision for collaborative knowledge culture (Levy 1999). As well as gathering, mastering, and deploying pre-existing information and concepts, members of a collective intelligence would additionally work with the collected facts and viewpoints to discover ways of thinking and coordinating. By focusing on the effect of crowdsourcing and collective intelligence in the decision-making process, especially in a game environment, we were able to explore the possibility of using games, not only as a learning tool for gamers, but as a means of learning from the gamers themselves.

To answer these questions, we designed, implemented, and tested a serious board game involving simulating the urban networks of the trade routes of the past (Figure 1). Using the actual topography, we abstracted several layers of GIS data for the game board design. In the game environment, we simulate the movement of trade, while the players as agents simulate the possible intercity networks. We recorded the traces of the players' collective movements using a ceiling-mounted video camera recording of the positions of the markers and players' behaviors. By tracking their movements and behaviors, we were able to collect user data for spatial analysis, game statistics, and user behaviors. The initial intended goal of this game was to identify possible locations of lost architectural heritage sites. The gamification of digital heritage, and using player participation in the form of the recorded observation was the research method of this study.

The central challenge of this research is the application of unconventional methods to quantify, qualify, represent, and experiment with these networks. Using this approach, the following questions are addressed: Can we reconstruct social networks of the past using gamers? Can unconventional computing help to uncover the Anatolian urban networks of

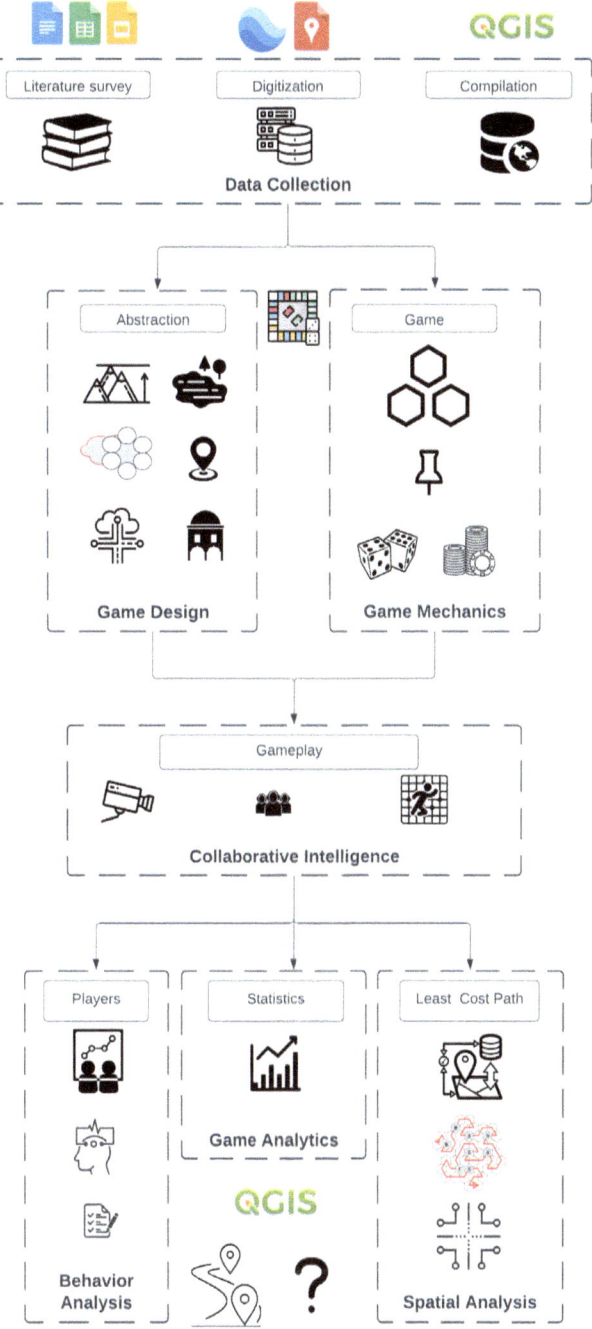

2

the past? Does the complex system of cities and civilizations correlate with the gamers' decision-making patterns? Using the players' decision-making skills, gaming can shed new light on the field of cultural heritage.

This research employs the collective intelligence of gamers to imitate the development of these ancient networks (Figure 1). As a case study in GIS, we collect and compile data from the ancient Silk Road routes, more specifically, the architectural evidence of these routes, namely, caravanserai, a roadside inn of ancient times. Based on the caravanserais' geographical coordinates, we used the GIS to overlay the following five layers: (1) base map from OpenStreetMap; (2) hexagon tiles on the board, in which all cells are equidistant; (3) terrain displaying the difference between flat valley floors, steep mountain climbs, and water elements; (4) today's traffic networks, referring to both the current route network and the density of the urban population, and finally (5) the location of caravanserais. The caravanserais, as nodes for safe-stay, are represented by solid hexagons. By allowing gamers to progress towards these, we aim to reveal the distribution networks of roads, paths, and traces. We monitored and documented gamers' pathfinding/pathmaking for comparison with computational simulations. Thus, the experiments and gamers' stigmergy provided predictions for lost urban networks of Anatolia.

STATE OF THE ART

At the core of this paper is an interdisciplinary approach at the intersection of collective learning, gamification, and computational tools.

Game as Stigmergy and Collective Intelligence

Stigmergy is a form of indirect communication and coordination in which agents modify the environment to pass information to their peers. Stigmergic interactions can occur variously in social species, robotics, web communities, and human societies (Heylighen 2016). This emerging phenomenon is highly relevant to the understanding of modern human society. Another concept, that of collective intelligence, similarly emerges from many individuals' collaboration, collective efforts, and competition, and is seen in consensus decision-making. Rather than simply gathering, mastering, and deploying preexisting information and concepts, members of collective intelligence work with the collected facts and viewpoints to discover new ways of thinking and coordinating.

Collaboration is an important aspect of architectural design. Kalay and Jeong (2003) make the connection between traditional project-based learning methodology based on individual design decisions, and the potential of collaborative design through multi-player games. Through their redesign of popular board games such as Scrabble and Monopoly, they were able to use gamification in collaborative design. Another example, Play the Koepel, is a multi-stakeholder game for developing collective scenarios as part of a city consultation process (Play the City 2020). Multiple stakeholders are invited to the game, designed specifically to promote interaction to bring together different ways of thinking for a possible interpretation of the city complex. Santacruz (2019) designed ScarCity Game, a board game and a pedagogical tool to involve architecture students in a series of experimental design sessions aimed at understanding the design process

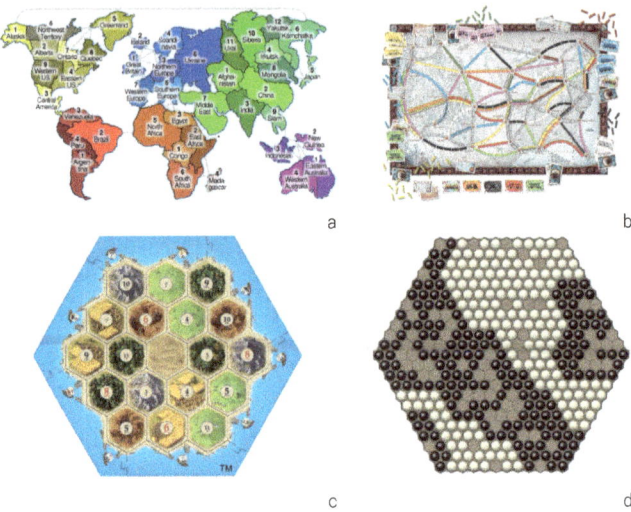

3 Levels of abstraction in games: (a) Risk, (b) Ticket to Ride, (c) CATAN, and (d) Stigmergy

of scarcity, and the true relation between craft and the digital. Game environments provide students with the medium to challenge conventional methods of teaching and learning in design.

Game design involves planning various aspects of the game, such as mechanics and dynamics. A game should be a challenging abstraction of reality and should involve following specific rules to achieve a goal. In the examples above, the game format was not only considered as a teaching aspect, but also aimed at providing an innovative and unconventional environment in which to solve scientific or design problems. An example of design problem-solving gamification is Ahlqvist and colleagues' game on resource management. This real geography component inspired Ahlqvist et al. (2011) to develop tools that others could use to create location-based learning games, making geography and GIS a fundamental component for game players. Efforts were made to integrate GIS to Massive Multiplayer Online Gaming (MMOG) to support the integrated modeling of human-environment resource management and decision-making (Ahlqvist et al. 2012). Games are also used to test specific innovations in close-to-real world conditions (Piette et al. 2021). Since the advent of Artificial Intelligence (AI), games have become testbeds not only because of their simplicity and popularity, but also due to their capacity to allow a clearer understanding of human perceptions of problems and decision-making processes, in comparison with those of machines.

Many games, in essence, are simulations of real world scenarios. Most well-known historical games use maps at various abstract levels; for example strategy games, such as chess or checkers using black/white board, or, in the case of backgammon, the board allows for a combination of strategy and luck. At the abstract level, Risk, a strategy board game, depicts the political map of the world, divided into six continents and 42 territories, where players occupy territories to eliminate others (Figure 3a). A well-known pathfinding game is Ticket to Ride, a cross-country train adventure in which the players build routes in between the cities, with more points scored for longer routes (Figure 3b). CATAN is a strategy game in which the aim is to trade, build, and settle an imaginary island, where players control their own civilizations and aim to spread across a modular hex board in competition for victory points (Figure 3c). At a higher level of abstraction, the game called Stigmergy is a borderless territory game for two, played on cells of hexagons, similar to Go, in which each player aims to gain possession of a certain number of positions, occupying these with stones of their color (Figure 3d).

Analysis Tools
Recent developments in emerging technologies have allowed the establishment of the phenomenon of digital heritage. These developments grew out of conventional computing, and the use of GIS in archaeological studies from the early 1990s, which represented progress towards a more fully macro-scale analysis of human societies and their environment. This approach included the representation of data in layers, the integration of statistical and spatial programs, and most importantly, the ability to work on 3D terrains. Thus, GIS has evolved as an invaluable aid for the heritage sector, allowing the incorporation of historic map data, physical details of the landscape, and known information about the past inhabitants. A significant example is the web application entitled The Stanford Geospatial Network Model of Roman World (ORBIS) (see ORBIS n.d.), simulating movement along principal routes of Roman road networks. This project enables the reconstruction of the time needed and the expenses associated with a wide range of different types of travel in antiquity.

Spatial analysis deals with the use of space. Network spatial analysis, as opposed to traditional planar spatial analysis, concerns events strongly constrained by their networks e.g. the locations of car crashes and fast-food shops on streets (Okabe and Sugihara 2012). Urban network analysis, although developed for the analysis of urban streets and networks, is also suited to railway networks, highway networks, or utility networks (Sevstuk and Mekonnen 2012). In the heritage context, network spatial analysis in GIS involves finding patterns of distribution of archeological finds, features, sites and monuments (Hodder and Orton 1976). The distribution map, able to show the totality of information, is one of the main instruments of archaeological research and exposition.

GIS facilitates agent-based simulation, which is an approach to modeling systems composed of individual, autonomous, interacting agents. There are a variety of simulation tools in GIS for analyzing the distribution of archaeological heritage along the network. Our focus is the "least cost path," allowing the determination of the most cost-effective route between a source and destination. Thus, least cost path analysis shows the optimal path between a source and destination in terms of cost function. Some of the previous works in the book *The Least Cost Analysis of Social Landscapes* (White and Surface-Evans 2014) focus on contexts such as hunter-gatherer zones, water transport trails, and prehistoric trail networks. Another prominent study takes the approach of juxtaposing GIS and archaeologically mapped ancient road routes for a region of ancient Italy, and reveals the map differences (Hodza and Butler 2022). Similarly, in these games, there are tools that can be used for analysis, such as the heatmap, density/location-based aggregated visualization of the users' movement, and GIS (Drachen and Canossa 2019).

Emerging technologies allow the better analysis of the user experience of space through systematic data collection via path tracking, bioinformatics, sonic analysis, and visual recording. However, computer vision provides another opportunity for behavior analysis and gesture tracking to recognize patterns. Moments in Time at MIT CSAIL (2019), is a research project dedicated to building a very large-scale dataset to help machine learning systems recognize and understand actions and events in videos (Monfort et al. 2019). In addition, specialized equipment such as a wearable electroencephalography (EEG) headset can present a framework for classifying the game player's expertise level.

CONTENT AND SCOPE
As humans traveled and traded with neighbors, exchanging goods, skills, and ideas, Eurasia became covered with intersecting routes, known today as the Silk Road. The Silk Road's inland routes were marked at intervals with caravanserais and guesthouses that provided accommodation for traveling merchant caravans. Extending along the Silk Road from China to Turkey, connecting east to west, these buildings not only provided a safe resting place, but also served to promote the exchange of cultures, languages and ideas, allowing the spread of news, but also, unfortunately, diseases. Caravanserais were carefully positioned a day's journey apart, which means every 30-40 km in well-traveled areas, approximately 6 (+/-2) hours' journey time. The diversity of paths, tracks, and roads changed not only in the long-term, but also seasonally, as weather conditions made river crossings and mountain passes impassable, forcing travelers to take different routes, often across wide valleys and steppes. We adopted the approach of identifying major nodes (large cities) along the Silk Road routes; identifying caravanserais as waypoints between these, and, rather than identifying

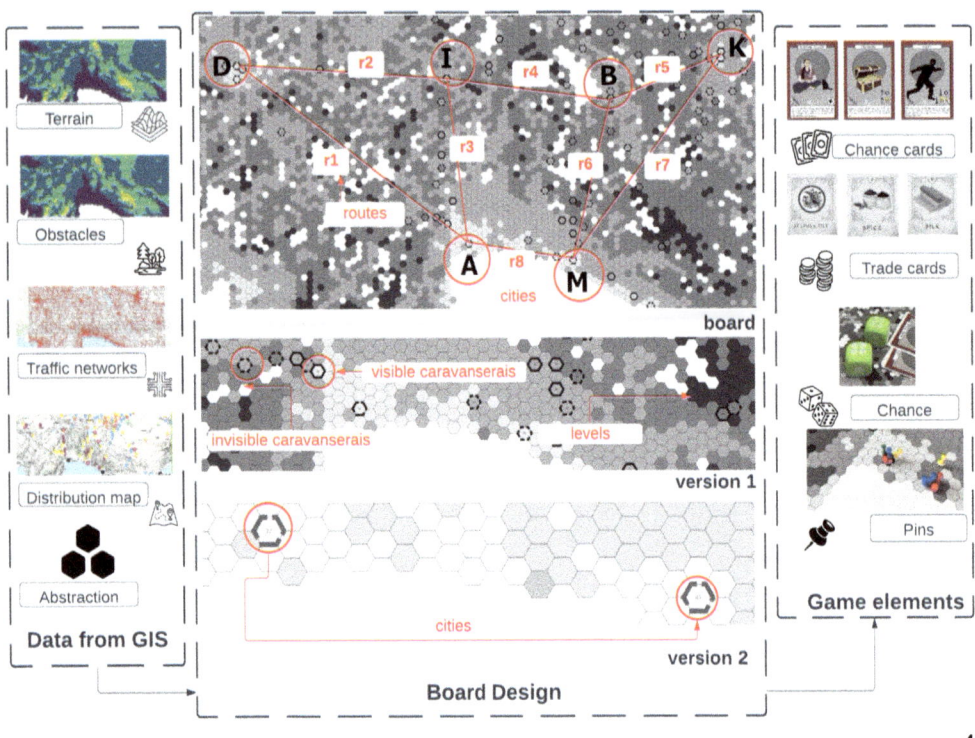

4 Data layers in (left) GIS; (center) board design, and (right) game elements

5 V.1 of the board, four main colors show players' paths, while black line shows the (a) optimal path; V.2 of the board, white line shows the (b) optimal path; (c) heatmap; and (d) modern intercity road networks and caravanserais plotted on the game board

potential specific roads, broadening out these routes between the nodes to represent the corridors of movement and impact. Selected as a case study, Antalya was a prominent ancient city on these roads, representing a secure transit point for goods, its harbors providing access to the maritime trade routes of the period. Another advantage was its proximity to networks of caravanserais. Studies in modern-day Antalya/Manavgat revealed that the city is at the intersection of eight caravan roads.

METHODS

Data Collection

To address the lack of a holistic approach in geospatial inventory of caravanserais of Anatolian Seljuk, we digitized and created a compilation of all the major cataloging studies. We created a structured database using maps, web inventories of travelers, books, a list of individual researchers, and a list of caravanserais. Subsequently, we reorganized this data, with each building being categorized according to the nearest cities, and made this information accessible in the Google Earth platform cloud service, allowing the team to work remotely on a common database. In the later stages, we exported the online database to Google My Maps, for public dissemination, and distributed it through social networks for travelers' guidance. This was a very early effort to crowdsource the data collection methodology.

Game Design

The game board granularity (size of the hexagon and number of elevation shades) has a considerable impact on the players' traversal. First, we designed the board, in which the various terrain differences were simplified, and represented by hexagon tiles of four shades of gray. We used GIS to overlay the multiple layers: base map from OpenStreetMap, hexagon tiles representing 5 km span, the terrain representing flat valley floors, steep mountain climbs and water elements, modern traffic networks referring to both the current route network and the geographical coordinates of caravanserais (Figure 4, left). The 5 km distance between two cells gives enough granularity to provide a reasonable level of topographic detail. Each player's movement is limited to six hexagons at the same level, represented by the shade of gray and one less movement per each level change. The six hexagon-move corresponds to 30-40 km in real world travel.

There are two versions of the board: V.1 refers to the board showing both visible and invisible caravanserais. Visible caravanserais refer to located and identified buildings in varying stages of preservation, while invisible caravanserais refer to an approximate location only, for example, those whose existence is implied in settlements whose name includes the word "han." V.2 refers to the board design with cities but without any caravanserais locations (Figure 4, center).

The game also employs playing cards, divided into chance and trade cards, which are used to determine players' movement on the board and their score (Figure 4, right). The visual and textual information on the cards is inspired by the architectural heritage of this period, utilizing contemporary symbols, instruments, characters, tales, food, clothes, artifacts, animals, etc. Subsequently, these elements were displayed in 2D graphics designed to visualize the daily life associated with the particular cultural heritage. The trade cards define the overall earned score, defined by the roll of dice, which determines how much money is won or lost. The trade cards are not related to the movement of the players. The chance cards may lead to a positive or negative outcome, leading the player either to proceed forward on the board and gain money, expressed as trade cards in the game, or to lose properties and move backward. Card dynamics add an element of flow, without disturbing players' path making decisions.

The game involves pathfinding through visiting caravanserais (waypoints) to reach the main cities (nodes). The aim is to score as many points as possible and return to the starting point as quickly as possible. The players are allowed one move at a time (six hex at the same level or one less movement per level change). When reaching a caravanserai, they stop, roll dice, and pick a trade card. If the caravanserai is invisible, they stop, but do not take a trade card. If they decide not to rest at a caravanserai, they pick a chance card, but this might bring negative outcomes, and should, if possible, be avoided. Players have to visit all cities. The winner is the first player to arrive back at Antalya, the starting point, and who has collected the most trade cards. The rules of the game motivate players to visit as many caravanserai as possible, and reach all of the cities in the fewest turns, while avoiding staying out in the open (cells without caravanserais). These rules provide the motivation for players to follow the actual historical trade routes.

Spatial Analysis

To test the historical hypothesis of the potential locations of trade routes, we collected the traces of gamers' movements in GIS. These traces show players' responses to the conditions reflecting the trading mechanics of the era, providing data for locating lost connections along the routes. The game is played six times to record the players' decision making under different conditions. The first four instances are played with all the caravanserais on the board, the fifth instance, with only visible caravanserais, and the final instance, with only cities visible. The aim of playing with different levels of information on the game board is to compare the deviation of paths from the optimal least-cost path. We conducted six game sessions of 24 participants, each session lasting approximately 90 to 120 minutes. After being instructed on the purpose of the game, a group of four players traced paths in between the cities. The first four sessions targeted the gameplay and game testing, on a board that includes the both visible and invisible caravanserais (V.1). In the first version of the board, four main colors show players' paths, while black lines show the optimal path (Figure 5a). The fifth

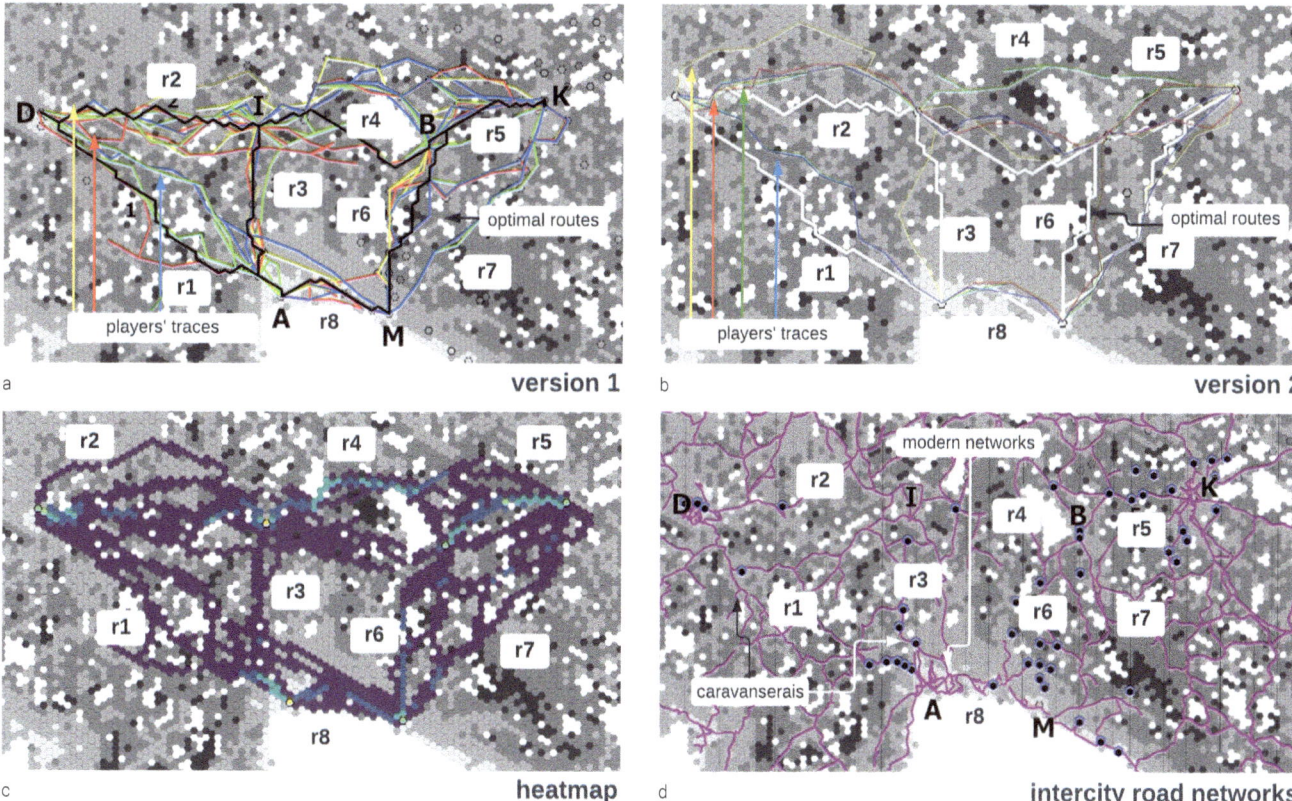

and sixth sessions were played on different board designs with no caravanserai data (V.2). In the second version of the board, four main colors show players' paths, while white line shows the optimal path (Figure 5b). A heatmap of all the traces of players' movements was created to show the cumulative paths (Figure 5c). Finally, modern roads, the location of caravanserais and the gamers' routes were plotted together, showing the relation between these (Figure 5d). Intercity road networks were obtained from OpenStreetMap (OSM) street networks of Turkey. We selected the roads with the type "highway," "primary," "secondary," and "trunk."

To compare the players' movements with the optimal topographic conditions, we used "least-cost path analysis" from the grids, which involved calculating optimal paths between cities using QGIS shortest path function. In the hexagon tiles, each cell has its own gray shade which also indicates its level. To calculate the least-cost path between two points, the movement cost of every cell along the route should be added (Figure 6). The granularity of the board selected for this project means that the spatial analysis is limited to topographical optimal path analysis.

Behavior Analysis

The students, as players, take on the role of agents simulating traders' behavior. Following the participatory design/co-design perspectives, the game was used as a research method rather than as a product. It is important to note that we additionally evaluated the users' behavior to assess the game design (Figure 7).

We recorded the players' behaviors and movements across the board game by using markers through video recordings. A camera focused on the game board recorded the players' movements. Each player used a distinctive pin marking their movement on the game board (Figure 7a). The locations of these were registered in the recordings using Python OpenCV libraries, converted into lines, and stored in QGIS. In addition, another camera was used to record the players' gestures (Figure 7b).

Moments in Time, co-developed by MIT and IBM, is a large-scale dataset for recognizing and understanding motion in videos (MIT CSAIL 2019). It was used to analyze players' behavior while playing the game. We compared the frequency of spoken communication in each session, detecting some frequent behaviors, such as "combing," "shrugging," etc. Figure 7b shows an example of action recognition of a three second video clip, and the prediction result showed some potential. The error in detecting the behavior was caused by the usage of facemasks, and could be enhanced by removing this visual obstruction.

RESULTS AND DISCUSSION

The game was designed to help the researchers to build heritage information. The initial results of the study based on gameplay, and also the statistics showed a need to divide the routes into eight fragments. Depending on the changes on topography, and the availability of modern traffic networks, each segment of the network (routes 1-8) displayed a variety of results. In this part, each segments' results are presented and discussed in relation to the gamers' decisions and optimal routes on both versions of the game.

The V.1 game results showed that along the routes with a major obstacle, a small majority of players chose the optimal route, especially in relatively uniform landscapes, but in segments with small obstacles of minor variations, a clear majority tended to choose the optimal routes. Exceptions were routes 6 and 7, where most diverted away from the optimal paths. These routes had the most frequent changes in topography and greatest uncertainty in the location of the caravanserais. These exceptions might indicate inconsistency in the method of abstraction of topography while designing the board. The 5 km distance of hexagons might have missed some natural features, as the average (overall height domain/4 levels of gray) cannot represent the scale of historical travel or human biodynamics with great accuracy. The V.2 game results proved that without any caravanserais information, players tend to follow paths further from the optimal routes, with an exception of routes 7 and 8, which have the fewest major obstructions.

Optimal routes partly coincide with players' decisions, but more completely coincide with existing modern traffic networks, with the exception of route 6, for which some

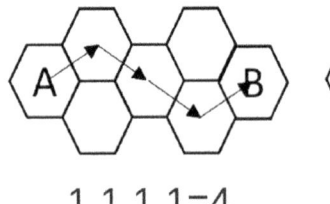
1,1,1,1=4

a

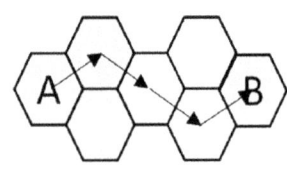
1,2,1,2=6

b

1,0.2,1,1=3.2

c

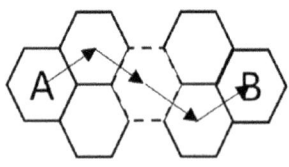
1,0.4,1,1=3.4

d

6

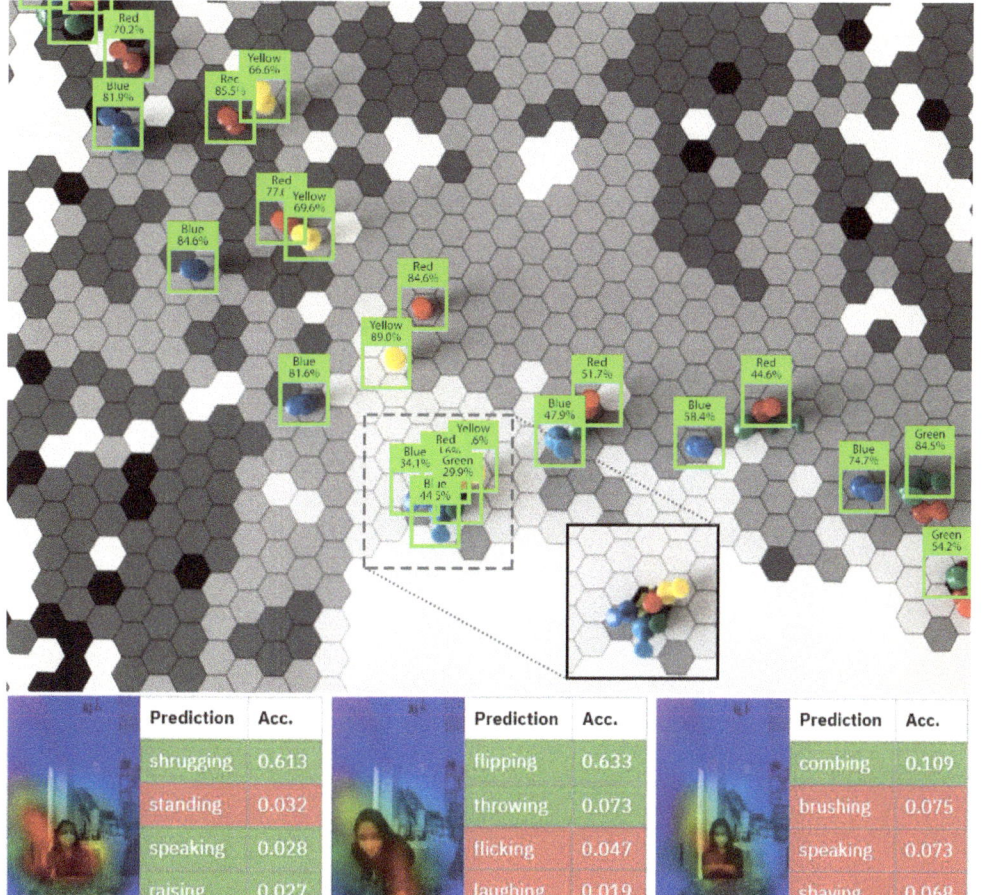

6   Movement costs for different cell configurations: The cost of movement along the same level grid cells is one for each cell from A to B, the cost of movement is 4 (a). If there is a change in the level of the grid cell, the cost of movement increases by one to reflect the need to climb up or down the slope in the topography, therefore the cost of movement from A to B is 6 (b). If there are caravanserais (bold cells) the cost of movement for that cell is 0.2, as a result the cost of movement from A to B is 3.2 (c). If there are hidden caravanserais (dashed cells), the cost of movement for that cell is 0.4, so the cost of movement from A to B is 3.4 (d).

7   Tracking players' pins (top) and their gestures (bottom)

caravanserais do not coincide with the modern roads. This suggests the possibility of another city or political center around route 6, which is already distinguished by its location next to steep mountains. This seems to justify the need for a higher resolution abstraction method in locations with greater topography changes.

CONCLUSION

Keeping in mind the rhetorical question of human rational and irrational decision-making processes, the study explored the potential of the game environment as an analysis tool. In addition to the use of gamers as agents to discover trade routes, the research also presents a novel perspective on the analysis of the past, to predict the actual and future development of cities and city networks. In other words, the aim was to understand the urban dynamics in the current era, which is characterized by constant change in city networks, due to climate change, immigration, and the recent pandemic period. The use of gamification allows understanding of the behavior of crowds in a visual and collaborative manner. A similar effort is the CityScope, a project conducted at the MIT Media Lab City Science group (2021), including tangible and digital platforms dedicated to solve spatial design and urban challenges. Another potential outcome of the project is to allow the integration of immersive aspects, and develop it into an AR game.

The game was designed and implemented as a board, around which the users can physically sit and interact, avoiding the need for an interface between individual players, the game environment and co-players, allowing them to collectively discuss, and agree or disagree with each others' moves. The competitiveness of the game required thinking about other players' moves and improving them to find the best strategy. The collaborative and competitive nature of the game helped reveal optimal paths in the topography. The removal of the caravanserais in V.2 showed that, in this case, players diverged farther from the optimal routes, which would provide evidence contradicting the players' own pathfinding ability. However, this result should be treated with caution. The sample set is limited to six games and 24 players, and warrants a more targeted summary, and therefore, it is likely that the determination of what is optimal, and thus efficacy of the game schema, needs further exploration.

The behavior of the users was analyzed through computer vision, to identify key moments in thinking and deciding. As our current focus is on the game rather than the student learning, no background information about the users was presented. However, for the next stages, gamer analysis could be improved by collecting more background information (age, gender, nationality).

In conclusion, this proof-of-concept project explored how games can be used to simulate ancient trade routes and how the players of the game can be used as a data source for research. Further studies could explore the benefits of transferring this game to a virtual platform, allowing automation of data collection from the users. In this virtual version, the targeted data set would be larger, leading to greater accuracy in determining the users' pathmaking. This large data set would allow us to use reinforcement learning methods, with humans as agents revealing emergent patterns.

ACKNOWLEDGMENTS

Prof. Varinlioglu's project titled "Stigmergy as a Tool for Uncovering Anatolian Urban Networks of the Past" is supported by the Fulbright Visiting Scholar Program and hosted by Prof. Nagakura at Massachusetts Institute of Technology, Department of Architecture, Computation Group. The game was designed and implemented during the MIT Design Heritage course 4.570 taught by Prof. Nagakura and Dr. Tsai. We would like to thank Yuxuan Lei, Rohit Priyadarshi Sanatani, Han Tu, Chunfeng Yang, Boya Zhou, Tamar Michaela Ofer, Yufei Chen, Rinako Sonobi, Barbara Felipe, Rafael Sousa, Nevin Sahin, Doris Qingyi Duanmu, Selin Sahin, Rafael Defreitas, Ozge Tekin, Mohammedwesam Amer, Antonio Alvarez Naveira for their contribution in game play, and Hilal Kaleli, Sarvin Eshaghi, Sepehr Vaez Afshar, Ezgi Yilmaz for their contribution in game design.

REFERENCES

Ahlqvist, Ola. 2011. "Converging Themes in Cartography and Computer Games." *Cartography and Geographic Information Science* 38 (3): 278–285.

Ahlqvist, Ola, Thomas Loffing, Jay Ramanathan, and Austin Kocher. 2012. "Geospatial Human-environment Simulation through Integration of Massive Multiplayer Online Games and Geographic Information Systems." *Transactions in GIS* 16 (3): 331–350.

Drachen, Anders, and Alessandro Canossa. 2009. "Analyzing Spatial User Behavior in Computer Games Using Geographic Information Systems." In MindTrek '09: Proceedings of the 13th international MindTrek Conference: Everyday Life in the Ubiquitous Era. Tempere, Finland. 182–189.

Heylighen, Francis. 2016. "Stigmergy as a Universal Coordination Mechanism I: Definition and Components." *Cognitive Systems Research* 38: 4–13.

Hodder, Ian, and Clive Orton. 1976. *Spatial Analysis in Archaeology*. Cambridge: Cambridge University Press.

Hodza, Paddington, and Kurtis A. Butler. 2022. "Juxtaposing GIS and Archaeologically Mapped Ancient Road Routes." *Geographies* 2 (1): 48–67.

Kalay, Yehuda E., and Yongwook Jeong. 2003. "A Collaborative Design Simulation Game." *International Journal of Architectural Computing* 1 (4): 423–434.

Lévy, Pierre. 1999. *Collective Intelligence: Mankind's Emerging World in Cyberspace*. Cambridge, MA: Perseus Books.

MIT CSAIL, Computer Science & Artificial Intelligence Laboratory. 2019. "Moments in Time Dataset." Accessed June 1, 2022. http://moments.csail.mit.edu.

MIT City Science Group. 2021. "CityScope." Accessed October 27, 2022. https://www.media.mit.edu/projects/cityscope/overview/.

Monfort, Mathew, Alex Andonian, Bolei Zhou, Kandan Ramakrishnan, Sarah Adel Bargal, Tom Yan, Lisa Brown, Quanfu Fan, Dan Gutfruend, Carl Vondrick, and Aude Oliva. 2020. "Moments in Time Dataset: One Million Videos for Event Understanding." In *IEEE Transactions on Pattern Analysis and Machine Intelligence* 42 (2): 502-508.

Okabe, Atsuyuki, and Kōkichi Sugihara. 2012. *Spatial Analysis Along Networks: Statistical and Computational Methods*. Hoboken, NJ: Wiley.

ORBIS. n.d. Stanford Geospatial Network Model of the Roman World. Accessed March 20, 2022. http://orbis.stanford.edu.

Piette, Eric, Matthew Stephenson, Dennis J. N. J. Soemers, and Cameron Browne. 2021. "General Board Game Concepts." In *2021 IEEE Conference on Games (CoG)*. Copenhagen: IEEE. 1–8.

Play the City. 2020. "Play the Koepel; City of Breda" project. Accessed May 22, 2022. https://www.playthecity.eu/playprojects/Play-the-Koepel.

Santacruz, Axel Becerra. 2019. "The Architecture of ScarCity Game: The Pedagogy of Scarce Design Process." In *The Routledge Companion to Games in Architecture and Urban Planning: Tools for Design, Teaching, and Research*, edited by M. B. Dodig and L. N. Groat. New York: Routledge. 76–91.

Sevtsuk, Andres, and Michael Mekonnen. 2012. "Urban Network Analysis. A New Toolbox for ArcGIS." *Revue Internationale De Géomatique* 22 (2): 287-305.

White, Devin A., and Sarah L. Surface-Evans, eds. 2012. *Least Cost Analysis of Social Landscapes: Archaeological Case Studies*. Salt Lake City: University of Utah Press.

Guzden Varinlioglu is an architect and academician who has dedicated her research efforts to the preservation and presentation of cultural heritage. Her interests in this area led to the establishment of the Digital Humanities Lab, VRLab, and a digital fabrication team in Turkey. She expanded her research horizons at institutions worldwide, through positions at TAMU and as a visiting scholar at UCLA. She is currently a Fulbright Scholar at MIT. Her designs, presentations, writing, and photography have garnered numerous awards.

Ozgun Balaban is an Adjunct Assistant Professor at Columbia GSAPP. He is a computational design researcher and an electrical engineer. His research interests are data-informed design and urban planning, machine learning in design, the use of game environments for design research, digital heritage, building information modeling, and GIS.

Daniel Tsai teaches design and computation at MIT. He also mentors student startup ventures in the MIT DesignX innovation accelerator. His cultural heritage research interest focuses on the architecture of Andrea Palladio. He is also an information systems consultant.

Charles Ngai-Hang Wu is a designer, engineer, and researcher based in Cambridge, MA. With a background in architecture and computational design at MIT, his current interest is on multi-model interactive visual systems and human-centered AI.

Takehiko Nagakura is an architect from Tokyo. At MIT, he teaches courses related to computer-aided design, and his research focuses on the representation and computation of architectural space and formal design knowledge. In 1996, he founded the Architecture, Representation and Computation group (ARC) at MIT, which he continues to lead. His recent projects include a series of digital heritage workshops that use photogrammetry, game engine, and XR tools to capture and represent historic sites such as Machu Picchu. He is key advisory member of Arcbazar.com, an online competition platform created to democratize the architectural design process through crowdsourcing.

Session Introduction

Advanced Production Methods II: Knitted Architecture

Tsz Yan Ng, Chair

CNC knitting has become a robust area of research in academia in the last ten years. So much so that this year's ACADIA conference includes a dedicated session: "Advanced Production Methods II: Knitted Architecture." The four papers in this session represent distinct applications and research approaches to CNC knitting, but they all harness a combination of three characteristics unique to knitting. The first is the customizability of the textile, that a single surface does not have to be homogenously structured. Through specification of yarn type (material) and knit structure (looping construction), a knitted textile could have varied behavior across a seamlessly knitted surface. This aspect of knits defies our normal conception of material logic, that across a single material, one can have localized behavior that contributes to and at the same time unique from the global behavior. In this respect, no researcher could escape the intrinsic nature to working with knits, where one must toggle between calibrating at the stitch level against the overall behavior of the entire surface. The second aspect unique to knits, and perhaps because of the customizability, is the ability to tune the knit's performance toward targeted goals. As demonstrated by Zilka and Underwood and Sinke et al., performance is integrated via computational methods/simulation. Through the calibration of parameters, the knitted textile is an engineered system, controlling for elasticity, formal geometries, and tensile strength. Such control enables applications at diverse scales, from the sleeve of a space suit by Tessmer et al. to ribs of a self-supported structure by the Newcastle research team. The third characteristic to knits is the capacity to embed and integrate other systems into the material with ease. This includes sensing systems knitted into the material as shown by Tessmer et al. for monitoring performance, or in the case of the Newcastle team, to allow mycelium to stiffen the softness of the knitted textile so that the textile could be leveraged for architectural applications. The papers herein highlight the keen awareness that knits, with its unparallel design capacity, is a material system that will have wide impact in the future for architectural production.

From Garment to Building

Dr. Leanne Zilka
RMIT University

Dr. Jenny Underwood
RMIT University

Research into the synergies, techniques, and application of a textile based approach to the fabrication of large scale elements in architecture

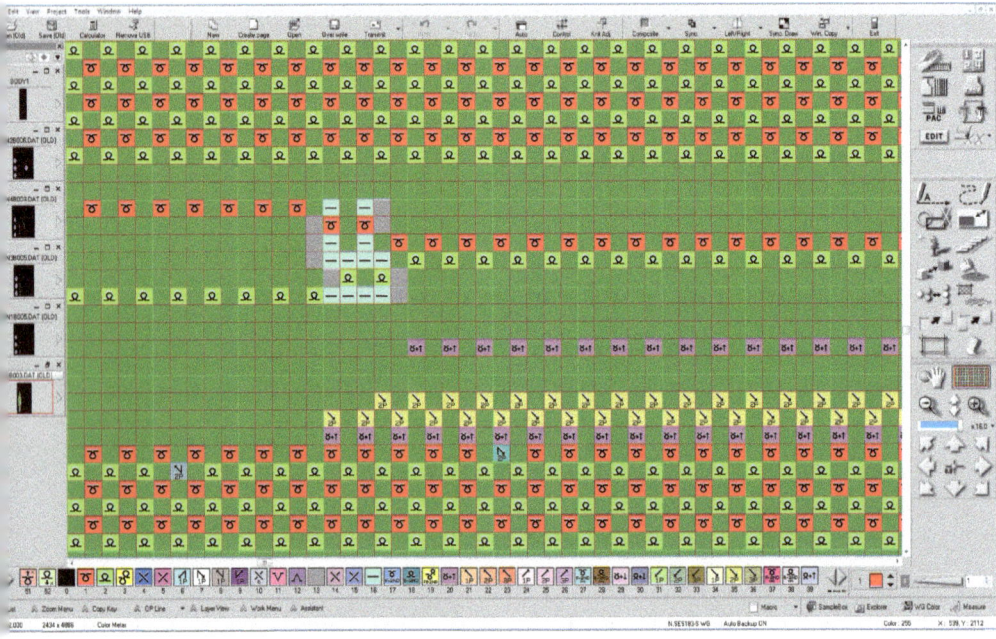

1. Screenshot of the pixelated code used in the Shima Seki knitting machine to produce the knitted forms in Figure 2, which have been digitally programmed to follow an architectural language

2. Installed knitted structures: Objects were 4 m high with widths varying between 0.3 m and 1.2 m

ABSTRACT

The research shown here revolves around an installation at the National Gallery of Victoria, Melbourne, Australia. It illustrates a novel approach to the design of building elements by combining the expertise embedded in the disciplines of architecture and textile design into a hybrid practice. We found that by using textile technologies to 'build' architectural elements we can rethink waste in construction, expand the materials available for use in architecture, as well as minimize time involved in the post-fabrication of complex forms. This paper describes an installation that demonstrates a direct translation between architectural form and whole garment knitting machines without the need for prototyping. The world of architecture and textile design have been closely linked through time, the first shelters were a mix of textiles that were able to effectively shed water and keep occupants dry. Since these primitive structures, there has always been a romantic link between the two disciplines, but they have diverged such that techniques, materials, and scale have become incompatible with each other. This project seeks to bridge the gap between the disciplines to offer alternative ways to think about 'skinning' a building.

CONTEXT OF RESEARCH
Textile at Scale

The work shown here translates the architectural model into fabrication that uses non-standard fabrication technologies, in order to enrich the fabrication language available to architecture. While the work produced is visually important and is part of an exhibition, the driving interest is in how to scale up knitted forms such that they can be used in fabrication or building of architecture.

To give some background as to where this research is located, it is important to distinguish the work between artists working with fibers who manually or even digitally knit their structures and the work by architects who are looking to develop a way of working with non-standard architectural fabrication technologies. The former is about the visual experience of the object and does not seek to develop a technique for use in the fabrication of building elements. That is not to say that they are inspirational and have many lessons in the development of a fiber-based architecture, but they are limited to an art practice that differs greatly from an architectural one. The latter researches techniques and methodologies, for use in the fabrication of architecture, that can be used regardless of the aesthetic or conceptual work produced.

This research is closely aligned with practices that include Metta Thomsen's work for the 2018 Venice Biennale Danish Pavilion.[1] This work uses advanced industrial knitting machines to create tensile forms with the use of Dyneema® yarns that have similar strength to carbon fibre but have the capacity to bend without breaking. Dyneema® yarns are commonly used in sailmaking or in tensile structures. The second practice of relevance is Zaha Hadid's collaboration with ETH on the KnitCandela[2] prototype that uses custom knitted formwork to create complex curvature in concrete shells. Hadid and Thomsen both employ whole garment digital knitting machines as a tool for use in the creation of customized architectural membranes that demonstrate high performance capacities that can offer an alternative way of thinking about tensile structures and expand on the work of Frei Otto and tensile membrane structures. In the case of Thomsen's work, alternative yarns that have the capacity to challenge the need for steel in construction—while showing capacity for venturous forms—points to a new direction in the use of 'floppy'[3] materials in architecture. The term floppy, refers to the qualities of materials in the knitted or woven category. The work looks to harness the qualities of fabrics when developing structural support for use shelter or enclosure. Floppy, a term unfamiliar to architecture allows us to traverse territories between architecture, fashion and textile design where the latter deals with the qualities of floppiness when developing garments or textiles.

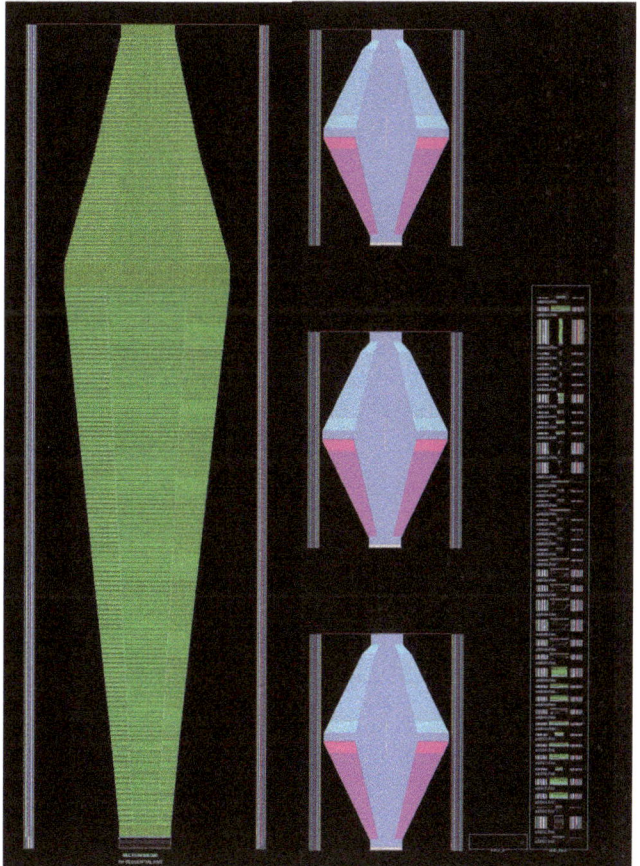

3 Large scale knitted skins coming off the 3D knitting machines

4 Digital model as understood by the 3D knitting machine

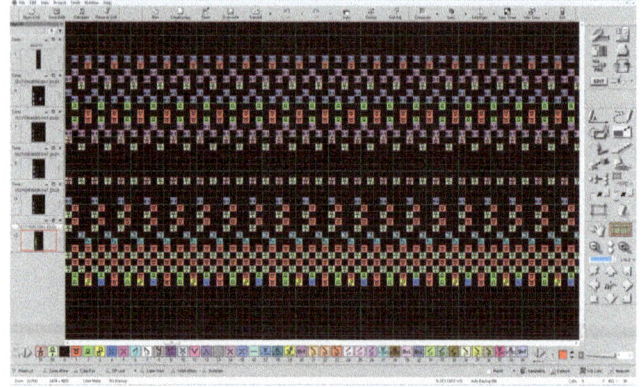

5 Knitted code showing a closer look at the knitted language of loop, drop stitch, fiber changes

6 Knitted form as it comes out of the Shima Seki whole garment knitting machine

7 Knitted from fitting over structure with limited sag, ready for ribs to articulate the form

8 Installed complex knitted forms

Expanding from Hadid's and Thomsen's work, the customized knitted structures shown here, demonstrate how we can add digital knitting machines as a tool for architecture that expands the capacity of buildings to harness advanced fiber technologies to better address climate, environment, and a buildings responsiveness. Looking at how knitted manufacturing techniques can help the production of complex curvature (Figures 1, 2) in the design and fabrication of architectural elements, our project adds to the development of a symbiotic relationship between the disciplines, where architecture can learn from the mastery of form seen in fashion and textile design, and fashion and textile design can benefit from the simulations possible via the digital model common in architecture. We are working towards a practice that directly translates from digital model to 3D knitted elements, such that little or no postproduction is required.

WHOLE GARMENT KNITTING AS A TOOL FOR ARCHITECTURE

Whole garment knitting machines can produce multiple garments with a range of fibers and patterns quickly and efficiently. Like most industrial machinery, they are designed for large scale production with minimal waste or need for manual labor. The set templates allow for some customization but are limited to the basic forms of sweater, pants, socks, vests, shorts, etc. In order to adapt for use in architecture, these templates are used as the basis of complex form creation through the careful mapping between the digitally generated 3D models in architecture and the language of the whole garment knitting programs. Within this translation there is an understanding of the behavior of knitted structures, which is vastly different from wovens. Whole garment knitting machines work on a pixel unit of measurement that does not relate to scale or metric/imperial units (Figures 3-6). Shapes are made through the dropping or picking up of stiches within the 1200 mm width of the knitting bed that, on the other hand, has unlimited length capacity of the knitted object. This 'shape making' is quite different from the architectural digital model, which describes lines and curves through metric or imperial units and angles.

There are several challenges in translating the architectural form to the language of the knitting machines. The first challenge includes working out a way to give the pixel a dimension such that direct translation from the 3D digital model can be seamlessly translated into a knitted object. Architectural models are visualized at a scale, but the language of textile design is scale-less; there is no incremental scale development of a garment, instead they are sampled directly at 1:1. This means the architectural scale needs to be understood prior to inserting this data into the whole garment knitting machines.

A second challenge is the issue of tension, which is an important consideration when working with knitted elements. The looped stitches naturally create stretch and needs to be controlled to avoid unwanted floppiness or looseness in the final knitted skin such that it can be precisely placed over a structure without the need for adjustment. Knitted structures look like they are just stretched over a structure but they need to be carefully mapped onto a form to ensure that an accurate fit is made. By working out a pixel-based script that has built-in restrictions conforming to the template parts of the whole garment knitting machine, we have translated an architectural model into the pixel language of the knitting machine, producing a complex form that is not a multitude of parts, but one that is ready to fix to a support structure with little post-production.

CONCLUSION

This project demonstrates the successful translation between the 3D digital architectural model and its fabrication using the whole garment knitting machines to produce architectural scaled elements containing complex forms. The benefits of this hybrid project taps into the efficiencies in the textile fabrication industry and explores the possibility of producing complex forms without the need for prototyping. The possibilities of pursuing a textile-based architecture means that buildings will be able to tap into the advanced fibers offered in textile design such as strong Dyneema® yarns, solar harvesting fibers, and other responsive fibres that can help address issues around climate.

When reflecting on the completed work shown in this paper we now know that complex curature can be coded to bridge between the architectural 3D model and the pixel based code of the whole garment knitting machine. However we need to work on how to create surfaces that adapt to the scale of a building but work within the 1200 mm width of the knitting machines. In addition to this the application of this knowledge to real problems facing our built environment is a key next step in the success of adding whole garment knitting technologies to the fabrication tools available to architecture.

ACKNOWLEDGMENTS

The research shown here is possible from the support of RMIT University, School of Architecture and Urban Design and School of Fashion and Textile Design, National Gallery of Victoria.

NOTES

1. Riccardo La Magna, Valia Fragkia, Philipp Langst, Julian Lienhard, Rune Noel, Yuliya Šinke Baranovskaya, Martin Tamke, Mette Ramsgaard Thomsen, "Isoropia: an Encompassing Approach for the Design, Analysis and Form-Finding of Bending-Active Textile Hybrids," *IASS Symposium 2018*, edited by Caitlin Mueller, Sigrid Adriaenssens (Boston, USA, 2018).

2. Mariana Popescu, Mattias Rippmann, Tom Van Mele, and Philippe Block, "Knitcandela: Challenging The Construction, Logistics, Waste and Economy of Concrete-Shell Formworks," *Fabricate 2020: Making Resilient Architecture*, edited by Jane Burry, Jenny Sabin, Bob Sheil, and Marilena Skavara (London: UCL Press, 2020).

3. Term referenced from Leanne Zilka, *Floppy Logic: Experimenting in the Territory Between Architecture, Fashion and Textiles* (Barcelona: Actar Publishers, 2020).

IMAGE CREDITS

Figure 1: ©Jenny Underwood, 2022
Figure 2, 6, 7, 8: ©Peter Bennets, 2022
Figure 3: ©Leanne Zilka, 2022
Figure 4, 5: ©Jenny Underwood, 2022

Dr. Leanne Zilka is a registered architect and academic based in Melbourne, Australia. Her architecture practice, ZILKA Studio and her academic position at RMIT University in the School of Architecture and Urban Design is a multidisciplinary one that brings together architecture, fashion, textile design, and material research and develops new ways to improve the built environment. By looking at materials not familiar to architecture, we can harness material and technology advances that can be used in buildings and cities to improve performance, function, and capacity that address pressing issues around climate change. Leanne's practice looks to land new technologies and materials in novel architectural propositions working in the digital fabrication realm, hacking into fashion fabrication technologies, such as whole garment knitting machines, to create architecturally scaled elements.

Currently investigating soft facades for the retrofit of multistory buildings, Leanne is working on prototypes that incorporate advanced fibers such as Dyneema®, solar fibers to harvest energy, irrigation tubing to support plant life and other external use yarns for use in customized knitted surfaces.

Leanne's research and practice has been recognized nationally and internationally through awards, a recently published book titled *Floppy Logic* by international publisher Actar, featured in the 2018 and 2021 Venice Architecture Biennale, 2019 Tallin Architecture Biennale, and most recently was part of the 2021 National Gallery of Victoria "Sampling The Future" exhibition.

Dr. Jenny Underwood is Associate Professor and Associate Dean of Fashion and Textiles Technology at the RMIT University School of Fashion and Textiles. In this role, she provides strategic and academic leadership for the fashion and textiles technology discipline and has overseen the creation of the Bachelor of Fashion and Textiles (Sustainable Innovation) degree.

Jenny's research is practice-based and interdisciplinary at the nexus of design, art, technology, and science. She is recognized internationally for her interdisciplinarity and as the leading expert in 3D seamless knitting. Her textile design practice brings together understandings of material behavior, parametric design approaches, digital fabrication, and physical prototyping within the context of architecture, speculative and critical design, and fashion. What emerges are material-led design strategies to create complex forms that are lightweight, flexible, and efficient at scale.

Jenny also researches and writes on fashion and textiles education for sustainability. Her current focus is on how relational connections between disciplines centered on place, materials, and the United Nation's Sustainable Development Goals that can support transformative learning experiences. She is also an experienced senior PhD supervisor, having supervised some of the first practice-based research PhDs within the field of fashion and textile design internationally.

BioKnit

Armand Agraviador
Newcastle University

Jane Scott
Newcastle University

Romy Kaiser
Newcastle University

Elise Elsacker
Newcastle University

Aileen Hoenerloh
Newcastle University

Ahmet Topcu
Newcastle University

Ben Bridgens
Newcastle University

Computation and material investigation in the design of biohybrid textiles towards architectural integration

ABSTRACT

This research paper evaluates a hybridized computational and material-based approach to form-finding in a biohybrid structure. It reports on the fabrication of *BioKnit*, a system that integrates knitted fabric, mycelium composite, and bacterial cellulose panels in the fabrication of a self-supporting installation that stands 1.8 m high with a 2.0 m diameter base. The form-finding opportunity of designing for growing materials that transition from soft to hard was explored using catenary geometry. This approach enabled the optimization of form during the fluid growth phase to produce an efficient free-standing structure in a final solidified state.

The paper discusses how catenary geometry was used to define parameters for knitting and mycelium, and how they were applied to the design of a 3D knit preform. In addition, the paper evaluates the success of the bespoke growth chamber fabricated for this research. The growth chamber was designed to support the hanging preform as a catenary vault during growing and to optimize mycelium growth via environmental controls. Findings of the research highlight the significance of computational methods to enable the design and construction of biohybrid textile systems that move from an assimilation of discrete material elements with defined boundaries to a cohesive technological approach.

1 *BioKnit* is a free-standing biohybrid knitted structure composed of fungal mycelium, bacterial cellulose, wool, and linen

INTRODUCTION

As researchers look to biohybrid strategies in the search for sustainable construction approaches, the possibility of breaking away from conventional methods to novel materials and processes is becoming increasingly apparent. The familiar vocabulary of construction can expand beyond the linear assemblages of blocks and frames that dominate our built environment and encompass the structures and systems more associated with the natural world.

This research paper evaluates a hybridized computational and material-led approach to form-finding in a biohybrid structure. The biohybrid system comprised mycelium composite integrated into a tubular knit fabric to create a homogenously grown architecture greater than the sum of its constituent materials. The overall knit system acts as both a permanent formwork and a textile surface that is assimilative rather than sacrificial (Figure 1). The soft components were transformed into an integrated architecture using a bespoke growth chamber designed to obtain catenary geometry via suspension of the form.

Through discussion of the design and fabrication of *BioKnit*, the following research questions are addressed:

- How can computational design coordinate with material investigation in the development of new biohybrid textile systems?
- How can the design and construction of biohybrid textile systems move from an assimilation of discrete material elements with defined boundaries to a cohesive technological approach?

Background

The drive to net zero necessitates both a transition to more sustainable construction practices and the development of alternative processes to replace the carbon intensive material systems currently in use. Biohybrid systems present an opportunity to rethink not only the way our buildings are made, transitioning from construction to growth (Dade-Robertson 2020), but also the form and materiality of our architectural spaces and surfaces. There is an opportunity to move towards structurally optimized geometries established by pioneers such as Gaudi (Heurta 2004), and this research seeks to redefine a catenary logic for materials that transition in state from soft and flexible to rigid and stiff, through the process of growth.

BioHybrid Material Systems

Biohybrids combine biological organisms alongside other materials such as polymers or ceramics (Gao and Maruyama 2014). Mycelium, the root network of fungi, has huge potential

2 Detail of mycelium/knit biohybrid: mycelium binds substrate together with the knitted fabric as it grows

3 Manual finishing of the *BioKnit* preform; here the biohybrid is in a soft state prior to growth

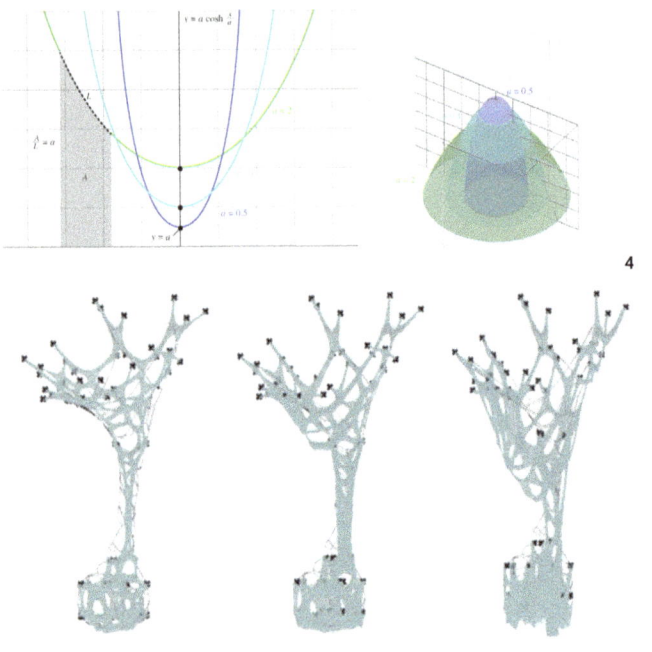
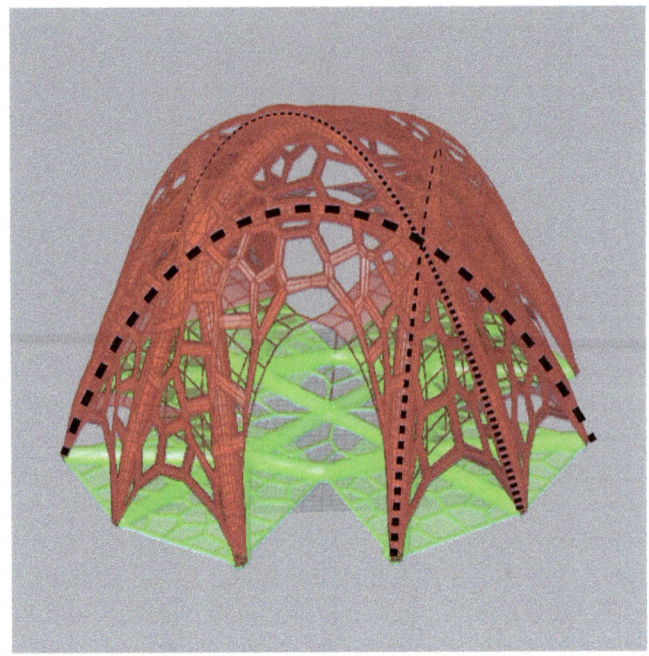

to act as a binder in the creation of bulk composite materials. Precedents have established scale (Benjamin, 2017), formability (Heisel et al. 2017), and compatibility with other material systems including textiles (Yogiman et al. 2020). Whilst mycelium can provide compressive strength (Lelivelt et al. 2015), the use of mycelium as part of a biohybrid system offers the potential to expand the functionality and applications for this biomaterial (Hoenerloh et al. 2022). Bacterial Cellulose (BC) is a flexible sheet material grown through a process of fermentation. It can be grown either as a pure culture or through a Symbiotic Culture of Bacteria and Yeast (known as a SCOBY). When dried, this biomaterial transforms into a leather-like translucent skin, with textile qualities that have predominantly been exploited in clothing and product design (Derme et al. 2016; Nguyen et al. 2018). The compatibility of different biomaterials growing together is challenging because of the different environmental conditions required, and the risk of contamination during growth (Hoenerloh et al. 2022); however, one method to integrate mycelium and BC at an architectural scale is to use a textile scaffold to guide the growth of mycelium and provide surface texture for adhesion of BC (Scott et al. 2022).

Knit as a Biohybrid Forming Agent

Knitted fabric is ideally suited for use in biohybrid systems because variable properties can be incorporated within a continuous material (Scott 2013). The sophistication of industrial knitting technology enables fabrics to be programmed as shaped panels, 3D tubes, or in preforms that combine both. At an architectural scale, knit has been integrated within composites as permanent formwork (Popescu 2018) and used as flexible, self-shaping molds (Ng et al. 2020). The adjustability of fabric parameters during design and fabrication improves the mechanical performance of mycelium by combining the tensile strength of knitted fabric with the compressive strength of mycelium composite to create a stronger, more robust composite material (Yogiman et al. 2020; Scott et al. 2022).

More fundamentally the integration of knit tectonics into architectural discourse has transformed the understanding of materials and materiality in architecture (Ramsgaard Thomsen and Bech 2012; Sabin et al. 2018; Ahlquist 2016). During knitting, the properties of a yarn are transformed through the knitting process to produce different textures and structures, and this directly impacts performance criteria such as tensile strength and extensibility. Knit programming is specified on a stitch-by-stitch basis therefore specific properties can be localized for material efficiency. This transformative process is a powerful conceptual tool for the development of biohybrid systems where material performance is transformed through the growth of a microorganism, and the biohybrid systems evolves through multiple states of flexible and rigid, wet and dry, and living and non-living (Figure 3).

Catenary Geometry

The catenary is defined as the curve that a slack length of idealized uniform chain or cable assumes under its own weight when held at its two end points (Gregory 1676; Hooke 1676). Inverted catenary curves have been employed in arches, vaults, and domes throughout architectural history due to their unique structural property of carrying their own weight, and

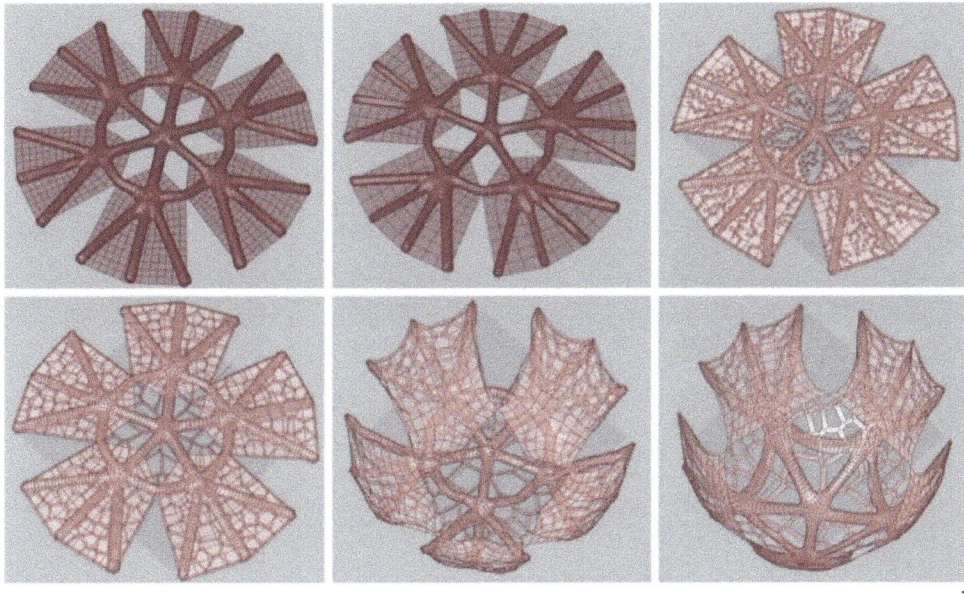

4 Catenary functions on Cartesian coordinates and inverses revolved around the y-axis

5 Example of partially suspended forms to occupy a vertical space, with sagging and tension observed for different initial parameters that affect the final form

6 Comparative initial (green) and final (red) geometries in which a resultant intersecting arch structure was observed (dashed lines)

7 Top: variations in topological nets with webbed flat panel, pre-simulation: (left to right) 6:3 branching, 5:4 branching, 5:3 branching with shortest-path vein substructure; Bottom (left to right): 5:3 branching with Voronoi vein substructure, early mid-stage of simulation, simulation approaching convergence

any uniformly distributed loading, in pure compression with no bending (Lluis et al. 2017). The catenary is described as a hyperbolic cosine function $y = a \cosh x / a$ (Figure 4), where both gravity and tension scale by the same factor assuming an inextensible chain. This means that the curvature is determined only by span and length.

Catenary curves are a function of our physical world; they manifest throughout it, beyond hanging webs and strings, in the section of a minimal surface of revolution taken by stretched films as well as some geometric oddities, such as the sectional path of a rolling polygon (Gil 2005). With traditional construction materials such as stone or brick, catenary curvature can only be achieved by setting up curved supporting formwork in a time-consuming process. For materials that are initially soft yet self-supporting, there is the possibility to achieve catenary curvature through suspension and later inversion. This process has recently been explored in float glass (Pawloswska 2020), and the opportunity to use catenary curves in BC development has also been identified (Turhan et al. 2021). Our research applies the logic to biocomposite construction to hybridize a living behavior and natural properties with a structurally optimal form to propose a self-organizing alternative approach to construction.

METHODS

The interdisciplinary nature of this research required multiple streams of research incorporating computation, biomaterial development (both mycelium substrate and bacterial cellulose), and knit design and technology. This paper focuses on the computational design strategies and how they were employed to facilitate design, fabrication, and assembly.

Underlying Design Requirements

Prior to modeling, the fundamental knitting requirements to produce a biohybrid knit scaffold for both mycelium substrate and bacterial cellulose were identified. These included:

- A tubular knitted fabric component to contain the mycelium substrate and create compressive arches within a vaulted form
- An open, textured fabric panel component for the bacterial cellulose adhesion
- A knitted preform composed of individual modules with minimal seaming points to reduce post-knitting assembly and the disruption to fabric performance generated by seams

Additional parameters (stitch density, knitting width) were controlled by the specification of the knitting machine (Shima Seiki SSR122). The stitch density is determined primarily by the machine gauge (gg), which specifies the number of needles per inch on the machine. Here a fine gauge fabric was knitted by a 12 gg machine. The maximum knitting width is determined by the width of the needle bed. The SSR122 has a needle bed width of 122 cm; however, the length is unlimited.

Modules were knitted from 2/30 nm merino wool and 1/30 nm linen. To improve fabric strength, yarns were doubled up for knitting, meaning that each feeder contained two ends of yarn. This changed the stitch density; with two ends achieving

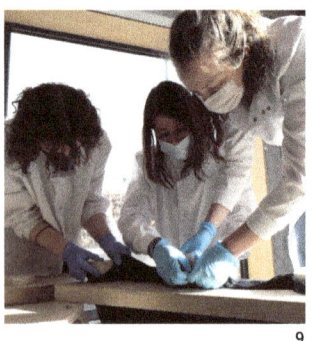
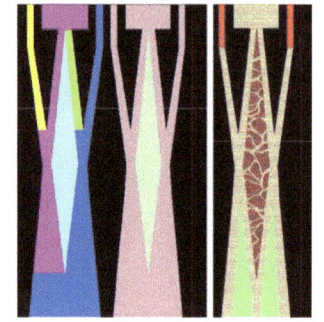

8 Knit Preform: testing layout of modules prior to assembly
9 Filling and construction whilst trying to maintain sterile conditions
10 Knit Preform Programme Data illustrating feeder allocation (left), yarn data (center), and structure (right)

a stitch density of 56 stitches/cm², compared to 48 stitches/cm², for a single end of yarn. Machine settings were adapted accommodate this, with the stitch length set extremely high at 70 (conventional knitting 25-45), and the take-down increased to help control the stitches as they were formed on the machine. Knitting with 2 ends of yarn had minimal impact on extensibility but increased the tensile strength of the fabric (Table 1).

Form-Finding

The initial challenge was to design a structural system that would contain and give form to the mycelium substrate despite the extensibility of a knitted fabric. In addition, the fluctuation in weight between the saturated, freshly inoculated mycelium substrate injected into the fabric, and final dehydrated composite material also posed a challenge for modeling.

Preliminary design studies explored predetermined starting geometries including parametrically defined geometries using Grasshopper (Rutten 2022). Using the Kangaroo2 plugin (Piker 2017), physical simulation was applied to these as mesh objects to observe their resting positions if made from materials of varying extensibilities (Figure 5).

However, the opportunity to work with the transition from flexible to stiff via growth enabled the design team to optimize the structure of the biohybrid using catenary logic. Subsequent form-finding studies employed the catenary curve as a hanging form, allowing gravity to define the resting geometry of a soft system with the intention of inverting it once it had undergone the transition from soft to stiff. This solution utilizes the soft material phase, rather than merely accommodating it as a challenge or limitation.

Simulations for Modular Topologies

To simplify the fabrication process, a branching topology was explored to set out a rotationally symmetrical tubular structure that could be divided into workable modules. The lines in this topology were intersected to form a 2D net, given variable thicknesses and joined to create a mesh object. It was to this geometry that physics simulations were tested using Kangaroo2 components for Grasshopper (Figure 7). Prior to being computed by a solver, the input parameters were translated into several simulation goals including:

- Anchor: Outermost vertical mesh points are kept in place to prevent translation of the entire system (i.e. moving vertically while maintaining starting shape)
- Load: Applied to all mesh points in a negative z vector to simulate gravity when suspended
- Length: Applied to all mesh edges with a target length

TABLE 1
Comparison of physical properties of 1 end versus 2 ends of 2/30 merino knitting plain knit on 12 gg Shima Seiki SSR122

2/30 nm merino	Stitch Setting	Tensile Strength (MPa)	Extension at break (%)	Stitch density (stitches/cm²)
1 end	26	0.76	67.32	48
2 ends	70	1.61	71.85	56

11 Growth chamber; once fabrication was complete the chamber was wrapped in thermal foil and humidity was maintained for the duration of the growth period

defined as a maximum factor of original length to emulate uniform extensibility with a strength input to control the effect of load
- Pressure: Applied to the mesh surface as an outward (normal) force vector to each face to maintain volume and emulate an internal filling of the tubes

This branching topology evolved into one that employed a net of intersecting arcs. This incorporated the arches produced by the intersecting tubes into the outset of the design with more structural intent. It also created an oculus at the apex of the dome instead of a junction, which solved potential issues of substrate concentration and liquid pooling at the suspension stage.

In the previous single-stage simulations, the flat lines of the base topology were piped into a tubular net to be hung with an exaggerated extensibility to allow enough slack for a desired catenary curve. A two-stage simulation process was employed (Figure 6). A first stage works exclusively on the tube paths to define the lengths and joint locations to be manufactured, while a second approximates the real-world behavior of such a structure only once the tube diameters are defined. The dimensions of the paths after the first simulation stage are used to input into programming for the knit machine. Meanwhile the volumes reported after the second stage identified how much substrate would be required to fill the final geometry with mycelium.

Material performance in physical knit analogues, such as extensibility and curvature in suspension, were tested in parallel to these simulations in order to compare with the digital output, creating a hybridized approach to observing behaviors in a structural system.

Knit Programming

The aim of knit programming was to integrate all structural requirements into a shaped preform that could be produced from a repeating module and assembled into a cohesive form ready for inoculation with mycelium. The underlying module incorporated shaped tubes of fabric intersected with panels of mesh (Figure 10). Programming was developed using the Shima Seiki Apex3 system. Within the programming interface the stitch structure, yarn specification, and color/material placement are specified within a single program. During pattern development the program is translated into knitting instructions suitable for the SSR122 machine.

RESULTS

The results focus on the assembly process and the growth stage with specific reference to the knitted modules and growth chamber design. The discussion evaluates how discrete material elements were assimilated into a cohesive technological approach and highlights the role of computation to predict, adapt, and respond to the requirements of a growing structure.

Knitted Preform Inoculation and Contamination Control

The prototype was composed of seven knitted modules. Each module measured 2 m x 1.2 m (widest point) to form a 7-sided polygon when laid flat (Figure 8). Critically the tubular parts of

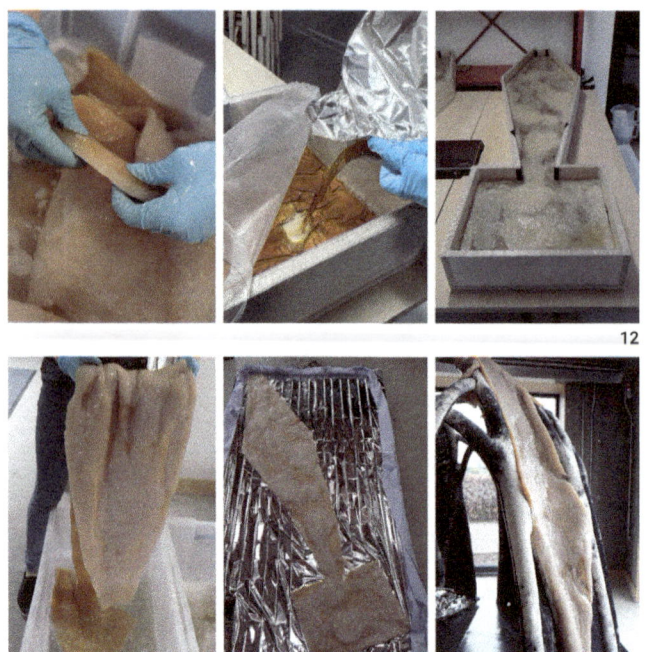

the structure extended beyond the shaped panel at each side and horizontally at the top. These tubes (4 on each module, 28 in total) created the seaming points to join the preform together. The modules were sewn together using polyester thread, leaving gaps in the seam at regular intervals to provide filling points for the mycelium substrate. This was important to maintain consistent, tightly packed mycelium substrate throughout the structure.

The mycelium substrate was specifically designed as a viscous paste for use with textiles to enhance the consistency and improve workability in soft molds. The mycelium was combined with a carefully proportioned mix of nutrients to create a substrate that is fast growing and achieved a dense and cohesive mycelial network around the fibers and the textile. Humidified beechwood sawdust fibers were inoculated with mycelium spawn in a sterile environment and pregrown for ten days at 28°C. A growth strategy that allowed the mycelium to colonize the substrate homogenously prior to the unsterile environment of the exhibition space. The paste was prepared in-situ by adding gelling agents and water, which increased the weight of the mixture considerably. For the mycelium to grow on the textile, all knitted modules were saturated with water and sterilized. A weight of approximately 200 kg was calculated as the total hanging load, including the mycelium paste and mid knit.

Growth Chamber with Integrated Hanging System

To produce a catenary vault, the filled preform was suspended throughout the 13-day growing period. During growth the mycelium hyphae grow through the substrate mix and knitted fabric, consuming nutrients and binding the materials together. To optimize the growth of mycelium composites, careful control of environmental conditions are required. For mycelium the principal controls are a temperature between 21-28°C, a relative humidity of 50-60%, and no light. These conditions are conventionally maintained under lab conditions; for a large-scale structure the design of the growth chamber was vital for successful growing.

The design approach selected for *BioKnit* brought together the environmental controls required for optimized growing with a structural framework that could support the 200 kg weight of the knit/mycelium preform as it grew. A bespoke framework was constructed from structural timber with holes spaced around the circumference of the top frame. The ends of the knit/mycelium tubes that would become the catenary arches of the prototype were secured within these holes (Figure 9).

The growth chamber was erected after the filled preform was attached to it (Figure 11). Whilst this production stage was potentially challenging due to the weight of the filled preform, the flexibility of the fabric acting as a soft mold enabled the preform to be manipulated into place, and the top level of the growth chamber lifted into place.

Growing Conditions

For the growing stage, the chamber was covered in layers of thermal reflective foil (Figure 10), which helped maintain a constant temperature of around 21°C and provided darkness. Humidity was actively increased to approximately 50% using a humidifier placed within the growth chamber. After 13 days the foil was removed, humidity reduced, and the structure was

12 Process of growing the bacterial cellulose skin in customized molds

13 Washing bacterial cellulose skin and placing on the mycelium-knit structure to dry

14 Detail illustrating the dried bacterial cellulose skin

15 *BioKnit* on exhibition in the OME, Newcatle University; in its final state the structure is freestanding at a height of 1.8 m

left to air dry for 22 days. After that time the structure had become stiff and solid. Although the mycelium was not heat-killed, the dehydrated fungal cells are unlikely to revive and continue growing, unless they would be humidified at 80-90% relative humidity.

Flipping the Vault

The final stage of fabrication was to rotate the structure 180° to form the catenary vault. Whilst there was concern that during rotation the structure would not be able to withstand the bending forces acting on the mycelium tubes, the combination of the compressive strength of the mycelium and the tensile strength of the knitted fabric meant that there were no problems during rotation. At this stage the prototype was entirely self-supporting (Figure 15).

Bacterial Cellulose Skin

As a secondary process, bacterial cellulose panels, grown to shape (200 cm x 50 cm) and weighing 14 kg (hydrated state), were applied to the surface of the self-supporting structure to form a skin over two of the modules (Figure 12 and 13). After application the bacterial cellulose dehydrated over five days (final thickness 2 mm) adhering to the textile where it contained mycelium composite but resisting adhesion to fabric surface alone (Figure 14).

DISCUSSION

The discussion evaluates how discrete material elements were assimilated into a cohesive technological approach and highlights the role of computation to predict, adapt, and respond to the requirements of a growing structure.

Knitted Preform

The knit structure was devised to act as both support and containment of the mycelium substrate while maximizing the oxygen interface to aid in growth. The fabric also operated as an adhesive interface for a cellulose skin, as well as providing a textured surface to the vaulted form.

The knitted modules performed extremely well generating the anticipated catenary vault, containing the mycelium substrate, and supporting excellent mycelium growth. This yarn also provided a good adhesive surface for bacterial cellulose. Optimizing the fabric by doubling the amount of yarn within the structure resulted in a knitted fabric that retained extensibility and enhanced tensile strength, both essential for structural performance. This could have been achieved by knitting on an alternative gauge of machine and further research will explore the role of stitch density and loop size on preform performance.

The dome-like catenary vault was chosen for the *BioKnit* prototype as the optimal structural form to utilize the phase change and take advantage of the compressive strength of the mycelium. It minimized bending in the slender columns and addressed the soft to hard state change without needing to create formwork. Parametric modeling was vital to determine the geometry and in particular the required fabric dimensions and volume of substrate.

Table 2 compares the effect of key parameters on the output geometry from the two-stage simulation and approximates the closest corresponding catenary arc (Figure 16).

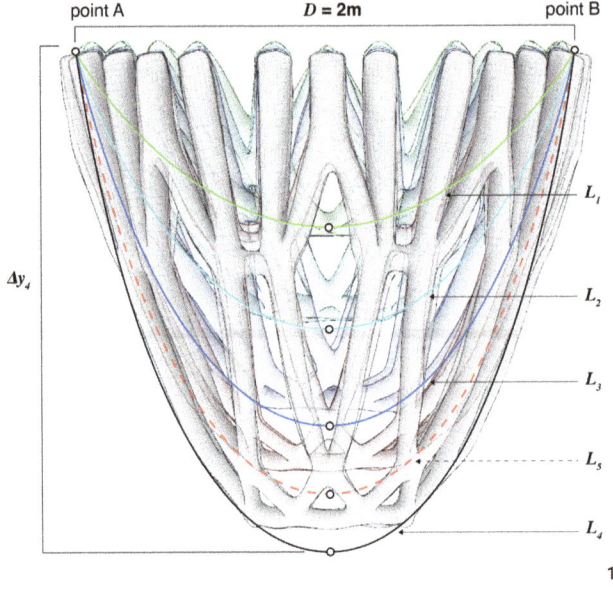

16

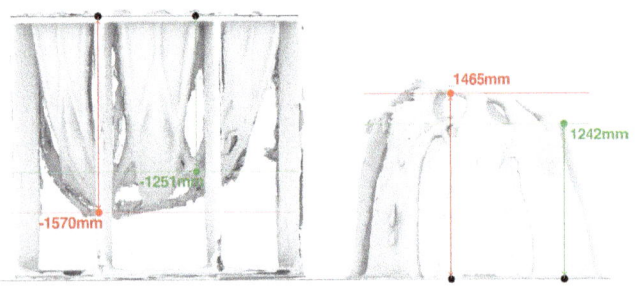

17

16 Model illustrating two-stage simulation with key input parameters for Grasshopper script and approximate corresponding catenary equation with elevation illustration

17 3D scan images and measurements before and after maturation

TABLE 2.
Table illustrating two-stage simulation key parameters for Grasshopper script and approximate corresponding catenary equation

		1	2	3	4	5
Script parameters	Path sink factor (sim 1)	1	20	50	100	100
	Extensibility factor (sim 2)	1.1	1.1	1.1	1.1	1.0
	Fill Volume* (m3)	0.21	0.27	0.32	0.37	0.30
Catenary equation	Arc length* L (m)	2.58	3.00	3.73	4.61	4.20
	Sag distance* Δy (m)	0.67	1.07	1.44	1.93	1.70

Notes: *Output parameters determined by path sink and extensibility factors (as well as unlisted constants); Diameter D set as a constant 2 m; Textile assumed to extend by max. 10% except in test 5 (red dashed) where inextensibility is tested

It should be noted that while the simulated morphologies closely approximate corresponding catenary curves, the branching topology of the structure means that each joint acts as a node and endpoint for an individual branch's catenary system. The overall aggregated effect is attributed to the pathfinding in the first simulation step; if the tubes were not fabricated to the lengths reported at this stage, the final vault would not have followed catenary curvature, even if its constituent arches had. Adding to this, the simulation assumes a uniform distribution of load which is difficult to achieve in injected substrate, which accounts for the dissymmetry in the prototype.

The catenary modeling can be compared to 3D scans of the structure during hanging (wet) and after flipping (dry) to illustrate the dimensional change that occurred (Figure 17). As the structure grew and consolidated the resultant height at the highest arch was reduced from 1570 mm to 1465 mm a reduction of 95 mm. Interestingly the height change measured at the lower arch from 1251 mm to 1242 mm indicates only a reduction of 9 mm. This could be explained by an uneven distribution of load at the highest point during hanging, or the tolerances and accepted variation in stitch length in the knitted preforms and differential packing of the substrate during assembly.

Weight Versus Efficiency in a Growing Structure

The *BioKnit* prototype weighed 200 kg during growth stage. As the structure air dried it dehydrated, and the final weight is estimated at 50 kg accounting for water loss. This transition from wet to dry was anticipated in the design stage and the catenary design modeled to anticipate the increased sag from the wet mycelium suspended in the soft structure. The weight of the wet mycelium did require a substantial timber frame for the growth chamber (Figure 10). This is potentially wasteful of materials; however, the growth chamber was designed to be disassembled such that the timber and fixings could be reused for other projects. Further work is required to engineer an alternative structural growth chamber that can accommodate the changing load of the preform without utilizing excessive materials in its construction.

CONCLUSION

The outcome of this research, the *BioKnit* prototype, could be viewed as the embodiment of a textile's lightweight formability, transformed into a self-supporting structure by the rigidity of mycelium composite and the structural efficiency of catenary geometry. Rather than growing small mycelium components in lab conditions and then joining them to make a larger structure, a key outcome of *BioKnit* is to demonstrate the challenges and benefits of growing a continuous mycelium structure. Whilst addressing challenges of large-scale

18 Detail illustrating mycelium/knit composite; the pattern of stitches is still visible in places along the knitted tube

fabrication and contamination, the benefit is a highly efficient, continuous structure with no connections, which tend to be weak points in the structure and/or require complex fabrication.

Whilst the successful production of the prototype does constitute a key finding of the research, the significance of the work also lies in the ability for computation to negotiate material synthesis rather than specify a predefined outcome and in this way reflects a logic of textile making. To coordinate with material investigation, computational design was used to specify conditions for growth rather than specify the precise dimensions of an outcome. Whilst the precise dimensions were not specified, the fabrication method (suspended in the growth chamber) ensured that a catenary dome was achieved without the need for complex curved formwork, and the result was a self-supporting structure with remarkably slender columns.

In this research computational design provides tools to encapsulate the interfaces of hybrid elements within primitive geometries to which proxies for real-world interactions are applied. Through programmed form-finding and physics simulation, any required material variation or modularity in the installation can be realized as part of a wider whole. The intention is that the resultant structure can be observed in the same abstraction as the technology it presents. While visual programming allows initially seamless form-finding to be modularized for fabrication and reconnected on installation, the aesthetic homogeneity that develops as the mycelium hyphae propagate gradually reflects those continuous digital forms. This research presents a precedent for a biohybrid system that has extended beyond the assimilation of discrete materials elements to a system that embodies the behavior of living organisms during growth towards the realization of a cohesive biotechnological approach.

ACKNOWLEDGMENTS

This research is funded by Research England's Expanding Excellence in England (E3) Fund as part of the Hub for Biotechnology in the Built Environment (HBBE).

REFERENCES

Ahlquist, Sean. 2016. "Sensory Material Architectures: Concepts and methodologies for spatial tectonics and tactile responsivity in knitted textile hybrid structures." *IJAC* 14 (1): 63–82.

Benjamin, David. 2017. "Living Matter." In *Active Matter*, edited by. S. Tibbits. Cambridge, Mass.: The MIT Press.

Dade-Robertson, Martyn. 2020. *Living Construction*. London: Routledge.

Derme, Tiziano, Daniela Mitterberger, and Umberto Di Tanna. 2016. "Growth based fabrication techniques for bacterial cellulose." In *ACADIA 16: Posthuman Frontiers; Data, Designers, and Cognitive Machines; Proceedings of the 36th Annual Conference of the Association for Computer Aided Design in Architecture (ACADIA)*, 488–495.

Gao Jueyuan, and Atsushi Maruyama. 2014. "Biohybrid Materials." In *Encyclopedia of Polymeric Nanomaterials*, edited by S. Kobayashi and K. Müllen. Berlin: Springer.

Gil, Juan. 2005. "The catenary (almost) everywhere." *Boletın de la Asociación Matemática Venezolana* XII (2): 251.

Gregory, David. 1676. "Catenaria." *Philosophical Transactions of the Royal Society* 19 (231): 637–52. https://doi.org/10.1098/rstl.1695.0114.

Heisel, Felix et al. 2017. "Design, Cultivation and Application of Load-Bearing Mycelium Components: The MycoTree at the 2017 Seoul Biennale of Architecture and Urbanism." *International Journal of Sustainable Energy Development* 6 (1): 296–303.

Huerta, Santiago. 2004. "Structural design in the work of Gaudi." *Architectural Science Review* 49 (4): 324–39.

Hoenerloh, Aileen, Dilan Ozkan, and Jane Scott. 2022. "Multi-Organism Composites: Combined Growth Potential of Mycelium and Bacterial Cellulose." *Biomimetics* 7 (2): 55.

Hooke, Robert. 1676. *A Description of Helioscopes and Some Other Instruments Made by Robert Hooke, Fellow of the Royal Society*. London: John Martyn.

Lelivelt, Rjj Robert, Gerald Lindner, Patrick Teuffel, and Hans Lamers. 2015. "The production process and compressive strength of Mycelium-based materials." In *First International Conference on Bio-based Building Materials*, 1–6. Clermont-Ferrand, France.

Lluis I Ginovart, Josep, Sergio Coll-Pla, Agustí Costa-Jover, and Mónica López Piquer. 2017. "Hooke's Chain Theory and the Construction of Catenary Arches in Spain." *International Journal of Architectural Heritage* 11 (5): 703–716.

Nguyen, Peter, Noémie-Manuelle Dorval Courchesne, Anna Duraj-Thatte, Pichet Praveschotinunt, and Neel S. 2018. "Engineered Living Materials: Prospects and Challenges for Using Biological Systems to Direct the Assembly of Smart Materials." *Advanced Materials* 30 (19): 1–34.

Pawlowska, Gosia. 2020. "Viscous Catenary." In *ACADIA 20: Distributed Proximities; Proceedings of the 40th Annual Conference of the Association for Computer Aided Design in Architecture (ACADIA)*, 170–175.

Piker, Daniel. *Kangaroo*. V2.42. McNeel. Mac. 2017.

Popescu, Maria, L. Reiter, A. Liew, T. Van Mele, R. J. Flatt, and P. Block. 2018. "Building in Concrete with an Ultra-lightweight Knitted Stay-in-place Formwork: Prototype of a Concrete Shell Bridge." *Structures* 14 (March): 322–332.

Ramsgaard Thomsen, Mette, and Karin Bech. 2012. "Suggesting the Unstable: A Textile Architecture." *Textile, The Journal of Cloth and Culture* 10 (3): 276–289.

Rutten, David. *Grasshopper*. V1.0 McNeel. Mac. 2022.

Scott, Jane. 2013. "Hierarchy in knitted forms: Environmentally responsive textiles for architecture." In *ACADIA 2013: Adaptive Architecture; Proceedings of the 33rd Annual Conference of the Association for Computer Aided Design in Architecture (ACADIA)*, 361–366. Waterloo, Canada.

Scott, Jane et al. 2022. "Knitted cultivation: Textiling a Multi-Kingdom Bio Architecture." In *Proceedings of ICSA 2022 5th International Conference on Structures and Architecture*. Aalborg, Denmark. 2–10.

Turhan, Gözde Damla, Varinlioglu Guzden, and Bengisu Murat, 2021. "An Integrated Structural Optimization Method for Bacterial Cellulose-Based Composite Biofilms." In *Towards a New, Configurable Architecture; Proceedings of the 39th eCAADe Conference*, vol. 1, edited by V. Stojakovic and B. Tepavcevic. Novi Sad, Serbia. 115–120.

Ng, Tsz Yan, Sean Ahlquist, Evgueni Filipov, and Tracey Weisman. 2020. "Active-Casting; Functionally Graded Knits for Volumetric Casting." In *ACADIA 20: Distributed Proximities; Proceedings of the 40th Annual Conference of the Association for Computer Aided Design in Architecture (ACADIA)*, edited by M. Yablonina et al. 546–555.

Yogiaman, Christine. et al. 2020. "Knitted Bio-Material Assembly." In *ACADIA 20: Distributed Proximities; Proceedings of the 40th Annual Conference of the Association for Computer Aided Design in Architecture (ACADIA)*, edited by M. Yablonina et al. 58–66.

IMAGE CREDITS

Figure 2: ©Dilan Ozkan
All other drawings and images by the authors.

Armand Agraviador is an architectural designer with expertise in Building Information Modeling, environmental design, visual programming, prototyping, and visualization. He has worked in several architectural practices across a range of project scales from furniture to master planning. In his role as research assistant in prototyping biological architecture in the Hub for Biotechnology in the Built Environment (Newcastle University, UK), Armand played a key role in computational design and visualisation of the BioKnit prototype.

Jane Scott is a NUAcT Research Fellow in the Hub for Biotechnology in the Built Environment at Newcastle University, UK, where she leads the Living Textiles research group. Her interdisciplinary research is located at the interface of programmable textiles, architecture, and biology. Before joining Newcastle, Jane was an academic at the University of Leeds and held a Visiting Research Fellowship in Biomimicry at Central Saint Martins. She completed her doctorate in Programmable Knitting, at the Textiles Futures Research Centre, Central Saint Martins. Jane was a member of the Board of Directors of ACADIA (2017-2021) and conference co-chair for ACADIA 2021, Realignments Toward Critical Computation.

Romy Kaiser is a bio-designer and researcher focusing on biomaterials, textile thinking, and fiber-based bio fabrication methods. Currently she holds a PhD position at the Hub for Biotechnology in the Built Environment, Newcastle (UK) working at the intersection between biology, textiles, and architecture as part of the Living Construction and Living Textiles Groups. Romy's research project *Textile Hosting* is investigating the scaffolding potential of knitted textiles for mycelium growth contributing to the field of bio fabrication.

Elise Elsacker is a postdoctoral researcher in the Hub for Biotechnology in the Built Environment at Newcastle University, working on Living Mycelium Materials. She is pioneering the development of materials in which fungal organisms maintain viability during their lifespan. Her current focus is to study the fungus's regeneration behavior leading to the material's self-healing upon damage. Her PhD, entitled "Mycelium Matters," in Engineering (2017-2021) at Vrije Universiteit Brussel (Belgium), is one of the first to characterize all principal factors affecting the biological and material properties of mycelium composites.

Aileen Hoenerloh is a PhD candidate at Newcastle University with an interest in bio-digital fabrication. Before starting her PhD in October 2019, she earned a Bachelor's Degree in Architecture in Germany and a Master's Degree in Computational Methods in Architecture from Cardiff University. In her Master's dissertation, she explored interactive manipulations on 3D clay printing using a remote-controlled actuator. Aileen's PhD research is focused on large scale, 3D bio-fabrication using bacterial cellulose.

Ahmet Topcu is an interior designer and researcher interested in biomaterial applications in the interior design and architecture context. He holds a Bachelor's degree in Interior Architecture from Karadeniz Technical University and a Master's degree in Interior Design from the University of Portsmouth. Currently, he is a PhD candidate at Newcastle University in the School of Architecture, Planning, and Landscape. His research investigates fungal-fungal interactions and mycelial growth to develop a novel biological form-making and fabrication strategy.

Ben Bridgens is a Senior Lecturer in Architectural Technology in the School of Architecture, Planning, and Landscape at Newcastle University, and a founding member of the Hub for Biotechnology in the Built Environment (www.bbe.ac.uk). Ben works at the interface of structural engineering, architecture, and design, critically examining sustainable technologies and considering the potential of low-tech, traditional approaches for construction and operation of the built environment.

Design-to-production Workflows for CNC-knitted Membranes

Strategies for design-to-production workflows for 3-dimensional non-developable CNC-knitted membranes

Yuliya Sinke
CITA / Royal Danish Academy

Mette Ramsgaard Thomsen
CITA / Royal Danish Academy

Drua Sif Simone Albrechtsen
CITA / Royal Danish Academy

Martin Tamke
CITA / Royal Danish Academy

1 Close up of *Zoirotia* CNC-knitted membranes complex surfaces

ABSTRACT

This paper discusses the design integration of knitted textiles into architectural workflows. The increasing interest in knit as a building material asks new questions of how knit can be designed, analyzed, specified, and fabricated at a large scale, with appropriate tolerances and verified performances. This paper presents a classification of two methods for fabricating 3-dimensional non-developable structural membranes with CNC-controlled industrial knitting machines: the graph method and the graded field method. It outlines the key differences in the two methods and discusses their individual benefits and limitations. It presents *Zoirotia* as a design case for the graded field method and describes a series of advances to this method as developed through the project.

INTRODUCTION

During the last decade research into knitted textiles for architectural application has gained increasing interest. Knitting technology presents a promising alternative to woven materials traditionally used in tensile textile design, as it allows for the construction of three-dimensional fabrics through the strategic placement and selection of stitches. Employed as alternative textiles substrates for structural membranes (Ramsgaard Thomsen et al. 2015; Ramsgaard Thomsen et al. 2019; Ahlquist 2015), formwork (Popescu 2019; Liu, Li, and Yuan 2020), and actuated structures (Scott 2013; Ramsgaard Thomsen and Karmon 2012; Ahlquist 2019), knit is understood as an innovative material system enabling complex geometry and bespoke fabrication while addressing the problem of waste in alternative textile systems such as a mass-produced weave (Ramsgaard Thomsen and Hicks 2008). By selecting high-performance yarns, such as high molecular weight polyethylene, aramid fibers, or high-performance polyester, the performance of the textile membranes can be designed in accordance with specific behavior demands.

This innovation relies on the creation of design-to-fabrication workflows that interface architectural design systems and CNC industrial knitting machines. CNC industrial knitting machines are developed for the garment industry and include specific workflows and software relying on the manual compiling of machine instructions. Knitting for architecture presents radically new demands to these workflows. Architectural knitted textiles depend on their three-dimensionality for performance, are large scale, have high demands to tolerance and customization, and need to incorporate material performance analysis that allow designers to control the geometry, strength and structure of the textile membrane. The acknowledgment that garment-based approach for knit design cannot be applied to an architectural scope has compelled the field to search for new approaches towards knit programming (Ramsgaard Thomsen et al. 2016; Popescu et al. 2017). In the field of architecture, two methods of design-to-fabrication automation are emerging. Both develop means of knitting three-dimensional non-developable geometries on flatbed knitting machines. Flatbed knitting machines are linear, therefore operating planarly during the production; however, addressing the control of the machine motion and knit structure will result in non-planar materials. The two methods, the graph method with a homogeneous knit structure and the graded field method with a discretized knit structure, respectively, use techniques of partial knitting or structural grading to create non-developable textile surfaces. This paper outlines the two methods, presenting their differences and individual limitations. It presents the *Zoirotia* design case and discusses the development of series of advances to the graded field method.

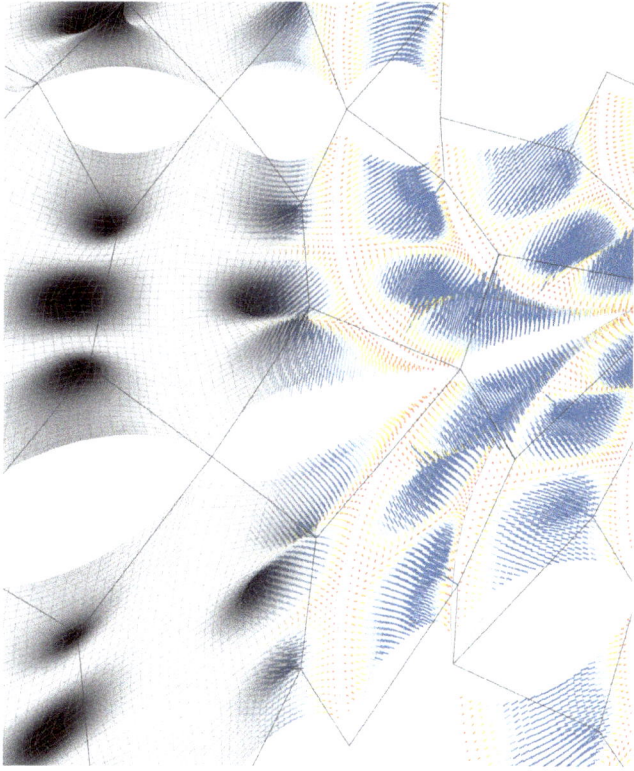

2 Frontal view of *Zoirotia*, 2022

3 Surface displacement visualization during the form-finding of *Zoirotia* membranes

Rounds 1,3,5: Knit.
Round 2: Kfb (2 sts.)
Round 4: Kfb, kfb (4 sts.)
Round 6: K2, yo,k2 (5 sts.)
Round 7: K2 (k1,yo,k1) in nxt st.
Round 8: K2tog, k1, incL (k1, yo, k1)
Round 9: K2tog, k2, incL (k1, yo, k1)
Round 10: K2tog, k3, incL (k1, yo, k1)

A B 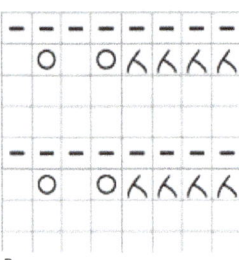 C 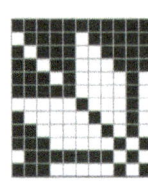 D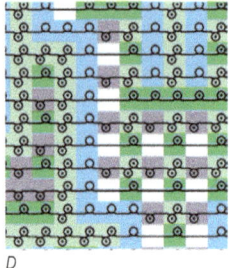

4 Examples of knitting instructions: (A) String-based descriptions; (B) Symbol-based descriptions; (C) Colored grid based; (D) Color and symbol-based for CNC-production

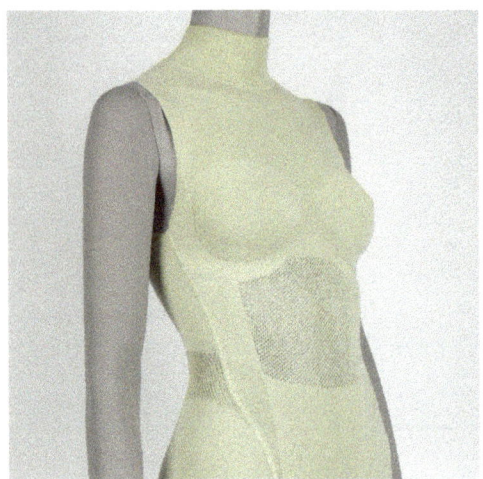 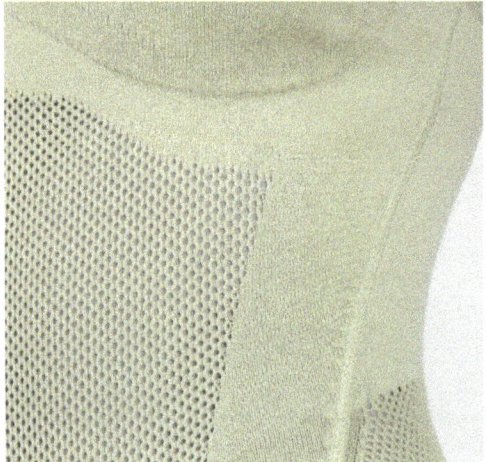 

5 Example of pre-set pattern from the STOLL pattern shop

INSTRUCTIONS AND DESIGN-TO-PRODUCTION WORKFLOW FOR 3-DIMENSIONAL NON-DEVELOPABLE KNITTED MEMBRANES

The tradition of knit instructions relies on linear descriptions in which the material assembly and structural logic of the material is declared sequentially, annotating needle-yarn manipulation row-by-row and following the order of fabrication. In manual knitting, these instructions are expressed as text- or symbol-based depictions organized either line-by-line or in a grid form (Figure 4). For CNC knitting, this underlying logic is continued. Here, the fabric material is encoded through the series of sequential instructions for the machine, existing within a two-dimensional drawing environment of the knitting machine software (M1Plus for STOLL or KnitPaint for SHIMASEIKI). These are represented in a form of time-needle schedules (Kaspar 2022; McCann et al. 2016), made of colored grid-organized symbols (Figure 4D). In these, the x-axis represents the needle space, while the y-axis represents the time. Through the color and symbols, these schedules allow for programming the engagement of the machine knitting carriage with the needles and the yarn carriers to form the stitches.

The preparation of knitting instructions requires a strong level of competence in understanding the relations between the pattern encryptions and their physical stitches equivalent to predict the fabric outcome and the volumetric behavior. The creation of new custom patterns is accompanied by extensive prototyping to evaluate the design, structure, and performance of the textile as well as its aesthetics, and if necessary, manually fine-tuning of the pattern configuration to the best outcome (Figure 5).

In contrast to industries where knitting is used for mass-production (garment, shoe or automotive), the architecture field has a greater demand for highly customized non-repetitive modules. This necessitates automated interfaces that can input file variations without labor-intensive manual compilation. Architectural membranes, furthermore, are inherently three-dimensional, using complex geometry to support structural performance. Structural membrane design implies volumetric design modeling through form-finding and simulation, to arrive at well-performing geometries that support pretensioning for structural stability. These result in highly three-dimensional surfaces, which need to be translated into a two-dimensional representation in order to be manufactured on flatbed knitting machines.

The time-needle schedule informs the two-dimensional machine instruction programming and can be replaced with a bitmap image, loaded to the native machine software. This pixel-based raster bitmap file plays the role of the linking interface to native machine programming tools

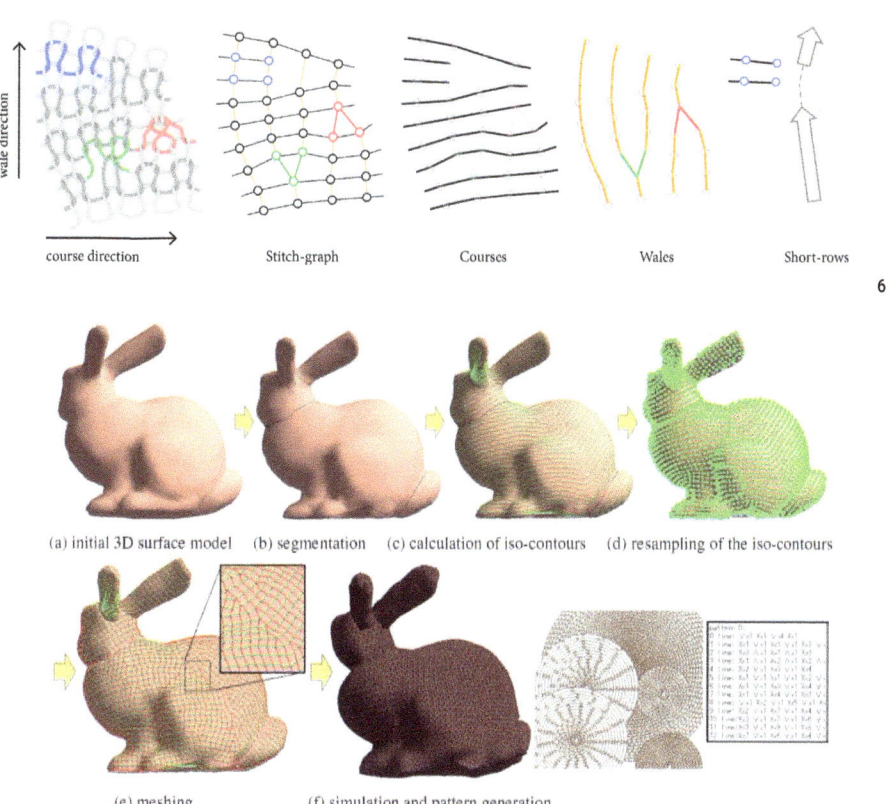

6 Partial knitting technique illustration; garment design workflows for on-demand machine knitting

7 Knitting a 3D model: graph method workflow

and can be prepared outside of the native knitting software with stronger links to the 3D shape data. The interfacing of the three-dimensional membrane geometry and its manufacturing two-dimensional bitmap file is understood through various ways of evaluating the membrane design three-dimensionally and its specification into an arranged pixel-based bitmap. The choice of methods used for volumetric shape evaluation is strongly linked to a knitting technique that is selected for the fabrication of membranes and affects the size of a design space.

TWO METHODS FOR SHAPE SPECIFICATION OF THREE-DIMENSIONAL KNITTING

To aid the fabrication of knitted complex non-developable surfaces, two methods for shape specification and manufacturing instruction representation are described further. These are graph and graded field methods. The following classification is defined by the paper authors and informed by existing workflow descriptions within the community of computer scientists and architects working with architectural knitting. The two methods achieve three-dimensionality through either accumulating selected stitches (partial knitting technique) or through the grading of the stitch structure (structural knit grading technique).

Graph Method

The graph knitting method employs a partial knitting technique to produce a three-dimensional fabric. By building on techniques coming from the field of computer graphics (Igarashi, Igarashi, and Suzuki 2008; McCann et al. 2016; Narayanan et al. 2018; Kaspar, Makatura, and Matusik 2019; Wu, Swan, and Yuksel 2019; Nader et al. 2021), the method allows the introduction of short rows, locally taking in and taking out stitches, resulting in the accumulation of material and hyperbolic bulging out of the plane, thereby forming a 3D surface (Figure 6). Focusing mainly on complex non-developable geometries to be made with the short-row technique, the fundamental discoveries of these laid a ground for further extension of the method to a larger scale of architectural elements and structures (Liu, Li, and Yuan 2020; Popescu et al. 2017; 2020). Although mentioned authors have some minor differences in their research aims and results, it is possible to define unified steps for describing the graph method, that lay fundamentally behind the method in our research.

To describe the three-dimensional surface in this manner, the graph method develops a series of steps to analyze and describe the surface along its courses (Figure 6).

The first step analyzes the textile form to define the start and end of the production cycle and determine the direction of knitting. Then the model is segmented and sliced into isocurves, which introduce the knitting rows (Figure 7). Each isocurve is sampled (divided) and populated with evenly distributed points, equal to the number of stitches per row. These points then are interconnected vertically to define pairing stitches. Unpaired

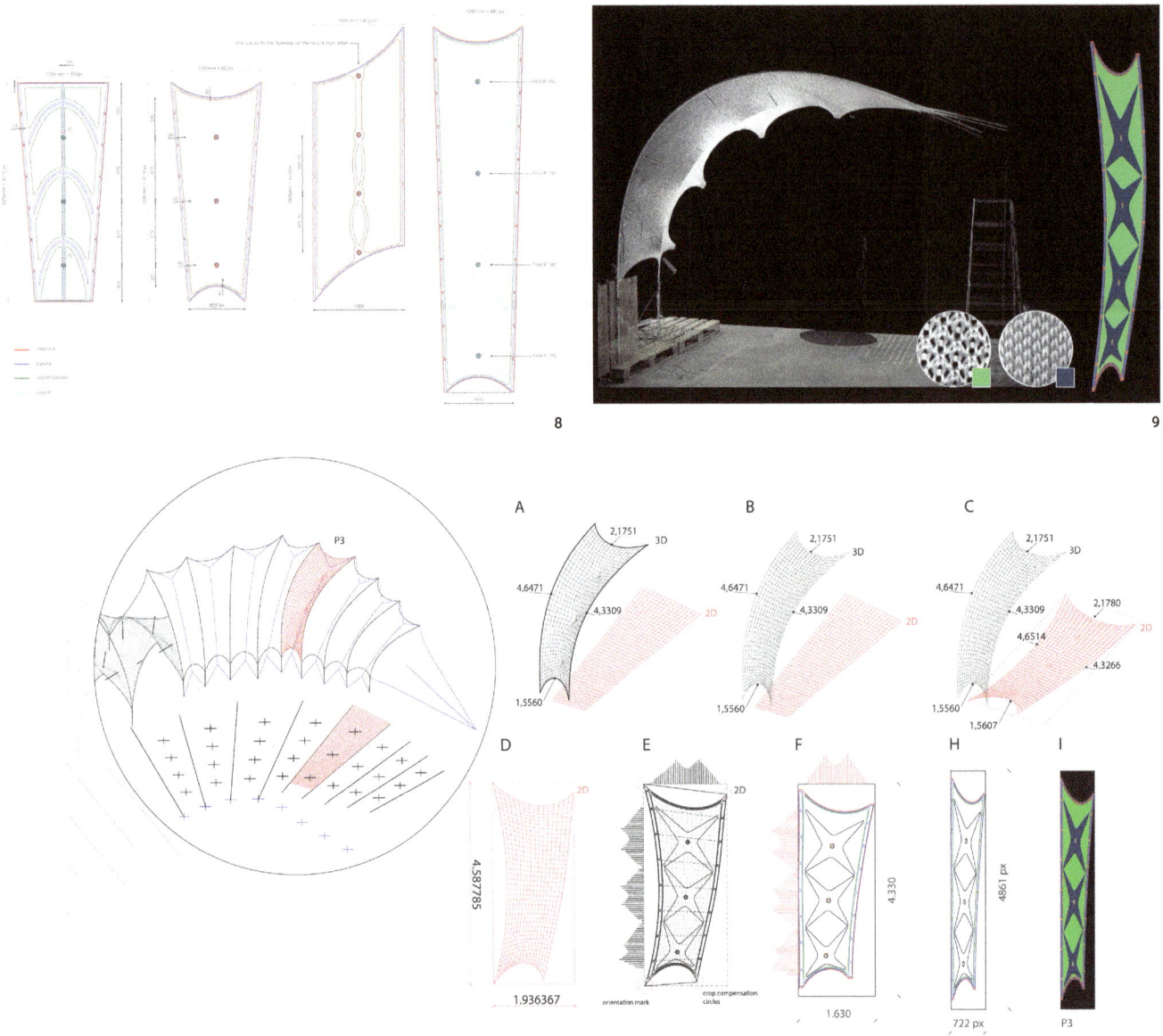

stitches would mark the boundary of the short row. Constructed links between the points form a surface network, that informs the build-up of the graph diagram, representing the relations between rows and columns. As a result, a grid-bound planarly arranged map of square pixels is constructed (bitmap) to be used for further machine instructions.

The graph method results in a three-dimensional textile with a homogeneous structure. As the stitch structure remains the same across the surface, the textile is undifferentiated in its initial structure. Further additions of yarn changes or structural changes can expand this expression.

Graded Field Method

The second method, developed at CITA uses the strategic grading of the textile surface by interchanging the stitch patterns. In knit the interchange between how the stitches are connected, the correlation between the front and the back bed (flatbed knitting), and the yarn tension create different locking points for the individual stitches. The locking point is the state in which the knit stitches are fully in tension. The grading of locking points is supported by the knitting machine's automated take-down systems that allow the differentiated tensioning of the textile under production. Unlike the graph method, in which the same stitch is used across the textile, the graded field method constructs each row out of varied stitches creating a heterogeneous pattern. These differ in their geometrical configuration, and therefore, in their structural performance, which results in the formation of the three-dimensional surface. Stitches with bigger and smaller locking point lengths are placed in relation to structural performance and surface displacement to enable the textile expansion under the load (Figure 9).

8   Varied graded field geometric shapes are iterated during the prototyping phase

9   Stitches with bigger and smaller locking point lengths, represented by green and dark blue colors, are used to achieve membrane surface expansion

10 Graded field method workflow (2008) in *Isoropia* project by CITA

11 Close up to the structurally graded membrane of *Zoirotia*, with the visible difference in the knitted structure composition

The graded field method defines the planarized boundary of the three-dimensional membrane shape. The areas of three-dimensional extension are then defined around their tension point. Through the prototyping, different field shapes were examined, and the most controlled and expanded shapes were chosen (Figure 8).

This allows us to define the graded field in which we place stitch types with bigger locking point lengths. In the first iterations of the method, the fields and their resulting membrane extension were defined through processes of prototyping and analyzed using the parametric calculation of the yarn usage per row. The method uses the automated generation of pixel maps as a common interface to the CNC knitting machine. Through the architectural design environment, the graded fields are defined as colored zones with specific RGB values and a BMP file with a predefined resolution is exported (Ramsgaard Thomsen et al. 2019). This BMP file is used to interface the knitting machine software through the assignment of specific stitch specifications and machine motion control (Figure 10).

EXTENDING THE GRADED FIELD-BASED METHOD: *ZOIROTIA* AS DESIGN CASE

The following presents the further development of the graded field method. Here, we depart from predefined field shapes to a structurally informed gradient field where the resulting three-dimensional extension is directly informed by a geometric analysis of the membrane performance. The method is developed through the design case *Zoirotia*, a large-scale installation developed for the ZKM Center for Art and Media in Karlsruhe, Germany. The method extends the graded field method by enabling a stronger performance integration and a more refined surface differentiation by introducing dithering strategies that allow to locally tune the transition of the membrane expansion properties. This is achieved through the development of an automated workflow allowing for the integration of complex structural grading informed by geometrical analysis with a high degree of detail for the interface file specification.

Zoirotia consists of 88 unique tensile CNC-knitted structural membranes, that are three-dimensional and pretensioned, which creates a non-developable membrane geometry. To achieve their geometry and structural performance, the membranes are graded using two stitch types with different locking point lengths. In *Zoirotia* we use a highly differentiated stitch structure (double jacquard and unravel double jacquard) that allows us to locally expand the membranes by up to 200% (Figure 11).

The method builds on the automated generation of pixel maps as an interface to the CNC knitting machine and extends this by further detailing its file generation capacity to enable highly detailed specifications that allow the dithered grading

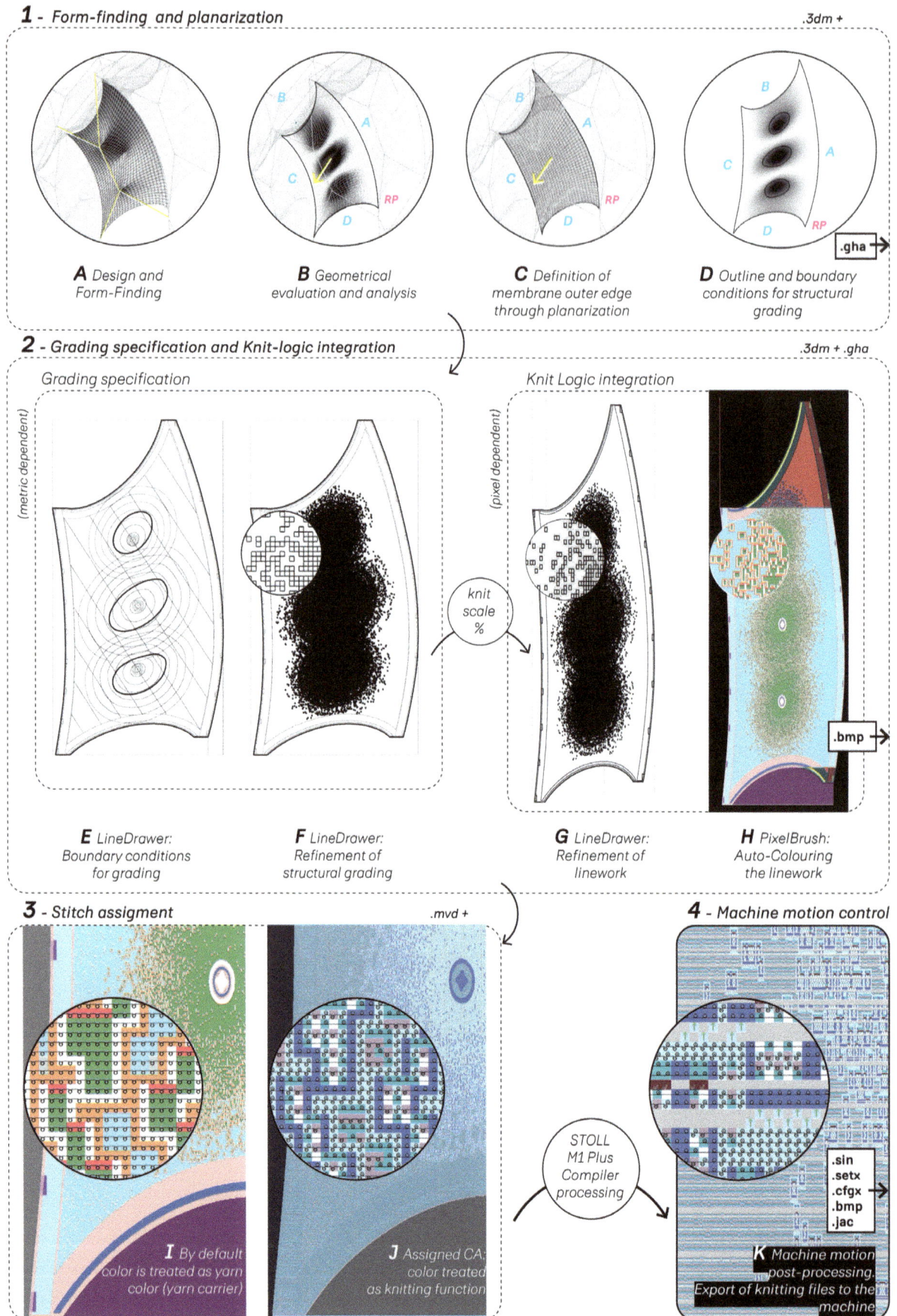

12 Extended workflow diagram for graded field method (2022) for structurally differentiated knitted membranes

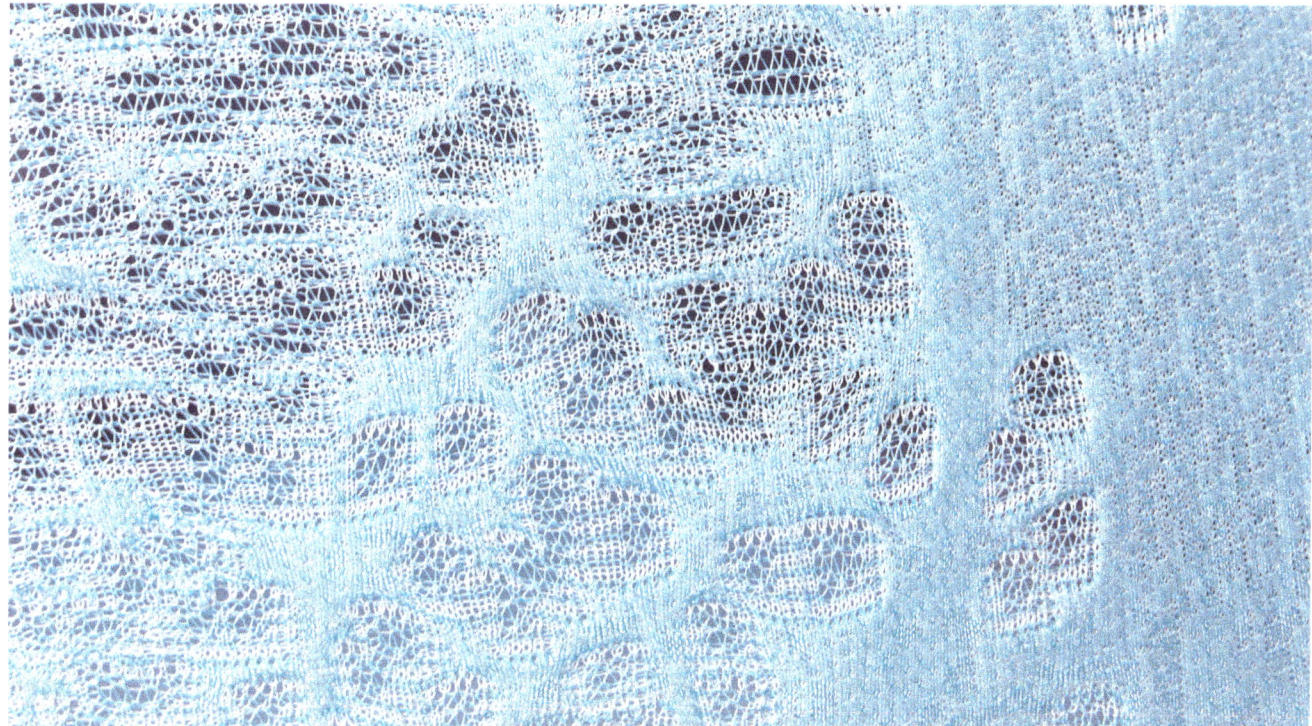

13 Close up of the surface differentiation with the double jacquard and unravel double jacquard stitches

of the membranes. It includes a stronger integration of knit logic that analyzes the structural integrity of the membrane and automates the adaptation to fabrication needs.

The method extends beyond the pixel map generation to create a bespoke parametric tool (Color Arrangement, heretofore referred to as CA) that assigns stitch functions to the files. This step allows to treat the color in the imported bitmaps as knit functions, rather than yarn carriers, as usually done by default in the STOLL software for knitted garments with double jacquard.

KNIT SPECIFICATION PHASES IN EXTENDED GRADED FIELD METHOD

The workflow process described below demonstrates an extension of the graded field method used to design and fabricate the surface graded membrane. For that, two custom parametric modules (LineDrawer + PixelBrush and CA STOLL module) are developed in collaboration between the design team and the machine operator. These tools allow for the automation of material specification and machine motion control. The workflow includes four specification phases moving from design to production through: 1) form-finding and planarization; 2) grading specification and knit logic integration; 3) stitch assignment; and 4) machine motion control specification (Figure 12).

Specification Phase 1: Form-finding and Planarization
The first phase sees the definition of the textile outline and its three-dimensional extension. The translation from 3D design to 2D field graphics linework takes place through the strategic planarization of the volumetric shape (Figure 12C). Here, the untensioned surface is evaluated against its final structural pretensioning, allowing us to calculate the displacement, and therefore, the needed surface expansion (Figure 12B). Here, a hierarchy of membrane measurements is established, defining the outer boundary of the membrane, and the outline shape of the graded field is planarized onto a 2D plane (Figure 12D). To inform this, we implement a monochrome displacement gradient map used to define the areas with a larger expansion value (unravel double jacquard stitch) (Figure 13).

Specification Phase 2: Grading Specification and Knit Logic Integration
Once the outer boundary of the membrane and outline shape of the graded field are generated (Figure 12D), a series of refinements of the membrane outlines are implemented (Figure 14A-J) and the programming of the graded field can begin.

To translate the monochrome gradient map into a dithering strategy structurally grading the membrane, we define a series of circular threshold boundaries that control the number of sequential unravel stitches in each row in turn defining the three-dimensionality of the membrane (Figure 12E). At first, this is treated as a binary grading in which the grading strategy is declared. After this, in a second process, we generate a series of boundary stitches defining the drop stitch and the transitional boundaries around the dithered

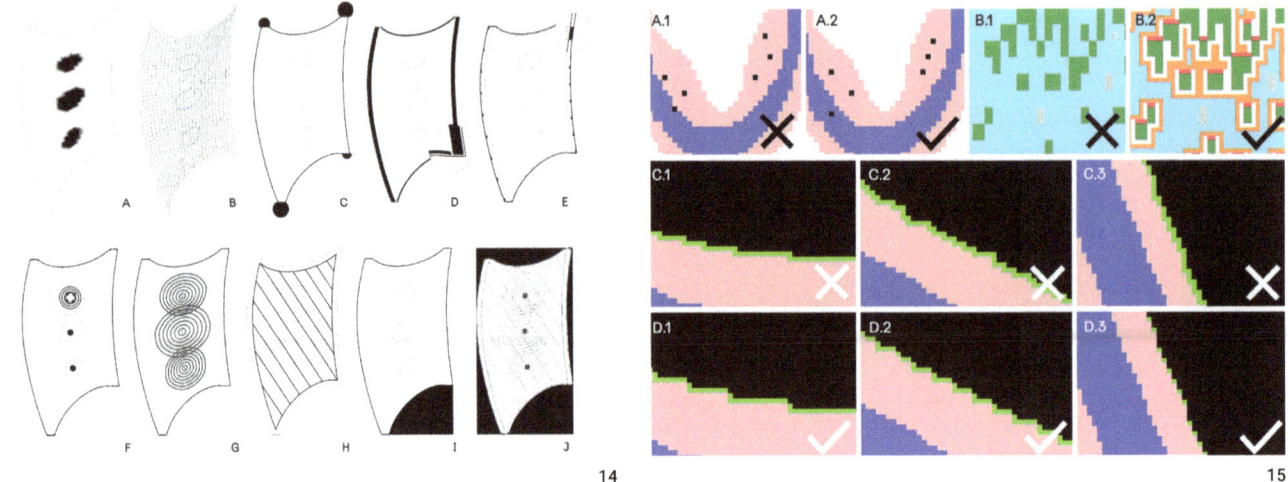

14

15

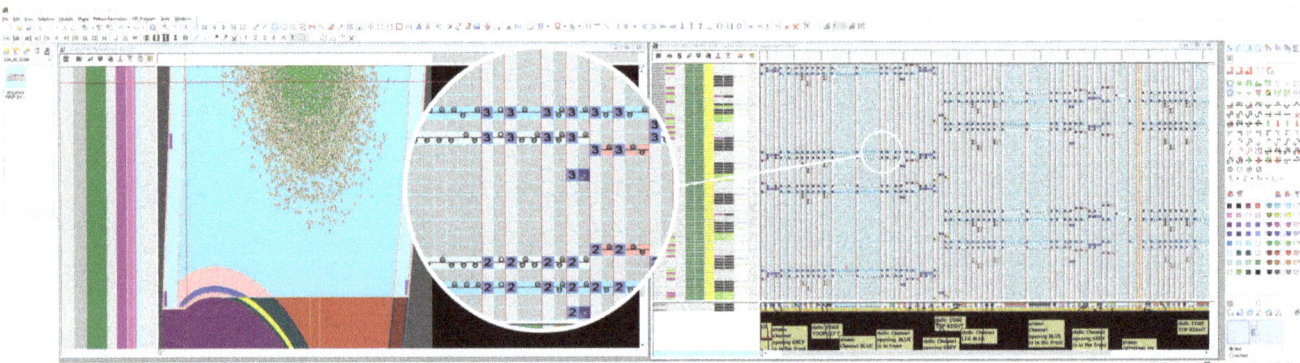

16

fields (Figure 12G). These are crucial for the unravel double jacquard texture integrity during the production (Figure 15, B.1-B.2).

The membranes are knitted to shape by decreasing or increasing the number of stitches at the edge of the membrane. It follows fabrication-driven rules for stitch stepping angles. To enable this we generate the placement of cast-on and cast-off functions at each edge of the membrane, where the stepping configuration is informed by the curvature of the edge outline in respect to the knitting direction (Figure 15, C-D).

Once all the refined field linework is generated, it is assigned an RGB color value and exported as an index-colored bitmap file. The detailed knit interface programming results in automatically generated highly-differentiated bitmap files, each characterized by total 52 indexed colors, corresponding to the number of assigned knitting functions (Figure 19).

Specification Phase 3: Stitch Assignment

The automatically generated bitmaps are imported into the STOLL native knitting machine software M1Plus), where the pixel map colors are associated to machine instructions. The use of a custom-developed parametric component in CA allows for the automated function assignment. The CA component contains small units of pattern codes that are programmed to be applied to specific color areas within the canvas (Figure 16). The colors, therefore, no longer correspond to the yarn carrier but to a knit function and act as an information package that can be unfolded by the program. Here, all information related to a needle actuation is programmed and machine-specific values can be edited within the global CA to match the chosen machine type (Figure 17). In *Zoirotia*, a universal CA module is created containing all possible knitting functions within all 88 bespoke membranes of the structure.

Specification Phase 4: Machine Motion Control

During the last phase of the workflow, the additional carriage and needle bed-related functions (effecting transfers and racking) are integrated into the pattern through a native post-processing compiler within STOLL M1 Plus software. It unfolds the stitch modules into compound elements (simpler needle motions) and expands the knitting rows based on the carriage drive direction (grey lines) (Figure 18). The completion of this phase relies fully on the structural integrity of the material grading and the knit logic of the support stitches around the dithered fields, where the pixel-dependent knitting features are refined so to make the program physically knittable.

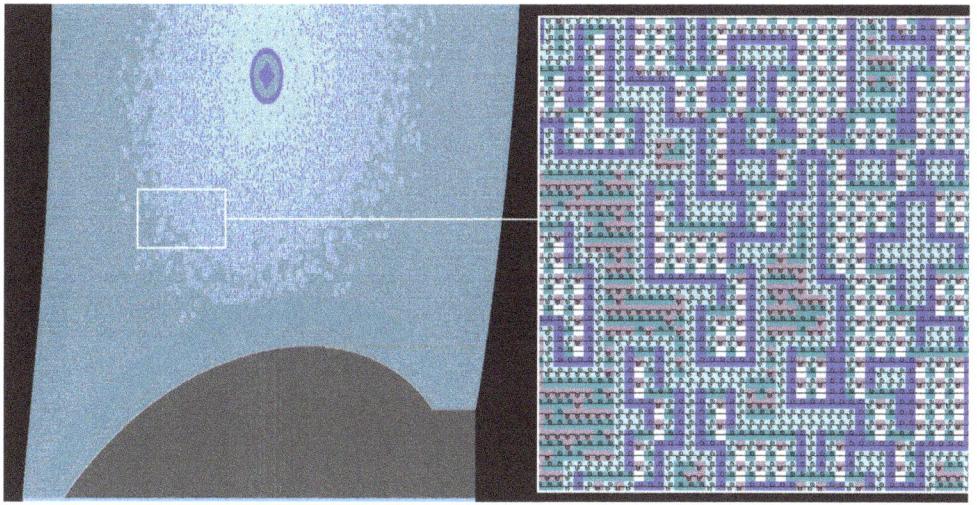

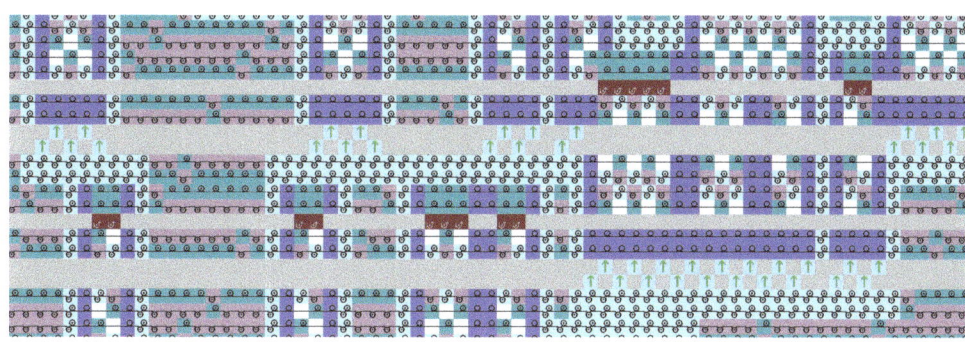

14 Grading knit specification defining the fields for distinct knitting features and structures: refinement of the outer boundary (C), introduction of detailing (D and E, F), subdivision of zones for structural grading (G), the definition of color grading (H), integration of waste yarn to support the knitting machine's automated take-down systems (I); (J) shows joined linework

15 Illustration showing the level of detail for structural grading of the membrane with a graded field approach

16 Color Arrangement (CA) interface within STOLL M1 Plus

17 Stitch assignment phase; close up of the color-symbol based assignments

18 Machine motion generation during the last phase specification by adding transfers and racking functions

RESULTS

This paper presents two methods for the fabrication of three-dimensional non-developable knitted membranes. Below these methods are evaluated through a set of criteria to that outline their benefits and limitations.

The main differentiation lies within the nature of their stitch structure: homogeneous in the graph method, and heterogeneous in the graded field method. This nature of surface composition has multiple implications in achieved topology, surface density, aesthetics, and material use.

Both methods allow for producing of double-curved shapes through the composition of stitches, either identical or varied. Here, the curvature degree of achieved surfaces varies based on the used knitting technique. Graph method, that uses partial knitting technique, allows for continuous surface addition and bending, which offers the opportunity for the production of helix, spiral, and hyperbolic surfaces. The graded field method works with the given width of the machine knitting row and operates within the space of stitch length modification. Here, the width between the needle beds and gauge of the machine defines the stitch expansion limits, which reflects in the smaller amplitude of the surface articulation.

The use of identical stitches in the graph method results in a homogeneous surface density, while in the graded field method, where variegated stitches are used, the surface shows variegated transparencies. The combination of double jacquard with unravel double jacquard stitch types in *Zoirotia* results in a very low-density surface in the places of surface expansion. However, it is possible to achieve firmer surfaces by combining of similar density stitch textures, while still maintaining a degree of surface expansion, like in the *Isoropia* project by CITA. The choice of varied stitch types defines the expansion property of the fabric and should be done carefully in response to the design and performance criteria.

The accuracy of both approaches is highly dependent on the loop dimensioning and requires preliminary prototyping to define these. However, the behavior prediction of structurally graded surfaces (graded field method) is challenging as should be prototyped at 1:1 scale rather than through an extracted smaller surface sample, which is acceptable for homogeneous surfaces. Therefore, the use of this method strongly relies on simulation tools in order to predict material behavior as 1:1 prototyping is not always feasible at the 1:1 architectural scale.

The graph method uses an initial design geometry without taking into consideration the internal material stresses and

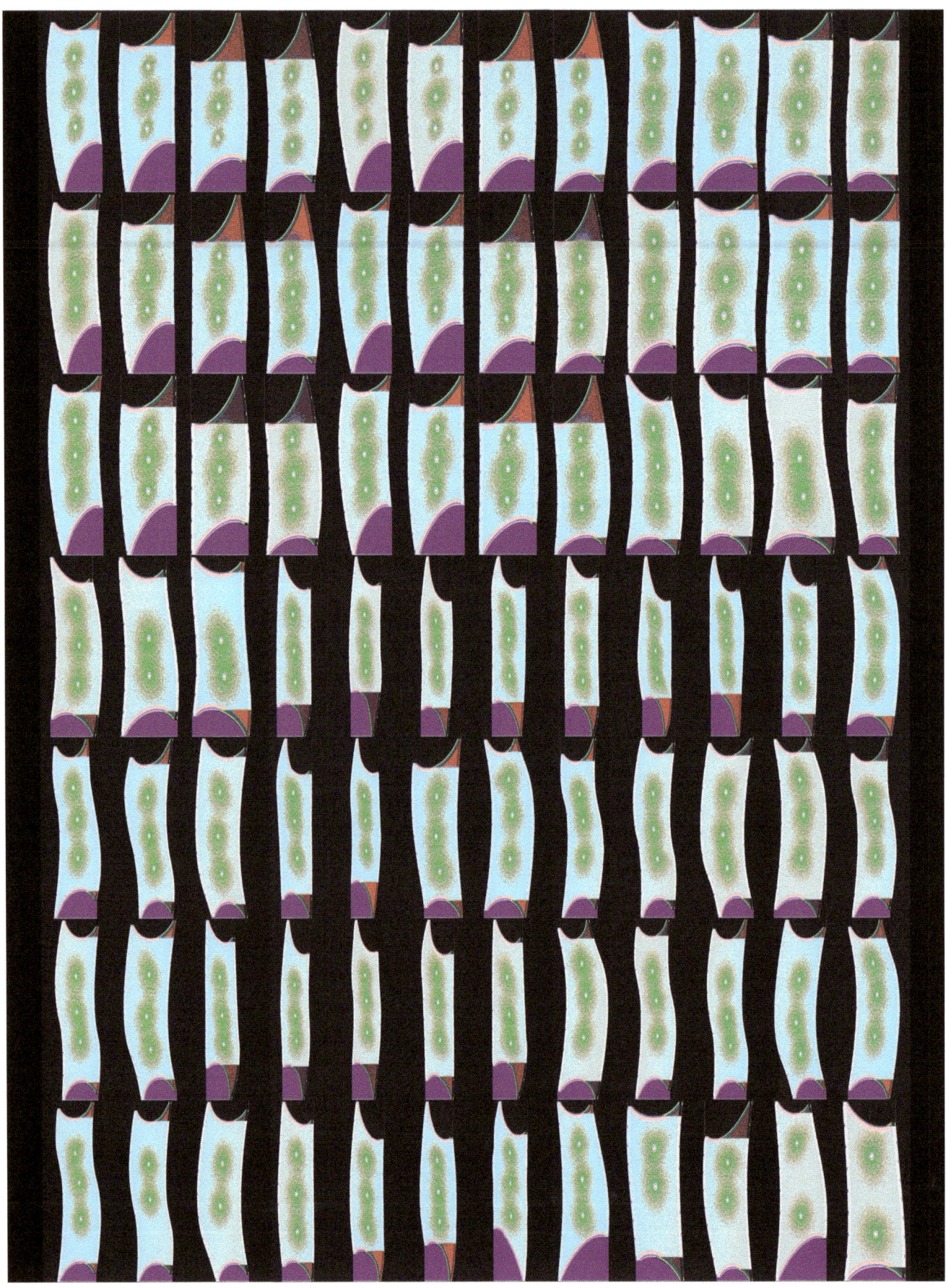

19 Layout of 88 unique bitmap files, generated for the fabrication of the *Zoirotia* structure

20 Interior space of *Zoirotia* made of highly bespoke knitted membranes

physical processes but as a geometric shape that is subdivided for fabrication. In difference, the graded field method integrates the surface tensioning, and therefore, its performance under load to define the material expansion.

For achieving three-dimensional shapes both methods essentially use a different amount of yarn material. The graph method uses more stitches to achieve three-dimensionality, which results in a denser and more materially intense surface. The graded field method uses modified stitches of different lengths. This allows the minimization of material use.

CONCLUSION AND FUTURE DEVELOPMENT

The presented methods are understood as complementary as each has its benefits and limitations. However, the combination of the methods could bypass the limitations of each, while joining their advantages for further expansion of the surface three-dimensionality, density, and performance control. Here, the application scenarios of knitted fabrics in architecture could be extended and new functional surface expressions could be imagined.

For example, the dense homogeneous knitted membranes as manufactured with the graph method and used for formwork could potentially benefit from adding functionally graded areas of lower density offering the opportunity for a tighter engagement between the formwork material and casting liquid material. The examples of architectural application with the graded field method, which are often tensile roofs and canopies, are often criticized for their low precipitation retention capacity, as fabrics let the rain through. To improve the surface weatherproofing ability, a surface water repellent casting layer could be introduced; however, it often requires a homogeneous, relatively dense surface that could be achieved by employing the graph method.

ACKNOWLEDGMENTS

This project could not have been possible without the commission of ZKM, Center for Art and Media in Karlsruhe, Germany for the exhibition *Bio-Media*, curated by Peter Weibel, Sarah Donderer, and Daria Mille, and of the yarn sponsorship by AMMAN Group. We would like to thank our collaboration partners Viola-Stils SIA, Kobleder GmbH, Karl Mayer Stoll Textilmaschinenfabrik GmbH, Julian Lienhard, and Dongyuan Liu from Tragwerksentwurf Uni Kassel. Additionally, we would like to acknowledge the effort and contribution of research assistants Simona Hnídková, Martynas Seskas, Nihit Borpujari, Viktoria Millentrup, and CITA Computation in Architecture students Muchen Yan, Jack Young, Chloe Liang Xiuling, Chih Wei Chan, Carl Hampus Vilhelm Carlström, Camila Martinez Alarcon, and Julie Amanda Aagaard Andersen.

REFERENCES

Ahlquist, Sean. 2015. "Integrating Differentiated Knit Logics and Pre-Stress in Textile Hybrid Structures." In *Proceedings of Design Modelling Symposium*, 101–11. Copenhagen: Springer International Publishing. https://doi.org/10.1007/978-3-319-24208-8_9.

Deleuran, Anders Holden, Ida Katrine Tinning Friis, Henrik Evers Leander, and Michel Schmeck. 2015. "Hybrid Tower, Designing Soft Structures." In *Modelling Behaviour*, edited by Mette Ramsgaard Thomsen, Martin Tamke, Christoph Gengnagel, et al. Cham: Springer. 87–99.

Igarashi, Yuki, Takeo Igarashi, and Hiromasa Suzuki. 2008. "Knitting a 3D Model." *Computer Graphics Forum* 27 (7): 1737-1743. https://doi.org/10.1111/j.1467-8659.2008.01318.x.

Kaspar, Alexandre, Liane Makatura, and Wojciech Matusik. 2019. "Knitting Skeletons: A Computer-Aided Design Tool for Shaping and Patterning of Knitted Garments." In *Proceedings of the 32nd Annual ACM Symposium on User Interface Software and Technology*, 53-65. New Orleans, USA. https://doi.org/10.1145/3332165.3347879.

Liu, Yige, Li Li, and Philip F. Yuan. 2020. "A Computational Approach for Knitting 3D Composites Preforms." In *Proceedings of the 2019 DigitalFUTURES*, 232–46. Singapore: Springer Singapore. https://doi.org/10.1007/978-981-13-8153-9_21.

McCann, James, Lea Albaugh, Vidya Narayanan, April Grow, Wojciech Matusik, Jen Mankoff, and Jessica Hodgins. 2016. "A Compiler for 3D Machine Knitting." *ACM Transactions on Graphics* 35 (4): 1–11. https://doi.org/10.1145/2897824.2925940.

Nader, Georges, Yu Han Quek, Pei Zhi Chia, Oliver Weeger, and Sai-Kit Yeung. 2021. "KnitKit: A Flexible System for Machine Knitting of Customizable Textiles." *ACM Transactions on Graphics* 40 (4): 1–16. https://doi.org/10.1145/3450626.3459790.

Narayanan, Vidya, Lea Albaugh, Jessica Hodgins, Stelian Coros, and James Mccann. 2018. "Automatic Machine Knitting of 3D Meshes." *ACM Transactions on Graphics* 37 (3): 1–15. https://doi.org/10.1145/3186265.

Popescu, Mariana Adriana. 2019. "KnitCrete: Stay-in-Place Knitted Formworks for Complex Concrete Structures." PhD thesis, ETH Zurich.

Popescu, Mariana, Matthias Rippmann, Tom Van Mele, and Philippe Block. 2017. "Automated Generation of Knit Patterns for Non-Developable Surfaces." In *Humanizing Digital Reality*. Singapore: Springer. 271–84.

Ramsgaard Thomsen, Mette, and Toni Hicks. 2008. "To Knit a Wall, Knit as Matrix for Composite Materials for Architecture." In *Ambience 08 International Scientific Conference: Proceedings 2008: Smart Textiles - Technology and Design.* 107–15.

Ramsgaard Thomsen, Mette, and Ayelet Karmon. 2012. "Listener: A Probe Into Information Based Material Specification." In *Ambience '11 Proceedings; …Where Art, Technology and Design Meet.* The Swedish School of Textiles. 158–163.

Ramsgaard Thomsen, Mette, Yuliya Sinke Baranovskaya, Filipa Monteiro, Julian Lienhard, Riccardo La Magna, and Martin Tamke. 2019. "Systems for Transformative Textile Structures in CNC Knitted Fabrics – Isoropia." In *Proceedings of the TensiNet Symposium 2019 Softening the Habitats.* 95–110.

Scott, Jane. 2013. "Hierarchy of Knitted Forms: Environmentally Responsive Textiles for Architecture." In *ACADIA '13: Adaptive Architecture; Proceedings for the ACADIA 2013 Conference.* Cambridge, Ontario. 361–66.

Wu, Kui, Hannah Swan, and Cem Yuksel. 2019. "Knittable Stitch Meshes." *ACM Transactions on Graphics* 38 (1): 1–13. https://doi.org/10.1145/3292481.

IMAGE CREDITS

Figure 1: ©Anders Ingvartsen, CITA
Figure 2: ©Anders Ingvartsen, CITA
Figure 5: ©STOLL Pattern Shop
Figure 6: ©Alexandre Kaspar, 2022
Figure 7: ©Yuki Igarashi, 2008
Figure 11: ©Anders Ingvartsen, CITA
Figure 13: ©Anders Ingvartsen, CITA
Figure 20: ©Anders Ingvartsen, CITA

All other drawings and images by the authors.

Yuliya Sinke is a PhD fellow at CITA, Center for IT and Architecture. She is conducting her research within the UN Sustainable Goal 12, Responsible Consumption and Production framework, where she is investigating the application of CNC-knitted membranes in architecture; she is interested particularly in the aspects of digital simulation and design models for graded textile properties.

Mette Ramsgaard Thomsen examines the intersections between architecture and new computational design processes. During the last fifteen years, her focus has been on the profound changes that digital technologies instigate in the way architecture is thought, designed, and built. In 2005 she founded the Centre for IT and Architecture research group (CITA) at the Royal Academy of Fine Arts, School of Architecture, Design and Conservation, where she has piloted a special research focus on the new digital-material relations that digital technologies bring forth. Investigating advanced computer modelling, digital fabrication, and material specification, CITA has been central in the forming of an international research field examining the changes to material practice in architecture.

Drua Sif Simone Albrechtsen is a dedicated Danish textile designer. She is recipient of a Master of Fine Arts degree with specialization in Textile Design (June 2016) and a Bachelor's degree in Fashion and Textile Design from Kolding School of Design in 2014. Drua designs textiles for both fashion and interiors within the high-end market and focuses upon woven and knitted fabrics.

Martin Tamke is pursuing design-led research on the interface and implications of computational design and its materialization. He joined the newly founded research center CITA in 2006 and shaped its design-based research practice. His focus is on new design and fabrication models and workflows with an emphasis on feedback from environment and process. His latest research focuses on bio-based material and is characterized by strong interdisciplinary links to computer science (Machine Learning and 3D sensing), structural engineering (simulation and ultralight hybrid structures), and material science (bespoke CNC-knit, engineered timber and additive fabrication with recycled and biobased materials).

3D Knit Spacesuit Sleeve

With Multifunctional Fibers and Tunable Compression

Lavender Tessmer
MIT Architecture

Ganit Goldstein
MIT Architecture

Guillermo Herrera-Arcos
MIT Media Lab / K. Lisa Yang Center for Bionics

Volodymyr Korolovych
MIT Institute for Soldier Nanotechnologies

Rachel Bellisle
Harvard-MIT Health Sciences and Technology

Cody Paige
MIT Aeronautics and Astronautics / MIT Human Systems Lab

Christopher Shallal
Harvard-MIT Health Sciences and Technology / K. Lisa Yang Center for Bionics

Atharva Sahasrabudhe
MIT Chemistry / Research Laboratory of Electronics

Hugh Herr
MIT Media Lab / K. Lisa Yang Center for Bionics

Svetlana V. Boriskina
MIT Mechanical Engineering

Dava Newman
MIT Media Lab / MIT Aeronautics and Astronautics

Skylar Tibbits
MIT Architecture / Self-Assembly Lab

1 3D Knit Spacesuit Sleeve prototype

ABSTRACT

Textiles were among the first engineered materials to be used in protecting the human body from the harsh environment of outer space. Throughout history, these advances in soft material and garment manufacturing techniques have inspired scientists and engineers in partnership with designers to create adaptive solutions for extreme conditions. This research establishes a multidisciplinary collaboration within a team of designers, engineers, and scientists to address current technical challenges associated with spacesuits' multi-layer fabric requirements.

This paper presents a novel approach to spacesuit fabrication and functionality using CNC knitting to enable precise material control throughout the three-dimensional structure, creating higher functionality in a seamless and minimal textile architecture. We have developed a 3D textile framework consisting of a computational design workflow, multifunctional fiber integration, and a highly customizable 3D layering method that can be adapted to the personalized dimensions of the body. This method includes designating regions for mobility, tunable compression, integrated sensing, and quick donning and doffing within a single sleeve prototype as a first step toward a novel approach for spacesuit fabrication. While this work has focused on the spacesuit application, we imagine future applications in other textile architectures and next-generation apparel with integrated monitoring for increased performance, environmental regulation, and improved comfort.

INTRODUCTION

Since the 1950s, spacesuit technologies have shown to be a continuous effort of negotiating between material capabilities, modes of fabrication, and the stringent requirements for survival in the extreme environment of outer space. The fabrication methods employed in spacesuit manufacturing have evolved with the technologies of material and textile production. Combined with manual fabrication techniques adopted from the garment industry, new technologies in soft materials influenced the spacesuit design solutions in earliest space expeditions, where new material layers functioned as reflective insulation, thermal lining, and gas-impermeable membranes (Thomas 2006; De Monchaux 2011). Following this history, we seek to propose a new approach to spacesuit construction utilizing the latest advances in material and textile manufacturing to create a seamless integration and minimal spacesuit construction (Figure 1). The extreme challenges of the space environment present a unique site for collaboration between science, engineering, and design, establishing technical requirements while imagining new adaptations of fabrication technologies. Through this cross-disciplinary team, we explore how CNC knitting, newly developed materials, and fiber-based sensors can address the unique technical challenges posed by spacesuits.

BACKGROUND

To protect the body from the vacuum of outer space, a spacesuit must provide breathable atmosphere and body surface pressure. Spacesuit designs differ in their technical strategies for maintaining pressure on the body surface in a vacuum environment. There is not a single solution to this problem, but rather different strategies that each negotiate a balance of trade-offs. The most common spacesuit type, and the only type used operationally in human spaceflight thus far, is gas pressurization. In this strategy, pressure is achieved by providing an airtight environment with non-permeable layers and gas pressure. Alternatively, another type of suit has used a porous membrane, usually a textile, to provide mechanical pressure directly to the body (Webb 1968). Our approach pursues tunable mechanical counter pressure (MCP) through knit fabric construction.

Background to Traditional Spacesuits

Extravehicular Activity (EVA) spacesuits are highly familiarized in media imagery from the first lunar landing to recent spacewalks from the International Space Station. Though time-tested and highly reliable, these suits are extremely cumbersome because they require an atmosphere of pressurized air around the inhabitant. The suits' airtightness further necessitates many layers and mechanisms for temperature and moisture regulation, since the body can no longer maintain these functions on its own in the absence of garment porosity (Thomas 2006). As a result, the necessary protective functions are performed by multiple layers of woven fabric in the suit's bulky assembly, and these materials are responsible for maintaining an impermeable membrane and regulating a steady interior thermal environment. Though the layers are individually thin and delicate, the airtight pressure requirements hinder astronaut mobility, and the motions of the joints are continuously at odds with the forces of pressure within the suit (Thomas 2006).

Background to Mechanical Counter Pressure (MCP) Suits

In contrast to this, MCP suits seek to alleviate the challenges of mobility and airtightness by providing mechanical pressure on the body through direct contact with the material. Though less well known compared to NASA's EVA suits, MCP suits have been in development since high-altitude flights were first made possible by advances in 1930s aviation technology, and MCP concepts were first explored for space flight applications in the 1950s (Figure 2a) (Webb 1968). MCP suits offer several proposed advantages over gas-pressurized versions, such as reduced bulk, improved mobility, and natural thermoregulation. These suits typically rely on elastic fibers which conform to the complex motion and geometry of the human body while maintaining the necessary pressure. Examples from NASA MCP Space Activity Suit prototypes in the 1960s and 1970s utilize elastic materials such as elastane and rubber cord in bobbinet and powernet textile constructions to provide mechanical pressure. These prototypes employed multiple layers of elastic mesh to generate the requisite level of elastic power to maintain pressure on the body (Figure 2b) (McFarland, Ross, and Sanders 2019). More recently, the MIT BioSuit™ research proposes a new version of the MCP suit

 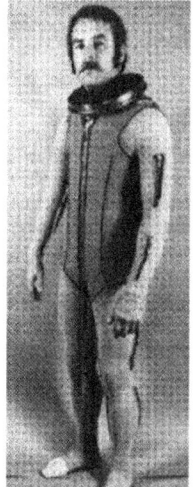 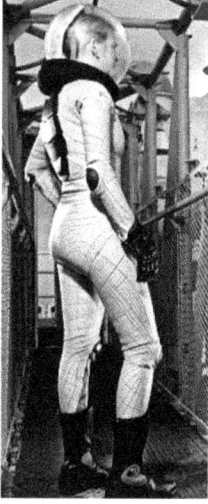

a b c

2 Evolution of Mechanical Counter Pressure (MCP) suits including: (a) the David Clark MC-3 partial pressure suit for high-altitude flight; (b) the 1960s Space Activity Suit; and (c) the MIT BioSuit™

concept, envisioning a future of lightweight, highly mobile, and individually adapted spacesuits (Figure 2c) (Newman 2007). The MIT BioSuit™ work has investigated methods to improve mobility, including Iberall's "lines of non-extension" (Iberall 1970) and skin strain mapping techniques (Obropta 2015, 2016; Wessendorf 2012), aiming to quantify skin movement to inform a "second-skin" garment.

Though MCP suits demonstrate a compelling alternative to the bulky traditional EVA suits, there remain multiple challenges that need to be addressed through alternative means of design and fabrication. Previous examples of MCP designs lack control of local fabric properties because they are assembled from layers of homogeneous sheet materials. In the early MCP suits that contained multiple layers of non-differentiated elastic mesh fabric, material that is easily compliant in thin layers becomes difficult to stretch when compiled into a high-power multilayer construction. As a result, convex bends in joints such as knees and elbows begin to experience resistance to mobility due to the thickness of the elastic fabric and inability to differentiate between circumferential and longitudinal strain in the material (McFarland, Ross, and Sanders 2019). Furthermore, the large amount of resistance in the elastic materials inhibits easy donning and doffing (putting on and taking off the suit), prompting the development of donning and doffing techniques and technologies (Anderson 2010; Holschuh 2012; 2013; 2015; 2016). Our project addresses these challenges through CNC knitting, which is capable of combining highly differentiated fabric properties into a single seamless assembly while maintaining the multilayered material functions required to support survival in space.

Background to CNC Knitting

CNC knitting is an ideal fabrication process to apply to these challenges because of its ability to combine multiple materials, designate localized fabric behavior, and integrate multilayered geometry together in one seamless textile panel— functions that are absent in homogenous textile sheets. Knit fabrics are composed of rows of interlocking loops, called "stitches," where various choices can be made by the designer to organize materials and properties of the fabric both across its faces and within its cross section. Other stitch types such as "tuck" and "float" can enable the placement of material within the fabric without looping (Figure 3) (Spencer 2001). The organization of stitches and stitch types controls the local knit structure, which can be leveraged to embed different fabric behaviors into different regions. In its simplest form, adaptive knit structure is ubiquitous in everyday garments and is visible in the ribbed cuffs of sleeves and socks, often differentiated from the plain body of the garment (Black 2012). There is widespread precedent for the application of knit structure both aesthetically and functionally, in garments

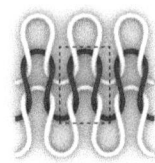 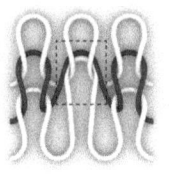 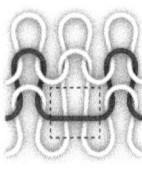

Front Stitch   **Tuck**   **Float**

3   Basic knit stitch types showing different configurations of inter-looped fiber

and in architectural research (Ahlquist 2013; Popescu et al. 2020; Tamke et al. 2021). Deployed into more advanced configurations, knit structure can be leveraged as a tool to affect behavior and the form of a surface. Furthermore, knit structure can be organized along with material selections to differentiate and fine-tune every area of the fabric. Our knitting strategy seeks to move beyond simple surface-based application of knit structure and material integration, demonstrating a multi-layer three-dimensional seamless assembly where the properties of each layer can be independently manipulated.

FUNCTIONAL REQUIREMENTS

The proposed spacesuit prototype responds to a range of functional requirements, including pressure on the body, ease of mobility, speed of donning and doffing, potential thermal and radiation protection, and the ability to integrate fiber-based sensing systems. We manage these requirements through a seamless knit fabric construction that can integrate multiple materials, layers, and functions that are tailored to different regions of the body.

Compression

Though many factors affect the necessary amount of pressure to survive in space, 24 kPa has been proposed as a target pressure for MCP suits (McFarland, Ross, and Sanders 2019). However, solutions to pressure suit design can be studied at lower pressures during initial prototyping. Our sleeve seeks to develop a path forward for 3D knitting in spacesuit pressure garments and to provide a minimum of 5 kPa to exceed the typical pressures (1.9-2.3 kPa) of flight socks (Belcaro 2003). Compression force is typically managed by manipulating the reduction factor of the elastic fabric relative to the corresponding circumference of the body. In contrast to a fluctuating reduction factor, our prototype applies a constant reduction factor across varying circumferences of the body and modulates the fabric power to maintain a uniform compression force. A reduction factor for the sleeve prototype was selected by testing a set of knit elastic cuff samples to determine the required amount of fabric power for different circumferences of the body (Figure 4). Furthermore, the surface of the body also contains many concavities, such as behind the knee or the inside of the elbow. These regions create additional challenges for fabrication, where pressure must be maintained through integrated padding to fill the concave areas. Here we demonstrate that this can be achieved through a combination of fiber materials and seamless knit construction.

Mobility

As a result of the high level of pressure applied to the body, enabling joint mobility while maintaining the required level of compression can present significant challenges. Our prototype seeks to provide at least 45 degrees of joint mobility in the elbow while maintaining the required mechanical pressure on the body. We aim to achieve this primarily through the manipulation of knit structure to provide a range of different elasticities embedded in different regions of the fabric through both elastic and non-elastic materials. In non-elastic layers, mobility is enabled through a region of higher surface area surrounding the outer elbow. In elastic layers, specific regions allow higher stretch in the longitudinal direction.

Donning and Doffing

The high level of pressure also creates challenges for the ease and speed of donning and doffing the suit. Simple garment closure mechanisms are not fully applicable in this scenario, often requiring significant assistance during the donning process. Rather than a single closure mechanism, we utilize the multilayered knit construction to separate the closure into different stages, distributing the compression forces into two closure sequences to speed the donning process to under one minute and enable an individual to put on the garment without assistance.

Integration with Sensing Systems

Finally, the system is designed to create a seamless integration of sensing systems to collect data about the performance of the suit (i.e., pressure) or the physical motions of the occupant. To match the mechanical and geometrical characteristics of the knit fabric, we use fiber-based sensors that demonstrate the potential for soft and stretchable sensing integration. The smart sensing is enabled by highly soft, flexible, and stretchable fibers fabricated through thermal drawing, a scalable and non-expensive method for microfabrication. Here, we propose a built-in network of channels and pockets into which soft stretchable fiber-based sensors can be integrated. Our prototype contains two circumferential sensors for measuring pressure at the bicep and forearm as well as a longitudinal sensor across the elbow joint which measures joint movements using strain.

METHOD

To test our approach, we fabricated a prototypical spacesuit sleeve. Containing a range of conditions within the arm region, the sleeve prototype allows us to evaluate our fabrication strategy centered on CNC knitting and our material, pressure, and sensor integration. We begin by establishing a pattern for an individual's unique body shape and determining the distribution of zones, layers and components, with a set of specific fiber materials that serve unique functions in the knit fabric. The pattern is then developed into a knitting file that addresses the requirements through fine-tuned placement of materials and knit structures. Finally, a two-layer closure is installed on the longitudinal edges of the knit panel and assembled with the sensing components.

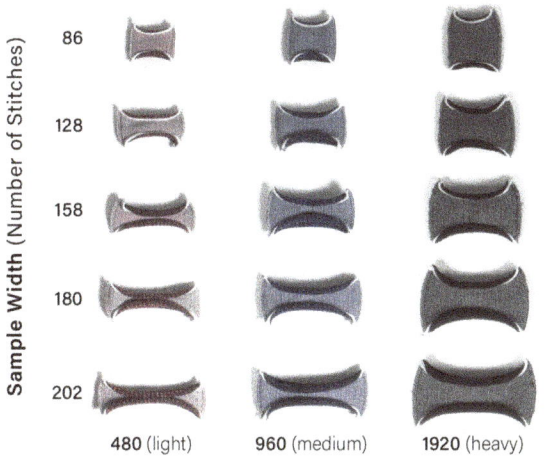
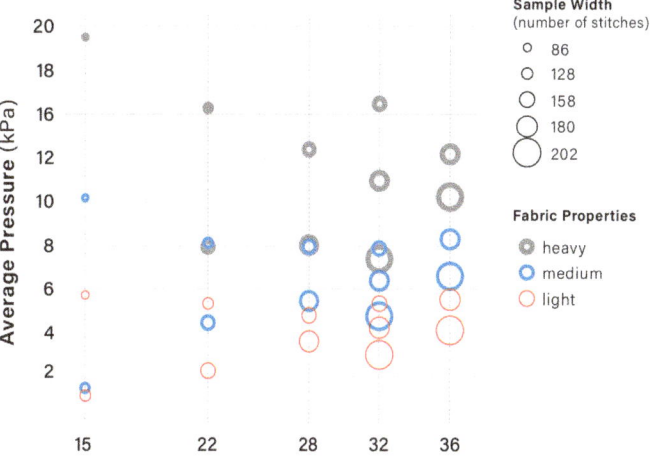

4 Knit cuff samples for collecting pressure data (left) and pressure measurements showing relationship between sample width, fabric properties, and reduction factor (right). As expected, pressure increases with a decreased sample width (i.e., an increased reduction factor), and pressure increases with heavier fabric properties, both due to increased fabric tension. These effects are consistent with the relationship defined by the hoop stress equation, which states that pressure is directly related to fabric tension and inversely related to cylinder radius (i.e., radius of the arm) (Schauss 2022).

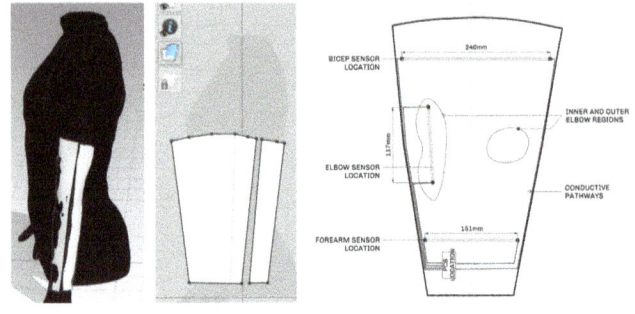

Situating the Requirements on the Body

We have developed a computational manufacturing approach focused on creating a customized fit of the knit spacesuit that is tailored to an individual's body scan. To create the custom sleeve, a person's scanned 3D arm was mapped into zones and unrolled into a 2D pattern (CLO Virtual Fashion 2022). The mapped pattern was then divided into segments to estimate mobility needs in the two primary regions, including the anterior and posterior elbow. The lower arm, elbow, and upper arm horizontal measurements defined the pattern reduction of the knit structure, while maintaining the scan's precise placement of the pattern. The pattern dimensions have unique zones on the arm, which are then converted to a file for knitting.

Knitting

The textile panel is composed of multiple layers that are knit simultaneously: a thin elastic layer against the skin, a thicker transverse layer of elastic fibers in the central layer, and a protective polyethylene layer on the exterior (Figure 6). Each layer is independently capable of a range of variability in the pattern design, enabling the single textile sleeve to adapt to the dimensions of the body and the functional requirements of the spacesuit allowing properties of the fabric to be differentiated across the structure (Figure 7). In addition to the in-plane variation, the layers are capable of separating and merging with each other in cross section, enabling the formation of interior channels, pockets, and integrated plackets that accommodate assembly components. The pattern was developed in STOLL M1 Plus software interface and produced on a STOLL CMS 330 HP-W TT Sport 12-10 knitting machine (STOLL 2021).

The central elastic layer is responsible for achieving the compression requirements of the sleeve. The prototype employs multiple spools, or "ends" of elastomeric nylon Lycra™, but the method can potentially employ any knittable elastic material. To increase and decrease fabric power in different areas of the sleeve, the knit pattern is configured to customize how many cumulative strands are placed into different areas of the fabric. Formed from a series of horizontal tuck stitches, the density of horizontal courses of elastic fiber can be increased or decreased based on the desired compression force and body circumference. The pattern can be manipulated to adjust the number of elastic strands that are placed inside each knitting row. As a result, the prototype demonstrates a gradient of elastic fiber density as the circumference shifts from its smallest at the wrist to its largest at the bicep, containing twice as many elastic fibers per unit area as the wrist (Figure 8).

Mobility requirements are addressed through the interaction of the innermost and outermost layers of the knit panel. On the exterior layer, extra courses of stitches create a region with a higher surface area compared to the surrounding fabric, creating pleats and folds that expand and collapse with motion of the elbow (Figure 9, left). On the innermost elastic layer, the elbow region is differentiated by adding twice as many rows of elastic stitches where the textile requires the greatest elasticity, increasing the potential for longitudinal

5 3D scan (left) converted to unrolled 2D pattern (right) with key regions for knit structure, locations of integrated sensors, and conductive pathways

6 Detail of three-layered fabric construction

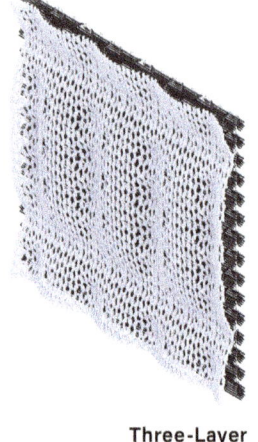

Three-Layer Fabric Construction

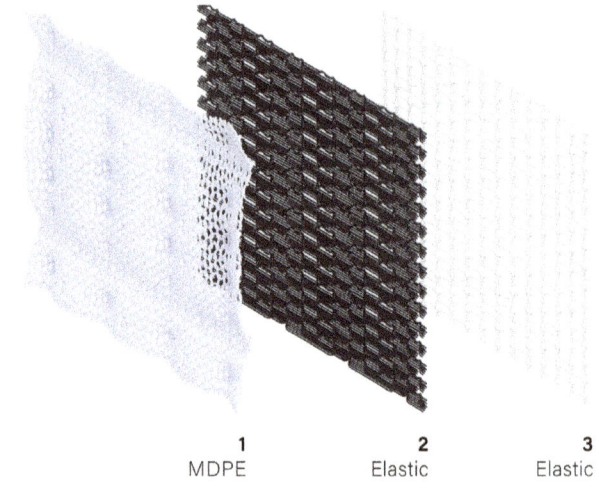

1 MDPE 2 Elastic 3 Elastic

strain (Figure 9, right). Conversely, surface area and elastic stretch are restricted in the interior elbow region by adding fewer stitches within the same area of the pattern.

The exterior layer of the knit structure creates a continuous protective shell of 892-denier Medium-Density Polyethylene (MDPE) multifunctional yarn. The exterior shell is integrated with the padded regions, which consist of a heat-responsive bicomponent fiber. This outer layer seeks to create a continuous protective barrier with radiation shielding capability while maintaining the seamless integration for compression and mobility requirements. Initially soft and pliable, the MDPE fiber is heat-set at 85°C after knitting, which rigidifies the material and reduces the large porosity in the knit structure. The exterior layer is connected to the inner elastic layers through intermittent tuck stitches reaching to the opposite face of the fabric, forming a rectangular "quilted" appearance (Figure 10. left). The rectangular pattern negotiates the surface area differences between the elastic inner layers and inelastic exterior layer, building enough slack throughout the

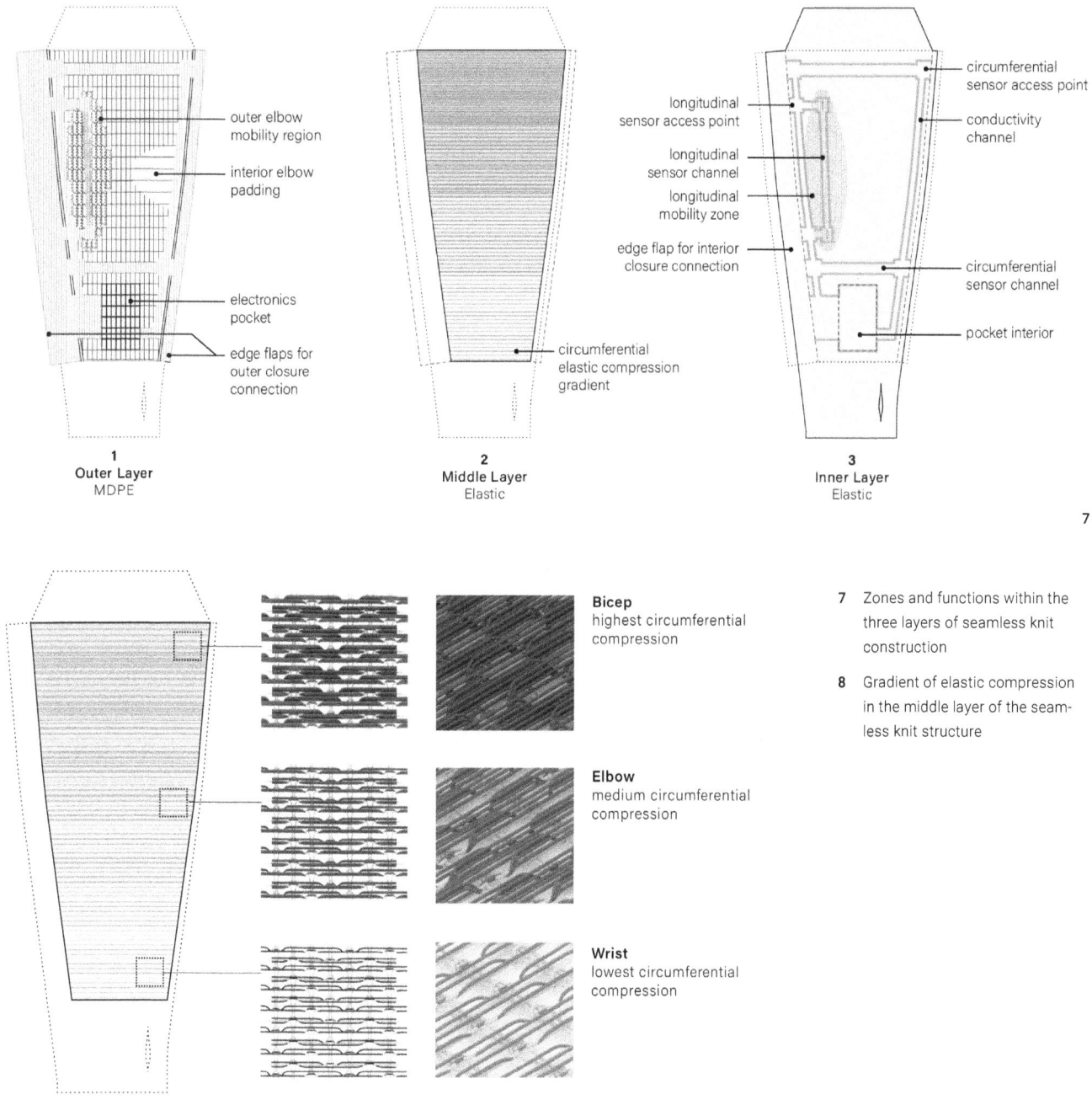

7 Zones and functions within the three layers of seamless knit construction

8 Gradient of elastic compression in the middle layer of the seamless knit structure

9 Knit zones enable longitudinal strain in the outer layer of non-elastifc material (left) and innermost layer of elastic material (right)

10 Detail of typical exterior layer condition (left) and integrated padding in concave elbow region (right)

11 Scanning electron microscopy (SEM) image of a 72-filament PE yarn (~7° tilting view); scale bar is 200 μm; linear density of yarn is 892 denier. The PE yarn was fabricated via a melt extrusion process by a filament line machine at Hills Inc. (Florida, USA). In a typical example, the pellets of medium density polyethylene resin were placed into the hopper with a screw rod that pushes them into the extruder barrel and then pumped through a die with 72 spherical holes. The extruded polyethylene fiber rapidly solidified in air at room temperature, forming a yarn with 72 filaments.

inelastic material system that all the materials are able to flex together (Figure 10, right). Padded areas are produced using a 72-filament, 290-denier, custom-made bicomponent fiber, containing both LLDPE and nylon polymers in its cross section. Before knitting, the material begins as a smooth, flat fiber, but responds to heat after knitting by becoming bulky and pillow-like. The fiber enables selective placement of padded areas directly in a knit fabric without any post-insertion or assembly.

Polyethylene is a versatile material with multiple functions related to space garments. PE molecules are composed of repeating units of two carbon atoms linked to four hydrogen atoms and have high hydrogen content to efficiently absorb and disperse harmful radiation (Narici et al. 2017; Barry 2005). PE is already being used to supplement radiation shielding of the sleeping quarters on the international space station. Existing research shows that polyethylene is 50% better than aluminum at shielding solar flares and 15% better at blocking cosmic rays (Narici et al. 2017). Padded areas that contain LLDPE material further enhance radiation protection properties of the sleeve. The multifunctional PE fabric platform also offers other unique advantages beyond shielding from ionizing radiation. These include stain-resistance, antibacterial properties, passive heat management, and low weight in the spacesuit industry and beyond (Holschuh et al. 2012; Boriskina 2019; Alberghini et al. 2021). As can be seen, the diameter of yarn is commensurate with the conventional sizing of polyester, nylon or hybrid stretchable yarns thus making PE yarn immediately capable of being integrated into standard textile industry processes (Figure 11) (Holschuh et al. 2012; Alberghini et al. 2021).

Finally, the multilayered CNC knitting approach can be leveraged to manipulate the geometric organization of the fabric. A network of interconnected channels was formed between the inner layer and the others, accommodating the strain sensors towards the skin and establishing the required pathways for electronic conductivity (Figure 12a). Junctures in the channel network are accessible through knit-in openings along the sides of the fabric panel, enabling the electronic components to be installed and maintained (Figure 12b). A pocket for

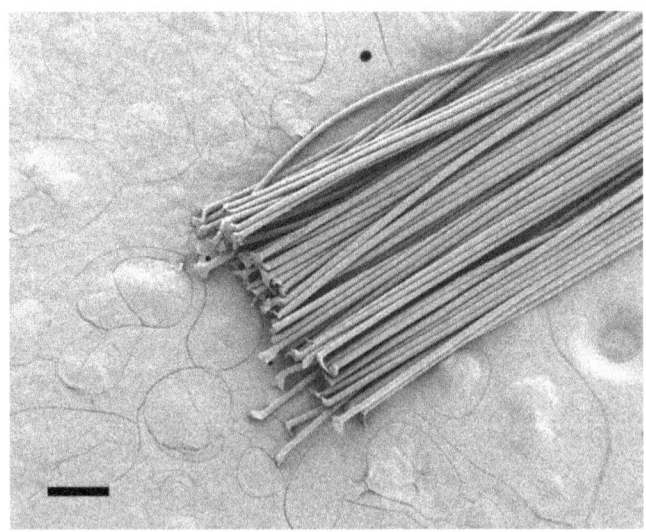

additional electronic components was formed by separating the exterior surface from the inner layers with interconnection points between the pocket region and the channel network (Figure 12c). On the longitudinal edges of the panel, the materials were arranged to form a highly elastic inner membrane belonging to the interior closure, and a sturdier outer layer for attaching to the secondary ratchet closure (Figure 12d).

Assembly

The multi-layer knit fabrication was created to demonstrate a seamless and minimal structure while meeting the needs of the functional requirements and the integrated built-in sensing system. The final assembly of the spacesuit prototype was done by first introducing the two separated closure mechanisms and secondly, by integrating electrical components to the mapped knitted channels.

By dividing the closure mechanism into two stages, the multilayer fabric architecture distributes pressure evenly over the arm while allowing the sleeve to be easily put on and off, spreading the compression forces over two closure sequences. The longitudinal inner membrane of the sleeve panel's edges was sewn onto a zipper. One side was made with a longer knitted length to allow the user to first zip the two edges into the arm in a "relaxed mode;" following this, the remaining pressure is applied with a magnetic boa ratchet system (Figure 13). The outside high-compression layer is sealed with an interwoven cord that was looped between closed-hook fasteners on both sides of the larger strip edges (Figure 13). The ends of the fastening cord were attached to two boa ratchets on upper and bottom parts of the sleeve, allowing faster donning with adjustable tightening along the sleeve.

Internal knitted channels with conductive wires were routed to a knitted pocket with a PCB board, allowing the fiber-based sensors to communicate and perform signal conditioning, processing, and wireless communication via an Arduino Nano 33 BLE Sense. To enable modularity within the integration of the fiber sensor, we introduced a conductive snapping mechanism where one side was connected to the conductive wire system of the PCB and the other to the fiber-based sensor. For the internal wiring of the sensing components, our prototype contained two different stretchable conductive wires that were assembled into the elastic channels of the sleeve. One version produced an insulated connection by threading a stainless-steel yarn into a silicone tube, while the other produced an uninsulated connection by knitting a steel yarn with an elastic yarn to produce an expandable conductive cord (Figure 14a, Figure 14b).

The sensors themselves are fibers, composed of Styrene-ethylene-butylene-styrene (SEBS) and filled with liquid gallium, allowing high stretchability (~100% strain) (Xu et al. 2017). The piezoresistive mechanism of these fibers is used to sense a variety of physical properties such as pressure and strain. Each fiber is encased in a thin layer of soft silicone (Smooth-On Ecoflex 00-30), creating a protective layer (1.2 mm wide by 0.4 mm thick) while maintaining the flexibility and stretchability requirements (Figure 15). The pressure and strain data for each sensor is communicated individually to the on-board system. Signals from these sensing fibers are

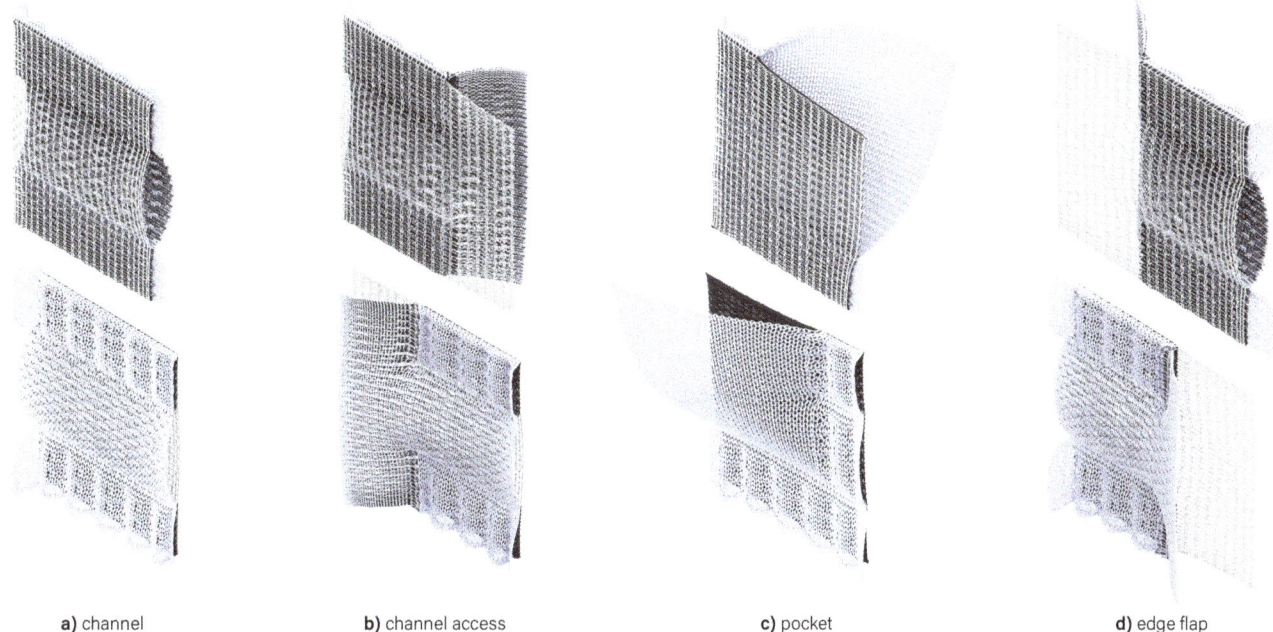

a) channel b) channel access c) pocket d) edge flap

12 Various conditions of surface-to-surface connections and openings produced throughout the multilayer knit panel.

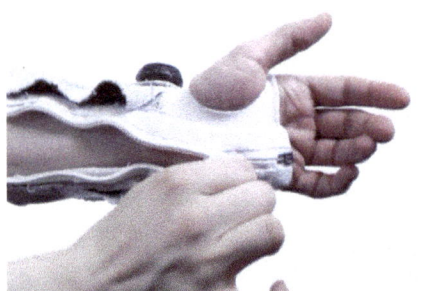 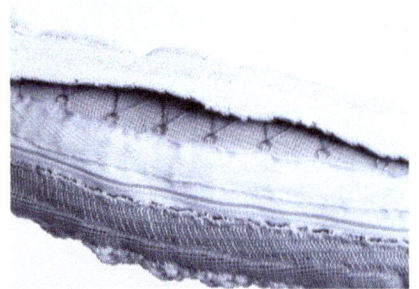 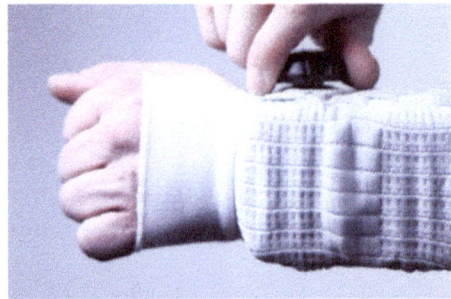

13 Process of donning the sleeve showing inner zipper (left), ratchet cord (center), and ratchet mechanism (right) for tightening outer closure

used to monitor the amount of constant pressure on the body, as well as the strain generated by the wearer (i.e. through joint movements).

RESULTS AND DISCUSSION
Evaluation of Results

A final complete prototype was assembled and tested to evaluate its success relative to our initial goals of compression, mobility, ease of donning and doffing, and integration with sensor systems (Figure 16). Overall, we believe the spacesuit sleeve shows significant variations in fabric behavior across the arm and demonstrates the overall viability of applying CNC knitting to achieve various regions of differing compression and mobility requirements. The sleeve pressure was evaluated by using the Pliance® pressure sensing system (Novel Electronics 2022). Sensors (20 mm diameter) were placed on the inside and outside of the elbow joint to measure the consistency of pressure in different regions as the joint is moved and flexed underneath the fabric, and the sleeve was able to achieve pressures ranging from 4-8 kPa, meeting the minimum pressure goal for the sleeve. Smaller prototype samples were also able to achieve pressures of nearly 20 kPa, showing the knit technique can potentially be effective for generating the full amount of pressure required for a vacuum environment. The multi-layer surface geometry was able to accommodate the installation of sensors and corresponding layout of conductive connections and pathways, and successfully accommodated access points for connecting and adjusting sensor components. We believe our workflow is effective in enabling the design and fabrication of a functional spacesuit prototype that is able to meet multiple requirements in a single seamless system. This result likely could not be achieved with other textile fabrication methods, and our prototype establishes a new way forward for applying CNC knitting and novel material characteristics to future spacesuit development.

Next Steps and Discussion

Though successful as a first step, our initial prototype has limitations, requiring further refinement and testing to move forward as a truly viable spacesuit. First, additional prototyping is needed to calibrate and fine-tune the range of behaviors that exist in the fabric, particularly in how they respond in motion. Currently, an excess of rigidity in the interior elbow and excess material in the exterior elbow results in an uneven distribution of pressure along the circumference of the elbow region, lacking consistency during motion of the arm. Further improvements in the sleeve performance can be achieved through iteration and calibration of fabric properties. While we were able to reduce donning and doffing time to under one minute, additional refinements to the fabric tension are needed in the two-layer closure to further facilitate ease of the donning method. Improvements can be achieved through further iteration and refinement to the knit structures and pattern layout.

To more thoroughly integrate high-performance materials, PE yarns can be engineered to perform a wide range of mechanical properties and replace all the commercial fibers used in the current sleeve prototype. Furthermore, improved radiation shielding and abrasion resistance can be achieved through doping of PE fibers with nanomaterials such as metal-based powders (Alberghini et al. 2021; Blachowicz and Ehrmann 2021).

Immediate next steps include increasing compression levels to meet target pressures above 20 kPa, followed by testing in a vacuum environment. Additionally, future work will include fabricating tests for additional areas of the body, moving

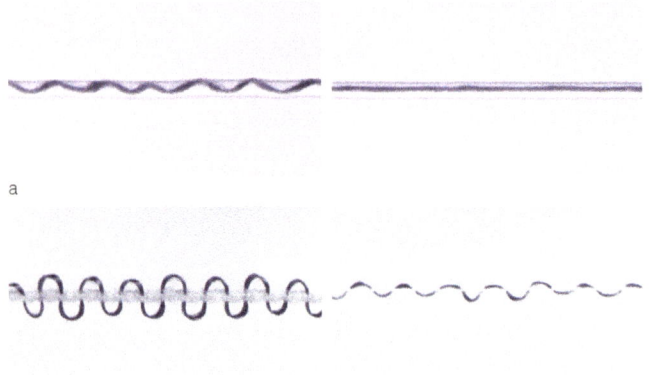

14

towards a full-body spacesuit prototype. Finally, our current testing procedures are heavily focused on evaluating pressure. While this was a primary goal in demonstrating the basic viability of our fabrication method, many other tests would need to be performed to prove that a prototype is prepared for space exploration. Ultimately, the ability to fabricate garments with customizable compression gradients and mobility zones has many potential applications not limited to space exploration; our continued work aims to translate this approach to other architectural textile applications in the near future.

CONCLUSION

The research presented in this paper proposes a new method for the design and manufacturing of spacesuits based on CNC knitting and a multi-functional three-dimensional textile architecture. Through a unique collaboration between designers, scientists, and engineers, we have demonstrated a novel process for creating customized spacesuits, based on a scan of an astronaut's body, and embedded with tunable compression, quick donning / doffing, and integrated sensing. These capabilities enable an approach toward a seamless and minimal spacesuit design without the reliance on bulky systems and complex layered assemblies. Our approach could offer better fit, increased comfort, greater range of motion and more information-rich interaction with the astronaut. The spacesuit sleeve prototype is one step towards this larger vision of fully integrating performance and functionality within an elegant, knitted architecture to allow for better human/space functionality and longer-term survival in the extreme environment of outer space.

ACKNOWLEDGMENTS

This work was partially supported by the DEVCOM Soldier Center through the US Army Research Office (W911NF-13-D-0001), the MIT Deshpande Center, the United States Navy and NERAMCO LTD (N6833521C0489), and the MIT Center for Art, Science & Technology (CAST). SVB and VK thank Brandon Henry, Michael Lampkin, Tony DeLaHoz, and Jeff Haggard from Hills Inc. for help with the PE yarns fabrication, the Dow Chemical Company for providing polymers, Maren Cattonar (NERAMCO), and Yijian Lin (Dow) for useful discussions. The team would also like to acknowledge the contributions of Gihan Amarasiriwardena (Ministry of Supply), Emelie Eldracher (MIT), and Don Haddad (MIT).

REFERENCES

Ahlquist, Sean, and Achim Menges. 2013. "Frameworks for Computational Design of Textile Micro-Architectures and Material Behavior in Forming Complex Force-Active Structures." In *ACADIA 13: Adaptive Architecture; Proceedings of the 33rd Annual Conference of the Association for Computer Aided Design in Architecture (ACADIA)*, 281-292. https://doi.org/10.52842/conf.acadia.2013.281.

Alberghini, Matteo, Seongdon Hong, L. Marcelo Lozano, Volodymyr Korolovych, Yi Huang, Francesco Signorato, S. Hadi Zandavi, et al. 2021. "Sustainable Polyethylene Fabrics with Engineered Moisture Transport for Passive Cooling." *Nature Sustainability* 4 (8): 715–24. https://doi.org/10.1038/s41893-021-00688-5.

Anderson, A., A. M. Hilbert, P. Bertrand, S. McFarland, and Dava J. Newman. 2014. "In-Suit Sensor Systems for Characterizing Human-Space Suit Interaction." In *44th International Conference on Environmental Systems*. Tucson, Arizona. http://hdl.handle.net/2346/59683.

Barry, Patrick L. 2005. "Plastic Spaceships: A 'Designer Material' Derived from Plastic Could Help Protect Astronauts on Their Way to Mars." NASA SCIENCE. August 25, 2005. Accessed Month 18, 2022. https://science.nasa.gov/science-news/science-at-nasa/2005/25aug_plasticspaceships.

Belcaro, G., M. R. Cesarone, A. N. Nicolaides, A. Ricci, G. Geroulakos, S. S. G. Shah, E. Ippolito, et al. 2003. "Prevention of Venous Thrombosis with Elastic Stockings During Long-Haul Flights: The LONFLIT 5 JAP Study." *Clinical and Applied Thrombosis/Hemostasis*. 9 (3): 197–201. https://doi.org/10.1177/107602960300900303.

Blachowicz, Tomasz, and Andrea Ehrmann. 2021. "Shielding of Cosmic Radiation by Fibrous Materials." *Fibers* 9 (10): 60. https://doi.org/10.3390/fib9100060.

Black, Sandy. 2012. *Knitting: Fashion, Industry, Craft*. London: V&A Pub.

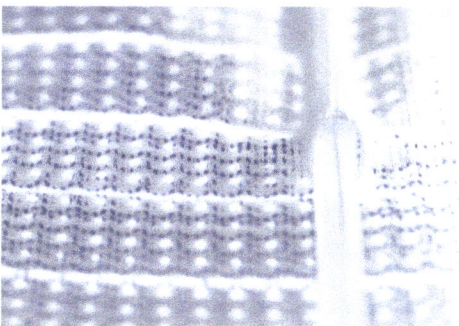
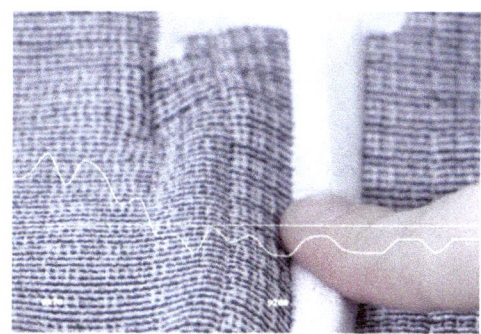

14 Stretchable conductors created from steel fiber threaded into (a) silicone tubing; and (b) by inserting as weft inlay into CNC-knit elastic

15 Details of soft sensors embedded in knit fabric

Boriskina, Svetlana V. 2019. "An Ode to Polyethylene." *MRS Energy & Sustainability* 6: 19. https://doi.org/10.1557/mre.2019.15.

CLO Virtual Fashion LLC. *CLO*. V. 6.2. 2022.https://www.clo3d.com/en/.

De Monchaux, Nicholas. 2011. *Spacesuit: Fashioning Apollo*. Cambridge, Mass: MIT Press.

Holschuh, Brad, and Dava Newman. 2015. "Two-Spring Model for Active Compression Textiles with Integrated NiTi Coil Actuators." *Smart Materials and Structures* 24 (3): 035011. https://doi.org/10.1088/0964-1726/24/3/035011.

Holschuh, Brad, and Dava Newman. 2016. "Morphing Compression Garments for Space Medicine and Extravehicular Activity Using Active Materials." *Aerospace Medicine and Human Performance* 87 (February): 84–92. https://doi.org/10.3357/AMHP.4349.2016.

Holschuh, Brad, Edward Obropta, Leah Buechley, and Dava Newman. 2012. "Materials and Textile Architecture Analyses for Mechanical Counter-Pressure Space Suits Using Active Materials." In *AIAA SPACE 2012 Conference & Exposition*. AIAA SPACE Forum. American Institute of Aeronautics and Astronautics. https://doi.org/10.2514/6.2012-5206.

Holschuh, Brad, Edward Obropta, and Dava Newman. 2015. "Low Spring Index NiTi Coil Actuators for Use in Active Compression Garments." *IEEE/ASME Transactions on Mechatronics* 20 (3): 1264–77. https://doi.org/10.1109/TMECH.2014.2328519.

McFarland, Shane M., Amy J. Ross, and Robert W. Sanders. 2019. "The 'Space Activity Suit' – A Historical Perspective and A Primer On The Physiology of Mechanical Counter-Pressure." In *49th International Conference on Environmental Systems*, Boston, MA. https://ttu-ir.tdl.org/handle/2346/84941.

Narici, Livio, Marco Casolino, Luca Di Fino, Marianna Larosa, Piergiorgio Picozza, Alessandro Rizzo, and Veronica Zaconte. 2017. "Performances of Kevlar and Polyethylene as Radiation Shielding On-Board the International Space Station in High Latitude Radiation Environment." *Scientific Reports* 7 (1): 1644. https://doi.org/10.1038/s41598-017-01707-2.

Newman, Dava, Marita Canina, and Guillermo Trotti. 2007. "Revolutionary Design for Astronaut Exploration — Beyond the Bio-Suit System." *AIP Conference Proceedings* 880 (January): 975–86. https://doi.org/10.1063/1.2437541.

Obropta, Edward W., and Dava J. Newman. 2016. "Skin Strain Fields at the Shoulder Joint for Mechanical Counter Pressure Space Suit Development." In *2016 IEEE Aerospace Conference*, 1–9. https://doi.org/10.1109/AERO.2016.7500744.

Novel Electronics Inc.. *Pliance-x 16 Expert*. V. 23.3.4. 2022.

Popescu, Mariana, Matthias Rippmann, Andrew Liew, Lex Reiter, Robert J. Flatt, Tom Van Mele, and Philippe Block. 2021. "Structural Design, Digital Fabrication and Construction of the Cable-Net and Knitted Formwork of the KnitCandela Concrete Shell." *Structures* 31 (June): 1287–1299. https://doi.org/10.1016/j.istruc.2020.02.013.

Schauss, Gabriella, Rachel Bellisle, Kothakonda Akshay, and Anderson Allison. 2022. "High Performance Mechanical Counter-Pressure Spacesuit Glove for Martian Surface Exploration." In *51st International Conference on Environmental Systems*, ICES-2022-191, 10–14. St. Paul, Minnesota.

Spencer, D. J. 2001. *Knitting Technology: A Comprehensive Handbook and Practical Guide*, 3rd edition. Cambridge: Woodhead Publishing.

STOLL. *M1Plus*. V 7.2.037. 2021.

Tamke, Martin, Yuliya Sinke Baranovskaya, Filipa Monteiro, Julian Lienhard, Riccardo La Magna, and Mette Ramsgaard Thomsen. 2021. "Computational Knit – Design and Fabrication Systems for Textile Structures with Customised and Graded CNC Knitted Fabrics." *Architectural Engineering and Design Management* 17 (3–4): 175–95. https://doi.org/10.1080/17452007.2020.1747386.

Thomas, Kenneth S. 2006. *US Spacesuits*. Springer-Praxis Books in Space Exploration. Berlin: Springer.

Webb, P. 1968. "The Space Activity Suit: An Elastic Leotard for Extravehicular Activity." *Aerospace Medicine* 39 (4): 376–83.

Wessendorf, Ashley M., and Dava Newman. 2012. "Dynamic Understanding of Human-Skin Movement and Strain-Field Analysis." *IEEE Transactions on Bio-Medical Engineering* 59 (12): 3432–38. https://doi.org/10.1109/TBME.2012.2215859.

Xu, Jie, Sihong Wang, Ging-Ji Nathan Wang, Chenxin Zhu, Shaochuan Luo, Lihua Jin, Xiaodan Gu, et al. 2017. "Highly Stretchable Polymer Semiconductor Films through the Nanoconfinement Effect." *Science* 355 (6320): 59–64. https://doi.org/10.1126/science.aah4496.

IMAGE CREDITS

Figure 2a: ©David Clark / NASA
Figure 2b: ©Paul Webb / NASA
Figure 2c: As follows below:
Inventor, Science & Engineering: Professor Dava Newman, MIT
Design: Guillermo Trotti, AIA, Trotti and Associates, Inc. (Cambridge, MA)
Fabrication: Dainese (Vicenza, Italy)
Photography: Dougas Sonders

All other drawings and images by the authors.

Lavender Tessmer is a PhD Student in Design and Computation in the Department of Architecture at MIT.

Ganit Goldstein is an MS student in Design and Computation in the Department of Architecture at MIT. She is a Research Assistant in the Self-Assembly Lab.

Guillermo Herrera-Arcos is a PhD student and Graduate Researcher at the MIT Media Lab in the K. Lisa Yang Center for Bionics.

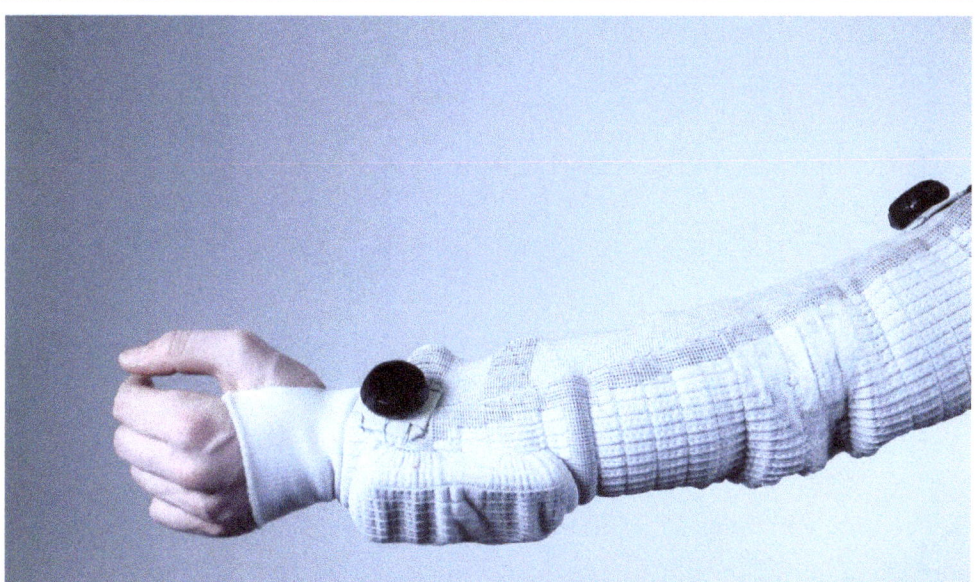

16 Completed and assembled sleeve prototype

Volodymyr Korolovych is a Research Scientist at the MIT Institute for Soldier Nanotechnologies (ISN).

Rachel Bellisle is a PhD Candidate in the Harvard-MIT Health Sciences and Technology Program and a Draper Scholar. She is a Graduate Researcher in the MIT Human Systems Lab (HSL) in the Department of Aeronautics and Astronautics.

Cody Paige is a PhD Candidate in the Department of Aeronautics and Astronautics at MIT and a Graduate Researcher in the MIT Human-Systems Lab (HSL).

Christopher Shallal is a PhD student in Medical Engineering and Medical Physics (MEMP) in the Harvard-MIT Health Sciences and Technology Program. He is a Graduate Researcher in the MIT Media Lab in the K. Lisa Yang Center for Bionics.

Atharva Sahasrabudhe is a PhD student in the Department of Chemistry and a Graduate Researcher in the Bioelectronics Lab in Department of Materials Science.

Hugh Herr is a Professor of Media Arts and Sciences at the MIT Media Lab and co-director of the K. Lisa Yang Center for Bionics at MIT.

Svetlana V. Boriskina is a Principal Research Scientist in the Department of Mechanical Engineering at MIT and Director of the META Research group.

Dava Newman is the Director of the MIT Media Lab and the Apollo Professor of Astronautics at MIT.

Skylar Tibbits is co-director of the Self-Assembly Lab and Associate Professor in the Department of Architecture at MIT.

Session Introduction

Advanced Materials

Ezio Blasetti, Chair

This session explores the state of the art in the relationship between Design and "Advanced Materials." The selected papers manifest how design methods and scope evolve with the introduction of new ways of looking at material processes.

Matter is active and seen through the lens of time: real time through its use and re-use at the construction site; simulation time in custom physics software; fabrication and curing time in the production of prototypes; circular and lag time in parametric materials.

The research opens up fundamental questions about architectural representation, negating conventional abstractions. The authors create unique digital representations as models and datasets of various resolutions that correspond to different methods of inquiry and control of materials. The intention is to interrogate matter with all of its irregularity and complex behaviour in order to forge a collaboration with artificial intelligence and robotic fabrication.

Computation is used as an augmentation of the agency of the designer as the decision-maker with a variety of performative, aesthetic and environmental criteria. Aesthetics are not reserved only for the final product or prototype, instead drive decisions at multiple steps in the digital models, simulations and physical experiments.

This research pushes against the limits of computation and robotic fabrication in order to simulate, predict and actuate matter in a range of scales. The papers are presented in a sequence that performs a long zoom in, extending the scope of design across different scales. The sequence begins at the building scale, zooms into a part, focuses on particle behaviour and stops just below the frequency of light at the scale of electron microscopy.

"Advanced Materials" offer multiple possibilities for hybridization. By establishing protocols of information exchange, authors, sensors, algorithms, robotic end effectors and materials enter in complex relationships. This produces hybrid processes and artefacts with a variety of human and non-human agencies registering their own haecceities on the design outcome. The designer is often found performing a delicate calibration that controls the hybrids and haecceities of each of the contributing elements and processes.

Inventory

Gil Sunshine

CAD for Medium Resolution Materials

1 A pile of "medium resolution" materials in the *Inventory* modeling space

ABSTRACT

With commands like "extrude," "trim," "revolve," and "array," contemporary CAD software contains analogues to the machine processes and abundances of industrial mass-production. This produces in the designer an affinity towards the predictable surface of standardized building materials. Furthermore, by instantiating geometries—that appear seemingly out of nowhere—with minimal apparent effort or cost, such software obscures the extraction, processing, transportation, and waste necessary to physically realize said geometries. This suggests that the only limits on the material world are processing power and file storage capacity, which poses a challenge as it is becoming increasingly apparent that circular construction is key to minimizing the deleterious environmental impacts of construction (Heisel 2021).

Inventory offers an alternative to contemporary CAD software, where the gap between digital models and physical constraints is vast. Rather than abstract commands that project forth a not yet existing material condition, *Inventory* is based on digital representations of specific pieces of material and processes for fabricating assemblies of parts. By the very nature of their being digital, these representations are necessarily approximations of their physical counterparts. They inhabit the space between the low resolution of pure geometric abstraction and high resolution of physical phenomena, and therefore, might be called "medium resolution" (Sunshine 2022). *Inventory* uses game engine physics to embed simulations of physical constraints in the digital modeling process. *Inventory* is a software interface for making architecture in a medium resolution world.

INTRODUCTION

During the industrial revolution, machines and techniques were invented and advanced to form materials into standardized shapes. This led to the development of the industrially mass-produced, standardized building materials used today and left an indelible mark on architectural thought. The architecture of modernism, especially the International Style, embodied the logics and efficiencies of the factory, while the endless linear and rotational processes of industrial production were quite literally imprinted on its geometries. To an unprecedented degree, architecture became a practice of blind trust in superficially dimensioned materials specified from afar.

This proved to be highly expedient in the case of the interwar and postwar periods in the United States, when swaths of the suburban landscape were populated with kit houses. Taken to the extreme in this context, the standardized material palette made it possible for buildings themselves to be standardized, calling into question the role of the architect entirely (Mumford 1930). Like the 2x4, an entire building could be mail ordered and delivered to site as a kit of parts, and to be pieced together by the builder with no architect in sight. Whether or not one is concerned with the status of the architect or with the aesthetic condition of sameness that the standardized material palette produced, it should also be noted that production of standardized materials is materially inefficient, at least in the case of the timber products from which the postwar suburban home was built. The dimension of the 2x4, which actual measures 1 1/2" x 3 1/2", does not respond to some existing condition in the dimensions of a growing tree to maximize its use, but rather is the result of the pressures of the housing market (Curtis 2018).

This intertwining of mass manufacturing and architecture has only intensified over time, finding its way into the tools of architect. The 3D modeling software used by the architect contains analogs to the machine processes and abundances of industrial production with commands like "extrude", "trim", "revolve" and "array". In Software Takes Command, Lev Manovich argues that this can be understood as the architectural logic developed around mass-produced housing encoded as software (Manovich 2013). Today, however, as we increasingly face the effects of unsustainable resource consumption and related disruptions to the building material supply chain, architecture must develop a broader material palette to include the found, the unwanted, the offcut and the wasted. This produces a new relevance for an architecture of underprocessed and irregular materials and demands a reconsideration of CAD tools that favor the predictability of the industrially produced surface.

In order to adapt to material irregularities, architects have adopted various 3D scanning techniques to produce digital representations of materials, which can then be manipulated in 3D modeling software to derive digitally informed fabrication and assembly processes. By the nature of their discrete sampling, these representations vary in their precision. What the architect encounters in the 3D modeling software is not the material itself in its infinite specificities, with its weight, moisture content and smell, but rather a surface representation, a mesh, composed of a large but finite set of points. This surface might be called "medium resolution."

If there exists a deeply ingrained connection between 3D modeling software used by the architect and the processes of industrial mass production, then in order to overcome the cognitive grasp of the standardized material palette, the 3D modeling software itself must be reconsidered. In general, 3D modeling software allows the designer to create and then alter a geometric representation. Though the geometry in a 3D modeling environment always references future actions and energy inputs upon construction, shape itself does not suggest process. Digitally, a sphere booleaned from a cube suggests a process of removal. Physically, this shape might be fabricated using an additive process like 3D printing.

This research develops an alternative to this model of design in the form of CAD software named *Inventory* that foregrounds the constraints of materials and fabrication processes. *Inventory* is an inventory of inventories that each contain medium resolution representations of specific pieces of material, sites, actions used to digitally fabricate joints between parts, and algorithmic patterns of interaction for assembling parts. *Inventory* uses game engine physics to allow the designer to model with some of the constraints of physical making. Unlike CAD software where geometric freedom is paramount, *Inventory* prioritizes the considerations of building in the physical world.

NONSTANDARD MATERIALS IN ARCHITECTURE

The widespread availability of 3D scanning technology has given way to a range of research concerned with the accommodation of nonstandard materials in CAD and digital fabrication. Common throughout this broad category of research is the use of 3D scanning to build a library of parts that can be digitally manipulated and assembled by computational means before being realized physically through digital fabrication workflows. Such research recognizes limitations of the standardized material palette and the latent potential—in terms of embodied carbon, structural performance, and aesthetics—that working with nonstandard materials provides.

The use of nonstandard timber components is of particular interest in that preserving the geometric irregularity of a tree section, especially the tree crotch, is directly tied to preserving its structural efficiency (Mollica and Self 2016; Amstberg et al. 2020; Von Buelow et al. 2018; Enns 2010). For the structure of the "Wood Chip Barn," designed and built by students and staff at the Architectural Association's Hooke Park, scans of tree forks are simplified to centerline geometries and oriented to match a goal geometry using an evolutionary optimization algorithm (Mollica and Self 2016). With full mesh scans oriented, bearing geometries are embedded within the parts to join them with additional support for tension and shear provided by steel elements. In this project, the stock is minimally altered, avoiding the need for additional spanning materials. Alternatively, the "Structural Upcycling" project from the Digital Structures Group, "LIMB" from a team of researchers from Michigan University, and "NATUREFRAME" by Johnathan Enns use smaller tree fork sections as structural nodes with standardized lumber infill.

Masonry construction is another area of research ripe for similar approaches to the use of nonstandard materials. "Digital Rubble" by Bastian Wibranek and Oliver Tessmann explores the use of 3D printed connections between irregular stones to allow for freestanding dry-stacked masonry structures (Wibranek and Tessmann 2019). The project uses evolutionary solvers to optimize the orientation of stone along a thrust line. Brandon Clifford and Wes McGee's "Cyclopean Cannibalism" uses rubble stone and concrete from building sites in the construction of a masonry wall prototype (Clifford and McGee 2018). This project offers an example where the constraint of complete contact between units in a dry masonry construction dictates that irregular stock must lose its original shape. That said, a recursive algorithm is used to fit the largest four-sided polygon within scanned pieces of stock, which has the effect of minimizing waste downstream in the process.

Inventory builds on existing research into the use of nonstandard materials in architecture, offering a general-purpose framework and interface for accommodating a range of materials and fabrication processes. Though the prototypes presented here are largely made from split wood sections, the process described can be applied to a range of materials.

PHYSICS-BASED MODELING

Inventory leverages game engine physics as a tool to aid in the complexities of modeling with nonstandard materials. Physics simulations have been used by designers for modeling architectural form with a range of divergent motivations. On the one hand, physics simulations offer strategies for optimizing form. The transition of Kangaroo from plug-in to built-in within Grasshopper is indicative of the popularity of such strategies. In addition to their use as form finding tools, physics simulations have found relevance as a compositional technique. MOS Architects' software experiments offer examples of purpose-built software interfaces where geometry collides, collapses, piles, spreads, drapes, etc. to produce forms (or non-forms) that seem to emerge out of incidental encounters between parts. MOS's Michael Meredith argues that such an approach allows for a more "situational" or "relativistic" approach to architecture as opposed to more deterministic models of form making (Meredith 2013). Similar physics-based compositional strategies are now pervasive in architectural practice (Ago 2019).

In this research, the physics simulation is used neither as an optimization technique nor as a purely compositional one. *Inventory* applies game engine physics to the modeling process in three key ways. First, rigid body physics prevents parts from overlapping with one another, emulating the physical reality that solid materials do not intersect. Second, gravity and attraction forces are used to organize parts, such that many parts do not need to be manually placed. Finally, *Inventory* takes advantage of "joints" in the Unity physics engine to constrain the movement and rotation of parts relative to one another (Unity Technologies 2020). In this research the physics simulation is used as tool to organize irregular parts and to plan future fabrication actions through the simulated joining of parts. In that parts (or "architectural particles," as Meredith calls them) are assembled into wholes, the approach taken here might be understood as more similar the one taken by MOS. Rather than the use of pure geometric forms, however, here the parts are 3D scanned mesh representations of specific pieces of material.

METHODS

The *Inventory* interface gives the user access to four primary inventories: parts, sites, actions, and protocols. These inventories can be browsed and an item selected, which will bring up a panel for said item. Any given item can itself be understood as an inventory, a container, for information and media that may be accessed through the item panel. For example, the panel for a part may contain links to video showing the harvesting of the material or information related to its embodied carbon (Figure 2). Item panels also contain relevant UI elements for adding parts and sites to the scene, or activating actions and protocols.

Parts

The parts inventory contains representations of specific pieces of material that can be added to the modeling environment. This instance of *Inventory* is primarily populated with pieces of split wood from logs sourced on Craigslist that have been

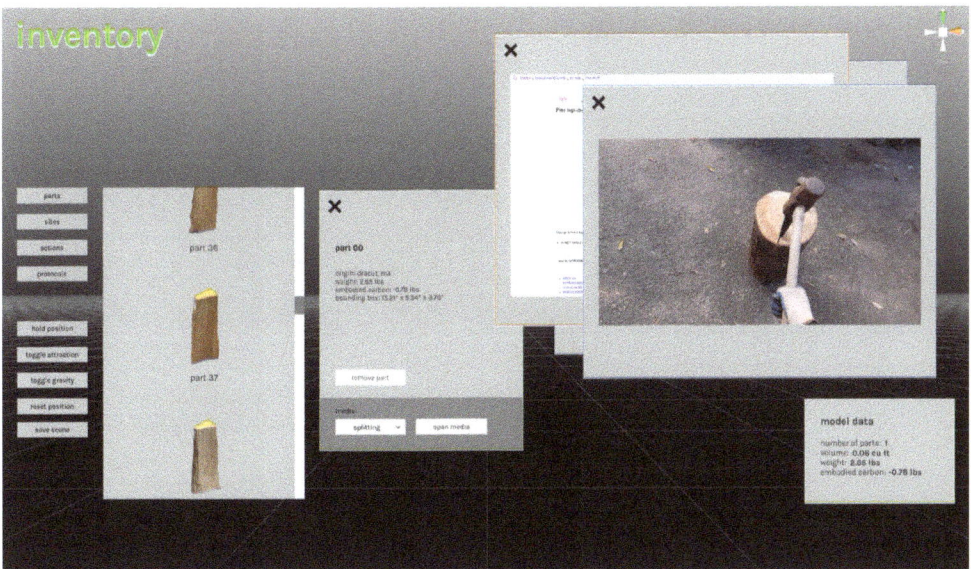

2 The *Inventory* UI showing browsable part inventory, part panel, model data panel, and media related to the sourcing and processing of the selected part

3 Parts collide with one another in the *Inventory* modeling space; the part on the right is selected as indicated by the transform gizmo

4 A scanned site in the *Inventory* modeling space

left behind from the felling of trees in Massachusetts. That said, all manner of matter, is subject to the medium resolution treatment. Specifically depicted here are short softwood log sections not suitable for use as lumber or as indoor firewood and, therefore, fated to end up in the chipper and left to rot, releasing their carbon to the atmosphere. Minimal processing of stock through splitting limits material waste and energy input, preserves the structural integrity of the wood's grain, and reduces distortion as the wood dries. Once standard practice for producing all manner of building materials, from structural members to roofing shakes, wood splitting today has been relegated to firewood production alone. However, the technique finds new relevance in a medium resolution world, where producing planarity through further processing of the stock offers no benefit. *Inventory* understands architecture as the temporary configuration of materials. Therefore, processing does not increase the value of a piece of material; it only limits its potential for further use. Stock may be taken up into an assembly one day, only to be returned to the inventory the next.

Parts are unique mesh representations of specific pieces of material "imported" from the physical environment to the digital. Pieces of material were 3D scanned using the photogrammetry software Metashape (Agisoft 2021). Using the Unity physics engine, parts are configured as rigid bodies with mesh collisions enabled (Unity Technologies 2020). Parts can be translated and rotated using Runtime Transform Gizmos (Octamodius 2021). In *Inventory*, there is no object snapping. Rather, objects collide, settling into one another. There is no coplanarity, only closeness (Figure 3). No "align bottom," only gravity.

Sites

Sites are mesh representations of physical locations. In a medium resolution world there is no distinction between the representation of parts and of sites (Figure 4). Both are meshes. The site is not the place where something is to be built or where materials are to be assembled. Instead, it is understood as part of the assembly itself.

Actions

Actions are digital abstractions of physical operations made on parts. "Dowel Part" for instance refers to making a hole in a piece of stock using a 6-axis operation in order to receive a dowel (Figure 5). "Add Wax" refers to adding a piece of polycapralactone wax, a reusable bioplastic, between two pieces of stock (Figure 6). These actions were developed through a series of physical prototypes and observation of the constraints that pieces of stock, dowels, polycaprolactone wax and rope put on one another in various configurations in order to join parts (Figure 7).

5

6

5 Drilling a hole in a piece of stock to receive a dowel

6 Applying softened polycapralactone wax to a joint

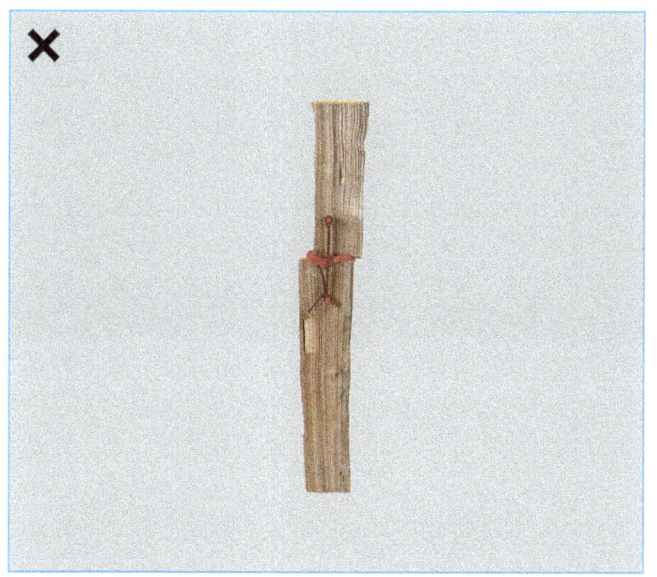 

 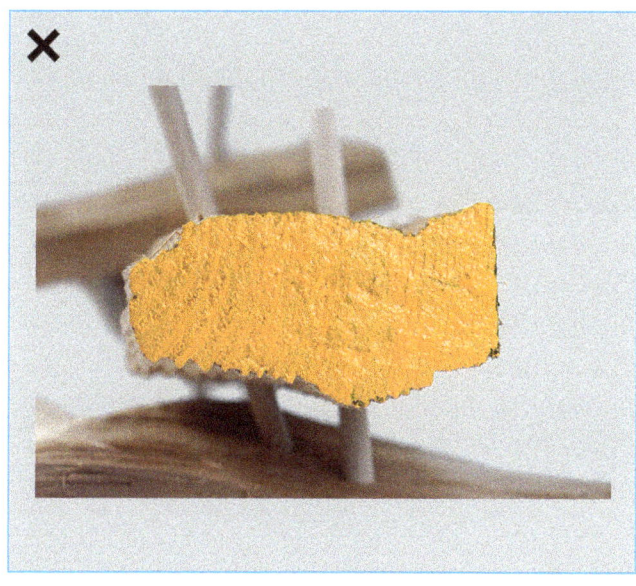

7 Overall views (left column) and corresponding details (right column) of prototype joints used to develop the fabrication actions abstracted in *Inventory*

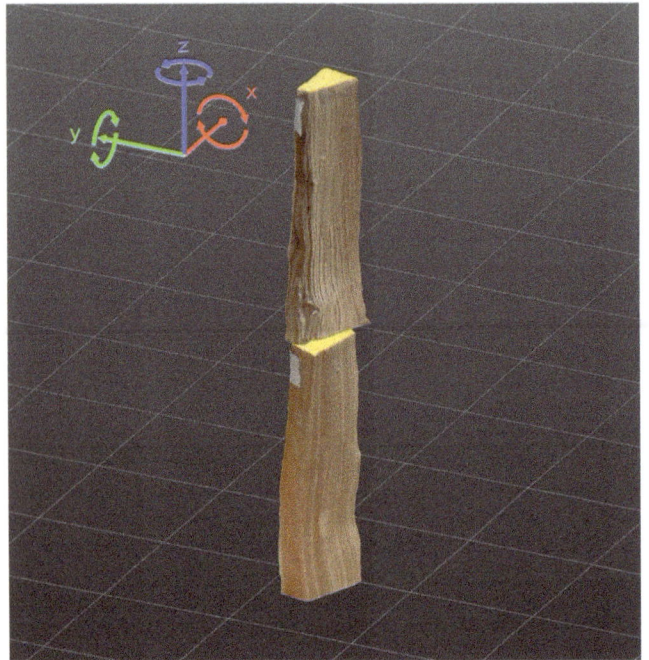
unconstrained joint

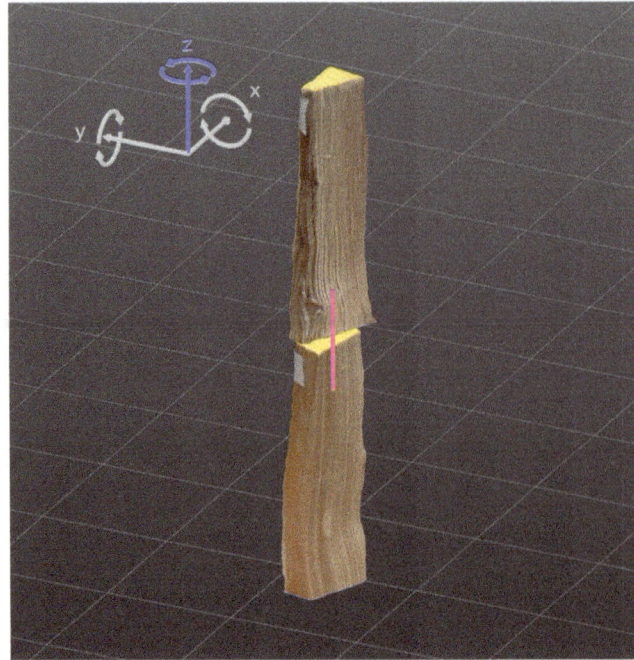
pin added

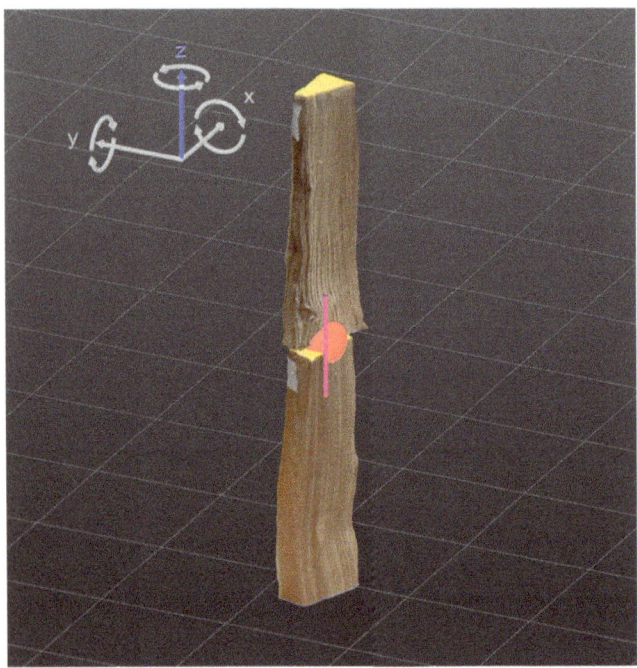
wax added

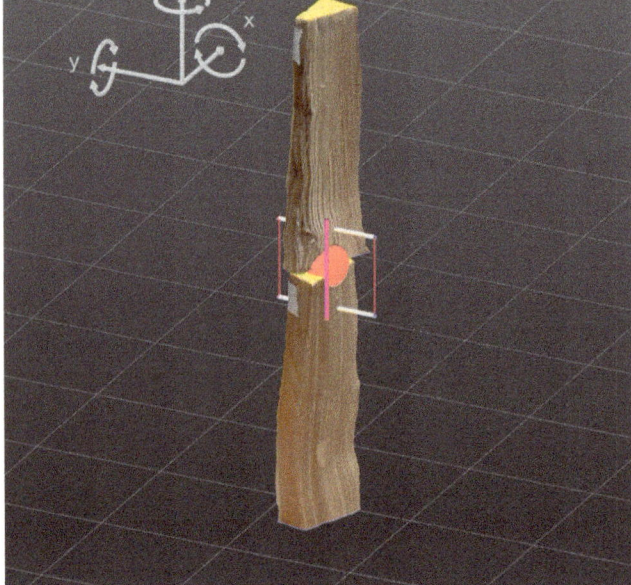
ties added

8   Sequence for modeling an unconstrained to fully-constrained joint between two parts in *Inventory*

 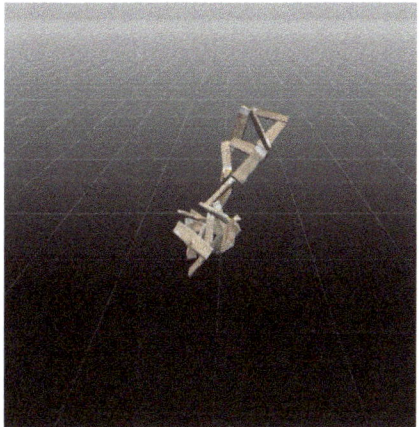 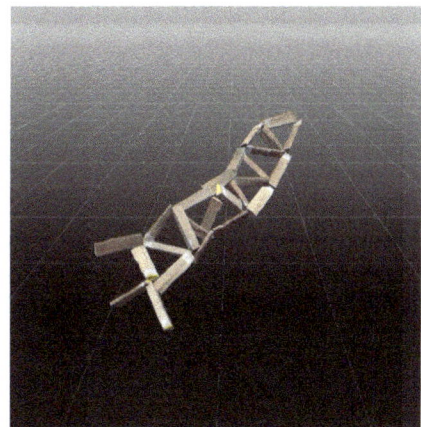

9   Parts organizing into a truss structure based on a protocol

10   Parts interacting with a site to produce a floor-to-ceiling column based on a protocol

Once physical joining methods were developed, they were abstracted into digital actions. Practically speaking, these actions apply joints to rigid bodies in the *Inventory* physics simulation. "Pin Two Parts," for example, asks the user to select two parts and add a pin, representing a wooden dowel, between them. If the z-axis is understood to be along the length of the pin, then once the joint has been activated it restricts motion between parts in the xy-plane and rotation around the x- and y- axes, but allows rotation around and movement along the z-axis (Figure 8). These are the same constraints that a dowel would place on two pieces of stock. In order to restrict rotation around the pin (z-axis), the user might select the "add wax" action, which is an abstraction of smooshing polycaprolactone wax between parts. The polycaprolactone wax, which is a reusable bioplastic, is not an adhesive, but by gripping onto the irregular surface of the stock, it restricts rotation between parts. In order to restrict motion along the z-axis and fully constrain the joint, the "add ties" action could be implemented. First, dowels would be added to each of the pieces of stock, roughly perpendicular to the initial pin (z-axis), and ties added between them. This offers just one example of the various ways in which several actions can be combined to create fully constrained joints.

Like the geometric representations of materials contained within *Inventory*, these digital actions might also be called medium resolution. They are not the actions themselves but fuzzy analogues with material and bodily implications. The actions do not accurately simulate their physical counterparts but rather extract their essential function in constraining rigid bodies.

Protocols

In addition to placing parts one by one, protocols may be used to algorithmically determine patterns of interaction between parts. Parts are assigned anchor points that determine points of attraction, and distance constraints are applied, pulling them together (Figure 9). In addition to attraction forces between parts, sites may also be assigned anchor points creating site-specific assemblies (Figure 10). Form is determined by the unrelenting drive of parts to be close to one another as the physics simulation runs, with the collision of rigid bodies preventing attracted parts from overlapping with one another.

Prototypes

Once a given organization has been achieved by a protocol, actions may be used to fix the position of parts relative to one

11 Fabricated column prototype

another, and the assembly fabricated. The physical actions that were abstracted into *Inventory* again become physically real, now choreographed by the information contained within the software. In this preliminary phase of research, modeled geometry was exported from *Inventory* and imported into Rhino such that KUKAprc could be used for path planning (Robert McNeel and Associates 2021; Association for Robots in Architecture 2021). Future research could develop greater integration of robot controls within *Inventory*.

In order to machine parts, a custom end effector was made to clamp parts of varying dimensions and irregular geometry (Figure 13). Once attached to the robot arm, the part is positioned close to a target graphic at a known location within the robot cell (Figure 13). The positioned part is scanned once more in order to locate it relative to the robot. Within Rhino, the geometry exported from *Inventory* is matched with the geometry of the scanned part attached to the robot arm. In the case of this research, the matching was done manually though this process could be automated using the Iterative Closest Point (ICP) method. With each of the prototypes presented here, the robotic operation was limited to drilling. Drilling preserves a piece of stock to a large extent, maximizing the potential for reuse. With the part attached, the robot arm is positioned relative to a drill press within the robot cell, and the drill bit is lowered manually. The adjustable bed of the drill press is used to support the part against the forces applied by the drill press given the limited payload of the robot arm used.

The first prototype explored the design and construction of a self-supporting column. The protocol for the column involved the end-to-end attraction of several collections of parts. Each of these collections was additionally drawn to an attraction point and a gravity force applied, causing strands of parts to hang to the floor (Figure 11). A second prototype explored the construction of a vault. Parts were attracted to form a collection of three strands all of which met at a "keystone" wood element. The parts at the other ends of each of the three strands were attracted to given points to a achieve a three-legged structure. With a force applied in the positive direction along the z-axis, i.e. opposite gravity, the vault form is achieved.

When there is no longer a need for a given assembly, or if a structural failure occurs, as was the case in the assembly of the vault structure, the parts can be returned to the inventory (Figure 12). However, these are no longer the parts in their original state. Embedded within them are the information and scars of their prior use. In this way, physical parts are inventories of their own histories. The broken parts of the vault assembly are removed and the remaining parts reassembled based on the existing joints they contain. The wax is melted down and reformed to match the new surfaces that meet. Broken parts can also be reinserted into the assembly to tie together the reconfigured vault elements producing a new form (Figure 14).

CONCLUSION

What this research offers is a proposal for how to design. *Inventory* composes architecture through the interaction of parts, sites, actions, and protocols. Each category is made up of digital abstractions of physical phenomena that encode and make apparent their physical impact. Rather than the model-forward practice of design that dominates today and that is predicated on trust in the availability of standard materials, *Inventory* establishes a dialogue between physical material conditions and digital models. As established by the beam prototype, in *Inventory* design and construction are not a one-way stream of information and materials ending as a file in a computer's trash and a 2x4 in a landfill.

It is not that our world should look like these prototypes or be made from these materials in particular, though in part it could be. Instead, it is about disentangling ourselves from the standardized material palette and our reliance on the predictable surface complicit in the failed present, in order to find ways forward. *Inventory* peels back some of the layers of abstraction authored into contemporary CAD software to allow the designer to be more specific and explicit about their interaction with the physical world.

As this research is expanded upon, greater integration of fabrication protocols, specifically robotic path planning and controls, would be an important addition to realize more

12 Assembly of vault prototype to the point of failure

13

13 Piece of stock attached to the robot arm via custom end effector. Target graphic can be seen in the background in red.

14 Beam fabricated from the reconfigured parts of the vault prototype.

14

fully *Inventory* as an interface between the designer and the physical world. Furthermore, as an open-ended platform for design this research anticipates the expansion of each of the inventories to accommodate a greater diversity of materials (parts), places (sites), fabrication processes (actions), and algorithmic design strategies (protocols).

ACKNOWLEDGEMENTS

The research presented herein was completed as part of a Master of Architecture thesis at the Massachusetts Institute of Technology. I would like to acknowledge the faculty who advised the thesis, Axel Kilian, Brandon Clifford, and Caitlin Mueller for their support and feedback throughout the process. I would also like to thank the architecture shop staff for their support in realizing the prototypes presented in this paper.

REFERENCES

Ago, Viola. 2019. "Compositional Physics and Other Diagrams of Force." *Log* 46 (Summer): 33–43.

Agisoft. 2021. *Metashape*. V.1.7.3.

Amstberg, Felix, Yijiang Huang, Daniel J.M. Marshall, Kevin Moreno Gata, and Caitlin Mueller. 2020. "Structural Upcycling: Matching Digital and Natural Geometry." In *Advances in Architectural Geometry 2020*, edited by O. Baverel, C. Douthe, R. Mesnil, C. Mueller, H. Pottmann, and T. Tachi. Paris: Presses des Ponts. 486–504.

Association for Robots in Architecture. 2021. KUKAprc.

Clifford, Brandon, and Wes McGee. 2018. "Cyclopean Cannibalism: A Method for Recycling Rubble." In *ACADIA 2018: On Imprecision and Infidelity; Proceedings of the 38th Annual Conference of the Association for Computer Aided Design in Architecture*, 404–13. Mexico City: ACADIA.

Curtis, Oliver J. 2018. "Nominal Versus Actual: A History of the 2x4." *Harvard Design Magazine* 45 (S/S): 40–41.

Enns, Jonathan. 2010. "Intelligent Wood Assemblies: Incorporating Found Geometry and Natural Material Complexity." *Architectural Design* 80 (6): 116–21.

Heisel, Felix. 2021. "Reuse and Recycling: Materializing a Circular Economy." In *The Materials Book*, edited by Ilka Ruby and Andreas Ruby. Berlin: Ruby Press. 156–60.

Manovich, Lev. 2013. *Software Takes Command*. London: Bloomsbury Academic.

Meredith, Michael. 2013. "After After Geometry." *Architectural Design* 83 (2): 96–103.

Mollica, Zachary, and Martin Self. 2016. "Tree Fork Truss: Geometric Strategies for Exploiting Inherent Material Form." In *Advances in Architectural Geometry 2016*, edited by S. Adriaenssens, F. Gramazio, M. Kohler, A. Menges, and M. Pauly. Zürich: VDF. 138–53.

Mumford, Lewis. 1930. "Mass Production and the Modern House." *Architectural Record* 66 (1): 13–20.

Octamodius. *Runtime Transform Gizmos*. V.1.3. 2021.

Robert McNeel and Associates. 2021. Rhinoceros. V.7.8.21196.05002.

Sunshine, Gil. 2022. "Medium Resolution." Master's Thesis, Massachusetts Institute of Technology.

Unity Technologies. *Unity*. V. 2020.3.8f1. 2020.

Von Buelow, Peter, Omid Oliyan Torghabehi, Steven Mankouche, and Kasey Vliet. 2018. "Combining Parametric Form Generation and Design Exploration to Produce a Wooden Reticulated Shell Using Natural Tree Crotches." In *Proceedings of IASS Annual Symposia, IASS 2018 Boston Symposium: Timber Spatial Structures*. 1–8.

Wibranek, Bastian, and Oliver Tessmann. 2019. "Digital Rubble: Compression-Only Structures with Irregular Rock and 3D Printed Connectors." In *Proceedings of the IASS Annual Symposium 2019, IASS 2019 Barcelona Symposium: Advanced Manufacturing and Non-conventional Materials*. 1–8.

IMAGE CREDITS

All images by the author.

Gil Sunshine develops software tools and digital fabrication processes that reconsider accepted modes of designing and making. Gil is a recent graduate of the Master of Architecture program at the Massachusetts Institute of Technology.

Parametric Matter

'Pushing' Updates into Materials and the Implications of Legacy and Lag

Dr. Adam Blaney
LICA, Lancaster University

Dilan Ozkan
HBBE, Newcastle University

Dr. Emel Pelit
MSI, Lancaster University

Dr. Mariana Fonseca Braga
LICA, Lancaster University

Dr. John G. Hardy
MSI, Lancaster University

Dr. Mark Ashton
MSI, Lancaster University

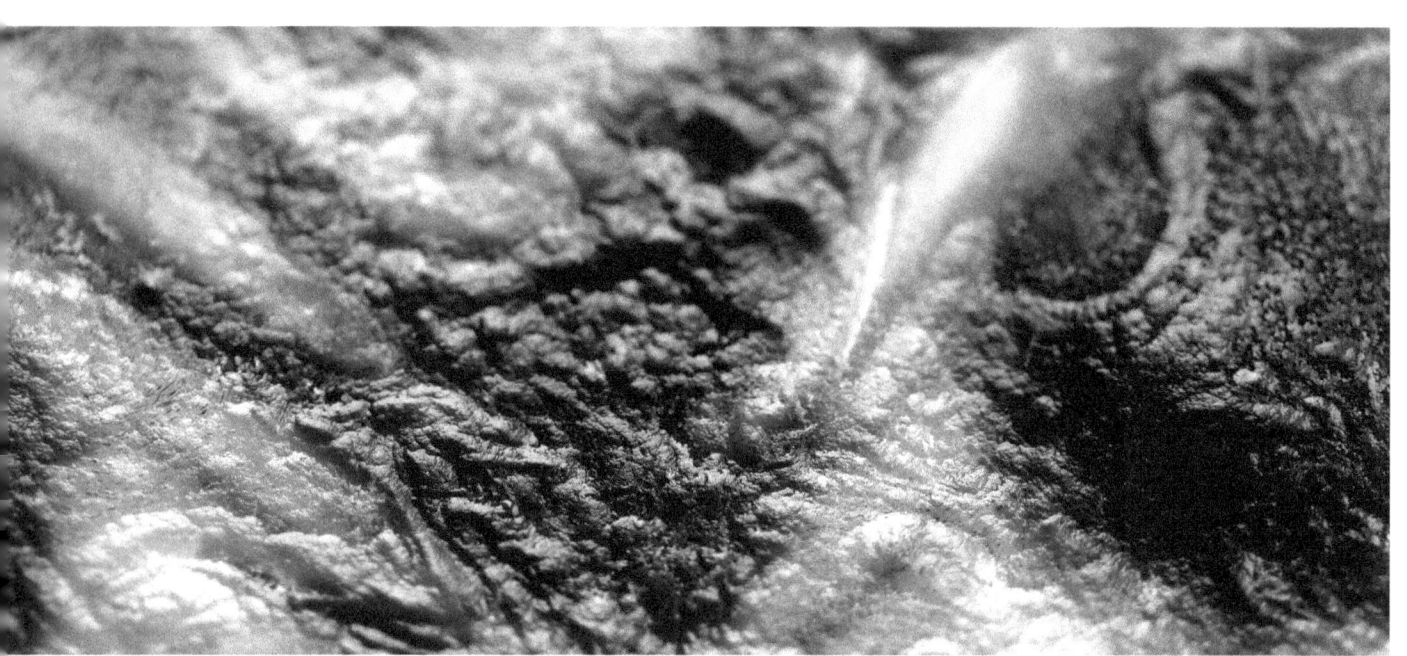

1A

ABSTRACT

This paper discusses an ongoing interdisciplinary research project that develops a design and fabrication approach termed; tunable environments. This is an explorative approach, which enables updates from a digital parametric interface to be 'pushed' into a 2D, 18 x 18 cm material sample, by modulating stimuli, so multiple properties can be updated/tuned at high resolutions. Our prototype explores how iterative updates can be achieved, which can be temporarily frozen in time. This opens up the idea of creating Parametric Matter/circular materials, which could reduce waste that can be attributed to typical linear processes. Additionally, highly bespoke, 'time-based' structures could be achieved. However, new implications for design and fabrication arise based on: time-lag of materials, a legacy of interactions, resetting materials as well as challenges of determining associations and desirable material properties.

1A Close up of material sample highlighting multiple properties that can be tuned at high resolutions

INTRODUCTION

Developments in digital design tools have enabled designs to be infinitely updated if they remain in their digital environments. However, typical design and fabrication processes are linear, which means that the final fabricated structures lose the ability to have multiple properties updated (shape, composition, color, texture). As a result, significant material waste, pollution, and resource depletion are generated when a design becomes outdated (aesthetics, capacity, environmental, etc.) or damaged. This inability to update a physical structure's properties is due, in part, to no discourse or associations being maintained between design parameters, material properties, and fabrication mechanisms over time. To address this issue, new design and fabrication approaches are required that are capable of leveraging material's computational abilities at high resolutions (granular/particles, molecular). In doing so, materials could be continually interacted with, finely tuned, and self-healed when damaged. Achieving these abilities build toward a notion of circular materials.

This paper presents initial findings from our feasibility project that aims to develop these types of updatable/circular materials at high resolutions. We present 2D material samples that can have multiple properties updated, at high resolutions, by modulating two stimuli (heat and magnetism). We term this design and fabrication approach "tunable environments" (Blaney et al. 2019). Employing tunable environments as a design and fabrication approach opens up the idea of circular material "abilities" (Blaney et al. 2021): materials can be self-healed when damaged, and updates can be 'pushed' into materials remotely. These abilities are made possible when a discourse is maintained between design tools and material properties. However, this study highlights new design implications when iteratively interacting with and 'programming' non-linear materials via stimuli, especially as the material samples are scaled up in size.

To further explore this area, firstly we contextualizes the design and fabrication approach and how it relates to existing research. Secondly, we present methodology, outlining our prototype set-up, how and why we will interact with materials via a simple digital interface, material sample development and how we recorded our results. In regards to our methodology, Research Through Design (Frayling 1993) is employed as a flexible approach (Gaver 2012) since the research carried out is explorative by nature and is guided by an overall aim: how can updates be pushed into materials to leverage material computation at high resolutions? Thirdly, we present our results to date from scaling up to an 18 x 18 cm material sample as an annotated portfolio (Bowers 2012). Finally, we discuss our key insights and our future intentions.

STATE OF THE ART

Physical materials demonstrate the ability to compute form when stress/stimuli (gravity, magnetism, tension etc.) are induced upon them. However, typical design and fabrication processes remove, or do not leverage these abilities, as materials are typically treated as static and or inert. Material computation has a rich history within architectural design processes to generate geometric forms such as Gaudi's catenary strings. These tension-based models create highly sculptural forms that can be translated into pure compression-based construction (Burry 2016). Additionally, Otto's form-finding experiments further demonstrate how various analog models/material platforms (soap bubbles, strings, polystyrene chips) can respond and reconfigure when a stimulus is induced to generate various design solutions, such as tensile and branching structural systems (Otto and Rasch 1995) as well as urban distribution models (Otto 2003).

To create highly adaptive, circular materials and continually customizable objects and structures, we believe material computation needs to be embedded within an object's or structure's material make-up, so that it can be leveraged on demand. Related research and developing stimuli-based design and fabrication processes can open up these abilities and enable continued interactions and high material resolutions.

Maintaining Discourse & Material Assembly

Persistent Modelling developed by Ayres (2011) demonstrates how modulatin a digital design representation (i.e. a digital parametric model) can be used to iteratively interact with and deform the global shape up to the material elastic limits (Ayres 2012) of predesigned metal components. Significantly, this approach enhances a structure's material capacities; Ayres demonstrates how a structure can change its shape to meet fluctuating design demands based on associations (Ayres et al. 2014). However, the implication of shape-changing abilities that are constrained to a material's elastic limit highlights opportunities for incorporating tunable stimuli with material processes capable of autonomous assembly. Principally, the idea of using stimuli to interact with materials that can be governed by a design representation opens up the design space for iteratively updating multiple material properties of a structure's material make-up investigated in this paper.

MIT's Self-Assembly Lab has developed approaches to programming matter that is capable of autonomous assembly and geometric reconfiguration/responses (Tibbits 2016). This approach pre-programs individual material units by designing their geometries and the material interfaces, which creates reconfigurable structures and achieves computational processes, such as self-error correction without embedding hardware (Papadopoulou et al. 2017). Interestingly, the role of external stimuli becomes increasingly apparent within this approach because to achieve autonomous assembly, the material units are supplied with random external energy, such as fluid agitation (Papadopoulou et al. 2017). Tolley and Lipson demonstrate that by 'tuning' the fluid agitation supplied to these types of material units, the assembly process, which is stochastic, can be speeded up (Tolley and Lipson 2010). This illustrates the beneficial role that modulating parameters of external stimuli can play in fabrication processes based on the assembly of individual material units, and how continued interactions can be achieved. Designing the geometry and interfaces of individual units as an approach to programming matter enables robust geometric assembly and reconfigurations, as well as visual evidence when the assembly process is complete. However, what if higher material resolutions and multi-material updates are desirable? Could tuning multiple properties lead to increasingly customizable/bespoke design solutions that can be finely tuned to a user's demands? We are interested in exploring how Persistent Modelling can be combined with material platforms capable of autonomous assembly could open up continued material interactions and higher material resolutions so that increasingly flexible systems can be created.

The concept of "programmable matter" first defined by Toffoli and Margolus (1991) outlines a vision of a universal material platform capable of high flexibility, scalability, and material resolutions. More recently, Ishi et al describe 'Perfect Red,' a speculative vision for programmable matter capable of performing actuation, sensing, and communication at molecular resolutions (Ishii et al. 2012). The ability to program and continually interact with matter at this resolution highlight how multiple material properties can be finely tuned so that a design solution can become increasingly bespoke based on user interactions. However, miniaturization becomes difficult when material units have hardware embedded into them to achieve sensing and actuation (Gilpin and Rus 2012). Various magnetically responsive material platforms have been developed that capture the original idea of programmable matter/ Perfect Red. For example, the global shape-change of slime-like materials in response to magnetic stimuli developed by Dickey (2017), which are capable of performing as soft robots (Wang et al. 2021). Within architectural applications, Goldman and Myers demonstrate how magnetic stimuli/fields can be used to extrude materials that, when frozen in time, create highly sculptural and detailed 3D forms (Goldman and Myers 2017). These magnetized material platforms provide a sound starting point. However, we aim to iteratively update multiple properties to increase their flexibility. To do this we explore

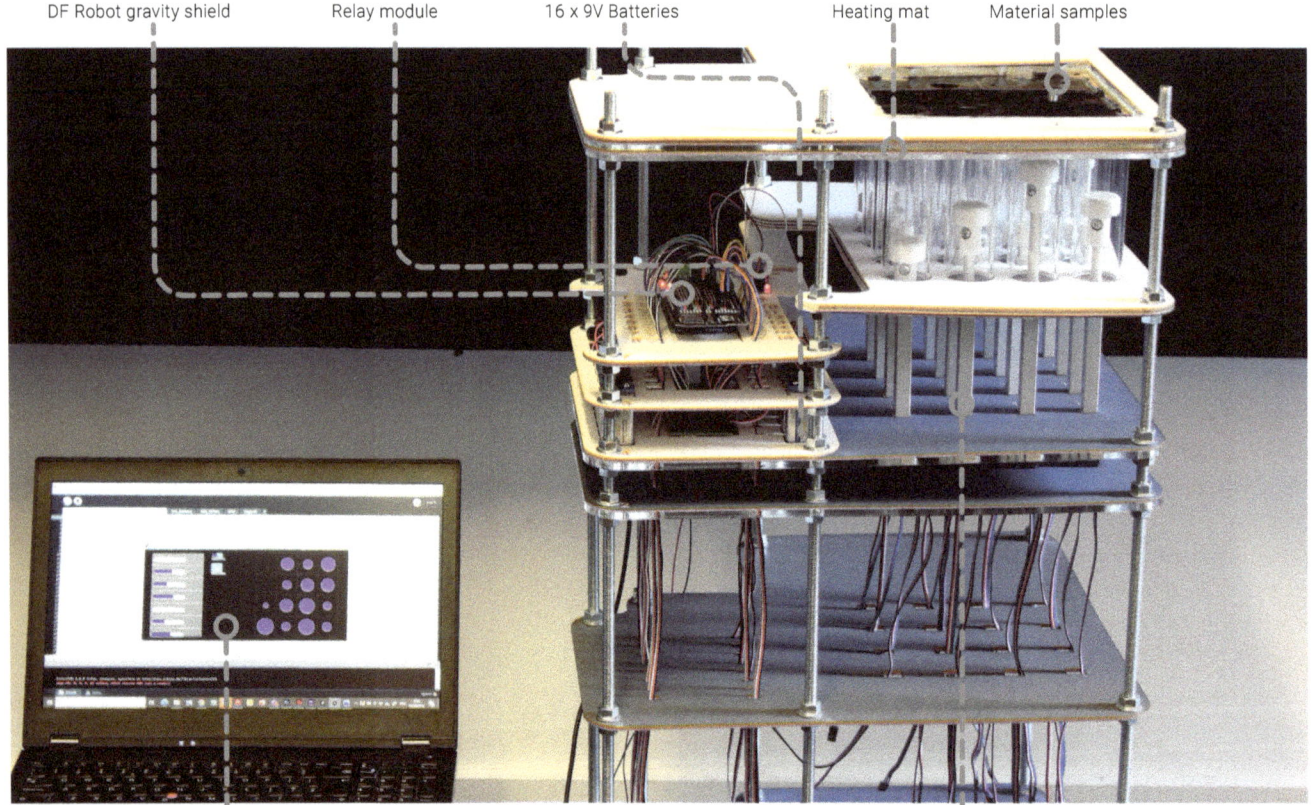

1B The prototype set up with a material sample interaction

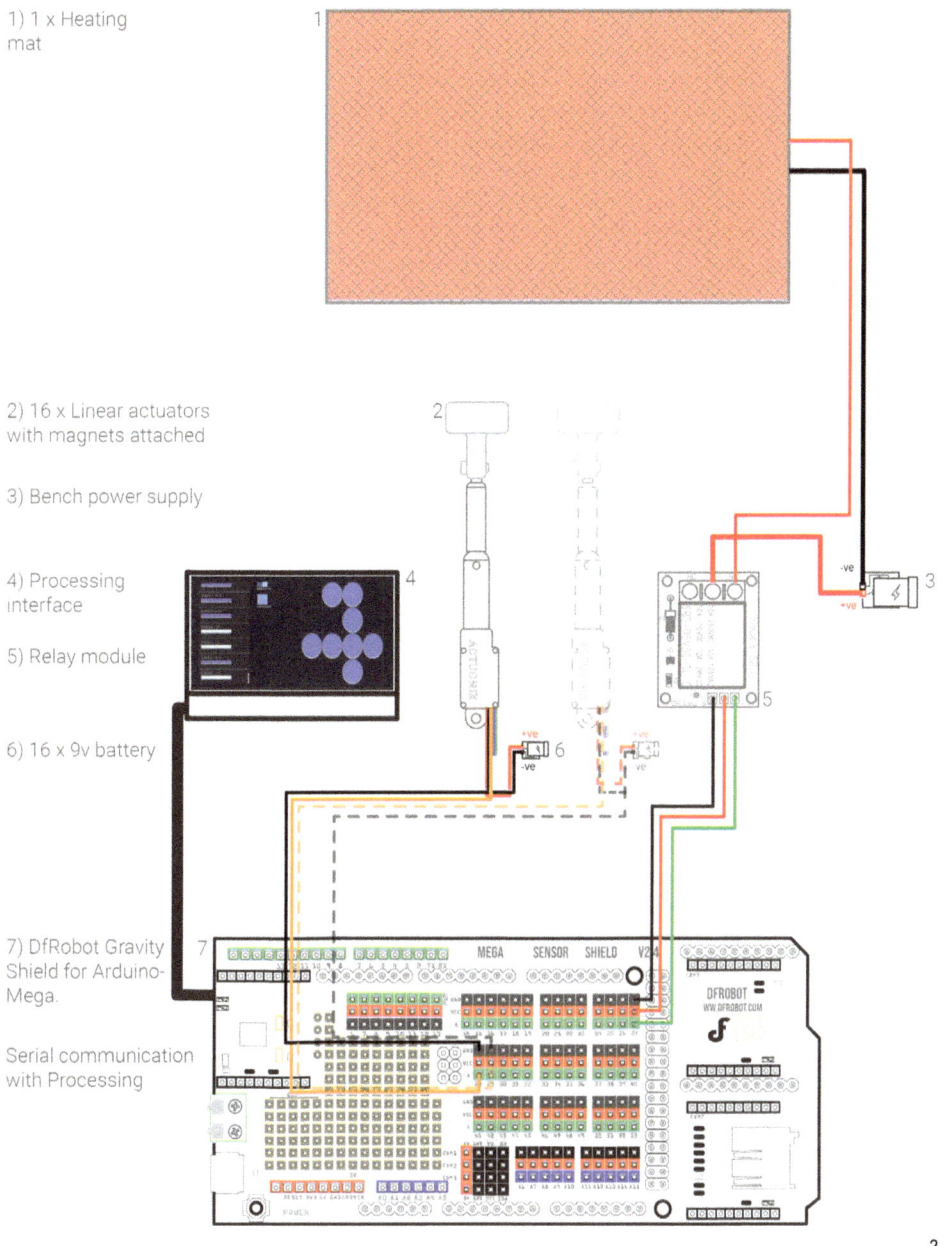

1) 1 x Heating mat

2) 16 x Linear actuators with magnets attached

3) Bench power supply

4) Processing interface

5) Relay module

6) 16 x 9v battery

7) DfRobot Gravity Shield for Arduino-Mega. Serial communication with Processing

2 System hardware diagram showing highlighted connections between components

how modulating stimuli can open up new potential and resolutions for programmable matter.

Previous research by the authors highlights how modulating stimuli based on digital design tools can update multiple material properties at high resolutions (molecular) when using chemical platforms (Blaney 2020). Additionally, previous research by the authors demonstrates how modulating stimuli can be applied to bio-material platforms to parametrize their properties (Ozkan et al. 2022). This highlights the flexibility of modulating stimuli and how it can interact with and programmed matter. However, determining reliable feedback between material properties generated with associated design parameters becomes an issue when not directly embedding sensors so that material resolution is not compromised (Blaney 2021). This current research incorporates state-changing magnetized materials and a multi-stimuli system to achieve iterative multi-material actuation at high-resolution and investigates their implications when developing design and fabrication processes for interacting with non-linear materials.

METHOD: DEVELOPING PARAMETRIC MATTER

To test and support the design challenges of developing circular materials that can have multiple properties updated, the authors designed and built a prototype set-up. The set-up is capable of inducing two stimuli, heat and magnetism, that are used to interact with state-changing, magnetized material samples.

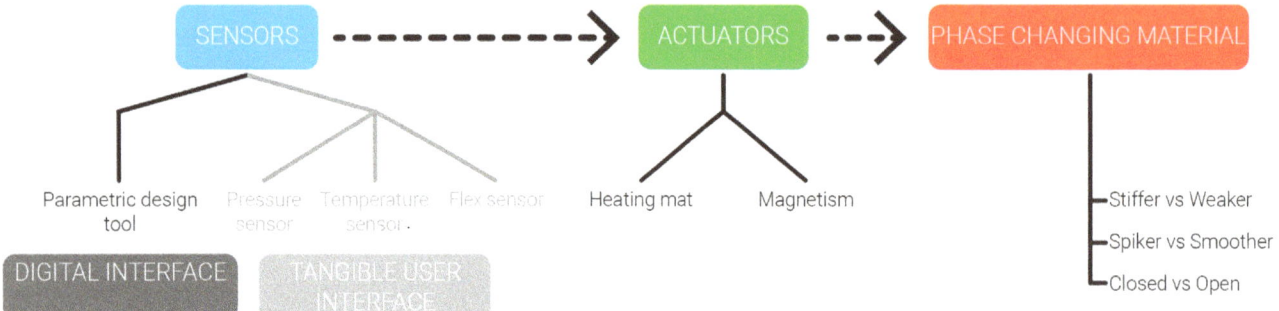

3 System relations highlight that there is no feedback between the digital interface and properties generated and specific design demands enable by further sensing

Prototype Set-up

The set-up modulates two stimuli: heat and magnetism. Figures 1 and 2 document the system's hardware and components that allow us to modulate the two stimuli. Importantly, there is no feedback mechanism at this stage between the design tools and material properties generated (Figure 3). Essentially, the system is an open-loop control system.

An ArduinoMega is used to control multiple actuators based on data sent from a basic parametric interface developed in Processing (Figure 4). A digitally controlled relay module is used to control a heating mat upon which the material samples were placed to melt them. Sixteen magnets are connected to sixteen linear actuators in a 4 x 4 grid, which was positioned directly below the heating mat. By connecting the magnets to the actuators, the strength of the magnetic force induced upon the material sample could be varied by changing the vertical position, which produces a dissipating effect. The combination of two stimuli and state-changing materials is used to explore how multiple properties can be iteratively updated across the sample's area—global shape-change, patterns, rigidities, volumes, porosity/openness, and surface texture—whereby updates can be 'pushed' from design tools into materials and temporarily frozen in time. In doing so, it becomes possible for samples to be taken out of their fabrication environment, interacted with, and then re-fabricated/updated or healed if they became damaged or outdated by placing them back in the fabricator.

'Pushing' Updates into Materials:
User Interface Development

A simple parametric interface is used to manually control the positions of sixteen magnets (Figure 4). This is done over serial communication between Processing to Arduino. On the interface, two magnets (in the top left corner) will be kept at position zero to act as a control area. The proximity of the other fourteen magnets will be changed randomly. Since the magnets have a dissipating effect on the samples, the further they are away, the less impact they have on the material. Varying this stimulus will be explored in future studies to understand the type of material responses that can be elicited and the impact of time. The closest a magnet can get to the material sample is 1 mm and the farthest away is 100 mm. For each sample, multiple iterations will be carried out by varying the positions of the magnets and solidifying or melting the sample by turning on or off the heat supplied to it.

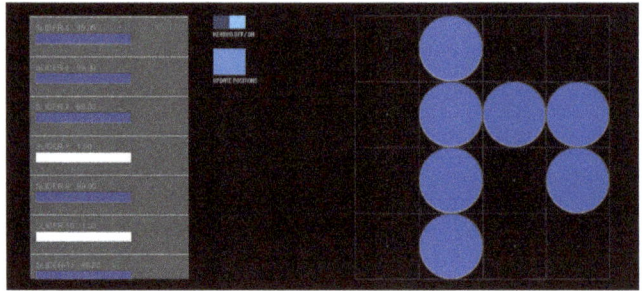

4 Simple parametric interface used to control material patterns developed in Processing

The re-melting and varying of the magnet's position to interact with the sample act as one iteration. Iterative interactions will be explored to understand their implications on design and fabrication processes.

Material Sample Development

The chemistry department produced multiple biodegradable and non-toxic material samples of Poly ε-caprolactone (PCL), which is a FDA approved semi-crystalline aliphatic polyester (Patricio et al. 2013; Asvar et al. 2017). In the design experiments, polycaprolactone with six different molecular weights (530, 900, 2000, 14000, 45000, and 80000 Da) and Iron (II, III) oxide (Fe_3O_4) with two different particle sizes (powder (P) <

5 μm and nanopowder (N) 50-100 nm) were used. The Fe_3O_4 powder and nanoparticles had good adhesion and dispersion of the particles in the polycaprolactone matrix. While the molecular weight affected material viscosity, the particle size of iron oxide gave ferromagnetic properties to the samples and affected their magnetic behavior. The samples became less viscous with heavier molecular weight and exhibited super magnetic behavior with smaller iron oxide particle size (Rezai et al. 2021; Wu et al. 2010).

Fe_3O_4-PCL composites were prepared by dissolving the PCLs (300 mg) in tetrahydrofuran (THF, 3mL) solvent using an ultrasonic cleaner at room temperature. Then nanopowder or powder-sized iron oxide (30 mg) was added and sonicated for an hour. Lastly, the mixture was cast, and the solvent was removed.

Figure 5 documents the various small-sized (approx. 2 x 2 cm) PCL samples that were developed (see also video reference information). The main criteria to select samples for further development and scaling up were: 1) mechanically robust when solid, and 2) enabling properties to be updated when in a liquid state, with a focus on global 2D shape-changes, material gradients, surface texture, volumes, and rigidities. In line with the desired criteria, the chemistry department further developed two final samples discussed in the paper.

The smaller sample, P-2 80 50:50, was composed of 250 mg of PCL with a molecular weight of 2,000, 250 mg of PCL with

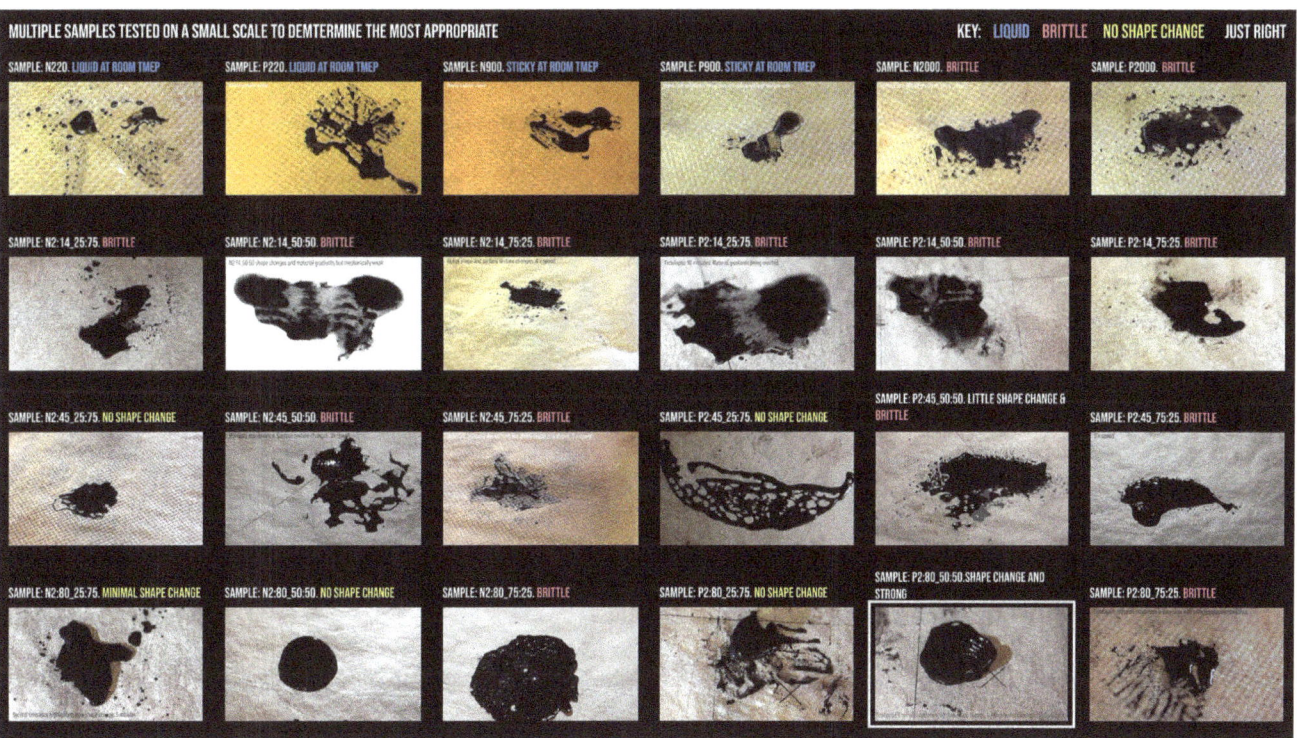

5   Video still of multiple samples tested on a small scale to determine the most appropriate samples to be scaled up; see NOTES section for video link.

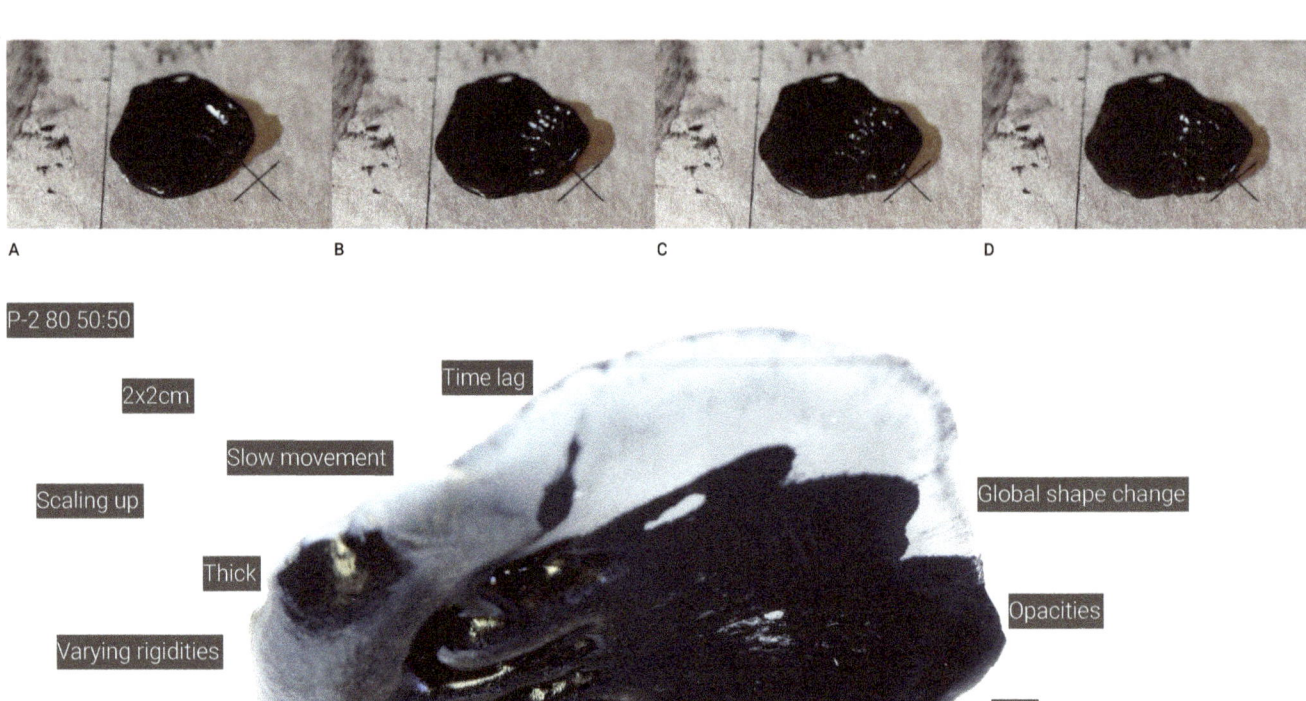

6 A-D) Time-lapse photos (11-minute duration) of 2x2 cm sample melting at 98°C and slow response to magnetic stimuli; E) Solidified 2x2 cm sample, annotations highlighting multiple properties achieved; F) Time lapse (12-minute duration) of 5x5 cm sample demonstrating expected movement to magnet locations; and G) Solidified 5x5 cm sample, annotations highlighting multiple properties achieved

a molecular weight of 80,000, and 50 mg of Iron (II, III) oxide powder size. The ratios (50:50) indicated the mass ratio of polycaprolactones with different molecular weights. The 18x18 cm sample P14 50:50, was produced on a larger scale, 25 grams of polycaprolactone with a molecular mass of 2,000, 25 grams of polycaprolactone with a molecular weight of 14,000, and 5 g of Iron (II, III) oxide powder size.

Recording Interactions and Responses

The changes in the material have been documented using photography, time-lapse, and videography. The camera used is a Canon EOS 600D with an f/2.8 Macro USM for macro images (fine detail). A 3.5-5.6 Zoom EF-S 60 mm is used for less detailed videography. Videography has been used to reveal real-time responses. Time-lapse photography is used to reveal slow and high-resolution responses. This is because time plays an important role in documenting various responses that occur at different rates. Importantly, documenting time using these strategies highlights the implications of developing design strategies to interact with materials.

RESULTS

The properties of the P-2 80 50:50 and P-2 14 50:50 samples and the performance of the set-up are now discussed.

Initial results from P-2 80 50:50 sample

The 2x2 cm sample response rate was evident but slow in sample P-2 80 50:50; it changed shape and relocated on the heating mat, as seen in Figure 6 A-D. Surface texture, volumes, rigidities, and mono-color gradients were observed while interacting with the sample.

As a next step, the sample area and volume were increased in size to 5x5 cm. The sample demonstrated global shape changes, as assumed, to magnet positions defined (Figure 6 F) as well as various surface textures, which were unexpected. This was due variations in heat throughout the sample, which created a thin 'skin' formed on the sample's surface as it was colder than the volumes in direct contact with the heating mat. However, this could still be associated with magnetic stimuli and material volumes as the greatest surface texture variations were evident where the magnetic stimulus is most significant.

Results from P-14 50:50 Sample

Iterations with sample P-14 50:50 demonstrated that multiple material properties could be updated at high resolutions. Figure 7C documents the multiple updates, such as global patterns, gradients, reflectance, surface texture, material height, volume and translucency. Comparatively, on the whole, P14 50:50 responded faster that P-2 80 50:50, but the finer patterns/gradients became more apparent via time-lapse photography i.e. longer durations. It was observed that the material becomes stiffer and more opaque on spots where more material is located. This was only tested manually. Additionally, the material gradient, from clear to black, impacts material stiffness. The ability to iteratively update material stiffness, shape-change, and textures could be applied to further customize medical prosthetics. The ability to update these properties is important due patients' physiological changes over time (Turner et al. 2022) and other fluctuating demands, like seasonal heat (Ghoseiri and Safari 2014); these fluctuations may lead to ill-fitting devices, discomfort, sores, and if they become damaged or unfit, the need for a new device.

It was witnessed that increased material volumes, which generated global surface bumps, reduce over time. Initially, the material would create large lumps where the magnets were positioned at 100% height, which would make them stronger. However, after some time, the material bumps reduce dramatically (see Figure 7). This highlights the significant implications of time windows and constraints for the designer to interact with the material to elicit and achieve certain material properties. There are trade-offs over time as certain material properties become more apparent (e.g. gradients, global patterns), whilst others diminish (e.g. bumps/volumes, surface textures), indicating that hierarchies need to be defined for given applications.

7 A) The initial state of the sample P-14 50:50; B) Simple parametric interface used to control patterns; C) The first iteration of global patterns demonstrates multiple properties; D) 18x18 cm sample demonstrating multiple properties, highlighted by annotations, across its area; E–H) Multiple material properties at high resolutions are highlighted, such as gradients, textures, reflectance, and opacity. See NOTES section for video link.

8 A–B) The induced patterns upon the sample before finalizing on the final fourth pattern; C) The final pattern is used to inform material properties; D) The photograph highlights significant 'legacy' properties from the first and two other prior interactions; E) Sample P-14 50:50 in its second iteration demonstrates multiple material properties that have been updated; F–G) Highlights multi-material properties updates at high resolutions.

9 A–D) Patterns tested to reset the sample; E) Material mixed manually to reset as a low-tech solution; F) Photographs documenting the sample becoming material reset in areas compared to the previous iteration; G) Area that was manually reset highlights a zone that is more thoroughly mixed compared to the stimuli reset areas.

10 Highlights of the different areas where the sample has been reset; close up photographs presenting the color-connection and transition of the sample P-14 50:50.

A

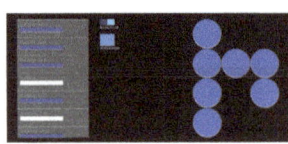

B

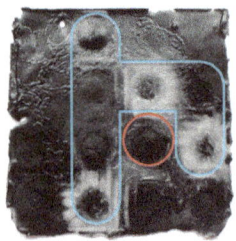

◯ 1st pattern update
◯ Legacy pattern

C

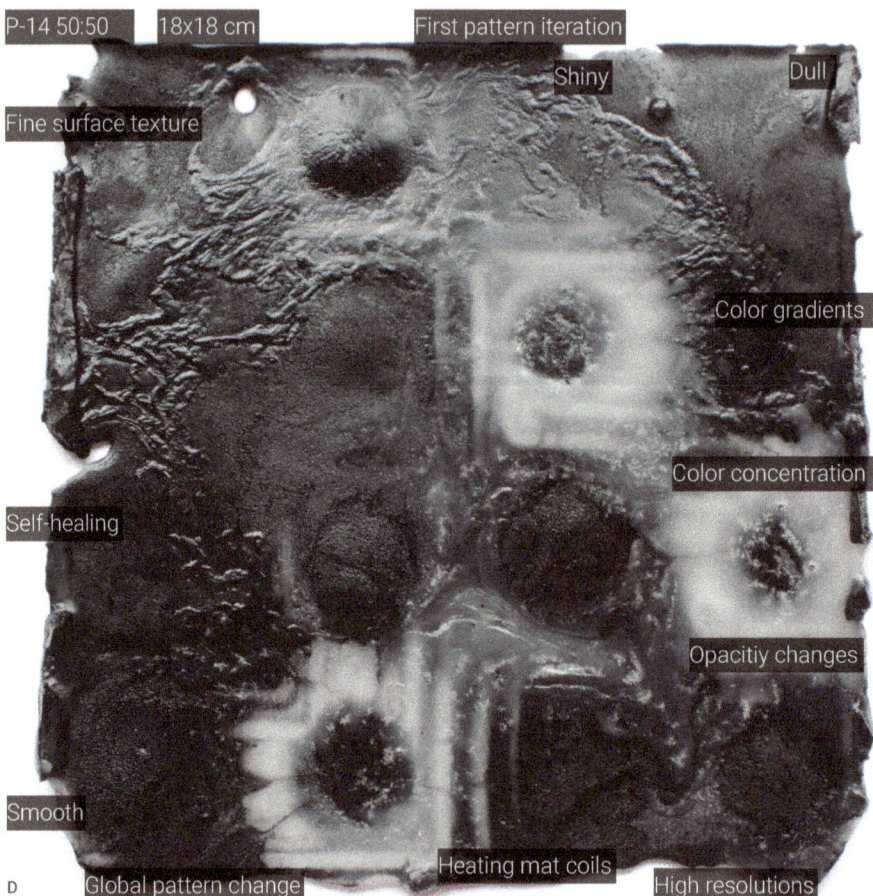

D

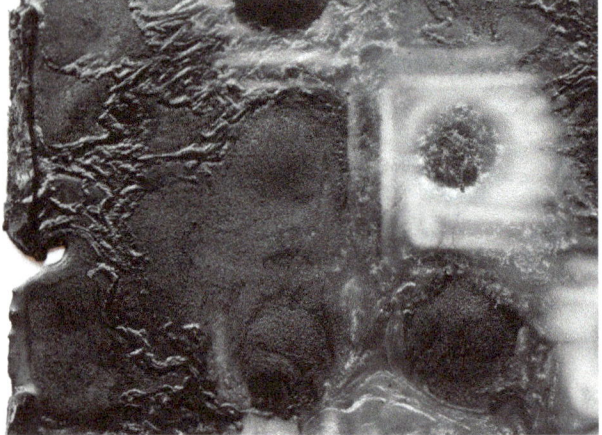

E

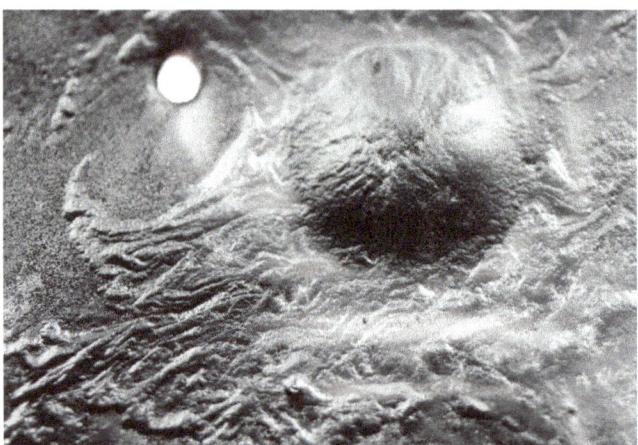

F

G

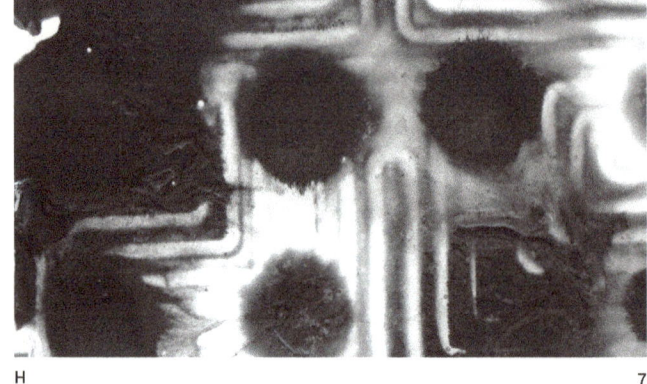

H

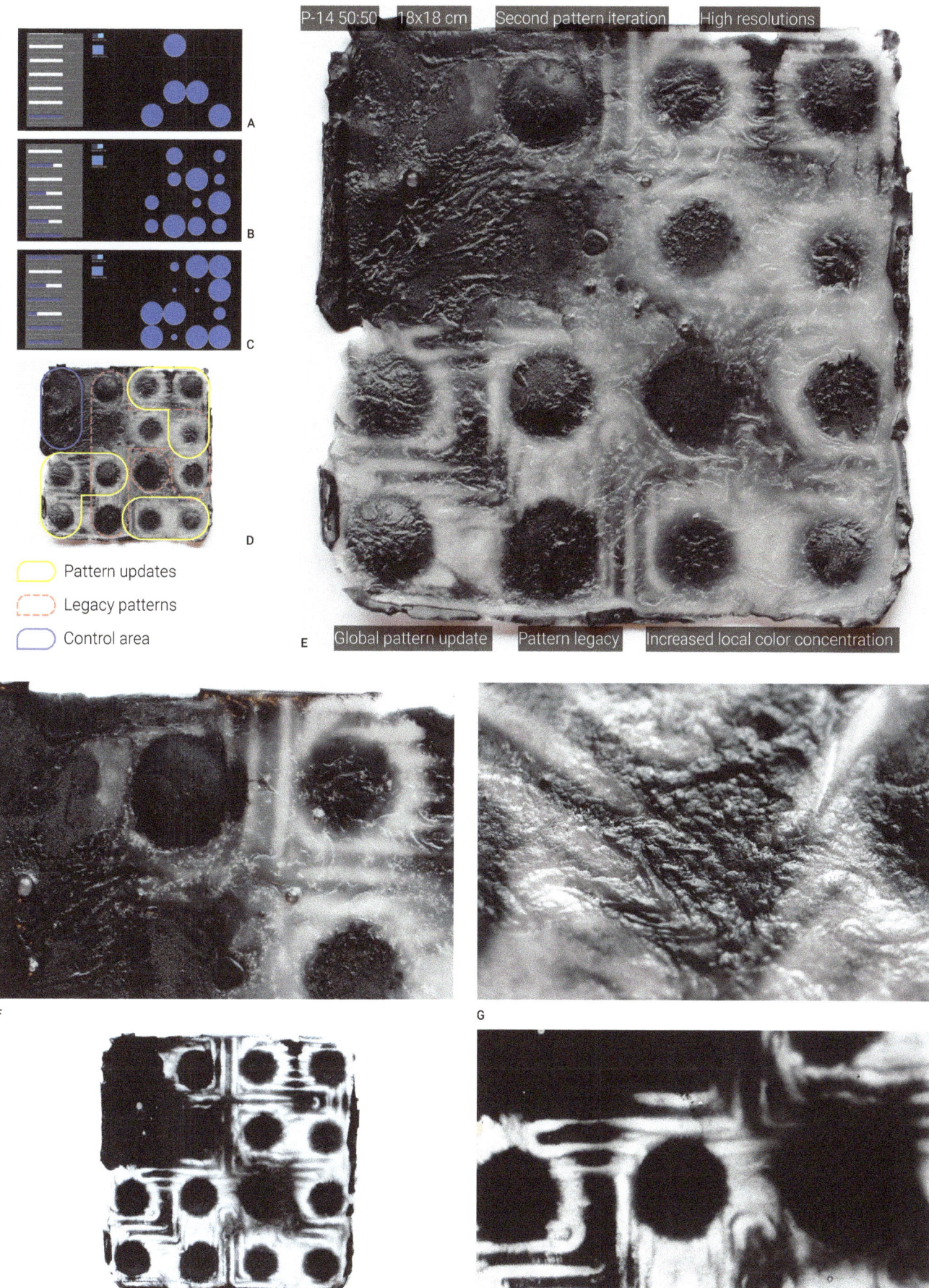

8

ADVANCED MATERIALS

Hybrids & Haecceities 217

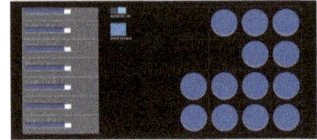 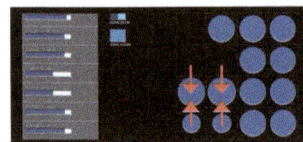 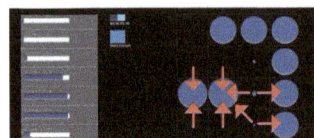 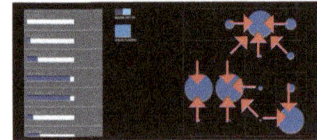

A   B   C   D

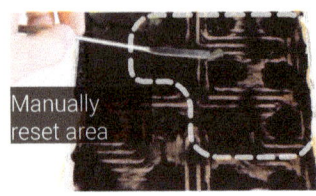   

E

Previous iteration   Sample reset   Manually reset area

F   G

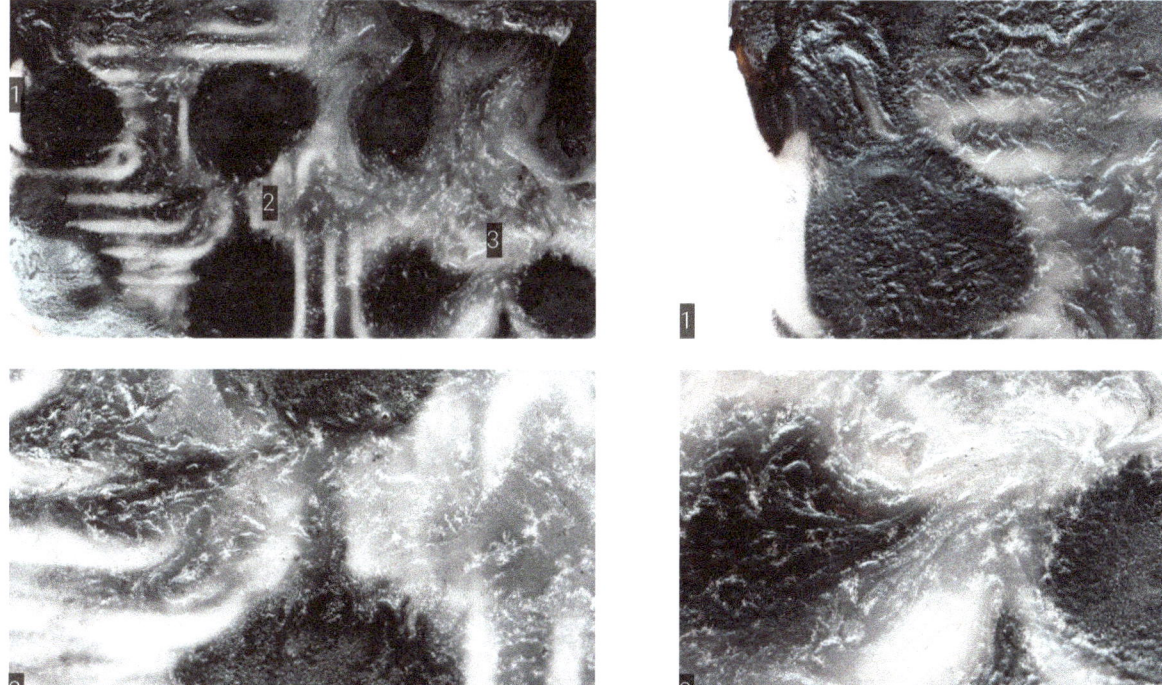

ADVANCED MATERIALS

Lastly, a second iteration was carried out where the sample was re-melted and magnet positions updated. This demonstrates that multiple properties can be iteratively updated based on digital design updates. However, this also highlighted that properties from previous interactions remain. This reveals implications of 'legacy' within the patterns (Figure 8D) and becomes more apparent if the neighboring magnets do not induce a strong enough effect on the materials. Due to this legacy, there is a need to reset the material (especially in larger samples) to a blank canvas. Resetting the sample gives it a clean slate, a uniform black color, which may help to not restrict future interactions. To test if resetting the material sample could be automated and if the magnetized particles can be evenly distributed across the sample, various magnet positions were tested.

Resetting Materials

To determine if the resetting process could be automated via stimuli, all magnet heights were set at 85% (15 mm away from the material sample) (Figure 9A). Then two magnets were lowered to 60% height (Figure 9B). This step was followed by lowering all neighboring magnets (Figure 9C-D); the assumed and witnessed direction of magnetic particle movement is illustrated using arrows. The stages of the material responses were photographed every three minutes while changing the magnet heights. The sample was most effectively reset when any directly neighboring magnet was lowered so it had no impact. This allows the black (magnetized) particles to reconnect to neighboring areas (Figures 9D and 11). To automate the process via stimuli, the shortest distances between neighboring magnets were reconnected via trails. To achieve more through mixing, magnets that tessellate together would be needed or an optimal proximity between them would need to be determined. Additionally, the top area (Figure 9E) was mixed by hand using a metal spatula to test manual resetting, which makes possible a more rapid homogenous mixing.

CONCLUSION

This work acts as a proof of concept and demonstrates that multiple material properties can be iteratively updated at high resolutions when using a two stimuli system combined with state-changing materials. It highlights potential alternatives to remanufacturing materials completely at end of life, as design and fabrication approach enables circular material production, i.e. materials that can be updated. However, multiple implications have been highlighted from this research when interacting with materials to push updates remotely via stimuli. These are:

- There are trade-offs over time as certain properties diminish others become increasingly significant; this highlights the need to define time-based hierarchies for various properties when they are desirable and for given applications.

- Due to slow material responses, time lag is created between digital updates and corresponding physical responses. As a result, in-situ and real-time responses are not possible with this set-up, currently. This highlights two challenges: 1) average data values would be required when iteratively updating material properties; and 2) the time lag, current set-up and process (placing materials back in the fabricator) lends itself to applications based on iterative updates/fine tuning, for example, in medical prosthetics/splints, track cycling skin suits, fashion/wearables.

- What is the role of material legacy? Does legacy compromise future interactions? How could legacy help to enhance bespoke design solutions?

- What digital design processes need to be developed/employed to create reliable interactions and properties being generated? How is feedback achieved without limiting material resolutions and the range of responses?

The simplistic digital interface used and lack of material behavior represented is a limitation of this research, but it was not the focus. Combined with what we term as time lag highlights issues of predictability when interacting with materials to elicit their computational abilities. Johns (2014) discusses the need for stochastic digital models when employing a robotics system and stimuli to interact with materials when they have indeterminate aspects. Additionally, the role of stochastic models has been discussed within distributed self-assembling materials process (Zykov and Lipson 2007) as well as biomaterials (Ozkan et al. 2022). This highlights a converging challenge for interacting with non-linear materials in order to maximize a greater range of responses but ensure robust responses that could open up other viable application areas.

Future work intends to develop a tangible user interface (TUI) to investigate how various sensor data can be used to interact with and inform material properties. The aim of developing a TUI is to understand how these design implications and system hierarchies impact more tangible/real-world applications that could lead to increasingly bespoke properties unique to user(s). However, this raises the challenge of determining what constitutes a desirable material response for various applications and becomes increasingly important as material properties become increasingly non-linear and involve multiple stakeholders.

Understanding this challenge and incorporating more sophisticated digital design processes could lead to increasingly bespoke structures and products that could address issues of waste attributed to linear design and fabrication processes. Importantly, the potential flexibility, continued interactions, and high-resolution achieved in these samples is made possible because they have been fabricated by leveraging and instilling the material's computational abilities within them.

NOTES

1. Figure 5 video link: https://vimeo.com/712053739
2. Figure 7 video link: https://vimeo.com/712054727

ACKNOWLEDGMENTS

The research was funded by Connected Everything ii (EP/S036113/1). We would like to thank the supportive and collaborative CEii network. Additionally, the collaborative nature between disciplines within the team of researchers contributed greatly to the results.

REFERENCES

Asvar, Z., E. Mirzaei, N. Azarpira, B. Geramizadeh, and M. Fadaie. 2017. "Evaluation of electrospinning parameters on the tensile strength and suture retention strength of polycaprolactone nanofibrous scaffolds through surface response methodology." *J. Mech. Behav. Biomed. Mater.* 75 (Nov.): 369–378.

Ayres, P. 2011. "Free-from Metal Inflation & the Persistent Model." In *Fabricate 2011: Making Digital Architecture*, edited by R. Glynn and B. Sheil. London: UCL Press.

Ayres, P. 2012. "Microstructure, Macrostructure and the Steering of Material Proclivities." In *Manufacturing the Bespoke: Making and Prototyping Architecture*, edited by B. Sheil. Chichester, UK: John Wiley and Sons Ltd. 220–237.

Ayres, P., K. Stoy, D, Stasiuk, and H. Gibbons. 2014. "Multi-scalar Shape Change in Pneumatically Steered Tensegrities: A Cross-disciplinary Interest in Using Material-scale Mechanisms for Driving Spatial Transformations. In *Fabricate 2014: Negotiating Design & Making*, edited by F. Gramazio, M. Kohler, and S. Langenberg. London: UCL Press.

Blaney, A. 2020. "Designing Parametric Matter: Exploring adaptive self-assembly through tuneable environments." PhD thesis, Lancaster University.

Blaney, A. 2021. "Material Units; Uploading information into matter via stimuli and the challenges of determining feedback." In *Towards a New, Configurable Architecture; Proceedings of the 39th eCAADe Conference*, edited by V. Stojakovic and B. Tepavcevic. University of Novi Sad, Novi Sad, Serbia.

Blaney, A., J. Alexander, N. Dunn, and D. Richards. 2019. "Designing Parametric Matter." In *IASDR 2019: Design Revolutions; International Association of Societies of Design Research Conference 2019*. Manchester, United Kingdom.

Blaney, A., D. Richards, A. Gradinar, and S. Michael. 2021. "Prototyping Circular Materials Based on Reprogrammable Matter." In *IASDR 2021: [_] with Design: Reinventing Design Modes; International Association of Societies of Design Research Conference 2021*. PolyU Design, School of Design, Hong Kong Polytechnic University.

Bowers, J. 2012. "The logic of annotated portfolios: Communicating the value of 'research through design'. In *DIS '12: Proceedings of the Designing Interactive Systems Conference*. Newcastle, UK. 68–77.

Burry, M. 2016. "Antoni Gaudí and Frei Otto Essential Precursors to the Parametricism Manifesto." *Architectural Design* 86 (2): 30–35.

Dickey, R. 2017. "Soft Systems: Rethinking Indeterminacy in Architecture as Opportunity Driven Research." In *Protocols, Flows, and Glitches; Proceedings of the 22nd CAADRIA Conference*, edited by P. Janssen, P. Loh, A. Raonic, and M. A. Schnabel. Suzhou, China. 811–820.

Frayling, C. 1993. "Research in Art and Design." *Royal College of Art Research Papers* 1 (1): 1–5.

Gaver, W. "What should we expect from research through design?" In *CHI '12: Proceedings of the SIGCHI Conference on Human Factors in Computing Systems*, edited by J. A. Konstan et al. Austin, Texas, USA. ACM. 937–946.

Ghoseiri, K. and M. R. Safari. 2014. "Prevalence of heat and perspiration discomfort inside prostheses: literature review." *Journal of Rehabilitation Research and Development* 51 (6): 855–868.

Gilpin, K. and D. Rus. 2012. "A distributed algorithm for 2D shape duplication with smart pebble robots." In *2012 IEEE International Conference on Robotics and Automation*. Minnesota, USA. 3285–3292.

Goldman, M. and C. Myers. 2017. "Freezing the Field: Robotic Extrusion Techniques Using Magnetic Fields." In *ACADIA 2017: Disciplines & Disruption; Proceedings of the 37th Annual Conference of the Association for Computer Aided Design in Architecture (ACADIA)*. Cambridge, Mass. 260–265.

Ishii, H., D. Lakatos, L. Bonanni, and J.-B. Labrune. 2012. Radical atoms: beyond tangible bits, toward transformable materials. *Interactions* 19 (1): 38–51.

Johns, R. 2014. "Augmented Materiality: Modelling with Material Indeterminacy." In *Fabricate 2014*, edited by F. Gramazio, M. Kohler, and S. Langenberg. London: UCL Press.

Otto, F. 2003. *Occupying and connecting; Thoughts on Territories and Spheres of Influence with Particular Reference to Human Settlement*. Stuttgart/London: Edition Axel Menges.

Otto, F. and B. Rasch. 1995. *Finding Form: Towards an Architecture of the Minimal*. Stuttgart: Axel Menges.

Ozkan, D., R. Morrow, M. Zhang, and M. Dade-Robertson. 2022. "Are Mushrooms Parametric?" *Biomimetics* 7 (2): 60–76.

Papadopoulou, A., J. Laucks, and S. Tibbits. 2017. "From self-assemblies to evolutionary structures." *Architectural Design* 87 (4): 28–37.

Patricio, T., and P. Bartolo. 2013. "Thermal stability of PCL/PLA blends produced by physical blending process." *Procedia Engineering* 59: 292–297.

Rezaei, V., E. Mirzaei, S.-M. Taghizadeh, A. Berenjian, and A. Ebrahiminezhad. 2021. "Nano Iron Oxide-PCL Composite as an Improved Soft Tissue Scaffold." *Processes* 9 (9): 1559.

Tibbits, S. 2016. *Self-Assembly Lab: Experiments in Programming Matter*. London, New York: Routledge.

Toffoli, T. and N. Margolus. 1991. "Programmable matter: concepts and realization." *Physica. D: Nonlinear Phenomena* 47 (1-2): 263–272.

Tolley, M. T., and H. Lipson. "Fluidic Manipulation for Scalable Stochastic 3D Assembly of Modular Robots." In *IEEE International Conference on Robotics and Automation*. Anchorage, Alaska. 2473–2478.

Turner, S., A. Belsi, and A. H. McGregor. 2022. "Issues faced by people with amputation(s) during lower limb prosthetic rehabilitation: a thematic analysis." *Prosthetics and Orthotics International* 46 (1): 61–67.

Wang, B., K. F. Chan, K. Yuan, Q. Wang, X. Xia, L. Yang, H. Ko, Y.-X. J. Wang, J. J. Y. Sung, P. W. Y. Chiu, and L. Zhang. 2021. "Endoscopy-assisted magnetic navigation of biohybrid soft microrobots with rapid endoluminal delivery and imaging." *Science Robotics* 6 (52): eabd2813.

Wu, A., P. Ou, and L. Zeng. 2010. "Biomedical Applications of Magnetic Nanoparticles." *NANO: Brief Reports and Reviews* 5 (5): 245–270.

Zykov, V. and H. Lipson. 2007. "Experiment Design for Stochastic Three-Dimensional Reconfiguration of Modular Robots." In *IEEE/RSJ International Conference on Robots and Systems, Self-Reconfigurable Robotics Workshop*. San Diego, CA.

IMAGE CREDITS
All drawings and images by the authors.

Dr. Adam Blaney is a lecturer at Lancaster School of Architecture. His research interests mainly focus on developing digital design and fabrication processes that aim to create physically responsive, adaptive and self-healing objects, devices and architectural structures. Additionally, based on principles from these prototypes, he is also interested in developing speculative visions of materially adaptive architectures and cities.

Dilan Ozkan is an architect and researcher who focuses on working with living systems. She aims to push the limits of traditional architectural production and bring different approaches by discovering new material making processes. Dilan completed an architectural design Master's at Pratt Institute in New York, where she was first inspired by the strange aesthetics of living organisms. Currently, she is a PhD student at Newcastle University. Within her research, she investigates nonlinear materials and working fungi as a biomaterial probe. She formed a study group called Mycology for Architecture to collaborate with other disciplines and share knowledge about fungi.

Dr. Emel Pelit is a postdoctoral research associate working in the field of organic chemistry. Her research interests focus on the synthesis of heterocyclic compounds and the study of their biological activities, and on conjugated polymers. In addition, she is interested in performing organic reactions in accordance with the principles of green chemistry, using methods such as aqueous media and sonochemistry.

Dr. Mariana Fonseca Braga is a hybrid designer with industry and academic experience in developing and implementing projects utilizing diverse design approaches to unlock innovation and tackle societal challenges with interdisciplinary teams. She is currently a postdoctoral research associate International in Imag-ination Lancaster at Lancaster University, having been awarded an MSc in Industrial Engineering and a PhD in Design with honors from Politecnico di Milano. Braga leads research in community-led enterprises, focusing on how these can feed into effective policy and technological design. Her recent publications encompass the value of design and challenges in organizations, and in tackling the UN SDGs.
https://www.linkedin.com/in/mariana-fonseca-braga

Dr. John G. Hardy was awarded his MSci and PhD in Chemistry in 2002 and 2007 from the University of Bristol and University of York, respectively. From 2006 to 2015 he undertook postdoctoral research in Chemical Engineering at the University of Strasbourg in France, in Bioengineering at the University of Bayreuth in Germany, in Biomedical Engineering at the University of Texas at Austin and the University of Florida in the United States, and in Pharmacy at Queen's University Belfast in Northern Ireland. He is currently a 50th Anniversary Senior Lecturer in the Department of Chemistry and Materials Science Institute at Lancaster University.

Dr. Mark Ashton received his PhD in Chemistry from Lancaster University in 2020 under the supervision of Dr. John G. Hardy. He currently holds the role of Technical Innovation Fellow at the Centre for Global Eco-innovation at Lancaster University. His PhD and subsequent postdoctoral research focused on the synthesis of stimuli responsive drug delivery systems with a specific interest in the use of non-invasive methods of controlling drug release. From 2015–2016 he spent time working at Procter & Gamble, USA, in the fabric care department and obtained his Master's degree (2017) after working with Professor Ross W. Boyle on the synthesis and development of porphyrin radio-sensitizers for use in cancer treatments.

Nanotectonica SEM-GAN

The Strange Materiality of Subvisible Bodies

Jonas Coersmeier
Pratt Institute
University of Pennsylvania

James Nanasca
Pratt Institute

Ivan Yan Man Hin
Pratt Institute

Ezio Blasetti
University of Pennsylvania
Pratt Institute

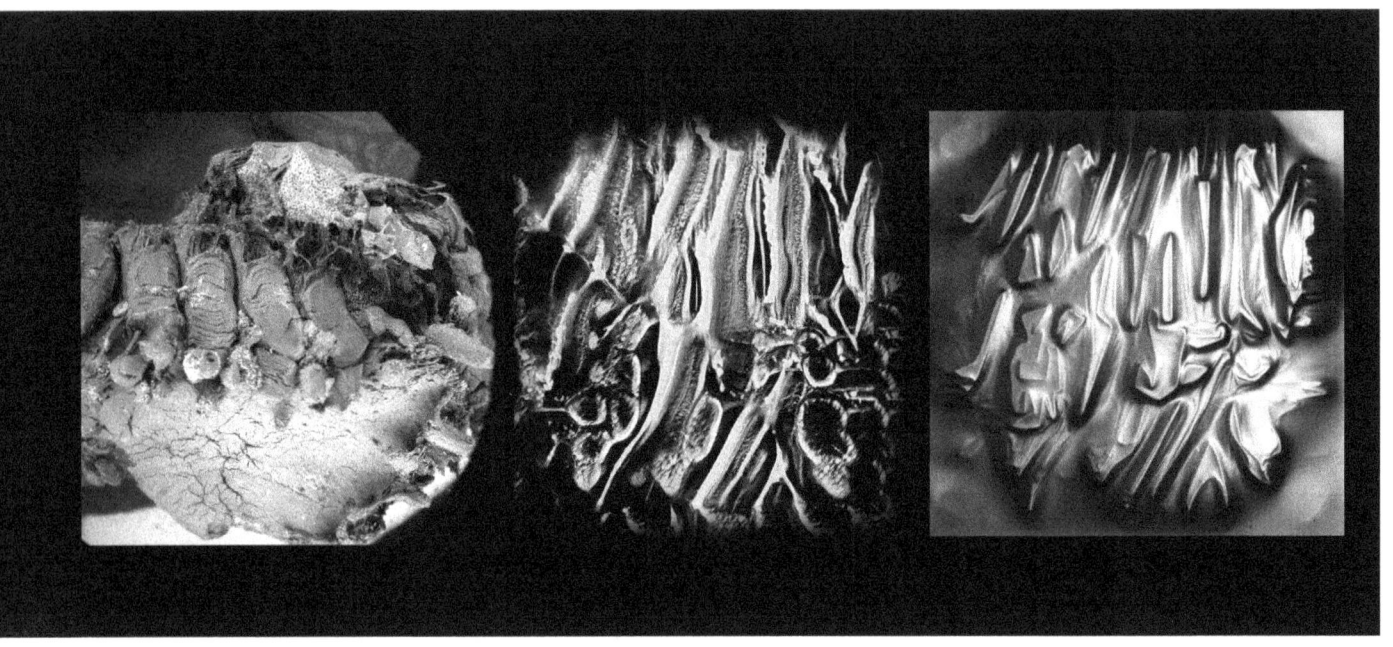

1   Left: Scanning Electron Microscopy (SEM) image

    Middle: Generative Adversarial Network (GAN) image

    Right: Robotic Incremental Metal Formed (RIMF) steel relief

ABSTRACT

Nanotectonica is an architectural research project that examines the convergence of nanotechnology and contemporary design tools. The present study, *Nanotectonica SEM-GAN,* focuses on two processes for image production, one based in the field of nanotechnology and the other in machine learning: Scanning Electron Microscopy (SEM) and Generative Adversarial Networks (GAN). It establishes commonalities of these routines as they pertain to aesthetics and design methodology, and it explores methods of spatializing and materializing images produced in their interaction. The study of transposing rich image material to three-dimensional geometry and material artifact is considered relevant not only to the particular study at hand, but also to the general problem of image-based machine learning techniques when applied in the spatial design disciplines. A third process, Robotic Incremental Metal Forming (RIMF), advances the aesthetic language of SEM-GAN through the sculptural method of the relief. Analogous to the electron beam probing minute bodies in nanoscopic imaging, the end effector impresses robust steel plates in robotic fabrication.[1]

The study elaborates a "strange materiality," identified in SEM imaging, enhanced through GAN, and materialized via RIMF. It refers to the inherent unfamiliarity of subvisible expressions that are made visible through the scanning electron microscope. SEM visuals are generally misperceived as black-and-white photography, yet they are not produced with light (photons) but with electrons (matter); hence, the strange materiality of the aesthetic quality—the infinitesimal manifestation of things rendered visible in total darkness. Materialism here refers to the mind-independent existence of objects under observation. Strangeness hints at expressing the vibrant agency of their invisible matter through appropriated aesthetic methods such as SEM and GAN.

INTRODUCTION

The study begins by discussing parallels in the operating processes of the Scanning Electron Microscope (SEM) and Generative Adversarial Network (GAN) applications, and it explores their potential for a non-deterministic and collaborative design method, here referred to as speculative design.[2] SEM and GAN applications both entail blind procedures, which momentarily suspend control by the human operator either through a certain degree of autonomy granted to the machine in training (GAN) or through indirect observation (SEM) [see *Blind Probing* section of this paper].

In addition to operational parallels, the study establishes aesthetic commonalities found in images produced by the SEM and GAN processes. It argues that both share a particular visual language with an intrinsic quality of "strange materiality." Images produced by the scanning electron microscope and those generated by adversarial networks express spatial effects through gradient pixel fields that depict shade and shadow and work without line graphics; both relate inputs that are not visible per se. However, GAN and SEM visuals differ in two fundamental ways. First, GAN applications process as well as produce digital images, whereas SEM produces images from subvisible physical specimens. Second, GAN sources and produces the entire digital spectrum of color, while the SEM operates without light and thus without color [see *SEM Operation*]. SEM grayscale images render smooth gradients into blurred fields and produce an often moody atmosphere, while GAN images evoke dreamlike scenarios; both depict strange worlds as they relate to an invisible source.

In *Nanotectonica*, strange materiality refers to the inherent unfamiliarity of subvisible expressions that are made visible through the Scanning Electron Microscope. The SEM visuals are generally misperceived as black-and-white photography, yet they are not produced with light (photons) but with electrons (matter). Hence, the strange materiality of the aesthetic quality—the infinitesimal manifestation of things rendered visible in total darkness by a direct matter-to-matter reading, and granted novel expression via GAN manipulation. Materialism here refers to the mind-independent existence of objects under observation. Strangeness hints at expressing the vibrant agency of their invisible matter through appropriated aesthetic methods such as SEM and GAN.[3]

While identifying inherent aesthetic commonalities produced by SEM and GAN operating separately, the study explores the two techniques in conjunction. Here SEM images function as source material for the GAN operation, and together they serve two parallel modes of inquiry. The first is a continuation of the aesthetic discussion (above) and attempts to amplify the established effects, i.e. the unfamiliarity of the subvisible

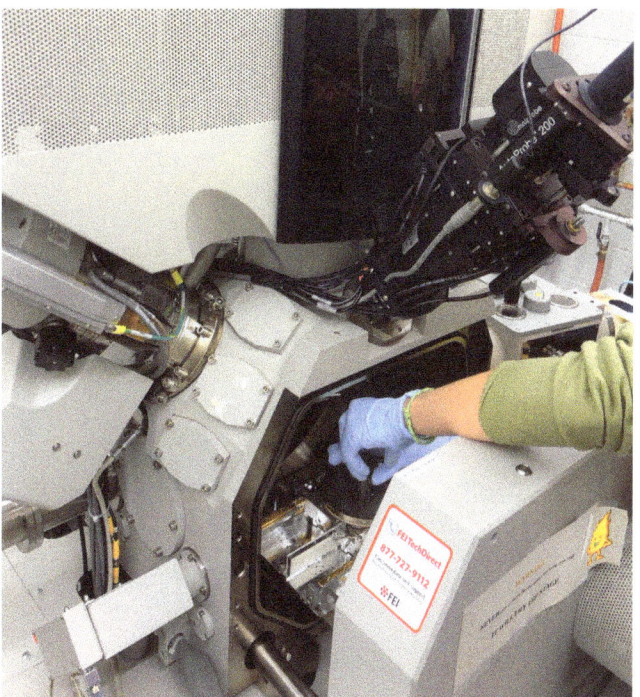

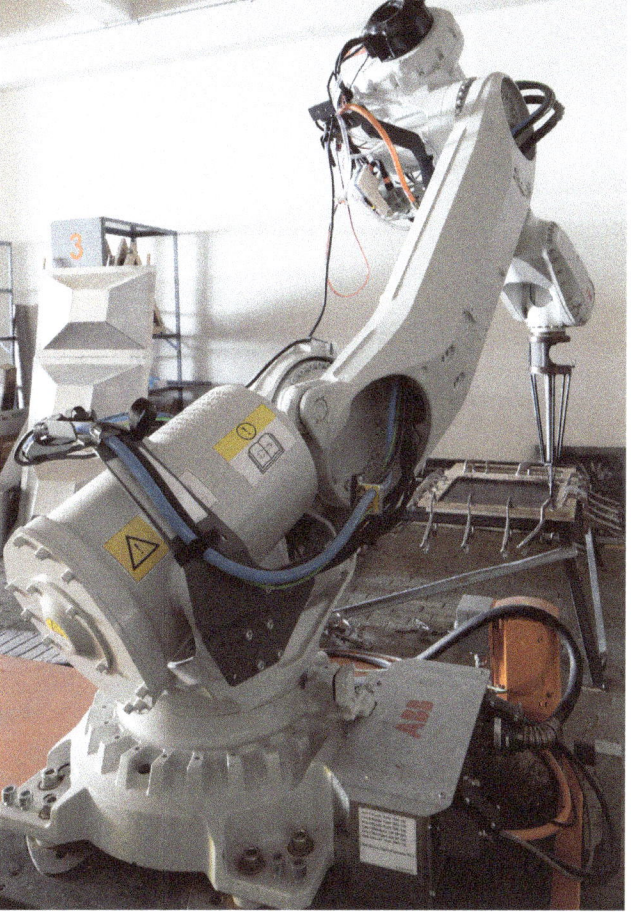

2 Scanning Electron Microscope: FEI, Helios NanoLab650; *Nanotectonica* at The New York Structural Biology Center, 2019

3 Robot: ABB IRB6700 *Nanotectonica* at The Consortium for Research and Robotics, hosted by Pratt Institute, 2022

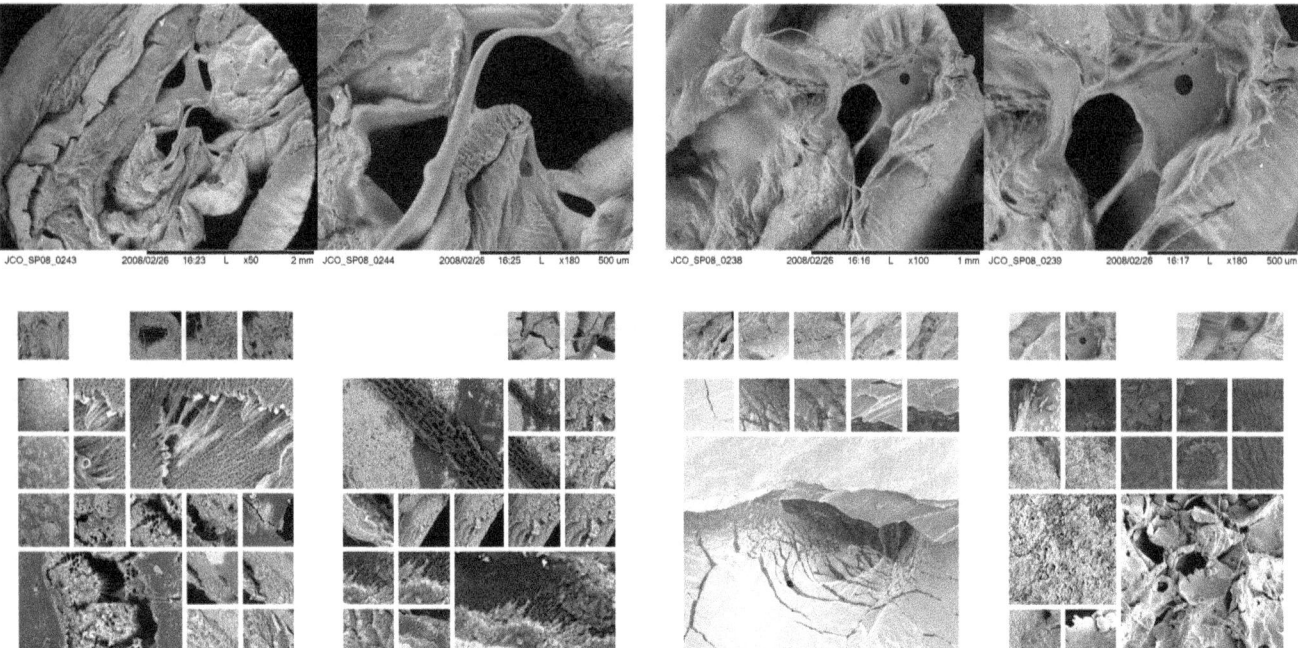

4

world is heightened by machine intelligence; GAN estranges the SEM image further.[4] The second utilizes the GAN as a taxonomic and archival operator of the SEM image collection Nanographia.[5] By training the GAN with a vast data set of SEM-produced image material, the study explores new modes of taxonomy making. Here, a non-human actor is invited to advance "open taxonomies" of the subvisible world [see *SEM-GAN Taxonomy*]. As a technical consideration, the study compares two different methods of training a GAN engine on the Nanotectonica SEM image library [see *Comparing Methods*].

Subsequently the study turns towards the problem of image spatialization and critically discusses various methods of transposing 2D image data to 3D geometry. With this Nanotectonica, SEM-GAN aims to contribute to a particular aspect of the discourse in the spatial design disciplines (architecture, and urban-, fashion-, and industrial design, etc.) relating to the use of image-based GANs. Initially the study takes three approaches to image spatialization: working with image stacks to generate monolith geometry; with multi-dimensional orthographic image projection to generate sculptural form; and with image displacement mapping to generate topographic relief. It concludes that sculptural relief is the preferred method for elevating the inherent aesthetic qualities of SEM-GAN [see *Image to 3D*].

Finally the study explores and advances Robotic Incremental Metal Forming (RIMF) as a method for fabricating sculptural reliefs derived from the SEM-GAN process. For relative flexibility in the robotic metal forming process, initial material tests were performed with aluminum panels. The fabrication aspect of *Nanotectonica SEM-GAN* currently focuses on light-gauge steel panels for RIMF, as the material expression of the resultant relief panels most directly corresponds to, and elevates the aesthetic qualities identified for SEM-GAN; in particular through its dark graphite qualities in matte finish [see *Metal Forming*].

BACKGROUND
Nanotectonica

The present study—*SEM-GAN*—is embedded within the larger design research project *Nanotectonica*, which examines the relationship between natural and architectural systems through the convergence of nanotechnology and contemporary design tools.[6] Nanotectonica is the encompassing term for the design research into the structures, aesthetics, and design ramifications of the Nanoscale, which originated as project-based investigations within our architectural practice around 2000, and then developed into a series of academic seminars beginning in 2007.[7] The nineteenth installment of this seminar is currently conducted as an advanced technology course Arch 720AP Nanotectonica directed by Jonas Coersmeier at Pratt Institute, School of Architecture, with participating graduate students James Nanasca, Man Hin Ivan Yan, and external consultant Ezio Blasetti.

A design research and production project that studies structures and organizations at multiple scales, *Nanotectonica* utilizes computational techniques to design, construct, and build novel material systems, intricate assemblies, and architectural artifacts.[8] The design research employs

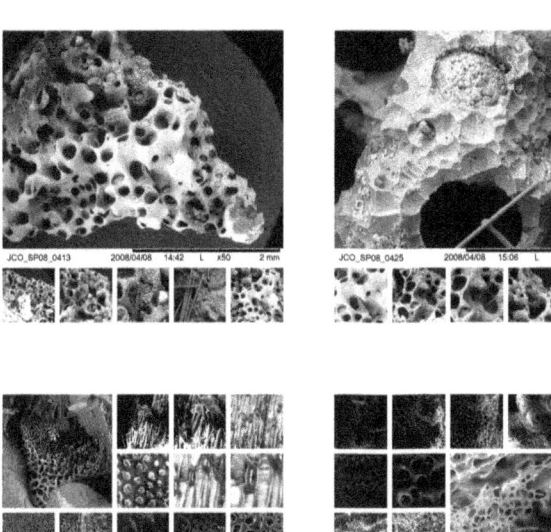
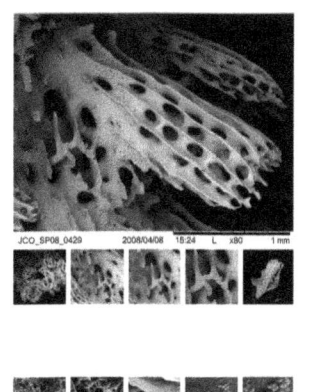
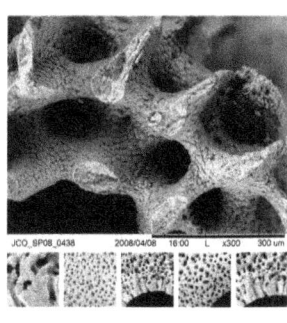
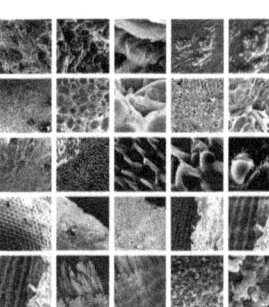
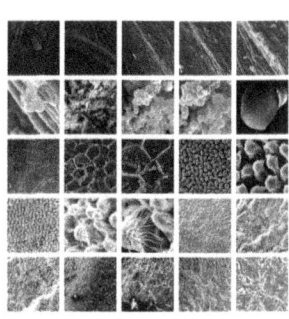

4 Selection of SEM images from the Nanographia archive

5 Selection of SEM images from the Nanographia archive

nanotechnology, specifically the scanning electron microscope (SEM), and digital tools of analysis for a deeper understanding of structures at various scales. The investigation is not limited to the phenotypic expressions, but seeks to decipher organizing and form-building principles. While the SEM is used as an instrument for the analysis of subvisible structures, it also serves as a model for a speculative design method, Blind Probing, which operates outside of the duality of the generative and determinative routines.[9]

Early findings of the Nanotectonica project were presented at ACADIA 2010 (Coersmeier 2010), during the conference at The Cooper Union and in the parallel exhibition at Pratt Institute. These early findings focused on a research, design, and fabrication process chain that entailed electron microscopy, parametric design, CNC flip milling, and fiberglass construction.[10]

SEM Operation (Technical Context)

Electrons have a shorter wavelength than visible light, and thus electron microscopes can detect smaller objects than optical microscopes. The Scanning Electron Microscope (SEM) images a sample by probing it with a focused beam of electrons that scans across its surface; in response, the sample emits secondary electrons which carry information about the properties of the specimen surface. This information is recorded and mapped into images that represent the surface morphology of the sample. Unlike other types of electron microscopes, the SEM has a significant depth of field, which allows it to produce two-dimensional imagery with three-dimensional visual qualities reminiscent of those achieved in photography. In the absence of light, secondary electron shadows sculpt spatial effects, rendered in grayscale pixel fields.

SEM Aesthetics

Nanotectonica embraces the SEM as a prolific machine for aesthetic production. The aesthetics of the SEM are based in part on the device's particular ability to produce spatial effects in the absence of light and shadow. While other types of electron microscopes generate flat images that evoke a sense of abstraction, SEM-based images hold an intrinsic quality of realism. Ever so close to black-and-white photography, these grayscale images often render smooth gradients into blurred fields and produce a kind of detached, moody atmosphere.[11] There is an uncanny quality to these images, which momentarily suspends the association with photography. The representational qualities of the SEM visuals enhance the inherent strangeness of the subvisible object, which itself is never seen directly and is shaped by unfamiliar forces.[12] SEM representation plays with the familiar and unfamiliar, describing an alien world in visually accessible terms. In New Landscape in Art and Science, Gyorgy Kepes (Kepes et al. 1956) describes how the gross world of regular sense perception can be connected to the subtle world by scientific instruments, and he establishes a relationship between images produced by these devices and those of contemporary abstract art. Images produced by the SEM often suggest just this relationship to artistic expression, in the form of strange materiality.

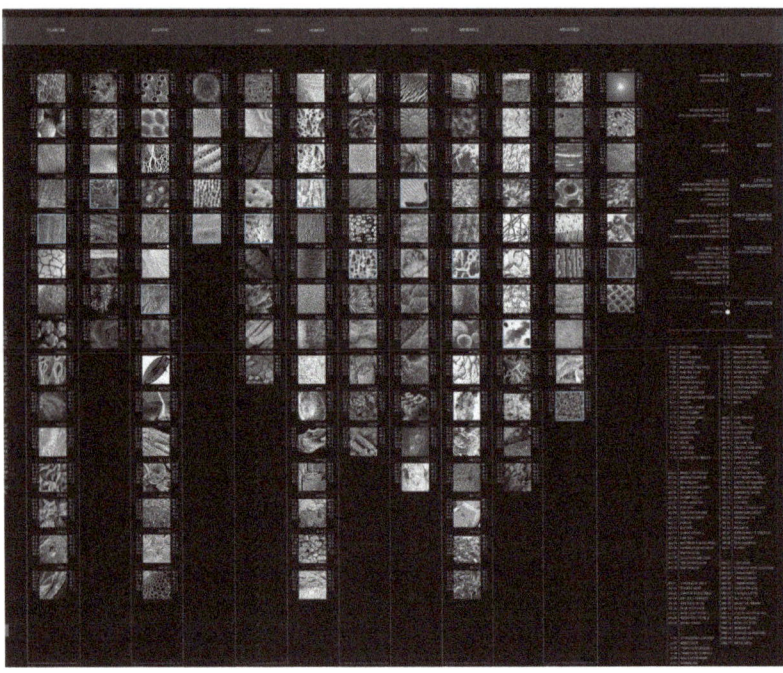
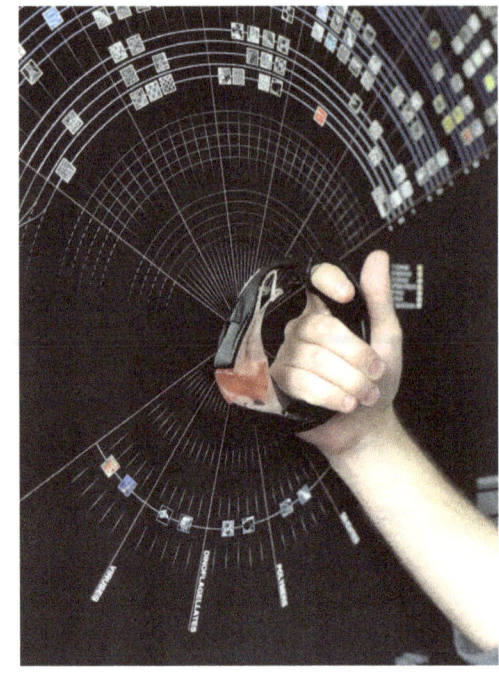

SEM Taxonomy

Nanotectonica explores the innovative potential of open taxonomies in the architectural design process. The formation of taxonomies is considered a creative act in itself, as it directs the search within an infinite space of possible structures; it gives texture to this space and provides momentary orientation. We refer to taxonomies here not in the sense of standard biological classifications, but as systems of structural and architectural commonalities crystallized in the sorting of rich nanographic material.[13] Accordingly, the terms used in these taxonomies do not adhere to biological nomenclatures, but refer to architectural expression.

This study approaches classification via taxonomy and not typology, despite the usual association of typology with architecture, and taxonomy with biology.[14] Taxonomy classifies according to observable and measurable characteristics, having no idealized point of reference (datum). Typology on the other hand refers to concepts rather than empirical cases, to idealized constructs rather than objects of reality. *Nanotectonica*'s model of design relates to speculative realities more than it does to ideal types, and so privileges the taxonomic approach over the typological.

SEM-GAN
SEM-GAN Taxonomy

The study explores the potential of GANs to advance open classifications for objects and structure of the subvisible world. The study explores new modes of taxonomy making by training the GAN with a vast data set of SEM-produced images and considering the generated images in terms of species relations. While the SEM catalog refers to a finite set of physical specimens, the images produced through machine learning processes suggest an infinite number of synthetic species that display similar formal characteristics as the source material. With GAN the definitions of these common attributes are generated through statistical analysis. Taxonomic identifiers are not limited to matrices of humanly crafted parameters but they are generated in an open field of commonalities. This advances the idea of open taxonomies and their speculative potential in the design process, as it liberates it further from the model of standard architectural classifications (typology).

GAN Technology (State of the Art)

StyleGAN is a style-based GAN (Generative Adversarial Network) architecture yielding data-driven unconditional generative image modeling (Karas et al. 2020). Improving on a previous iteration of StyleGAN, Karas et al. further develop NVIDIA's StyleGAN into StyleGAN2 that refines the GAN architecture's overall image quality and control of image generation. At the time of conducting this research, StyleGAN3 was released by NVIDIA, but was not utilized due to the proliferation of StyleGAN2, and readily available scripts from the likes of Jeff Heaton (Heaton 2022) and companies like RunwayML (Runway AI, Inc. n.d.).

In the last five years there has been an unprecedented interest in architectural design research with GAN and machine learning tools. These techniques were first introduced in speculative design studios taught by architects such as Karel Klein in 2018, and their use became ubiquitous in academic

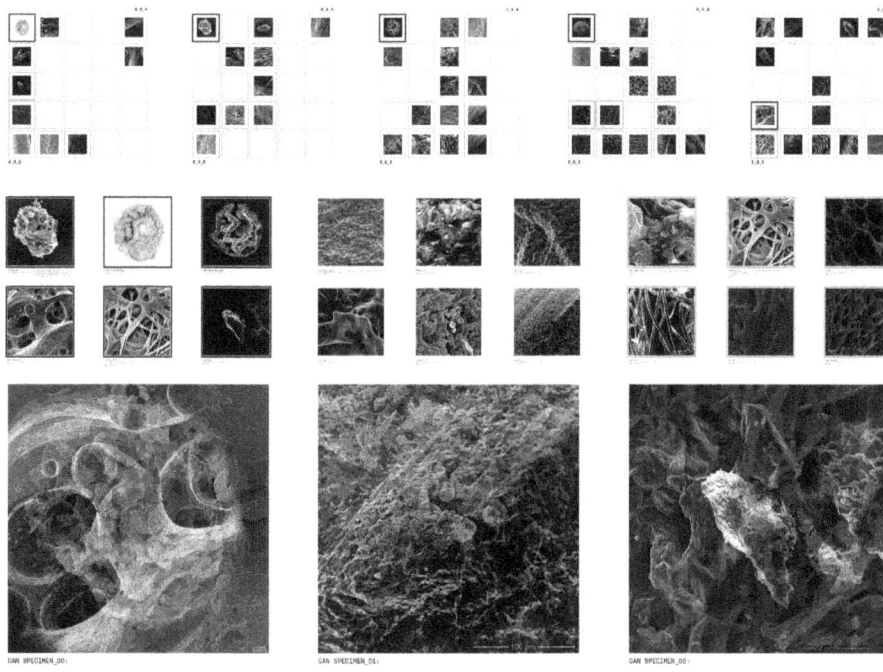

8 SEM-GAN taxonomy *Prototype*: images from the Nanographia archive are arranged according to formal commonalities, represented specimens manually hybridized

6 SEM Taxonomy, indexical
7 SEM taxonomy, radial

research in a very short period of time. Early works that speculate on the creative design potential of these tools include the academic studios and practices of Matias del Campo, Kyle Steinfeld, Daniel Bolojan, Garbiel Esquivel and others. At the time of writing this paper, we are witnessing a massive acceleration in the production of machine learning applications. Tools like Nightcafe, Wombo, Midjourney, and DALLE that work primarily with text or images as input have millions of active users everyday. The research presented in this paper is also part of this acceleration. To our knowledge, no other researcher is working specifically with images or datasets from the nano scale for purposes of machine learning for design.

Comparing Methods

The study compares two different methods of running a StyleGAN2 engine trained on the same SEM Nanotectonica archive library. With a myriad of Python scripts available, the study tested one from Jeff Heaton at University of Washington St. Louis (Heaton 2022), and RunwayML's web-based StyleGAN2 interface. RunwayML's StyleGAN2 allowed the options of pre-trained discriminator datasets of faces, cats, vehicles, etc. Training the Nanotectonica archive with RunwayML's pre-trained vehicle discriminator dataset resulted in more "SEM-like" aesthetics with discernable objects and fields that the study pursued.

The differences between Heaton's StyleGAN2 script, and RunwayML's StyleGAN2 output are seen in stark contrast below (Figure 11). Heaton's method resulted in field-like conditions almost exclusively, as well as producing a green tint from grayscale SEM images. The pretrained discriminator from RunwayML generated more recognizable "SEM-GANs", leading the research to select RunwayML as its StyleGAN engine of choice. While not the sole subject of the paper, there is much to be gained in Nanotectonica research by training our own discriminator of SEM images to then allow a StyleGAN to more successfully generate SEM-GAN imagery.

The Synergetic SEM-GAN Workflow

The investigation and production of the SEM-GAN images quickly found credence in a hybrid workflow. Human command over the images gave way to machine driven outputs, with said digital outputs guided by human intuition and selection. This method expressed in itself the logic proposed by Kepes regarding art and science [see *SEM Aesthetics*] in which the human eye injects artful finesse into the project, while the computational procedure remained objective in the process. This partnership fundamentally anchored the methodology of how we began to perceive these new creatures, as not one or the other, but a product of human-machine interactions, which generated SEM-GAN images of strangely foreign yet familiar worlds.

IMAGE TO 3D

State of the Art (Existing Research)

The problem of deriving 3D artifacts from image material is as old as the (spatial) design disciplines themselves. Since machine intelligence technology first arrived in design via image-based GANs, this problem has been given a particular status in recent disciplinary discourse. The research presented in this paper builds upon previous work from one of its authors with focus on the computational generation of 3D

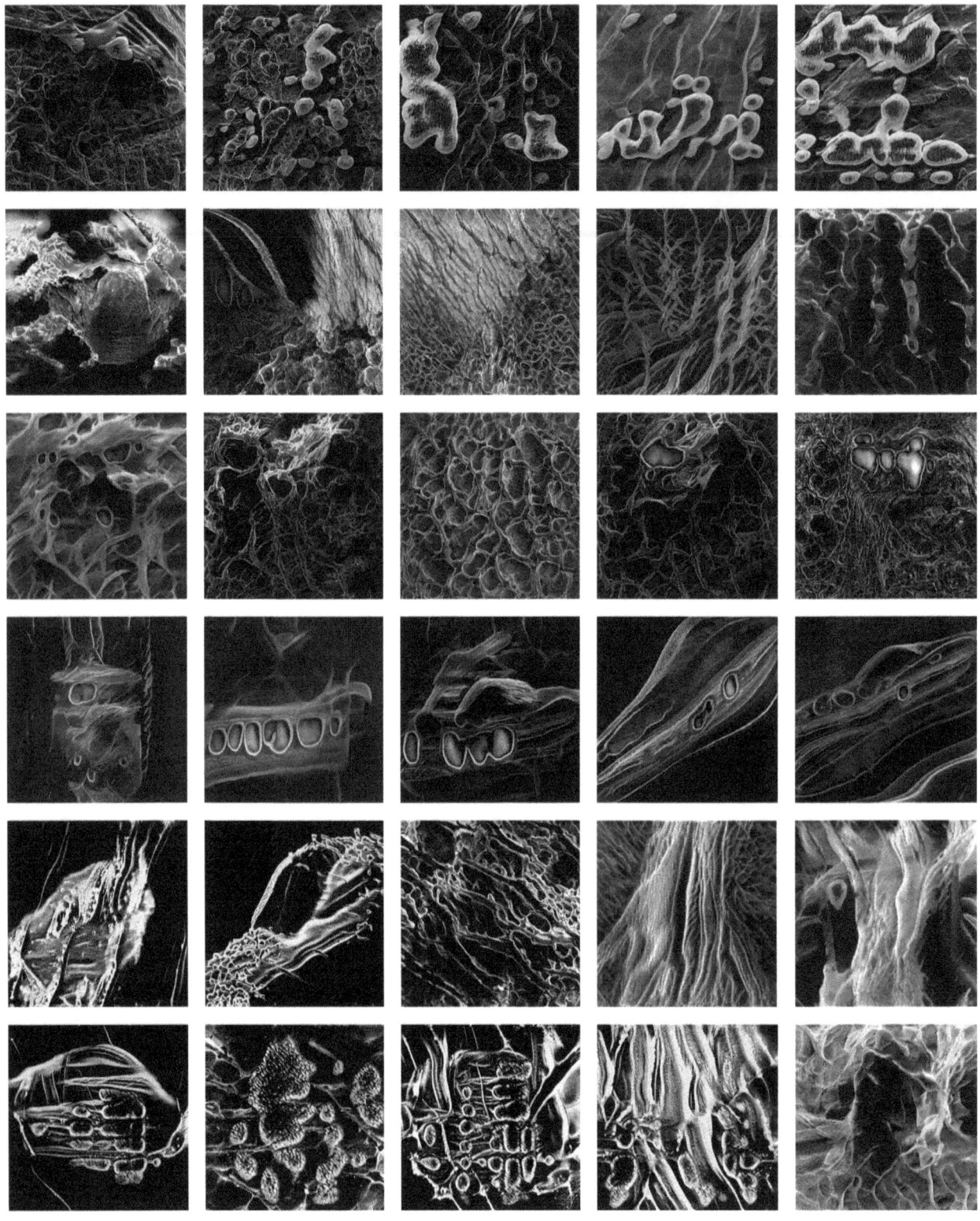

9 Nanotectonica SEM-GAN: GAN images generated from Nanotectonica SEM Archive Nanographia

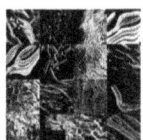

10 Nanographia SEM images are scored and organized on a matrix of structural commonalities

11 (left) Nanotectonica SEM archive dataset trained using Heaton's Google Collab StyleGAN2; (right) Nanotectonica SEM archive dataset trained using RunwayML StyleGAN2

form from image based machine learning tools.[15] Early work on specific machine learning techniques for the encoding and generation of 3D form includes the work of Hiroharu Kato from the University of Tokyo[16] (Kato et al. 2018). Today's advanced digital tools, such as those discussed by Chan from Stanford University and NVIDIA, allow for the "[u]nsupervised 3D generation of images and 3D shapes using collections of single-view 2D photographs…" (Chan et al. 2021). Tools like these have been making their way into architectural research. As an example, a recent project by Kyle Steifeld for the Venice Biennale titled "Artificiale Releivo" attempts to train a GAN model with 3D information embedded in semi-transparent depth map representations. The more accessible tools in 2D-to-3D generation reside in mesh and NURBS modeling softwares, as those softwares are relatively user friendly.

Three Methods for Image to 3D

A readily available and contemporary method of turning 2D images into 3D is the use of voxelization softwares. This initial study leaps from image to 3D by transposing unprocessed SEM-GAN imagery into mesh geometry.[17] Stacking images along a guiding curve (which in this initial case is oriented to the z-axis) allowed for the 3D geometry to be generated based on light and dark values of the stacked SEM-GAN images. The slicing of images along the z-axis results in an eroded column-like structure (Figure 13 left). While aesthetically interesting, this 3D generation deviated from the research agenda and qualities of the SEM imagery and was not further pursued.

Another attempt was made of translating an image to 3D via image projection into a voxelized field (Figure 13 center). Compared to the stacked method, the projection allows a network of light and dark values of three images along the x-, y-, and z-axis to generate geometry based on guiding curves. Like the stacked method, this guiding geometry of two circles and a vertical curve produced a column-like structure but with a less eroded effect than the stacked method. The column-like forms and eroded aesthetics were discontinued to pursue methods that would produce geometry and aesthetics closer to the source material of the generated SEM-GANs.

Still using voxelization software, the study pursued the extrusion of light and dark values of the chosen SEM-GAN image into a displacement-mapped mesh geometry. Some images translated into mesh geometry better than others depending on the values of adjacent dark and light pixels. A large difference between adjacent pixel values resulted in "sharp" or "spiked" meshes (Figure 13 right), rendering the geometry incompatible with fabrication. The method of voxelized displacement-mapped meshes began to produce geometry closer to the aesthetic qualities discussed in the research, so other methods furthering displacement-mapped geometry were pursued.

Displacement Method: Monocular Depth Estimated SEM-GAN Images

Focussing on the displacement geometry, the research looked at turning SEM-GAN images into a NURBS surface instead of a voxelized mesh geometry. Filtering the images through a Monocular Depth Estimated machine learning engine via RunwayML produced smoother and less "spiked" geometry. Figure 12 displays the translation of the chosen SEM-GAN image, filtered through the Monocular Depth

	SEM-GAN	HEIGHT FIELD	TOOLPATH	INITIAL PRESS	SECOND PRESS
00					
01					
02					
03					
04					
05					

12 SEM-GAN specimen and operation matrix

 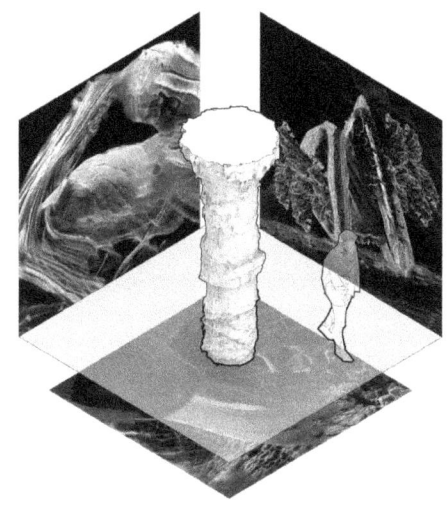 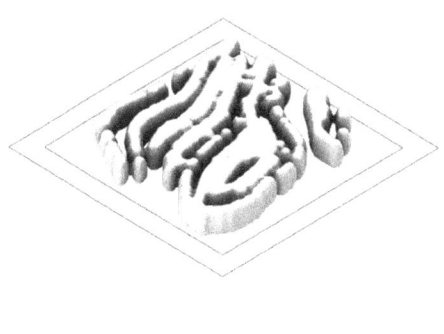

13 (left) Sliced SEM-GAN 3D object; (center) Projected SEM-GAN 3D object; (right) Displacement SEM-GAN 3D object

estimating software, and ultimately translated into a surface that retains the surface continuity of the source image. A clear lineage of image-3D is established and primed to turn into a physical artifact.

The relief became the most effective process in capturing the aesthetic qualities of the SEM-GAN images. Relief is also the most conducive of the chosen fabrication method of robotic incremental metal forming (RIMF), enabling the setting of a toolpath derived from the Monocular Depth Estimated SEM-GAN Surface and hand tailored to achieve the greatest possibility of success. This focus allowed us to set the main agenda as a 2.5D image production, revealing the most direct and authentic/accurate translation of the image from a 2D plan into the intended 3D object.

METAL RELIEF
RIMF State of the Art
Robotic Incremental Metal Forming (RIMF) is a well documented fabrication method of turning flat sheet metals into formed pieces, first pioneered by Schrafer and Schraft (2005). Ammar Kalo and Michael Jake Newsum (Kalo and Newsum 2014, 2) take RIMF further by pressing non-planar components to aggregate an assembly of self-similar parts. More recent research done by Cui et al. (2022) rigorously details methods of finite element analysis and 3D scanning that can be deployed to minimize geometric inaccuracies with RIMF. The goal of this study is not primarily to expand upon the work in RIMF or StyleGAN separately, but to explore design potential and aesthetic richness in the interaction of these processes.

Toolpathing and End Effector
Toolpathing to press the SEM-GAN imagery is generated by contouring the surface using the outputs from the Monocular Depth Estimated Images. Kalo and Newsum (Kalo and Newsom 2014) illustrate that a spiral toolpath allows for constant engagement of the end effector and alloy(s) being pressed. Other tool path generations that were not spiraled resulted in visible seams once pressed into the alloys. A constant acute angle of 30 degrees measured planarly from the inside of the pressed form was found to be the most reliable angle to press the relief (Figure 16). Neglecting to adhere to this prescribed angle led to the tearing of the steel [see *Material*] as the material would be stretched inconsistently.

The end effector traces the path of the spiralized toolpath, acting as a displacer of the steel sheet. Similar to how the computer using the monocular estimated depth method will read the light and dark values of pixels and assign them corresponding heights, the end effector and robot work in a choreographed routine stretching the sheet. The current 1 foot and 11¼ inches long end effector is constructed of a steel frame that interfaces with ABB 6700 baseplate and a hardened carbide tip to engage with the stock. Using a 2 ft by 2 ft sheet, the end effector was able to press both steel and aluminum into 2¾-inch depths while maintaining the alloy's integrity. In this current research the longer end effector, which allows for deep reliefs, was not utilized, but future research could leverage the reach of the end effector by pressing larger stock (Figures 12, 13).

Aligning with the research of Kalo and Newsum (2014), this study also performs multiple toolpath operations on a single

14

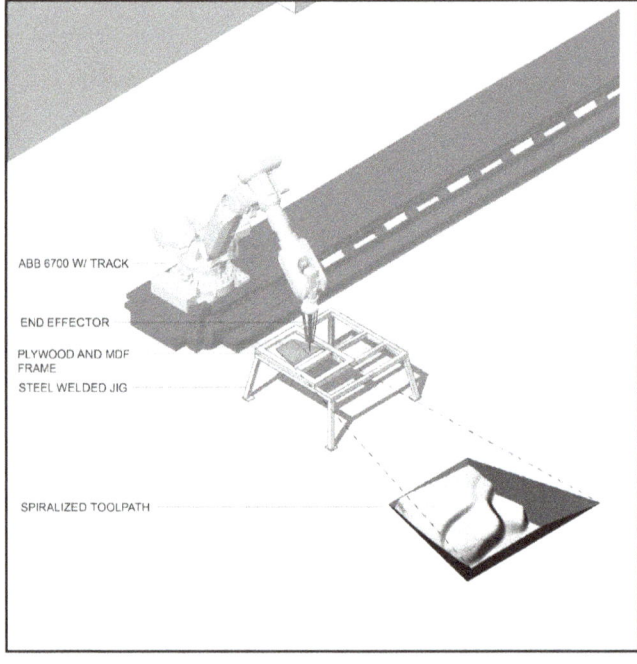

15

sheet with the aim of producing detail-rich steel panels representative of the SEM-GAN images. The research found that select panels prompted a "second act" to enhance resolution. Specimens 03-05, served as good candidates for "second acts" due to highly varied and high contrast images. The first spiralized toolpath resulted in a global press, wherein the general 3D shape was pressed by visualizing some low resolution detail of the SEM-GAN. The "second act" allowed for the inclusion of localized and higher resolution details to be pressed. Select geometries local to a given specimen were manually culled, translated into spiralized toolpath geometry, then pressed into a higher resolution steel form.

Material

Aluminum at 14 gauge and 1/16-inch thick, and steel at 22 gauge and 1/32-inch thick were tested in this research project. Aluminum afforded more malleability when compared to steel, and allowed for more successful pressing of the toolpath generated with minimal risk of tearing the sheets. However, the aesthetic qualities found in steel aligned more directly with the research agenda, and SEM imagery (Figures 22-25), so ultimately steel was pursued and tested on large (24 in x 24 in, 22 ga sheets) and small (12 in x 12 in, 22 ga sheets) scales. Testing allowed us to visually quantify the stress limits of the steel through observation of its physical responses when pressed, revealing the material's breaking points. The material choice led to the codifying of aesthetic properties that drew from SEMs. Specifically steel's inherent aesthetic qualities referenced the language that emerged under the scanning electron microscope, with highlighted bright spots and deep graphite coloring on matte fields.

Human–Robotic Fabrication

The synergetic SEM-GAN workflow previously detailed was extended into the fabricating portion of the paper's research, where we found a fine balance between trusting the robotic arm to perform as programmed, but made manual adjustments through human hands and mounting points to address the stress symptoms exhibited by the pressed alloy. The choreography quickly reinforced the idea of a partnership in which the operating agents made an action and the other reciprocated with a response—a production method that relied on the strengths of the two partners to push and pull against the mounted alloy in order to guarantee the greatest chance of success in pressing the relief.

The robotic arm possessed the proficiency and payload to stretch the anchored metal sheet, but was constrained by the programmed input and its sensors that rejected the shifting and pinching moments of the material as it was stressed to

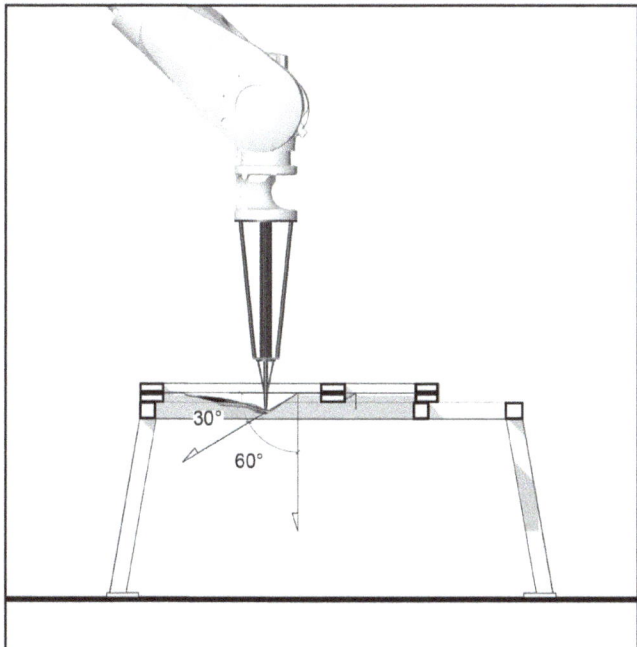

14 ABB IRB6700 with clamped steel sheet setup
15 Diagram of setup [see Figure 14]
16 Diagram of Robotic Incremental Metal Forming (RIMF) section [see Figure 17]
17 RIMF in operation

its limit. The human, on the other hand, lacked the ability and power of the robotic arm, but played tandem to the process through monitoring the audible and visible stressing of the steel that is evident to human senses but inaccessible to the robot's digital sensors. Addressing the live conditions by applying more machining lubricant or shifting the mounting clamps allowed for greater tolerance and shifting of the material and ensured the greatest chance for the routine to be completed without material failure.

This partnership codified the SEM-GAN as a collaborative; with a focus upon subvisible structures (SEM) hybridized through machine learning (GAN) training that is directed towards a three dimensional and aesthetic production as a workflow of human-machine interface (Figure 19).

CONCLUSION

Nanotectonica SEM-GAN develops a strange materiality, first identified in subvisible (physical) expressions imaged by the Scanning Electron Microscope, and carried through the process of Generative Adversarial Networks, to the physical artifact of metal relief. It addresses the three processes—SEM, GAN and RIMF—in interaction. While the study advances specific techniques for each one, it principally contributes to a cross-process knowledge base as it pertains to design methodology and material aesthetics.

The study elaborates a model for human-machine-material collaboration in design research and production. It ascribes design agency to all three actors, who perform an open dance rather than a scripted routine.[18] With this the study critically discusses the algorithmic project in architectural discourse, which typically sets up a dichotomy between the generative and the compositional design method.[19] *Nanotectonica SEM-GAN* aims at disrupting these categories and offers an integrated model for design.

The general problem of deriving 3D artifacts from image data is approached by assessing three methods of spatialization along two criteria—the immediacy of a method and

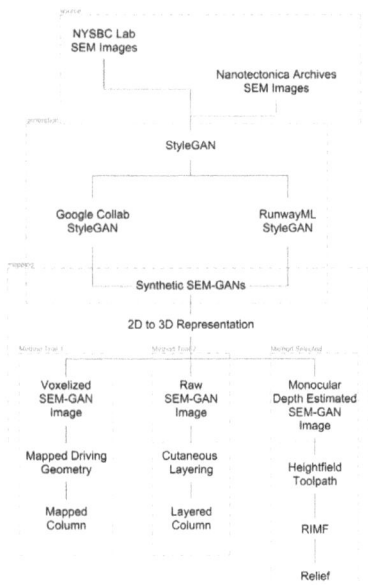

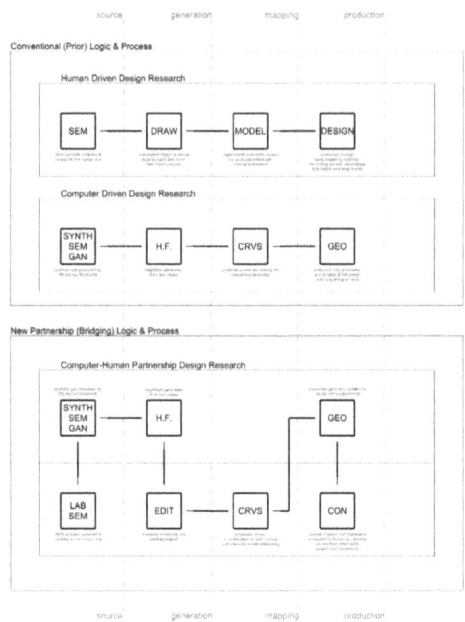

18 *Nanotectonica SEM-GAN* research map

19 Comparing linear design model with the human-robotic integrated model

the relevance of an output geometry to a given material and fabrication system.[20] The direct translation from image to relief via the method of displacement stays close to the source data, while it generates topographic relief rather than tectonic artifact. As this study is invested in exploring a particular aesthetic language migrating across various media, the immediacy of this method proved to be significant.[21] In a next stage the study will explore this proximity in practical terms, by developing a direct routine for robotic material displacement based on SEM-GAN image data.[22]

Future *Nanotectonica* research seeks to capture three dimensional material information in parallel to training GAN discriminators in 2D and 3D. Recent advances in computer graphics and microscopy, as well as the democratization of fabrication tools and autonomous processes, will allow *Nanotectonica* to explore the multidimensional qualities of matter further. In microscopy, the Focused Ion Beam-Scanning Electron Microscope (FIB-SEM) allows for the three dimensional tomography of matter. In computer graphics, Neural Radiance Fields (NeRF) provide three dimensional reconstructions with only few 2D image inputs to produce volumetric representations. In the field of robotic fabrication, multi-material 3D printing and weaving will expand the geometric and tectonic translations between mediums that this study engages with. Future design research will include some of these techniques to explore new models of engaging with the strangeness of subvisible matter.

ACKNOWLEDGMENTS

The research for this paper was supported by Pratt Institute School of Architecture, where the *Nanotectonica* seminar is offered every Spring Semester. The Consortium for Research and Robotics, hosted by Pratt Institute, has provided the robotic facilities. Research on the electron microscope was made possible through institutional and industry sponsors by the New York Structural Biology Center, LPI Inc., and the Hitachi Corporation. The authors would like to thank Gisela Baurmann for ever-present inspiration, Gökhan Kodalak for ongoing conversation and guidance, Adam Elstein for professional photography, and Patrick Rutan for research assistance.

NOTES

1. In 1665 Robert Hooke published *Micrographia: or Some Physiological Descriptions of Minute Bodies Made by Magnifying Glasses. With Observations and Inquiries Thereupon.*

2. "Collaboration" here refers to the interaction between human designer and machine. The term 'speculative design' was "popularized by Anthony Dunne and Fiona Raby as a subsidiary of critical design. The aim is [..] to design proposals that identify and debate crucial issues that might happen in the future. Speculative design is concerned with future consequences and implications of the relationship between science, technology, and humans" (attributed to Dunne and Raby 2014, from Wikipedia article "speculative design" as accessed June 2022).

3. Our philosophical influences include new materialists such as Manuel DeLanda and Jane Bennett. Our aesthetic references draw from art historical categories of the "strange" (Herbert Grabes), the "weird"

(Mark Fisher), and the "uncanny" (Sigmund Freud and Anthony Vidler)

4. Russian formalist Viktor Shklovsky coined the term "defamiliarization" in his 1917 essay *Art as Device* in which he writes: "The purpose of art is to impart the sensation of things as they are perceived and not as they are known. The technique of art is to make objects 'unfamiliar'...."

5. "Nanographia" is the term used for the Nanotectonica archive of original Scanning Electron Microscopy images produced since 2007. It comprises 1,665 images produced in various electron microscopy labs including the New York Structural Biology Center, the Interdisciplinary Nanostructure Science and Technology, University Kassel, Germany, and the Nanotectonica SEM lab at Pratt Institute [cf. *Micrographia* (Hooke 1665)]

6. Nature is capitalized to indicate that the term itself is subject of the larger research project's inquiry. Nature is considered an artificial construct that refers to various and changing concepts. *Nanotectonica* discusses these concepts as they pertain to ecological thinking and building, and the architectural mandate in the midst of a global climate crisis. It points at the problem of distinguishing Nature from technology, investigates a new understanding of living systems, and offers an integrated reading of the term "Natural structures." *Nanotectonica* critically discusses ideas of bionics and biomimicry, and rejects scientific and design methods that idealize and reduce Nature to an empirical field for investigation. In parallel *Nanotectonica* conducts historical studies that refer to a lineage of naturalists, microscopists, and engineers that have advanced ecological thinking and building.

7. Architectural practice by Gisela Baurmann and Jonas Coersmeier. Academic seminars at Pratt Institute, Undergraduate Architecture (2007-2012,) Graduate Architecture GAUD (2013-2022;) University Kassel School of Architecture, Digital Design Department Jonas Coersmeier (2008-2009.)

8. Design research in *Nanotectonica* refers to three linked modes of inquiry: the first invests in the concept of design itself, an ontology of design. It discusses the problem of the "creative act" in its relation to media and methods, and offers a design methodology as testing ground for this discourse. The second engages in project-based research production. This includes historical references and cross disciplinary sources for cognitive and material models in support of a specific design agenda. The third entails original research production, the work with the Electron Microscope, which simultaneously is the most concrete and speculative form of research here: concrete as it borrows its technical routine and device from the natural sciences and produces tangible (visible) results; speculative as it turns away from objectifying and recording nature and instead proposes the multi-dimensional and interactive operations (the blind folded dance) with the electron beam as a model for the moment of design. As such it offers answers to disciplinary questions posed in the first mode, the concept of design (Coersmeier 2020).

9. Blind Probing: "The work on the electron microscope provides access not just to the world of subvisible structures, but through its unique operating procedure to the obscure moment of design innovation. It can help externalize and thus prepare for theorizing this moment. [..] The process is blind in two respects: Firstly, it happens in the dark; light does not enter the scene, but an electron beam like a white cane scans the probe space. Secondly, the exploration is conducted without an overview or perceptual reference to the specimen. In a process of constant reorientation, local scans only gradually assemble a sense of object gestalt...." (Coersmeier 2020).

SEM Method: "In *Nanotectonica* the Scanning Electron Microscope (SEM) does not embody the purely analytical routine of the scientific method. Instead, it operates as a model for design, both as a conceptual model for the moment of design innovation, as well as a practice model for speculative design sensibility. The former refers to the non-deterministic character of the blind search. In this model the search is conducted in a vast space of design potential, that comprises immanent yet unrealized forms and ideas. The search is not indiscriminate, as design intention structures the space, nor is it globally directed, as the intention acts like the electron beam locally and in real time. The latter model is a design trainer and refers to the actual work on the scanning electron microscope. We conduct electron microscopy laboratory sessions in order to gain first-hand experience in operating the SEM. While the work on the machine is initiated by the desire to explore subvisible structure and to produce images of a particular aesthetic quality, it serves as a training exercise that helps develop a light touch for design speculation. The work in the SEM-lab induces an instantaneous flow of mediate interaction with material, a state of focused distraction conducive to design" (Coersmeier 2020).

10. See *ACADIA 10: LIFE in:formation, On Responsive Information and Variations in Architecture, Proceedings of the 30th Annual Conference of the Association for Computer Aided Design in Architecture (ACADIA). New York 21-24 October, 2010.*

11. In some instances they feature sharp-edge, high-contrast depictions of specimens and evoke the strange illuminant effect common in astrophotography. Highlights are blown out by bursts of locally charged electrons rather than solar radiation.

12. Morphologies of the subvisible are less subject to gravitational force than those of the visible world. Electromagnetic force produces different forms.

13. Taxonomies of architectural expressions: the associations formed in this process are unconstrained by established species' relations, and occasionally they run in parallel but often counter to them.

14. At the end of the eighteenth century Jean-Nicolas-Louis Durand set out to systematize architectural knowledge. He developed a theory of 'type,' a kind of science for architecture, we now call typology (Durand 1799-1801).

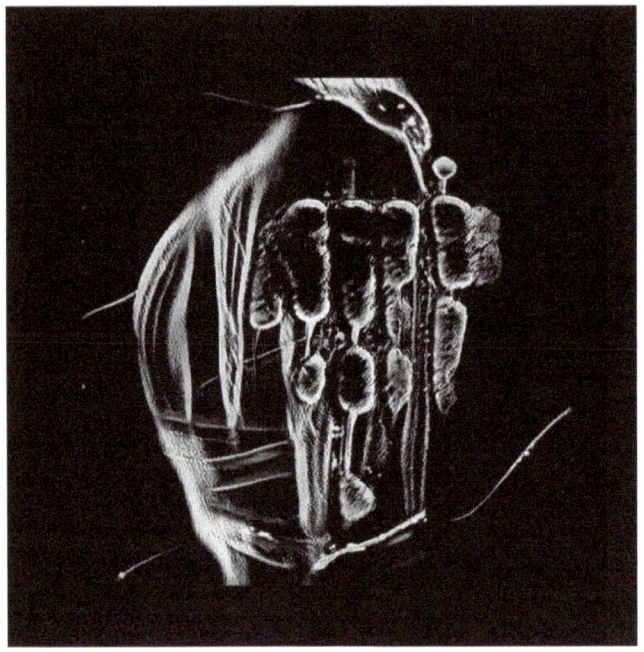 

20 (left) SEM-GAN Specimen 05; (right) Global RIMF relief

15. Early experiments in translating the 2D output of styleGAN and cycleGAN into 3D form were conducted by Hang Zhang in the research studio of Ezio Blasetti and Cecil Balmond in Spring of 2019 using similar computational techniques: Serial Stack, Multiview. The research was later published in a paper titled "3D Architectural Form Style Transfer Through Machine Learning" by Hang Zhang and Ezio Blasetti.

16. The researchers explore the potential of the integration of a mesh renderer into neural networks in their paper "Neural 3D Mesh Renderer" (Kato et al. 2018).

17. *Monolith*, a plug-in for Grasshopper3D, and is described by its author, Andy Payne as a "voxel-based modeling editor...and three dimensional voxel based image processing with the aim of allowing a very fine level of control over volumetric material distributions..." (Payne 2017).

18. During the process of imaging (SEM), the observed specimen responds to the electron beam by changing its surface topography and thus the recorded image; during the phase of artifact production (RIMF), the behavior of the pressed steel plate induces adjustments in the robotic fabrication setup in real time. In each case the material disrupt and direct human-machine interaction. The Robot and GAN are considered autonomous systems and ascribed (artificial) design intelligence.

19. We first explored the convergence of nanotechnology and contemporary design tools at a time when the idea of generative architecture re-emerged in the context of digital technologies. We refer to this moment as the algorithmic project in architecture, and we discuss it critically in this design research. It is argued that the search for generative design methods, along with the critique of compositional and allegedly more deterministic methods, had been part of architectural discourse since the early twentieth century. "Self-generation has been a consistent goal in architecture for over a century" (Mertins 2004). Since the advent of the algorithmic project at the end of the past century, architectural effects of the generative method have been widely privileged over compositional qualities, and the two have been considered incompatible. Compositional qualities have been associated with a higher degree of direct, top down engagement by the designer, operating at the level of design expression, while generative qualities have been seen as the result of operations at the scripted substrate of the design engine.

20. Testing three methods of image-to-3D—image stacking, 3D projection, and displacement mapping—the study determines that displacement is the most relevant for the purpose of this study. "Immediacy" here refers to the direct translation of image to topography, as well as to the conceptual proximity of the geometric method to the chosen fabrication method (RIMF.)

21. In the context of additive or subtractive fabrication methods, or projects that foreground the logic of tectonic assembly, other methods of moving from image to 3D, including the two alternatives studied here—stacking and projection—may be more relevant.

22. Displacement mapping translates grayscale value to changes along the normals of a surface, analogously the robotic arm impresses topographical changes to the metal surface. At the current state the study identifies this conceptual proximity of displaced geometry and displaced material, while the practiced translation from one to the other remains mediated, as it entails several steps of 3D geometric authoring and the standard practice of robotic toolpathing.

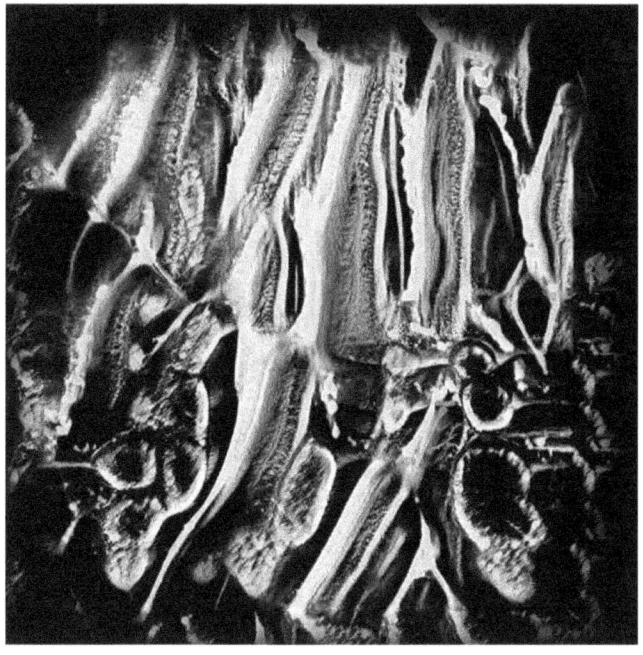 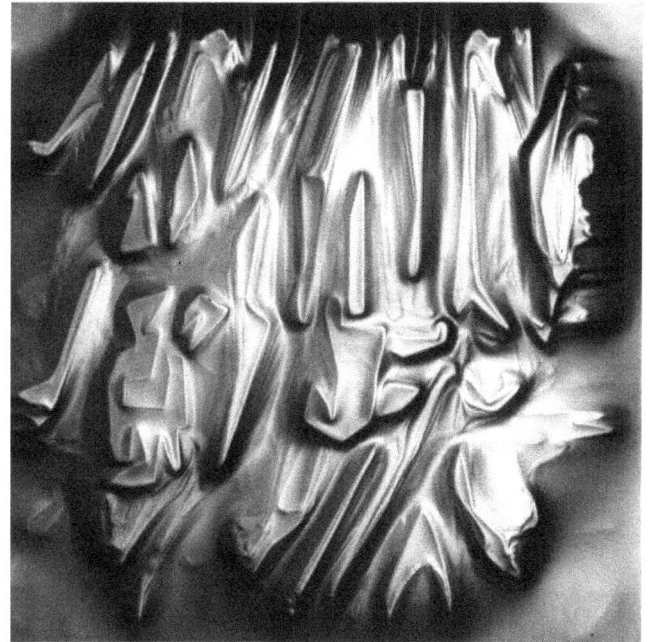

21 (left) SEM-GAN Specimen 03; (right) Localised RIMF relief

REFERENCES

Bennett, Jane. 2010. *Vibrant Matter: A Political Ecology of Things*. Durham: Duke University Press.

Chan, Eric R., Connor Z. Lin, Matthew A. Chan, Koki Nagano, Boxiao Pan, Shalini De Mello, Orazio Gallo, et al. 2021. "Efficient Geometry-Aware 3D Generative Adversarial Networks." *arXiv*. Accessed April 27, 2022. https://arxiv.org/abs/2112.07945.

Coersmeier Jonas, and Donovan N. Leonard. 2008. "Nanotectonica: Architectural Design Studio and TableTop SEM." *Microscopy Today* 16 (5): 40–43 https://doi.org/10.1017/S1551929500061769.

Coersmeier, Jonas. 2010. "Nanotectonica." In Evolutive Means: Exhibition of the 2010 ACADIA Conference, The Cooper Union, New York, Oct. 21 - Nov 19, 2010.

Coersmeier, Jonas. 2013. "Nanotectonica." In *Architecture in Formation: On the Nature of Information in Digital Architecture*, edited by Pablo Lorenzo-Eiroa and Aaron Sprecher. New York: Routledge.

Coersmeier, Jonas. 2020 "Nanotectonica (Draft)." Issuu. April 2021. https://issuu.com/coersmeier/docs/nanotectonica_draft.

Cui, Quian, Sidharth Suhas Pawar, Mengxi He, and Chuan Yu. 2022. "Forming Strategies for Robotic Incremental Sheet Forming." In CAADRIA 2022: POST-CARBON, Proceedings of the 27th International Conference of the Association for Computer-Aided Architectural Design Research in Asia (CAADRIA), vol. 2, 171–180.

DeLanda, Manuel. 2015. "The New Materiality." *Architectural Design* 85 (5): 16–21. https://doi.org/10.1002/ad.1948.

DeLanda, Manuel and Graham Harman. 2017. *The Rise of Realism*. Malden, MA: Polity Press.

Deleuze, Gilles, and Felix Guattari. 1988. *A Thousand Plateaus: Capitalism and Schizophrenia*. London: Athlone Press.

Dunne, Anthony, and Fiona Raby. 2014. *Speculative Everything: Design, Fiction, and Social Dreaming*. Cambridge, Mass.: The MIT Press.

Durand, Jean Nicholas-Louis. 1799-1801. *Recueil et parallèle des édifices de tout genre, anciens et modernes : remarquables par leur beauté, par leur grandeur, ou par leur singularité, et dessinés sur une même échelle*. Paris: l'Imprimerie de Gillé fils.

Fisher, Mark. 2017. *The Weird and the Eerie*. London: Repeater.

Grabes, Herbert. 2004. *Einführung in die Literatur und Kunst der Moderne und Postmoderne: Die Ästhetik des Fremden*. Basel: A. Francke Verlag.

Heaton, Jeff. 2022. "Jeff-Heaton-DL/t81_558_class_12_01_ai_gym. Ipynb at Master · Karthy257/Jeff-Heaton-DL." *GitHub*. April 29, 2022. https://github.com/karthy257/Jeff-Heaton-DL/blob/master/t81_558_class_12_01_ai_gym.ipynb.

Kalo, Ammar, and Michael J. Newsum. 2014. "Bug Out Fabrication: A Parallel Investigation using the Namib Darkling Beetle as a Biological Model and Incremental Sheet Metal Forming as a Fabrication Method." In *ACADIA 14: Design Agency; Proceedings of the 34th Annual Conference of the Association for Computer Aided Design in Architecture (ACADIA)*, 531–538. Los Angeles, October 23-25, 2014.

Kalo, Ammar, and Michael J. Newsum. 2014. "Performing: Exploring Incremental Sheet Metal Forming Methods for Generating Low-cost, Highly Customized Components." In *Fabricate: Negotiation Design and Making*, edited by Fabio Gramazio, Matthias Kholer and Sike Langenberg, 166-173. Zurich: gta Verlag.

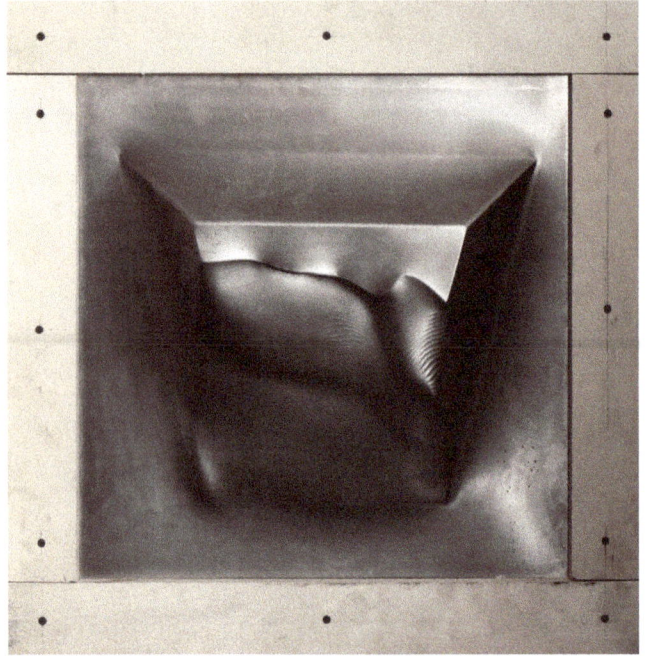

22 Specimen 00, RIMF steel relief secured with metal fasteners in birch plywood

23 Specimen 05, RIMF steel relief secured with metal fasteners in birch plywood

Karras, Tero, Samuli Laine, Miika Aittala, Janne Hellsten, Jaakko Lehtinen, and Timo Aila. 2020. "Analyzing and Improving the Image Quality of Stylegan." In *2020 IEEE/CVF Conference on Computer Vision and Pattern Recognition (CVPR)*, 8107-8116. Online, June 16-18, 2020. https://doi.org/10.1109/cvpr42600.2020.00813.

Kato, Hiroharu, Yoshitaka Ushiku, and Tatsuya Harada. 2018. "Neural 3D Mesh Renderer." In *The IEEE Conference on Computer Vision and Pattern Recognition (CVPR)*, Salt Lake City, Utah, June 18-22, 2018.

Kepes, Gyorgy, John E. Burchard, and Jean Arp. 1956. *The New Landscape in Art and Science by Gyorgy Kepes*. Chicago: Theobold & Co.

Kodalak, Gökhan. 2020. "Spinoza and Architecture: The Air of the Future." *Log* 49.

Mertins, Detlef. 2004. *Bioconstructivism, Lars Spuybroek NOX: Machining Architecture*. London: Thames & Hudson.

Mildenhall, Ben, Pratul P. Srinivasan, Matthew Tancik, Jonathan T. Barron, Ravi Ramamoorthi, and Ren Ng. 2020. "NeRF: Representing Scenes as Neural Radiance Fields for View Synthesis." Paper presented at *European Conference on Computer Vision*, virtual, August 23-28, 2020.

Payne, Andrew. 2017. "Monolith 0.3.6415.29557." Monolith. https://www.food4rhino.com/en/app/monolith.

Runway AI, Inc. n.d. *RunwayML, StyleGAN2 Interface*. Accessed 2022. https://app.runwayml.com/.

Shklovsky, Viktor. 2017. *Viktor Shklovsky: A Reader*, translated by Alexandra Berlina. New York: Bloomsbury.

Schafer, Timo, and Rolf Dieter Schraft. 2005. "Incremental Sheet Metal Forming by Industrial Robots." *Rapid Prototyping Journal* 11 (5): 278–86. https://doi.org/10.1108/13552540510623585.

Vidler, Anthony. 1992. *The Architectural Uncanny: Essays in the Modern Unhomely*. Cambridge, Mass.: The MIT Press.

Zhang, Hang, and Ezio Blasetti. 2020. "3D Architectural Form Style Transfer Through Machine Learning." In *RE: Anthropocene, Design in the Age of Humans; Proceedings of the 25th CAADRIA Conference*, vol. 2, 659–668. Chulalongkorn University, Bangkok, Thailand, August 5-6, 2020.

IMAGE CREDITS

Figures 4, 5: ©Coersmeier, Nanotectonica, Pratt Institute.

Figure 6: ©Coersmeier, Nanotectonica, Pratt Institute, Spring 2019, Graduate students Francisco Gallegos, Leonardo Martinez.

Figure 7: ©Coersmeier, Nanotectonica, Pratt Institute, Spring 2019, Graduate student Thomas J. Diorio.

Figures 1 [Right], 3, 17, 20 [Right], 21 [Right], 22-27: ©Adam Elstein, August 12, 2022.

All other drawings and images by the authors.

24 Specimen 04, RIMF steel relief secured with metal fasteners in birch plywood

25 Specimen 03, RIMF steel relief secured with metal fasteners in birch plywood

Jonas Coersmeier is a German architect in New York, who runs the office Büro NY with partner Gisela Baurmann. He is a professor at Pratt Institute, where he teaches architecture and urban design and conducts research on new materialism in design: Material Urbanism, Nanotectonica. At the University of Pennsylvania, he teaches the design studios LoLux that develop architectural models for a more equitable housing future. He studied at MIT, Columbia University, and TU Darmstadt, and he is a registered architect in Berlin, where he conducts the interdisciplinary program Pratt Berlin.

James Nanasca completed training as an architect at Pratt GAUD (2022) and SAIT (2012). He has also developed geographical sensibilities through studying Urban Planning at the University of Calgary (2019). Currently the Director of Design Technologies at New Bedford Research and Robotics, James carries expertise in computational and generative design, CNC machining, multi-material design and fabrication, multi-media research, and robotic fabrication.

Ivan Yan Man Hin is a MArch Graduate of the GAUD at Pratt Institute (2022), where he has served as a member of the Graduate Student Council and worked as a research assistant under Haresh Lalvani at the Center for Experimental Structures. Originally from Hong Kong, he received his honours BA from the University of Toronto's DFALD (1T9). He has been recognized in the National and New York Society of American Registered Architects Design Awards for his graduate cap studio work and, in 2021, he was made a Fellow of the Taconic Fellowship in the Pratt Center for Community Development.

Ezio Blasetti is a registered architect in Europe (TEE-TCG), and his academic and professional research focuses on the application of advanced technologies in all phases of architectural design, from the initial composition to robotic fabrication. He is a founding partner at Maeta Design, an architectural design and research firm based in New York and Athens. In 2009 he co-founded ahylo architects, an architectural design and construction practice, and "apomechanes," an intensive design lab on algorithmic processes and fabrication. He is a Senior Lecturer at the Stuart Weitzman School of Design at the University of Pennsylvania and a visiting faculty at Pratt Institute.

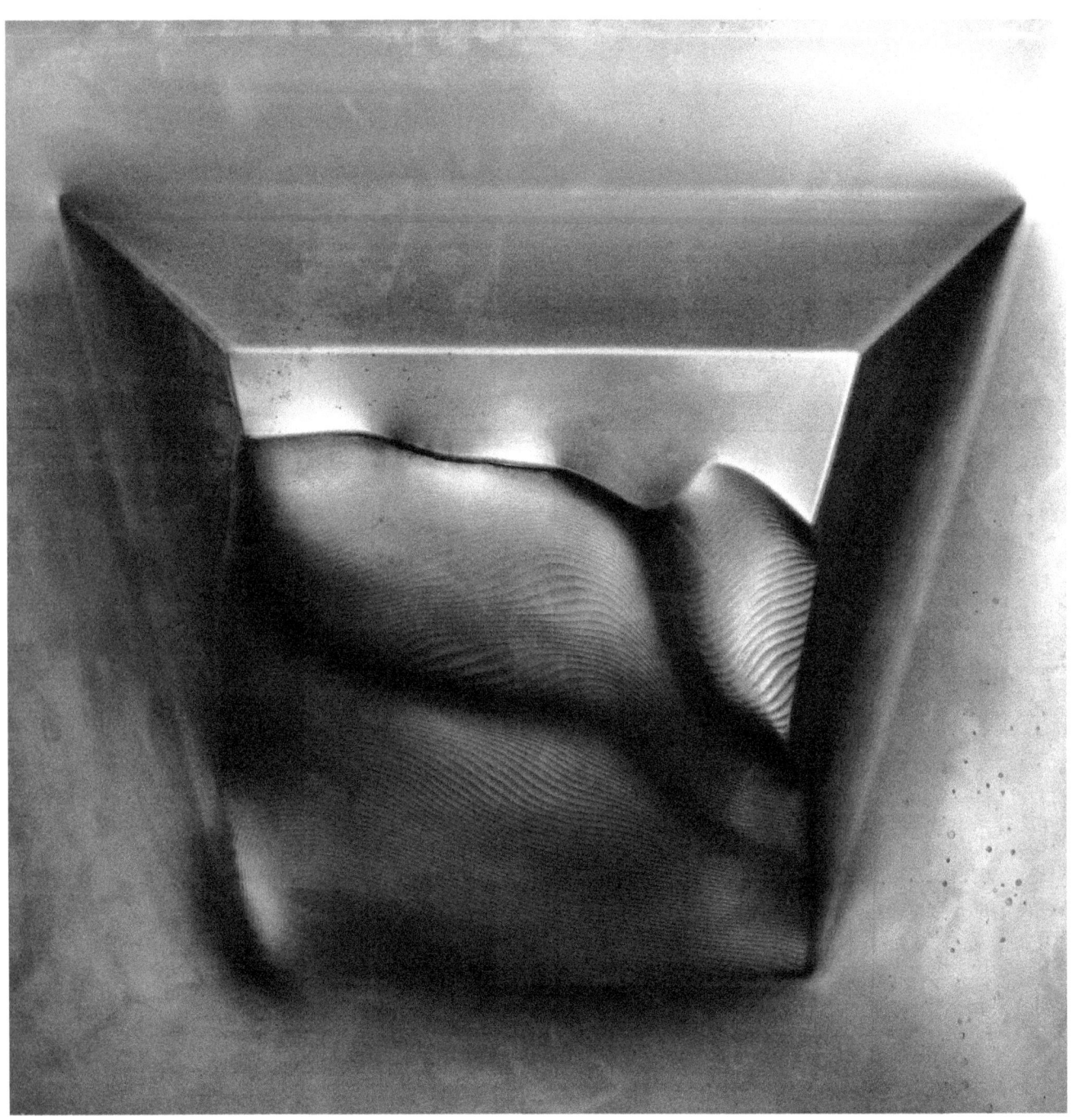

26 RIMF steel surface, Specimen 00

27 RIMF steel surface of Specimen 03

Introducing Bespoke Properties to Slip-Cast Elements

Designing a Process for Robotically Controlled Rotational Casting

Davis Dunaway
University of Pennsylvania

Dan Rothbart
University of Pennsylvania

Layton Gwinn
University of Pennsylvania

Dr. Nathan King
University of Pennsylvania/Virginia Tech Center for Design Research

Robert Stuart-Smith
University of Pennsylvania/UCL

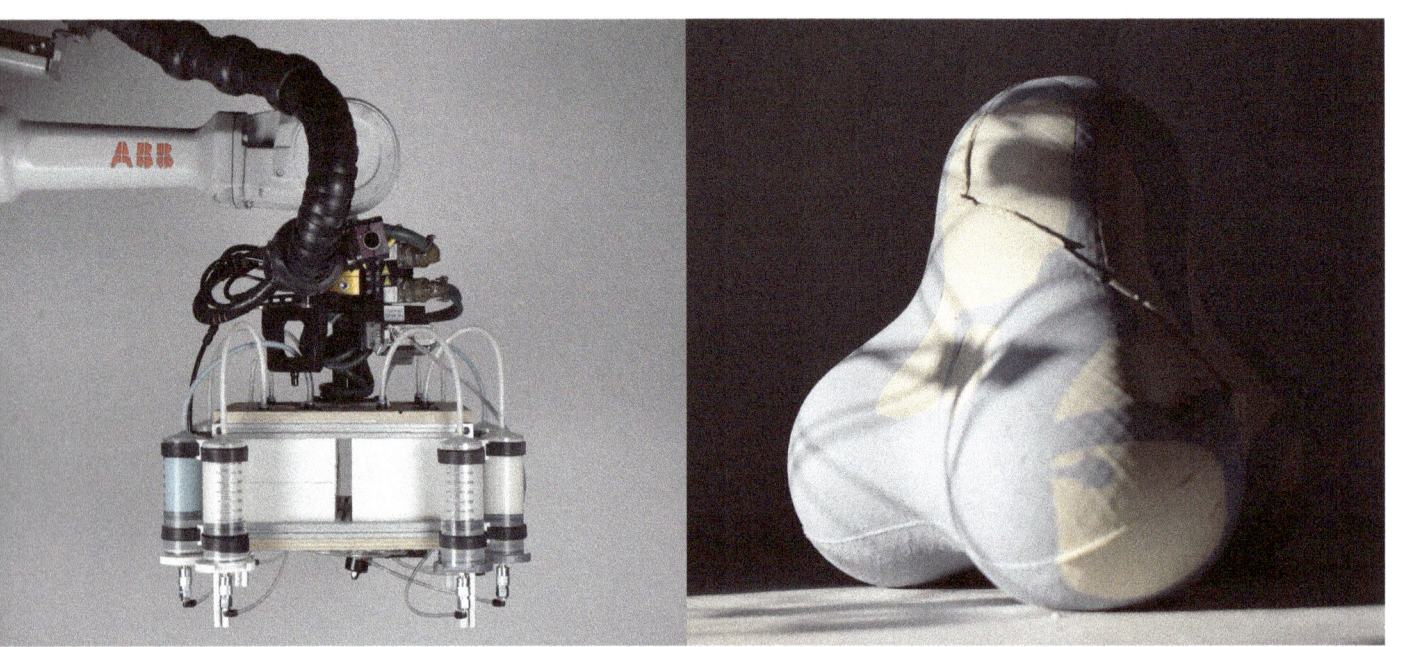

1A (left) The completed end effector for preforming robotically controlled slip casting, allowing the robot to control the injection of slip into the mold and for the molds to be held in virtually any orientation; (right) An example of a slip cast part that demonstrates a desirable marbling pattern

ABSTRACT

The industrial use of slip casting is niche but highly recognizable. The phase-changing nature of the clay slip makes the process ideal for the production of complex, standalone geometries, such as those needed by the sanitary ware and fine porcelain industries. Slip casting, however, currently lacks the ability to produce meaningful visual variation between components without the need for an entirely new mold. This research explores a novel technique for creating bespoke, slip-cast artifacts through the use of 6-axis robotic motion. By incrementally injecting different amounts of colored slip into the mold while it is rotated, we are able to achieve variable color, pattern, and structure. Because of the highly precise nature of the robotic motion, this variation can be repeated with a high degree of accuracy. In addition, the incremental injection of slip also allows us to achieve a full cast with a minimal amount of slip, removing the draining process of traditional slip casting entirely. The level of control this process might give a designer is explored through a series of tetrahedral components that demonstrate the types of marbling that can be achieved. This work borrows heavily from the field of plastic rotational molding, as numerous parallels can be made between the two processes' flexibility and parametrization. By drawing on this neighboring field, we hope to bring new variables into the world of mass-manufactured slip-cast ceramics in the form of controllable color and pattern.

INTRODUCTION

At its core, slip casting is a simple and elegant process. The process consists of two parts, a liquid clay body often referred to as slip and a plaster negative of the desired geometry to serve as a mold. To create an artifact, slip is poured into the mold until it is completely filled. Over a short period of time, the plaster draws moisture out of the slip causing it to harden into a leather like shell, completely taking the form of the interior of the mold. The remaining slip is then drained and the mold is removed, creating a hollow version of the desired part. The slip's ability to go from liquid to solid is what makes it unique as a clay body. It is this phase-changing nature that makes slip casting a go-to process for the production of complex ceramic geometries by both artisans and industry (Bechtold, Kane and King 2015).

There are currently two basic ways of producing variation between slip-cast parts that come from the same mold. The first is to increase the wait time before drainage. This allows the plaster mold to absorb more moisture from the slip resulting in a final part that has thicker walls. While this does provide critical control over the structure of the part, the variation is not at all externally visible. The second way to control variation is to change the composition of the slip. This can provide control over the overall color of the final part but requires every variation to have its own unique clay body which often makes it unsuitable for providing a wide degree of variation. Because the amount of control over the visual qualities of a slip-cast part is minimal, we often see the process used to create identical artifacts, both structurally and visually, in mass quantity. We, however, believe that through the use of industrial robotics, an augmented version of the slip casting process can produce results with a near infinite number of visual variations. More specifically, we propose that by incrementally injecting different amounts of colored slip into a mold that is rotated by a 6-axis robotic arm, we can achieve variable color, pattern, and in some cases controllable aperture on the surface of the part.

This hypothesis was evaluated rigorously using both human and robotic tests to ensure that the proposed process is both functional and feasible to perform (Figure 1B). These tests took the form of casting a series of tetrahedral components. We successfully found that we were able to achieve a wide range of constantly repeatable patterns, demonstrating that this process does, in fact, give a new level of control to the designer of slip-cast parts.

Additionally, we found that an auxiliary benefit of the process is that we can drastically reduce the amount of slip needed to produce a successful part. Because the mold is being rotated while slip is injected, the interior of the surface can become coated with slip without needing to fill it entirely. This removes the need for a drainage step as we can fill the mold with the minimum amount of slip needed to get a full coating. This means that no slip is wasted in the process.

BACKGROUND

Overview of Traditional Slip Casting

Slip casting's use is both widespread and time tested. It can be seen used in European ceramics dating back to the 18th century, and its use in Chinese ceramics even earlier (Paxton and Fairfield 1980). Industrial use of slip casting can be traced back to mid-14th century China during the Ming Dynasty, where the town of Jingdezhen used the process to mass produce the iconic blue-and-white porcelain pottery that is still common today (Canby 2009).

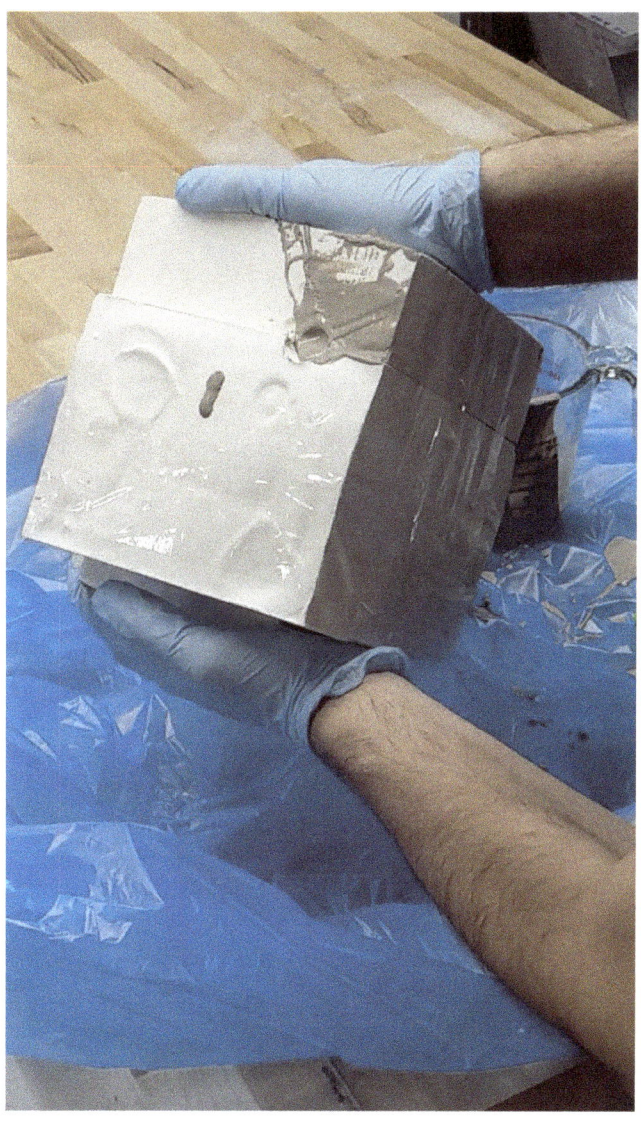

1B A manual test of the robotically-assisted rotational casting process: two colors of liquid slip were poured into a plaster mold and rotated intermittently in order to bring about a controlled marbling effect in the slip cast artifact

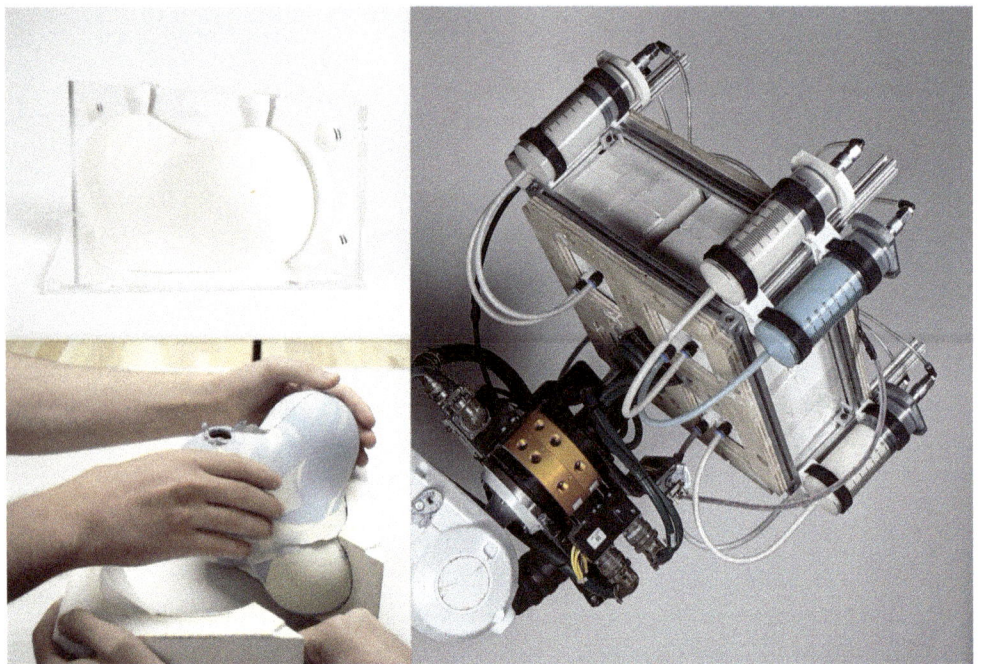

2   Composite image of the major parts of the robotically-assisted rotational casting process: (top left) A 3D-printed PLA and laser cut acrylic positive for the production of a plaster mold; (right) Custom end-effector for injecting multiple colors of slip into a series of plaster molds while the robot is changing their position and orientation; (bottom left) A completed slip cast being removed from the plaster mold, a process that often requires a delicate touch

In the modern industrial context, the efficiency of the process has been increased. The plaster molds have a relatively short lifespan, and creating those negatives can often be the most arduous part of the slip casting process. Because of this, larger operations now use stainless steel formworks to produce plaster molds quickly and efficiently. Given a large enough number of plaster molds, identical slip-cast elements can be produced virtually continuously making it ideal for mass production.

Slip casting is also commonly used by artisans due to its simplicity and cheap material costs. In fact, while the core of the slip casting process has remained relatively untouched by mass manufacturers, artisans have been able to explore more experimental changes. While not exactly slip casting, we can see the use of colored slip to achieve decorative surface marbling as early as the mid-19th century (Erickson 2003). Artists such as Peter Pincus and Jenny Rijke have taken this idea of using multiple colors of clay body and have applied it to slip casting. We took note of Jenny Rijke's work in particular, as it achieves the marbling effect by using a pre-swirled multi-colored slip in the casting process. This technique can create variation in color and pattern, but the result is always different as it is not repeatable from part to part.

Current Research of Casting Compatible Materials

Slip casting has largely been unexplored within the field of architectural robotics. Because of this, our research primarily builds off of work that uses other casting materials such as concrete, plaster, and plastics. Concrete, in particular, is the material of choice for a number of projects that explore methods for generating bespoke results through robotic fabrication. We have seen the use of flexible fabric formworks in pre-cast and on-site concrete, the use of robotic manipulation to produce thin shell structures via rotational molding, and the use of 6-axis robotic arms to manipulate a formwork over the length of column-like elements, varying their sectional geometry serially with a high degree of control (Hawkins et al. 2016; Tessmann and Mehdizadeh 2019; Lloret-Fritschi et al. 2020). When analyzing the last of these examples in particular, we can begin to see a direct relationship between robotic motion and fabricated artifact. This relationship gives credence to the use of robotics, specifically the use of a 6-axis industrial arm, in low volume settings.

This specific type of relationship between robot and artifact has been given the name of Design Robotics and is formally defined as an argument for the use of robotics that "bridges the gap between primarily artistic endeavors and the construction automation research of the building industry" (Bechtold and King 2013). While exploring this idea in the field of architecture is not new, applying these design principles to casting-based processes is still in its infancy, as overcoming the overwhelming influence the mold has on the final part has proved challenging. The research previously mentioned attempts to overcome the mold in different ways, most notably through the use of flexible formwork that can be manipulated robotically (Lloret-Fritschi et al. 2020); however, this is currently not an option for the production of ceramics as the material characteristics of the plaster mold are essential in facilitating the phase change of the slip (Bechtold, Kane and King 2015). For inspiration on how to tackle this problem, we looked to existing processes that existed outside of both the field of architectural robotics as well as ceramic production as a whole. This led us to a practice commonly used in the production of plastic components called rotational molding. Recent increased parametrization and automation of material add-ins to rotational-molded plastic parts started to show glimpses of how mold-based production methods might be made more bespoke (Gupta and Sangani 2020).

METHODS

Plaster Molds

While the creation of plaster molds in low volume settings is incredibly well documented, our specific geometry and need for highly accurate dimensions forced us to be creative in how we produced our molds. We created custom formwork out of 3D-printed PLA and laser cut acrylic that would then be used to produce our four-part plaster mold (Figure 2). In order for the formwork to be reusable, it was critical that all of the materials be resistant to water. When casting the plaster molds, we used the typical plaster-to-water ratio of slip casting using USG number 1 plaster; however, some of the casts took an extended period of time to dry before being usable. Batches of molds were put either into an oven at 150° or in direct sunlight for extended periods of time in order to aid the drying process; however, neither process significantly decreased the overall drying time during our testing period.

Robotic End Effector

We developed an end effector that would allow for a robotic arm to manipulate the molds freely as well as fully automate the injection of slip into each mold (Figure 2). The end effector was designed to carry four molds simultaneously. The loading and unloading of the carriage was heavily considered. A human operator only has to release four bolts, unload and load the plaster molds, and rescrew the same four bolts before the robot can begin another round of casting. For automatic slip injection, we opted for the use of a pneumatically powered system of plungers to feed slip into the molds.

A custom 3D printed backing was designed for a 500 ml syringe that allowed the syringe to be connected to compressed air via an industrial quick connect fitting. The optimal pressure for operating these syringes was 30 psi. The end effector was fitted with eight of these syringes, one for each color of slip for each mold, and were connected in groups of four to allow independent actuation of each color of slip when injecting. Because these backings were a semi-permanent addition to the syringes, we had to devise a way to refill the syringes with slip after use. We used a small venturi module to generate a vacuum that could pull the plunger back down causing the syringe to refill. While this pneumatic loading method did show promise as a relatively low cost solution, we found that the optimal pressure was not achievable in our lab setting, as we were limited to 80 psi total (20 per syringe) and even at optimal pressure the flow rate of slip was inconsistent. Ideally, the system would be designed for use with a pneumatic cylinder or a linear actuator to more accurately control the slip. For our final tests, we manually loaded slip into each mold in order to ensure the quantity of slip was accurate.

Robotic Toolpaths

We also developed and tested a series of robotic toolpaths. These aim to consistently coat the entirety of the mold with slip when run in combination. The order in which the toolpaths are layered define the creation of unique patterns that are always replicable (Figure 3). Each of these toolpaths used different methods to coat the molds. Some focused on moving quickly from corner to corner, while others focused on

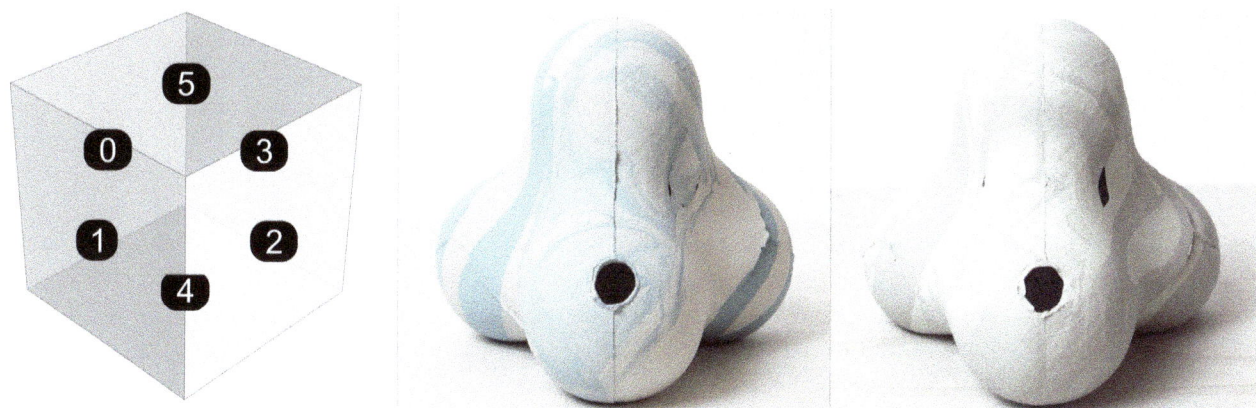

3   Composite image describing the face classification of plaster molds when rotating them by hand alongside two separate casts which produced the same marbling patterns using the same rotational motion while casting. Both of these casts introduce one mason stain of slip into the plaster mold, moving the mold such that sides 0, 1, 2, and 3 faced downward, and repeating with the second color. When side 5 is faced downward after 400 ml of slip is introduced and swirled in that rotation, the pictured effect is achieved.

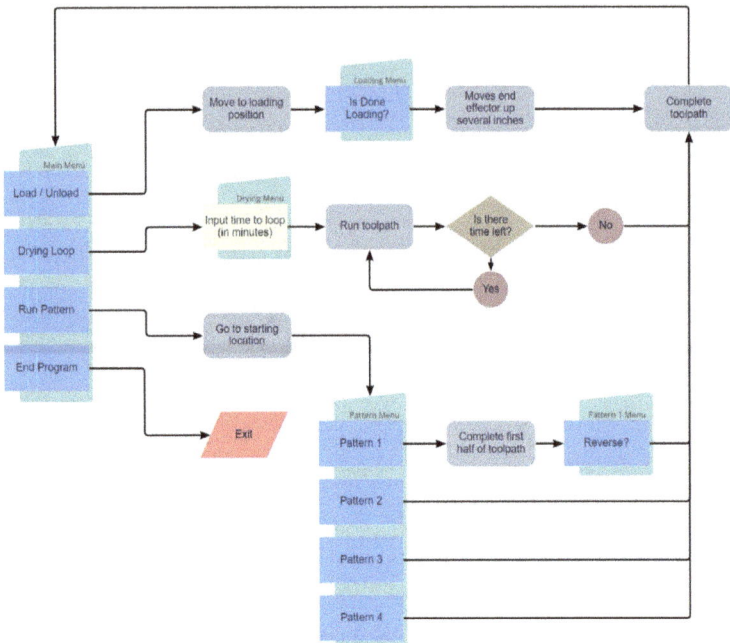

4   Flowchart depicting the organizational structure of the RAPID code script that allows the user to select from a number of different toolpaths necessary for performing the rotomolding process. This includes the four swirling patterns, a repetitive drying loop, and the loading and unloading routine.

slow and continuous rotation. In addition to the coating toolpaths, we also produced toolpaths for loading and unloading the end effector, as well as a looping drying cycle that could be run continuously after the coating process. Because of the large number of different procedures that we needed to execute in a given production run, it was important to create a master program that allowed us to trigger manually the desired function quickly and efficiently from the robot's teach pendant. We created a small RAPID program that prompts the user with the different possible toolpaths and ensures that nothing can be run in an order that would cause issues for the robot (Figure 4).

Preparation for Experiments
One of the most important variables to control was our clay body. Early tests were conducted with Leuders Casting Slip—a stoneware casting slip that comes pre-mixed in a two-gallon bucket—due to its availability from a local ceramics shop. The majority of our tests, however, were instead conducted using Laguna Fine Porcelain NS-125 Very White Liquid Slip. The clay body was mildly altered before use by the addition of mason stain in order to achieve the different colors of slip necessary for our process. The mason stains we used were Mason Color Works colors—Peacock 6266, Robin's Egg 6376, and Gunmetal 6591—all in a ratio of 1/4 lb mason stain to 1 gallon of slip. For manual tests where an exact quantity of slip was needed, a 500 ml food-grade syringe was used that allowed us to accurately add slip in increments of 10 ml. When casting, the plaster molds were fastened together using standard packing tape. We opted for tape over large rubber bands or other common fastening methods due to the fact that our molds consisted of four parts. The tape allowed the assembled molds to have a small amount of play until we were confident in the mold's alignment. To precisely track the motion in the initial manual tests, such that we could translate them to robotic motion and ensure repeatability of the results, we restricted most of the rotations of the plaster molds to the six faces of the molds' cubic bounding boxes. After casting, pieces were left in their molds for an additional 30 minutes before removal. Once removed, the seams on the cast would be trimmed by hand, and the mold would be placed on a sheet of drywall to dry. Each mold was limited to two casts per day in order to prevent over-saturating the plaster with moisture.

RESULTS AND DISCUSSION
Results from Early Experimentation
Three different geometries were explored during the initial hand tests before choosing a final geometry. The tested geometries were a regular sphere, a metaballed cubic corner (referred to as cornerpod), and a metaballed tetrahedron (referred to as tetrapod) (Figure 5). These three different geometries were designed with varying spatial packing and interlocking capabilities in mind. This was done with the aim of potentially informing future research into architectural applications such as facade screen assemblages (Figure 6).

The cornerpod was explicitly modeled to allow for the interlocking of two or more pieces while maintaining an open aperture in some areas. The tetrapod serves as the counterpart to this, as it was modeled without a specific organizational packing structure in mind. All three of these molds were first subjected to the traditional method of slip casting. Each mold was filled entirely with stoneware casting slip and left for varying amounts of time before drainage. When waiting longer than 30 minutes before draining, all molds produced highly unstable results, often collapsing under their own weight. At shorter wait times, both the sphere and tetrapod proved feasible, while the cornerpod often failed. This was because the mold would trap air inside itself leaving uncoated surfaces. Additionally, slip often became trapped in the thin neck between nodes resulting in an almost solid barrier. This prevented the cast from draining fully and often resulted in partial collapse of the cast after being removed from the mold. While this particular geometry did not result in any successful casts, it seemed to support our motivation for wanting to rotate the mold during the casting process as adding slip in small increments rather than all at once allowed for all surfaces of the mold to become coated in slip without needing to worry about the release of trapped air. While we believe we could now cast this geometry successfully using our new production method, it proved too difficult to explore via hand tests and was dropped in favor of the other two molds.

Our next series of tests involved filling the molds with a fraction of their interior volume's worth of slip and proceeding to rotate the mold in a regular fashion. We sought to determine a sufficient relationship of interior volume, volume of slip used, and rotational motion used to achieve a full coat of the mold shape. For these tests we switched over to a porcelain-based slip as it is the standard for most industrial slip casting processes. Our original motivation for this test was to achieve varying thickness in different parts of the cast; however, this proved exceedingly difficult to control with any sort of reliability in both molds. We did, however, make two unexpected discoveries. The first was that while we were unable to consistently control thickness, we were able to constantly control which parts of the mold were coated in slip and which parts were not. This led to the creation of apertures within the individual pieces that warrant further research. The second was that we were occasionally able to coat the entirety of the mold without the need for draining excess slip. This was achieved by adding a specific volume of slip relative to the part's overall volume to the mold and rotating it onto all six sides. It is worth noting that the percent volume of slip that the sphere and the tetrapod molds required in order to achieve this result varied, as the sphere mold required about 50% slip volume to the tetrapod's 15%. This relationship requires more research, as we were unable to derive a formula for the ratio due to the relatively small sample size of our tests.

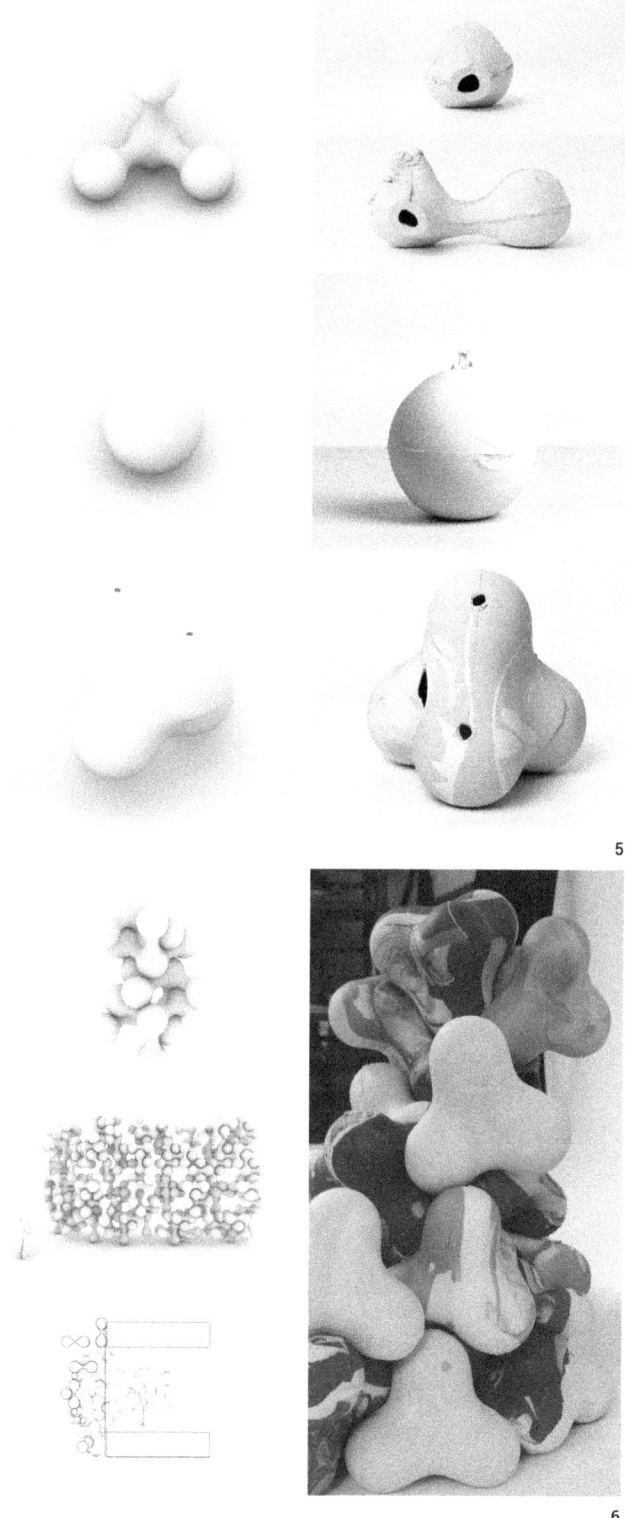

5   Renderings of the component shapes selected for the robotically-assisted slip casting experiments: a form with a cubic shape that resulted in failure due to the difference in weight between its spherical and its connecting elements, a spherical form used to test a casting process using only a fraction of the amount of slip, and a form with a tetrahedral base for use in a robotic cast

6   Renderings of facade screen application of cornerpod, physical aggregation of components displaying color and formal opportunities of the tetrahedral component

Table 1: Notable tests of volume ratios - spherical component (550ml volume)

Test Number	Volume of Slip Injected (ml)	Rotation Specification	Notes	Image of Result
1	550	Sides 1,2,3,4,5,6: 2.5 minutes ea.	Unstable, high drainage, collapsed under own weight	
2	225	Sides 1,2,3,4,5,6: 1.5 minutes ea.	Stable, full coat achieved, fairly delicate	
3	112	Sides 1,2,3,4,5,6: 1.5 minutes ea.	Stable, does not achieve a full coat	

Table 2: Notable tests of volume ratios – tetrahedral component (2543ml volume)

Test Number	Volume of Slip Injected (ml)	Rotation Specification	Notes	Image of Result
10	532	Color 1, 0-1-2-3, Color 2, 0-1-2-3, repeat. Then 1 minute on each side, then 30 minutes of drying	Good result, some drainage	
12	355	Color 1, 0-1-2-3, Color 2, 0-1-2-3, repeat. Then 1 minute on each side, then 30 minutes of drying	Not a full coating	
14	400	Color 1, 0-1-2-3, Color 2, 0-1-2-3, repeat. Then 1 minute on each side, then 25 minutes of drying	Most successful; ready for replication on robot	

In current industrial practice, the step to remove the drainage in the slip casting process would not be incredibly impactful, as slip is entirely recyclable; however, this newfound lack of drainage does open the doors to several new methods of creating bespoke artifacts from a single mold that were previously not feasible. In our research, this took the form of the simultaneous use of multiple colors of slip. Normally, this would be impractical as the drained slip would be an unknown mixed color and, therefore, be unable to be recycled, but the lack of a drainage step mitigates this entirely. Additionally, the marbling effect that results from swirling two different colors of slip is a perfect design characteristic to directly showcase the robotic movement that lead to its creation.

Our last series of hand tests focused on achieving a uniformly coated cast that consistently produced no excess slip. For these tests, we decided to focus purely on the tetrapod as its form lent itself to the injection of two different colors of slip without immediate mixing. A new mold was designed that allowed for two separate injection sites at the top of the mold, one for each color. Unlike the previous tests, slip was injected incrementally between rotational movements rather than all at once. For the tetrapod mold, we found that injecting 400 ml of slip in 50 ml increments consistently resulted in a uniform coating without the need for drainage. We achieved different patterns of swirling by changing the order in which we rotated the mold between injections.

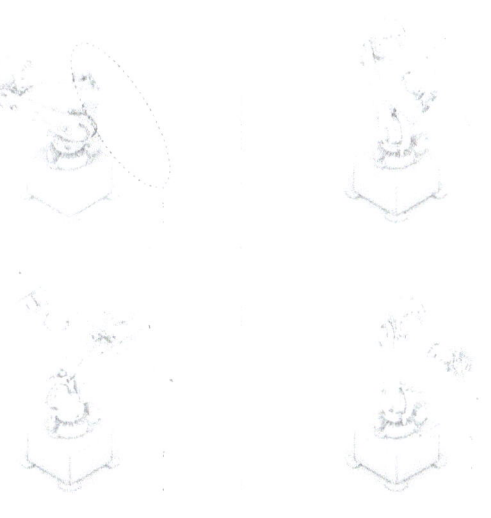

7   A composite image depicting four toolpaths used in the robotically-assisted rotational casting process: four plaster molds are rotated 90, 180, or 270 degrees while moving along the lines notated by the arrows. In the image, the IRB 4600 industrial robot arm is covered by canvas and construction bags as a safeguard against leaking slip.

8   Two composite images comparing separate robotically-assisted slip-cast marbling patterns. Examples of repeatibility inherent to our process, the components with the same coloration were cast completely separate from one another. The placement of all major color blocks is almost identical and some more intricate features have striking similarities.

Results from Robotic Tests

For these tests, we attempted to keep as many variables as possible constant in order to look solely at the effects that the toolpaths had on the marbling and see to what degree it was repeatable. To do this, the molds were loaded identically to our most successful hand test, 200 ml of each color loaded alternating 50 ml increments via syringe between each toolpath. Despite the ability for our end effector to carry four molds, we opted to only test with two as this would reduce the time that the mold remained stationary for the filling process.

Like the rotational motion tests done by hand, we were able to see a direct relationship between the robotic toolpath and the achieved artifact through the created marbling pattern. Each test ran a series of eight toolpaths constructed from the four present in our RAPID interface. As mentioned previously, the four toolpaths use different techniques to systematically coat the mold (Figure 7).

In order to determine if the marbling patterns were repeatable, we ran the same combination of toolpaths multiple times across multiple days. While not identical, the relationships in the marbling show a high degree of similarity (Figure 8).

We believe that at least a portion of the remaining variation comes from the varying injection angle of the slip when loading manually. This is something that could easily be minimized in a more industrial context. In theory, any number of feasible toolpaths could be designed and run in any sequence in order to generate new patterns, giving the designer almost complete control over the marbling effect that occurs on the surface of the part.

CONCLUSION

In this paper, we have proposed a method for augmenting the traditional slip casting process with industrial robotics in order to afford a designer more control over the visual qualities of the final artifact. We have developed this process with the principles of Design Robotics in mind in order to ensure that what we are creating truly warrants the use of robotics. Marbling effects are achievable by hand; however, the addition of the robotic motion makes the process repeatable and significantly less wasteful. Both of these qualities are a necessity in bringing the practice of marbling to high volume production settings.

In parallel with our research, we began to explore potential architectural applications that would necessitate the need for a large quantity of parts produced in this way. Attempts were made to explore the aggregation potential of our tetrapod components by pressing them into each other before firing (Figure 9).

Almost all of these attempts resulted in tearing, cracking, slumping or other critical failures in the components that prevented more than a small number of pieces from being joined together. While our research into this particular aspect was not thorough enough to completely rule out the possibility of pre-firing aggregation, the process was clearly telling us that it was better suited for the creation of individual artifacts. Architects are often the ones driving Design Robotics style research, and the current lack of an immediately clear architectural use that would not be better achieved with concrete or other castable materials likely contributes to the fact that slip casting is an underexplored process. All this said, slip casting still excels at the creation of stand alone artifacts, which is why, despite not commonly seeing it used in an architectural setting, it is still a valuable process that deserves the attention that other materials and processes are getting in Design Robotics research. The technique proposed in this paper only further enhances its abilities, as it now introduces controllable variables of color and pattern into a process that typically only produces exact copies of an artifact. One can begin to imagine the implications that this process could have on an assembly line as controlled marbling could be added to any number of ceramic artifacts without any increase in time (Figure 10).

9   Photograph of an aggregation of robotically-assisted rotational cast components, fused to one another during the firing process

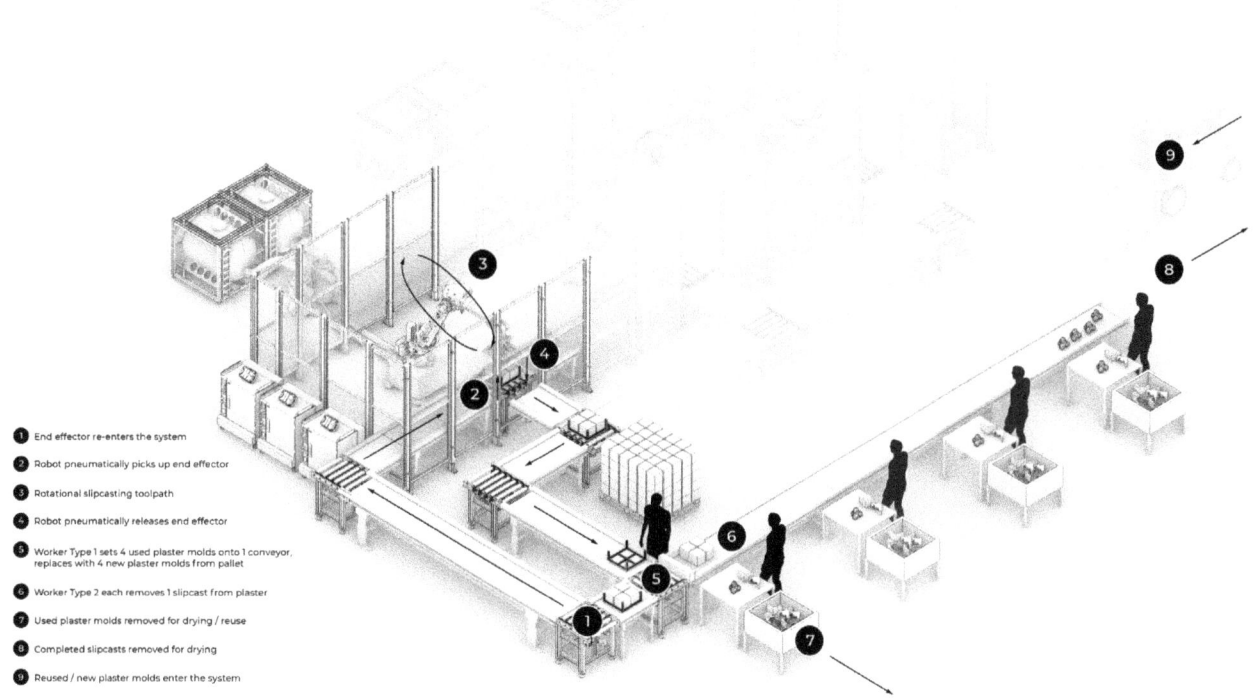

1. End effector re-enters the system
2. Robot pneumatically picks up end effector
3. Rotational slipcasting toolpath
4. Robot pneumatically releases end effector
5. Worker Type 1 sets 4 used plaster molds onto 1 conveyor, replaces with 4 new plaster molds from pallet
6. Worker Type 2 each removes 1 slipcast from plaster
7. Used plaster molds removed for drying / reuse
8. Completed slipcasts removed for drying
9. Reused / new plaster molds enter the system

10 Diagram of a hypothetical proposal to industrialize this ceramic rotomolding process: changes to the end effector such as a system for automatic attachment to the robot and locating the slip off the robot for easier refilling would allow it to run continuously. Humans are stationed to perform the delicate task of removing the slipcast part from the mold.

While this process has proved highly promising in the initial stages of research, there is still much to be done. More variations of robotic toolpaths must be explored in order to gain a better understanding of how different types of motion contribute to different types of visual effects. A catalog of these relationships must be created in order to allow a designer to effectively utilize them in the design process. Additionally, a strict equation or, at the very least, methodology must be developed for determining how much slip should be used in a given mold. This is needed in order to allow new molds to be quickly implemented into the process without the need for trial and error.

While not critical to its cementing as a process, the possibility of creating aperture on the surface of a part is also something that should be further explored. This would allow for control over the actual geometry of the part which could open up brand new possibilities for designers using the process. We believe that with this additional research, this process has the potential to become a serious method for the production of ceramic artifacts. It adds the ability to create meaningful and repeatable visual variation in parts that come from the same mold, something that current slip casting techniques, both industrial and artisan, are unable to do. In a world where the bespoke is becoming ever more sought after, that ability makes the process valuable and worth future research endeavors.

10 A series of robotically-assisted rotational cast components with varying degrees of aperture, as a result of an incomplete coat of the plaster surface

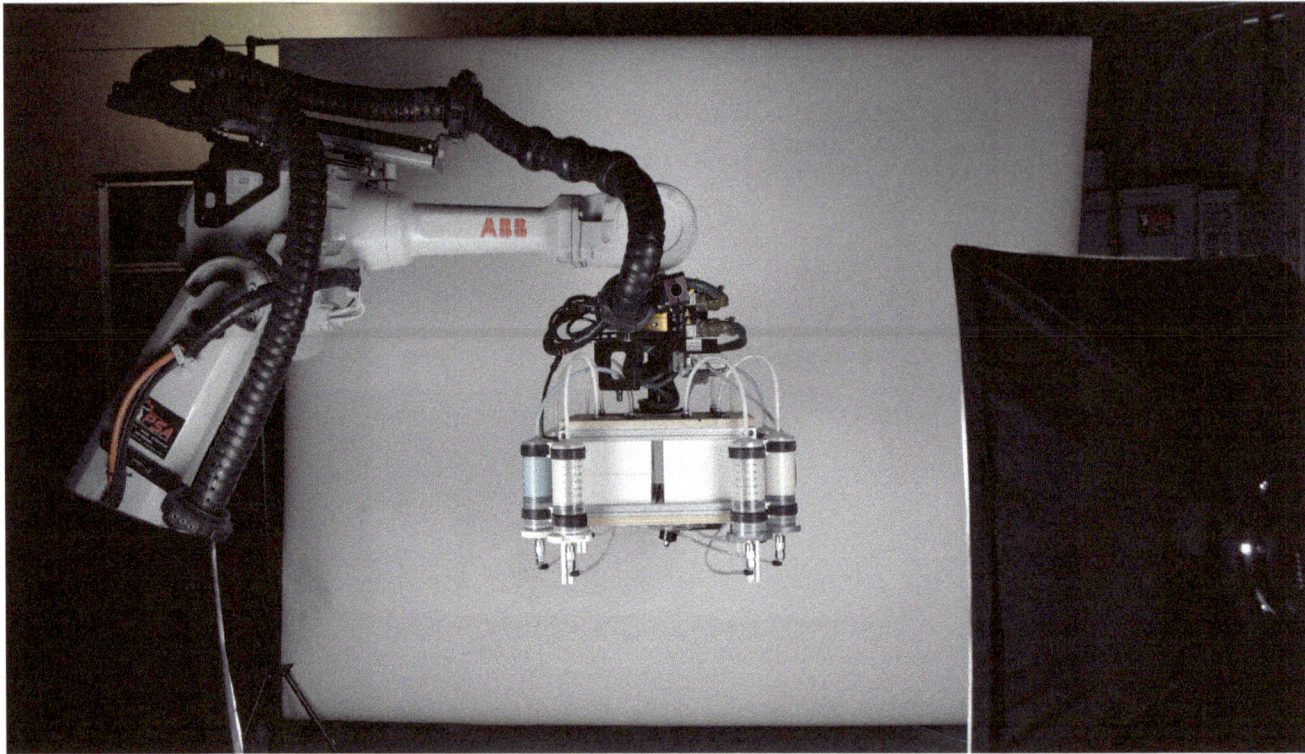

11 Photo of the completed end effector being tested for the first time

ACKNOWLEDGMENTS

This research was primarily performed in the Master of Science in Design: Robotics and Autonomous Systems program, at the Stuart Weitzman School of Design, University of Pennsylvania. A 16-week project integrated across four courses, this work owes a lot to the input of Jeffrey Anderson, Billie Faircloth, Nathan King, and Robert Stuart-Smith in their courses Advanced RAS Programming, Scientific Research & Writing, Experimental Tooling, and Material Agencies, respectively. Each of these faculty, as well as David Forero, had valuable contributions to the work. This paper was written as a part of the Scientific Research & Writing course and the pneumatic end-effector was developed for the Experimental Tooling and Material Agencies courses.

REFERENCES

Bechthold, Martin, Anthony Kane, and Nathan King. *Ceramic Material Systems: in Architecture and Interior Design*. Berlin, München, Boston: Birkhäuser, 2015.

Canby, Sheila R. *Shah 'abbas: The Remaking of Iran*. London: British Museum Press, 2009.

Erickson, Michelle. 2003. *Swirls and Whirls: English Agateware Technology. Ceramics in America*. Milwaukee, WI and Hanover, NH: Chipstone Foundation.

Bechthold, Martin, and Nathan King. 2012. "Design Robotics." In *Rob | Arch 2012*, edited by S. Brell-Çokcan and J. Braumann. 118–30.

Gupta, Nikita, P. L. Ramkumar, and Vrushang Sangani. 2020. "An Approach toward Augmenting Materials, Additives, Processability and Parameterization in Rotational Molding: A Review." *Materials and Manufacturing Processes* 35 (14): 1539–56.

Hawkins, Will J., Michael Herrmann, Tim J. Ibell, Benjamin Kromoser, Alexander Michaelski, John J. Orr, Remo Pedreschi, et al. 2016. "Flexible Formwork Technologies; A State of the Art Review." *Structural Concrete* 17 (6): 911–35.

Lloret-Fritschi, Ena, Timothy Wangler, Lukas Gebhard, Jaime Mata-Falcón, Sara Mantellato, Fabio Scotto, Joris Burger, et al. 2020. "From Smart Dynamic Casting to a Growing Family of Digital Casting Systems." *Cement and Concrete Research* 134 (August): 106–17.

Paxton, John, and Sheila Fairfield. 1980. *Calendar of Creative Man*. London and Basingstoke: Macmillan.

Tessmann, Oliver, and Samim Mehdizadeh. 2020. "Hollow-Crete." In *DMSB 2019, Impact: Design With All Senses*, edited by C. Gengnagel, O. Baverel, J. Burry, M. Ramsgaard Thomsen, and S. Weinzierl. Cham: Springer. 474–86.

Valenstein, Suzanne G. 1989. *A Handbook of Chinese Ceramics*, rev. and enl. ed. New York: The Metropolitan Museum of Art.

IMAGE CREDITS

All drawings and images by the authors.

Davis Dunaway received his Bachelor of Design from Carnegie Mellon University and is a recent graduate of the Master of Science in Design, Robotics and Autonomous Systems from the University of Pennsylvania.

Dan Rothbart holds a Master of Science in Design, Robotics and Autonomous Systems from the University of Pennsylvania and a Bachelor of Architecture from Rensselaer Polytechnic Institute. He currently works as a Project Engineer at A. Zahner Company, Kansas City, Missouri.

Layton Gwinn received his Bachelor of Arts in Architecture from Clemson University and is a recent graduate of the Master of Science in Design, Robotics and Autonomous Systems from the University of Pennsylvania. He is currently employed at Quarra Stone Company LLC as a Design Engineer in their QLAB.

Dr. Nathan King is the Co-Director of the Center for Design Research at Virginia Tech, an Instructor at the University of Pennsylvania and Harvard University, and leads the Autodesk Research organization focusing on the Industrialization of Construction.

Robert Stuart-Smith is Director of the Master of Science in Design, Robotics and Autonomous Systems program at the University of Pennsylvania, Director of the Autonomous Manufacturing Lab (AML) at the University of Pennsylvania (Architecture), and Co-Director of the AML at University College London (Computer Science).

Designer Agency in 3D Packing of Irregular Material Stock

Patricia Dueñas Gerritsen
Massachusetts Institute of Technology

Emily Wissemann
Massachusetts Institute of Technology

Jose Luis García del Castillo y López
Harvard University

1   Example array of irregular material stock

ABSTRACT

This paper presents a flexible computational method to simulate architectural design elements out of nonstandard and irregular sets of material stocks. This approach contributes to ongoing efforts to reduce material processing in common reuse strategies by developing a framework for human-machine design collaborations. Our three-dimensional packing method accepts as inputs user preferences with the goal of creating a framework for design that integrates creative decision-making, enabling architects to simulate and conceptualize alternative design schemes for material reuse. As part of this research, we propose the simulation of sets of parametrically derived objects of irregular geometry and size in the form of digitally generated stones, children's toys, fragmented bricks, and cracker crumbs as proxies for post-demolition building materials. Through a series of case studies, we demonstrate the potential of a heuristic combinatorial algorithm to respond to design preferences.

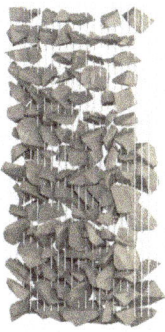   

2 Final packing assembly representations with imagined connectors and mortar strategies of steel tension elements, plastic tubes, mortar, and peanut butter

INTRODUCTION

The building industry is responsible for almost 40% of global carbon emissions and therefore has a significant role to play in mitigating climate change (Global Alliance for Buildings and Construction 2021). As building performance standards increase and environmental regulations become stricter, embodied energy, or the sum total of the energy associated with construction materials and processes, represents an increasing share of this percentage. The effects of carbon emissions that are encompassed by the metric of embodied energy is immediate, unrecoverable, and more difficult to track. In fact, most building industry professionals do not calculate the embodied carbon emissions of their projects.

The importance of material reuse as a primary method to reduce overall global energy consumption led to the identification of grouted masonry materials commonly used in construction. This subset of masonry is particularly difficult to recycle and thus does not enter conventional recycling streams like that of plastic, glass, or metal, which can be melted and reconstituted. Existing recycling processes for masonry rely on the manual sorting of less damaged units that can be directly reused, while the remaining material is discarded (Hansen 1992). If irregular masonry—such as grouted groups or debris extracted from prior construction—could be used directly, energy-intensive procedures of recycling, disposal, and new construction could be circumvented. Expanding the direct processing of masonry materials beyond the reclaiming of individual salvageable units requires an acceptance of alternative assembly processes for sets of irregular materials.

In this research, we explore the potential creative use of nonstandard material stocks in construction efforts. Whereas a plethora of digital tools to model abstract 3D form exist, tools to model, design, and iterate through assemblies of pre-existing material stock are lacking. The availability of a tool that gives a user agency in reuse scenarios could encourage designers and architects to utilize preexisting or reclaimed materials. A cultural shift away from new, standardized material stocks as a default choice could both have new aesthetic implications, as well as, a considerable impact on reducing carbon emissions.

To do this, it is necessary to understand how irregular material stock, comprised of fragmented objects, might pack into any desired larger system. The reassembly of broken, small-scale objects into a new form is both a computational problem and a design challenge. Rather than aiming for an optimized solution that might reduce gaps between parts, we developed a framework that provides users the possibility of designing assemblies through the customization of their patterning parameters. The spacing of gaps, order of units, degrees of rotation, number of units, and bounding geometry are the design parameters that through the iteration and permutation generate varied aesthetic outcomes.

Our approach is to create an open-ended system that can input large material stock collections of irregular objects as proxies for construction debris and give these items a second life by reassembling them within a range of permutations. While the assembly of irregular materials is commonplace in vernacular constructions globally, our proposed framework gives architects and designers the ability to visualize a range of aesthetic outcomes. This visualization tool is a step towards the acceptance of nonstandard construction assemblies within the discipline. In this paper, we analyze prior research done on designing with reclaimed construction materials, contribute a framework for a user-controlled combinatorial system of material reuse, and outline the current limitations and future possibilities of our approach.

3

BACKGROUND

Three-dimensional packing is a complex extension of two-dimensional (2D) and one-dimensional (1D) packing problems that are commonly encountered in logistics and Operational Research for the transport and shipping industry (Khairuddin et al. 2020). Although 1D packing is known to be NP-Complete, three-dimensional questions are NP-Hard (Martello, Pisinger and Vigo 2000), which means first-fit and optimized solutions exist only through approximation algorithms (Johnson 1974).

Packing has been the subject of study in an extensive body of research initiated in the 1970s (Johnson and Demers 1974).

Three-dimensional bin packing is an optimization problem that has widely been studied through heuristic approaches of aggregation of dimensioned boxes inside of regular rectangular containers of known dimensions (George and Robinson 1980). Recent studies have since adapted Genetic Algorithms (GA) in order to optimize solutions with varied applications. One approach to the problem is to produce increasingly smaller containers optimized to their contents (Khairuddin et al. 2020). Opposite approaches of optimization exist where GAs simulate contents that "grow" from multiple sides of a set container to maximize the contents that fit in the container (Lim and Ying 2001). Our research is a variation of previous

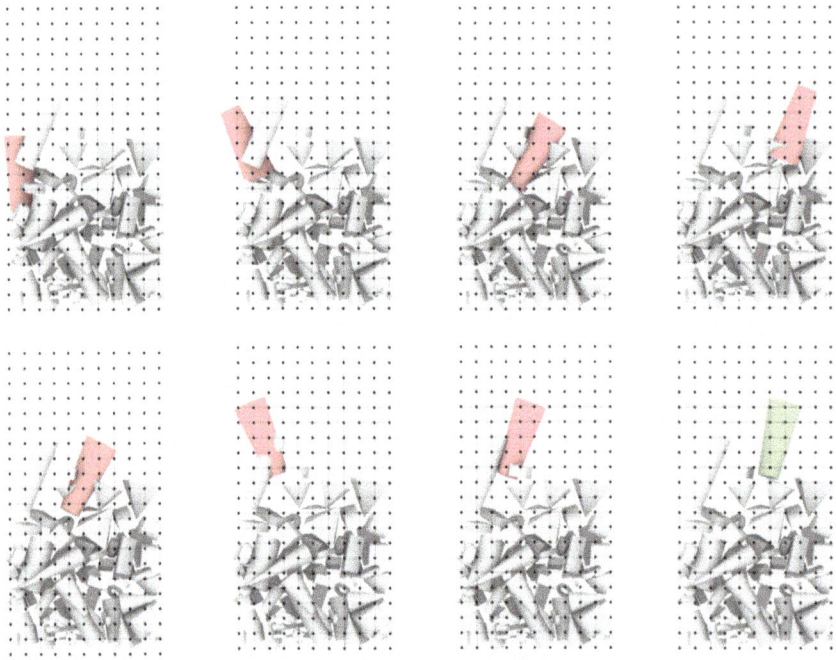

3   The input of different collections by increasing size

4   Algorithm procedures where green is the placed object and pink demonstrates failed placement attempts due to intersections, removed from the final assembly output

4

5 Material stock collections of stones, children's toys, fragmented bricks, and cracker crumbs that serve as proxies for a real-use scenario of 3D scanned objects.

work towards a framework of user-guided "bin packing" for architectural facade construction. The classical first-fit three-dimensional packing problem is further complicated here with the use of true-to-life irregularities of recycled and reused building stock.

Rubble and scraps have been integrated into building processes and documented extensively since the medieval period, though such strategies are rarely used in contemporary construction (Campbell 2013). However, this strategy is gathering increasing attention in the form of architecture that reclaims local material through irregular assemblies and vernacular techniques, such as in the Ningbo Museum by Wang Shu (Chau 2015). The capacity to algorithmically generate an optimized arrangement of scrap material for new structures out of neglected waste has been studied using pattern recognition to map a 2D container (Nolte and Witt 2016).

Contemporary architectural design processes have long incorporated computer-aided design through parametric frameworks as means of generative design and for optimization of structural and thermal performance. Recent advances in robotic fabrication methods have opened up opportunities for research that examines the application of computer-aided sorting and thus the ability to transform irregular, non-uniform stock into new forms (Raspall 2015).

In contrast, our approach incorporates user judgment in order to pair human-machine protocols, stimulates aesthetic outcomes, and demonstrates how user-controlled parameters can act as a tool for designer agency. Parametric modeling and optimized simulations can be highly valuable in the early stages of the design process, rather than as ex-post-facto checks against finished design schemes (Bradner, Iorio, and Davis 2014). Advances in the integration of simulations into early-stage design processes as well as fast-feedback tools and interfaces (Tsigkari and Chronis 2013) allow the opportunity to pair parametric simulation with informed design responses.

METHODS

The basis of this research is the unsorted and irregular material stock of deconstruction and demolition processes. The material stock collections employed as case studies for this research are digital representations of post-demolition brick masonry, stones, children's toys, and broken saltine crackers—all chosen as representations of stock collections of irregular size and geometry (Figure 2). Such representative geometries visualized in this research could be extended to represent post-deconstruction and demolition slate or concrete slabs, timber offcuts, or concrete masonry units. The resulting visualizations present objects as suspended in space in order to prioritize the varied arrangements produced from manipulated parameters. Each real-world material would have a corresponding connection and support construction technique. Advancing technologies of three-dimensional scanning and digital robotic fabrication could allow for these digital representations to be replaced by true-to-life collections of reused objects. Digital mesh objects could then be representative of actual objects, to be tagged, scanned, and sorted computationally before assembly through robotic means (Raspall 2015).

In order to visualize the reconstruction of reused material, this research simulates the packing of objects through a heuristic approach that accepts any sized collection of material stock and assembles the objects into a user-defined bounding geometry (Figure 3).

Assemblies are generated by filling the bounding geometry with objects from the material stock collection until either the collection is exhausted or the bounding geometry is full. This procedural approach works on the basis of an underlying coordinate system, testing the placement and rotation of each object along points on the coordinate system until a

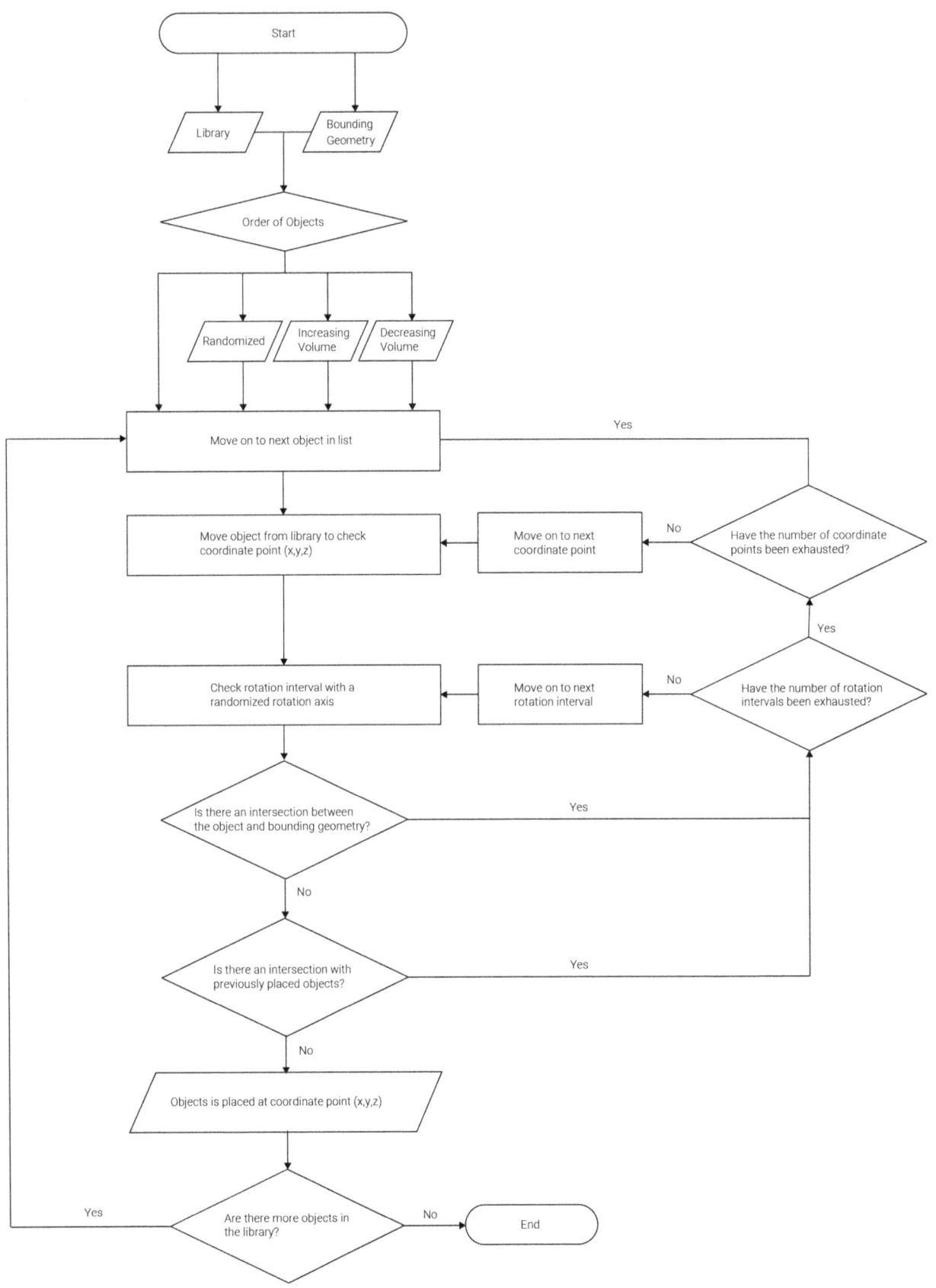

6 Flowchart describing the order and priority of user-controlled parameters

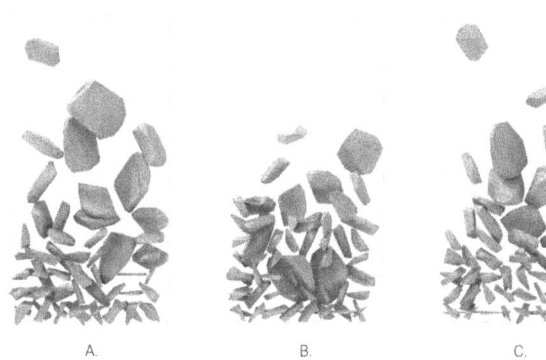

7 The order in which objects are input creates a design bias in the assembly distribution

location is found where there are no intersections with either the bounding geometry or any previously placed objects (Figure 4). The underlying structure of this approach is adaptable given user input through a set of parameters, allowing designers to generate iterative assemblies based on changes in packing density, degree intervals of rotation, and the pre-sorting of objects.

Material Stock Collections

To test the packing algorithm, four collections of material stock, each with 50-200 unique objects, were developed: stones, children's toys, fragmented bricks, and cracker crumbs. The stones, bricks, and crackers were scripted in a parametric modeling tool to generate large collections of irregular material stocks. Each stone was created by extending randomized planes to slice a flattened spherical geometry. The bricks and crackers were made by utilizing a Delaunay mesh at randomized locations and orientations to slice whole brick and cracker forms. The platonic shape collection, or children's toys, were generated at varying scales through polysurface object commands in a 3D modeling tool. These collections all serve as proxies for real discarded or salvaged material that could be 3D scanned and integrated into the same workflow.

Three-dimensionally scanned objects could be reoriented and arranged computationally before or as a step in the packing algorithm. The orientation and order of the material stock collections are important because they create a design bias in the output assembly. This is apparent in Figure 5 where the units have been laid out in particular patterns: the stones are grouped roughly by size; the children's toys are organized by geometry in columns and size in rows; the fragmented bricks are arranged with their original dominant length consistent across the collection; and the crackers are laid out on their larger area dimension. The effects of these differing initial arrangements can be seen in the case studies discussed later in this paper.

The core packing algorithm is scripted in a C# programming plug-in using the parametric modeling tool Grasshopper in the popular 3D computer graphics and computer-aided design application Rhinoceros 3D. It employs the RhinoCommon library for mesh and NURBS geometry operations. The case study results are rendered in post-production with material swatches and render maps using V-Ray©, a plug-in for Rhinoceros.

Packing Algorithm Overview

Order, rotation, and spacing are defined parametrically in order to give the designer agency to iterate through different packing arrangements. Once the material stock collection and bounding geometry are established, the packing algorithm goes through a particular order and priority of user-controlled parameters (Figure 6). The first optional user-controlled parameter is the initialization of the ordering of the material stock collection items, which can be randomized, or based on properties of the objects, such as volume in increasing or decreasing order. After the order is established, if the user chooses to alter it from the initial state, each object is placed at the first coordinate point where it does not intersect with other objects or the bounding geometry. The coordinate points are determined based on the spacing parameter decided on by the user. While placing these objects at particular locations, the algorithm also tests rotational angle configurations at degree intervals decided by the user along randomized rotation axes.

User-Controlled Parameter Case Studies

The user-controlled parameters of order, spacing, and rotation were tested with the stone, children's toy, and fragmented brick material stock collections, respectively. While the bounding geometry is user-generated, for purposes of these case studies it remains constant and orthogonal.

As mentioned previously, the order in which collection objects are sorted by the algorithm has a dramatic effect on the packing outcome. In Figure 7, the order of objects

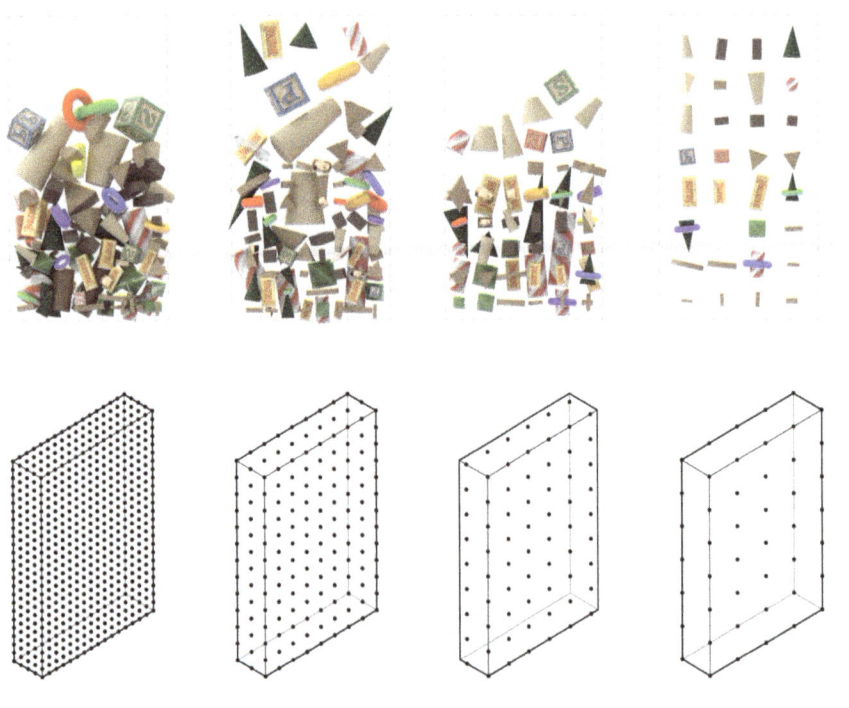

8   Four different spacing interval values within bounding geometry constraints generate different packing densities

9   Detail images of tightly packed assemblies where no intersections between objects were found

has been assigned in four ways: manually determined by user (A), processed through randomization (B), arranged by increasing volume (C), and arranged by decreasing volume (D). The most significant difference can be seen between the increasing and decreasing volume orders. Increasing volume sorting creates a gradient from small to large stones towards the top of the bounding geometry. In the decreasing volume order, large stones were placed first, leaving room for smaller stones to be placed in the same starting locations without intersections occurring with previously placed objects. While the decreasing volume order produces the tightest packed assembly, the other order options produce alternative aesthetic outcomes that may be desirable depending on the application.

The spacing parameter dictates at what x, y, and z coordinate point intervals locations are tested within the bounding geometry. In order to understand the effects of subtle manipulations, the bounding geometry for all of these case studies was set to a 1 x 4 x 6 model unit rectangular prism. In Figure 8, the range of values shown is from 0.25 to 1 model units with each example increasing by 0.25 intervals from left to right. As the spacing interval number increases from 0.25 to 1, the number of points for the placement of the objects decreases at a much quicker rate. From left to right, the quantity of points that the packing algorithm computes is as follows: 2210, 351, 108 and 70.

At the closest spacing, objects are shown as almost resting on each other and the torus geometry allowed two objects

10 By constraining the angle rotation intervals, a design bias in the direction of the objects can be achieved

10

to be placed at the same coordinate point (Figure 9). The spacing parameter primarily determines the density or how tightly packed objects in an assembly are. The denser assemblies allow for a larger quantity of objects from a material collection to be used. At smaller spacing intervals, there is also a greater exploration of different rotational configurations of objects.

This parameter case study suggests that gaps in this packing algorithm are not a problem to be solved, but rather an opportunity to be intentionally designed. Skewers, strings, dowels, grout, and a variety of other techniques can be considered as potential connectors. The type of connector has an effect on the degree to which certain variables can be modified within a design framework. Discrete connection elements allow for a drastic changes in density and spacing, while workable materials, such as mortar (or peanut butter), lend themselves to experiments with assembly order and rotation.

The rotation interval parameter determines at what angles the algorithm checks if an object fits at a particular location. The smaller the interval, the more tightly packed the results because objects are checked at an increasing number of angles. The rotations are applied to the original orientation of the object in the material stock collection. In Figure 10, angle rotations include 45, 90, 180, and 270 degrees from left to right. The 90-degree rotation produced orthogonal results similar to a traditionally constructed brick wall. All of the other rotation intervals present more dynamic and non-orthogonal results that suggest novel methods of assembling masonry.

RESULTS AND DISCUSSION

This framework acts as a simulation tool for irregular assemblies, however, construction procedures, such as manual or computer-aided sorting through tagging and labeling, are currently outside the scope of this research. The digital material stock representations created for this research may be specified to a particular stock of materials through the use of 3D scanning or photogrammetry technologies. Further explorations on the scalability of 3D scanning, analyzing, and managing stock collections within industrial material processes are needed.

In addition, future research could focus on an interactive system of design. For example, a computational strategy could be developed that responds to designer input, such as the manual selection and placement of a specific object. Designer preferences could then be assigned as inputs and fed back into the initial algorithm. Such an approach would expand on the proposed tactics of human-machined design collaboration and contribute to the emerging field of mixed reality and robotic fabrication.

This research assumes non-structural applications of the resulting assemblies. Next steps could include considerations on how to incorporate structurally defunct material parts into a load-bearing system. Physics-based forces, such as mass and center of gravity, could be built into the algorithm to influence assemblies and produce realizable structural configurations. Presently, the creation of a rainscreen or shading system may be a good first test of the algorithm for a built project. Similar to how this research presents case study assemblies with proxy objects of irregular material stocks, the built project would place an emphasis on the spaces between upcycled material units. Appropriate connectors and supporting elements would be designed based on the scenario. Further work could also be done on developing a larger list of user-controlled parameters that affect the output assemblies.

CONCLUSION

This research reenvisions the manual process of sorting and configuring stacked wall assemblies to create a parametric framework for designers while addressing the pressing need to convert waste into a resource. The simulation of design parameters for sorting, spacing, and orientation in packed assemblies expands the capacity of a designer or architect to address the aesthetic implications of utilizing irregular material stocks for building. The framework and tool introduced here could significantly improve design freedom in the use of recycled materials. In addition to discovering new design possibilities, when implemented in early phase design explorations, this framework could play a role in creating new spaces and methods of fabrication.

ACKNOWLEDGMENTS

This research was conducted as part of the course GSD-6338: Introduction to Computational Design, at the Harvard Graduate School of Design with collaborator Sam Pires. The authors would like to thank the entire teaching team: Jose Luis García del Castillo y López, Xiaoshi Wang, Indrajeet Haldar, Jessica Chen, and Dongyun Kim for their support throughout the semester. The authors would also like to thank Samier Merchant for his advice and contributions.

REFERENCES

Bradner, Erin, Francesco Iorio, and Mark Davis. 2014. "Parameters Tell the Design Story: Ideation and Abstraction in Design Optimization." *SimAUD '14: Proceedings of the Symposium on Simulation for Architecture and Urban Design*. 1–8.

Campbell, James W. P. 2013. "The Supply of Stone for the Rebuilding of St Paul's Cathedral 1675-1710." *Construction History* 28 (2): 23–49.

Chau, Hing-wah. 2015. "The Aesthetics of Reuse: the Materiality and Vernacular Traditions of Wang Shu's Architecture." In *Unmaking Waste 2015 Conference Proceedings: Transforming Production and Consumption in Time and Place*. 358–364.

George, John A. and David F. Robinson. 1980. "A Heuristic for Packing Boxes into a Container." *Computers & Operations Research* 7 (3): 147–156. https://doi.org/10.1016/0305-0548(80)90001-5.

Global Alliance for Buildings and Construction. 2021. "2021 Global Status Report for Buildings and Construction." United Nations Environment Programme. Accessed May 5, 2022. https://globalabc.org/sites/default/files/2021-10/GABC_Buildings-GSR-2021_BOOK.pdf.

Hansen, Torben C., ed. 1992. *Recycling of Demolished Concrete and Masonry*. London: CRC Press. https://doi.org/10.1201/9781482267075.

Johnson, David S. 1974. "Approximation Algorithms for Combinatorial Problems." *Journal of Computer and System Sciences* 9 (3): 256–278. https://doi.org/10.1016/S0022-0000(74)80044-9.

Johnson, David S., Alan Demers, Jeffrey D. Ulman, Michael R. Garey, and Ronald Graham. 1974. "Worst-case Performance Bounds for Simple One-dimensional Packing Algorithms." *SIAM Journal on Computing* 3 (4): 299–325.

Khairuddin, Uswah, Nasuh Razi, M. Abidin, and R. Yusof. 2020. "Smart Packing Simulator for 3D Packing Problem Using Genetic Algorithm." *Journal of Physics: Conference Series*, vol. 1447. https://doi.org/10.1088/1742-6596/1447/1/012041.

Lim, Andrew and Wang Ying. 2001. "A New Method for the Three-dimensional Container Packing Problem." In *IJCAI'01: Proceedings of the 17th International Joint Conference Artificial Intelligence*, vol. 1, 342–347.

Martello, Silvano, David Pisinger, and Daniele Vigo. 2000. "The Three Dimensional Bin Packing Problem." *Operations Research* 48 (2): 256–267.

Nolte, Tobias, Andrew Witt, Mike Degan, Jason Tucker, Cody Glen, Claire Kuang, and David Hamm. 2016. "Mind the Scrap." Certain Measures, Installation video, Collection Centre Pompidou. Accessed May 5, 2022. https://certainmeasures.com/MINE-THE-SCRAP.

Raspall, Felix. 2015. "Design with Material Uncertainty: Responsive Design and Fabrication in Architecture." In *Modelling Behavior*, edited by M. Thomsen, M. Tamke, C. Gengnagel, B. Faircloth, and F. Scheurer. Cham: Springer. 315-327. https://doi.org/10.1007/978-3-319-24208-8_27.

Tsigkari, Martha, Angelos Chronis, Sam C. Joyce, Adam Davis, Shuai Feng, and Francis Aish. 2013. "Integrated Design in the Simulation Process." In *SimAUD '13: Proceedings of Symposium on Simulation for Architecture and Urban Design*. 1–8.

Patricia Dueñas Gerritsen is a researcher and designer. She is currently a Master's student in the Department of Architecture and Planning at the Massachusetts Institute of Technology and holds a BA in History of Art and Architecture from Brown University. In her work, she is committed to reducing the negative impacts of building. Her research interests include reclaimed material use, vernacular building practices, architecture of additions and alterations, and the distribution of agency in design and construction.

Emily Wissemann is researcher, artist, and designer. They are currently a Master's student in the Department of Architecture and Planning at the Massachusetts Institute of Technology. They hold a BA in Studio Art from Bard College. Their research interests include information theory, digital and physical storage systems, archives and value.

Jose Luis García del Castillo y López is an architect, computational designer, and educator. His work focuses on the development of digital frameworks that help democratize access to creative technology for designers and artists. Jose Luis is a registered architect and holds a Doctorate in Design and a Master in Design Studies on Technology from the Harvard University Graduate School of Design, where he is currently Lecturer in Architectural Technology at the Material Processes and Systems Group (MaP+S). He also leads ParametricCamp, an online platform for open knowledge in computational design.

Session Introduction
Field Notes

Melissa Goldman, Chair

The Field Notes began as a format at the first online conference in 2020 as a place for experimental and nascent ideas, projects, and quandaries within the set of more polished papers and projects. During an unsettled time with limited access to our usual spaces and tools of work, Field Note authors found new ways to research, leading to new ways to publish findings. Meant to be provocative, Field Notes could take a wide range of topics as well as form.

Our set of Field Notes tells a collection of stories through provocations. The authors ask us to take focused views of a process, a relationship, a detail, or a platform, and to use this understanding of the local to reform a new view of the global. The group includes a manifesto that invokes the microcosm and the detail, workshops with craftsfolk that celebrate fabrication methods and materials as well as communication platforms, a project that reframes the graphic norm, and deep dives into details and data which present new ways to think about the global wicked problems that cross borders both physically and virtually. We traverse scales of ecotones, design graphic languages, explore expertise in craft, and launch into the Metaverse.

The panel discussion of the Field Note session at ACADIA 2022 led us into what each author found hopeful about their calls for change. We spoke about the hopeful conversations they have had crossing disciplines and platforms to connect globally while simultaneously gaining a better understanding of local materials, cultures, and ecosystems. The authors also recognized the difficulties and messiness of starting these conversations and relationships, whether sitting down at a transdisciplinary table dealing with the health of cities, bringing new knowledge into a team with generations of expertise, or trying to make sense of virtual avatars with no sense of gravity.

Each of these stories reveal a messy process. The Field Notes as a format sheds light on this process, giving a platform to a different kind of publication of works-in-progress. We hope to keep the space of these types of conversations in many ACADIA conferences to come.

Alternative Typographic Histories

Arabic Script as a Driver of Language Display Technology

Levi Hammett
VCUArts Qatar

Fatima Abbass
VCUArts Qatar

Hind Al Saad
VCUArts Qatar

Mohammad Suleiman
VCUArts Qatar

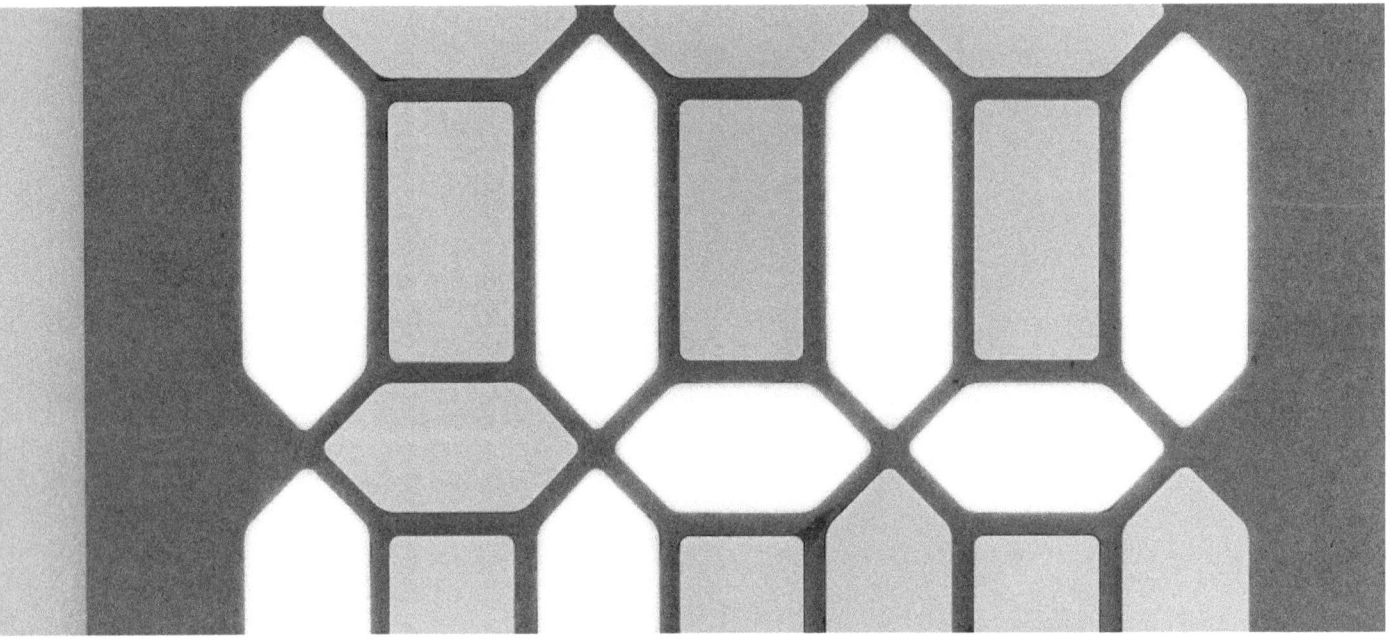

1 Close up shot of the installed segmented display showing the Arabic letter "Seen"

The evolution of printed language has been driven by technological innovation aimed at increasing our ability to communicate over time and space. The twentieth century saw an exponential increase in typographic diversity created to facilitate and adapt to new technological innovations. However, the locality of these innovations have been situated primarily in the west where the Latin typographic script is the predominant form of written language. The influence of these developments created significant pressure to reform connected scripts, such as written Arabic and Persian, whose communities were not in a position to influence the design of these technologies or have a leading voice within these new typographic domains (Parhami 2019).

Technology is often applauded for its ability to disrupt established systems through innovation, efficiency, and its tendency for rapid adoption within a given domain. The transformative nature of technology can have negative implications, particularly when technologies propagate beyond the context for which they were developed without adequate critical discourse. When we look at the recent developments of the Arabic and closely related Persian scripts, the promise of efficiency and increased productivity created significant incentives for engineers and type designers to conform to western developed frameworks for typographic rendering, which required significant aesthetic compromise of the formal conventions of the Arabic script (Nemeth 2019). Viewed broadly, there is a historic tendency for mechanical reduction of the formal structure of typographic characters toward a paradigm of unification, which has in many cases reduced the visual harmony of scripts outside of the standard Latin character set (Boutros et al. 2009).

2

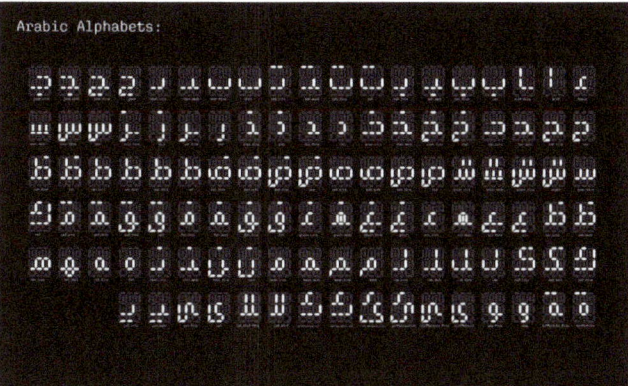

3

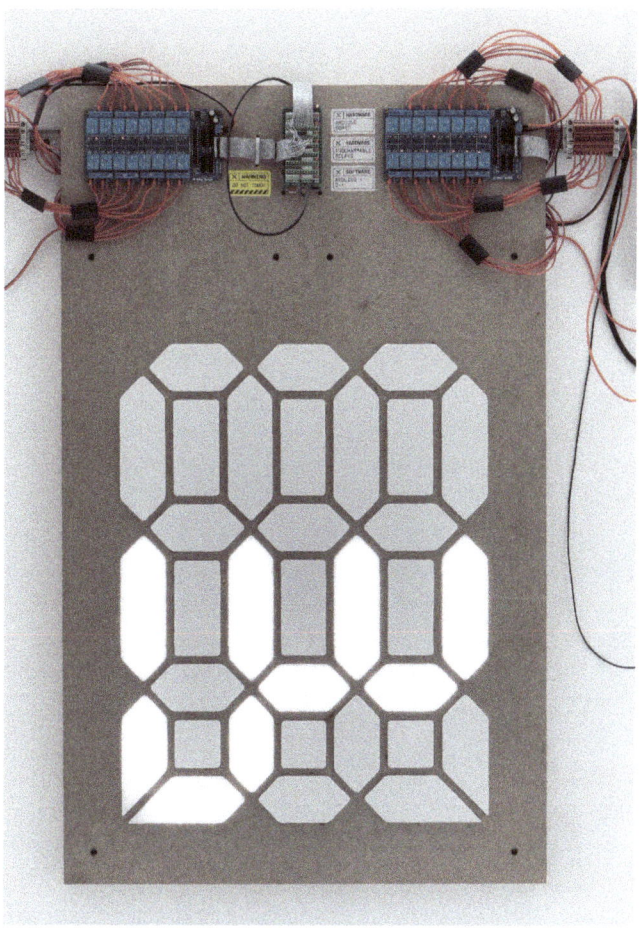

4

2 31-segment grid design of the segmented LED display

3 Design of the full Arabic character set

4 Installed segmented display measuring approximately 63 cm x 100 cm

This body of work explores these themes through the development of a series of anachronistic Arabic typeface designs that reimagine an alternative timeline where the Arabic language played a leading role in the development of specific typographic display technologies. These typefaces have been designed within the limitations of obsolete typographic technologies and reimagined as computational installations, informed by and building on existing research related to structural developments in Arabic script (Azad et al. 2004; Osiur et al. 2008; Beg and Hassan 1987).

The first phase of artistic research in this body of work was a large scale illuminated display constructed around a novel segmented grid designed to display a single Arabic character shown in Figure 2. The design of the grid is able to render the entire Arabic alphabet which includes over one hundred unique forms as demonstrated in Figure 3. The design artifact employs a microcontroller together with electromagnetic relays that enable the programming of custom-designed and fabricated LED segments shown in Figure 4. The segments can be individually illuminated with precise timing able to display animated transitions between letters and sequenced to display words and longer texts.

This speculation of alternative typographic histories is an attempt to re-examine the evolutionary path of the Arabic script within inherited technological constraints. The research attempts to highlight key moments of typographic regression resulting from technology that was out of step with the type-making process (Nemeth 2017). In so doing, this work aims to uncover new pathways for the development of the Arabic script in order to add new perspectives to the co ntemporary type design discourse.

REFERENCES

Azad, Abul Kalam, Rezwana Sharmeen, and S. M. Kamruzzaman. 2004. "Universal Numeric Segmented Display." In *Proceedings of the 7th International Conference on Computer and Information Technology (ICCIT-2004)*. 887–892. Dhaka, Bangladesh.

Beg, M., and Wasim Ahmad. 1987. "Dot Matrix Alphanumeric Display System for Arabic." *IEEE Transactions on Consumer Electronics* CE-33 (1): 47–50.

Boutros, Mourad et al. 2009. *Talking About Arabic*. New York: Mark Batty Publisher.

Nemeth, Titus. 2017. *Arabic Type-Making in the Machine Age*. Leiden; Boston: Brill.

Parhami, Behrooz. 2019. "Evolutionary Changes in Persian and Arabic Scripts to Accommodate the Printing Press, Typewriting, and Computerized Word Processing." In *Proceedings of TeX Users Group Conference* 40 (2): 179–186. Palo Alto, California.

Rahman, M. O., M. I. Shafique, E. Scavino, A. Hussain, and H. Basri. 2008. "The design of a complete uniform segmented display unit for Arabic alphanumeric characters." In *2008 International Symposium on Information Technology*, vol. 4. 288–293.

IMAGE CREDITS

All other drawings and images by the authors.

Levi Hammett is a designer and Associate Professor exploring the synthesis of computational processes and traditional crafts with an emphasis on the development of culturally constructive graphic objects.

Fatima Abbas is a graphic designer whose work pushes the boundaries of Arabic type design by rethinking the fundamental structure of the language to imagine new typographic forms.

Hind Al Saad is a computational artist and teacher, who explores automation systems within digital code and analog printmaking processes to create emergent graphical forms.

Mohammad Suleiman is an architect and Assistant Professor who investigates spatial mapping and data-driven design workflows for the existing built environment.

Straddling the Boundary

Julianna Cano

A Tiny Manifesto

1   Folly #5 - Lasker Columbarium Aerial ©Julianna Cano, 2021

The architectural project often begins with a broader idea, representative of the parti: schematic in resolution and macro in scope, increasing in detail over time. Design phases reflect this progression, starting with a concept and ending in documentation. The urban is often considered before the architectural, and the building always precedes its details. Urban planning initiatives tend to dominate cities, imposing a universal development strategy and enforcing building standards to meet sustainability and efficiency goals, often ignoring local conditions.

Moreover, the architectural object has become a reactive tool that responds and conforms to environmental standards rather than initiating a position based on the subtleties of its immediate surroundings. The building industry has become increasingly standardized, making buildings performative machines rather than thoughtful architecture. The convergence of architecture and environmentalism has resulted in an unsettling imbalance: architecture has become subservient to an environmental program rather than serving as an equal contributor towards the production of our material world. Mark Foster Gage echoes this sentiment, stating that "the form is legitimized, in this case, through its ability to solve the abstract problem set at the beginning—not from the resultant form itself…. We are reinforcing a system where architecture is being legitimized not for what it is, physical architecture, but only what it can do or can successfully refer to" (Gage 2011, 18–19).

Simultaneously, environmentalists and philosophers have condemned American "exceptionalism" in favor of global interconnectedness and equity. Eco-critical discourse has argued for an erasure of national boundaries to help tackle the global climate crisis. Alan C. Braddock highlights the shift towards a "transnational perspective [that is] broadly ecological insofar as it extends—like global warming or a migratory bird—beyond such boundaries" (Braddock 2015). Transnational thinking emphasizes the erasure of national boundaries and figure-ground relationships in a time of cultural dissemination (Braddock 2015). Both the disciplines of architecture and environmental sciences have shifted towards a globalist mindset in an attempt to overcome the anthropocentric and hierarchical constructs that are deeply rooted in society. A macrosociological method in how we address cultural and ecological incongruities is promising in theory; however, we risk repeating mistakes of the past when we impose a universal solution that ignores the heterogeneity of peoples, regions, and ecological bodies.

This manifesto argues that any action to erase or simply ignore boundary conditions will in most cases trigger the opposite effect: it will reinforce them. The condemnation of boundaries, which can be defined as ideological, political, ecological, or geographical, to name a few, does not eliminate the dichotomies that underpin their separation. The dissolution of boundaries implies that the whole is greater than the sum of its parts, and therefore, fails to acknowledge the complexity of those parts. Moreover, it stunts the organic and often messy hybridization of divergent practices and agents. We as designers must be cognizant of this ideology and its potential influence over how we address cultural and geographical nuances associated with a given project or site. The unintentional consequences of a transnational or globalist approach towards how we approach environmental concerns related to our built environment is a green light (no pun intended) for increased standardization, diluted architectural interventions and disseminated monotonous cities.

2 Opaque Wall, Hadrian's Villa, Tivoli (formerly Tibur), Italy

3 Porous Wall, Oil Factory at Brisgane, Algeria

In a 1982 lecture given by Michael Graves titled "The Sense of Boundary in Architecture," he reaffirms this through the analogy of the wall stating that, "[i]t is essential I think to make understood to us, the society, that space is not homogenous, that indeed we need the separations and boundaries of place to place, to understand our individual and particular realms…. Because it is that boundary that we ultimately make the place of passage. The idea that one first makes the wall to separate two things and then finds a way to combine them again" (Graves 1982). Similar to Graves's comparison of two wall conditions—one opaque and one porous—the boundary can also manifest itself in different ways. It can stay rigid and reinforce the separation between two entities, or it can become permeable and entice leakage, slippage, or "accidental" comingling. It is our willingness to engage with the boundary that will determine its evolution and ultimately our fate.

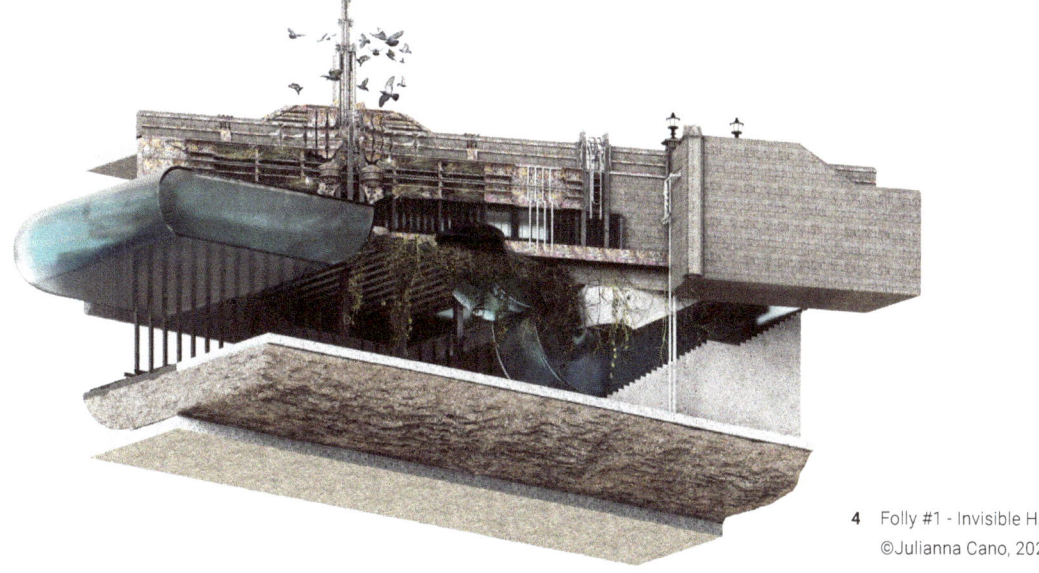

4 Folly #1 - Invisible Harvester
©Julianna Cano, 2021

All this to say, we must avoid falling back into cultural exceptionalism at all costs. In *A New Philosophy of Society*, Manuel DeLanda proposes an assemblage theory to overcome reductionist reactions to the classic micro-macro problem. He briefly summarizes that individualism can be understood as the micro-reductionist position and collectivism, the macro-reductionist position (DeLanda 2006, 4–5). Exceptionalism and transnationalism seem to echo this dichotomous trap: operating on the assumption that an alternative method cannot exist. Like DeLanda suggests, we should avoid extremist positions. Instead, we should aim for the gooey middle; an approach that supports an inherent multiplicity of ecologies, communities, architectures, and interactions. We should build a world where microcosms thrive.

It is important for us as architects to reframe how we engage with the boundary conditions that both separate and shape space and culture. These conditions can no longer be seen as adversaries to dismantle or avoid but as existing conditions to work directly with. The architect must straddle the very boundary that they wish to dissolve. Timothy Morton (Morton 2019) reaffirms this position and emphasizes the need to not wish away problems.

> "Wholes subscend their parts, which means that parts are not just mechanical components of wholes, and that there can be genuine surprise and novelty in the world, that a different future is always possible. It is good to regard things...as physical beings, not simply as fictions that would disappear if we just stopped believing in them. But what kind of physical being are they? If they are subscendent it means that we can change them, if we want."

The project titled *Teeter-totter Wall*, by architecture studio Rael San Fratello does not regard the boundary as fictitious, and instead takes the liberty of occupying it. This project is a series of seesaws that penetrate the international border wall between the United States and Mexico, providing the opportunity for citizens of both countries to interact. The boundary in this case becomes the site and the project demonstrates that any action on one side has a direct impact on the other. Another example of a boundary typology might be ecological. An 'ecotone' supports this idea that the boundary is neither here nor there but a place in its own right. According to William deBuys "[t]he area where two or more distinct habitats adjoin is called an Ecotone. Because it is a border zone where multiple sets of resources and opportunities become available, an ecotone tends to support greater biological diversity than either of the systems it mediates between" (deBuys 2013, 156). The boundary becomes a sense of place, and its divisiveness becomes diversified.

It is no longer appropriate to think large scale and ignore discrete and often delineating conditions. The gooey middle calls on us to be diligent. For architecture to reaffirm its value in contemporary culture, we must focus on identifying these boundary conditions, both contested and benevolent, before we get to work. Once we begin, we must focus our efforts on scale and resolution. Architects and environmentally conscious designers must embrace a working process that operates at the tiniest of scales in order to interrogate the often-invisible boundary and address the complexities of the nonhuman and the marginalized. The tiny architectural object has the agility to straddle both the ideological and geographical boundary condition and focus on the subtleties and complexities unique to its locality.

A micro approach is intimate, discrete, and complex and inverts our current design process. We should begin at a high resolution, with a fragment and build outward. We should think ontologically big but physically small. We should begin with a tree, not a landscape, a room not a house, a chimney not a roof, a window, not a wall. Or even a bike rack not a park, a pillar not a threshold, a greenhouse not an industrial farm or a bird loft not a zoo. Aldo Rossi understood the importance of creating "small things, because the possibility of great ones has been historically precluded" (Rossi 1981, 83).

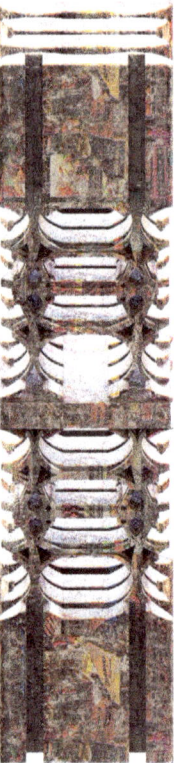

5   Park Entrance Pillar ©Julianna Cano, 2021

Starting small and building outward enables us to loosen our anthropocentric grip and engage in a new consensual relationship with our environment. Not only does this approach create empathy, it eliminates the opportunity for normative solutions. With this approach, physical architectural development would no longer be measured by footprint but instead by its impact on the microcosms that the architecture initiated and supported. In doing so, these discrete physical interventions would no longer serve larger institutionalized agendas and would be free to develop from their own component parts. This freedom gives rise to new architectural potential that has long been stunted by predefined programmatic constraints and assumptions.

6   Folly #2 - Nested Greenhouses ©Julianna Cano, 2021

Operating at the tiniest of scales allows the architect to fit into the nooks and recesses of our material world and observe everyday experiences with greater intimacy. The tiny architectural object serves as a mediator, disruptor, adapter, or witness to the binaries that the boundary separates. It is adaptable to its immediate surroundings and does not impose a diluted universal solution. The minute architectural intervention celebrates the particularities of its immediate context through its ability to respond and hybridize with it. The tiny architectural object is conceived by its details and evolves from the epicenter of chaos, the boundary. Its power lies in its ability to engage with the invisible and the forgotten, guiding us towards a new way of seeing. We, the co-authors of these tiny interventions, must seek out these boundaries, get on our knees, touch our cheeks to the soil, and stare into the dwellings of aphids if we are to learn anything.

REFERENCES

Braddock, Alan C. 2015. "From Nature to Ecology: The Emergence of Ecocritical Art History." In *A Companion to American Art*, edited by J. Davis, J. A. Greenhill, and J. D. LaFountain. Chichester West Sussex, UK: Wiley Blackwell. 447–67.

deBuys, William. 2013. "Ecotone." In *Home Ground: A Guide to the American Landscape*, edited by B. Lopez and D. Gwartney. San Antonio, TX: Trinity University Press.

DeLanda, Manuel. 2006. *A New Philosophy of Society: Assemblage Theory and Social Complexity*. London: Continuum.

Gage, Mark Foster. 2011. "Introduction." In *Aesthetic Theory: Essential Texts for Architecture and Design*. New York; London: W. W. Norton & Company.

Graves, Michael. 1982. "The Sense of Boundary in Architecture." London, England: Pidgeon Digital.

Morton, Timothy. 2019. *Humankind: Solidarity with Non-Human People*. London, UK: Verso. 102.

Rossi, Aldo. 1981. *Aldo Rossi: A Scientific Autobiography*. Cambridge, Mass: MIT Press.

IMAGE CREDITS

Figure 1: ©Julianna Cano, Thesis: Spectral Minutiae, Folly #5 - Lasker Columbarium Aerial, 2021

Figure 2: Tivoli, Wall of the Poikile of Hadrian's Villa (Villa Adriana), 2019, ©Patrick Kunec, via Wikipedia, Creative Commons Attribution-Share Alike 4.0 International

Figure 3: Oil Factory at Brisgane, Algeria, via https://www.leguidetouristique.com/ruinesbr/huilerie-de-berzguene-brisgane

Figure 4: ©Julianna Cano, Thesis: Spectral Minutiae, Folly #1 - Invisible Harvester, 2021

Figure 5: ©Julianna Cano, Thesis: Spectral Minutiae, Park Entrance Pillar, 2021

Figure 6: ©Julianna Cano, Thesis: Spectral Minutiae, Folly #3 - Nested Greenhouses, 2021

Julianna Cano is a designer at Snøhetta in New York City and a graduate of the Master of Architecture program from the University of Pennsylvania where she was the recipient of the Faculty Prize. Her thesis, "Spectral Minutiae" produced under the advice of Ferda Kolatan, focused on society's outmoded relationship with nature and interrogated New York City's Central Park through the implementation of a series of post-humanist follies. The thesis aimed to push the boundaries of speculative futures through surrealist imagery while critically examining current ecological issues surrounding urban communities.

Collateral Computation

Misri Patel
University of Tennessee

Catechizing Craft and Paradoxes of High Tech

1 Excerpt from a virtual conversation sharing on-site processes with off-site participants

Technological advancements in the building industry and fabrication have led to an abundance of custom geometries and unprecedented precision of parts. In contrast, hundreds of craft traditions that form the second largest workforce in developing countries after agriculture struggle to compete against machine-made products. Although numerous design projects have attempted to address and integrate the dichotomy between them, the dialogue between the makers from "then" (traditional craft) and "now" (technological craft) seems to be largely focused on reviving craft. Most commonly, the studies have focused on learning from the craft on-site and reinterpreting the creative processes in a state-of-the-art lab context. While this practice has enhanced the knowledge of makers from now, one can argue that comparatively, makers from then have benefitted marginally.

Traditional craftspeople with fewer resources and limited or no access to new technology have grappled with adapting to the changing building landscapes. This disparity in knowledge transfer has been further strengthened throughout the pandemic. Concurrently, architecture and design schools have tested and implemented innovative workflows to impart technological fabrication knowledge. The process of instructing advanced fabrication workflows challenges the need to be in close proximity to fabrication facilities.

Building on this replicable prototype of distance learning, can we extend the pedagogical model to include craft-based communities that have accumulated knowledge and skills through

2 Introductory site visit to Lonavala, India

3 Introductory site visit to Nashik, India

4 Preliminary tests using 3D printed PLA as a substrate for adobe slurry on site

generations? With the new generation of these communities attending schools and colleges remotely, can we as educators eliminate the gap to facilitate dialogue between researchers and artisans?

This pilot study elaborates on this statement and begs the question, can we as designers and makers in the 21st century have an equitable, mutually dependent relationship between rural craftsperson and users of cutting edge technology? By contrast, can one see the reinvigorating qualities of craft and guiding principles accompanying modern ways of making? Emanating from questions around architectural models as a tool for speculation, the setup initiates discussions between a brick-making community in rural India and a computational designer in Knoxville, Tennessee. Prior to setting up a working model, it was crucial to identify and understand the inherent conditions that challenge the working of inclusive collaboration. This approach allowed the participants to objectively evaluate the site conditions, like access to a personal computer, internet, and a mode of communication.

The widespread use of personal devices and virtual conferencing platforms during the global pandemic has created a favorable scenario. The pilot study leverages this characteristic to create an outline of a predictable pedagogical model.

The collaboration relied on an introductory site visit subsequently followed by a series of virtual conversations. The preliminary site visit allowed the team to understand traditional techniques of rammed earth production and identify both limitations in the process and potentials in the forms. In addition to this, a desktop 3D printer was used on-site to expose the artisans to advanced additive manufacturing techniques. Instead of expecting the artisans to learn and master the machine, the mentoring session allowed them to be aware and cognizant of the technology. The goal of this session was to establish a relationship with the community and invite the artisans to be active contributors to the project. The speculative prototypes focus on three key resources: traditional material knowledge, processes in craft, and advanced manufacturing techniques.

5   Excerpt of a collaborative session between participants testing shipped 3D printed formwork

The traditional model of construction by using rammed earth has been revived recently as a sustainable building method, but the typical expectation is that it would always have flat faces, thus reinforcing the ubiquitous qualities of sheet stock and its widespread use among contemporary building systems. This study investigates a conceptual framework for the digital practice of rammed earth architecture, which positions digital fabrication and manufacturing as an integral part of the computational design process. In particular, the prototypical study places assembly—the act of putting together discrete elements—as the main generative driver to bridge the gap between craft (age-old ways of making) and advanced digital fabrication techniques (3D printing) to construct novel building blocks.

Subsequent virtual discussions speculated on why, how, and what initiatives can convene a system of knowledge transfer that leverages craft practices with communities based in Nashik, India, and New Mexico, United States. The current stage foregrounds a reciprocal relationship by continuing the dialogue and making on- and off-site. This thematic domain aimed to convert the interactions into applicable tacit knowledge.

The result of this experiment posed a few challenges including the community's access to a desktop 3D printer, the economic burden of getting vocational training, and understanding the know-how of the technology. To aid this, 3D printed prototypes were developed in conjunction with the craftspeople during the introductory site visit. Subsequently, additional studies were printed off-site and shipped locally to test on-site. The combined knowledge of indigenous slip casting and digitally crafted 3D printed PLA is used as formwork to create scaled prototypes.

This intervention proved to be pivotal in knowledge sharing and allowed younger artisans to guide artisans with greater experience. The awareness created in the community actively cultivated the desire to be further involved in the sequential fabrication processes. In addition to this, the participants extended the provocations: Does the status as a maker come through a different set of materials, mediums, and tools? Does the development of new knowledge and processes have to be mutually exclusive?

Given our current climate and social contexts, craft, the assimilation of culture and other foreign influences need to

6   Craftman's prototype using a 3D printer and rammed earth on site

be merged with cutting-edge research in advanced building material systems. This model can be adapted to recognize and pursue entrepreneurship for minorities. Over and above, the recent lockdown conditions presented a challenge that turned into an opportunity to engage with participants (local workers, students, and tutors) from across the globe that articulates a wider conversation on pedagogy and research methodology.

ACKNOWLEDGMENTS

This project is funded by School of Interior Architecture, CoAD, University of Tennessee, Knoxville. The author would like to acknowledge the tireless intellectual and logistical support of local craftspeople, Shantaram, Bharti, and Tukaram. Special thanks to Catie Newell, Milagros Zingoni, Dishant Patel, and Ada Tolla for their generous support and guidance.

REFERENCES

Byars-Winston, Angela, and Maria Lund Dahlberg, eds. 2019. *The Science of Effective Mentorship in STEMM*. Washington, DC: National Academies Press. https://doi.org/10.17226/25568.

Chickering, Arthur W., and Zelda F. Gamson. 1991. "Appendix A: Seven Principles for Good Practice in Undergraduate Education." *New Directions for Teaching and Learning* 1991 (47): 63–69. https://doi.org/10.1002/tl.37219914708.

Mangual, Isaac, Garret Wood-Sternburgh, Larisa Sherbakova, Misri Patel, and Ryan Craney. 2022. "Digital Digital Fabrication Fabrication: Remote Collaborative Teaching and Learning in Advanced Fabrication." In *ACSA 110th Annual Meeting*, Virtual Conference, May 18-20.

IMAGE CREDITS

All drawings and images by the author.

Misri Patel is an architect and researcher from Mumbai, India. She currently serves as a Visiting Assistant Professor of Practice and IA fellow at the School of Interior Architecture, CoAD, University of Tennessee. She was the 2019-2020 Ballard Fellow at CoAD, Lawrence Technological University and gained professional experience at sP+a, Mumbai and LOT-EK, New York. She received a Master of Science in Digital and Material Technologies from Taubman College, University of Michigan and BArch from NMIMS BSSA. Prior to joining UTK, she worked as a research lead at Taubman College and visiting lecturer at the University of Pennsylvania's Stuart Weitzman School of Design.

Nemagari-no-Takumi Workshop

Mixed-reality Crafting with Unwieldy Logs

Nicholas Bruscia
University at Buffalo,
State University of New York

Daiki Kanaoka
FabCafe Tokyo

Hideaki Asaoka
Hidakuma

Kotaro Iwaoka
Hidakuma

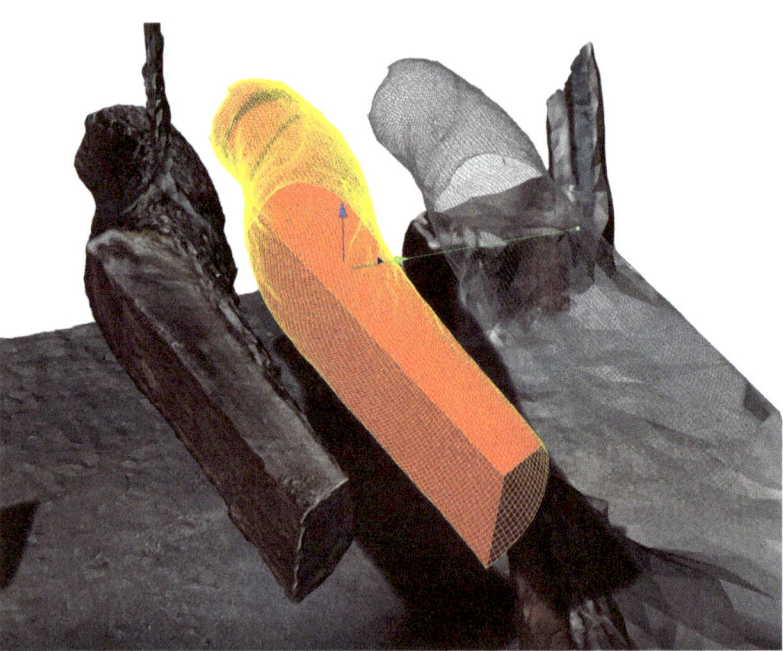

1 Mixed-reality chainsaw cutting and 3d scan verification

Following the second world war, Japan's Ministry of Agriculture and Forestry began an aggressive planting campaign to address the country's reconstruction. To enhance the productivity of national forests, huge amounts of cedars were planted due to their quick and resilient growth, replacing the native broad-leaf species in many areas. Simultaneously, Japan's cultural connection to cedar as a building material waned as the country continued to westernize, and eased regulations on imported lumber reduced the country's self-sufficiency rate to 18% by the turn of the century (Sugimoto 2007).

Over time, vast amounts of abandoned tree plantations have grown too dense, blocking sunlight for smaller plants that contribute to healthy biodiversity. This has resulted in increased risk of landslides, increased pollen, and changes to the ecosystem as excessive nitrogen in the soil due to human activity washes into rivers and ponds causing algal blooms as the smaller plants are not present to absorb it (Ru and Chiwa 2021). The forests surrounding Hida, a town in Gifu, Japan are facing these challenges. This region is known historically as a center for Japanese carpentry and is home to a wide variety of traditional crafts that remain in practice today. These mountains are also steep and receive heavy snow, often bending younger trees at their trunk under the load. The tree will continue to grow with a curve, making its lumber difficult to harvest and process within standard industrialized processes. Adding to the region's legacy, when rice was the tax, and since Hida is too mountainous to grow enough to contribute, local carpenters were instead commissioned to build temples and shrines throughout Japan (Iimori 1912).

2 Nemagari in a forest near Hida, Japan (Hidakuma 2022)

3 3D scan data of nemagari trees, Hida, Japan (Hidakuma 2022)

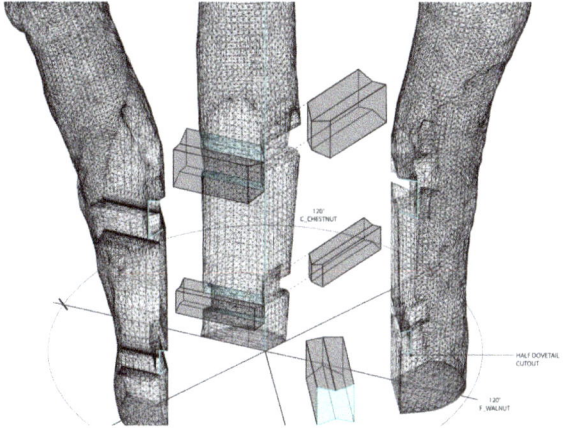

4 Refined meshes from 3D scans prepared for MR-guided chainsaw cutting

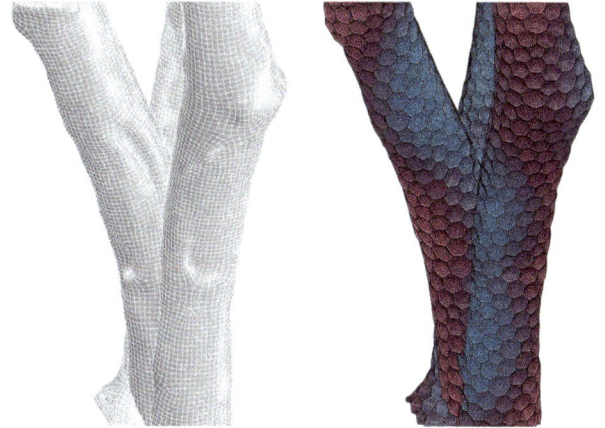

5 Refined, textured meshes from 3D scans prepared for MR-guided sculpting

"Bent-Root" Timber

Within this context, this project reflects a recent collaborative and international workshop that connected faculty and students based in the USA with craftspeople and consultants that reside and work in Hida, Japan. The team developed mixed-reality (MR) fabrication workflows to utilize large-scale *nemagari* (bent-root) timber. Nemagari is typically avoided, but when harvested out of necessity, they are reduced to chips or firewood due to their shape. The material remains underutilized as the trees' natural geometry is considered a downgrade in standardized milling operations (Amtsberg et al. 2020).

The MR workflow is intended to enhance the skill of the craftsperson with real-time holographic guides, enabling efficient processing of large and unwieldy logs into architectural elements that make use of their natural curvature which may otherwise be a hindrance to manual, CNC, and robotic fabrication techniques. Construction from an inventory where no two parts are exactly alike necessitates new tools for visualizing the design intent and for accurately cataloging the exact shape of the logs (Lok and Bae 2022). For example, LiDAR scanning was used to verify accuracy prior to and following each chainsaw cut, so that the holographic guides may be adjusted as necessary.

Nemagari Crafting Workshop

The workshop attempted to bridge technological epistemologies, manual and digital, and seamlessly connect faraway, interdisciplinary collaborators. Results shown are a work in progress, pointing toward numerous ways to fine-tune the process and toward other novel uses for MR crafting. The primary toolset consisted of digital modeling and simulation in Rhino/Grasshopper, mixed reality with Microsoft HoloLens 2, and Fologram. Team efforts were divided into four trajectories, each with specific visualization and calibration criteria: structure, balance/weight distribution, texturing, and weaving/thatching.

The "structure" project was the focus of the three-day workshop that resulted in a branching assembly of large nemagari, each approximately 3 meters in length. The 3D scan data for each log was combined and modified into a three-part assembly, requiring two long chainsaw cuts guided by intersecting holographic planes or a full holographic twin of the desired part calibrated to each log as per the chainsaw operator's request (Figure 1). A 3d scan of each cut log was compared to the digital model and necessary adjustments were made to ensure the cut logs align when assembled. Oversized butterfly joints were intended to hold the logs

6 MR chainsaw cutting and assembly

7 Interactive holograms showing a collaborative MR-guided sculpting process

together into a column-like assembly, the joinery of which would be done using hand tools also guided by calibrated holograms. Future work includes tool calibration to track the exact location of the chainsaw blade providing real-time feedback of the blade's position relative to a target cut surface or desired cut depth, and the implementation of MR chainsaw cutting and hand tooling in a completed structure.

With an appreciation for the cultural relationship to trees exemplified in various *matsuri* (civil or religious festivals), the "balance" project focused primarily on ways in which holographic visualizations could demonstrate how a large and heavy log could be carefully balanced from an off-center fulcrum point. Holographic interactive simulations attempted to describe the intended awe of a precariously balanced tree while the fabrication process remained open to interpretation. Future work includes testing a variety of visualization techniques that reveal invisible factors, i.e. log density and moisture content, and implying weight and balance within interactive holograms.

Taking a more granular approach, the "texture" project drew inspiration from *naguri*; timber processing techniques that leave behind unique markings along the surface of the log. Mesh modeling and calibrated holographic visualizations were initially intended to guide the same traditional hand tools that are historically associated with naguri, but are equally useful to carving processes using power tools. Color gradients applied to the 3D scan data converted into customized meshes dictate the intention to vary the depth of texture—a direct result of tool use that can only be translated by experienced hands. The speed, pressure, and strike angle of chisels, axes, and electric grinders alike are controlled by the interpretation and skill of the craftsperson. Future work includes material prototyping, customizable holographic user interfaces, and collaborative hybrid sculpting.

Acknowledging calibration as an important part of MR fabrication, the "weave/thatch" project investigates the combination of the two. The study here is multi-faceted; first, to calibrate a bending simulation to real-world material parameters by overlaying a hologram modeled in K2Engineering/Kangaroo2 over an actively bent wooden lath. When aligned, the numerical inputs are less arbitrary and the simulation can more accurately reflect the bending of the material being used. Second, the project reveals the geometric rules of *kagome* basketry by projecting holographic nodes of specifically arranged triangles. By connecting the dots, a novice unaware of how the topology determines the form can learn about the mathematical concepts underlying traditional weaving techniques. Future work includes developing MR guided and novel thatching sequences for non-traditional structures.

ACKNOWLEDGMENTS

The authors would like to thank our mixed-reality chainsaw operators Kazunori Yanagi (Hida) and Wade Georgi (Buffalo) for their expertise and valuable insight on future improvements to this body of work, and Yanagi Mokuzai for supplying the timber and hosting the fabrication

8 Workshop prototype installed at Yanagi Mokuzai, Hida, Japan and its digital twin of raw 3D scan data and refined meshes

and assembly. Sincere gratitude to the students involved in the workshop for their enthusiastic and clever work. Thanks to the Department of Architecture at the University at Buffalo, Hidakuma, FabCafe Tokyo, and FabCafe Hida for supporting this work-in-progress.

REFERENCES

Amtsberg, Felix, Yijiang Huang, Daniel Marshall, Kevin Moreno Gata, and Caitlin Mueller. 2020. "Structural Up-cycling: Matching Digital and Natural Geometry." In *Advances in Architectural Geometry 2020*. Ecole de Ponts ParisTech and Universite Gustave Eiffel. https://thinkshell.fr/wp-content/uploads/2019/10/AAG2020_25_Amtsberg.pdf.

Iimori, Rokujiuyen. 1912. *Hida no Takumi Monogatari (The Story of a Hida Craftsman)*, translated by F. V. Dickins. London and Glasgow: Gowans & Gray, LTD.

Lok, Leslie, and Ji Yoon Bae. 2022. "Timber De-Standardized 2.0 – Mixed Reality Visualizations and User Interface for Processing Irregular Timber." In *Post-Carbon: Proceedings of the International Conference for The Association for Computer-Aided Architectural Design Research in Asia (CAADRIA) 2022*, Hong Kong.

Sugimoto, Ari. 2007. "Balancing Environment and Economics with Compressed Cedars – Hida Sangyo's Challenge." *Think Daily*, vol. 31, January 15, 2007. Accessed September 7, 2022. http://www.thinktheearth.net/thinkdaily/report/2007/01/rpt-31.html#page-1.

Yang, Ru, and Masaaki Chiwa. 2021. "Low nitrogen retention in a Japanese cedar plantation in a suburban area, western Japan." *Scientific Reports* 11 (1): 5335. https://doi.org/10.1038/s41598-021-84753-1.

IMAGE CREDITS

Figure 2, 3: Hidakuma 2022
Figure 4, 6, 8 (left): Chudy, Heiser, Nicpon 2022
Figure 5, 7: Clay, Marsh, Mohammadyar 2022

All other drawings and images by the authors.

Nicholas Bruscia is an assistant professor in the Department of Architecture at the University at Buffalo, where he is also a researcher in the Sustainable Manufacturing and Robotic Technologies fabrication factory and the Center for Architecture and Situated Technologies.

Daiki Kanaoka is the CTO and Fabrication Engineer at FabCafe Tokyo, where he applies a wide range of manufacturing knowledge to collaborative projects with partners around the world.

Hideaki Asaoka is a designer and fabrication master at FabCafe Hida with a deep knowledge of Hida forests, history, and carpentry.

Kotaro Iwaoka is the CEO at Hidakuma, founder and director of FabCafe, and lecturer at the Arts Information Center of the Tokyo National University of Fine Arts and Music.

Composite Mies

Nick Safley
Kent State University CAED

Image Joints and the Joints They Image

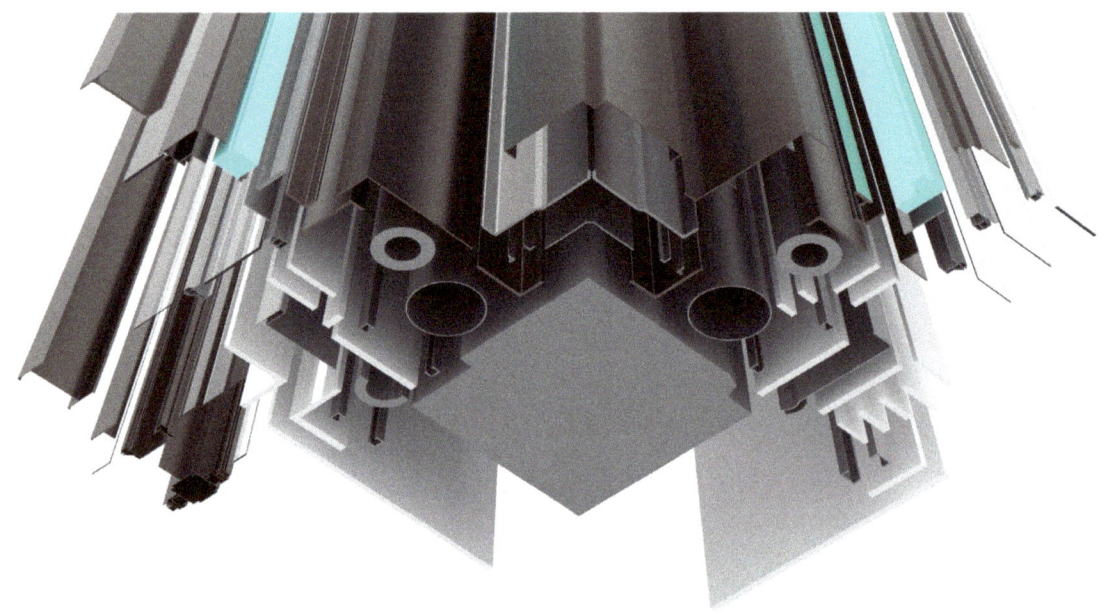

1 Composite Miesian corner detail

IMAGE JOINTS

Architectural detail drawings existed historically as a clear connection between representation and construction. They formed isolated spaces for discussion of intent and control around building. Unlike other historical modes of working upon details, such as tonal renderings, technical linework drawings, idealized material artifacts using geometric measurements, and a reductive color pallet. Pragmatically, linework detail drawings allowed construction data to be compressed for easy transport and distribution to those building and shaping materials from those designing and forming geometry. Today information for construction is less in need of compression as computers continue to gain processing power (Carpo 2017). This condition leaves the detail drawing in a strange position as a disciplinary tool without a pragmatic necessity, ill-equipped to address contemporary discourse, and in danger of atrophy. Current image discourse sidesteps explicit discussions of representation, and in this discussion technical details lose their agency due to their focus on geometric abstraction and communication (Linder 2012). Computationally, vector-based graphics have allowed greater drawing precision and small file size, but this reduction has also distanced the detail from image discourse and cultural relevance as a speculative tool. The detail has become a "how" drawing as opposed to a "why" drawing. This project addresses this condition by operating upon technical linework details, such as those of Mies Van de Rohe, in raster-based software. Once overlaid with pixels these technical drawings can be manipulated with raster-based tools and techniques opening the underlying initial detail to new processes and tools.

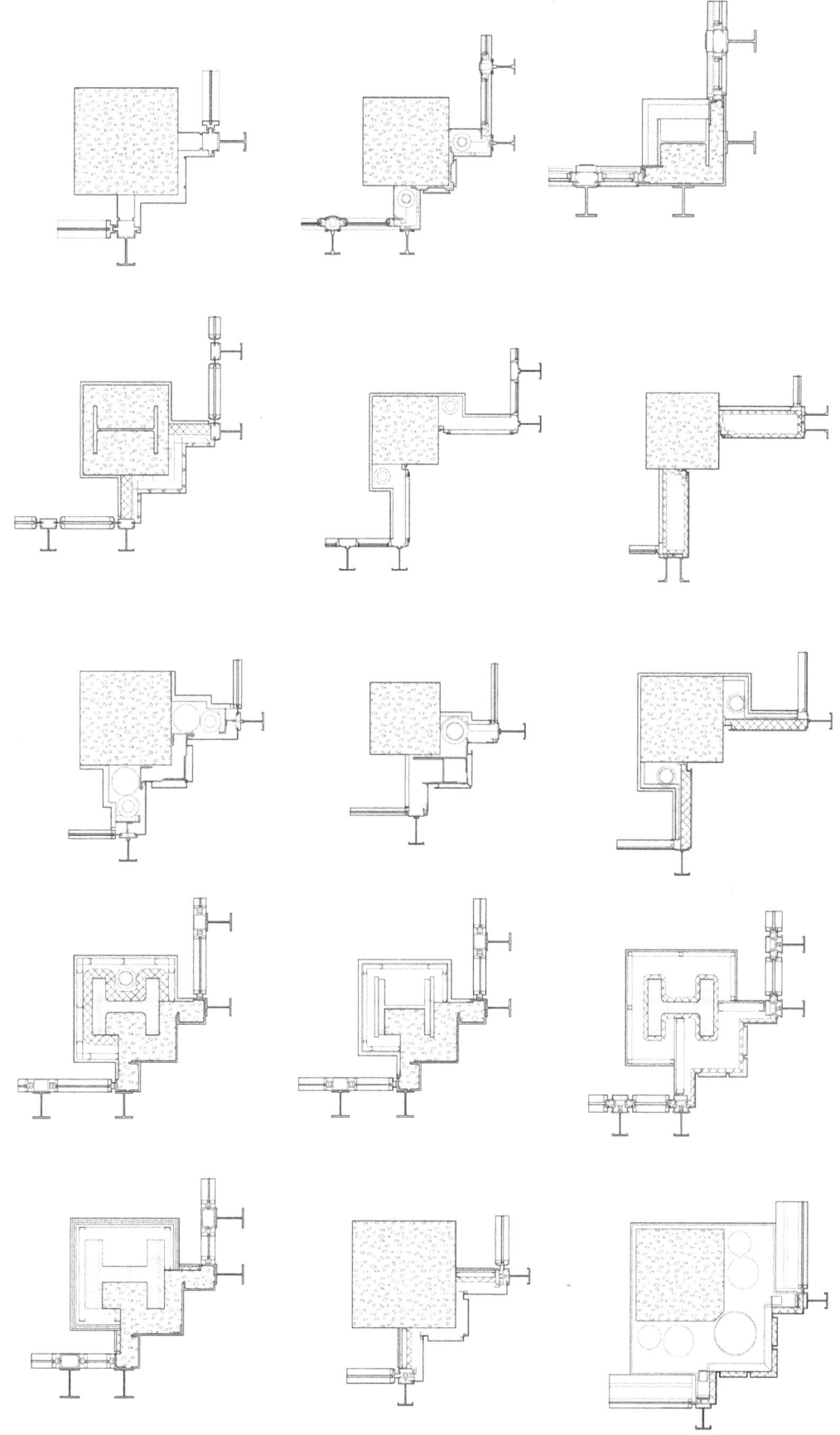

2 Mies van der Rohe corner details, redrawn by seminar students as adapted from John Winter (Winter 1972)

3   Twelve Boston physicians and their composite portait (Bowditch 1892)

Students executed this speculative project during the fall semester of 2021. The seminar course was held virtually and examined material details and digital methods of connecting. Students received a series of corner details from Mies van der Rohe as JPEG files with a consistent pixel ratio and resolution. These details were initially drawn by John Winter and published in the February 1972 issue of *The Architectural Review* and downloaded from the architecture blog Socks Studio (Winter 1972; Lucarelli 2018). The initial details express Mies's working through an outside corner condition in different construction materials, in specific climates, and under different building code contexts. Each student curated a selection of 4-5 details from the collection for a particular construction element or architectural character (Figure 2).

While the initial details were constructed as ink on mylar drawings, the digital file format afforded a platform to reconceptualize the construction detail using image-based techniques. Images of the details were input into Photoshop and composited using the photomerge tool, native to the software. This tool evaluates the input images for color and tone continuity by evaluating the pixels of each image and creating seamless collages of the input images. Unforeseen connections between the initial details and illogical constructions between materials emerge due to pixel-based values in each image being compared to those in other images. Interestingly contrasting with Mies's perspectival collages, which used the more traditional methods of juxtaposing, this process is only possible with recent image processing software (Vassallo and Herreros 2016). Once the new seamless collages were compiled, the raster image of the composite detail was translated into vector drawings using Rhino. Once translated back into vector-based geometry, the details were extruded to a specific height and given photorealistic VRay rendering materials (Figure 5-7). The new details contain steel and aluminum profile extrusions, glazing areas and frames, concrete and steel structural elements, and gypsum board with supporting light framing. Outside of the initial material construction logic of the details, the new material assembly follows an image-based composite of pixel values that create seams as surface joints between the images. Once extruded, the seams manifest as three dimensional joints whose construction logic is derived from the composite image. These joints form nonidealized construction assemblies, often duplicating elements, such as I-beam profiles, and create novel corner conditions in contrast to their source material.

Drawing upon the history of composite photography outlined by Mark Linder's analysis of Francis Galton, the project uses

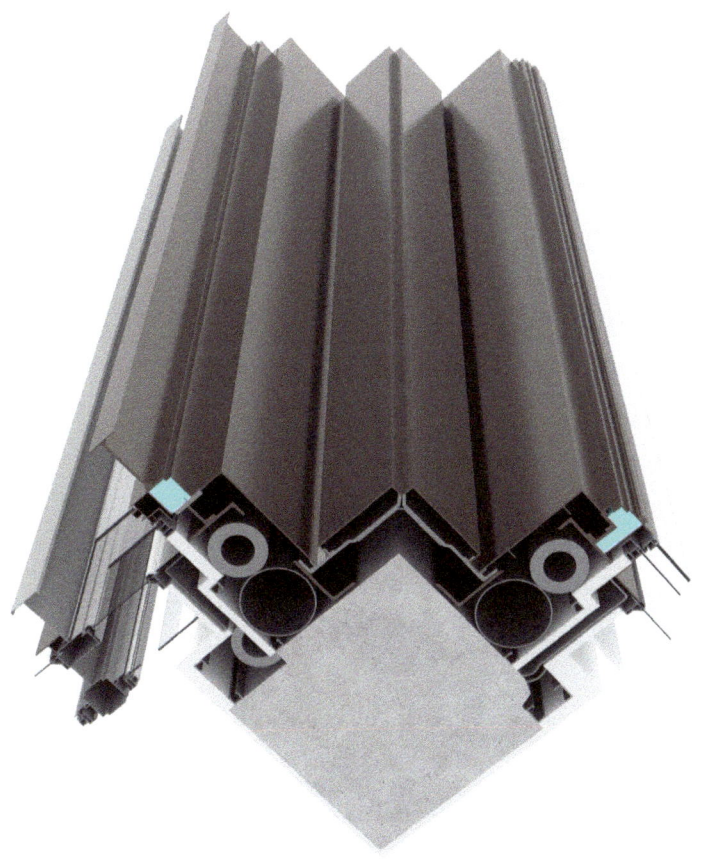

4   Composite corner detail created by photomerging existing details. The detail contains construction information from the initial disciplinary drawings but has been connected to separate corners using rastered digital information and the photomerge tool image to create a new image.

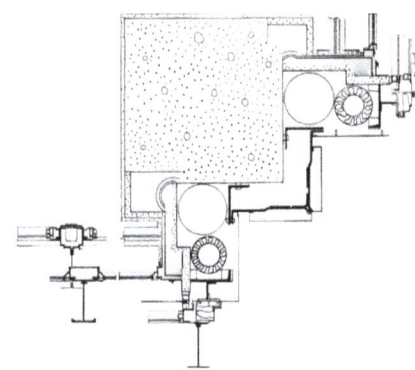

5   Composite image of Mies corner details combined automatically using the Photoshop's photomerge tool

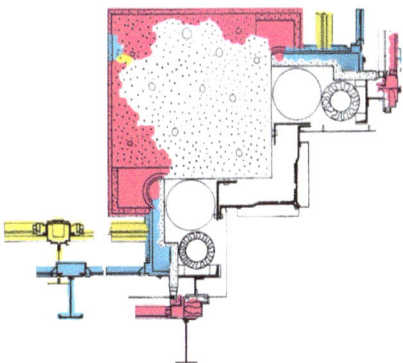

6   Various individual images color coded to show their extents and how they adjoin images, which are adjacent

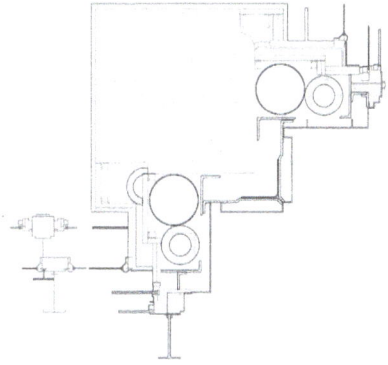

7   A vector-based technical drawing was traced from the composite raster details image to allow extrusion

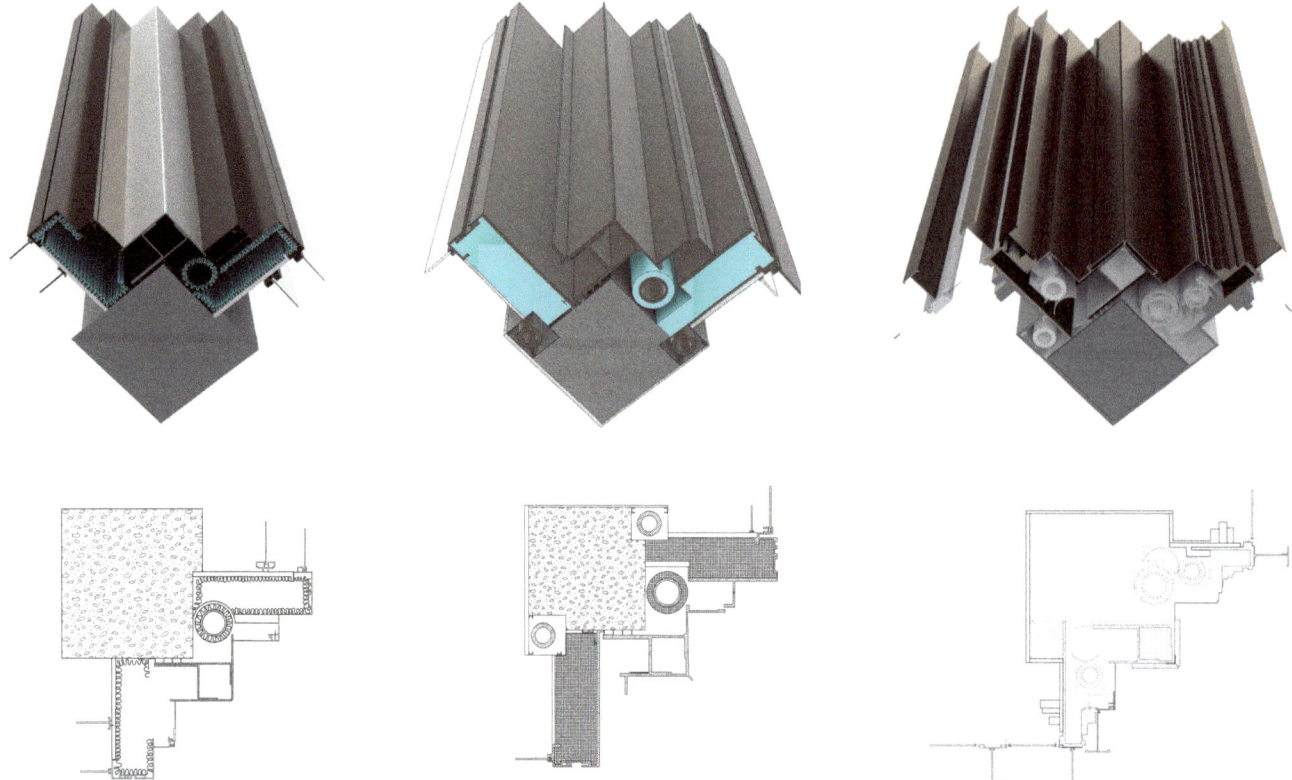

8 Student composite details created from photomerged Miesian corners

the photomerge tool as a technical device operating upon images to speculate upon an architectural material organization (Figure 3). Importantly, this work does not seek beauty by averaging detail traits or the moralization of a generalized type of detail, like Francis Galton (Linder 2019). Instead, the new technical joints or seams between images privilege the image as a digital material composed of data in place of the physical material controlled by the initial linework detail drawings.

Artist Rachel Harrison's *Voyage of the Beagle* from 2007 is another precedent that creates composite images and locates the project within contemporary visual culture. Harrison uses portraits of human and non-human faces, or face-like compositions, placed next to one another in a gallery setting to solicit connections from the viewing subject. This work forces the viewer to create a connection between each successive image. Once each image in the show has been viewed, each visitor has a unique composite portrait held in their minds without technical apparatus and, more importantly, without a statistical averaging effect. The detail composites formed in Photoshop for the seminar are explicitly technical, like Galton's. Still, they seek an impact similar to Harrison's, where the result of the process does not produce a singular artifact. Each of the seminar's students created a Miesian outside corner with a great degree of variability in outcomes and open interpretation of the material compositions created.

Drawing has been a corollary to methods of construction where lines contain information destined for material construction. In this relationship, a human author is required to translate expert knowledge, connecting material and representation. However, when considered as raster-based approximations of the drawing's geometry, the agency of the detail moves away from processes native to the human and opens the detail to manipulations not native to the material world but informed by its technical image. This project engages a post-digital outlook on our current cultural condition. It places the digital and material worlds in a non-hierarchical relationship that strives to reinvigorate the architectural detail with agency and speculative power through its image.

ACKNOWLEDGMENTS

Thanks to Professor Mark Linder for his willingness to discuss my work publicly in the fall of 2020 as part of the Kent State CAED Faculty and Friends Lecture series. Without his inspiration and teaching this project would not have been conceived. The work presented here was completed in a seminar during spring of 2021 at Kent State University. The proposed details pictured above were created with the generous talent and effort of Oriel Behboudnia, Bruce Wolf, Ryan Cashman, Zook Crain, Shelby Dolan, Amanda Harrer, Devanshi Jariwala, Summer Nairn, Caleb O'Bryon, Zach Petrus, Haley Scott, Kristyn Svetlak, and Yuting Chang.

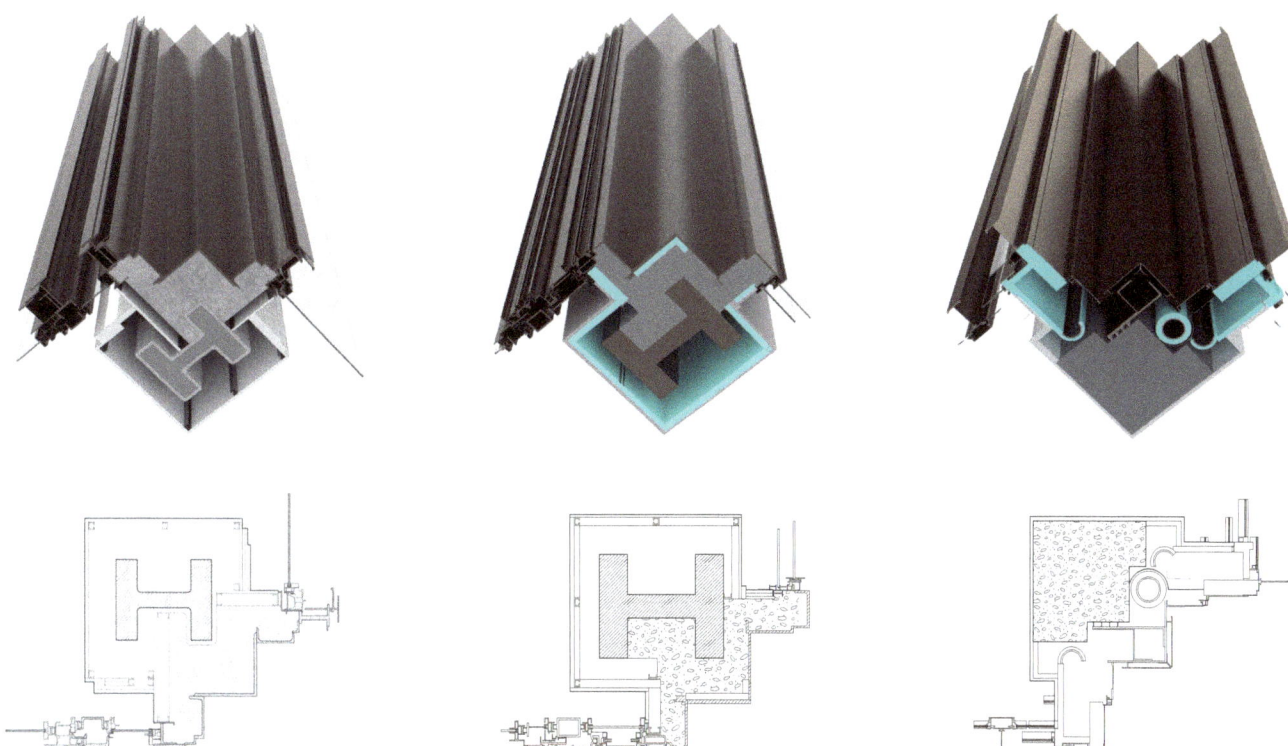

REFERENCES

Bowditch, H. P. 1892. "Twelve Boston physicians and their composite portrait." Harvard Medical Library, *OnView: Digital Collections & Exhibits*. Accessed November 11, 2022. https://collections.countway.harvard.edu/onview/items/show/6212.

Carpo, Mario. 2017. "The Second Digital Turn: Data-Compression Technologies We Don't Need Anymore." In *The Second Digital Turn: Design Beyond Intelligence*. Cambridge, Mass.: The MIT Press. 9–19.

Linder, Mark. 2012. "Images and Other Stuff." *Journal of Architectural Education* 66 (1): 3–8.

Lucarelli, Fosco. 2014. "Corner Solutions of Mies Van Der Rohe's Towers (John Winter, 1972)." *SOCKS Studio* (blog). Accessed February 9, 2018. https://socks-studio.com/2014/10/07/corner-solutions-of-mies-van-der-rohes-towers-john-winter-1972/.

Linder, Mark. 2019. "Episodes in the Emergence of Imaging Practices." In *Instabilities and Potentialities: Notes on the Nature of Knowledge in Digital Architecture*, edited by Chandler Ahrens and Aaron Sprecher. New York, NY: Routledge. 17–32.

Vassallo, Jesús, and Juan Herreros. 2016. "Seamless: Digital Collage and Dirty Realism in Contemporary Architecture." In *Seamless: Digital Collage and Dirty Realism in Contemporary Architecture*. Zürich: Park Books. 175–78.

Winter, John. 1972. "The Measure of Mies." *The Architectural Review* 900 (Feb): .

IMAGE CREDITS

Figures 2, 8: Kent State Seminar Students, 2021

Figure 3: H. P. Bowditch, 1892, Harvard Countway Library of Medicine

All other drawings and images by the author.

Nick Safley is a designer whose work focuses on architectural character, materiality, and fabrication. He teaches courses on construction and coordinates Kent State University's undergraduate Integrated Design Studio. Safley received a BArch from the University of Oklahoma and a MArch from the University of Michigan.

Experimentations in Neuroscience for Architecture

An Interdisciplinary Approach for Designing Healthier Environments

Kristine Mun
Academy of Neuroscience for Architecture

Biayna Bogosian
Florida International University

1   A work-in-progress immersive installation focused on the relationship between stress levels and green spaces in dense urban environments

Creating an intimate and well-tempered (Banham 1984) relationship of bodies in space is an endeavor of an architect, as our bodies and the environment are intricately intertwined (Figure 2). Literature in social sciences has revealed that many people living in densely populated urban contexts suffer from depression, loneliness, and other stresses brought on by the natural and built environments. Furthermore, recent studies in neuroscience have shown that living with such stressors can "alter our neural processing of acute social stress" (Lederbogen et al. 2011). As curators of socio-spatial dynamics, architects understand that the built environment affects the occupants by its form, structure, material, and composition of spaces. The building's performative qualities, such as atmospheric lighting and ephemeral qualities of materials, are the deliberations of architects and urban designers.

Integrating computational methods in design and construction has intensified the discipline's focus on quantifying buildings and occupants. However, when quantifying the human body, the metrics are often related to the physical dimensions (proportions, loads, heat, cold) and not psychological or psychophysical (stress, perception, emotion). This Field Note builds on the authors' experience and research focused on a 'Neuroscience for Architecture' approach to designing healthier environments.

Neuroscience for Architecture

Neuroscience, including cognitive behavioral science and computational neuroscience, addresses fundamental questions about human emotions, memory, intelligence, and so forth to understand how we interface with external environments. Although the study of the brain

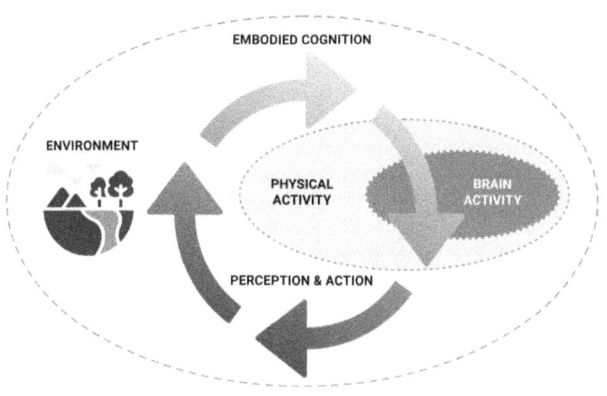

2   A model for Embodied Cognition

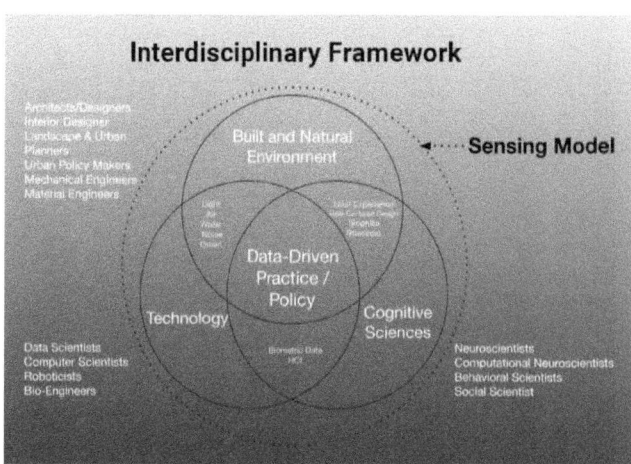

3   Diagram of authors' Interdisciplinary Framework for Neuroscience for Architecture Research

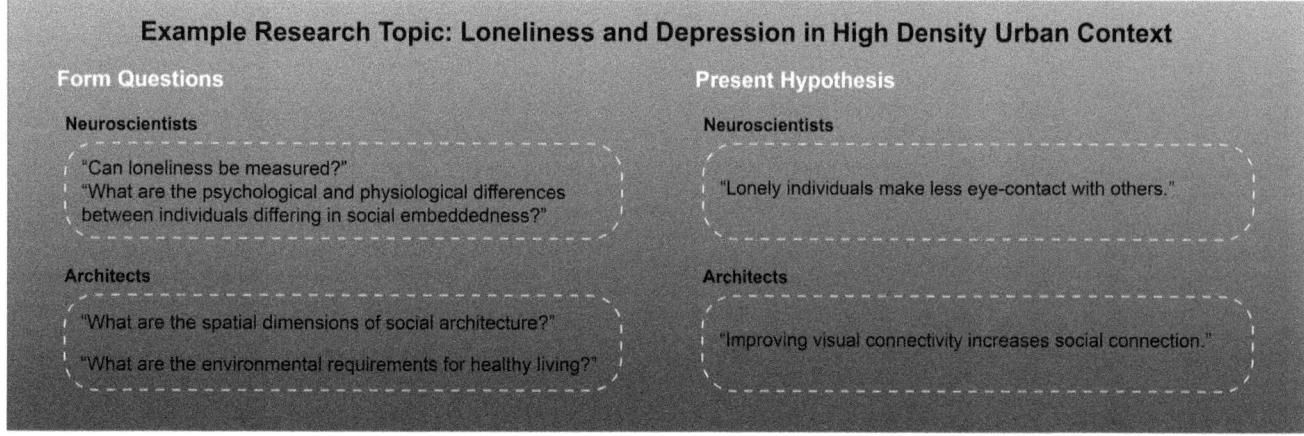

4   Examples of framing interdisciplinary questions and arriving at possible hypotheses

dates back before Vitruvius's time, the field of neuroscience as an explicit discipline was introduced in 1966 at Harvard Medical School (DeFelipe et al. 1992).

More recently, in 2003, the AIA San Diego chapter established the Academy of Neuroscience for Architecture (ANFA) after architect John Eberhard, and neuroscientist Fred Gage began their collaboration at the Salk Institute. Over the years, the interdisciplinary network of academics and professionals exploring Neuroscience for Architecture (NfA) has grown rapidly and internationally (Robinson et al. 2015). For example, neuroscientist Satchin Panda and architect Fred Marks examine circadian rhythm and the impact of interior light on our body's performance (Panda et al. 2015).

Additionally, with advancements in computer science, neuroscience is regarded as an important field for inspiring and validating Artificial Intelligence research (Seth 2021). This momentum, along with developments in neurotechnology, has made neuroscience the new frontier of the 21st century (Figure 3).

Methodology and Tools

NfA research adapts scientific methodology for collecting and analyzing data about the brain and behavior (Figure 4). A successful interdisciplinary collaboration aims to (1) understand the technical terminologies of each discipline, for instance, the term "space" is defined differently (Gepshtein 2001); (2) conjointly develop the scope of the problem; and (3) agree on tools and techniques of experimentation.

To apply a NfA approach aiming for healthier environments, we identified and tested a series of environmental parameters, such as artificial light and noise, related to form, color, scale, and navigation in various urban settings. Experiments were conducted using Virtual Reality (VR) and Augmented Reality (AR) head-mounted displays coupled with biometric sensors, as well as traditional quantification methods like surveys. Collected data were analyzed using statistical methods to determine trends and identify outliers.

The experiments we devise have largely depended on the tools and methods that are readily available. For example, machines

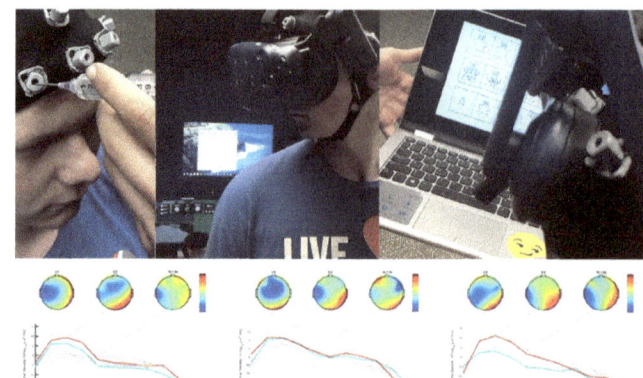

5 On-site experiments with mobile multi-modal biometric setup

6 (top) 24nodal EEG Cap, eye-gaze tracking, and HRV in VR measuring arousal and attention; (bottom) EEG patterns displaying frequencies correlated with time intervals of viewing

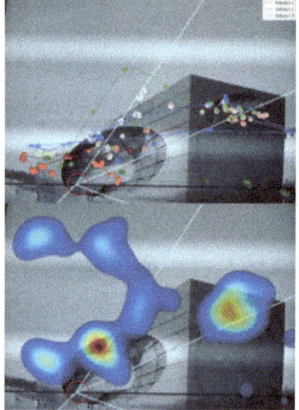
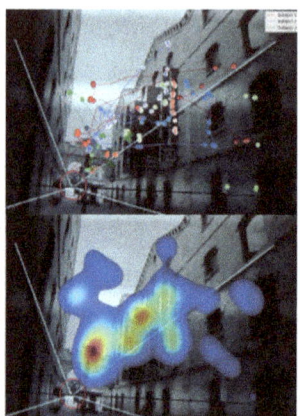
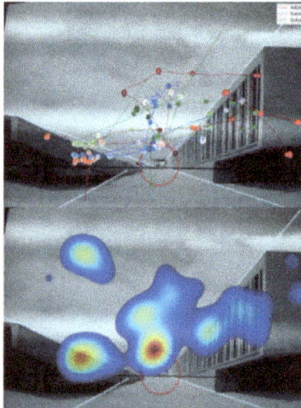

7 Eye-gaze tracking of saccades (top) and duration (bottom) to analyze field of view in various urban settings

such as fMRI, EEG, and EKG monitoring devices in neuroscience research laboratories are incredibly costly. Still, more affordable multi-modal mobile devices allow for replicability by architects who might not have access to neuroscientific labs (Figures 5-8).

Conclusion and Future Directions

Advancements in ubiquitous computing and the proliferation of the Internet of Things (IoT) are turning our built environment into a live information platform. Additionally, the state-of-the-art biometric sensing technology has the potential to correlate human bodies to sensor-integrated environments further. The prospect of the NfA is immense but comes with significant challenges, such as finding collaborators, securing funding, and conducting science-based experiments. However, the necessity to humanize our built world in various environmental crises requires architects and city planners to take on this challenge. If we begin to ask the right design questions, engage in interdisciplinary research, and create data-driven spaces based on wellness parameters, the NfA approach can provide critical data for a healthier and more inclusive future.

ACKNOWLEDGMENTS

This research was supported by the SOM Foundation's Humanizing High Density Research Prize as well as Prof. Eduardo Macagno's N-LEAD workshop at UCSD. A complete list of credits and research is available at the SOM Foundation website, https://somfoundation.com/fellow/bogosian-mun/.

REFERENCES

Banham, Reyner. 1984. *Architecture of the Well-tempered Environment*. Chicago: University of Chicago Press.

Box-Steffensmeier, Janet M., Jean Burgess, Maurizio Corbetta, Kate Crawford, Esther Duflo, Laurel Fogarty, Alison Gopnik et al. 2022. "The future of human behavior research." *Nature Human Behaviour* 6 (1): 15–24.

DeFelipe, Javier, and Edward G. Jones. 1992. "Santiago Ramón y Cajal and methods in neurohistology." *Trends in Neuroscience* 15 (7): 237–246.

Gepshtein, Sergei. 2001. "The perceptual organization of visual space-time," PhD dissertation, University of Virginia Charlottesville.

Hinton, Andrew. 2014. *Understanding Context: Environment, Language, and Information Architecture*. Sebastopol, CA: O'Reilly Media.

8   (left) On-site experiments with mobile multi-modal biometric setup; (right) Sensing cognitive attention of street luminance in VR using 360 capture; Experimenting with object detection and visual saliency

Lederbogen, F., P. Kirsch, L. Haddad, F. Streit, H. Tost, P. Schuch, S. Wüst, J. C. Pruessner, M. Rietschel, M. Deuschle, and A. Meyer-Lindenberg. 2011. "City living and urban upbringing affect neural social stress processing in humans." *Nature* 474 (7352): 498–501.

McCullough, Malcolm. 2013. *Ambient Commons: Attention in the Age of Embodied Information*. Cambridge, MA: The MIT Press.

Panda, Satchidananda, and Frederick Marks. 2015. "Circadian Lighting for Health." *ASHRAE Journal* 57 (11): 84–87.

Robinson, Sarah, and Juhani Pallasmaa, eds. 2015. *Mind in Architecture: Neuroscience, Embodiment, and the Future of Design*. Cambridge, MA: The MIT Press.

Seth, Anil K. 2021. *Being You: A New Science of Consciousness*. New York: Penguin Random House.

IMAGE CREDITS

Figure 2: Image adapted from Hinton (Hinton 2014, Fig. 4-4).
Figures 5, 6, 7: Courtesy of Prof. Eduardo Macagno and Dr. Kristine Mun's N-LEAD workshop at UCSD.

All other drawings and images by the authors.

Dr. Kristine Mun is an architect, experimental designer, professor, and the Director of SensoryArchitectures. She earned her PhD from the Architectural Association and a MArch from Cranbrook Academy of Art. Her interest lies in how machines and bodies form a/synthesis in our environment towards an "Empathic Architecture-Architecture that Feels." She has taught at USC, AADRL, University of Minnesota, University of Brighton, and Pratt Institute. She was formerly the Head of Neuroscience for Architecture Program at the NewSchool of Architecture & Design. She currently practices in Los Angeles and is an Advisory Council member of the Academy of Neuroscience for Architecture.

Biayna Bogosian is an interdisciplinary researcher and educator focused on innovation in design and technology within a broader environmental context. She explores data-driven and participatory approaches to improve the built environment. Biayna is an Assistant Professor at Florida International University, Director of FIU's Immersive Media for Environmental Explorations Lab, and Associate Program Director of FIU's Robotics and Digital Fabrication Lab, where her research is supported by several NSF and NASA grants. In addition, Biayna is a PhD candidate in Media Arts and Practice at the University of Southern California and holds a MS in Advanced Architectural Design from Columbia University. Since 2021, Biayna has been on ACADIA's Board of Directors and co-chaired the ACADIA 2021 Conference, "Realignments: Toward Critical Computation."

Field Guide to Meta-Architecture

Cameron Nelson

Lessons for Architects in the Era of Web3

1   Screenshot from a crit in SL, site of Archi21, a EU-funded project to study experimental design education and language

Architecture, as a discipline, is enjoying a front row seat to the cosmological big bang of the Metaverse. Countless designers have flooded into a new forum for 3D worldbuilding, with a significant portion of new content attributable to moonlighting architects. This trend is partly catalyzed by accessible game engine technology, increased emphasis on technical literacy in the field, and the rise of remote work and the gig economy as alternatives or supplements to traditional career paths. Educators, practitioners, students, and amateurs curious about this space should be aware of earlier "virtual worlds" like Second Life in order to contextualize the aesthetic and social contributions of the "Metaverse's" latest incarnation. This article also calls upon architects to help imagine a more digital-native architecture of the Metaverse, beyond mere imitation of the real.

A SECOND LIFE FOR SECOND LIFE

Architecture is enjoying a front row seat to the cosmological big bang of the Metaverse, whatever that is. In 2022, designers of virtual worlds, regardless of background, are dubbed "architects," and their ranks are replete with many real-life architects too, from undergraduates with a side-hustle to giants like Bjarke Ingels Group and Zaha Hadid Architects (Schulman 2022). Everyone is getting in on the hype, some bullish, some skeptical. But we've been here before.

Rewind to June 23, 2003, when Linden Lab launched a 3D virtual world called Second Life. Many in Architecture saw its potential. College studios were taught, and entire virtual design firms were founded within this virtual world (Newitz 2006; Wong 2007). However, despite unprecedented design freedom, architecture in SL averaged out to a bland facsimile of suburbia, endless hedges of low-poly, cookie-cutter mansions.[1] Most architects lost interest and the early Metaverse was relegated to obscurity.

Fast forward nineteen years: the Metaverse is given a new lease on life thanks to Facebook's dramatic rebranding and the furor over cryptocurrencies and Web 3.0. If the Metaverse is now experiencing a renaissance, it is with an emphasis on the prefix "re-": a secondary iteration, an offspring, a mutation. What is different, this time around, for architecture?

A NOMAD IN THE METAVERSE

This winter, picture me: not long out of graduate school, living out of my car and semi-aimlessly wandering the southwest, picking up remote gig work where I can as Covid wreaks havoc across west Texas, where I am laying low. An advertisement leads me to Mona, a Metaverse platform that has launched a "build-a-thon" virtual world competition (Dezeen 2022). Nearly broke, but equipped with a laptop and a degree in computational design, I toss in my hat. Between a growing trend of digital nomadism, the pull of the gig economy, and the all-encroaching tide of crypto, I suspect fragments of this story might resonate broadly in our post-Covid culture.

I built my world while camped in coffee shops or libraries, and when these establishments closed for the night, my studio would move to the front seat of my car, my laptop propped on some luggage, screen set to the lowest brightness to make the most of its battery life. I extensively used free or open source tools like Blender and Unity. Functionally, Mona is a pipeline from Unity to the Metaverse, simple enough that even non-technical architecture students can bring their existing designs into the Metaverse in an afternoon, without paying a dime. Whereas it once required expensive plugins like Enscape or a good amount of scripting to experience your CAD designs immersively, that is no longer the case.

2   Ana-Kata in Unity: rendering virtual environments within the game engine blurs the line between representation and experience

3   User treepledreamers explores Ana-Kata during a virtual open-house

Between the low barrier to entry, overlapping skill sets, and the economic and physical pressure to move to remote and/or gig work, it should come as no surprise that the Mona Discord channel, and numerous other Metaverse forums that I scoped out as an accidental ethnographer, was heavily populated with young architects. Which also perhaps explains why so much of the Metaverse today looks like a Google image search result for "parametricism" or a speculative rendering one might see left tacked up on the walls of a college studio. It's a step up from copy-pasting suburbia, but still highly derivative of prevailing aesthetic trends coming from an academic setting.

COPY-PASTE AESTHETICS

This raises the question, why do Metaverse buildings look like buildings at all? When limits of the body, senses, gravity, or structural stability are all thrown out the window, why do we still have stairs, doors, elevators, and the like? More than merely formal and aesthetic, the physio-virtual parallels extend to program as well. Many of the virtual spaces I have witnessed are proxies for real-world typologies: galleries and museums, nightclubs, concert venues, a classroom, a town square. There are at least two possibilities: one, as Patrick Schumacher suggests, is because "if you're flying around in some abstract geometry universe, you wouldn't understand where you are and what's going on," not without familiar structures to guide you (Schulman 2022). Another, is that our creativity is lagging behind technology, and gradually these familiar forms will melt away into new, improbable forms we simply haven't imagined yet.

A few weeks later my submission "Ana-Kata," itself intended to embody a kind of transit hub in an as-yet disembodied Metaverse still glued together by urls and hyperlinks, garnered a design award and a healthy sum of Ethereum. I begin to understand why so many students have been taking up Metaverse architecture as a secondary source of income.[2] The fact that the judging panel included representatives from *Dezeen* indicates the intention of Metaverse architecture to be taken seriously as a player in art and design. Metaverse platforms are actively wooing the discipline.

PLAYTESTING ARCHITECTURE

Ana-Kata is later featured by Mona in a virtual open-house, and I get to witness for the first time a group of people (read: avatars) inhabiting one of my architectural designs. Inhabiting, not looking at renderings, or leaning over models, but actually jumping and running and climbing on the walls. While I look around at strange characters in outlandish "skins," I notice a couple of avatars slowly clambering their way up the central spire. I had never considered that one would be able to climb this menacing, levitating spike at the center of the world, and yet here they were climbing it. Not five minutes spent in Ana-Kata and these two anonymous visitors had already discovered an unintentional use of my architecture. In the real world, for fear of bodily harm, most buildings would never be subjected to such rigorous playtesting, but in the Metaverse, iteration and experimentation is cheap. This, if nothing else, convinces me that we don't need familiar architectural elements to guide our behavior; rather, it is only a matter of time before unexpected behavior will guide us, as architects, to design for use-cases we might never have dreamt.[3]

NOTES

1. It is hard to say why SL replicated suburbia, but Tor Lindstrand, Professor at the Royal Institute of Technology, Stockholm, and head of virtual studio LOL Architects, likens it to MTV Cribs.

2. At SXSW Interactive 2022 Mona co-hosted the "NFT House," featuring talks by crypto-artists including an ambitious undergraduate student from Los Angeles who had rebranded himself as a metaverse architect, apparently to some commercial success. Stories like his are becoming increasingly common.

3. If it seems unlikely these new use-cases will feed back into real-life architecture, I caution that the feedback may be subtle and indirect but no less influential for all that. A fine example is an online marketplace like Amazon: virtual marketplaces may minimally affect the way we design for retail, at least at present, but the displacement of retail zones in favor of remote fulfillment warehouses is a very real, very spatial effect of behaviors enabled by virtual reality.

REFERENCES

Au, Wagner. 2022. "Metaverse: "The" Or "A", Capital Or Lower Case: A Naming Convention Proposal." *New World Notes* (blog). Accessed June 1, 2022. https://nwn.blogs.com/nwn/2022/05/metaverse-definition-naming-convention.html.

Dezeen. 2022. "Call For Entries To Mona's Renaissance Metaverse Competition." Accessed June 1, 2022. https://www.dezeen.com/2022/01/10/mona-renaissance-metaverse-competition-design-virtual-reality/.

Newitz, Annalee. 2006. "Your Second Life Is Ready." *Popular Science*. Accessed June 1, 2022. https://www.popsci.com/scitech/article/2006-09/your-second-life-ready/.

Ring, Heather, and Tor Lindstrand. 2007. "Architecture's Second Life." *Archinect*. Accessed June 1, 2022. https://archinect.com/features/article/47037/architecture-s-second-life.

Schulman, Pansy. 2022. "Inside The Metaverse: Architects See Opportunity In A Virtual World." *Architectural Record*. Accessed June 1, 2022. https://www.architecturalrecord.com/articles/15625-inside-the-metaverse-architects-see-opportunity-in-a-virtual-world.

Wong, Kenneth. 2007. "First-Hand Architecture In Second Life." *Cadalyst*. Accessed June 1, 2022. https://www.cadalyst.com/aec/first-hand-architecture-second-life-tech-trends-feature-3676.

Cameron Nelson holds a BA in Mathematics and Architecture from Yale University and an MS in Matter Design Computation from Cornell University. Their research touches on topics of programmable matter, simulation, and algorithms.

Session Introduction

Computational Methods for Structural Design

Jonas Coersmeier, Chair

The title of this session "Computational Methods for Structural Design" is composed of three rather flexible terms—computation, structure, and design—in combination with several possible meanings: it may refer for instance to either the Design of structures or the Structure of design. And if it refers to the design of structures, are we looking at Physical building structures, Urban structures, or Data-structures—block chain structures even? It was Cecil Balmond, here at Penn, who pointed out that Structures Design is not limited to the redirection of physical forces but may elaborate connective patterns of information of all kinds. The papers in this session address all of the above.

At the Institutions represented here—University of Pennsylvania, PennState, University of Stuttgart, and UT Austin—a networked understanding of structural analysis and design has been developed with ground-breaking achievements, even before the advent of the digital in architecture. We may (for example) think of the intricate physical models developed by Robert Le Ricolais here at Penn, as a means of removing mass in structures, or to find answers for his simple and brilliant question: Where to place the holes? Or we may think of the physical computation models developed by Frei Otto in Stuttgart and the pioneering Light Weight Structures they informed.

Today we get an insight into the state-of-the-art in structural design even beyond early digital, topological, and parametric models, as the authors involve cutting-edge Machine Learning and Agent-based systems. Perhaps we can call it the algorithmic project in structural design 3.0.

The order of presentation is loosely arranged according to the size of the structures addressed in them—from urban to textile fabrics. However, the papers all address multiple scales: the first paper specifically addresses part to whole relations or "part-relationships" between architecture and its city, from prefab building components to urban-form and policy.

Hao Zheng and Masoud Akbarzadeh introduce a web-based structural design tool that implements the workflow of Graphic Statics and machine learning of the dragonfly wing structures. In previous research, the authors developed a workflow to use Graphic Statics to analyze natural structure and machine learning models to generate the topology and geometry of a dragonfly wing structure. In their paper, in order to help all designers to access and benefit from their method, they developed and present a user friendly web-based tool.

Daniel Koehler's paper asks "What if one cannot rely on, or build on, [existing] data?" when exploring emergent characteristics in buildings, settlements, and the city. Koehler proposes a workflow to simulate entire data sets, and a case study that exemplifies how quickly one can generate and analyze a set of buildings at the resolution of BIM modeling. The work suggests that synthetic data sets could become a feature of daily design workflows, promoting new kinds of typological thinking.

Mathias Maierhofer and Achim Menges write that "significant material savings can be achieved by providing load-bearing structures with the ability to actively...adapt to forces and deformations." They describe an agent-based design environment, previously developed to explore adaptable truss topologies beyond established structural typologies, and aim to develop strategies to negotiate design intent with the adaptability objectives of such agent system. Their paper addresses these challenges through the example of elementary objectives common in early architectural design.

The work of Farzaneh Oghazian, Sam Moradzadeh, and Felecia Davis shows the potential of different Grasshopper plugins for simulating and analyzing knitted textile structures, elaborating the form-finding and analysis of such structures using Kangaroo2, K2Engineering, and Kiwi3D!

A Web-based Interactive Structural Pattern Generation Tool with Graphic Statics and Machine Learning of Dragonfly Wings

Hao Zheng
Masoud Akbarzadeh

Polyhedral Structures Laboratory, Stuart Weitzman School of Design, University of Pennsylvania

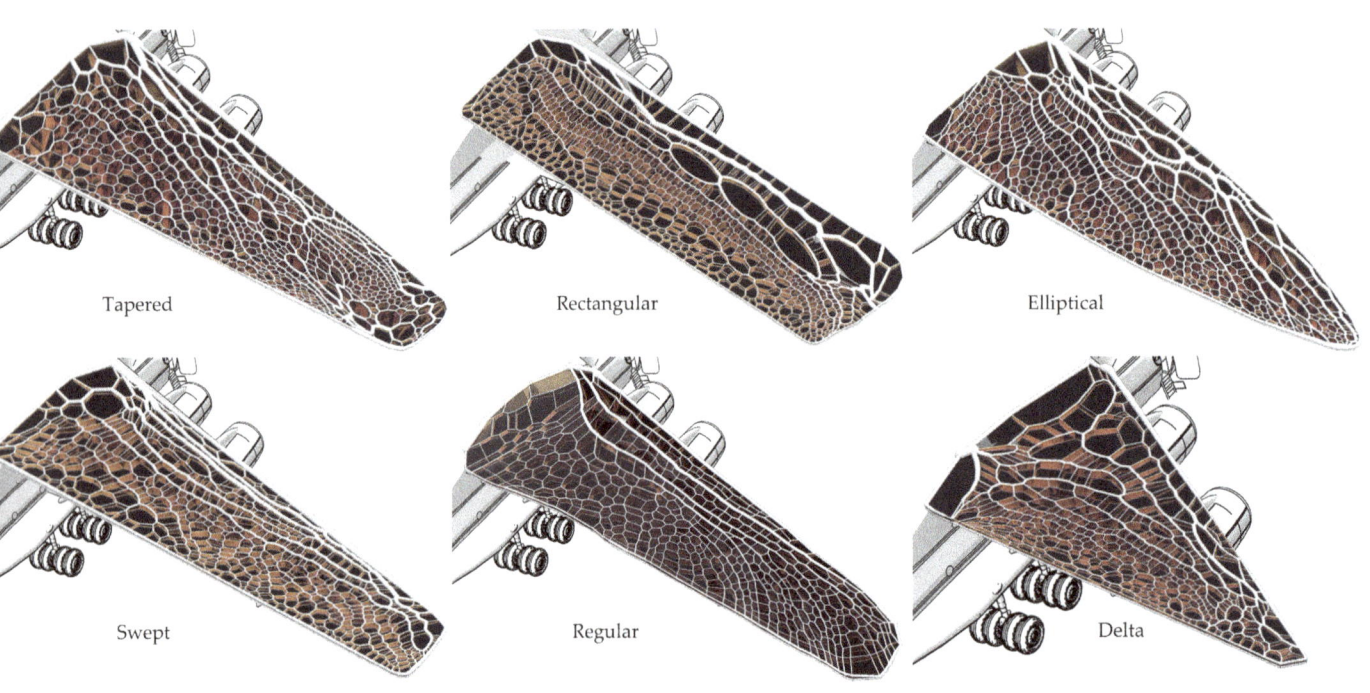

1 Airplane wing structures generated by our tool

ABSTRACT

Designers, engineers, and scientists have always drawn inspiration from nature. The dragonfly wing is among the many natural structures that have intrigued many researchers to study its geometry and performance as a bioinspired design. In previous research, we developed a workflow to use Graphic Statics to analyze the dragonfly wing structure and machine learning models to generate the topology and geometry of a dragonfly wing structure. However, the current workflow involves multiple geometric algorithms and the implementation of complex machine learning models, making it is difficult for designers to follow and use. Therefore, in this paper, we introduce a web-based tool that implements the workflow. It includes (1) the input control panel from the user to define the constraints of the structure; (2) the backend server that proceeds the generation of the structure with Graphic Statics and machine learning models; (3) the output control panel to allow the interaction of the result for frontend display; and (4) the file manager to store and restore the generated result. On the web page, designers can easily input the boundary and parameters of a wing/cantilever structure and generate a funicular structural form in our cloud server. Analytical results such as Minkowski Sum and FEM analysis are shown to the user. Finally, the user can export the STL model for other purposes such as aerodynamic analysis or digital fabrication.

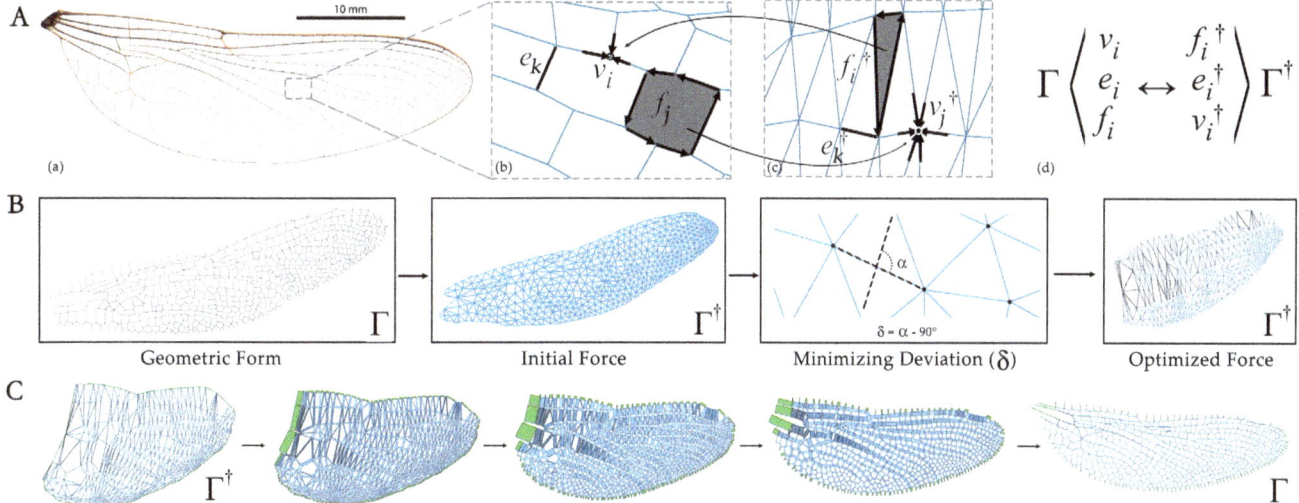

2   Previous research on the Graphic Statics of dragonfly wing structures: (A) The form and force diagrams; (B) The iterative solution of Graphic Statics; (C) The Minkowski Sum between the force diagram and the force diagram

INTRODUCTION

Graphic Statics on Natural Structures

The geometry of a structural system with axial-only internal loads can be related to its force equilibrium using two geometric diagrams as proposed by J. Clerk Maxwell (Maxwell 1864): (a) the form diagram that represents the form of the structure, the length of members, and the locations of supports and applied loads; and (b) the force diagram that consists of closed polygonal faces and shows the equilibrium of forces in each node of the structure. The numbers of the edges of the two diagrams are equal, and the magnitude of the force in each member of the structure is proportional to the lengths of the corresponding edge in the force diagram. This method is known as Graphic Statics (Akbari, Akbarzadeh, and Bolhassani 2019; Akbari et al. 2020; Akbari, Lu, and Akbarzadeh 2021; Akbari et al. 2022) and has been used previously to describe the force equilibrium of convex-only, natural networks such as a spider web (Whiteley et al. 2013). In the case of a dragonfly wing, the internal 2D pattern bounded by the boundary edges mainly consists of convex polygons. Thus, the method of graphic statics can be used to analyze the static equilibrium of forces in the system (Figure 2A).

The geometry of the force equilibrium for the network of the wing is found using iterative methods. The wing's network will be referred to as the form diagram from which the topology of its dual force diagram is extracted. Each polygon in the form diagram is reciprocal to a vertex in the force and vice versa. Subsequently, the edges of the force diagram are rotated iteratively to become normal to the edges of the form diagram. The difference measured from the right angle of 90° and the angle of the two corresponding edges is defined as the deviation. The iteration process minimizes the value of the deviation to derive a solution of the force diagram within the predefined tolerance (Figure 2B). The PolyFrame plugin (Nejur and Akbarzadeh 2021) was used for this iterative operation which is a free plugin for Rhinoceros software (Robert McNeel & Associates 2018)..

Machine Learning of Topology

Based on the above research, in the following phase, we developed two methods using Machine Learning models capable of generating the entire structural form of the wing from a user-input boundary with an intermediate product of the force diagram. The implemented machine learning techqnues inlcude Generative Adversarial Network (GAN) to generate patterns as images, and Artificial Neural Network (ANN) to predict structural properties as vectors. In the first method (Figure 3A), to build the training and testing dataset for the machine learning models, geometries in different stages are transformed into images. Inspired by the identification of the main veins by (Hoffmann et al. 2018), a similar method is developed to extract the main path of the force diagram of the dragonfly wing. Then, image-to-image machine learning models (Isola et al. 2017) are used to learn the mapping between each stage of the dragonfly wing data and generate the force diagram from the form boundary as an image. In addition, a vector-based machine learning model is trained to predict the edge lengths of the form diagram using a dataset of the edge lengths extracted from the dragonfly wing geometries. Therefore, the four machine learning models can predict all information needed to generate the structural form of the wing.

However, in the first method, the geometries of a force diagram need to be reconstructed from the image manually

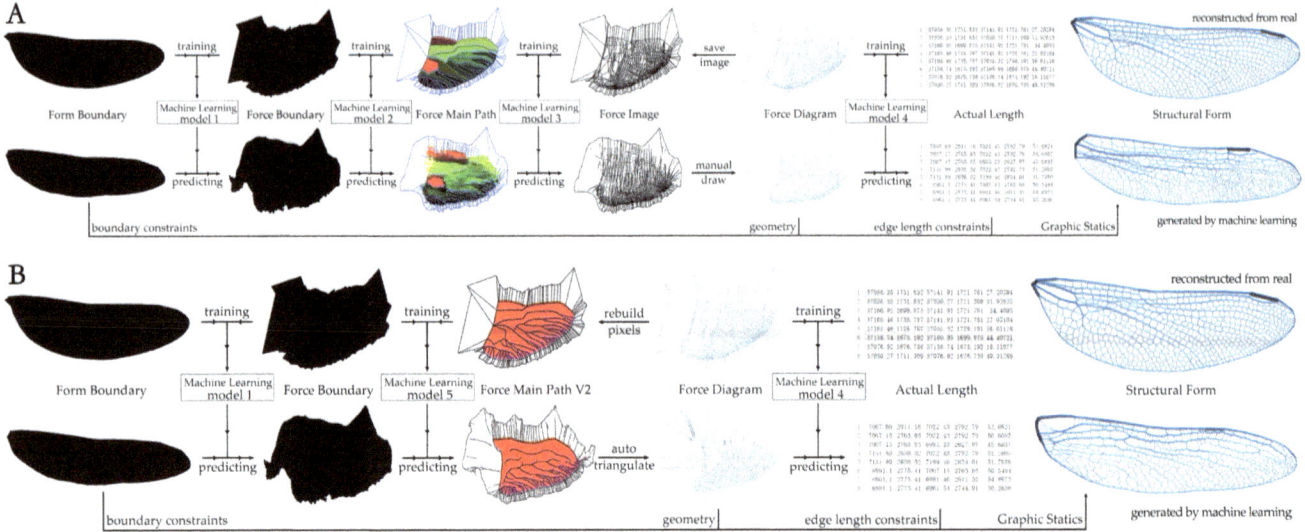

3 Previous research on the machine learning of the topology of the dragonfly wing structures: (A) Machine learning models with a manual process to draw the force diagram and generate the structural form; (B) Machine learning models with an automatic process to directly generate the structural form

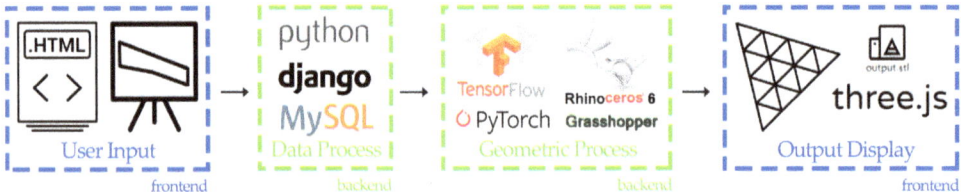

4 The proposed workflow of the web tool

to maintain precision. To automatically generate the force geometry, we develop the second method that represents the vertex information as the pixel values in the main path image and triangulates the regions in the main path according to the recognized vertexes (Figure 3B). Thus, the force geometry can be directly generated and proceeded into the structural form. The first method produces more visible information for a human to understand, while the second method generates abstract information for the machine to rebuild the geometry.

Problem Statement and Objectives

With the above previous research (Zheng, Hablicsek, and Akbarzadeh 2021; Zheng and Akbarzadeh 2022), we have successfully developed the workflow of generating funicular wing/cantilever structures with Graphic Statics and machine learning. However, technically speaking, implementing the workflow locally on a computer is particularly difficult for a design without much computational and programming knowledge.

Therefore, to better help designers access our method, we develop a web-based tool that accepts user inputs and feedback generated structures. The web tool is implemented as an online resource and open for designers to visit as a web page. Meanwhile, a local server proceeds the input data, generates the output structure, and sends the model file to be displayed on the web page. With this web tool, designers can easily adjust the input parameters and boundary conditions, and obtain the structure model online, without going through the complex local computing process.

To be specific, Figure 4 shows the proposed workflow of the web tool, which contains the frontend and backend. It follows four steps: 1) The user inputs the boundary and parameters in the web page, and HTML and Javascript (Eich 2020) transform the data from Canvas (Fulton 2011) into a formatted database; 2) The server stores and proceeds the data (Holovaty 2022; Widenius 2022), and updates an indicator file; 3) The geometric components (Robert McNeel & Associates 2018; Rutten 2020) and the machine learning components (Google 2019; Paszke 2018) in the server work together to generate the structural model as CSV file, and send it back to the web page; 4) the web page (Cabello 2022) displays the model, and allows the user to generate and download the STL file.

METHODOLOGY

User Input

First, the web page should receive the input boundary and parameters from the user. Figure 5 shows the icons, names, ranges, and buttons in the "Input Control Panel." The boundary is defined as a closed polygon of several vertexes. The user can select the number of vertexes and adjust the positions of each vertex by dragging them on the canvas. The boundary

Icon	Name and range	Icon	Name and range
	Input Boundary (canvas input)		Subdivision Density (0.01-1.00)
	Sharpness (0.00-1.00)		Length Constraint Multiplier (0.01-2.00)
	Boundary Constraint Multiplier (0.01-9.99)		Iterations (10000-30000)
	Total Length of the Wing (10-50000)		Machine Learning Model (selection)
	Default Settings (button)	Submit	Submit (button)
help	Instruction on how to use (button)	about	Developer information (button)

5 The control parameters and functional buttons in the user input panel and the compute panel

vertexes are transformed into numeric values based on their coordinates. Noted that, the most left curve in the boundary is the place where the structure is anchored. Therefore, additional marks are made as slash lines to represent the anchor.

Besides, the "Input Control Panel" also includes parameters that are defined as numeric values. A set of six parameters can be input from the user and sent to the server. To be specific, "Subdivision Density" defines the density of the structural members. By increasing it, more members will be generated with a longer time cost. "Sharpness" defines the upper and lower bounds of the length constraint for each edge. Increasing it will give more freedom to the geometric generation process in graphic statics. "Length Constraint Multiplier" defines the relaxation of the edge length constraints. Increasing it will cause more rectangular cells than circular cells. "Boundary Constraint Magnitude" defines the magnitude of the boundary constraint to the form. Increasing it will make the structure attach closer to the boundary. "Iterations" defines the number of iterations in the geometric generation process in graphic statics. Increasing it will generate a more accurate structure but with a longer time. "Total Length of the Wing" defines the size of the generated model in millimeter. Last, the user can select the machine learning model of different species, the default is set as the dragonfly wing model.

Noted that, the web page will restore the input parameters when the user successfully submitted the last time, thus the user can more easily adjust the parameters. There is a button "Default Settings," by clicking it, the input parameters will be set as the default values. If a new user does not understand the meaning of each parameter, he/she can move the mouse cursor to the button of each parameter to see its name and click the "help" or "about" button to see the detailed instructions.

Backend for Geometric Processes
When finishing adjusting the six input parameters, the user can click the "Submit" button in the "Compute" panel to send the first set of input parameters to the server. Figure 6 shows

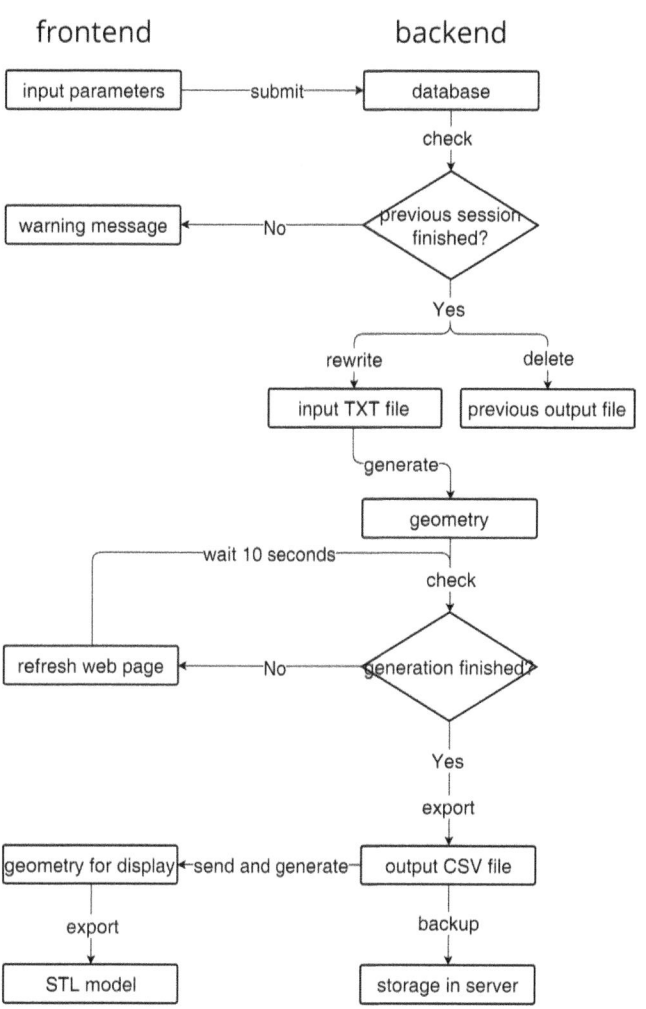

6 The workflow in the backend and the frontend of the server after the user submits the input parameters

the algorithm in the backend and the frontend of the server after the user submits the input parameters. First, in order to avoid duplicated submissions from multiple users before the server completely responds to the current request, an indicator value in the server is firstly loaded by the web page. If it indicates that a request is being proceeded, a warning message will be shown to the user when he/she clicks the "Submit" button, and no values will be submitted to the server. Besides, the entire generation process in the server usually takes three to seven minutes. The web page will be automatically refreshed every 10 seconds until it receives the result from the server.

When the server finishes the generation process, it will send back a CSV file to the web page, which contains the geometric information of the generated structure, the graph information for implementing the Minkowski Sum, and the numeric information of the FEM analysis. The geometric information contains the coordinates of the start and end points of each

Icon	Function	Icon	Function
	View the generated model		Remove the generated model from the scene
	View the Minkowski Sum		Remove the Minkowski Sum geometry from the scene
	Show the external forces		Hide the external forces
	Show color-coding		Hide color-coding
	Turn on the FEM result under the load of self-weight		Turn off the FEM result under the load of self-weight
	Turn on the FEM result under the point load on the farthest vertex		Turn off the FEM result under the point load on the farthest vertex
	Material of the Wing (selection)	FEM	Request the recomputation of the FEM results from the server

7 The functional buttons in the output control panel and the FEM control panel

Icon	Function	Icon	Function
	Export the current scene as an STL model	1	Restore the pre-generated sample 1 from the server
	Download the vector-based data file to save the result	2	Restore the pre-generated sample 2 from the server
	Upload the vector-based data file to restore the result	3	Restore the pre-generated sample 3 from the server

8 The functional buttons in the file manager panel

edge, as well as its corresponding force magnitude. The graph information stores the connectivity matrix of the form and the force diagrams. The FEM analysis result contains the deformation magnitude for each edge in the structure.

Frontend for Display

Next, when the web page receives the data file, the user can choose a different display mode in the "Output Control Panel" and "FEM Control Panel" (Figure 7) to turn on or off the generated model and the Minkowski Sum. Also, the user can turn on or off the FEM results under the load of self-weight or under the point load in the farthest vertex.

In the case of the normal display mode, the web page regenerates the structural members according to the information from the file and the second set of the user input parameters of the minimum radius and the maximum radius. An additional transparent box geometry is shown to indicate the anchor of the structure. The user can control the camera with the mouse in the main display window to better view the generated model. The generated 3D model is displayed on the web page with pre-set lighting environment. However, to reduce the computational load from the local device, the shadow is represented as a series of static geometries on the ground with gray lines. The color setting for the main geometry keeps constant with that in PolyFrame. In the display control panel, the user can also change to turn on or off the display of the external forces.

In the case of the Minkowski Sum mode, the web page reads the user input of the Minkowski Sum indicator (MSI) and calculates the corresponding status in the form-to-force transformation. The graph information in the feedback file contains the following items: (1) the coordinates of the vertexes in the form diagram; (2) the index of neighbor cells of each cell in the force diagram; (3) the index of the shared edges in the neighbor cells of each cell in the force diagram; (4) the index of edges in each cell; and (5) the coordinates of the start and end points of each edge in the force diagram.

By scaling the cells in the force diagram with the MSI value and moving them to each corresponding vertex in the form diagram, each edge in the force diagram will become an area with thickness. Therefore, with a gradually-changed MSI value from 0 to 1, the areas shift from the edges in the form diagram to the edges in the force diagram, thus showing the transformation between the form and force. Still, the user can turn on or off the external forces in the Minkowski Sum mode.

For the FEM analysis, Karamba (Karamba 2022) is used to calculate the deformation of edges based on the user input of the span and the material of the structure. Applicable materials include steel, wood, concrete, and aluminum, and the material property is embedded in Karamba. We provide two types of loading: (1) self-weight loading for all vertexes; and (2) point loading for the farthest vertex to the anchor with the magnitude of half of the self-weight. In the CSV data file, the FEM analysis part includes the following information for each edge: (1) coordinates (x, y, z) for the start and end points (deformation in z-axis included); and (2) color-coding value (R,G,B). The user can adjust the multiplier in the frontend to increase or decrease the deformation magnitude, and view the color-coded edges and color scales. When the user changes the material setting or the structural thickness, a recomputation request can be sent to the server and the FEM results will be updated in round ten seconds.

Server-free Data Files

In addition, to better help users restore the previously generated results, in the "File Manager" panel (Figure 8), users can download the data as a CSV file and save it on their local computer. If the user input an email address when submitting a request to the server, when the computation is finished, the server will send an email to the user with the CSV data file attached. By uploading the file to the web page, users can restore the input parameters, the output structure, the Minkowski Sum geometry, and the FEM analysis results. Restoring the previous result only requires the local computer from the user, thus it is an offline process and does not require a connection to the server. Also, the user can export and download the STL file for 3D printing or other purposes.

UI Design of the Web Page

Therefore, with this web page implementation (http://www.

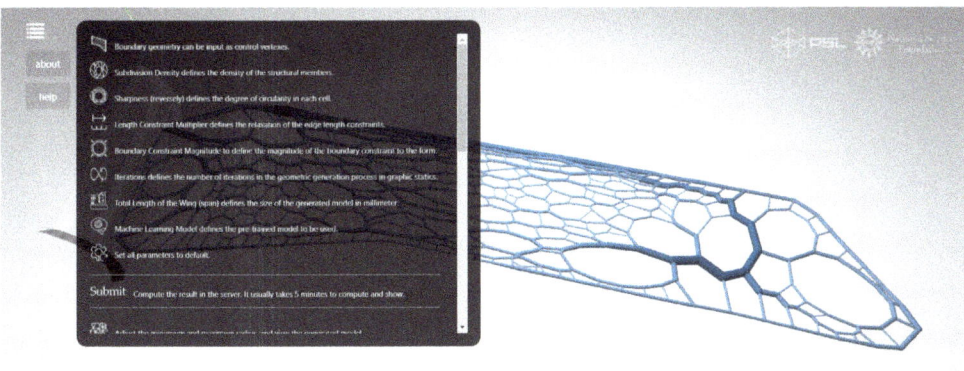

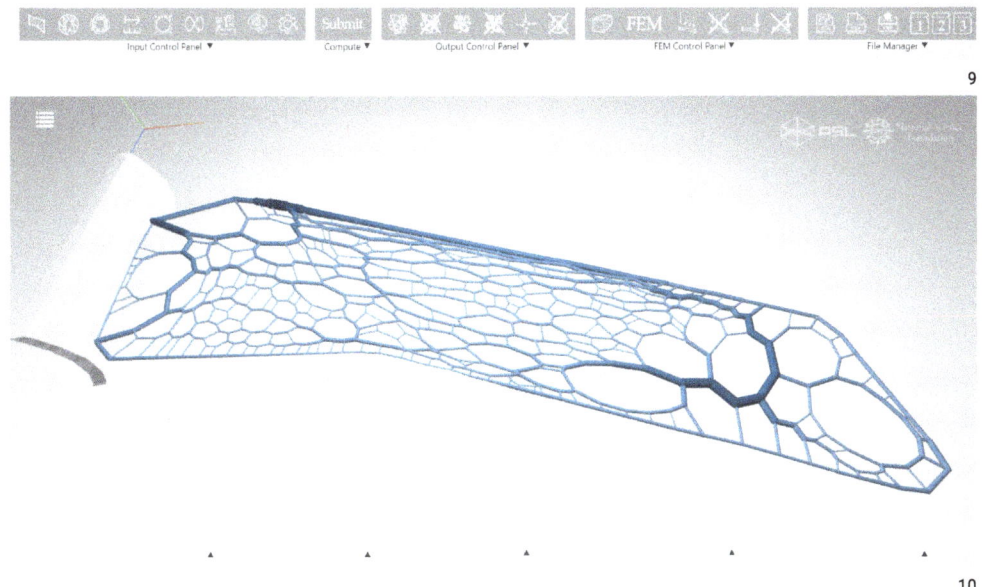

9 The implemented web page with panels unfolded
10 The implemented web page with panels folded

ai-gs.com/frontend/DFW-GH.html) (PSL 2022), users even without much knowledge of machine learning and graphic statics could easily generate lightweight and high-performance structures within given boundaries. Figure 9 shows the web page with control panels unfolded, and Figure 10 shows the web page with control panels folded. The user can freely decide to fold or unfold each panel to better balance the UI and the model display.

APPLICATIONS

In this section, several cases are generated and shown using our web tool. The user can either submit multiple requests to our backend server to generate structures with different input-related parameters or adjust the output-related parameters in the frontend to view and export results.

Form-finding with Various Boundaries

In the first case (Figure 11), structures with different input boundaries are generated. The user can select the number of control points in the input boundary, and drag the points to adjust their positions. Even if the user inputs an invalid boundary such as crossing curvatures (Figure 11 bottom right), our geometric script will merge it into a pixel-based black-and-white image, and send the image to machine learning models. In addition, the user is not required to input a wing-like boundary. As long as the anchor is on the left, any cantilever structure can be generated with the input boundary, which contains the features of dragonfly wings.

Precise Control of Generated Structures

The second case shows one example of controlling other input parameters such as the subdivision density (Figure 12). Among the input parameters, the subdivision density is the most important one since it directly controls the complexity of the generated structure. A smaller subdivision density can significantly simplify the structure while keeping the features of dragonfly wing patterns. In our recent research, we found that simplifying the structure to a certain degree would increase the structure performance, and make it easier to fabricate in the real world. Thus, the user can consider the fabrication ability and adjust the subdivision density. Other parameters such as sharpness and iterations can also greatly affect the generated result, but changing the input parameters requires a re-computation from our server and it usually takes around five minutes to respond.

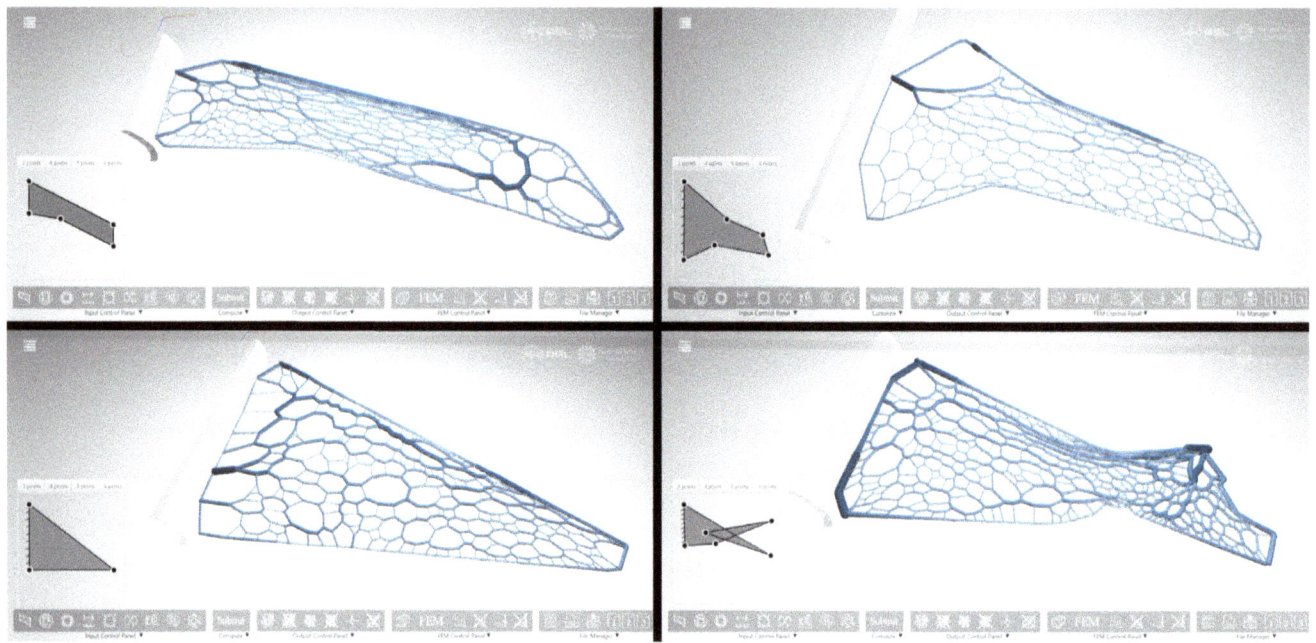

11 The generated structures with different user-input boundaries

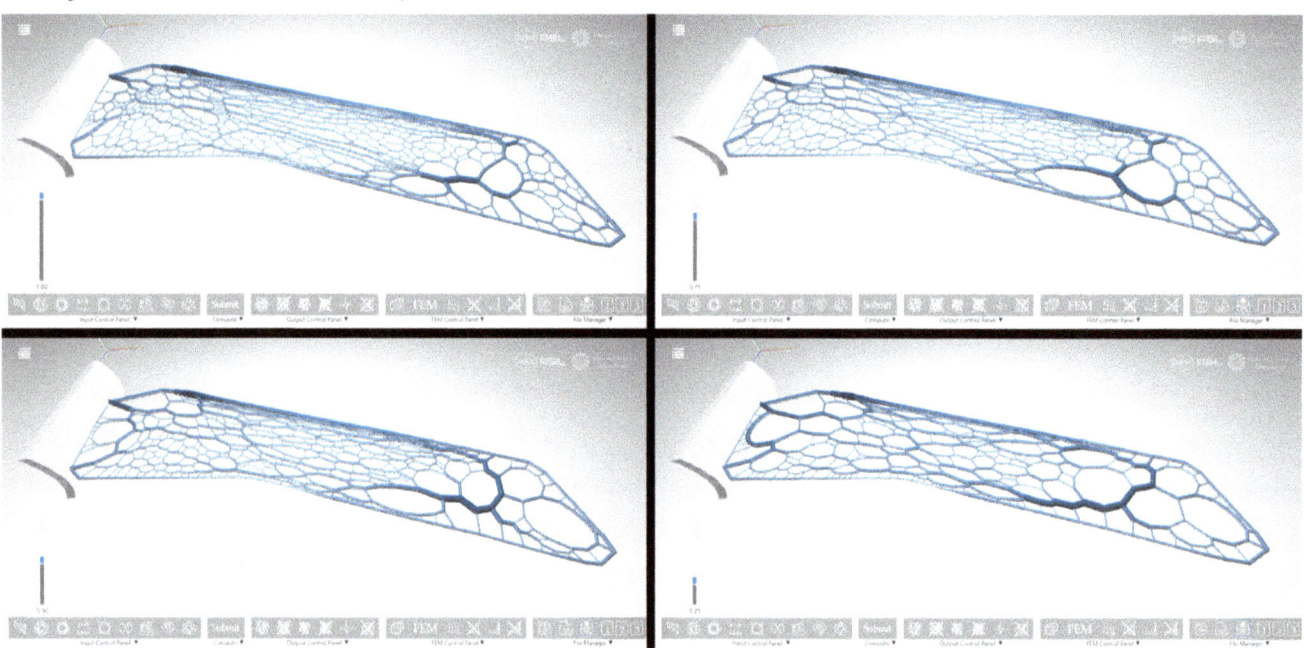

12 The generated structures with different subdivision densities: (top left) SD=1.0; (top right) SD=0.75; (bottom left) SD=0.5; (bottom right) SD=0.25

Minkowski Sum Display

Besides changing the input parameters, the user can also adjust the output display modes and the related indicators to show or hide the generated structures and the analytical results. Figure 13 shows the case of Minkowski Sum in different stages with different values of the indicator. The indicator defines position in percentage of the current Minkowski Sum in the form-to-force transformation. The result is closer to the form diagram with a smaller value of the indicator, while it is closer to the force diagram with a larger value of the indicator. The user can adjust the value of the indicator and view the smooth transformation from the form to the force.

Finite Element Method Analysis

Also, Figure 14 shows the FEM results under different loading scenarios and materials. As mentioned, in the geometric mechanism of our server, we provide the FEM results of the deformation under the self-weight or a point load. The user can decide to show none/one/both of them by clicking the corresponding buttons in the output control panel, or hide the main structure to better compare the FEM results. The color-coded scale is also shown on the right of the web page when the corresponding FEM result is shown. The scale includes the minimum and maximum values of the percentage of deformation compared with the span, and the real values of

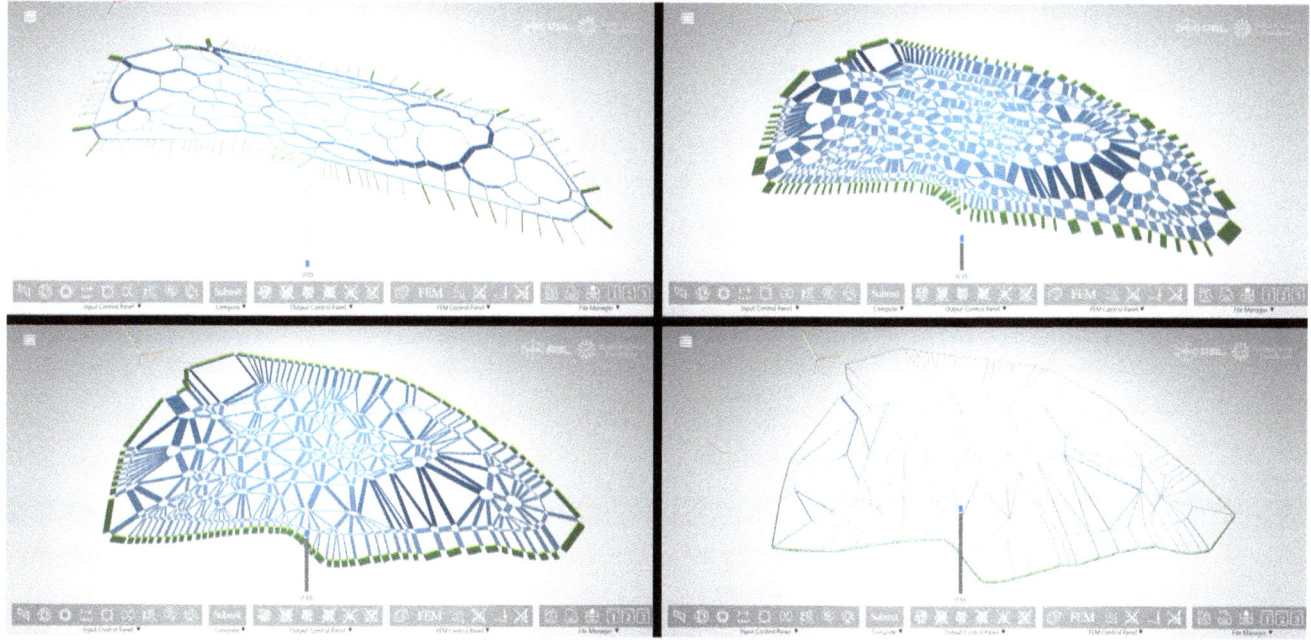

13 The generated Minkowski Sum with different stage indicators: (top left) MSI=0.05; (top right) MSI=0.35; (bottom left) MSI=0.65; (bottom right) MSI=0.95

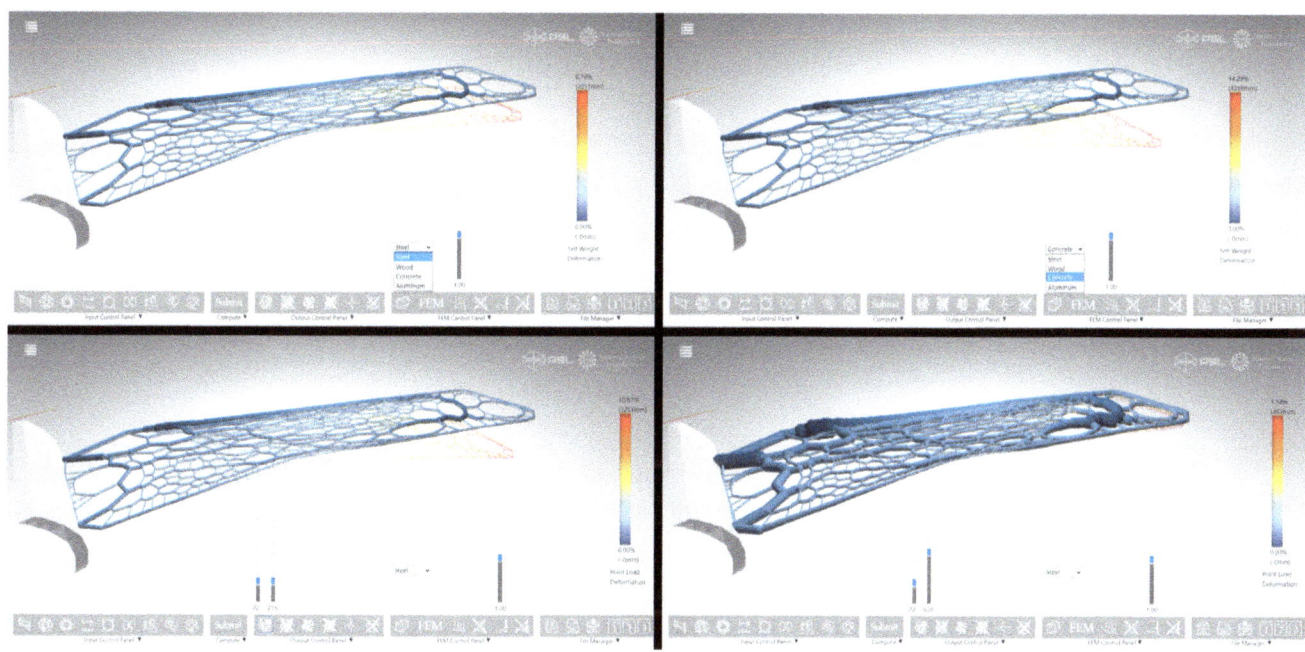

14 The FEM analysis results with different loading scenarios, materials, and structural thickness: (top left) self-weight load with steel material; (top right) self-weight load with concrete material; (bottom left) point load with steel material; (bottom right) point load with steel material and larger structural thickness

the deformation in millimeter. When keeping the deformation multiplier constant, the user can also directly compare the FEM results from different structures.

Exported Models

The final case shows the application of our web tool in exporting the result to other platforms for various purposes. In the normal display mode, the user can adjust the minimum and the maximum radius to control the range of the thickness for edges, and export the structure as an STL model (Figure 15). The exported STL model can be imported to a variety of platforms, including modeling software for further analysis and 3D printing software for digital fabrication. Therefore, our web tool completes the logic of loop by accepting the user input and exporting the generated result back to the user.

CONCLUSION AND DISCUSSION

This paper introduces a web-based structural design tool that implemented the workflow of Graphic Statics and machine learning of the dragonfly wing structures. It helps designers generate funicular cantilever structures without any installation on the local computer, by inputting the boundary and

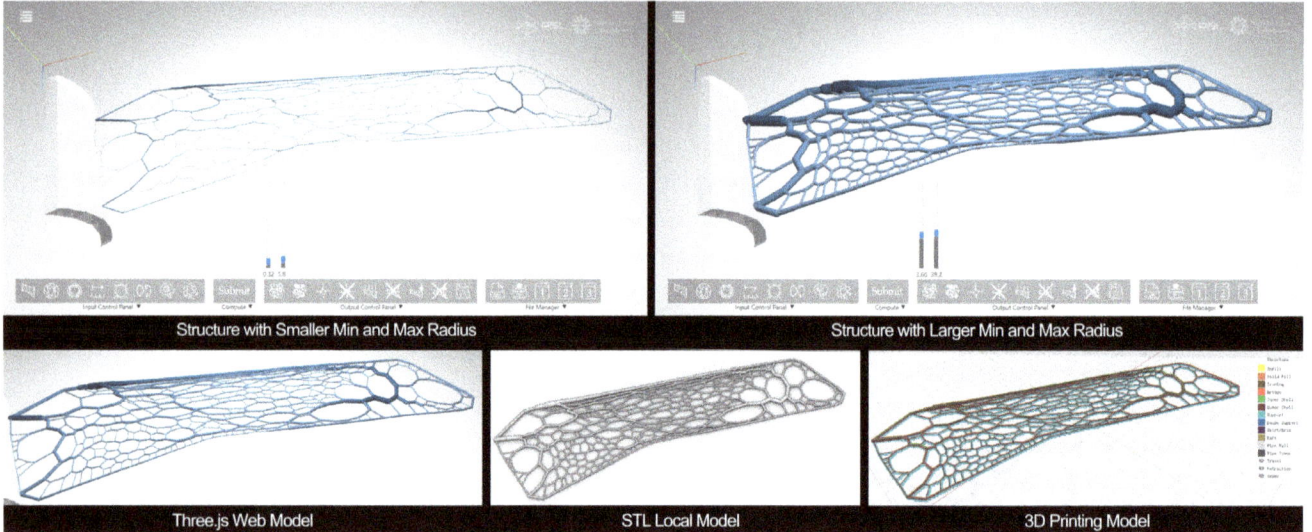

15 (top) Structures with different minimum and maximum radius; (bottom) Exported STL model and its 3D printing preview

control parameters directly on the web page. The backend server computes the topology and geometry of the structure, and the frontend web page displays the generated structure, Minkowski Sum, and FEM analysis results. The user can also export the STL model for further purposes, or download and upload the CVS data file to retrieve the previous results.

In addition, this web-based implementation method can not only be used in serving our workflow of generating structures, but also be used in any local computational process involving the geometric generation and machine learning predictions in various design fields, such as architectural geometry generation (Huang and Zheng 2018; Ren and Zheng 2020), urban feature prediction and plan generation (Zheng and Yuan 2021; He and Zheng 2021), and structural evaluation and generation (Zheng, Moosavi, and Akbarzadeh 2020). Especially in the situation that the workflow involves the data flow in multiple platforms and scripts, this implementation method can reduce the risk of errors raised by compatibility issues in local computers, thus serving the input and output without installation requirements to the user.

To further develop our web tool, the future research includes the following aspects. In the UI design, free-drawn boundaries will be allowed for users to input any closed area as input images. The intermedia outputs such as the force diagram will be allowed to export for further manipulation of design purposes from users, thus users can freely apply the force diagram to different architectural and structural design cases. In the system design, future improvement includes: (1) parallel computing to serve a larger number of users; (2) user login system and online data storage; (3) precise structural analysis options; and (4) implementation of the machine learning models from other natural structures. Besides, with the generated structural forms, further materialization experiments can proceed, depending on the usage of the structures.

REFERENCES

Akbari, M., Y. Lu, and M. Akbarzadeh. 2021. "From design to the fabrication of shellular funicular structures." In *ACADIA 21: Realignments: Toward Critical Computation; Proceedings of the 41st Annual Conference of the Association for Computer Aided Design in Architecture (ACADIA)*. Online.

Akbari, Mostafa, Masoud Akbarzadeh, and Mohammad Bolhassani. 2019. "From Polyhedral to Anticlastic Funicular Spatial Structures." In *Form and Force: Proceedings of the IASS Annual Symposium 2019*. Barcelona, Spain.

Akbari, Mostafa, Armin Mirabolghasemi, Hamid Akbarzadeh, and Masoud Akbarzadeh. 2020. "Geometry-based structural form-finding to design architected cellular solids." In *SCF '20: Proceedings of the 5th Annual ACM Symposium on Computational Fabrication*.

Akbari, Mostafa, Armin Mirabolghasemi, Mohammad Bolhassani, Abdolhamid Akbarzadeh, and Masoud Akbarzadeh. 2022. "Strut-Based Cellular to Shellular Funicular Materials." *Advanced Functional Materials* 32 (14): 2109725.

Cabello, Ricardo. *Three.js*. V. 0.151.3. GitHub. PC. 2022.

Eich, Brendan. *Javascript*. V. 2020. Brendan Eich. PC. 2020.

Fulton, Steve. *HTML Canvas*. V. 5. Steve Fulton. PC. 2011.

Google. *TensorFlow*. V. 2.0. Google. PC. 2019.

He, Jingyi, and Hao Zheng. 2021. "Prediction of crime rate in urban neighborhoods based on machine learning." *Engineering Applications of Artificial Intelligence* 106 (Nov): 104460.

Hoffmann, Jordan, Seth Donoughe, Kathy Li, Mary K. Salcedo, and Chris H. Rycroft. 2018. "A simple developmental model recapitulates complex insect wing venation patterns." In *Proceedings of the National Academy of Sciences* 115 (40): 9905–9910.

Holovaty, Adrian, and Simon Willison. *Django*. V. 4.2. PyPI. PC. 2022

Huang, Weixin, and Hao Zheng. 2018. "Architectural Drawings Recognition and Generation through Machine Learning." In *ACADIA 18: Recalibration; Proceedings of the 38th Annual Conference of the Association for Computer Aided Design in Architecture*. Mexico City, Mexico.

Isola, Phillip, Jun-Yan Zhu, Tinghui Zhou, and Alexei A Efros. 2017. "Image-to-image translation with conditional adversarial networks." arXiv preprint.

Maxwell, J. Clerk. 1864. XLV. "On reciprocal figures and diagrams of forces." *The London, Edinburgh, and Dublin Philosophical Magazine and Journal of Science* 27 (182): 250–261.

Nejur, Andrei, and Masoud Akbarzadeh. 2021. "Polyframe, efficient computation for 3d graphic statics." *Computer-Aided Design* 134 (May): 103003.

Paszke, Adam, Sam Gross, Soumith Chintala, and Gregory Chanan. *PyTorch*. V. 1.0.0. PyTorch. PC. 2018.

Polyhedral Structures Laboratory (PSL). 2022. "Wing Structure Generator, Machine Learning + Graphic Statics." Accessed April 28, 2022. http://www.ai-gs.com/frontend/DFW-GH.html.

Preisinger, Clemens. *Karamba3D*. V. 2.2.0. Karamba3D. PC. 2022.

PSL. 2022. "Wing Structure Generator | PSL." accessed 04/28. http://www.ai-gs.com/frontend/DFW-GH.html.

Ren, Yue, and Hao Zheng. 2020. "The Spire of AI - Voxel-based 3D Neural Style Transfer." In *Proceedings of the 25th International Conference on Computer-Aided Architectural Design Research in Asia (CAADRIA)*. Bangkok, Thailand.

Robert McNeel & Associates. *Rhino*. V. 6. Robert McNeel & Associates. PC. 2018.

Rutten, David. *Grasshopper*. V. 1.0.0007. Robert McNeel & Associates.

Whiteley, Walter, Peter F. Ash, Ethan Bolker, and Henry Crapo. 2013. "Convex polyhedra, Dirichlet Tessellations, and Spider Webs." In *Shaping Space*, edited by M. Senechal. New York: Springer.

Widenius, Ulf Michael. *MySQL*. V. 8.0. Oracle. PC. 2022

Zheng, Hao, and Masoud Akbarzadeh. 2022. "The Dragonfly Wing Project." *Architectural Design* 92 (3): 132–133.

Zheng, Hao, Marton Hablicsek, and Masoud Akbarzadeh. 2021. "Lightweight Structures and the Geometric Equilibrium in Dragonfly Wings." In *Proceedings of International Association for Shell and Spatial Structures Annual Symposia (IASS)*. Guildford, UK.

Zheng, Hao, Vahid Moosavi, and Masoud Akbarzadeh. 2020. "Machine Learning Assisted Evaluations in Structural Design and Construction." *Automation in Construction* 119 (Nov): 103346.

Zheng, Hao, and Philip F Yuan. 2021. "A Generative Architectural and Urban Design Method through Artificial Neural Networks." *Building and Environment* 205 (Nov): 108178.

IMAGE CREDITS
All drawings and images by the authors.

Hao Zheng graduated from the PhD program at the University of Pennsylvania, specializing in machine learning, digital fabrication, mixed reality, and generative design. He holds a Master of Architecture degree from the University of California, Berkeley, and Bachelor of Architecture and Arts degrees from Shanghai Jiao Tong University. Previously, Hao worked as a research assistant at Tsinghua University and UC Berkeley with a concentration on the robotic assembly, machine learning, and bio-inspired 3D printing. His teaching experience includes workshop tutor at Tongji University, lecturer at the University of Pennsylvania, and teaching fellow at Shanghai Jiao Tong University.

Masoud Akbarzadeh is a designer with academic background and experience in architectural design, computation, and structural engineering. He is an Assistant Professor of Architecture focusing on Structures and Advanced Technologies and the Polyhedral Structures Laboratory (PSL) director. He holds a DSc from the Institute of Technology in Architecture, ETH Zurich, where was a research assistant in the Block Research Group. His main research topic is Three-Dimensional Graphical Statics, a novel geometric method of structural design in three dimensions. In 2020, he received the National Science Foundation CAREER Award to extend the methods of 3D/Polyhedral Graphic Statics for Education, Design, and Optimization of High-Performance Structures.

Building Synthetic Data Sets or How to Learn from Future Architectures?

Daniel Koehler
University of Texas at Austin /
School of Architecture

1 Selection of simulated buildings from a synthetic data set

ABSTRACT

Machine learning models learn from data-emergent characteristics of entire sets, also known as hyper-localities. We anticipate that such hyper-local features, or what we call "communal physics," can lead to new forms of contextual-driven building typologies. However, what if one cannot rely or build on data? In this paper, we propose a workflow to simulate entire data sets. Simulating synthetic data can induce design speculation to machine learning applications. Leaning on density studies for modernist settlements, we propose an approach that mixes ratios of sets to generate buildings quickly. A case study exemplifies how quickly one can generate and analyze a set of buildings at the resolution of BIM modeling. We conclude that synthetic data sets could become a feature of daily design workflows due to being computationally inexpensive and easy to adapt. Synthetic sets can complement data for design beyond the scale of a building, or its common context, like weather, or pollution. That promotes new kinds of typological thinking, the testing, and adaptation of a building strategy to a diverse range of contexts. In that way, virtual but physical environments allow us to evaluate inter-building-related criteria and their impact on urban forms.

INTRODUCTION

Machine Learning (ML) applications build insights from data by their capability to converge plural values. More than the statistical capability to regress future events from given data, ML applications contribute new insights by learning emergent characteristics of entire sets. Applied to space, Nadav Hochman and Lev Manovich describe those as hyper-localities that replace traditional notions of place through topological features of intersubjective data (Hochman et al. 2014). Simple example: the value of a home today is less valued by the actual parts of the house than by a home value index, a complex and often non-transparent amalgam calculated by real estate platforms that rate schools, commuting, market availability, seasonal market shifts, etc. (Zillow Research 2019). Once learned, hyper-localities are reliable resources to an extent that Alex Pentland elevated Machine Learning to Social Physics (Pentland 2015). Understood critically and coined as hyper-capitalism by the economist Thomas Piketty, trained hyper-features have rewritten or opened entire new markets. First layer insights of ML applications, the weighting of a data set by building regression or classification models, have very little meaning for the spatial organization of a building. At the current use, plans are ranked in the difference between them by external labels, like number of bedrooms but not through the inherent representational features of walls, entrances etc. As highlighted by Stanislas Chaillou (Chaillou 2021), the capability of ML models to learn hyper-local characteristics might be its actual contribution to architecture. Hyper-localities render qualities of space. One finds first computational descriptions in Space Syntax approaches (Hillier and Hanson 1984) or Stanislas Chaillou's ML applications (Chaillou 2019). Both use the comparison of inherent graph-based features, like connectivity, depth of access, or orientation for qualitative assertions. Buildings involve diverse communal decision-making processes, from planning, financing, and construction, to their use and life cycle. Hyper-local features extended beyond formal criteria could offer alternative ways to describe a building as a kind of communal physics. Moving into Web 3.0 with its distributed, participatory applications, we anticipate that such features, consciously analyzed and applied, will lead to new forms of plan-making and building typologies. Such a data-driven approach could lead to urgently needed new building typologies as a direct design response to a hyper-local, contextual-environmental understanding of a building's performance.

However, what if one can hardly learn from existing data? Databases at an architectural resolution are sparse for several reasons. Architectural notations, like plans, are expert representations that hardly appear on social media, or other web-based platforms from which large databases source today. Even within the field, architectural forms of representation differ depending on the context, graphics, and meaning.

2 The Three-Nine, a fully simulated building proposal for Austin. Based on the data of the case study, three architecture machines arranged 382,659 building parts into variations on a nine-square grid resembling the daily spatial use of 500 people situated in downtown Austin.

For example, annotating a stair with an arrow in plan can indicate its upward or downward direction depending on the country in which one draws the plan. Due to the ambiguity between different architectural drawings, the first data-driven applications appeared in large architecture practices with a sufficient, coherent oeuvre of work to train networks. An aesthetic milestone is the DeepHimmelblau Research by Coop Himmelblau, which uses photos from build work to interpolate the compositional design space between those and inherent to the practice oeuvre (Coop Himmelb(l)au 2021). However, to not only automate but also expand its own design space, Coop Himmelb(l)au would need access to additional data sets from other offices.

Theoretically, notations from most buildings exist in the public realm. Most communities archive planning material. However, those are difficult to access; older building plans must first be digitized with expensive equipment in a time-consuming process. Also, access to building information often needs the consent of multiple parties involved. Therefore, we find public databases only in cases of urgent necessity. For example, most fire departments today maintain digital databases with escape plans allowing an early briefing of firefighters during driving to an emergency.

The lack of coherence and legal or economic constraints render it more likely that large-scale databases of buildings faster evolve outside the building industry using non-classical representations, like photogrammetry or 3D scanning. Scan-to-BIM approaches are a well-established practice, especially for plannings in existing contexts (Laing 2015). Global three dimensional data sets evolved through photogrammetry and platform-driven data collection in recent years. The real estate platform Zillow generates floor plan layouts from uploaded photos (Li 2020). In late 2022, the Google Maps immersive view feature will enable fully textured, three dimensional interior walkthroughs on standard cell phones (Daniel 2022). With such technical advances, it is quite possible that semantically formatted 3D data will be available for most buildings in the near future. But, how democratic will access to such data be? Scans can be made today on common smartphones but mostly with limited temporal and data access. From an ontological perspective, we should also ask: How adaptable or insightful are those datasets to the rapidly changing contexts that cities today are? Housing prices were once an entry data set to ML regression models, but no one could have predicted the surge over the last two years (Kaggle 2014). Remote work and adjusting supply chains led to new programmatic mixtures and building types, like Cloud Kitchens (Colpaart 2019). Carbon emissions should approach net-zero in 2050, but how to learn carbon-neutral structures from a city that is not net-zero proves challenging (IPCC 2022). The recent economic, social, and environmental challenges question whether we could or should even learn from existing data.

One possible solution is the simulation of data sets of building configurations is the use of synthetic building data. Synthetic data can play a crucial role in leveraging access to ML in architecture and other design fields. Adapting McGough's general list of advantages (McGough 2022), synthetic data in architecture can help with generating building sets at: 1) negligible resources, and 2) without the need to acquire and label data. Generating data sets of fully three dimensional building ensembles needs very little time today, comparable to the processing time of architectural visualizations. Such a speed enables data-driven hypothesis-tested design workflows. Analog to a sketch or diagram, the design hypothesis prospects a possible design space, though unlike a sketch, it generates a speculative data set that can be tested against further criteria. Subsequently, and previously out of reach for most architects, synthetic data enacts designers to integrate data-dependent valuations directly into an iterative design process. In that way, the design workflow reciprocally appreciates Human & AI explorations, where on the one hand, ML valuations weigh the designer's judgments, and on the other hand, convergent insights open new design opportunities (Wilson 2018). Like in the following case study (Figures 1, 2), simulated within a game engine, the building ensembles are physically viable, tokenized at the resolution of a BIM model, and reversible between three dimensional geometry and graph-structured data. Custom data fields and color-coded images allow the labeling of architectural features. That means designers generate labels by a deterministic process that ensures the labels are correctly assigned, and more importantly, architects design labels with an architectural connotation in the first place. Synthetic building sets can be massive and semantically well-structured, both preconditions for most ML approaches today.

State of the Art

Simulating data is, in most fields, a well-established method on an industrial scale. As a simulation platform, Unity provides a great overview of how its game engine is used to simulate landscapes to train self-driving cars, interactive shop interiors to test security protocols, or medical procedures to train robotic interaction (Unity 2022). With interest in communal spaces' physical features, we propose a literal approach to simulation, using the simulation for a combinatorial assembly of interactive building parts following agent-based modeling. Simulating the spatial layout of a building directly as an assembly of building parts has several advantages. First, the building figuration can be sourced from a BIM library of building components to promote modular fabrication. Second, the simulation of an assembly links in reverse also to its disassembly. Design methods incorporating a timeline can also map fabrication processes relating to reusability and extend to the life-cycle design of a building. Computationally, the method has the advantage that the simulation can be connected to the well-established reinforcement learning (Hosmer et al. 2020), or the promising models of federal learning (Qinbin 2021), or distributed learning approaches (Verbraeken 2019) by applying models for social decision-making processes to a timeline-based construction of space.

Method

This paper proposes an additional workflow to automated building design[1] by building first synthetic data set of buildings through iterative simulation of building assemblies made of discrete automata, referred to as discrete parts (Retsin 2019) and composed through mereological relationships (Koehler 2019) Technically, we build on a design workflow that Unity proposed to simulate synthetic data sets for object detection (Navarro 2021).[2] In their case, the simulation randomly positions a thtree dimensional, textured asset in front of a randomized background. The setup allows for semantic labeling of content using the perception package (Unity 2022). The overall aim is to quickly generate a set of images and textual descriptions at a probabilistic even distribution to train for further ML applications.

With the long-term focus on designing data sets for learning hyper-local features, in our case, each set entry should vary compositionally. Therefore, the simulation randomizes compositional parameters, like the quantity, ratio, position, orientation, or linkage between building parts. This method links to a history of design strategies by literally shuffling programs and formal distributions through data analysis. The method finds precedents in the work of Ildefons Cerdà, Ludwig Hilberseimer, Eckard Schultze-Fielitz, Yona Friedman, MVRDV (Koehler 2017). Closing the loop between analysis and proposition, those architects built on a method named "mixed-settlement." Directly referring to the shuffling of built form, "mixed-settlement" investigates the combinatorial synergies between different building types. By mixing the ratios and proportions of types and their arrangements, researchers derived urban design principles that resulted in the most livable environments over time. This study extends the method of typological mixing through the workflow to simulate synthetic data sets. Center to the simulation is a stackable assembler agent that configures spatial parts, like rooms, assembles assemblers of spatial parts to building parts, and so forth. Types are seen here not as static, closed units but as reconfigurable spatial arranged set of tokens. In that way, a building arrangement is available as a flat structure of building parts or as clustered set following traditional segmentation of room, flat, house, or block (Figure 6).

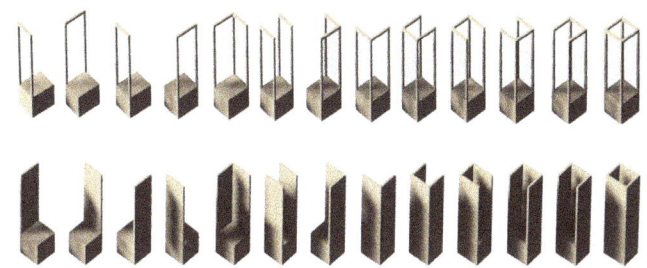

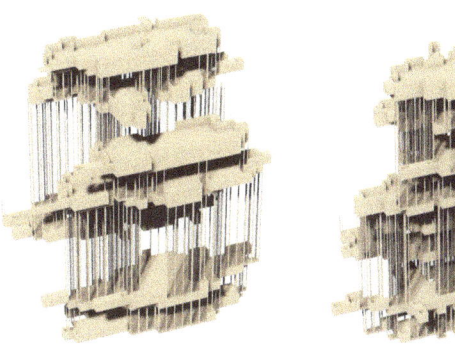

3   Responsive Part computed from a selection of prefab components using a state machine design pattern

4   Screenshots of an adaptive, procedural envelope

5   Building ensemble and its voxelated massing, used to sample data; here are indexed light intensity in green and sun insolation in blue

The workflow consists of the following steps. First step is the design and import of specific BIM models at the resolution of prefabricated building parts. Following a blockchain data structure, those BIM parts are treated as "tokens," and an assembly as a "block" is saved as a list of links between assembled tokens. Saved in JSON format allows simple interoperability between platforms, where an assembly can be restored by reading and transforming the token IDs and their relative position and orientation in space. Data related to the overall assembly is treated likewise as a specific token and is stored at a dynamic position in a block.

In a second step, tokens are combined with spatial parts using an assembler agent pattern. Those assemblers build rooms,

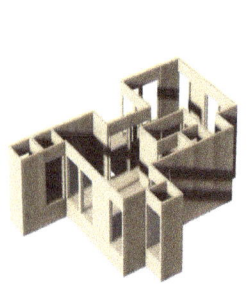  

6   Nested scales of Assemblers: (from left to right) responsive part, spatial segment, house configuration, block ensemble

platforms, and other spatial segments from a manageable number of imported BIM models, usually described as building parts, like walls, doors, and floors. They can be trained with reinforcement learning using Unity's ML-Agents or their part-to-part linkage constrained using a Wave Function Collapse Algorithm (Boris 2020). They integrate physics behavior, from non-overlap of parts to simulation of structural performances, as part of the assembly. The assembly can also follow a state machine pattern that allows the integration of procedural behaviors (Figure 3). Procedural methods are beneficial to condition envelopes, closure, handrails towards voids like those shown in Figure 4 or passages between rooms, etc. In this particular case study, the state machine pattern shown in Figure 3 consists of two nested agents. The parent agent uses a procedural method to compute an envelope by simply checking for void space to its sides, and in case placing a façade or handrail element. Adjacency to other building blocks groups the block to a ML-Agent trained for room and plan layout generation for a limited area. The agent would then switch sides, or the block between open, framed, filled.

Each assembler composes an absolute number of different parts at a specific ratio. The characteristic of a composition and proportion of a spatial segment is defined by ratios and quantities of the number of different parts available, their ratio, and the number of absolute parts used per segment. Those values represent a domain of possible inputs for probable outputs by shuffling building parts. In a third step, one describes more extensive arrangements by recursively combining smaller assemblers. A repeatable pattern of nested assemblers assembles a building (Figure 6). The simulation itself consists of complementary methods to iterate, evaluate, and store building ensembles once generated. The data-set simulation follows a Monte-Carlo-Simulation approach that uses random sampling to represent a design space. In the beginning, we generate the inputs—namely the number and ratios between the building parts used by the assemblers— randomly from a probability distribution over the domain. Then, in representing the spatial diversity of a set, we avoid Gaussian distributions by using not built-in random methods but by calculating first uniform distributions to take into account the combination of all parameters and the desired size of the simulated set. For a statistically meaningful sampling of a set, we typically simulate in excess of 3,000 assemblies.

Once an assembly is completed, it is stored via JSON files following a blockchain data structure. Further evaluations can be added simply as a separate token to that data structure. As the building assembly is clustered into groups of discrete building parts, some values, like mass and area, can be easily derived by used building materials or by the program, respectively. For comparative evaluation, we suggest the further tessellation into a voxel space. That allows for quantifying mass-void distributions, or as in Figure 5, local light intensities by sampling raytracing. The simulated data is storable and then available at a 1:1 BIM resolution. Additionally, evaluations or labeled content can be saved. The simple file format with a reversible assembly allows also to quickly rebuild and analyze output in another software at a later point.

Results and Discussion

As shown in Figure 7, Monte Carlo simulations can also be used to spread measured values of non-parametrically

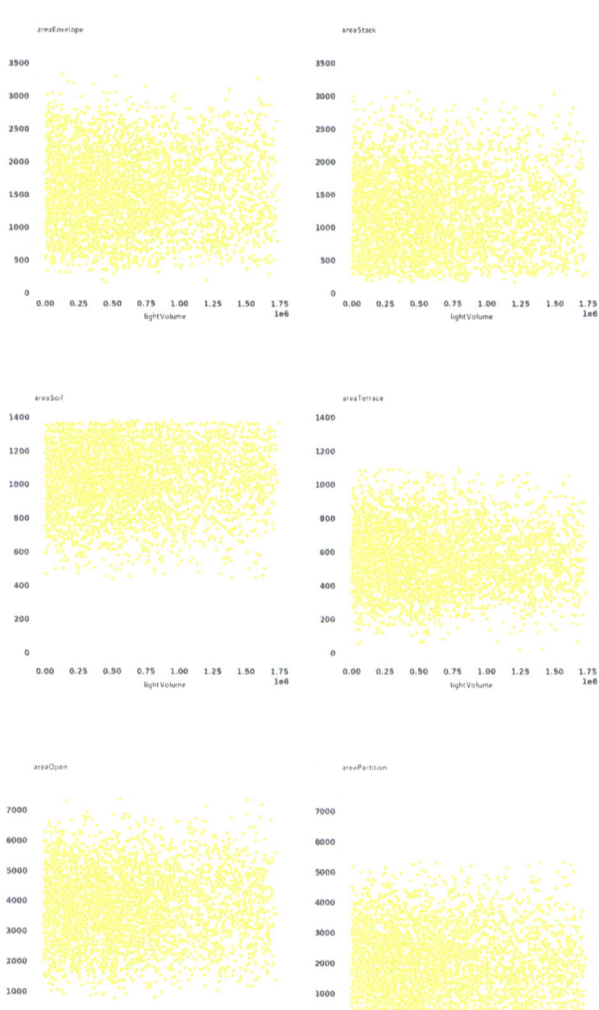

7 Benchmark graphs mapping ratios of light intensity to programmatic areas

controlled properties. Thus, the generated set itself can be viewed as an alternative to standard optimization workflows. One can easily navigate through the datasets using basic simple sorting or clustering methods, here with Python Panda library, as seen in Figure 8. That allows to quickly conclude the performance of single properties and in combination with multiple values.

The approach of shuffling assemblies offers ways to overcome often seen limitations of bottom up approaches, like wave function collapse, that describe entire arrangements only through the constrained combinatorial action between part-to-part linkages. Scaled up, this quickly leads to unmanageable interconnections for designers. By stacking different assemblers, interconnections can be better manipulated, swapped or recombined.

The presented case study focuses on the workflow of simulating data sets, and the above-mentioned potential to discuss buildings with alternative topological characteristics through ML models: their "Communal Physics" can only be part of further texts. Technically, there are limitations with the integration of external libraries. For example, Unity's ML-agent models cannot be nested, they are difficult to combine, and require considerable resources. Also, most libraries use object pooling patterns, and instantiate parameters before the start of a simulation. Thus, entire scenes must be reinstantiated in each iteration, requiring significantly more resources. We see our case study only as one possible application among many to build synthetic data sets for an architectural realm. The shown bottom-up approach has computational limitations, too. By increasing the number of parts, the number of dependencies and computation time increases exponentially. The mereological partitioning into building groups helps to flatten resources but is compared in processing time not competitive to top-down approaches, like diffusion models. With the recent success of those, we prospect that one direct application in rendering simulated building representations will be to enrich the image databases with architectural themes to train diffusion models.

CONCLUSION AND OUTLOOK

It is remarkable how fast one can generate a dataset. In the case study presented here, the simulation, evaluation, visualization, and storage take seconds (house) or minutes (high-rise), depending on the combinatorial complexity. Those numbers refer to using a single laptop, although such simulations could be easily scaled using Unity's cloud computing services. With today's pricing of GPU hours, the simulation of a set of 3,000 houses equates to approximately five dollars. Comparable to the rendering time and expenses of animations, it is foreseeable that data-rich design, similar to architectural visualizations, become an everyday aspect of a standard design process. By that, each building design turns into a design space of buildings.

The scale and performance of simulating data sets promises to break the dependence on existing data. It also bridges data with speculative design thinking. Thus, in the long run, we speculate that designing with statistical methods will not lead to universally adaptable products but to precise contextually developed approaches. Easy access to data-rich design will allow building-specific models for a variety of materials, regulations, construction methods, spatial, socio-programmatic, and environmental influences. The models will not only be more precise, but also more robust. In a simulation environment, one can simulate building part or contextual data to its extreme to so-called "black swan" events that are rare but significant, like breaking structures or extreme weather conditions.

Synthetic data can also drastically reduce the need for real-world training data by simulating contexts. Leveraged access to

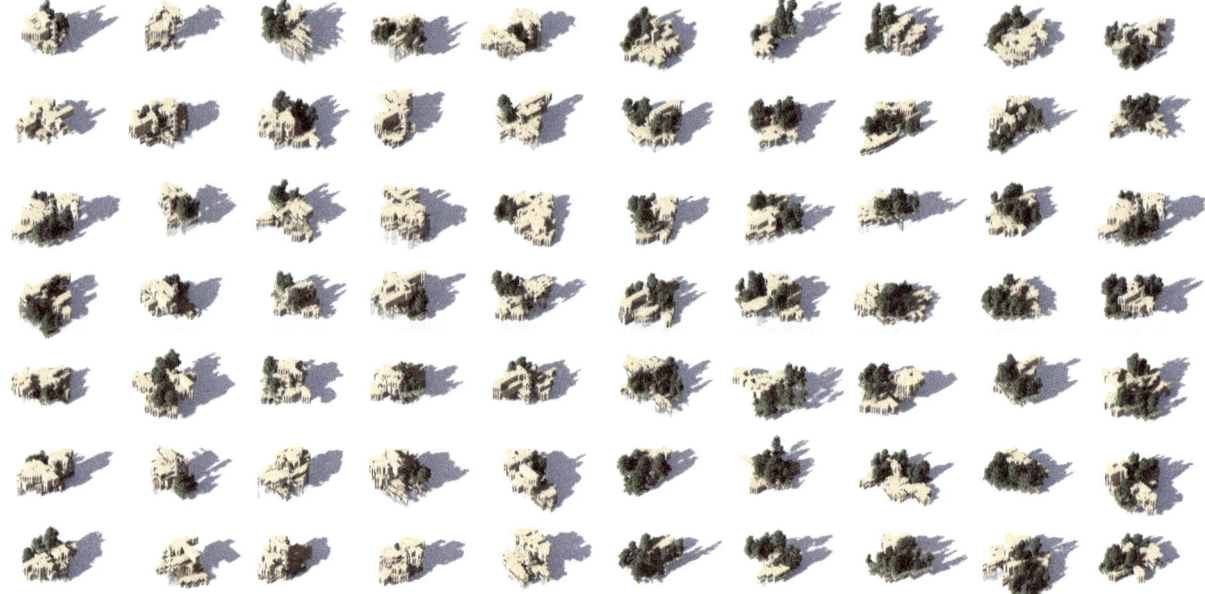

8   Matrix from two subsets; (left half) sorted set of maximum sun insolated volume to massing volume; (right half) sorted set of min 35% (sun volume to massing) plus maximum of integrated deep soil area for large vegetation

contextual data can predict a higher design sensibility to a building's environmental condition. Data simulation, as highlighted by one of the anonymous peer reviewers of this paper, allows the simulation of complementary data to traditional building information. It promises the opportunity to include "non-standard and non-building data into a compiled solution set for architecture." Building on Manovich's concept of the extra values hyper-localities convoluted from data, simulated sets would be most valuable in complementing data for design beyond the scale of a building, or its common context, like weather or pollution. That promotes new kinds of typological thinking, the testing, and adaptation of a building strategy to a diverse range of contexts. In that way, virtual but physical environments allow us to simulate and quantify housing trends and their impact on urban forms. Further, we can anticipate planning with inter-building-related criteria—like heat islands, which have often devastating communal implications—and we are only slowly begin to understand by learning from data (Plumer et al. 2020). With careful consideration toward unintended biases, synthetic data simulation can be an approach to include data insights into a design practice for just environments. Simulations are ideal for exploring multiple sets of parameter spaces. Subsequently, simulated datasets provide a source for a data-driven comparison of the physical impact of urban policies on urban form, and can become a valuable complementary contribution to the politics of cities.

ACKNOWLEDGMENTS

This work was supported by Good Systems, a research grant challenge at the University of Texas at Austin.

NOTES

1. For a recent overview of automated floor plan generation, see Weber et al. (2022).

2. Our work builds on Navarro's approach by translating it into an architectural realm. Please refer to that paper for a detailed technical description (Navarro et al. 2021).

REFERENCES

Boris. 2020. "Wave Function Collapse Explained." *BorisTheBrave* (blog). April 13, 2020. Accessed May 5, 2020. https://www.boris-thebrave.com/2020/04/13/wave-function-collapse-explained/.

Chaillou, Stanislas. 2019. "AI & Architecture. An Experimental Perspective." *Towards Data Science* (blog). February 24, 2019. Accessed December 12, 2019. https://towardsdatascience.com/ai-architecture-f9d78c6958e0.

Chaillou, Stanislas. 2021. *Artificial Intelligence and Architecture: From Research to Practice*, 1st ed. Boston: De Gruyter.

Colpaart, Ashley. 2019. "Everything You Need to Know About Cloud Kitchens (aka. Ghost Kitchens) in 2020." *The Food Corridor* (blog). December 5, 2019. Accessed December 5, 2019. https://www.thefoodcorridor.com/2019/12/05/everything-you-need-to-know-about-cloud-kitchens-aka-ghost-kitchens-in-2020/.

Daniel, Miriam. 2022. "Immersive View Coming Soon to Maps—plus More Updates." *Google* (blog). May 11, 2022. Accessed May 11, 2022. https://blog.google/products/maps/three-maps-updates-io-2022/.

Hillier, Bill, and Julienne Hanson. 1984. *The Social Logic of Space*. Cambridge, UK: Cambridge University Press.

Coop Himmelb(l)au. 2021. "Deep Himmelblau." Accessed January 12, 2022. https://coop-himmelblau.at/method/deep-himmelblau/.

Hochman, Nadav, Lev Manovich, and Mehrdad Yazdani. 2014. "On Hyper-Locality: Performances of Place in Social Media." In *Proceedings of 2014 International AAAI Conference on Weblogs and Social Media (ICWSM)*.

IPCC. 2022. "IPCC Sixth Assessment Report: Climate Change 2022: Mitigation of Climate Change." *United Nations, Intergovernmental Panel on Climate Change*. https://www.ipcc.ch/report/ar6/wg3/.

Kaggle. n.d. "Housing Prices Competition for Kaggle Learn Users." Accessed December 12, 2021. https://kaggle.com/competitions/home-data-for-ml-course.

Koehler, Daniel. 2017. "The City as an Element of Architecture: Discrete Automata as an Outlook beyond Bureaucratic Means." In *Shock! Sharing of Computable Knowledge; Proceedings of the the 35th eCAADe Conference*, vol. 1, edited by Antonio Fioravanti. 523–532.

Koehler, Daniel. 2019. "Mereological Thinking: Figuring Realities within Urban Form." *Architectural Design* 89 (2): 30–37.

Laing, R., M. Leon, J. Isaacs, and D. Georgiev. 2015. "Scan to BIM: The Development of a Clear Workflow for the Incorporation of Point Clouds within a BIM Environment." In *WIT Transactions on The Built Environment* 149: 279–89.

Li, Yujie. 2020. "Zillow Floor Plan: Training Models to Detect Windows, Doors, and Openings in Panoramas." *Zillow Tech Hub* (blog). August 6, 2020. Accessed August 6, 2020. https://www.zillow.com/tech/training-models-to-detect-windows-doors-in-panos/.

Li, Qinbin, Zeyi Wen, Zhaomin Wu, Sixu Hu, Naibo Wang, Yuan Li, Xu Liu, and Bingsheng He. 2021. "A Survey on Federated Learning Systems: Vision, Hype, and Reality for Data Privacy and Protection." In *IEEE Transactions on Knowledge and Data Engineering* 1-1.

McGough, Mason. 2022. "PeopleSansPeople: Generating Synthetic Data of Virtual Human Beings in Unity." *Toward Data Science* (blog). March 9, 2022. Accessed March 9, 2022. https://towardsdatascience.com/peoplesanspeople-generating-synthetic-data-of-virtual-human-beings-in-unity-a1847a56895c.

Navarro, Anthony, and James Fort. 2021. "Supercharge Your Computer Vision Models with Synthetic Datasets Built by Unity." *Unity* (blog). April 19, 2021. Accessed April 19, 2021. https://blog.unity.com/technology/supercharge-your-computer-vision-models-with-synthetic-datasets-built-by-unity.

Pentland, Alex. 2015. *Social Physics: How Social Networks Can Make Us Smarter*. New York, NY: Penguin Books.

Plumer, Brad, Nadja Popovich, and Brian Palmer. 2020. "How Decades of Racist Housing Policy Left Neighborhoods Sweltering." The New York Times. August 31, 2020. Accessed September 1, 2020. https://www.nytimes.com/interactive/2020/08/24/climate/racism-redlining-cities-global-warming.html.

Retsin, Gilles. 2019. "Toward Discrete Architecture: Automation Takes Command." In *ACADIA 19: Ubiquity and Autonomy; Proceedings of the 39th Annual Conference of the Association for Computer Aided Design in Architecture*. 532–41.

Thaman, Alex, Souranil Sen, Shounak Mitra. 2021. "Data-Centric AI with Unity Computer Vision Datasets." Unity (blog). December 10, 2021. Accessed December 12, 2021. https://blog.unity.com/technology/data-centric-ai-with-unity-computer-vision-datasets.

Hosmer, Tyson, Panagiotis Tigas, David Reeves, and Ziming He. 2020. "Spatial Assembly with Self-Play Reinforcement Learning." In *ACADIA 2020: Distributed Proximities; Proceedings of the 40th Annual Conference of the Association for Computer Aided Design in Architecture*, vol. 1, edited by B. Slocum, V. Ago, S. Doyle, A. Marcus, M. Yablonina, and M. del Campo. 382–393.

Unity Blog. n.d. "Real-Time 3D" (topic). https://blog.unity.com/topic/real-time-3d.

Unity Technologies. 2022. *Perception Package* (Unity Computer Vision). C#. Unity Technologies. https://github.com/Unity-Technologies/com.unity.perception.

Verbraeken, Joost, Matthijs Wolting, Jonathan Katzy, Jeroen Kloppenburg, Tim Verbelen, and Jan S. Rellermeyer. 2019. "A Survey on Distributed Machine Learning." *arXiv*. http://arxiv.org/abs/1912.09789.

Weber, Ramon Elias, Caitlin Mueller, and Christoph Reinhart. 2022. "Automated Floorplan Generation in Architectural Design: A Review of Methods and Applications." *Automation in Construction* 140 (August): 104385.

Wilson, H. James, and Paul R. Daugherty. 2018. *Human + Machine: Reimagining Work in the Age of AI*. Boston: Harvard Business Review Press.

Zillow Research. 2019. "Zillow Home Value Index Methodology, 2019 Revision: What's Changed & Why." *Zillow* (blog). December 19, 2019. Accessed August 1, 2020. https://www.zillow.com/research/zhvi-methodology-2019-highlights-26221/.

IMAGE CREDITS
All drawings and images by the author.

Daniel Koehler is an urbanist, architect, and assistant professor for architecture computation at the University of Texas at Austin. Prior to joining UT Austin, Daniel conducted research at the Bartlett in London and Innsbruck University, where he wrote his PhD dissertation "The Mereological City," a study on the part-relationships between architecture and its city in the modern period. Daniel's work has been exhibited in Prague, Milan, Venice, Graz, Montreal, London, Austin, and is part of the permanent collection of the Centre Pompidou in Paris. His current research focuses on the implications of artificial intelligence, on the design practice of cities and their architecture.

Methods for Integrating Architectural Design Intent into the Agent-based Design of (Adaptive) Truss Structures

Mathias Maierhofer
ICD, University of Stuttgart

Achim Menges
ICD, University of Stuttgart

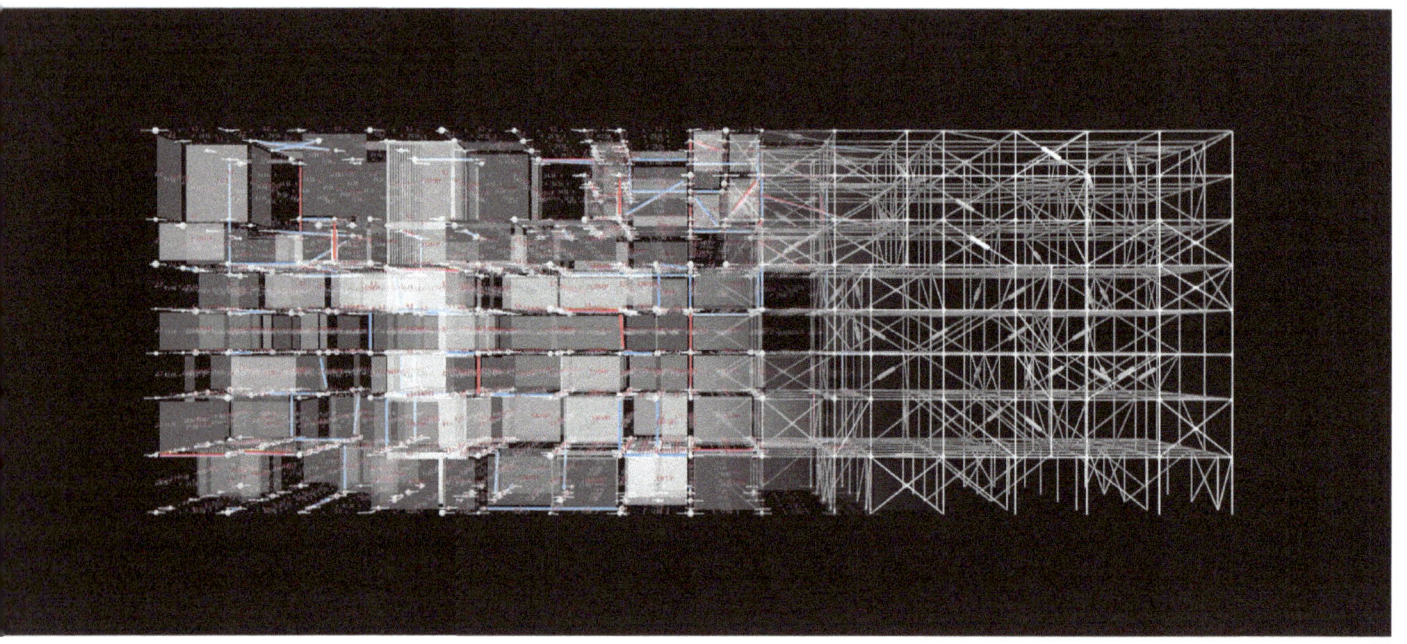

1A Exploration of structurally adaptable and architecturally meaningful truss topologies

ABSTRACT

Significant material savings can be achieved by providing load-bearing structures with the ability to actively manipulate (i.e., adapt) forces and deformations resulting from adverse loads. In order for this approach to be effective, structures must exhibit a high degree of adaptability. In the case of adaptive truss structures, adaptability relates to truss topology and the redundancies therein. Given a highly non-linear relationship between adaptability and topology, an agent-based design environment was previously developed to explore particularly adaptable truss topologies beyond established structural typologies, as such exhibit limited adaptability.

However, when introducing structural adaptation to the architectural domain, design methods cannot explore adaptability in isolation but must also acknowledge a myriad of architectural requirements. The scope of this research is, therefore, to develop strategies for translating design intent into constraints and instructions to be eventually negotiated with adaptability objectives by the agent system. The key challenge here is to ensure that these constraints and instructions are specific enough to be sufficiently integrated, and at the same time flexible enough not to compromise the solution space and exploration of adaptable topologies. This paper addresses these challenges through the example of elementary objectives common in early architectural design.

INTRODUCTION

Load-bearing structures are typically designed for extreme yet rarely occurring loads. Therefore, much of the construction material used remains underutilized during most of a building's service life. Contrary to the mere reliance upon material, adaptive structures utilize sensor-actuator mechanisms to detect and actively manipulate (i.e., homogenize) forces and deformations resulting from adverse loads (Sobek 2016).

Homogenization, in this context, involves the induction of artificial forces and deformations that counteract those acting on the system. In doing so, structures can be designed for more frequent and less severe influences. This allows for material savings of up to 30% and a doubling in service life compared to optimized passive structures (Ostertag et al. 2020). From an environmental perspective, this constitutes a significant reduction in embodied energy that outweighs the small amounts of operational energy required for infrequent actuation (Senatore, Duffour, and Winslow 2019). Structural adaptation thus presents a promising strategy to meet the rapidly increasing demand for construction activity in line with global decarbonization goals (UNEP 2020).

Motivation

The development of adaptive structures presents a multifaceted design challenge that goes beyond the mere equipment of structures with sensors and actuators. Above all, the prerequisite for a structure to effectively manipulate forces and deformations is a high degree of adaptability (Geiger et al. 2020a). In the case of adaptive truss structures, adaptability directly relates to truss topology and the distribution of redundancies therein (Geiger et al. 2020a). Specifically, it describes the degree to which actuation (i.e., a minimal change of length) of certain members can influence the load-bearing behavior of all other members.

Though the quantification of the nontrivial relationship between topology and adaptability is still under investigation, Steffen et al. (2020) indicate that existing structural typologies exhibit only limited adaptability. Relying on design knowledge, intuition and methods for conventional structures might thus lead to inadequate adaptive structures (Geiger et al. 2020b).

This raises the question of what means are required to identify, validate, and ultimately integrate adaptability objectives in the design of adaptable structural topologies that are yet to be explored. In addition, the dissemination of structural adaptation as a future building technology is ultimately contingent upon architects being able to design for and with it. Consequently, the means to be developed must acknowledge the peculiarities of architectural design. The research underlying the present work thus addresses the question of what methods and tools are necessary to support architects in the exploratory design of adaptive buildings, so that non-intuitive adaptability requirements can be incorporated with utmost agency, and from the outset of a design process when far-reaching design decisions are made (Bogenstätter 2000).

BACKGROUND

Agent-based Modeling and Simulation

Agent-based Modeling and Simulation (ABMS) presents as a promising strategy to address the above question (Maierhofer and Menges 2019; Maierhofer et al. 2020). ABMS is a bottom-up modeling and simulation method typically used for gaining insights into the dynamics of complex adaptive systems (Macal and North 2010). Gaining insights, in turn, means understanding the relationship between system behaviors on the micro-level and the respective (i.e., emergent) phenomenon on the macro-level (Bonabeau 2002). In ABMS, systems are therefore modeled from the perspective of their constituent units. These units are conceived of as autonomous decision-making entities (agents) that exhibit distinct behaviors (intentions) informed by external stimuli (feedback). The agent's self-organization over time, in turn, is a function of their continuous interaction with each other and the environment in which they are situated.

With its characteristics of emergence and self-organization, the relevance of ABMS to architectural design lies in its ability to integrate, i.e., negotiate, a variety of (potentially conflicting) design objectives and to explore the ways in which they manifest at the building level. Agent-based modeling thus accommodates the divergent and convergent nature of architectural design, "which oscillates between increasing variation, as part of an explorative search process, and selection, as part of homing in on the design solution" (Schwinn 2021).

Moreover, compared to other design integration and optimization methods such as black-box algorithms (Wortmann 2017), agent-based models are inherently interactive. This not only applies to the interaction between agents, but also to the interaction between the agent model and designers. By setting behavior parameters, manipulating the environment, or the agent itself (Groenewolt et al. 2018), designers can actively participate in the exploratory process, where the immediate response of the agent model provides instant design feedback.

The primary challenge when using ABMS in architecture is the definition of what an agent represents, and the abstraction of global design objectives to low-level agent behaviors (Heath and Hill 2010). Schwinn et al. (2014), for example, who define agents as the individual segments of segmented timber shells, employ locomotive behaviors to negotiate geometric, structural, and fabrication-related requirements. Overviews of further ABMS approaches in architecture and construction can be found in Stieler et al. (2022), Schwinn (2021) and Gerber et al. (2017).

Agent-based Truss Model

In the present work, similar to Schwinn et al. (2014), we conceive of agents as building elements. Specifically, we define agents as truss nodes, whereas an explicit relationship between two agents represents a truss member (Figure 1B-a). Agents exhibit two types of behaviors that inform their connectivity and position in space, and consequently the topology and geometry of the truss structure they describe:

- Topology-oriented behaviors through which agents iteratively vote for members to be removed from or added to the truss structure, and
- Position-oriented behaviors which result in translation vectors that contribute to a desired change in position (see Reynolds 1999)

The above behavior types are informed by information stored in two corresponding types of agent environments: a topological environment that reflects the connectivity of the truss structure, and a geometric environment that represents the Euclidean space in which the truss structure exists and to which agents are constrained (e.g., the building site).

The topological environment is computationally described as a non-manifold and multidimensional mesh using the AHF data structure (Dyedov et al. 2015). It is composed of vertices (agents), edges (truss members) and faces (truss fields) (Figure 1B-b). Truss fields are used to determine an agent's topological neighborhood, comprising truss nodes and elements associated with truss fields incident to the agent (Figure 1B-d). The geometric environment, in turn, is defined by a three-dimensional Euclidean geometry that is discretized into two-dimensional planes which correspond to floor slabs and constrain agents to horizontal movements (Figure 1B-c).

About This Paper

As part of ongoing research with engineers from structural mechanics, topology-oriented behaviors are investigated as a means to incorporate structural requirements and adaptability objectives and to validate respective evaluation metrics. Key to this investigation are adaptability analysis routines that are recomputed at each iteration. The corresponding results for each truss member are stored within the topological environment, from where they are accessed by agents to inform their decision of whether or not to cast a 'remove' or 'add' a vote. Votes, however, can only be cast for members that represent bracings and exist within the agent's neighborhood (Figure 1B-d).

The research presented here augments the above investigation with architectural objectives. The goal is to ensure that truss topologies resulting from the exploratory process are not only adaptable and structurally sound, but also meaningful architecturally and functionally.

A methodological prerequisite for achieving this goal is the identification of adequate strategies for translating design intent into constraints and instructions to be negotiated with adaptability objectives by the agent system. The key challenge here is to ensure that these constraints and instructions are specific enough to be sufficiently integrated, and at the same time flexible enough not to compromise the solution space and exploration of adaptable topologies. Moreover, to accommodate for the iterative and open-ended nature of architectural design exploration, strategies for integration must also allow the for on-the-fly reformulation or adjustment of objectives.

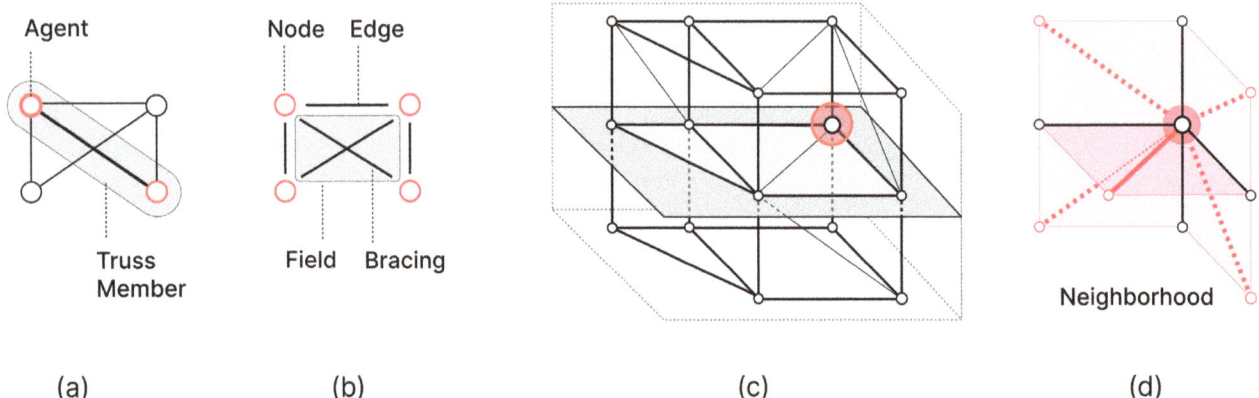

1B Truss agents (a), topological model (b), topological and geometrical environment (c), and topological agent neighborhood (d)

This paper addresses these challenges through the example of elementary objectives common in early architectural design. The aim is to establish a catalog of strategies for their co-integration with adaptability objectives and thus a sound methodological foundation for future research on implementation and validation.

AGENT-BASED ARCHITECTURAL DESIGN METHODS

Though agents constitute inherently autonomous decision-making entities, there are several ways in which designers can pro-actively engage in agent-based design processes beyond the overall definition of model constructs. In other words, providing building elements with agency does not mean jeopardizing but rather augmenting designer agency.

However, contrary to the accustomed modes of geometric or parametric modeling, designing with ABMS does not mean to explicitly define, model ,or associate geometries but to define and encode objectives and constraints which, in turn, inform the (inter-)actions of agents. Here, given the two types of behaviors and environments, encoding (architectural) design intent involves answering the following questions:

- What is the objective to be encoded and how can it be described?
- Does the objective involve a geometric or topological response?
- What is the feedback that informs the agent's response, and where is it stored, i.e. accessed by the agent?
- What is the mapping from feedback to response?

Truss Initialization

A key concept to be addressed in addition to translating design objectives to agent model constructs is that of the base structure. Given that architectural design briefs typically specify basic programmatic requirements, e.g., areas and adjacencies of certain functions, designers must be able to translate such requirements into a preliminary truss topology from which the agent model is initialized and the exploratory process is started. In doing so, the solution space can be tailored to non-negotiable objectives, and design exploration can be focused on requirements that are more flexible or ill-defined.

With regard to the agent-based truss model, the proposed method for obtaining a base structure involves three steps. First, designers define zones, i.e., programmatic units, for each floor slab based on a variable grid system that allows both for specifying the locations and floor areas of zones relative to one another (Figure 2a). Second, potential truss node positions are extracted from the user input and evaluated with regard to their potential impact on the user input (Figure 2b). Third, truss elements and fields are derived from the set of nodes (Figure 2c), considering basic truss design parameters, such as maximum length for elements and a node count limit equal to four for truss fields.

It is worth noting, however, that the goal of this process is not to fully satisfy architectural, structural or adaptability requirements, as these are subject to the exploratory process, but to provide a basic structural model at a topological resolution that is adequate for the respective design task.

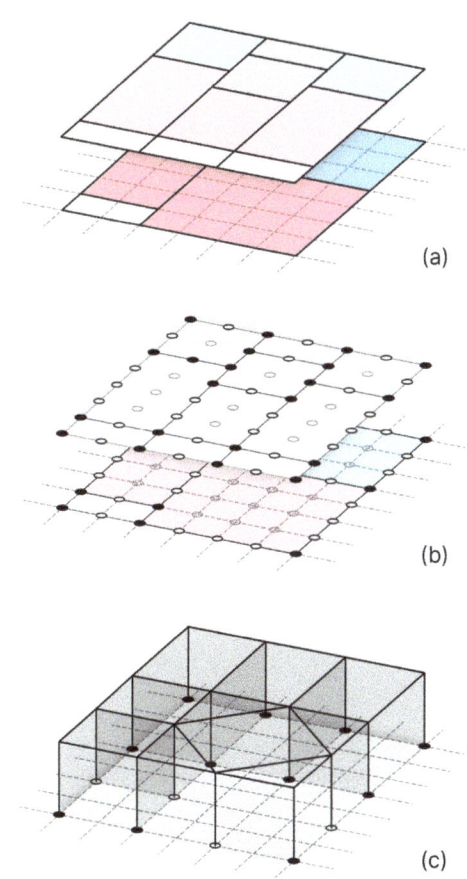

2 User-defined configuration of programmatic zones (a), their abstraction to potential truss nodes with varying priorities (b), and the resulting base truss mesh (c)

Site-related Objectives

In addition to programmatic requirements, architectural designs are typically constrained by the boundary conditions of the building site. These include the maximum building volume (footprint and height), minimum distances to neighboring buildings or building lines and vision axes as defined by the development plan. From a design perspective, the massing of a building might be additionally specified by global design decisions related to the urban context or the function of a building, involving desired pathways, plazas, or setbacks to provide sufficient natural lighting. In order for such requirements to be accommodated both geometrically and topologically, agents must be provided with instructions and constraints that inform their positioning in space and decision of whether or not to establish truss elements in certain areas.

A straightforward approach to informing agents about site-related objectives is to encode information into the Euclidean environment. To do so, designers define areas within the three-dimensional instance of the geometric environment that are to be avoided or populated by agents and thus truss elements (Figure 3a), using geometric input such as points, curves or volumetric geometries, combined with scalar values for prioritization. In a second step, this information is translated to vector fields and mapped to the corresponding two-dimensional instances of the environment (i.e., floor slabs), thereby defining areas of attraction and repulsion (Figure 3b).

From the agent's point of view, the vector field of its associated floor slab contributes to the desired movement vector at a given iteration i. However, agents are only aware of attraction and repulsion vectors within a user-defined vision radius that is declared for the corresponding position-oriented behavior (Figure 3b). This behavior, here referred to as site behavior, computes a movement vector v, where $\widehat{v_f}$ denotes the unitized average of all vectors within the agent's vision, w the weight of the behavior and f a scaling factor.

$$v_f = \tfrac{1}{n} \times \textstyle\sum_{i=1}^{n} v_{fi} \quad \Big| \quad v = w \cdot (\widehat{v_f} \cdot f)$$

Truss-related Topological Objectives

From an architectural perspective, the topological peculiarities of adaptable truss structures present both a design opportunity and challenge. The opportunity lies, above all, in the exploration of novel structural topologies beyond those inherent to established structural typologies. The challenge, in turn, is to architecturally account for topological conditions that do not unfold in well-known and intuitive ways.

In other words, if not constrained, truss members might occur in areas where they interfere with architectural spaces (Figure 4). This might involve spatial interferences (e.g., elements that cross and thus jeopardize functional areas), functional interferences (e.g., elements that prohibit access to certain areas) or visual interferences (e.g., elements that obstruct views).

To allow designers to pro-actively address unfavorable impacts of truss elements, an approach based on the concept of zones (see "Truss Initialization") is pursued. Zones represent user-defined sets of parameters and objectives that are assigned to constructs in the topological environment. In the case of a zone with topological constraints, e.g., a "no-bracing zone," topology-oriented behaviors (those focused on structure and adaptability) would avoid adding elements in truss fields labeled as "no-bracing."

However, a zone can also be defined more globally by associating programmatic areas with specific objectives, which are then applied to all truss fields labeled accordingly. For example, a "living-room" zone applied to all living rooms in the building, that requires at least one adjacent façade field to remain free from bracings, e.g., in order to install balconies.

Another example would be a circulation-specific zone, where fields in the floor slab are to remain free from diagonals to install staircases or elevators. These examples also illustrate the potential range of priorities, given that unobstructed views might be compromised in some cases in favor of adaptability, whereas vertical circulation cannot be jeopardized.

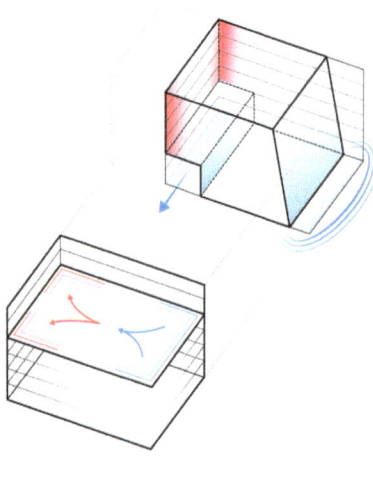

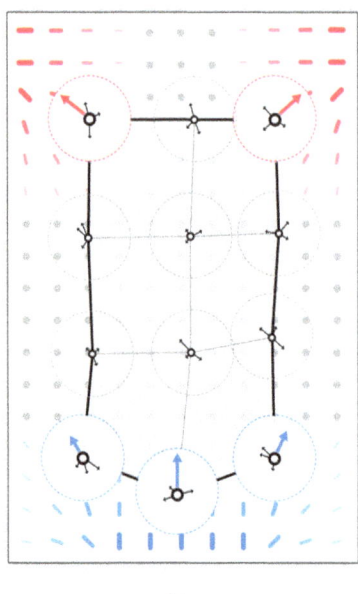

3

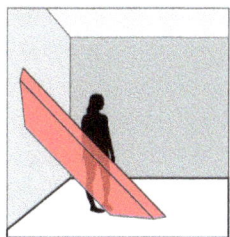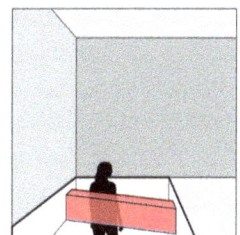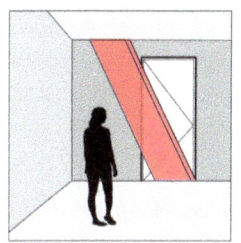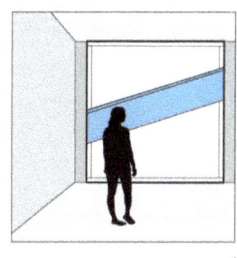

Truss-related Geometrical Objectives

The concept of zones can also serve as a means to encode geometrical objectives into the topological model. These include desired floor areas, angles between truss members as well as the lengths of truss members (Figure 5d). This way, functional areas at first rather roughly defined during the initialization of the truss topology can be further specified in view of specific design intentions. Three corresponding geometry-oriented agent behaviors—all of which are informed by the topological environment—are proposed.

The goal of the area behavior is to maintain a desired target floor area assigned to a truss field in the topological environment (Figure 5a). For agents that are incident to those fields, it computes a translation vector v from the unit vector $\hat{v}_c$, which moves the agent away from or towards the field's centroid, depending on the delta Δ of the desired and current area. The magnitude of v, in turn, is determined by delta Δ, the behavior weight w and the scaling factor f.

$$v_c = CA \quad | \quad \Delta = a_x - a \quad | \quad v = w \cdot (\hat{v}_c \cdot f \cdot \Delta)$$

The goal of the angle behavior is to maintain a desired angle between two adjacent truss members within a field (Figure 5b). For agents that are incident to both truss members, it computes a translation vector v from the unit vector $\hat{v}_n$, which is based on the addition of the two vectors defined by the two truss members. The magnitude of v, in turn, is determined by the delta Δ of the desired and current angle, the behavior weight w and the scaling factor f.

$$v_n = AA_i + AA_j \quad | \quad \Delta = n_x - n \quad | \quad v = w \cdot (\hat{v}_n \cdot f \cdot \Delta)$$

The goal of the length behavior is to maintain desired truss member lengths (Figure 5c). For agents that are incident to those truss members, it computes a translation vector v from the unit vector $\hat{v}_e$, which moves the agent away from or towards the other agent incident to the truss member. The magnitude of v, in turn, is determined by the delta Δ of the desired and current length, the behavior weight w and the scaling factor f.

$$v_e = A_iA \quad | \quad \Delta = l_x - l \quad | \quad v = w \cdot (\hat{v}_e \cdot f \cdot \Delta)$$

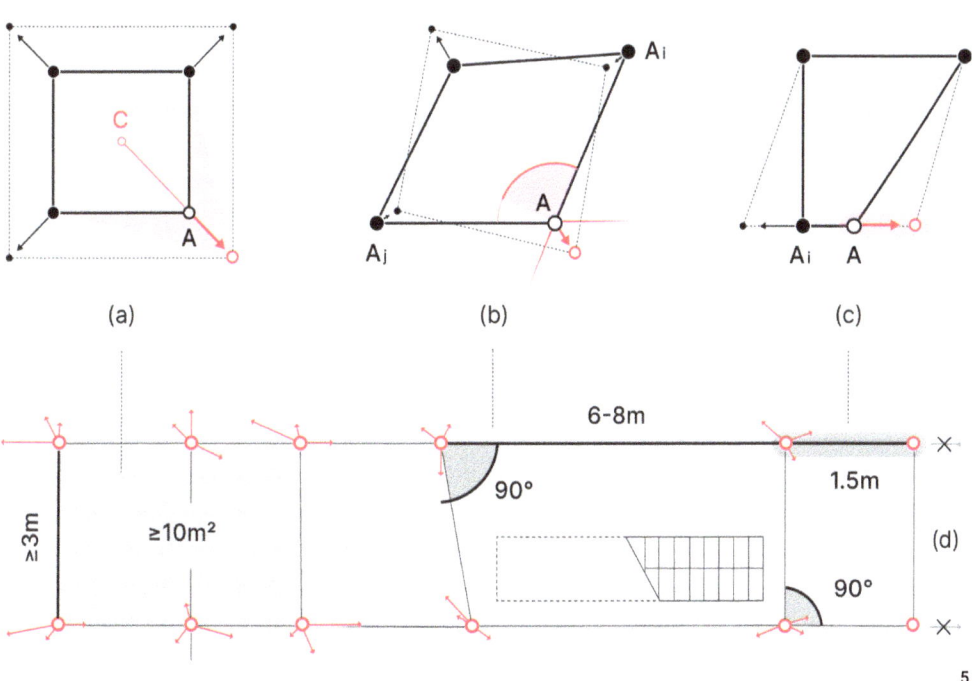

3 Site environment: user-defined areas to be avoided or populated by agents (a), translation of user input to vector fields (b)

4 Adverse spatial, functional, and visual impacts of truss members on architectural spaces

5 Area (a), angle (b), and length (c) behaviors; encoding of geometrical objectives in the topological agent environment (d)

Designer Interaction

While certain architectural requirements are well-defined and can be encoded at the beginning of a design process, many objectives are fluid, fuzzy, or only become apparent over the course of several iterations. The proposed methods for integration and negotiation are, therefore, conceived to be receptive to on-the-fly adjustments and designer interaction. We categorize the modes of interaction beyond defining behaviors and environments as indirect and direct.

Indirect interaction involves, assigning, adjusting, or extending parameter sets related to design objectives (e.g., floor area); as well as behavior prioritization by controlling behavior parameters through which the behavioral response to certain objectives (e.g., vector scaling factors) or the relationship between behaviors (e.g., behavior weights) can be tuned. These types of interactions are considered to be indirect since they merely affect the actions of agents and not their actual state.

Direct interaction, on the other hand, involves explicit designer interventions in the agent model. These include dragging agents to desired positions, fixing agents at desired positions, or adding and removing truss elements. In doing so, designers can counteract unfavorable agent actions or express intentions that are subjective, cannot be quantified or described explicitly.

CONCLUSION AND DISCUSSION

This research addresses agent-based modeling of adaptive truss structures from the perspective of architectural design. Specifically, it investigates the methodological implications for architects when exploring adaptable truss topologies with ABMS.

In view of the constituents and characteristics thereof, the paper proposes agent-oriented strategies for translating design intent to agent behaviors and environments, by focusing on a few objectives inherent to early architectural design processes. These include boundary conditions inherent to the building site, the architectural program defined in the design brief, as well as the corresponding geometrical or topological requirements. To accommodate the iterative and exploratory nature of early design processes, the proposed methods allow for objectives to be encoded and adjusted on-the-fly.

This work lays the methodological groundwork for designing architecturally meaningful adaptive structures, by providing a methodology that is receptive to both architectural and adaptability-specific design requirements. Therefore, it contributes to the "designability" of adaptive structures, and consequently, to the dissemination of structural adaptation as a relevant technology for low-emission construction and operation. It also contributes to the field of ABMS in general, as it showcases its potentials for architectural design exploration and integration. Lastly, it contributes to the broader field of computational design in architecture, as it bridges the gap between research and practice through addressing real-world design requirements along with the development of novel building systems and corresponding design methods.

However, to reflect the actual boundary conditions in practice, the scope of investigation must be expanded towards further aspects of design, such as materiality, building physics or construction. This could be addressed through additional agent constructs, or dedicated communication protocols to augment the agent model with other state-of-the-art modeling environments such as Building Information Modeling (BIM).

Another key question that remains open is that of design feedback. Even though the response of the agent model to user-defined constraints and objectives provides some degree of feedback to the designer, it does not provide information as to what a designer could possibly change to yield a more favorable response. This would be particularly relevant in the case of conflicting design goals that cannot be mediated by agents, or to inform direct designer interactions. Developing means for agents to communicate the origin or even potential solutions for conflicts would thus allow for a more productive and genuine collaboration between human and agent designers.

ACKNOWLEDGMENTS

This work was funded by the Deutsche Forschungsgemeinschaft (DFG, German Research Foundation), Project-ID 279064222 – SFB 1244 (A02). The authors are grateful for the generous support.

REFERENCES

Bogenstätter, Ulrich. 2000. "Prediction and Optimization of Life-Cycle Costs in Early Design." *Building Research & Information* 28 (5-6): 376–86. https://doi.org/10.1080/096132100418528.

Bonabeau, Eric. 2002. "Agent-Based Modeling: Methods and Techniques for Simulating Human Systems." In *Proceedings of the National Academy of Sciences* 99 (Suppl 3): 7280–87. https://doi.org/10.1073/pnas.082080899.

Dyedov, V., N. Ray, D. Einstein, X. Jiao, and T. J. Tautges. 2015. "AHF: Array-Based Half-Facet Data Structure for Mixed-Dimensional and Non-Manifold Meshes." *Engineering with Computers* 31 (3): 389–404. https://doi.org/10.1007/s00366-014-0378-6.

Geiger, F., J. Gade, M. von Scheven, and M. Bischoff. 2020a. "A Case Study on Design and Optimization of Adaptive Civil Structures." *Frontiers in Built Environment* 6. https://doi.org/10.3389/fbuil.2020.00094.

Geiger, F., J. Gade, M. von Scheven, and M. Bischoff. 2020b. "Optimal Design of Adaptive Structures Vs. Optimal Adaption of Structural Design." *IFAC-PapersOnLine* 53 (2): 8363–69. https://doi.org/10.1016/j.ifacol.2020.12.1604.

Gerber, D. J., E. Pantazis, and A. Wang. 2017. "A Multi-Agent Approach for Performance Based Architecture: Design Exploring Geometry, User, and Environmental Agencies in Façades." *Automation in Construction* 76: 45–58. https://doi.org/10.1016/j.autcon.2017.01.001.

Groenewolt, A., T. Schwinn, L. Nguyen, and A. Menges. 2018. "An Interactive Agent-Based Framework for Materialization-Informed Architectural Design." *Swarm Intelligence* 12 (2): 155–86. https://doi.org/10.1007/s11721-017-0151-8.

Heath, B. L., and R. R. Hill. 2010. "Some Insights into the Emergence of Agent-Based Modelling." *Journal of Simulation* 4 (3): 163–69. https://doi.org/10.1057/jos.2010.16.

Macal, C. M., and M. J. North. 2010. "Tutorial on Agent-Based Modelling and Simulation." *Journal of Simulation* 4 (3): 151–62. doi:10.1057/jos.2010.3.

Maierhofer, M., and A. Menges. 2019. "Towards Integrative Design Processes and Computational Design Tools for the Design Space Exploration of Adaptive Architectural Structures." In *Proc. of the 1st Int. ICETAD Conference*. Toronto, Canada. 113–20.

Maierhofer, M., M. Ulber, M. Mahall, A. Serbest, and A. Menges. 2020. "Designing (For) Change: Towards Adaptivity-Specific Architectural Design for Situational Open Enviornments." In *Proceedings of the 38th eCAADe Conference*, vol. 2. Berlin, Germany.

Ostertag, A., M. Dazer, B. Bertsche, F. Schlegl, S. Albrecht, P. Leistner, A. Gienger, J. Wagner, C. Tarín, and O. Sawodny. 2020. "Reliable Design of Adaptive Load-Bearing Structures with Focus on Sustainability." In *Proceeding of the 30th European Safety and Reliability Conference and 15th Probabilistic Safety Assessment and Management Conference*. Venice, Italy. 4703–10.

Reynolds, C. W. 1999. "Steering Behaviors for Autonomous Characters." In *Proc. of Game Developers Conference*. 763–82.

Schwinn, T. 2021. "A Systematic Approach for Developing Agent-Based Architectural Design Models of Segmented Shells: Towards Autonomously Learned Goal-Oriented Agent Behaviors." PhD diss., University of Stuttgart.

Schwinn, T., O. D. Krieg, and A. Menges. 2014. "Behavioral Strategies: Synthesizing Design Computation and Robotic Fabrication of Lightweight Timber Plate Structures." In *Proceedings of the 34th ACADIA Conference*. Los Angeles, California.

Senatore, G., P. Duffour, and P. Winslow. 2019. "Synthesis of Minimum Energy Adaptive Structures." *Structural and Multidisciplinary Optimization* 60 (3): 849–77. https://doi.org/10.1007/s00158-019-02224-8.

Sobek, W.. 2016. "Ultra-Lightweight Construction." *International Journal of Space Structures* 31 (1): 74–80. https://doi.org/10.1177/0266351116643246.

Steffen, S., S. Weidner, L. Blandini, and W. Sobek. 2020. "Using Influence Matrices as a Design and Analysis Tool for Adaptive Truss and Beam Structures." *Frontiers in Built Environment* 6. https://doi.org/10.3389/fbuil.2020.00083.

Stieler, D., T. Schwinn, S. Leder, M. Maierhofer, F. Kannenberg, and A. Menges. 2022. "Agent-Based Modeling and Simulation in Architecture." *Automation in Construction* 141:104426. https://doi.org/10.1016/j.autcon.2022.104426.

UNEP. 2020. "2020 Global Status Report for Buildings and Construction." Accessed August 20, 2022. https://globalabc.org/sites/default/files/inline-files/2020%20Buildings%20GSR_FULL%20REPORT.pdf.

Wortmann, T. 2017. "Opossum - Introducing and Evaluating a Model-Based Optimization Tool for Grasshopper." In *Proceedings of the 22nd CAADRIA Conference*. Hong Kong, China. 283–92.

IMAGE CREDITS

All drawings and images by the authors.

Mathias Maierhofer is a research associate at the Institute for Computational Design and Construction (ICD) at the University of Stuttgart, where he is also involved in teaching graduate students within the 'Integrative Technologies and Architectural Design Research' (ITECH) M.Sc. Programme. He holds a Bachelor's degree in architecture from Vienna University of Technology as well as a Master's degree with distinction from the ITECH Master program at the University of Stuttgart.

Achim Menges is a registered architect in Frankfurt and professor at the University of Stuttgart, where he is the founding director of the Institute for Computational Design and Construction (ICD) and the director of the Cluster of Excellence on Integrative Computational Design and Construction for Architecture (IntCDC). In addition, he has been Visiting Professor in Architecture at Harvard University's Graduate School of Design and held multiple other visiting professorships in Europe and the United States. He graduated with honors from the Architectural Association, AA School of Architecture in London.

Form-finding of Architectural Knitted Tensioned Structures

An Investigation on Form-finding and Analysis of Knitted Tensioned Structures in Architecture

Farzaneh Oghazian
Pennsylvania State University

Sam Moradzadeh
Pennsylvania State University

Felecia Davis
Pennsylvania State University

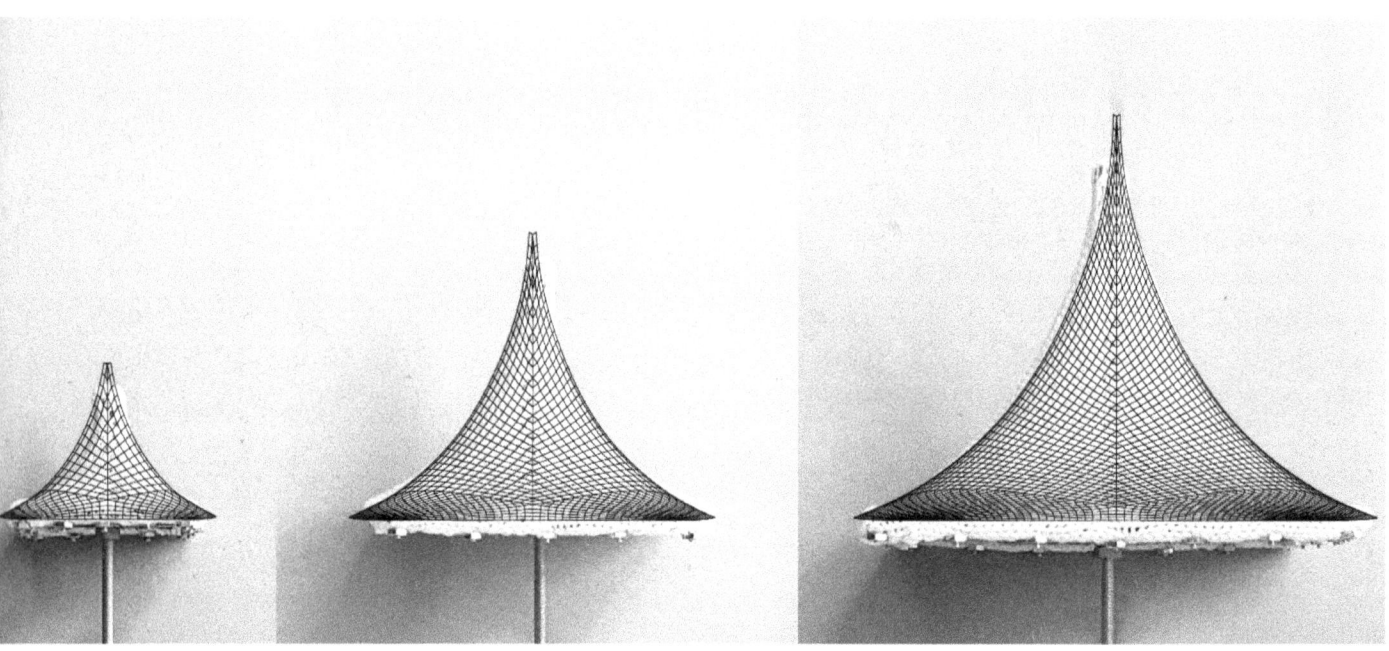

1   Comparison of physical and digital models at different scales

ABSTRACT

This study discusses the form-finding process of tensile structures and specially knitted tensioned structures. The performance of different plugins in Grasshopper is investigated for behavioral simulation of architectural knitted tensioned structures. Whereas multi-directionality and flexibility are the main characteristics of knitted textiles and allow these materials to be used to develop more complex architectural shapes, such characteristics also make digital simulation of knitted textiles more challenging. We explore the extent to which available tools can assist during the design process. Firstly, the theoretical aspects of the form-finding of tensile structural systems are discussed. Secondly, the form-finding and analysis of knitted textile structures are explored using Kangaroo2 along with K2 Engineering and Kiwi!3D. The performance of these tools is judged by comparing the digital form-found model against the actual physical model behavior and shape. Results shows that Kangaroo2 is a faster tool than Kiwi!3D when the actual stitch distribution is applied to generate the initial shape; additionally, tuning the length of the mesh edges to replicate the physical model behavior is easier in Kangaroo2. On the other hand, Kiwi!3D is a faster tool if a rough initial simulation for gaining insights into overall form. While Kangaroo2 yields a more detailed and accurate form-found shape, for the FE structural analysis it needs to be coupled with K2 Engineering, or the output of the form-finding from Kangaroo2 can be used as an input for FE analyses using Kiwi!3D. Rather than competing with each other, these software environments are different problem solving approaches; the process requires alternating between these plugins to achieve the project goal. The research shows that in using materials such as knitted textiles to develop architectural forms, successful digital modeling can be obtained by form-finding through the material instead of form-finding onto the material.

INTRODUCTION

Tensile, especially textile, structures are one of the main categories in building construction. For many years, lightweight structures made with technical woven systems have been one of the dominant methods of developing such designs. Within the last decade, however, other techniques of textile development, such as knitting, have found their way into architecture. Recent improvements in CNC knitting and computational design have increased the viability of knitted fabrics in architecture and enhanced the research in this field of study (Tamke et al. 2021; Thomsen et al. 2016; Ahlquist and Menges 2013; Sabin 2013).

Several potentials of knitted textiles, such as flexibility and seamless knitting, allow for the development of complex architectural shapes and composite systems. Additionally, knitted fabrics made of organic yarns enable architects to apply these materials to create sustainable structures. However, the successful implementation of knitted textiles in architecture requires developing a digital model that accurately represents the material's behavior (Oghazian, Brown, and Davis 2022).

Literature shows that tools already developed for simulating the tensile behavior of the fabrics are essentially focused on the two-directional characteristics of the woven fabrics. Although these two main directions can also be recognized in knitted textile structures—course and wale—knitted textiles are multi-directional materials. Studies have already been conducted showing the comparison or integration of multiple tools during the design process of developing different types of tensile structural systems. For example, the behavior of rods in bending-active forms is compared in three plugins, Kangaroo, Kiwi!3D, and SOFiSTiK, by Bauer et al. (2018). Magna et al. (2018) implemented different tools such as Kangaroo, Kiwi!3D, and SOFiSTiK at various design stages for bending-active structures, including rods and fabric. In another study by (Cuvilliers et al. 2018), two common algorithms of Kangaroo1 and Kangaroo2 are compared regarding the reliability, speed, and accuracy of the form-finding process by defining numerical calculation from physical model measurements and comparing them with the form-found models.

However, few studies have been conducted to understand the potential of the available computational tools for simulating the behavior of 3D knitted architectural tensioned structures. This research focuses on comparing existing solvers to demonstrate each one ability and limits and suggesting improvements for simulating the behavior of tensioned knitted structures. The selected approaches are Kangaroo2 (Piker 2018), K2Engineering (Brandt 2019), and Kiwi!3D (2020), and for future investigation, we will also add SOFiSTiK. The chosen tools are not competing with one another and could be beneficial at different stages of the design process. Alternative software solutions in the field of knitted tensioned structures that are commercially available or presently under development are not covered here since they are either in beta testing or not fully functional.

FORMFINDING OF TENSILE STRUCTURES, COMMON TOOLS, AND POTENTIALS

In general, the current design process of the tensile structures includes three main steps: the first step is form-finding (defining initial shapes and boundary conditions); the second is adding material properties, loading, and analysis; and the third step is generating cutting patterns (Figure 2) and sometimes compensation (Puystiens et al. 2016).

The form-finding step includes creating an initial shape based on the boundaries' condition and prestress in the textile. The output of the form-finding step is an equilibrium form. The next step is to analyze the form-found model considering loading effects, membrane stresses and displacements, material, and geometric properties of the flexible fabric. If computed stresses and displacements are not within allowable limitations, the form-finding process should be repeated, modifying boundary conditions and initial prestresses (Dutta and Ghosh 2019).

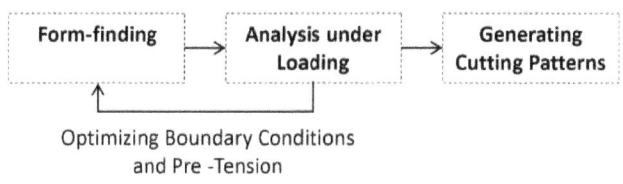

2   Methodology for the design of membrane structures (Puystiens et al. 2016)

The most popular form-finding methods are transient stiffness, force density, and dynamic relaxation (Lewis 2008). Iterative computation for producing shapes in static equilibrium is the primary goal of the three approaches (Figure 3):

- The Transient Stiffness method is dependent on the small displacements and rotations of elements. To discretize the surface in this method, lines in two directions intersect in a node. Stiffness resulting from the prestress changes the nodal forces and consequently alters the surface geometry (Lewis 2008).

- Force Density uses surface discretized as a system of branches. The boundary conditions are assumed, and the only factor affecting the structure's shape is the force density value (Lewis 2008). Force density is defined as force per unit of the length of the discretized membrane network.

Form-Finding Method	Definition	Features	Surface Discretization Method
Transient Stiffness	Assumption of small displacements and rotations based on the forces	No material properties are required for the form-finding step The ratio of force to length or stress to surface area is central in this method (Veenendaal and Block 2011)	Lines in two directions that intersect in a node
Force Density	The boundary conditions are assumed. The value of the force densities determines the shape of the structure	No material properties are required for the form-finding step Only geometric stiffness is required for the form-finding (Dutta and Ghosh 2019)	Branch of elements
Dynamic Relaxation	Equilibrium state in the structural system over time under the influence of the damping	Analogy with motion	Triangular mesh of points for constant tension membranes and the forces are independent of the shape of the triangular meshes

3   Three main approaches for form-finding in tensile structures

- In the Dynamic Relaxation method, the mass of discretized continuum is supposed to be concentrated at the surface's nodes. A small number of arithmetic operations is required because computations for nodes will be in turn instead of analyzing all the nodes simultaneously. This method is more efficient than the Transient Stiffness method. In Dynamic Relaxation method, equilibrium will be obtained over time and under the damping influence (Dutta and Ghosh 2019). This method uses the motion approach, which means residual forces are converted to the nodes' velocities, and mass determines the acceleration (Veenendaal and Block 2011).

It is apparent that in all of these approaches, the problem of form-finding is geometric and thus material-independent (Dutta and Ghosh 2019). Digital platforms for the design and analysis of tensile structures adopt one of these form-finding approaches. As presented in Figure 4, most software applications freely available to architects and architecture students are plugins embedded in Rhino and Grasshopper, such as Rhino Membrane, Kangaroo2, K2Engineering, and Kiwi!3D. Some other software applications require subscription and are mainly used for structural analysis and construction of tensile structures, such as K3-Tent, Tensyl, Easy, Touch Cad, and RFEM (DLUBAL n.d.).

All software applications mentioned in Table 2 are designed for cable-net and tensile structures using technical woven composite fabrics. How we can adopt the current platforms to simulate knitted textiles is the central question of this research.

FORMFINDING OF TENSIONED KNITTED TEXTILES AND CHALLENGES

Knitted textiles are new materials in architecture. As discussed in the previous section, not many studies have been conducted to simulate these materials' structural behavior and shape. Different processes for the simulation of these materials should be considered compared to common woven tensile structures because, in knitted textiles, the textile and overall form will be shaped simultaneously. Therefore, the properties of the knitted textile will inform the shape and size of the initial as well as the final form that will emerge after post-tensioning.

Regarding simulations of the behavior of the knitted materials in architecture, springForm, developed by Ahlquist and Menges (2013), is a tool that uses physics-based simulation in processing for simulating knitted textiles shapes. In a study by Schmeck and Gengnagel (2016), a grid with hexagonal cells that substitutes stitches in a knitted textile structure for a planar shape is studied using Kangaroo2 for the form-finding.

Software/Method/Structures	Initial Shape	Form-Finding	Analysis	Cutting Pattern
Rhino Membrane/ Natural Force Density Method/ Tensile and cable net structures, Woven textiles	Mesh			No
Kangaroo2/ Dynamic Relaxation Method/	Mesh			No
K2Engineering/ Dynamic Relaxation Method/ Cable nets and Grid shells	Mesh			No
Kiwi3D/ Finite Element Methods and Isogeometric Analysis (IGA)/	NURBS			No
K3-Tent (LLC)	Yes	Yes	No	Yes
Tensyl (Shepherd 2007)	Yes	Yes	Yes	Yes
Easy (GmbH)/ Force Density Method/ lightweight structure design (Ströbel and Singer 2005) [Images from Easy website]	Yes	Easy.Form	Easy.Stat	Yes Easy.Cut
Touch Cad (Lundström-Design)	Yes	No	Yes	Yes
RFEM (DLUBAL) [Images from RFEM website]				Yes

4 Common plugins for the form-finding process and designing tensile structures

In another study by Oghazian, Farrokhsiar, and Davis (2021), a 3D knitted shape is used as an initial shape, and through manipulation of the length of the mesh edges, the methods for inputting the material characteristics are explored. The table in Figure 5 illustrates some of these studies in the literature.

METHODOLOGY

This section represents the form-finding and analysis of a conical knitted tensioned structure as a case study using Kangaroo2 in conjunction with K2 Engineering and Kiwi!3D. The conical shape is selected as one of the most common tensioned structure models we could knit seamlessly using our lab's simple hand-held knitting machine (Silver Reed SK 840, standard gauge). Figure 7 and 8 show a series of conical models made with wool yarn. The first section of Figure 4 also shows the simulation results for another common form in tension structures known as hypar.

In all of the plugins, the main steps of form-finding are: defining the initial shape, determining the boundary conditions, applying forces, and inputting knitted textile material characteristics. The potential and limitations of the tools for simulation of the 3D knitted tensioned structure are highlighted in the following sections.

Reference	Software/ Interface/ Logic	Pictures
(Ahlquist and Menges 2013)	springFORM/ Processing/ Spring particle-based Knitting	
(Schmeck and Gengnagel 2016)	Kangaroo2/ Grasshopper/ Spring particle-based Knitting	
(Oghazian, Farrokhsiar, and Davis 2021)	Kangaroo2/ Grasshopper/ Spring particle-based Knitting	

5 Software and plugins that are designed or used in literature studies for analyzing knitting behavior

Kangaroo2 and K2 Engineering

The Dynamic Relaxation (DR) approach for solving structural problems was initially created in the 1960s (Day 1965) and has since gained widespread acceptance in various areas. DR is often related to stiffness-independent membrane shape discovery in the built environment (Barnes 1999) or with the distribution of architectural grids across curved surfaces (Williams 2001). Instead of using the usual force-based technique, where added forces at nodes produce accelerations, the Kangaroo solver works based on the configuration of the vertices with various degrees of freedom. While Kangaroo1 uses Dynamic Relaxation, Kangaroo2 logic is based on the Projective Constraints method (Cuvilliers et al. 2018). Hooke's Law is used in Kangaroo for determining axial stiffness and bending stiffness is based on the Barnes/Adriaenssens model (Barnes, Adriaenssens, and Krupka 2013).

Since Kangaroo accepts mesh as an input for the initial shape, one advantage of using such a tool is that the mesh can replicate the initial form that should be knitted with each mesh face as a substitute for a stitch. Also, the sizes and number of these mesh faces can be controlled so that the shape and size of this initial shape correspond to the actual material properties. As it is illustrated in Figure 3, the initial mesh of a cone is simulated, and the mesh faces are distributed to correspond to the knitting patterns. Then the form-finding settings should be determined. One advantage of the Kangaroo2 is that the initial and final positions of the anchors can be different. It is crucial because in simulating materials like knitted textiles that possess elasticity obtained from their unique structure, the stretchability of the material and extent to which the material could be stretched can be simply simulated by assigning points from the initial mesh as an initial anchor state and points from the final destination. Another advantage of the Kangaroo2 is that a length factor can be assigned to the mesh edges that represent the stretchability of different sections of the model. In working with knits as a flexible and multi-directional material, stitches will behave differently in different areas of the model. Assigning different length factors to these various areas allows having better control over the length change of the meshes to correspond to the actual material behavior. More details about the model's simulation and the automated process can be found in the authors' previous papers (Oghazian, Farrokhsiar, and Davis 2021; Oghazian, Brown, and Davis 2022).

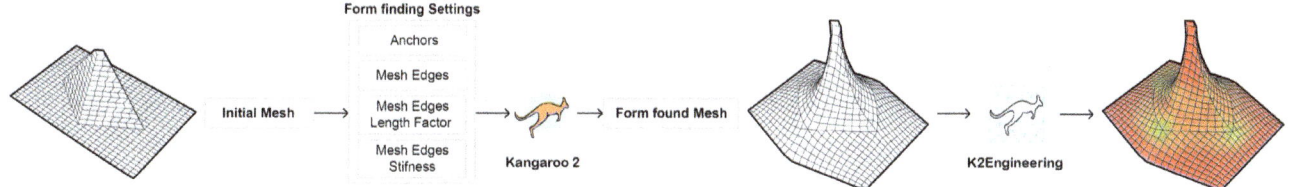

While Kangaroo can perform the form-finding process, K2Engineering can be used for the further structural analysis of the form-found shape (Figure 9). K2Engineering for Kangaroo is a plugin developed by Cecilie Brandt-Olsen that contains a set of customized Kangaroo2 Grasshopper components that can be used for calibrating goals regarding the structural performance of the tensioned fabric structures (Brandt 2019). Both form-finding and analysis can be performed within the Grasshopper environment using a combination of Kangaroo2 and K2Engineering. Figure 10 shows a series of simulations under the self-load of the knitted textiles of different E modulus. While at the form-finding step, the length factor of the meshes is the essential element in determining the overall form, at the analysis step, the features such as the E modulus determine how the structure performs under the self-weight or external loads.

Kiwi!3D

Kiwi!3D is another plugin for Grasshopper/Rhino that uses Isogeometric Analysis (IGA) techniques developed by Hughes et al. (2005) for finite element analysis. The advantage of this approach is that it uses Non-Uniform Rational B-Splines (NURBS) as the basis of the analysis. It is essential to consider because most form-finding tools perform on mesh as the initial shape, while in Kiwi!3D control points play a role in form-finding; therefore, remeshing and element discretization is not required. While this is an advantage in a working environment such as Rhino, it would become problematic—when simulating knitted textiles—to substitute each stitch with a simpler shape to have better control of the stitch itself and the combination of stitches that define the overall form. Each stitch can, however, be simulated as one brep, and the joined B-reps can be fed into the form-finding algorithm. One manipulation required at this point would be to change the polynomial degree of individual B-reps to one so that each B-rep edge can be treated as a line. One drawback of this method is that it would be computationally more expensive than using meshes in Kangaroo. Thus Kiwi!3D would be a good method if simulating the stitch distribution is not the case. Otherwise, Kangaroo performs much faster. Figure 11 illustrates a simulation process that takes the output of Kangaroo2 simulation for further non-linear analysis of the model under self-load in Kiwi!3D. Another advantage of IGA during the design process is the independence of the parametrization for boundary conditions such as loads, supports, and coupling entities. This allows the user to seamlessly integrate

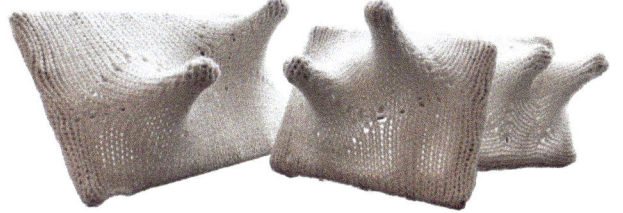

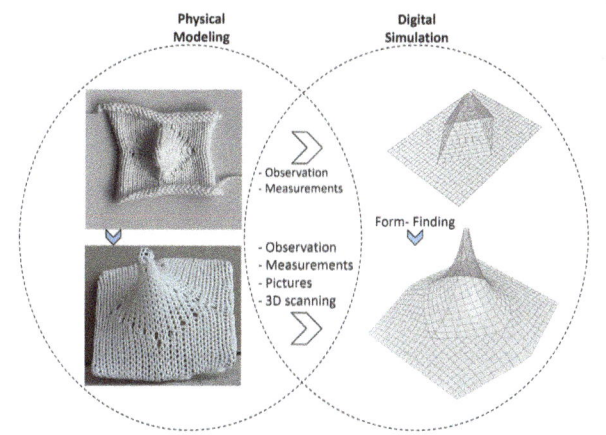

6 Form-finding with Kangaroo and displacement analysis with K2Engineering

7 A series of conical knitted samples

8 Single conical shape used as a case study of the simulations

9 Correlation between physical and digital simulation as well as initial and final form-found mesh

different design elements and have better assessments without losing or needing to estimate the stresses and displacements (Bauer et al. 2018).

PHYSICAL MODELING/DIGITAL SIMULATIONS

To have a reliable model, the digital simulations should always be compared against the physical models. Since the dimensions and shape of the initial mesh are according to the actual knitted sample, if we could predict the length change of meshes at different sections of the models, then we can get a form-found model that represents the actual material behavior. Figure 9 shows the interrelated connection between the physical and

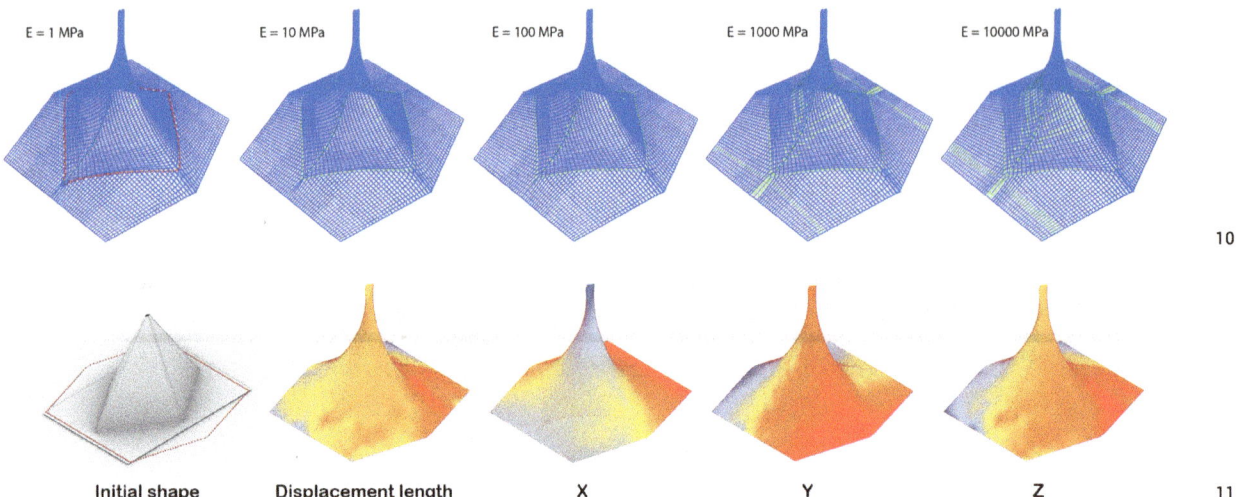

10 Analysis with K2Engineering; E modulus is the variable, and colors show the axial force

11 Kiwi!3D analysis with different force factors

digital model as well as the initial and final form-found mesh. Additionally, the simulations are done for a series of models of the same shape but at different scales to understand the behavior change across the scales. As we scale up, the number of stitches in the actual physical model increases, as does the number of mesh faces in the digital model. This increases the computational process time. One solution would be to pack a series of mesh faces into one mesh face. How reliable to model would be is the topic of another ongoing research.

CONCLUSIONS AND CONTRIBUTION

This research study shows the potential of different Grasshopper plugins for simulating and analysis of knitted textile structures. The main contribution of this study is elaborating the form-finding and analysis of knitted tension structures using Kangaroo2, K2Engineering, and Kiwi!3D, emphasizing the material behavior and shape of the knitted model. The results show that Kangaroo 2 is a faster tool in case the stitches' exact distribution is to be applied during the simulation process. Having a mesh with such a mesh face configuration is essential because it allows the users to have control over the individual stitch. This becomes more helpful, especially when dealing with knitted textiles with a gradient pattern. Additionally, having a mesh that represents the actual stitch distributions allows the designer to get more accurate feedback after the analysis step.

The main takeaway of this study is that these tools suggest a new way of practicing that incorporates software that allows for multiple behaviors from a 'material,' a craft, rather than a design being imposed on an abstract 'material' (hylomorphic and Albertian) (Ingold 2009). With the drafted algorithms using the aforementioned plugins, we now have a way to be in conversation with the formation, thus conducting form-finding through the material, not form-finding onto the material.

REFERENCES

Ahlquist, S., and A. Menges. 2013. "Frameworks for Computational Design of Textile Micro-Architectures and Material Behavior in Forming Complex Force-Active Structures." In *ACADIA 2013 Adaptive Architecture; Proceedings of the 33rd Annual Conference of the Association for Computer Aided Design in Architecture*. 281–292.

Barnes, Michael R. 1999. "Form finding and analysis of tension structures by dynamic relaxation." *International Journal of Space Structures* 14 (2): 89–104.

Barnes, Michael R., Sigrid Adriaenssens, and Meghan Krupka. 2013. "A novel torsion/bending element for dynamic relaxation modeling." *Computers and Structures* 119: 60–67.

Bauer, Anna M., Philipp Längst, Riccardo La Magna, Julian Lienhard, Daniel Piker, Gregory Quinn, Christoph Gengnagel, and Kai-Uwe Bletzinger. 2018. "Exploring Software Approaches for the Design and Simulation of Bending Active Systems." In *IASS 2018: Creativity in Structural Design*.

Brandt, Cecilie. 2019. K2Engineering. V. 1.0.0. Released on 2019-04-27. https://grasshopperdocs.com/addons/k2engineering.html.

Cuvilliers, Pierre, Justina R. Yang, Lancelot Coar, and Caitlin Mueller. 2018. "A comparison of two algorithms for the simulation of bending-active structures." *International Journal of Space Structures* 33 (2): 73–85. https://doi.org/10.1177/0266351118779979.

Day, A. S. 1965. "An introduction to dynamic relaxation (Dynamic relaxation method for structural analysis, using computer to calculate internal forces following development from initially unloaded state)." *Engineer* 219: 218–221.

DLUBAL. n.d. *RFEM*. Product page. https://www.dlubal.com/en/products/rfem-fea-software/what-is-rfem.

Dutta, Subhrajit, and Siddhartha Ghosh. 2019. "Analysis and Design of Tensile Membrane Structures: Challenges and Recommendations." *Practice Periodical on Structural Design and Construction* 24 (3). https://doi.org/10.1061/(asce)sc.1943-5576.0000426.

GeoS Center LLC. n.d. *K3 Tent*. https://k3-tent.com/what-is-it/.

Hughes, Thomas J. R., John A. Cottrell, and Yuri Bazilevs. 2005. "Isogeometric Analysis: CAD, Finite Elements, NURBS, Exact Geometry and Mesh Refinement." *Computer Methods in Applied Mechanics and Engineering* 194 (39–41): 4135–4195.

Ingold, T. 2009. "The textility of making." *Cambridge Journal of Economics* 34 (1): 91–102. https://doi.org/10.1093/cje/bep042.

Kiwi!3D. 2020. *Kiwi!3D*. https://www.kiwi3d.com/.

La Magna, Riccardo, Valia Fragkia, Philipp Längst, Julian Lienhard, Rune Noëlb, Yuliya Šinke Baranovskaya, Martin Tamke, and Mette Ramsgaard Thomsen. 2018. "Isoropia: An Encompassing Approach for the Design, Analysis and FormFinding of Bending-Active Textile Hybrids." In *Proceedings of IASS Annual Symposia, IASS 2018 Boston Symposium*.

Lewis, W. J. 2008. "Computational form-finding methods for fabric structures." *Proceedings of the Institution of Civil Engineers; Engineering and Computational Mechanics* 161 (3): 139–149. https://doi.org/10.1680/eacm.2008.161.3.139.

Lundström-Design. n.d. *TouchCAD*. https://www.touchcad.com/index_tc.html.

Oghazian, F., N. Brown, and F. Davis. 2022. "Calibrating A Formfinding Algorithm For Simulation of Tensioned Knitted Textile Architectural Models." In *Proceedings of the 27th International Conference of the Association for Computer-Aided Architectural Design Research in Asia (CAADRIA)*, vol. 1. 111–120.

Oghazian, F., P. Farrokhsiar, and F. Davis. 2021. "A simulation process for implementation of knitted textiles in developing architectural tension structures." In *Inspiring the Next Generation; Proceedings of the International Conference on Spatial Structures 2020/21*.

Piker, Daniel. 2018. *Kangaroo2*. https://grasshopperdocs.com/addons/kangaroo-2.html.

Puystiens, Silke, Maarten Van Craenenbroeck, Lars De Laet, Danny Van Hemelrijck, Wim Van Paepegem, and Marijke Mollaert. 2016. "Integrated analysis of kinematic form active structures for architectural applications: Design of a representative case study." *Engineering Structures* 124: 376–387. https://doi.org/10.1016/j.engstruct.2016.06.038.

Sabin, Jenny E. 2013. "myThread Pavilion: Generative Fabrication in Knitting Processes." In *ACADIA 13: Adaptive Architecture; Proceedings of the 33rd Annual Conference of the Association for Computer Aided Design in Architecture (ACADIA)*. 347–354.

Schmeck, Michel, and Christoph Gengnagel. 2016. "Calibrated Modeling of Knitted Fabric as a Means of Simulating Textile Hybrid Structures." *Procedia Engineering* 155: 297–305. https://doi.org/10.1016/j.proeng.2016.08.032.

Shepherd, Paul. n.d. *Tensyl*. https://people.bath.ac.uk/ps281/software/tensyl/.

Ströbel, Dieter, and Peter Singer. 2005. "Computational Modelling of Lightweight Structures: Formfinding, Load Analysis and Cutting Pattern Generation." In *Textile Roofs: The 10th International Workshop on the Design and Practical Realisation of Architectural Membrane Structures*. Berlin, Germany.

Tamke, Martin, Yuliya Sinke Baranovskaya, Filipa Monteiro, Julian Lienhard, Riccardo La Magna, and Mette Ramsgaard Thomsen. 2021. "Computational knit–design and fabrication systems for textile structures with customised and graded CNC knitted fabrics." *Architectural Engineering and Design Management* 17 (3-4): 175–195.

technet GmbH. n.d. *Easy*. https://www.technet-gmbh.com/en/products/easy/.

Thomsen, Mette Ramsgaard, Martin Tamke, Ayelet Karmon, Jenny Underwood, Christoph Gengnagel, Natalie Stranghoner, and Jorg Uhlemann. 2016. "Knit as bespoke material practice for architecture." In *ACADIA 16: Posthuman Frontiers: Data, Designers and Cognitive Machines; Proceedings of the 36th Annual Conference of the Association for Computer Aided Design in Architecture*. 280–289.

Veenendaal, Diederik, and Philippe Block. 2011. "A Framework for Comparing Form Finding Methods." In *Taller, longer, lighter: Meeting growing demand with limited resources*, vol. 98. IABSE ETH Zurich.

Williams, Chris JK. 2001. "The analytic and numerical definition of the geometry of the British Museum Great Court Roof." *Mathematics & Design 2001*, edited by M. Burry, S. Datta, A. Dawson, A. J. Rollo. Geelong, Victoria, Australia: Deakin University. 434–440.

IMAGE CREDITS

Figure 2: Redrawn after Puystiens et al. (2016); figures in the tables are referenced according to the sources used; images in Figure 3 are redrawn from associated sources.

All other drawings and images by the authors.

Farzaneh Oghazian is a PhD architecture candidate and researcher in the Stuckeman Center for Design Computation at Penn State. Her research focuses on developing computational methods to enhance the implementation of knits as materials for architectural application.

Sam Moradzadeh is a Master's student and researcher in Design Computing. His research focuses in the intersection of Human-Centered Design, Extentended Reality, and design research.

Felecia Davis is an associate professor of architecture in the Stuckeman School at Penn State and the director of the Computational Textiles Lab (SOFTLAB) in the Stuckeman Center for Design Computing.

Session Introduction

New Ecologies I: Biomaterials / Biotech

Laia Mogas-Soldevila, Chair

These texts unfold the complexities of biomaterial-driven design and fabrication at all scales of the architectural discipline. Underwater landscapes are scrutinized, and their sedimentation dynamics redesigned in favor of coral reef restoration; new grid shell logics are derived by embracing baboo growth eccentricities in favor of performance heterogeneous structures; to make walls, cellulose and other soft waste materials are combined into pastes to manufacture modules able to interlock despite their brutal shrinkage dynamics; column-making mycelium is guided to achieve intricate surface geometries by growing within printed lost formwork, and finally; extruded fibrous hydrogels restore wooden window frames while augmenting aesthetics with in-situ pigment diffusion dynamics.

In all projects, the role of material heterogeneity is refreshingly celebrated as a design driver instead of traditionally discarding it as a nuisance. The multiagent dynamics of oceanic sediment deposition directly inform new solid-to-porous geometries that redirect geological shape to protect corals from humans and their effect in the environment. Intrinsic non-regular biological geometric properties that make the section of bamboo vary throughout its length are direct inputs to develop a digital design workflow. Cellulose pastes' hygroscopicity and shrinkage are carefully harnessed to derive cyclopean stacking as materials harden becoming compression stable. New solutions arise when facing micro-fibrillated cellulose's segregation within its carrier hydrogel and when discovering mycelium's bonding to certain 3D printing materials.

The generation of new data is a crucial aspect in these investigations as biomaterial-aware digital design and fabrication is still in its infancy and new datasets globally devised by researchers become bodies of knowledge to push the discipline further. Contemporary digital design platforms are already able to account for and accurately simulate complex phenomena such as fluid dynamics, finite element analysis, energy transmittance, computer-aided-machining etc.; however, they yet do not have the ability to design with the properties of natural materials. The authors unveil specifically relevant new information on evaporation-driven self-shaping, beams' section eccentricity, differentially distributed porosity, time-based digestion, or pigmentation during hydrogel drying. Importantly, they can encode it within digital design to fabrication workflows or are able to parametrize it for others to do so in a joint effort towards an inevitable era of biomaterial-driven tools.

Finally, works hint at the creation of a new aesthetic. One that results from bringing the organic into the digital world; from guiding unruly, curly, meandering, soft, and flowy matter into dry robust structures and systems. In collaboration with biology, our built environment will soon embrace currently uncommon phenomena of digestion, flow, deformation, wicking, shrinkage, or decay. Efforts presented next, contribute to its much needed cultural and disciplinary acceptance.

RePrint

Digital Workflow for Aesthetic Retrofitting of Deteriorated Architectural Elements with New Biomaterial Finishes

Rebecka Rudin
Malgorzata A. Zboinska
Department of Architecture
and Civil Engineering/
Chalmers University of Technology

Sanna Sämfors
Paul Gatenholm
Department of Chemistry and
Chemical Engineering,
Chalmers University of Technology

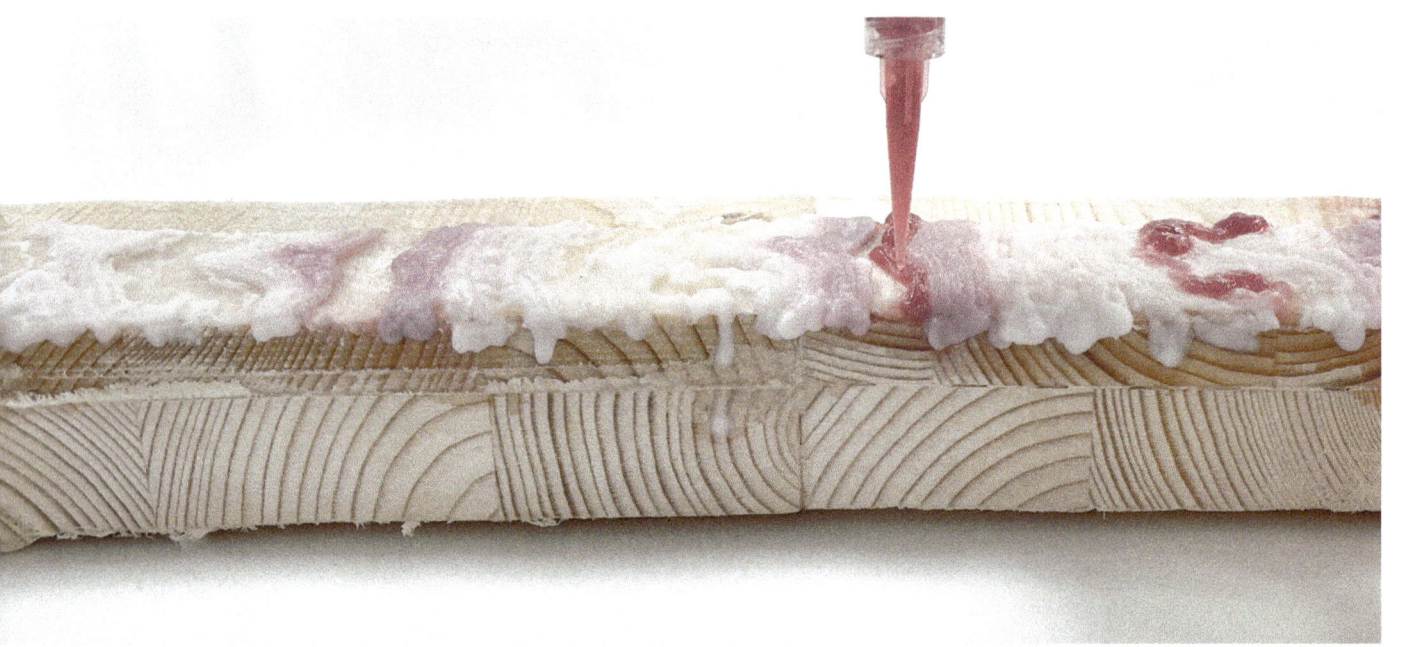

1A Robotic 3D printing of a regenerative biobased coating, applied onto a digitally reproduced model of a historic window casing

ABSTRACT

Digital fabrication offers new opportunities for revitalizing aged buildings in the time of craft expertise decline and higher demands regarding the sustainability of employed materials. Precise reproduction of architectural elements with digital 3D reconstruction methods, such as photogrammetry, and their repair using agile robotic 3D printing involving new environmentally friendly materials can save time and resources, leading to more circular design and manufacturing.

This study presents digital workflows for architectural restoration, based on the concept of aesthetic retrofitting of deteriorated wooden architectural elements through the application of surface finishes from a novel biomaterial—microfibrillated cellulose hydrogel upcycled from forestry waste. The workflows were established through experimental digital design and reproduction of wooden architectural details in an existing historical building, and executed within an integrated digital framework combining photogrammetry, 2D graphics processing, computational design, and robotic 3D printing. The investigation sought to demonstrate the potential of microfibrillated cellulose as a material suitable for applications in renovation and conservation. Further, the intention was to elucidate the role of digital tools as new media of restoration that enable to uplift cultural assets in an alternative way—by allowing the embedding of aesthetic features that convey the contemporaneity of remedial interventions. Aiming to contribute to current work in experimental preservation, the study offers a novel approach in which deteriorated architectural elements are endowed with a new materiality that follows the new logic of circularity in contemporary design and construction.

INTRODUCTION

In the time of depleted resources, where the construction sector generates large quantities of environmentally harmful waste, researchers advocate for new solutions involving renewable, biobased materials that can make the shift toward more circular and resource-efficient architectural design. Among such proposals, those that harness the potentials of existing architectural elements through repair, redesign, and reuse are prominent. A good example is *The Re-Use Atlas*, issued by the Royal Institute of British Architects, that presents a large and diverse collection of experimental projects following these approaches (Baker-Brown 2017). Theoretical discourse on the subject is also visible in a growing number of critical texts on reuse and appropriation (Brilliant and Kinney 2016; Reeser Lawrence and Miljački 2018).

Related to these contexts, this study synergizes knowledge from digital restoration, novel sustainable material developments, computational design, and robotic fabrication to investigate how deteriorated architectural elements could be preserved and creatively reimagined by retrofitting them with new, 3D printed surface finishes from an architecturally unexplored yet resource-efficient biomaterial—microfibrillated cellulose (MFC) hydrogel derived from forestry waste (Figure 1B). The work forms the first stage of a larger investigation exploring hybrid interfacing of aged wood with a new material that also originates from the forest yet has a profoundly different materiality.

Our initial trials of applying pure MFC hydrogel onto various wooden surfaces indicated limitations in the types of textures it is able to adhere to without other additives (Figure 2). We also observed significant changes in the material's appearance upon drying, such as alterations of form and color. Consequently, the first stage of the study was limited to retrofitting wooden architectural elements outside the confines of a stringent conservation practice, permitting more liberal interventions. In further studies, however, we aim to establish material compositions adapted to specific types of degraded wood, to facilitate application in traditional conservation as well.

The focus of the presented work has arisen from the ambition to critically reflect on the possibility of a hybrid, extended restoration practice that would concern all aged elements, regardless of historic or cultural status. Most buildings contain degraded components, and such elements will always require careful assessment of aesthetic, architectonic, cultural, and historic values. We argue that this evaluation would benefit from dual perspectives—of a historic preservationist on the one hand, and an architect designing new interventions on the other. Therewith, the customary foci of both preservation and architectural design could be broadened by placing particular emphasis on the intention of the interventions, allowing for hybrid agendas that preserve and recreate while also enriching material artifacts.

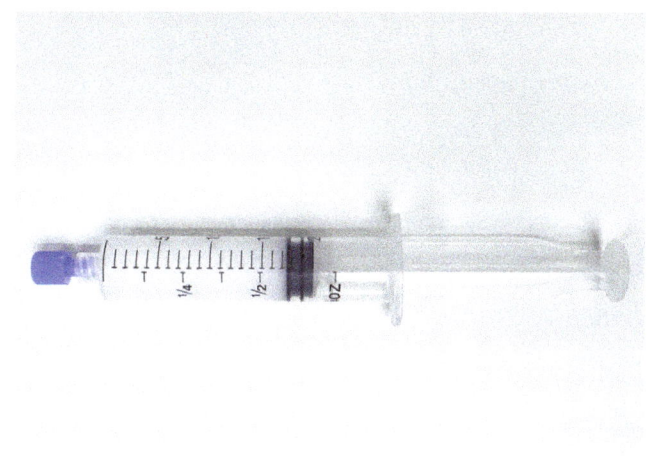

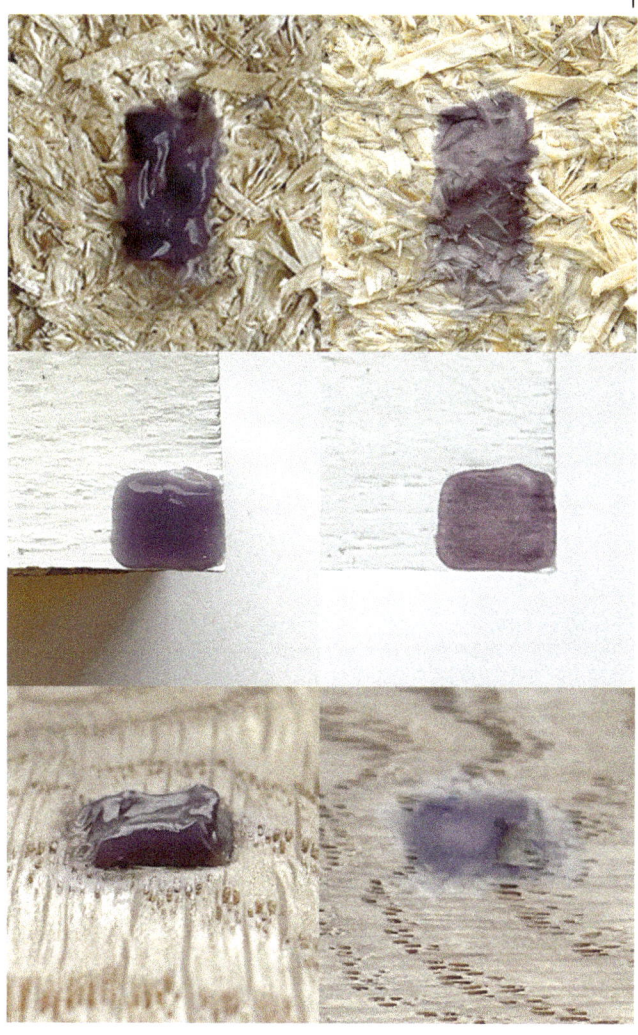

1B Unpigmented microfibrillated cellulose hydrogel, loaded into a syringe

2 Excerpt from initial trials of coating various underlays with MFC hydrogel, in wet (left column) and dry (right column) state; from top to bottom, on chipboard, acrylic-painted wood, and oiled oak

If architectural values of an element considered for repair or restoration are insufficient, then an extended, hybrid conservation practice could serve to negotiate a new materiality that highlights the contemporaneity of the intervention while protecting any noteworthy historical or cultural layers. This could offer a prolonged life span to numerous aged architectonic elements that would otherwise be regarded as construction waste. Instead of unsustainable strategies of landfill disposal, combustion, replacement with new products, or repair using synthetic paints, resins and other chemicals, this study argues for retaining degraded elements within buildings by retrofitting them with contemporary, bio-based material layers and additions, in line with emerging concepts of circularity and rational resource use in construction.

STATE OF THE ART

Digital restoration is a field that deals with historical reconstruction and preservation using latest digital technologies. It is represented by practices such Factum Arte in Spain and the Institute of Digital Archaeology in the UK, as well as by numerous academic studies by research institutions across the globe. The primary focus of mainstream digital restoration is faithful reconstruction and creation of accurate replicas. Hence, the current methods, tools and materials comprise an established, standardized repertoire, which is essential for maintaining the required high quality and accuracy of the restorative work.

An analysis of recent literature indicates that the methodology of digital restoration relies on a common routine. It encompasses four linearly progressing steps of element reproduction, prototyping, post-processing, and fitting on-site, each supported with off-the-shelf digital tools employed according to state-of-the-art protocols of use (Bonora et al. 2021). The first step involves high-precision 3D reproduction of an artifact using established methods, such as laser scanning or photogrammetry (Aicardi et al. 2018). The second step embraces digital processing, fitting, and sculpting of the reconstructed 3D model of the artifact in a 3D modeling environment to achieve high resemblance in relation to the physical piece. The third step encompasses standardized toolpath generation through 3D model slicing in dedicated software, followed by planar, layer-by-layer CNC milling or 3D printing of an object's physical replica (Xu, Ding, and Love 2017). Usually, it is desirable to manufacture a copy with no machine traces, which necessitates a fourth step of post-processing, involving manual procedures of smoothing, sanding, and coating (Higueras, Calero, and Collado-Montero 2021).

The materials employed in mainstream digital restoration, for artifacts recreated using digital manufacturing techniques such as 3D printing, often feature mixes of synthetic polymers with natural ingredients, e.g., ceramics, wood fibers, and stone powder, whereas the materials applied manually as the final finish layer range from traditional ones, based on natural ingredients, such as ceramic putty, to synthetic ones, e.g. acrylic paint, latex, and resins (Acke et al. 2021). Hence, plant-based hydrogels have not yet been established as materials for architectural restoration. Instead, prior studies demonstrated their use for 3D printing of new, standalone architectural elements (Duro-Royo, Mogas-Soldevila, and Oxman 2015; Malik et al. 2020).

Interestingly, alongside the mainstream work, there also exists an experimental and critical trajectory in digital restoration that aims to challenge the creation of precise historical replicas by questioning their authenticity and materiality. For instance, Factum Arte's three reproductions of a historic sculpture 'Venus Victrix,' showcased at the Venice Architecture Biennale in 2016, were intentionally presented in three different materials—glass, wax, and resin, and fabricated using different techniques—traditional casting into molds, modern 3D printing, and a combination of the two. In this way, the project has shown that certain features, such as fabrication tool traces and specific properties of the chosen reproduction material, should be regarded as constitutional traits of the replica, profoundly impacting its visual expression and historic interpretation. A similar intention was expressed when reconstructing the Notre Dame Cathedral in Paris after the fire in 2019. Because genuine Parisian stone and traditional crafting knowledge are no longer available, a proposal was put forth to recreate the cathedral's stone sculptures using 3D printing with a new material comprising powdered limestone and ashes collected from the fire. It was argued that, through this, the bygone materiality of the building could be closely reproduced while more truthfully displaying its new character (Geboers and Baldassari 2019).

These alternative approaches, although indicating an interesting turn of restoration and preservation toward more liberal stances, rely nonetheless on the already mentioned state-of-the art methods, tools, and materials. The aim is still to achieve a certain degree of precision and similarity to the original. While these objectives of preservation are, and will be, valid in many contexts, the question of authenticity and materiality of the replica remains open. Hence, further architectural knowledge is needed to more comprehensively reveal if and how the negotiations between the known and emergent, existing and new materials, as well as the digital media employed to process them, could lead to an acceptable yet wider range of departures and deviations from the original state of the restored artifact.

With this in mind, our study aimed to elucidate this knowledge and inspire further developments in research and practice of digital restoration. Instead of aiming for faithful restoration of an original surface state, we have sought to show the potentials of endowing it with a new appearance. This was achieved by employing a novel bio-based material, microfibrillated cellulose

3   Eighteenth-century tower 'Götiska tornet' in Stockholm, Sweden (left), with an inventory of deteriorated wooden elements (middle) and an incomplete wooden window casing selected for the experimental restorative intervention (right)

hydrogel, having sustainable properties radically different from those of the traditional and often environmentally non-neutral materials used in preservation.

In terms of digital workflows and tools, a novel element contributed by our study is the introduction of non-standard, bottom-up machine 3D printing toolpath design explorations, enabled by the introduction of parametric tools and custom robotic 3D printer programming into the restoration toolkit, to replace the stage of machine toolpath generation using off-the-shelf software relying on automated digital 3D model slicing. This results in a new workflow that deviates the typical linear, four-stage path mentioned earlier. Namely, it features two loops of exploration, encompassing toolpath design iteration and mockup production, to enable more extensive design investigations of the new materiality to be introduced in the restored artifact.

DESIGN RESEARCH INVESTIGATION
Intervention Site and Materials

The site chosen for design experimentation in this study is a historical, eighteenth-century tower 'Götiska tornet' in Stockholm, Sweden, having unfinished, deteriorated wooden interior cladding. A part of such cladding, comprising a window casing with profiled pilasters and panels, raw and uncoated, was selected for our experimental restorative intervention (Figure 3).

The restorative intervention was done with microfibrillated cellulose hydrogel Exilva F 01-L, a circular, sustainable material comprising 2% cellulose fibrils and 98% water. Originally translucent, in our study it was additionally pigmented using water-soluble food colorants. This specific type of a hydrogel is suitable for pressure-actuated 3D printing because it is thixotropic. Accordingly, when pressure is applied, the viscosity of the material decreases, encouraging flow and therewith enabling extrusion. Once the pressure is removed, the material returns to its original viscous state, which allows for it to maintain the desired shape after deposition.

Custom Robotic Extrusion System

The 3D printing of new coatings from MFC hydrogel was done using a pressure-actuated extrusion system, custom developed specifically for the project. The system was commissioned on an industrial robot KUKA Agilus KR10-1100-SIXX.

The end-effector of the robot comprised of 200 ml dispensing syringes, equipped with plastic plungers and mounted into custom-designed, 3D printed clamps, providing ease of material reload (Figure 4). Based on initial 3D printing experiments, the syringe nozzle diameter was 0.6 mm, and the material deposition speed was 0.015 m/s at a constant pressure of 0.25 bar. The global air pressure in the system was provided from an air compressor. Local pressure control in each dispensing syringe was enabled by pressure regulators operated via manual gauges as well as 3/2-way solenoid valves activated through digital I/O communication with the robot controller (Figure 5).

The robot work cell included an industrial robot fastened to an aluminum frame table, onto which a 70 x 120 cm printing bed was mounted (Figure 6). The bed was constructed from a 2 cm thick MDF board, fastened to the main table using two 4 x

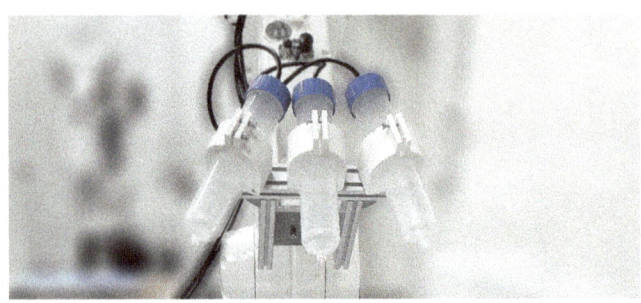

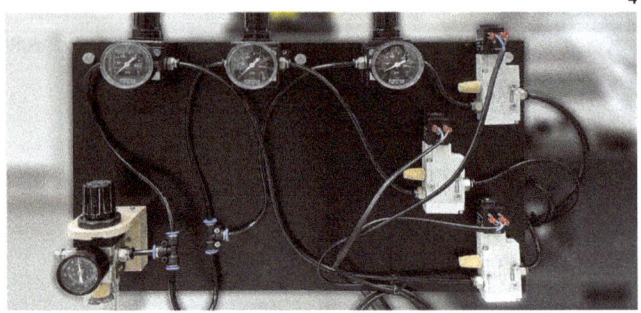

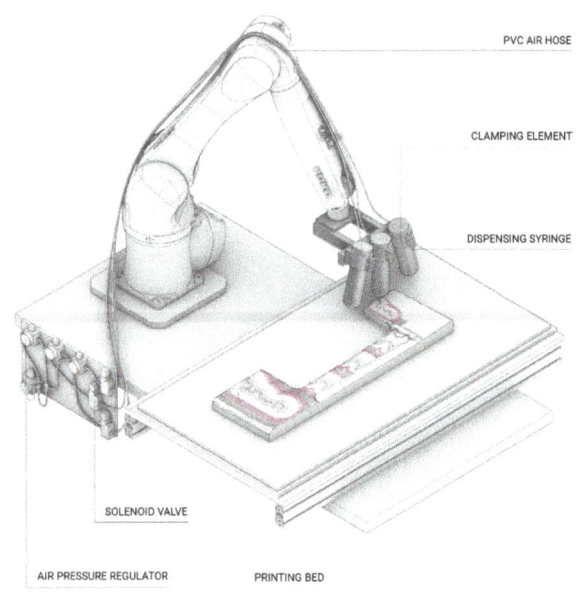

8 cm aluminum profiles. The bed leveling was enabled through a system of six screws and washers placed at the mounting supports between the aluminum profiles and the MDF board.

RePrint: A Digital Workflow for Aesthetic Retrofitting of Deteriorated Architectural Elements

The workflow established in the design experiments was based on a progression from digital reproduction of an existing detail, through iterative mockup prototyping enabling extensive exploration of toolpath designs and their aesthetic implications, up to a restorative intervention through robotic 3D printing of a new surface finish from microfibrillated cellulose hydrogel (Figure 7).

The first step of the workflow involved the digital reproduction of the chosen building element through photogrammetry. To enable this, the element needed to be captured in photographs, taken according to the standard protocols for capturing data for photogrammetry, using a digital camera, in our case an Apple iPhone 12 camera. The photographs were then loaded into the photogrammetry software, i.e., a mobile application Polycam (Polycam Inc. 2022) that generated a texture-mapped, open 3D mesh model of the element in OBJ file format.

The second step of the process embraced post-processing of the mesh to match the original dimensions of the element, patch mesh holes, and create a closed mesh exportable to an STL file format required for the next stage of the process. In our study, the mesh processing operations were done in a 3D modeling software Rhinoceros 3D (Robert McNeel & Associates 2018).

The third step involved two interlinked, looped phases of design explorations, digital and physical, in which various options for the restorative intervention were investigated. To enable this, a scaled physical mock-up model of the digitally reproduced window casing, created as two halves, was first fabricated via CNC milling in birch wood, carried out on a SainSmart Genmitsu 3018-PROVer machine. The process then proceeded with parametric explorations of robot toolpath designs defining how the new surface finishes would be 3D printed. In our case, the toolpath designs were based on raster and vector graphics processing, done in Illustrator (Adobe Inc. 2022) and allowing to create curves that could be further explored and fine-tuned parametrically using Grasshopper (Rutten 2021). Once the toolpath designs were complete, the KUKA|prc add-on was employed for visual simulation of the robotic process, as well as to generate a robot program in the native SRC file format, permitting to 3D print on the mockup models.

In the final exploration stage, the designed surface finish patterns were 3D printed onto the CNC-milled scaled physical mockups to better understand their new materiality. The mockup explorations formed a knowledge base for the last stage of the process, in which one toolpath design would be chosen for fabrication in full scale on the actual element.

Robot Toolpath Design Strategies

The robot toolpath designs relied on a simplified method of projecting 2D toolpath curves onto a 3D surface of the digitally captured object, to create material deposition paths that follow the outline of the existing element in z-axis direction while securing a constant distance of the extrusion nozzle from the surface being printed on (Figure 8). Hence, the machine setup presented herein assumes no extruder rotation and, consequently, a geometric limitation of the robot toolpath to surfaces

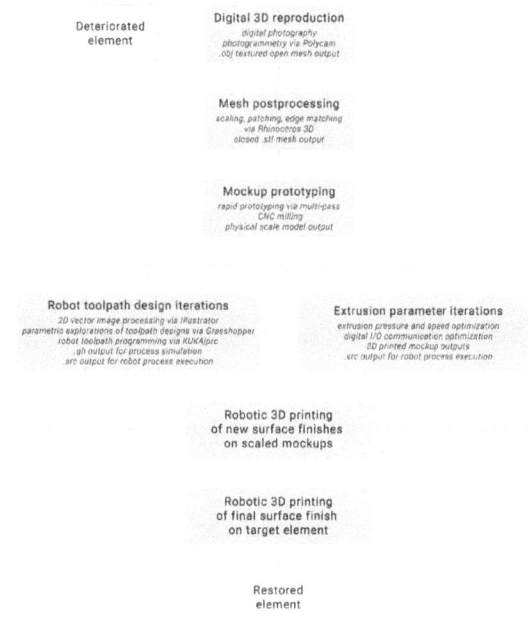

4   Custom-developed robot end-effector, enabling pressure-actuated extrusion of microfibrillated cellulose hydrogel

5   Pressure regulation and control system connected with an industrial robot controller via digital I/O communication

6   Arrangement of the work cell for robotic 3D printing with the hydrogel

7   Digital workflow underpinning the RePrint approach (left) and outputs of the main process stages preceding the production of microfibrillated cellulose coatings on physical mockups: digital 3D mesh reproduced via photogrammetry (top right) and a scaled physical mockup digitally manufactured via CNC milling (bottom right)

that are close to planar. In further research, a 3D printing method for convex and concave elements will be established by introducing rotation of the robot's end-effector, to be able to follow more steeply shaped geometries of the retrofitted pieces. This further investigation will also allow to establish the maximum angle range limits of the extruder tilt in relation to local surface curvatures as these parameters will profoundly affect material adhesion for variously shaped elements.

To demonstrate various directions for restoration from the conceptual standpoint, two aesthetic retrofitting strategies and accompanying toolpath designs were developed. The first strategy related to the encountered appearance and surface qualities of the retrofitted architectural elements, taking into account the noticeable weathering features. Accordingly, the toolpath was designed to highlight surface discolorations left by excessive exposure to rainwater. The imperfections were acknowledged in a new surface layer that followed the discoloration outlines. This was achieved via color hue tracing of a raster image of the retrofitted element in Illustrator to create vector paths following the boundaries of the discolorations (Figure 9). Thereafter, the vector paths were imported into Rhinoceros 3D and parametrically processed with Grasshopper.

The material hues in this strategy were inspired by those found in the Baroque interior of the Royal Palace in Stockholm, Sweden, specifically, in Queen Hedvig Eleonora's Bedchamber, featuring blue and navy tones. The color scheme was designed as congruous with the original staining pattern of the weathered wood while introducing hues from the precedent historic interior. Accordingly, the intervention featured dark blue tones corresponding with the colder and darker colors in the existing weathered wood and light blue as well as white hues relating to the warmer and brighter weathering tones. The design intention was to introduce a material layer that celebrates and preserves the past maculation traces while introducing a new materiality.

The second toolpath design strategy offered a new pattern and color distribution logic to the existing surface. The robot toolpath was generated by creating and editing a fractal Koch curve. The fractal curve was chosen to conceptually and geometrically link to the historical period in which the tower was built, the Baroque. From the standpoint of geometry, the organic aesthetic of the fractal curve was meant to relate to Baroque's characteristic curved, furcating rocaille motifs (Figure 10).

Conceptually, the proposed patterning was inspired by the theories of modern philosopher Gilles Deleuze, which in an architectonic way interpret the theories of Baroque mathematician and philosopher Gottfried Wilhelm Leibniz. In the seminal work *The Fold: Leibniz and the Baroque*, Deleuze pictorially describes different manners of folds—the folds of the soul and the pleats of matter—that make up different kinds of textures (Deleuze 1993). Relating to these notions, the proposed fractal-based patterning and coloration logic of the surface aimed to convey the tower's historical origins.

The color scheme of this intervention introduced a palette of reds, again inspired by colors found in the precedent Royal Palace interior, specifically, the white-burgundy tones in the Main Staircase. To relate to the furls and folds of matter described by Deleuze, the colors were designed in gradient transitions from translucent white, to light red, red, and burgundy. Burgundy was used to accentuate the most complex fractal curves of the pattern and

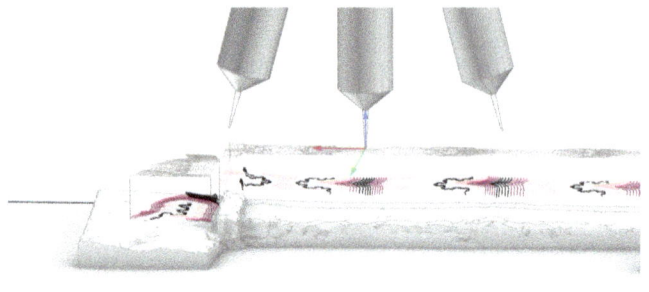

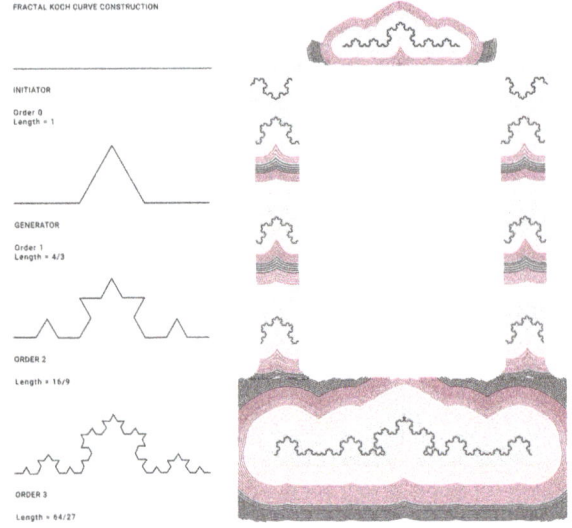

8 Generation of a non-flat toolpath, shown as colored 3D curves, derived through planar projection of 2D curves, shown in grey, onto a 3D mesh surface to be printed on

9 Toolpath design following the weathering pattern of the existing wooden element through vectorized color hue tracing of a raster image

10 Toolpath design inspired by Baroque rocaille motifs, derived by parametric variation of fractal Koch curves and their processing in a vector graphics software

highlight the edges of the lowermost part of the window casing. The other color gradients provided a smooth color transition while accentuating the furls and folds of the new surface.

RESULTS AND DISCUSSION
Material Effects and Aesthetic Expressions

Two properties of microfibrillated cellulose hydrogel have profoundly affected the 3D printing process and the final appearance of the new material layers. The first property is the hydrogel's non-homogenous character, caused by a locally uneven distribution of the cellulose fibrils in the polymer network. This has caused variations in the width and height of the deposed material strands. The generous differentiation of the strands introduced intricate features of textural variation and an irregular aesthetic expression (Figure 11). Such material-specific, unique effects would not have emerged if the deposition had proceeded in perfectly uniform, continuous layers.

The second influential property of the hydrogel was its high water content. Microfibrillated cellulose hydrogel left in ambient conditions shrinks due to water evaporation. Hence, starting at approximately one hour and for the next 36 hours post printing, the hydrogel layer underwent geometric transitions upon water loss. The transitions were also influenced by the gradual absorption of water from the newly printed layer of the hydrogel by the wooden surface of the mockup.

The speed and effects of absorption and drying over time were not uniform. The thinner layers dried more quickly, which has led to local surface ruptures due to material strains caused by strong bonding with the wooden underlay. The water in the thicker layers of the material was absorbed to a lesser extent and these zones dried slowly. As those zones started to shrink at the end of the drying period, geometric shape transformations and local erasure of the initial toolpath traces occurred, creating new patterns that in some areas entirely deviated from the original toolpath (Figure 12).

The material coloring has also undergone changes over time during bonding and drying. The non-pigmented parts of the hydrogel that originally had a translucent white tint became fully transparent after drying. The material with small quantities of pigment became translucent, maintaining the intensity and tone of its color. The hue of the material with the largest content and darkest tone of the pigment was amplified, exhibiting much darker tones than directly after 3D printing.

The encountered material effects could be further embraced in different ways. On the one hand, they could be regarded as design opportunities, following Manuel DeLanda's new materialism philosophy, which argues that the material world has an inherent agenda independent from our intentions and that

11 (left) Initial aesthetic outcomes of architectural element retrofitting with microfibrillated cellulose hydrogel (fully wet prototypes, 30 minutes after 3D printing); (right) Final aesthetic outcomes of architectural element retrofitting with microfibrillated cellulose hydrogels (fully dried prototypes, 36 hours after 3D printing)

we should allow this agenda to affect the materiality of our constructs instead of attempting to fully control it (DeLanda 2015). An opposite path would be to generate a more predictable design outcome. This could be achieved through closer collaboration with material scientists, who could fine-tune the material composition to achieve specific material effects.

Contribution Summary and Future Research

The study has sought to contribute to the existing body of work in digital restoration, by expanding the scope of its typical digital workflow with a looped design exploration phase, enabling aesthetic versioning, evaluation, and acquisition of a deeper understanding of new material interventions before they are applied onto actual architectural elements. In our scaled conceptual mockups, we have also shown how customized parametric toolpath design of new material layers allows to carefully combine them with existing ones, yielding new material hybrids and aesthetic haecceities. These hybrids and haecceities prompt the reemphasis of the importance and value of careful interfacing of historical layers, contemporary materials, and new manufacturing processes.

The work also contributes to architectural restoration discourse by proposing a new approach and a material solution driven by incentives of sustainability and circularity. Through this, it aligns with emergent voices in experimental restoration discussed in the 'State of the Art' section of this article, relating to authenticity and sustainability. The lack of skilled artisans, the impossibility to access the original material, and the unsustainability of the material acquisition process can often motivate the need for an alternative approach harnessing novel biomaterials and building components that would conventionally be treated as waste. This permits to retrofit a wider range of damaged or aged elements with a contemporary touch left by today's digital fabrication techniques.

Simultaneously, the proposed approach of introducing new materials to an existing historical context has limitations regarding the range of cultural artifacts it can be applied to. Careful considerations should still be in place to determine if and what fragments of architectural objects can be handled in this way, what to preserve, which historic layers to reveal and recreate, and which interventions to conceal. In some cases, the character introduced by a new material and digital manufacturing technique might misalign with the existing context, outweighing the sustainability benefits and certain authenticity aspects. In such cases, a more traditional preservation approach could still have priority.

Future research includes continued in-depth explorations of bio-based material interventions, to probe a wider range of architectural effects and material combinations. Another interesting aspect to explore from the computational perspective would be to include machine vision techniques in our workflow, to guide the fabrication process with real-time data inputs regarding the geometry and texture of the treated objects. This could lead to novel, highly integrated workflows that further boost both the creative and the pragmatic aspects of digital restoration.

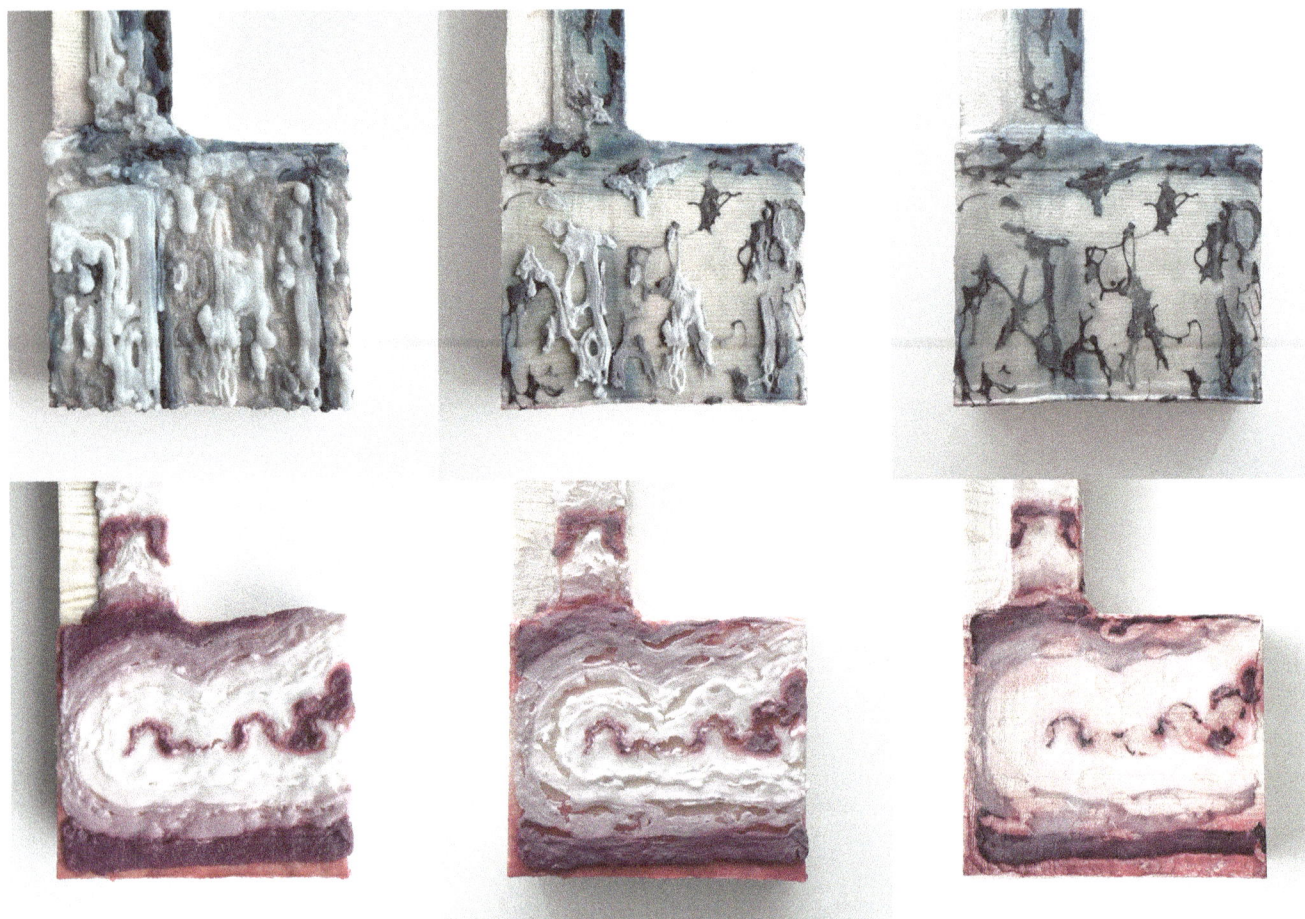

12 Aesthetic and geometric transitions of the new material layers from microfibrillated cellulose hydrogel seen in detail over time for the two different toolpath variants (from left to right: 3, 18 and 36 hours post 3D printing)

CONCLUSION

The work has demonstrated an alternative architectural restoration approach based on aesthetic retrofitting of existing deteriorated elements through digital design and robotic additive manufacturing with a new sustainable biomaterial. By exemplifying, through prototypes and customized digital workflows, how a novel material can be 3D-printed on degraded wood as a conceptually motivated aesthetic coating, the study aimed to cultivate familiarity with and stimulate further discussion and research on alternative digital preservation and restoration methods, highlighting potentials and issues raised by introducing new material hybrids to a historic architectural context.

Today, all design interventions, including those in historic settings, are expected to be respectful toward the natural environment. In that sense, digital restoration engaging with novel sustainable biomaterials offers a move toward more circular treatment of the built environment where respectful mending and care of existing artifacts and spaces is done using latest innovations in material science and new digital technologies. This offers opportunities for zero-waste manufacturing and an extended scope of restoration of existing architectural elements, underpinned by notions of circularity and resource efficiency.

If a structure or artifact cannot be restored with original materials and crafting skills, then it is legit to consider whether it would be more truthful to engage contemporary materials and digital tools. Exposing the nature of artifacts conceived with those tools allows to add a contemporary layer to the history of an object. This aligns with Beatriz Colomina's and Mark Wigley's reflection that design not only conveys the physical results of our ideas, but also the intangible techno-cultural structures and work processes invented throughout human history (Colomina and Wigley 2016). Therefore, it can be argued that utilizing digital tools for restoration while paying close attention to environmental and cultural aspects adds a valuable layer of history to an object, telling the story of our present attempts to redefine the current design systems into more sustainable and resourceful ones.

ACKNOWLEDGMENTS

This study underpins the MSc thesis "RePrint" by Rebecka Rudin, supervised by Malgorzata Zboinska at Chalmers University of Technology, within a research project "Nanocellulose in Architecture: Aesthetic Applications through Robotic 3D Printing," funded by Adlerbertska Foundation and Chalmers Area of Advance Material Science. Authors thank Karl Åhlund for assisting in the extruder development and the sustainable biorefinery Borregaard for providing microfibrillated cellulose.

REFERENCES

Acke, Lien, Kristel De Vis, Stijn Verwulgen, and Jouke Verlinden. 2021. "Survey and literature study to provide insights on the application of 3D technologies in objects conservation and restoration." *Journal of Cultural Heritage* 49 (May-June): 272–288.

Adobe Inc. *Adobe Illustrator*. V. 26.1. Adobe Inc. PC. 2022.

Aicardi, Irene, Filiberto Chiabrando, Andrea Lingua, and Francesca Noardo. 2018. "Recent trends in cultural heritage 3D survey: The photogrammetric computer vision approach." *Journal of Cultural Heritage* 32 (Jul.-Aug.): 257–266.

Baker-Brown, Duncan. 2017. *The Re-Use Atlas: A Designer's Guide Towards a Circular Economy*. London: RIBA Publishing.

Bonora, Valentina, Grazia Tucci, Adele Meucci and Bernardo Pagnini. 2021. "Photogrammetry and 3D Printing for Marble Statues Replicas: Critical Issues and Assessment." *Sustainability* 13 (2): 680.

Brilliant, Richard and Dale Kinney. 2016. *Reuse Value: Spolia and Appropriation in Art and Architecture from Constantine to Sherrie Levine*. Oxon: Routledge.

Colomina, Beatriz and Mark Wigley. 2016. *Are We Human? Notes on an Archaeology of Design*. Zürich: Lars Müller Publishers.

DeLanda, Manuel. 2015. "The new materiality." *Architectural Design* 85 (5): 16–21.

Deleuze, Gilles. 1993. *The Fold: Leibniz and the Baroque*. Minneapolis: University of Minnesota Press.

Duro-Royo, Jorge, Laia Mogas-Soldevila, and Neri Oxman. 2015. "Flow-based fabrication: An integrated computational workflow for design and digital additive manufacturing of multifunctional heterogeneously structured objects." *Computer-Aided Design* 69 (2015): 143–154.

Geboers, Eric, and Matteo Baldassari. 2019. "Rebuilding Notre Dame: A Phoenix Rising from the Ashes." Accessed April 14, 2022. https://medium.com/@eric_geboers/rebuiling-notre-dame-a-phoenix-rising-from-the-ashes-f087bf89f5ed.

Higueras, Maria, Ana Isabel Calero, and Francisco Jose Collado-Montero. 2021. "Digital 3D Modeling Using Photogrammetry and 3D Printing Applied to the Restoration of a Hispano-Roman Architectural Ornament." *Digital Applications in Archaeology and Cultural Heritage* 20 (March): e00179.

Malik, Shneel, Julie Hagopian, Sanika Mohite, Cao Lintong, Laura Stoffels, Sofoklis Giannakopoulos, Richard Beckett, Christopher Leung, Javier Ruiz, Marcos Cruz, and Brenda Parker. 2020. "Robotic extrusion of algae-laden hydrogels for large-scale applications." *Global Challenges* 4 (1): 1900064.

Robert McNeel & Associates. *Rhinoceros 3D*. V. 6.0. Robert McNeel & Associates. PC. 2018.

Polycam Inc. *Polycam*. V. 2.2. Polycam Inc. Apple iPhone 12. 2022.

Reeser Lawrence, Amanda, and Ana Miljački. 2018. *Terms of Appropriation: Modern Architecture and Global Exchange*. Oxon: Routledge.

Rutten, David. *Grasshopper*. V. 1.0.0007. Robert McNeel & Associates. PC. 2021.

Xu, Jie, Lieyun Ding, and Peter E.D. Love. 2017. " Digital reproduction of historical building ornamental components: From 3D scanning to 3D printing." *Automation in Construction* 76 (2017): 85–96.

Rebecka Rudin has an MSc in Architecture and Urban Planning from the Department of Architecture and Civil Engineering at Chalmers University of Technology in Sweden and a background in Art History from Uppsala University. She is interested in how digital technologies and new biomaterials can be employed to reappropriate the built environment, which she seeks to explore in her practice as an architect.

Malgorzata A. Zboinska is an Associate Professor in Digital Design, Fabrication and New Media Art and Creative Leader of the Robotic Fabrication Laboratory at Chalmers University of Technology. Her research expertise bridges new media theory, digital design, computation and robotic fabrication with new sustainable material applications, circularity and resource efficiency in architecture and construction.

Sanna Sämfors is a biomaterial scientist with a background in biotechnology engineering and analytical chemistry, and a researcher affiliated to Chalmers University of Technology and the Wallenberg Wood Science Center in Sweden. In her current work, she develops new wood-based biomaterials for different 3D bioprinting applications.

Paul Gatenholm is a Full Professor of Biopolymer Technology at Chalmers University of Technology. His research is focused on the relation between the structure and material properties of polysaccharides, especially their assembly into hierarchical structures. The work has resulted in more than 370 publications and international patents for novel biomaterial applications in various products and industry sectors.

Integrated Design Strategies for Multi-scalar Biopolymer Robotic 3D Printing

Gabriella Rossi
Ruxandra-Stefania Chiujdea
Laura Hochegger
Ayoub Lharchi
Paul Nicholas
Martin Tamke
Mette Ramsgaard Thomsen

CITA/Royal Danish Academy
Institution

1 Demonstrator wall showcasing an assembly of forty-four 3D printed biopolymer components

ABSTRACT

Additive manufacturing technologies have the potential to initiate changes in architecture's material culture and move us towards a bio-based paradigm. Robotic 3D printing can propose new design languages, logics and tectonics specific to wet biopolymers. In this paper we present strategies and workflows for cellulose-based biopolymer 3D printing. We propose a digital design framework informed by the fabrication system and guided through human design input. The workflow stabilizes the material at the scale of the toolpath, the component, and the wall assembly, by integrating joinery and cross-bracing together with the component geometry. We showcase the feasibility of a large-scale dry-assembly of 3D printed biopolymer components. The demonstrator wall allows us to evaluate our workflows and discuss the challenges and implication of bringing biomaterials in our built environment.

INTRODUCTION

Rising awareness of architecture and construction's impact on the climate crisis through carbon pollution and material overconsumption (Allen et al. 2014) pushes us towards a fundamental rethinking of the way we design and produce our physical spaces. Not only should we rethink our building systems, but also the design logics that feed into them, and the material practices that underpin them. This implies a radical re-evaluation of the way we understand architecture, its production, and its objective. Decades of research into restorative and regenerative design principles have led to new political frameworks of operation such as the European Green Deal (European Commission, Directorate General for Research and Innovation 2021) and other green initiatives. A new model for material consumption would entail shifting away from the threatened and finite materials of the geosphere, and towards the abundant and cyclical materials of the biosphere. This bio-based agenda is building momentum in architecture and design, with attention placed on timber glulam construction, bamboo and rattan, natural fibers, and an emerging class of bio-polymer composites (Hebel and Heisel 2017). This allows us not only to build in a bio-integrated manner, but also to think about our built environment as a carbon sink (King 2017).

Pragmatically, there is a marked knowledge gap curbing this transition. On the one hand, the contemporary construction industry, its supply chains, design workflows and construction systems, are designed for materials that are static, firm, homogenous, and stable. This relies on centuries of accrued knowledge and efforts to standardize material behavior and durability through certified industrial fabrication processes. On the other hand, working with bio-based materials requires a fundamentally different paradigm for they are shaped by growth cycles, and so are characterized by heterogeneity and anisotropy (Pradhan et al. 2021). They are furthermore dynamic as they age (such as in creep), degrade, and respond to environmental triggers such as temperature and humidity (swelling and shrinking). Advances in computational design can close this gap by proposing new data-rich design workflows and smart fabrication pipelines that can account for and work with the complexity of bio-based materials (Thomsen and Tamke 2022). The goal of this research is to propose ways where bio-materials can interface and complement existing built environment in novel hybrid ways.

In this paper we present strategies and workflows for cellulose-based biopolymer 3D printing. Through the production of design prototypes and a full-scale demonstrator, we explore how 3D printing can be instrumentalized to propose new applications of bio-based materials in architecture. Cellulose is the most abundant organic compound on Earth (Pattinson and Hart 2017). However, biopolymers are unruly materials; they are less stable and less durable than their petrochemical equivalents (Nagalakshmaiah et al. 2019). Rather than operating within a schema of material conservation by optimizing design for minimal material usage, strength, and durability, working within a bio-based material paradigm asks us to shift our design logics to embrace a new architectural language defined by shorter life spans, heterogenous properties and emerging design aesthetics. The demonstrator presented in this paper is an experiment allowing us to create and test an interior bio-printed element, whose lifespan and performance is impacted by slow temperature and humidity fluxes within an inhabited environment, while protected from more aggressive degradation processes such as precipitation and UV. We report on various experiments that explore methods of material stabilization, at the intersecting scales of the toolpath, the component, and the wall assembly. We combine geometric design aspects and parametric workflows with fabrication and material system constraints to bring an unruly material into architectural tolerance through a new digital tectonic expression specific to digitally designed, robotically-produced biopolymer prints.

BACKGROUND

Curing of Large-scale Robotic 3D Prints

Large-scale robotic 3D printing has been gaining ground in architectural research and industry. Differently from small-scale rapid prototyping applications which usually involve high resolution slicing, and dense infills with aim to reproduce the element at high fidelity (Chua and Leong 2014), 3D printing at large scale entails a deeper understanding of the material system. On the one hand, efforts are made to tune the extruder system to the specific material that is being extruded, and on the other hand, the printed material's composition and physical properties are tuned for best extrusion flow and print stability. Large scale concrete 3D printing for instance requires specific mixes rich in plasticizer to prevent buckling during the print, and a designed pumping system for flow regulation from mixer to extruder (Gosselin et al. 2016). Similar to cast concrete, the curing of the print is a chemically induced exothermic reaction lasting several weeks. This allows to print dense prints with very thick beads. In the case of large-scale plastic 3D printing, the curing is immediate as the material flows outside of the extruder nozzle thanks to the usage of fans that cool the material below its liquid flow point. This allows to create both sliced surfaces (Schork et al. 2021) as well as spatial trusses (Soler, Retsin, and Garcia 2017). In these material systems, the curing is independent of the geometry of the print. This is not the case with 3D printing of natural materials, such as earth, clay, or bio-composites. Here, the materials undergo a two stage fabrication process: an initial rapid forming followed by a slower evaporative hardening phase.

Most bio-based extrudable materials are water-based slurries. During the slow and long curing phase, the water content will evaporate and allow the print to dry and gain strength. Water, although problem causing, is a necessary component in the mix, as it allows to reach an extrudable viscosity (Campos, Cruz, and Figueiredo 2020; Rossi et al. 2022a). Therefore, geometric strategies for toolpath design and curing control become a crucial aspect of biomaterial 3D printing. In current work, localized curing control is commonly implemented through fans. In the "Ocean Pavilion" project (Mogas-Soldevila et al. 2015), the material is printed as a thin flat sheet, topped by a computer-controlled evaporation system, composed of one hundred fans, which precisely controls the hydration through computerization and drives the self-folding behavior of the sheet. Alternatively, for smaller scale prints, the fan grid is positioned on the extruder itself, thus concurrently stabilizing the ongoing print and initiating the curing process (Dritsas, Hoo, and Fernan 2022). Toolpath control and strategic digital design is particularly present in clay and earth printing, where the mass of the printed elements is too large to be influenced by fans. For example, wall sections have been specifically designed with vertical ventilation shafts for evaporative cooling performance of the building (Chronis et al. 2017). These shafts, while designed for the usage phase of the wall, also contribute to the curing phase, promoting an equal drying of the wall which prevents shear cracking and failures. With biopolymers and bio-receptive composite 3D printing we see two tendencies. The first is to use parametric spatial lattices that offer possibilities for lightweight aerated porous panels to be tiled and attached to a substructure (Chiujdea and Nicholas 2020). The second is the usage of space filling curves (Dristas et al. 2020) or reaction diffusion algorithms (Goidea, Floudas, and Andréen 2020) to generate layered column structures. Both strategies allow to maximize the surface to volume ratio, which promotes ventilation during the evaporative drying.

While these presented projects constitute pioneering efforts in the field of biopolymer 3D printing, the results showcase examples of standalone monolithic objects, with a simple stackable assembly. This lack of consideration for dry assembly joint solutions for 3D printed elements exists across multiple efforts within the large-scale 3D printing community. Here, the difficulty in printing overhangs and interlocking geometries with acceptable tolerances are seen as key limitation of the fabrication process (Shaker et al. 2021). Our research goes beyond state-of-the-art by extending considerations of standard assembly and joinery prevalent in timber construction to 3D printed biopolymer elements. This is achieved through the design decision to rotate our non-standard components 90 degrees with respect to the print bed. In this way, we work with the width and breadth of the print rather than its thickness. Furthermore, we are able to harness the structural capacities of the bead-oriented fibers that are embedded in the material. This introduces biopolymers to the territory of prefabricated, transportable, and maneuverable large-scale assemblies. Our proposal of a wall assembly demonstrator alludes to possibilities of using these printed biomaterials as retrofitting systems that interact with, and improve qualities of existing built environment—a matter of high priority in the EU context (Uihlein and Eder 2010).

MULTI-SCALAR METHODS FOR STRUCTURALLY STABLE BIOMATERIAL 3D PRINTS

In this research we explore the 3D printing of a cellulose-based biopolymer slurry, and we present associated design strategies and fabrication workflows. Our approaches suggest a shift in design logics to embrace the abundance and heterogeneity of biomaterials and develop novel design aesthetics and program potentials for them. Our material recipe is developed in-house (Rech et al. 2021). It blends cellulose flock, wood flour, glycerol, xanthan gum, calcium chloride, and water at 72% of the total weight ratio. This water evaporates during a 15-day post-printing period enabling the material to harden and gain strength, but also causing the material to shrink. To best understand the unruliness of our material and be able to bring it to architectural scale, we develop a Material Monitoring Framework to study its behavior. Our results (Rossi et al. 2022a) showcase that geometry as the critical driver for surface evaporation. Our findings showed that open geometries, which expose more surface area to airflow, dry more evenly, while denser geometries are more prone to warpage. This became a driving consideration for the design of the demonstrator components.

In this paper we focus on design tools, strategies, and potentials of 3D printing large scale biomaterial assemblies. We have developed an interactive parametric model that generates print toolpaths integrating structural and assembly features into the design language of the components, based on designer input. The model ensures material stabilization at the scale of the toolpath, the component, and the wall assembly. The model combines geometric design aspects with fabrication understandings and material system constraints to bring an unruly material into architectural tolerance. We detail various aspects of the model across increasing scales.

Print Stability: The Loop

One of the main challenges of printing with wet materials is that they should be able to bear their own weight during the printing process, therefore avoiding buckling and print failure. In the concrete 3D printing industry, this problem is solved with the addition of plasticizers, thick print beads, multiple shells, and print delays. We seek a geometrical solution to the

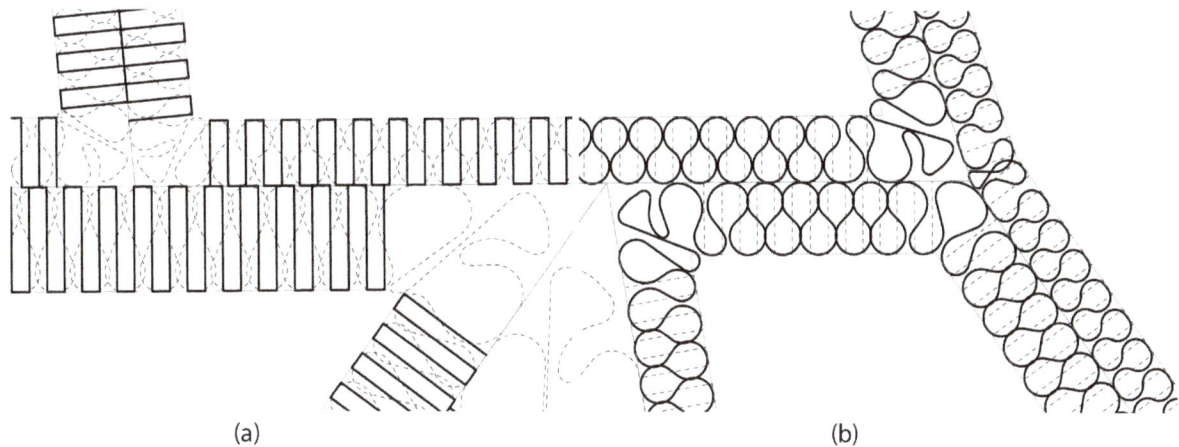

problem since our recipe is to be maintained 100% organic and biodegradable, and thicker beads or multiple shells would impede the proper drying of the print (this would cause the formation of a crust, and rot in the center). Curvature-induced stabilization is achieved through a half circular toolpath we term "The Loop." The Loop allows us to gain cross-sectional moment of inertia without compromising the aeration and ventilation. The sizing of the loops is informed by the fabrication setup. The algorithm (Figure 2) operates on the basis of the print nozzle size, which informs both the base subdivision parameter (1.5x the nozzle diameter) as well as the overlapping parameter (0.5x the nozzle diameter). Since our material is fiber-oriented, the correct overlap parameter is crucial to ensure aeration without delamination.

Component Stability: Cross Bracings and Joints
While the sizing of the loops is informed by the fabrication setup, their depth is informed by the load to which the component is subjected. Our algorithm uses a simple load approximation given by the specific weight of the material and the gravity and self-weight load network. The higher the load, the thicker the cross-sectional frame of the component. The algorithm operates using a data tree structure which allows to add extra features to selected edges without having to manually manipulate the loops, which would be cumbersome. Instead, the algorithm searches for affected edges based on the manual designer guideline input, manipulates the underlying polyline, and replaces the edge branch in the data tree. This method allows us to add different features sequentially using the same integrated loop language, in a simple plug-and-play flexible method: cross bracings, edge thickeners, in-plane joints, and transversal joints (Figure 3). For example, to generate cross bracings that guarantee the stability of the cross-sectional frame against torsion, the designer draws a simple line within the component, the algorithm finds the affected parallel edges, pulls the closest perpendicular loops to the drawn bracing ridge line, and regenerates the

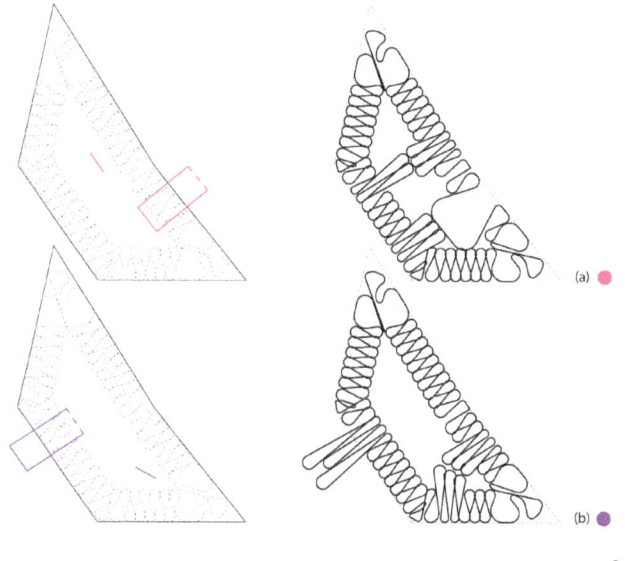

2   Loop generation algorithm: (a) A polyline is generated between the original outline and the offset outline. The spacing is informed by the fabrication parameters, and the polyline is then expanded, manipulated during the design process, and the curved printing toolpath (b) is generated in the final stage.

3   Example of designer input on the base loop (left) and integrated component design with joints and cross-bracings (right). The cross bracings are informed by little dashes that pull together the affected edge loops, while the joints are defined by rectangles reaching between the neighboring components. The dashed edge points to the male side of the joint: in (a) it falls outside the component, and therefore, a female joint is generated, and in (b) it falls inside the component and a male joint is generated.

edge. For in-plane joints, a similar search is conducted. The designer draws a rectangle on the common edge between two components where the joint should be hosted, using the connectivity graph; the male and female components are identified, and a protruding male joint is generated by pulling the loops, whereas an accommodating female joint is generated by deleting the loops. Inner corners of the components

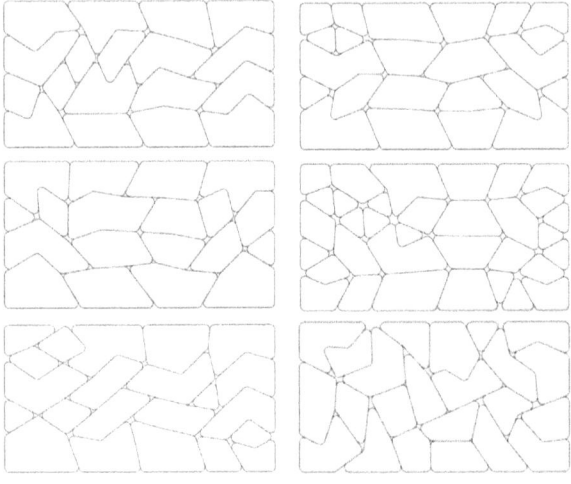
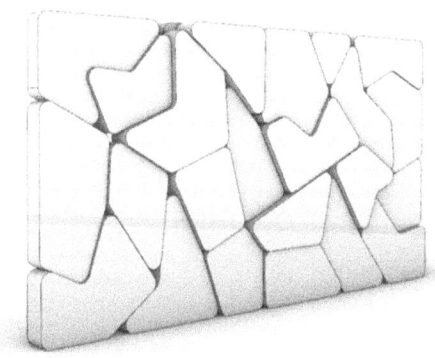

4

are reinforced using a tripartite loop, as preliminary tests have showed that they constitute a weak point in the component if left without reinforcement.

Structure Stability: Macro-scale Assembly

The algorithm operates with an input of component outline. From our structural characterization testing, we know that it behaves better in compression than in tension (with a density of 527.06 kg/m3, and an approximately 140 MPa modulus). To ensure cyclopean compression-based stacking, we therefore, design a tessellation system based on a triangular module. The components are generated using an agent-based system that takes into consideration geometrical constraints that stem from the fabrication system, for instance, component maximum and minimum area, maximum and minimum length of edge and angle with neighboring edge, and maximum and minimum number of edges per components. The resulting tessellations (Figure 4) are evaluated based on the load network explained above. Discontinuities in the load path inform where the in-plane dovetail joints should be located. Finally, the assemblies are tapered in thickness towards the top of the structure.

EXPERIMENTS AND RESULTS

Multi-material Prints

Our recipe has the capacity to change its properties by varying the fiber used. We have so far experimented with replacing the cellulose paper flock with linen fibers, cotton fibers, ground bark, and seagrass. Each different fiber lends the mix not only different properties in term of color and texture, but also structural performance. This opens the possibilities for multi-material prints where the recipe is topologically graded to respond to functional criteria. Figure 5 shows two multi-material prints where the two material beads have been printed adjacently and, through the shrinkage that occurs while drying, have been fused into each other.

Multi-scalar Prints

Our developed toolpathing algorithm can be adapted to print at multiple scales since it is nozzle-size based. Our material presents viscosities that are compatible with cement extrusion hardware. For the printing of the small-scale prototypes, we have used an ABB1600 robot, a custom in-house end-effector extruder fitted with a 6 mm nozzle and an auger screw for flow stabilization, fed with a pressurized 10 L acrylic tube at 2.5 bar. For the printing of the wall demonstrator components, we used the same extruder fitted with a 11 mm nozzle and fed by a Mai 2Pump Pictor. Further experiments are currently being carried out for large-scale components using a concrete 3D printing gantry system fitted with 30 mm extruder nozzle (Figure 6).

Demonstrator Wall

As a proof of concept of internal partitions made of 3D printed cellulose components, we have fabricated a wall assembly spanning 3 meters long by 1 meter wide by 1.8 meters tall (Figures 1, 7). It is composed of 44 polygonal components and tests a corner configuration through a T-shaped composition. The connection between the two orthogonal elements is achieved using a dovetail joint. The joint pieces are printed with staggered heights to ensure three dimensional interlocking. The components are rotated 90 degrees with respect to the print bed. This allows the fiber orientation embedded in the print bead to be used for structural performance. The roughness between the component outlines and the friction within the male/female joint hold the structure standing. This

4 (left) Tessellation iterations produced by the agent-based algorithm; (right) final tessellation volumetrically staggered

5 Examples of multi-material prints using different fibers: cellulose flock (gray), cotton (blue), and bark (brown). Different techniques are tested: interlayer material switch, and side-to-side bead switch

6 (left) Demonstrator component print in our robot lab with a 40 x 40 cm area; (right) scaled up component printed on the gantry system spanning 200 x 90 cm

shows that prefabricated biopolymer components can be integrated together using dry assembly, which can be further stabilized with mechanical fixings if needed. The components express a new tectonic language that integrates frames, bracings, and joints. It explores notions of opacity across the wall. The components are light-weight, and the wall can be assembled by two people in one hour.

DISCUSSION

The demonstrator wall is a successful example showcasing the feasibility of a large-scale dry-assembly of 3D printed biopolymer components. The research points to a series of key considerations in working with biopolymer composites. In scaling up from small probes and prototypes to the demonstrator, substantial performance and fabrication driven changes challenge the design workflow. Firstly, some challenges are due to production and drying logistics. For instance, while the hopper pump allows us to print without a capped volume limit, material must be constantly prepared and fed to it. Using an industrial bread mixer allows us to mix 10 L batches at a time, yet a print in the wall uses an average of 25 L of material. Material can be prepared a few days in advance, but it must be stored in airtight bags, and we have found that mixing material batches of different ages in the same print can cause problems in print consistency. Moreover, the prints must be moved from the print bed to the curing rack while wet; this is when they are most fragile and prone to buckling, and also most heavy. In order to mitigate this, we designed a stretcher system allowing for their transport by two operators. The curing room must be kept at high temperature (27 degrees) and low humidity (35-40%) and be fitted with fans to ensure constant airflow and quick extraction of the water from the material. This process can be quite lengthy; we have found that the densest components (the bottom row of the demonstrator) required 3-4 weeks of drying time, while the lightest pieces (the top row of the demonstrator) were cured within 7-10 days. We have also found that loading the components before they are completely dry can cause delamination and failure.

Secondly, other challenges are inherent to the material system itself. We are aware that as the material dries and loses water, the geometry shrinks. This shrinkage has been considered by calibrating the oversizing of the printed components so that the dried geometry interlocked together. However, we have found that while the tolerances on the component outlines were sufficient for fitting, the orthogonality of the component edges and the tolerance around the interlocking joints was challenged. Here we found that the taller the component is, the higher are the chances of the beads in their wet state to buckle under their own weight and bulge out of plane. This constituted a problem given the need to rotate our component and stack them against each other's planar edges. As

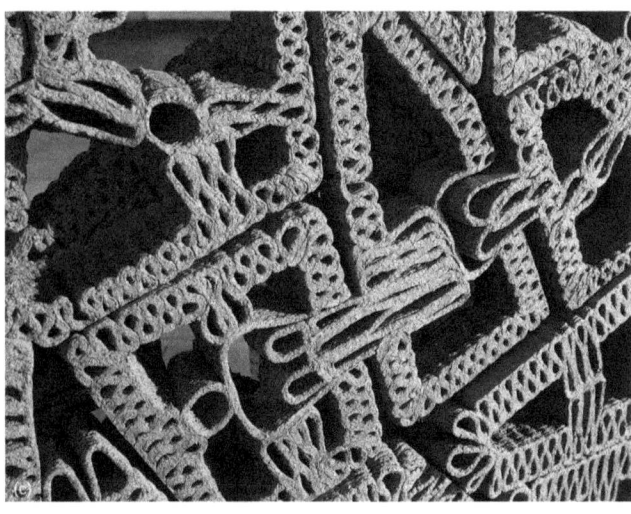
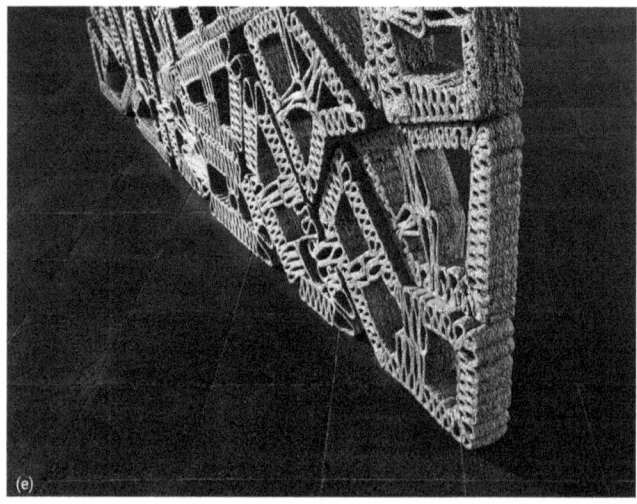

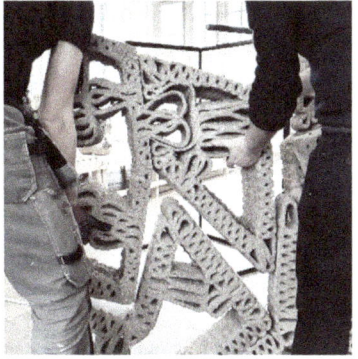

7 (a) Frontal view of the demonstrator wall; (b, c) detail shots of the components cross-bracings and joints; (d) rear view of the corner solution; (e) textured side edge of the wall

8 (left) Joint refining using a hand-held jigsaw; (right) two interlocking pieces fit together through an in-plane joint and subsequently placed in the interlocking assembly

our material can be easily post-processed with standard woodworking tools (Figure 8), this print artifice could be easily corrected, and our assembly logic remained valid. To track these deformations, we have monitored the components during their curing using an Opti-Track setup and have compiled a dataset to train a machine learning algorithm for warpage mitigation. We developed methods for geometric data encoding, tolerance-informed data augmentation and statistical modeling, which are reported in Rossi et al. 2022b.

CONCLUSIONS

In this paper we have presented integrated design and fabrication strategies for multi-scalar biopolymer robotic 3D printing that are showcased through the production of a room-scale demonstrator composed of dry-assembled cellulose biopolymer components. The proposed digital workflows are informed by the fabrication system and are guided through human design input. The outcomes bring together joinery and cross bracing in an integrated aesthetic and tectonic logic that is more suited to design with unruly biomaterials than to 3D printing.

From a broader perspective, this project has allowed us to demonstrate the possibility of printing and drying larger cellulose components that is enabled by the geometrical strategy we have adopted and that maintains both structure and aeration in the components. Both the scale of the elements and the reversibility of the dry assembly are compatible with a retrofitting architectural paradigm and allow for a responsive maintenance regime and replacement of parts. The ability of the elements to carry a load equivalent to that of an interior wall reinforce this potential. Immediate realizations are that these biomaterials have the potential to regulate occupancy, humidity, and sound, and thus to improve the quality of existing built space. We will explore these aspects in future work.

In future work, we will also examine the improvement of the tolerance control of the components from wet to dry. The research presented here runs in parallel with a machine learning track that investigates how to predict the deformation of the original print during drying and ultimately to inform the design of objects made from biopolymers. Here, our goal is to differentiate the tolerance control around the joint and edges. Furthermore, we plan to expand the 3D printing process to non-flat printing beds using conformal techniques.

ACKNOWLEDGMENTS

This project is funded by Independent Research Fund Denmark (DFF) PROJECT NUMBER 9131-00034B "Predicing Response," in collaboration with Anders Egede Daugaard and Arianna Rech (Denmark Technical University) and John Harding (University of Reading).

The authors thank Computation in Architecture master students Konrad Sonne, Cheng Sin Ariel Lim, and Ee Pin Choo for their help with the demonstrator production.

IMAGE CREDITS

Figure 1, 7: Anders Ingvarsten ©CITA / Royal Danish Academy, 2022
Figure 2, 3, 5, 6, 8: Gabriella Rossi ©CITA / Royal Danish Academy, 2022
Figure 4: Ruxandra Chiujdea ©CITA / Royal Danish Academy, 2022

REFERENCES

Allen, Franklin, Jere R. Behrman, Nancy Birdsall, Shahrokh Fardoust, Dani Rodrik, Andrew Steer, and Arvind Subramanian. 2014. *Towards a Better Global Economy: Policy Implications for Citizens Worldwide in the 21st Century*. Oxford University Press. https://doi.org/10.1093/acprof:oso/9780198723455.001.0001.

Campos, T., P. J. Cruz, and B. Figueiredo. 2020. "Paper in Architecture: The Role of Additive Manufacturing." In *Industry 4.0–Shaping The Future of The Digital World*, 1st ed., edited by P. J. da Silva Bartolo, F. Moreira da Silva, S. Jaradat, and H. Bartolo. 167–72. Boca Raton, FL: CRC Press/Balkema Book. https://doi.org/10.1201/9780367823085-30.

Chiujdea, Ruxandra Stefania, and Paul Nicholas. 2020. "Design and 3D Printing Methodologies for Cellulose-based Composite Materials." In *Anthropologic: Architecture and Fabrication in the Cognitive Age; Proceedings of the 38th ECAADe Conference*, vol. 1. TU Berlin. 547–554.

Chronis, Angelos, Alexandre Dubor, Edouard Cabay, and Mostapha Sadeghipour Roudsari. 2017. "Integration of CFD in Computational Design." In *ShoCK!, Sharing Computational Knowledge!; Proceedings of the 35th ECAADe Conference*, vol. 1. Sapienza University of Rome. 601–610.

Chua, Chee Kai, and Kah Fai Leong. 2014. *3D Printing and Additive Manufacturing: Principles and Applications (with Companion Media Pack). Rapid Prototyping*, 4th ed. World Scientific Publishing Company. https://doi.org/10.1142/9008.

Dritsas, Stylianos, Yadunund Vijay, Samuel Halim, Ryan Teo, Naresh Sanandiya, and Javier G. Fernandez. 2020. "Cellulosic Biocomposites for Sustainable Manufacturing." In *Fabricate 2020: Making Resilient Architecture*, edited by J. Burry, J. Sabin, B. Sheil, and M. Skavara. London: UCL Press. 74–81. https://doi.org/10.2307/j.ctv13xpsvw.14.

Dritsas, Stylianos, Jian Li Hoo, and Javier G Fernan. 2022. "Sustainable Rapid Prototyping with Fungus-like Adhesive Materials." In *Post-Carbon; Proceedings of the 27th CAADRIA Conference*. Sydney. 263–72.

European Commission, Directorate General for Research and Innovation. 2021. *European Green Deal: Research & Innovation Call*. Publications Office of the European Union. https://data.europa.eu/doi/10.2777/33415.

Goidea, Ana, Dimitrios Floudas, and David Andréen. 2020. "Pulp Faction: 3d Printed Material Assemblies through Microbial Biotransformation." In *Fabricate 2020: Making Resilient Architecture*, edited by J. Burry, J. Sabin, B. Sheil, and M. Skavara. London: UCL Press. 42–49.

Gosselin, C., R. Duballet, P. Roux, N. Gaudillière, J. Dirrenberger, and P. Morel. 2016. "Large-Scale 3D Printing of Ultra-High Performance Concrete; A New Processing Route for Architects and Builders." *Materials & Design* 100 (June): 102–9. https://doi.org/10.1016/j.matdes.2016.03.097.

Hebel, Dirk E., and Felix Heisel. 2017. *Cultivated Building Materials: Industrialized Natural Resources for Architecture and Construction*. Berlin, Boston: Birkhäuser. https://doi.org/10.1515/9783035608922.

King, Bruce. 2017. *The New Carbon Architecture: Building to Cool the Climate*. Gabriola Island, BC, Canada: New Society Publishers.

Mogas-Soldevila, Laia, Jorge Duro-Royo, Daniel Lizardo, Markus Kayser, Sunanda Sharma, Steven Keating, John Klein, Chikara Inamura, and Neri Oxman. 2015. "Designing the Ocean Pavilion: Biomaterial Templating of Structural, Manufacturing, and Environmental Performance." In *Future Visions: Proceedings of the International Association for Shell and Spatial Structures (IASS) Symposium*. Amsterdam.

Nagalakshmaiah, Malladi, Sadaf Afrin, Rajini Priya Malladi, Saïd Elkoun, Mathieu Robert, Mohd Ayub Ansari, Anna Svedberg, and Zoheb Karim. 2019. "Biocomposites." In *Green Composites for Automotive Applications*. Duxford, UK: Elsevier. 197–215. https://doi.org/10.1016/B978-0-08-102177-4.00009-4.

Pattinson, Sebastian W., and A. John Hart. 2017. "Additive Manufacturing of Cellulosic Materials with Robust Mechanics and Antimicrobial Functionality." *Advanced Materials Technologies* 2 (4): 1600084. https://doi.org/10.1002/admt.201600084.

Pradhan, R. A., S. S. Rahman, Ahmed Qureshi, and Aman Ullah. 2021. "Biopolymers: Opportunities and Challenges for 3d Printing." In *Biopolymers and Their Industrial Applications: From Plant, Animal, and Marine Sources, to Functional Products*, edited by T. Sabu, S. Gopi, and A. Amalraj. Amsterdam: Elsevier. 281–303. https://doi.org/10.1016/B978-0-12-819240-5.00003-1.

Rech, Arianna, Ruxandra Chiujdea, Claudia Colmo, Gabriella Rossi, Paul Nicholas, Martin Tamke, Mette Ramsgaard Thomsen, and Daugaard Anders Egede. 2021. *Predicting Response: Waste-Based Biopolymer Slurry Recipe for 3d-Printing (CelluloseFloc_v01)* [Data set]. Zenodo. https://doi.org/10.5281/ZENODO.5557218.

Rossi, Gabriella, Ruxandra Chiujdea, Claudia Colmo, Chada ElAlami, Paul Nicholas, Martin Tamke, and Mette Ramsgaard Thomsen. 2022a. "A Material Monitoring Framework: Tracking the Curing of 3d Printed Cellulose-Based Biopolymers." In *ACADIA 21: Realignments: Toward Critical Computation: Proceedings of the 41st Annual Conference of the Association for Computer Aided Design in Architecture*. Online.

Rossi, Gabriella, Ruxandra Chiujdea, Laura Hochegger, John Harding, Paul Nicholas, Martin Tamke, and Mette Ramsgaard Thomsen. 2022b (forthcoming). "Statistically Modelling the Curing of Cellulose-Based 3d Printed Components: Methods for Material Dataset Composition, Augmentation and Encoding." In *Design Modelling Symposium Berlin 2022: Towards Radical Regeneration*. Online.

Schork, Tim, Ninotschka Titchkosky, Chris Bickerton, Dagmar Reinhardt, Michael Bennett, David Pigram, and Mohammed Makki. 2021. "The Geometry of Air: Large-Scale Multi-Colour Robotic Additive Fabrication for Air-Diffusion Systems." *Construction Robotics* 5 (1): 49–61. https://doi.org/10.1007/s41693-021-00054-z.

Shaker, Aya, Nour Khader, Lex Reiter, and Ana Anton. 2021. "3D Printed Concrete Tectonics: Assembly Typologies for Dry Joints." In *ACADIA 21: Realignments: Toward Critical Computation: Proceedings of the 41st Annual Conference of the Association for Computer Aided Design in Architecture*. Online.

Soler, Vicente, Gilles Retsin, and Manuel Jimenez Garcia. 2017. "A Generalized Approach to Non- Layered Fused Filament Fabrication." In *ACADIA 2017: Disciplines and Disruption; Proceedings of the 37th Annual Conference of the Association for Computer Aided Design in Architecture*. Cambridge, MA. 562–571.

Thomsen, Mette Ramsgaard, and Martin Tamke. 2022. "Towards a Transformational Eco-Metabolistic Bio-Based Design Framework in Architecture." *Bioinspiration & Biomimetics* 17 (4): 045005

Uihlein, Andreas, and Peter Eder. 2010. "Policy Options towards an Energy Efficient Residential Building Stock in the EU-27." *Energy and Buildings* 42 (6): 791–98. https://doi.org/10.1016/j.enbuild.2009.11.016.

Gabriella Rossi is a PhD Fellow at CITA, Royal Danish Academy. Her research focuses on Machine Learning as an emerging modeling paradigm, and how it can change architectural practice. She focuses on architectural datasets, harvested or simulated, and potential applications for complex material behavior, robotic fabrication, and design performance. She teaches the master program Computation In Architecture at the Royal Danish Academy, and the AI and Architecture module at the MaCAD program at IaacC.

Ruxandra Stefania Chuijdea is a research assistant at CITA, Royal Danish Academy. Her research focuses on digital modeling and fabrication across conventional and emerging materials focusing on 3D printing with biocomposites and metal sheet forming using industrial robotics. She was a resident at Autodesk BUILDspace, investigating assembly and robotic forming of curved metal sheets. She assists teaching the Master's program Computation in Architecture.

Laura Hochegger is a research assistant at CITA, Royal Danish Academy. She holds a Civil Engineering Master from BOKU Vienna. During her studies she focused on structural engineering and resource-efficient building construction. While working on her Master's degree in economics, she works at a construction company where she looks for new technologies and digital solutions that contribute to a sustainable future of the construction industry.

Ayoub Lharchi is PhD candidate and research associate at CITA, Royal Danish Academy. He has a deep interest in complex geometries, advanced digital fabrication, and assembly techniques. He is a registered Architect and Computational Designer.

Paul Nicholas is an Associate Professor at the Royal Danish Academy and leads the international Master's program Computation in Architecture. He holds a PhD in Architecture from RMIT University, has practiced with Arup consulting engineers, and has taught in Australia, China, and Europe. His research explores new material practices linking design, fabrication, and materiality with focus on machine learning, biomaterials, adaptive robotic fabrication, and multiscalar modeling.

Martin Tamke is Associate Professor at CITA, Royal Danish Academy. He pursues pursuing design-led research on the interface and implications of computational design and its materialization with feedback from environment and process. His current focus is on machine learning and 3D sensing of CNC-knitted, engineered timber and biobased materials.

Mette Ramsgaard Thomsen is Professor, Head, and Founder of CITA at the Royal Danish Academy, where she focuses on the profound changes that digital technologies instigate in the way architecture is thought, designed, and built. Recent work examines new design principles for bio-design and how processes of renewable, regenerative, and restorative resource thinking leads to sustainable design practice.

3D Printed Formwork for Mycelium Bound Composites

Matthias Leschok
DBT/ ETH Zurich

Benjamin Dillenburger
DBT/ ETH Zurich

1 Polymer 3D printed formwork for mycelium-based composites;- close up of fruiting bodies on the *MycoChair*

2 Close up of the dissolving PVA formwork of *The Living Column*

ABSTRACT

Mycelium-bound composites (MBC) are gaining increased interest in replacing high-embodied energy building materials. This trend is propelled by material scarcity and awareness of increasing construction waste. MBC can be grown locally using agricultural waste and is fully biodegradable. Furthermore, they are lightweight and showcase great thermal performance. Despite those benefits, MBC suffers from low structural strength, hindering their broader architectural application.

For wider applicability, MBC needs to be coupled with a structurally informed design that can be facilitated by bespoke geometries. 3D printed formwork offers the creation of bespoke elements at no extra cost. However, there is little to no research on 3D printed formwork for MBC.

This paper presents investigations on 3D printed formwork for MBC using different polymers. The method is demonstrated via two large-scale prototypes, and major findings are critically discussed. It is shown that different polymer materials can be used for growing MBC, and that the choice of material affects the final component. The research aims to promote the creation of low-embodied energy and fully biodegradable building components by exploring 3D printed formwork for mycelium-bound composites.

INTRODUCTION

Growth-based fabrication as an alternative to standard construction processes is gaining increased attention (De Belie et al. 2018; Derme and Mitterberger 2016; Bitting et al. 2022; Oxman et al. 2013). Material scarcity and non-recyclability of common construction materials are propelling this development. Global construction and demolition waste is expected to reach 2.2 billion tons by the year 2025, nearly double the value of 2012 (Transparency Market Research 2022). Furthermore, buildings are major contributors to greenhouse gas emissions (GHG), and the grey energy of building materials accounts for up to 25% of the life cycle of GHG. Finding alternative, low-embodied energy carbon materials is one of the major challenges in the upcoming decades. Growth-based fabrication offers a solution to these challenges by providing bio-based and biodegradable materials.

There are different growth-based fabrication processes relevant to architecture. For example, bacteria can be used to create self-healing concrete (De Belie et al. 2018) or to grow cellulose membranes (Derme and Mitterberger 2016). Silkworms are used to establish a biologically-driven strategy for material optimization and to spin the "Silk Pavilion" (Oxman et al. 2013). However, mycelium is one of the most promising growth-based fabrication processes for creating components relevant to architecture due to its ease of cultivation and scale. Mycelium is the root structure of a fungus (hyphae) and is similar to the roots of a tree. Mycelium-based Composites (MBC) do not require energy-intensive production processes. Furthermore, MBC does not rely on resource extraction or long-distance transportation, as they can be grown locally, for instance, with agricultural waste or by-products (Angelova, Brazkova, and Krastanov 2021). At the end of their life they can be biodegraded (Van Wylick et al. 2022).

However, research has shown that MBC has reduced compressive strength compared to well-established construction materials. Therefore, it is concluded that a successful application of MBC needs to be coupled with a structurally informed design and an appropriate digital fabrication method to create these bespoke geometries (Bitting et al. 2022).

Polymer 3D printing arises as a prominent method to digitally fabricate molds for MBC and has proven its viability in the application of concrete formwork. However, there is little to no research on the application of 3D printed formwork for MBC. This paper describes investigations on 3D printed formwork using three different thermoplastic materials to create MBC. The methods are demonstrated via two large-scale prototypes and major findings are critically discussed. The aim of the research presented is to promote the creation of low-embodied energy and fully biodegradable building materials by exploring 3D printed formwork for mycelium-bound composites.

STATE OF THE ART

The relevant state of the art for this research is organized into three topics: (1) types of mycelium materials, (2) mycelium architecture, and (3) polymer 3D printed formwork.

Mycelium

There are two main categories for mycelium based materials:
- Mycelium-bound Composites (MBC)
- Pure Mycelium Material (PMM)

2

MBC is created using a substrate usually consisting of a lignocellulosic material like sawdust, hemp, flax, and water, providing a matrix for the mycelium to grow in (Elsacker et al. 2020). The substrate provides nutrients and humidity to the living organism forming a breeding ground during its growing phase. The mycelium creates a dense bio-polymer matrix.

PMM, on the other hand, is grown directly in a given container without the presence of a lignocellulosic substrate. This process requires a liquid mycelium culture and a soft scrim with a nutritive substrate (Karana et al. 2018).

Until now, MBC has been used mainly as packing materials or for architectural applications, whereas PMM finds its application in the textile, footwear, and paper-making industry. This publication focuses solely on MBC.

Application of MBC in Architecture

MBC can be found in or as commercially available building products like thermal insulation boards (Ecovative n.d.). The company Biohm provides insulation panels with thermal conductivity of as low as 0.024 W/m*k (Biohm n.d.). In parallel to good thermal insulation, MBC can also provide sound insulation properties. The Italian-based company Mogu provides acoustic panel and flooring systems based on mycelium (Mogu n.d.).

Besides these commercial products, MBC is investigated for their architectural potential in several prototypes, projects, installations, or pavilions. Those can be further categorized as prefabrication-reliant MBC applications and in-situ growth.

In past research, MBC has been shaped with abrasive robotic wire-cutting to serve as stay-in-place formwork for concrete (Elsacker et al. 2021). The formwork of the prototypical concrete slab facilitates material savings (bespoke shape) while, in parallel, providing a layer of thermal insulation.

MBC in the form of bricks is used in the *Hy-Fi* tower (UrbanNext 2014) and the 2012 *Mycotecture* project (Mok 2018). Both projects use MBC as discrete building blocks that are prefabricated and assembled on site. *Mycotecture* is an arched structure made entirely from mycelium bricks, and it is considered to be the first prefabricated architectural structure using MBC. The *Hy-Fi* tower is considerably larger, a 13-meter tall structure assembled using ten thousand MBC bricks for an outdoor installation for the MoMA PS1. Lastly, *MycoTree* presents MBC as a part of a load-bearing structure (Heisel et al. 2017). The formwork of the bespoke components is fabricated using laser-cut plastic sheets.

Compared to the aforementioned projects the following are fabricated in situ. *Monolithic Mycelium* is a large-scale (2.5 x 2.5 x 2.5 meter) pavilion made from 800 kg of substrate. The pavilion is shaped by an interior (lost) formwork that stabilizes the structure and a laminated plywood formwork that is removed after an initial growing period (Dessi-Olive 2019). MBC is used as a building envelope in the *The Growing Pavilion* in combination with other bio-based building materials (The Growing Pavilion n.d.).

Polymer 3D Printed Formwork

Polymer 3D printing (3DP) (often referred to as FDM or FFF) describes a process where a thermoplastic material is heated up in a hotend and deposited layer by layer to form a 3D object.

Polymer 3DP has been used to create bespoke formwork for concrete elements. Its application ranges from stairs to slabs (Jipa, Barentin et al. 2019; Jipa, Giacomarra et al. 2019), to columns (Burger et al. 2020; Leschok and Dillenburger 2019, 2020; Doyle and Hunt 2019), to form structural elements within building envelope prototypes (Loh, Qu, and Leggett 2020). The materials used for creating polymer formwork are diverse and range from biodegradable, to recyclable, to water dissolvable.

Challenge

For a successful application of MBC in architecture, digital fabrication methods for shaping need to be investigated. The goal of these investigations is to enable the fabrication of bespoke designs that can mitigate the low-compressive strength of MBC. Polymer 3D printed formwork provides an opportunity to create those bespoke shapes. However, there is little to no research on 3D printed formwork for MBC.

3D printed molds/formwork for MBC must include material-specific constraints and provide a prosperous environment for the fungus. The formwork must be non-toxic, withstand sterilization, and ideally be biocompatible with the mycelium. Furthermore, the formwork needs to support the MBC during its growth for more than a week.

The research that follows here investigates 3DP formwork to create bespoke MBC. It examines different polymers and provides guidelines and use-cases for each material.

FORMWORK FOR MYCELIUM

The method of fabricating an MBC in 3DP formwork includes the following steps and can be described as follows (Figure 4):

As a first step, the substrate and the 3DP formwork need to be prepared. This includes, but is not limited to, harvesting substrate, mixing with secondary ingredients (like gypsum or flour), and sterilization or pasteurization. As an alternative, ready-to-use substrates, e.g., those provided by Ecovative or local mushroom farmers, can be used. The formwork needs to be designed and printed using a chosen polymer. Before it can be filled with substrate, the formwork also has to be sterilized, often with chemicals like bleach or alcohol.

The preparation step is followed by the growth process, which includes, but is not limited to, (a) filling of the 3DP formwork with the substrate, (b) inoculation with mycelium, and (c) the incubation phase (Ghazvinian and Gürsoy 2022). Depending on the chosen substrate, the inoculation can be skipped as mycelium is already included in the substrate. After the initial incubation phase, depending on the chosen polymer, the formwork gets removed, stays in place, or is already dissolved.

To retrieve an MBC, the material needs to be consolidated. Consolidation is usually facilitated by controlled drying. The drying process stops the growth of the mycelium and, once fully dried, results in a lightweight MBC component. If the formwork is removed after consolidation, the formwork can be reused to create another MBC, washed and recycled to 3D print a new formwork, or decomposed. After the MBC's lifetime, the element can be decomposed or crushed and become a substrate anew. The formwork is printed using off-the-shelf FFF

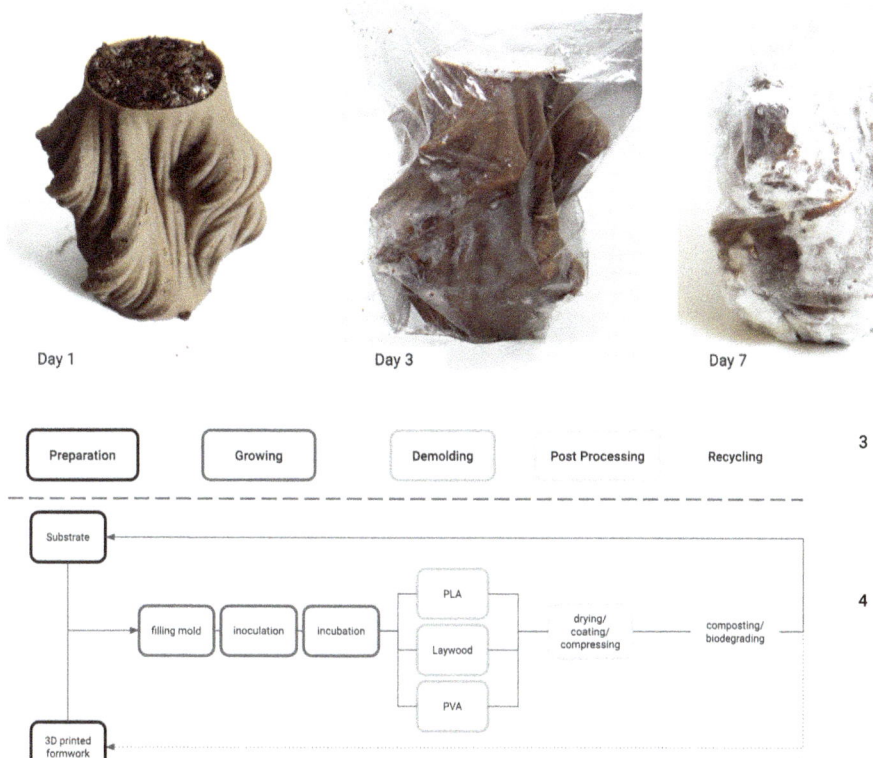

3   Incubation of lignocellulose formwork: Day 1, no visible mycelium; Day 3, mycelium skin grows on top; Day 7, mycelium clusters appear on the formwork; Day 14, fruiting bodies grow in the open area and through the formwork

4   Cradle-to-cradle scheme of 3D printed formwork for MBC creation. Phases: Preparation, growing, demolding, postprocessing, and recycling. The formwork can be shredded and reprinted depending on the chosen polymer material.

printers with a build volume of 300 x 300 x 600 mm. The tested materials in the following experiments are PLA, Laywood, and PVA. They can be found off-the-shelf and are either bio-based or biodegradable:

Polylactic acid (PLA) is a bio-based and biodegradable polymer commonly created from corn starch. First polymerized in 1845, it was commoditized in the 1990s (Auras et al.). Today it is amongst the most popular polymers for FDM/FFF 3D printing.

Laywood (LW) is a polymer composite with a high cellulose content (approx. 40%). Initially developed in 2012 the material became commercially available under the name Laywood™ (3ders 2015). Today, different companies provide composites with biological additives (bamboo, algae, etc.).

Poly Vinyl Alcohol (PVA) is a water dissolvable thermoplastic that is non-toxic and fully biodegradable. It is most commonly used in wood glue or the packing film industry (Hallensleben, Fuss, and Mummy 2015). For FDM/FFF 3D printing, it is used as a support material for overhangs and cantilevered areas.

APPLICATIONS

The following section summarizes project-based experiments. The sample size and geometries are not consistent; however, they are incremental in scale.

Lignocellulose Polymer Composites

3D printable "wood-based" composites gained attention in architecture for their ability to change shape (Correa et al. 2015; Özdemir et al. 2022). Their material composition is of interest for the application as formwork for MBC. Those materials consist of a high cellulose content (approx. 40%) and a polymer as a binding agent. As substrate and formwork share the same base bio-material (lignocellulose), the mycelium is expected to not repel the 3DP but create a stay-in-place formwork.

A formwork piece of approximately 10 x 10 x 20 cm is 3D printed and filled with a pre-inoculated substrate as described in the section *Formwork for Mycelium*. Figure 2 documents the incubation process over a time of 14 days. As expected, the non-enclosed area of the formwork (gas exchange area) starts to grow a mycelium skin after three days. After seven days, the mycelium grows through the lignocellulose formwork, and spots of hyphae appear. The growth continues but never covers the entire formwork, nor does the mycelium densify as in the area not enclosed by the formwork.

Due to this observation, the dehydration phase is not triggered. The authors wanted to observe if the mycelium continues to grow through and on top of the 3DP formwork. After 14 days, fruiting bodies grow on top and on the formwork itself. However, further densification of the hyphae is not observed.

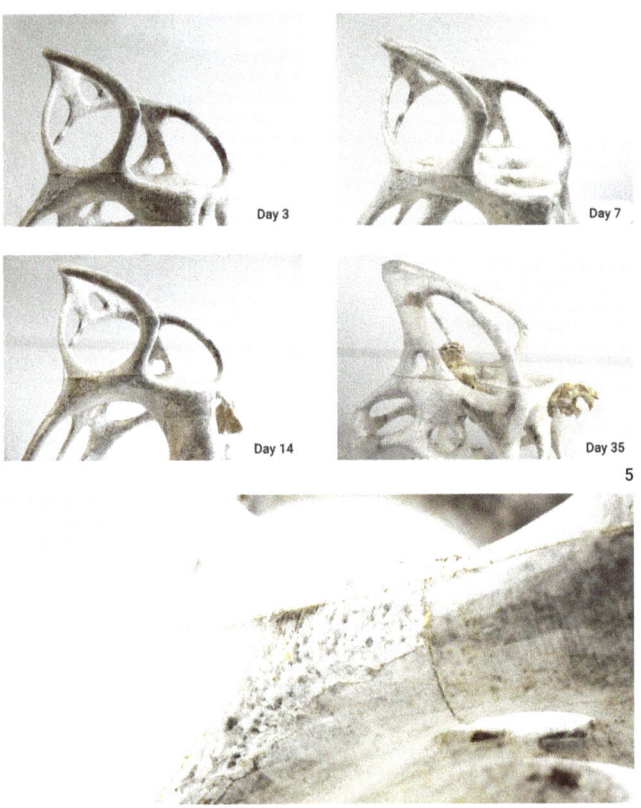

5 Incubation timeline: Day 3, slight mycelium growth, no visible infections; Day 7: growth continues; Day 14: Final state of the mycelium, fruiting bodies appear

6 Printing errors become gas exchange areas, resulting in a mycelium skin

7 Consolidated MycoChair

MycoChair

The *MycoChair* investigates if PLA is a suitable material for 3D printing formwork for MBC. PLA is of interest as it is easy to print, inexpensive, and biodegradable. PLA has been recently used for several architectural applications. Those range from formwork for concrete to directly 3D printed building components (Naboni, Breseghello, and Kunic 2019) or facade prototypes (Mungenast 2017; Taseva et al. 2020).

The design traces the shape of a chair with a set of curves. These define a volumetric mesh representation of the object using isosurfacing through Chromodoris (Newham 2022). The 3D model is segmented into 45 pieces to fit the 3D printers' built space. The formwork is printed as single shell objects with a thickness of 0.5 mm and a resolution of 0.2 mm. The segments are oriented, so that they do not need support structures. However, due to the topology of the segments, they cannot be printed without local failures and partially open areas. Those openings can be beneficial as mycelium produces CO_2 during its growing phase, so constant gas exchange is necessary for healthy growth.

After 3D printing, the formwork is sterilized, and the first set of formwork pieces is preassembled. A 3D printing pen is used to extrude PLA and weld the elements together, creating non-structural connections. After the first set of pieces is filled, the adjacent elements are attached, and the filling/compressing procedure continues.

The *MycoChair* is covered with a foil to keep the substrate humid and protect it during incubation. After three days, the mycelium starts to cover the substrate, and no infections are visible from the outside (Figure 5). The formwork's local defects/open patches start to be covered with a dense mycelium skin after seven days (Figure 6).

After two weeks, a first fruiting body grows and the foil is removed to start the drying process (Figure 1). As the consolidation did not progress as expected, and the substrate is in a fragile state, so removing the formwork by cutting and breaking is not possible. Therefore, to propel the consolidation process the formwork is penetrated using a drill.

After 30 days, the fruiting bodies quit growing and the chair is finally consolidated and ready to be used (Figure 7). Fruiting bodies reduce mass within the MBC itself and can result in cracks or disconnected elements within a complex/thin tubed geometry and should therefore be avoided. Strategies to consolidate MBC in PLA formwork need to be investigated.

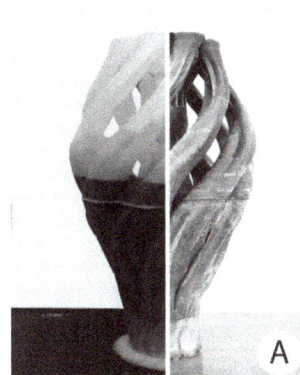
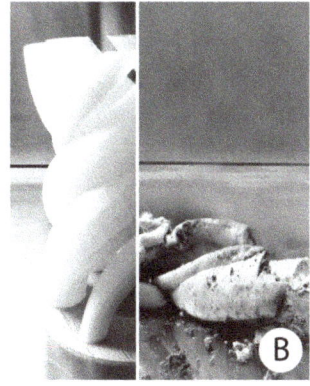
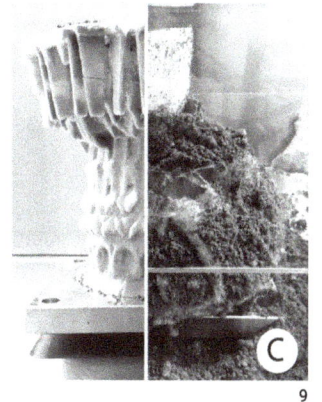

8   3DP formwork made from PVA: (left) directly after filling the samples; (right) samples after seven days of growth

9   Comparison of 3DP PVA formworks: (A) for concrete; (B) and (C) for MBC; MBC PVA formwork fails due to long incubation time and high water content

The Living Column

The Living Column investigates if 3D printed PVA can be used to fabricate bespoke MBC. PVA is a dissolvable thermoplastic known in the field of tissue engineering for its excellent biocompatibility (Kumar and Han 2017). Due to its water solubility, PVA has been investigated as formwork for concrete (Doyle and Hunt 2019; Leschok and Dillenburger 2020; Meibodi 2020). The authors see great potential in using PVA as formwork material for MBC as it potentially acts as a nutrient and can be dissolved, allowing consolidation.

An initial set of elements (60 to 80 mm) are printed with PVA filament, sterilized, and filled with substrate. These experiments showcase promising results as the fungus grows through the 3DP formwork and covers it entirely (Figure 8). A set of larger experiments followed these initial experiments. As shown in Figure 9, thin-shell 3DP formwork made from PVA can be used to cast bespoke concrete elements. However, reproducing a similar geometry or cantilevering structures for MBC is challenging. Concrete reaches a sufficient strength within hours, but MBC has a long incubation time. The moisture content in the substrate significantly weakens the PVA and results in collapse.

A two-component formwork system is introduced to support the PVA during the growth period. Two materials are used, PVA as a primary shell and a secondary, non-dissolvable scaffold. The authors investigated two approaches to fabricating such formwork: 1) dual printing, and 2) distributed printing.

A 3D printer with two printing heads is needed for the first approach. For the second, multiple 3D printers are needed. As dual printing is usually associated with longer printing time and a higher failure rate, the final prototype is printed using multiple 3D printers.

The Living Column is designed with a custom mesh subdivision script. It is segmented into seven different pieces, of which two need to be split into four separate elements. The segmentation is mainly driven by the reduction in print time. To reduce fabrication time even further, the thickness of the formwork is decreased depending on its location in the final structure, from 3 mm at the bottom to 0.8 mm at the top of the column. The supporting outer structure of the column is printed out of PLA.

The 2.5-meter tall column is assembled in one working day, packing 150 liters of substrate into the 3DP formwork (Figure 10). Once assembled, it is covered with a foil, similar to the *MycoChair* project. After seven days, we observe that the column was infected with mold. There may be several reasons for this: a) an uncontrolled environment (access to the room can be controlled/monitored), or b) the type of substrate, as a pasteurized substrate has a higher infection rate than the autoclaved substrate. After consolidation, a 3D scan of the column documents the fabrication challenges for large-scale mycelium components (Figure 11). The final column showcases shrinkage and deformation. The initially observed behavior of mycelium penetrating the formwork was not reproducible for

10 Top view of the column after filling

11 Close up of the living column; lower segment with a thickness of 3 mm did not dissolve as the upper area (shown in Figure 1); infections are visible

12 (left) 3D model and 3D column scan showing deformation and shrinkage; (right) close-up column after demolding

the large-scale column. The upper part of the formwork (thinnest walls) dissolves before the actual dehydration process starts. However, it keeps the MBC compacted long enough to grow successfully. The middle area is partially dissolved (Figure 2). The lower formwork segments with a thickness of 3 mm are not dissolved, nor does the mycelium grow through it (Figure 11).

RESULTS AND OUTLOOK

From the experiments conducted it was observed that:

- All tested polymers can be sterilized and does not interfere with the growth of mycelium.
- Gas exchange areas should be integrated to facilitate homogenous growth.
- Thin members (< 20mm) might result in cracks in the MBC and should be avoided.

LW—

- Laywood can be used as a stay-in-place formwork for MBC if it is designed to interlock with the substrate physically.
- The MBC grows through the formwork. This effect can be improved by printing certain areas in the formwork with reduced thickness (non-structural areas).
- Design for physical bonding: Laywood could be implemented as a stay-in-place formwork for MBC. However, the formwork needs to integrate geometrical features to improve the physical connection between MBC and formwork.

PLA—

- The MBC does not connect or grow through the formwork.
- Design for disassembly: PLA formwork prevents the MBC from drying/consolidating. The MBC needs to be demolded for consolidation. As an alternative, the entire formwork needs to be printed in an open mesh manner (Figure 6).
- Demolding in a non-consolidated state is critical and should be considered during the formwork design, e.g., by introducing a two-piece mold per segment.

PVA—

- Depending on the wall thickness, the mycelium grows through the PVA formwork (Figure 14).
- Depending on the geometry and scale, the formwork needs to be supported with a secondary scaffold during incubation.
- Depending on the wall thickness, the formwork dissolves and does not need to be manually removed.

This paper presents a first set of experiments using 3D printed formwork for MBC. It gives first insights into material behavior and limitations. Future research needs to tackle more mycelium-specific aspects of functional integration. 3D printing has shown several functional integration strategies for the application of concrete formwork. However, for MBC, these functions need to be redefined. The authors believe that the following aspects are of relevance.

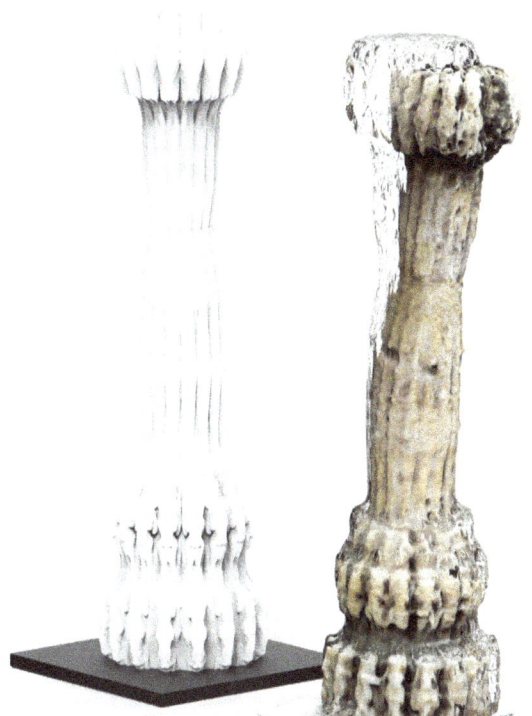

12

As envisioned in Figure 13, the formwork needs to be able to host natural reinforcement (e.g., bamboo poles) and introduce water/nutrient channels for homogenous growth. These can be integrated directly into the formwork design without creating cumbersome manual labor. Furthermore, the design must distinguish structural and semi-permeable areas for controlled gas exchange. This can be achieved by changing printing parameters without causing manual post-processing. This is especially important for scaling up the size of components. Mycelium grows in the direction of oxygen; therefore, components cannot be simply scaled up without considering growth-relevant parameters.

In addition to that, other polymer materials need to be tested and evaluated. The range of available polymers or polymer blends is growing. Especially interesting is a material that is specifically made for enabling the growth of biological materials, i.e. Lay Filaments (3ders 2015). As polymer 3D printing offers the possibility of printing with multiple materials, a multi-material formwork can be envisioned. Areas that are structural (stay-in-place) can be combined with areas that dissolve on their own to achieve consolidation or bio-welding.

CONCLUSION

The research showcases that 3D printing can be a viable option for creating formwork for MBC. Three different materials have been tested and two larger prototypes are presented. Using bio-based printing materials like PLA, LW, or PVA, in a quasi-waste-free production process is in line with the sustainability benefits of MBC. 3D printed formwork extends the design freedom for MBC; therefore, the latter can be shaped to help overcome the structural deficits of the material. The high amount of manual labor needed to create formwork for bespoke structures (e.g., concrete or composites) is their main cost driver. For MBC, custom designs or structural informed designs can be of great relevance to extend their applicability. 3DP offers a cost-effective solution to this problem. Compared to other digital fabrication methods, such as laser cutting in the *MycoTree* project, 3DP drastically reduces the manual labor involved.

3DP formwork was proven to be an appropriate digital fabrication method to create bespoke geometries made from MBC. To fully harvest the potential for MBC, these findings need to be now combined with a structurally informed design. The proposed 3DP formwork method is cheap, easily accessible, and scalable. Therefore, the presented research can have a relevant impact on propelling the successful application of MBC for architectural applications. Mycelium-bound composites coupled with bio-based or biodegradable 3D printed formwork have the potential to create hybrid structures that can help to decarbonize the construction sector due to their low-embodied energy and full recyclability.

ACKNOWLEDGMENTS

The authors would like to thank a number of partners and collaborators whose dedication helped to fulfill the research described in this paper. From the Karlsruhe Institute of Technology (KIT), our sincere gratitude goes to Prof. Dr. von Both, Dr. Koch, Florian Rothermel, and Nicklas Dorsch. Furthermore, we would like to thank Andrei Jipa

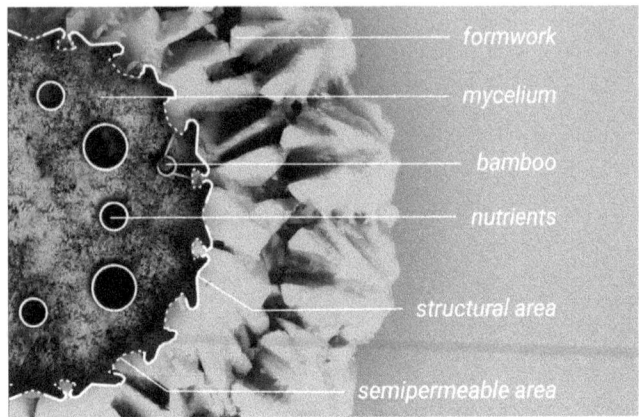

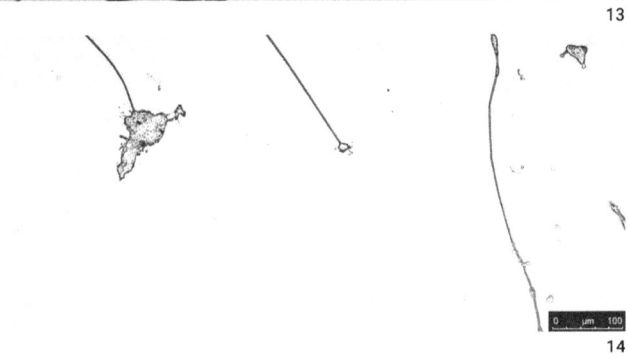

13 Vision of an integrated formwork system for MBC, including guides for natural reinforcement and nutrient channels; differentiated printing parameters distinguish between structural and gas-exchange areas

14 Microscopy indicates that the hyphae (thin black lines) enclose the PVA and grow through and around it

and Rena Giesecke for their guidance during the thesis research, to Marirena Kladeftira for editing the paper, and to Jan Wohlfarth from the Department of Health Sciences and Technology for microscopy and analysis of the sample. This research was partially supported by the NCCR Digital Fabrication, funded by the Swiss National Science Foundation (NCCR Digital Fabrication Agreement #51NF40-141853).

NOTES

Pasteurised substrate (provided by Kernser Edelpilze GmbH) was used for the column, autoclaved (provided by Biomycotec) for the rest of the experiments. The substrate from Biomycotec is sterilized, and the other is pasteurized. Sterilized substrates reduce the risk of infections/mold creation, pasteurized substrates on the other hand can speed up the growth of mycelium, but bare higher risks of infections/mold. To mitigate the risks of infections, the pasteurized substrate was "pre-grown" for seven days before being filled into 3DP molds.

REFERENCES

3ders. 2015. "Kai Parthy is back with LAYWOOD-FLEX, a flexible version of LAYWOOD 3D printer filament." 3ders.org (blog), May 13, 2015. Accessed April 26, 2022. http://www.3ders.org/articles/20150513-kai-parthy-is-back-with-laywood-flex-a-flexible-version-of-laywood-3d-printer-filament.html

Angelova, G. V., M. S. Brazkova, and A. I. Krastanov. 2021. "Renewable Mycelium Based Composite, Sustainable Approach for Lignocellulose Waste Recovery and Alternative to Synthetic Materials: A Review." *Zeitschrift Für Naturforschung C* 76 (11–12): 431–42. https://doi.org/10.1515/znc-2021-0040.

Auras, R., L.-T. Lim, S. E. M. Selke, and H. Tsuji, eds. 2010. *Poly(Lactic Acid): Synthesis, Structures, Properties, Processing, and Applications*. Hoboken, NJ: John Wiley & Sons.

Biohm. n.d. "BIOHM | The Future Of Home | London." Accessed April 23, 2022. https://www.biohm.co.uk/.

Bitting, S., T. Derme, J. Lee, T. Van Mele, B. Dillenburger, and P. Block. 2022. "Challenges and Opportunities in Scaling up Architectural Applications of Mycelium-Based Materials with Digital Fabrication." *Biomimetics* 7 (2): 44.

Burger, J., E. Lloret-Fritschi, F. Scotto, T. Demoulin, L. Gebhard, J. Mata-Falcón, F. Gramazio, M. Kohler, and R. J. Flatt. 2020. "Eggshell: Ultra-Thin Three-Dimensional Printed Formwork for Concrete Structures." *3D Printing and Additive Manufacturing* 7 (2): 48–59.

Correa, D., A. Papadopoulou, C. Guberan, N. Jhaveri, S. Reichert, A. Menges, and S. Tibbits. 2015. "3D-Printed Wood: Programming Hygroscopic Material Transformations." *3D Printing and Additive Manufacturing* 2 (3): 106–16.

De Belie, N., E. Gruyaert, A. Al-Tabbaa, P. Antonaci, C. Baera, D. Bajare, A. Darquennes, et al. 2018. "A Review of Self-Healing Concrete for Damage Management of Structures." *Advanced Materials Interfaces* 5 (17): 1800074

Derme, T., D. Mitterberger, and U. Di Tanna. 2016. "Growth Based Fabrication Techniques for Bacterial Cellulose: Three-Dimensional Grown Membranes and Scaffolding Design for Biological Polymers." In *ACADIA 16: Posthuman Frontiers, Data, Designers, and Cognitive Machines; Proceedings of the 36th Annual Conference of the Association for Computer Aided Design in Architecture*. 488–495.

Dessi-Olive, Jonathan. 2019. "Monolithic Mycelium: Growing Vault Structures." In *Construction Materials & Technologies for Sustainability. 18th International Conference*. Nairobi, Kenya.

Doyle, S. E., and E. L. Hunt. 2019. "Exploring Additive Manufacturing for Reinforced Concrete." In *ACADIA 19: Ubiquity and Autonomy, Proceedings of the Association for Computer Aided Design in Architecture*. University of Texas, Austin. 178–187.

Ecovative. n.d. "Ecovative, We grow better materials." Accessed April 23, 2022. https://ecovative.com/.

Elsacker, E., A. Søndergaard, A. Van Wylick, E. Peeters, and L. De Laet. 2021. "Growing Living and Multifunctional Mycelium Composites for Large-Scale Formwork Applications Using Robotic Abrasive Wire-Cutting." *Construction and Building Materials* 283 (May): 122732.

Elsacker, E., S. Vandelook, A. Van Wylick, J. Ruytinx, L. De Laet, and E. Peeters. 2020. "A Comprehensive Framework for the Production of

Mycelium-Based Lignocellulosic Composites." *Science of The Total Environment* 725 (July): 138431.

Ghazvinian, A., and B. Gürsoy. 2022. "Mycelium-Based Composite Graded Materials: Assessing the Effects of Time and Substrate Mixture on Mechanical Properties." *Biomimetics* 7 (2): 48.

Hallensleben, M. L., R. Fuss, and F. Mummy. 2015. "Polyvinyl Compounds, Others." In *Ullmann's Encyclopedia of Industrial Chemistry*. Weinheim, Germany: Wiley-VCH Verlag. 1–23.

Heisel, F., J. Lee, K. Schlesier, M. Rippmann, N. Saeidi, A. Javadian, A. Reza Nugroho, T. Van Mele, P. Block, and D. E. Hebel. 2017. "MYCOTREE Design, Cultivation and Application of Load-Bearing Mycelium Components: The MycoTree at the 2017 Seoul Biennale of Architecture and Urbanism." *International Journal of Sustainable Energy Development* 6 (1): 296–303.

Jipa, A., C. Calvo Barentin, G. Lydon, M. Rippmann, M. Lomaglio, A. Schlüter, and P. Block. 2019. "3D-Printed Formwork for Integrated Funicular Concrete Slabs." In *Proceedings of the IASS Annual Symposium 2019 – Structural Membranes 2019*. Barcelona, Spain.

Jipa, A., F. Giacomarra, R. Giesecke, G. Chousou, M. Pacher, B. Dillenburger, M. Lomaglio, and M. Leschok. 2019. "3D-Printed Formwork for Bespoke Concrete Stairs: From Computational Design to Digital Fabrication." In *Proceedings of the ACM Symposium on Computational Fabrication*. Pittsburgh Pennsylvania: ACM. 1–12.

Karana, E., D. Blauwhoff, E.-J. Hultink, and S. Camere. 2018. "When the Material Grows: A Case Study on Designing (with) Mycelium-Based Materials." *International Journal of Design* 12 (2): 119–136.

Leschok, M., and B. Dillenburger. 2019. "Dissolvable 3DP Formwork; Water-Dissolvable 3D Printed Thin-Shell Formwork for Complex Concrete Components." In *ACADIA 19: Ubiquity and Autonomy, Proceedings of the Association for Computer Aided Design in Architecture*. University of Texas, Austin. 188–197.

Leschok, M., and B. Dillenburger. 2020. "Sustainable Thin-Shell 3D Printed Formwork for Concrete." In *Impact: Design With All Senses*, edited by C. Gengnagel, O. Baverel, J. Burry, M. Ramsgaard Thomsen, and S. Weinzierl. Cham: Springer International Publishing. 487–501.

Loh, P., M. Qu, and D. Leggett. 2020. *Future Prototyping: Catalogue: Architecture, Art, Engineering, Culture, Fashion*. Melbourne, Australia: Melbourne School of Design, the University of Melbourne.

Mogu. n.d. "Homepage." Accessed April 23, 2022. https://mogu.bio/.

Mok, Kimberly. 2018. "Mycotecture: Building With Mushrooms? This Inventor Says Yes." Treehugger (blog), October 11, 2018. Accessed April 23, 2022. https://www.treehugger.com/mycotecture-mushroom-bricks-philip-ross-4857225.

Mungenast, M. 2017. "Fluid Morphology; 3D-Printed Functional Integrated Building Envelope." In *Proceedings of the 12th Conference of Advanced Building Skins*. Bern, Switzerland. 109–24.

Naboni, R., L. Breseghello, and A. Kunic. 2019. "Multi-Scale Design and Fabrication of the Trabeculae Pavilion." *Additive Manufacturing* 27 (May): 305–317.

Newnham, Cameron. *Chromodoris* (plugin). V. 0.91. 2022. https://github.com/camnewnham/ChromodorisGH.

Oxman, N., J. Laucks, M. Kayser, and C. D. Gonzalez Uribe. 2013. "Biological Computation for Digital Design and Fabrication." In *Computation and Performance (International Conference of eCAADe)*. Delft: eCAADe and Delft University of Technology. 585–594.

Özdemir, E., L. Kiesewetter, K. Antorveza, T. Cheng, S. Leder, D. Wood, and A. Menges. 2022. "Towards Self-Shaping Metamaterial Shells: A Computational Design Workflow for Hybrid Additive Manufacturing of Architectural Scale Double-Curved Structures." In *Proceedings of the 2021 DigitalFUTURES*, edited by P. F. Yuan, H. Chai, C. Yan, and N. Leach. Singapore: Springer Singapore. 275–285.

Snooks, R. 2018. "Sacrificial Formation." In *Towards a Robotic Architecture*, 5th ed., edited by A. Wit and M. Daas. Novato, CA: ORO Editions.

Taseva, Y., N. Eftekhar, H. Kwon, M. Leschok, and B. Dillenburger. 2020. "Large-Scale 3D Printing for Functionally-graded Facade." In *RE: Anthropocene, 25th International Conference of CAADRIA*. Hong Kong. 183–192.

The Growing Pavilion. n.d. "Home - The Growing Pavilion." Accessed April 23, 2022. https://thegrowingpavilion.com/.

Transparency Market Research. 2022. "Construction Waste Market Share, Research Insights by 2025." Accessed April 27, 2022. https://www.transparencymarketresearch.com/construction-waste-market.html.

urbanNext. 2014. *Hy-Fi*. Accessed April 23, 2022. https://urbannext.net/hy-fi/.

Van Wylick, A., E. Elsacker, L. L. Yap, E. Peeters, and L. De Laet. 2022. "Mycelium Composites and Their Biodegradability: An Exploration on the Disintegration of Mycelium-Based Materials in Soil." *Construction Technologies and Architecture* 1: 652–659.

IMAGE CREDITS

Figures 10, 11 ©Catherine Leutenegger.
All other drawings and images by the authors.

Matthias Leschok is a PhD researcher at Digital Building Technologies, ETH Zurich. He holds a MAS in Architecture and Digital Fabrication from ETH Zurich and a Dipl. Ing. in architecture from KIT. He is also co-founder and COO of the startup SAEKI Robotics AG.

Benjamin Dillenburger, PhD is Professor at the Institute of Technology in Architecture, ETH Zurich. He is leading the research group Digital Building Technologies (DBT), which investigates computational design and digital fabrication with a focus on large scale additive manufacturing in architecture.

Developing a Digital Design Workflow for Nexorade Bamboo Structure

Jonas Hauptman
Ramtin Haghnazar
Sara Saghafi Moghaddam

School of Design, College of Architecture, Arts and Design, Virginia Tech

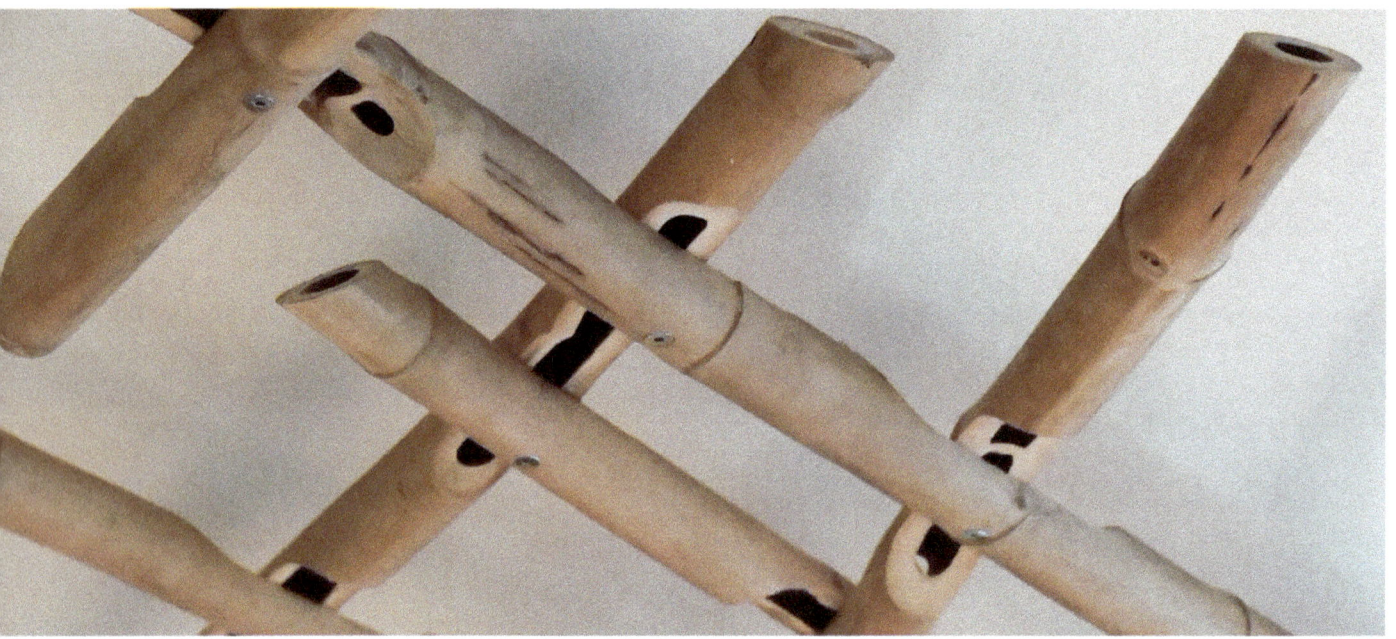

1 Bamboo research faculty apartment ceiling, Virginia Tech

ABSTRACT

This paper presents a case study integrating generative design and bamboo culm geometries. Our goal is to improve the application of biological materials in a responsive Computer-Aided Design (CAD) process. We argue that fabrication-aware design tools are currently inadequate because they do not embed intrinsic biological geometrical properties into the design or fabrication processes. Although research with CAD tools has advanced the development of generative and fabrication-aware design processes for the built environment, most of these tools are developed for off-the-shelf materials that are typically unitized, industrialized, and homogeneous. This has made it difficult to find and create computational tools for naturally variable biological materials. While employing eccentric biological materials such as bamboo imposes an added layer of complexity on the design-to-fabrication process, it may also offer more sustainable material application and expand the frontiers of design and fabrication research methods. The methods explored in this paper are deployed to realize freeform Nexorade structures (FNS) that are explicitly tailored to individual bamboo culms (BC); each of these has been measured to explore the potential that material eccentricity may be a district benefit rather than a detriment to the quality and efficiency of a design.

TERMINOLOGY

Terms and Abbreviations used in this paper:

Term	Short Description	Full Description
1. IB	Iron Bamboo	*Dendrocalomus strictus*, a tropical and subtropical clumping bamboo species native to Southeast Asia, heretofore identfied in this paper as Iron Bamboo or IB.
1. Culm	Live culm shoot	A living bamboo stalk still growing and supported by a rhizome.
2. R-Stock	Raw stock	A cut and treated sections of culm ready for analysis and fabrication are often cut to standard stock lengths and graded or binned into groups of approximate diameters.
3. M-Stock	Node measured stock	The raw stock that has been named with an ID tag and measured node to the node for its lengths.
4. B-Lib	Bamboo library	A library of bamboo poles that have been weighed and measured. Measurements include the base outer diameter (OD) at the widest end of the pole, and a list of dimensions calling out the bamboo node to node distances for each pole. This data is tied to a unique alphanumeric identifier assigned to each raw bamboo in the library. The ID is also labeled on the actual raw bamboo stock.
5. Elements	Theoretical elements	The theoretical lines that aggregate to describe a Nexorades structure.
6. T-Comp	Theoretical component	A revolved and closed NURBS polysurface that approximates a real multi-nodal segment of bamboo from M-Stock with an approximate diameter and node distribution but with less accuracy than a scan, useful in matching theoretical elements from stock to place them with the best utilization in the Nexorade structure.
7. M-Comp	Milled component	A scanned, indexed, and CNC-cut bamboo part to match a desired part of the Nexorade assembly.
8. Mesh	Typical mesh	Triangular tessellation of faces represented by edges and vertices points approximates a more complex surface or form.
9. N-Line	Nexorade line	A vector defined by two endpoints and 4 near intersection points needed to aggregate it with many similar parts to form the Nexorade theoretical models.
10. Surface	Base surface	A manual CAD input made by the user designer to specify the form intention and to be used by the tool to generate the FNS. Created as a NURBS Surface.
11. Face	Mesh face	A flat Polygon/Face described by a minimum of 3 edges and 3 points.
12. B-Node	Bamboo node	Diaphragm found intermittently along the length of a bamboo culm.
13. G-Node	Geometry node	A point of intersection(s) of lines, also known as a vertex.

INTRODUCTION

This paper describes a novel generative design tool that was developed to assist users in creating FNS for measured and serialized bamboo. The tool incorporates individual culm geometry and paves the way for waste reduction and a closer alignment of material with form and pattern. Bamboo holds great promise as a natural building material that responds to the emerging sustainability concerns in the AEC industry. As a rapidly growing and renewable agricultural resource, bamboo consumes fewer resources, grows faster than other sources of woody biomass, and holds enormous potential as a carbon bank, able to sequester carbon both in its rhizomatic root structure and in permanent bamboo constructions. Bamboo shares some mechanical characteristics with lumber and with steel. These properties make it appropriate for innovation in building systems for rapidly urbanizing tropical zones where bamboo flourishes.

Presently, bamboo is not widely used in the building industry for multiple commonly cited reasons. These include cultural perceptions, a lack of know-how within contemporary building industries, a lack of codes and standards, as well as challenges with managing its complexity as a non-standard and eccentric material (Hidalgo López 2003; Steffens et al. 2000; Schumann et al. 2019; Hauptman et al. 2019). Of particular interest to us is the lack of established systems and strategies in the design, engineering, and construction processes for utilizing the full capability of materials with various geometrical shapes and other properties. As a structural building element, bamboo culm faces many challenges, including the need to identify its intrinsic geometric variability and to incorporate this into formal design and fabrication techniques. Although digital tools offer advantages for custom fabrication, most design-to-fabrication processes relies on the specifications of the standard, industrially manufactured material units (i.e., stone slabs, milled lumber boards, plastic pipes, or plywood). The variables afforded by the generic component are predictable and formalized regardless of whether it is assembled from standardized units or derived from subtractive processes. The current materially confined domain of digital design to the fabrication process highlights the necessity to investigate alternate tool-making approaches that address the complexity of biological materials. While retaining the inherent aesthetic benefits of biological materials, the advancement in accessibility in manufacturing can yield more materially sustainable solutions.

This study documents a CAD tool to design and fabricate FNS using *Dendrocalomus strictus*—heretofore identified in this paper as Iron Bamboo or IB. Rhinoceros 3D (Robert McNeel & Associates 2020) Grasshopper (GH) (Rutten 2020) definition is used to develop a tool that aids in the design and fabrication process by searching, identifying, and matching specified needed Theoretical Components (T-Comp.) with actual IB from previously measured bamboo stock. These actual elements are assigned to the appropriate position in the structure based on the bamboo's unique geometrical properties. Compared to conventional procedures, this method creates a symbiotic relationship between the natural material and the specific geometry required, thus resulting in near-zero waste and producing elegant and robust structural IB elements. This integration of generative design, non-industrialized material geometry, and analysis elevates the importance of understanding the anomalies of natural materials in design and digital fabrication.

Bamboo

IB reaches harvestable maturity in less than five years. It also offers uniquely high solidity for bamboo at approximately 95% solid, and when compared to structural softwood (2x4), it has similar stiffness and strength in half the height and cross-sectional area (Hauptman et al. 2019).

As a biomaterial, bamboo is reasonably regular compared to a hardwood tree branch. Branches contain a greater variety of types, including more rapid tapering and more kinks and curves from forking at comparable lengths and diameters to IB. To further clarify the amount of variation in IB the following quantifications are provided. From a set of 200 culms if measuring the distance along 15 nodes therefore 14 segments they vary from 119 inches to 213 inches. Furthermore, not all node to node distances are equal in fact they vary from 0.5 inches to 15 inches apart if evaluating all 2800 segments. Finally each culm thickness varies approximately 30 percent across the sample, and base thickness of all culmsset ranging from 1.3 to 2.6 inches.

For these reasons, bamboo has been chosen as the material to explore. These characteristics allow us to explore fabrication methods and vernacular gestures in ways that many other hollower, thinner-walled bamboo species or other wood raw bio-materials do not. Finally, in terms of size, IB is ideal for

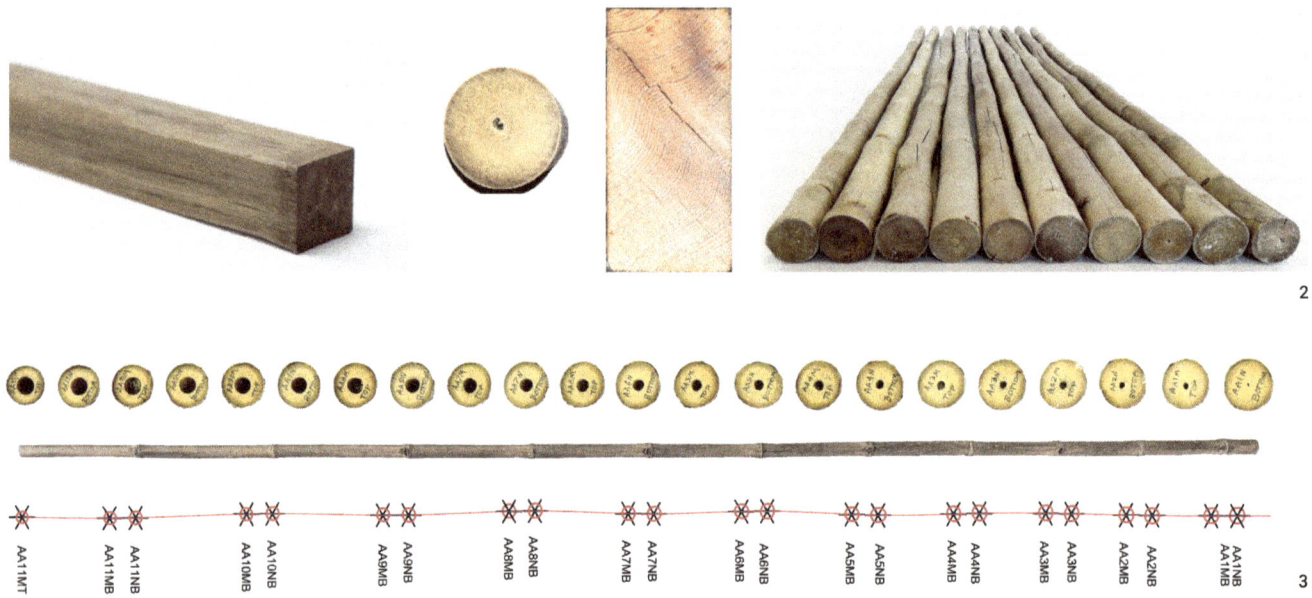

2 (left) IB ripped to square seemingly solid shape; (center) IB and 2x4 cross-section at 1:1 scale; (right) Group of IB stock 6-foot lengths to demonstrate natural variation

3 A single 12-foot IB culm cut near all nodes to express common ranges of wall thickness

our research because it is one of the smallest construction/timber bamboos, measuring only an average of 50 mm at its largest OD.

Reciprocal Frames

The three-dimensional reciprocal frame structures (RFS), also known as Nexorade systems, are composed of linear flat or inclined elements that support each other across their span (Baverel 2000). With no structural hierarchy, they form a closed unit, and this system provides a self-supporting spatial configuration capable of spanning multiple lengths with an individual unit (Popovic Larsen 2008). Advances in digital design, fabrication, and architectural prototyping expand the research opportunities for reciprocal frames as a modular, lightweight, and low-cost structural system (Apolinarska 2018). Using straight elements, the RFS enables the exploration of innovative curved 3D complex structures (Popovic Larsen 2014).

Furthermore, these structures offer a potential approach for lowering member connection complexity (Sénéchal, Douthe, and Baverel 2011). RFS is composed of short members—all joints and members are identical in symmetrical configurations—and the assembly of their components is entirely dependent on precisely identifying their geometric qualities (Popovic Larsen 2014).

In this research, bamboo was coupled with a Nexorade structure approach because bamboo is known to have challenges in joining, and Nexorade makes it easy. Moreover, Nexorades allow short-length parts, making indexing, work holding, and fabrication more accessible, manageable, and accurate. Finally, Nexorades elements' length can be extended without any change in the structured points of connection, and this offers a considerable and unique benefit. Using this system for a variable material like bamboo makes for the most reliable joints, with all penetration occurring only in between nodes and never in a region of the cut culm that is left open without a terminating bamboo node.

Computational Design and Digital Fabrication

Advances in CAD/CAM processes have enabled the development of complex assemblies and freeform structures in the building industry. This allows for the exploration of materials with intrinsic geometrical and biological complexity that traditional methods could not address (Lorenzo and Mimendi 2020).

This paper presents the design to the fabrication process of an FNS utilizing BC as an eccentric material with inherent variations. The form-finding process for this seemingly simple structure is complex. In one study, genetic algorithms were used to construct the form-finding, and in another study,

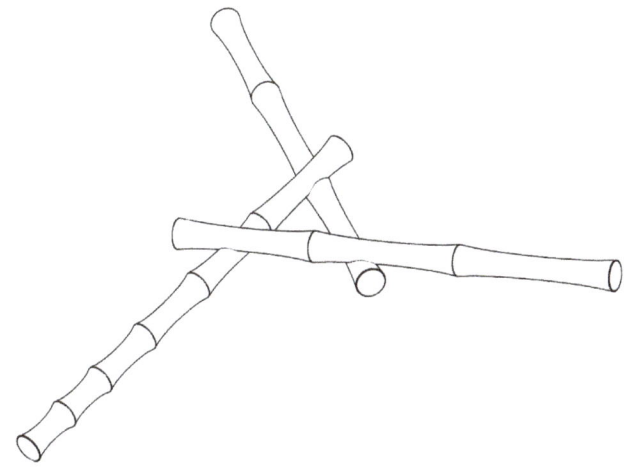

4 Illustration of a reciprocal array of three Theoretical Components (T-Comp)
5 Assembly of actual IB CNC-milled components
6 Element on the right represents the proper B-Node to cut relationship; element on the left has the incorrect hole to B-node relation

fictitious mechanical behavior was used to solve the dynamic relaxation problem (Baverel and Nooshin 2007; Douthe and Baverel 2009). This method has now been adapted for use in a Grasshopper workflow. Non-linear algebra and analytical solutions using regular polyhedron rotations were also used to describe dynamic relaxation (Mesnil et al. 2018; Sénéchal, Douthe, and Baverel 2011). Each of these methods requires computational design for its analysis and to enhance accuracy.

Besides the complex behavior of the Nexorade system, its joints must be positioned at particular angles in the structure. Each element meets four additional components at four joints that are placed at different coordination and polar angles in space. Four-axis computer numerical controlled (CNC) milling is the fabrication method used for joints in this research, and it requires data from a CAD tool.

CAD tools have been leveraging computational design methods and allowing new exploration opportunities in design. Designers now have a broader range of methodologies, which influence the generation of fabrication information (Haghir et al. 2021). Computers and simulation technologies have enabled users to explore entirely new areas of knowledge through tailored design tools. These tools encourage studying and combining various types of data to develop and evaluate ideas from different perspectives, which eventually assist designers in making more informed decisions (Johns and Foley 2014; Oliyan Torghabehi 2020; Rastegar et al. 2021).

Current CAD tools lack an approach that allows the integration of design and fabrication complexity with the variability presented by natural materials. This research addresses this challenge by developing a tailored design tool and embedding specific and serialized bamboo geometry into the process. The eccentricity of bamboo makes it impossible to industrialize it in twentieth century terms; instead, it requires a twenty-first century data-driven approach.

METHODOLOGY

We are proposing a novel digital design workflow (DDW) to aid designers in creating FNS that symbiotically accommodate both the natural geometrical qualities of IB and human design intent (MacDonald, Schumann, and Hauptman 2019). Typically, designers use standard materials to create and build novel and complex forms. As noted in the literature, the majority of bamboo projects do not employ digital design or fabrication tools and are often constructed traditionally, relying on carpentry know-how rather than predetermined design variables to ensure reliability; a considerable amount of redundancy is usually included to compensate for a significant lack of information about the capacity of bamboo. To use bamboo in FNS, where its wider material potential is fulfilled, designers need a DDW that allows them to integrate material considerations early in the design process.

The process begins by measuring each bamboo culm to discover its physical dimensions. This step is not required in conventional or industrial material design and fabrication, and ever since the industrial revolution, this has been considered a drawback. However, in a technological age where the Internet of Things has democratized sensors, microcomputers, and the storing and exchange of data, new efficiencies are possible, and it makes sense to begin the process this way.

As a next step, a DDW helps the designer generate structures based on the convergence of both human needs and the biological potential derived from analyzing the existing bamboo inventory. Then the required bamboo will be matched to its best-suited location, and the ultimate product will be within a range of efficient alternatives. This can result in a near-zero-waste solution and still reflect the initial design intention. The process presented in this paper attempts to define an algorithm for such a design method and implement the logic via Grasshopper to create a useful DDW. The tool first gets the input data containing the bamboo dimensions, then the design parameters, and finally, it assists the designer in generating structures by creating a BIM model, fabrication data, and any other essential outputs for project realization.

These include both manufacturing instructions like G-codes and the solid model elements useful for architectural and construction drawing. With such data, both finite element analysis and life cycle assessment can be developed, as the design data is highly representative of the constructed project. The following steps summarize the generative process:

- Search M-Stock against all needed parts
- Assign M-Stock to Theoretical Elements

7   Axis CNC milling machine with IB actual pre-milled component chucked up

- Check visualization for improper fits of Theoretical Elements
- Assign final locations and identifiers to all M-Stock that will be utilized
- Cut the appropriate multi-nodal section from the M-Stock
- Generate joinery holes and surface milling CAM data
- Generate a four-axis toolpath including pecked holes for bolts and lap joint cuts to make mating surfaces
- Generates assembly instructions

Biological Approach

Our approach is to explore several facets of the process simultaneously to create a symbiotic solution that respects natural bamboo's geometrical potential while delivering design intent that addresses a human need. In a world of surplus brought on by mass extraction and industrialization, designers have grown accustomed to standard materials and use them rapidly, but not necessarily efficiently in a material sense.

For instance, a designer might use a standard timber, or a rectangle board as opposed to a natural raw tree limb to create a line of structure. Perhaps because of this standardized process in an industry that must meet both site and human needs that are variable, the construction industry in the United States is responsible for around 30 percent of the total weight of building materials supplied to a construction site ending up in landfills (Osmani 2011). As a result of this standardizing process, renewable and environmentally friendly materials, such as bamboo, have been disregarded. The natural qualities of bamboo are more diverse than those of lumber, making it impossible to convert them into identical profiles. As the AEC community becomes increasingly concerned about the shortage of resources and the consequences of human actions on the environment, we seek an approach that aims at a natural rather than an industrialized norm for material processing. This research offers an example of how humans can use a natural standard of materials, such as bamboo, to rethink what is meant by efficiency.

Although our study's scope does not encompass all the variables related to bamboo geometry, it does examine how some variables impact the design process. As a rule of thumb in fabrication, cutting bamboo should be limited to specific locations along its length because it leaves segments prone to splitting, crushing, and rot. Therefore, it is necessary to locate a bamboo with an internodal dimension that can fit the particular position in the final structure.

BCs are often not perfectly straight and have a meandering centerline, so their individual shape is important in applying them to a structural system. Finally, because they are not available in standard or infinitely variable sizes, it is good to cluster elements into different types and use thicker elements near the ground where gravity is less of a concern and lighter and thinner elements in the upper regions of a structure to lower the weight that needs to transfer from upper regions of a structure.

In our process, we capture data about the size, internode length, shape, and weight to capture each potential material unit's unique anatomy and capacity. We then use this information to influence the design. The most important property in this research is the internodal length, which is vital when the tool wants to replace the original Theoretical Element with a real piece of bamboo (Measured Element) from the existing Bamboo Library.

System Design

An important feature of Nexorade structures is their simplicity of assembly. This is because they have no hierarchy of elements, and there are no beams or columns but simply arrays of interconnected elements with no hierarchy. Our experimental structures will be made by simply bolting bamboo elements without the need for highly complex joints or hubs commonly needed to assemble other types of bamboo structures. However, Nexorades are complex to design, especially when applied to natural materials. Thus, the researchers decided to streamline the variables to more easily manage them in a DDW.

Bamboo culms are milled and drilled by the CNC to manage the diameter differences and kinks in axes. Although the

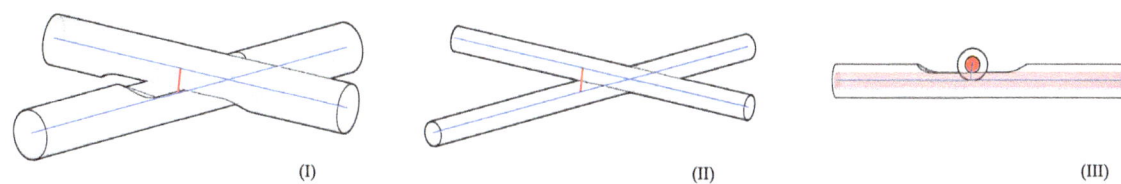

8   The ER concept illustrations: the red line represents the constant distance and axis between the pairs of bamboo elements in FNS, and regardless of individual bamboo OD, is always in that position; a joint volume in relative position to the axis is subtracted from the element to guarantee the equal distance of the axis anywhere in the structure.

bamboo diameters are variable, they are within a range of tolerance. To manage this variation, a method was developed in the system design, termed here as the effective radius (ER). In this way, all bamboos are considered the same in radius at their joint location. In other words, the distance between the axis of the two connected bamboo elements is always the same, and the difference in the diameter only affects the depth at which to cut the joint. Thus, the distance between the imaginary axis of the element, the wire model, and the joint faces of the bamboo after milling is always constant and equal to the ER, as illustrated in Figure 8. It is important to know that the ER must be smaller than the physical radius of the available bamboos in the inventory.

Digital Design Workflow

Considering the aforementioned criteria and strategy, it is helpful to have a digital design tool in 3D modeling software. The authors extract the design logic from the criteria, strategies, and limitations and implement them in a DDW to help the designers easily realize an FNS.

In order to design an FNS in this way, it is needed to have an algorithmic method to solve the problem. In complex problems, there are substantial geometrical interconnected challenges to manage. In other words, when a designer tries to solve a geometrical issue in a part of a structure, a new problem is created on the other side of it; that is the nature of the complex problems in FNS. To find a suitable result, it is necessary to query the design parameters repeatedly and test the solutions. Only an algorithmic design tool can do it in real-time and let the designer generate many alternatives to reach a suitable solution.

This tool has a set of inputs containing essential properties of the bamboo culms, such as length, the distance between nodes, OD of each bamboo, design parameters, and a base surface. The result will be a BIM model containing a geometry model for each element, fabrication data containing cutting data, CNC G-codes, and assembly instructions.

THE ALGORITHM

The algorithm is a logical procedure implemented in GH to convert the input data to the final BIM model containing any essential information for fabrication and assembly. In this section, this procedure is described step by step to determine how this tool works and can help designers to design FNS efficiently.

Inputs

Our tool considers different bamboo variables; first, the bamboo shape properties including the acceptable effective radius (ER) as illustrated in Figure 8-II; also, as seen in Figure

9 Illustrates how a kink in bamboo can make for an infeasible part even when the ER and the OD are a potential match

10 (I) Base surface; (II) Randomized vertices on base surface; (III) Attractor point to produce non-uniform logic; (IV) Generating the sphere, where the radius is defined by the attractor point; (V) Soft sphere packing to reach the desired triangular mesh density; (VI) Final weighted mesh

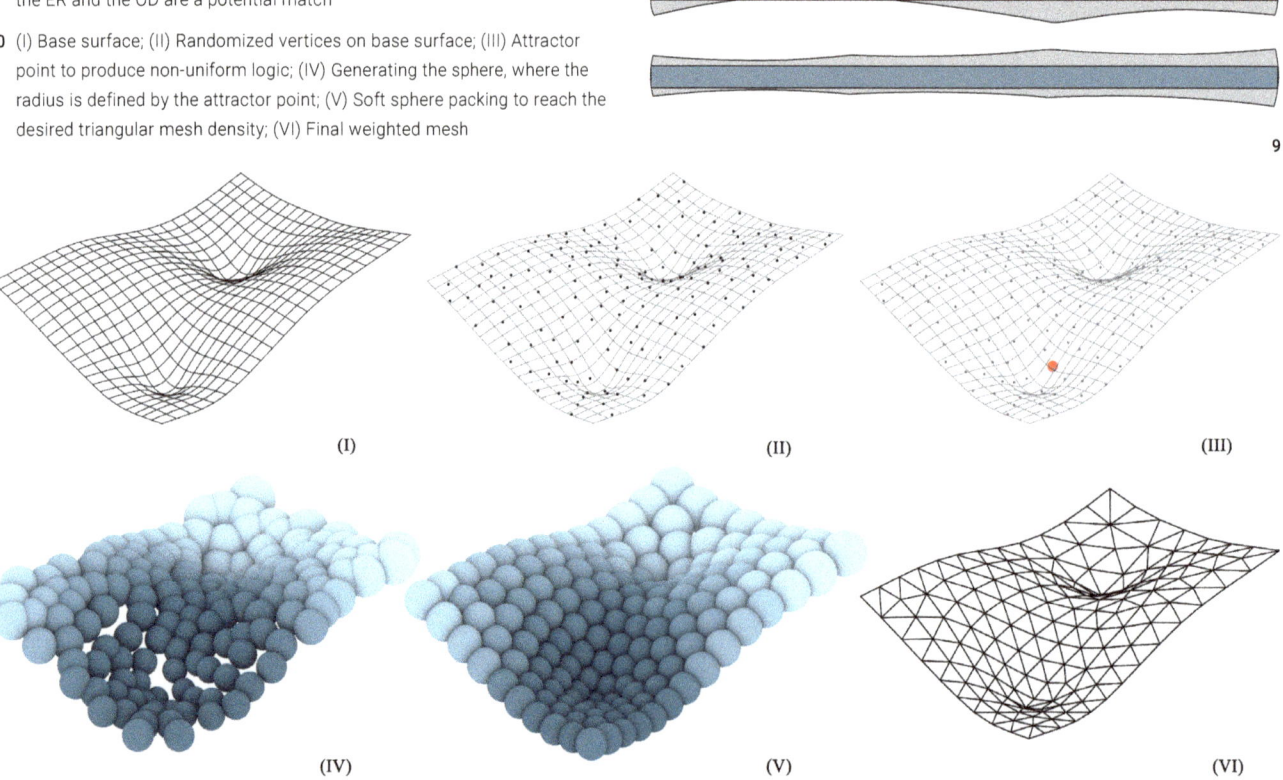

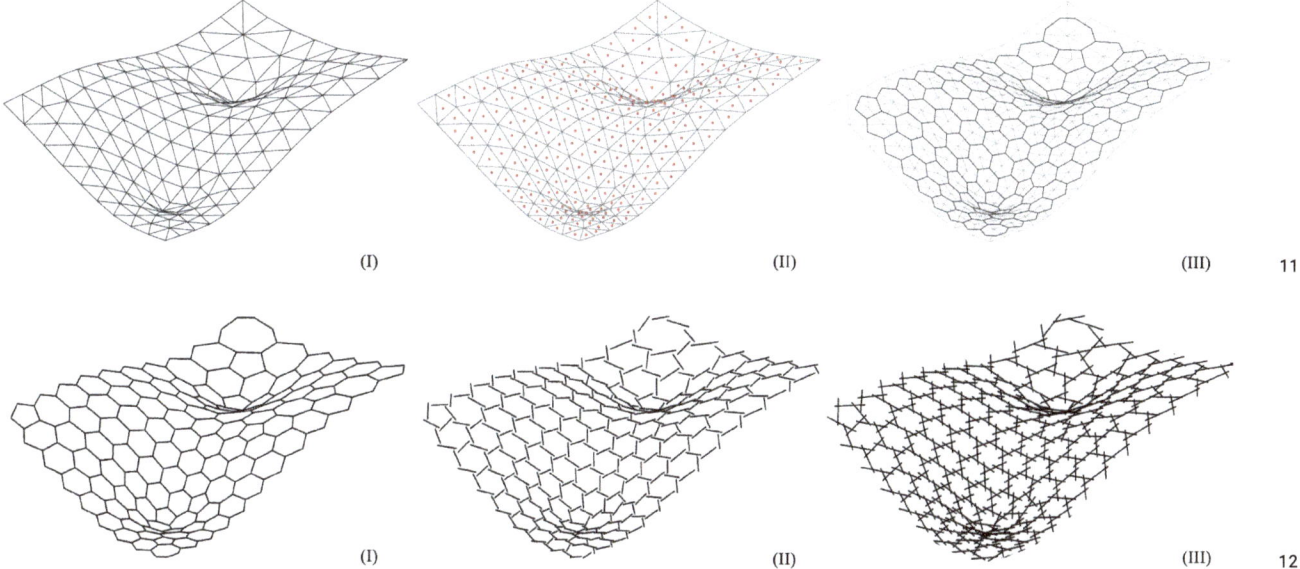

9, the blue illustration is an acceptable section of R-Stock while the red one is not. This is because the minimum ER can violate the physical stock even of an acceptable OD if the bamboo is substantially kinked.

Second, the main surface can be a NURBS or mesh representing the design model form. Although the code has an embedded part of generating the form by a form-finding method, any kind of geometry is acceptable depending on the designer's needs.

Third, point count and desirable cell size are the final two categories of inputs that define sphere radius in the soft sphere packing step. With attractor points, it is possible to change the density of the FNS.

Step 1: Pattern Generator
In order to generate the FNS, the user supplies a base surface. Next, our tool generates a field of randomly defined mesh vertices. After which a soft sphere packing algorithm, using Kangaroo3D (Piker 2017), is used to regularize the mesh, and an attractor point is used to define the radii of the spheres and to create mesh variability.

The triangulated mesh is not suitable for the generation of a Nexorade pattern because it creates overcrowding of intersections. Thus, the code uses a dual pattern algorithm to convert from a triangular mesh to a mostly hexagonal tessellation.

Step 2: Wire Modeling
The result of the previous step is only a tessellation and requires conversions to become a Nexorade wire model. There is a rotation of each element from its center point to avoid three elements colliding at a point; this allows them to rest on each other two at a time. To do so, the code first finds

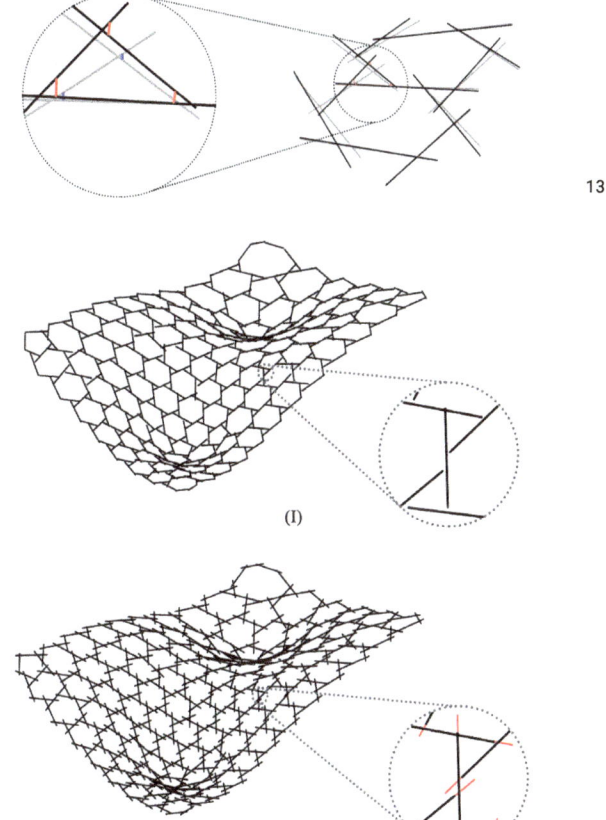

11 (I) Triangulated mesh; (II) Center of each triangular face; (III) Hexagonal tessellation

12 (I) Main pattern; (II) Rotating; (III) Extending

13 The relaxation function moves the elements in a way that all of the red strings become equal/balanced in length

14 The red extensions occur at both of the outer intersections, and all these are set to user-defined equal lengths

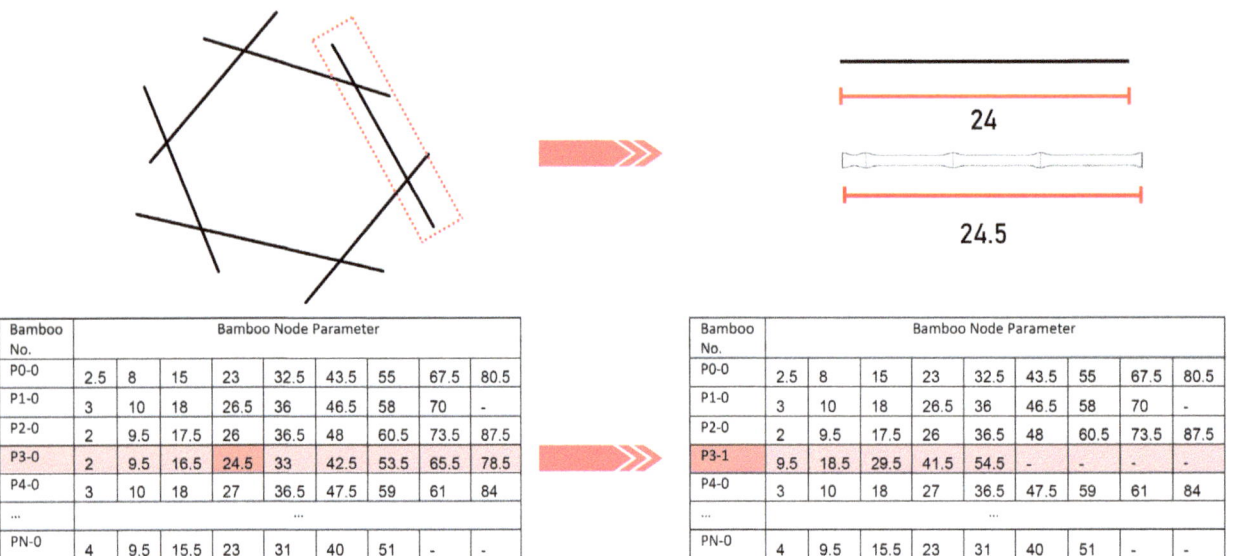

15

the midpoint of each edge, then finds the tangent plane of the base surface at that point, and then uses this information for the rotation plane in the 3D space. Then, the Theoretical Elements must be extended to have an intersection with adjacent elements, which is demonstrated in Figure 12.

Although the wire model approximates the Nexorade pattern, the distance between the elements is not yet defined and is variable. To reach the position of each element in a way that the distance between the element and four other connected elements is the same as twice the ER, the tool runs a physics-based relaxation procedure. This part works to solve the theoretical stretchable string between all of the connected elements by considering the resting length of the strings as twice the ER.

After this step, the axes of elements are in their final locations, but their full lengths are not extended to be accurate. The code can now compute the intersection position based on the final axis locations. There is no actual intersection in the current state of the wire model; the intersection is considered the point at which two linear elements are closest in 3D space. There is always an extension needed to resolve the Nexorade wire model; the code extends linear elements to a minimum specific dimension past the intersections (Figure 14). This is the final wire model of FNS.

Step 3: Pairing Bamboo Library with Theoretical Elements
As described in the design strategy, the bamboo should be cut only on B-Nodes. Thus, the elements cannot be the exact lengths as defined by the wire model, but instead must be equal to a specific multi B-Node to B-Node linkage from the

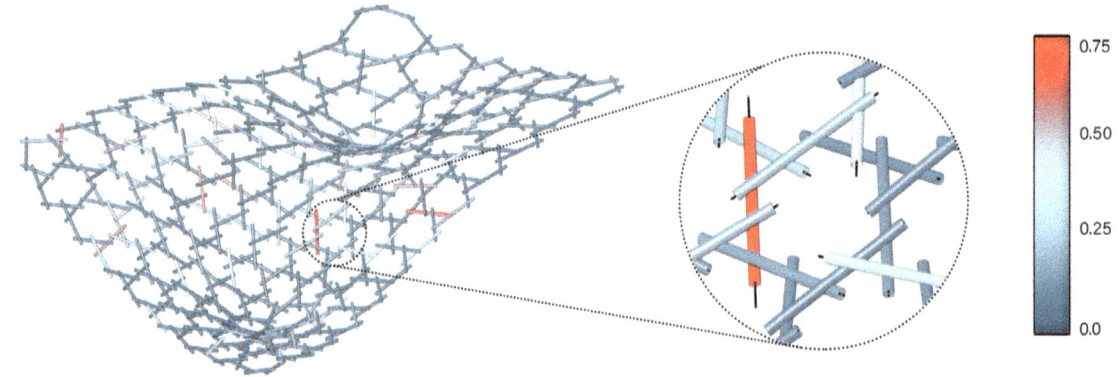

16

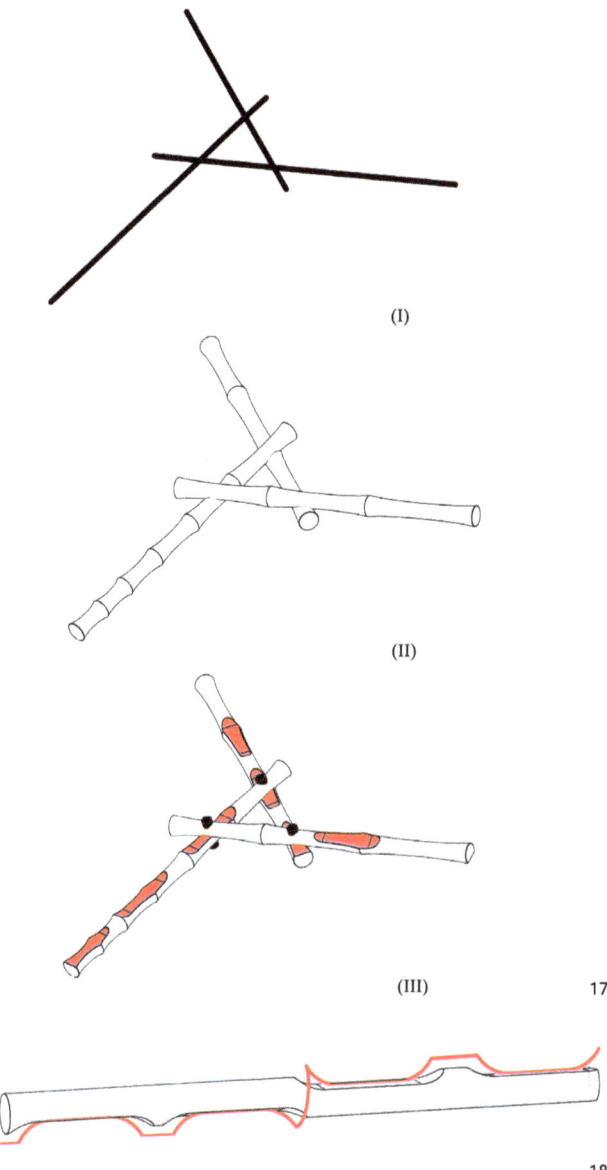

M-Stock of the B-Lib. To find the best-fit bamboo element, the code sorts all needed Theoretical Elements, starting from the largest needed one. It searches among the B-Lib to find the best fit and assigns that entity as the replacement for that Theoretical Element. Then this entity is deleted from the B-Lib, thereby updating the available resources. The algorithm finds replacements for all possible elements and continually updates the remaining inventory. As long as enough bamboo has been allocated and the longest parts are within tolerance, a complete solution is generated.

In this part, the code visualization shows the difference between the Theoretical Element of the wire model and the replaced elements to find the extra length of each element of M-Stock. As is shown in Figure 16, the red pipe is a poor fit, and the ice blue ones have replacements that are a close fit to the associated Theoretical Elements.

Step 4: BIM and Fabrication Data
As the B-Lib contains the B-Node to B-Node distances, the code can generate an approximation model of physical bamboo which we call the T-Comp. A series of subtracted volumes are removed from the model to create the final geometry. The code finds the locations and direction of the volume by using the springs from the relaxation procedure because the springs vectors used for the subtraction are surface normals and thus perpendicular to the final mating lap joint surface.

After the digital subtraction process, the final model of the bamboo structure is ready, and the G-code for the four-axis CNC milling can easily be generated based on this data. The overall stock to be removed is illustrated with the red path below minus the perpendicular bolting holes not shown in Figure 18 but integral to the computational tool and the resulting physical structure.

THE CONCLUSION

This research develops a strategy through which to apply natural materials like bamboo in complex FNS. The main difference is that BCs, unlike industrialized materials are challenging to integrate into systematized methods of design and fabrication because of their inherent variation. This research explores this variation as an opportunity toward a more integral, elegant, and sustainable design approach. It provides the designer with a tool to minimize the inherent complexities associated with natural variation and remove the risk of this novel approach to the design outcome. Furthermore, because the approach accommodates multiple solutions, it also supports fabrication changes and future needs for upkeep via part substitution. Although our approach requires new forms of material awareness, it promises a safe and reliable process

15 Finding the best-fit part of the bamboo from the inventory and updating the remaining inventory information after picking the element

16 Graphical analysis tool that displays the extension amount in the replacement step to the user; for example, the red one is the worst-case scenario, in which the inventory's best-fitting bamboo part is nine inches longer than the Theoretical Elements and outside of the preferred tolerance, and all blue elements are within the targeted range.

17 (I) Theoretical Elements; (II) T-Comp; (III) M-Comp BIM model containing the joint subtraction and the bolt

18 Fabrication data can be created by the Grasshopper tool, represented as a final surfacing 4D path

without wasteful redundancies of industrial materials. Moving design and fabrication into a symbiotic state with the environment replacing the brawn of industrialized materials with the brains of computational and biological systems.

The tool presented in this paper collects essential user information analyzed, digitized and codified bamboo into fabrication inventory (Bamboo Library). Providing the designer with various opportunities and efficiencies to deliver dynamic, elegant, stable, and efficient freeform bamboo Nexorade structures. Thus, the designer can generate several alternatives instantly and observe the effect of the changes in real-time. In this process, the tool finds the best-fitted part of the BC in the inventory and places it in the proper place in the structure, updating the structure with new data. Finally, it generates a high detail model containing every element of the structure in a visually realistic mode in a way that the designer can see the joint shape and condition, joint place in comparison with bamboo node, bamboo final shape after CNC, and even each bolted connection. Moreover, this tool generates all essential information for fabrication containing the CNC tool paths and g-code, and a table of dimensions and information to pre-cut, serialize and index the bamboo.

ACKNOWLEDGMENTS

The work presented in this paper was produced with support from the American Institute of Architects' Upjohn Research Initiative Grant and the School of Architecture + Design at Virginia Polytechnic Institute and State University.

REFERENCES

Apolinarska, Aleksandra Anna. 2018. "Complex Timber Structures from Simple Elements: Computational Design of Novel Bar Structures for Robotic Fabrication and Assembly." PhD thesis, ETH Zurich.

Baverel, Olivier, and Hoshyar Nooshin. 2007. "Nexorades Based on Regular Polyhedra." *Nexus Network Journal* 9 (2): 281–98.

Baverel, Oliver L.S. 2000. "Nexorades: A Family of Interwoven Space Structures." PhD thesis, University of Surrey, Space Structure Research Centre.

Douthe, Cyril, and Olivier Baverel. 2009. "Design of Nexorades or Reciprocal Frame Systems with the Dynamic Relaxation Method." *Computers & Structures* 87 (21–22): 1296–1307.

Hauptman, Jonas, Katie MacDonald, Kyle Schumann, Daniel Hindman, and Tom Hammett. 2019. "Structural Performance of Faced Calcutta Bamboo (Dendrocalamus strictus) for Use in Joined Structural Assemblies." In *4th International Sustainable Buildings Symposium*, edited by A. B. Gültekin. Dallas, Texas: IntechOpen. 257–263.

Hauptman, Jonas. 2021. "Building with Biology," CITYX Venice Italian Virtual Pavilion, May 22, 2021, YouTube video, 00:03:07, https://youtu.be/5QrjJJAQze4.

Haghir, Saeid, Ramtin Haghnazar, Sara Saghafi Moghaddam, Danial Keramat, Mohammad Reza Matini, and Katayoon Taghizade. 2021. "BIM Based Decision-Support Tool for Automating Design to Fabrication Process of Freeform Lattice Space Structure." *International Journal of Space Structures* 34 (3):164–179.

Hidalgo López, Oscar. 2003. *Bamboo: The Gift of the Gods*. Bogota, Colombia S.A.: The Author.

Johns, Ryan Luke, and Nicholas Foley. 2014. "Bandsawn Bands." In *Robotic Fabrication in Architecture, Art and Design 2014*, edited by W. McGee and M. Ponce de Leon. Cham: Springer International Publishing. 17–32. https://doi.org/10.1007/978-3-319-04663-1_2.

Lorenzo, Rodolfo, and Leonel Mimendi. 2020. "Digitisation of Bamboo Culms for Structural Applications." *Journal of Building Engineering* 29 (May): 101193. https://doi.org/10.1016/j.jobe.2020.101193.

MacDonald, Katie, Kyle Schumann, and Jonas Hauptman. 2019. "Digital Fabrication of Standardless Materials." In *ACADIA 19: Ubiquity and Autonomy; Paper Proceedings of the 39th Annual Conference of the Association for Computer Aided Design in Architecture*, edited by K. Bieg, D. Briscoe, and C. Odom. Austin, Texas: Acadia Publishing Company. 266–275.

Mesnil, Romain, Cyril Douthe, Olivier Baverel, and Tristan Gobin. 2018. "Form Finding of Nexorades Using the Translations Method." *Automation in Construction* 95 (November): 142–54. https://doi.org/10.1016/j.autcon.2018.08.010.

Oliyan Torghabehi, Omid. 2020. "Generative Reciprocity: A Computational Approach for Performance-Based and Fabrication-Aware Design of Reciprocal Systems." PhD diss., University of Michigan.

Osmani, Mohamed. 2011. "Construction Waste." In *Waste: A Handbook for Management*, edited by T. M. Letcher and D. A. Vallero. Amsterdam: Elsevier Academic Press. 207–218.

Popovic Larsen, Olga. (2008). *Reciprocal Frame Architecture*. Amsterdam: Elsevier.

Popovic Larsen, Olga. 2014. Reciprocal Frame (RF) Str Exploratory. *Nexus Network Journal* 16 (April): 119–134. https://doi.org/10.1007/s00004-014-0181-0.

Piker, Daniel. *Kangaroo3D*. V 2.42. Windows OS. 2017.

Rastegar, Raha Motamed, Sara Saghafi Moghaddam, Ramtin Haghnazar, and Craig Zimring. 2021. "'TAL-CAT' a Computer-Aided Tool Prototype to Quantify User Experience in Design Workflow: A Case Study of Teamwork Assessment in Primary Care Clinics." In *XXV International Conference of the Iberoamerican Society of Digital Graphics*. Sao Paolo: Editora Blucher. 147–60. https://doi.org/10.5151/sigradi2021-79.

Robert McNeel & Associates. *Rhinoceros 3D*. V 7.14. Robert McNeel & Associates. Windows OS. 2020.

Rutter, David, Grasshopper 3D. V 1.0.0007. Robert McNeel & amp; Associates, Seattle, WA. Windows OS. 2020.

Schumann, Kyle, Jonas Hauptman, and Katie MacDonald. 2019. "Addressing Barriers for Bamboo: Techniques for Altering Cultural Perception." In *ARCC Conference Repository* 1 (1). https://www.arcc-journal.org/index.php/repository/article/view/664.

Sénéchal, B., C. Douthe, and O. Baverel. 2011. "Analytical Investigations on Elementary Nexorades." *International Journal of Space Structures* 26 (4): 313–20. https://doi.org/10.1260/0266-3511.26.4.313.

Steffens, Klaus, Jean Dethier, eds. 2000. *Grow your own house: Simón Vélez und die Bambusarchitektur,* bilingual ed. Weil en Rheim, Germany: Vitra Design Museum.

IMAGE CREDITS

Figures 1, 5: ©Chiravi Patel
Figures 2, 3, 6, 7: ©The VT Bamboo Research Group
Figures 4, 8-18: ©Yasaman Ashjazadeh

All other drawings and images by the authors.

Jonas Hauptman is a Co-Founder of the VT BioDesign Research Group. His research explores various materials including recycled plastics and bamboo. He has won several prestigious awards including the 2004 University of Michigan's Oberdick Fellowship, and a 2018 AIA Upjohn award. He is a member of the International Bamboo and Rattan Organizations Construction Task Force and faculty advisor for a 2021 Musk Foundation, Carbon XPrize team. His work has been featured in the media including *Architect*, *Metropolis*, *Dwell*, *The New York Times*, and others. Finally, recently his video "Building with Biology" (Hauptman 2021) was included in the 17th Venice Architecture Biennale.

Ramtin Haghnazar obtained a Bachelor of Architecture, Master of Architecture Technology, and PhD in Computational Design at the University of Tehran. He is a Co-Founder of Dahi Studio, a research group in the Computational Design and Digital Fabrication field that helps other architects design and build complex and freeform buildings. His projects were built in several countries including Canada, Qatar, Turkey, Azerbaijan, and Iran, and he won several national and international prizes, including Architizer A+ Award and Memar Award, and was nominated in some other competitions. In 2018, Dahi Studio established a new Digital Fabrication Laboratory entitled Digital Craft House in cooperation with the University of Art to develop new construction methods under the supervision of Ramtin and his partners.

Sara Saghafi Moghaddam is a practicing architect and researcher, currently pursuing her doctoral studies in architecture and design research at Virginia Tech. As a graduate researcher and instructor, her research interest employs the integration of extended reality (XR) and design knowledge into the initial phase of the design process and its impact on informed design decisions and design pedagogy. Prior to this, during her Master's studies, she worked on digital fabrication methods and biomaterials. Sara holds a BArch from the University of Tehran, a MArch from Milan Polytechnic University, and an MS in Computational Design from Georgia Tech.

Tactical Sedimentation of Architectural Reef System

Hybrid simulation framework for nearshore underwater landscapes

Colleen Duong
Carnegie Mellon University

Dana Cupkova
Carnegie Mellon University

Azadeh Sawyer
Carnegie Mellon University

Marantha Dawkins
University of Virginia

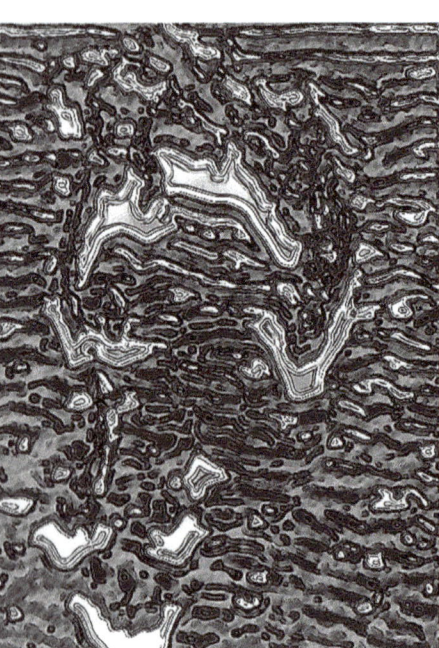
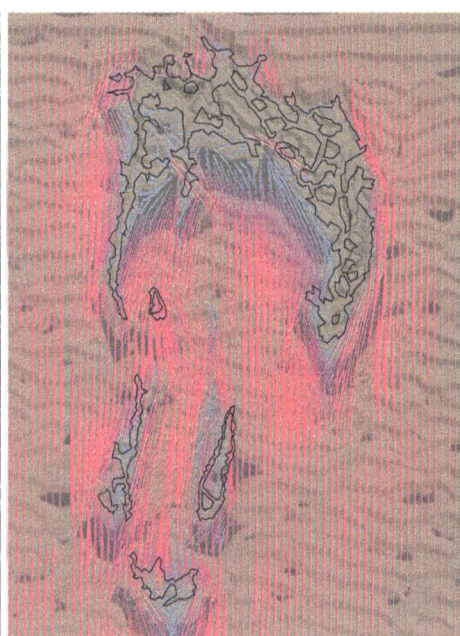

1 Analysis workflow of overlaying digital CFD and sediment movement simulations to evaluate potential areas for safe sediment accumulation to enhance coral growth

ABSTRACT

Coral reefs are rapidly dying due to climate change and anthropogenic activities. Because these sensitive ecosystems are critical to ocean health, new approaches for designing synthetic reef systems have emerged in the last 50 years to sustain and promote coral diversity. However, despite their success, these studies lack the larger-scale and higher-level ecological analysis that accounts for anthropogenic threats to these ecosystems. Without considering how contemporary near-shore environments are hybrid, novel landscapes, artificial reefs are not designed for shared ecologies. This study proposes a novel simulation framework that expands the existing analytical modeling methods, allowing us to visualize and test underwater eco-spatial phenomena within dynamic systems to better identify a design space for intervention with the goal of mitigating the conventional human-reef relationship through tactically choreographing sedimentation. This paper focuses on developing a method of modeling and analysis linked to the ecological characterization of coral species and simulation sequences allowing us to: 1) study diverse, underwater landscapes that are not easily visible or accessible, and 2) project complex environmental change and sedimentation over time. This method works across tools, scales, and media to develop a computational ecological approach to designing sensitive habitats. Our simulation sequence proposes an overlay of (a) CFD analysis with (b) computational sand dune formation and (c) physical experimentation using a simulated sand and water table to study the sediment response to morphological intervention. The goal is to identify zones of intervention within the dynamic underwater landscape that encourage strategic increase or decrease of sediment build-up, nurturing coral health. This method creates a strategic, responsive framework for a site-specific coral reef typology that enables a deeper understanding of dynamic environments.

INTRODUCTION

Coral Reefs are known as the "rainforests of the sea" because of the diversity of life that is found within them. They are one of the most diverse, valuable ecosystems in the world, supporting more species per unit of area than any other marine environment (NOAA n.d.). They exist in a variety of shapes and sizes with various morphologies formed throughout time, providing different needs for different species: a place for food, reproductive space, space to care for young, and shelter. There are also many benefits to humans that come with coral reefs; reefs serve as a natural marine barrier to protect coastal communities from high-impact waves during natural disasters, provide food and medicinal resources, and support the economy by increasing job opportunities.

Despite their importance, coral reefs are dying at a rapid pace. Over 50% of the world's coral reefs have died in the last 30 years. Up to 90% are predicted to die within the next century (IUCN 2021). Warming waters, pollution, acidification, overfishing, sedimentation, and physical destruction are some of the causes of coral bleaching and death. Climate change is one of the greatest global threats against coral reef ecosystems. As temperatures continue to rise, disease and coral bleaching become more common in coral reef ecosystems. Undisturbed, it takes approximately 9 to 12 years for a coral reef to recover from a coral bleaching event.

Other than climate change, one of the leading causes of the rapid decline of coral reef health is human activity. Anthropogenic activities are major threats to coral reefs (NOAA n.d.). Pollution, overfishing, harmful fishing practices, collecting live coral for the aquarium market, and mining coral for building materials are some of the many ways that people damage reefs around the world. For example, deep water trawling, which involves dragging a fishing net along the sea bottom, can tangle up and kill reef organisms. Also tourism can create direct, harmful contact with coral reefs when tourists do not understand how to be careful with these delicate ecosystems.

A known method to combat coral reef death is the production of artificial reefs (Hilbertz 2009; POSCO 2000; Moffitt 1989; Faridah 2015). These artificial reefs help to restore fisheries and to mitigate the effects of resource exploitation and destructive practices like trawling. Many artificial reef material case studies around the world have been successful in multiple ways, whether from their strength, the biodiversity they attract, or their effects on coastal communities. However, if they are not installed with sensitivity to the larger underwater dynamic system or designed with a deeper knowledge of the effects of their form and material, they can be destructive to existing reefs. Understanding the underwater landscapes requires complex, often hybrid approaches. More

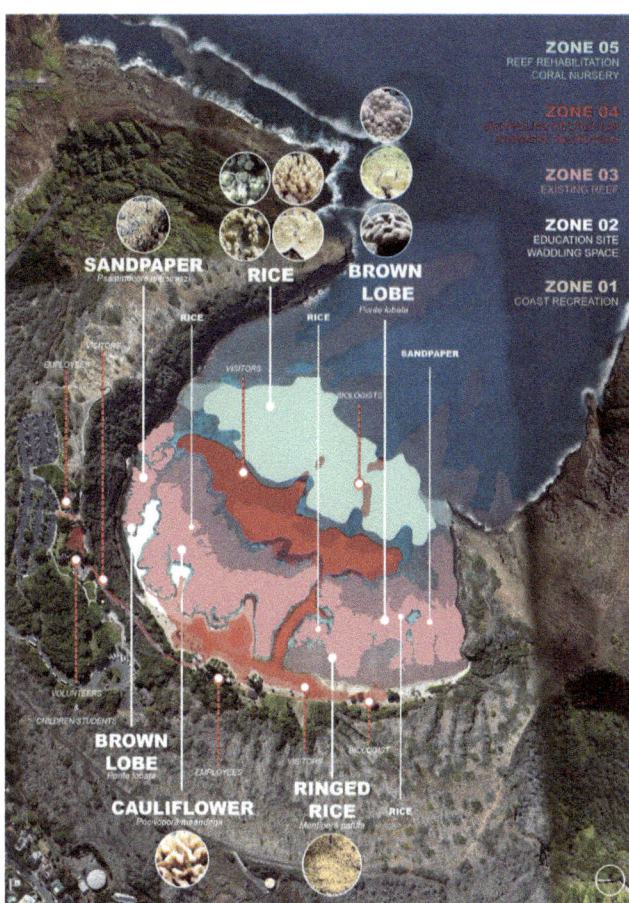

2

3

2 Qualitative data gathered on Hanauma Bay looking into the existing coral species and their location on site, human visitor circulation and density, and types of marine creatures that reside in the Bay

3 An analysis on different zonal conditions (education spaces, recreational spaces, and rehabilitative/nursery spaces) that can reside in the Bay based on the gathered qualitative data

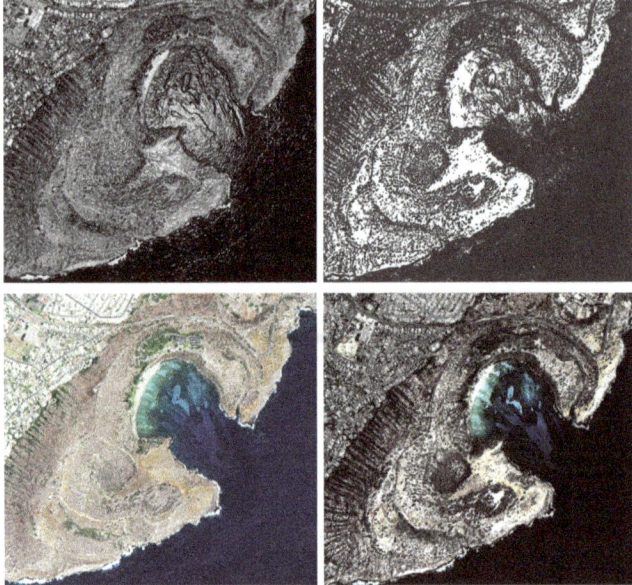
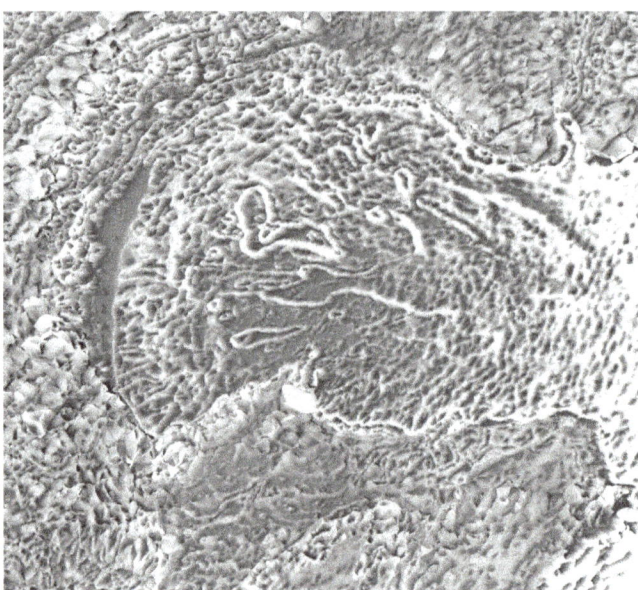

4 A series of filtered rastered Google Earth satellite images of the site

5 Final 3D mesh generated using depth mapping of coral reef in Hanauma Bay allowing for identification of coral species referenced by field data

6 Re-sketching site organization based on simulated studies focused on computational fluid dynamics of water and sediment movement across the bay to create new pathways for multi-species.

standard numerical/physics-based modeling in nearshore environments is particularly difficult given the turbulence of wave action in the transition zone between land and ocean (Wang 2005). Climate change has increased the stochasticity of these sensitive thresholds and unpredictable human use has introduced further complications. These studies on artificial reefs seem promising. However, many of these studies omit the analysis or acknowledgment of the anthropogenic threat to these coral reefs (Kleypas 2001). This is, in part, because of the lack of non-intrusive methodologies that allow for more nuanced modeling and experimentation within such a sensitive environment to understand the complex interactions between natural forces, humans, and reefs.

THE CASE STUDY SITE: HANAUMA BAY

Hanauma Bay in Honolulu, Hawai'i, contains an endangered coral reef ecosystem. This location is a popular tourist destination, which is a large source of income for the Hawaiian economy. However, since this beach was opened to nonlocals, the coral reef ecosystem has rapidly decayed because of both climate change and the impacts of recreational activities like swimming, diving, and snorkeling. Several studies have tried to determine "acceptable limits of human disturbance" to allow visitors to continue visiting without degrading the reef's health. Despite these measurements, humans still impact the coral reef's health, but at a much slower pace, keeping these reefs from fully recovering (Tsang and Stefanak 2020). During this study, dynamic entanglement between all elements and actants on the case study site was mapped to identify the areas of intervention that allow healthy coexistence.

LACK OF DATA: UNDERWATER MORPHOLOGY MODELING PROCEDURE

Initially, past research was collected and mapped to understand the types of visitors—both human and nonhuman—that reside in the bay (Figure 2). The aim of this data collection was to create multiple zonal conditions on site, mapping multi-species connections (Figure 3). These include zones focused on human recreation, rehabilitation, and education, respectively.

The study began by digitally modeling the existing site conditions and coral reefs on site, focusing on the brown lobe, rice, sandpaper, and cauliflower coral species. However, no online database exists for mapping underwater topographies of habitats. Many small islands that house coral reefs, like Hawai'i, lack digital databases due to their small size and secluded locations. There is also no standard method for modeling dynamic over-time relationships between coral reefs and recreational landscapes. To respond to this gap, we propose a method to construct a hybrid digital model for coral reef typology using the scarce online resources.

Resourcing topographical information of Hanauma Bay via CityEngine (ArcGIS n.d.)—which directly utilizes ArcGIS data—an initial 3D model of Hanauma Bay was constructed. Then, it was hybridized by adding additional bathymetric data to model the ocean floor. A cohesive surface model of Hanauma Bay nearshore topography was developed that could be used as a substrate for depth- and texture-mapping of the coral reef morphology.

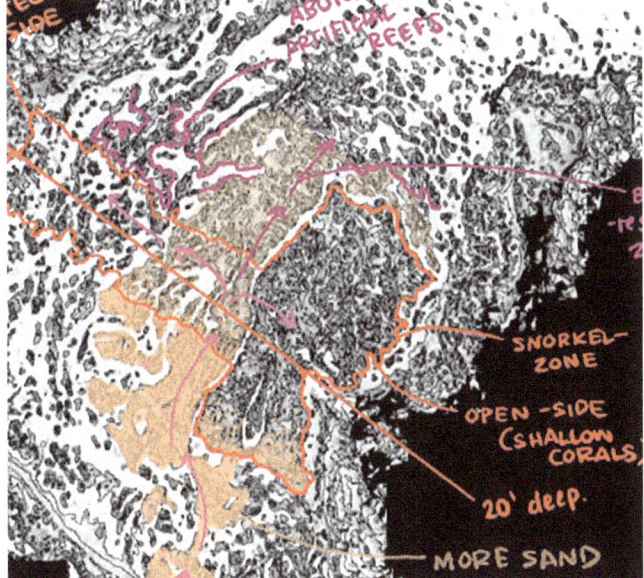 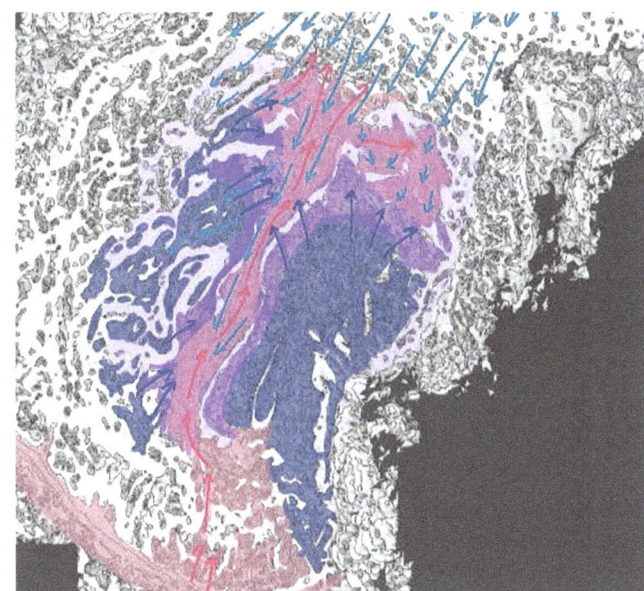

6

To create a specific digital coral morphology, several Google Earth satellite images of the site were enhanced to visualize the shapes of coral textures (Figure 4). Then, this modified image was applied to the 3D site model through a depth-mapping procedure (Figure 5).

Using depth map modeling, a series of 2D rasterized data maps were enhanced and superimposed directly onto the 3D mesh model of Hanauma Bay to spatialize various coral reef ecosystems in the bay (Figure 5). The accuracy of this new 3D mesh morphology directly relates to the resolution of depth-based image channels, which can be partially amplified, even from lower-resolution satellite images. Existing contemporary 3D hydrodynamic models such as CMS Wave/CMS Flow and Delft3D can solve continuity and Navier Stokes equations, or more simply, accurately represent the physics of this context. However, these models 1) rely on robust 3D data sets, which are not available for many sites like Hanauma Bay, and 2) oversimplify the fate of landscapes that have to account for human use. Our proposed workflow can enhance the accuracy of this modeling method with higher-resolution satellite imagery. Using the proposed workflow, this study focuses on identifying the dynamic phenomena of sedimentation patterns relative to diverse coral speciation.

Identifying Strategies for Entanglements: What to Simulate? Species Identification

Combining sketches with the newly-developed coral reef morphological model allows for choreographing various human and nonhuman activities to support dynamic formations while allowing the entanglement of pathways for humans and marine creatures to be protected by both artificial and natural coral reefs (Figure 6).

Sedimentation is known to hurt corals; it can lead to suffocation due to suspended sand blocking sunlight or sand settling on corals (NOAA n.d.). However, corals have developed methods to protect themselves from high levels of sedimentation such as secreting protective shields (NOAA n.d.) or evolving into a funnel shape to create "sedimentation traps" (Riegl 1996) to protect other corals that grow on its surface. If designed carefully, sedimentation can actually be accumulated in a way that benefits coral reefs. Due to rising sea levels, corals have been unable to grow quickly enough to gather sufficient sunlight, causing them to starve (Sanborn 2017). Increased sedimentation in certain areas could elevate growing beds on the seafloor as well as encourage vertical coral growth, allowing these coral species to be closer to the sea surface and have more access to sunlight.

These areas of decreased or increased sedimentation can be shaped through artificial reefs placed strategically around the bay in response to the existing natural reef. This can allow us to shape pathways for humans and pathways for marine creatures to safely interact with one another while setting up a methodology for simulation and experimentation in further understanding the conditions of the natural environment acting upon the coral reef in response to safe, tactical areas of increased sedimentation.

Separation of Coral Form for Analysis

To perform these test analyses, the site was divided based on prior research that identified portions of the coral reef differently affected by human activity (Severino 2019). The northern portion is a coral reef located closer to shore with a more dense and resilient form. The southern portion is a coral reef that is more scattered and fragile forming. By using this method, smaller, more typical segments of coral

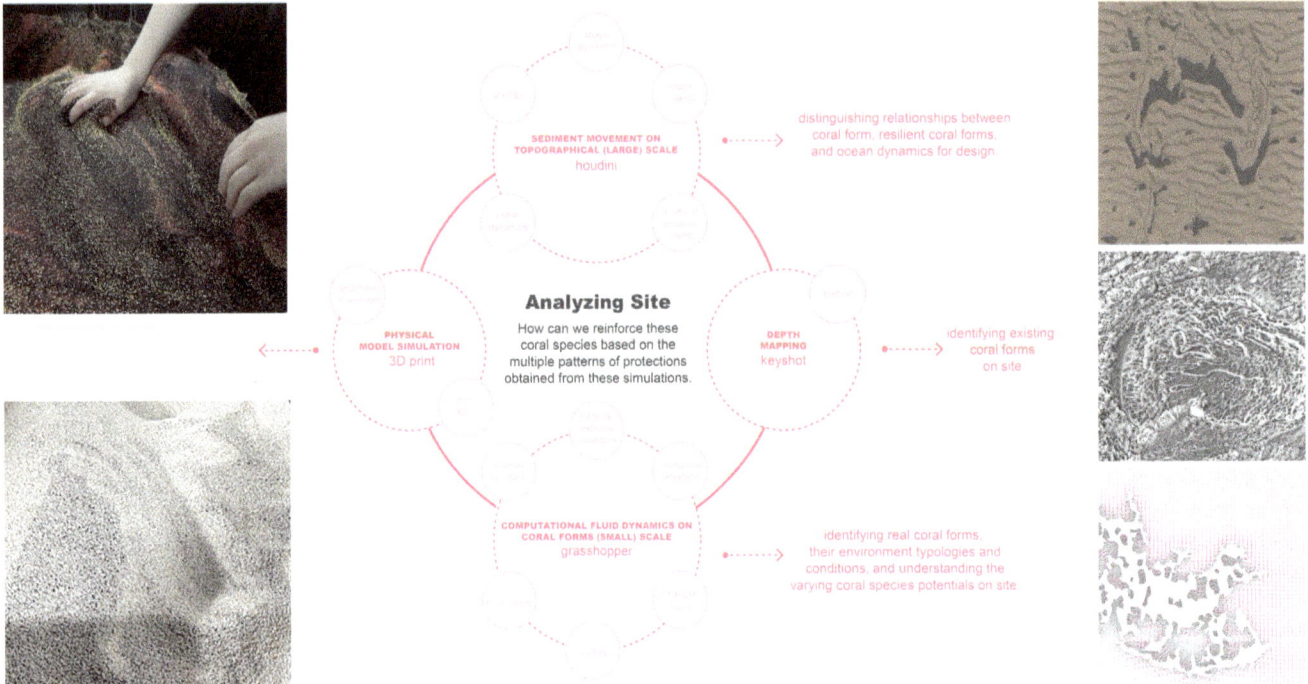

7   Circle of analysis and simulations focusing on constantly circulating through various simulation (fluid dynamics, sediment movement) and media (digital, physical) to create a constant flow of information that speaks to one another

reef forms could be studied to evaluate how they are affected by natural forces.

This project utilizes architectural methods of modeling underwater morphologies based on accessible satellite images. For higher accuracy of the coral reef topography model, the satellite data was enhanced to intensify the depth channel. This is a partially speculative procedure that is juxtaposed with a more rigorous, scientific method of analysis such as fluid dynamics and sediment movement. Furthermore, the gathered data was expanded and simulated this phenomena into a potential design integration of an artificial reef system that could be implemented on the site.

DIGITAL AND PHYSICAL SIMULATIONS

To begin understanding the conditions of the coral reef in Hanauma Bay, a series of simulations was conducted on the coral reef at various scales, including simulating the natural flows (fluid and sediment movement) and the media (digital and physical simulation) (Figure 7). The goal of this simulation series was to understand the sediment accumulation and disbursement phenomena that begin to reinforce particular coral species based on the multiple patterns and scales of protection as well as to understand how digital and physical simulations compare.

Computational Fluid Dynamics: Water Flow

Computational fluid dynamics (CFD) in Rhino-Grasshopper-Butterfly was used to examine the movement of water forces on the existing coral form (Simscale n.d.). The CFD simulations ran through the two large coral reef sections on site using an average wind speed (pink lines) of 10.3 mph and a low wind speed (blue lines) of 3.4 mph (Figure 8). The blue lines created a shadow of the overall coral form that highlighted where the slowest water movement occurred.

These CFD simulations show a "formal shadow," which was created by the water hitting the coral (Figure 8). This shadow contains the area with the slowest moving water. These shadows potentially allow for an area that could be built up to enhance coral growth. Additionally, these areas could be redesigned, adding structures to carefully increase water speed in specific areas. Corals are known to thrive in fast currents, so this analysis can locate intervention areas to increase water speeds in specific areas of the bay, especially zones focused on rehabilitation.

Sediment Movement

Sand sedimentation simulations focused on observing how sediments collect in various areas surrounding a coral reef and looked for potential areas in which it could carefully accumulate, and encourage vertical growth in the coral reef. Vertical coral growth is a series of evolutions that corals have undergone to avoid trampling, specifically on sites that are heavily populated by humans. Hanauma Bay has an abundance of vertical corals because of the high human activity that occurs. However, sediment accumulation must be thoughtfully choreographed; too much sand could lead to coral suffocation. We

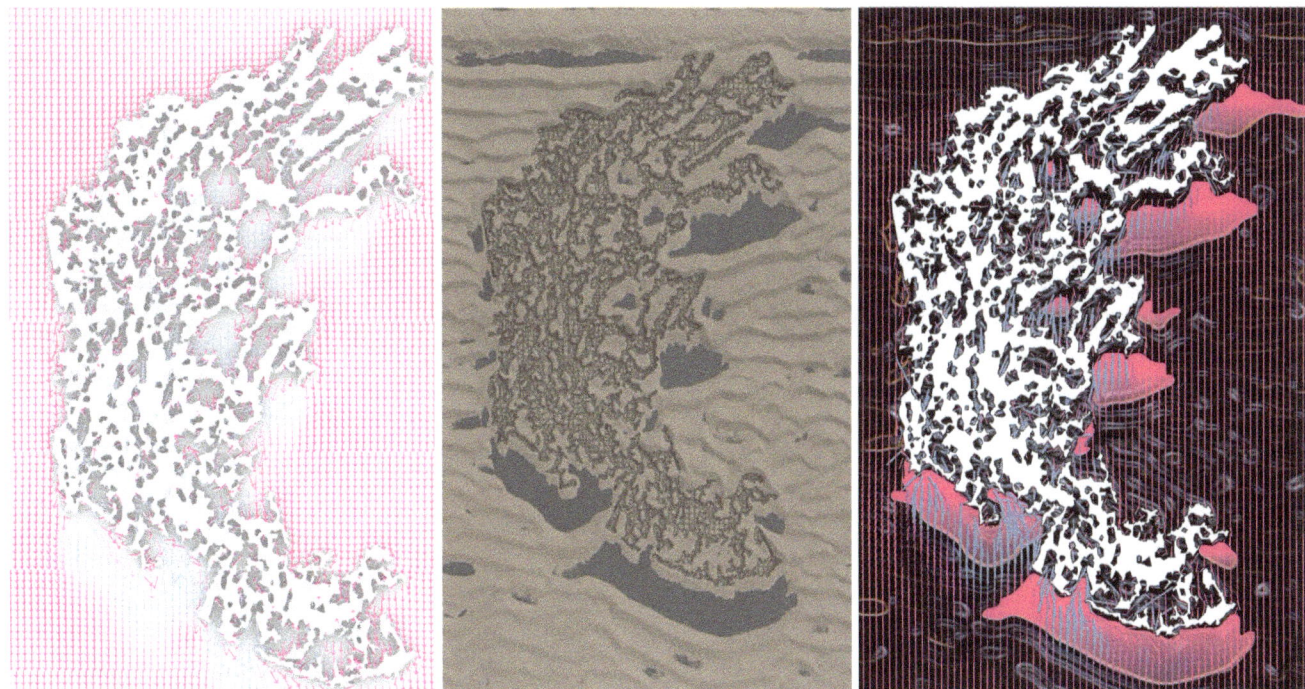

8 Workflow of conducting CFD simulations and sediment movement simulations then overlaying the two results on top of one another to locate potential areas for safe sediment accumulation

must design carefully to maximize the benefits of sediment accumulation, encouraging vertical coral growth.

Houdini-dune solver was used for this analysis. Dune solver (LP) (Meeker 2021) is an asset for Houdini created by Barrett Meeker. It is a height field and terrain tool that simulates the ways sand dunes are formed from wind, sand depth, and other factors.

These experiments visualize a trend of locations with little to no sand that could be redesigned for safe sediment accumulations. These sandless areas are potential locations for careful sediment accumulation to encourage coral reef growth on higher grounds.

Applying Shadows: Overlaying Digital Simulations

The CFD and sediment simulations were overlaid to investigate the relationship between the two sets of data and to discover potential areas for land building or speeding up and slowing down water to enhance coral growth (Figure 9). Using these simulations, potential areas were identified for high sand densities informed by both the coral form and the open ocean floor (Figures 9, 10).

The simulation overlap revealed more complex sediment movement and water flow patterns, identifying shadows from the coral form. These shadows are potential areas for an artificial reef placement to enhance coral growth. More nuanced simulation on a smaller scale could also reveal formal characteristics for soft artificial reef barrier designs.

Such soft artificial reef barriers help accumulate sand at an angle driven by the water direction, allowing for corals to grow vertically. This strategy would bring the coral reef closer to the ocean's surface, allowing them to better collect sunlight while also raising awareness of their location to any visitors nearby to avoid trampling. Also, these barriers must be large enough to safely separate humans and the fragile coral reefs to allow them to grow undisturbed.

In addition to informing these self-forming, localized interventions, the shared simulation effects reveal the natural forces that occur in interaction with the coral reefs; these forces can help support the production of new lands through sediment accumulation, thus encouraging a positive relationship between multiple species as well as a more biodiverse environment.

Physical Simulations with Mixed-media Sediments

The physical simulations with the water table (Figure 11) and engineered sand (Figure 12) focused on sediment movement by comparing the results of the digital and the physical simulations. The overlaps between the two sets of simulations contributed to identifying more nuanced sedimentation phenomena and aided in the process of designing artificial reefs that can accumulate amounts and forms of sediment. Initially, several simple shapes were tested in the water table to establish a baseline for the comparison of digital and physical simulations (Figure 13). Digital simulations have a more uniform set of information due to the constant water speed and flow that moves across the entire surface, while physical

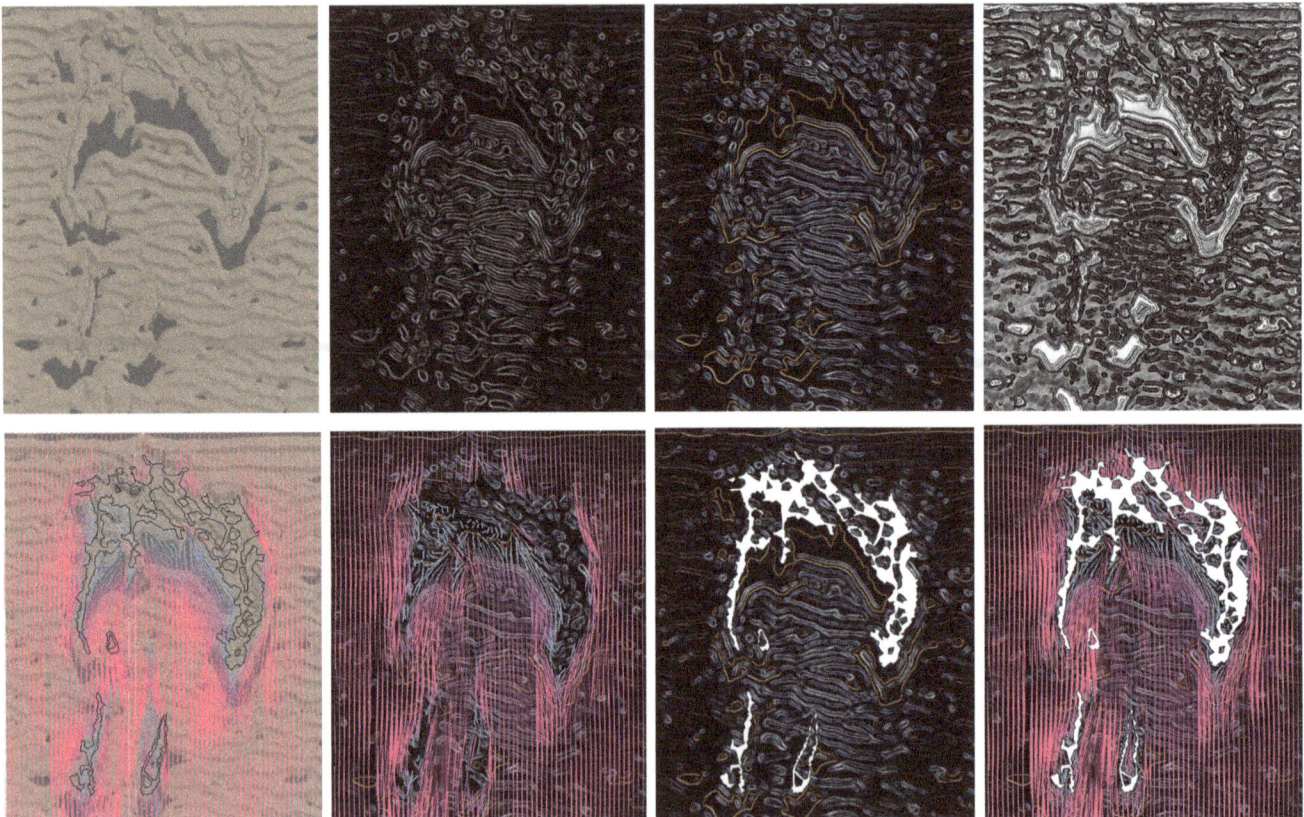

9 Enhancement and overlaying of the two conducted simulations, CFD water flow and sediment movement, to bring out the patterns of information shown

10 Analyzing coral reef shadows as potential spaces for sediment collection and growth from the overlaid simulation results

simulations create a more concentrated flow because of the single-point water source.

Overlaps of information show the shadow of sediment accumulation (Figures 14). The darkest areas shown in the simulation (outlined in red) show the locations with the largest potentials for sediment accumulation. The brightest areas (outlined in green) showcase where the accumulated sediments could grow over time. This information could help create a standard for different types of artificial reef forms, shapes, and placements on the seafloor to respond to this data set for safe accumulations of sediments for coral growth.

"Shadows" represent potential areas for increased or decreased sedimentation in relation to the existing natural reef topology. These shadows identify areas of safe sediment accumulation or areas that may have too much sedimentation for optimal coral reef growth. Sedimentation can be regulated in areas where there is more existing sediment build up (darker portions of the simulations) to reduce coral drowning caused by increased sedimentation. Increased sedimentation can be designed in spaces that have less sedimentation as

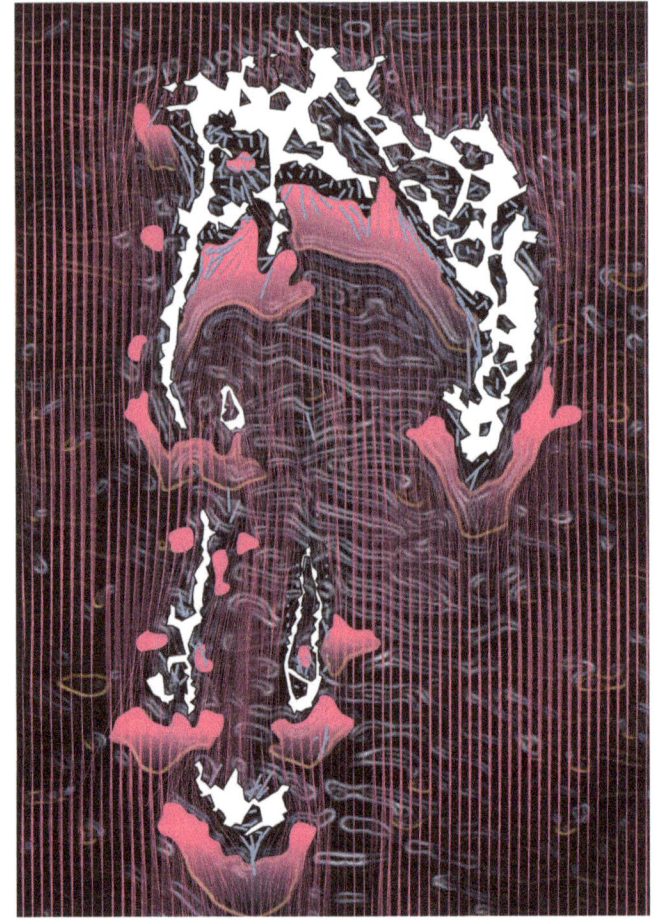

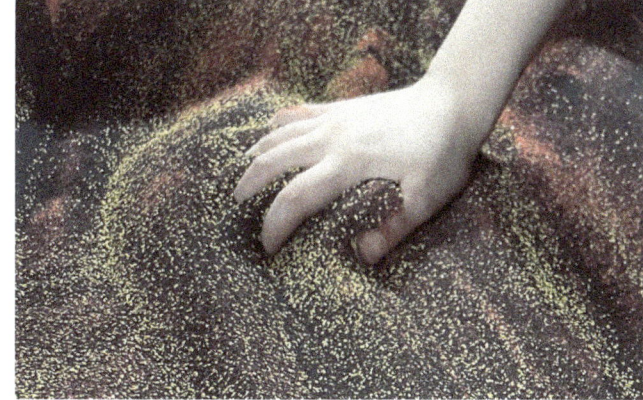

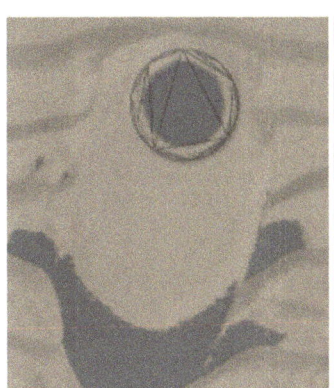
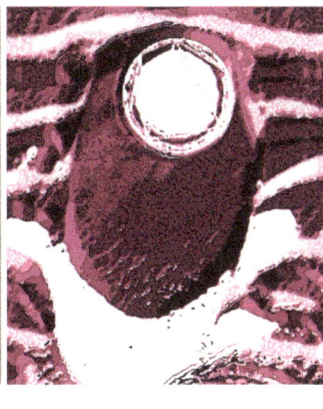
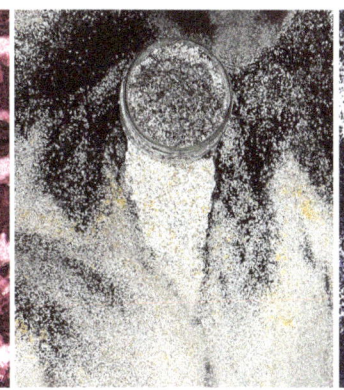

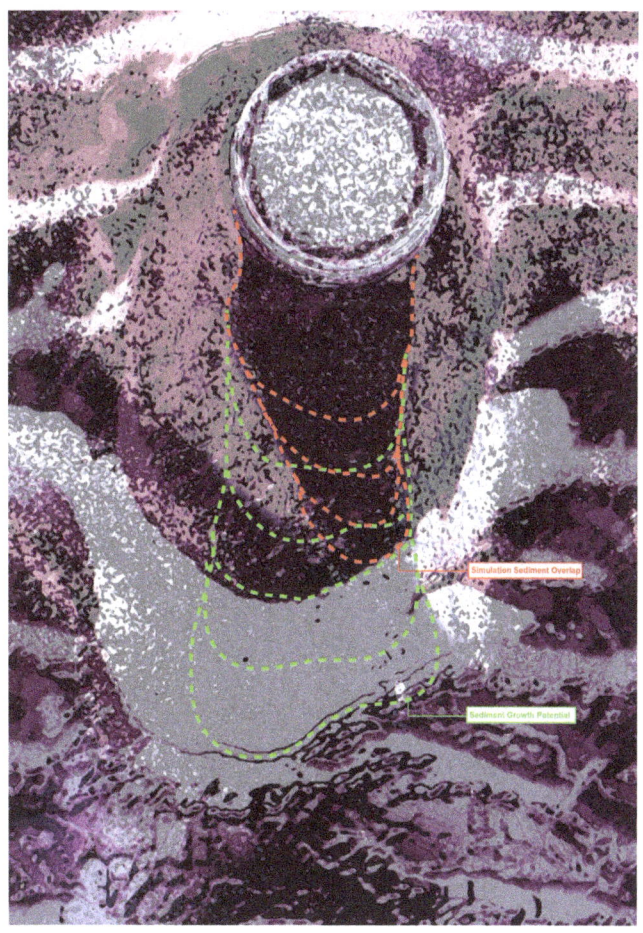

11 Sand table set up for physical simulation testing using two (36 in x 24 in) utility tubs, one solar hot water heater circulation pump with DC power supply adaptor (low noise, 3 m head, 8 LPM, 2.1 GPM), two (1/2 in ID x 3/4 in OD x 25 ft long) clear PVC flexible vinyl tubing, four plastic barbed hose fittings (3/8 in adapter hose ID, 1/4 in NPT male), four plastic pipe fitting (straight connector, 1/4 in NPT female threaded pipe), and three screw tops that filter water.

12 Emriver (Emriver n.d.) color-coded media sand mixture. The color-coded media allows for easier visualization of the sediments being transported according to their grain size. The sizes are approximately 1.4 mm for the yellow particles, 1.0 mm for white, 0.8 mm for black, and 0.4 mm for red.

13 Physical and digital simulation results enhanced for further analysis

14 An overlap of physical and digital simulations, looking at various areas and levels (dark vs. bright) of overlap to determine spaces safe for sediment accumulation

a result of the coral form (the lightest portions of the simulation) to elevate seabeds and raise up corals so they can gather more sunlight.

The simulations identify a potentially interesting phenomenon that is not typically considered. The simulations allow us to map a multitude of scenarios for sediment accumulation between coral morphology and water behavior. However, identifying these patterns should not be thought of as a 'final result,' but rather an alternative approach to generating a coral growth strategy in the context of human engagement.

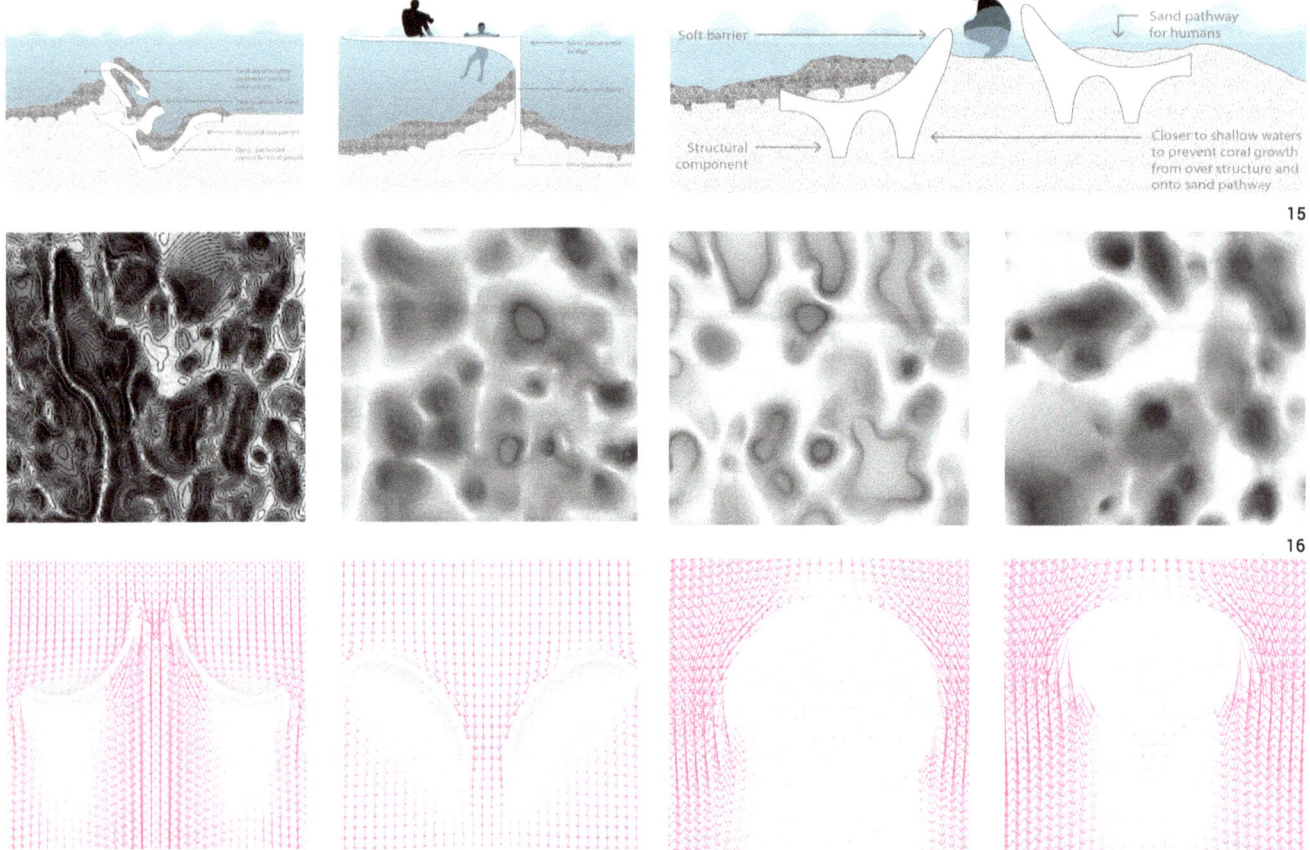

15 Proposed designs of artificial reef barriers and bridges focused towards developing new lands to create pathways safe for human-nonhuman interactions, enhancing coral growth, creating visual and physical barriers, and the encouragement of vertical coral growth while grounding itself deep into the sea floor to increase resilience against strong natural forces created in underwater conditions.

16 3D depth mapping rugosity models of coral species that reside in Hanauma Bay (brown lobe, rice, sandpaper, cauliflower)

17 Additional studies on forms that create water tunnels with increased water speeds, water barriers with slower moving water, and water vortexes to trap and clean out sediments.

RE-SITUATING SIMULATION RESULTS ON SITE
Designing New Lands

The results of the various simulations (digital fluid dynamics, digital sediment movement, physical sediment movement) were combined and superimposed onto the Hanauma Bay coral reef model to study the potential for the formation of new land.

By gathering the collective data from previous experiments, new circulation patterns were observed that overlapped existing coral reefs, human pathways, and areas of potential new land. These aspects of sedimentation suggest new potential zones for education, recreation, and rehabilitation. Thus, artificial reef designs can be created that are focused on more specific activity types in specified zones.

Design Integration and Testing

Utilizing the information gathered from the simulations and observations, specific trends were identified that support a more diverse design of artificial reefs and land building. The goal is to understand and adapt to the speed and direction of water movement relative to specific coral morphologies. The aim is to strategically move and collect sediment to support the reef ecosystem.

Two sectional design typologies were developed for artificial reefs in this study: a barrier and a bridge. A barrier is a localized structure embedded into the sea floor that accumulates a soft pile of sediment around it. Besides accumulating sand, it functions as a physical and visual barrier to protect fragile corals and encourage vertical coral growth. On the other hand, the floating bridge extends to the surface of the water and utilizes surface rugosity and porous textures to increase biofouling (Figure 16).

The sectional strategies of a barrier and a bridge should be coupled with planar forms that perpetuate sediment movement in the horizontal plane. A series of potential formal strategies emerged based on the simulation results (Figure 15), including forms that support coral growth, protect corals, accumulate sediments, and allow for safe cohabitation between humans and nonhuman species. This is a hybrid formal framework that utilizes adaptive modularity; these artificial reefs are shaped bespokely and are placed strategically across the bay as well as around the coral reef to speed up, slow down, or redirect water currents (Figure 17).

RE-IMPLEMENTING DESIGN ONTO THE SITE
The Zones
We propose a series of overlapping zones on site: an educational zone, a recreational zone, and a rehabilitation zone (Figure 17). This design moves away from the traditional idea of mixed-use programming used in architecture, and begins to create hybrid dynamic fuzzy zones for multi-species engagement. Moving away from the typological segregation of public and private spaces in terms of marine creatures and coral reef ecologies for humans, this project prioritizes creating safe spaces for multiple species to interact without entirely separating them. The idea of 'private' is centered around creating specific areas that are directed towards rehabilitation and research rather than exclusion. This creates 'safe public spaces' for humans, coral, and marine species to interact. These proposed strategies inspire us to rethink how these artificial reefs could construct self-maintaining, regenerative architectures for sensitive landscapes. There is an abundance of different architectural typologies with the potential to further enhance coral reef growth.

RESULTS
As a tourism-heavy site, Hanauma Bay would benefit from an architectural reef-artificial system intervention. This system allows the rehabilitation of natural coral reefs, attracting increased biodiversity as well as allowing humans and marine species to safely cohabitate. To do this, a novel method of combining quantitative and qualitative data was created to enhance design intervention in dynamic underwater landscapes.

By developing a method of modeling this hybrid topography, one can begin to understand the natural energetic forces that act upon an underwater ecosystem like a coral reef through both digital and physical tests. The digital model can be used to analyze forces such as sediment movement, water speeds, and flows that move across the site. Then, this information can be expanded and analyzed to compare with the physical simulation studies done on sediment movement and the analysis of how these forces act upon simple shapes and forms other than the complexity of a coral reef. This information leads to a more developed model for designing artificial reefs that respond to the needs of the specific ecosystem that is in recovery.

From these simulated studies, new ideas were discovered for controlling these energetic forces by accumulating sand to create new lands as well as slowing down or speeding up water flows to grow coral reefs. It has been previously shown that fast-moving water greatly benefits the growth of coral reefs while sedimentation has been known to have detrimental and potentially positive effects on coral reef growth. If controlled properly and carefully, sedimentation can be effectively used to encourage vertical coral reef growth, which would be tremendously beneficial in an area that is highly populated by humans; this would lead to corals being less susceptible to death from trampling. These raised areas also allow for corals to respond better to rising sea levels, supporting increased growth at higher elevations and increased sunlight collection; they would collect more sunlight quicker than they can naturally grow to match the rapidly rising sea level. Though subsequent in situ physical testing is necessary to validate the simulation results collected in this experiment, this study identifies a design strategy for cross-species zonation. The methods developed in this study can be used to identify potential areas for safe sediment accumulations as well as areas that would most benefit from fast- or slow-moving water based on the various coral species present.

As a result of this study, a diverse set of design interventions was created that focus on a floating bridge typology and a soft barrier typology. These two reef architecture types propose the added benefit of textured surfaces to encourage biofouling, porous spaces to create a difference between interior and exterior surfaces for growth, and the ideas of accumulating different amounts of sand or controlling the speeds of water through the use of modularity. Such artificial reefs would be strategically placed across the bay to enable hybrid activities within each zone: the educational zone, the recreational zone, or the rehabilitative zone; the goal is to create an environmentally friendly landscape that enables regenerative relationships between multiple species.

The modeling method using depth map texture modeling with satellite imaging offers instant visualization of existing habitat while massing and patterns can be updated over time as new growths emerge. The limitation of this cyclical feedback is that the textural maps must be periodically remodeled to reflect shifts in topographies in relation to natural forces and/or human recreation and use. This method has greater accuracy with evolving data accessibility. Additionally, it can be expanded with additional scientific information. It also allows for a more integrated design approach across multiple dynamic forces, bringing together physics with the human and habitat considerations.

CONCLUSION
This study proposes re-evaluating methods for creating responsive hybrid architectures that encourage positive

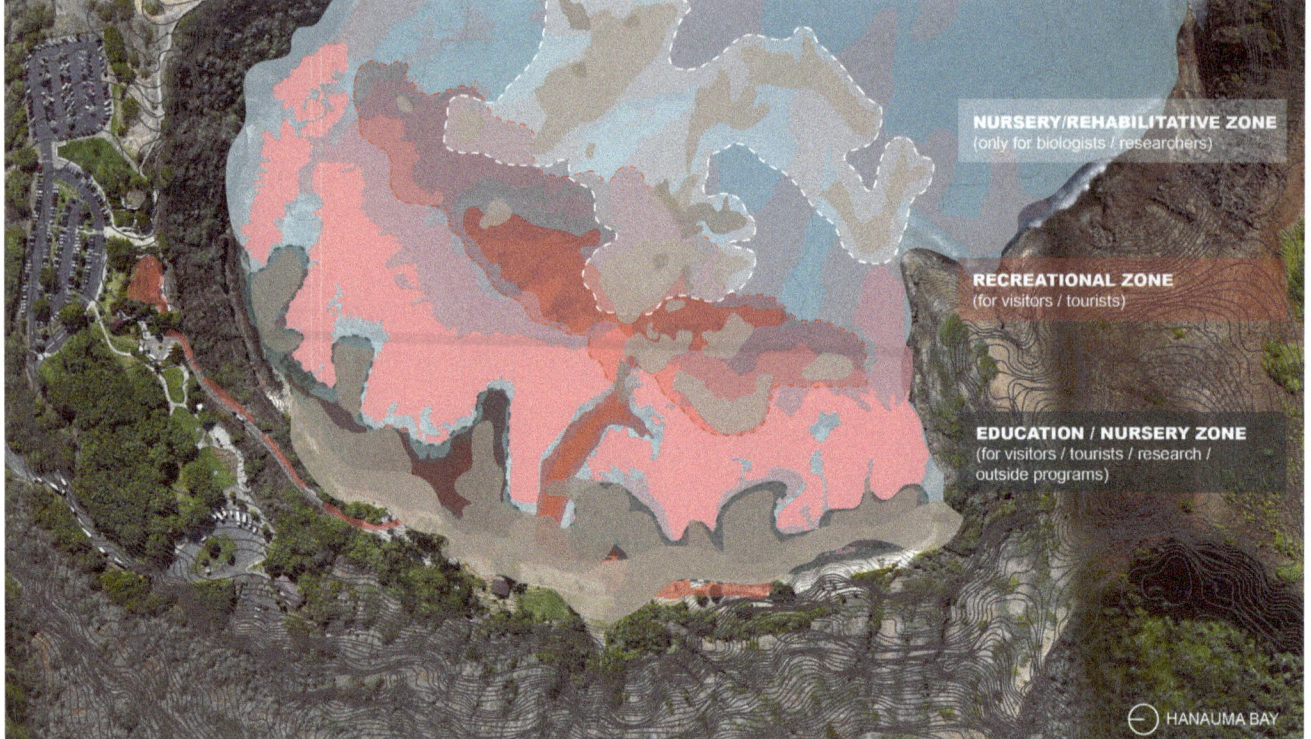

18 Overview of the different types of zones, activities, and visitors that take place in Hanauma Bay with the implementation of artificial reefs building up new lands

interactions between multi-species environments within dynamic underwater and nearshore landscapes. By studying site-specific coral reef species and their morphologies, new representations emerge that help us uncover data-rich dynamic phenomena for tactical sediment accumulation and land building. Proposing a workflow that uses accessible means of digital drawing, modeling, and simulation to visualize dynamic behaviors relative to larger ecological impacts is critical in the current context of design for climate change. Focused on identifying the area for intervention in outlining a potential design brief for aquatic architectures, a method was proposed looking into an overlap between digital and physical procedures for emerging design strategies. In this study, methods for creating a series of digital and physical simulations was developed to inform the design process based on limited site data. By comparing the results from the digital and physical studies, a better understanding between the distinct formal languages of the natural forces that act upon these sensitive landscapes and of sedimentation can help redefine the strategic placement of design interventions with a more tactical approach to ecological design.

ACKNOWLEDGMENTS

This study was developed within the Master of Science in Sustainable Design (MSSD) program, led by Dana Cupkova at the Carnegie Mellon School of Architecture. The work has been partially funded by Carnegie Mellon University Graduate Small Project Help Fund (GuSH). Special thanks to Bradley Cantrell and Akshay Mehra for their insight and feedback.

REFERENCES

ArcGIS. n.d. *ArcGIS CityEngine*. Accessed March 11, 2022. https://www.esri.com/en-us/arcgis/products/arcgis-cityengine/overview.

Brock, Richard E. 1994. "Beyond Fisheries Enhancement: Artificial Reefs and Ecotourism." *Bulletin of Marine Sciences* (55): 1181–1188.

Brown University. 2010. "Faster Water Flow Means Greater Diversity of Invertebrate Marine Life." *News from Brown University* (press release). November 17, 2010. https://news.brown.edu/articles/2010/11/currents.

Cantrell, Bradley. 2019. "Choreographing Topography." Dumbarton Oaks. 2019. https://www.doaks.org/research/mellon-initiatives/mellon-initiative/midday-dialogues/cantrell-2019-02-06.

Coastal Inlets Research Program. n.d. *Coastal Modeling System (CMS)*. Accessed May 1, 2022. https://cirp.usace.army.mil/products/cms.php.

Deltares. n.d. *D-Waves*. Accessed May 1, 2022. https://www.deltares.nl/en/software/module/d-waves/.

Emriver. n.d. "Emriver Modeling Media." Accessed Month 15, 2022. https://emriver.com/.

Fisher, Lucy V. and Andrew R. Barron. 2019. "The recycling and reuse of steelmaking slags—A Review." *Resources Conservation and Recycling* 146 (July): 244–255.

International Union for Conservation of Nature (IUCN). 2021. "Coral reefs and climate change." *IUCN Issues Brief*. Accessed October 20, 2021. https://www.iucn.org/resources/issues-brief/

coral-reefs-and-climate-change.

Kleypas, J. A., R. W. Buddemeier, and J. P. Gattuso. 2001. "The future of Coral reefs in an age of global change." *International Journal of Earth Sciences* 90 (2): 426–437.

Meeker, Barrett. *Dune Solver Asset*. Version. Orbolt. https://www.orbolt.com/asset/LuckyPause::dunesolver. 2021.

Moffitt, Robert B., Frank A. Parrish, and Jeffrey J. Polovina. 1989. "Community structure, biomass and productivity of deepwater artificial reefs in Hawaii." *Bulletin of Marine Science* 44 (2): 616–630.

National Oceanic and Atmospheric Administration (NOAA). n.d. "Coral Reefs: Rainforests of the Sea." Accessed October 19, 2021. https://oceanservice.noaa.gov/ocean/corals/.

Perry, C. T., S. G. Smithers, P. Gulliver, N. K. Browne. 2012. "Evidence of very rapid reef accretion and reef growth under high turbidity and terrigenous sedimentation." *Geology* 40 (8): 719. https://doi.org/10.1130/G33261.1.

Piller, W. and B. Rigl. 2003. "Vertical versus horizontal growth strategies of coral framework (Tulamben, Bali, Indonesia)." *International Journal of Earth Sciences* 92 (4): 511–519.

POSCO. 2020. "POSCO Uses Steel Slag to Create a Sea Forest and Save the Marine Ecosystem." POSCO Newsroom (press release). June 3, 2020. https://newsroom.posco.com/en/posco-uses-steel-slag-to-create-a-sea-forest-and-save-the-marine-ecosystem/.

Riegl, Bernard, Carlton Heine, and George M. Branch. 1996. "Function of funnel-shaped coral growth in a high-sedimentation environment." *Marine Ecology Progress Series* 145: 87–93.

Rodgers, K.S. et al. 2017. "Patterns of bleaching and mortality following widespread warming events in 2014 and 2015 at the Hanauma Bay Nature Preserve, Hawai'i." *PeerJ* 5 (May): e3355. https://doi.org/10.7717/peerj.3355.

Sahari, Faridah. 2015. "Cockle Shell As An Alternative Construction Material For Artificial Reef." In *International Conference on Creativity and Innovation for Sustainable Development*. International Islamic University of Malaysia, Kuala Lumpur.

Sanborn, Kelsey L. et al. 2017. "New evidence of Hawaiian coral reef drowning in response to meltwater pulse-1A." *Quaternary Science Reviews* 175 (Nov.): 60–72. https://doi.org/10.1016/j.quascirev.2017.08.022.

Severino, S. and K. Rodgers. 2019. "Hanauma Bay Biological Carrying Capacity Survey 1st Annual Report: May 2018 - May 2019." University of Hawai'i: Hawai'i Institute of Marine Biology, Coral Reef Ecology Laboratory/Coral Reef Assessment and Monitoring Program.

Severino, S.J.L., K.S. Rodgers, A. Graham, A. Tsang, Y. Stender, and M. Stefanak. 2021. "Hanauma Bay Biological Carrying Capacity Survey 2020/21 Annual Report." University of Hawai'i, Hawai'i Institute of Marine Biology: Coral Reef Ecology Laboratory/Coral Reef Assessment and Monitoring Program. Honolulu: Parks and Recreation.

SimScale. n.d. "What is CFD | Computational Fluid Dynamics?" *SimScale Documentation (SimWiki)*. Accessed March 5, 2022. https://www.simscale.com/docs/simwiki/cfd-computational-fluid-dynamics/what-is-cfd-computational-fluid-dynamics/.

Van Woeik, R., Y. Golbuu, and R. Roff. 2015. "Keep up or drown: Adjustment of western Pacific coral reefs to sea-level rise in the 21st century." *Royal Society Open Science* 2 (7): 150181.

Wang, P. 2005. Nearshore Sediment Transport Measurement. In *Encyclopedia of Coastal Science*, edited by M.L. Schwartz. Dordrecht: Springer. https://doi.org/10.1007/1-4020-3880-1_224.

IMAGE CREDITS
All drawings and images by the authors.

Colleen Duong is a Master's graduate of the Carnegie Mellon University, School of Architecture. A native of Hawai'i, she grew up with the saying "mālama i ka 'āina," meaning to respect and care for the land, which she took with her through her education. Colleen's interests are in designing architectures that respond to the environment, focused specifically on the implementation of natural resources like water, and developing new stories and ideas that speak to the relationships and experiences between humans, nonhumans, and architectural space. Currently, she is working as an Architectural Designer in San Francisco.

Dana Cupkova holds an Associate Professorship at Carnegie Mellon School of Architecture and directs Epiphyte Lab, Architectural Design & Research Collaborative focused on shaping environments between architecture and landscapes. Situated at the intersection of technology and ecology, Dana's work is centered on issues of design and environmental stewardships, while engaging computationally driven processes, circular materiality, embodied energy, and advanced manufacturing frameworks, with a particular interest in thermodynamics, and construction waste streams.

Azadeh Sawyer is an Assistant Professor at the Carnegie Mellon University School of Architecture. Azadeh holds a PhD in Building Technology from the University of Michigan, where she was also a Rackham Predoctoral Fellow. She has a Master of Science in Architecture from the University of Michigan, Master of Design Studies in Sustainable Design from Harvard University Graduate School of Design, and is the 2011 recipient of the Harvard Daniel L. Schodek Award for Technology and Sustainability. Her research focuses on building facades, daylighting, energy performance, and evaluation through simulation and immersive virtual reality.

Marantha Dawkins is a landscape architect and PhD student at the University of Virginia School of Architecture. With a focus on long-term and large-scale climate resilience, her research and teaching primarily explores how landscape form and process can inform nature-based infrastructure.

Session Introduction

Performative Design: Structural and Material Synthesis

Shelby Doyle, Chair

This session brings together five projects that developed and applied computational tools to simulate and assemble complex, nonstandard materials, with special attention to flexible and bending-active materials and structures. Strategies presented in "Performative Design: Structural and Material Synthesis" include multi-material bending active structures, non-standard materials such as bamboo, bespoke 3D printed structural nodes, actuated textiles, robotic self-assembly, curved-creased folding formworks, mixed reality models, and 3D scanned non-uniform reused lumber.

Though these projects span a variety of materials and approaches, they also share an ecological agenda of reducing material waste and maximizing the potentials of thin, light, and flexible materials as construction strategies. During the conference, the session panel discussion resulted in a thoughtful conversation around the following questions.

What is the relationship between assembly and computational labor in 'digital' projects that rely on extensive material manipulation through packing, shipping, construction, and disassembly? Is automation the necessary next step for this work or is integrating human construction labor essential to a sustainable future? What is the final scale of this work? Are these projects full-scale or are they scaled models of larger systems? How does the relationship between material and assembly change if these projects are scaled up and become too large to be assembled by hand? And how does assembly method (robotic assembly, self-assembly, and human-assembly) change how designers approach the design and simulation of project potentials?

LOOPS

A Mobile, Shape-Changing Architectural System: Robotically-Actuated Bending-Active Tensile Hybrid Modules

Valentina Soana
Yichao Shi
Tongyao Lin
The Bartlett School of Architecture, University College London

Yiting Ma
Ling Dai
University College London

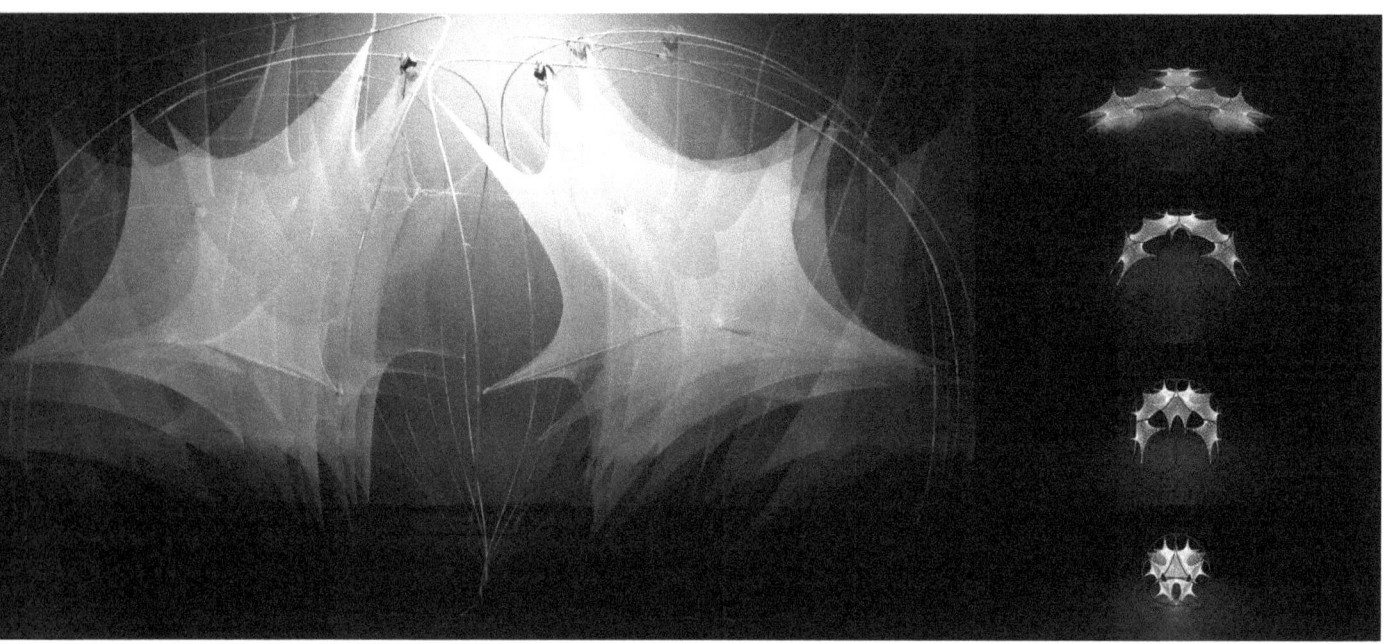

1 LOOPS self-forming and shape-changing system. The image on the left shows an overlay of multiple states of the robotic prototype operating. The images on the right side show a simulation of four states of a LOOPS system comprised of nine units, four cables and four robotic actuators.

ABSTRACT

LOOPS is a mobile and shape-changing architectural system that achieves multiple states through robotically controlled elastic material deformations. LOOPS is part of a wider research agenda on elastic robotic structures (ERS). ERS are lightweight, adaptive and can perform multiple behaviors with material and actuation efficiency, leveraging the capability of elastic materials to undertake large deformations (Lienhard 2014). This approach provides an alternative to conventional rigid body kinetic systems that rely on multiple parts and connections (Schleicher 2016). Due to the challenges of controlling large deformations of continuously operating systems, elastic kinetics at architectural scale remain underexplored (Soana et al 2020). Recent ERS research has sought to address this by proposing approaches that combine lightweight structure design methods used in architecture and engineering with the integration of robotic solutions.

LOOPS is the first modular ERS that was conceived to operate at large scale and that can provide a wide range of design and behavior options. It comprises an aggregation of robotically actuated bending active tensile hybrid (BATH) units. Novel approaches had to be developed in response to the challenge of designing and controlling global deformations of an interacting network of elastically deformable modules. The research focused on the development of design, fabrication, and control methods, demonstrated through a series of architectural systems, and of fully operating robotic prototypes. The work is relevant beyond its technical contribution. It proposes a new vision of the design potential of ERS as a building system; and, more broadly, a new vision of movable, intelligent structures.

INTRODUCTION

This paper presents design, fabrication, and control methods of LOOPS: a mobile and shape-changing architectural system that can achieve multiple states of equilibrium through robotically controlled elastic material deformations (Figure 1).

LOOPS is part of a wider research agenda concerned with the development of elastic robotic structures (ERS). ERS can change state in response to multiple factors (human, material, structural, environmental). The research aim is to develop novel adaptive lightweight systems, combining design and fabrication approaches of lightweight elastic structures with the integration of robotic solutions.

Elastic systems are inherently adaptive, given their capacity to undergo large deformations in response to different load conditions (Lienhard 2014). The advantage of elastic kinetic systems, relative to traditional rigid body approaches, is their ability to achieve a wide range of states with greater material and actuation efficiency, since their motion results from material behaviors (Schleicher 2016). ERS research aims to exploit this adaptive potential. To date, due to the challenges of controlling large deformations of continuously operating systems at architectural scale, elastic kinetic systems have only been deployed in small structures or in adaptive systems that display limited changes of shape. Recent ERS research has sought to address this by focusing on the conceptual and technical development of single units. Example systems include Self-choreographing Network (Maierhofer and Soana et al 2018) and ELAbot (Soana el al 2020). LOOPS builds on this agenda by advancing technical development and exploring a wider range of shapes, topologies and behaviors, within a larger architectural system.

LOOPS is the first modular ERS system based on the aggregation of robotically actuated bending active tensile hybrid (BATH) modules. In BATH structures equilibrium and global shape emerges from interaction between tensile (form active) and elastically bent (bending active) elements (Lienhard 2014; Alquist and Menges 2013). By continuously changing the connection between the tensile and bending active elements, multiple shapes of equilibrium can be achieved.

LOOPS research sought to establish a more generic design approach for robotic BATH at architectural scale, proposing: 1) the development of a parametric design approach in a calibrated simulation environment, enabling the generation of multiple shapes and behaviors; 2) methods for the design, fabrication, and control of robotic LOOPS for self-forming, shape changing and mobile behaviors; and 3) the implementation of a previous ERS cyber-physical control network, enabling the real-time control of LOOPS in response to changing conditions. The work is relevant beyond its technical contribution. It proposes a new vision of the design potential of robotic BATHs as a building system; and, more broadly, of movable, intelligent structures and potential application scenarios.

Adaptivity in Architecture

Traditionally, adaptive architecture has been pursued to enable spatial reconfiguration for multiple uses and functions (Kronenburg 2007). More recently, focus on adaptive architecture has been motivated by the aim to increase efficiency and material optimization, while reducing energy consumption in construction (Barozzi et al. 2016). Within this context, adaptive lightweight structure research that seeks to generate changes of state, leveraging material and structural behaviors (Sobek et al. 2006), is particularly relevant. This approach aims to leverage material behaviors to enhance performances of structures and building systems, during both construction and operation.

Recent technological advancements in simulation, sensor and actuation systems call for revision of the adaptive architecture debate of the 1960s. Those visions were spurred by developments in cybernetics, artificial intelligence and information technology; a provocative response to architecture's rigid, inflexible articulation of space (Kolarevic 2009). Among others, in 1964, Archigram members proposed their ideas on adaptive, deformable architecture. Later that decade, Charles Eastman developed the concept of adaptive-conditional architecture, an automated system that self-adjusts based on feedback from spaces and users (Kolarevic 2009). Nicholas Negroponte proposed the soft architecture machine, a sensory, actuated, performative assemblage of spatial and technical systems stimulated by users' interactions and their behavior (Negroponte 1975). The broader aim was to establish a radical new vision of intelligent architectural systems. More recent research has also investigated continuously operating architectural spaces (Bier 2018). However, the majority of adaptive buildings rely on mechanized rigid systems that exhibit limited actuation and intelligence. For example, kinetic façades that aim to optimize energy consumption (Decker and Zarzycki 2014). Jean Nouvel's Institut du Monde Arabe (1989) was the first significant, large-scale building with an adaptive envelope (Kolarevic 2009), exemplifying the limitations of mechanically operated designs (Decker and Zarzycki 2014), which depend on multiple electronically controlled moving parts.

In order to change state, adaptive systems need to perform a partial or full motion. Conventionally, movable systems are based on the principle of rigid body mechanics (Schleicher 2016), while compliant systems offer the opportunity to achieve motion through elastic deformations (Liehnard 2014).

Elastic Structures in Architecture and Engineering
Examples of lightweight elastic structures can be found in the distant history of vernacular architecture, where traditional materials, such as bamboo, were employed to achieve self-stabilizing structural systems through the elastic deflections of their members (Lienhard 2014). From the research of Frei Otto and his multidisciplinary team at the University of Stuttgart in the late 1960s, emerged novel design methods to control fundamental relationships between the form of a structure and the forces acting on it. This work produced multiple lightweight projects in which elastic behaviors were implemented to achieve large span structures with minimum material and high structural efficiency. Among these projects are the German Pavilion in Montreal and the 1972 Olympic Stadium in Munich, where Frei Otto inspired a new research agenda in lightweight architecture (Otto 1988). These achievements and subsequent projects established the field of lightweight structures in architecture and engineering.

Elastic structures cannot be designed using conventional geometrical approaches, as their form emerges from the equilibrium of forces and material behaviour (Suzuki 2020). In recent years, multiple research institutions such as ICD and ITKE at the University of Stuttgart, and CITA in Denmark, have sought to address the challenges of predicting and simulating non-linear elastic behaviors, through a series of cutting-edge research projects. The outcome of these efforts produced novel theoretical, numerical, and practical methods for the design of lightweight elastic systems in architecture and engineering. The work on BATH structures is particularly relevant. Projects such as the CITA Tower, a lightweight structural system constructed by stacking BATH modules (Holden Deleuran et al. 2015) and the textile hybrid M1 (Ahlquist and Lienhard 2013) established new methods and visions for the design and realization of these systems.

From Elastic Kinetics to ERS
The ability of elastic systems to undertake deformation without having to change their topology or degree of static determinacy (Lienhard 2014) provides an alternative to conventional kinetic systems, which achieve motions through mechanical rigid bodies (Khoo and Salim 2011). In elastic systems kinetic energy can be stored as strain energy in the deflected member and released when necessary (Schleicher 2016). Recent projects that explore the kinetic potential of elastic deformations for architectural structures range from room scale systems such as Flexing Room (Kilian 2018) to the adaptive façade of Thematic Pavilion EXPO by Soma (SOMA 2012), to medium scale prototypes such as the work on kinetic tensile systems conducted at Universiteit Brussel (Brancart, Laet, and Temmerman 2016). Examples are limited, both in number and range of behaviors, reflecting the technical challenges of designing, controlling, and fabricating elastic kinetic systems. Chiefly these derive from a lack of appropriate prediction and control methods. Elastic kinetics are difficult to model given their non-linear geometrical behavior, which depends on stress conditions and material properties. While the work of Seiichi Suzuki (2018) and Sean Ahlquist (2013) significantly advances simulation design approaches, actual material behavior cannot be fully predicted ex ante for continuously operating systems that can achieve large changes of shape. In the context of a highly deformable system operating in uncontrolled environment, one simply cannot compute all physical conditions (internal, external) acting on the system at a specific moment in time (Soana et al. 2020).

ERS integrate simulation methods with robotic planning and control solutions, so that control values can be constantly generated, evaluated and updated based on the negotiation between simulation and physical criteria (Maierhofer et al. 2018). ERS offer potential benefits in construction, where this approach could be used to self-form elements, reducing material, fabrication and construction effort and the need of many external supports, such as scaffolding. Once a building is operational, it can respond to changes in internal and external conditions. From the perspective of human-space experience, ERS enables architecture that is not rigid, determined, stable, but that leaves more space for human engagement, interaction and playfulness.

Previous research projects focusing on unit-scale ERS prototypes, such as Self-choreographing Network and ELAbot, have produced a series of techniques and tools. LOOPS extends these methods to: a) enable the design of larger modular systems; b) explore multiple behaviors such as shape changing and crawling; c) improve the efficiency of the cyber-physical control network by integrating a human tracking system; d) improve prediction and robotic planning through machine learning; and e) offer operation and design scenarios for new movable, adaptive systems interacting with humans.

METHODS
The LOOPS design process entailed development of the 1) material-actuation system; 2) robotic system; and 3) cyber-physical control networks and behaviors.

Material-Actuation System Design
LOOPS is based on the aggregation of multiple ELAbot modules. In ELAbot, the module comprises an actuated BATH loop, where one of the connections between the tensile surface and the bending active loop consists of a cable of variable length. The change of length affects the tensioning of the tensile surface, inducing global bending deformation

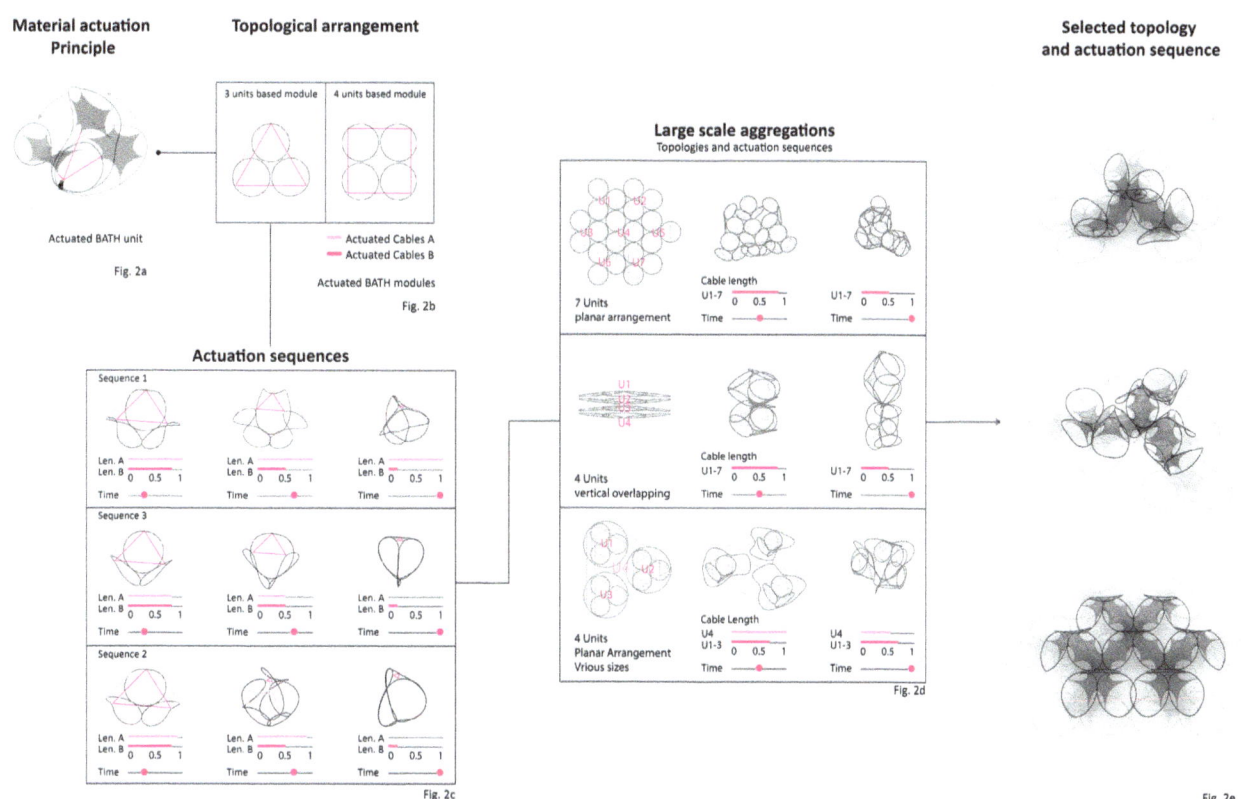

2 Overview of LOOPS design process in the parametric simulation environment. The figure shows: 2a) the material actuation logic of the robotic BATH unit, based on changing the length of the connection cable between the tensile surface and the bending active loop; 2b) the topological arrangement of 3 and 4 robotic BATH units into modules; 2c) actuation studies of one module, demonstrating how global deformations of the same topology can vary based on the actuation sequence; 2d) topological variations of large scale aggregations at three different states; 2e) three examples of different topologies in 2D and 3D state.

of the whole module (Soana et al. 2020). Based on this principle (Figure 2a), LOOPS explored the design potentials of a network of ELAbot. In LOOPS, global deformations are a result of multiple actuated modules interacting with each other. The actuation principle integrates ELAbot system with additional actuating cables connecting the units. Since the shape of a BATH system depends on material and geometrical parameters, the LOOPS design process considers multiple aspects simultaneously: number and arrangement of modules, material proprieties, local and global actuation logics, and actuation sequence (Figure 2).

The design emerges through a form-finding process, made possible by means of numerical simulations and physical prototyping. Given the complexity of simulating non-linear material behaviors, numerical simulations of BATH systems are usually calibrated by physical experiments (Lienhard 2014). LOOPS is a kinetic BATH where the form-finding process had to take place at multiple states. The main challenge was to find an equilibrium between material proprieties and actuation sequence, to enable the system to be flexible and stiff at the required times. Multiple tests focused on defining parameters affecting shapes and stability. Material behaviors were conducted iteratively and constantly between physical and numerical modelling. These parameters were: a) the diameter, length, and material of the bending active rods; b) the cutting pattern and elongation behaviors of the textile surfaces; c) the position, diameter and material of the actuated cable(s).

Simulation studies were conducted in Kangaroo 2.0 interactive physics engine (Piker 2013) within the Rhino/Grasshopper environment. Kangaroo simulation parameters were calibrated with material properties and results coming from physical experiments, while topological parameters of the network of modules could vary through a series of custom c# components. This process created a design tool that enabled rapid iteration of different topologies and actuations, predicting their shapes and behaviors. Within this context two main behaviors were explored: 1) self-forming and shape changing; and 2) crawling.

The self-forming behaviors enabled LOOPS to change from two- to three-dimensional states, and vice versa. Aggregation

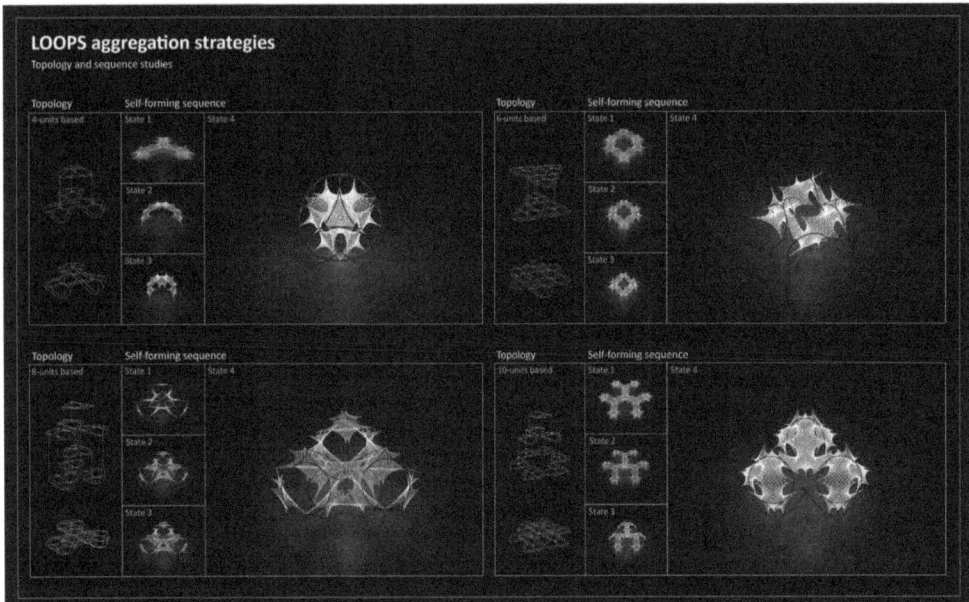

3   The images show simulations of four self-forming LOOPS. Multiple outcomes and behaviors were produced by varying the number of modules, topological arrangements, actuation position and sequence.

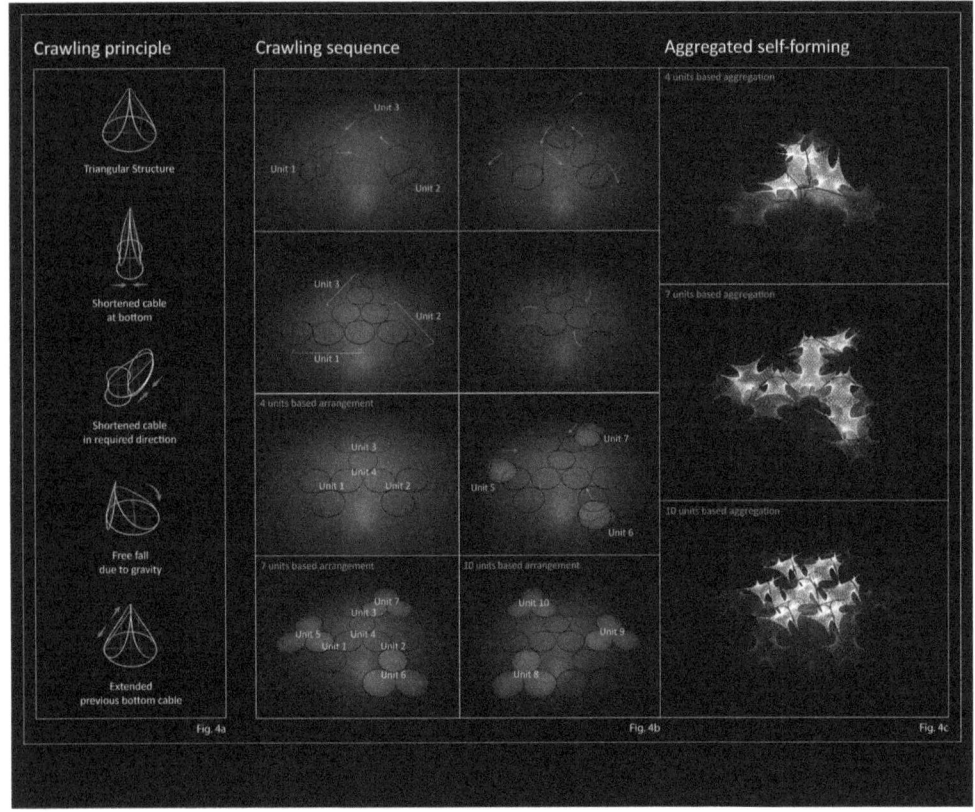

4   The images show simulations of LOOPS crawling principle (4a); a sequence of images of crawling units aggregating into a larger system (4b); and a self-forming and shape-changing sequence of larger systems (4c)

studies started testing a network of three and four loops, connected by variable length cables and arranged circularly in plane. Global changes were affected by the stiffness of the bending rods and the position and sequence of the actuation cable(s). Initial material studies were conducted with bending active loops and cables. Combinations of three and four loops performed best, in terms of avoiding redundant complicity and excessive material consumption, while maintaining stability. Once the actuation logic was tested, multiple shapes and strengths of membrane constraining the bending active loop were tested. These parameters affected global stiffness and provided different spatial features and aesthetics. The variety of cutting patterns and connection conditions generated multiple design outcomes. Self-forming units of three

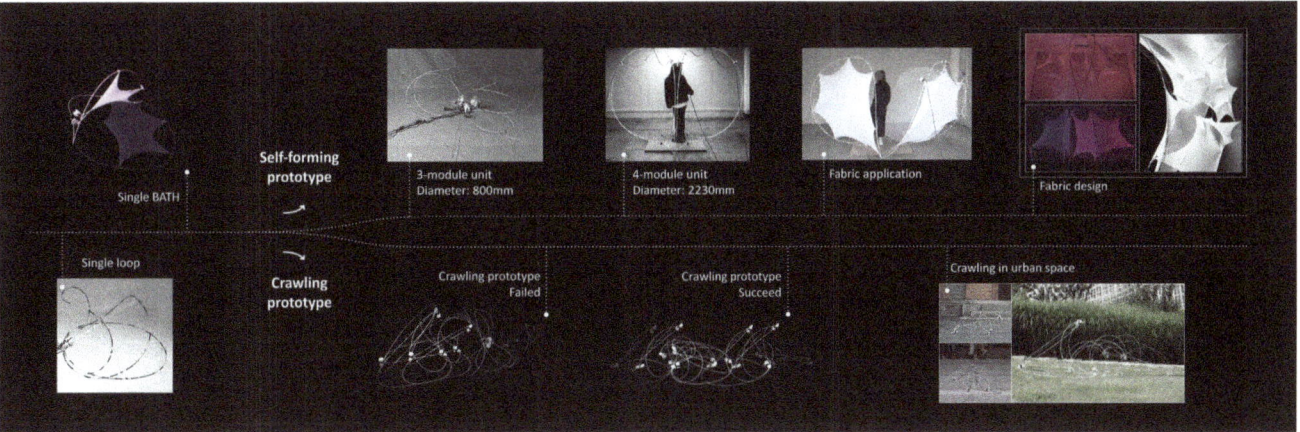

5 Overview of all the robotic prototypes

and four modules were further aggregated, in pursuit of yet larger scale. Aggregation and actuation patterns were tested iteratively in the parametric-based model, integrated within Kangaroo. Modules were either aggregated horizontally or vertically, generating different outcomes (Figure 3).

Crawling behavior emerged from the interaction between structural deformation, friction, and gravity. The module consisted of three robotic BATH units, each comprising two perpendicular cables to ensure the initial double curvature and structural stiffness. The three units were connected to each other to form a triangular structure. To achieve deformation of the module and produce global motion, the vertical cable of each unit was actuated by changing its length. The actuation sequence of each cable was key to achieving crawling behavior. Multiple actuation sequences and combinations were tested and evaluated. The successful sequence is demonstrated in Figure 4a. Continuous crawling was achieved by repeating this process. To simulate continuous crawling behavior within Kangaroo, custom C# based components that could automatically compute the length sequence of each cable, integrating forces such as gravity and friction, were developed. The modular parametric approach enabled testing of multiple crawling and self-forming sequences (Figure 4).

Robotic System

LOOPS' robotic system was designed for two behaviors: self-forming and crawling. Multiple scales and materials were tested (Figure 5).

LOOPS followed a two-step development process: 1) design, fabrication, and assembly of the physical system; and 2) mechanical design and feedback control of the actuation system. Fabrication of the physical prototypes was informed by design and simulation studies. The robotic system was an aggregation of 3 and 4 BATH units made of glass fiber rods (GFR) and tensile surfaces, connected by cables and actuated through multiple motor-driven spooling mechanisms that allowed cable length to be controlled. Fabrication studies focused on determining material properties of the rods, textiles and cables, to enable the system to deform in a controlled manner while avoiding collapsing at different scales.

The actuation mechanism comprised a dynamixel smart-servo capable of sensing and a 3D-print spool. The network of actuators was controlled by a custom feedback control system operating in ROS (Robot Operating System), which implemented the subscribing and publishing logic to enable the processing and exchanging of data between multiple devices. Within this environment, custom control methods, based on the Dynamixel SDK software development kit for ROS, were written to control and coordinate the network of actuators. The motor actuation sequence depended on the length variation sequence from behaviors that had first been designed in the simulation, then translated into a sequence of motor values.

The self-forming prototype was made of three and four BATH modules connected by robotically controlled variable cables. Initial studies were undertaken with a combination of three actuated BATH loops, each with a diameter of 80cm and restrained by two orthogonally arranged cables, in order to develop the basic control logic between multiple actuators and deformation (Figure 6a). For each module, the vertical cable was actuated by a motorized spool mechanism. To calibrate the control values between digital and physical environments, the displacements of actuated physical prototype were measured and compared to the simulation under various motor values (speed, torque, position). Once the multi-actuator control logic was developed, experiments aiming for stability at a human scale were carried out (Figure 6b).

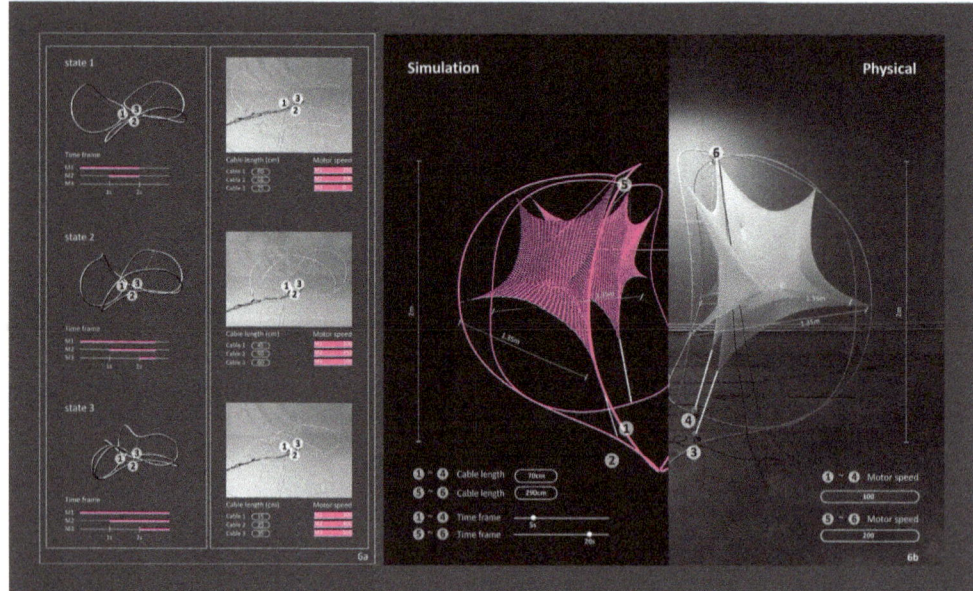

6   Overall development of the self-forming and shape changing robotic LOOPS. Images on the left (6a) show the first robotic system made of three units. The system was operating through the cyber-physical network. The images show a real-time comparison between simulated and physical states. The image on the right (6b) shows both the simulated and physical state of the human scale robotic system. Values on the left and right show cyber-physical data at that specific state.

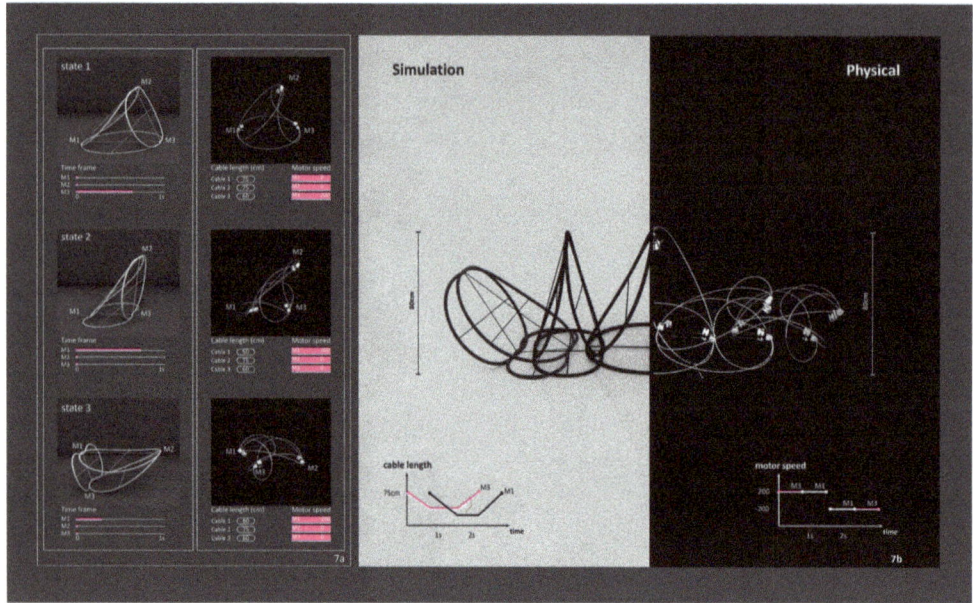

7   Overall development of the crawling robotic LOOPS. Images on the left (7a) show the crawling system operating through the cyber-physical network. The images show a real-time comparison between simulated and physical states. The image on the right (7b) shows the simulated and physical state of crawling LOOPS. Values on the left and right show cyber-physical data at that specific state.

The topology of the crawling prototype was defined by simulation studies. The system comprised three robotic BATH units connected by hinge joints. Multiple rods were tested to find appropriate dimensions able to support and balance the weight of the actuators. As noted above, actuation sequences were then translated into motor values. Control challenges included the calibration of motor values (such as speed) to alternate global rotations of the system to states of equilibrium (Figure 7).

Cyber-Physical Behaviors
Given the difficulty of controlling highly deformable elastic systems operating continuously, LOOPS behavior was embedded within a cyber-physical network where control values were processed in real-time, negotiating between digital and physical feedback. This network was developed based on previous work on ERS systems (Soana et al. 2020) that enabled real-time data computation between: 1) the simulation in Grasshopper/Kangaroo; 2) the robotic control in ROS (Robot Operating System); and 3) the User Interface (UI) in Unity game engine (Figure 8).

The main technical novelties were: 1) updating the control logic to enable the coordination of multiple actuators; 2) running the simulation environment within the UI to optimize data exchange velocity; 3) integrating human movement feedback

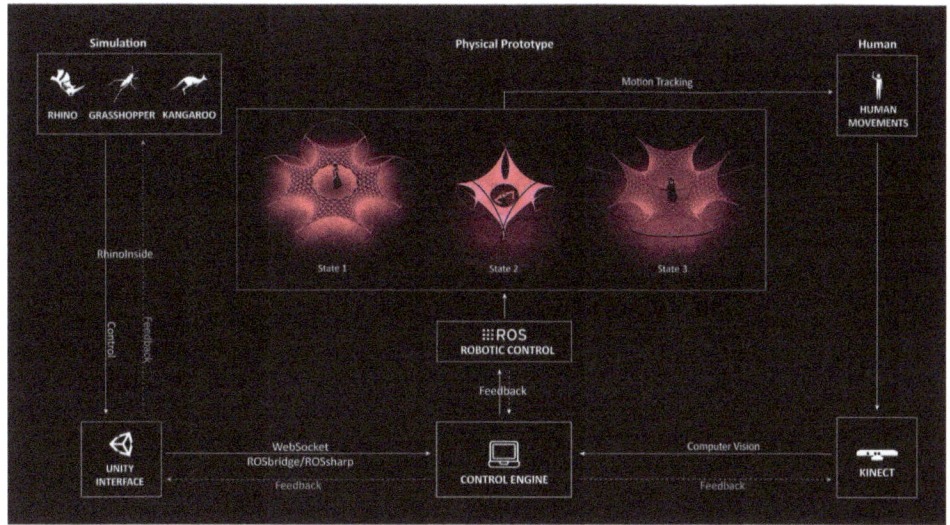

8 Cyber-physical control network. Robotic control values were computed based on feedback and negotiation of data from the simulation, physical and UI environments.

9 LOOPS self-forming and shape-changing cyber-physical behavior. The images show three states of LOOPS robotic system and the data exchange and feedback between the simulation, UI, and robotic control environments. Each physical state of the robotic prototype corresponded to a series of actuator states (control values), computed through a negotiation between design intentions, users' preferences, human motions, and robotic feedback.

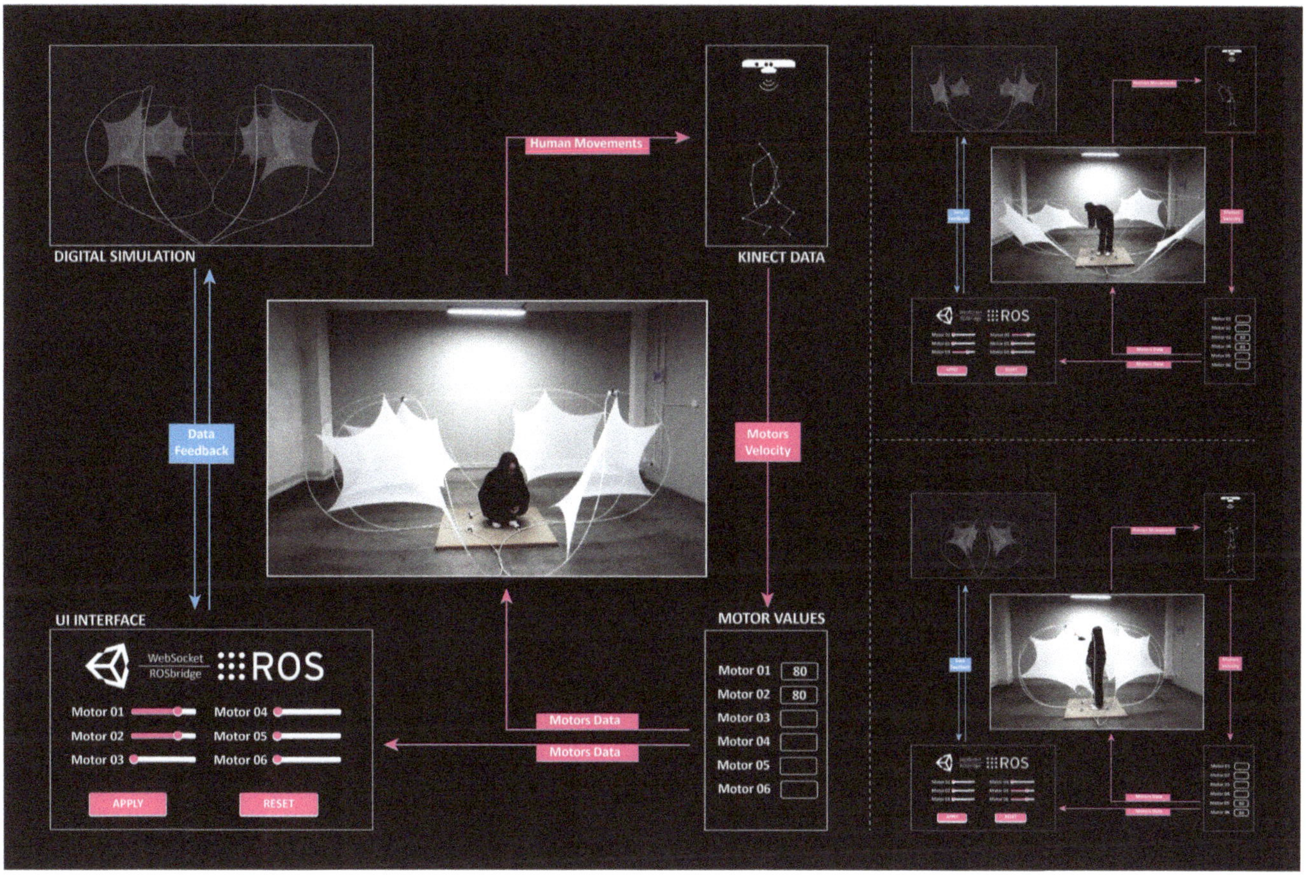

through an external tracking system; and 4) developing a machine learning process to compute crawling behavior. Kangaroo simulation ran inside Unity through Rhino.inside. It was communicating to the robotic control via ROS_Sharp and ROS_Bridge plugins connecting Unity and the ROS network. Another novelty was the integration of real-time tracking that captured human motions with the Kinect sensor.

This control logic integrated predetermined sequences with real-time user preferences and human motion detection. By integrating the Unity interface with ROS and the Grasshopper simulation, the physical prototype and digital simulation were synchronized. A predetermined sequence of changing values could be set to actuate the robotic system. Users could also control the system via the UI. In parallel, the Kinect sensor

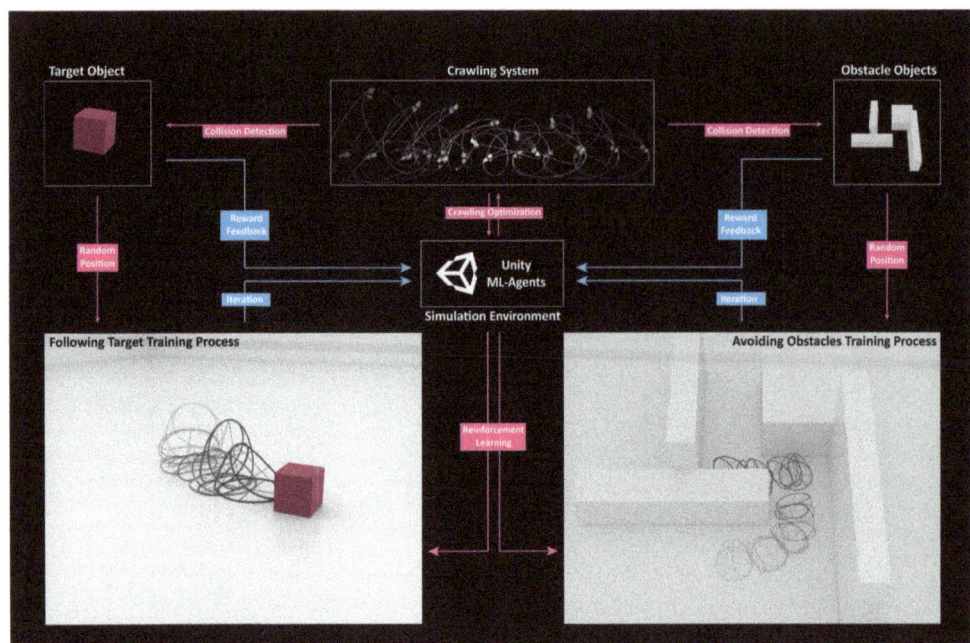

10 Machine Learning optimization process of LOOPS crawling system

detected human gestures and stored them as variables. Custom computational methods were developed to enable the integration and negotiation of these values, and their conversion into actuator control variables (Figure 9).

The control logic of the crawling behavior was computed through the robotic system calibration phase described above. Two optimization approaches were also developed. The first aimed to compute optimal crawling control values, to enable the unit to reach changing targets while avoiding obstacles (Figure 10). The process was developed in Unity through a reinforcement learning system based on Unity ml-agents.

The second aimed to output control values of a swarm of crawling units. This approach was based on Boids and Cellular Automata algorithms. With Boids flocking algorithm, multiple crawling units could troop onto their destination, avoiding collision with obstacles or other units (Knievel and Krueger 2022). Each unit was conceived as one agent in a multi-agent simulation environment developed by ELAbot. The goal of the multi-agent system was to determine the actuation sequences necessary to achieve target configuration and position.

RESULTS

The outcome of LOOPS research is demonstrated through the proposal of a series of architectural structures and two operating robotic systems. Design methods developed within the calibrated simulation environment enabled the generation of multiple robotic structures in different scenarios. This new mobile, adaptive system was conceived as a lightweight temporary structure that offered a wide range of spatial and volumetric experiences. Within the parametric environment the designer could input simple volumetric goals in relation to the project brief such as site, function and area. Different topologies and actuation strategies could then be generated, evaluated and selected based on initial goals (Figure 11).

In the final studies multiple LOOPS were explored with the aim of creating systems that could self-build without the need for additional construction substructures, such as scaffolding, while providing different spatial experiences over time (Figure 12).

Beyond the creation of shape-changing structures, technical development achievements enabled the researchers to speculate and propose a novel provocative vision for mobile architectural systems. The work was motivated by the envisioning of architectural robots that could crawl around our cities and self-aggregate into larger moving structures.

In parallel to the digital studies, two fully working robotic prototypes were tested. One crawling robot was built and operated in the urban streets of Shanghai. The crawling prototype attracted the attention of many passersby, providing an early glimpse into how architectural robots could start to occupy our cities (Figure 13).

Two self-forming prototypes were built as part of two specific scenarios. The first was built in Shanghai. The system tracked the movement of its user (indoors) and changed shape based

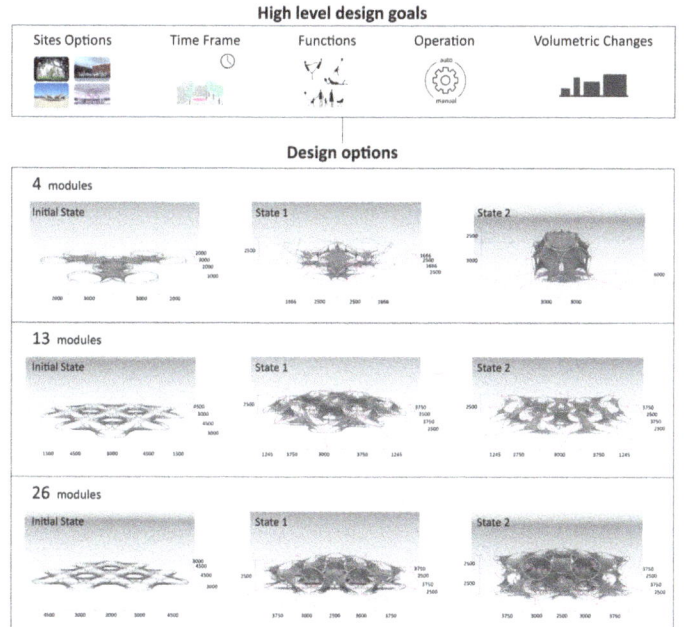

Design Options # Modules	Number of Actuators	Initial Footprint (m²)	State	Footprint (m²)	Height (m)	Features	Max. Capacity
4 modules	24	54	1	37.5	2.5	1 Storey, 1 internal space, 6 external spaces	18
			2	36	5.5	1 Storey, 1 centralised internal space	18
13 modules	78	162	1	112.5	2.5	1 Storey, 7 divided internal spaces, 9 external spaces	56
			2	112.5	2.5	1 Storey, coherent internal spaces, 9 external spaces	56
26 modules	156	162	1	112.5	2.5	1 Storey, 7 internal spaces, 9 external spaces,	56
			2	112.5	5	2 Storeys, 14 internal spaces in total, 9 external spaces	112

11 LOOPS architectural system design process. The first step was to identify high level design goals translated into volumetric goals. Based on these goals a series of LOOPS topologies could be generated in the parametric simulation environment. Each option could be analyzed based on initial goals, to determine the most suitable system. The preferred option could then be determined by the topology with fewer modules and actuation required to achieve spatial and volumetric goals at multiple states.

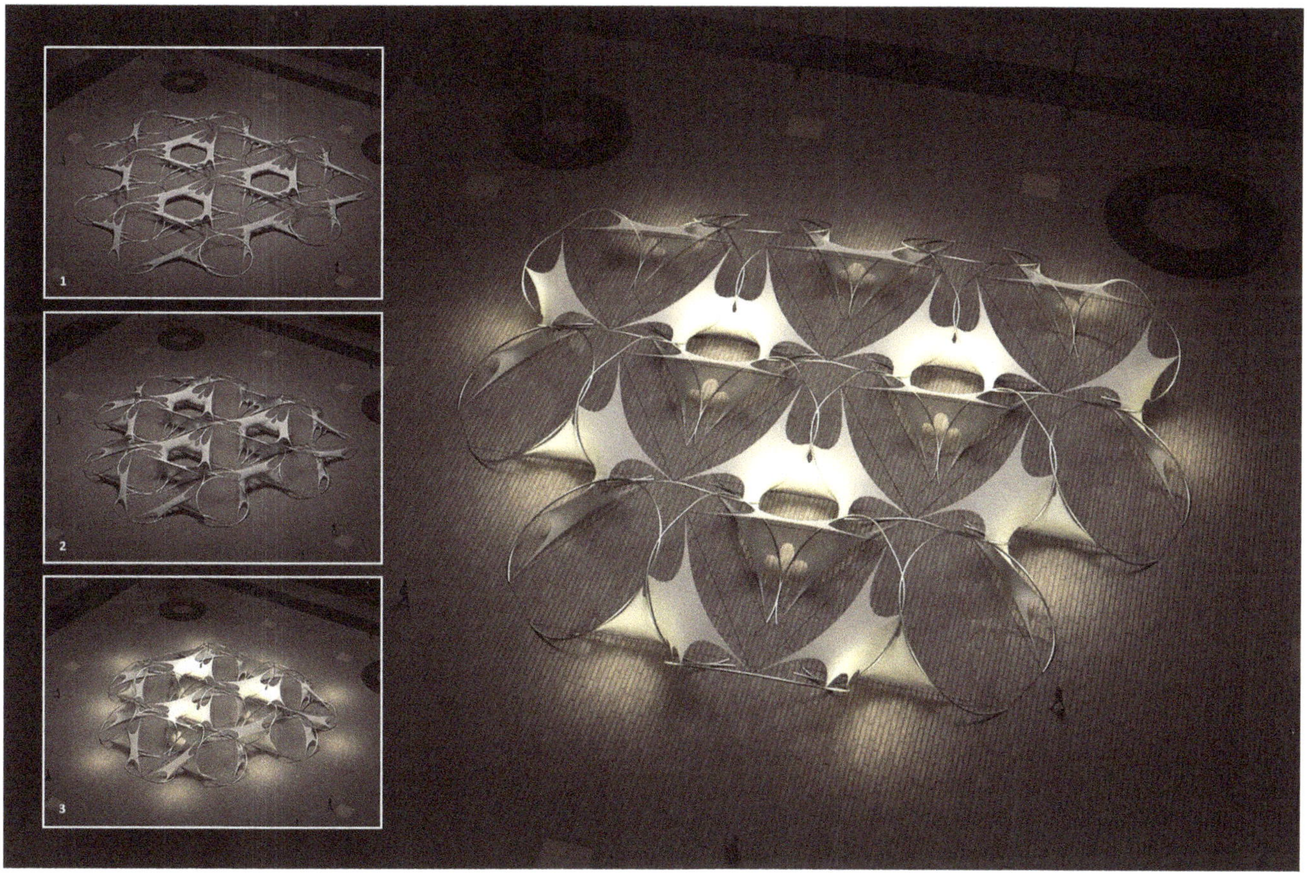

12 LOOPS system architectural vision; the images show potential LOOPS robotic systems in different environments

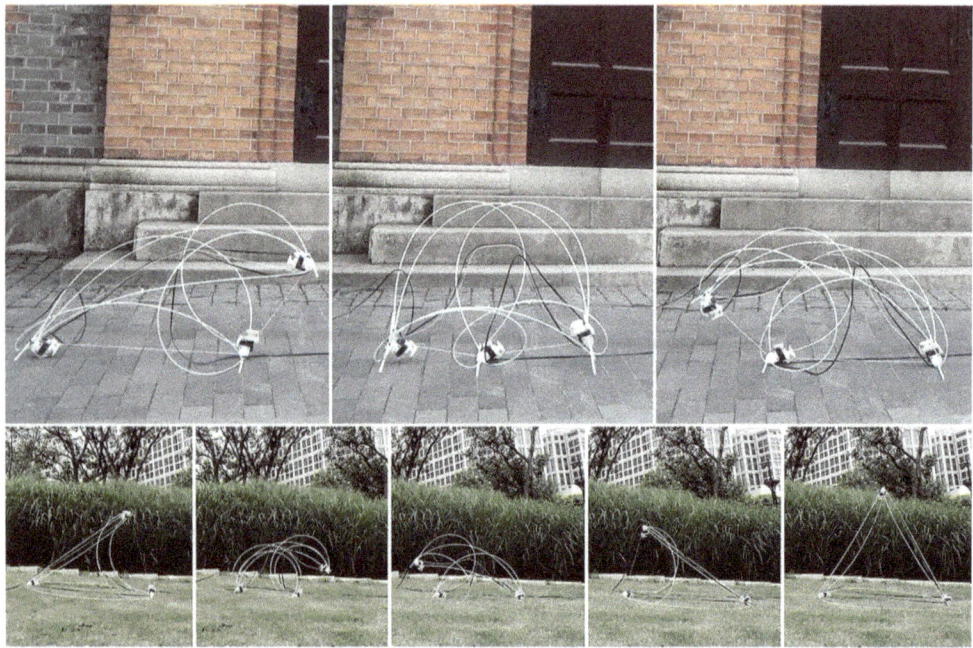

13 Images of LOOPS crawling robotic system in urban environment.

14 Images of LOOPS self-forming and shape-changing robotic system indoor (14a) and in House Block Exhibition (14b).

14a

14b

on human gesture recognition (Figure 14a). The same system was disassembled and shipped to House Block take-over in London, demonstrating that it could be easily transported and re-assembled by following basic instructions. It then performed in House Blocktake-over as part of a public event (Figure 14b).

The work questions how humans could start interacting with architectural robots. LOOPS control platform responds to this question with a speculative answer (Figure 15).

Humans could own architectural robotic modules, giving them a platform to collaborate directly with other users, to build emerging structures in an urban environment. Within the "Search" function, the nearby LOOPS could be displayed depending on each user's location sharing intentions. Users could nominate themselves as a convener and call others to aggregate the LOOPS into larger scale structures. The users and LOOPS modules that responded to the call could be shown on the "Friends" page. In the "Design" section, allied users could select and regulate the site, arrange, and forecast the aggregation results. After a site was chosen, the participating LOOPS unit could aggregate to the site based on the Boids Flocking Algorithm. Users could organize these discrete units by inputting parameters, generating various patterns of aggregation arrangements and behaviors. The structure could then operate within the cyber-physical environment, integrating real-time feedback to perform its behavior.

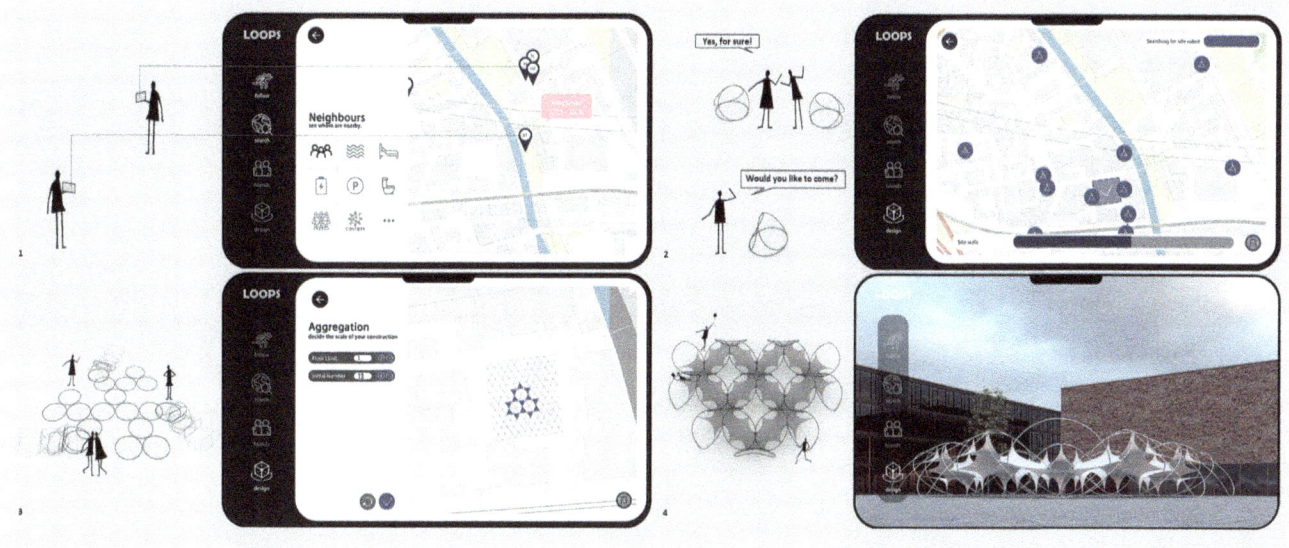

15 The images show the concept for LOOPS control platform, demonstrating how users "owning" LOOPS crawling modules could connect, gather and interact with other users to generate a temporary robotic LOOPS system in the urban environment.

CONCLUSION

The development of the application platform was a first attempt to speculate on the future of robotic architectural modules that could self-assemble in continuously changing structures. Multiple complex parameters and solutions remain to be considered, before LOOPS can operate in the built environment. In parallel to mechanical and structural advancements, future works will focus on the development of participatory decision-making processes to determine LOOPS behaviors. The most challenging and interesting aspect of this will be to continue speculating on how control values are generated. On the one hand LOOPS' cyber-physical system enables real-time evaluation of its current state based on material performances. On the other hand, other variables, such as human preferences and environmental feedback, could be collected. The integration of multiple sensor data will increase the level of complexity of control strategies negotiating between structural, material and actuation optimization, while integrating design external preferences and inputs. Given the complexity of solving the technical and conceptual questions around LOOPS, the work will continue at the intersection of multiple disciplines, such as robotics, architecture and engineering.

ACKNOWLEDGEMENTS

The presented work was developed as part of the Bpro, AD program Research Cluster 2 2020/2021. The authors would like to thank the tutor Georgia Kolokoudia, the skill tutors Christos Chatzakis, Emmanouil Dimitrakakis, and Harvey Stedman for their support.

REFERENCES

Ahlquist, Sean, and Achim Menges. 2013. "Frameworks for computational design of textile micro-architectures and material behavior in forming m complex force-active structures." In *ACADIA 13: Adaptive Architecture; Proceedings of the 33rd Annual Conference of the Association for Computer Aided Design in Architecture*. Cambridge, Ontario.

Ahlquist, Sean, and Julian Lienhard. 2013. "Physical and Numerical Prototyping for Integrated Bending and Form-Active Textile Hybrid Structures." In *Rethinking Prototyping: Proceedings of the Design Modelling Symposium*. Berlin.

Barozzi, Marta, Julian Lienhard, Alessandra Zanelli, and Carol Monticelli. 2016. "The Sustainability of Adaptive Envelopes: Developments of Kinetic Architecture." *Procedia Engineering* 155: 275–84. https://doi.org/10.1016/j.proeng.2016.08.029.

Bier, Henriette, Alexander Liu Cheng, Sina Mostafavi, Ana Anton, and Serban Bodea. 2018. "Robotic Building as Integration of Design-to-Robotic-Production and -Operation". In *Robotic Building*, edited by H. Bier. Springer Series in Adaptive Environments. Cham: Springer. 97–120.

Brancart, Stijn, Lars De Laet, Niels Temmerman. 2016. "Deployable Textile Hybrid Structures: Design and Modelling of Kinetic Membrane-restrained Bending-active Structures." *Procedia Engineering* 155: 195–204. https://doi.org/10.1016/j.proeng.2016.08.020.

Decker, Martina, and Andrzej Zarzycki. 2014. "Designing Resilient Buildings with Emergent Materials." In *Fusion: Proceedings of the 32nd eCAADe Conference*, vol. 2, edited by E. M Thompson. Newcastle upon Tyne, England. 179–184.

Holden Deleuran, Anders, Michel Schmeck, Gregory Charles Quinn, Christoph Gengnagel, Martin Tamke, Mette Ramsgaard Thomsen. 2015. "The Tower: Modelling, Analysis and Construction of Bending Active Tensile Membrane Hybrid Structures." In *Proceedings of the International Association for Shell and Spatial Structures (IASS): Future Visions*. Amsterdam, The Netherlands.

Khoo, Chin Koi and Flora Dilys Salim. 2011. "Designing elastic transformable structures: Towards soft responsive architecture." In *Proceedings of the 16th International Conference on Computer Aided Architectural Design Research in Asia (CAADRIA)*. The University of Newcastle, Australia. 143–152.

Kilian, Axel. 2018. "The Flexing Room Architectural Robot. An Actuated Active-Bending Robotic Structure using Human Feedback." In *ACADIA 18: Recalibration, On imprecision and infidelity; Proceedings of the 38th Annual Conference of the Association for Computer Aided Design in Architecture*. Mexico City, Mexico. 232–241.

Knievel, Christopher, and Lars Krueger. 2022. "Sensor Object Plausibilization with Boids Flocking Algorithm." *arXiv*. https://doi.org/10.48550/arXiv.2203.08036.

Kolarevic, Branko. 2009. "Exploring Architecture of Change." In *ACADIA 2009: Building a Better Tomorrow; Proceedings of the 29th Annual Conference of the Association for Computer Aided Design in Architecture*. 58–61.

Kronenburg, Branko. 2007. *Flexible: Architecture That Responds to Change*. London: Laurence King.

Lienhard, Julian. 2014. "Bending-active Structures: Form-Finding Strategies using Elastic Deformation in Static and Kinetic Systems and the Structural Potentials Therein." PhD diss., University of Stuttgart. http://dx.doi.org/10.18419/opus-107.

Maierhofer, Mathias, Valentina Soana, Maria Yablonina, Seiichi Suzuki Erazo, Axel Körner, Jan Knippers, and Achim Menges. 2018. "Self-Choreographing Network." In *ACADIA 19: Ubiquity and Autonomy; Proceedings of the 39th Annual Conference of the Association for Computer Aided Design in Architecture*, edited by K. Bieg, D. Briscoe, and C. Odom. Austin, Texas. 654–663.

Negroponte, Nicholas. 1975. *Soft Architecture Machines*. Cambridge, Mass.: The MIT Press.

Otto, Frei Paul. 1988. *Gestaltwerdung. Zur Formentstehung in Natur, Technik und Baukunst (Arcus)*. Cologne: R. Müller.

Piker, Daniel. 2013. "Kangaroo: Form Finding with Computational Physics." *Architectural Design* 83 (2): 136–37. https://doi.org/10.1002/ad.1569.

Schleicher, Simon. 2016. "Bio-inspired Compliant Mechanisms for Architectural Design." PhD diss., University of Stuttgart. http://dx.doi.org/10.18419/opus-123.

Soana, Valentina, Harvey Stedman, Durgesh Darekar, Vijay M. Pawar, and Robert Stuart-Smith. 2020. "ELAbot". In *ACADIA 20: Distributed Proximities; Proceedings of the 40th Annual Conference of the Association for Computer Aided Design in Architecture*, vol. 1, edited by B. Slocum, V. Ago, S. Doyle, A. Marcus, M. Yablonina, and M. Del Campo. Online + Global. 340–349.

Sobek, Werner, Patrick Teuffel, Agnes Weilandt, and Christine Lemaitre. 2006. "Adaptive and Lightweight." In *International Conference on Adaptable Building Structures (ICABSA)*. Eindhoven, Netherlands. 6–38.

SOMA Lima. 2012. "One Ocean, Thematic Pavilion EXPO 2012." *ArchDaily*. Accessed March 10, 2022. https://www.archdaily.com/236979/one-ocean-thematic-pavilion-expo-2012-soma.

Suzuki, Seiichi, and Jan Knippers. 2018. "Digital Vernacular Design. Form-finding at the edge of realities." In ACADIA 18: Recalibration, On imprecision and infidelity; Proceedings of the 38th Annual Conference of the Association for Computer Aided Design in Architecture. Mexico City, Mexico. 56–65.

Suzuki, Seiichi. 2020. "Topology-driven form-finding: interactive computational modelling of bending-active and textile hybrid structures through active-topology based real-time physics simulations, and its emerging design potentials." PhD diss., University of Stuttgart. http://dx.doi.org/10.18419/opus-10865

IMAGE CREDITS
All drawings and images by the authors.

Valentina Soana is a roboticist, designer, and researcher working at the intersection of architecture, engineering, and art. Her research focuses on the development of adaptive lightweight robotic structures that leverage material properties to achieve multiple states. She is a Lecturer in Architectural Design and Architectural Computation at The Bartlett School of Architecture (UCL), where she leads a research cluster a design studio. She is also a doctoral researcher at the soft haptics and robotics lab, part of UCL Robotics at the Department of Mechanical Engineering (UCL). She received a Master of Architecture from KTH Stockholm and a Master of Science (ITECH) from the University of Stuttgart, both with distinction. Valentina has worked as an architectural designer, researcher, and academic at architectural firms and universities in the US, Europe, and New Zealand. Her work has been published and exhibited internationally, most recently at the Venice Architecture Biennale.

Yichao Shi is a MS researcher at the School of Architecture, College of Design at the Georgia Institute of Technology. He received a MArch degree awarded by the Bartlett School of Architecture at University College London with distinction. His research focuses on integrating design computation and the development of cyber-physical system for autonomous architectural structures.

Tongyao Lin recently graduated with a Master in Architectural Design degree from University College London. Her research interests include kinetic architecture, bending-active structures, and computational design methods.

Yiting Ma holds a Master of Architecture from University College London and a Bachelor of Fine Arts from Donghua University. Her research focuses on the adaptive architecture practices incorporating robotic system and elastic materials.

Ling Dai is a postgraduate at The Bartlett School of Architecture at University College London and an architect at Atkins. His research focuses on robotic material systems and kinetic architecture.

Digital Bamboo

A Study on Bamboo, 3D Printed Joints, and Digitally Fabricated Building Components for Ultralight Architectures

Marirena Kladeftira
Digital Building Technologies/ ETH Zurich

Matthias Leschok
Digital Building Technologies/ ETH Zurich

Eleni Skevaki
Digital Building Technologies/ ETH Zurich

Davide Tanadini
Structural Design/ ETH Zurich

Patrick Ole Ohlbrock
Structural Design/ ETH Zurich

Pierluigi D'Acunto
Structural Design/ Technical University of Munich

Benjamin Dillenburger
Digital Building Technologies/ ETH Zurich

ABSTRACT

This paper presents a novel construction system that integrates natural and artificial components through the case study of the Digital Bamboo. A reversible non-standard structure is made of unprocessed bamboo poles connected with 3D printed joints and covered by lightweight 3D printed shading panels. The system combines multiple technologies to prefabricate all parts of the structure, which are controlled with a chain of computational tools.

Bamboo has gained increased attention in the construction industry due to its fast growth, natural tubular shape, and its ability for carbon sequestration, while it is locally available in different parts of the globe. However, joining bamboo becomes especially challenging because it is characterized by natural deviations and dissimilar mechanical characteristics amongst products.

In this paper, we portray how combining additive manufacturing and robotic fabrication methods allows the development of novel integrative systems for such non-standard materials. In parallel, we describe a computational workflow that allows the negotiation and control of the multiple methods applied, but also acts as an interdisciplinary collaboration platform between architects and engineers. The developed methods showcase the degree of agility that is necessary to tackle the increased complexity of such projects and are presented here through the development and execution of a building demonstrator.

1 View of the center of the pavilion portraying the two connecting systems and the attached shading panels

2 Digital Bamboo in exhibition at the Venice Biennale 2021

3 Digital Bamboo in exhibition at the ZAZ Bellerive

2

3

INTRODUCTION

The investigation of low-embodied-carbon materials and carbon sinks for the production of building components is gaining interest as a response to the large contribution of the construction sector to the climate crisis. Existing research on the embodied carbon per unit mass of construction materials (Cabeza et al. 2021) pinpoints the use of biomaterials, such as bamboo, in order to drastically reduce the embodied CO_2 footprint of future buildings.

Bamboo is a fast-growing natural material with a short harvesting cycle (Moses et al. 2015) that can sequester carbon dioxide throughout its life cycle. Its tubular shape and reinforcing rings along its length define the longitudinal orientation of its fibers that contribute to its high strength-to-weight ratio. This natural shape makes bamboo an excellent candidate for replacing energy-intensive metals in space frame construction, where loads are mainly transferred axially. However, the natural variability of unprocessed bamboo poles hinders their use with existing connection systems. Furthermore, proprietary connection systems are designed for specific materials like metal tubes or timber, hence the connection typology is primarily defined by those. In parallel, they limit the possible geometric configurations one can build, as their connections define largely the complexity and topology of structures.

The inherent variability of unprocessed bio-materials can be easily embedded in connection systems that are manufactured with 3D printing technologies (3DP). Developments in 3DP of metal alloys and high-strength thermoplastics create opportunities for fabricating bespoke structural connections in a simple supply chain. In parallel, reports indicate the wider adoption of additive manufacturing (AM) technologies by the construction industry in the near future (Wohlers et al. 2018). Based on these facts we can conclude that AM presents opportunities for manufacturing connections for standard-less bio-sourced materials like bamboo.

Bespoke Joints for Non-standard Structures

Research on customized connections for bespoke spatial structures is not a novelty in architecture. Architects like B. Fuller (Fuller 1954), K. Wachsmann (Sumi and Burkhalter 2018), F. Haller and others have developed numerous typologies of joints during the second half of the 20th century when proprietary systems started to dominate the market as initiated by the Mero system (Mengeringhausen 1974). Until recent years, most bespoke prefabricated connections were manufactured by casting metal alloys in unique molds or by welding steel plates forming a 3-dimensional node. Systems like the Ball Mero Node have simplified the fabrication of such connections, however the more complex the structure, the more the logistics, cost and lead time increase. Additionally, although the system is somewhat flexible in the joint creation, its use is limited to compatible materials and possible configurations.

There is a growing interest to harvest the flexibility of AM for manufacturing bespoke connections. In 2014 Arup created the first 3DP prototypical connector for a tensegrity structure (Galjaard et al. 2015). Subsequently, other researchers explored the opportunities of AM for space frames using metal Binder jetting (Raspall et al. 2019), as well as AM polymers with extrusion-based processes (Crolla et al. 2017). Although first investigations in the field address material reduction and design freedom, opportunities emerged to define new systems that include biomaterials like bamboo. These challenge the practices of traditional bamboo joinery developed over centuries in southern Asian and American regions, an overview of which was portrayed by Dunkelberg (Dunkelberg 1992). In 2018 AM plastic connections were fabricated to perfectly fit scanned bamboo sections for a freeform canopy (Amtsberg and Raspall 2018). Similar research investigated 3DP bio-composite materials for bamboo joints (Paola and Mercurio 2020; Wassenhove et al. 2021), while a recent study focused on the connection element as the geometric space between reciprocal bamboo poles (Qi et al. 2021).

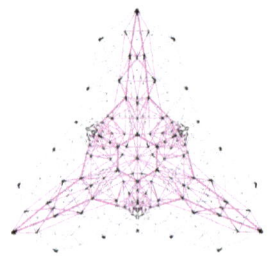 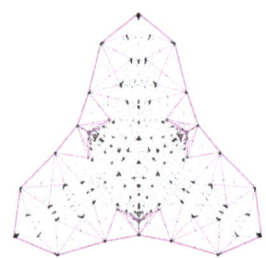 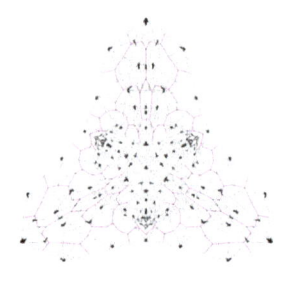

4   The three parts of the structure: primary, secondary truss, and tertiary structure

5   Top view of the structure and the axes of symmetry

Using AM to create connections for bamboo challenges the idea of overlapping poles seen in traditional connections, while it allows for more precision during built-up relying less in craftsmanship to deliver better results. Many joints that connect poles end-to-end require a chemical connection between the bamboo and the joint, while those that use a mechanical connection follow conventional mechanisms used for industrial materials.

In the sections to follow we will present a new reversible connection system that was specifically developed for bamboo and leverages AM technologies. Furthermore, we will show how its function can be augmented to encapsulate more layers of the structure, and finally how a digital process chain can facilitate the development of such novel systems for local architecture. These findings are presented in the case study of a bespoke pavilion with different digitally fabricated components that are all part of one digital design-to-production chain.

DIGITAL DESIGN AND MATERIAL SYSTEMS
Form Finding and Anatomy of the Structure

The Digital Bamboo is a space frame that covers an area of 40 m² and has a mass of only 200 kg. It is composed of 930 bamboo poles connected with 381 3DP joints. The geometry of the pavilion is rotationally symmetric, consisting of three cantilevering wings supported by three spatial columns (Figure 2). Each of the three cantilevers further displays a reflection symmetry on the vertical plane that separates the part in two halves (Figure 5). The symmetry of the pavilion serves multiple purposes on the aesthetic, structural, and fabrication level.

The geometry of the pavilion is derived from an equilibrium-based form-finding process that considers the self-weight of the structure as the dominant load case. The spatial truss outlining the form is triangulated in three dimensions, forming polyhedral cells to define the topology of the space frame. The load-bearing structure is composed of three distinct parts (Figure 4) that interact hierarchically (Tanadini et al. 2022):

- The primary truss consists of the main load-bearing part that spans from the columns to the tip of the cantilevers.
- The wide-spanning cantilevering wings are formed with a secondary three-dimensional triangulated truss.
- The tertiary structure, which is composed of planar, non-load-bearing elements.

Due to the limited tensile capacity of the connecting system described in the following section, the primary truss is post-tensioned with six steel ropes that are positioned along the reflection symmetry axes. Bamboo poles of circa 20 mm in diameter were chosen to create a filigree triangulated structure with high geometric stiffness.

Connecting Systems

The bamboo poles are joined by a novel system of bespoke connections that are specifically designed for AM. Most custom connections produced with AM are manufactured in metal (M3DP) or a thermoplastic material such as polylactic acid (PLA) through a layer-based extrusion process (FFF). However, these methods are, in our opinion, unsuitable for connecting bamboo. M3DP can be used to produce mechanically strong joints but excessively so compared to the properties of bamboo. FFF can produce lightweight parts which are, however, mechanically weak, anisotropic and often delaminate.

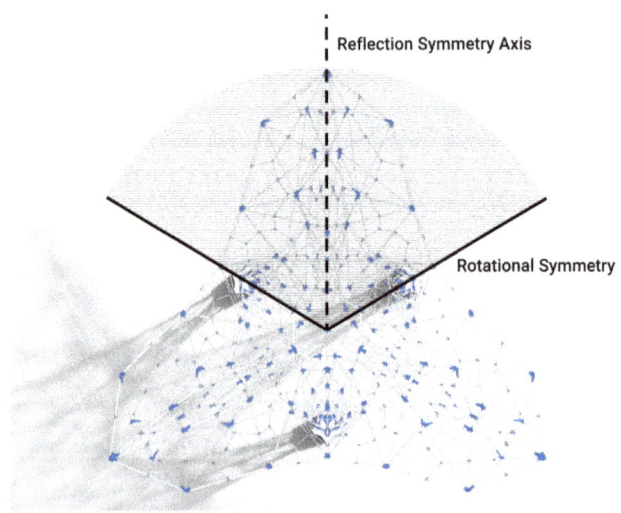

Digital Bamboo Kladeftira, Leschok, Skevaki, Tanadini, Ohlbrock, D'Acunto, Dillenburger

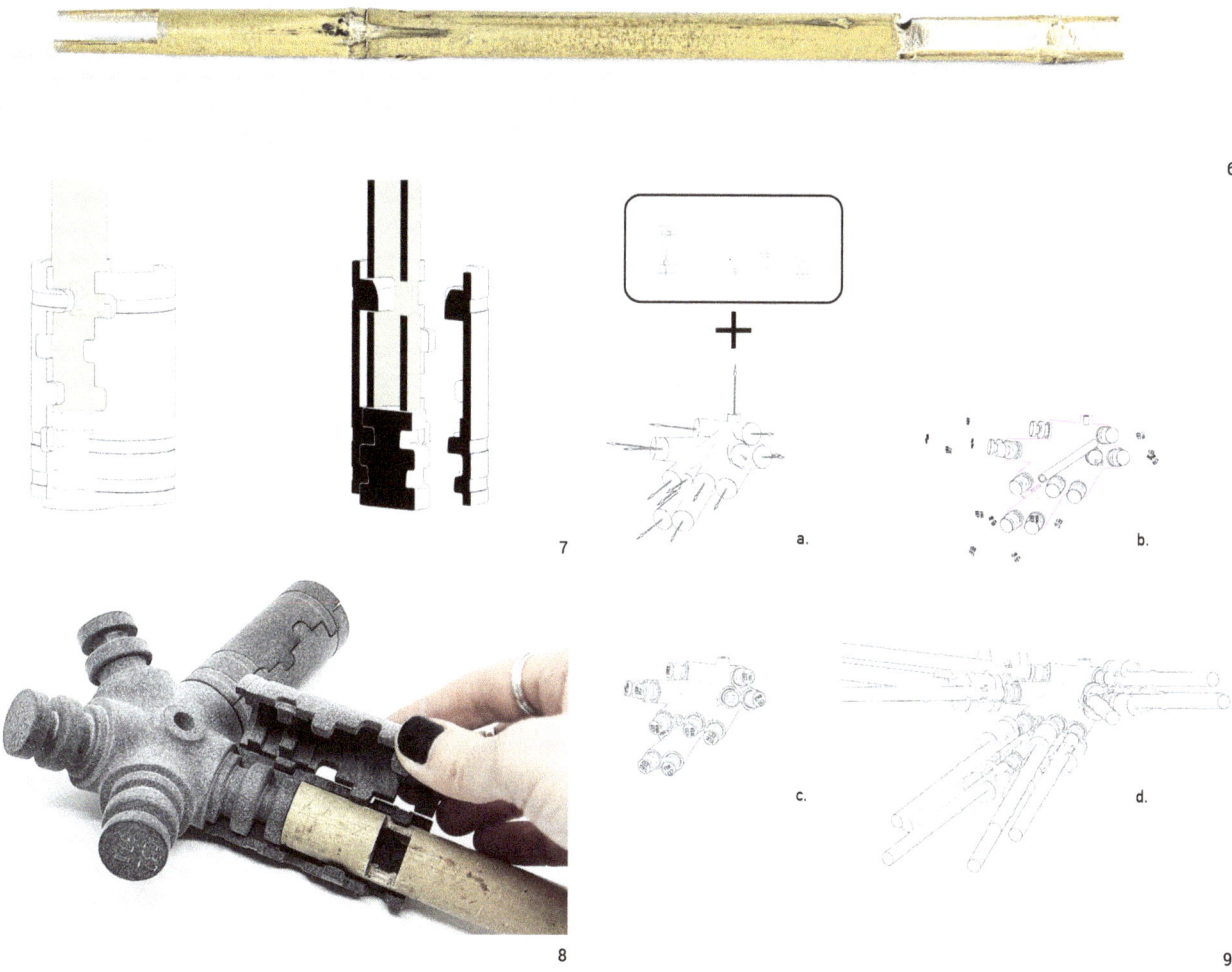

6   Split of bamboo fibers due to excess stress: failure of the notch on the left, and the pin on the right

7-8 The connecting system, or "notch," between bamboo and 3D printed connection

9   a) individual definition of parts based on a parametric model that allows easy iterations and adjustments, b) orientation in space according to defined planar interfaces between parts, c) combination of all geometries via modification of the distance values, and d) generation of a mesh file for fabrication

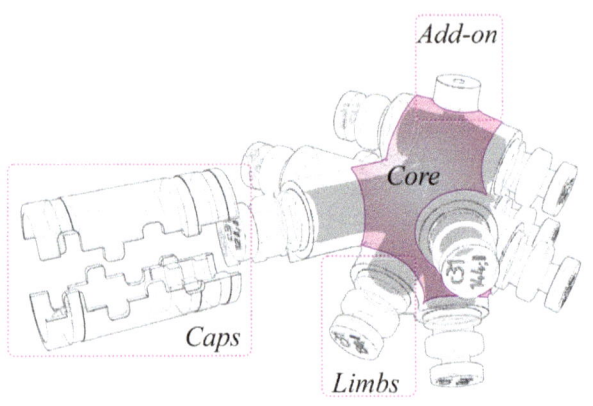

10 The core, limbs, add-on skin, caps, and assembly info are integrated into the joint

11 The central connection connects 18 bamboo poles and 6 post-tension cables; due to structural demands, it is manufactured in steel with DMLS

The species of bamboo used was unknown and therefore also its mechanical properties. We performed tensile tests on bamboo poles with a pin connection utilizing a simple testing station with 0.2 kN incremental loading. Based on the tests' results (1.4 kN), we chose to use the MultiJet Fusion (MJF) printing process because it delivers slightly better tensile strength than the bamboo. Thus, we ensure that the material is adequate, not over performing and does not represent the weakest point in the structure. The material printed is polyamide 12 (PA12) with characteristic tensile strength of 48 MPa, while lightweight with a specific weight of 1.01 g/cm³. MJF is a powder-based method that guarantees isotropic performance of printed parts and is used for the first time in architectural applications, to the best of our knowledge.

A number of requirements made the design and development of the connection system specifically challenging:

- Bamboo is a natural material with variable cross-sections along the longitudinal axis and varying diameters among the different poles.
- We intended to create a fully-reversible connection between the nylon and the bamboo, as the structure is a temporary installation.
- The joints should not only fulfill their structural role in the space frame, but also incorporate additional elements, in this case, the anchorages for the post-tension cables, the connectors to the shading elements, and the bundling of multiple bamboo poles.

Each 3D printed node generates a notch to evenly transfer stresses between the nylon and the bamboo (Figures 7, 8). Unlike typical pins or friction-reliant connections, the notch acts similar to a pin but distributes the forces over a larger area. Therefore, the longitudinal splitting of fibers is prevented (Figure 6) delivering better performance in tension than a pin connection. The rounded shape of the notch and the high tolerance between bamboo and connection, in conjunction with the triangulation of the structure ensures that there is no friction or bending in the connection. After tensile testing on more than 100 poles, the performance of the notch was determined according to the fifth percentile (1.8 kN) of the measured ultimate strength. The result was subsequently used to perform further structural analysis of the structure. Based on the analysis we identified the weak points, where a single pole is replaced by a bundle of three, adopting the technique as commonly used in traditional bamboo structures. The bundles highlight the force flow through the space frame in a continuous manner and architecturally outline the pavilion and the inner compression ring.

Each joint consists of the core, the limbs, add-on features, the caps, and 3D printed assembly information (Figures 9, 10). The core is defined by a custom algorithm that analyzes the topology of the structure and, based on the material profiles specified, calculates the minimum geometric region so that the bamboo poles do not collide. The limbs stem from the core in the directions where the joint meets the bamboo. They feature the necessary logistic information (enumeration,

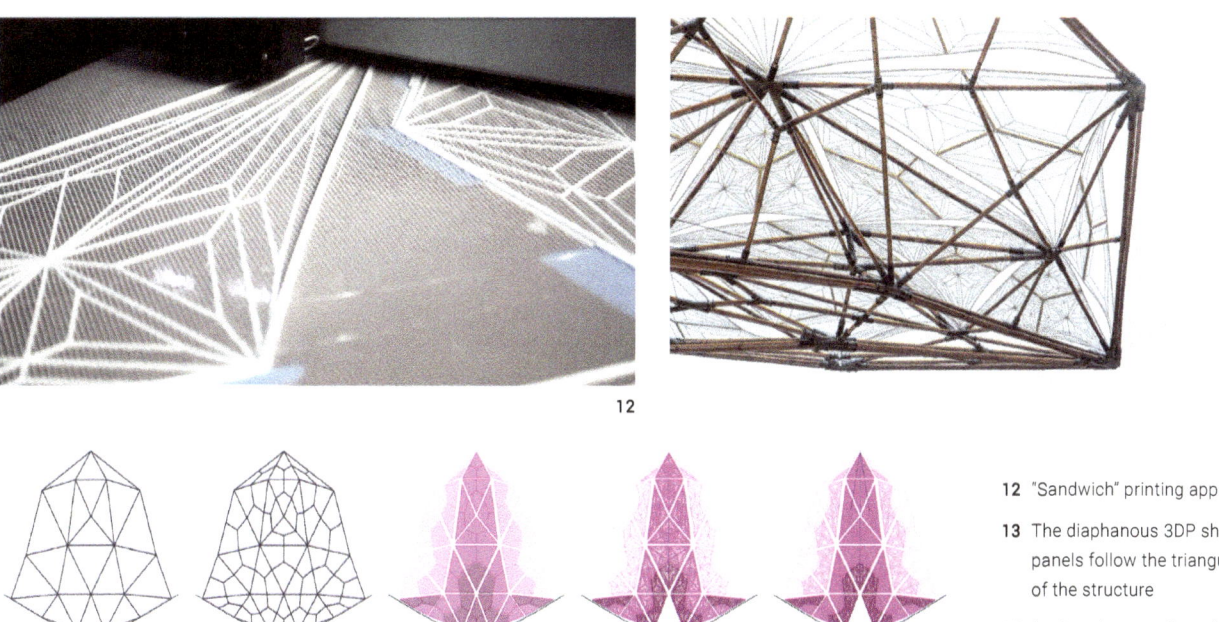

12 "Sandwich" printing approach

13 The diaphanous 3DP shading panels follow the triangulation of the structure

14 Assigned groups based on the distance of the center and the symmetry axes. rules are applied according to the color groups

location, etc.) for the assembly. The add-on features include the insertion and fastening details for the post-tensioning cables (7 mm channels along the axis of the cables), and the connections to the shading panels (threaded tubes perpendicular to the planes of the latter). The caps are standardized parts that surround and connect the limbs of the joint with the bamboo and are fastened with zip ties. The joints are hollow with a minimal wall thickness of 4 mm, which was identified to be adequate through multiple tensile tests of the connection system.

Since each node is composed of several parts, the configuration of which is different for each specific location in the structure, the design of these parts is not trivial with conventional mesh-based methods (Kladeftira et al. 2021). To tackle this problem, we employed volumetric modeling as an effective method for designing complex connections. This approach allows for an easy and precise combination of 3D shapes while avoiding the arduous process of Boolean operations. The interface of the computational method was developed within the Rhino/Grasshopper environment. The topology of the joints was created with a custom script based on the COMPAS framework (Mele et al. 2021) and the geometry generation was performed with a customized workflow utilizing Axolotl (Bernhard and Dillenburger 2019). The computational method and detailed development of the connection system are fairly complex and described in depth in a dedicated companion article.

The nylon-based connection system fulfills the structural demands of all connections in the pavilion except for one special joint located in the center of the structure. This joint connects 18 bamboo poles and 6 post-tension cables and follows the same principle as the rest of the joints but it is fabricated in steel with Direct Metal Laser Sintering (Figure 11).

Skin

The pavilion is covered by rigid panels that provide shade. This shading layer follows a panelization derived from the top-layer triangulation of the bamboo truss (Figures 3, 13). Following the lightweight character of the pavilion, the 75 bespoke panels that cover the structure are digitally fabricated by combining a mesh Lycra textile and an extruded thermoplastic in a new process called add-on 3D printing (add-on 3DP). This process allows the production of ultralight panels, as the textile has a weight of 173 g/m² and the 3DP plastic is applied only for forming/framing and stiffening in selective areas.

The fabrication method impacts the aesthetics of the shading panels, as one can clearly distinguish between the textile and printed plastic. Therefore, the stiffening ribs are designed in a computational process that follows the vocabulary of the space grid and serves the structural integrity of the panels. We applied mesh grammar rules that subdivide the frame of the individual panels according to area, location and patterning schemes in the steps shown below:

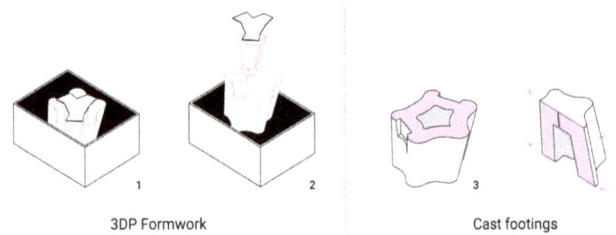

3DP Formwork   Cast footings

15

1. For every vertex *v* of the starting triangular face, create a new face between vertex *v*, the midpoints of the edges that are incident to *v* and the centroid of the face. With this scheme, every triangle is now split into three quads (Figure 14 a-b).
2. Assign groups to the resulting quad faces based on their distance to the center of the structure (Figure 14c). The number of subdividing iterations varies for every group: the outermost are subdivided once, while the innermost are complete after 4 subdivision steps (Figure 14d).
3. In order to reduce the wind load, introduce cut-outs inside the sub-triangles of the pattern, which vary in size and frequency depending on their position inside the structure (Figure 14e). In the digital environment, this operation is as simple as culling the last face generated by the termination rule.
4. Add a connection detail to each original vertex *v* that includes a hole for a screw connection to the corresponding joint.

The panels were printed flat and bent to an inverted catenary shape during assembly. The result is a three-dimensional filigree, diaphanous skin with graded porosity, which makes for a playful and adapted shadow pattern (Figure 13).

To fabricate the panels a sequential printing process was developed to create a "sandwich" material system where the textile is placed and locked between printed layers of plastic. The mesh-like porous fabric is used for stronger adhesion between the printed layers (Figure 12), as the plastic penetrates the fabric and adheres directly to the first layers. Two layers of PETG are printed, the fabric is placed and more layers are printed on top. The first PETG layer of the upper part is printed with a small offset to the initial print path in order to avoid collisions with the textile. In order to compensate for the increased starting height, the print speed is reduced during the first top-layer.

Foundation

A steel plate acts as a non-invasive foundation for the pavilion. Its weight ensures the anchoring of the space frame to the ground as wind loads would cause up-lifting of the ultra-light structure. Bespoke concrete footings act as transition elements to connect the structure to the steel plate. Any tolerance resulting from an uneven placement of the steel plates or from the assembly of the structure is mitigated by those.

The design of the footings reflects the shape of the base connector (the lowest 3D printed connection) that joins together 19 bamboo poles and the post-tensioning cable. The post-tension cables are anchored inside the concrete feet and inserted before the latter are fixed to the steel base. The orientation of the cable inside the concrete is guaranteed by a 3DP guiding tube embedded in the concrete. The concrete feet are cast into a removable polymer 3DP formwork made from biodegradable plastic (PLA) (Figure 15). An inner cavity in the mold system helps to reduce material, but also provides easy access to anchor the post-tensioning cable inside the footing.

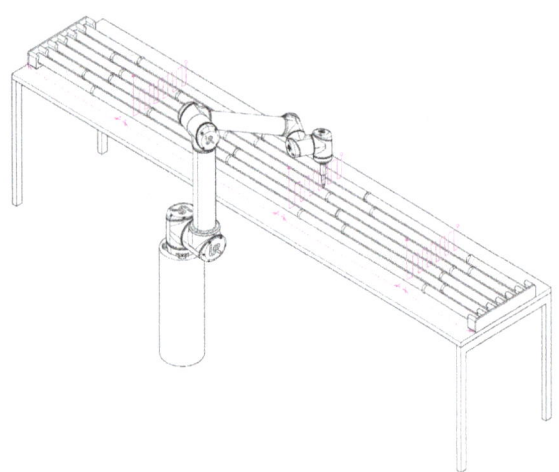

16

Part	Processes	Steps	No of Items	Production Time (days)	Assembly Time (hours)
Bamboo	Robotic Marking/ Cutting/ Notching	3	930	1 / 2 / 5	-
Joints	3DP MultiJet Fusion / 3DP DMLS	1	380 / 1	12	24
Foundation	3DP formwork / Casting	2	3	4	4
Skin	Add-on 3DP / Cutting	2	75	15	8

TABLE 1. Overview of the different manufacturing processes and production logistics

15 3DP formwork is submerged into sand for casting and after the hydration of concrete is removed by applying heat
16 The cuts needed are nested with an algorithm to make the most of the full length of the bamboo poles and marked with a collaborative robot
17 Workflow and data flow diagram for the Digital Bamboo
18 Assembly sequence

Digital Bamboo Kladeftira, Leschok, Skevaki, Tanadini, Ohlbrock, D'Acunto, Dillenburger

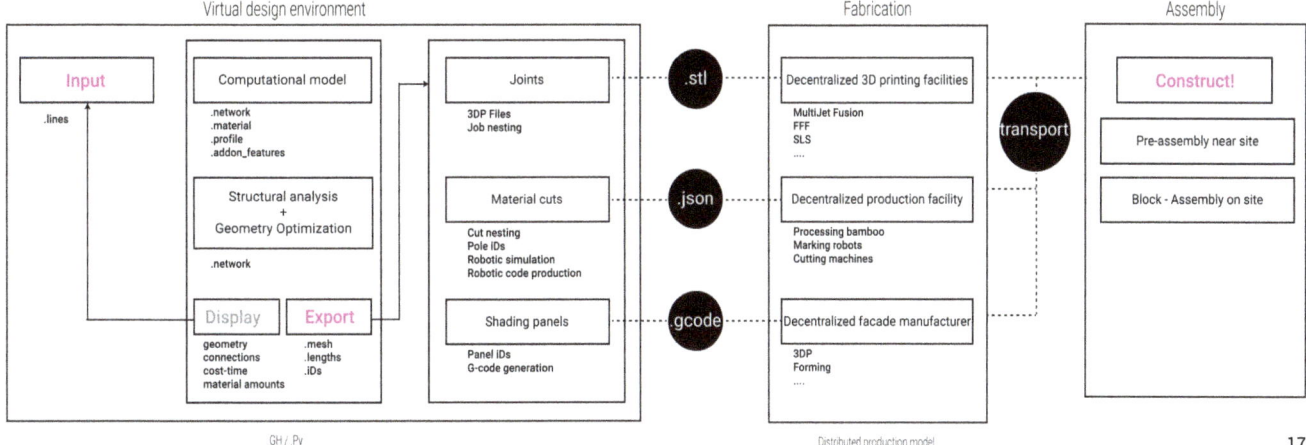

PRODUCTION AND ASSEMBLY LOGISTICS
Production Logistics

The computational tools developed ensure a smooth design-to-production workflow (Figure 17). The use of multiple digital fabrication methods to manufacture all the parts of the structure allows an efficient production in a short time while maintaining high precision. The most crucial part of the assembly process is the matching of joints and bamboo poles. The joints are equipped already in the design stage with a labeling system in the limbs that features the joint ID and the pole ID that corresponds to the specific limb. The system is simple and, therefore, can be assembled by anyone like a puzzle: starting from any one joint you match the IDs embedded and start growing outwards until an entire module is complete.

All the nylon joints are commissioned and produced with the MJF technology in an external facility. Each print job takes approximately 9 to 12 hours to complete and about 2 hours to retrieve and clean the excess powder from the parts as show in Table 1. The production can be distributed to multiple printers for faster production. With each job multiple parts can be printed at a time without compromising the quality expediting the production for large structures.

Overall, the production time with AM is reduced compared to conventional manufacturing where each joint is produced individually. However, the true benefit lies in the significant reduction of human labor and post-processing. All parts of the digital manufacturing chain are automated: the parts take into account the constraints of the AM methods (wall thickness, maximum dimensions, aspect ratio, resolution) and are sent to the fabrication facility nested and ready to print. In the remote location, where printing takes place, someone loads the print job and subsequently unloads the parts to the cleaning basket. Then, the joints are shipped to site.

In order to efficiently cut over 900 bamboo poles in their respective length, reduce imprecision and avoid human errors, we leveraged the precision of a collaborative robot to mark all the poles within 10 hours. We created a robotic marking station consisting of a UR10, a custom table that can host six 3-meter-long bamboo poles and a 3D printed pen-holder end-effector (Figure 16). Six poles are marked simultaneously

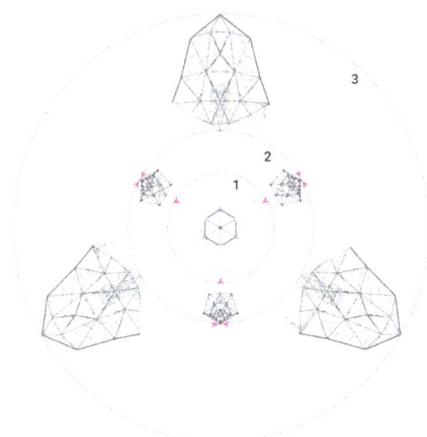
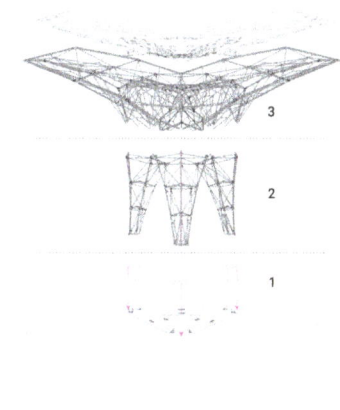

19 The Digital Bamboo exhibited at the courtyard of the ZAZ Bellerive Museum in Zurich in 2020

by the robot. In parallel, a human collaborates by labeling the individual elements with a serialized tag ID system. The bamboo is then cut at the marks using a chop saw. Two notches are produced on each bamboo at both ends in order to match the detail of the 3DP connections. For structural reasons a constant inner core thickness is ensured (8 mm after notching). The notch is not a typical connection type and requires minimal processing with machinery to be carved out. However, all notches are identical and performed at a similar distance to the end of the pole. Therefore, this step can easily be automated even with low cost machines made to carve wood-like materials (MacDonald et al. 2019). As the connection system is rotation-free, the orientation of the notches on one end does not need to align with the other end of the pole, as one can rotate the bamboo to find the most convenient orientation for assembly.

Discretization and On-site Assembly Sequence

The assembly of the structure happens in a two-step process. Six distinct modules (3 columns and 3 cantilevers) are pre-assembled on-site. In a second step, they are mounted on the base with the help of a small crane. Then the structure is post-tensioned. This strategy minimizes the volume that is transported as all the materials are shipped before any assembly takes place. On-site time is reduced as the larger modules are light and can be mounted very quickly. The second step lasts only half a day with the help of a small crane (Figure 18). The process is repeated vice versa for disassembly.

RESULTS, DISCUSSION AND FUTURE WORK

The notion of a digital design-to-production process is becoming increasingly relevant in architectural research and practice if we are to achieve a more efficient and sustainable model of construction that harvests the opportunities of digital fabrication. The Digital Bamboo brings forward a new vocabulary of space-frame structures merging innovative manufacturing methods with a bio-sourced material. It has unique aesthetics that speak of an amalgam of the digital and the vernacular.

The structure shelters a space of 40 m^2 with a weight of only 5 kg/m^2. In existing buildings, material quantities typically range from 500 to 1500 kg/m^2 (Wolf 2014). Although these quantities include many other layers like services and facade systems, they are at the lower end a hundred times heavier than the proposed system. Other researchers investigating lightweight building systems with bio-sourced materials enabled by digital fabrication have managed to reduce the material weight of a similar pavilion structure to a little over 30 kg/m^2 (Alvarez et al. 2020), more than six times heavier than Digital Bamboo. The ultralight character of the Digital Bamboo relies not only on the choice of bamboo but also on the development of the structural, connection system and the materiality of the connections. The ability to digitally fabricate all the components of the structure guarantees the minimization of excessive material use and the application of each material only where it is truly needed.

The resources and fabrication time are kept to a bare minimum. Space frames are one of the most efficient ways to scale up AM by fabricating only the connections that comprise a small fraction of the entire structure. In parallel, high-embodied-carbon materials are used sparsely only where needed as shown with the single stainless-steel node and the add-on 3DP polymer to stiffen and form the textile panels. In fact, for the production of the whole pavilion, there was almost zero waste except for the three millimeter-thick 3DP formwork shells for the concrete foundation that can be recycled. The cut-outs of textile and bamboo were scarce as the pieces needed for the pavilion were nested via an optimization algorithm ensuring almost no leftover pieces. The steel base underneath the structure was cut from excess material that would have been discarded.

There are significant benefits to the 3DP connection system that was developed. All the logistic information is embedded in the 3DP parts, which are designed to facilitate human assembly. During the assembly process several people with no expertise or experience of the system helped intuitively in the build-up of the structure by matching the information on the different pieces like a big puzzle. As complexity increases in the construction industry and digital fabrication methods are increasingly more sophisticated, we believe that creating smart systems such as this one will help in their wider adoption.

Furthermore, the system can be entirely disassembled as all parts are attached mechanically with non-permanent fastenings. This was an overarching principle during the development phase so as to ensure that the pavilion can be transported in minimal volume and re-assembled elsewhere but also that all materials can be recycled or reused at the end of life of the structure. The material used for the joints was chosen because of its strength and isotropic behavior. Alternatives such as bioplastics are gaining popularity, however, current bio-composite filaments are printable with FFF processes and thus do not deliver the required strength while they have anisotropic behavior. This would add additional constraints as joints are highly three-dimensional objects and it would affect both the design and the fabrication of the elements. This project focuses on the ability to disassemble all the distinct materials that can therefore be treated accordingly at the end of life.

We see great potential in this combinatorial scheme of multiple digital fabrication technologies. We believe that the future of architecture and construction will be shaped by the new technologies available to the point that their coexistence in future buildings will be seamless. In this respect, we also envision that the presented construction system can be enhanced by other technologies in the prefabrication state, like the use of augmented and mixed reality devices for precision and quality control during assembly or the use of mobile robotic units and UAVs in a prefabrication cell to create bespoke prefabricated space frame modules. These technologies can be integrated in the process by adding tracking and localization tags in the printed joints the geometry of which can be modified according to assembly constraints at no manufacturing expense.

CONCLUSION

This paper presented a holistic approach for the development, fabrication, and assembly of a novel lightweight structure that is entirely based on digital fabrication technologies (Figure 19). This is possible through interdisciplinary collaboration between specialized architects and structural engineers that use computational design and analytical tools from conception to realization in order to control and harvest the benefits of digital fabrication. The methods presented describe the design and rationale for all the digital layers of the structure, as well as their fabrication with suitable digital manufacturing methods.

Digital Bamboo provides an insight into how we can build with bio-based materials without compromising precision, quality, and aesthetics. It showcases how one can design with uncertainties and imprecision that are inherent to standardless materials and use this knowledge as a driving force for innovation towards sustainable and scalable construction.

ACKNOWLEDGEMENTS

The authors acknowledge the students of the MAS DFAB '19-20 for their contribution to the initial design studies, Yael Ifrah, our collaboration partner HS HI-Tech, and our sponsors that enabled the research: HP Inc., SGSolution AG, AF Fercher AG, Holcim, Abuma Gmbh, and ZAZ Bellerive. We are thankful for the exhibition hosts ZAZ Bellerive and ECC Italy. The research was supported by the ETH Zurich #IPG 02-112019 and the NCCR Digital Fabrication, funded by the Swiss National Science Foundation (NCCR Digital Fabrication Agreement #51NF40-141853).

NOTES

The bamboo used for this research was left untreated against degradation because of the temporary nature of the structure. We acknowledge that if used untreated for an uncovered permanent structure it might affect its longevity. In such cases additional impregnation with chemical means or other methods that prevent degradation because of environmental, external factors or insects might be appropriate.

REFERENCES

Alvarez, Martín, Hans Jakob Wagner, Abel Groenewolt, Oliver David Krieg, Ondrej Kyjanek, Daniel Sonntag, Simon Bechert, Lotte Aldinger, Jan Knippers, and Achim Menges. 2020. "Buga Wood Pavilion." *Architecture Research Building* April: 150–61.

Amtsberg, Felix, and Felix Raspall. 2018. "BAMBOO^3." In *Learning, Adapting and Prototyping, Proceedings of the 23rd International Conference of the Association for Computer-Aided Architectural Design Research in Asia (CAADRIA) 2018*. Hong Kong. 245–254.

Bernhard, Mathias, and Benjamin Dillenburger. 2019. *Axolotl*. https://www.food4rhino.com/app/axolotl.

Cabeza, Luisa F., Laura Boquera, Marta Chàfer, and David Vérez. 2021. "Embodied Energy and Embodied Carbon of Structural Building Materials: Worldwide Progress and Barriers through Literature Map Analysis." *Energy and Buildings* 231: 110612.

Crolla, Kristof, Nicholas Williams, Manuel Muehlbauer, and Jane Burry. 2017. "Smartnodes Pavilion." In *Protocols, Flows and Glitches: 22nd International Conference on Computer-Aided Architectural Design Research in Asia (CAADRIA)*. 467–76.

De Wolf, Catherine. 2014. "Material Quantities in Building Structures and Their Environmental Impact." MSc Thesis, Massachusetts Institute of Technology. 92. http://dspace.mit.edu/handle/1721.1/91298.

Dunkelberg, Klaus. 1992. *IL 31, Bambus-Bamboo: Bambus Als Baustoff; Bauen Mit Pflanzlichen Stäben*. Stuttgart: Krämer.

Fuller, Richard Buckminster. 1954. Building Construction. US Patent 2,682,235, filed December 12, 1951, issued June 29, 1954.

Galjaard, S., S. Hofman, N. Perry, and S. Ren. 2015. "Optimizing Structural Building Elements in Metal by using Additive Manufacturing." In *Proceedings of the International Association for Shell and Spatial Structures (IASS) Symposium 2015, Future Visions*. Amsterdam.

Kladeftira, Marirena, Matthias Leschok, Eleni Skevaki, and Benjamin Dillenburger. 2021. "Redefining Polyhedral Space Through 3D Printing." *AAG Advances in Architectural Geometry 2020*: 306–29.

MacDonald, Katie, Kyle Schumann, and Jonas Hauptman. 2019. "Digital Fabrication of Standardless Materials." In *ACADIA 19: Ubiquity and Autonomy; Paper Proceedings of the 39th Annual Conference of the Association for Computer Aided Design in Architecture*. 266–75.

Mengeringhausen, Max. 1974. Process and apparatus for the production of connectors for space frame works or the like. US patent 3826584a, filed January 31, 1972, issued July 30, 1974.

Moses, A, O Simon, and E Oke. 2015. "Comparative Analysis Of The Tensile Strength Of Bamboo And Reinforcement Steel Bars As Structural Member In Building Construction." *Comparative Analysis of The Tensile Strength Of Bamboo and Reinforcement Steel Bars as Structural Member in Building Construction* 4 (10): 47–52.

Paola, Francesco Di, and Andrea Mercurio. 2020. "Design and Digital Fabrication of a Parametric Joint for Bamboo Sustainable Structures." *Advances in Intelligent Systems and Computing* 975: 180–89.

Qi, Yue, Ruqing Zhong, Benjamin Kaiser, Long Nguyen, Hans Jakob Wagner, Alexander Verl, and Achim Menges. 2021. "Working with Uncertainties: An Adaptive Fabrication Workflow for Bamboo Structures." In *Proceedings of the 2020 DigitalFUTURES. Singapore: Springer*, edited by P. F. Yuan, J. Yao, C. Yan, X. Wang, and N. Leach. 265–79. https://doi.org/10.1007/978-981-33-4400-6_25.

Raspall, Felix, Carlos Banon, and Jenn Chong Tay. 2019. "AIRTABLE. Stainless Steel Printing for Functional Space Frames." In *Computer-Aided Architectural Design Research in Asia (CAADRIA) 2019*, vol. 1. 113–122.

Sumi, Christian, and Marianne Burkhalter. 2018. *Konrad Wachsmann and the Grapevine Structure*. Zurich: Park Books.

Tanadini, Davide, Patrick Ole Ohlbrock, Marirena Kladeftira, Matthias Leschok, Eleni Skevaki, Benjamin Dillenburger, and Pierluigi D'Acunto. 2022. "Exploring the potential of equilibrium-based methods in additive manufacturing: the Digital Bamboo pavilion." In *Proceedings of the IASS 2022 Symposium*.

Van Mele, Tom et al. *COMPAS: A framework for computational research in architecture and structures*. V. 1.0. ETH Zurich. 2021.

Wassenhove, Romain van, Lars De Laet, and Anastasios P. Vassilopoulos. 2021. "A 3D Printed Bio-Composite Removable Connection System for Bamboo Spatial Structures." *Composite Structures* 269 (December 2020): 114047.

Wohlers, Terry T., Ian Campbell, Olaf Diegel, and Joseph Kowen. 2018. "Wohlers Report 2018: 3D Printing and Additive Manufacturing State of the Industry; Annual Worldwide Progress Report." Washington, DC: Wohlers Associates.

IMAGE CREDITS

Figure 11,13,19: ©Andrei Jipa
All other drawings and images by the authors.

Marirena Kladeftira is a Doctoral candidate at the chair of Digital Building Technologies, ETH Zurich and affiliated with the NCCR - Digital Fabrication, where she is leading the research on 3D printed joints for non-standard structures. She holds a Dipl. Arch. Eng. from the NTU of Athens and an MAS in digital fabrication from ETH Zurich.

Matthias Leschok is a PhD researcher at Digital Building Technologies, ETH Zurich. He holds a MAS in Architecture and Digital Fabrication from ETH Zurich and a Dipl. Ing. in architecture from KIT. He is also co-founder and COO of the startup SAEKI Robotics AG.

Eleni Skevaki is a Research Assistant at the chair of Digital Building Technologies, ETH Zurich. She studied Architecture at the NTU of Athens and ENSA Paris Malaquais before joining the MAS DFAB at ETH Zurich. She has participated in various large-scale research projects that employ digital fabrication for concrete and other materials.

Davide Tanadini is a lecturer and PhD student at the Chair of Structural Design at ETH Zurich. His research focuses on applications of the theory of plasticity and graphic statics on timber structures and timber joints, and their implementation in digital fabrication and robotic assembly. Davide graduated from ETH Zurich as a civil engineer in January 2018.

Patrick Ole Ohlbrock holds a degree in Civil Engineering since September 2013. In 2020, he obtained a PhD with distinction from ETH Zurich. He is currently a Postdoctoral Researcher and Lecturer at the Chair of Structural Design at the ETH Zurich and in parallel works as a consulting structural designer.

Pierluigi D'Acunto graduated in Building Engineering-Architecture from the University of Pisa in 2007 and received a Master of Architecture from the AA School of Architecture in London in 2012. In 2018, he completed his PhD with distinction at ETH Zurich. Pierluigi is currently Assistant Professor of Structural Design at the Technical University of Munich.

Benjamin Dillenburger is Professor at the Institute of Technology in Architecture at the Department of Architecture, ETH Zurich. He is leading the research group Digital Building Technologies, which investigates computational design and digital fabrication with a focus on large scale additive manufacturing in architecture.

Curved-crease Folding of Bending-active Plates as Formwork

A Reusable System for Shaping Corrugated Concrete Shell Structures

Lotte Scheder-Bieschin
BRG ETH Zurich

Tom Van Mele
BRG ETH Zurich

Philippe Block
BRG ETH Zurich

1   Unfolding of the curved crease-folded bending-active formwork

ABSTRACT

The critical environmental impact of the concrete construction industry demands material-efficient structures and construction methods applicable to low-tech and high-tech contexts. Thin, structurally-informed shells with corrugations as stiffeners are material-efficient solutions as they gain their strength through their non-standard geometry. However, their bottleneck lies in their costly and wasteful formwork systems. This research introduces curved-crease folding (CCF) of bending-active plates as a flexible, lightweight, and reusable formwork system for shaping corrugated concrete shell structures. CCF is extended to an initially closed configuration that unfolds initially-planar bending-active strips into a 3D formwork when actuated on-site. The curved creases control the shape and structurally stiffen the formwork shaping a concrete shell structure with stiffening corrugations.

The paper concentrates on the system design covering theoretical, computational, and fabrication aspects. The primary focus for the computational methods lies in implementing and extending the reflection method for the initially closed CCF—for the materialization method in textile hinge solutions for the curved creases. The approach is demonstrated with a small-scale proof-of-concept prototype. The proposed system offers a material-efficient, self-supporting formwork solution that can be flat-packed for transport, rapidly erected on-site through the actuation of the CCF mechanism, and reused after concreting and decentering. The proposed formwork's geometry is not sensitive to stiffness variations, as it is constrained by the CCF. Furthermore, the CCF makes the formwork independent of advanced machine technology, thus allowing for the construction of complex customized shapes also in low-tech contexts.

INTRODUCTION

Facing the critical impact of the AEC sector on the climate crisis, resource depletion, and waste production, we must decarbonize not only building structures, but also their construction methods. Moreover, for a global impact, sustainable construction solutions must be broadly applicable and accessible not only in high-tech but also low-tech construction contexts. Innovation is particularly important for the widespread concrete construction industry with its large impact on global CO_2 emissions (Lehne and Preston 2018). Rethinking its conventional design and construction methods that rely on redundant material placement in standard slab and beam typologies offers great potential. Instead, thin funicular shell structures with corrugations as stiffeners offer material-efficient solutions as they gain strength through their structurally-informed geometry (Block et al. 2020). However, shaping such non-standard concrete structures poses a bottleneck. Conventional formwork systems for custom shapes are typically high in cost, material, and waste and limited to high-tech machining, such as CNC-milled timber or foam with GFRP coating (Kudless et al. 2020) or plywood plates mounted onto waffle substructures (Peri Group 2020).

Alternative formwork solutions proposed in research range from additive manufacturing to flexible formworks (Jipa et al. 2019; Veenendaal et al. 2011). Flexible formworks base their efficiency on the structurally-informed geometry of their form-active structural system. Such systems are made from fabric shuttering together with a tensile cable-net falsework (Mendez et al. 2019; Popescu et al. 2020) or a bending-active falsework (Cuvilliers et al. 2017; Scheder-Bieschin et al. 2022). Shuttering and falsework are installed in consecutive steps and are not walkable.

The research presented aims to introduce a flexible formwork system of combined formwork and falsework. It proposes using curved-crease folding (CCF) of bending-active plates as a formwork system for shaping corrugated concrete shell structures. The combination of CCF and active bending is a bilateral mechanism where one actuates and amplifies the other. Planar strips are connected alongside curved-crease hinges and fold or unfold into a globally doubly-curved geometry serving as formwork (Figure 2a, b, c). The curved creases control the shape and structurally stiffen the formwork, consecutively shaping the concrete shell structure with stiffening corrugations (Figure 2d).

The proposed system would offer a material-efficient formwork solution that can be flat-packed for transport, quickly erected on-site, immediately walkable, and reusable. The primary material is plywood, typical for conventional formworks. Since its sophistication lies in the geometric principle of the CCF rather

2a

2b

2c

2d

2   Curved-crease-folded bending-active formwork in its closed state (a); half-unfolded state (b); and fully-unfolded state (c); and the resulting corrugated concrete shell structure (d) shown on the proof-of-concept prototype

than in advanced machine technology, it allows for custom shapes in low-tech contexts. The design development of such a construction system must be approached holistically, integrating aesthetic, structural, and fabrication opportunities and constraints.

This paper focuses on the novel system design with its underlying concepts and proof-of-concept workflow from design to fabrication. The core of the system design is the CCF's shape control and stiffening for the formwork and resulting shell. Further, this research introduces an extended CCF formulation that starts from a closed configuration and unfolds into a 3D geometry. In the proof-of-concept pipeline, the paper primarily addresses the computational methods of the extended, initially closed CCF configuration in the COMPAS framework (Van Mele et al. 2022), as well as the materialization challenge of the curved creases with a textile hinge strategy. The process is demonstrated through a small-scale prototype.

BACKGROUND

Curved-crease folding (CCF) is a popular research stream for its intriguing design space and sophisticated computational modeling methods. Huffman (1976), Duncan and Duncan (1982), and Fuchs and Tabachnikov (1999) established theorems on the behavior of CCF based on differential geometric analysis and introduced the method of reflection for modeling the CCF based on basic geometric principles. More recently, there have been many approaches to advanced computational simulation and optimization of CCF (Kilian et al. 2008; Bhooshan et al. 2015; Rabinovich et al. 2019). To simulate mechanical behavior, finite element analysis (FEA) simulation with the commercial software SOFiSTiK allows computing internal bending stresses for curved plates based on the unstressed state so that the system equilibrates with non-linear, third-order analysis away from the ideal target towards its deformed equilibrium shape (Bellmann 2017).

CCF finds application in non-structural architectural installations as kinetic façades or as foldable utility objects because of its simple fabrication of elegant shape from developable strips, its amplified actuation behavior, and its flat-packed transportability (Bhooshan et al. 2015; Choma 2021; Körner et al. 2016; Frommeld 2011). Bhooshan et al. (2015) utilized CCF as molds for shaping concrete for small prefabricated nodal segments assembled into a rib structure. These examples are materialized with thin sheets. Recent examples extend the use of CCF with thicker plates to structural applications (Maleczek et al. 2020; Basnak et al. 2020). At such scales, the plates are considered bending-active, and the curved creases are materialized with hinges between non-continuous plates.

Active bending is a method to elastically bend slender planar elements, such as splines or plates, into curved shapes without formworks (Lienhard 2014). It offers easy deployment, advantageous packaging, and lightweightness. It allows constructing globally doubly-curved structures from planar plates with reduced waste, like in the historical example of Fuller (1959) and recent research (Schleicher and La Manga 2016). Bending-active structures suffer the dilemma of flexibility for forming and stiffness for structural performance to withstand external loads. Stiffening strategies are built-in tension cable elements (Takahashi 2016) or could be curved-crease folds resulting in corrugated sections that increase the moment of inertia as in conventional folded plate structures.

Curved folded plate structures have been constructed either from actively or passively bent separate plates whose edges are then interlocked in the curved state (Buri et al. 2011; Correa et al. 2016; Robeller et al. 2014) or by the assembly of the plates in their flat state and then actively bending and folding, actuated and shape-controlled by the curve-folding mechanism (Maleczek et al. 2020). The latter is fast, practical, and reversible. However, the foldability poses geometric limitations in the design space. This folding mechanism requires flexible hinges along the curved folds, such as textile hinges (Maleczek et al. 2020; Basnak et al. 2020). The separate plates enable starting

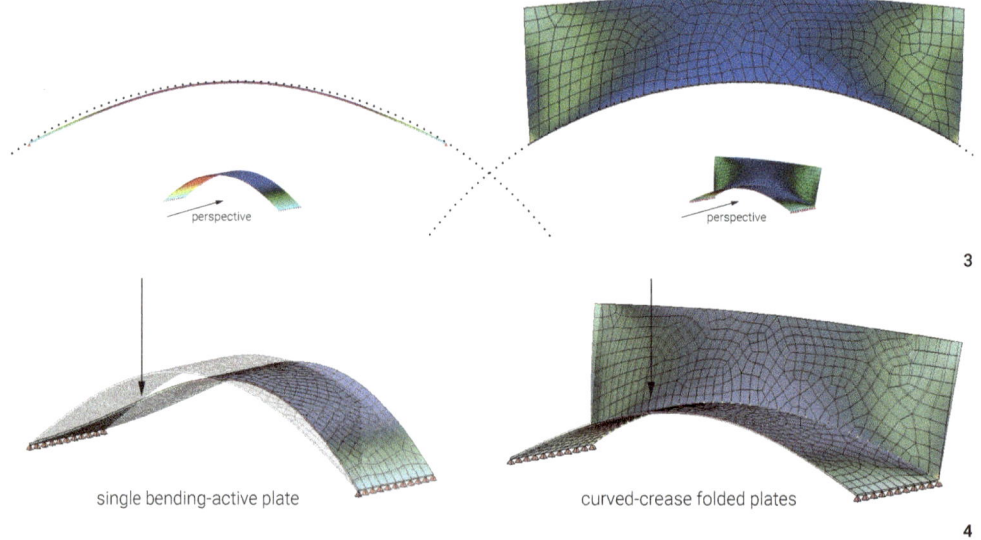

3   Shape control comparison of a single bending-active plate and curved-crease folded plates with its crease curvature designed for constant bending curvature to a circular arc (dotted line) with FEA

4   Stiffness comparison of a single bending-active plate to curved-crease folded plates by deformation under asymmetric wet-concrete load with FEA

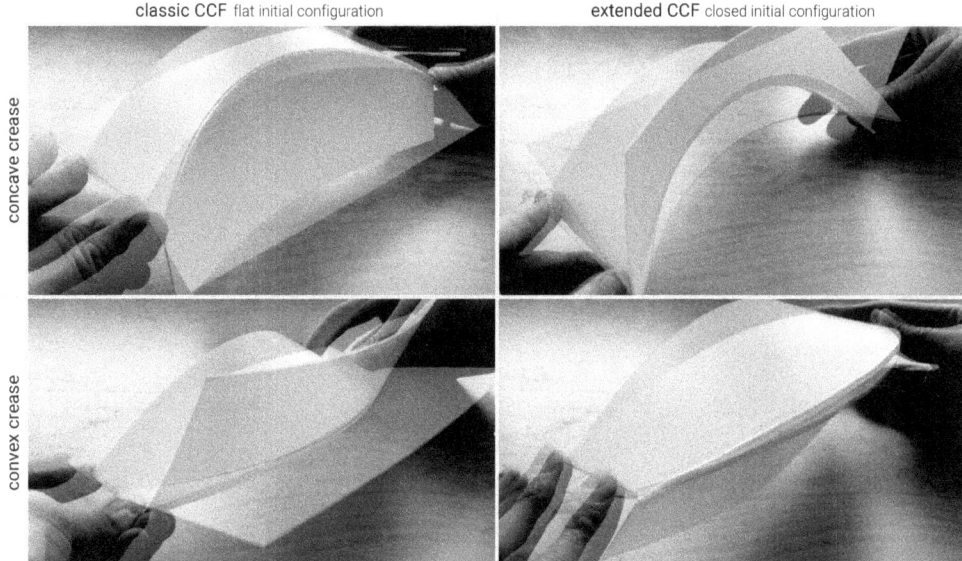

5 The four configurations of CCF in its classic and extended version with concave and convex creases showcased in paper models overlaying the starting and actuated positions

from a closed configuration as identified in a single crease in the prototype of Basnak et al. (2020). For flat crease patterns, the folding of nonzero crease patterns to avoid self-collision is of particular interest (Ku and Demaine 2016).

With the discussed potentials and limitations, the applicability of an extended CCF mechanism with bending-active plates and textile hinges as self-supporting formwork and falsework are investigated in the presented research.

SYSTEM DESIGN WITH UNDERLYING PRINCIPLES

The system design of the formwork and the resulting shell relies primarily on the shape control of CCF and the stiffening effect of CCF that translates to the concrete shell. Built-in restraining cables achieve further stiffening and actuate the system's unfolding. The system designs are based on two types of CCF, the classic version and an extended variation that offer flat-packed formworks that unfold like an accordion.

Shape Control by Curved Creases of the Formwork

The curved creases control the equilibrium shape of the bending-active plates. Naturally, a single plate of constant stiffness actively bends into an Euler Elastica curve (Lienhard 2014). In CCF, the curvature of a curved crease controls the curvature of its adjacent plates; thus, a wide range of shapes can be achieved. In the example of Figure 3, the crease's curvature is computed with the reflection method such that the curvature of the plate satisfies a perfect arc. It is demonstrated by a form-finding simulation with the FEA-software SOFiSTiK, modeling the mechanical bending resistance of the plates. The shape control only holds true in the proximity of the curved crease. Thus, the plates must be controlled by creases on both sides.

Stiffening by Curved Creases of the Formwork and Shell

A further advantage of the CCF is the stiffening effect on the formwork. The formwork must carry asymmetric loading of the wet-concrete weight. Compared to a single plate, the

6 Closed flat-packed formwork (left) and the globally doubly-curved CCF formwork (right) after unfolding like an accordion, shown on prototype design geometry

adjacent plate acts as a restraining diaphragm wall such that deformations decrease and stability increases dramatically, as demonstrated with the FE simulation (Figure 4).

The CCF shape results in a corrugated shell in concrete and the stiffening scheme is translated to the shell structure. The corrugations increase the static height, i.e., the moment of inertia of the section, and if the concrete shell is form-found around an envelope of thrust lines for asymmetric loads, it can result in a funicular, compression-only concrete structure, not in need of structural reinforcement (Block et al. 2020).

Concept of Classic CCF and the Initially Closed CCF

Curved-crease folding starts from a flat, continuous sheet or thicker plates cut and re-joined with a hinge. During actuation, as the adjacent plates fold towards each other, they bend with single curvature in opposing directions. As an extension, this research introduces an extended CCF that starts from a flat-packed, initially closed configuration. The plates are non-continuous and mirrored along their creases. Contrary to the classic CCF, the plates unfold away from each other and the system opens up. The plates' curvature is oriented away from each other for a concave crease and towards each other for a convex crease. Figure 5 shows the four basic types of the classic and extended CCF with concave and convex creases, respectively.

The classic CCF offers the advantage of being flat and the downside of a large format and is thus considered more suitable for prefabrication. In contrast, the extended CCF offers the advantage that its flat-packed pile is practical for transportation and opens up like an accordion into 3D geometry (Figure 6), hence considered more suitable for in-situ applications.

Further, introducing the extended CCF broadens the constrained design space of the classic CCF. However, it demands modified computational modeling methods and poses the challenge of the hinge materialization in the closed, obstructed state.

COMPUTATIONAL MODELING METHODS
Computational Workflow

The proposed construction system requires a computational modeling framework that simultaneously integrates the geometric, structural, and fabrication constraints for both the formwork and the resulting shell. The computational methods are implemented in COMPAS, an open-source framework for research in AEC (Van Mele et al. 2022). The design workflow is shown based on a design that serves for the proof-of-concept prototype. The workflow (Figure 7) commences with the funicular form finding of the concrete shell that defines the target geometry for the formwork. The design exploration toward this target geometry for the CCF formwork is implemented with the geometric reflection method. The CCF geometry is verified and analyzed with a form-finding simulation through a COMPAS interface with SOFiSTiK. Finally, the cutting pattern for the textile and plate strips is generated and exported as fabrication data from COMPAS. This paper focuses on the reflection method for the closed CCF; the other workflow steps will be discussed in a complementary paper.

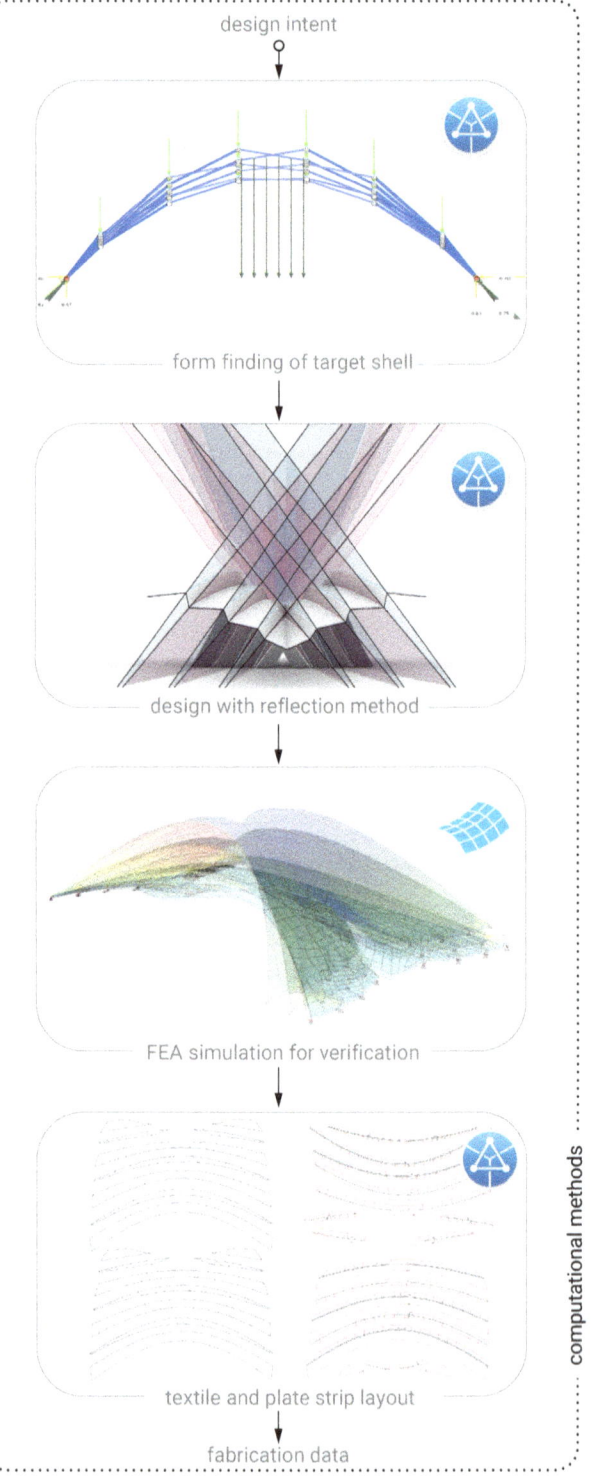

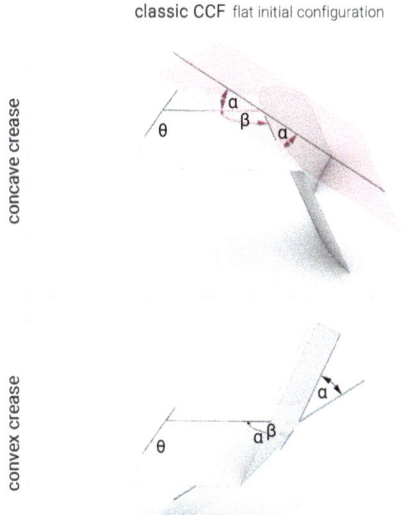
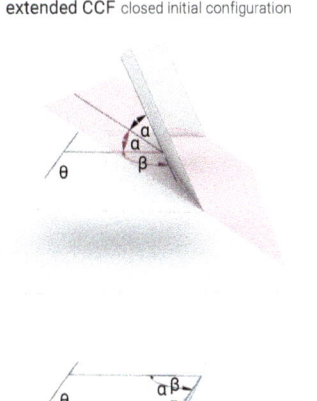
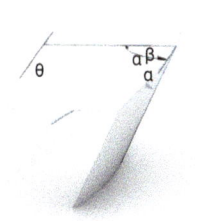

7 Computational modeling workflow

8 Reflection method for all four configurations of CCF with the ruling line, osculating plane (pink for concave, blue for convex), α reflection angle, β sector angle, θ dihedral angle

9 Design rules for all four configurations and their flipped state with a labeling convention of "+" for positive and "−" for negative curvature

Method of Reflection

The theorems for the reflection method and the differential geometry for the classic CCF can be applied with modification to the closed CCF. They define the interrelationships among angles associated with creases and curvature definitions of the developable surfaces. Figure 8 shows all four configurations for cylindrical surfaces with parallel ruling lines. The ruling lines are where all points on the surface share the same tangent plane and indicate the direction of zero Gaussian curvature. These are reflected along osculating planes with the reflection angle α. The sector angle β that defines the crease curvature correlates with the dihedral angle θ that defines the plate bending curvature. In the classic flat case, the continuous plate, which is a reflection of the initial strip, is reflected along the osculating plane. In contrast, in the extended closed case, the identical plate is reflected along the osculating plane as they are directly mirrored. The reflection method is limited to planar creases, resulting in a constant reflection angle over the entire crease; however, the angle normal to the crease varies unless for special cases.

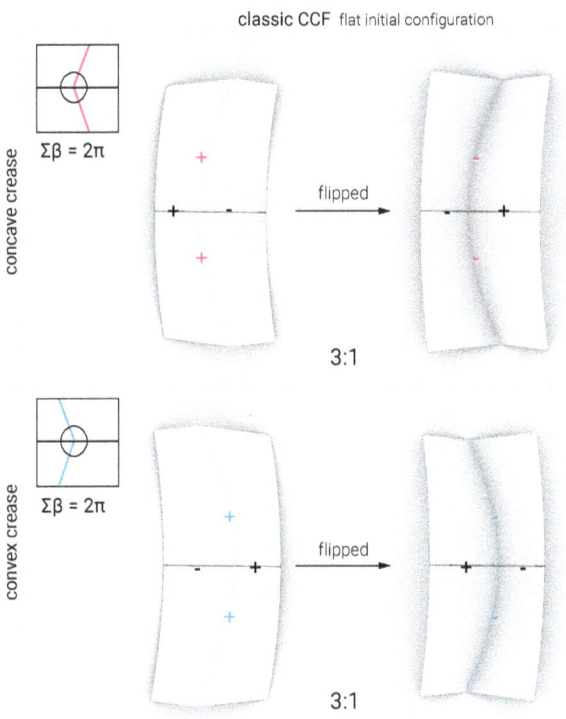
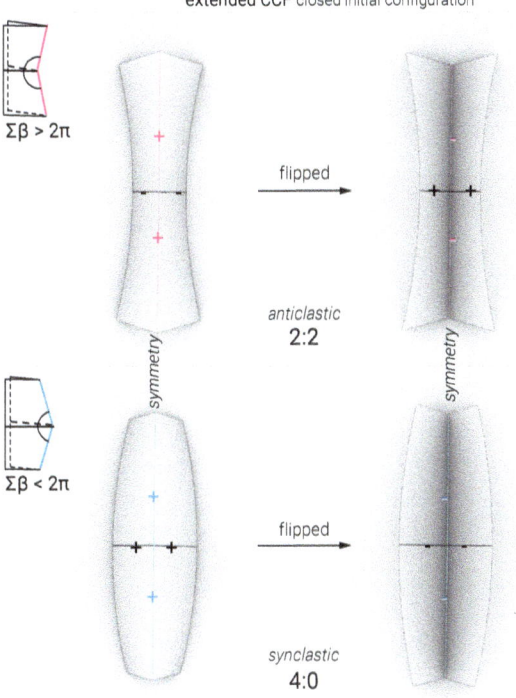

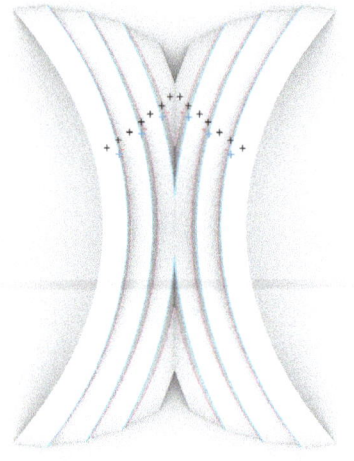

10 Application of the rules to the prototype design with osculating planes of convex (blue +) and concave (pink −) creases and curvature of plates (black +)

10

The reflection method is implemented for both the classic and the extended CCF in the COMPAS framework with planar quad meshes. Further, the implementation includes the simulation of the folding and unfolding process.

Design Rules with the Reflection Method

In the classic CCF of developable continuous plates, the sum of all plane sector angles Σβ remains constant at exactly 2π at all times during folding (Figure 9 small diagram). However, in the extended CCF, the plates are cut out and re-joined with a sum of all plane sector angles Σβ greater than 2π for the concave crease and smaller than 2π for the convex crease. These sector angles also remain constant at all times during unfolding.

Huffman (1976) introduced a labeling convention with "+" for positive and "−" for negative curvature. This is useful for analyzing the resulting shape and the possibilities of combining multiple strips. In the classic CCF, the crease and ruling lines must always be in a ratio of 3 + to 1 − and vice versa. This results in shapes around the crease that are neither anti- nor synclastic, as it still is formed from a continuous developable surface. This research extends the convention for the closed CCF; for the concave crease, the ratio is 2 + and 2 − alternating, and for the convex crease, the ratio is either 4 + or 4 -. This leads to anticlastic and synclastic curvatures, respectively, aligning with the sum of sector angles (Figure 9 render diagrams). In the plan view, the closed CCF is symmetric along its crease due to the reflection of the identical plate, whereas the classic CCF is antisymmetric due to the reflection of the reflected side.

11 Materialized open and closed CCF with classic (a) and extended (b) initial configuration, and cable-actuated folded (c) and unfolded (d) 3D geometry

12 Textile hinge strategies with one-sided, sandwiched, and two-sided/wrapped laminated fabric (blue) with sewing seam at the textile hinge position (pink), with normal (n) or tangential (t) acting force in glue interface (shown for one side only, but symmetrical)

13 Integrated cable attachment loop into textile hinge during unfolding

11

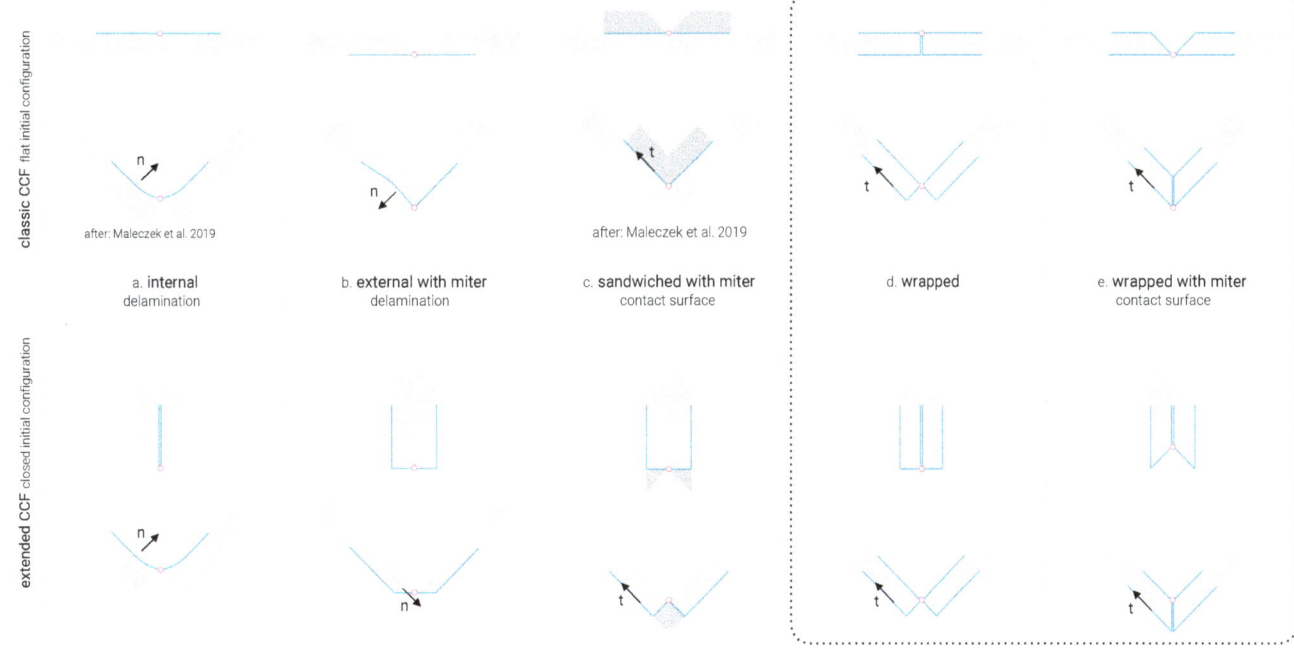

12

| a. internal delamination | b. external with miter delamination | c. sandwiched with miter contact surface | d. wrapped | e. wrapped with miter contact surface |

These rules are applied to the computational design of the physical prototype (Figure 10). The design consists purely of extended CCF creases and alternates in ridges along concave creases and valleys along convex creases. Its global shape is defined by variable osculating plane inclination and distance, where each orientation affects the subsequent creases. The osculating planes are rotated around parallel axes and translated such that the boundary and rise curve inwards, resulting in a globally doubly-curved geometry. In elevation, this results in variable corrugation widths wider towards the supports and almost acute towards the ends.

MATERIALIZATION METHODS

The aim of this research is to develop a practical and reusable formwork system applicable to a wide range of contexts. The main challenge is the materialization of the curved hinges, using broadly available materials, with a fabrication workflow that is feasible in low-tech as well as high-tech contexts, adjustable to the availability of labor or technology. This work predominantly focuses on a low-tech approach to oppose the common state-of-the-art high-tech manufacturing strategies for custom shapes. It is demonstrated on the proof-of-concept prototype that materializes the computationally designed proposal.

Textile Hinges

The curved creases are materialized with textile hinge strips laminated to plates. The hinge strategy is developed to prevent delamination, obtain a hinge position compatible with a nonzero thickness model, and achieve a joining strategy for the closed CCF configuration. As the inner side of the plates in the closed CCF is not accessible during the prefabrication,

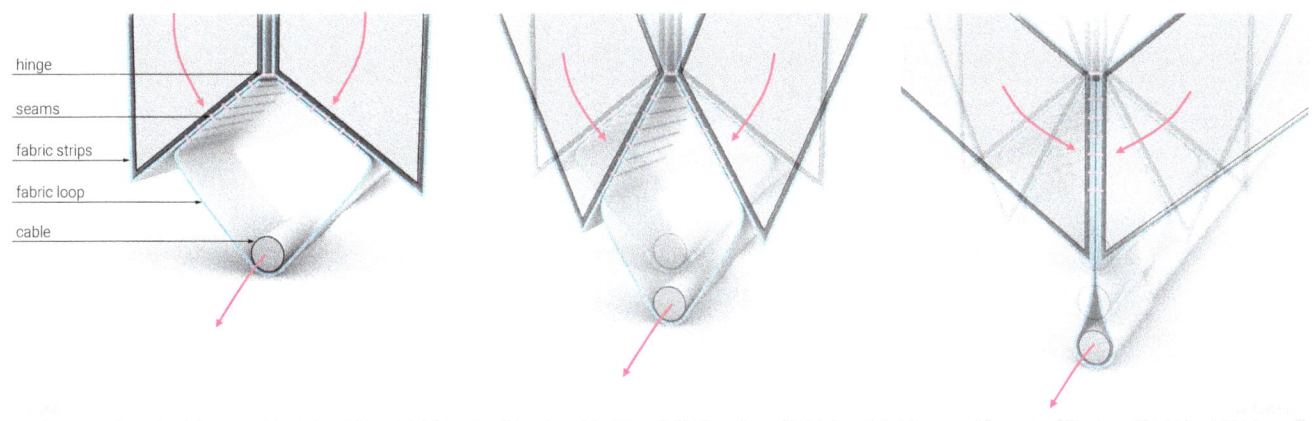

13

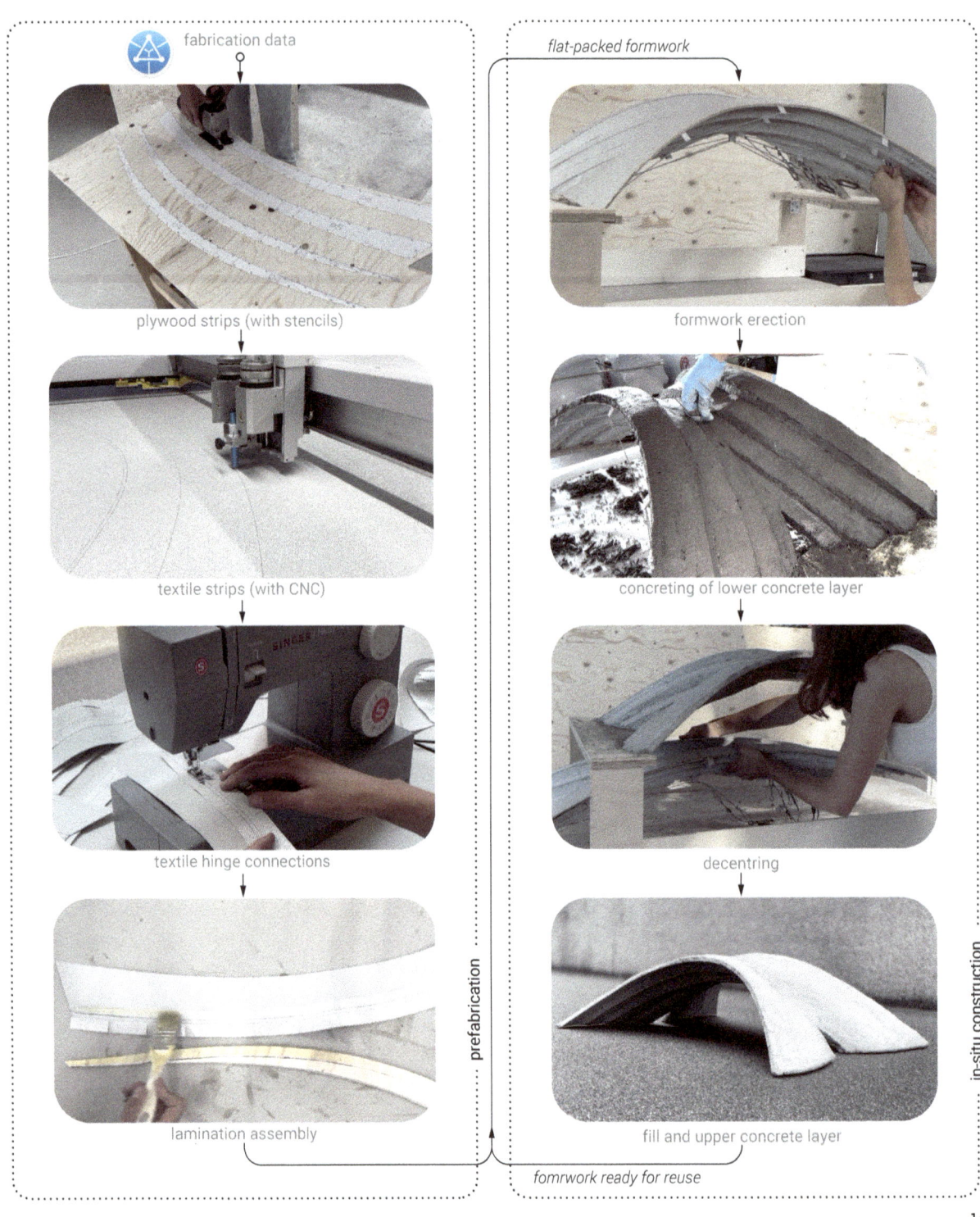

a joining strategy must be chosen that does not require tool access as for mechanical fasteners. Instead, lamination with a bonding agent allows connecting a textile to the plates by applying external pressure without the need for internal access between the plates and textile.

A textile hinge laminated to a plate only on one side is at risk of peeling off when subject to forces normal to the plate (Figure 12a, b). To prevent this failure, Maleczek et al. (2019) demonstrated that this can be avoided by clamping the textile with an additional slat (Figure 12c). The limitation of this solution is that it requires miter joints and that the slats increase the sectional height decreasing the minimal allowable curvature. Building upon this strategy, this research proposes laminating textile to both sides of the plate connected at the hinge position with a sewn seam, sandwiching the plate in the middle.

As a result, the forces normal to the plate are redirected to the opposite side and taken by tangential forces in which a laminated connection performs best (Figure 12d, e).

In the closed CCF, the hinge pole must be positioned at the contact surface to the adjacent plate so that the plates can freely rotate around it when the CCF system unfolds. When materialized with a miter cut, the rotation is locked to the angle where contact forces block the CCF unfolding. This offers additional stiffness to the curve folded formwork. Figure 11 shows the applied strategy of the wrapped hinge with miter joint (Figure 12e) for both the classic and extended CCF.

The custom-sewn textile hinges allow integrating features such as accommodating attachment loops for the actuation cables. These loops can be sewn into the seam simultaneously with the connection of two adjacent textile strips. If the loop is connected at the inside of the miter joint, the longitudinal tensioning of the cable to unfold the formwork supports the unfolding as it pulls together the abutting faces without inducing delaminating forces normal to the surface (Figure 13).

Formwork Material Choice

In the physical prototype, the bending-active strips are spruce plywood plates of 5 mm thickness. Plywood is commonly used for conventional formworks. Furthermore, it is particularly suitable for active bending with its lightweight property of approximately 2.5 kg per m^2 and its ratio of flexural strength to stiffness greater than 2.5^{E-3}. The orthotropic material is oriented such that the stiffer direction is along the ruling direction and the weaker direction along the bending direction. The orthotropic layout is particularly compatible with cylindrical surfaces where the rulings are parallel, which is not the case for conical or tangent developable surfaces.

For this prototype, the textile hinge strips are materialized as one-sided PVC-coated woven polyester textiles. The strips are designed as wide as to cover the plywood plates entirely to simultaneously serve as a protection layer for the plywood from the concrete and offer a smooth demoldable surface on the PVC-coated side. The non-coated side is crucial for adherence to the plates.

Fabrication Workflow

The fabrication workflow comprises the prefabrication of the formwork and the in-situ construction (Figure 14). The workflow is demonstrated on the small-scale physical prototype, testing the system design and investigating the logistical steps of the process.

In the prefabrication, first, the plywood and textile pieces are cut and labeled based on the fabrication data from the design process. Depending on the context and availability of technology, these prefabrication tasks can either be executed by labor with paper stencils or using CNC machines. Second, the textile strips of both sides of a hinge are connected with a sewing machine, integrating the cable attachment loops. Third, the plate and textile strips are assembled with adhesive connections using a high-strength bonding agent. Lastly, the cables are attached by threading through the loops along the hinge lines and anchored to the plates at their ends. The resulting prefabricated element is compact, flat-packed, and relatively lightweight, practical to be transported to the site also on larger scales.

The erection of the in-situ formwork is actuated by turnbuckles in the contracting cables that pull the supports towards each other and actuate the bending of the plates together with the unfolding of the creases. Through this bilateral mechanism, the formwork system deforms from a compact folded-up into an articulated 3D globally double-curved geometry (Figure 1). Onto this articulated formwork, the concrete is directly applied. The concrete is a regular mix with a low water-cement ratio of 0.37 for high viscosity and small aggregates for workability. The developable geometry of the strips offers the advantage that the concrete can be shaped following the ruling lines with a straight tool. This step creates the lower layer of the double-layered sandwich concrete shell. Before or after

14 Fabrication workflow, from prefabrication to construction

15 Section of the concrete sandwich shell with low-strength, low-embodied carbon fill

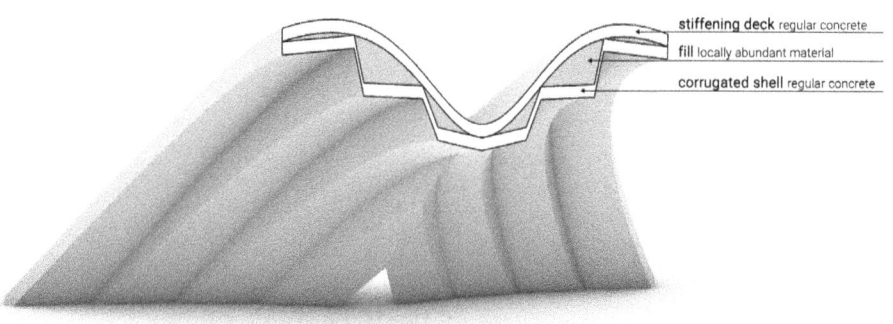

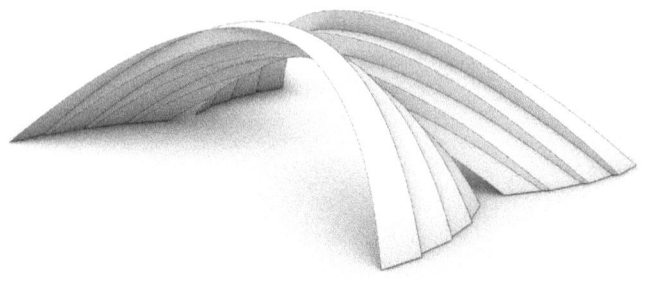

decentering, a low-strength, low embodied-carbon fill and upper concrete layer are applied, resulting in a double-layered sandwich shell (Figure 15). The decentering of the removable formwork is carried out by first lowering the formwork slightly to avoid collision with the concreted shell and then refolding it. The formwork could be immediately reused for another concrete shell.

RESULTS AND DISCUSSION

The proposed computational and materialization methods were demonstrated with a proof-of-concept digital model and physical prototype, revealing the potentials and limitations of bending-active CCF for formwork systems for efficiently constructing corrugated concrete shells. The resulting physical prototype with a span of 1 meter is congruent in shape to the form-found digital model (Figure 16). This demonstrates that the computational design process and the fabrication methods suit the proposed formwork system. The system design allowed for the exploration of a custom geometry in the design process as well as its realization without the need for advanced technology.

The materials used in the prototype are low-cost and broadly available. The quality of the plate must not be high, as the geometry is controlled by the robust CFF system. For example, a plate strip of the physical prototype was assembled from two plates joined in the middle. Its shape displayed a kink along its connection line when bending the plate independently. However, when later integrated into the CCF prototype, no kink could be observed as the strip was well constrained by the curved creases to its neighboring plates that impose the curvature. However, in future work, other material choices, such as GFRP for CCF like in the work of Körner et al. (2016) or Choma (2021), could be investigated to offer a robust and mono-material formwork system for reusability in high-tech contexts.

The formwork was load tested with a tenfold (25 kg) of its self-weight (2.5 kg) and supported the wet-concrete weight with minor deflections. This demonstrates the stiffening effect of the folded plates and the restraining cables. In contrast to textile formwork systems, this stiff formwork could be immediately walkable also at larger scales. The formwork system is the self-supporting, load-bearing falsework and the concrete-shaping shuttering in one. This could be a further advantage over the textile formwork systems consisting of separate falsework and shuttering as it could dramatically simplify the on-site logistics.

The purpose of the small-scale physical prototype was for proof of concept. Future research on system development, structural design, and prototypes will focus on upscaling the

16 Computational geometry (a) in comparison to the physical prototype (b) of the first concrete shell layer

17 Design possibilities for alternating convex and concave initially-closed creases (extended CCF) with parallel translation or rotation of reflection planes

system to make it applicable to the construction industry. Critical limitations lie in the implications on the fabrication and the structural integrity, including admissible forces for the textile hinges and stability of the plates while unfolding and supporting the wet-concrete weight. Extra stiffness in the non-bending direction could be achieved by extra stiffeners, as in Basnak et al. (2020), or by subtractive processing of thicker plates along the rulings. In addition to the actuation cables, gravity could drive the actuation. Further, fabrication at larger scales must consider the limited available sizes of plates by either joining multiple plates as the stiffness variation does not impact the CCF-controlled shape or by dividing the spanning direction with curved creases resulting in more intricate layouts.

Such intricate layouts demand further development of computational methods. These could also enable complex geometries with parallel translation and rotation of the osculating planes, inclinations of the planes around varying axes, combining the classic and the extended CCF, including cone and tangent surfaces, and, lastly, non-planar creases. This would enlarge the design space of the system dramatically. However, the potential collision of the formwork with the concrete shell during decentering limits the range of geometries. Already introducing the extended CCF with planar creases enabled a wide range of custom shape opportunities (Figure 17).

CONCLUSION

The proposed formwork system for shaping corrugated concrete shells allows for a high degree of prefabrication, flat-packed transport, and rapid on-site deployment through the actuation of the CCF mechanism and the integration of falsework and shuttering. It is walkable, reusable, and self-supporting. The system requires neither high material consumption nor high technology as its structural strength and fabrication strategy rely on its geometrical constraints. Its independence from advanced machine technology allows for the construction of complex customized shapes also in low-tech contexts. If scaled up, the proposed construction system could find applications in corrugated shell structures for vaulted floors, roofs, bridges, and funnel-shaped columns. This would bring CCF out of the research realm toward an application that addresses the construction industry's critical challenges in developing and industrialized construction contexts.

ACKNOWLEDGMENTS

This research was supported by the ETH Architecture & Technology Doctoral Fellowship 2019 - 2022. The author would like to thank Rupert Maleczek, Shajay Bhooshan, and Axel Körner for inspiring discussions on curved-crease folding, and Prof. Jan Knippers for his valuable advice.

REFERENCES

Basnak, O., M. Klammer, G. Rihaczek, R. La Manga, A. Körner, J. Knippers, and A. Menges. 2020. "timbRfoldR: Making Bending Active Systems Accessible in Planning, Fabrication, and Assembly." ITECH Thesis Project, University of Stuttgart. Accessed May 10, 2022. https://www.itech.uni-stuttgart.de/itech-thesis-projects/2020-timbrfoldr/.

Bellmann, J. 2017. "Active bending starting on curved architectural shape." In *8th International Conference on Textile Composites and Inflatable Structures; Structural Membranes 2017*, edited by K.-U. Bletzinger, E. Oñate, and B. Kröplin.

Bhooshan, S., V. Bhooshan, A. Shah, H. Louth, and D. Reeves. 2015. "Curve-folded form-work for cast, compressive skeletons." In *Proceedings of the Symposium on Simulation for Architecture & Urban Design*. 221–228.

Block, P., T. Van Mele, M. Rippmann, F. Ranaudo, C. Calvo Barentin, and N. Paulson. 2020. "Redefining structural art: Strategies, necessities and opportunities." *Structural Engineer* 98 (1): 66–72.

Buckminster Fuller, R. 1959. Self-strutted geodesic plydome (Patent No. 654166).

Buri, H., I. Stotz, and Y. Weinand. 2011. "Curved Folded Plate Timber Structures." In *35th Annual Symposium of IABSE / 52nd Annual Symposium of IASS / 6th International Conference on Space Structures: Taller, Longer, Lighter - Meeting Growing Demand with Limited Resources*. London, UK.

Choma, J.. 2021. Foldable composite structures. United States. Patent No. US10994468B2, filed April 11, 2019, and issued May 4, 2021.

Correa, D., O. D. Krieg, and A. Menges. 2016. "Robot-Made: Double-Layered Elastic Bending." ICD Workshop, University of British Columbia. Accessed May 13, 2022. https://www.icd.uni-stuttgart.de/teaching/workshops/double-layered-elastic-bending-for-large-scale-timber-structures.

Cuvilliers, P., C. Douthe, L. du Peloux, and R. Le Roy. 2017. "Hybrid Structural Skin: Prototype of a GFRP Elastic Gridshell Braced by a Fiber-Reinforced Concrete Envelope." Journal of the International Association for Shell and Spatial Structures 58 (1): 65–78. https://doi.org/10.20898/j.iass.2017.191.853.

Duncan, J. P., and J. L. Duncan. 1982. "Folded Developables." *Proceedings of the Royal Society of London; Series A, Mathematical and Physical Sciences* 383 (1784): 191–205.

Frommeld, M.. 2011. "Folding Boat." Accessed May 13, 2022. https://www.maxfrommeld.com/folding-boat.

Fuchs, D., and S. Tabachnikov. 1999. "More on Paperfolding." *The American Mathematical Monthly* 106 (1): 27–35. https://doi.org/10.1080/00029890.1999.12005003.

Huffman, D. A. 1976. "Curvature and Creases: A Primer on Paper." *IEEE Transactions on Computers* C–25 (10): 1010–1019. https://doi.org/10.1109/TC.1976.1674542

Jipa, A., C. Calvo Barentin, G. Lydon, M. Rippmann, G. Chousou, M. Lomaglio, A. Schlueter, P. Block, and B. Dillenburger. 2019. "3D-Printed Formwork for Integrated Funicular Concrete Slabs." In *Proceedings of the IASS Annual Symposium 2019 – Structural Membranes 2019: Form and Force*, edited by C. Lázaro, K.-U. Bletzinger, and E. Oñate.

Kilian, M., S. Flöry, Z. Chen, N. J. Mitra, A. Sheffer, and H. Pottmann. 2008. Curved folding. *ACM Transactions on Graphics* 27 (3): 75. https://doi.org/10.1145/1360612.1360674.

Körner, A., A. Mader, S. Saffarian, and J. Knippers. 2016. "Bio-Inspired Kinetic Curved-Line Folding for Architectural Applications." In *ACADIA 16: Posthuman Frontiers; Proceedings of the 36th Annual Conference of the Association for Computer Aided Design in Architecture*. 270–279.

Ku, J. S., and E. D. Demaine. 2016. "Folding flat crease patterns with thick materials." *Journal of Mechanisms and Robotics* 8 (3): 1–6. https://doi.org/10.1115/1.4031954.

Kudless, A., J. Zabel, C. Naeve, and T. Florian. 2020. The Design and Fabrication of Confluence Park. In *Fabricate 2020*, edited by J. Burry, J. Sabin, B. Sheil, and M. Skavara. London: UCL Press. 28–35.

Lehne, J., and F. Preston. 2018. *Making Concrete Change; Innovation in Low-carbon Cement and Concrete*. Chatham House Report.

Lienhard, J.. 2014. "Bending-Active Structures: Form-finding strategies using elastic deformation in static and kinematic systems and the structural potentials therein," PhD diss., University of Stuttgart. https://doi.org/http://dx.doi.org/10.18419/opus-107

Maleczek, R., G. Stern, A. Metzler, and C. Preisinger. 2020. "Large Scale Curved Folding Mechanisms." In *Impact: Design With All Senses; Proceedings of the Design Modelling Symposium, Berlin 2019*, edited by C. Gengnagel, O. Baverel, J. Burry, M. Ramsgaard Thomsen, and S. Weinzierl. Cham: Springer Nature. 539–553. https://doi.org/10.1007/978-3-030-29829-6.

Méndez Echenagucia, T., D. Pigram, A. Liew, T. Van Mele, and P. Block. 2019. "A Cable-Net and Fabric Formwork System for the Construction of Concrete Shells: Design, Fabrication and Construction of a Full Scale Prototype." *Structures* 18 (April): 72–82. https://doi.org/10.1016/j.istruc.2018.10.004.

Peri Group. n.d. "Freeform Formwork." Peri Group, Products (webpage). Accessed June 2, 2022. https://www.peri.ltd.uk/products/formwork/special-formwork/freeform-formwork.html.

Popescu, M., M. Rippmann, A. Liew, L. Reiter, R. J. Flatt, T. Van Mele, and P. Block. 2021. "Structural design, digital fabrication and construction of the cable-net and knitted formwork of the KnitCandela concrete shell." *Structures* 31: 1287–1299. https://doi.org/10.1016/j.istruc.2020.02.013.

Rabinovich, M., T. Hoffmann, and O. Sorkine-Hornung. 2019. "Modeling curved folding with freeform deformations." *ACM Transactions on Graphics* 38 (6): 1–12. https://doi.org/10.1145/3355089.3356531

Robeller, C., S. S. Nabaei, and Y. Weinand. 2014. "Design and Fabrication of Robot-Manufactured Joints for a Curved-Folded Thin-Shell Structure Made from CLT." In *Robotic Fabrication in Architecture, Art and Design 2014*, edited by W. McGee and M. Ponce de Leon. 67–81. https://doi.org/10.1007/978-3-319-04663-1

Scheder-Bieschin, L., K. Spiekermann, M. Popescu, S. Bodea, T. Van Mele, and P. Block. 2022. "Design-to-fabrication workflow for bending-active gridshells as stay-in-place falsework and reinforcement for ribbed concrete shell structures." In *Design Modelling Symposium Berlin 2022: Towards Radical Regeneration*, edited by C. Gengnagel et. al. Cham: Springer Nature.

Schleicher, S., and R. La Magna. 2016. "Bending-Active Plates: Form-Finding and Form-Conversion." In *ACADIA 16: Posthuman Frontiers; Proceedings of the 36th Annual Conference of the Association for Computer Aided Design in Architecture*. 260–269.

Takahashi, K., A. Körner, V. Koslowski, and J. Knippers. 2016. "Scale effect in bending-active plates and a novel concept for elastic kinetic roof systems." In *Proceedings of the IASS Annual Symposium 2016; Spatial Structures in the 21st Century*, edited by K. Kawaguchi, M. Ohsaki, and T. Takeuchi.

Van Mele, T., G. Casas, R. Rust, B. B. Lytle; L. Chen, et al. 2022. COMPAS: A computational framework for collaboration and research in architecture, engineering, fabrication, and construction. V. 1.16.0 [source code]. http://compas.dev. https://doi.org/10.5281/zenodo.2594510.

Veenendaal, D., M. West, and P. Block. 2011. "History and overview of fabric formwork: using fabrics for concrete casting." *Structural Concrete* 12 (3): 164–177. https://doi.org/10.1002/suco.201100014.

IMAGE CREDITS

All drawings and images by the authors.

Lotte Scheder-Bieschin (formerly Aldinger) is a PhD Researcher at the Block Research Group. She has a background in structural engineering and received a Master of Science with distinction from the ITECH program at the University of Stuttgart with a thesis on tailoring the self-formation of membrane-actuated composite gridshells. She had leading roles in the ICD/ITKE Research Pavilion 2016/17, BUGA Wood Pavilion 2019, Urbach Tower in 2019, and KnitNervi Pavilion 2022. Her PhD research focuses on bending-active formwork systems for concrete shells with an integrative computational, structural design, and fabrication approach driven by the motivation toward more sustainable construction practices.

Dr. Tom Van Mele is co-director and head of research at the Block Research Group (BRG), ETH Zurich, and lead developer of COMPAS, an open-source computational framework for research and collaboration in architecture, engineering, and construction. Tom studied architecture and structural engineering at the Vrije Universiteit Brussel, where he received his PhD in 2008. Tom joined the BRG in 2010 and was appointed Senior Scientist at ETH in 2018. His technical and computational developments form the backbone of multiple flagship projects, including the Armadillo Vault (2016), Striatus Bridge (2021), and the NEST HiLo unit (2021).

Dr. Philippe Block is full professor at the Institute of Technology in Architecture (ITA), ETH Zurich, where he leads the Block Research Group (BRG) with Dr. Tom Van Mele and is Head of the Institute. Philippe is also Director of the Swiss National Centre of Competence in Research (NCCR) - Digital Fabrication. He studied architecture and structural engineering at the Vrije Universiteit Brussel and at the Massachusetts Institute of Technology, where he earned his PhD in 2009. He has received numerous awards, including the Rössler Prize for most promising young professor from ETH Zürich and the Berlin Arts Prize for Baukunst.

HoloWall

A Mixed Reality-Informed Hollow Core Wall Assembly

Leslie Lok
Cornell University

Jiyoon Bae
Cornell University

1   HoloWall showcasing the thickening and thinning of layers and porosity

ABSTRACT

HoloWall is a wall assembly that integrates mixed reality (MR) protocols with nonuniformly sized lumber to develop a customized hollow-core cross-laminated timber (HCCLT). The performance-driven design workflow leverages the MR technology and tiling automation of nonuniform wood boards to guide material processing and fabrication of a customized HCCLT prototype. This paper proposes to expand the usage and the viability of customized HCCLT as a structural component. Upcycling locally salvaged wood elements, the prototype develops a material language of lamination that peels away in calibrated gradients to generate structural and visual porosity. By engaging with the computational environment and the physical making process through the MR workflow, users are able to explore an accessible design streamline.

To test this hybridized design approach, the paper investigates the following objectives: 1) utilize locally sourced salvaged wood boards in response to the emerging deconstruction practice to minimize material waste; 2) leverage MR-aided process for the utilization of nonuniform wood materials and the fabrication of prototypes; 3) implement design parameters such as structural and program requirements for customization; 4) develop a semiautomated wood board packing protocol for nonuniform materials; and 5) propose a performance-driven load-bearing HCCLT wall at housing scale as a further step of architectural application. This MR-informed process facilitates the reuse of discarded wood wastes for the customized HCCLT assembly and provides an intuitive workflow for expert and nonexpert users.

INTRODUCTION

In the building life cycle, construction and demolition waste (CDW) from the building industry is one of the major contributors to carbon emissions. Wood accounts for 6 to 7% of total CDW (Cochran and Townsend 2010). Even though active CDW management has been practiced globally in the construction sector, it is estimated that approximately 35% of wood CDW is landfilled (Menegaki and Damigos 2018). Furthermore, the wood CDW greatly affects the project cost. Thus, in the construction process, the industry has employed a sustainable waste management hierarchy consisting disposal, recovery, recycling, reuse, minimization, and prevention (Nagapan et al. 2012). Meanwhile, when buildings reach their end of life, demolition is typically regarded as an inexpensive and for site clearing. Building demolition produces a markedly larger volume of waste than construction. According to the United States Environmental Protection Agency, demolition accounts for 90% of total construction and demolition debris (Tolaymat et al. 2017).

By contrast, deconstruction is a promising solution that replaces demolition with upcycling of construction materials. It is usually defined as the disassembly of structures to retain the economic value of salvaged building materials. The process is expected to divert the maximum amount of building parts from the waste stream, which immediately provides economic and environmental benefits (Chini and Bruening 2003). Nevertheless, this practice is being employed in limited socioeconomic conditions due to requirements of time, budget, regulated policy, skilled workers, and advanced tools, which consequently makes this practice less viable at the local level.

In response, this paper proposes an integrated design approach within a local material economy, starting from utilizing locally sourced salvaged wood resources to hybridize an existing concept of timber construction with emerging mixed reality (MR) technology. The research aims to suggest a bespoke circular construction process to develop the MR-informed design protocol. It expands the notion of deconstruction and material upcycling by allowing for a mass-customized design approach at local scale using salvaged wood elements.

To test this approach, this paper investigated the following objectives through a full-scale HoloWall installation (Figure 1) by leveraging a custom MR framework: 1) the MR technology was utilized for processing irregularly shaped salvaged wood elements with specific holographic guidance (Figure 2); 2) design parameters such as structural performance, wind deflection, and architectural programs were embedded for customizing the structural and visual porosity of hollow-core cross-laminated timber (HCCLT) panels; 3) the implementation of a semiautomated tiling protocol for the nonuniform wood elements; 4) MR- and augmented reality (AR)-aided fabrication and installation process for the HoloWall prototype; 5) proposes a performance-driven load-bearing HCCLT wall as a further step of architectural application. By leveraging the MR- and AR-aided protocol throughout research

2 Waste wood from a local collapsed barn that was used for the prototype

phases, users are expected to explore the efficient and user-friendly design pipeline that could contribute to minimizing wood wastes.

STATE OF THE ART
Cross-Laminated Timber

Even though wood has been one of the most preferred building materials due to its sustainability, affordability, and abundant availability, challenges still exist. The material properties of wood, including shape, diameter, cellular structure, and moisture content, vary widely depending on the species (Mallo and Espinoza 2015). Because of these heterogeneous characteristics, only limited wood species have been highly valued and processed as timber products in the processes of mass timber harvesting. This consequently causes the abandonment of a large amount of low-valued harvested trees (Luppold and Bumgardner 2003).

Considering this aspect, cross-laminated timber (CLT) is a promising building construction component. Developed and standardized in Europe, CLT has been globally employed as a well-recognized engineered timber product (Brandner et al. 2016). CLT panel brings several benefits in different aspects. First, environmentally, it utilizes low-grade timbers that have small diameters or irregular shapes in comparison to traditional mass timber elements. It can be revalued as a building material for an architectural application. This contributes to better forest management by minimizing the material waste from harvesting (Mallo and Espinoza 2015). Second, structurally, CLT structures are built with layers of wood members laminated orthogonally to each other. These adjacent layers perform as reinforcement, carrying the load in multiple directions (Steiger and Gulzow 2010). This structural characteristic allows CLT to be used as load-bearing wall, plate, or roof structures. Third, and last, the CLT system enables the construction process to be simple and efficient due to its prefabricated nature.

Haus Gables, designed by Jennifer Bonner, is a two-story single residence in the US built with CLT. All structural components, including exterior and interior walls, floors, and roofs, consist of CLT panels. The panels were prefabricated, custom cut, transported to site, and assembled in fourteen days (Bonner 2018). Thus, mass standardization and prefabrication simplify the on-site installation and reduce construction costs. (Mallo and Espinoza 2015). Even though CLT structures are intended to perform as various structural components, they are more commonly employed as only load-bearing slabs and roof structures. Moreover, CLT is mass produced as a generic wood product, which represents an overlooked opportunity to embed design parameters such as structural optimization and programmatic integration.

Hollow-Core Cross-Laminated Timber

In the U.S., the application of cross-laminated timber (CLT) has grown considerably due to the benefits addressed above. CLT panels consist of multiple layers of orthogonally bonded dimensional lumbers. From an engineering perspective, the multiple layering panels aim to generate structural depth. However, only the outer layers are fully utilized for structural stability in a one-way system and the core layers obtain limited structural benefits and work merely as a spacer between adjacent layers (Mayencourt and Mueller 2019). Moreover, the regularity and solidity of the manufactured panels limit the incorporation of potential design considerations. To address these aspects, the research questions how material usage in CLT can be minimized but still retain performative and aesthetic benefits. Some studies suggested an alternative way of producing CLT by selectively removing core timber elements, making hollow-core cross-laminated timber (HCCLT) (Huang et al. 2021). For example, SPLAM (SPatial LAMinated Timber) is a timber pavilion exhibited for the 2021 Chicago Architecture Biennial, led by a collaborative design team of SOM and professors Tsz Yan Ng and Wes McGee at the University of Michigan. The pavilion adopted customized CLT framing panels that minimize material usage by 46% compared to the timber usage of conventional CLT construction. It serves as an outdoor education facility and gathering space. (Skidmore, Owings & Merrill 2021).

A customized HCCLT system has advantages over generic CLT panels: HCCLT reduces material usage by selectively omitting core timber elements; secondly, it enables users to explore customized porosity for design; lastly, design iterations can be informed by various parameters such as structural performance and program requirements. This performance-driven design process informs the panel thickness, layer density, and porosity. Therefore, this paper investigates a MR-aided design process for customized HCCLT walls with locally sourced salvaged wood resources.

Mixed Reality Technology Informed Design and Construction

Advanced computational technologies have enabled designers to explore complex geometric and structural forms. However, the physical making process requires highly skilled labor, expert-level knowledge, and fabrication infrastructure. In this aspect, mixed reality (MR) technology allows the design complexity to be hybridized with competent physical making processes that might align with a more bottom-up approach. Recent investigations in MR-aided workflow has provided designers with visual protocols to manage the design and construction process. For instance, Augmented Feedback, research developed by Chinese University of Hong Kong, employed the holographic environment to fabricate and

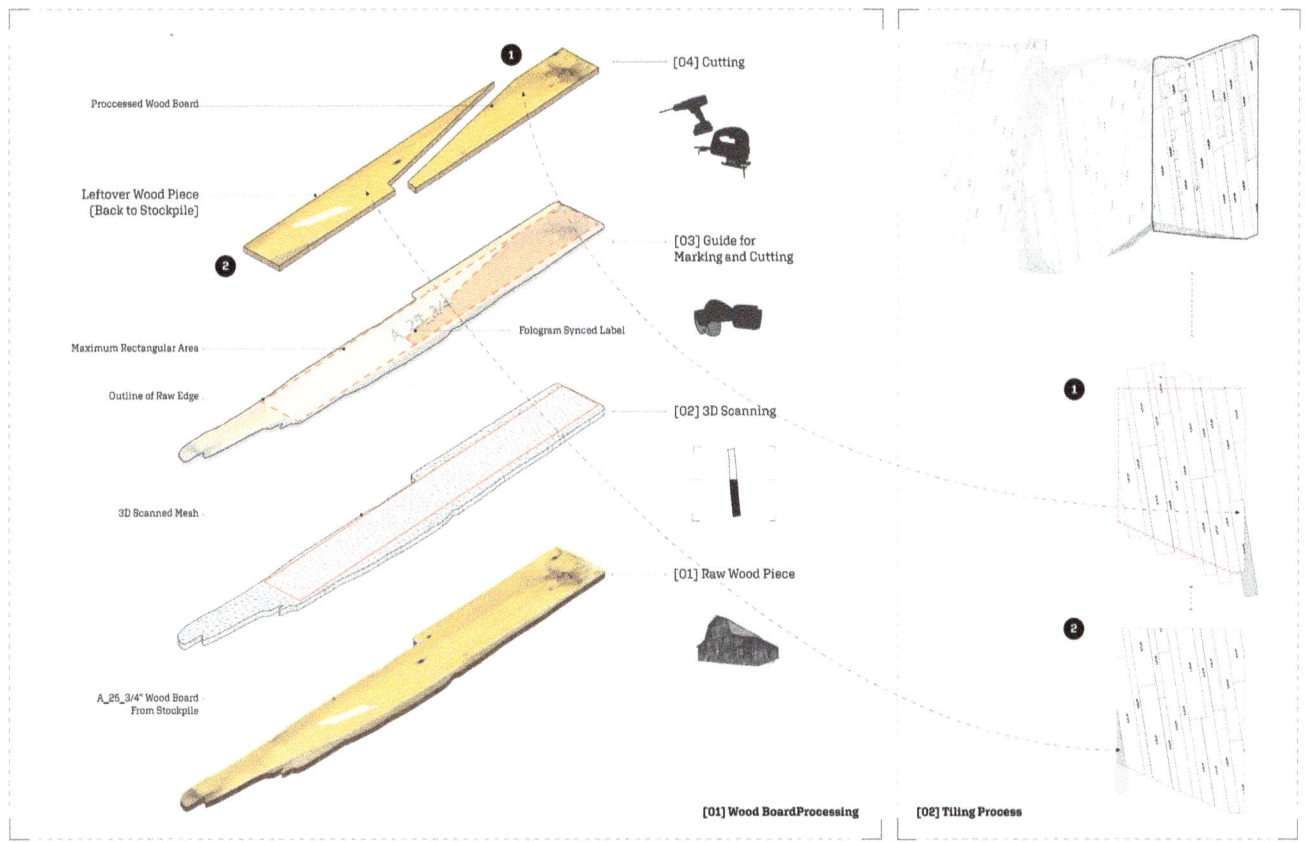

3 Workflow for material processing and semiautomatic tiling protocol

install a bamboo pavilion. The research demonstrated that holographic guidance enabled participants to better understand the design complexity and construction protocols. Therefore, it significantly increases productivity and achieves design accuracy by minimizing on-site measurement (Goepel 2021). Additionally, Timber De-standardized leverages MR technology to analyze and to utilize irregularly shaped logs with structural feedback loops for a spatial timber structure. This research delivered a MR-informed approach in which nonstandard tree log structures can be designed and fabricated without using advanced robotic technology. This pipeline provided users with new opportunities to intuitively configure wood structures with nonuniform elements within a user-friendly MR environment (Lok et al. 2021). The HoloWall research explores the reassembling of salvaged construction materials by hybridizing a customized wood construction methodology with emerging MR technology in response to current environmental and socioeconomic contexts. In this paper, the HoloWall prototype was developed and installed at a specific location to validate its feasibility as a parameter-driven design as well as its MR- and AR-guided material processing, fabrication, and assembly. A 1:1 scale prototype is intended to provide the outdoor environment with seating space where users can be sheltered from the wind and interact with surrounding context through visual openings.

METHODS
Research Workflow
The paper aims to describe an MR-informed design pipeline that develops a customized HCCLT load-bearing wall using locally sourced salvaged wood materials. The research is structured in the following phases (Figure 3): 1) source the material from a local collapsed barn and 3D scan each elements for digital cataloging; 2) transform salvaged wood boards into variably sized orthogonal elements with new material boundaries by computationally calculating the maximum usable rectangular area; 3) integrate design parameters such as structural performance and architectural program requirements; 4) processed wood members are automatically tiled on the panels' outer layers in the CAD environment, and a manual recalibration is performed to optimize the packing pattern by eliminating gaps; 5) the research team leverages AR/MR guidance to laminate unique wood elements and to assemble on-site installation by following intuitive holographic guides. Lastly, the paper proposes potential architectural application of HoloWall as a load-bearing system in two-story housing studies.

Material Sourcing, Cataloging, and Processing
The primary objective for this specific MR-guided material processing is to provide a user-friendly working interface to

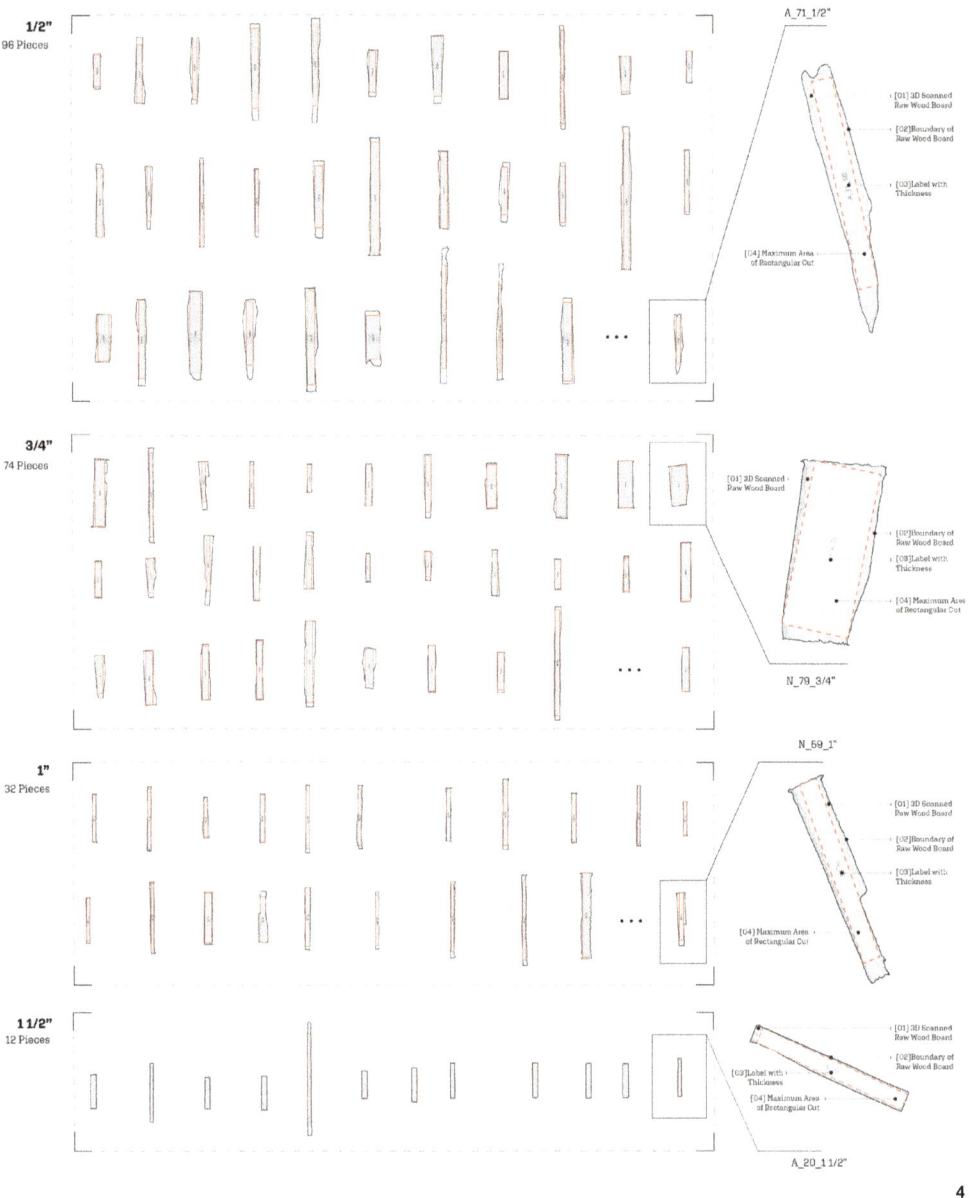

4   A portion of the 3D-scanned digital stockpile and visual reference of usable material boundaries
5   Design iteration of HoloWall prototype and design parameters
6   CFD analysis of HCCLT wall section based on different wind directions
7   CFD analysis with dominant wind directions at installation site
8   Structural analysis for optimizing hollowness and layer thickness
9   Voided volume of customized HCCLT structure

upcycle locally sourced wood wastes. Therefore, the visual guidance projected through the Hololens plays a key role in allowing users to intuitively manage materials' irregularity. The salvaged wood members are mostly irregular in their profile, including edge condition, size, and thickness. To analyze and process the stockpile, wood pieces are 3D scanned and converted into mesh data to create a digital stockpile (Figure 4). New material boundaries of the maximum usable rectangular area are computationally calculated using a custom Grasshopper script. Additionally, individual elements are digitally labeled and sorted according to categories of thickness including 1/2", 3/4", 1" and 1 1/2". These digital references are transmitted to the MR environment for guiding the physical material processing.

Performance-Driven Design and Parameters

CLT panel brings several environmental and structural benefits. However, it has been typically used as slab or roof structures, and its application for shear wall is less common in comparison. Therefore, this paper explores a customized HCCLT to optimize the material usage and expand its architectural application through the prototype, HoloWall, a public installation at a specific site. By selectively reducing core timber to optimize the void volume, the HoloWall saves construction materials and leverages the cavity as a design parameter. The structure adopts multiple design parameters to customize the HCCLT panels, such as structural performance, wind deflection, and view interaction to inform the panel thickness and porosity (Figure 5).

The form and environmental functionality of HoloWall are inspired by the traditional windbreak wall. Windbreak walls have been used to deflect wind for outdoor structures and the protection of livestock. Referencing the traditional windbreak

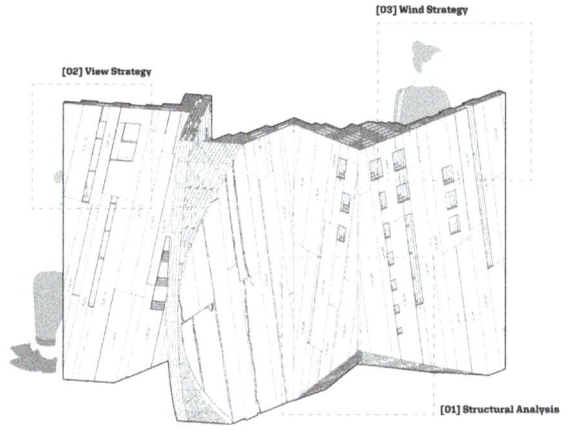
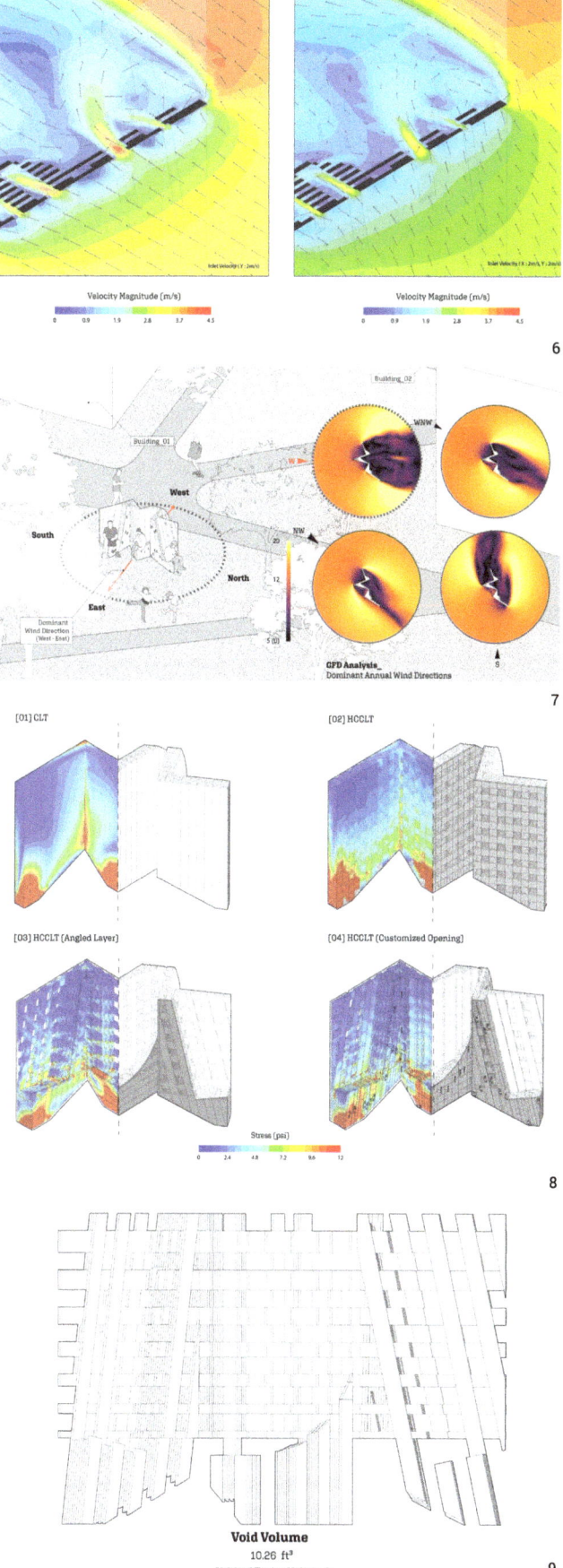

shelters in an agricultural landscape, airflow deflection is another design driver that informs the porosity and position of HoloWall on-site. The following computational fluid dynamic (CFD) tests show that the velocity and vector of airflow are highly influenced by inlet airflow direction (Figure 6). The design of the lamination pattern creates calibrated openings that are oblique to the wind direction, this allows the customized hollow gradients to act as filters for dissipating blows of wind and to prevent low pressure build-up behind the wall.

The CFD analysis informs the placement and organization of wall panels on site in order to maximize the wind sheltering area. The annual wind rose data reveal that the dominant wind directions are northwest and south throughout the year. Therefore, the wall is faceted into five segments angled 45 to 90 degrees to the west, and the seating area is integrated on the east side. The following CFD analysis results illustrate that the wall configuration can sheltered visitors from various dominant wind directions at the site (Figure 7).

The compositive layering with divertive angles and sizes of the cavity generates unique porosity. These openings allow for air penetration and for users' visual interaction with the surroundings. The layering strategy involves the thinning and thickening of lamination to further calibrate porosity for view factors as a design consideration. The lamination thickens on the east side to incorporate seating for user interactions. The work seeks to provide a space where dialogue can be shared by individuals seated across the wall where light and sound are filtered through the perforated layers. The research team analyzed the structural performance of design iterations to optimize the layer thickness and to study the impact of porosity on structural stability. Static analyses for each iteration, including CLT, regular HCCLT, and customized HCCLT, are evaluated using the cloud-based analytic tool Simscale (Simscale GmbH 2013). The following simulation results suggest that the omission of core elements does not significantly compromise the structural stability. Furthermore, the

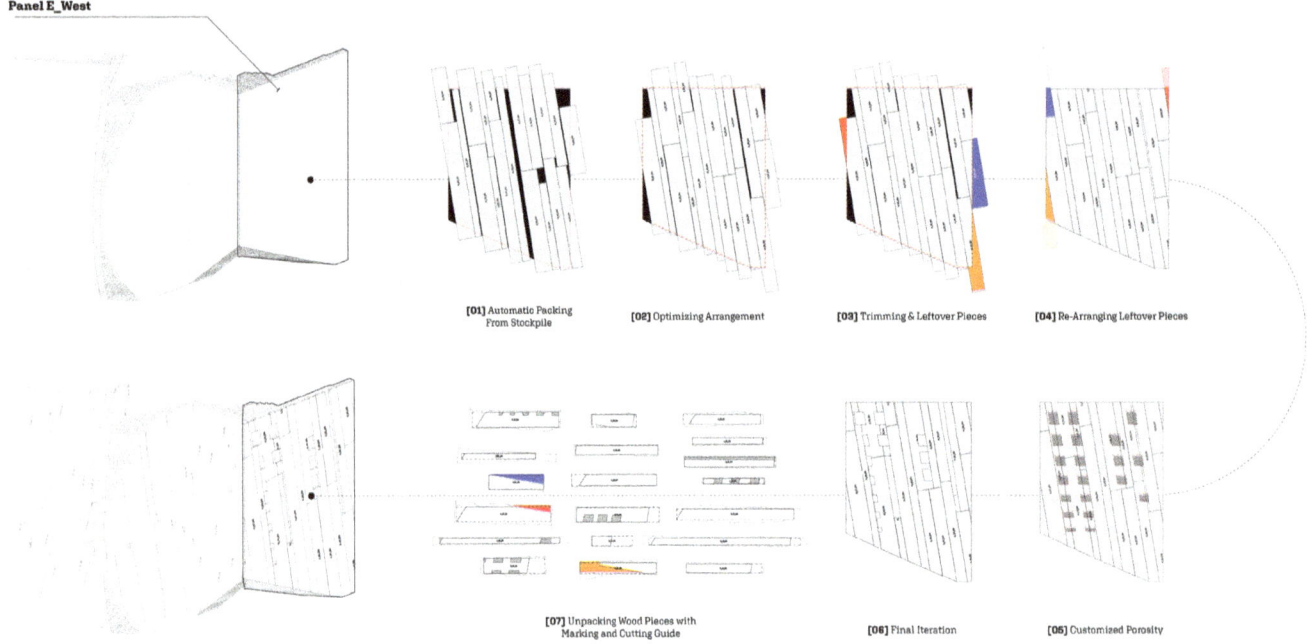

10 Semiautomated tiling process for wood pieces

analysis reveals that there is no apparent difference in pressure distribution between CLT and the customized HCCLT structure (Figure 8).

It is suggested that HCCLT structures can eliminate more than 30% of material (Huang et al. 2021). As a prototype, the HoloWall extracted 30% of the entire volume (10.26 sf) in the final iteration (Figure 9). Based on the structural analysis, it achieved significant material saving while embedding architectural programs like seating and openings that interact with the surrounding context.

Computational Design Protocol for Material Tiling

HoloWall consists of multiple wood elements that are irregular in dimension. Therefore, efficient material management is a key for users to work with the wood elements' nonuniformity. This research phase explores a computational design protocol that enables designers to automatically track the labeled unique wood components. Additionally, the custom Grasshopper script packs irregular wood boards onto any geometries of panel boundary. This automatic tiling protocol culls used wood boards from the digital stockpile to prevent the duplication of material usage.

The specific semiautomatic packing protocol follows these processes (Figure 10): 1) computationally pack wood components from the digital stockpile to the outer layers of the HoloWall panel utilizing a custom Grasshopper script with a plug-in called Packrat (Yconst 2012); 2) manually move the digital wood elements to optimize the automatic packing process by eliminating gaps between materials; 3) trim wood materials within the boundary line of each panel; 4) reuse residual trimmed materials to fill gaps from the initial packing process; 5) customize porosity, which deflects wind and creates openings for users' visual interaction. The completed packing layout will be then transmitted to holographic guidance in the MR environment.

Mixed Reality–Aided Fabrication and Installation Process

The research utilizes MR technology to provide users with visual guidelines for both fabrication and installation. Upon being digitally recalibrated to maximum usable area and labeled, the wood boards are transmitted to the MR interface with specific visual graphics that represent different material usage. Then the irregularly shaped wood boards are marked, trimmed, and labeled by following the holographic instructions.

Specifically, 3/4-inch-thick standardized plywood is used for the inner layers, and nonuniform salvaged wood elements are used for the outer layers. Thin solid curves represent the wireframe of original wood geometries. Dashed lines with transparent surfaces illustrate the usable material boundaries of the maximum area for trimming. Red solid surfaces indicate trimmed residual portions to reuse for panel filling. Lastly, green solid surface represents customized openings. By following these holographic references, users can operate

within a bespoke MR workspace in real time to measure and mark physical materials without referencing any drawing instructions. The augmented boards are visualized in a linear assembly at an oblique angle to the working table. These holographic instructions are integrated with Fologram's (Newnham et al. 2017) parameter sliders allow users to view upcoming boards and to navigate through the labeled pieces for processing (Figure 11). This MR-guided material processing enables both expert and nonexpert users to intuitively and efficiently mark and trim material with ease.

In the next lamination process, users can first dry laminate each layer to check geometric and dimensional accuracy. Subsequently, each wood component is laminated along the holographic gluing position (Figure 12). After completing the lamination of one layer, users can transition to the upper layer by tapping the holographic UI arrow. The inner core is coded by panel type, layer number, and element order from bottom to top for horizontal layers and left to right for vertical layers. The research team dry-laminates and glues wood boards with the holographic references that display the wood elements' arrangement for each layer and gluing positions (Figure 13). The intuitive user interface (UI) allows users to shift back and forth from layer to layer of a HoloWall panel by tapping the up and down arrows within the MR environment.

In the final installation process, MR and AR technology was utilized to position the structure at with a predetermined

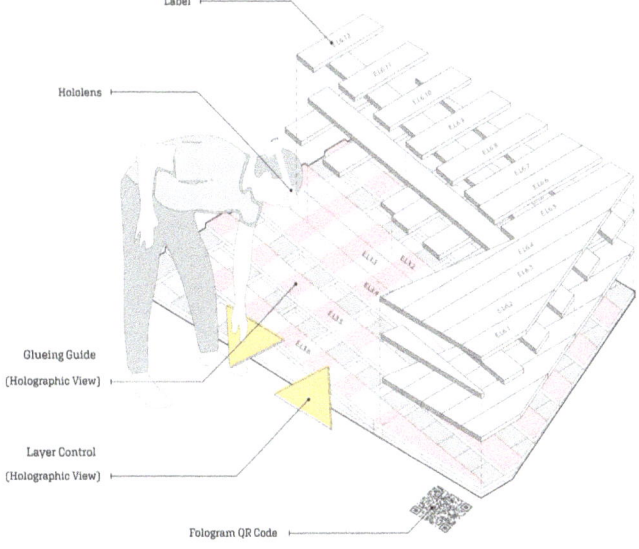

11 Mixed-reality guided marking and cutting process
12 Mixed-reality guided laminating and gluing process
13 Custom mixed-reality user-interface (UI) for lamination with board labels and up and down arrows to shift between layers

14 On-site installation of HoloWall prototype utilizing AR view

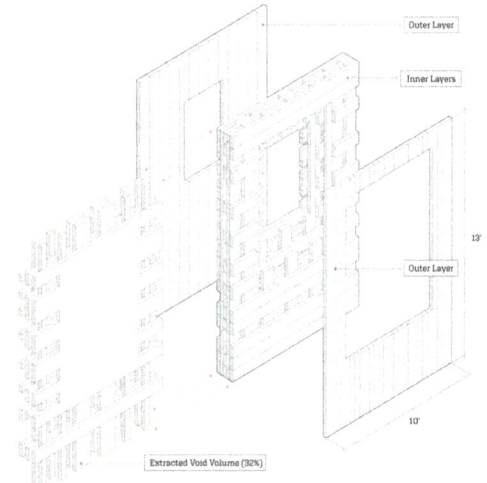

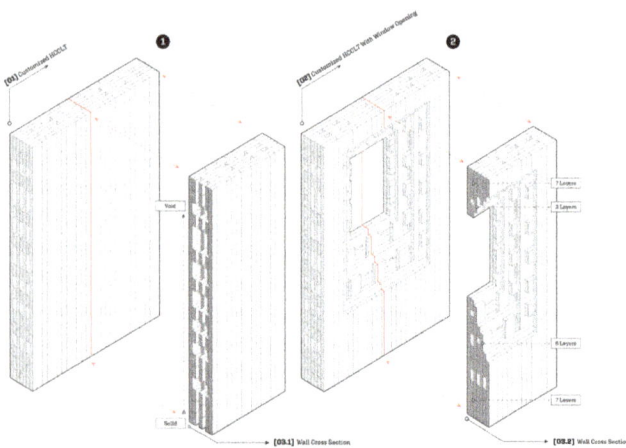

15 Customized layers of HoloWall panel

orientation (Figure 14). The team used holographic view to position the anchoring steel plates and precisely align the openings for the anchoring connection. This workflow could be similarly applied to custom installations or construction processes.

Customized HCCLT Load-Bearing Wall

With the MR-driven research pipeline developed for the HoloWall, the paper further investigates a possible architectural application of the HCCLT panel as a load-bearing wall system by deploying the customized panels within a generic two-story housing configuration. This process aims to test customized HCCLT load-bearing walls as a structural component that supports the dead load of the housing building. Built upon the investigation of the HoloWall, the variation of layering strategies can be informed by design parameter such as structural performance. The typical layup of CLT is decided by a given application. CLT panels are bonded orthogonally with an odd number of layers, from 3 to 11 layers. Three-ply CLT is used for walls and secondary construction. A 5-ply CLT panel is a standard layup used for walls and floors that span less than 5.5 m. The maximum number of layers of 11-ply CLT panel are used for floors or bridge decks (Brandner et al. 2016).

Considering the above-mentioned practices, a 10 ft wide by 13 ft tall, 7-layer panel (Figure 15) was used as a starting base. The research team customized each wood element's distance and layer thickness based on the structural analysis. The result reveals that the bottom portion of HCCLT receives more pressure than the upper portion. Moreover, the pressure is minimized with additional layers. Therefore, the exploration customized spacing between each horizontal and vertical layer in response to the structural analysis. In particular, the bottom portion becomes relatively solid in comparison to the upper part. Likewise, areas with the least pressure have increased void volume to generate structural or visual porosity. This process extracted approximately 30% of the volume compared to the same thickness of CLT structures.

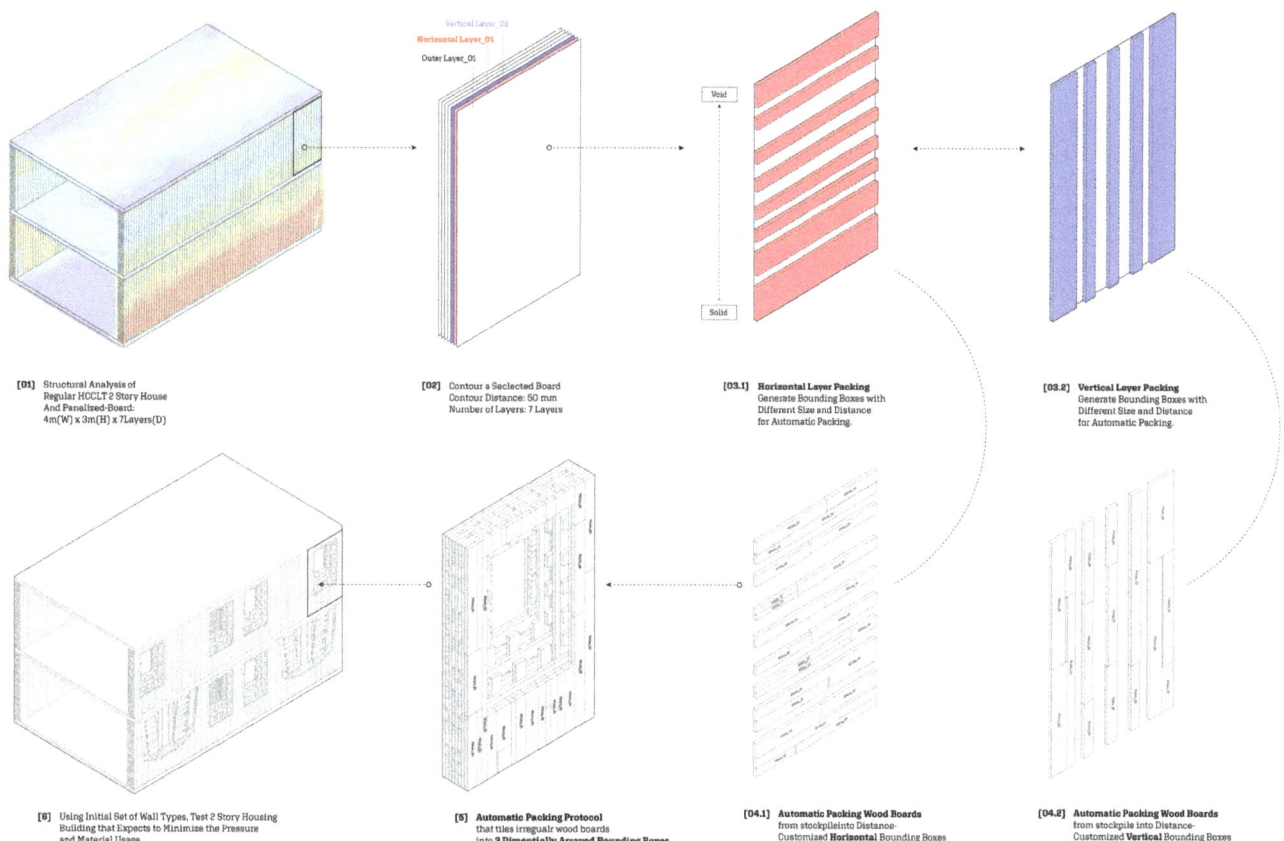

16 3D Packing process for two-story customized HCCLT load-bearing walls

The HoloWall load-bearing wall also leverages MR technology for its packing and lamination process as a prefabricated product. In this phase, the 2D packing process is extended to a 3D packing protocol. The distances between wood components are rearranged corresponding to the structural analysis, which informs the manipulation of the bounding box. Subsequently, the wood boards from the digital stockpile are automatically tiled into each bounding box along with the boards' labels. These three-dimensionally arrayed bounding boxes allow designers to efficiently pack and manage the irregularly shaped wood components for the further MR-guided fabrication process (Figure 16). This research phase introduces initial studies of customized HCCLT load-bearing wall types that could be further developed and refined a specific wall types to accommodate wall openings and building system integration.

DISCUSSION

This paper explores the customized HCCLT wall informed and enabled by mixed reality (MR) technology. The locally sourced salvaged wood pieces are utilized to illustrate an alternative way of upcycling wood as a construction material for the installation, HoloWall. MR technology played a significant role throughout the research streamline. First, the material processing leveraged MR guidance to recalibrate the new material boundaries of maximum usable area of irregularly shaped salvaged wood boards. This MR-guided material processing enables us to minimize material waste and renders the process accessible to nonexpert users. Second, the MR-aided fabrication process contributed to providing users with an intuitive working environment. Third, the proposed HCCLT load-bearing wall as a possible architectural application might also generate impact for further mass customization with approximately 30% material saving.

The immediate next step of this research is to continue developing the physical outcomes of customized HCCLT load-bearing wall types. In conjunction, the research team will mainly aim to refine the study of wall types, considering extended building types, connection detail, and additional program requirements. Building upon the initial set of wall designs, the future research will also explore the expanded MR-guided process to visualize the catalog of new wall type collection, its added holographic components and protocols while maintaining a customized UI for both expert and nonexpert users.

CONCLUSION

Considering the enormous amount of available wood reallocated from construction and demolition waste that would

17 Close-up view of the HoloWall installation showing customized voids between layers

18 Side view the HoloWall installation showing various degrees of porosity

19 HoloWall sited within the Arts Quad at Cornell University

have been discarded and landfilled, mixed reality technology provides an alternative solution to recycle and manage the discarded wood resource. Therefore, this research investigated MR-guided design processes using nonuniform wood material to develop the customized HCCLT installation, a HoloWall prototype (Figures 17, 18, 19). From material preparation to the fabrication process, the MR-informed protocol provided users with an intuitive hybridized work environment for material processing and fabricating. The user-friendly MR environment enabled both expert and nonexpert users to easily understand the materials' irregularity and their usage. As a further architectural application, the proposed HoloWall load-bearing wall system is expected to open up the possibility of the HCCLT system as a structural component by integrating emerging technology like MR and AR processes.

ACKNOWLEDGMENTS

HoloWall is a research project by the Cornell Rural-Urban Building Innovation (RUBI) Lab. Full-scale prototype team members: Nina Koscica, Sahir Choudhary, Jiyoon Bae, Sahil Adnan, Bushra Aumir, Asbiel Samaniego. This project is partially funded by the Cornell Council of the Arts and the Aref and Manon Lahham Excellence Fund for Sustainability.

REFERENCES

Bonner, Jennifer. 2018. "Haus Gables." Single Family Residence. Atlanta, Georgia. Accessed March 10, 2021. https://jenniferbonner.com/01-Haus-Gables.

Brandner, R., G. Flatscher, A. Ringhofer, G. Schickhofer, and A. Thiel. 2016. "Cross Laminated Timber (CLT): Overview and Development." *European Journal of Wood and Wood Products* 74: 331–351. https://doi.org/10.1007/s00107-015-0999-5.

Chini, Abdol R., and Stuart F. Bruening. 2003. "Deconstruction and Materials Reuse in the United States." *International Electronic Journal of Construction, The Future of Sustainable Construction* (Special Issue).

Cochran, K. M. and Timothy G. Townsend. 2010. "Estimating Construction and Demolition Debris Generation Using a Materials Flow Analysis Approach." *Waste Management* 30 (11): 2247–2254.

Goepel, Garvin, and Kristof Crolla. 2021. "Augmented Feedback: A case study in Mixed-Reality as a Tool for Assembly and Realtime Feedback in Bamboo Construction." In *ACADIA 21: Realignments, Toward Critical Computation; Proceedings of the 41st Annual Conference of the Association for Computer Aided Design in Architecture*.

Huang, Haoyu, Xiaoqi Lin, Junhui Zhang, Zhendong Wu, Chang Wang, Brad Jianhe Wang. 2021. "Performance of the Hollow-Core Cross-Laminated Timber (HC-CLT) Floor under Human-Induced Vibration." *Structures* 32: 1481–1491.

Lok, Leslie., Asbiel Samaniego, Lawson Spencer. 2021. "Timber De-Standardize." In *ACADIA 21: Realignments: Toward Critical Computation; Proceedings of the 41st Annual Conference of the Association for Computer Aided Design in Architecture*.

19

Luppold, William, and Matthew Bumgardner. 2003. "What is Low-Value and/or Low-Grade Hardwood?" *Forest Products Journal* 53 (3): 54–59.

Mallo, Maria Fernada Laguarda, and Omar Espinoza. 2015. "Awareness, Perceptions and Willingness to Adopt Cross-Laminated Timber by the Architecture Community in the United States." *Journal of Cleaner Production* 94: 198–210.

Mayencourt, Paul, and Caitlin Mueller. 2019. "Structural Optimization of Cross-Laminated Timber Panels in One-Way Bending." *Structures* 18: 48–59.

Menegaki, Maria and Dimitris Damigos. 2018. "A Review on Current Situation and Challenges of Construction and Demolition Waste Management." *Current Opinion in Green and Sustainable Chemistry* 13: 8–15.

Nagapan, Sasitharan, Ismail Abdul Rahman, Ade Asmi, Aftab Hameed Memon, and Imran Latif. 2012. "Issues on Construction Waste: The Need for Sustainable Waste Management." In *IEEE Colloquium on Humanities, Science and Engineering (CHUSER 2012)*. Kota Kinabalu, Malaysia. 325–330.

Newnham, Cameron, Nick van den Berg, and Gwyllim Jahn. *Fologram*. V.3.5. Fologram. Rhino 3D. 2017.

Simscale GmbH. *Simscale*. Web browser. 2013.

Skidmore, Owings & Merrill. 2021. "SPLAM Timber Pavilion." Chicago. Accessed December 21, 2021. https://www.som.com/research/splam-timber-pavilion/.

Steiger, Rene, and Arne Gulzow. 2010. "Validity of Bending Tests on Strip-Shaped Specimens to Derive Bending Strength and Stiffness Properties of Cross-Laminated Solid Timber (X-lam)." In *Final Conference of COST Action E53, The Future of Quality Control for Wood & Wood Products*. Edinburgh: Edinburgh Napier University, Forest Products Research Institute / Centre for Timber Engineering.

Tolaymat, T., M. Krause, J. Smith, and T. Townsend. 2017. "The State of the Practice of Construction and Demolition Material Recovery." U.S Environmental Protection Agency, Washington, DC, EPA/600/R-17/231.

Yconst. *Packrat*. V. 0.6. Food4Rhino. Grasshopper. 2012.

IMAGE CREDITS
All drawings and images by the authors.

Leslie Lok is Assistant Professor at Cornell University and directs the Rural-Urban Building Innovation (RUBI) Laboratory. The lab investigates new building methods that couple digital construction technologies with natural and non-standardized material for the design of adaptable buildings in rural-urban contexts. She is also a co-founder at HANNAH, an experimental design practice that utilizes innovative forms of construction to advance building design.

Jiyoon Bae is a recent graduate of Master of Science in Matter Design Computation from Cornell University. He received his Master of Architecture degree at Syracuse University. He is currently a PhD student at the University of Pennsylvania Stuart Weitzman School of Design.

Tailoring Bending Behavior

3D Printing-Based Variable-Density Materials Design for Bending-Active Structure

Lei Gong
Tongji University

Xinjie Zhou
Tongji University

Hua Chai
Tongji University

Junguang Liu
Tongji University

Philip F. Yuan*
Tongji University

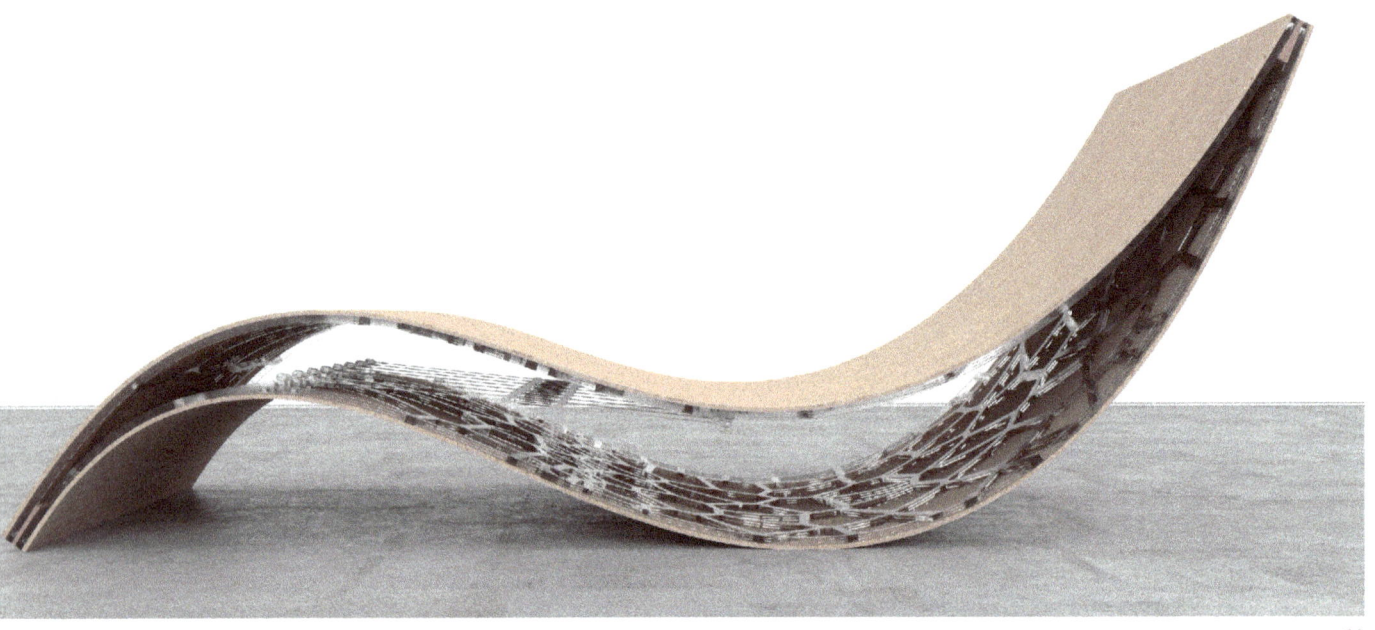

1A Rendering of the bending chair

ABSTRACT

Bending-active structures have drawn considerable attention in the past decades of research and practice. However, most existing bending active structures are made of homogeneous materials with constant bending properties, making it difficult to achieve complex design intentions. This paper presents a novel hybrid material design strategy that enables the realization of curved active structures with complex geometries. This hybrid material consists of birch plywood and 3D printed PETG. The bending behavior of the hybrid material can be adjusted by changing the density of the 3D-printed part.

First, the variable density material is designed according to the target shape of the bending-active structure. A form-driven approach based on Graphic Statics is introduced to find the EI distribution of materials that can realize the desired bending shape under a specific restraining system. Then the distribution will be translated into the density variation of 3D additive printing pattern density. Next, bending experiments with mixed materials of different densities were conducted to verify the effectiveness of variable density materials. Unlike previous methods that blindly searched for forms that bending-active structures could form, this method achieves the target form by tailoring the density of the material.

This paper attempts to combine decoration and structure by using variable density patterns to provide non-uniform structural properties to the hybrid material. This method does not require molds and can be used for façade and wall designs.

INTRODUCTION

Bending active structures are defined as a class of lightweight structural elements that utilize the elastic deformation of straight or flat elements, usually sheets or rods, to achieve curved geometries (Guerguis et al. 2020). One classic example of pure bending-active structures is Buckminster Fuller's *Plydome* (Figure 1B). Its construction principle approximates the basic geometry of a sphere with a polyhedron made of identical plywood, which is fixed together by bending the plates at their corners (Bruetting et al. 2017).

Previous studies have explored multiple ways to control bending-active behavior. *Berkeley Weave* (Figure 2) and *timbRfoldR* (Figure 3) utilize large monolithic materials to form large bending-active structures (Schleicher et al. 2016; Fleischmann et al. 2011), while *Double-Layered Lightweight Load-Bearing Structures* (Figure 4) use relatively small size components (Stanojevic et al. 2019). These structures are made of homogenized materials, and the different forms are achieved by the constraints on the boundaries between the components. In addition to the above methods, the bending behavior can be controlled by designing different 3D printing patterns (La Magna 2017) and the thickness of the material (Boulic et al. 2020). Compared with the boundary control method, controlling the material properties to achieve the design goal is more precise in controlling the final form and provides more design possibilities.

Based on graphic statics principles, Boulic and Schwartz explore a design method for Bending-active structures that derives material properties from form (Boulic et al. 2020). According to this approach, the constraint system and the varying bending stiffness (EI) can be used to control the shape of the active bending element (Boulic et al. 2017). However, the study still used homogeneous materials, thus limiting the form of bending-active structures.

Many studies have investigated the topology optimization methods based on variable density materials (Naboni et al. 2019; Do et al. 2021), and the method of variable density is also used to control the bending-active behavior (Rodriguez et al. 2022). The variable density approach not only improves the performance of components and saves materials, but the rich mechanism of its formation also provides more possibilities for architectural aesthetics.

Based on this, this paper attempts to combine decoration and structure by using variable density patterns to provide non-uniform structural properties to the hybrid material. Wood is one of the most suitable materials for bending-active structures, and 3D printed plastics can easily print patterns of various densities, so this paper combines wood and 3D printed plastics to build a composite variable density material.

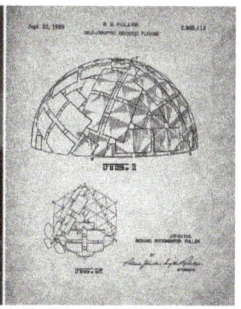

1   Buckminster Fuller's *Plydome*

2   *Berkeley Weave* installed at the courtyard of UC Berkeley's College of Environmental Design (CED).

3   ITECH M.Sc. Thesis Project 2020: *timbRfoldR*

4   *Double-Layered Lightweight Load-Bearing Structures*

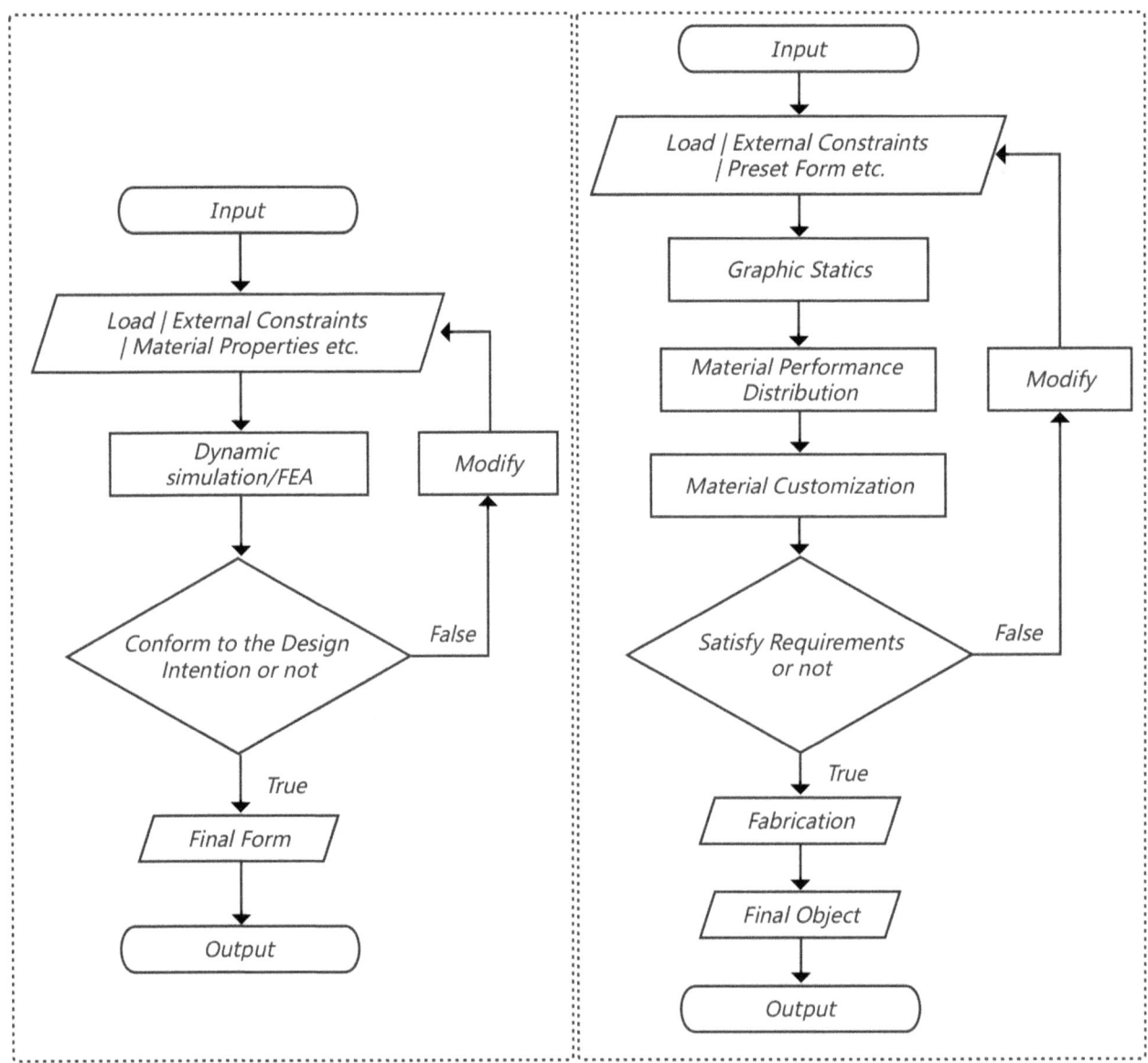

5   Two different bending-active structure design strategies

By combining other materials, the bending properties of the wood itself will be enhanced (Johns et al. 2000), and at the same time, multi-material structures offer new opportunities for the integration of architectural decoration and structure (Snooks 2022).

METHODOLOGIES
Theoretical principle

The traditional form-finding approaches do not offer direct control over the consequential equilibrium shape. To control the target form precisely, tedious and inefficient iterative attempts must be taken to achieve the initial design intention. The main difference between form-finding and form-driven approaches lies in whether the reverse process is realized (Boulic et al. 2018). The final design form is used as the input parameter, and the corresponding material performance parameter is used as the output value (Figure 5).

Essentially, the curvature of a spline and its bending stiffness are related. The form-driven approach is based on two strategies, which can also be combined, as follows: the bent equilibrium geometry can be obtained either through adapted bending stiffness or through additional restraining forces.

Regarding the form-driven approach, the initial target bending curves are discretized. In the equilibrium process of graphic statics, any mechanic force acts directly on node points. Therefore, the bending moment generated by Bending is transformed into a pair of shear forces applied to the node.

The shear forces are respectively (Boulic et al. 2017):

$$S_{i,i-1} = (EI)_i \frac{2\sin\alpha_i}{l_{i-1} \cdot l_{i-1,i+1}} ; S_{i,i+1} = (EI)_i \frac{2\sin\alpha_i}{l_i \cdot l_{i-1,i+1}}$$

Where (Figure 6):

$S_{i,i-1}$ and $S_{i,i+1}$ refer to the magnitude of the two pairs of shear forces applied at node i.

α_i refers to the magnitude of the complementary angle formed by the two lines adjacent to node i.

l_{i-1} refers to the length of the line between node i-1 and node i.

$l_{i-1,i+1}$ refers to the length of the line between node i-1 and node i+1.

$(EI)_i$ refers to the flexural stiffness at node i.

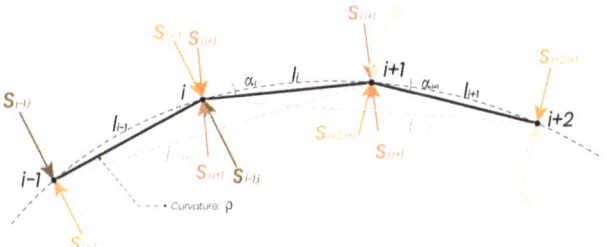

6   Bending member's internal force representation

After the equilibrium of force diagram, the EI distribution adapted for the given loads can be calculated. The intervention of graphic statics makes the solving process intuitive and readable, which eventually feeds back a series of discrete EI data to guide the customization of material properties.

Material Behavior
Several scholars have investigated the bending properties of multi-material sheets (Chattopadhyay et al. 1996; Zhang et al. 2008). However, because of the complexity of the pattern, the cross-section is different at each location, so the weighted average method (Junqing et al. 2022) is taken in this paper to approximate the EI, and it is assumed that the different materials are tightly coupled with each other. The calculation formula is as follows:

$$EI = \sum_{i=1}^{n} E_i \left(\frac{b_i h_i^3}{12} + (y_i - y_0)^2 \times b_i h_i \right)$$

Where:

E_i, b_i and h_i represent the modulus of elasticity, width and height of the ith part of the section.

y_0 refers to the position of the neutral axis of the entire section.

y_i refers to the position of the neutral axis of the ith part of the section.

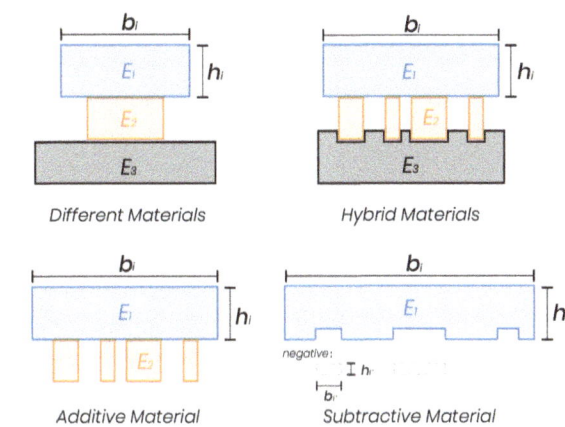

7   Multiple compounding methods

From the above equation, it can be seen that the flexural stiffness can be adjusted by changing the size or elastic modulus of the cross-section. However, changing the bending modulus of a material is rather challenging to implement, even if theoretically possible. An alternative solution is to consider the macroscopic combination of hybrid materials; the most typical example of it is the reinforced concrete materials, which are reinforced with steel to enhance the tensile properties of pure concrete structures to meet higher usage requirements, which inspired us to adopt the methodology of hybrid materials.

STRATEGIES
Implementation of Variable Section Moment of Inertia
The method of adjusting the moment of inertia of the section is essential to control the shape of the active bending system. This paper puts forward several schemes to adjust the section moment of inertia and analyzes the advantages and disadvantages of each scheme (Figure 7).

Material modification operations from the cross-sectional moment of inertia are divided into two main categories: the first, manipulation of the material thickness, and another, manipulation of the material width.

Manipulation of the thickness of the material:

- The thickness of the material can be adjusted by digital milling technology to meet the requirements. This method is efficient, has low construction difficulties, and can be performed without complex equipment. However, the subtractive behavior will inevitably cause material waste. In large-scale construction, the raw material size is too large, and the processing cost is magnified exponentially.

- Additive manufacturing offers another way. The thickness issue fits well with the construction logic of additive technologies, where the thickness is obtained through laminar stacking.
- However, Additive manufacturing is too dependent on the quality of printing parameters.

Manipulation of the width of the material:

- The advantage of additive processes, especially robotic 3D printing technology, lies in shaping the thickness of the material rather than the width. To manipulate the width of the geometry, a sequential planar extrusion is required.
- Subtractive manufacturing makes it easy to change the width of the material. However, there is still the problem of material waste.

Control the equivalent width (Figure 8) by the material density:

- The method achieves overall width control through material density, which takes inspiration from spatial structures such as honeycomb and cellular structures (La Magna et al. 2018). Solids and voids are arranged according to a pattern to construct a tunable equivalent cross-sectional width.
- The subtractive manufacturing strategy deals with material density by removing some parts of the material. The remaining solid parts can be involved in calculating cross-sectional moments of inertia as a unified whole. When the pattern density is high, only a small amount of subtraction is needed to achieve the effect. However, it is not applicable to the case of low pattern density. Additive manufacturing, on the other hand, does not need to worry about this problem.
- Summarizing several section moment of inertia customization schemes mentioned above, we can conclude that additive manufacturing has significant advantages in material efficiency; regarding process parameters and technical difficulty, subtractive manufacturing is more feasible. However, for the form-driven approach, the ability to fit the flexural stiffness along the axial direction as accurately as possible is the key to the whole process, so the means based on additive manufacturing to adjust the equivalent width becomes the solution adopted in this paper.

Additive printing prototype

Under a certain number of printing layers, geometric objects with different section moments of inertia can be manufactured by printing patterns with different densities. Similarly, the same effect can be achieved by varying the number of layers printed while keeping the print pattern density constant. Therefore, additive manufacturing needs to weigh the above two printing prototypes when facing specific problems.

The former is easier to print. Its disadvantage is that it may make the pattern too dense or too sparse. Because once the print thickness is constrained, the pattern of the structure may appear very dense and very sparse in parts. In the form-driven approach, the bending stiffness result of a specific shape is difficult to predict. So the material modification operations can only passively approximate the solution data to ensure that there will be no unacceptable deviation or error between the curvature in the practical and ideal state. The latter ensures the uniformity of the pattern, however, it is not easy to print. Multiple layers need to be printed to make target objects with different thicknesses. If the print path is not continuous, the robot must start and often stop, which greatly increases the probability of failure.

In a word, any single additive printing prototype has its limitations. Combining the advantages of both ways, printing variable density patterns in the horizontal direction and printing multiple layers in the vertical direction, we can accurately match the bending stiffness value of the target geometry.

Comparison of Different Printing Patterns

The mechanical properties of additive printing materials result from the joint action of multiple factors, in which the design of the pattern plays a dominant role. Therefore, the use of a two-dimensional pattern design to reflect the one-dimensional linear bending stiffness distribution is the focus of material modification in this paper.

Based on the ideal state, the bending mechanical properties of

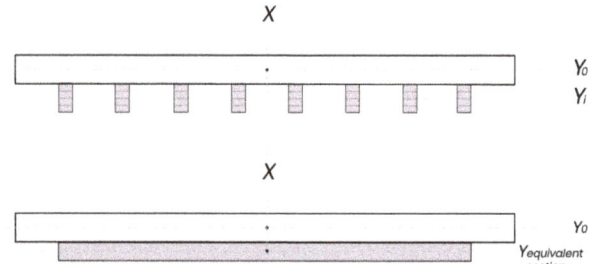

8   Equivalent width

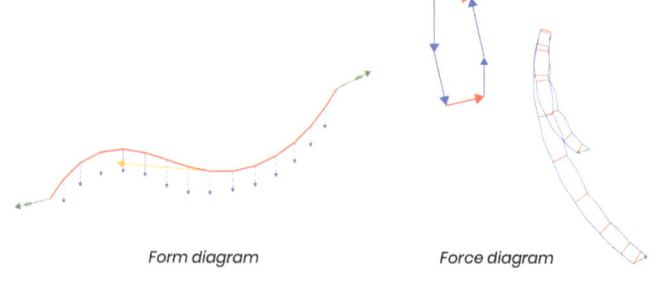

9   Graphic statics calculation

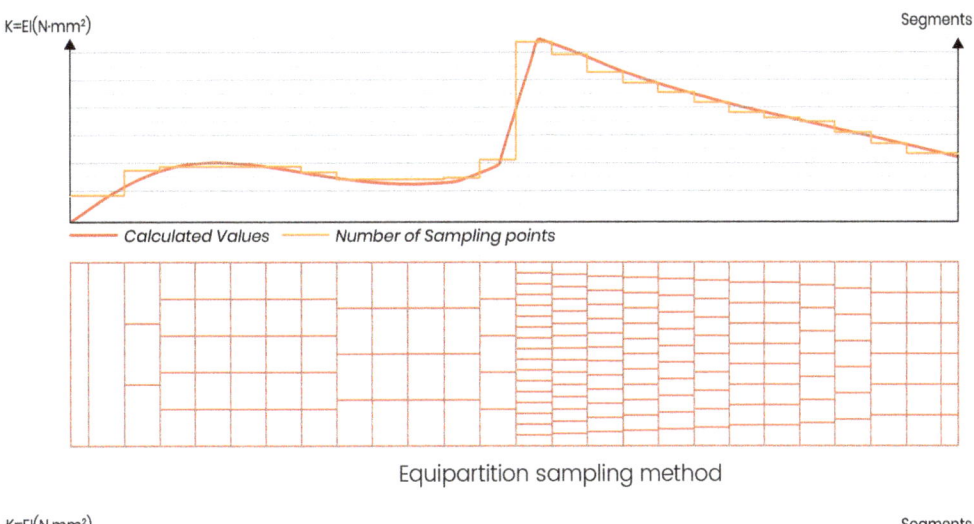

Equipartition sampling method

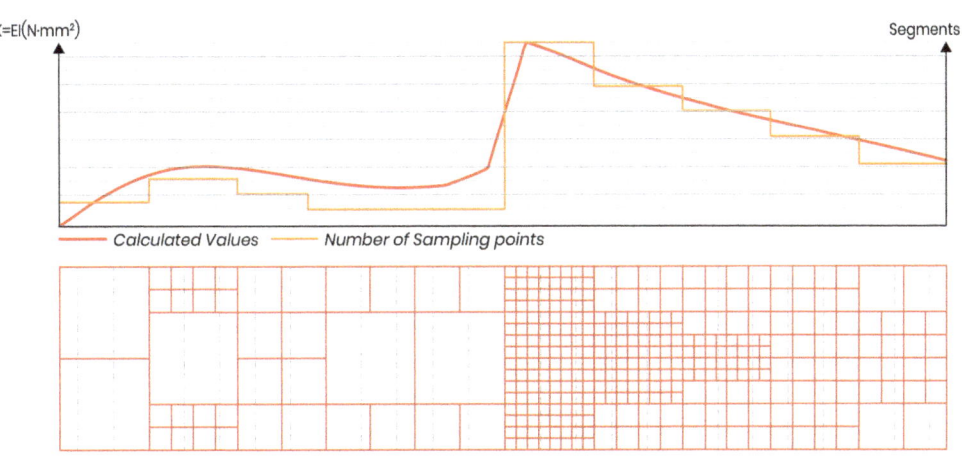

Subdivision sampling method

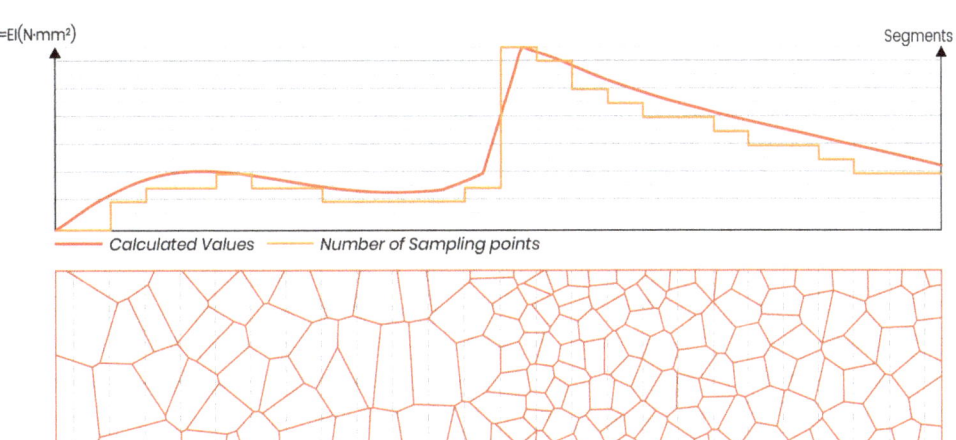

Voronoi point random sampling method

10 Calculated flexural stiffness distribution and sampling points of three different patterns

the printing components generated by three different printing patterns in the actual printing state are investigated in this paper. Figure 9 shows the form diagram and force diagram of the target curve curvature under ideal conditions through graphic statics calculation.

As shown in Figure 10, three strategies are used to generate sampling points of variable density. The first strategy uses the method of equal division, where the design area is divided equally according to the direction of gradient change, and each interval is adjusted for density using equal division. The second strategy uses the subdivision method, constantly subdividing panels to meet density requirements. The third strategy uses voronoi points random sampling method to adapt to changes in density by adjusting the probability of generating points at different locations, and ultimately uses Voronoi graphs to make them printable. The distribution of all above three patterns is based on the same area to achieve density change.

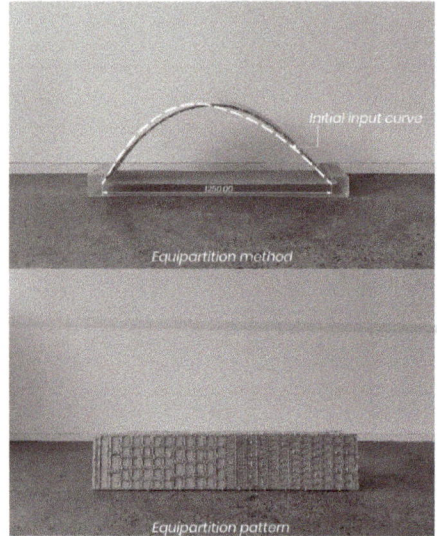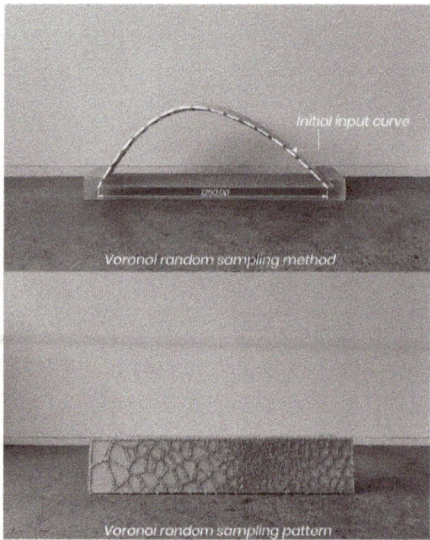

11 Bending behaviors of three different patterns

12 Bending-active hybrid structure with Voronoi pattern

We printed three test samples. These samples were all made of 5 mm thick birch plywood laminated with 9 mm thick PETG printed layers, 300 mm wide and 1500 mm long. The wood and printed layers were secured with a rope between them. Then they are all fixed on supports of 1250 mm long (Figure 11).

The final results show that variable density patterns using random sampling methods (Figure 12) can achieve the design goals more accurately. So the final case of this paper will be designed in this way.

CASE STUDY

The above multiple groups of experiments have demonstrated that the pattern variable density design method based on graphic statics could be applied to the construction process of non-uniform curvature structures. As a consequence, a matched design and fabrication process based on reverse engineering thinking and graphic statics is proposed (Figure 13).

A chair with non-uniform curvature is taken as the final design objective. Firstly, the curvature of the chair outline contour profile is derived from the point cloud data of ergonomics; Secondly, the axial bending stiffness data obtained from the graphic statics algorithm is remapped into the distribution density data of the cellular pattern, and the continuous printing path of the robot is generated; Finally, the final fabrication and assembly procedure was completed through the onsite KUKA kr90 r2900 robots (Figure 14).

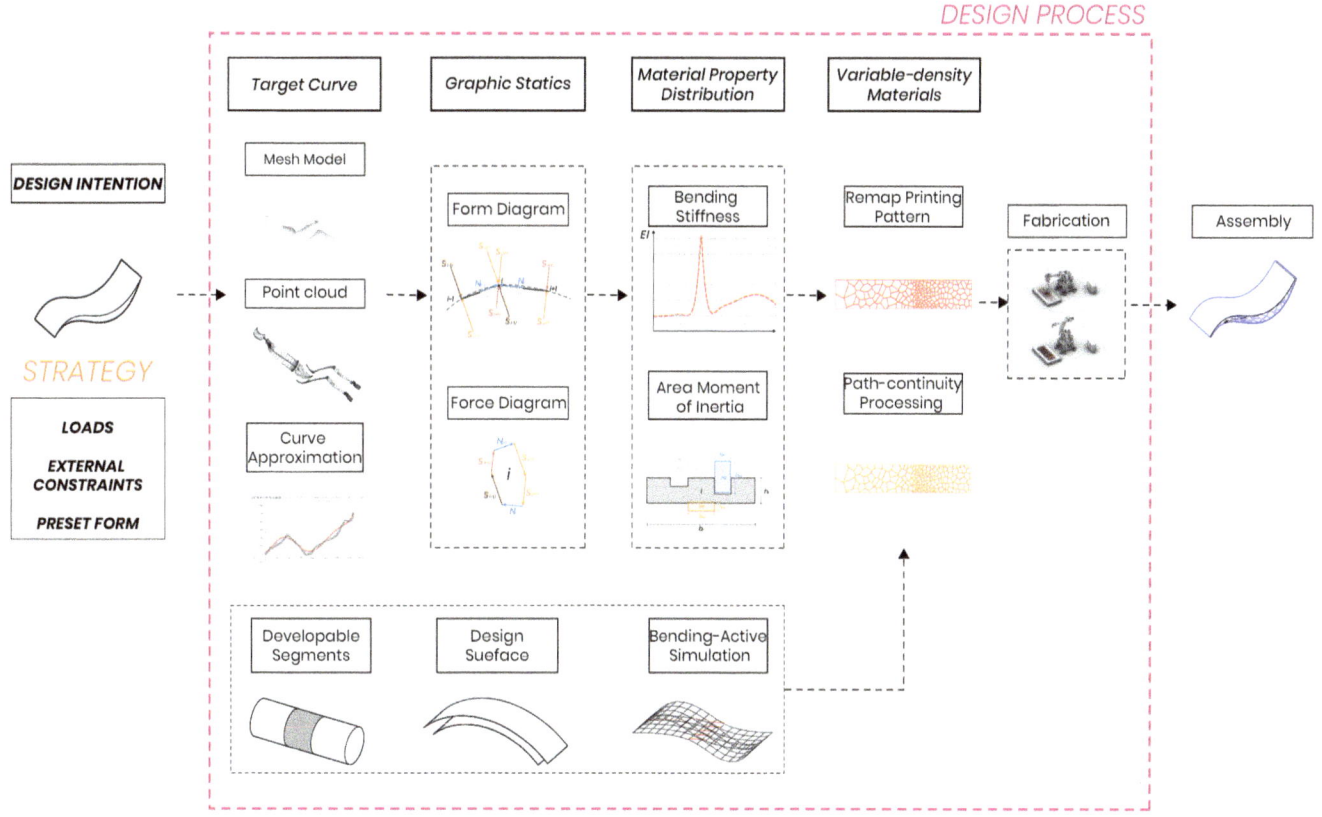

13 The matched design potential and process

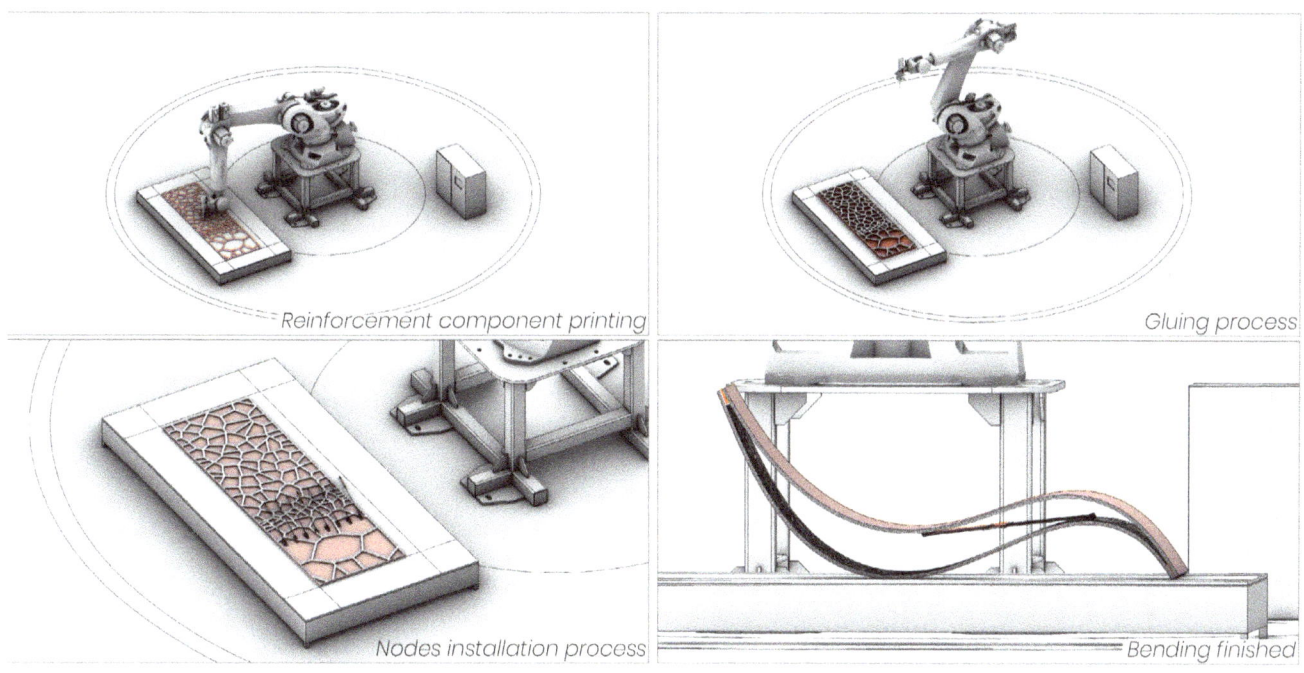

14 Digital fabrication procedures

PERFORMATIVE DESIGN: STRUCTURAL AND MATERIAL SYNTHESIS

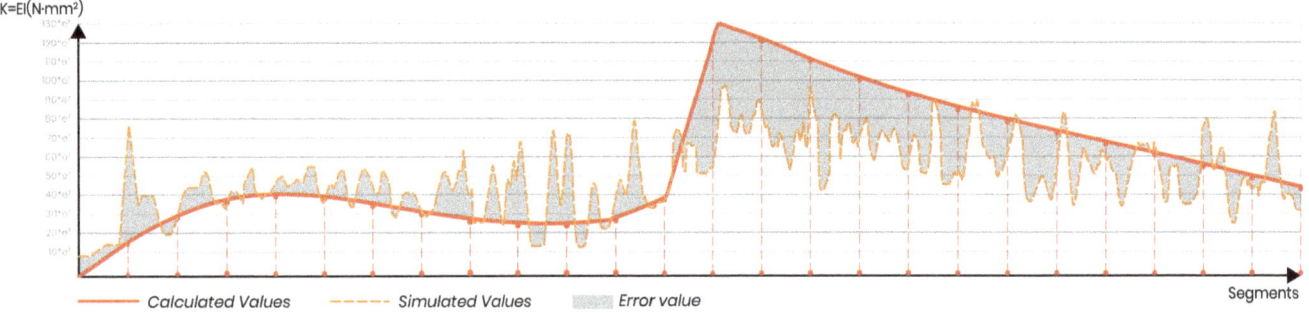

15 Error and deviation

A row of cables was added to the structural system to achieve the basic shape of the chair and the form is determined by both the plywood panels and the 3D printed layers of variable density (Figure 1A).

DISCUSSION

In this paper, we propose a hybrid material with variable density. Through experiments, we found that the actual effect of the Voronoi model based on random sampling points is better than the effect of the subdivision method and the equalization method, and the morphology formed by this method bending is more in line with expectations.

However, the way we perform point sampling in Voronoi has limited ability to fit the EI design values. The centroid of the cellular unit—the independent variable for printing pattern density—cannot precisely describe the expected mechanic data. It can only be obtained by the integer multiple of the flexural stiffness of the unit section made from 3D printing technology, thus, generating errors. As is shown in Figure 15, curve 1 represents the EI value in the ideal state, curve 2 represents the EI value in the actual state, and the gray area represents the error between the two. The properties of the modified material should be as consistent as possible with the real situation.

Even though some measures have been actively taken at each step of the workflow to optimize and reduce errors, the hybrid materials made of plywood and PLA prints inevitably could not achieve the required flexural stiffness distribution in the ideal case due to the technical parameters of additive printing and the inherent characteristics of the material.

At the same time, the error accumulation resulting from the approximation of the calculation process is not negligible. The section of the additive printing trajectory is a rounded rectangle rather than a rectangle. The error caused by how the cross-section moment of inertia is solved also makes the final bending morphology prediction difficult.

CONCLUDING REMARKS

The paper is trying to propose an original workflow to construct a bending-active structure system through graph statics and a hybrid material system. We take chair design and manufacturing as an example to demonstrate the workflow. There are several techniques used in the workflow. The form-driven strategy based on graphic statics is used to calculate the distribution of the bending stiffness, and the variable-density materials derived from gradient 3D printing pattern are used to match the output value of the specific bending-active objective.

This paper attempts to combine decoration and structure by using variable density patterns to provide non-uniform structural properties to the hybrid material. This approach will also provide new possibilities for façade and wall design

ACKNOWLEDGMENTS

The presented case studies were supported and funded by National Natural Science Foundation of China (Grant No.U1913603), Ministry of Housing and Urban-Rural Development (Grant No.2021-R-085) and Shanghai Science and Technology Committee (Grant No.21DZ1204500).

A special thanks to Fab-Union for its support to the course.

REFERENCES

Boulic, Léa, Pierluigi D'Acunto, Federico Bertagna, and Juan José Castellón. 2020. "Form-Driven Design of a Bending-Active Tensile Façade System." *International Journal of Space Structures* 35 (4): 174–90. https://doi.org/10.1177/0956059920931021.

Boulic, Léa, and Joseph Schwartz. 2017. "Graphic Statics Principles for the Design of Actively Bent Elements Shaped with Restraining Systems." In *Proceedings of the IASS Annual Symposium 2017; Interfaces: Architecture Engineering Science*. Hamburg.

Boulic, Léa, and Joseph Schwartz. 2018. "Design Strategies of Hybrid Bending-Active Systems Based on Graphic Statics and a Constrained Force Density Method." 2018. In *Proceedings of IASS Annual Symposia*

Bruetting, Jan, Axel Körner, Daniel Sonntag, and Jan Knippers. 2017. "Bending-Active Segmented Shells." In *Proceedings of the IASS Annual Symposium 2017*. Hamburg.

Chattopadhyay, A., and H. Gu. 1996. "Exact Elasticity Solution for Buckling of Composite Laminates." *Composite Structures* 34 (3): 291–99. https://doi.org/10.1016/0263-8223(95)00150-6.

Do, Quang Thang, Cong Hong Phong Nguyen, and Young Choi. 2021. "Homogenization-Based Optimum Design of Additively Manufactured Voronoi Cellular Structures." *Additive Manufacturing* 45 (September): 102057. https://doi.org/10.1016/j.addma.2021.102057.

Fleischmann, Moritz, and Achim Menges. 2011. "ICD/ITKE Research Pavilion: A case study of multi-disciplinary computational design." In *Computational Design Modelling, Proceedings of the Design Modeling Symposium Berlin*, edited by C. Gengnagel, A. Kilian, N. Palz, and F. Scheurer. Berlin; Heidelberg: Springer. 239–248.

Guerguis, Maged S., Alex Stiles, and John-Michael Worsham. 2020. "Hybrid Bending Active Systems: A Novel Application of Carbon Fiber in Lightweight Structures." In *Proceedings of IASS Annual Symposia, 2020*. International Association for Shell and Spatial Structures (IASS). 1–12.

Johns, Kenneth C., and Simon Lacroix. 2000. "Composite Reinforcement of Timber in Bending." *Canadian Journal of Civil Engineering* 27 (5): 899–906.

Junqing, Zhao, Zhen Yubao, Zhou Peng, Wang Jun, and H. U. Hengshan. 2022. "Methods for Determining the Neutral Axis Position and Stress Analysis of Composite Beam[1]." *Mechanics in Engineering* 44 (1): 184–187.

La Magna, Riccardo. 2017. "Bending-Active Plates: Strategies for the Induction of Curvature through the Means of Elastic Bending of Plate-Based Structures," PhD diss., University of Stuttgart: Institut für Tragkonstruktionen und Konstruktives Entwerfen.

La Magna, Riccardo, and Jan Knippers. 2018. "Tailoring the Bending Behaviour of Material Patterns for the Induction of Double Curvature." In *Humanizing Digital Reality*, edited by K. De Rycke, C. Gengnagel, O. Baverel, J. Burry, C. Mueller, M. M. Nguyen, P. Rahm, and M. Ramsgaard Thomsen. Singapore: Springer Singapore. 441–52. https://doi.org/10.1007/978-981-10-6611-5_38.

Naboni, Roberto, Luca Breseghello, and Anja Kunic. 2019. "Multi-Scale Design and Fabrication of the Trabeculae Pavilion." *Additive Manufacturing* 27 (May): 305–17. https://doi.org/10.1016/j.addma.2019.03.005.

Rodriguez, Emmanuel, Georges-Pierre Bonneau, Stefanie Hahmann, and Mélina Skouras. 2022. "Computational Design of Laser-Cut Bending-Active Structures." *Computer-Aided Design* 151 (Oct): 103335.

Schleicher, Simon, and Riccardo La Magna. 2016. "Bending-Active Plates: Form-Finding and Form-Conversion." In *ACADIA 16: Posthuman Frontier; Proceedings of the 36th Annual Conference of the Association for Computer Aided Design in Architecture*. 260–269.

Snooks, Roland. 2022. "Behavioral Tectonics: AgentBody Prototypes and the Compression of Tectonics." *Architectural Intelligence* 1 (1): 1–14.

Stanojevic, Djordje, and Kenryo Takahashi. 2019. "Strip-Based Double-Layered Lightweight Timber Structure." In *Proceedings of IASS Annual Symposia 2019*. International Association for Shell and Spatial Structures (IASS). 1–8.

Zhang, Da-Guang, and You-He Zhou. 2008. "A Theoretical Analysis of FGM Thin Plates Based on Physical Neutral Surface." *Computational Materials Science* 44 (2): 716–20. https://doi.org/10.1016/j.commatsci.2008.05.016.

IMAGE CREDITS

Figure 1B: © R. Buckminster Fuller, 1957
Figure 2: © Simon Schleicher and Riccardo La Magna, 2016
Figure 3: © Gabriel Rihaczek, Maximilian Klammer, and Okan Basnak
Figure 4: © Héctor Pineda, 2020; image courtesy of Centro de Estudios Superiores de Diseño de Monterrey CEDIM

All other drawings and images by the authors.

Lei Gong is a Master of Architecture student of Tongji University.

Xinjie Zhou is a PhD student studying Architecture at Tongji University.

Hua Chai is a postdoctoral researcher at the College of Architecture and Urban Planning (CAUP) of Tongji University. His research focuses on the computational design and robotic fabrication of timber structures, with a special attention on advanced robotic production technologies for complex timber building systems.

Junguang Liu is a Master of Architecture student of Tongji University.

Philip F. Yuan is professor and doctoral supervisor of the College of Architecture and Urban Planning (CAUP) at Tongji University, and served as Thomas Jefferson Professor at the University of Virginia (2019), visiting professor at the Massachusetts Institute of Technology (2019), and the Royal Melbourne Institute of Technology (2021).

Session Introduction

New Ecologies II: Responsive/Adaptive Design Methods

William Braham, Chair

The projects presented in this session focused on adaptive design methods, mostly rule-based approaches using different materials and fabrication techniques. The exception is the project by Marengo and Neri, which uses an algorithm based on circuit theory to enhance ecological connectivity within urban areas, producing a kind of route-finding recommendation for vegetation to connect. The rest of the session involves different design and fabrication of physical prototypes, three papers using robotic fabrication techniques, and a fourth using digitally-designed formwork to create customized and optimized rammed earth walls.

The interesting questions raised by the group involves the role of the constraints imposed by the choice of material and the method of construction. The project by Gupta and Cupkova begins with the goal of enlarging the formal possibilities of rammed earth, accepting the materials and methods of construction, which focus their innovations on the generation of formwork and its optimization for self-shading, with sequential stacking of formwork for greater flexibility and a proposal for process feedback with computer vision evaluation.

The project by El MAhdy, El-Rahim, and Atassi explores 3D printing of shaped, hollow wall elements with clay, configured to enhance wind-driven ventilation to cool the assembly. Prototypes are tested at small scale, then enlarged to full-size, revealing many features that had to be refined as the scale effected structural and fluid dynamic behavior. A second project using additive robotic printing by Zidek, Aman, Li, and Aghaeir-Meibodi uses thermoplastic to build prototypes of a layered wall assembly to support green plants, their soil beds, and a watering feeding network. The capabilities of thermoplastic, especially structural adhesion under stress resulted in a series of design refinements to avoid failure points. The final robotically fabricated project by Marcus, Kim, and Reichert is a computationally generated pattern printed on a film applied to the inner layer of glass in a public walkway. The nature of the project allows for multiple full scale mockups tested on site and continual refinements before the final production.

Discretizing Low-tech Adaptive Rammed Earth Formwork

Geometric Figurations for Thermally Performative Earthen Wall Panels

Pragya Gupta
Carnegie Mellon University

Dana Cupkova
Carnegie Mellon University

1 Rammed earth wall section prototype, 18 in x 18 in

ABSTRACT

Rooted in a hybrid material and climate-based approach to design, this study proposes a computational design framework for low-tech rammed earth adaptable formwork that allows for variable surface figuration, related to thermal and aesthetic design parameters. Built as vertical panel prototypes, as in-situ vertical construction, this study proposes to couple thermal performance with sequenced constructability of varied surface geometries through an adaptable repetitive kit-of-parts formwork that can be constructed with limited advanced manufacturing capabilities. Addressing the need for low-impact sustainable construction in vast areas of the world where thermal mass, coupled with earthen construction, would radically contribute to the reduction of carbon footprint, this study builds upon the extensive cultural and ecological history of rammed earth in those regions. Providing a novel methodology for shaping in-situ vertical rammed earth design to construction, this study utilizes ray-tracing simulation that sets constraints for geometric constructability. Our goal is to address the perception of earthen construction as a low-tech natural anachronistic material and expand its design repertoire while addressing a climate-sensitive approach to shaping architectural systems. Here, ray-tracing analysis is used to determine the geometric capacity for continuous load bearing of earthen material being rammed vertically, while enhancing its thermal mass effectiveness through surface geometry. This method expands on the design vocabulary of the earthen construction beyond highly simplified shapes, or material color variation, and also differs from other methods that use complex horizontally laid formwork to erect panels vertically post-curing. This study aims to embrace and build upon vernacular practices of rammed construction, with added design variability, rooted in local circular material practice.

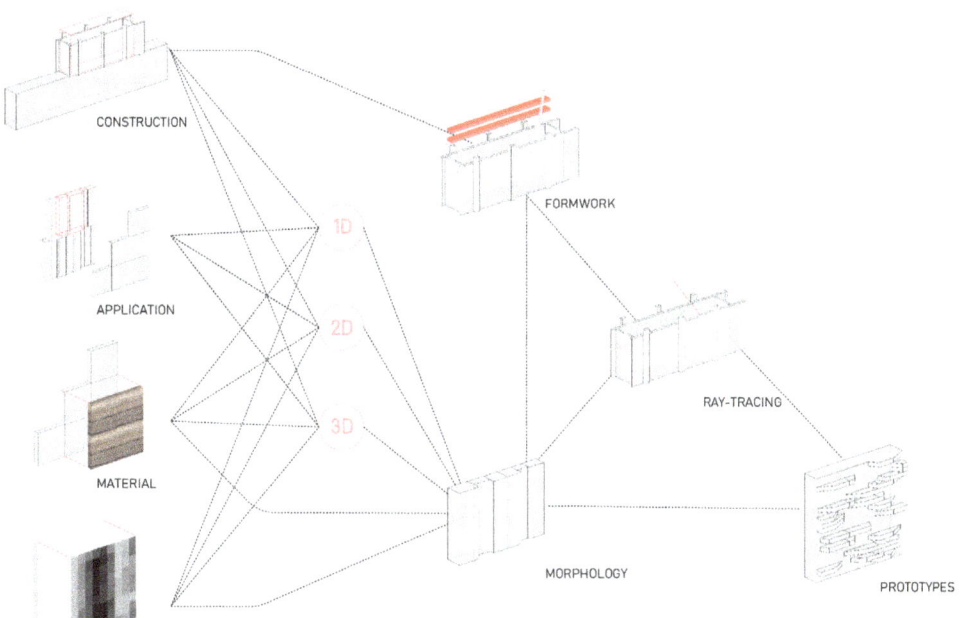

2 Variables constraining the process of rammed earth formation: diagram of digital workflow fabrication and physical prototype construction, setting up a systematic study to determine geometric freedom of variable forms for rammed earth panels based upon current construction technology, performance criteria, and material properties

INTRODUCTION

Rammed earth construction is an ecologically benign alternative to prevalent non-biodegradable construction materials, with the possibility of utilizing locally sourced materials. It is a healthy, non-toxic material that requires minimal industrial processing, while providing healthier indoor air quality and thermal comfort. This study explores the effectiveness of vertical rammed earth construction in hot and arid climates, and aims to identify digital-to-analog workflows to introduce greater design variability in rammed earth. Digital fabrication technologies focusing on additive manufacturing, robotics, and the design engineering of cementitious materials have advanced significantly in recent times (Asprone et al. 2018). For the most part, those studies expand on automation through the introduction of robotic manufacturing, morphological explorations, bespoke fabrication of earthen systems, or direct Life Cycle Analysis to validate engagement of earthen construction regarding carbon footprint reduction (Reddy and Kumar 2010). Building upon this greater knowledge framework, this research lies at the intersection of digital simulation capabilities and a climate-based approach to shaping architectural materials, in providing a low-cost and low-impact construction methodology applicable in vast areas of the world—where thermal mass coupled with earthen construction would radically contribute to the reduction of carbon footprint. This study explores surface figuration for rammed earth construction through the introduction of ray-tracing simulation and a reconfigurable adaptive formwork design, and builds on previously validated research focused on calibrating the thermal exchange rate of vertical surface geometries to improve both the aesthetic and thermodynamic performance of passive heating and cooling systems in buildings (Cupkova and Azel 2015) and a thermal simulation study comparing the performance of rammed earth assemblies with mainstream wood and concrete assemblies, that account for both heat resistivity and capacity (Gupta et al. 2020). This workflow combines geometric figuration of thermal mass with analytical digital tools for the formation of novel surface tectonics, coupled with low-tech constructability (Figure 2).

SITUATING RAMMED EARTH CONSTRUCTION

Vernacular architecture has its materiality deeply rooted in social and cultural contexts, connected to local material instrumentality, climate, and labor systems (Figure 3). Presently, we must consider the geopolitical context of architectural production, together with climate impact and redirect current design practices towards more historically aware and socially nuanced modes of production. Contemporary rammed earth construction and paneling are being reinvented as part of the sustainable manufacturing endeavor, with projects focused on industrializing panelization and pre-casting (REW, n.d.; Sirewall 2019). Traditionally, rammed earth construction was used in many countries, including present-day China, Europe, India, Africa, and the Middle East (Gramlich 2013). The earliest mention of rammed earth construction is by the Roman author Pliny the Elder, in approximately 79 A.D. Earliest rammed earth remnants are often found in hilly areas, since the natural soil composition consists of a combination of sand, silt, and clay (Jaquin 2011). More recently, rammed earth construction was used extensively in the Ladakh region, that extends over parts of India and Nepal (Chayet, Jest, and Sanday 1990). Many contemporary approaches that embrace vernacular construction methods intend to elevate earthen buildings within advanced manufacturing space, thus expanding on

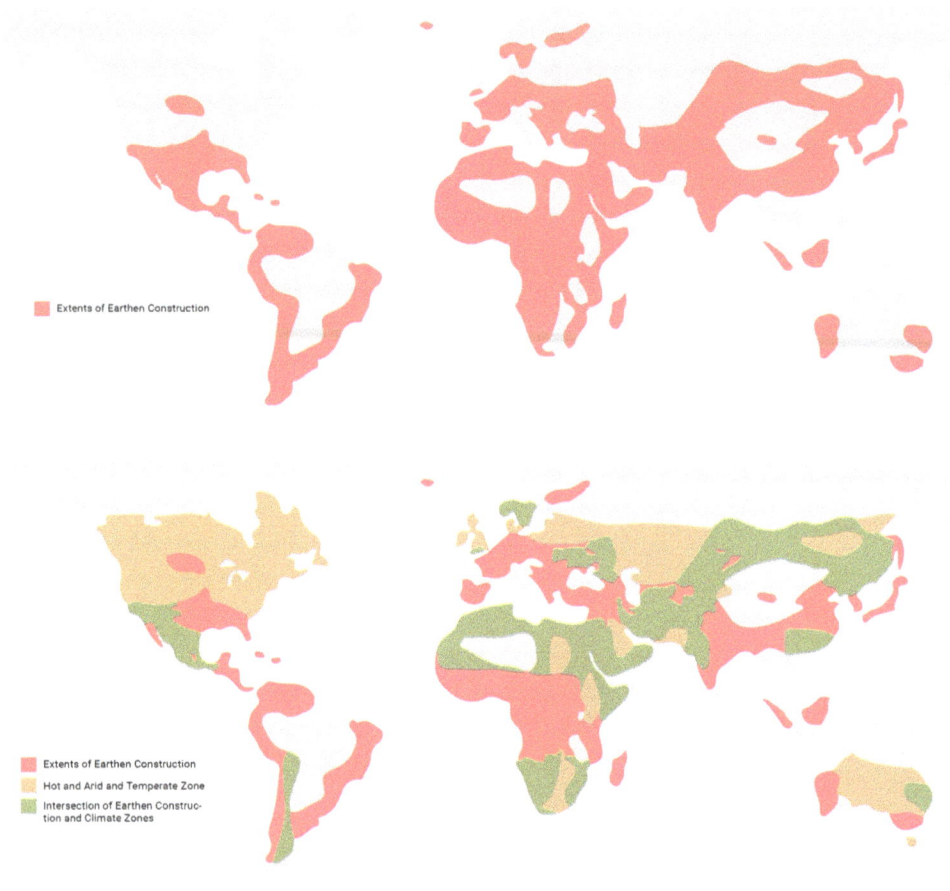

3   Geographical impact of rammed earth: historical and current earth construction extends across the globe (Data source: Auroville Earth Institute)

4   Geographical impact of rammed earth overlaid with hot and arid and temperate climate zones

5   Discretization logic is based on the spatial geometry of ramming the earth that sets up the relationship between ray-tracing as a method to introduce variability and discretization of the formwork

aesthetic language through the use of technological advances such as 3D printing (Chiusoli n.d.; Gibson 2019) and robotic deposition (Chaltiel et al. 2020; Derme and Mitterberger 2020). These experiments propose novel types of architectural forms, tightly coupled with their automation frameworks.

Our interest lies in the expressive characteristics of the rammed earth figuration, while combining low-tech constructability and thermal characteristics. Historically, rammed earth construction techniques differ across regions, and their characteristics are evident in the expression of architectural surfaces. Differentiation of form and aesthetics are predominantly affected by construction methods and climate needs. In India, rammed earth was built mainly in parts, wherein each subsequent wall section was constructed by lifting the formwork on top of the previous one, thereby giving the buildings a distinct characteristic where the walls step inwards as they grow vertically. Another method within the same region is the use of formwork that supports the next layer of the construction of each subsequent layer of rammed earth, leaving the gap-like openings in the walls as part of the architectural language.

Furthermore, considering that buildings contribute to approximately a third of global greenhouse gas emissions, refocusing on instrumentalizing design and construction methods with rammed earth—especially within hot and arid climates—is critical. The building construction industry contributes to 10% of the total global greenhouse gas emissions, while the emissions from operational energy have increased to a record high of 28% from all global energy-related CO_2 emissions (UNEP 2020). With such a high impact on global energy consumption and greenhouse gas emissions, it is imperative that we find solutions that would minimize the carbon footprint of novel construction over the entire life-cycle of the building—construction, operation, and demolition. In addition to rammed earth construction being an environmentally benign alternative to current construction materials, the use of rammed earth as a thermal storage device has the potential to reduce the energy associated with heating and cooling a building while reconnecting with vernacular practices of architecture.

MATERIAL AND DESIGN TECHNIQUE APPROACH

Vernacular techniques of earthen construction give the initial impetus and set the constraints for translation of the traditional and native knowledge of earthen systems, through the operation of digital simulation in rethinking the fabrication of variable geometries. This array of material experiments expands on the range of approaches to material resilience and aesthetics, as well as add foundations to the future

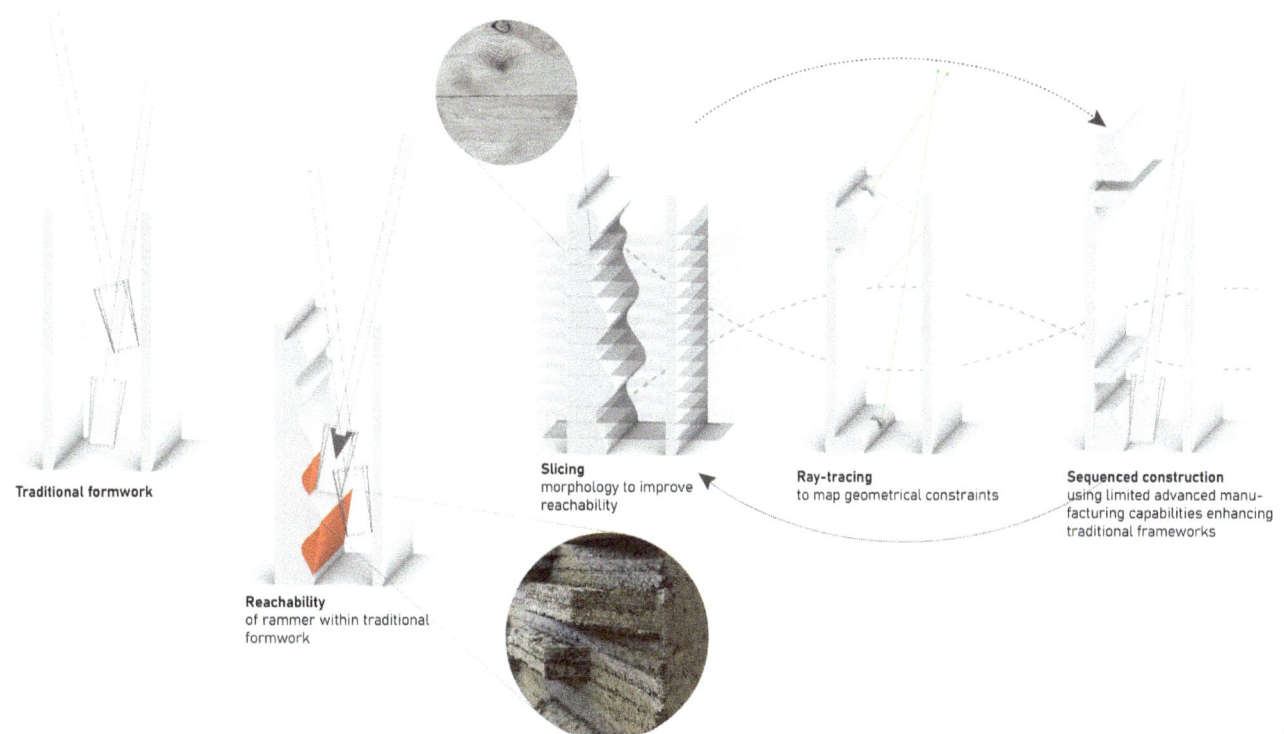

5

development of distributed technologies, in lieu of centralized industrialization within construction, while relying on the process of ramming related to a range of geometric space that allows for deviation from vertical wall surfaces (Figure 5).

CONSTRUCTION FORMWORK GEOMETRY FACTORS

Understanding material characterization as the main design framework lies at the center of the ecological approach to shaping architectural systems. Our approach to understating the characterization of rammed earth is based on combining thermal, structural, and material logic into an aesthetic framework that builds deeper intuition regarding possible formal expressions within the affordances of its making. This approach to design computation, simulation, and construction situates architecture within the hyperlocal approach to making. To adopt and utilize the hygrothermal response of rammed earth construction, a more holistic material-based design approach is required.

Thermal Mass Performance

Thermal inertia can be utilized as recoverable energy storage to maintain thermal comfort. The thickness of thermal mass and its diffusivity dictate the time required to reach a steady-state condition. The time taken by the wall assembly to transfer heat from the outside surface to the inside surface is known as 'thermal lag'. The energy storage in a wall is dependent on the heat storage capacity of the material, its surface area and thickness, and the temperature differential through the assembly's cross section. Maximizing the surface area and minimizing the thickness can, therefore, be an optimal use of building material as heat storage, to which figuration geometry can aid in offsetting thermal load (Cupkova and Promoppatum 2017). Using thermal geometry rulesets to affect the thermal mass was the starting point of this study.

The surface geometries, with differing degrees of smoothness and roughness, are developed based on incident solar radiation studies (Figure 6). Solar insolation values were used to manipulate the external surface of rammed earth panels to achieve a greater surface area-to-mass ratio. A range of morphologies were tested as their thermal performance varies based upon their particular geometric taxonomy related to surface-to-volume ratios in thermal mass, and can be used by designers to tune the performance—based upon their specific contextual requirements.

Typically, heat transfer calculations conducted using simulation software (or manually) are conducted for steady-state heat transfer, wherein the temperature on either side of the envelope assembly is assumed to remain constant. However, this does not represent the actual conditions experienced by the building, wherein the external and internal temperatures fluctuate with each time step. A series of surface geometries based on this ruleset was generated to test the formal expression gesture relative to fabrication limitations.

An 18 ft x 18 ft south-facing panel in a hot and arid climate zone was modeled with a 4-inch grid. An annual solar insolation study was performed using Radiance (Radiance 2016).

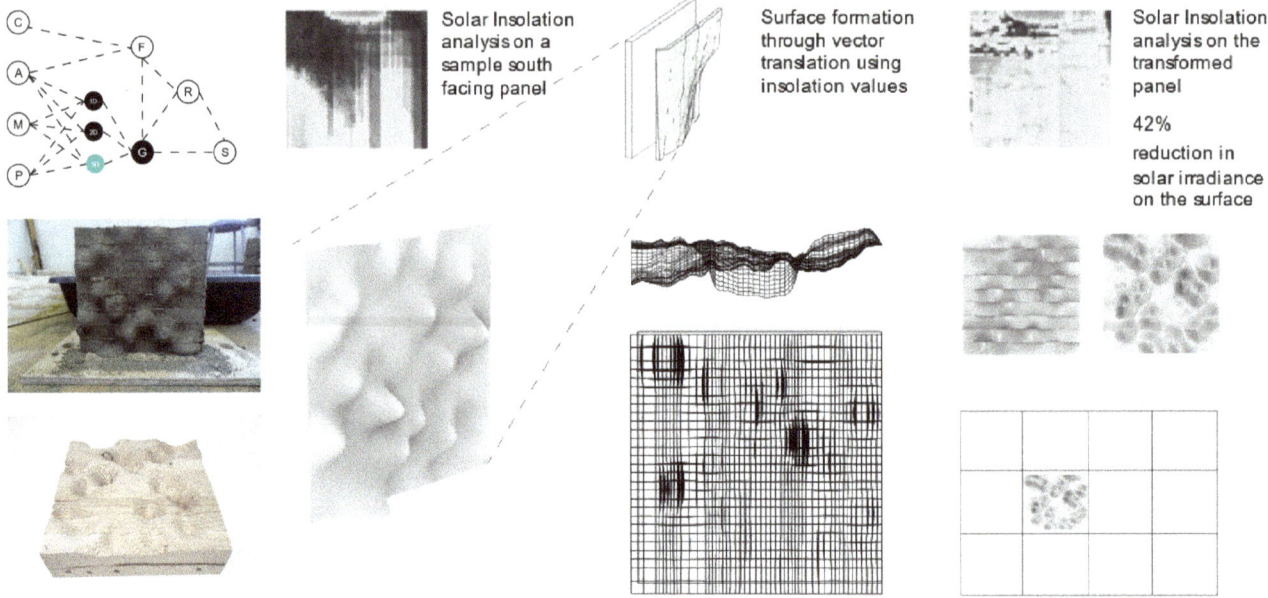

The sensor data was consequently employed to manipulate the surface by performing a vector translation based on the values. A smoothing algorithm was applied to the surface. The solar insolation study was repeated on the morphed panel. The result was a 42% reduction in annual solar insolation on the panel due to self-shading and non-uniform distribution of incident solar radiation along the curvature of the surface (Aronescu and Appelbaum 2021).

The negative of the panel was used to perform ray-tracing simulation and verify the limits of the geometry for physical prototyping of the formwork. The result is an iterative process between surface morphing, slicing height, and ray-tracing simulations (Figure 5).

Material Composition

Rammed earth construction primarily consists of wooden formwork that is filled in layers with a lightly moist mixture of sand, silt, and approximately 20% clay. Each layer is consequently tamped using a wooden tamper to almost half of its original volume. This process of tamping the earth is morphologically similar to the formation of sedimentary rocks, and has the characteristic layering visible in the walls. The formwork is then removed and moved horizontally to construct the next section, or 'lifted' vertically. The wall is then allowed to dry slowly naturally. The tightly tamped soil achieves a high density and, therefore, acts as a thermal mass. The walls are usually constructed as load-bearing walls due to their high compressive strength and are historically 12 to 18 inches thick.

The ideal mixture for rammed earth construction is a combination of different particle sizes. Gravel provides high compressive strength and structural stability to the mixture. Clay particles act as adhesives, binding the materials together. However, the total clay content in the rammed earth mixture must be controlled to enable tamping and avoid shrinkage. The rammed earth mixture has low plasticity achieved by only slightly dampening the earth mixture with less than 8% water content. This also ensures that there is almost negligible shrinkage. High clay content increases the plasticity index and, therefore, must be lower than 20% for most common rammed earth mixes. The soil composition most suitable for rammed earth construction consists of 40-60% sand and gravel content by mass, 20% clay content, and 20-30% silt by mass. The plasticity index should be lower than 30 (Gramlich 2013).

According to field tests conducted on the soil that was excavated from a nearby farm, the mixture used for this study had a ratio of 2 parts rich clay, 2 parts clay, 0.5 parts sand, and 0.1 parts lime, which amounts to roughly 20% clay content, 50% sand, 20% silt, and 4% lime. To evaluate the mixture (Figure 7), field tests were performed. Sedimentation tests were conducted to derive the initial clay recipe, and the drop test and ribbon tests were used to test the plasticity and homogeneity of the mixture.

Structural Performance and Fabrication

Rammed earth has high compressive strength and can withstand considerable vertical loading. The layers are perpendicular to the direction of vertical loading, reducing the probability of developing cracks and shearing, even with the articulated surfaces that this study focused upon. The current research in advanced form-making with rammed earth is concentrated on a horizontally laid construction that is then

6 Solar insolation study used to determine thermal heat delay through surface geometry and consequent vector translation, to set up ray-tracing simulation for fabrication setup

7 Sedimentation field test, clay mixing, and drop test to develop the soil mixture

erected vertically; this is a departure from the traditional load transfer profile of rammed earth technology, where the layers are perpendicular to the dominant force direction and could likely lead to structural performance reduction. By changing the layer direction, the lateral shear resistance of the wall is reduced. This study proposes a methodology that augments the traditional load profile and construction technique (Figure 8).

RAY-TRACING LOGIC AS CONSTRAINT FOR DISCRETIZATION OF FORMWORK

Rammed earth construction has historically lacked variability in surface figuration due to its constructability. The aesthetic quality of a rammed earth structure is typically limited to its surface finishing and layering, or artistic sculpting. Our interest is in developing a technique that allows for repeatability, future automation, and greater variability of vertical construction while simultaneously varying both sides of the wall (Figure 9).

The current formwork design and construction technique limitations might have also possibly been the defining factors in the familiar monolithic rammed earth wall design although two critical members contribute to this design vocabulary. Firstly, the formwork needs to be rigid, stable, and withstand the construction forces. Secondly, to achieve a dense and well-compacted wall, the rammer must reach each section of the rammed earth module uniformly and wield even tamping pressure—roughly perpendicular to the layer direction.

This resulted in rigid design outcomes that could only be manipulated in one direction. Once novel geometries were evaluated and attempted, it was clear that under these constraints, the possibility of geometric manipulation would be highly limited.

This study proposes a discretized reusable formwork system that collapses thermal approach and structural loading, while setting up the limits that address both. It allows for a vertical construction with variable or repeatable surface patterns on both sides, thus responding to thermal performance for solar shading and heat storage, as well as usability, aesthetics, and other factors (Figure 8).

We developed a modular formwork logic that can be fitted as an inlay within the conventional formwork to develop novel continuously varying geometries throughout construction (Figure 12). Ray-tracing serves as an analysis to set the limits to the original geometry for constructability through ramming. The digital model is sliced into repeatable vertical sections. A ray-tracing algorithm, which verifies the ramming constraints, is applied based upon variables such as slicing dimensions, wall width, and module height relative to the external formwork (Figure 11). The internal formwork consists of individually cut timber or plywood pieces. The digitally generated morphology is consequently milled or sawed into the internally placed module. This increases the reach of rammer providing the ability to design and construct complex geometries that were otherwise not possible with conventional formwork.

Ray-Tracing Geometry

To map the reachability of the rammer based on a maximum variation of discretized formwork logic, a computational model was developed to conduct a ray-tracing simulation. The vector reachability is indicative of the ability to achieve

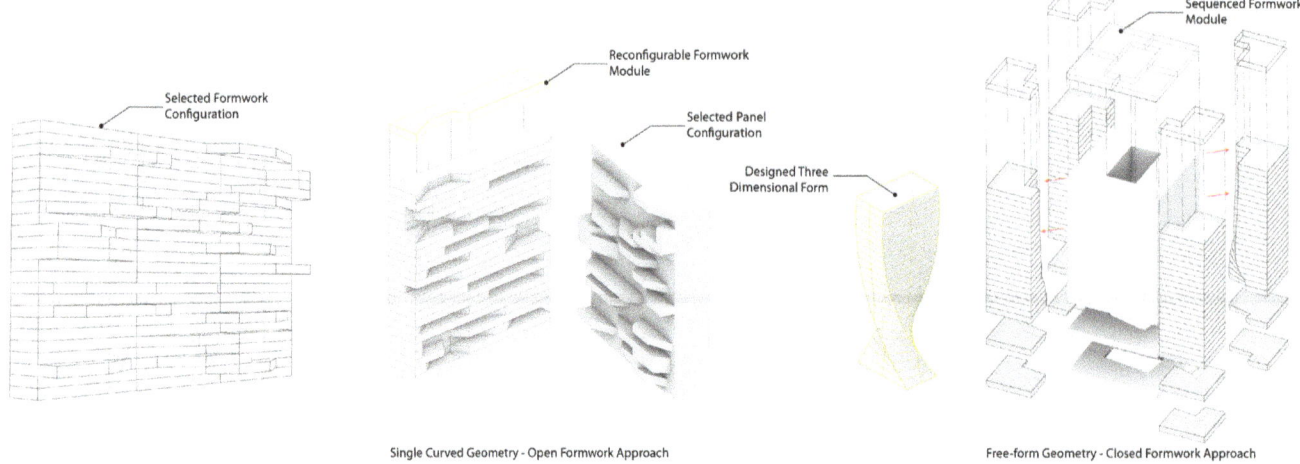

8   Thermal variation of the structural rammed earth system is limited to maintain loading of the vertical force, with incremental variations at each layer of the formwork

9   Diagram of geometric logic as a constraint for discretized formwork

10 Diagram of the ray-tracing methodology set-up

11 Geometry range and constructability of surface figuration, while maintaining structural compression load

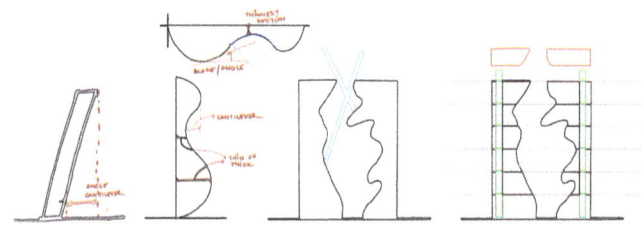

even compression by direct access of the rammer to each surface node (Figure 10).

The ray-tracing simulations were performed sequentially for the bottom and top modules within the formwork model, for a series of differing wall widths of rammed earth, ranging from 4 to 12 inches. This methodology helps evaluate the reachability of the rammer and the limits of variability. This is an iterative process combined with an appraisal of individual module/vertical slice height dimensions within the desired form. With thinner slices, we can achieve greater reach. Based on this initial study, we developed the following design guide represented in Figure 11.

STACKABLE FORMWORK

The digital model is sliced into manageable sections with repeatable heights (Figure 12). Each section can be milled (or sawed) based on its profile within the internally placed module. This method increases the reach of rammer, providing the ability to design and construct complex geometries that were otherwise not possible with conventional formwork. The formwork can be manufactured using any accessible wood, timber, or plywood pieces.

The milled formwork contains negative spaces that are filled with rammed earth, allowing for ramming throughout the surface for maximum effectiveness. Parallel with increased surface figurations, the corners of rammed earth walls are generally more susceptible to breaking and chipping. This became a key consideration in the design of formwork and setting up physical prototypes for testing. Slicing of the formwork provides sequential access of the rammer to reach all areas of the soil mixture. The ray-tracing algorithm helps to identify the slicing constraint and can be used to test multiple permutations of the discretized kit.

GEOMETRIC PROTOTYPES

The heights of the slices within these prototypes were digitally optimized using ray tracing and based upon widely available cost-effective timber. The 2x4 timber was prepared for milling, including drilled holes for steel rods that serve as stabilizers and guides for the pieces to slide into the already-widely used traditional formwork.

Subsequently, the modules are clamped together to create the dimensions of the prototypes and are milled on a CNC machine, based upon the digital model. This entire process creates an iterative feedback loop, starting with a panel design, understanding solar implications, setting the limits through the ray-tracing methodology, and then retesting the formwork using physical prototyping.

Study One: Single-Curved Geometries

The first tests focused on simple surface geometry manipulations to achieve either single curved or rectangularly extruded variations of the surfaces (Figure 14). Two degrees

10

11

of smoothness were tested during physical prototyping to verify a wider range of morphological typologies.

The geometry was developed through multiple iterations, derived from geometric constraints relative to solar logics, aesthetics, and ray-tracing procedures that inform fabrication. The formwork slicing was controlled to develop a reconfigurable kit of parts that would achieve even greater variability of form, as the formwork is lifted vertically and horizontally.

Figure 12 illustrates the completed formwork insert using threaded metal rods for sliding the discretized pieces into place, in any desired order. The potential of ad-hoc reordering of the slices is most evident in Figure 15, showing rectangular geometry with variable surface articulation during phased construction, reusing the same set of discretized slices with variable repetition.

Study Two: Double-Curved Geometries

Double curve surface figuration in this study (Figure 13) is a result of the solar insolation analysis, used to explore its limitation on constructability within the identical methodology. Introducing doubly curved surface variations requires an additional step in ensuring that the rammer projections are reachable. This geometry often requires tamping at an angle in addition to vertically loaded tamping. During the first few prototyping sessions, it was observed that the ramming process could cause portions of soil (that are not within direct reach) to coalesce, which put additional limitations on geometry, reactive to material characterization.

Study Three: Three-Dimensional Forms with 2D Formwork

Sequential discretized formwork opens the possibility for even more complex rammed earth profiles. Figure 17 depicts an experiment in the sculpting of the 3D form—across all three axes—using repeatable two-dimensional formwork. The height of the sliced module determines the fidelity of the outcome relative to the computational model. In this case, a 1/2 inch piece of plywood was used to create the layers. The plywood was planarized and then CNC milled to accommodate the shape synthesis and four holes for guide rods. The slices fit the conventional formwork framework, similar to the prior methodology, but each layer is additionally sliced into four pieces to aid formwork removal.

FABRICATION CHALLENGES

Each step is critical and can have a significant impact on the final panel. It was noted that reducing the module sizes below the limits identified in the ray-tracing simulation can result in a lack of adhesion with the cantilever geometries. Another point of importance is the removal of formwork. Due to the intricacies of these geometries, it is important to remove the formwork within 12 hours, or when the wall dries by approximately 30-50%. If the formwork is left for too long and the earth mixture dries within it, the removal process results in breakages.

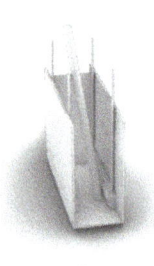
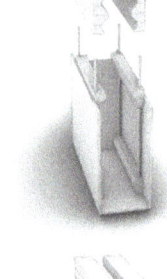
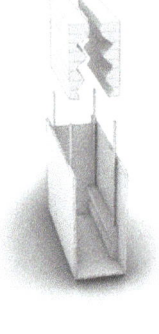
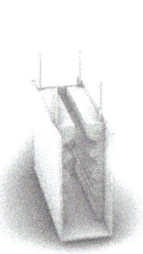

12 Sequential insertions of discretized formwork pieces during the ramming process

13 Double curved geometry does not completely conform with the ray-tracing, resulting in a constructability tolerance test

14 Prototype of single curved surface geometry in rammed earth

CONCLUSION AND PROJECTIONS

This study systematically evaluated the in-situ vertical constructability of varied surface geometries related to the thermal performance of rammed earth, leveraging digital technologies and fabrication methodologies. By introducing design variability in rammed earth construction, there is a possibility of further improving its performance and consequently conferring greater design freedom. Rammed earth has significant ecological benefits over other prevalent construction technologies. It is low-cost and reduces the energy spent on production and transportation, as easy availability can result in significant operational energy savings and reconnects the architectural practices to vernacular and local roots. By undergoing systematic geometrical exploration, the limitations and opportunities of the material can be explored to further improve its design and applicability for in-situ vertical adaptation. Through employing existing formwork techniques with a few minor additions, the discretization of sliding formwork can be leveraged to introduce even greater design freedom.

Additionally, to increase future automation, the author proposes the use of adaptive formwork that can be manipulated in series, possibly by a robotic arm. The proposed prototype consists of the outer hard shell, which is similar to traditional formwork and an inner flexible layer that is sequentially manipulated to achieve the designed variability. The author proposes the use of sand as an infill material between the flexible and rigid segments, to reduce deformation due to lateral forces. Furthermore, the adaptation or manipulation of nodes must be performed similarly to the current stackable formwork—layer by layer. This will increase the range of reachability by the tamper.

Another critical conversation surrounding adaptive formwork is controlling the deformation. The author proposes an iterative methodology using adaptive formwork to address this issue. Using computational design techniques, especially with the use of geometric comparison and feedback via computer vision/point-cloud-based scanning, the node manipulation can be adjusted to design for the deformation. Optimum node spacing can also be determined to further obtain the desired geometry. Such an iterative process can result in beautiful geometric evolution of the material, while simultaneously reducing the need for fabricating custom formwork. Another area that is vastly understudied is the introduction of automation within rammed earth procedures. While pneumatic rammers have considerably reduced the labor requirement for manual tamping, using robotic fabrication can greatly benefit the future shape-ability of rammed earth. Leveraging digital technologies for rammed earth construction can have a significant impact on the adoption of environmentally benign and regenerative materials, while simultaneously enhancing design freedom.

ACKNOWLEDGMENTS

This research project was developed in the 2018-19 academic year within the Master of Science in Sustainable Design (MSSD) program, led by Dana Cupkova at the Carnegie Mellon School of Architecture. Special thanks to Professor Stephen Lee, the thesis project secondary co-advisor, for his advice relative to manufacturing processes; and Lola Ben Alon, the thesis tertiary co-advisor, for her insight regarding soil composition and earthen construction techniques. The work was partially funded by Carnegie Mellon University Graduate Small Project Help Fund (GuSH), as well as CMU MSSD discretionary graduate program funding.

13

REFERENCES

Aronescu, Avi, and Joseph Appelbaum. 2021. "Solar Radiation on a Parabolic Concave Surface." *Energies* 14 (April): 2245. https://doi.org/10.3390/en14082245.

Asprone, Domenico, Ferdinando Auricchio, Costantino Menna, and Valentina Mercuri. 2018. "3D Printing of Reinforced Concrete Elements: Technology and Design Approach." *Construction and Building Materials* 165 (March): 218–31. https://doi.org/10.1016/j.conbuildmat.2018.01.018.

Auroville Earth Institute. n.d. "Earthen Architecture in the World Introduction - Earth-Auroville.com." Accessed April 3, 2019. https://earth-auroville.com/maintenance/uploaded_pics/0-earth-world-intro-en.pdf.

Brooks, Rodney A. 1990. "Elephants Don't Play Chess." *Robotics and Autonomous Systems* 6 (1): 3–15.

Chaltiel, Stephanie, Maite Bravo, Diederik Veenendaal, and Gavin Sayers. 2020. "Drone Spraying on Light Formwork for Mud Shells." In *Design Transactions: Rethinking Information Modelling for a New Material Age*, edited by B. Sheil et al. London: UCL Press. 150–57. https://doi.org/10.2307/j.ctv13xprf6.30.

Chayet, Anne, Corneille Jest, and J. Sanday. 1990. "Earth Used for Building in the Himalayas, the Karakoram and Central Asia - Recent Research and Future Trend." In *6th international conference on earthen architecture*. Las Cruces, Mexico, France: HAL SHS. 29–34.

Chiusoli, Alberto. n.d. "3D Printed House TECLA - Eco-Housing - 3D Printers | WASP." 01 21. Accessed October 19, 2019. https://www.3dwasp.com/en/3d-printed-house-tecla/.

Cremers, Jan. 2011. "Energy Saving Design of Membrane Building Envelope." In *International Conference on Textile Composites and*

14

Inflatable Structures: Structural Membranes 2011, edited by E. Oñate, B. Kröplin and K.-U.Bletzinger. Barcelona: CIMNE. 148–155.

Cupkova, Dana, and Nicolas Azel. 2015. "Mass Regimes: Geometric Actuation of Thermal Behavior." *International Journal of Architectural Computing* 13 (June): 169–94. https://doi.org/10.1260/1478-0771.13.2.169.

Cupkova, Dana, and Patcharapit Promoppatum. 2017. "Modulating Thermal Mass Behavior Through Surface Figuration." In *ACADIA 2017:*

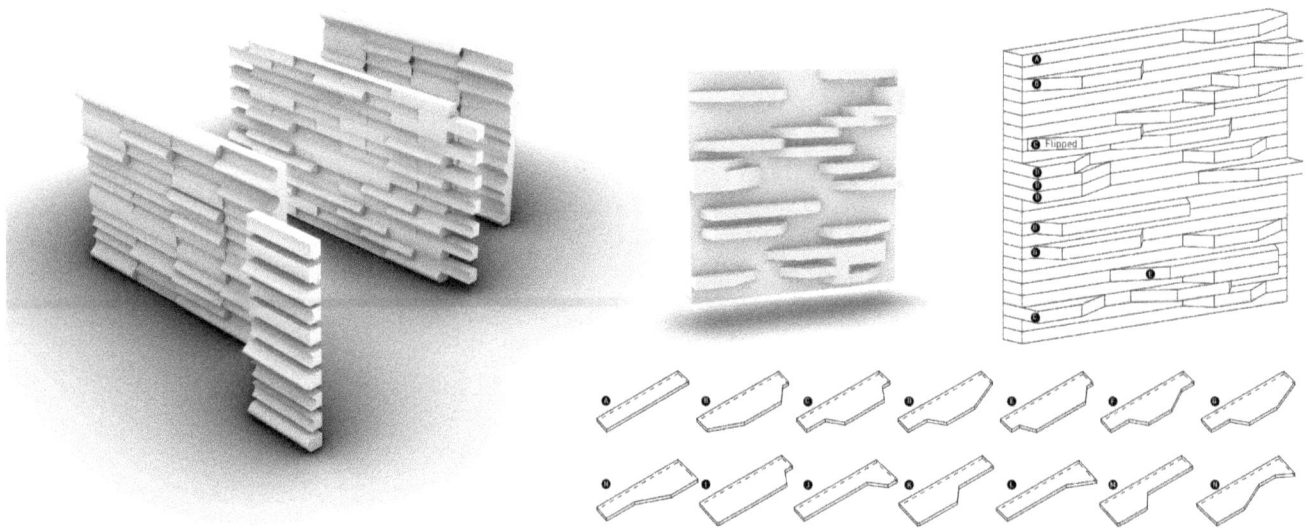

15 Rammed earth prototype with a limited set of discretized formwork parts to create a variable surface effect

16 Single curved prototype related to negative formwork configuration

17 Three-dimensional rammed earth form constructed using two-dimensional repetitive discretized formwork

18 Author's prototype for adaptable formwork that can increase the possibility of future automation for ramming of variable surfaces: each threaded rod serves as a node that can be robotically modulated to control an inner flexible layer. Sand will be poured between the rigid outer shell and the inner flexible membrane to control deformity during ramming.

Disciplines & Disruption; Proceedings of the 37th Annual Conference of the Association for Computer Aided Design in Architecture. 202–211. https://doi.org/10.52842/conf.acadia.2017.202.

Derme, Tiziano, and Daniela Mitterberger. 2020. "Digital Soil: Robotically 3D-Printed Granular Bio-Composites." *International Journal of Architectural Computing* 18 (June): 147807712092499. https://doi.org/10.1177/1478077120924996.

Fox, Michael, and Miles Kemp. 2009. *Interactive Architecture.* New York: Princeton Architectural Press.

Gibson, Eleanor. 2019. "Rael San Fratello 3D Prints Mud Structures as Prototypes for Low-Cost Construction." *Dezeen.* October 3, 2019. Accessed September 29, 2019. https://www.dezeen.com/2019/10/03/mud-frontiers-rael-san-fratello-3d-printed-low-cost-construction/.

Gramlich, Ashley. 2013. "A Concise History of the Use of the Rammed Earth Building Technique Including Information on Methods of Preservation, Repair, and Maintenance," M.S. diss., Historic Preservation, University of Oregon. https://scholarsbank.uoregon.edu/xmlui/handle/1794/12982.

Gupta, Pragya, Dana Cupkova, Lola Ben-Alon, and Erica Cochran Hamee. 2020. "Evaluation of Rammed Earth Assemblies as Thermal Mass through Whole Building Simulation." In *2020 Building Performance Analysis Conference and SimBuild co-organized by ASHRAE and IBPSA-USA.* 618–25. https://www.proquest.com/openview/713c137976685285c6a6d1b94959e697/1?pq-origsite=gscholar&cbl=5014767.

Hasdell, Peter. 2009. "Pneuma: An Indeterminate Architecture, or Toward a Soft and Weedy Architecture." In *Design Ecologies: Sustainable Potentials in Architecture*, edited by L. Tilder and B. Bolstein. New York: Princeton Architectural Press. 92–113.

Jaquin, Paul. 2011. "A History of Rammed Earth in Asia." In *International Symposium on Innovation & Sustainability of Structures in Civil Engineering.* Xiamen University, China. https://web.statler.wvu.edu/~r-liang/ihta/papers/11%20FINAL%20Paul%20Jaquin_paper_workshop.pdf.

Radiance-Online. 2016. "Radiance." Accessed September 14, 2019. https://www.radiance-online.org/.

REW. n.d. "Rammed Earth Works - Pre-Cast Rammed Earth." Blog. Accessed October 2, 2019. https://www.rammedearthworks.com/blog/2018/1/12/interior-rammed-earth-panels-thinner-prefab-panels-and-a-cool-blue-hue.

Sirewall. n.d. "SIREWALL System; Structural Insulated Rammed Earth." Accessed September 30, 2019. https://sirewall.com/sirewall-system/.

UNEP. 2020. "2020 Global Status Report for Buildings and Construction: Towards a Zero-Emission, Efficient and Resilient Buildings and Construction Sector." Accessed December 20, 2020. https://wedocs.unep.org/handle/20.500.11822/34572;jsessionid=A6F-31460C35B8EBE9156B28EAB3C9627.

Venkatarama Reddy, B. V., and P. Prasanna Kumar. 2010. "Embodied Energy in Cement Stabilised Rammed Earth Walls." *Energy and Buildings* 42 (3): 380–85. https://doi.org/10.1016/j.enbuild.2009.10.005.

IMAGE CREDITS

Figure 3, Data Source: https://www.earth-auroville.com/maintenance/uploaded_pics/0-earth-world-intro-en.pdf

All drawings and images by the authors.

Pragya Gupta graduated with Master of Science in Sustainable Design from Carnegie Mellon School of Architecture in 2019. She is a firmwide Design Computation Team Leader and Design Performance Analyst at NBBJ, Seattle. Her work intersects high-performance building design and computational practices to create spaces that are healthy, sustainable, and bring people closer to Nature. She works with project teams to use data as a tool to drive and evaluate design ideas, and is an advocate for zero carbon design.

Dana Cupkova holds an Associate Professorship at Carnegie Mellon School of Architecture and directs Epiphyte Lab, Architectural Design & Research Collaborative. Engaging issues of environmental stewardship in design, Dana's work is situated at the intersection of built environment and ecology, focused on computational methods, materiality, embodied energy, and advanced manufacturing frameworks, with a particular interest in thermodynamics and construction waste streams.

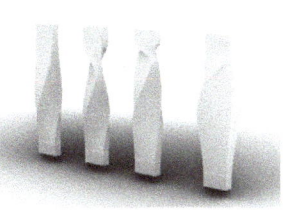

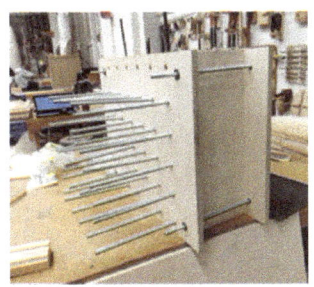

Robotic Fabrication of 3D Printed Clay Opening as a Passive Cooling System

Deena El-Mahdy
The British University in Egypt /
Iaac, The Institute for Advanced
Architecture of Catalonia

Marwa Abd ElRahim
Iaac, The Institute for Advanced
Architecture of Catalonia

Adel AlAtassi
Iaac, The Institute for Advanced
Architecture of Catalonia

1 Testing 3D printed opening through a smoke test

ABSTRACT

Natural ventilation in buildings is one of the most important passive cooling strategies to provide better thermal performance. This paper presents the assessment of a 3D printed opening with a dual function of cooling and heating through a cavity wall using clay. Two prototypes are printed using a desktop clay printer for the small scale, followed by using a robotic Kuka arm for the large scale. The design of the opening aimed at achieving passive cooling by utilizing natural ventilation, which is inspired by wind catchers and Trombe wall concepts. A computational fluid dynamic (CFD) simulation test is done to examine the airflow through the inlet and outlet opening through the space, followed by a smoke test to validate the resulting data. The results recorded a difference in the structural performance and material behavior between both small and large scales considering the thickness of the nozzle, the weight of the materials, and the drying process.

INTRODUCTION

Passive cooling systems that rely on natural ventilation have been used for centuries, especially in the hot arid climate region (Nalamwar, Parbat, and Singh 2017; Cruz-Salas, Castillo, and Huelsz 2014). This system uses natural forces such as wind and air to cool the outdoor air which have an impact on spaces' thermal comfort. Similar traditional systems can be seen in the windcatchers, courtyards, solar chimneys, and Trombe walls (Dabaieh and Elbably 2015; (Obeidat, Kamal, and Almalkawi 2021). The increased demand for energy required for heating and cooling calls for the need to revisit these traditional ventilation strategies and reintegrate them in the most efficient ways. The appearance of computational design and digital fabrication, including 3D printing (3DP), allows for the reintegration of traditional cooling systems in optimized ways.

The application of 3D printing can vary from being fully functional to purely aesthetic. The precision and flexibility of 3DP allow the easy generation of customized and complex forms without the need for formwork and wall finishing, which correspondingly can lead to passive ventilation solutions. The potential of 3DP in architecture can be seen in the ability to print on-site and resposiveness to the climatic conditions which reduce waste, time, materials, and labor cost.

Several materials have been used for 3D printing in construction such as concrete, earth, steel, fibers, and plant-based materials. Printing with concrete is more popular in construction, yet, with the use of cement, the process remains a source of high carbon emissions which harm the environment. The main driver of this research was the lower environmental impact of clay compared to that of concrete. However, one of the main challenges of 3DP with earthen material is the longer drying time. The flowability of clay through the printer, depends on a sufficient amount of water. Since clay depends on air drying to harden, its drying time is typically longer and depends on climatic conditions compared to the drying time of concrete, which is faster. Another challenge is the potential shrinkage after drying—due to the high-water content of clay—that results in cracking.

Clay has been used in vernacular architecture in different countries, especially in Africa, Asia, and South America. It is known for its good thermal and fire resistance, good insulation and heat absorption, and for retaining humidity (Khosravi and Mahdavi 2021). The thickness of clay walls usually varies between 40 to 80 cm to prevent heat from reaching inside the space and keeping the wall humid. Despite the competitive thermal performance of clay, it is known for its weak properties in compression, which leads to limitations in building heights. To further the use of clay in 3D printed large-scale construction, the research presents the assessment of computational 3DP opening through a wall using clay, intended to respond to the high-temperature fluctuation in hot, arid climates by providing cooling in summer and heating in winter.

Different openings are tested to reach the optimized inlet and outlet geometries that provide the best airflow circulation through the space. The design of the wall considers providing natural ventilation that is inspired by the passive cooling system of the Trombe wall. Two prototypes are printed with: 1) a small-scale using a desktop clay printer, and 2) a large-scale using a robotic Kuka arm. A simulation test of computational fluid dynamics (CFD) is done to investigate the airflow circulation through the printed inlet and outlet openings to the space. A smoke test is done on the 3D printed opening to validate physically the results of the CFD analysis test. The results record a difference in the performance and material behavior between the two scales, considering the thickness of the nozzle, the weight of the materials, and the drying process as factors.

STATE OF THE ART

Clay has been integrated into several 3D printed construction projects that are done on a small scale, more than 1:1 scale. The research titled "Heat distribution, 878 J|KG_C" investigated the integration of a heating system within a 3D printed wall (Dubor, Marengo, and Ros-Fernández 2019), and it succeeded to achieve the required thermal performance by replacing the conventional technique with 3DP clay. The project implemented passive ventilation strategies taking into consideration the wall cavity, infill pattern, and geometry of the opening as the main parameters that affected the design. Another project was done using 3DP clay as an inspiration from the birdhouse (Sevostianov, Li, and Riaz 2017). Different patterns, texturing, openings, and orientations were implemented to provide a self-shading system on the wall surfaces. The shading resulting from the computational form provided more cooling through the reduction of the direct solar radiation on the surface. Another project titled *Digital Adobe* resulted in a 3D printed clay wall inspired by the Trombe wall (Chang et al. 2018). The project employed parametric modeling and computational evaluation methods in the design to generate optimized forms regarding the structure and climatic conditions. The bump pattern on the surface of the wall generated self-shading that provides cooling in summer and heat absorption in winter. The *Digital Adobe* wall includes openings at the bottom for the inlet, while the outlet opening is located at the top. The wall was assembled on-site using 3D printed small-scale blocks printed off-site. Based on these previous projects, our research proposes a new approach to integrate openings within 3DP clay walls on-site as a passive dual element for both cooling and heating. The main determinant of the opening design is the desired airflow performance—geometry follows airflow.

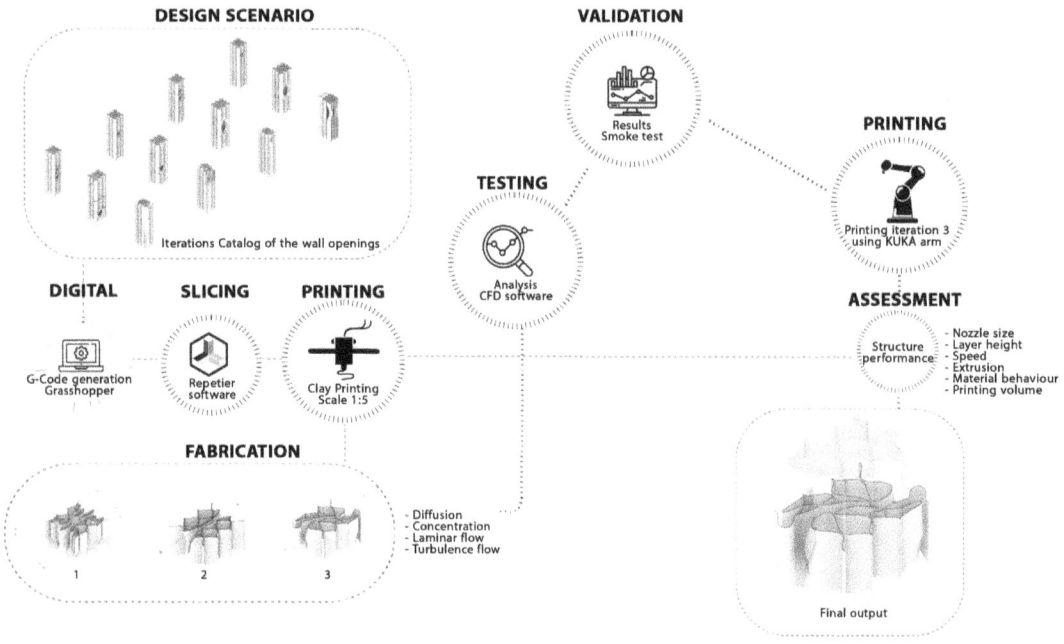

2

METHOD AND MATERIAL

In this paper, airflow pattern and behavior through 3D printed clay wall elements were investigated by employing a combination of both physical and digital tests. The methodological framework includes four main phases to examine the airflow behavior and its application to 3D printed walls, as shown in Figure 2. The first phase includes studying the basic concepts of airflow to understand the geometrical parameters that control it from outside to inside the space. Subsequently, in the second phase, a set of 3D printed clay pieces with geometrical variations were digitally simulated and printed at a small scale. The third phase includes the larger scale printing using a robotic Kuka Arm, scaling up the best design resulting in the most desirable airflow behavior. Printing the same geometry at two different scales allowed for the observation of the change in material behavior and the printing speed, therefore, the structural performance evaluation. Finally, the last phase includes the smoke test—currently in progress—which is done on the 3D printed unit to assess and verify the airflow speed through the inlet and outlet in relation to the time.

Catalog of 3D Printed Wall and Simulation

Based on the results from preliminary previous research on the possibilities of integrating openings within 3D printed walls, and on the predictions of the airflow paths through walls, a catalog with different openings and geometries was designed. The designs were later printed at a small scale to evaluate their stability and printability. Additionally, simplified models of the printed geometries were simulated through Autodesk CFD software. The openings were evaluated based on their diffusion and concentration of air and the type of flow that resulted, whether laminar or turbulent. Consequently, the opening that achieved the desired airflow pattern was printed at a larger scale.

Wall Design

The design of both inlet and outlet openings was modeled in Rhinoceros software, and the Grasshopper plugin was used to generate the G-code to be prepared for slicing (Robert McNeel & Associates 1995; Rutten 2007). Repetier-Host software (Littwin and Littwin 2011) was used for slicing the geometry and connecting the desktop printer with a laptop. While different parameters were taken into consideration: width and length of the cavity, inclination angle, and form of the openings. The location of the inlet and outlet openings were placed facing each other to allow more airflow acceleration. The dimensions of the wall were 60 cm thick, and 100 cm wide by 200 cm long for the opening, and an inclination angle of 30 degrees. The angle was selected based on research done by Erbay, Doğan, and Öztürk (2017), where a louver angle of 28.5° performed better in thermal conductivity. Based on Kosutova et al. (2019), the inclination angle between 30° and 45° will prevent airflow from entering smoothly where only a smaller free area is available. The relationship between the inclination angle, thermal performance, and temperature are identified for further research.

Fabrication of 3D Printed Wall at Small and Large Scale

For the small-scale prototype, a modified Creality Ender 5 printer with a nozzle size of 2 mm diameter was used. An air

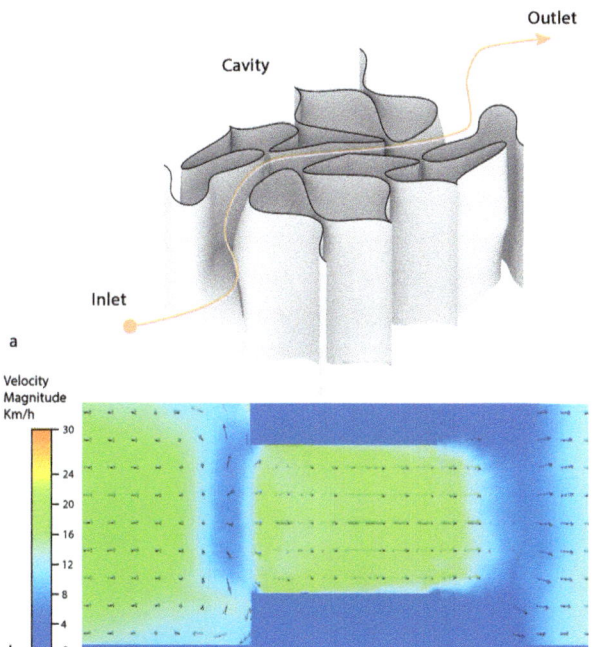

2 Methodological framework

3 a) Section through the cavity and the opening; b) CFD analysis test of the opening

compressor was connected to push the clay into the cartridge to the extruder. The scale of the printed opening piece was 1:5 with dimensions of 12 cm x 12 x 18 cm. The total print length was 14.1 meters with a layer height of 1.5 mm.

The large-scale prototype was printed with a 6-axis Kuka robotic arm with a payload of 60 kg and a nozzle diameter of 5 cm. WASP pump was used to mix and supply the material to the nozzle. The prototype was printed in scale 1:3 with dimensions of 20 cm x 20 cm x 30cm, with a total print length of 118.4 meters and a layer height of 5 cm.

Smoke Test
A physical smoke test was applied to the small-scale prototype. Smoke was released and directed through the opening to assess the airflow pattern and speed. The results were considered as a validation of the CFD analysis test.

RESULTS AND DISCUSSION
Design and CFD Analysis Test
The design of the opening in Figure 3a shows the behavior of the airflow through the cavity. The inlet opening is inclined to cause airflow suction, where a smaller area accelerates the air velocity and, in parallel, increases the speed, while the cavity in the middle of the wall is wider to cool the warm air—in case it is warmer than the inside.

The same inclination is applied to the outlet opening through the space to ensure speeding up of the airflow. The result of the CFD analysis test in Figure 3b shows that the inclination angle of the inlet opening by 30 degrees allows more laminar flow with an increase in the airflow. The opening acts as an air shaft that pulls the air inside the space, which is similar to the Venturi effect, while the variation in the input and outlet opening size creates different pressure zones, and therefore, more diffusion within the space will occur.

3D Clay Printing and Smoke Test
The geometry of the small-scale prototype as shown in Figure 4a is tested with smoke to validate physically the data of the opening after the CFD analysis. The smoke test records a high air speed in a short time compared to other geometries caused by the inclination angle as shown in Figure 4b.

Structural Performance in Different Printed Scales
It is essential to consider the local as well as the global center of inertia when designing for 3D printing. While printing, the local center of gravity shifts throughout the layers; therefore, it is important to ensure having enough material to counterbalance this shift and achieve a stable structure (Foroughi et al. 2019). Another consideration for 3DP, especially with clay, is that the structural properties of the material change based on the water content.

The prototype is printed all at once on the small scale, while the larger scale requires a slower speed with short pauses in between printing as shown in Figure 5 to ensure that the layers below are dry enough to hold the new layers. Additionally, the larger print scale has higher water content in the mixture, resulting in a longer drying time. It is observed that the material behaves differently on a larger scale due to its weight, where it becomes sensitive to buckling. As a solution to

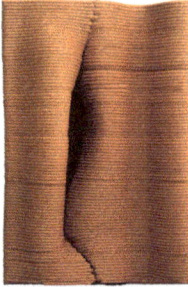

4 a) 3D printed clay inlet and outlet opening; b) Smoke test experiment through the opening

5 Large scale of the opening using robotic Kuka arm

prevent the buckling in the middle of the prototype, where compression occurred at the base of the geometry, the larger the area at the base, the more stable the geometry that can support any inclination and cantilever above. During scaling up, the volume of the geometry should consider the weight of the materials, which will result in different behavior and forces compared to the smaller scale. A relation between the height, the area, and the volume of the prototype are suggested for further research. The focus of the research was to assess airflow in relation to geometry; however, ensuring a structurally stable geometry is key to having a printable opening on the large scale. Therefore, the design was adjusted to eliminate extreme cantilevers that were printable on the small scale but problematic on the large scale due to the higher load of the material. A detailed evaluation of the structural performance will be pursued in further research.

CONCLUSION

The paper highlighted the possibility of printing an opening using clay as a material. The inlet and outlet positioning as well as the opening inclination angle were studied. The stability varied with the scale of the prototype due to the different weights of the materials as well as the water content in the mixture. A difference in the material behavior was recognized due to the difference in the scale. The printed wall can act as a dual system for both cooling and heating based on the form and the opening scale. Further research is to experiment and validate the data through a numerical study. In addition, the relationship between the inclination angle and thermal performance is left for further research. As the work is in progress, the architectural vision and the future work suggested is the application of a 3D printed passive system on a 1:1 scale within an enclosure to test the climatic response in parallel with the human thermal comfort inside the space. Another ongoing test is on making the opening operable to adapt and control the cooling and heating of the climate.

ACKNOWLEDGMENTS

This research was part of the work developed at IAAC 3D Printing program 2021/2022. The authors would like to thank the directors of the 3DPA program 2021/2022 Edouard Cabay and Alexandre Dubor at IAAC for their guidance and teaching. Special thanks to Lili Tayefi and faculty members who assist in providing feedback and follow-up during the research project: Guillem Baraut, Vincent Huyghe, Elisabetta Carnevale, Ashkan Foroughi, Eduardo Chamorro, and faculty assistant Francesco Polvi for assisting in managing printing on KUKA arm.

REFERENCES

Chang, Ya-Chieh et al. 2018. "Digital Adobe." *Institute for Advanced Architecture of Catalonia, IAAC* (blog). Open Thesis Fabrication Program. Accessed January 31, 2023. https://iaac.net/project/digital-adobe/.

Cruz-Salas, M. V., J. A. Castillo, and G. Huelsz. 2014. "Experimental Study on Natural Ventilation of a Room with a Windward Window and Different Windexchangers." *Energy and Buildings* 84: 458–65.

Dabaieh, Marwa, and Ahmed Elbably. 2015. "Ventilated Trombe Wall as a Passive Solar Heating and Cooling Retrofitting Approach; a Low-Tech Design for off-Grid Settlements in Semi-Arid Climates." *Solar Energy* 122: 820–33.

Dubor, Alexandre, Mathilde Marengo, and Pablo Ros-Fernández. 2019. "Experimentation, Prototyping and Digital Technologies towards 1:1 in Architectural Education." In *JIDA '19: VII Jornadas sobre la Innovación Docente en Arquitectura*. Barcelona: Universitat Politècnica de Catalunya. https://doi.org/10.5821/jida.2019.8381.

Erbay, Latife Berrin, Bahadır Doğan, and Mehmet Mete Öztürk. 2017. "Comprehensive Study of Heat Exchangers with Louvered Fins." In *Heat Exchangers; Advanced Features and Applications*, edited by S S. Murshed, M. M. Lopes. London: IntechOpen. 61–92. https://doi.org/10.5772/66472.

Foroughi, Ashkan, Ipsita Datta, Bhakti Vinod Loonawat, Nusrat Tabassum, Ozgur Cengiz, Pavlina Kriki, Payam Salahinezhad, Shahram Cawsi Randeria Y Du, I-Fan Liao, Yuchen Chen, and Yingxin Du. 2018. "Structural Toolkit." *IAAC Blog*, December 3, 2018. Accessed January 31, 2023. https://www.iaacblog.com/programs/structural-toolkit/.

Khosravi, Shive Najaf, and Ardeshir Mahdavi. 2021. "A CFD-Based Parametric Thermal Performance Analysis of Supply Air Ventilated Windows." *Energies* 14 (9): 2420. https://doi.org/10.3390/en14092420.

Kosutova, Katarina, Twan Hooff, Christina Vanderwel, Bert Blocken, and Jan Hensen. 2019. "Cross-Ventilation in a Generic Isolated Building Equipped with Louvers: Wind-Tunnel Experiments and CFD Simulations." *Building and Environment* 154 (May): 263–80. https://doi.org/10.1016/j.buildenv.2019.03.019.

Littwin, Marcus, and Roland Littwin. Repetier-Host. 2011. Hot-World GmbH & Co. KG. 22\www.repetier.com.

Nalamwar, Mahesh R., Dhananjay K. Parbat, and D. P. Singh. 2017. "Study of Effect of Windows Location on Ventilation by CFD Simulation." *International Journal of Civil Engineering and Technology* 8 (7): 521–31.

Obeidat, Bushra, Hammam Kamal, and Amal Almalkawi. 2021. "CFD Analysis of an Innovative Wind Tower Design with Wind-Inducing Natural Ventilation Technique for Arid Climatic Conditions." *Journal of Ecological Engineering* 22 (2): 86–97.

Robert McNeel & Associates. *Rhinoceros 3D*. Robert McNeel & Associates. 1995. www.rhino3d.com.

Rutten, David. Grasshopper. Robert McNeel & Associates. 2007. www.grasshopper3d.com.

Sevostianov, Philipp, Quan Li, and Sheikh Rizvi Riaz. 2017. "OTF Birdhouse Workshop." *Institute for Advanced Architecture of Catalonia, IAAC* (blog). Accessed January 31, 2023. https://www.iaacblog.com/programs/otf-birdhouse-workshop/.

IMAGE CREDITS

Figure 4, 5: ©Deena El-Mahdy, December 2021.

All other drawings and images by the authors.

Deena El-Mahdy is an Assistant Professor in the Architectural Engineering Department at the British University in Egypt. She holds BSc, MSc, and PhD degrees from Cairo University. Her research interests are digital fabrication processes, parametric design, computational design, and additive manufacturing techniques in both education and construction. Her interest in digital fabrication and 3D printing led her to follow a postgraduate program at the Institute for Advanced Architecture of Catalonia (IAAC) related to 3D printing with clay in 2021/2022, where she was awarded the Collete scholarship.

Marwa Abd Elrahim is a Sudanese architect and researcher based in Abu Dhabi. She holds a BSc in Architectural Engineering from the University of Sharjah and a postgraduate degree in 3D Printing Architecture from the Institute for Advanced Architecture of Catalonia (IAAC) 2021/2022. Marwa enjoys exploring the intersection between architecture, art, and technology. Her research focuses on utilizing digital and traditional design and fabrication methods to address environmental and social sustainability.

Adel AlAtassi is an architect and urban designer who lives and works in Hamburg, Germany. A native of Homs, Syria, he is an architect and technology enthusiast who has worldwide experience. He won several prizes in the field of architecture and digital methods, and finally he was awarded a scholarship to study in Barcelona at the Institute for Advanced Architecture of Catalonia 2021/2022, where he completed the postgraduate program in 3D Printing Architecture.

Integrative Green Building Envelope

Large Scale Robotic Additive Manufacturing

Juliette Zidek
University of Michigan

Laurin Aman
University of Michigan

Xinran Li
University of Michigan

Jumaanah Alhashemi
University of Michigan

Mania Aghaei Meibodi
DART, University of Michigan

1   Image of the full-scale IGBE system prototype

ABSTRACT

The urban heat island (UHI) effect, corresponding pollution, and a lack of biodiversity in urban areas pose serious threats to human health that will significantly increase with the rapid urbanization predicted in the next several decades. Green building envelope systems that incorporate plants and vegetation on the building exterior offer one solution to reduce UHI effect and air pollutants while increasing building performance through passive cooling, wind buffering, and added insulation.

While conventional green wall systems are often materially inefficient due to a lack of integration with functional building envelope layers, this research investigates the potential of large-scale 3D printing with robotic pellet extrusion (RPE) to produce a novel integrative green building envelope system. The developed envelope system fosters the growth of a self-watering, diverse plant ecology to passively cool buildings, filter contaminants from polluted air, and attract diverse forms of wildlife. This paper presents the conceptualization, design, and prototypes of the Integrative Green Building Envelope (IGBE) system. The fabrication and computational methods developed through this research resulted in the successful 3D print of a geometrically complex, full-scale prototype with multiple layers that can be adapted and applied to many different architectural typologies and locations.

INTRODUCTION

Urban areas face a host of existential environmental crises brought on by anthropogenic climate change. Urban Heat Island (UHI) effect and corresponding pollution pose serious threats to human health and well-being that will only magnify with the increasing frequency and intensity of extreme weather events (Masson-Delmotte et al. 2021). UHI effect is a phenomenon by which urban areas experience elevated temperatures in comparison with rural surroundings. This difference in atmospheric temperature can reach 12°C at night, while surface temperatures are found to be 15°C higher during the day (U.S. Environmental Protection Agency 2008). Caused by the high thermal absorption of construction materials, heat entrapment from urban canyon effect, heat generation from human activities, and a lack of green space in cities, UHI adversely affects urban communities through increased energy demand, thermal pollution of sensitive ecosystems, the formation of ground-level ozone and other air pollutants, as well as heat-related illness and mortality (U.S. Environmental Protection Agency 2008). UHI is estimated to cause an additional 1.1 million heat-related deaths in the United States each year while ambient air pollution is responsible for an additional 4.2 million deaths globally (Lowe 2015; World Health Organization 2018). The impact of these problems will be exacerbated by the rapid urbanization needed to accommodate two-thirds of the world's population by 2050. There will be a significant increase in UHI effect as more buildings are constructed and urban areas densify (United Nations, Department of Economic and Social Affairs, Population Division 2019).

Green building envelope systems that comprise vegetation on the building exterior offer one solution to reduce UHI effect and its negative impact on human health (Lehmann 2021). Plants cool their surrounding environments through evapotranspiration and shading, reducing ambient temperatures during extreme heat events and improving building performance. Green envelope systems also filter harmful pollutants and contaminants from the air, absorb carbon, increase building longevity, and provide insulation and wind protection to further reduce energy consumption (Perini and Rosasco 2013). In addition, urban ecosystems can benefit from the habitat and sources of food such envelope systems provide for birds and various invertebrates (Mayrand et al. 2018).

In cities, building facades offer expansive areas to cultivate plant life with green wall systems. Conventional green wall systems, however, are typically designed as superimposed structures that lack integration with building systems, resulting in material inefficiencies that increase their life cycle global warming potential (Ottele et al. 2011). These systems comprise mats (often non-recyclable felt) or modular planters with a uniform depth of growing substrate that limits plant

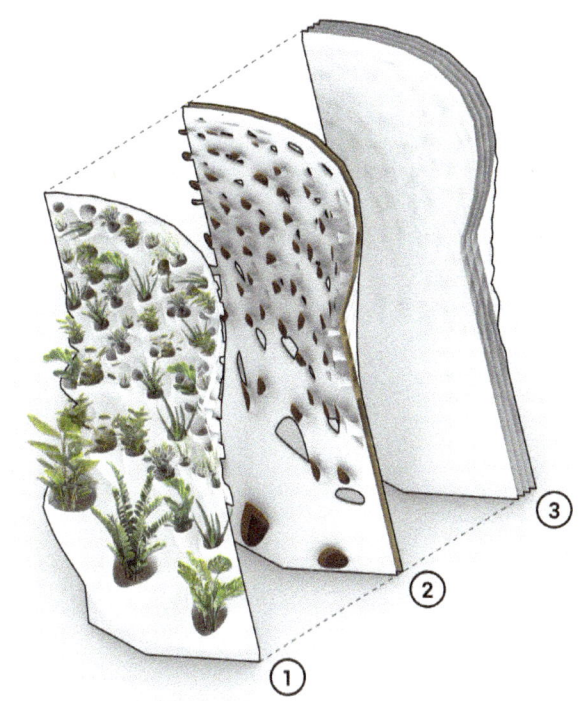

2   Multi-functional layers of the IGBE system include: 1) the exterior shell comprising plant pods of various sizes; 2) the capillary irrigation layer; and -3) multiple interior air layers

size, variety, and support of biodiversity (Convertino, Vox, and Schettini 2021). The environmental burden and high cost of mat systems, in particular, are largely impacted by the high installation cost of pre-vegetated panels and annual maintenance requirements, including frequent panel, plant, and irrigation pipe replacement (Perini and Rosasco 2013).

Large-scale 3D printing through fused filament fabrication (FFF) and fused granulate fabrication (FGF) has the potential to profoundly change the way we conceptualize, design, and build envelope systems by allowing for the seamless integration of many functions. FFF and FGF are additive manufacturing (AM) processes through which material is heated, extruded through a nozzle, and deposited to form objects layer by layer. FFF uses filament, while FGF uses granular material as feedstock. These fabrication methods allow for the fast manufacturing of complex components. In particular, printing with thermoplastics, which are lightweight and versatile, enables the invention of high-performance envelope solutions with energy saving benefits. Material research of thermoplastics, especially bio-based plastics, is a rapidly expanding field that promises to greatly reduce the carbon footprint of the construction industry (Bhatnagar 2019). Together, AM and thermoplastics can radically transform the production and performance of building envelope systems.

Recent research has demonstrated the interest of the Architectural Engineering and Construction industry in the

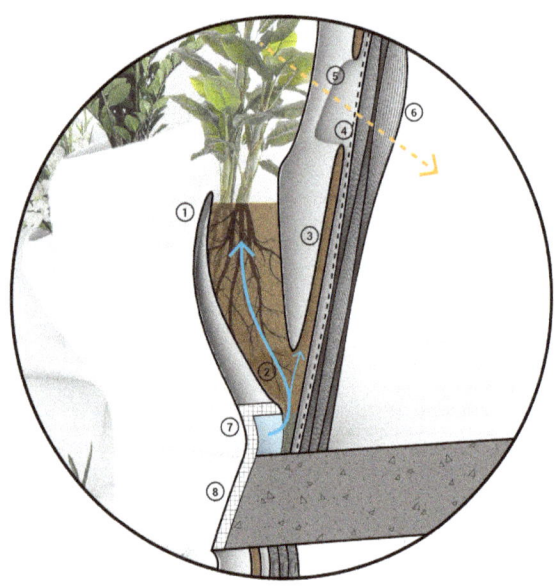

1. PLANT POD
2. CAPILLARY CHANNEL
3. CAPILLARY IRRIGATION LAYER
4. INTEGRAL VAPOR BARRIER
5. VOID FOR LIGHT AND AIR
6. AIR LAYERS
7. PERIMETER WATER SUPPLY
8. INSULATED SLAB EDGE

3   Schematic section perspective and detail of the IGBE system

thermoplastic 3D printing of building envelopes. For example, the Additive Manufacturing and Integrated Technology (AMIE) project, a research collaboration between Skidmore Owings and Merrill, LLP and the U.S. Department of Energy's Oak Ridge National Laboratory demonstrates the use of Big Area Additive Manufacturing (BAAM) technology to design and fabricate a high-performance building enclosure that integrates vacuum insulated panels, self shading windows, structural and electrical systems, photovoltaic panels and an innovative energy sharing platform (Guerguis et al. 2017). Students at ETH have 3D printed a functionally graded facade system with non-orthogonal discretization, gradient infill structures, and integrated snap-fit connectors (Taseva et al. 2020). Branch Technology, a Tennessee-based fabrication company, uses spatial 3D printing of ABS plastic and carbon fiber, a process they have patented as "Cellular Fabrication" (C-Fab), to build mass-customized rain-screen panels in addition to other bespoke building elements (Shelton 2017). The full potential of polymer 3D printing, however, remains unexplored. The application of novel 3D printing technologies will require a complete reimagining of construction practices, design processes, and the entire conceptualization of building envelope systems.

STATE OF THE ART

A new alliance between nature and construction is proposed to address the growing environmental issues that humans and urban ecosystems face. This research presents a novel green wall typology designed as an integral component of a multi-layer 3D printed building envelope system. Such a system combines the many functions of conventional building systems including structure, insulation, air and vapor barriers, and exterior cladding with components that support the cultivation of plant life. The proposed envelope system is composed of: a) an exterior shell entailing plant pods that vary in size; b) a capillary irrigation layer (CIL); c) channels that connect each plant pod to the CIL; and d) multiple, separate interior layers that provide thermal insulation (Figure 2).

This research advances the existing research on 3D printing thermoplastic envelopes and green building infrastructure by:

1. Developing 3D printing methods based on robotic pellet extrusion (RPE) for large-scale geometrically-complex parts,
2. Conceptualizing an integrative green building envelope (IGBE) system,
3. Developing a computational design method for the seamless integration of diverse plant species within a building envelope system, and
4. Developing tool pathing methods and control systems that improve the speed and efficiency of RPE 3D printing for construction.

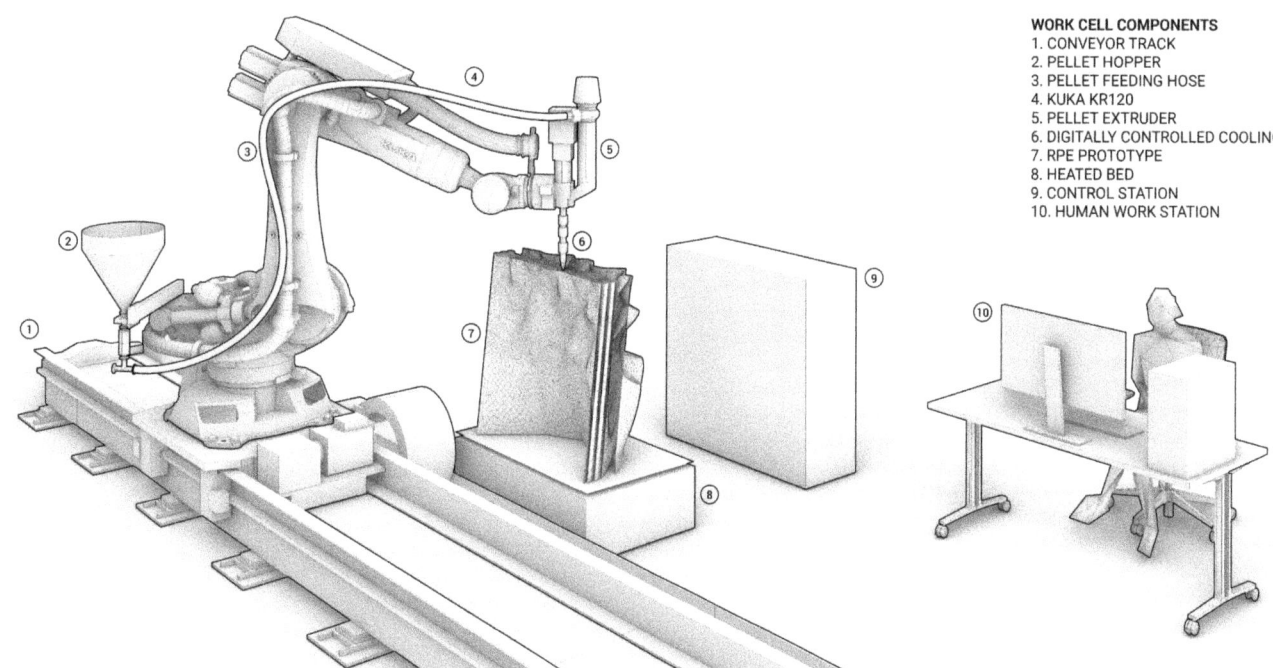

4 The construction scale RPE work cell

Achieving holistic integration of various envelope functions results in geometric complexity that can only be produced through 3D printing. A computational model based on Voronoi logic is developed to partition the exterior layer of the envelope into many unique plant pods with channels that connect each plant pod to the CIL.

Production of complex topologies through robotic extrusion demands toolpath strategies that account for shallow overhangs, minimizing or eliminating support material, and optimizing the route between discontinuous polylines both within a single layer and between successive layers. The toolpath is defined by contour curves that outline the designed form. A complex geometry with interior porosity and voids, as in this research, demands outlines of the form in the exterior and interior of the designed part. Therefore, contouring the overall mesh geometry to generate 3D printing data results in sets of open and closed polylines (with connected start and endpoints) per layer. These polylines are the source for the toolpath generation process and they contain all of the information needed to generate tool paths.

PETG is used as the 3D printing feedstock. While PETG is an affordable material that can produce durable and hard-wearing components, it is derived from non-renewable resources and causes environmental harm at various points along its supply chain. For this project, PETG is proposed as a proxy for an alternative, renewable thermoplastic. Numerous types of non-petroleum-based plastics are being utilized in architectural 3D printing applications. Projects such as the *3D Printed Urban Cabin*, located in Amsterdam, and the temporary facade of Europe Building designed by DUS Architects for the Dutch EU presidency in 2016 are printed with materials from renewable biomass sources (DUS Architects, n.d.). Technical advancements to the fabrication setup can allow future exploration of alternative materials. Through prototyping, the toolpath strategies, fabrication settings, computational model, and the geometry of the components were adjusted and refined. The developed envelope system, computational and fabrication methods in this research are generalizable and scalable.

METHODS
Conceptualization

The exterior cladding is devised for plants to inhabit pods in an upright orientation as they would inhabit the earth's ground (Figure 3). This approach is similar to conventional modular living wall systems that provide greater soil depth than mat systems and allow for more diverse plant habitation, greater plant longevity, and easier maintenance (Dover 2015; Jim 2015). Unlike conventional modular green walls, however, these integral plant pods vary in size. Proper container size is critical to the survival and growth of cultivated plants; containers either too small or too large can harm plant health, leading plants to become root-bound or develop root rot. Additionally, root depth and morphology vary among plant species, necessitating a variety of pod sizes to foster plant diversity. Support of diverse flora allows a unique tailoring of species specific environmental performance; for instance, plant species with small leaves and high foliage density have been found to exhibit greater cooling effects, woody plants have been found to store carbon more effectively, and leaf

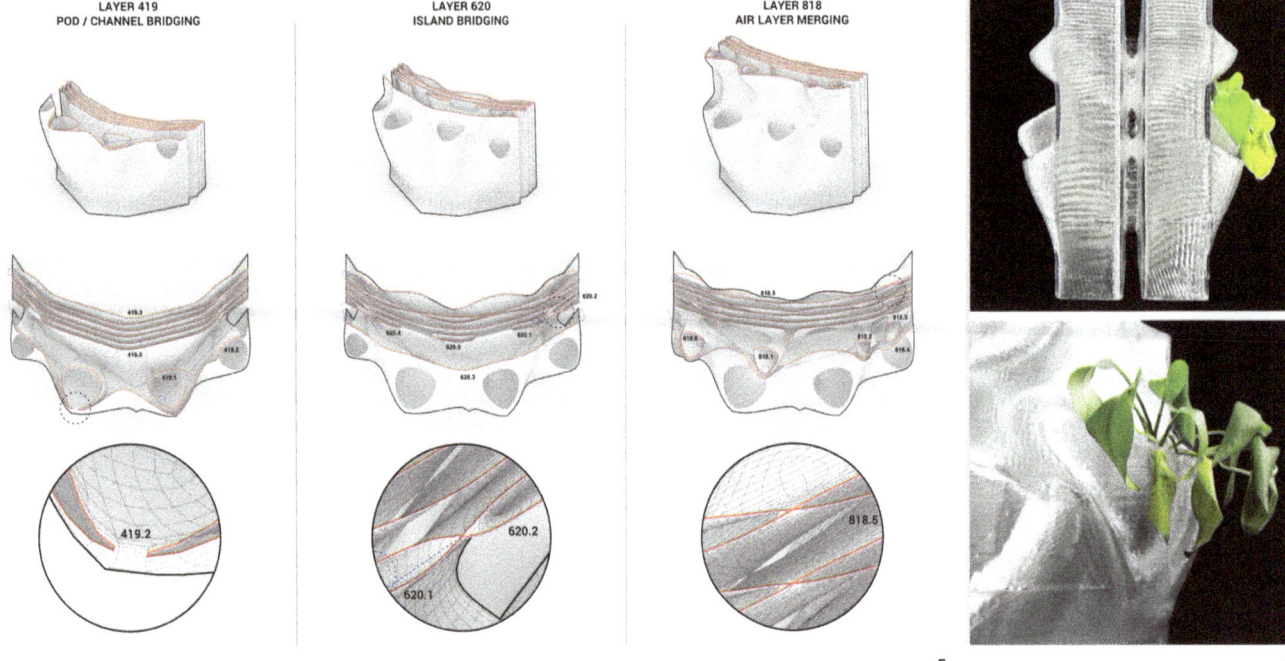

5   Enlarged tool paths generated at critical bridging and merging layers

6   Early prototypes testing plant pod geometry and print settings for speed and print quality

morphology has been found to affect the capture of airborne particulate matter (Charoenkit and Yiemwattana 2017; Ysebaert et al. 2021).

Here, pod size is parametrically adjustable through the computational design of the enclosure system. Each computationally generated pod is unique in its form and volume. In the proposed IGBE system, pods are largest in scale at the base of the envelope and decrease in size with height, giving clearance to plants for vertical growth and access to sunlight.

The plant pods are connected by interior channels to the CIL, which contains a substrate to transport water upwards through capillary action. These channels act as a pathway for plant roots to absorb water. A capillary irrigation system is proposed as an alternative to complex drip irrigation networks that are costly to maintain and prone to failure (Jim 2015). At each building floor, a perimeter water channel supplies water at the base of the CIL. This water channel can provide plants with water that has been collected from rainwater or supplied from the building's water system. Voids for light and ventilation perforate the CIL at each plant pod. While in summer these voids would be shaded by foliage, winter sun could passively heat buildings in temperate climate zones if perennial plants are used.

In the IGBE system, sand is proposed as the capillary substrate given its relatively large particle size. Rudimentary tests were conducted to prove that the sand filled irrigation layer holds water. Plastic is hydrophobic and a PETG test cylinder did not wick water. When the test cylinder was filled with sand, water wicked all the way to the top of the cylinder, and the sand was fully saturated. Capillary rise depends on the porosity of the infill material. Sand has the potential to rise 1-8 feet; silt 12-16 feet; and clay 12-20 feet (Thomas 2020).

Adjacent to the CIL are sandwiched layers that provide thermal insulation. The close proximity of the layers creates air pockets, and air is a heat insulator. Previous research in 3D printed insulation has demonstrated that more pockets result in a higher thermal resistance and smaller pockets reduce the thermal conductivity of encapsulated air (Grabowska and Kasperski 2020). Thirty air-filled layers would be necessary to achieve the required R-20 value for Climate Zone 4A, according to Lars Junghans, University of Michigan Associate Professor of Architecture (in conversation with authors, November 18, 2021). Argon and other noble gasses such as krypton and xenon could be used in place of air to improve the thermal resistance of these layers. Because vegetation has been found to increase the thermal insulation capacity of exterior wall systems, the R-value of the proposed enclosure system could also grow as the density and foliage of cultivated plants increases (Pötz 2016).

In the proposed IGBE system, three layers are sandwiched at a spacing of 25 mm. The extruder end-effector as well as the fabrication speed are factors that influenced the number of layers in the full-scale prototype. A future challenge is to achieve thirty layers without increasing the thickness of the overall facade, which can be achieved with advancements to the fabrication setup, most importantly designing an end-effector nozzle to print at various resolutions.

Construction Scale RPE Set Up and Challenges
A construction scale robotic pellet extrusion (RPE) refers to a method of FGF deposition using a pellet extruder conveyed by an industrial robotic arm. The fabrication set up for this research includes an industrial robotic arm with payload of 120 kg and a 2.5 m maximum reach, accommodated with an inhouse built pellet extruder, automatic pellet feeding system, digitally controlled cooling systems, and a 1236 mm by 762 mm heated bed (Figure 4).

The RPE process comes with certain challenges and limitations including: a) limited overhangs without support material; b) bridging islands without support material; c) adhesion between the first printed layer and the heated bed; d) layer adhesion and print stability as the height of the print increases; and e) unintended material deposition during traversal of the extruder. In this project these challenges were addressed respectively through computational modeling and fabrication setups that: a) integrate parameters to limit overhang angles; b) provide cooling and air flow at bridging moments; c) generate a dense raft beneath the first printed layer and increase extrusion rates of the first several layers to facilitate adhesion; d) tangentially merge internal layers to add stability; and e) optimize traversal path while refining traversal settings. At shallow cantilevers and moments of unsupported horizontal printing, manual techniques were also employed to ensure the success of the print such as manual reduction of print speed and addition of external support.

Form Generation from Fabrication Criteria
The generation of toolpaths for FGF processes always utilizes polyline geometries. Through planar slicing of a 3D geometry, each slice, or layer, contains a collection of 2D open and closed polylines, or islands. In this project, toolpaths are generated at 1 mm increments. The complex topology of the building envelope entails continuous internal cavities and passages for functional integration which produce collections of islands per layer. Islands each require starting and stopping the print with lead in and lead out motion parameters that affect fabrication time. While sufficient time for cooling can reduce sagging at overhangs and bridging moments, an overall reduction in fabrication time is desirable for efficiency and improved layer adhesion (Roschli et al. 2018). To reduce the number of islands and shorten fabrication time with toolpath continuity, alternating edges of the interior air layers were enclosed prior to slicing (Figure 5).

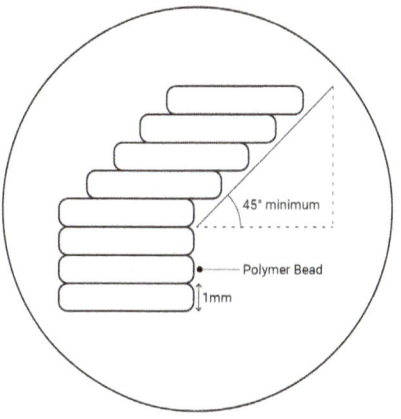

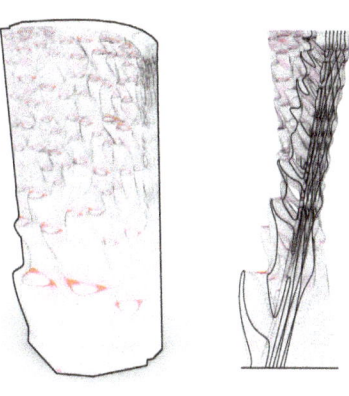

ORIGINAL MODEL

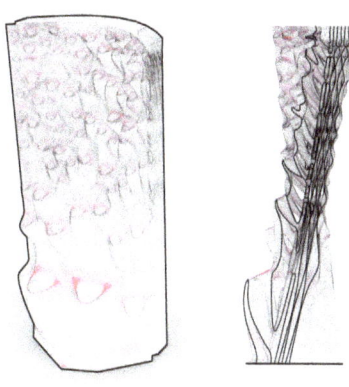

ADJUSTED MODEL

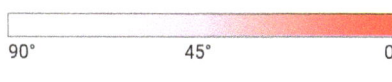

OVERHANG ANGLE FROM HORIZONTAL

7   The minimum cantilever angle from horizontal is shown above; the angle of each mesh face in an earlier version is compared with the printed version of the full-scale prototype

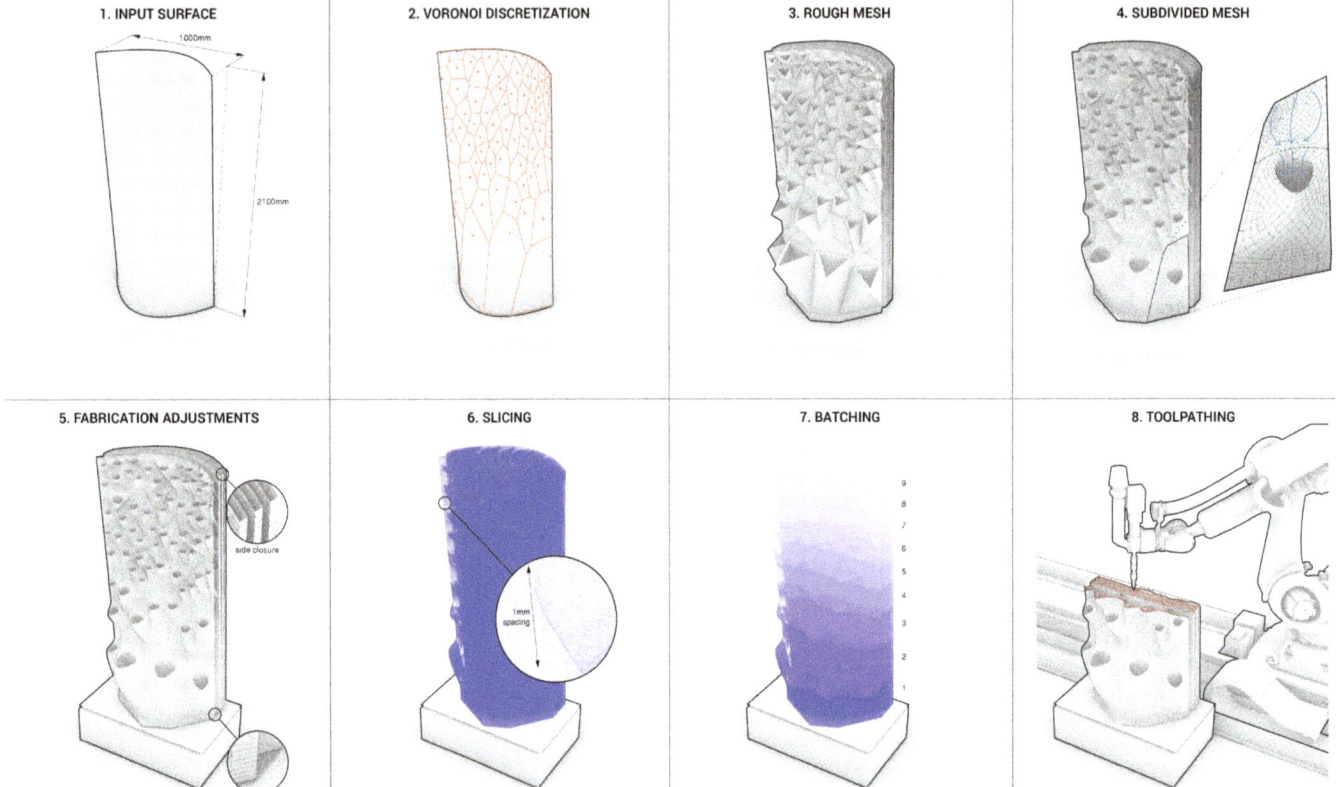

8 Steps of the computational design methodology

As islands increase in size in successive layers, they merge with each other. At the moment of merging, each island toolpath must intersect tangentially. This is considered the lamination layer. A perfect tangent in the lamination layer is critical to the watertightness and stability of the 3D printed part. A successful build-up of the printed part is determined by the overhang angle of islands in successive layers, as well as the amount of cooling delivered by the fan. Minimal unsupported distance between islands and high cooling decreases the chance of sagging in the lamination layer. Through prototyping, it was found that a minimum overhang angle of 45° from horizontal was required to reduce sagging and delamination between layers. This data was integrated in the computational model to guide the geometric design of the enclosure system (Figure 6).

The full adhesion of the printed geometry to the heated bed during the fabrication process is critical to the success of the print. Vibration of the robotic arm increases with height, which can lead to shaking of the print. Proper bed adhesion and the design of tangentially merging interior layers ensure the stability of the print as it grows in height. The temperature of the heated bed and surface area of the first several print layers contribute to the print adhesion. To prevent detachment of the printed part from the heated bed throughout the printing process, the temperature of the bed was set to 105 °C and a zig-zag raft with 4 mm thickness was introduced at the base of the prototype geometry.

Computational Design

The computational methodology of the IGBE system is designed as an adaptable framework to enable site-specific design strategies and involves the generation of design geometry and fabrication data. The computational design stages include: 1) inputting a given freeform surface; 2) discretizing the surface via a Voronoi algorithm; 3) generating a rough mesh of interconnected geometry from Voronoi cells; 4) subdividing the rough mesh with the Catmull-Clark algorithm; 5) modifying the mesh for fabrication constraints including a) a raft for bed adhesion, and b) layer enclosure for toolpath continuity; 6) slicing the mesh to generate contours; 7) batching the contours; and 8) generating and simulating toolpaths with printing parameters per batch (Figure 8).

To best accommodate a diversity of plant species within the building enclosure system, a computational model based on the Voronoi diagram is developed. This model allows parametric partitioning of any given exterior surface into regions close to the size required for a given set of diverse plants. For each plant, there is a corresponding region, the Voronoi cell. The bottom portions of these cells form the plant pods and their connection to the CIL, resulting in one continuous mesh. The top portions of the Voronoi cells are articulated as concave surfaces that are designed to direct precipitation to plants.

To integrate thermal performance, the developed computational model generates interior layers that can be controlled parametrically. The overall thickness of the envelope, spacing between layers, and tangential merging of layers can be computationally adjusted in response to design and fabrication considerations such as thermal resistance, daylighting, material efficiency, and print speed. The reduction in spacing between layers as they merge allows for variation in translucency and patterning across the facade. This merging between layers also provides structural stability during the fabrication process so that the full height prototype can be produced without the need for infill support, reducing the overall weight of the 3D printed part, material used, and printing time.

The computational design of the envelope was refined multiple times in response to a number of fabrication constraints after prototyping. For example, the geometries of plant pod rims were angled downward at a sufficiently steep angle to allow for successive layer deposition based on the observed minimum overhang angle. Plant pod and channel geometries were steepened and elongated as well. Interior voids were rescaled and shaped to limit the unsupported distance at bridging moments.

Toolpath Generation

The tool path generation is developed as an automated process. The complex and interconnected geometry of the envelope entails islands that require turning on and off the extruder as well as the unique naming of separate toolpaths on each layer. An algorithm was developed to automatically name these islands based on K-Nearest Neighbor (KNN) to optimize the travel path of the robot, and therefore minimize the time needed to fabricate a piece.

Besides toolpath generation, print settings for material deposition were developed and refined through prototyping. These print settings include: lead-in and lead-out path and parameters, material extrusion rate, robotic toolhead velocity, traversal velocity, extruder activation and deactivation locations, and cooling activation and deactivation locations (Table 1). Lead-in and lead-out settings were refined to minimize excess material deposition at the end of toolpaths and during traversal, while also balancing the reduction in speed necessary for such quality control measures. The introduction of helical lead-ins/outs resulted in neat and watertight seams with high resolution. Print speed, extrusion rate, and cooling settings were refined as well.

Fabrication and Prototyping

A series of prototypes at multiple scales were printed to investigate the relation between the geometric properties of the

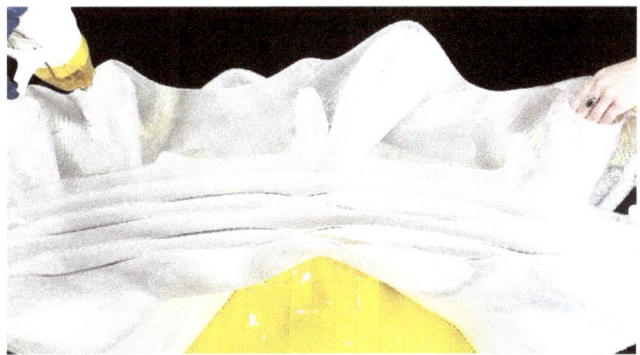

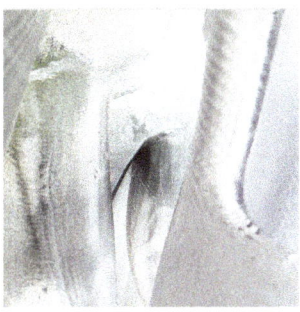
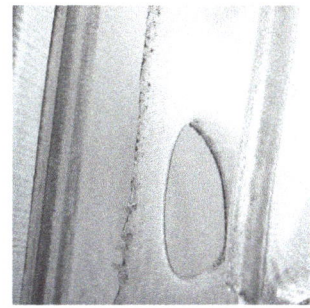

9   Fabrication images of the full-scale prototype highlighting the seamless connection of channels and interior layers

envelope design and the fabrication settings, constraints, and criteria (Figure 6). Based on the prototypes across these scales, strategies for computational design and fabrication methods were refined.

RESULTS

The resulting 1:1 scale prototype (1018 mm x 678 mm x 1022 mm) demonstrates the advantages of FGF 3D printing thermoplastic for integrative and green building envelopes. The 3D printed building envelope prototype entails a highly complex topology with an array of unique plant pods that form interconnected channels and multiple interior layers. The prototype weighs 58.9 kg and was 3D printed over the course of 64 hours (Table 2). The size of plant pods vary in several domains; here the largest pod (174 mm x 181 mm x 488 mm) measured 7,845 cm^3 and the smallest (70 mm x 67 mm x 295 mm) measured 710 cm^3. The prototype plant pods were filled with plants of various species and sizes to qualitatively evaluate the utility and design of the IGBE system. The planted envelope attracted pollinators as shown in Figure 12. Layers of translucent polymer shaded by plants and growing substrate allow diffuse light to permeate the envelope prototype.

Design and development of a robust raft was necessary to ensure the stability of the prototype throughout the 3D printing process. The 4-layer thick raft effectively secured the print to the heated bed throughout the entire printing time. The interconnected geometry and merging layers also provided adequate stability and support for a successful print. No additional support was required. The toolpath was refined to allow excess extrusion material to collect outside the printed geometry rather than along the toolpath. While early prototypes required manual care throughout printing to monitor excess material deposition, these changes to the toolpath made such observation unnecessary.

The limits of both the angle and the height were pushed while prototyping. Portions of the envelope with overhang angles shallower than the 45° minimum managed to print effectively, though minor sagging and delamination occurred at these moments (Figure 9). Human-robot collaboration to support these shallow areas was the cheapest and most time efficient method to ensure a successful print without any 3D printed support. In future prototypes, shallow cantilever limitations could be resolved through further geometric adjustments in the computational model. The full-scale prototype and earlier experiments established rulesets for the design geometries and toolpath settings, which will be useful to future advancement of this research.

10 Photograph of the full-scale planted IGBE prototype
11 Top view of the prototype revealing the inter-connected interior geometry
12 Photograph of a pollinator drawn to the planted prototype when brought outdoors
13 Photograph of the full-scale IGBE prototype

CONCLUSION

This project successfully demonstrates that RPE of thermoplastic can be used to establish a hybrid approach to architectural design and construction, integrating natural and building systems within a 3D printed building enclosure. The proposed IGBE system offers a strategy to remediate UHI effects and atmospheric pollution in urban areas while supporting an increase in biodiversity and contributing to the development of resilient urban ecologies. The fabrication and computational methods developed through this research resulted in the successful 3D print of a geometrically complex, full-scale prototype with multiple layers that can be applied and adapted to many different architectural typologies and locations. Speed and material efficiency of the RPE fabrication method have been studied through this research, though further research is needed to calculate and optimize energy use of the entire process and to apply such optimization at an industrial scale.

This research will continue in the following areas to advance the study of integrative green building envelope systems: development of the fabrication system to allow for the high speed production of thinner plastic layers; performance modeling and testing of the IGBE system for passive cooling effect, thermal resistance, air filtration, daylighting, and longevity; performance testing of the capillary irrigation layer to investigate the feasibility of a self-watering irrigation system; a thorough study of specific plant species, their unique performance profiles and habitation requirements; further study on the circular economic model of thermoplastic and bio-plastic building envelopes; and further study on bio-plastic 3D printing and its performance in the context of building envelope fabrication.

TABLE 1

Print Settings

Lead In	Style	Arc Tan Ramp
	Distance (mm)	10
	Overlap Distance (mm)	100
	Speed (m/s)	0.08
Lead Out	Style	Arc Tan Ramp
	Distance (mm)	10
	Overlap Distance (mm)	100
	Speed (m/s)	0.01
Extrusion	Rate	1.1
Path	Speed (m/s)	0.08
Traverse	Speed (m/s)	0.4
Extrusion	Activation Delay (mm)	0
	Activation Distance (mm)	-3
	Deactivation Delay (mm)	0
	Deactivation Distance (mm)	-3
Cooling	Activation Delay (mm)	0
	Activation Distance	20
	Deactivation Delay (mm)	0
	Deactivation Distance (mm)	0

TABLE 2

Results	Printed Portion (Segments 1-3)	Full Height (Segments 1-9)
Material	Thermoplastic PETG	Thermoplastic PETG
Dimensions (L x W x H)	1018 mm x 678 mm x 1022 mm	1018 mm x 745 mm x 2175 mm
Total Tool Path Length	9,427 m	19,842 m
Volume	47,135 cm^3	99,210 cm^3
Weight (using 1.25 g/cm3)	58.9 kg	124.3 kg
Printing Time	64 hours	135 hours
No. Mesh Faces	62,347	381,370
No. Curves	5,106	17,320

ACKNOWLEDGMENTS

The design and fabrication of the prototype was conducted as part of the Material Engagement Course taught by Dr. Mania Aghaei Meibodi at the Master of Science Digital and Material Technologies program at the University of Michigan. We would like to thank Prof. Lars Junghans for his consultation regarding the thermal performance of air layers in building envelope systems; Christopher Voltl for photographing the full scale prototype; and Alireza Bayramvand for his assistance preparing fabrication files. The transfer of the toolpath data to the robotic controller used Super Matter Tools, a toolpathing interface accommodating the Kuka robots at the Taubman College Digital Fabrication Lab.

REFERENCES

Bhatnagar, Navodita, ed. 2019. *Bio-Based Plastics: Materials and Applications*. N.p.: Arcler Education Incorporated.

Charoenkit, Sasima, and Suthat Yiemwattana. 2017. "Role of specific plant characteristics on thermal and carbon sequestration properties of living walls in tropical climate." *Building and Environment* 115 (January): 67-79. http://dx.doi.org/10.1016/j.buildenv.2017.01.017.

Convertino, Fabiana, Giuliano Vox, and Evelia Schettini. 2021. "Evaluation of the cooling effect provided by a green facade as a nature-based system for buildings." *Building and Environment* 203 (June). https://doi.org/10.1016/j.buildenv.2021.108099.

Dover, John W. 2015. *Green Infrastructure: Incorporating Plants and Enhancing Biodiversity in Buildings and Urban Environments*. N.p.: Taylor & Francis Group.

DUS Architects. n.d. "DUS Architects – Public architecture and design that consciously influences everyday life." Accessed August 21, 2022. https://houseofdus.com/.

Grabowska, Beata, and Jacek Kasperski. 2020. "The Thermal Conductivity of 3D Printed Plastic Insulation Materials—The Effect of Optimizing the Regular Structure of Closures." *Materials* 13 (19): 4400. https://doi.org/10.3390/ma13194400.

Guerguis, Maged, Leif Eikevik, Lucas Tryggestad, Andrew Obendorf, Philip Enquist, Brian Lee, Arathi Gowda, Brian Post, Kaushik Biswas, and Justin Shultz. 2017. "High performance 3D printed façade with integrated energy: built works and advancements in computational simulation." In *Proceedings of the 12th Conference on Advanced Building Skins 2017*. Bern, Switzerland.

Jim, C. Y. 2015. "Greenwall classification and critical design-management assessments." *Ecological Engineering* 77 (April): 348–362. http://dx.doi.org/10.1016/j.ecoleng.2015.01.021.

Lehmann, Steffen. 2021. "Growing Biodiverse Urban Futures: Renaturalization and Rewilding as Strategies to Strengthen Urban Resilience." *Sustainability* 13 (5): 2932. https://doi.org/10.3390/su13052932.

Lowe, Scott A. 2015. "An energy and mortality impact assessment of the urban heat island in the US." *Environmental Impact Assessment Review* 56 (October): 139–144. https://doi.org/10.1016/j.eiar.2015.10.004.

Masson-Delmotte, Valerie, Panmao Zhai, Anna Pirani, Sarah L. Connors, Clotilde Pean, Yang Chen, Leah Goldfarb, et al., eds. 2021. "IPCC, 2021: Summary for Policymakers." In *Climate Change 2021: The Physical Science Basis; Working Group I Contribution to the Sixth Assessment Report of the Intergovernmental Panel on Climate Change*. Switzerland: Intergovernmental Panel on Climate Change. 3–34.

Mayrand, Flavie, Philippe Clergeau, Alan Vergnes, and Frederic Madre. 2018. "Vertical Greening Systems as Habitat for Biodiversity." In *Nature Based Strategies for Urban and Building Sustainability*, edited by Katia Perini and Gabriel Perez. Oxford, UK: Elsevier Science.

Ottele, Marc, Katia Perini, A.L.A. Fraaij, E.M. Haas, and R. Raiteri. 2011. "Comparative life cycle analysis for green facades and living wall systems." *Energy and Buildings* 43 (12): 3419–3429. https://doi.org/10.1016/j.enbuild.2011.09.010.

Perini, Katia, and Paolo Rosasco. 2013. "Cost-benefit analysis for green facades and living wall systems." *Building and Environment* 70 (December): 110–121. https://doi.org/10.1016/j.buildenv.2013.08.012.

Pötz, Hiltrud. 2016. *Groenblauwe Netwerken Handleiding Voor Veerkrachtige Steden Green (Blue Grids Manual for Resilient Cities)*, translated by G. Forno. Delft: atelier GROENBLAUW.

Roschli, Alex, Katherin T. Gaul, Alex M. Boulger, Brian K. Post, Phillip C. Chesser, Lonnie J. Love, Fletcher Blue, and Michael Borish. 2018. "Designing for Big Area Additive Manufacturing." *Additive Manufacturing* 25 (11): 275–285. https://doi-org.proxy.lib.umich.edu/10.1016/j.addma.2018.11.006.

Shelton, Ted. 2017. "Cellular Fabrication: Branch Technology, 2014–Present." *Technology | Architecture + Design* 1 (2): 251–253. https://doi.org/10.1080/24751448.2017.1354636.

Taseva, Yoana, Nik Eftekhar, Hyunchul Kwon, Matthias Leschok, and Benjamin Dillenburger. 2020. "Large-scale 3D printing for functionally-graded facade." In *Anthropocene: Proceedings of the 25th International Conference of the Association for Computer-Aided Architectural Design Research in Asia*, vol. 1. 183–192.

Thomas, Geo. 2020. "Effects of Capillary Water on Structures and Foundation | Preventing damage." *EngineeringCivil.org*, July 16, 2020. Accessed June 4, 2022. https://engineeringcivil.org/articles/civil-engineering/effects-of-capillary-water-on-structures-and-foundation-preventing-damage/.

United Nations, Department of Economic and Social Affairs, Population Division. 2019. *World Urbanization Prospects: The 2018 Revision (ST/ESA/SER.A/420)*. New York: United Nations.

U.S. Environmental Protection Agency. n.d. "Heat Island Effect | US EPA." Accessed August 21, 2022. https://www.epa.gov/heatislands.

World Health Organization. 2018. "Ambient air pollution." *World Health Organization*. https://www.who.int/teams/environment-climate-change-and-health/air-quality-and-health/ambient-air-pollution/.

Ysebaert, Tess, Kyra Koch, Roeland Samson, and Siegfried Denys. 2021. "Green walls for mitigating urban particulate matter pollution—A review." *Urban Forestry & Urban Greening* 59 (April): 127014. https://doi.org/10.1016/j.ufug.2021.127014.

IMAGE CREDITS

Figure 10, 12, 13: ©Christopher Voltl, DART Laboratory, Taubman College, University of Michigan, May 13, 2022.

All other drawings and images by the authors. All images and diagrams are copyright of DART Laboratory, Taubman College, University of Michigan.

Juliette Zidek, AIA (she/her) is an architect, researcher and computational designer whose work explores new methods of sustainable design and construction through the use of emergent technology. Formerly a fellow of the MS Digital and Material Technologies program at the University of Michigan's Taubman College of Architecture and Urban Planning, Zidek currently practices as a Senior Designer at Studio Gang Architects in Chicago, IL.

Laurin Aman (she/they) is an architectural researcher, designer, and fabricator who is passionate about crafting materials through rigorous prototyping practices that engage digital technologies. They are a graduate and former fellow of the MS Digital and Material Technologies program at the University of Michigan's Taubman College Architecture of Urban Planning. Currently, Laurin is a Research Assistant and Teaching Aide at the University of Michigan, an Adjunct Professor at Lawrence Technological University, and a member of *Alibi Studio.

Xinran Li (she/her) is an architect. She received her Bachelor's degree in 2020 in architecture and her Master's degree of the MS Digital and Material Technology in 2022 at the University of Michigan. Her areas of interests include parametric design, digital fabrication, intelligent buildings and green buildings. Now she works in AIplanetwork in Shanghai as an architect, focusing on parametric design, digital construction, and finishing projects with BIM.

Jumaanah Alhashemi (she/her) is a multidisciplinary designer, artist, and fabricator based in Dubai and Abu Dhabi. She received her Master of Science in Digital and Material Technologies from the University of Michigan where she was also a research fellow. She is currently the Assistant Director of Research Visualization and Fabrication Services located in the Experimental Research at New York University Abu Dhabi.

Dr. Mania Aghaei Meibodi is Assistant Professor of Architecture and Chair of Digital Architecture Research and Technologies (DART) Laboratory at Taubman College of Architecture and Urban Planning at the University of Michigan.

Lines of Flight

Facade Design for Multispecies Migrations

John Kim*
Macalester College

Adam Marcus*
California College of the Arts

Molly Reichert*
University of Minnesota

*All authors contributed equally to this research

1 View of south facade of Lines of Flight, Human

ABSTRACT

Lines of Flight, Human is a large-scale architectural facade design completed for the new Minneapolis Public Service Building that examines the history of human migration to the region and its impacts on human and non-human life. The project employs a range of computational techniques to spatialize cultural data about human migration and dispossession in the design of a pattern that meets bird-safety standards for buildings. The work is situated within theoretical discussions of site-specificity, land acknowledgment, the politics of immigration, procedural art, and best practices for bird mortality reduction. The paper discusses the project as a case study in data spatialization by reviewing the design process, which employed custom algorithms to integrate data-driven and recursive workflows that integrate cultural data, communicative design, and bird-safe parameters. Discussion of the research reflects on the project's implications for future design of communicative facades, the limitations and biases of data-driven design techniques, and the potential to expand these workflows to engage with other aspects of building performance.

INTRODUCTION

Lines of Flight, Human is a large-scale architectural facade design completed for the new Public Service Building at City Hall Plaza in Minneapolis (Figure 1). The project, a custom-printed film pattern applied to the building's glass curtain wall, extends along 150'-0" of the second floor facade and 400'-0" of a public skyway bridge that connects to the county office building across the street. The facade's design leverages a range of computational techniques and workflows to meet three objectives: to spatialize historical and cultural data related to the site's history; to serve as a dynamic and engaging public artwork for both people within the space and pedestrians on the street level; and to meet design standards for reducing bird mortality. In this regard, computational design becomes a platform for synthesizing and expanding the architectural facade's latent capacities for data visualization, public engagement, and environmental performance.

The project examines the history of human migration to the region now known by its settler-colonial name, Minnesota (subsequently referred to as "Minnesota")[1] and its impacts on human and non-human life. *Lines of Flight, Human* references immigration's role—in the forced dispossession and genocide of Indigenous communities—with a visual land acknowledgment of Indigenous place names. The project also redresses human migration's consequences for non-human life by reducing bird mortality with bird-safe design. The facade pattern spatializes demographic data as a series of nested bands that weave over and under one another to create an interconnected network across the facade. Each band corresponds to a different constituent population that has immigrated to the region and serves as a kind of timeline, mapping the history of that respective population's migration to the state. The variable dash pattern of each loop intensifies in density and lineweight to indicate moments of greater migration over time. Between and among the bands is a field consisting of encoded Indigenous place names found within the region. The resulting pattern produces multiple readings for multiple constituents at multiple scales: as a bold super-graphic from the street, as an intricate, lacy pattern for the thousands of people who walk through the space each day, and as a bird-safe deterrent for migrating avian species. As a case study in data spatialization (Marcus 2014; Marcus et al. 2017; Marcus 2022), the project demonstrates how computational techniques can be leveraged to synthesize meaningful cultural data, dynamic perceptual qualities, and ecological sensitivity.

STATE OF THE ART
Migration and Land Acknowledgment

Lines of Flight, Human was conceived in the midst of heightened anti-immigration sentiment in the wake of the 2016

2 Concept sketch showing woven bands extending across the skyway facade

presidential campaign and election in the United States. The project responds to this political context by embedding histories of immigration and dispossession into the materiality of the building's facade. Reflecting the persistence of settler-colonialism and how the United States has become a country of immigrants, *Lines of Flight, Human* is an illustration of the national origins of immigrants to Minnesota from 1850-2016, a 160 year span. The artwork draws on data sourced from the Integrated Public Use Microdata Series (IPUMS), the world's largest individual-level population database (IPUMS, n.d.). Analyzing this enormous dataset provided a complex information base for the artwork's design, which recalls Charles Minard's *Map of World Migration*, a visualization of the numbers and destinations of emigrants from Europe, Africa, China, and South Asia (Minard 1862).

The history of immigration to this country is partly a story of forced migrations. In the nineteenth century, Scandinavian immigrants fled the failed revolutions of 1848 to resettle in Minnesota, and in the 1990s, Somali refugees resettled to Minnesota to escape destructive civil wars. According to the U.S. census and refugee-support agencies, Minnesota has the highest number of refugees per capita nationwide. With 2% of the nation's population, Minnesota has 13% of its refugees (Shaw 2018).

With the ongoing migration crises in the Mediterranean, the war in Ukraine, and similar conflicts in other places around the world, the topic of forced migrations continues to draw extensive public attention. *Lines of Flight, Human* echoes works found in recent art exhibitions on the topic, such as *When Home Won't Let You Stay*, an exhibition of works by contemporary artists in response to migrations and displacement of people worldwide. The exhibition borrows its title from a poem by Warsan Shire, a Somali-British poet who gives voice to the experiences of refugees. Reena Saini Kallat's wall piece, *Woven Chronicle*, shown in the exhibition, is a cartographic visualization that references the global flows and movements of travelers, migrants, and labor. "Kallat uses electrical wires—some of which are twisted to resemble barbed wire—to create the lines, which are based on her meticulous research of transnational flows" (Cantor Arts Center, n.d.).

Immigration to the region led to the genocide and forced dispossession of Indigenous peoples that inhabited this land for tens of thousands of years. *Lines of Flight, Human* aspires to recognize this history and draws from art works that suggest a visual land acknowledgment statement—a public recognition that North American lands were the traditional homeland of Indigenous peoples. Edgar Heap of Bird's series, *Native Hosts* (Caldwell 2016), for example, contains the settler name of a place written backwards. Spelled forward is the name of the Indigenous community that inhabits the land. Andrea Carlson's recent work, *You Are on Potawatomi Land* (Chicago Riverwalk, n.d.) is a site-specific project along the Chicago River that powerfully proclaims the Indigenous history of place.

Generative and Procedural Art

Lines of Flight, Human draws inspiration from the long history of generative and procedural artists who use code-based processes to create novel formal and aesthetic qualities. These include the work of Sol LeWitt, whose large-scale drawings differ from traditional murals in their ability to be recreated by others as a kind of executable program. "When an artist uses a conceptual form of art, it means that all of the planning and decisions are made beforehand and the execution is a perfunctory affair," wrote LeWitt in 1967. "The idea becomes a machine that makes the art" (LeWitt 1967). While LeWitt operated in an entirely analog mode, early pioneers of computational art such as Vera Molnar, Manfred Mohr, and Frieder Nake extended this ethos into digital workflows that blended rule-based processes with the computer's capacity to embed randomness, recursion, and iteration into the generation of two-dimensional drawings (Molnar 1975; Mohr 1971; Nake 2009). This project extends such ways of working to the architectural scale of a building's facade. It considers how procedural workflows might yield not only novel aesthetic possibilities, but also new syntheses with data-driven representation, public engagement, and building performance.

Bird Mortality Reduction

The human inhabitation of the Mississippi River basin has had destructive impacts on the non-human life that coinhabits and migrates through this space (Kim 2021). Bird populations, in particular, are increasingly threatened by the construction of buildings in flyway zones, migratory pathways, and habitats (Loss, Will, and Mara 2015). Buildings kill birds at an astonishing rate: an estimated 365 to 988 million birds die annually in the United States from building strikes (Loss et al. 2014).

Located blocks away from this project's site, the U.S. Bank Stadium in Minneapolis is a prime example of a publicly funded building designed without consideration of bird collisions. With its enormous reflective surface, the glass stadium is responsible for many bird deaths, estimated at over 100 birds per year (Loss et al. 2019). This was the context in which the new Minneapolis Public Service Building was conceived—as a building where bird safety was prioritized in the design.

There are a number of strategies for mitigating bird-strike mortality in building design (Klem 1990). The basic principle is to incorporate patterns on transparent fenestration to increase opacity and registration. This patterning signals to birds that a building is present. Frit patterns and other applied films typically use standardized repeating patterns. With advances in computational design and digital printing techniques on glass, bird-safe facade patterns present an opportunity to create unique immersive artworks that resonate at multiple scales and for multiple species.

METHODS

Recursive Tiling and Form Generation

The algorithmic design process for the pattern of *Lines of Flight, Human* integrated data-driven and recursive workflows. This allowed us to synthesize historical data, communicative design, and bird-safe parameters. The first step evolved out a conceptual framing of the project as a series of woven bands extending across the facade as a representation of the intertwined and diverse populations that constitute the history of human migration to Minnesota (Figure 2). To translate this concept to a working parametric model in the Grasshopper platform (Robert McNeel & Associates 2022), we developed a recursive generative algorithm predicated on simple, gridded patterns of lines and arcs. When tiled, the pattern produces unique, horizontally continuous curves, while also minimizing the appearance of repetition. The curves consist of straight segments and arcs, adhering to an underlying grid spacing that ensures formal consistency while allowing for a significant degree of variation. Using the Anemone plugin (Zwierzycki 2015), the algorithm selects a single pattern module from the catalog of manually designed options (Figure 3) and then uses a random seed to iteratively cycle through the catalog for the next sequential pattern module. For each iteration, it also tests rotated and mirrored versions of the selected pattern module, and then evaluates if the placed module tiles continuously with the previous module. A blend of manual and automatic processes, this logic is inspired by Sebastien Truchet's patterns of curves that seem random yet are generated from a limited set of repeated and rotated tiles (Pickover 1989). In this case, we drew pattern modules manually. The algorithm then rearranged these pattern modules, generating a range of unique tiling patterns across the facade from which we selected the final design (Figure 4).

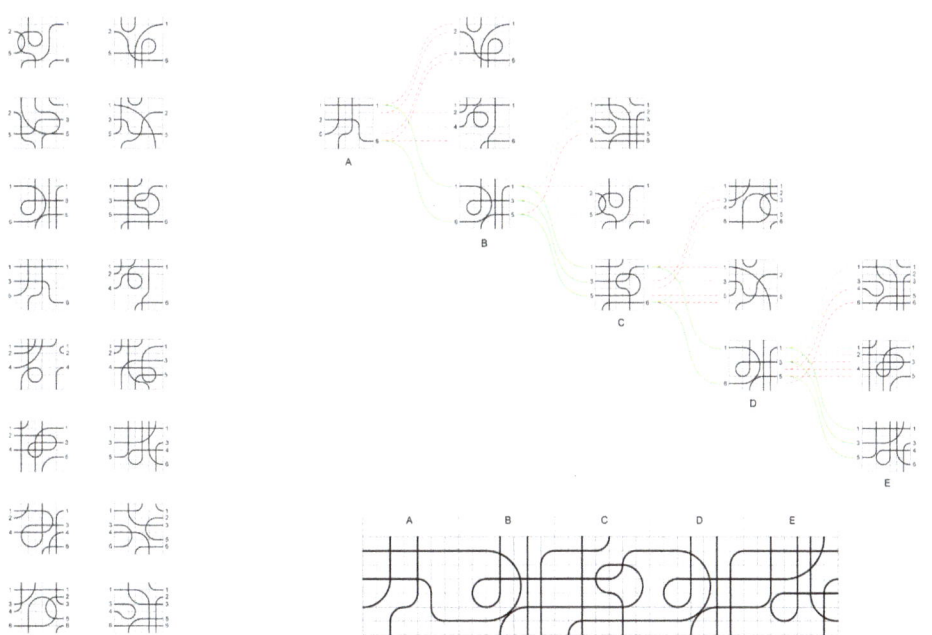

3   Catalog of gridded pattern modules (left) and algorithmic workflow of matching seams to ensure continuous tiling (right)

4   Iterative testing of algorithmically-generated tiling configurations of the pattern modules

5   Each of the 87 curves is analyzed by length and mapped to a specific population based on the quantity of immigrants over time. The diagram uses line thickness to represent length and shows where two populations, from Germany and Peru, are located in the facade.

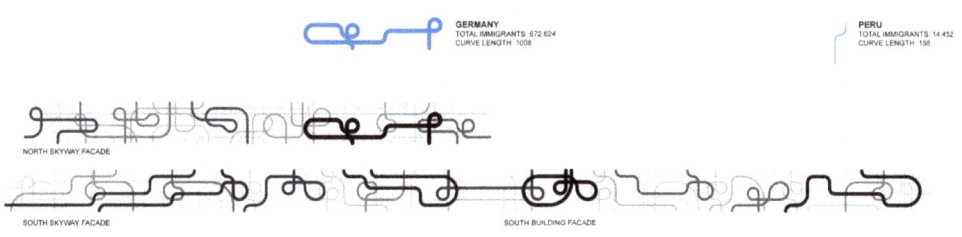

NEW ECOLOGIES II: RESPONSIVE/ADAPTIVE DESIGN METHODS

Hybrids & Haecceities

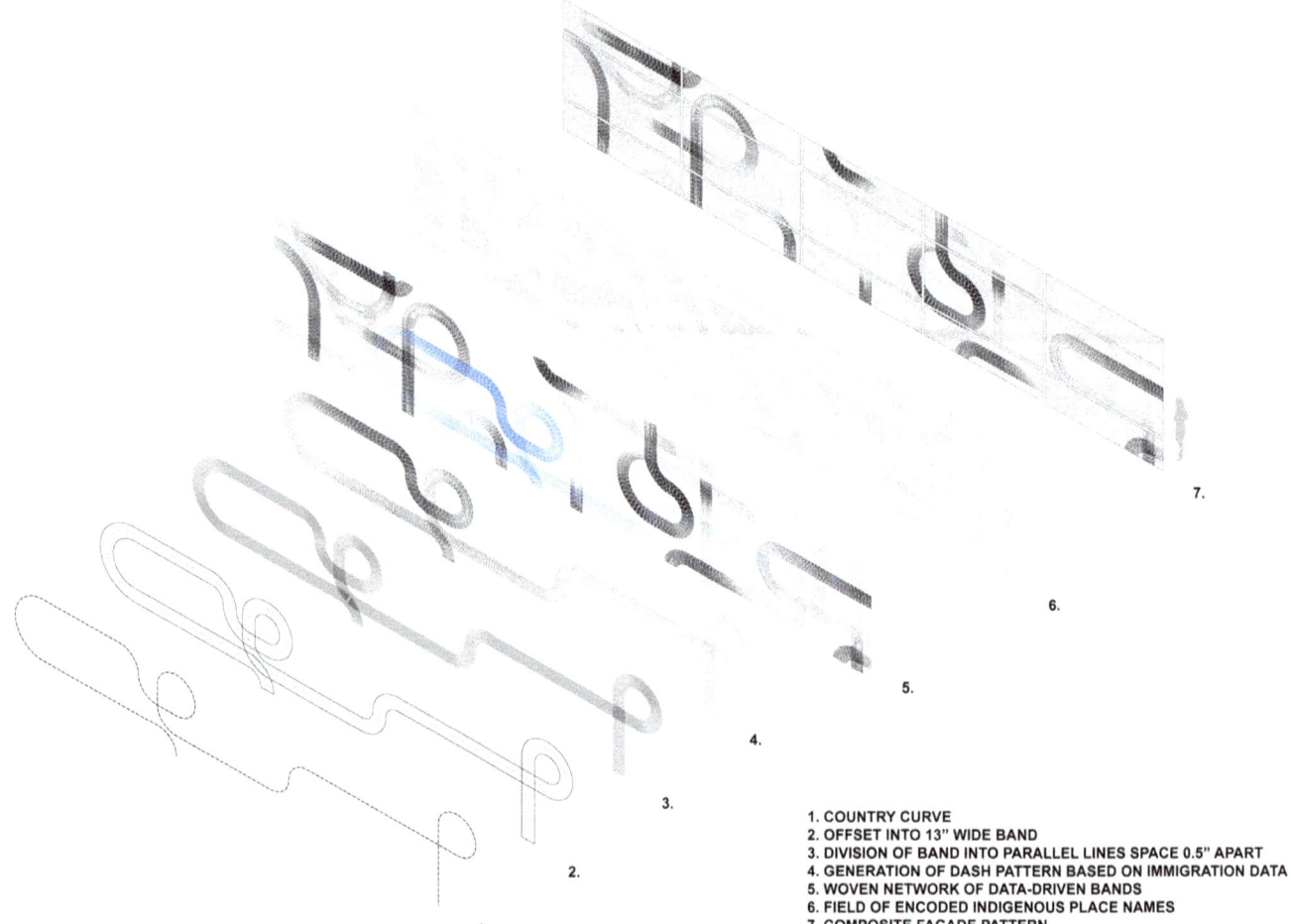

1. COUNTRY CURVE
2. OFFSET INTO 13" WIDE BAND
3. DIVISION OF BAND INTO PARALLEL LINES SPACE 0.5" APART
4. GENERATION OF DASH PATTERN BASED ON IMMIGRATION DATA
5. WOVEN NETWORK OF DATA-DRIVEN BANDS
6. FIELD OF ENCODED INDIGENOUS PLACE NAMES
7. COMPOSITE FACADE PATTERN

6

Data Mapping: Immigrant Population

The next step consisted of quantitatively analyzing the outcome of the tiling script and mapping census data by population to each resulting curve by length. IPUMS census data was sorted by both year and country of immigration. Each of the 87 curves produced by the tiling script was matched with a specific immigrant population, with the length of each curve corresponding to the total number of immigrants from that population to Minnesota.[2] As an example, the longest curve on the facade is Germany, because the largest number of immigrants to Minnesota over time have come from there (Figure 5).

Data Mapping: Time

Each curve was offset repeatedly to produce a 13-inch wide band consisting of 25 curves spaced 0.5 inches apart. These curves were broken into variable dash patterns, in which the dash length, spacing, and lineweight gradate based on the temporal mapping of the population data along each band (Figure 6). Each band is understood as a decennial timeline, with the start point representing 1850 and the end point representing 2016 (Figures 7, 8).[3] For each ten-year census interval, the immigrant population from each country is represented as a percentage of that interval's total number of immigrants. Bolder and more pronounced dashes correspond to periods of increased immigration from that country. Throughout the design process, a variety of full-scale prints and mockups were produced to calibrate optimal scale, resolution, and legibility of the dash patterns from multiple vantage distances (Figure 9).

Immigration and Dispossession as Figure-Ground

The field between the woven bands consists of parametrically generated dash patterns. The direction of these patterns respond to the curvature of the bands, but the composition of the dashes is driven by a different data set. The field of dashes uses the 8-bit ASCII standard (Mackenzie 1980) to encode Indigenous place names within the region as strings of zeros and ones, which are then converted to a running series of short and long dashes (Figure 10). The result is a dynamic field that is both abstract and also rich with information and cultural meaning. The field complements the census data bands to produce an interplay of figure and ground (Figure 11).[4]

Bird Safety Analysis

The final step of the design process analyzed both band and field dashes to ensure that the overall pattern conforms to

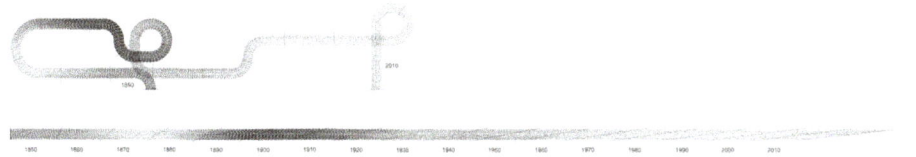

6   Pattern generation process diagram

7   The band representing immigrants from Germany shown in isolation (top) and in unrolled form (bottom). The longer and bolder dashes represent moments of increased immigration from Germany, as indicated in the period from 1890-1920.

8   Catalog of all unrolled bands, sorted by size of immigrant population

9   Full-scale printed mockups, both on paper and clear film, were produced during the design phase to optimize the pattern's legibility from multiple distances.

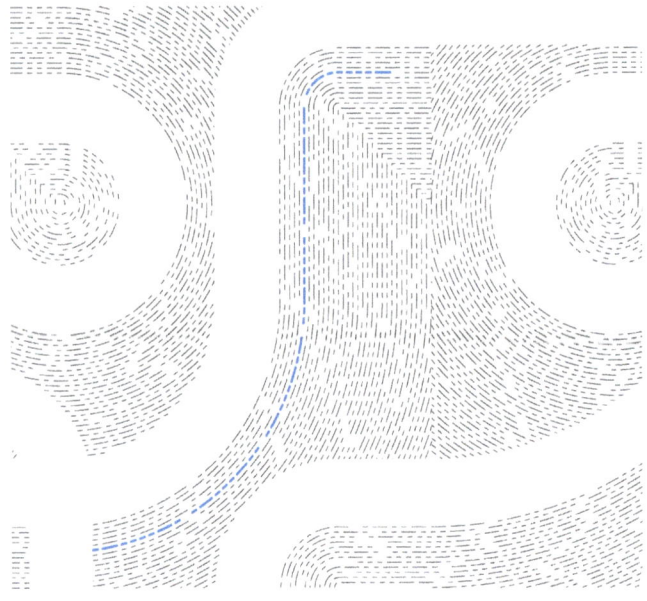

| 01000010 | 01000100 | 01000101 | 00100000 | 01001101 | 01000001 | |
| B | D | E | [space] | M | A | ... |

10

11

12

SKYWAY - NORTH ELEVATION

SKYWAY - SOUTH ELEVATION

BUILDING - SOUTH ELEVATION

13

10 The field of dashes consists of Indigenous place names encoded as a series of long and short dashes, which correspond to a binary zeros and ones. This example isolates one curve that contains characters from Bde Maka Ska, the largest lake in Minneapolis.

11 Detail photograph shows the interplay of the data-driven bands and the field of encoded Indigenous place names

12 0.25 inch diameter dots are placed algorithmically along field dashes to conform to minimum spacing requirements of bird safety standards (left); detail photograph shows integration of dots with pattern (right)

13 Elevation drawing of *Lines of Flight, Human*

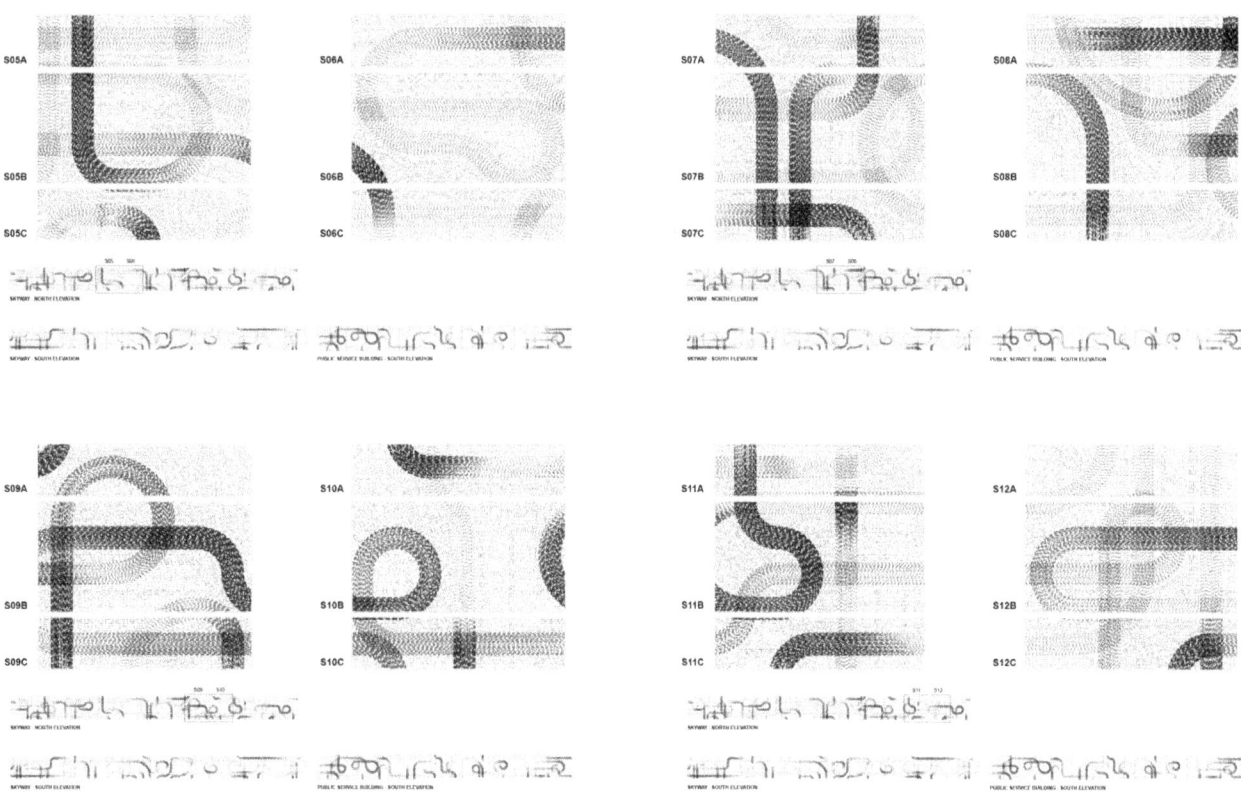

14 Examples of parametrically generated shop drawings

bird safety standards. Adopted by numerous municipalities and advocacy groups, the "two-by-four" rule defines best practices for designing glass patterns to prevent bird mortality (Klem 1990; San Francisco Planning Department 2011; Audubon Minnesota 2010). It recommends a pattern made up of elements at least 0.25 inches wide and arranged in a density of no more than two inches horizontally, and four inches vertically. The density of the band patterns already conformed to these standards, but the field dashes, as thinner lines, required additional pattern elements to satisfy the minimum density. We developed an algorithm to analyze the dashes' spacing and lineweight, to search the field for where additional graphic density is needed in order to meet bird safety requirements, and then to place 0.25 inch diameter dots at these locations. Rather than placing the dots arbitrarily along a "two-by-four" inch grid (as is typical with bird-safe facades), the algorithm searches the field and places dots at the ends of the dashes where necessary to meet the spacing standards. In this manner, the dots are optimized in their placement despite having a random appearance (Figure 12).

Fabrication Drawings

Developing these workflows in a single parametric model provided a means to test different iterations of data mapping, geometric scales, and densities of the dash pattern with relative ease. The integrated nature of the model facilitated the automation of shop drawing and fabrication file production. This eliminated much of the manual labor often associated with these tasks and allowed flexibility to fine tune the overall pattern (Figure 13) while maintaining the ability to easily output the fabrication files (Figure 14).

RESULTS AND DISCUSSION

Site-Specificity and Revisiting the Communicative Surface

Lines of Flight, Human demonstrates that site specificity can be a compelling way to add meaning and capacity for public engagement in the design of facades. Incorporating the complicated histories and legacies of a site can contribute to "re-mediating" viewers' understandings of place, such that they can gain new insight and relationality to it (Kim 2016). As anti-immigration sentiment continues to grow and become mainstream, it is important to remind the public that the United States, since its colonization, has become predominantly a country of immigrants. This history must be told alongside an account of the forced dispossession and genocide of Indigenous communities. One way of recognizing this history is land acknowledgment statements. They are often spoken, and thus ephemeral. By embedding histories of immigration and displacement into the materiality of the facade's design, as *Lines of Flight, Human* does, architecture itself provides a lasting record. This revisits theories of architecture's communicative capacities (Venturi, Scott Brown, and Izenour 1972;

15 View of south facade of *Lines of Flight, Human*

Jacob 2012) and demonstrates how computational design can be leveraged to deepen architecture's engagement with site and memory.

Assessment and Limits of Data Spatialization

Although *Lines of Flight, Human* draws on Minnesota demographic data and is illustrative of this history, the project is not a comprehensive data spatialization of human migration to the region. It is constrained by the availability, accuracy, and bias of the IPUMS data. In addition, subjective decisions were made in the design process that impacted how the data was utilized.

First, census data is collected every ten years, but beginning in 2000, the IPUMS database recorded immigration data annually. In order to make this consistent with the decennial data from 1850 to 1990, we opted to average the annual data from 2000-2009. The design phase for this project began in 2018, and at that time we only had access to data through 2016. We also averaged the partial set from 2010-2016 to represent the most recent decade.

Second, given the size and extents of the building's facade, there were limitations on how many bands could fit without compromising graphic legibility. Through iterative band generation testing (outlined above), we settled on 87 bands, as that number provided optimal graphic density and figure-ground balance. The 87 bands are countries with the highest numbers of immigrants to Minnesota, representing 97.8% of total recorded immigrants since 1850.

Third, when working with historical data about countries that spans centuries, one discovers that many nation-states no longer exist (for example, Prussia). We considered consolidating nonexistent countries with the country that currently occupies overlapping geographical areas (i.e. replace Prussia with Germany), but we decided that this would be subjective and inaccurate, as historical borders do not always align with present-day ones. There are two exceptions to this: Yugoslavia and Czechoslovakia. These two former countries are represented as individual bands, as significant numbers of people from these countries immigrated to Minnesota.

Fourth, the census data does not include a specific field for country of origin. In lieu of this information, we drew our data from the "birthplace location" field. There are potential inaccuracies in this approach, as it does not reference the

16 Interior view of *Lines of Flight, Human*

country from which a person last immigrated or felt culturally connected to, items that are unavailable in the census data.

Any process that involves the collection, analysis, and visualization of data must account for biases that exist in the available record (Crawford 2016; Criado Perez 2019; Benjamin 2019). In this project, there were multiple sources of bias. First, U.S. census data was only available from 1850, which excludes information about migration prior to this date. Second, our efforts to acknowledge Indigenous place names introduced bias as translation and digitization are deeply entangled with the legacies of colonialism. Indigenous names are first translated into anglicized text, and this text is encoded as zeros and ones, following ASCII protocols. Like all character encodings that enable the electronic transmission and storage of text, ASCII is a descendant of telegraph code (Mackenzie 1980), a technology that facilitated westward expansion and settler colonialism in the United States.

Future Opportunities for Computational Design

This project opens up consideration of additional opportunities for integrated workflows that blend manual and computational processes. First, facade patterns provide shading and can reduce solar heat gain inside the building, creating efficiencies in mechanical cooling and energy consumption. This was not considered in *Lines of Flight, Human*, as it was commissioned after the design of the curtain wall, which was already engineered to meet energy code requirements. A pattern that was optimized for shading capacity could involve an additional layer of variation in its design. Second, bird species fly at varying heights and from different directions. With additional research into bird mortality reduction, it may be possible to enhance the facade's pattern so that it is responsive to its vertical and directional placement on the building. Finally, with the wider availability of color and transparency in films applied to glass curtain walls, there will be opportunities to enrich pattern designs.

CONCLUSION

As a government building to house a variety of Minneapolis's city employees and services, the Public Service Building was designed to serve the city's residents. What public does the building serve? Who are the city's residents? *Lines of Flight, Human* sought to answer these questions by leveraging computational processes to explore histories of immigration to this region as a way to engage viewers with a design that references the site's history and the building's public role. When examining the history of immigration, however, one cannot overlook the history of genocide and forced dispossession of Indigenous communities that call this region home. *Lines of Flight, Human* embeds a visual land acknowledgment alongside the history of immigration of diverse groups to the region currently known as Minnesota.

The project's computational design integrates generative, data-driven, and analytical workflows to synthesize conceptual, communicative, and bird-safe parameters. This results in a design that is graphically bold, visually dynamic, and responsive to context (Figures 15, 16). Too often, the design of bird-safe glass frit and printed film applications is an afterthought, making use of repetitive, off-the-shelf patterns that foreclose opportunities to engage the public with a site's unique characteristics. *Lines of Flight, Human* suggests novel pathways for facade design that blend public engagement and computational design with a sensitivity to non-human inhabitants with whom we share the environment.

ACKNOWLEDGMENTS

Design: Futures North: John Kim, Adam Marcus, Molly Reichert, Daniel Dean

Design Visualization: Pete Pham

Data Consultant: Alicia Johnson

Building Design: Henning Larsen Architects, MSR Design

Fabrication: The Vomela Companies

Photography: Corey Gaffer Photography, Futures North

Client: City of Minneapolis

NOTES

1. Minnesota is the traditional homeland of the Dakota, Ojibwe, and other Indigenous peoples. The state's name is derived from the Dakota phrase for the region, Mni Sota Makoce, the Land Where the Waters Reflect the Clouds.

2. It is important to note that this process, while rigorous in its analysis of the migration data, nevertheless is incomplete in offering a perfectly accurate representation of human migration to Minnesota over time. This is discussed further in the Results & Discussion section.

3. Note that as this project began the design phase in 2018, we had access to population data only through 2016. The data for the ten-year period of 2011-2020 is thus partial and only extends through 2016. See the Results & Discussion section for further discussion of this.

4. The potential biases and limitations with visual land acknowledgments is discussed further in the Results & Discussion section.

REFERENCES

Audubon Minnesota. 2010. "Bird-Safe Building Guidelines." Accessed May 17, 2022. https://mn.audubon.org/sites/default/files/05-05-10_bird-safe-building-guidelines.pdf.

Benjamin, Ruha. 2019. *Race After Technology: Abolitionist Tools for the New Jim Code*. Cambridge: Polity.

Caldwell, Ellen. "The Cheyenne artist who is challenging the silenced history of Native Americans." *JSTOR Daily*, January 21, 2016. Accessed May 17, 2022. https://daily.jstor.org/cheyenne-artist-challenging-silenced-history-native-americans/.

Cantor Arts Center, Stanford University. n.d. "When Home Won't Let You Stay: Migration through Contemporary Art." Temporary Exhibition, February 5, 2021–May 30, 2021. Accessed May 17, 2022. https://museum.stanford.edu/exhibitions/when-home-wont-let-you-stay-migration-through-contemporary-art#Kallat.

City of Chicago. 2021. "You are on Potawatomi Land." Public Art on the Chicago Riverwalk. Accessed May 17, 2022. https://www.chicago.gov/city/en/sites/chicagoriverwalk/home/public-art.html.

Crawford, Kate. "Artificial Intelligence's White Guy Problem." *New York Times*, June 25, 2016. *Accessed May 17, 2022.* https://www.nytimes.com/2016/06/26/opinion/sunday/artificial-intelligences-white-guy-problem.html.

Criado Perez, Caroline. 2019. *Invisible Women: Exposing Data Bias in a World Designed for Men*. New York: Abrams Press.

Integrated Public Use Microdata Series (IPUMS), n.d. "U.S. Census Data for Social, Economic, and Health Research." Accessed May 17, 2022. https://usa.ipums.org/usa/.

Jacob, Sam. 2012. "The Communicative Mode of Architecture." MAS Context 14 (Summer). Accessed May 17, 2022. https://www.mascontext.com/issues/14-communication-summer-12/the-communicative-mode-of-architecture/.

Kim, John. 2021. "The Fourth Coast, Revisited." *The Anthropocene Review* 8 (3): 241249.

Kim, John. 2016. *Rupture of the Virtual*. St. Paul, MN: Dewitt Wallace Library.

Klem, Jr., Daniel. 1990. "Collisions Between Birds and Windows: Mortality and Prevention." *Journal of Field Ornithology* 61: 120-128.

LeWitt, Sol. 1967. "Paragraphs on Conceptual Art." *Artforum* 5 (10): 80.

Loss, Scott R., Tom Will, Sara S. Loss, and Peter P. Marra. 2014. "Bird–building collisions in the United States: Estimates of annual mortality and species vulnerability." *The Condor* 116 (1): 8–23. https://doi.org/10.1650/CONDOR-13-090.1.

Loss, Scott R., Tom Will, and Peter P. Marra. 2015. "Direct Mortality of Birds from Anthropogenic Causes." *Annual Review of Ecology, Evolution, and Systematics* 46: 99–120. https://doi.org/10.1146/annurev-ecolsys-112414-054133.

Loss, Scott R., Sirena Lao, Joanna W. Eckles, Abigail W. Anderson, Robert B. Blair, and Reed J. Turner. 2019. "Factors influencing bird-building collisions in the downtown area of a major North American city." *PLoS ONE* 14 (11): e0224164. https://doi.org/10.1371/journal.pone.0224164.

Mackenzie, Charles E. 1980. *Coded Character Sets, History and Development. The Systems Programming Series.* Reading, MA: Addison-Wesley Publishing Company, Inc. Accessed May 17, 2022. https://textfiles.meulie.net/bitsaved/Books/Mackenzie_CodedCharSets.pdf.

Marcus, Adam. 2022. "Arbor: Tectonic Contingencies and Ecological Engagement." In *ACADIA 21: Realignments: Towards Critical Computation; Proceedings of the 41st Annual Conference of the Association for Computer Aided Design in Architecture*, edited by K. Dörfler, S. Parascho, J. Scott, B. Bogosian, B. Farahi, J. Grant, J. L. García del Castillo y López, and V. A. A. Noel. Los Angeles: ACADIA.

Marcus, Adam, Molly Reichert, John Kim, and Daniel Dean. 2017. "Meander: Data Spatialization and the Mississippi River." In *2015 TxA Emerging Design + Technology Conference Proceedings*, edited by K. Bieg. Austin: TxA. 100–121.

Marcus, Adam. 2014. "Centennial Chromagraph: Data Spatialization and Computational Craft." In *ACADIA 14: Design Agency; Proceedings of the 34th Annual Conference of the Association for Computer Aided Design in Architecture*, edited by D. Gerber, A. Huang, and J. Sanchez. Los Angeles: ACADIA. 167–176.

Minard, Charles Joseph. 1862. "Carte figurative et approximative représentant pour l'année 1858 les émigrants du globe, les pays dóu ils partent et ceux oú ils arrivent." 51 x 59 cm. Library of Congress. https://lccn.loc.gov/98687134.

Mohr, Manfred. 1971. "Artist's Statement." Manfred Mohr Computer Graphics - Un Esthétique Programée. Musée d'Art Moderne de la Ville de Paris.

Molnar, Vera. 1975. "Toward Aesthetic Guidelines for Paintings with the Aid of a Computer." *Leonardo* 8 (3): 185–189.

Nake, Frieder. 2009. "The Semiotic Engine: Notes on the History of Algorithmic Images in Europe." *Art Journal* 68 (1): 76–89.

Pickover, C. A. "Picturing randomness with Truchet tiles." *Journal of Recreational Mathematics* 21: 256-259.

Robert McNeel & Associates. *Grasshopper for Rhinoceros 7.0.* V. 1.0. Robert McNeel & Associates. PC. 2020.

San Francisco Planning Department. 2011. "Standard for Bird-Safe Buildings." Accessed May 17, 2022. https://sfplanning.org/sites/default/files/documents/reports/bird_safe_bldgs/Standards%20for%20Bird%20Safe%20Buildings%20-%2011-30-11.pdf.

Shaw, Bob. "Minnesota has the most refugees per capita in the U.S. Will that continue?" *Pioneer Press*, January 13, 2018.

Venturi, Robert, Denise Scott Brown, and Steven Izenour. 1972. *Learning From Las Vegas*. Cambridge, MA: The MIT Press.

Zwierzycki, Mateusz. *Anemone*. V. 0.4. PC. 2015.

IMAGE CREDITS

Figures 1, 15: ©Corey Gaffer Photography, 2022.
All other drawings and images by the authors.

John Kim is a Professor of Media and Cultural Studies at Macalester College in St. Paul, Minnesota and a partner in Futures North. A theorist and practitioner of new media, John has published widely on media art, the history of the computer interface, Augmented Reality, and our changing experience of the material world. As a practitioner of new media, John has exhibited interactive art, sculpture, video games, and software internationally, including MassMOCA, House of World Cultures, ISEA, the Walker Art Center, and Northern Spark.

Adam Marcus is a registered architect and educator whose work operates at the intersection of technology, ecology, and public engagement. He is a partner in Futures North, a public art collaborative dedicated to exploring the aesthetics of data, and he directs Variable Projects, an independent architecture and research practice. He is an Associate Professor of Architecture at California College of the Arts in San Francisco, where he directs the Architectural Ecologies Lab. From 2011 to 2013 he was the Cass Gilbert Design Fellow at the University of Minnesota School of Architecture, and he has also taught previously at Columbia University in New York and the Architectural Association Visiting School in Los Angeles.

Molly Reichert is a designer, artist, and an educator whose passion for transdisciplinary collaboration and craft fuels her approach to design and education. She is a partner in Futures North, an award-winning public art collaborative dedicated to exploring the aesthetics of data. Molly currently teaches in the architecture department at the University of Minnesota. Her teaching combines digital design and fabrication methodologies with community engaged processes. Previously she taught in the architecture departments of University of California, Berkeley, San Jose State University, and was a founding faculty member of the professional Architecture degree program at Dunwoody College of Technology.

Computer-aided Ecological Connectivity

For Urban Design Within Climate Change Adaptation

Mathilde Marengo
IAAC, Advanced Architecture Group

Iacopo Neri
IAAC, Advanced Architecture Group

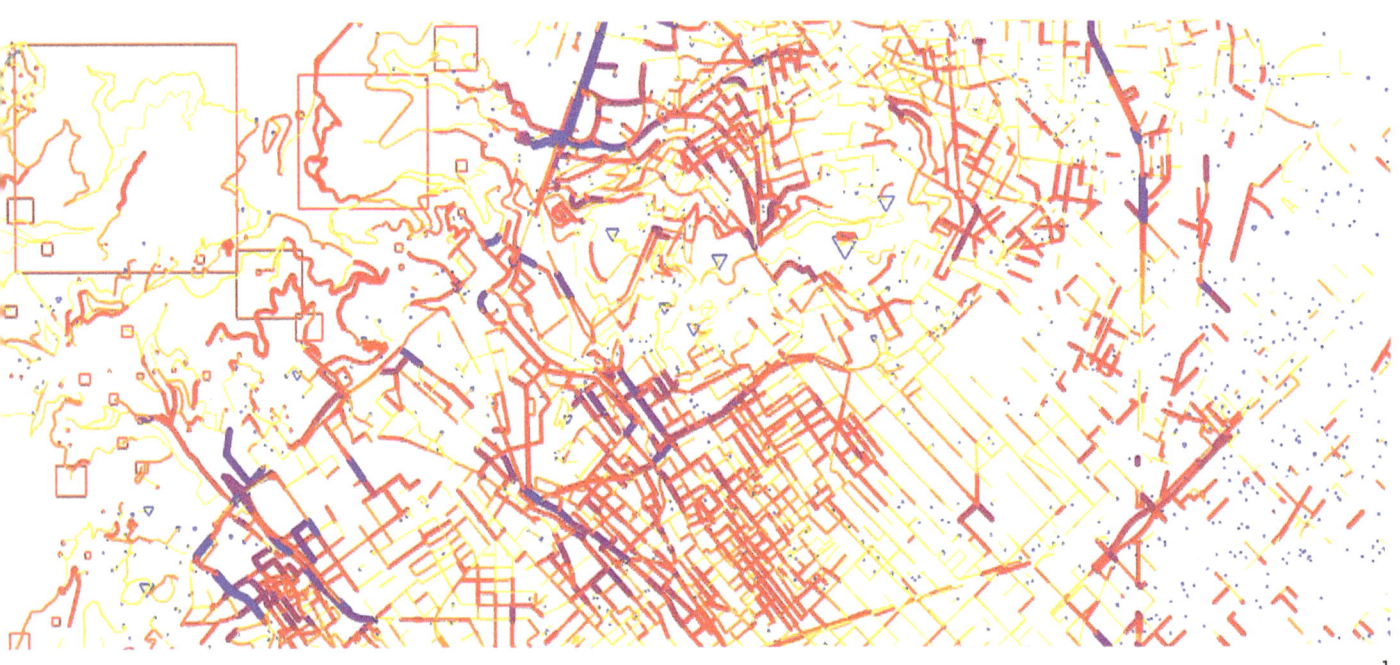

1 A detailed view of the priority map of the city of Barcelona

ABSTRACT

Today, cities represent one of the main threats to fine ecological balances. One of the issues that arise from this condition is landscape fragmentation. In the realm of urban design and planning, this condition raises the essential question of how nature and ecology can have an operative voice within the design process. The pathways to reach such objectives are not yet clear. Fostering widely accessible environmental and urban datasets, a design support methodology is developed to enable designers to design cities together with nature as an active partner, responding to the issues provoked by landscape fragmentation by enhancing ecological connectivity towards climate change adaptation. The body of work that follows discusses an experimental computer-aided methodology for design and planning processes that detects and amplifies potential and beneficial connections between urban parks and metropolitan forest patches, using as media roads and streets and their existing vegetation.

Technically, it introduces a computational protocol to gather and process georeferenced data necessary to run through CircuitScape a set of structural connectivity analyses in urban contexts. In order to overcome the barrier of local data accessibility, the methodology exploits Google Earth Engine to cross-read OpenStreetMap data with local NDVI imagery, allowing to calibrate the connectivity analysis on current ecological performances. The outputs of the analyses highlight key road segments where vegetation should be reinforced, and therefore, identify specific areas or streets of interest for intervention.

INTRODUCTION

Cities, today, represent the main threat to fine ecological balances. In this sense, one of the more impending environmental challenges we face today, most often caused by urban or mobility network growth, is landscape fragmentation. Landscape fragmentation consists in the breaking up and consequent disintegration of larger areas of natural land cover considered as continuous ecosystems "into smaller, more isolated patches, independent of a change in the total area of natural land cover" (Mitchell et al. 2015). The degradation, fragmentation and disconnection of natural habitats on land [...] has resulted in the alteration and isolation of habitat important for movement of organisms and for the maintenance of ecological processes present in previously connected landscapes [...]" (UNEP 2019). This disruption in structural ecological connections presents serious threats to the environment's capacity to provide vital ecosystem services, resulting in a decrease in habitat resilience in the general context of climate change. In order to visualize and explore the extent of landscape fragmentation within Europe's Functional Urban Areas (FAU)—cities and their commuting zones—the European Environment Agency has developed a dashboard visualizing 2018's data. Here we can observe that in the European Union the average per country of areas under high or very high fragmentation is 27%, and the average per city being a staggering 94% (EEA 2021).

Within the realm of urban design and planning this condition raises the essential question of how ecology can have an operative voice. The pathways to reach such an objective are not yet clear. The body of work that follows discusses an experimental computer-aided methodology for design and planning processes that detect and amplify potential and beneficial connections between urban and forest areas, fostering environmental data in order to design cities that engage with nature as an active partner in the process. In this regard, the proposed data-driven methodology uses computational logics exploited in environmental studies to predict ecological patterns for design and validation purposes.

CIRCUIT THEORY
FOR URBAN ECOLOGICAL CONNECTIVITY
The Case Study of Barcelona

The City Council of Barcelona proposes green infrastructure reinforced through ecological connectivity to mitigate landscape fragmentation. Taking this case as an initial testbed, and considering the widely available local data, the methodology is defined to model the structural connectivity between peripheral forest patches and urban parks, borrowing the logic of electric circuit theory (McRae et al. 2008). With the intention of evaluating road infrastructure performance to foster structural connectivity, graph modeling is adopted as a computational representation of the city. Graph-based modeling is an approach to reduce the complexity of the urban fabric to its most relevant geometrical features of nodes and edges, additionally informed by a set of weights—in this case: the width of the street, and the vegetation present—to consequently run purpose-specific analyses.

In this manner, graphs allow for a spatial simplification of the city, still maintaining the properties of its urban layout and offering a structure to store and relate geographical

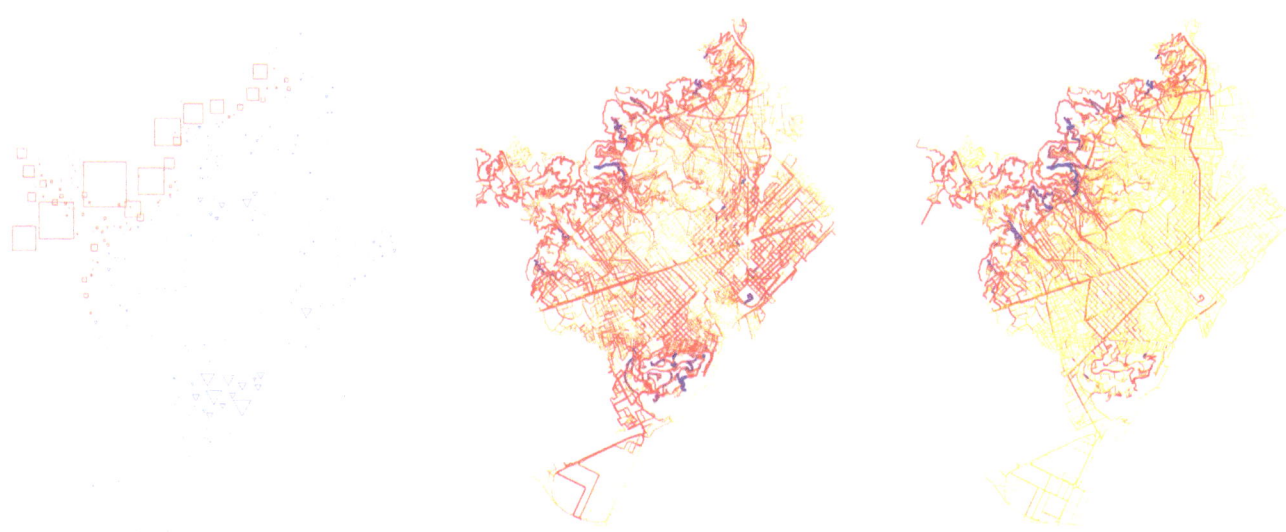

2 Source-Ground map with respective ecological connectivity charging points (forest patches and urban parks) visualized as the area parameter (left), first scenario map with average current values (center), and second scenario map with average flow values (right) evaluated on two sets of simulations each, respectively: forest to parks, and parks to forest.

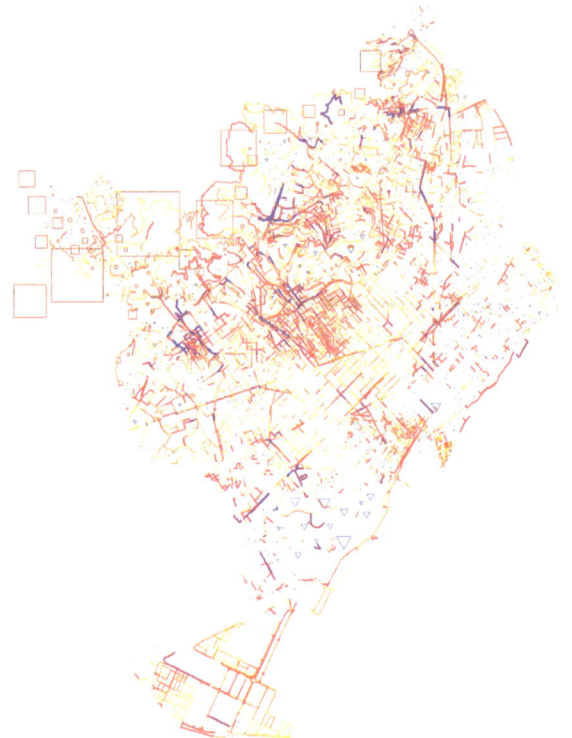

3 Priority map for the city of Barcelona, or the conversion potentials weighted on the connectivity analysis. The darker the color the more beneficial impact a greening intervention on the ecological connectivity as a whole.

axis, which is used as a proxy for ecological conductance: the denser and healthier the vegetation, the stronger the connection. Consequently, the potential conductance is extracted as the gap between the current NDVI value and the maximum achievable in consideration to the space available—streetscape—for planting.

For the purpose of this study, the connectivity simulations are run in graph-based advanced-mode, alternatively using the park and forest nodes as grounds and sources, with variable charges as per the NDVI cumulative values detected, which are then transposed onto the modified street network. The latter is evaluated under two scenarios: the first with the values of its conductance as its present NDVI values - the zonal statistics cumulative sum -, and the second as its potential maximum NDVI, meaning all the detections in the zonal statistics return to the maximum value of 1.

Finally, the process averages the four simulations into two scenario maps, predicting connectivity at the street level from the outer patches to the inner ones, and vice-versa, while anticipating current and potential scenarios (Figure 1).

Once the scenario maps are computed, a final cross-reading can be performed to identify the most relevant areas where vegetation should be enhanced to support the system's overall structural connectivity. By evaluating the difference between current and potential scenarios, it is possible to identify the street-specific conversion value, where the higher the values are—meaning that the difference between current and potential ecological connectivity is greater—the more relevant the activation of that segment is, and the more impact it will have on the system as a whole. In this manner, the methodology proposed contextualizes each road segment within the broader urban ecosystem, providing a connectivity metric that is not only locally relevant to its particular segment, but also to the entire urban ecosystem.

The priority map (Figure 2) can, therefore, be used as a base for strategic design and planning processes where, for each segment, a specific nature-based project solution should be identified in relation to a higher resolution of information, beyond that already embedded in the process, consequently becoming an essential tool for a fruitful ecological transition.

information (Agryzkov et al. 2017). For this reason graphs and networks, as computational objects, are used to evaluate city dynamics and have been successfully used by many scholars since the advent of computers. The most notable example consists in Space Syntax, which foresees pedestrian flows, assuming correlations between how people move in streets and the morphology of the street network itself (Hillier and Hanson 1984). On the other hand, concerning the modeling of landscape or ecological connectivity, many are the examples that prefer an image-based representation to obtain this functional simplification of the system (Dickson et al. 2019). This being said, the proposed pipeline adopts a graph representation over a raster-based approach considering its focus within the urban context and the importance of integrating the street layout with a higher degree of specificity. This adoption allows the methodology to support decision-making for urban environments, specifically within the possibilities of traditional planning practices and implemented street by street.

In order to create these connections, streets have been used as connectors, weighted by their actual and potential conductance, according to their spatial and environmental attributes. The latter is extracted from a high resolution Normalized Difference Vegetation Index layer (Open Data Barcelona 2022) and reflects the presence and health of vegetation along each

Towards a Replicable Methodology

The potential to replicate the study in urban contexts different from the one of Barcelona inevitably opens up challenges related to data availability and data interoperability. For this reason, a computational pipeline written in the Python programming language was created to collect, process, and visualize data inherent to city-specific performances that enhance ecological connectivity between the urban parks and their broader metropolitan forest ecosystems, and that consequently tackle the issue of landscape fragmentation within their urban cores. The outputs of this pipeline are designed to identify, by means of comparison, similar urban strategies for different cities, possibly triggering a beneficial exchange of knowledge among municipalities confronted with similar challenges. In order to reproduce the analysis for any city, independently of local data availability, two main global databases are exploited:

- OpenStreetMap (OSM): to geocode the location and municipal extension tool, and to retrieve all the necessary vectorial data layers (OpenStreetMap Wiki Contributors 2022).
- Google Earth Engine (GEE): to cross-read the vectorial data with local NDVI imagery and to aggregate this extra data to each geometrical feature, finally calibrating the connectivity analysis that follows on current ecological performances.

Prior to the connectivity analysis and in parallel to the databases exploited, the pipeline is structured in an initial quantitative phase focusing on loading, and geometrically processing, the data relative to the street layout and to the spatial boundaries of the parks, green pockets, and forest patches. A second qualitative phase adds to the previous step and focuses on allocating geometry-specific performances: their potential to enhance the connectivity based on the presence and health of greenery detected from the satellite images. Similar to the case of Barcelona, and necessary for the second phase, is the relative data of the street widths, which is calculated from the data available and collected in OSM.

Dynamically accessing OpenStreetMap was in great part achieved through the OSMnx Python library, which additionally provided a flexible toolkit to clean and prepare the street network. More precisely, the methodology queried the OSM database to load the vectorial extensions of the urban and metropolitan green areas according to the keys-values of *leisure: park* and *leisure: forest* falling within the municipal boundaries of the city under study. These were geocoded with the *geometries_from_place()* function, while similarly loading the street layout filtered on the primary, secondary, tertiary, residential, and pedestrian categories (Figure 3).

The filtered street layout represents a crucial step for the methodology as it implicitly executes a series of topological operations, converting the simple multilinestring OSM geometries of the streets into a graph (Boeing 2017), and therefore preparing the geometrical base needed for Circuitscape to run the connectivity analysis. In addition to the graph creation, the park and forest areas are linked to the graph with the closest

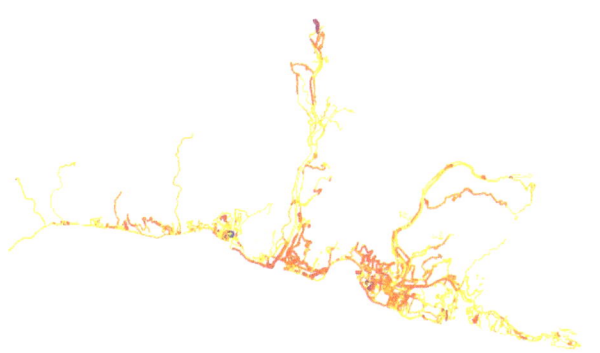

4 Priority map for the city of Genoa obtained with the discussed Python pipeline exploiting OSM and GEE

node to their centroids, calculated with GeoPandas and the *nearest_node()* function within the *distance* utilities of OSMnx. Overall, the methodology uses the GeoPandas and Pandas Python library for standard geometric and data manipulations, such as those relative to the projection between different Coordinate Reference Systems when necessary, or those required to fill in the missing data of the street widths. This information is currently scarcely available in OSM, as it is reported for the 0.29% of the total streets present in the database (OSM Taginfo 2022). Despite this shortage and with the expectation that more data will be crowdsourced in the future, the methodology aims at site-specificity regarding the street widths per city, calculating this criterion with the modal values per street category, as downloaded from OSM (Table 1). Finally, when no data is available, the methodology refers to generic widths that are imposed, as exemplified in Table 1 in the section *Width*.

Once the street graph is constructed and improved with the missing data of the streets' width, the methodology focuses on accessing Google Earth Engine to add the property relative to the coverage of urban greenery. In this regard, multiple zonal statistics are calculated in GEE to locally sample the values retrieved from a global NDVI image layer, in order to enrich the graph. This NDVI layer was computed as the normalized difference of *band 4* and *band 3* of the Landsat

Street typology	Extracted Mode Widths						Width
	Berlin	Milan	Madrid	Paris	Amsterdam	Barcelona	
Primary	8	8	7	12	5	12	10
Secondary	6	20	6.1	5	3.5	8	8
Tertiary	6	9	10	6	5.5	5	8
Residential	6	4.7	4.8	3	4	6	5
Pedestrian	5	3	4	4	7	10.5	5

TABLE 1. Mode widths per street typologies for the cities of Berlin, Milan, Madrid, Porto, Amsterdam, and Barcelona

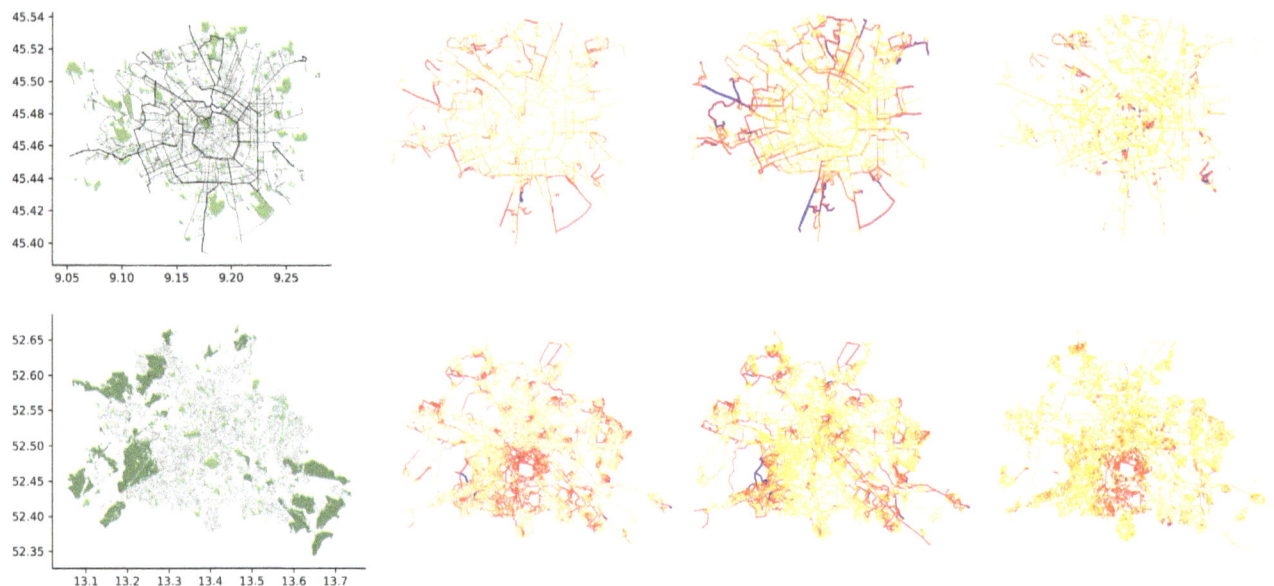

5 Cartographical outputs for the cities of Milan and Berlin (in order from left to right): vectorial base layers downloaded from Open Street Map of forest patches, urban green pockets and street graph, scenario map with average current values, scenario map with average flow values, and priority map

7 5-year TOA percentile composites layer (USGS 2022), and consequently queried with the *ee.ReduceRegions()* function to map the sum values on each road segment—edge of the graph—as well as on each green area—node of the graph, and source and ground for the connectivity analysis. In the case of the edges, a prior buffering operation was executed in order to attribute width-specific areas to the linear geometries. Finally, the conversion between GeoPandas dataframes and GEE features is performed through the eeconvert Python library.

After the topological processing and the NDVI data augmentation, the methodology has proven viable to run connectivity analyses, similarly to the case study of Barcelona, in several urban contexts (Figures 4, 5).

CONCLUSION

Through the methodology presented we begin to unpack the complexity related to understanding ecological connectivity and its relationship to morphological phenomena, through computational logics. Running the connectivity analysis for the city of Barcelona, both with local and OSM and GEE data, permits to evaluate the accuracy and trade-offs implicit when moving the scope of research from one city to the divers conditions of cities across the globe. The street width value is observed as playing a central role and, therefore, requires a level of resolution that, in this version of the methodology, could call for an integration of local data. This being said, the NDVI analysis has proven to maintain a good level of accuracy with respect to its local data counterpart. This, therefore, allows to determine factors related to ecological connectivity with sufficient precision, giving indications as to the identification of relevant pathways for ecology to connect and thrive, although not in relation to all site-specific morphological conditions, giving a minimum bias at this stage in the results. This can be counteracted in future iterations with the refining of datasets in relation to the specific street sections, either by providing an indication for specific local data requirements, or by exploring alternative higher resolution data mining processes, involving machine learning algorithms, within the current pipeline.

REFERENCES

Agryzkov, Taras, Jose Luis Oliver, Leandro Tortosa, and Jose Francisco Vicent. 2017. "Different types of graphs to model a city." In *WIT Transactions on Engineering Sciences*, vol. 118. Southampton, UK: WIT Press. 71–82. https://doi.org/10.2495/CMEM170081.

Boeing, Geoff. 2017. "OSMnx: New Methods for Acquiring, Constructing, Analyzing, and Visualizing Complex Street Networks." *Computers, Environment and Urban Systems* 65 (September): 126–39. https://doi.org/10.1016/j.compenvurbsys.2017.05.004.

Dickson, Brett G., Christine M. Albano, Ranjan Anantharaman, Paul Beier, Joe Fargione, Tabitha A. Graves, Miranda E. Gray, et al. 2019. "Circuit-Theory Applications to Connectivity Science and Conservation." *Conservation Biology: The Journal of the Society for Conservation Biology* 33 (2): 239–49. https://doi.org/10.1111/cobi.13230.

EEA. 2021. "Landscape fragmentation in Europe." Accessed May 10, 2022. https://www.eea.europa.eu/data-and-maps/data/data-viewers/landscape-fragmentation-in-europe.

Hillier, Bill, and Julienne Hanson. 1984. *The Social Logic of Space*. Cambridge: Cambridge University Press. https://doi.org/10.1017/CBO9780511597237.

McRae, Brad H., Brett G. Dickson, Timothy H. Keitt, and Viral B. Shah. 2008. "Using Circuit Theory to model Connectivity in Ecology, Evolution, and Conservation." *Ecology* 89 (10): 2712–24. https://doi.org/10.1890/07-1861.1.

Mitchell, Matthew G. E., Andrés F. Suarez-Castro, Maria Martinez-Harms, Martine Maron, Clive McAlpine, Kevin J. Gaston, Kasper Johansen, and Jonathan R. Rhodes. 2015. "Reframing Landscape Fragmentation's Effects on Ecosystem Services." *Trends in Ecology & Evolution* 30 (4): 190–98. https://doi.org/10.1016/j.tree.2015.01.011.

Open Data Barcelona. 2022. "Vegetable Cover of the City of Barcelona, Based on the NDVI Index, seen from the Sky." Accessed May 10, 2022. https://opendata-ajuntament.barcelona.cat/data/en/dataset/cobertura-vegetal-ndvi.

OpenStreetMap Wiki Contributors. 2022. "Width | Keys | Taginfo." Accessed May 10, 2022. https://taginfo.openstreetmap.org/keys/width#overview.

UNEP, United Nations Environment Programme. 2019. *Frontiers 2018/19: Emerging Issues of Environmental Concern*. https://wedocs.unep.org/20.500.11822/27538.

USGS. 2022. "Landsat 7 5-Year TOA Percentile Composites | Earth Engine Data Catalog." Accessed May 10, 2022. https://developers.google.com/earth-engine/datasets/catalog/LANDSAT_LE7_TOA_5YEAR.

IMAGE CREDITS
All drawings and images by the authors.

Mathile Marengo is an Australian, French, and Italian architect, with a PhD in Urbanism, whose research focuses on the Contemporary Urban Phenomenon, its integration with technology, and its implications on the future of our planet. Within today's critical environmental, social, and economic framework, she investigates the responsibility of designers in answering these challenges through circular and metabolic design. She is Head of Studies, Faculty and PhD Supervisor at the Institute for Advanced Architecture of Catalonia's Advanced Architecture Group, an interdisciplinary research group investigating emerging technologies of information, interaction, and manufacturing for the design and transformation of cities, buildings, and public spaces.

Iacopo Neri has been involved in academia since 2015, researching computational design and geospatial analysis. His research lies at the intersection of architecture, computer science, and urban planning. He has been involved in teaching activities at the University of Florence, The Polytechnic University of Milan, the Ecole des Ponts - ParisTech, and at IAAC - Institute for Advanced Architecture of Catalonia, where he is currently faculty of computational design and member of the AAG - Advanced Architectural Group's Computational Design Research Team.

Session Introduction

Environmental Performance and Simulation

Billie Faircloth, Chair

The researchers in Session 12, Environmental Performance and Simulation, emphasize the importance of modeling, simulating, prototyping, and measuring the built environment. They demonstrate how buildings, spaces, and surfaces affect people and their surroundings and explore design frameworks to help us understand heat flow, daylight access, and acoustics. Two studies address the potential of urban-scale passive cooling, climate change adaptation, and resilience. These are presented to off-set the urban heat island effect through greening empty land parcels in Philadelphia, Pennsylvania (Tian et al.), and to examine the capacity of existing multi-family housing to meet thermal comfort needs in the Republic of Cyprus (Ozarisoy and Altan). A study on daylighting and building massing investigates potentially competing demands between standards for daylight access and high-density planning (Xiao et al.). Enhancements to the acoustic performance of interior partitions are prototyped through novel assemblies and surfacing methods. An approach for addressing the poor acoustic performance of mass timber panels integrates Helmholtz resonators (Nguyen et al.), while the quest for acoustic control pairs plywood with kerfing techniques and novel applications of a design-simulation-fabrication-testing workflow (Borhani et al). During the panel, the authors talked about how their work might be extended through environmental policy and building standards. They discussed the importance of a mixed-methods approach and the challenges and opportunities of ground-truthing and validation. They emphasized the necessity for interdisciplinarity, skill-building, experimentation, and quality control.

Passive Cooling Strategies for Thriving in a Changing Climate

A Prototype Retrofit Housing Energy Modeling and Simulation Framework

Bertug Ozarisoy
Middle East Technical University
Northern Cyprus Campus

Hasim Altan
Prince Mohammad Bin Fahd University

1 Methodological framework of the building energy modelling stages for the present study

ABSTRACT

The United Nations Sustainable Development Goal 7 calls for universal access to affordable, reliable, sustainable and modern energy for all, and is expected to influence the near future trends in many countries across the European Union. The retrofitting of buildings is an important milestone in the evolutionary development of energy-efficient residential buildings, yet a significant proportion of the south-eastern European social housing stock is inadequate in this area. This paper investigates the thermal performance of 288 flats in three different nationally representative collective housing archetypes in the southeastern Mediterranean island of Cyprus, where the climate is subtropical (Csa) and partly semi-arid (Bsh), as designated in the Köppen climate classification system. The participants' experiences and thermal sensation votes were assessed to predict individual aspects of adaptive thermal comfort, and the relevance thereof on overheating, and in situ measurements—including indoor air temperatures, thermal imaging survey, recorded building-fabric-element heat fluxes, on-site environmental conditions monitoring, and review of household energy bills to accurately determine actual energy use—were collected. The results indicated a lack of diurnal temperature variations within the sample flats, which suggests that internal operative air temperatures remained relatively high throughout the day and night; indoor air temperature ranged from 28.5°C to 36.5°C, and there was a difference of +5°C between the actual and the simulated/predicted operative air temperatures.

INTRODUCTION

Understanding the importance of different variables in energy calibration studies that constitute a disciplinary or integrated framework in decision-making processes is essential; it is important to review all extant findings related to socio-technical variables that affect the housing energy sector. The conclusions of the present study may impact the manner in which future longitudinal field studies intended to detect heat loss in building envelopes are conducted, specifically with the use of in situ measurements and analyses of household energy bills, as shown in Figure 1. The proposed step-by-step methodological workflow will demonstrate the manner in which the building energy simulation (BES) can be undertaken to minimize the risk of discrepancies between actual and predicted energy use with a comprehensive illustration of a thermal imaging survey.

According to the Energy Performance of Buildings Directives (EPBD) and their mandates, every EU member state should develop building diagnostic methods to evaluate the energy performance of existing housing stock (López-González et al. 2016; Salvalai et al. 2015; Villca-Pozo and Gonzales-Bustos 2019). A 2020 report on assessing the energy efficiency of Cypriot housing stock revealed a shortfall in energy policy and regulations when conceptualizing an effective design methodology for the BPE studies in addition to a technical failure to assess the energy performance certificates (EPCs) of each building typology (Mavrigiannaki et al. 2020). The report also indicated that the Republic of Cyprus (RoC) housing stock should be retrofitted to achieve energy-efficiency targets and reduce CO_2 emissions by 2030.

To date, no field investigations or pilot studies have been undertaken to fill this knowledge gap; the present empirical research study sets a starndard and can be applied to a research and development framework in EU countries, under the scope of the EPBD, to improve energy-efficiency regulations at the conceptual and national levels. As such, this paper's contribution to the body of knowledge related to energy calibration studies is the potential development of feasible retrofitting strategies and energy governance regulations; moreover, the present study will fill the energy efficiency knowledge gap and implement measures within an energy-policy framework.

In this exploratory case study, infrared radiometer thermography (IRT) surveys and in situ measurements of environmental conditions were concurrently conducted with on site with participant households along with a distributed questionnaire survey; further, household energy bills were obtained from the Cyprus Electricity Authority (CEA) to verify the data derived from the BES analysis. This is the first study to use energy-calibration studies as diagnostic tools to examine the existing post-war social housing stock in the southeastern Mediterranean basin. This conventional building diagnostic method was adopted to verify building thermal properties and the impacts thereof on overheating risks and occupant thermal comfort. The aim was to obtain human-based results in the BES to prove that the end result of energy calibration analysis is not based in forecasting scenarios, but rather on actual household information related to occupancy patterns and home energy performance factors embedded in the BES to design effective retrofitting strategies that will improve the thermal efficiency of the existing housing stock.

Review of Available Energy Simulation Tools and Workflows

Several studies employed the IES software to assess the energy performance of archetype buildings and create an overview of household energy use and associated CO_2 emissions for energy policymaking decisions (Pasichnyi et al. 2019; Stojiljković et al. 2015). In a study conducted by Kristensen et al. (2018), Dutch residential buildings were selected as sampling strategy of existing housing stock. The IES software is widely used for commercial purposes, and a few exemplar research case studies utilised the IES software suite as part of their methodologies. Mahdavi (2020) insisted that the IES software should be widely used by scholars to demonstrate the universal validity of this BES platform in large-scale residential projects to eliminate research bias and uncertainty related to the results obtained from DTS analyses; this research gap should be addressed (Ball et al. 2020).

The IES software offers engineers a versatile range of energy modeling techniques when they are undertaking a calibration analysis to validate available data sources (Choi 2017). The platform provides effective tools to construct an actual building geometry and assign building thermal properties, occupancy profiles, ventilation schedules for each room to undertake various numeric experiments related to BPEs (Yang et al. 2020). The IES software consists of standardized occupancy profiles and infiltration rates for naturally ventilated buildings, and users are able to adapt an analytical energy model to their own purpose-built design and simulation input parameters, such as occupancy profiles, window-opening schedules, clothing value, and metabolic rates, to assess household energy performance.

BACKGROUND

Many field studies conducted by scholars from different climate zones across the globe have investigated correlations between households' habitual adaptive behaviors and their sociodemographic variables (Shipworth et al. 2010; Pelenur 2010; Pelenur and Cruickshank 2013). These studies highlight that household behavior is one of the factors that can affect the energy efficiency gap (EEG). Pelenur (2013) suggests that the forces behind the EEG can be further identified through comprehensive research on both the psychological and sociological aspects of households. These are factors that require further investigation in terms of their relationship to the EEG, and this is a knowledge gap that other scholars have failed to fill with previous multidisciplinary studies, as outlined in Figure 2. After thoroughly reviewing other pilot studies, the present study finds that researching the EEG can be integrated with the socio-technical-systems (STS) conceptual framework in order to provide reliable research outputs in energy policy. The present study, therefore, reviews previous scholars' work to discuss the strength of EEG theory.

DIRECT BENEFITS
(Expected positive impacts)
- Reduced energy bills
- Warmer home; increased thermal comfort
- More consistent heat
- Improving lighting
- More control over temperature
- Improved in-home comfort level

INDIRECT BENEFITS
(Unexpected positive impacts)
- Enhanced sense of safety
- Sense of investment in the property
- Improved property value
- Reduced environmental exposures
- Reduced health risks including stress and anxiety
- Improved tenant-landlord relations - less complaints

NEGATIVE CONSEQUENCES
- Damage to property as a result of upgrades
- New equipment not functioning properly
- Fears of increased rent as a result of building improvements and energy efficiency upgrades
- Less control over heat due to landlord remote control and keeping temperatures low

UNATTENDED ISSUES
(beyond the scope of energy efficiency)
- Poor building maintenance/"Sick building syndrome"
- Building safety concerns
- Poor relationships with landlords
- Over/under-heating
- Chronic economic hardship including economic energy insecurity
- Limited energy literacy including more effective/appropriate energy coping strategies

2   Strengths and constraints of identifying the EEG and implementation solutions

The importance of identifying the EEG began at policy level with the regulatory innovation introduced by Directive 2010/31/EU (Ascione et al. 2020; Bertoldi and Mosconi 2020). This policy regulation is also known as the Energy Performance of Buildings Directive (EPBD). The EPBD has become a mandatory policy regulation to assess the energy performance of buildings through a multi-decade process (Ozarisoy and Altan 2021). After the implementation of the EPBD, most scholars have changed their research focus to conducting comprehensive research studies aimed at improving the energy efficiency of existing housing stock. The outcome of EPBD implementation has had a positive impact on integrating the concept of the EEG as an energy policy tool.

Copiello (2017) indicates that there is a strong correlation between the rationale of the EPBD and the subject matter of the EEG. Copiello stresses that the EPBD has had a linear effect on improving energy efficiency awareness across households. Ozarisoy and Altan (2022) indicate that EPBD implementation, when considering the EEG, could provide an effective mechanism for establishing policy tools and supporting non-governmental organizations (NGOs) to further develop energy subsidization schemes. Ma et al. (2012) noted that the presence of the EPBD mandates has opened a new pathway for an evolutionary process of methodological approaches in the building engineering field. Tian (2013) highlights that the implementation of the EPBD has enforced households' need to apply for the EPCs for their buildings, which in turn has enabled people to better understand the energy performance of their buildings. This is because the concept of the EEG has a direct impact on understanding energy efficiency and assessing the energy performance of existing housing stock.

The present study seeks to corroborate the empirical model by integrating the findings from the questionnaire survey and energy calibration analyses to feed the BES procedures. It considers an adaptive thermal comfort theory mentioned in literature by integrating regression forecasting findings on an energy simulation platform with those measured in the case study multiunit residential buildings (MURBs). The degree of overheating in each occupied space is discussed and evaluated in terms of impact of different orientations and floor levels. Building energy performance is then calibrated by running a dynamic thermal simulation (DTS) to develop a reliable assessment of the overall energy end uses for space conditioning to optimize occupant thermal comfort. It also discusses the uncertain input parameters for the BES that quantitative modeling adopted for the calibration of dynamic thermal simulation findings in conjunction with occupants' socio-demographic characteristics, occupancy patterns, household size, and recorded environmental conditions.

METHOD AND TOOLS

This section presents the research methodology by explaining the rationale for the present study and the conceptual framework that was developed to address the EEG. The present study employed a mixed methodology of qualitative and quantitative data analyses to examine the influence of the STS approach through a building modeling simulation based on the archetype residential buildings to develop energy policies that are in accordance with the EPBD objectives.

Representative Archetypes

In this empirical study, key criteria in the representative case study building selection were that the sample was representative of the pre- and post-war medium- or high-rise residential tower block developments were built in three distinctive construction eras in the 1970s, 1990s, and 2010s in

TABLE 1 The parameters for selecting prototype multiunit residential buildings (MURBS) as base case scenario

Building	Construction year	Location	Climate Characteristics	Building Fabric	External walls U-value [W/(m²K)]
MURB - 1	1970	Suburban	Mediterranean climate *Csa* - coastal	Single/double-leaf hollow brick walls	1.23 W/(m²K)
MURB - 2	1990	Urban - city centre	Humid subtropical climate *Cfa* - Inland	Single-layer brickwork	1.6 W/(m²K)
MURB - 3	2010	Urban	Transition between *Cfa* and *Csa* climates - Semi-mountainous	Massive/perforate brickwork	2.1 W/(m²K)

terms of location, building climate characteristics, construction style (architecture, fabric thermal performance, form and energy consumption), as shown in Figure 3. Additionally, it is necessary to have sufficient existing building data and scope for survey and in situ measurements to allow a comprehensive study. Three case study buildings were selected from the coastal city of Famagusta, Cyprus in accordance with these criteria. The selected buildings are listed in Table 1.

Figure 4 shows classification of housing stock by construction period. In Cyprus, 36.5% of housing stock was built between 2000 and 2008. Cypriot housing stock is relatively newly built, in comparison to the other EU countries. Despite the high amount of newly built housing, the lack of regulatory bodies to implement the EPBD has led to thermally uncomfortable indoor conditions in many households. As shown in Figure 4, the sample fraction initially aimed for was 73.2% of the total selected building stock; however, the collective housing developments that were successfully recruited for the study constitute 36.7% of the total, which is still a relatively reasonable fraction. These buildings were chosen by the Board of Housing, Building and Planning in cooperation with Statistics by the State Planning Organization as statistically representative of the Mediterranean island of Cyprus residential building stock.

Questionnaire Survey

This research adopts a methodology that includes a questionnaire survey in three representative residential tower building prototypes in the coastal city of Famagusta, Cyprus. A thorough review of the study was conducted, including feedback obtained during the pilot study. The systemic retrofit policy framework was then optimized before implementing low-tech passive design strategies on the most ill-performing prototype MURBs. In addition, surveys (semi-structured interviews) were conducted among the residents on aspects of the use of the building in the summer, together with the socio-demographic profile of the occupants in order to generalize the sampling size of the questionnaire survey as shown in Figure 5.

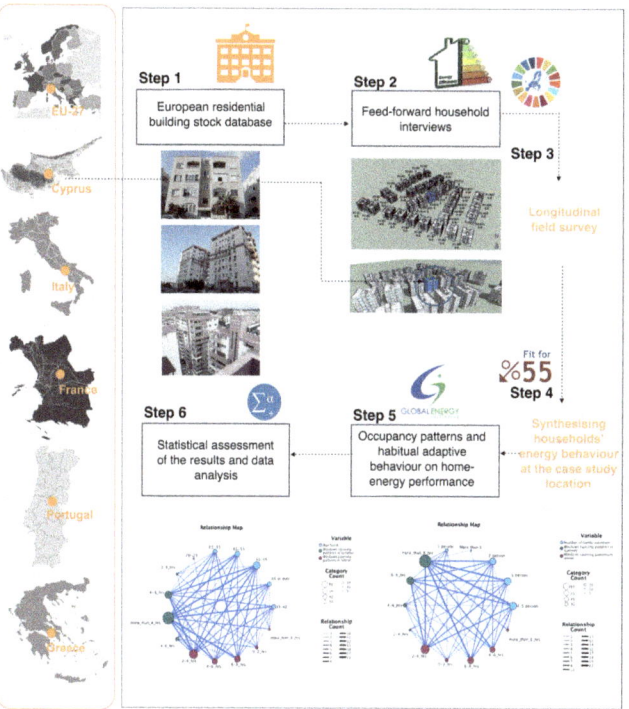

3   The method of design used to demonstrate extrapolation of archetype buildings by adopting the STS conceptual framework for effective policy making decisions in energy use

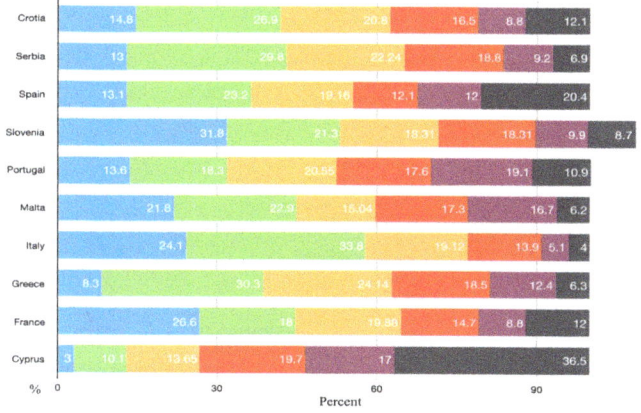

4   Construction period of housing stock in Europe

INTERVIEWED APARTMENT UNITS

PROTOTYPE 1 LORDOS RESIDENTIAL TOWER BLOCK DEVELOPMENT - 1970s

NORTH-WEST FACING APARTMENT UNIT	FLOOR PLAN LAYOUT	DETAILS	WINDOWS OPENING SCHEDULES	HEATING USE TYPES	COOLING USE TYPES
	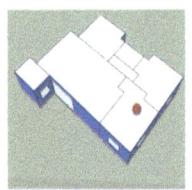	APARTMENT UNIT: LEVEL 9 - 2+1 OPEN PLAN KITCHEN - YEARS OF LIVING: 3 OCCUPANCY PROFILE: SINGLE RESPONDENT AGE: 36 EDUCATION: DEGREE IN ARCHITECTURE FULL TIME JOB - THURSDAY/WEEKEND WORKING FROM HOME HEATING WAS NOT IN USE CURTAINS/SHUTTERS OPENED HEATING TYPES: A/C - EXTERNAL GAS HEATER COOLING TYPES: A/C			

PROTOTYPE 2 GOVERNMENT'S SOCIAL HOUSING UNITS - 1990s

NORTH FACING APARTMENT UNIT	FLOOR PLAN LAYOUT	Energy Analysis Heating	WINDOWS OPENING SCHEDULES	HEATING USE TYPES	COOLING USE TYPES
	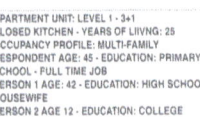	APARTMENT UNIT: LEVEL 1 - 3+1 CLOSED KITCHEN - YEARS OF LIVNG: 25 OCCUPANCY PROFILE: MULTI-FAMILY RESPONDENT AGE: 45 - EDUCATION: PRIMARY SCHOOL - FULL TIME JOB PERSON 1 AGE: 42 - EDUCATION: HIGH SCHOOL - HOUSEWIFE PERSON 2 AGE 12 - EDUCATION: COLLEGE STUDENT PERSON 3 AGE 10 - EDUCATION - PRIMARY SCHOOL STUDENT HEATING WAS NOT IN USE CURTAINS CLOSED ENCLOSED BALCONY TERRACE CURTAINS OPENED HEATING TYPES: A/C - EXTERNAL GAS HEATER COOLING TYPES: A/C - EXTRACTOR FANS WEEKEND IS NOT OCCUPIED			

PROTOTYPE 3 ALASYA PARK RESIDENTIAL TOWER BLOCK DEVELOPMENT - 2010s

SOUTH FACING APARTMENT UNIT	FLOOR PLAN LAYOUT	Energy Analysis Heating	WINDOWS OPENING SCHEDULES	HEATING USE TYPES	COOLING USE TYPES
		APARTMENT UNIT: LEVEL 3 - 2+1 OPEN PLAN KITCHEN - YEARS OF LIVING: 5 OCCUPANCY PROFILE: SINGLE RESPONDENT AGE: 34 EDUCATION: POSTGRADUATE DEGREE FULL TIME JOB - WORKING HOURS:16:00-23:00 HEATING WAS NOT IN USE BEDROOM 2 WINDOWS OPENED HEATING TYPES: A/C - EXTERNAL GAS HEATER COOLING TYPES: A/C - EXTRACTOR FANS WEEKEND IS NOT OCCUPIED			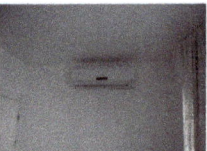

PROTOTYPE 3 ALASYA PARK RESIDENTIAL TOWER BLOCK DEVELOPMENT - 2010s

SOUTH FACING APARTMENT UNIT	FLOOR PLAN LAYOUT		WINDOWS OPENING SCHEDULES	HEATING USE TYPES	COOLING USE TYPES
		APARTMENT UNIT: LEVEL 1 - 3+1 OPEN PLAN KITCHEN - YEARS OF LIVING: 5 OCCUPANCY PROFILE: MULTI-FAMILY RESPONDENT AGE: 50- EDUCATION: PRIMARY SCHOOL - FULL TIME JOB PERSON 1 AGE: 47 - EDUCATION: PRIMARY SCHOOL - HOUSEWIFE PERSON 2 AGE 23 - EDUCATION: UNDERGRADUATE DEGREE - FULL TIME JOB PERSON 3 AGE 19- EDUCATION - UNDERGRADUATE STUDENT HEATING WAS IN USE COOKER WAS IN USE KITCHEN DOOR OPENED HEATING TYPES: EXTERNAL GAS HEATER COOLING TYPES: EXTRACTOR FANS		N/A	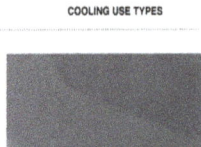

5 Interviewed apartment units in three different prototype MURBs

In order to collect subjective data from the building occupants for cooling energy use and comfort evaluation in the specified locations of the representative MURBs, a standardized questionnaire survey design was developed. The questionnaire includes 28 questions, which adopt a combination of open-ended, partially close-ended and predominantly closed-ended questioning approaches (see the questionnaire survey at https://repository.uel.ac.uk/item/8q713). The questionnaire proforma was designed to gather occupants' socio-demographic characteristics and its impact on domestic energy use. This method could also be utilized in examining how occupants' energy use and internal gains, e.g. from domestic appliances, can play a key role during the overheating risk assessment of buildings.

Building Energy Modeling and Simulation

To provide sufficient resolution for the analysis of occupants' thermal comfort it was deemed necessary to use the DTS model. The Integrated Environmental Solutions Virtual Environment (IES-VE) suite was selected as the most appropriate application for this purpose. In terms of validated performance, IES software interface is understood to meet a number of international standards including CIBSE *TM52 The limits of thermal comfort: avoiding overheating* (2016) in European buildings and is also accredited for use to European standard *EN 15251:2007 Indoor environmental input parameters for design and assessment of energy performance of buildings addressing indoor air quality, thermal environment, lighting and acoustics*. It is also necessary that IES-VE offers a number of features collectively that were found to be

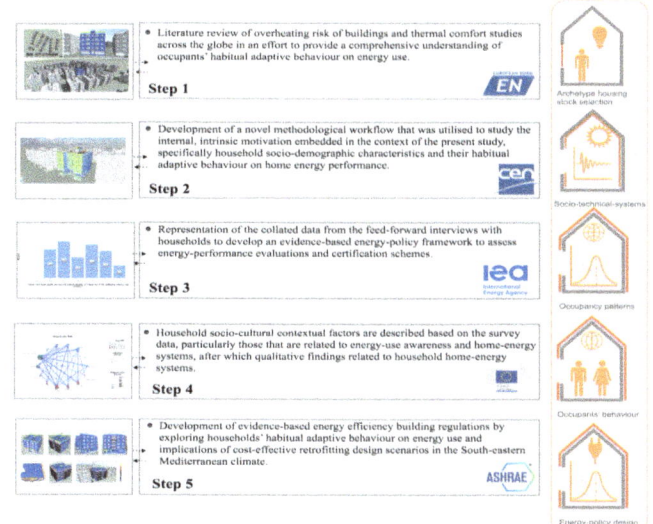

6   Stages of development of the STS conceptual framework of the present study

7   Meta-analysis of building envelope in three different prototype MURBs

8   Conceptualisation stages for the representation of reported parameters

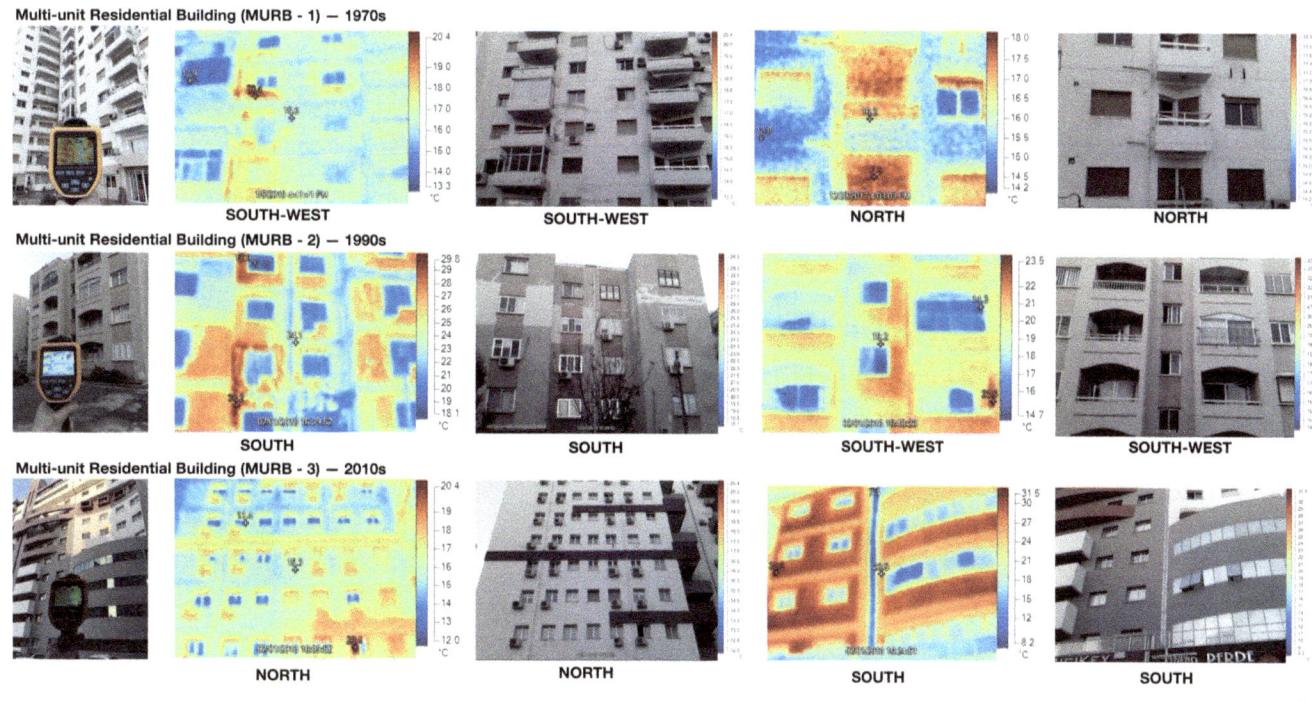

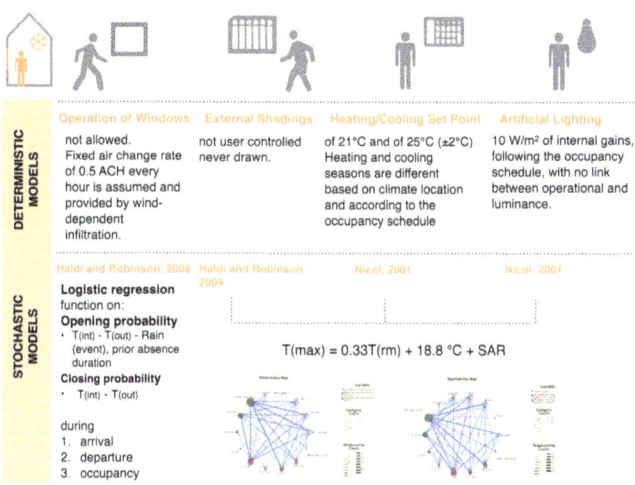

beneficial to the analysis. These included the following: close reproduction of the existing building geometry, detailed breakdown of the energy results by end use and zone, and ability to external interpolate the model settings (construction and zone profiles) to measure both the quasi-steady state and dynamic thermal scenario analysis. It should be noted that archetype buildings were modeled in IES's software suite, including the extensive modeling of the surrounding environment, to predict the impact of the mutual shading factor, high albedo surface frequency and vegetation of the neighborhoods.

It is also of interest to consider that in combination with the DTS components of the IES-VE software suite, it is possible to assess the energy performance of material changes concurrently. To assess the energy performance of prototype MURBs, thermal templates were constructed in the IES-VE software suite's Apache-Sim interface. These templates define the space conditioning systems (Apache Systems) and gain variation profiles for zones within the building. Energy simulations of the thermal behavior of the base-case prototype MURBs were conducted for conditions of typical meteorological year (TMY) of the coastline city of Famagusta, Cyprus. The framework for the methodology includes: first, architectural and other technical data from the privately-owned construction companies' archives for representative case study buildings were collected and analyzed.

- Step 1. The building geometry of base-case MURBs were constructed for its initial existing state and every floor with correspondent thermal zones and subdivisions.
- Step 2. Solar Analysis: To measure the risk of overheating of the MURBs via embedding IES-VE's solar analysis of Sun-Cast interface in order to identify external environmental factors such as heat gains through the number of hours' sunshine was gathered to calibrate the initial results from the modeling. Simulation was only performed on the same facing (south-west) MURBs, output intervals were measured daily with a simulation period of one year.
- Step 3. Evaluation of the existing cooling energy consumption and architectural measures in the scope of passive cooling design strategies (PCDS) developed for energy policy decisions.
- Step 4. Evaluation of the overheating risk criteria of occupied spaces in line with considering the CIBSE TM52 benchmarks. The in situ measurements were undertaken for each room separately in order to consider occupants' energy use patterns of a particular room.
- Step 5. Determination of the worst-case scenario based on the analysis of results obtained in previously mentioned steps and evaluation of; indoor comfort parameters (operative air temperature and relative humidity); heat losses and heat gains of buildings' components; annual final energy demand for cooling and total system energy use.

Thermal Imaging Survey

The study investigates the usefulness of infrared thermography for a quick diagnostic tool of the thermal performance of three representative prototype MURBs. The case study MURBs were surveyed and infrared thermal imaging was conducted with a thermal camera (FLUKE TiS20) at two times every day in winter period, in the late evening, and in the early morning, for avoiding the possible mistakes from direct solar radiation, as shown in Figure 7.

One of the main reasons is that infrared thermography is a potent technique for determining quick thermal conditions of the existing buildings and structures. In this study it is presented as a first diagnostic tool in order to identify the most ill-performing prototype MURBs in a worst-case scenario for the implementation of low-tech passive strategies at the final phase of the study. Figure 7 illustrates the thermal imaging timeline for investigating heat losses and assessing overheating risk of a building were undertaken between December 25, 2017 and December 1, 2018. The thermal survey was employed on three representative prototype MURBs by investigating three distinctive construction eras. MURBs selection was also based on convenience sampling, given the extended access required internally and externally for these experiments during the field studies. Following the survey period, all images were uploaded and assessed using the Fluke Connect Smart-view (beta version). This software enables image adjustment so that each has the same temperature span, including average temperature, minimum and maximum temperature of a measured environment. This allows us to understand the transient heat losses to be viewed more accurately.

Parametric Statistical Analysis

Statistical Package for Social Sciences (SPSS) software tool version 28.0 was used for the analysis of the data collected from the field studies, which were exported into spreadsheets. Separate analyses were conducted according to the interviewed three representative prototype MURBs and specified locations within the buildings, as shown in Figure 8.

Pearson correlations were computed to assess the correlation between pairs of variables. Correlations between multiple

Multi-unit Residential Building (MURB - 1) – 1970s

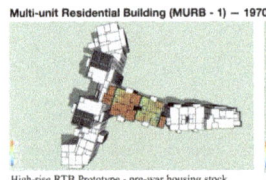
High-rise RTB Prototype - pre-war housing stock

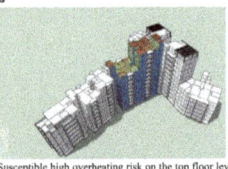
Susceptible high overheating risk on the top floor level flats

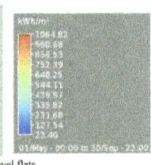

Multi-unit Residential Building (MURB - 2) – 1990s

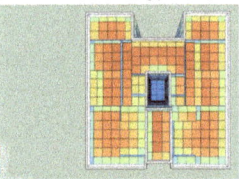
Medium-rise RTB Prototype - post-war housing stock

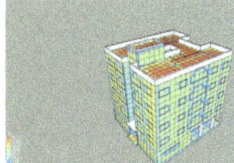
High solar radiation factor on the top floor level flats

Multi-unit Residential Building (MURB - 3) – 2010s

High-rise RTB Prototype - newly-built housing stock

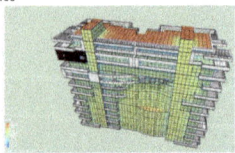
Overheating risk factor on the large glazed opaque surfaces

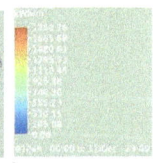

9 Solar irradiation factor and its impact on building envelopes

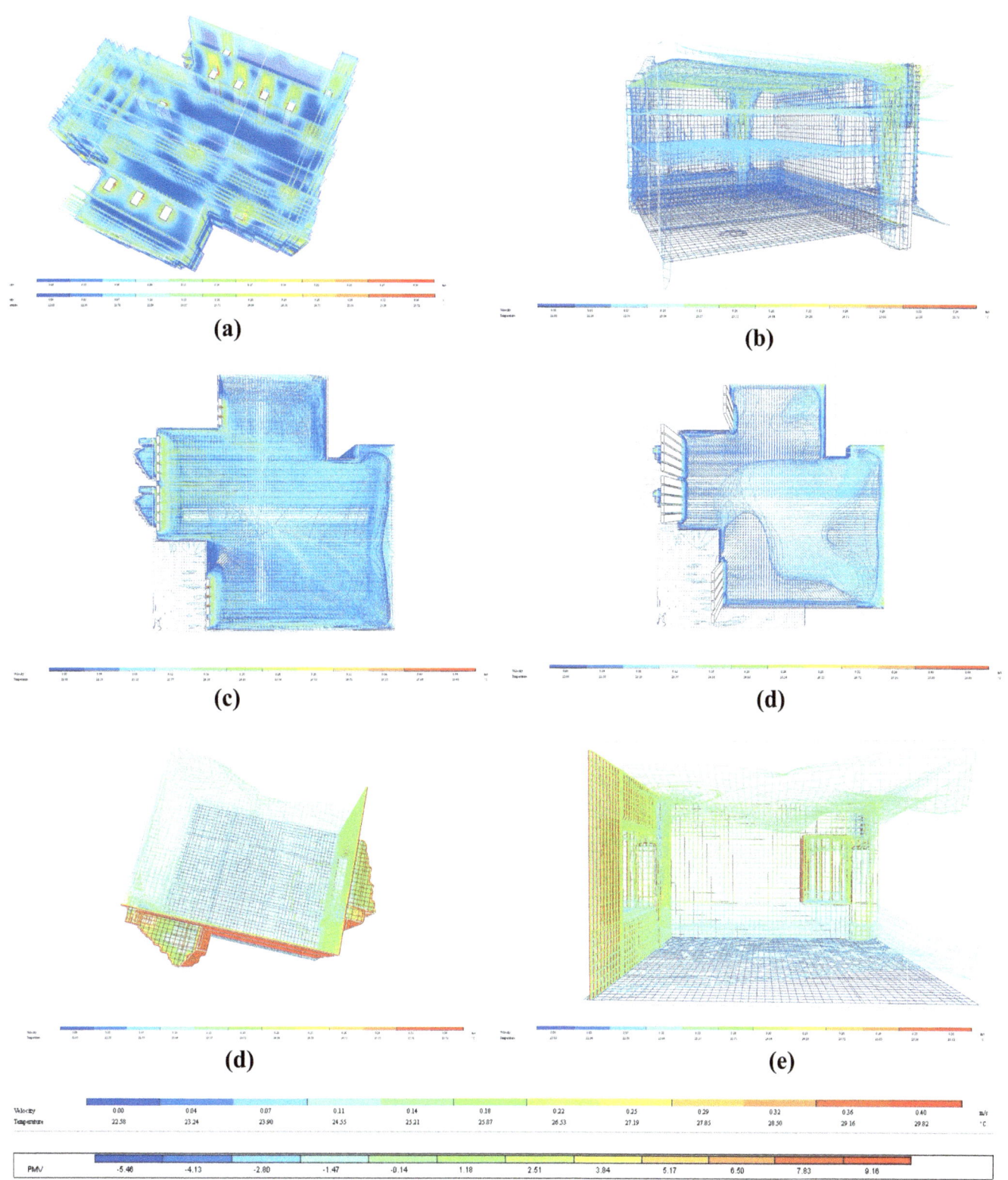

10 Computational fluid dynamics (CFD) analysis of a representative living room

ENVIRONMENTAL PERFORMANCE AND SIMULATION

parameters were collected from the questionnaire surveys in order to investigate the relationship between different parameters. The significant level of the analysis will be set to be 0.05 co-efficient for the statistical analysis. This means that the results were statistically significant when p-value <0.05. First, the descriptive analysis articulated the interview findings to report the occupants' behaviour in energy use, then the findings were formulated by the correlation analysis methods in order to evaluate the correlation between different parameters. For this, Pearson (two-tailed) correlation analysis was conducted.

RESULTS AND ANALYSIS

This section illustrates the quantitative and qualitative findings of the infrared radiometer thermography (IRT) survey and the building energy simulation (BES) studies to develop an evidence-based building energy performance evaluation for the southeastern Mediterranean climate of Cyprus. In the following sections, the results of the IRT survey and an analysis of data collected from on-site monitoring, in situ physical measurements, household energy bills, and building energy modeling are detailed to implement an evidence-based framework in domestic energy use policy.

Building Fabric Thermal Performance

The SunCast software interface tool was implemented in the building modeling simulation to assess the amount of solar radiation that was absorbed by any given external surface of the prototype MURBs, based on the orientation thereof and the effects from adjacent buildings. The SunCast simulation module was used to validate the qualitative and quantitative analyses of the survey findings that were obtained from the thermal imaging survey. Figure 9 shows the maximum solar radiation and mean values of the three analyses that were adopted for the worst-case scenario of the south-facing MURBs.

Only three external surfaces were exposed in Figure 10, and all three exhibited different heat gains throughout the year due to poor insulation in the exposed wall, with noted exacerbations in the summer, which created overheating risks. Upper floor flats demonstrated the greatest risk of overheating due to the impact of the U-values of the building envelopes and the solar panels for the hot-water tanks that were placed on top of the original surface; for this reason, all bedroom spaces in the upper- and intermediate-floor flats experienced a greater likelihood to overheat, compared to the CIBSE TM52 overheating criteria (Ozarisoy 2022). It was determined that the living rooms of these flats were also susceptible to overheating, but this was because of different factors: the rooms had significant window-opening ratios with no shading, and the spaces all faced either south or south-east and were, therefore, exposed to high-intensity sunlight throughout most of the day; the external walls, which were constructed from brick and exterior rendering without insulation, were also exposed to high solar-heat gains. A combination of these factors led

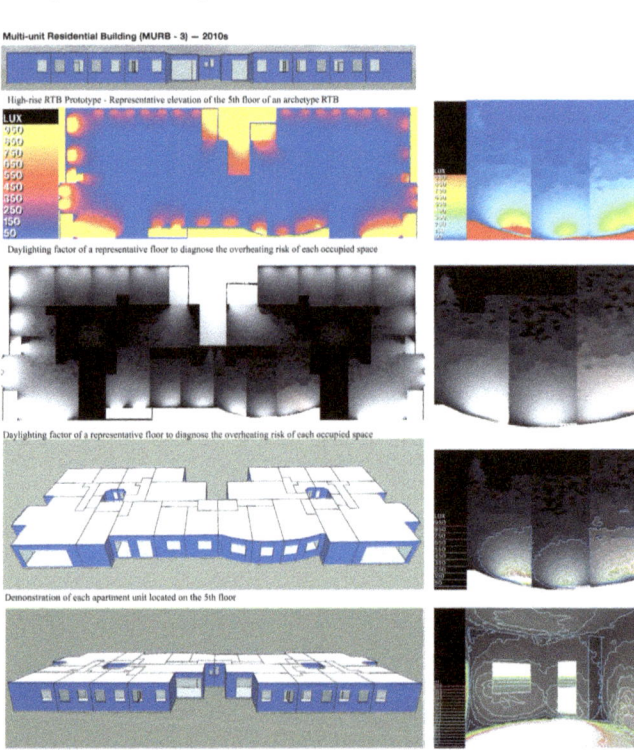

11 Daylighting analysis of representative flats in MURBs 1 and 2

12 Daylighting analysis of representative flats in MURB 3

TABLE 2A Results for TM52 Criteria 2 and 3 for the study year (ΔT = room temperature minus the maximum adaptive temperature for overall means/averages across representative flats)

Flat Location	Living Room	Bedroom 1	Bedroom 2	Bedroom 3
TM52 Criterion 2: Daily Weighted Exceedance (°C per hour)				
First Floor	51	45	43	37
Second Floor	60	48	62	69
Upper Floor	60	47	61	69
TM52 Criterion 3: Maximum ΔT (°C)				
First Floor	5	5	4	4
Second Floor	8	9	7	8
Upper Floor	11	10	7	8

TABLE 2B Overheating risk assessment criteria

	Assessment Criteria	Acceptable Deviations
Criterion 1	Percentage of occupied hours during which ΔT (ΔT = Top-Tmax rounded to the nearest whole degree) is greater than or equal to 1 °C	Up to 3% of occupied hours
Criterion 2	Daily weighted exceedance (We) in any one day > 6 °C.h (degree hours)	0 days
Criterion 3	Maximum temperature level (Tup) ΔT > 4 °C	0 hours

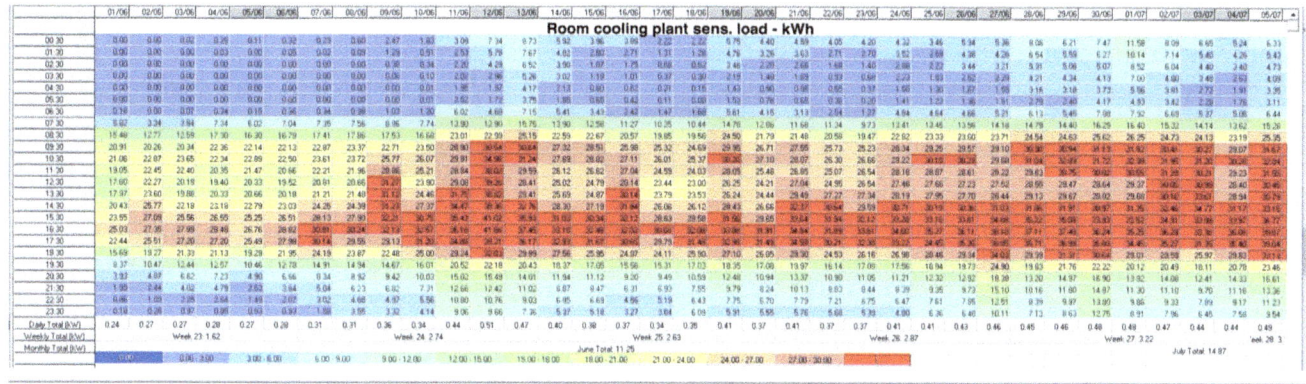

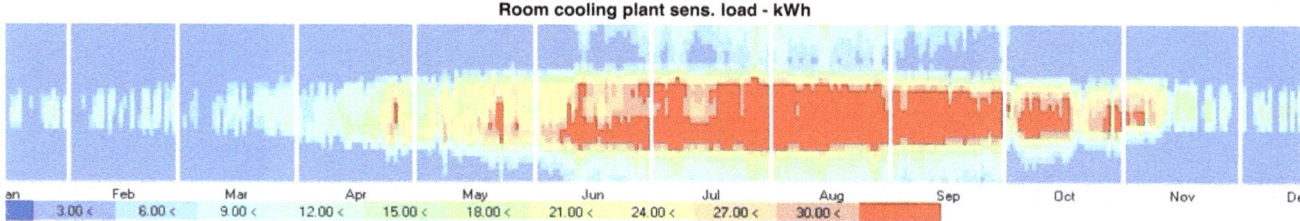

13 Overheating risk assessment criterion of representative apartment units

to overheating issues and significant occupant discomfort, especially in the summer. It should be noted that Figure 10 illustrates both the air velocity measurement and indoor air temperature to predict occupants' TSVs with minimum error margins. In this experimental analysis, the CFD period was chosen between May and September in the summer period for the northern hemisphere to represent the worst-case scenario conditions while assessing overheating risk of archetype MURBs in the South-eastern Mediterranean climate. Notably, the windows were open with 30% parameter which were identified average opening preference by the occupants.

Overheating Risk Assessment

Figure 11 presents overall observations to demonstrate the heat vulnerability of the base-case MURBs. The scope is to test how the existing representative floor plan and window dimensions allow direct solar heat gain to the interior space. Temperature readings from approximately 2830 images were analysed using the forward-looking infrared radiometer (FLIR) analysis tool to diagnose the thermal performance of all prototype MURBs for the worst-case scenario development. These assessments were performed during the 2017–2018 winter months, and all of the survey data were used to model the base-case building and validate the building energy simulation findings (Bayomi et al. 2021); the on-site field observation thermal images and photographic documentation and the quantitative temperature recordings of the building fabric systems also validated the findings related to the households' socio-demographic characteristics.

Building envelope surface temperatures in the southwest oriented MURBs (as shown in Figure 11) ranged between 5.4°C to 13.9°C, and this structure exhibited heat loss through the external wall; the thermographic image was taken on January 8, 2018 between 06:30–07:54AM, when the air-temperature

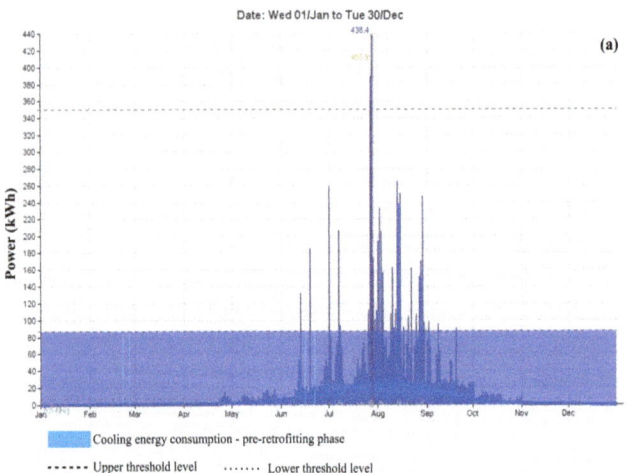

14a Total electricity consumption of representative base-case MURBs, pre-retrofitting

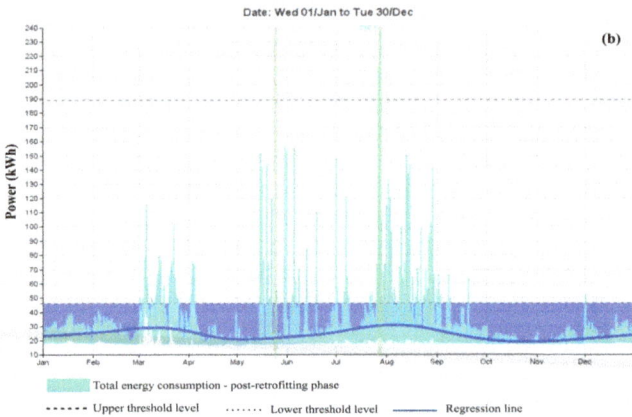

14b Total electricity consumption of representative base-case MURBs, post-retrofitting

was recorded at 43°C. The detected thermal anomalies were possibly due to the small kitchens in these units; motivated by limited floor area and inadequate meal-preparation, most of the surveyed households had refurbished their kitchen spaces, and these modifications led to significant heat loss through the wall surfaces. The photograph in Figure 12 shows the daylighting illuminance of a representative floor at the newly built prototype MURB 3.

As can be seen in Table 2(a), the CIBSE TM52 criteria was assessed according to the total number of days in the calendar year in which the exceeded degree-hours of 6°C per hour, while that specific zone was occupied. The highest ΔT-values for Criterion 3 indicate that all the simulated rooms in the first-, intermediate-, and upper-floor flats failed to achieve the criteria. Bedroom 1 in each of the sample flats exhibited the highest peak ΔT with a value of 10°C due to the north and southeast orientations thereof. A pragmatic way of quantifying the effects of thermal comfort is defined in CIBSE TM52 for new buildings; that is, major refurbishments and adaptation strategies should conform to Category II in BS EN 15251, as shown in Table 2(b) (British Standards Institution 2007; CEN 2007; CIBSE 2017).

As shown in Figure 13, the flat on the intermediate floor was again found to perform better than the flat on the upper floor. The living room of the first-floor flat was observed to be the critical zone within the sample flats, because it failed to achieve Criterion 3. This differentiation in the performance of the flat on the first floor can be attributed to the location of the rooms: the unit with three exposed walls has a reasonably reduced capacity to provide thermal comfort within the adaptive comfort limits. The living rooms in the first- and upper-floor flats exceeded 4°C by four and nine hours each year, respectively.

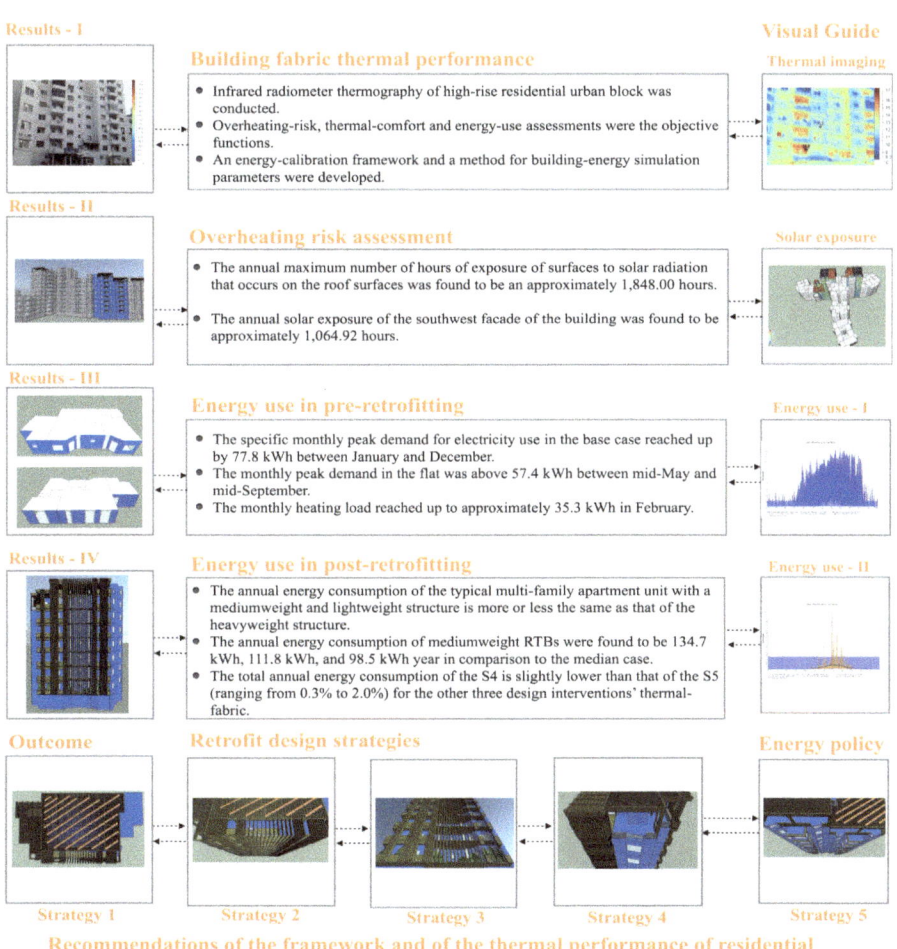

15a Schematic illustration of passive cooling design strategies implemented in prototoype MURBs

Notably, significant heat loss was detected where the front façades (i.e., the living room spaces) of the south-, southeast- and southwest-facing MURBs met the corner of the construction junction. Most of the south-oriented flats showed signs of significant thermal loss in the winter, and it appeared as if these MURBs also demonstrated a greater risk of overheating in the summer. The on-site measurement method allowed the worst-performing MURBs to be identified so that further calibration studies could be conducted in the building modeling phase of the present study.

Building Energy Simulation

In addition to the thermal imaging survey and in situ measurements, the energy performance of the sample rooms was investigated using dynamic thermal simulation modeling with a well-established suite of the IES-VE simulation software that included ModelIT, SunCast, ApacheSIM, MacroFlow, and VistaPro. The expected energy performance of the prototype MURBs and the overheating potential thereof were simulated between May and September 2021. Simulating, investigating, and optimizing the energy performance of different building component structures revealed significant differences in cooling loads in the base-case MURBs. The results of the simulation of the existing performance for the representative units indicated that the largest share of heat loss came from air infiltration, uninsulated exterior walls and windows, which led to a high annual energy demand for cooling measures.

The results of the simulations were analyzed to better understand existing energy use conditions and to calibrate the energy consumption patterns, especially the cooling demands of the representative first-, second- and top-floor flats. In terms of examining energy consumption as it relates to specific heat loss, the prototype flat consumed 438.4 kW of energy a per week during the pre-retrofitting phase and 237.1 kW of energy a per week during the post-retrofitting phase through its implementation of PCRDS onto the existing building envelope, as is shown, respectively, in Figures 14(a) and (b).

Passive Cooling Systems

This study presents an analysis of the PCRDS development framework to demonstrate an evidence-based integrated design approach for energy use, as shown in Figures 15(a) and (b). The proposed solution for passive-design retrofitting

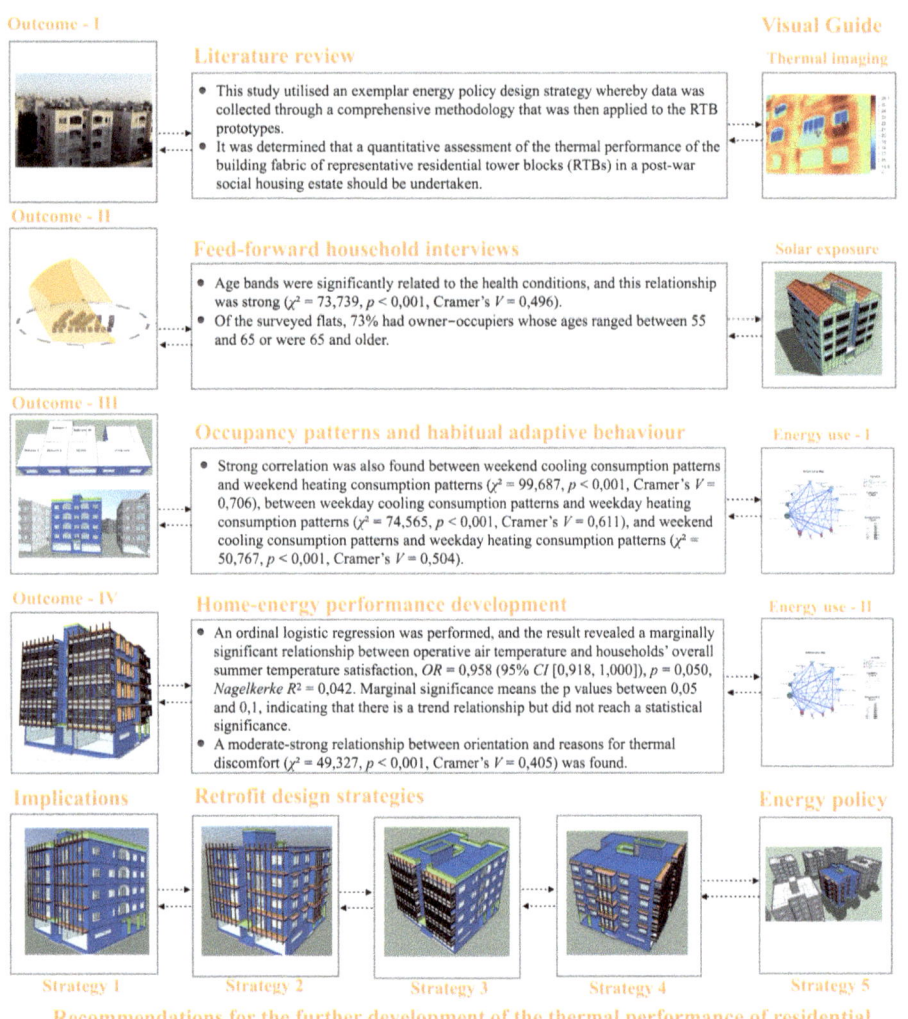

15b Schematic illustration of passive cooling design strategies implemented in prototoype MURBs

of the building envelope was the installment of a thick, thermal-insulated clay-tile external-facing system, the replacement of windows and door glazing (from single- to double Low-E glazing) and the use of timber-framed shading elements.

The presented scenarios were reviewed and studied globally, including the use of energy efficient building systems and local construction codes, and these evaluations yielded improvement models that were especially suitable for the specific region. The retrofitting measures of the design alternative solution were as follows:

EXTERIOR WALLS: The existing outer layer was removed, 245 mm of new insulation was affixed to an inner layer and a new outer concrete layer, and external clay tile cladding was installed (new U-value of 0.95 W/m2K).

ROOF: Old roof mastics and insulation were removed, and 340 mm of new insulation and a new asphalt mastic cover were installed (new U-value of 0.80 W/m2K).

BASE FLOOR: Additional external insulation was added (new U-value of 0.94 W/m2K), and all existing single-pane windows, balcony openings, and internal doors were replaced with double-pane windows (new U-value of 1.39 W/m2K).

Figure 16 illustrates the assigned construction properties of the base-case and the six strategies that were applied in the simulation.

In Figure 17, the cooling energy consumption is shown for the living zones (living room and bedroom 1) on the worst-case (level 10) sample flat. The initial observation is that the flat units greatly exceed the benchmark of 15kW/h. The flat on the typical floor is shown to have worse thermal performance than the one on the first floor. The worst performing flat is the top floor (three exposed external walls) corner flat. The living room in this unit is the worst performing zone with a performance that exceeds the benchmark by over eight times at a value of 60.4 kW/h. It is worth noting that Bedroom 1 exceeds the benchmark by over seven times with a value of 75.3 kW/h.

Multi-unit Residential Building (MURB - 1) — 1970s

S1 - Wind catcher systems attached to each space   S2 - Horizontal louvers for shading purposes

Wind-catcher systems

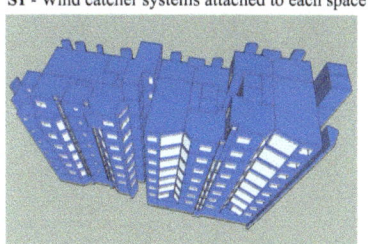

S1 - Overview of wind catcher systems   S3 - Renewables installed on the roof terrace

Multi-unit Residential Building (MURB - 2) — 1990s

S4 - Volumetric fenestration systems   S5 - Hybrid perforated shading systems

Multi-unit Residential Building (MURB - 3) — 2010s

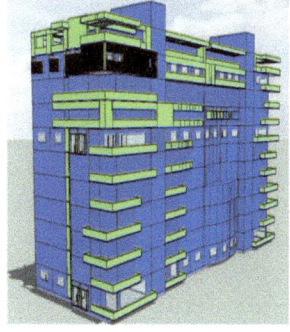  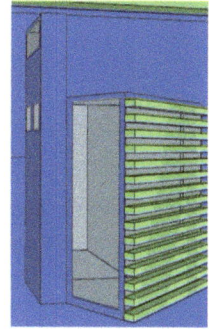

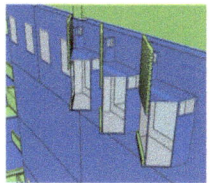 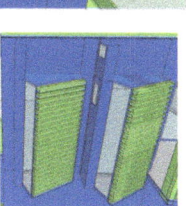

S6 - Integration of kinetic skin to optimise daylight and reduce overheating risk of each space

16 Overview schematic illustration of passive cooling design strategies developed

17a Total energy consumption of a representative apartment unit of MURB 1, after implementing Strategy 1 (wind catcher systems)

17b Total energy consumption of an upper level flat of MURB 2, after implementing Strategy 5 (hybrid perforated shading systems)

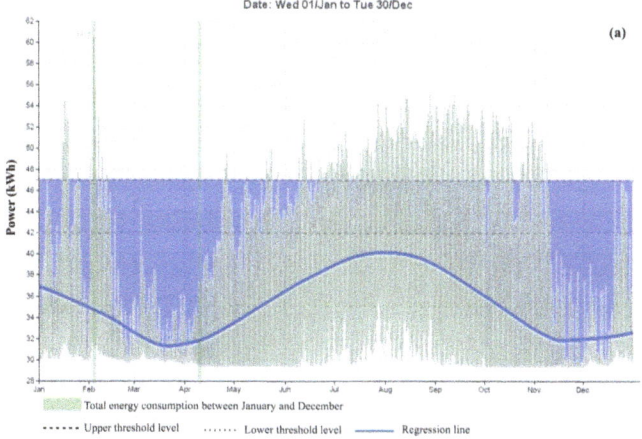

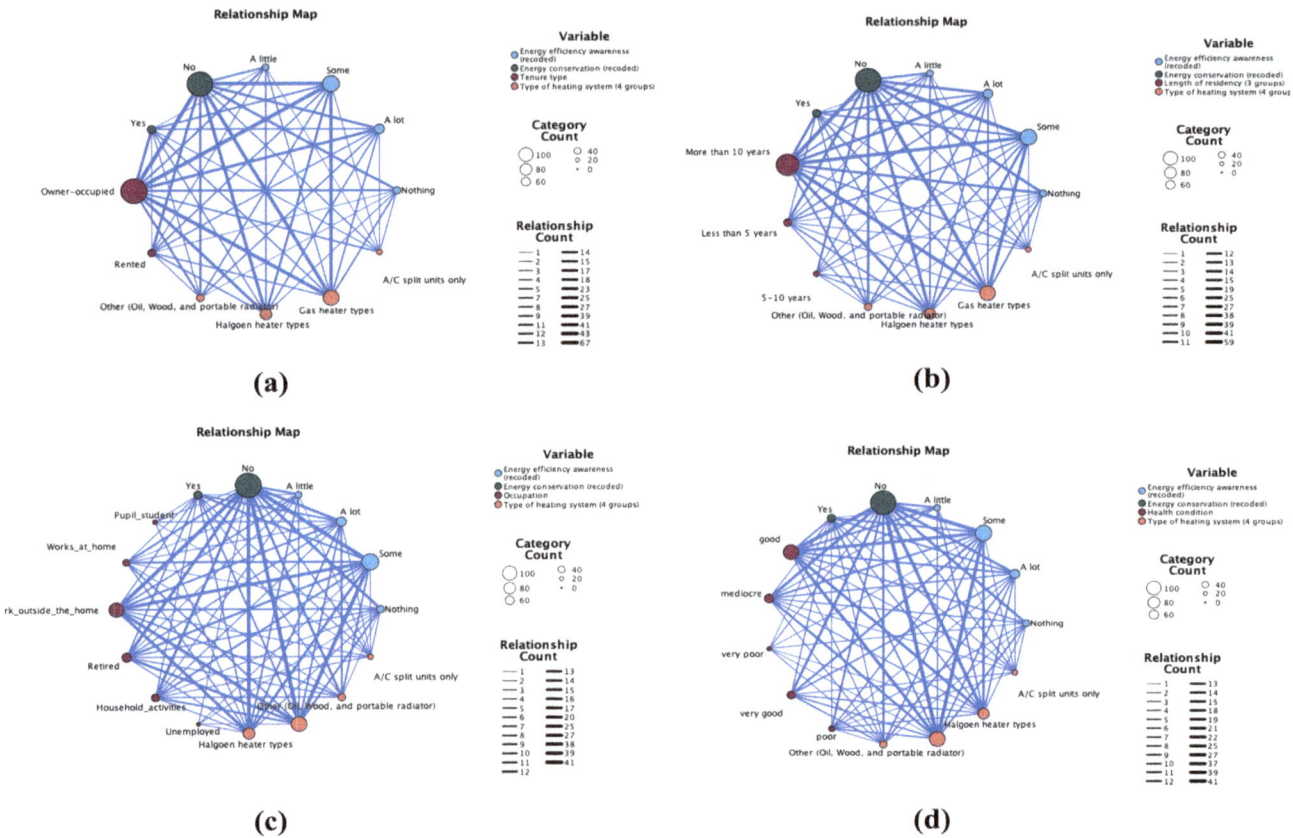

18 Selected socio-demographic characteristics of respondents: (a) tenancy status; (b) length of residency; (c) employment status; (d) health condition

The study elucidated the potential applicability of passive cooling design strategies in various retrofitting interventions to improve the energy efficiency of existing residential buildings. Based on this study, the passive design principles were shown to result in significant reductions in energy consumption and optimized thermally comfortable indoor air for the flat occupants. This was an important finding that needs to be further explored by a robust optimization study, which will provide a wider domain to assess and optimize the risk of overheating and better understand occupants' thermal comfort when seeking to enhance 'night cooling' effects in MURBs in this southeastern Mediterranean climate. The results indicate that indoor air temperatures in Famagusta, Cyprus follow a consistent pattern throughout the month of August. Indoor air temperatures in the sample units range from 28.5°C to 36.5°C throughout the day and night; this lack of diurnal temperature variation suggests that internal operative air temperatures (OTs) remain relatively high and do not induce cooling at night. Furthermore, the external building fabric, uninsulated roof and three exposed wall surfaces were found to be key determinant factors due to the high U-value of the building properties, the surface area and the amount of solar gain exposure, all of which resulted in high heat transmittance into and out of the upper floor flats and had a significant effect on the OTs of all the flats. One of the significant findings of this study was that cooling energy consumption decreased by 81% after all six passive design strategies were implemented. These conclusions will create the prerequisites and the background information that is needed for the development of a novel methodological framework and a ground-breaking epistemological design approach in the area of development and design, energy policymaking and the drafting of subsidization schemes and targeted actions to improve the energy efficiency of existing housing stock.

Households' Habitual Adaptive Behavior on Energy Use
This section presents the in-field findings on the effect of occupancy patterns and habitual household behavior on home energy performance. Household sociocultural contextual factors are described based on the survey data, particularly those that are related to energy use awareness and home energy systems, after which are presented qualitative findings related to household feedback that are structured according to the conceptual framework.

To investigate the energy saving awareness of each household, the respondents were asked close-ended questions to determine whether they frequently checked their electricity meter to monitor how much energy they had consumed, as shown in Figure 18. It was shown that 82% of the surveyed households reported that they did not check how much energy they consumed, while 18% of the households reported that they frequently checked their electricity meter. The households were also asked about the type of awareness they had related to energy savings. Additionally, it can be seen that 20% of the respondents reported that they always took energy-saving methods into consideration, while 50% of the respondents infrequently thought of energy-saving methods; notably, 15% of the respondents never considered energy-saving methods. Based on the above-cited results, it can therefore be concluded that the degree of awareness related to energy-saving measures is not only due to the occupant's level of education, but also to their income level; only the households who took these into considerations were mostly rental tenants and migrant workers with young children who took all possible energy-saving measures into consideration.

DISCUSSIONS

The empirical model that was created for this study addressed the development of an evidence-based STS conceptual framework that could explore the influence of building thermal properties on occupant thermal comfort; the results revealed relatively warm global temperatures that ranged between 24.5°C to 37.0°C. According to the CIBSE TM52 Criterion 1, the upper thermal-comfort limit was 33°C, yet the on site monitoring recorded a maximum outdoor temperature of 38.7°C, and the in situ measurements revealed the indoor air temperatures that ranged between 22.5°C and 35.0°C, which was well above the acceptable threshold limits. There was a strong positive correlation ($r=0.588$, $p<0.01$) between weekday heating and cooling consumption patterns, because this was when the occupants were at home for the greatest amount of time and desired optimum thermal comfort. Conversely, there was a strong negative correlation ($r=-0.621$, $p<0.01$) between type of occupancy and window-opening patterns in the summer. A moderate positive correlation ($r=0.215$, $p<0.05$) was found between the occupants' TSVs and indoor air temperature, while a moderate negative correlation ($r=-0.325$, $p<0.01$) was found between the occupants' TSVs and outdoor air temperature. According to the occupants' TSVs, 28.5°C was the 'neutral' temperature, and the upper thermal-comfort limit in warm indoor conditions was 31.5°C, which suggests that occupants in hot, dry climates who experience thermally uncomfortable indoor environments in the summer are able to tolerate warmer conditions than people in high and medium altitudes. The findings of the present study provide evidence that a methodological approach plays a decisive role in the calibration of building energy performance. As such, this study predominantly lies within the STS approach, which highlights the importance of upgrading the energy efficiency of buildings with the use of an appropriate methodology and implementing effective retrofitting interventions in the decision-making process that are influenced by human actions. The effective energy consumption reduction measures for social housing estates developed in the course of this study will further benefit from a conceptual level analysis and prioritization in accordance with the climate characteristics of the regional context.

CONCLUSIONS

This study engaged in an in-depth investigation into the thermal performance of buildings in Famagusta, Cyprus and the building fabric thermal performance in representative MURBs. This paper aimed to evaluate the overheating risk for these structures, while also considering summertime cooling energy demands and potential ways to lessen these demands through the development of a novel methodological workflow to assess the building diagnostics; an energy audit was conducted by integrating the usefulness of the IRT thermal-assessment technology. It can be observed that there was a lack of diurnal temperature variations within the sample flats, which means that the internal operating temperatures remained relatively high throughout the day and night, and ranged from a minimum of 28.5°C to a maximum of 36.5°C; this was not significant enough to induce night cooling. It was also concluded that the external building materials, uninsulated roof and three exposed wall surfaces were a key determinant factor because of the high U-value, surface area and level of exposure to solar gains; this then induced high heat transmittance into and out of the first-floor flats, which had a significant effect on the operative temperatures of these flats. This study will add significant value to ongoing and future efforts to achieve energy savings by integrating the IRT process into the retrofitting decision-making process for buildings. Finally, this study can and will serve as an exemplar when examining comparable building typologies from similar construction eras across Europe.

ACKNOWLEDGMENTS

The authors would like to acknowledge the Graduate School, School of Architecture, Computing & Engineering (ACE), London, United Kingdom. Data Access Statement: The datasets generated during and/or analyzed during the current study are publicly available at https://doi.org/10.15123/uel.8q713.

REFERENCES

Ascione, F., N. Bianco, T. Iovane, G. M. Mauro, D. F. Napolitano, A. Ruggiano, and L. Viscido. 2020. "A real industrial building: Modeling, calibration and Pareto optimization of energy retrofit." *Journal of Building Engineering* 29 (May): 101186. https://doi.org/10.1016/j.jobe.2020.101186

Ball, B. L., N. Long, K. Fleming, C. Balbach, and P. Lopez. 2020. "An open-source analysis framework for large-scale building energy modeling." *Journal of Building Performance Simulation* 13 (5): 487–500. https://doi.org/10.1080/19401493.2020.1778788.

Bayomi, N., S. Nagpal, T. Rakha, and J. E. Fernandez. 2021. "Building Envelope Modeling Calibration using Aerial Thermography." *Energy and Buildings* 233 (February): 110648. https://doi.org/10.1016/j.enbuild.2020.110648.

Bertoldi, P., and R. Mosconi. 2020. "Do energy efficiency policies save energy? A new approach based on energy policy indicators (in the EU Member States)." *Energy Policy* 139 (April): 111320. https://doi.org/10.1016/j.enpol.2020.111320.

Choi, J. H. 2017. "Investigation of the correlation of building energy use intensity estimated by six building performance simulation tools." *Energy and Buildings* 147 (July): 14–26. https://doi.org/10.1016/j.enbuild.2017.04.078.

Copiello, S., and E. Donati. 2021. "Is investing in energy efficiency worth it? Evidence for substantial price premiums but limited profitability in the housing sector." *Energy and Buildings* 251 (Nov): 111371. https://doi.org/10.1016/j.enbuild.2021.111371.

Haldi, F., and D. Robinson. 2008. "On the behaviour and adaptation of office occupants." *Building and Environment* 43 (12): 2163–2177. https://doi.org/10.1016/j.buildenv.2008.01.003.

Kristensen, M. H., R. E. Hedegaard, and S. Petersen. 2018. "Urban-scale dynamic building energy modeling and prediction using hierarchical archetypes: A case study of two Danish towns." In *Proceedings of BSO 2018: 4th Building Simulation and Optimization Conference*. 11–12.

López-González, L. M., L. M. López-Ochoa, J. Las-Heras-Casas, and C. García-Lozano. 2016. "Energy performance certificates as tools for energy planning in the residential sector. The case of La Rioja (Spain)." *Journal of Cleaner Production* 137 (Nov): 1280–1292. https://doi.org/10.1016/j.jclepro.2016.08.007.

Ma, Z., P. Cooper, D. Daly, and L. Ledo. 2012. "Existing building retrofits: Methodology and state-of-the-art." *Energy and Buildings* 55 (Dec): 889–902. https://doi.org/10.1016/j.enbuild.2012.08.018.

Mahdavi, A. 2020. "In the matter of simulation and buildings: some critical reflections." *Journal of Building Performance Simulation* 13 (1): 26–33. https://doi.org/10.1080/19401493.2019.1685598.

Mavrigiannaki, A., K. Gobakis, D. Kolokotsa, A. L. Pisello, C. Piselli, R. Gupta, M. Gregg, M. Laskari, M. Saliari, M.-K. Assimakopoulos , and A. Synnefa. 2020. "Measurement and verification of zero energy settlements: Lessons learned from four pilot cases in Europe." *Sustainability* 12 (22): 1–16. https://doi.org/10.3390/su12229783.

Nicol, F. 2011. "Temperature and adaptive comfort in heated, cooled and free-running dwellings." *Building Research & Information* 45 (7): 730–744. https://doi.org/10.1080/09613218.2017.1283922.

Ozarisoy, B. 2022. "Energy effectiveness of passive cooling design strategies to reduce the impact of long-term heatwaves on occupants' thermal comfort in Europe: Climate change and mitigation." *Journal of Cleaner Production* 330 (Jan): 129675. https://doi.org/10.1016/j.jclepro.2021.129675.

Ozarisoy, B., and H. Altan. 2022. "Significance of occupancy patterns and habitual household adaptive behaviour on home-energy performance of post-war social-housing estate in the South-eastern Mediterranean climate: Energy policy design." *Energy* 244 (Part B): 122904. https://doi.org/10.1016/j.energy.2021.122904.

Ozarisoy, B., and H. Altan. 2021. "Developing an evidence-based energy-policy framework to assess robust energy-performance evaluation and certification schemes in the South-eastern Mediterranean countries." *Energy for Sustainable Development* 64 (Oct): 65–102. https://doi.org/10.1016/j.esd.2021.08.001.

Pasichnyi, O., J. Wallin, and O. Kordas. 2019. "Data-driven building archetypes for urban building energy modelling." *Energy* 181 (Aug): 360–377. https://doi.org/10.1016/j.energy.2019.04.197

Pelenur, M. J., and H. J. Cruickshank. 2012. "Closing the Energy Efficiency Gap: A study linking demographics with barriers to adopting energy efficiency measures in the home." *Energy* 47 (1): 348–357. https://doi.org/10.1016/j.energy.2012.09.058.

Pelenur, M. 2013. "Retrofitting the domestic built environment: Investigating household perspectives towards energy efficiency technologies," unpublished PhD Thesis, University of Cambridge, United Kingdom. Accessed February 2, 2021. https://doi.org/10.17863/CAM.14069.

Salvalai, G., G. Masera, and M. M. Sesana. 2015. "Italian local codes for energy efficiency of buildings: Theoretical definition and experimental application to a residential case study." *Renewable and Sustainable Energy Reviews* 42 (Feb): 1245-1259. https://doi.org/10.1016/j.rser.2014.10.038.

Stojiljković, M. M., M. G. Ignjatović, and G. D. Vučković. 2015. "Greenhouse gases emission assessment in residential sector through buildings simulations and operation optimization." *Energy* 92 (Part 3): 420–434. https://doi.org/10.1016/j.energy.2015.05.021

Tian, W., Y. Heo, P. de Wilde, Z. Li, D. Yan, C. S. Park, X. Feng, and G. Augenbroe. 2018. "A review of uncertainty analysis in building energy assessment." *Renewable and Sustainable Energy Reviews* 93 (Oct): 285–301. https://doi.org/10.1016/j.rser.2018.05.029.

Yang, S., D. Zhou, Y. Wang, and P. Li. 2020. "Comparing impact of multi-factor planning layouts in residential areas on summer thermal comfort based on orthogonal design of experiments (ODOE)." *Building and Environment* 182 (Sept): 107145. https://doi.org/10.1016/j.buildenv.2020.107145.

Villca-Pozo, M., and J. P. Gonzales-Bustos. 2019. "Tax incentives to modernize the energy efficiency of housing in Spain." *Energy Policy* 128 (May): 530–538. https://doi.org/10.1016/j.enpol.2019.01.031.

IMAGE CREDITS
All drawings and images by the authors.

Dr. Bertug Ozarisoy is an architect and expert in building energy modeling in Cyprus. He holds a BSc (Hons) in Architecture, MSc in Architecture, MA in Architecture, Cultural Identity and Globalisation (University of Westminster, UK) and a PhD in Environmental Design Engineering (University of East London, UK). His research focuses on understanding the theory between architecture and energy policy design in conjunction with exploring the impact of passive cooling systems on domestic energy use and households' thermal comfort. He is interested in environmental design and the development of novel design applications throughout his architectural practice. He has focused on developing a novel methodological framework to assess overheating risk of buildings to tackle the detrimental effects of long-term heatwaves experienced as a result of climate change in Europe. He is an author of the *Handbook of Retrofitting High Density Residential Buildings: Policy Design and Implications on Domestic Energy Use in the Eastern Mediterranean Climate of Cyprus*, which is published by Springer. Currently, Dr. Ozarisoy is a postdoctoral researcher at the Middle East Technical University (METU) Northern Cyprus Campus, where he teaches two post-graduate courses entitled 'Deep Energy Retrofit' and 'Environmental Design and Engineering' as part of his postdoctoral research in the Sustainable Environment and Energy Systems (SEES) program.

Prof. Dr. Hasim Altan is Professor of Sustainable Design and Architectural Engineering in the College of Architecture and Design (COAD) at Prince Mohammad bin Fahd University (PMU) in Khobar, Saudi Arabia. He is a Chartered Architect (RIBA) and a Chartered Engineer (CIBSE) with over twenty years of academic and practice experience in the field of Architecture, Engineering, and Construction (AEC) in the Built Environment in UK, Europe, Middle East, and North Africa (MENA) regions. He is a founding member of the International Network on Zero Energy Mass Custom Home (ZEMCH), which so far has organized nine international conferences, several design workshops, and numerous technical visits. As well as having supervised 17 successful PhD theses (11 as first supervisor), and over 80 Masters (MSc, MArch and MEng) dissertations, he has published over 300 refereed international journal and conference papers, as well as technical reports, edited books, chapters, and editorials in related fields.

High-Density Building Form Generation Considering Daylight Performance

With a Daylight Performance Optimization Method Integrating Genetic Algorithm and Cellular Automata

Jun Xiao [1,2]
Yubo Liu [1,2]
Qiaoming Deng [1,2]

1. School of Architecture, South China University of Technology

2. State Key Laboratory of Subtropical Building Science, China

ABSTRACT

High-quality and uniform daylighting is important for frequently occupied buildings such as classrooms and large offices in which people have stable, uniformly distributed seats. It is difficult to achieve required daylighting in high-density contemporary buildings that receive daylight by side windows. Further, practioners of contemporary design cannot correctly predict the daylight performance in a complex-form building with the mixed use of side windows and skylights.

In this case, aiming for a high-density building designed with high-quality daylighting, this article develops a building form generation program for daylight performance optimization integrating Cellular Automata (CA) and Genetic Algorithm (GA), tools that can provide a global daylighting optimization through the balance of competition and concession of agents. The CA model provides randomness and structural restriction, while GA provide optimization and convergence by evaluating and selecting a series of CA models. The model applies designed CA rules on daylighting and a mathematic proxy model for daylight performance in GA fitness calculation. Validation and calibration are included to gather statistics of building floor parameters, which are necessary for this program to provide more precise predictions. The program is examined on the site located in southeast China. As a result, the proxy model maintains a low predictive error with R2 higher than 0.67 and 1/100 computational time compared to real-time simulation, and the high-density generative result provide 69% eligible rooms under the British daylighting standard.

1A Habitat 67, view from Montreal's Port (Legault 2021)

LIST OF SYMBOLS

Symbol	Unit	Explanation
i		Simulation square No.
N_i	unit(s)	Total number of simulation cells
N_{pa}	unit(s)	Total number of light paths
L_t	lux	Indoor illuminance of a specific time
t_d	hours	Daylight hours
$F_{time,\%}$	%	Fraction of time for which given value of illuminance is exceeded
$F_{plane,\%}$	%	Fraction of the reference plane for target illuminance level
Avg_p	%	Average $F_{time,\%}$ in all sensor grid of approximation
Avg_s	%	Average $F_{time,\%}$ in all sensor grid of simulation
Med	%	Medium of $F_{time,\%}$ in all sensor grid of simulation/approximation
SEE		Standard Error of Estimate
RMS		Root-mean-square for $F_{time,\%}$ in all grid cells of simulation/approximation
FI	%	Fitness of an offspring, number (percentage) of qualified rooms

1. INTRODUCTION

Daylight performance is a critical issue that is closely related to building design. In the consideration of spatial organization, contemporary multi-story buildings mostly prefer to receive daylight by side windows. However, side windows would cause nonuniform daylighting since the space near the window will be too bright and others inside may be too dark. The skylight can provide such uniform lighting for indoor space (Russell 2004). According to our practice in the school buildings, rooms on the lower floor can hardly meet high-quality lighting requirements considering illuminance and uniformity only by side windows.

Uniform lighting is helpful to spaces like classrooms, especially in cities with large populations. Students sit in organized positions, often for long periods of time, without consideration of daylight conditions. Non-uniform, poor daylighting in classrooms, therefore, can have negative impacts on students' vision health and wellbeing.

On the other hand, new urban construction is always required to preserve a relatively high plot ratio through increasing the land use intensity, making it harder to ensure the daylighting of a contemporary multi-story building since skylights can only be placed on the roof. However, a more complex building form can increase roof area and allow more skylights on the lower floors. Moshe Safdie's design for Habitat 67, a modular residential building (Figure 1A) for the Montreal World Expo in 1967, is an example of such a complex building form. It creates an interesting spatial experience and diversified neighborhood. At the same time, it has more uniform daylight combining side windows and skylight.

The contemporary workflow of daylighting design with design-simulation loops is not efficient for a complex, high-density spatial organization with unpredictable daylight performance. Designers can only complete daylight analysis after the first iteration, with most analytical decisions relying on the building form without rational logics, hints, or further form-making iteration for daylighting optimization. Further complicating the workflow, each iteration of the redesign-simulation loop takes more than a day.

In this context, parametric designers suggest applying the agent-based models to simulate the complex system of building underlying the guidelines of performance-based design to improve design efficiency and accuracy. Agent-based

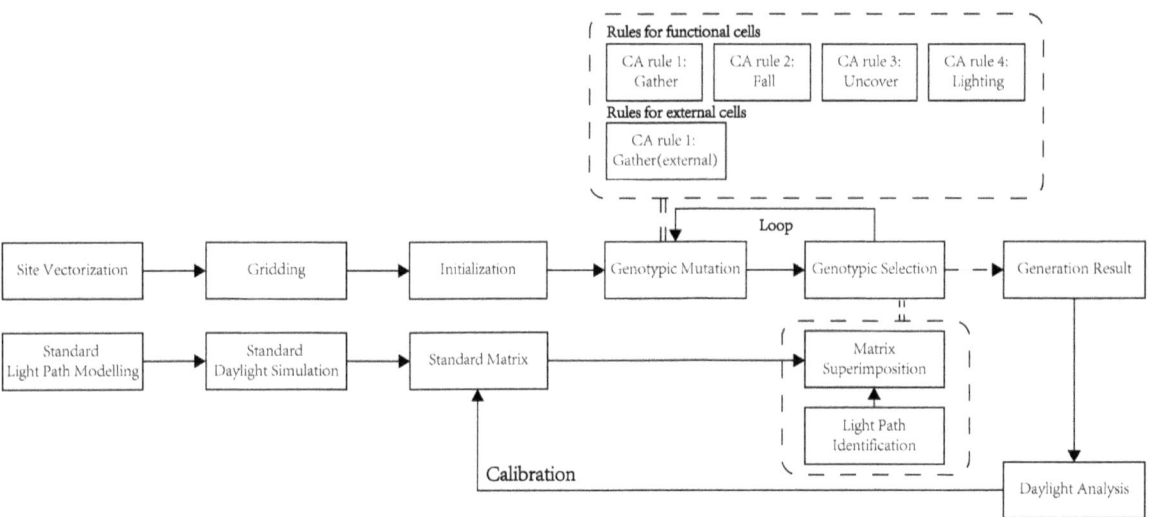

2 Program flow chart

model is a bottom-up design method that simulates the interactions among individual agents simultaneously with the setting of their relationship (Macal and North 2010). Among various kinds of agent-based models, Cellular Automata (CA) is an original agent-based model that directly regards the grid cells as agents. Since agents are spatially connected to rooms, CA has the potential to simulate complex building forms with modular rooms design considering building performance.

This article proposes to develop a complex building form generation program integrating CA and Genetic Algorithm (GA) aiming for high-density building design while considering high-quality daylighting like that of Habitat 67 (Figure 2). The program is composed by a loop of GA: 1agent) the environment is vectorized, grided, and pre-evaluated; a series of standard daylight simulations are completed and translated to matrix; 2) 500 sets of cells are initialized based on site issues and input parameters; 3) a completed GA loop is applied to optimize the sets of cells mutated following CA rules, evaluated under a quick-access approximated proxy model based on simulation matrix, and selected based on the fitness of evaluation, inbred to the next generation with 500 sets; and finally 4) output when the fitness meets the threshold.

This CA model can work on more complex rules than Elementary Cellular Automata (ECA) and is more efficient and universal than the filter methods. In the case of complex CA rules to simulate daylighting performance interaction between agents, to ensure optimization and convergence for CA, a set of CA models are evaluated, selected, and optimized in each GA population by the fitness function. A proxy model in place of real-time daylight simulation is added in fitness calculation, following building standards on daylighting to obtain the best daylight performance. This article also provides a comparison of the approximation method to daylight simulation to analyze and validate. A case study is developed on a site in southeast China, whereby a calibrated function for the method is developed based on first-time running model. The final generated result is tested by daylight simulation.

This research establishes the GA and CA model based on simulation result; therefore, it is reasonable and verifiable than those based on the abstract evaluation. It provides a better prediction for complex systems' daylight performance in the design process; furthermore, it can ensure convergence and optimization of the CA model and can have a practical, universal, and more efficient workflow than existing architectural CA methods.

2. STATE OF THE ART

CA used to be 2-dimensional in the computation model. The set of nearest cells of the tested agent, called its "neighborhoods," generally present as three types including Von Neuman Neighborhoods, Moore Neighborhoods, and Margolus Neighborhoods all of which can be extended to more dimensions (McIntosh 2008). The functions for agents' actions to follow call CA rules when mutations happen. With the agents chasing the individual best performance, the agent-based model can reach an optimized balance after competition and concession between individuals. Published research on architectural CA describe standard 2D CA models to yield a feasible workflow for building design (Watanabe 2002; Herr and Kvan 2007).

In comparison with 2D traditional CA, 3D CA model can be more flexible and feasible for complex spatial organization. Several methods have been developed to enhance convergence. For example, controller cells can be added into the model like circulation cells as the building core (Khalili Araghi and Stouffs 2015). Dillenburger et al. (2009) describe a 2.5D alternative CA model built from downstairs to upstairs to lower the CA model's complexity. Guo and Li (2017) apply a series of rules for the cells to detect neighbors below, considering structural issues and daylighting. In this research, it is difficult to quantitatively validate the advantages of the generated result, limiting the complexity of rules and efficiency of iteration and preventing the real-time simulation from taking part in rule setting.

Improvements of the CA method have been developed to enhance the rationality of the model under the limitation of convergence. One is the adaptation of ECA patterns (Zawidzki 2010) and the filter method, mostly working on façade design (Zawidzki 2015; 2017). Other improvements develop a secondary selection on their results (Kim 2013), include energy consumption and glare issues into the evaluation (Ayoub 2018), and has been repeated by several researchers (Fathy et al. 2015).

Other research tries to find assistance for complex CA models to help them be optimized and convergent like the combination with CA and Genetic Algorithm (GA) or Evolutionary Algorithm (EA). This method can adjust and optimize designed CA rules by GA before execution (Momeni and Antipova 2020) or apply GA as assistance to get convergence and optimization for designed CA rules (Ye et al. 2020). Other research integrating EA and 2.5D CA has also been explored (Dillenburger et al. 2009).

3. METHODOLOGY

This method is used to generate models for high-density building forms with high-quality daylighting as a guideline for performance-based design. Daylight simulation results are added to the program to build a quick-access approximation to daylighting in the genotype selection process which guides the convergence of the CA model.

The entire process of the program can be explained as follows: (1) in each population the program has 500 models that are given the initial conditions in a set of 25 models; (2) the program applies 100 transformations of the form for each model following four CA rules in the genetic mutation step; (3) the fitness of 500 models is evaluated under a quick-access daylighting prediction method; and (4) the best 5% models are maintained to set up the initial condition for the next population.

This study consists of four parts:

1. A generation algorithm combining CA and GA logic (Section 3.1)
2. A quick-access daylight evaluation as the fitness function in GA (Section 3.2)
3. Validation and Calibration of the evaluation method by running it on the first-time running model; a floor-level parameter (also a 6x6 grid) is created according to the Standard Error of Estimate (SEE) to fulfill the generation algorithm (Section 4 of this paper).
4. A case study located in southeast China is established under our building form generation program (Section 4 of this paper).

3.1 GA Mutation Setting: CA Model and Rules

In this research, GA is used to ensure optimization and convergence of the CA model. There are several definitions of GA models, as follows.

CELLS: Each room in the building mass is considered as one cell, either activated or inactivated.

CHILDREN: In this program, the genotypes are the state of cells (agents), therefore one child is one CA model composed of agents.

POPULATION: A set of children run through an entire GA loop including genotype inbreed, genotype mutation, fitness calculation, genotype selection calls a group of Population.

MUTATION: One change on a child is called a mutation. Each mutation includes hundreds of transformations of the models following CA rules.

FITNESS: The score of a child that will decide whether the children is to be maintained, inbred, or deleted.

MAINTENANCE: Top models ranked by fitness can be maintained for the next population, and in the next loop they will not have a mutation and directly attend the fitness calculation in order to avoid reversal iteration of the program.

INBREEDING: Another set of top models (bigger than the range of maintenance) will be cloned several times as the initial condition of children for the next population.

This program sets 500 mutations, 500 children, 5% maintenance, and 20% inbreeding in our GA program to get a better algorithm efficiency.

Facing the need for daylight simulation, this experiment chooses to set our CA model in a three-dimensional space. With the help of GA, the chaotic property of 3D Cellular can be converged and optimized under several simple rules. The aim of genotypic mutation, following the CA rules, is to add randomness under fundamental control on structure and space. The program sets each cell as an 8 meter by 8 meter square with a height of 3.5 meters; all faces that do not connect with other cells are automatically allocated a horizonal window—0.35 WWR (Window to Wall Ratio) for skylights facing south, 0.5 WWR for side windows—for us to simulate a maximal daylight

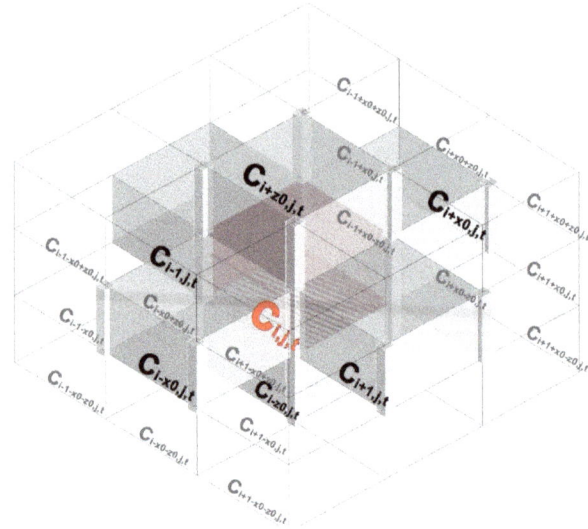

3 Neighborhood definition

TABLE 1 CA rules with 2 kinds of cells

No.	Probability	Description
Functional Cells		$C_{i,j,t} = 1$
1	80%	Gather: demolish* if $S_1 < 2$
2	100%	Block: 100% demolish* if $S_1 > 6$
3	100%	Fall: demolish* if $C_{i-z0,j,t} = 0$ & $S_1 < 3$
4	100%	Uncover: demolish* if $C_{i-z0,j,t} = -1$
5	100%	Daylighting: demolish* if light paths are less than 1
External Cells		$C_{i,j,t} = -1$
1	1-Block time%	Gather: demolish* if
Empty Cells		$C_{i,j,t} = 0$
		None.

*Demolish: inactivate the functional/external cell and randomly activate another one which is an empty cell. external cells are forced on the first floor.

*Block time means the percentage of time when the position of the cell is shaded by the surrounding building.

*All rules' probabilities are tested with their convergence and randomness.

indoor. The program keeps the same WWR of 0.5 for all windows and 0.3 WWR for skylights.

This program extends the classic Von Neumann CA (Macal and North 2010) into 3D space with two new neighbors at the top and bottom (Figure 3). The program contains 3 states of cells and 6 rules (Table 1), and all basic rules are extended from the classic 2D CA model into a 3D space.

For each GA population, 500 mutations under CA rules are applied to each child. Generally, in the mutation $t \in [0,499]$, each cells $i \in [0, N_t]$ in children $j \in [0,99]$ named $C_{i,j,t} = \{-1,0,1\}$ would check the state of its neighbor sets. $C_{i,j,t} = 0$ means the cell is inactive, $C_{i,j,t} = 1$ means the cell is a functional cell, $C_{i,j,t} = -1$ means the cell is an external cell. The neighbor sets of $C_{i,j,t}$ is:

$$L = \{C_{i-1,j,t}, C_{i+1,j,t}, C_{i-x0,j,t}, C_{i+x0,j,t}, C_{i-z0,j,t}, C_{i+z0,j,t}\}$$

Sets of neighbors' states S_1, S_2 will be calculated to decide the action $C_{i,j,t}$ should take to iterate into $C_{i,j,t+1}$:

$$S_1 = \sum_c^6 IFF[L(C) = 1,1,0]$$

$$S_2 = \sum_c^6 IFF[L(C) = -1,1,0]$$

External cells are used to control building density and represent the shading of surrounding building. All functional cells will not be reactivated above them. Its demolished action is controlled by percentage of blocked time based on a direct sunlight time simulation of the site. For example, if sunlight on the cell is blocked in 75% of time annually, the External Cell on this location will have only 25% of a possibility to be moved away.

For functional cells, a designed rule is added to ensure their daylighting. If none of their windows are presented in the standard lighting condition in Figure 5, the cell will be 100% demolished. The searching for the light path is implemented by the depth-first algorithm (Peng et al. 2017), and the program uses the same algorithm with a recorder list in fitness calculation to traverse all light paths for superimposition.

3.2 GA Fitness Function: Daylight Evaluation

To establish our fitness function, this article looks into published standards and research about daylighting. Illuminating Engineering Society of North America (IESNA 2013) provides different methods include Spatial Daylight Autonomy (sDA) and Annual Sunlight Exposure (ASE) for ranking indoor daylight performance. Its standard is mostly based on $sDA_{300/50\%}$. IESNA suggests $sDA_{300/50\%}$ meet or exceed 55% as "Nominally Accepted Daylight Sufficiency."

Li and Tsang (2008) suggest average Daylight Factor (DF) and daylight glare index (DGI) as the parameter. Nabil and Mardaljevic (2006) suggests using sDA to replace DF as daylight illuminance

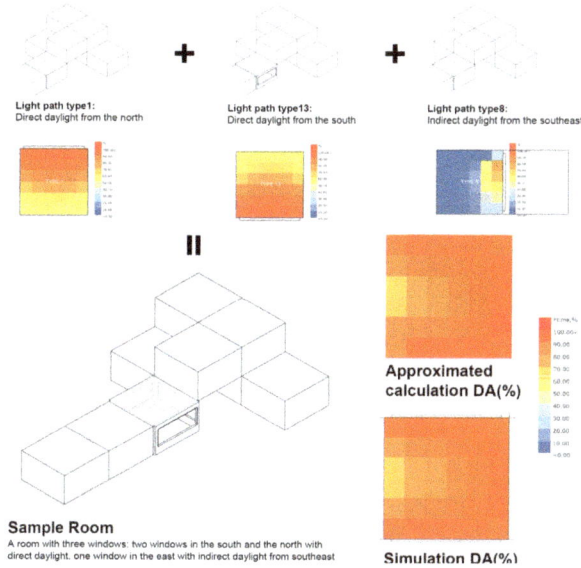

4 A case of superimposition with different windows/light sources

paradigm. The same daylight standard can be found in most of the research about daylighting and CA (Ayoub 2018), where the solar heat gain due to sunlight is inevitable and should be considered when designing energy-efficient façades. This research explores the potential contribution of utilizing monotonous-free Elementary Cellular Automata patterns as climate-adaptive shading systems, to be applied on buildings' façades in order to mitigate the undesirable impacts by excessive solar penetration in cooling-dominant climates. It also presents a new approach for evaluating the daylighting performance and energy demand for the dynamic shading systems at the early stages of design. Grasshopper is exploited for parametric modeling of Elementary Cellular Automata patterns. The methodological procedure is realized through two main phases. The first evaluates all 256 Elementary Cellular Automata possible rules to elect the ones with random patterns, and to ensure an equitable distribution of the natural daylight in internal spaces. The computational simulations are then conducted in the second phase using DIVA-for-Rhino and Archsim to evaluate the performance of the elected Elementary Cellular Automata patterns that are applied as dynamic shadings, based on the newly developed hourly-based metrics: Hourly Daylight Illuminance (HDI300/HOY. In this context, this program decides to set our original standard as $sDA_{300/50\%}$ which can be described as below:

$$F_{plane,\%} = \frac{\sum_{i=1}^{N} IFF[(F_{time,\%})_i > 50\%, 1, 0]}{N} *100\% \quad N=36$$

The method integrating GA and CA will improve the running time of the algorithm. Therefore, it is impossible to directly use real-time simulation in our program. To improve the efficiency, this article provides a quick-access, approximation method based on the pre-simulated result instead of real-time daylight simulation. In the CA model, the agents' lighting conditions, with several windows open, are composed of a series of "standard conditions." Each standard condition has only one way (one sunlight direction and one window open) for the daylight to come into the room. Those standard conditions are regarded as the light paths.

Light paths are distinguished in 30 styles with four directions and skylight; indirect daylighting is also defined by its window, including 4 directions, skylight, and 25 different indirect daylighting styles. All light paths run a daylight simulation in a 6 by 6 sensor grid before the generation program and get the statistical results with Daylight Autonomy which is presented as $F_{time,\%}$ in this article.

We develop a formula to superimpose the standard sensor grid to approximately calculate the agent's daylight performance in $F_{time,\%}$. The program calculates the spatial Daylight Autonomy (sDA) of each agent based on that $F_{time,\%}$ grid for daylight evaluation. For superimposition of light path sensor grid, the program uses the following formula to calculate final approximation 6 by 6 sensor grid for $F_{time,\%}$:

$$(F_{time,\%})_i = 1 - \prod_{pa=1}^{N_{pa}}[1-(F_{time,\%})_{i,pa}]$$

This calculation is based on the probability of radiated time for a sensor point. Daylight from multiple window opening sides can be approximately divided according to windows. Take an example of a room that has a window facing south, a window facing north, and a window facing east which gets indirect daylight from the southeast (Figure 4). If the sensor point can receive acceptable radiation $F_{pa}\%$ annually by the north window, $F_{pb}\%$ by the south window, and $F_{pc}\%$ by the east window, the total probability for it to get acceptable daylight should be:

$$F_p\% = (1-F_{pa}\%)*(1-F_{pb}\%)*(1-F_{pc}\%)$$

This presents errors in several aspects: (1) the secondary reflection from the window (originally as an entire wall) on another side is ignored; and (2) indirect sunlight from more than 8 meters away is ignored. This has a significant difference for daylighting from the skylight. After getting all approximated DA grids, our fitness function counts up the spatial acceptable percentage of cells which meet the threshold of $F_{time,\%} > ü$, presented in $F_{plane,\%}$.

$$F_{plane,\%} = \frac{\sum_{i=1}^{N} IFF\left[1-\prod_{pa=1}^{N_{pa}}[1-(F_{time,\%})_{i,pa}] > 50\%, 1, 0\right]}{N}*100\%$$

$$FI \aleph \quad IFF\left[F_{plane,\%} \quad 55\%, 1, 0\right]$$

We mandatorily regard all standard cells as first-floor cells in the first-time running and calibrate the method using

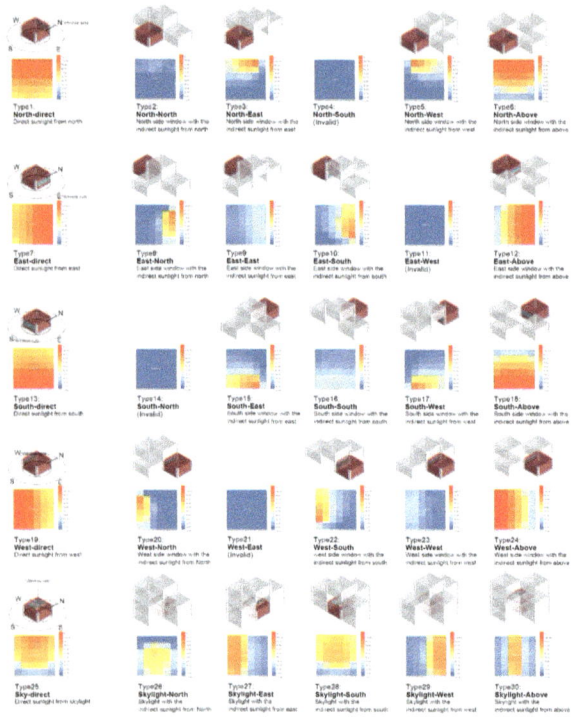

5

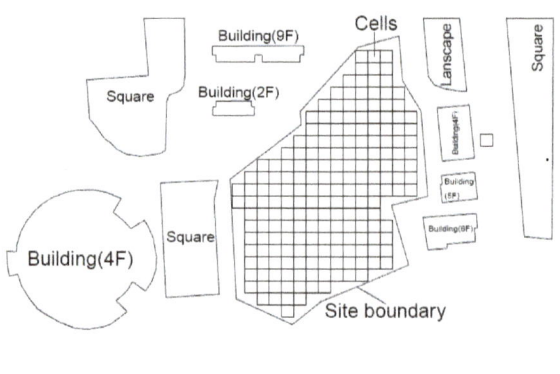

6.1

6.2

first-time running result. The post-simulation (Table 2) of light path (Figure 5) use the Radiance engine (Kharvari 2020) that has high flexibility to attend any step on computational design (Futrell, Ozelkan, and Brentrup 2015).

TABLE 2 General setting of daylight simulation

Opaque material	reflectivity	0.7
	roughness	0.2
Transparent material	reflectivity	0.31
	Transparency	0.69
Simulation parameter	Location	Guangdong. Guangzhou.592870
	Grid	6 x 6
	-ab	2
	-ad	5000
	-lw	2e-05

4. RESULT VALIDATION AND CALIBRATION
4.1 Example Site

The program is practiced on an example site located in Southeast China. It has an irregular site boundary surrounded by residences about 20 meters high. The program sets a relatively high plot ratio of 2.0, 2.5 and 3.0 in this area, with a 21 meters limitation of the building height. Site and first-time running building mass are presented in Figure 6.

In the first-time running model, the prediction errors are mostly caused by the regardlessness of different daylighting of different building levels. Therefore, this research develops the floor-level parameters set through validation and calibration. The program first applies our approximation method to get an approximated series of $F_{time,\%}$ grids for all the rooms; then uses Honeybee to simulate another series of $F_{time,\%}$ grids, which is on behalf of real performance.

4.2 Error Analysis

The Standard Error of Estimate (SEE) is an error definition that first comes from liner then polynomial regression. It can more correctly describe the discreteness of prediction with identification for extreme data. It is the square root of the average squared deviation:

$$SEE = \sqrt{\frac{\sum^{N}[(F_{time,\%})_{predition} - (F_{time,\%})_{simulation}]^2}{N}}$$

*N is the total number of simulated squares

The SEE analysis with predicted and simulated grides by floor on all cells of the building is completed and presented in Figure 7. For our three testing models (with the plot ratio of 2.0, 2.5, and 3.0), the overall SEEs are 27.38, 23.52, 24.94. The average SEE for three cases together is 25.12.

Most of the errors happen on the first two floors (Figure 8) for a six-story building. Errors occur mainly for three reasons: (1) it has failed to detect low indirect daylighting and has

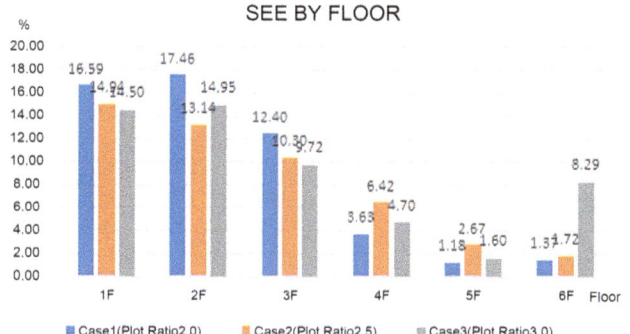

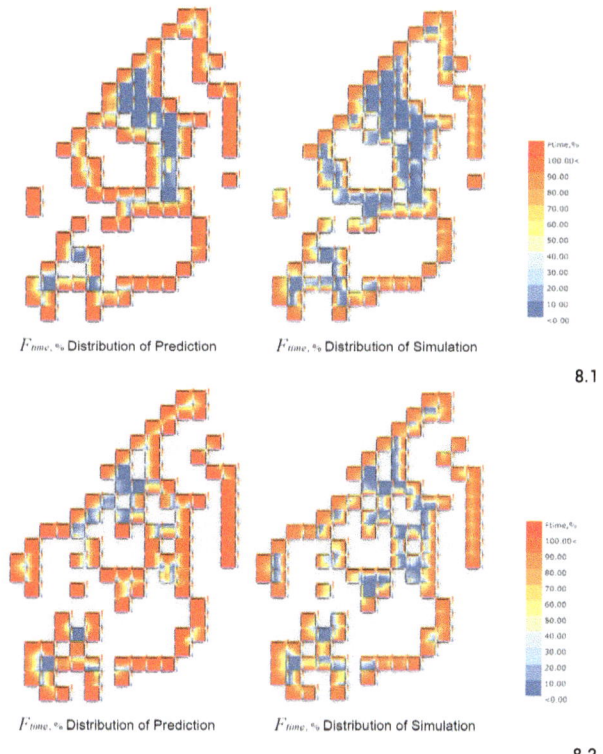

$F_{time,\%}$ Distribution of Prediction $F_{time,\%}$ Distribution of Simulation

8.1

$F_{time,\%}$ Distribution of Prediction $F_{time,\%}$ Distribution of Simulation

8.2

5 Light path and annual simulation result, presented in DA300
6.1 Case study site and griding
6.2 First-time generative result
7 SEE for 3 cases with different plot ratio
8.1 DA comparison between prediction and simulation for first floor
8.2 DA comparison between prediction and simulation for second floor

overestimated the daylight performance of skylight; (2) some rooms with side windows present higher daylight performance in prediction than the simulation; (3) contrast of an individual room is relatively higher than simulation.

There are two explanations to the errors: (1) according to our daylight prediction method, sunlight from a distance over 16 meters (two grid cells) will be ignored; and (2) errors on side windows and errors on contrast are mostly caused by building floor differences, suggesting a calibration to average and contrast on the results.

4.3 Calibration with Floor-level Parameter

Floor-level parameters are developed to calibrate the average and contract of $F_{time,\%}$ distribution. In our research the contrast can be represented by Root Mean Square Error (RMS):

$$RMS = \sqrt{\sum_{i=1}^{N}[(F_{time,\%})_i - Avg]^2 / N} \quad N = 36$$

The program transfers the RMS and average from simulation and prediction result into the baseline calibration formula:

$$(F_{time,\%})_{new} = \left[F_{time,\%} - Avg_p\right] * RMS_s / RMS_p + Avg_p + Med_s - Med_p$$

To avoid extreme numbers (especially the sensor grid of a dark room), the program also takes medium instead of average to match the light of approximation to simulate. All parameters for calibration are listed by floor in Table 3.

TABLE 3 Floor parameters set

	Floor 1	Floor 2	Floor 3	Floor 4	Floor 5	Floor 6
ü$_s$	21.69183	19.60209	17.11041	8.58694	1.74371	4.0862
ü$_p$	17.44286	16.14199	15.86402	6.74735	1.060228	5.266299
ü$_p$	0.680423	0.743702	0.775155	0.928021	0.989071	0.84088
ü$_p$	0.876463	0.889187	0.91976	0.979428	0.998476	0.915417
ü$_s$	0.598506	0.764384	0.824907	0.862017	0.914321	0.865006

According to the histogram of the predicted values (Figure 9), different levels of lighting intensity have a different deviation to simulation. The program multiplies to fix the predicted to the simulation according to its initial level:

$$(F_{time,\%})_{new} = F_{time,\%} + \lambda_i \left[\frac{RMS_s - RMS_p}{RMS_p} * \left(F_{time,\%} - Avg_p\right) + \left(Med_s - Med_p\right)\right]$$

λ_i is calculated from the difference of histogram and the average $F_{time,\%}$ by floor in prediction and simulation (Table 4).

A comparison of SEE and first-floor deviation is established after sending the calibrated grid to the approximate function (Figure 10). Over 90% of cells obtain the distribution with a

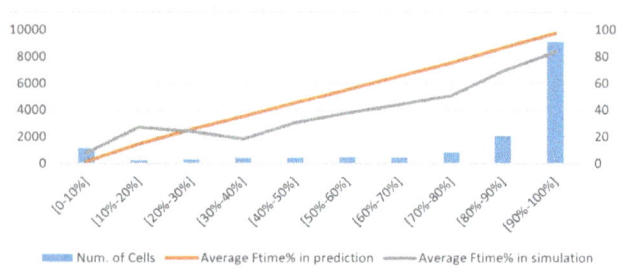

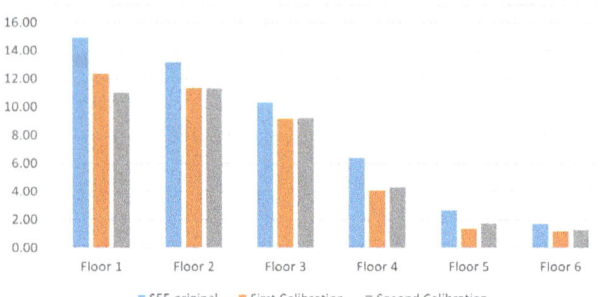

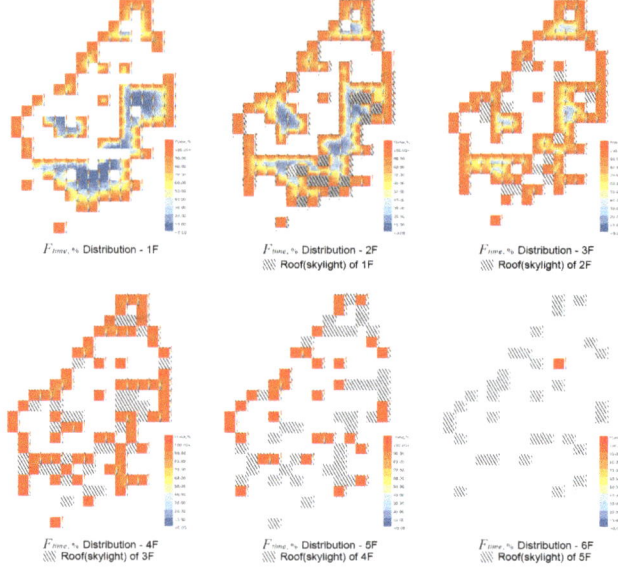

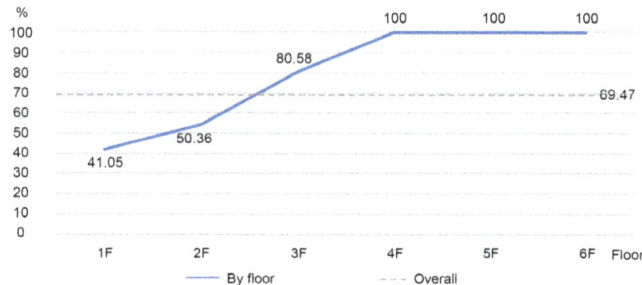

9 Histogram and difference of DA300
10 SEE before and after calibration
11 Daylighting analysis for Calibrated Result
12 Eligible rooms

deviation lower than 10%, which is acceptable in programming design, and the accuracy of prediction reach to 97% on the upper floor with a set of relatively high building density and plot ratio. Therefore, in the case of building lower than 15 meters and four floors, this method can provide a high-quality approximation, which can also strikingly reduce hardware and runtime requirements in CA and GA programs for architectural design.

TABLE 4 Histogram and difference of $F_{time,\%}$

	[0-10%]	[10%-20%]	[20%-30%]	[30%-40%]	[40%-50%]	[50%-60%]	[60%-70%]	[70%-80%]	[80%-90%]	[90%-100%]
Count	1142	251	301	408	417	473	461	834	2048	9109
Average in prediction	1.41	14.28	25.35	35.00	45.01	55.20	65.41	75.05	86.23	97.20
Average in simulation	7.91	27.16	24.07	18.19	30.39	37.73	44.01	51.06	69.06	83.58
λ	0.27	0.54	0.05	0.70	0.61	0.73	0.89	1.00	0.72	0.57

After calibration, a second-running of the program stops at the 250th gender. Our model reaches 66.2% of FI in an exceedingly early stage at about 6000 mutations (60th GA genders). Then, the rest of 5% preserve control groups are gradually optimized to 66.2% after about 25000 mutations (250th GA Genders). Thus, 66.2% is the upper limit of this generation process. 69% of spaces according to the simulation, and 66% of spaces according to the approximation meet the standard of $ü_{300/50\%}$ >55% which can be regarded as an acceptable room (Figure 11).

Further, the reliability and efficiency of the proxy model is tested compared to a typical model for daylight simulation using Artificial Neutral Network (ANN) (He et al. 2021). Error for sDA is presented in MSE and R2, and simulation/prediction computation time is presented in ms/grid. It also compared with Radiance real-time simulation (Table 5).

TABLE 5 Comparative study on methods (He et al. 2021)

Method	Error on sDA		Duration(ms/grid)	
	R2	MSE	Annual	Point-in-time
CNN+GAN	0.886(train)	0.020(train)	0.062	0.062
	0.650(test)	0.076(test)		
Mathematic Proxy	0.673	0.032	0.019	0.019
Simulation	1.000	0.000	4.581	1.145

5. CONCLUSION AND DISCUSSION

This research integrates CA and GA to provide a generation algorithm work on complex high-density building form design considering high-quality daylighting. GA provides convergence and optimization on daylighting to the CA model, by applying daylight simulation results on the fitness calculation process in GA. In this case, daylighting design of complex building form can be guided the outcome of this generation program that is validated by real simulation data. In the case study, our generation result can ensure about 69% of eligible room to meet the requirement of $ü_{300/50\%} > 55\%$, under the setting of building density of 55%, six-story building with a plot ratio of 2.5. For the reference in real constructions which have similar density with 1.5 plot ratio, the general building design can only provide about 25% to 30% of eligible areas with uniform daylighting based on the standard of China.

This research also needs improvement in several aspects. Firstly, the research needs a better solution on surrounding buildings and the environment. In the original design of our CA model, External Cells are set to ensure building density and to represent the shading of a neighbor building. However, cells' actions still show high randomness, in other words, the model on external cells lacks in optimization. Adding an independent fitness function on external cells in GA should be the best way to improve this issue.

Secondly, this program inclines to sacrifice the first-floor space to get a maximal number of eligible rooms (Figure 12). To ensure daylighting for first-floor rooms is relatively harder than in upper level rooms. By massive trial of GA, the probability of asymmetry gradually grows into inclination. Besides, the first-floor rooms are always attributed to lower daylight in a normal building unless the designers bring a solution to this problem. It suggests that other improvements in design should be added into our model to help it exceed the 70% threshold for FI.

Designers can solve this problem in public buildings by setting artificial lighting. If the inclination should be calibrated, the program can have several improvements: (1) set mandatory regulations to control the rooms' density by floor; (2) decrease the predicted sDA of lower floor to preserve more cells on lower floor; and (3) iterate the third CA rule (Table 2), the "Fall" rule to prevent cells dropping to the lower floor.

Thirdly, the calibration method, aiming for the floor-level parameters, inevitably underestimates the complexity of lighting on the lower floor. That is because the calibration is an abstract mathematical translation towards the predicted sensor grid. Although it shows well consistency in statistics of data for approximation and simulation, their $F_{time,\%}$ grids still have differences caused by light attenuation because of distance from the window and bright light reflection. In general, daylighting of an indoor room cannot be easily judged by light average and contrast.

Two improvements can be tried on the model. Firstly, the light paths can be elaborated considering the level issue, function, and even changeable cell size. Secondly, this model can use machine learning instead of mathematical calibration to get the floor-level parameters. Because the learning sample can be directly generated by our agent-based model, theoretically they can be built into a loop to work automatically.

ACKNOWLEDGMENTS

This research is supported by National Natural Science Foundation of China (No. 51978268 & No. 51978269) and State Key Lab of Subtropical Building Science (No. 2019ZA01).

REFERENCES

Ayoub, M. 2018. "Integrating illuminance and energy evaluations of cellular automata controlled dynamic shading system using new hourly-based metrics." *Solar Energy* 170 (April): 336–351.

British Standards Institution, European Committee for Standardization, & DIN Deutsches Institut für Normung. 2019. *Daylight in Buildings*.

Celik, T. 2014. "Spatial Entropy-Based Global and Local Image Contrast Enhancement." *IEEE Transactions on Image Processing* 23 (12): 5298–5308.

Dillenburger, B., M. Braach, and L. Hovestadt. 2009. "Building design as an individual compromise between qualities and costs: A general approach for automated building generation under permanent cost and quality control." In *Joining Languages, Cultures and Visions - CAADFutures 2009, Proceedings of the 13th International CAAD Futures Conference*. Montreal: Presses de l'Université de Montréal. 458–471.

Fathy, F., Y. Mansour, H. Sabry, S. Abdelmohsen, and A. Wagdy. 2015. Cellular automata for efficient daylighting performance: Optimized façade treatment." In *14th International Conference of IBPSA; Building Simulation 2015, BS 2015, Conference Proceedings*. Hyderabad: International Building Performance Simulation Association. 2705–2711.

Futrell, B. J., E. C. Ozelkan, and D. Brentrup. 2015. "Optimizing complex building design for annual daylighting performance and evaluation of optimization algorithms." *Energy and Buildings* 92 (1): 234–245.

Guo, Z., and B. Li. 2017. "Evolutionary approach for spatial architecture layout design enhanced by an agent-based topology finding system." *Frontiers of Architectural Research* 6 (1): 53–62.

He, Qiushi, Ziwei Li, Wen Gao, Hongzhong Chen, Xiaoying Wu, Xiaoxi Cheng, and Borong Lin. 2021. "Predictive Models for Daylight Performance of General Floorplans Based on CNN and GAN: A Proof-of-Concept Study." *Building and Environment* 206 (December): 108346.

Herr, C. M., and T. Kvan. 2007. "Adapting cellular automata to support the architectural design process." *Automation in Construction* 16 (1): 61–69.

Illuminating Engineering Society of North America (IESNA). 2013. *IES LM-83-12 IES Spatial Daylight Autonomy (sDA) and Annual Sunlight Exposure (ASE)*. New York: IESNA.

Khalili Araghi, S., and R. Stouffs. 2015. "Exploring cellular automata for high density residential building form generation." *Automation in Construction* 49: 152–162.

Kharvari, F. 2020. An empirical validation of daylighting tools: Assessing radiance parameters and simulation settings in Ladybug and Honeybee against field measurements. *Solar Energy* 207: 1021–1036.

Kim, J. 2015. "Adaptive façade design for the daylighting performance in an office building: The investigation of an opening design strategy with cellular automata." *International Journal of Low-Carbon Technologies* 10 (2015): 313–320.

Legault, R. 2021. "The Making of Habitat 67: A Tense Pas de Deux between Moshe Safdie and August Komendant. *Journal of the Society for the Study of Architecture in Canada* 46 (1): 30–50.

Li, D. H. W., and E. K. W. Tsang. 2008. "An analysis of daylighting performance for office buildings in Hong Kong. *Building and Environment* 43 (9): 1446–1458.

Macal, C. M., and M. J. North. 2010. "Tutorial on agent-based modelling and simulation." *Journal of Simulation* 4 (3): 151–162.

McIntosh, H. 2008. "Discrete Tools in Cellular Automata Theory Introduction." *Journal of Cellular Automata* 3 (3): 181–186.

Momeni, E., and A. Antipova. 2020. "Pattern-based calibration of cellular automata by genetic algorithm and Shannon relative entropy." *Transactions in GIS* 24 (6): 1447–1463.

Nabil, A., and J. Mardaljevic. 2006. "Useful daylight illuminances: A replacement for daylight factors." *Energy and Buildings* 38 (7): 905–913.

Peng, W., N. Peng, K. Ng, K. Tanaka, and Y. Yang. 2017. Optimal depth-first algorithms and equilibria of independent distributions on multi-branching trees. Information Processing Letters, 125, 41–45.

Russell, P. L. 2004. *Guide for Daylighting Schools*. Troy, New York: Lighting Research Center, Rensselaer Polytechnic Institute.

Treado, S., G. Gillette, and T. Kusuda. 1984. "Daylighting with windows, skylights, and clerestories." *Energy and Buildings* 6 (4): 319–330.

Watanabe, M. S. 2002. *Induction Design: A Method for Evolutionary Design*. Basel: Birkhäuser.

Zawidzki, Machi. 2010. "A Cellular Automaton Controlled Shading for a Building Facade." In *Cellular Automata*, edited by S. Bandini, S. Manzoni, H. Umeo, and G. Vizzari. Berlin, Heidelberg: Springer. 365–372.

Zawidzki, Machi. 2015. "Emergence in the modular shading system based on cellular automaton." *Advances in Building Energy Research* 9 (2): 280–292.

Zawidzki, Machi. 2017. "Cellular Automaton-Based Shading System (CASS)." In *SpringerBriefs in Architectural Design and Technology*. Singapore: Springer. 15–61.

The Potential of Mitigating Urban Heat Island with Vacant Lands in Philadelphia

Hui Tian
Stuart Weitzman School of Design
University of Pennsylvania

Jiali Yao
Stuart Weitzman School of Design
University of Pennsylvania

Shimin Tu
Stuart Weitzman School of Design
University of Pennsylvania

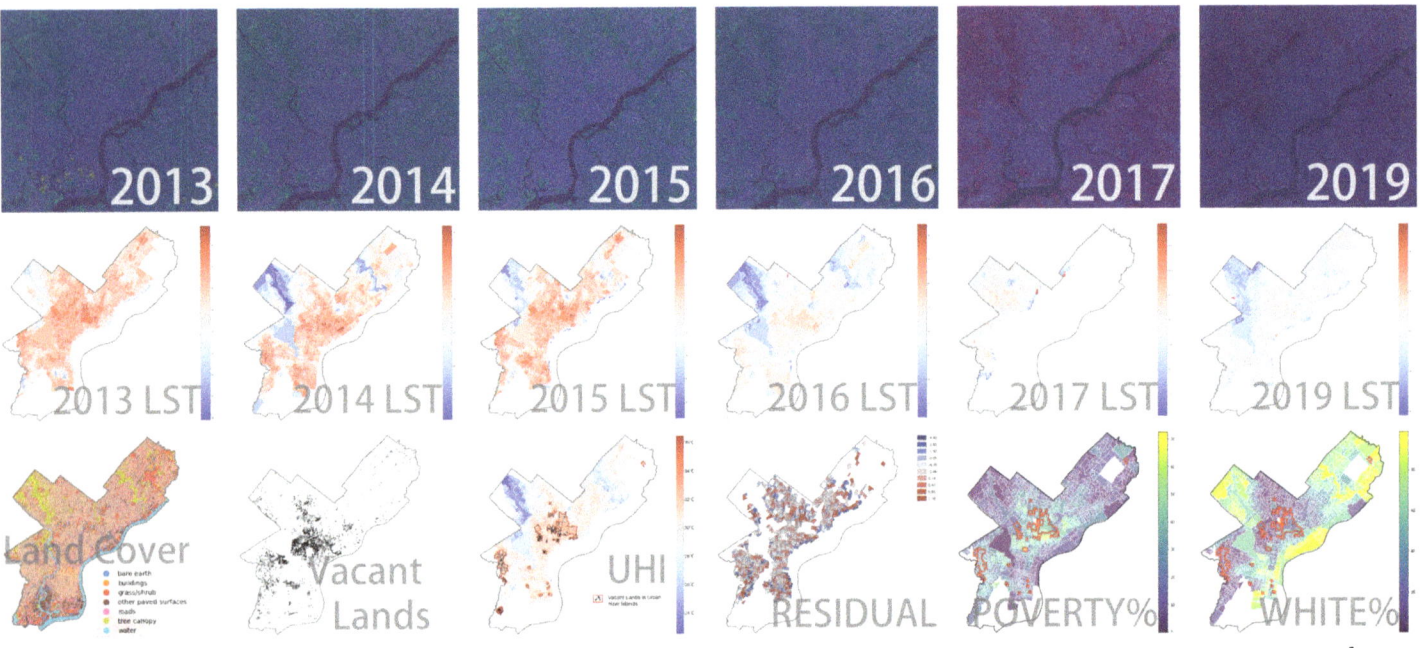

1 The raster datasets for training machine learning models and processing analytics

ABSTRACT

Under the context of climate change, the urban heat island (UHI) is a challenging problem in Philadelphia as the number of days with extreme heat every year keeps increasing. Taking into account limited green space but a considerable amount of vacant lands in Philadelphia, we would test the cooling effect of greening vacant lands in UHI by exploring the quantitative relationship between land covers and Land Surface Temperature (LST) with novel machine learning technologies. Firstly, we summarized Landsat 8 raster datasets to average LST on two geometry systems: Census Block Group (CBD) and vacant lands in Philadelphia. Then different types of land covers were spatially joined in geometries as predicting features to fit five types of regression models with LST as the target variables. The Geographically Weighted Regression (GWR) performed best to predict LST in the CBG unit with the highest regression coefficient R^2 of 0.79. Limited to the granularity of Landsat images and low proportion of vacant lands in most CBG areas, the GWR model was not able to capture apparent temperature changes when simulating all land covers in vacant lands to tree canopies. However, observed LSTs of vacant lands are cooler than the ones of surrounding CBGs, by an average of ~1.9°C, with a range of 0°C to 8.7°C, which may be attributed to their high vegetation cover. This study revealed a quantitative relationship between land covers and average LST with the GWR algorithm. It also proved existing vacant lands in UHI exerted cooling effects of ~2.1°C within their boundaries, compared to surrounding CBG. In terms of positive correlation between LST and social vulnerability in UHI, considering percentage of non-white, poverty, and vacant houses, strategical planting should be designed at micro-scale with simulation tools to reach a better cooling effect in the future.

INTRODUCTION

Problem Statement

Heat islands are urbanized areas that experience higher temperatures than outlying areas. The phenomenon usually occurs when natural land cover is replaced with dense concentrations of pavement, buildings, and other surfaces that absorb and retain heat (Hulley 2012). Urban heat island (UHI) effect increases energy costs, air pollution levels, and heat-related illness and mortality (Hammer et al. 2020). Under the context of climate change and global warming, urban areas will suffer higher temperature and harsher heat in UHI. It is a challenging urban problem that not only impacts sustainability and human health, but also reflects social, racial, and economic inequalities associated with disproportionate lack of green spaces in cities.

Background

Multiple studies of urban heat island indicate green space can mitigate urban heat island effect by creating cooling buffer zones. For instance, Chow's study indicates that green space air temperature can be 1–3°C, and sometimes even 5–7°C, cooler than surrounding built-up areas (Chow et al. 2010). Park's research in South Korea found that small green spaces can reduce air temperature of an urban block, with the polygonal type having a better cooling effect (Park et al. 2017). Pearsall analyzed the socio-spatial pattern in land surface temperature and vacant land across Philadelphia, and indicated the opportunities of converting vacant land to green space to reduce urban heat island effect (Pearsall 2017). Recent research from Mitz also explored the relationship between urban landscape and surface temperature in a case study of Philadelphia (Mitz et al. 2021). However, little research conducted a quantitative analysis of the mitigating effect of greenery vacant lands.

Project Goal

Against the backdrops of existing literature, this study aimed to examine the quantitative relationship between land cover and LST, as well as predict the cooling effect of converting built-up areas to green spaces with machine learning technologies. Philadelphia, a former major industrial center in the United States, contains vacant land as large as approximately 428 acres in the city (City of Philadelphia 2016). Under the context of climate change, extreme heat events have increased significantly every year in urban areas of Philadelphia. Though green spaces could cool down LST compared to paved areas, there are limited spaces and resources to develop new parks and recreational open spaces across the city. Thus, we would like to test the cooling effect of greening vacant lands in UHI by exploring the quantitative relationship between land covers and LST.

In addition, we explored the correlation between average LST and the socioeconomic status in CBG units, considering percentage of non-white, poverty, and vacant houses, and potential of utilizing vacant lands to serve for neighborhoods which have a higher, on average chance of being more sensitive to UHI effect than others.

METHOD

Project Workflow

The workflow of this study includes five processes: data collection and wrangling, exploratory data analysis, training machine learning models, comparing predicted and observed LST in original and simulated status, and discussion of social vulnerability and LST in UHI areas (Figure 2).

Remote Sensing Data Collection

- Landsat 8 satellite images with clouds less than 5% of Philadelphia were downloaded in resolution 40 m/pixel with dates of 20130820, 20140807, 20150725, 20160828, 20170730, and 20190922 (United States Geological Survey n.d.).
- American Community Survey (ACS2019) was downloaded from the U.S. Census Bureau API (United States Census Bureau n.d.).
- Vacant Property Indicators - Land (2016) was downloaded from Open Data Philly (City of Philadelphia 2016).
- Philadelphia Land Cover (2018) was downloaded from Open Data Philly (City of Philadelphia 2018).

Land Surface Temperature Calculation

The LST is generated with the algorithm of mapping LST from Landsat 8 thermal infrared sensor Band4 (Red), Band5 (NIR), and Band10 (Avdan and Jovanovska 2016).

After converting Landsat 8 images to LST maps (Figure 3), we spatially joined LST to geometry of the census block group and vacant lands in Philadelphia. Since census data were unavailable at a finer scale, we employed Census Block Group (CBG) as a unit to represent average land surface temperature in neighborhoods. The lowest LST of a CBG was 15°C, which looked normal. The extreme temperatures higher than 50°C were removed as outliers, which took up less than 15% in data. We removed the Landsat 8 images of 20170730 and 20190922 from datasets, because most data of the former one was identified as outliers and the average temperature of the later one has high deviations from the rest satellite images. The average LST of CBG in Philadelphia was 30.38°C, with a range of 22.54°C to 36.47°C. On the other hand, we also aggregated the average LST of vacant lands as 28.74°C, with a range of 21.36°C to 31.91°C, for later comparison.

Land Cover Vectorization

The raster data of Philadelphia Land Cover identifies seven categories: tree canopy, grass/shrub, bare earth, water,

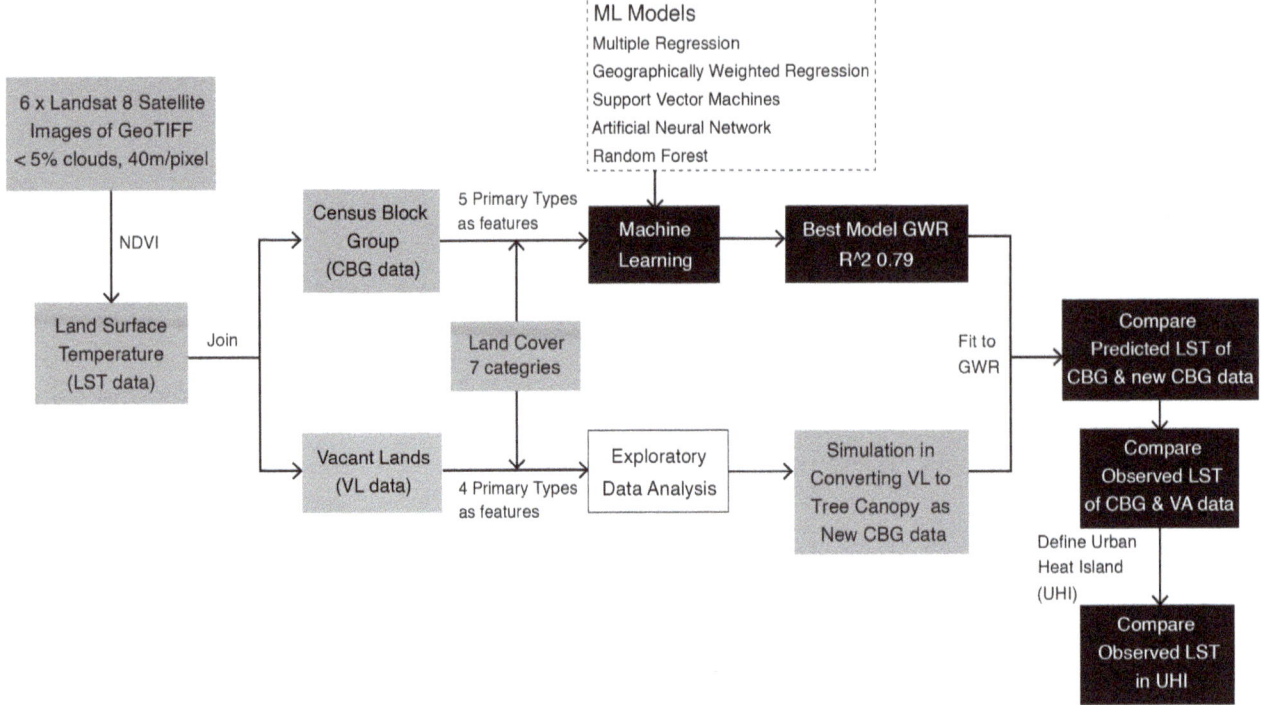

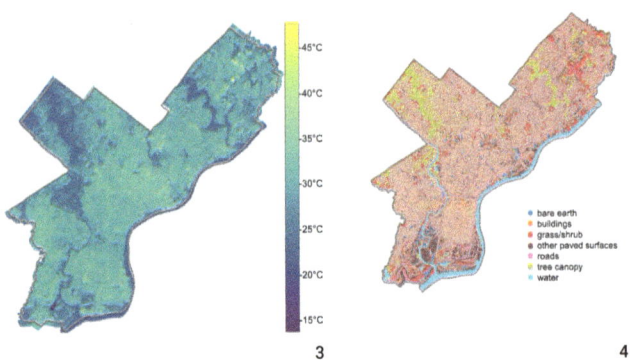

2 Workflow of technology
3 Land Surface Temperature map of Philadelphia on August 28, 2016
4 Land Cover dataset of Philadelphia in 2018

buildings, roads, and other paved surfaces (Figure 4). We converted it to vector polygons in ArcGIS Pro (version 2.8.6), and spatially joined them to CBG and Vacant Lands geometries. Looking at Figure 3 of LST and the land cover map, we found that urban heat islands always appear in areas with high density of buildings and low coverage of tree canopies.

Prepare Data to Train Models

DEFINE FEATURES

We have two datasets: the CBG data includes average LST and areas of different land covers in each CBG geometry; the vacant lands data covers average LST and areas of different land covers in each vacant land geometry. With the CBG dataset, we prepared the percentage of land covers as continuous features (independent variables): percent of tree canopy, grass/shrub, buildings, roads, other paved surfaces, bare earth, and water. The average temperature of each census block group was used as an observed outcome (dependent variables). On the other hand, we did exploratory analysis on vacant lands dataset and found the primary land covers were tree canopy, grass/shrub, buildings, and other paved surfaces. Then we simulated all land covers in vacant lands to tree canopies and employed the change to CBG data as a new CBG data for later simulation with the best fit model.

REMOVE OUTLIERS

We checked the distribution of CBG areas and found it was a significantly skewed right distribution. With Interquartile

Range Method, we removed outliers by defining limits on the sample values that are a factor 1.5 of the IQR below the 25th percentile or above the 75th percentile.

PEARSON CORRELATION

The Pearson correlation analysis measures the strength of the linear relationship between two variables. According to the Pearson correlation coefficient (r), we found that percentages of tree canopy, grass/shrub and water were negatively correlated with LST, while percentages of buildings, roads, other paved surfaces, and bare earth are positively correlated with LST. Due to the low percentage of bare earth, and water in CBG units, we excluded them in the features to train models (Figure 5).

MULTICOLLINEARITY ANALYSIS

Multicollinearity occurs when two or more independent variables are highly correlated with one another in a regression model. This means that an independent variable can be predicted from another independent variable in a regression model. Since several features are highly correlated, we used Principal Component Analysis (PCA) to address the problem.

PRINCIPAL COMPONENT ANALYSIS (PCA)

PCA is a linear dimensionality reduction technique which converts a set of correlated features into a series of uncorrelated features from a high to a low dimensional space. We used two principal components to replace the original five independent variables in regression models to prevent Multicollinearity. Figure 6 indicates that Multicollinearity did not occur between two principal components.

MACHINE LEARNING ALGORITHMS

We used supervised machine learning algorithms to build regression models, since the output variable (LST) is a continuous and known value. The common algorithms used for supervised learning are Linear regression, Random Forest, Support Vector Machines (SVM), and Neural Networks, which codes were referred from Scikit-learn: Machine Learning in Python. Though the output result of temperature is continuous value, SVM and ANN can be used for either a regression or classification model. Geographically weighted regression (GWR) is a spatial analysis technique that takes non-stationary variables into consideration (e.g., climate; physical environment characteristics). To better visualize local R^2 and predict LST in each CBG, we used GWR function in ArcGIS Pro.

DEFINE URBAN HEAT ISLAND

Urban Heat Islands are on average 1°C to 3°C warmer than their rural surroundings during the daytime (Bradford et al. 2015). According to the Philadelphia Heat Vulnerable Index (Hammer et al. 2020), the hottest 10% of differences in average surface

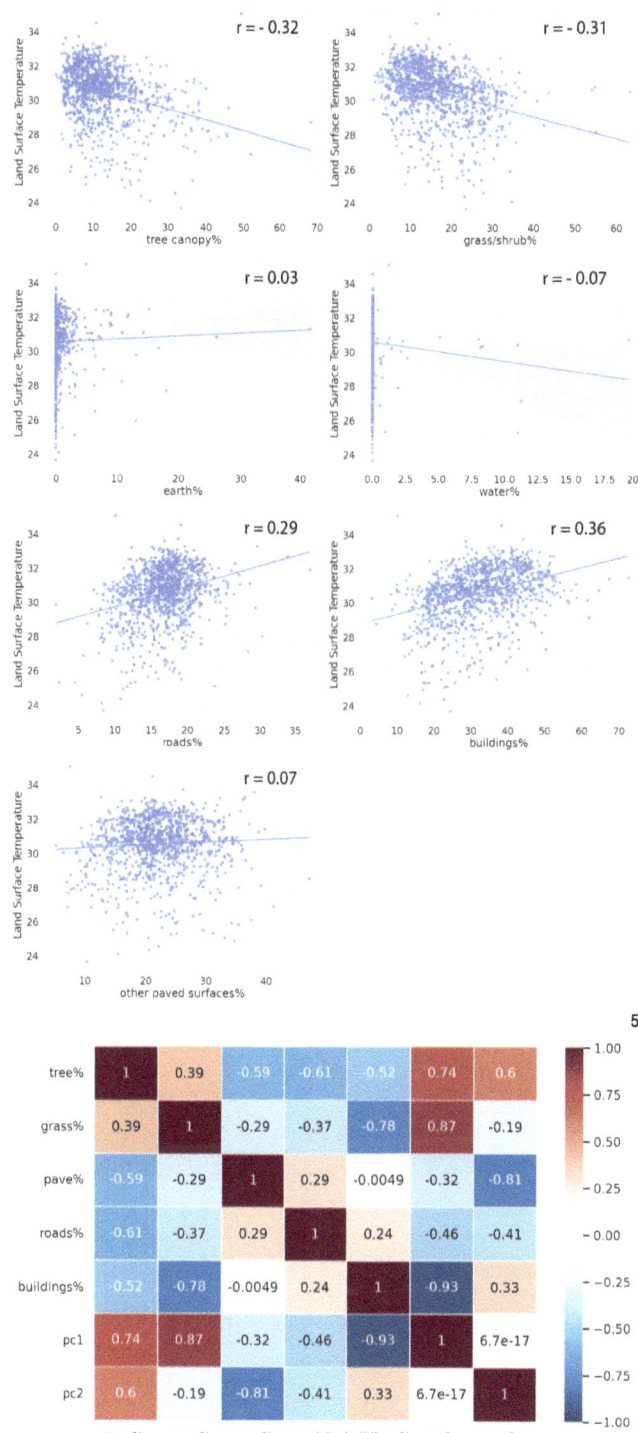

5 Pearson correlation between predicting features and average LST

6 Multicollinearity analysis of predicting features and principal components

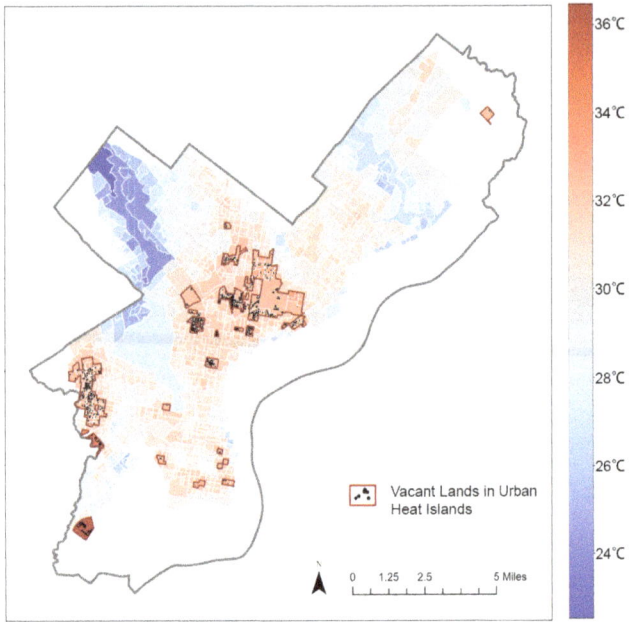

7   Vacant lands within urban heat islands on Choropleth map of Land Surface Temperature

temperature (the average LST of all CBGS was subtracted from the LST of each census block group) as the areas of hottest surface temperatures. Therefore, we referred to this method to define UHI in our study. After sorting the averaged LST of each census block group, the threshold temperature of the hottest 10% is 32.2°C (90°F). The vacant lands dataset records 28451 plots in polygon format in Philadelphia, and 2262 of them were within UHI (Figure 7).

Results and Discussion

The census block group data with land covers and mean temperatures (1113 observations) were split into Training set (70%) and Test set (30%). We compared the results from each model in Table 1. According to several metrics of each model, we evaluated their performances. Since we were predicting the temperature, unlike house price prediction, the mis-prediction on Celsius would be close to the actual temperature. MSE is more appropriate to evaluate the performance as an error metric, which could amplify the prediction error.

We used Google Colab with Python 3 Google Compute Engine backend, which provides RAM of 12 GB and a disk space of 358 GB, to train machine learning models. Since it is not a complex dataset with 1013 samples and fiver predicting features, the longest execution time is 1 minute 48 seconds for ANN model, while the shortest time is 21 miliseconds for Multivariable regression.

MULTIPLE REGRESSION

The coefficient of determination R^2 of this model is as low as 0.16, which means the relationship between percentages of land covers and land surface temperature is not linear. In spite of using PCA to solve multicollinearity, the result did not change much compared to using percentages of land covers. We did not choose it as the best model.

RANDOM FOREST

A random forest is a meta estimator that fits a number of decision tree classifiers on various subsamples of the dataset and uses averaging to improve the predictive accuracy and control overfitting. We included several sets of tuning parameters in the parameter grid and used three-fold cross validation to fit the model. The best hyperparameters were computed as n_estimator of 30, which represents the number of trees in the forest, and max_depth of 2, which represents the depth of each tree in the forest. The R^2 of the best RF model is presented as the best test score as 0.2. We did not choose Random Forest as the best model after comparing all metrics.

SUPPORT VECTOR MACHINES (SVM)

SVM aims to find one or more hyperplanes that separate the

Models w/ Best Parameters Metrics	Multiple Regression	Random Forest (max_depth=2, n_estimators=30)	SVM (C=1, gama=1, kernel=rbf)	ANN (batch_size=15, epochs=100)	GWR (gwr_bw=34)
R^2	0.16	0.2	0.13	0.18	0.79
Mean Absolute Error (MAE)	1.1	1.08	1.09	1.1	0.52
Mean Sqaure Error (MSE)	2.27	2.11	2.29	2.16	0.54
Mean Absolute Percentage Error (MAPE)	3.76%	3.63%	3.70%	3.69%	1.74%
Accuracy	96.24%	96.37%	96.00%	96.31%	98.26%.

Table 1. Result Comparison of Machine Learning Models

classes. We tested different parameters of C to control error and gamma to decide the curvature level in the parameter grid with 'rbf' kernel, finding best C as 1 and gamma as 1. The R^2 of the best RF model is presented as the best test score of 0.13.

ARTIFICIAL NEURAL NETWORK (ANN)
ANN is a deep learning model, as the extreme simplification of human neural systems. We built an input layer, a hidden layer, and an output layer in the models with units of 5, kernel initializer as normal and activation as "relu." The R^2 of the best ANN model is presented as 0.18. ANN would work better when dealing with image, audio, and text, so we did not choose it.

GEOGRAPHICALLY WEIGHTED REGRESSION (GWR)
GWR is supposed to take the spatial component into consideration, like the centroid and the according bandwidth is essential to the regression (Fotheringham et al. 2002). The latitude and longitude of the centroid in each census block group were considered as two features to add into the dataset along with 5 types of percentages of land covers. The ArcGIS Pro identified the best bandwidth of 34, presenting the max distance at which any feature has at most 34 neighbors. Like in Multiple Regression, R^2 is a measure of goodness of fit. The R^2 of the GWR model is 0.79, indicating 79% of variance in the dependent variable explained by the predictors, which is significantly improved compared to other models.

GWR generates parameter estimates for every census block group and the local residuals can be visualized in Figure 8, where spatial autocorrelation was not noticeable. Residuals are the predicted y values that are subtracted from the observed y values, which are the smaller, the closer the fit of the GWR model to the observed data. With the spatial autocorrelation tests, we calculated the global Moran's *I* as 0.013, suggesting that GWR residuals had no spatial autocorrelation and predicted results were reliable.

By comparing all error metrics and R^2 among five machine learning algorithms, GWR was the best model to predict the relationship between average LST and percentage of land covers in each census block group (Figure 9).

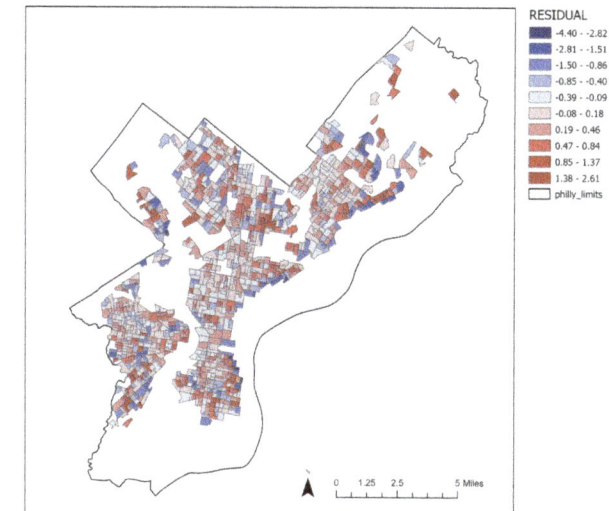

8

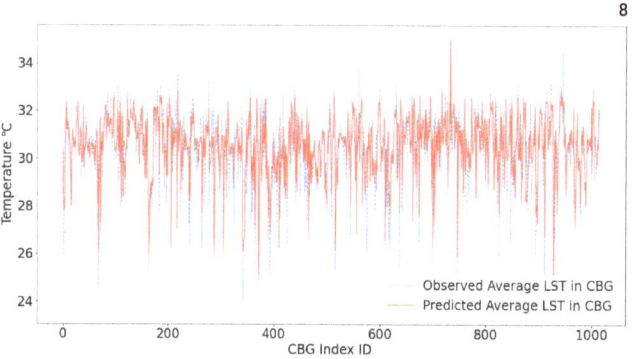

9

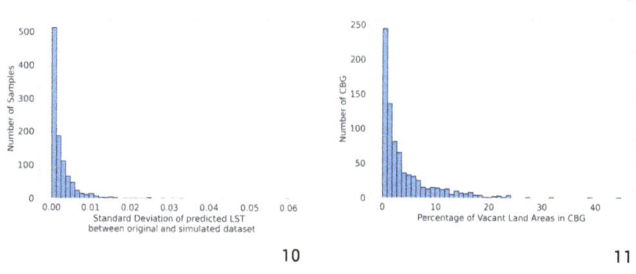

10   11

8   Choropleth map of the GWR regression residuals

9   Comparison of Observed LST and Predicted LST of census block groups

10 Histogram of standard deviation of Predicted LST between original and simulated CBG datasets

11 Histogram of percentage of vacant lands in census block groups

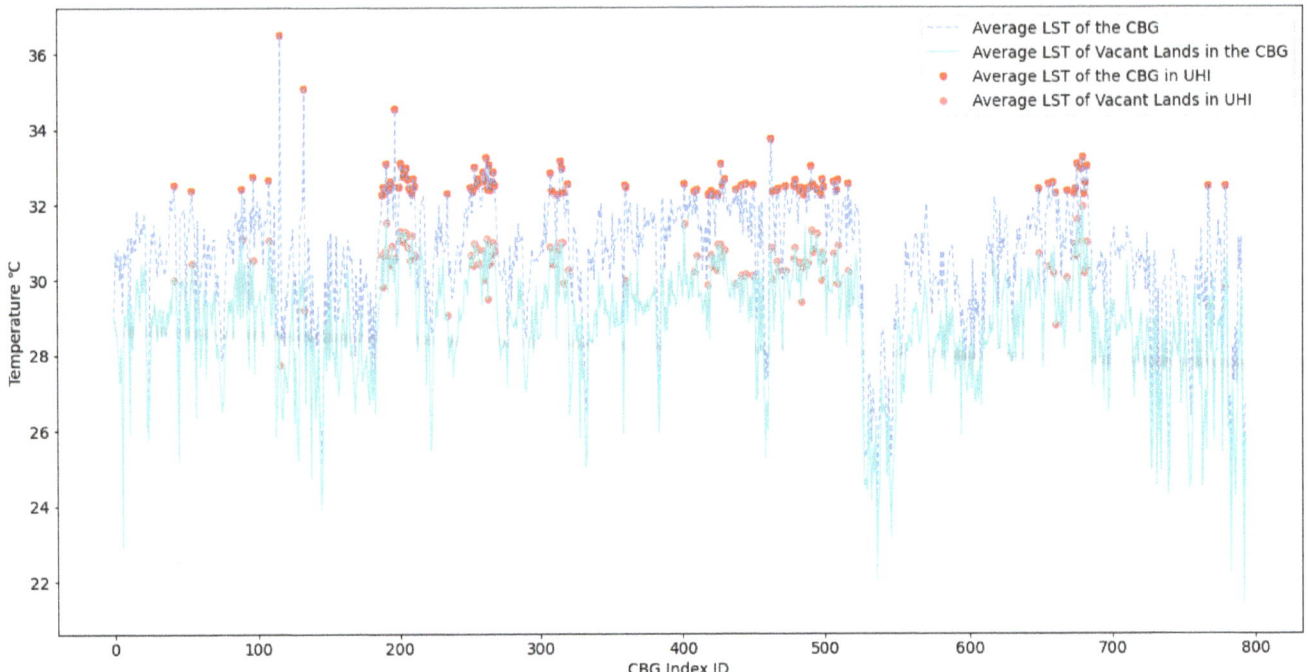

12 Observed LST of vacant lands and census block groups (CBG) on which they are located

SIMULATING LAND COVER CHANGES IN VACANT LANDS

In the dataset of vacant lands, the main land covers included grass/shrub (47%), tree canopy (26.3%), other paved surfaces (13.8%), and buildings (4.4%). Due to the limited areas of each vacant land, they always lacked a variety of land covers compared to the ones in a census block group. Since there are many zeros in the independent variables, the vacant lands dataset is not suitable to train the model. Therefore, we simulated land cover changes of vacant lands to tree canopies and employed them to the census block groups to which vacant lands belong. Then we tested their impact on predicted LST of each CBG with this new CBG data. The difference between new predicted and observed LST (new CBG data set) is very small, by an average standard deviation of 0.002 (Figure 10). Considering the granularity of Landsat raster datasets and low proportion of vacant lands in most CBG areas (Figure 11), the GWR model was not able to capture temperature changes in simulation of greening all vacant lands to tree canopies.

EXPLORATIVE ANALYSIS OF VACANT LANDS IN URBAN HEAT ISLANDS

Though simulating land cover change in vacant lands did not prove much of a cooling effect, we made important discoveries in analyzing existing conditions of vacant lands. The observed average LST of vacant lands are cooler than the ones of surrounding CBGs, by an average of ~1.9°C, with a range of 0°C to 8.7°C, which may be attributed to their high percentages of tree canopies (~26%) and grass/ shrubs (~47%) coverage. In Urban Heat Island (UHI) areas, the cooling effect of vacant lands reaches to ~2.1°C on average, compared to the surrounding CBG (Figure 12). Therefore, we believed vacant lands had exerted cooling influence with their high coverage of shrubs and tree canopies. According to Wong's research (Wong et al. 2021), the cooling effect beyond green spaces decrease with size, especially small green spaces (less than 0.1 ha) may not even beyond their boundaries. Considering of previous simulations of land cover change in vacant lands, we thought greenery vacant lands could reduce the LST efficiently within their boundaries, instead of mitigating the average LST in larger scale, like census block group.

SOCIAL VULNERABILITY

As urban heat islands can lead to heat-related illness and mortality, we also attempted to find places which have a higher, on average chance of being more vulnerable to heat exposure. Looking at Figures 13 and 14, we found the group with a relatively high percentage of poverty living in the urban heat islands have a higher risk of heat exposure. According to the Pearson correlation coefficient, the average LST of CBG is positively correlated with the percentage of income below poverty (r of 0.34), and the percentage of vacant properties (r of 0.25). On the contrary, there is negative correlation between the average LST of CBG and the percentage of white occupants with the coefficient of −0.32 (Figure 15). As the presented LST of vacant lands are cooler than surrounding CBGs by an average of ~2.1°C above, utilizing vacant lands strategically should be considered by policy makers and social service providers, especially for vacant lands in UHI. If legal circumstances allow, residents could be invited to use some vacant lands as public green spaces and benefit from

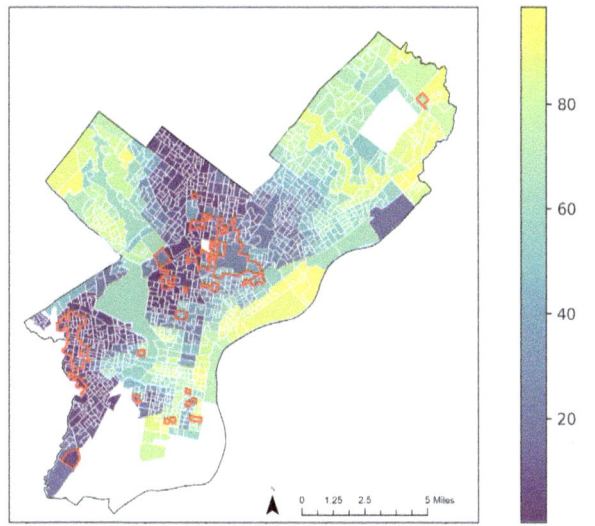
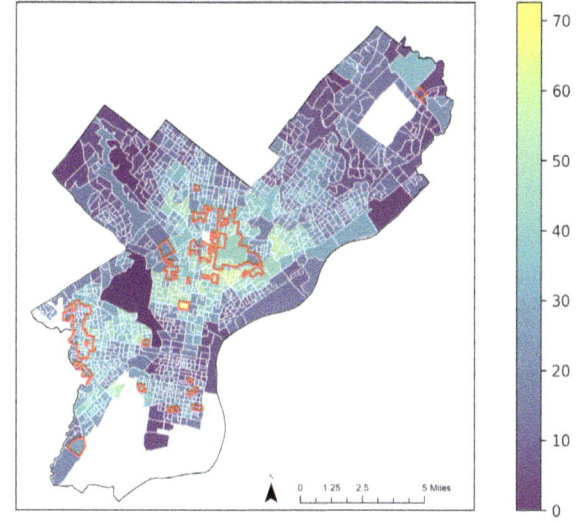

13 Choropleth map of percentage of white race

14 Choropleth map of percentage of income below the poverty line

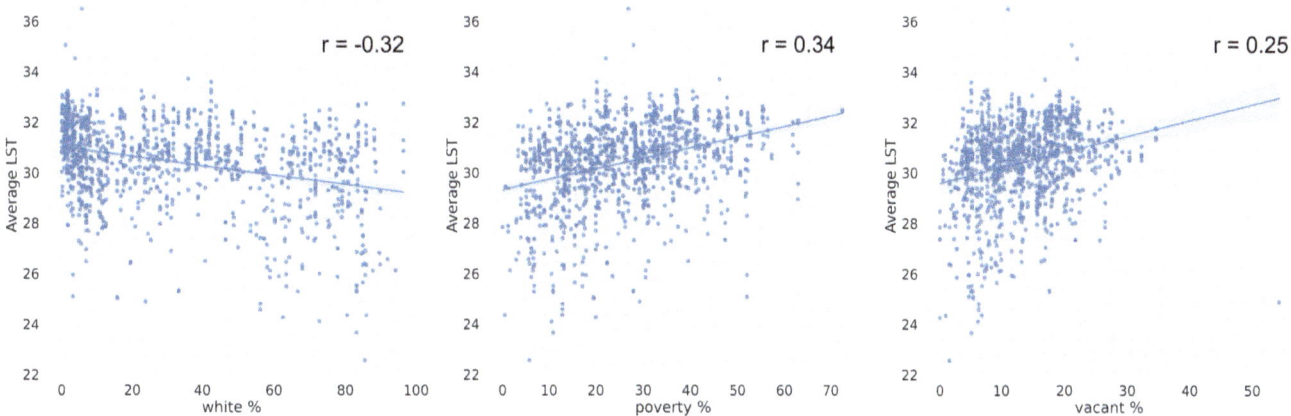

15 Pearson correlation between socioeconomic characteristics and average LST

their cooling effect. In addition, greenery vacant lands should be quantified on a micro scale with urban spatial modeling tools to maximize cooling effect in future research.

CONCLUSION

This study introduced a novel approach to applying machine learning methods to explore a quantitative relationship between land covers and average Land Surface Temperature (LST). It achieved three key contributions: (1) Geographically Weighted Regression performs best among five machine learning algorithms to interpret relationship between land covers and average LST of census block group with highest R^2 and lowest MSE; (2) current vacant lands are cooler than surrounding census block groups by an average of ~1.9°C due to high vegetation coverage; (3) it provided a method to identify social vulnerability with extra heat and explored potential cooling strategy with utilization of vacant lands there. Though simulating land cover changes in vacant lands did not impact average LST in census block group scale, the observed LST of vacant lands are cooler with a range of 0°C to 8.7°C than surrounding census block groups. In addition, LST of vacant lands in UHI are cooler than surrounding CBGs by an average of ~2.1°C, whose mitigating effect could benefit residents by opening vacant lands to public green spaces legally.

Several limitations should be noted. First, remote sensing technology provides a raster dataset in a larger scale, proving difficult for the simulation model in capturing subtle changes in LST. On-site measurements of LSTs should be made for higher granularity in future research. Second, social vulnerability to heat exposure is a complex topic, which may be related but not be limited to socioeconomic and demographic characteristics, like age, education level, language barrier, poverty, race, and social isolation. Due to the size of the paper, this study only discussed poverty and race. Lastly, some vacant lands are not available for public use due to legal issues. Strategically greening and using vacant lands for cooling effect in Urban Heat Islands still need broad collaboration among police makers, researchers, city planners, and stakeholders.

ACKNOWLEDGMENTS

We would like to thank Dr. Guray Erus (University of Pennsylvania), for his guidance in the data wrangling and machine learning process. We also want to thank Dr. Nick Hand (Office of the City Controller in Philadelphia), for his advice in the workflow of this research. Finally, we appreciate Professor Allison Lassiter (University of Pennsylvania), who inspired us to explore potentials of vacant lands as an urban resource.

NOTES

1. UHI: Urban Heat Island
2. CBG: Census Block Group
3. LST: Land Surface Temperature
4. GWR: Geographically Weighted Regression
5. ANN: Artificial Neural Network
6. SVM: Support Vector Machines
7. PCA: Principal Component Analysis

REFERENCES

Avdan, Ugur, and Gordana Jovanovska. 2016. "Algorithm for Automated Mapping of Land Surface Temperature Using LANDSAT 8 Satellite Data." *Journal of Sensors* 2016: 1–8. https://doi.org/10.1155/2016/1480307.

Bradford, Kathryn, Leslie Abrahams, Miriam Hegglin, and Kelly Klima. 2015. "A Heat Vulnerability Index and Adaptation Solutions for Pittsburgh, Pennsylvania." *Environmental Science & Technology* 49 (19): 11303–11311. https://doi.org/10.1021/acs.est.5b03127.

City of Philadelphia. n.d. *Philadelphia Land Cover Raster 2018.* Imagery dataset. Accessed May 25, 2022. https://www.opendataphilly.org/dataset/philadelphia-land-cover-raster.

City of Philadelphia. n.d. *Vacant Property Indicators: Land 2016.* Shapefile dataset. Accessed May 25, 2022. https://www.opendataphilly.org/dataset/vacant-property-indicators.

Chow, Winston T. L., Ronald L. Pope, Chris A. Martin, and Anthony J. Brazel. 2010. "Observing and modeling the nocturnal park cool island of an arid city: horizontal and vertical impacts." *Theoretical and Applied Climatology* 103 (1–2): 197–211. https://doi.org//10.1007/s00704-010-0293-8.

Esri Inc. *ArcGIS Pro.* V. 2.8.6. Esri Inc. 2022.

Fotheringham, Alexander Stewart, Chris Brunsdon, and Martin Charlton. 2010. *Geographically Weighted Regression.* Chichester: Wiley.

Google Inc. *Google Colab with Python 3.* V. 3.8.10. Google Inc. 2022.

Hammer, Jason, Dominique G. Ruggieri, Chad Thomas, and Jessica Caum. 2020. "Local Extreme Heat Planning: an Interactive Tool to Examine a Heat Vulnerability Index for Philadelphia, Pennsylvania." *Journal of Urban Health* 97 (4): 519–528. https://doi.org/10.1007/s11524-020-00443-9.

Hulley, Mike. 2012. "5 - The urban heat island effect: causes and potential solutions" In *Metropolitan Sustainability: Understanding and Improving the Urban Environment; Woodhead Publishing Series in Energy,* edited by Frank Zeman. Cambridge, UK: Woodhead Publishing. 79–98. https://doi.org/10.1533/9780857096463.1.79.

Mitz, Erik, Peleg Kremer, Neele Larondelle, and Justin D. Stewart. 2021. "Structure of Urban Landscape and Surface Temperature: A Case Study in Philadelphia, PA." *Frontiers in Environmental Science* 9: 592716. https://doi.org/10.3389/fenvs.2021.592716.

Park, Jonghoon, Jun-Hyun Kim, Dong Kun Lee, Chae Yeon Park, and Seung Gyu Jeong, 2017. "The influence of small green space type and structure at the street level on urban heat island mitigation." *Urban Forestry & Urban Greening* 21: 203–212. https://doi.org/10.1016/j.ufug.2016.12.005.

Pearsall, Hamil. 2017. "Staying cool in the compact city: Vacant land and urban heating in Philadelphia, Pennsylvania." *Applied Geography* 79: 84–92. https://doi.org/10.1016/j.apgeog.2016.12.010.

United States Census Bureau. n.d. *American Community Survey Data 2019.* Accessed May 25, 2022. https://www.census.gov/programs-surveys/acs/data/data-via-api.html.

United States Geological Survey. n.d. *Landsat-8 image 2013-2019.* GeoTIFF dataset. Accessed May 25, 2022. https://www.eos.com/landviewer.

Wong, Nyuk Hien, Chun Liang Tan, Dionysia Denia Kolokotsa, and Hideki Takebayashi. 2021. "Greenery as a mitigation and adaptation strategy to urban heat." *Nature Reviews Earth & Environment* 2 (3): 166–181. https://doi.org/10.1038/s43017-020-00129-5.

IMAGE CREDITS

Figure 1: ©USGS, 2013-2019

All other drawings and images by the authors.

Hui Tian holds a Master of Urban Spatial Analytics degree from the University of Pennsylvania, a Master of Landscape Architecture from the Louisiana State University, and a Bachelor of Landscape Architecture from China Agricultural University. She specialized in improving urban resilience and meditating climate change with data-driven approaches. Her research area includes urban green infrastructure, brownfield remediation, and saltwater intrusion prediction.

Jiali Yao holds a Master of Urban Spatial Analytics degree from the University of Pennsylvania and a Bachelor of City Planning from the Wuhan University of Science and Technology. Her research interests include real estate, smart city, and urban spatial analytics.

Shimin Tu holds a Master of Urban Spatial Analytics degree from the University of Pennsylvania and a Bachelor of Management degree from Nankai University, specializing in studying urban spatial topics with quantitative approaches. Her research interests include transportation planning, public policy analysis, and machine learning.

Resonant Hexagon Diffuser

Designing Tunable Acoustic Surfaces by Combining Sound Scattering and Helmholtz Resonators

John Nguyen
University of Toronto

Philipp Cop
University of Toronto

Nicholas Hoban
University of Toronto

Brady Peters
University of Toronto

Ted Kesik
University of Toronto

1 Hex CLT absorber and surface scatter assembly for airport lounges

ABSTRACT

Wood as a building material has been used in architecture for millennia. In recent years, technological advances have led to new timber products, mass timber, to be used in bigger and more complex building construction projects. The surge in mass timber buildings being constructed introduces unique acoustical challenges as mass timber is more permissible for sound to travel across floors, ceilings, and walls, especially for lower frequencies. In order to address these acoustical challenges, the absorption qualities of Helmholtz resonators and surface diffusion of scattering surfaces are leveraged by combining the two systems in an integrated structure using the tectonics of mass timber construction. Both the absorption behavior and the diffusive characteristics are tunable via geometric methods; this allows for the control of a gradient acoustic behavior to produce a surface ranging across frequency and sonic behavior. This paper investigates the potential of Helmholtz resonators to be used in combination with sound scattering surfaces to achieve optimal performance in cross laminated timber (CLT) panels through the use of a hexagonal pattern as the underlying design strategy. Prior design studies have researched Helmholtz resonators or sound scattering surfaces for architectural purposes, but none combined both sound design strategies to leverage the well depth requirements of the scatters as Helmholtz resonant cavities. The results from our investigation indicate that it is a functional approach towards creating prefabrication assemblies with optimized acoustical properties.

INTRODUCTION

Architects have been fascinated by resonant acoustic absorbers since the early days of western architecture. Early documentation of resonant absorbers can be found in Vitruvius' *The Ten Books on Architecture* where he describes how empty bronze vases were placed underneath rows of Roman and Greek theater seating to create better acoustic performance (Vitruvius 1960). Vases found built into the ceiling and walls throughout many European churches were thought to have been placed there for their acoustic properties (Crunelle 1993). However, the efficacy and application of these vases as acoustic absorbers is a topic debated amongst researchers (Kanev 2020). Experiments conducted by Bruel (2002) inside Danish churches, Mijic and Sumarac-Pavlovic (2004) in old Serbian churches, and Zakinthinos and Skarlatos (2007) inside an old Byzantine church concluded that their respective vases did not significantly contribute to better acoustic performance of the spaces. In contrast, Godman (2007) states that the reason many of these spaces do not perform as intended is due to how vases were cemented in place instead of being placed freely as Vitruvian examples indicated. A relationship between geometrical proportions and performance of acoustic resonators exists allowing for the dimensions of absorbers to be adjusted to tune for specific resonant frequencies (Cox and D'Antonio 2009). Although resonant absorbers are a powerful tool, their absorption spectrum and bandwidth are limited to the restrictions of neck friction losses (Lee, Nomura and Lizuka 2019). Resonant absorbers are the most powerful tool when dealing with low frequencies (Everest 2001). In order for broadband absorption to be achieved, multiple resonators need to be coupled together (Godbold 2008). New advances in the realm of digital modeling and simulation tools are making accessibility to study resonators much more feasible in the realm of architecture (Cop, Nguyen, and Peters 2021). Large scale Helmholtz prototypes can now be performance tested at original scale. This type of investigation has previously not been possible to test at scale as mainstream physical impedance tubes are limited to diameters ranging from 16 mm to 100 mm (Brüel & Kjær n.d.; Placid Instruments n.d.). Software packages such as Comsol Multiphysics Acoustic Modules can simulate complex geometries at 1:1 scale through virtual impedance tubes. Recently released Comsol Multiphysics software updates to improve native meshing capabilities has improved the granularity level for simulations at finer geometrical scales. This has made the study of broadband absorbers with shared resonator volumes much more accessible. In this research investigation, various types of Helmholtz resonator hexagonal assemblies are simulated for single band frequency studies that are then used to inform broadband absorption arrangements. This resulted in the creation of hexagonal cluster assemblies at 25 Hz intervals where the depth created on the flipped surface side was utilized as a scattering surface.

BACKGROUND

Helmholtz Resonators in Architecture

A resonant absorber functions through the science of vibrations where sound waves are pressured into cavities through air flow (Cox and D'Antonio 2009). The earliest form of documentation for the concept of the resonant absorber dates to the time of architect Vitruvius. Yet this remains a point of contention amongst researchers as limited forms of documentation remain (Crunelle 1993). According to the publication by Arns and Crawford (1995) the majority of vase type resonators are located throughout European churches and cathedrals numbering in the hundreds; St. Petronio Cathedral in Bologna, Italy contains four hundred resonators cavities (Crunelle 1993) and St. Nicholas Church in Pskov, Russia contains 300 pots inserted into wall cavities. Kanev (2020) and Crunelle (2011) state that the plans for how to arrange the location of resonant absorbers would have been a secret known only to the designated guilds responsible for overseeing their construction. Zakinthinos and Skarlatos (2007), in their investigation of resonators in an old Byzantine church, arrived at the conclusion that the vases did not provide noticeable acoustic improvements to the space, as their analysis of the resonator's effects revealed that the limited quantity of resonators in the church had an unobservable difference. This may be the reason for the discontinuation of resonant absorbers throughout Europe. The study of churches through Serbia by Mijic and Sumarac-Pavlovic (2004) arrived at similar conclusions where the resonant vases frequencies were too low to be excited by the human voice and that the quantity of resonators was too limited for any noticeable acoustic difference. They concluded that the building of acoustic resonators was a result of orally transmitted tradition rather than any proper understanding of their acoustic qualities or functions.

Acoustic Absorption of Helmholtz Resonators

In the field of acoustics, there exist two main types of acoustic resonators which can absorb acoustic waves through resonance: membrane absorbers and Helmholtz resonators. The membrane type absorbs acoustic energy by converting the acoustic pressure wave into mechanical oscillation of a membrane (Frommhold et al. 1994). Membrane absorbers are commonly implemented as large panels suspended from ceilings or walls to absorb low frequencies in a room. Likewise, Helmholtz resonators are also used to provide low frequency absorption but over broader ranges (Everest 2001). The Helmholtz resonator is named after Hermann von Helmholtz, who was the first person to calculate the resonant frequency where the most absorption could be obtained (Long 2006). The Helmholtz resonator can be configurable as many forms, as it is simply an enclosed volume of air with a small opening neck. It functions like an air spring where the sound wave enters the opening neck of the resonator, thereby pushing the air into the larger cavity, causing the compression of air. At resonance, the period of high and low pressure of a propagating pressure wave interacts with the geometric proportions of the resonator in such a way that an oscillating system is formed. Air moves faster in the neck of the resonator, operation as a mass on the spring of the larger resonator body. This oscillating system takes acoustic energy out of the room and convers it mostly into heat energy due to increased friction at the neck of the resonator. This process forms an oscillating system with a specific resonance frequency (Cox and D'Antonio 2009). The most common forms of Helmholtz resonators are empty bottles, string instruments, or perforated wood, or gypsum board. The control of absorption properties and their relation to reverberation time is one of the key criteria in architectural acoustics still relevant today (Sabine 1922). Helmholtz resonators may also be used for dampening specific eigenmodes of a space, which develop at low frequencies. A noticeable shortcoming of Helmholtz resonators is the ability to only absorb waves over very narrow frequency bands. A way to overcome this issue is to utilize absorbing material such as fleece, foam, or mineral wool within the resonator cavity (Fuchs 2013). It has been shown that Helmholtz resonators can play a key role in providing a solution to architects when designing for low frequency absorption scenarios (Cop, Nguyen, and Peters 2021).

The Helmholtz resonator equation is (Long 2006):

$$f_0 = \frac{c}{2\pi}\sqrt{\frac{S}{V(l+\delta)}}$$

For circular holes, the error adjustment is (Long 2006):

$\delta \cong 0.8d$, where

f_0	is the Design Frequency
c	is the Speed of Sound in Air
S	is the Area of the Opening
d	is the Diameter of the Opening
V	is the Volume of the Resonating Chamber
l	is the Length of the Neck

Sound Scattering Surfaces

The terminology for scattering and diffuse reflections are often used interchangeably throughout the field of acoustics. In this paper, Cox and D'Antoni's (2009) definition is used, where scattering refers to the measurable quantity of sound that is scattered away from the incoming specular reflection direction. The past decade has seen a rise in interest from researchers and designers towards leveraging computational power for the exploration of sound scattering surfaces (Peters et al. 2019), as scattering surfaces are beneficial not only to traditional acoustical spaces such as concert halls and performance venues (Haan and Fricke 1997), but also in regular classrooms and meeting rooms (Choi 2013). A key criterion for any acoustic surface is the sound scattering coefficient, which is used to define the quality of sound in a space (Cox and D'Antonio 2009). The scattering coefficient is the ratio of scattered sound waves to the total reflected sound (Peters et al. 2019) and can be extracted from the attenuation curve provided by the exponential-power formula and measured in a room with uneven distribution of absorption wall qualities (Lavrova and Kanev 2020). The definition of sound scattering surfaces, associated diffusion coefficients and methods of measurement have been internationally agreed upon under ISO (2012) due to its importance. The scattering coefficient is also very important in providing accurate simulation results for computational simulations conducted (Peters et al. 2019). A relationship exists between the depth of a surface geometry in relation for the optimal sound frequency scattering to occur (Cox and D'Antonio 2009). Any surface detailing can have an impact on the scattering performance (Peter and Olesen 2010).

Sound Scattering equation (Cox and D'Antonio 2009):

$$S = \frac{\alpha_{spec} - \alpha_s}{1 - \alpha_s}$$

S	is the Scattering Coefficient
α_{spec}	is the Apparent Specular Absorption Coefficient
α_s	is Absorption Coefficient

Absorption Coefficient equation (Cox and D'Antonio 2009):

$$\alpha = 1 - |R|^2$$

α	is the Absorption Coefficient
$|R|$	is the Magnitude of the Pressure Reflection Coefficient

Mass Timber

The origins of utilizing mass timber as a construction material date to the period of the Mesopotamians during 3400 BC (NRC 2021). During this time, wood strips were glued together at opposing angles, which is similar in form to current day mass timber products called Glued Laminated Timber (GLT) or Cross Laminated Timber (CLT). In recent time, technological advances in assembly methods and manufacturing have allowed mass timber products to innovate rapidly to a point where numerous high rises are currently under construction or being proposed (Kesik and Martin 2021). The key development to this new mass timber technology is that it consists of a family of products that can be interconnected to other smaller mass timber elements, allowing for larger systems to

DESIGN	PARAMETRIC SCRIPT	ACOUSTIC SIMULATION		FABRICATION
Concept Development	Create Parametric Model using Rhino 7 - Grasshopper	Simulate 3D Thermoviscous and Pressure using COMSOL	Simulate 2D Scattering using AFMG Reflex and custom FDTD script in Processing	CLT Milling using Powermill exported for Kuka Robotic Arm

Parameter Optimizations ← ← ←
Design Adjustments ←

2 Workflow diagram

be formed. The benefit of this approach is that the primary load-bearing structure is not limited to just solid wood but can be combined with engineered lumber products. The label *engineered lumber* refers to composite materials where the dominant material is based on soft woods (Kesik and Martin 2021). The use of composite materials allows architects to design for more idealized forms consisting of complex geometries for structural elements such as beams, columns, arches, roofs, floors, and walls. In recent years the demand for more sustainable building materials has seen increased interests as the environmental costs of concrete and steel constructions projects have become more apparent as made evident by the Paris Climate Agreement to call for a carbon free environment by 2050. In order for this deadline to be met, architects need to actively use building materials with lower embodied carbon emissions. The use of carbon sequestering materials such as mass timber is a beneficial alternative option as global forests store a large quantity of the planets carbon and any building constructed out of mass timber continues to retain it too. It is widely known that the building industry is a significant contributing factor to climate change (Dangel 2017), which is why mass timber is such an appealing material to address the problem. Many architects have already begun to design entire high-rises using primarily mass timber as the main building material (Orta et al. 2020). This strategy of using entire mass timber buildings will introduce unique acoustical challenges as mass timber assemblies are not as dense as concrete, allowing for the transmission of sound to travel across floors, ceilings, and walls. This is especially true for lower frequencies, which provides an opportunity for the integration of Helmholtz resonators and sound scattering surfaces to be implemented as a solution.

METHODS

The methods used in this investigation builds upon our experience and knowledge gathered in prior and ongoing research related to Parametric Acoustics (Peters et al. 2010), Surface Scatter (Peters et al. 2019), Simulation Techniques (Peters and Nguyen 2021), Helmholtz Resonators (Cop, Nguyen, and Peters 2021), and Acoustic Metamaterials (Cop, Nguyen, and Peters 2022).

Parametric Modeling

3D modeling was conducted in Rhinoceros 3D (McNeel et al. 2010), where data derived from acoustic simulation software was used to inform Grasshopper scripts for parametric geometry creation that could be generated for the sought-after acoustic performance qualities. Controllable parameters were hex depth, diameter, neck width, neck depth, increment size, and cluster quantity. The use of a parametric model allows for quicker and precise generation of options using manual slider adjustments or in combination with genetic algorithms plugins such as Galapagos (refer to Figure 2 for workflow integration).

Ray-based Acoustic Simulation

Room acoustic performance simulations are divided into two categories: wave-based and ray-based (Siltanen et al. 2010). The ray-based approach abstracts sound travel as linear reflections and is computed using the geometric method. This method is the most common approach in which architectural acoustic performance is predicted and used in most commercial room acoustic simulation software packages. A shortcoming to the ray-based approach is that it cannot be used to simulate for the performance of resonators and requires the complete values for all absorption coefficients. This makes the ray-based approach more suitable to study scattering surfaces for room acoustics.

Wave-based Acoustic Simulation

Wave-based simulation methods utilize numerical modeling to predict solutions for the wave equation as it is unable to solve it directly. The main numerical methods for acoustic simulation are: the finite element method (FEM), the boundary element method (BEM), and the finite-difference time-domain method (FDTD) (Cox and D'Antonio 2009). Numerical techniques involve the discretization of a space which makes for a computationally intensive calculation. In this study, the FEM software COMSOL Multiphysics (Littmarck and Saeidi 2022) was used to model the acoustic wave interaction with the Helmholtz resonator geometry, while a customized FDTD script using the Processing(p3) language was used to animate the wave propagation against the scattering surface, and the BEM software AFMG Reflex (Ahnert and Feistel 2011) was used to conduct two-dimensional scattering studies.

Virtual Impedance Tube

Impedance is the resistance to flow of acoustic wave propagation through a medium; this resistance to flow can be used to understand a material's reflection and absorption coefficient. In architectural acoustics practice, the impedance of a material sample is analyzed in a physical impedance tube. An impedance tube consists of a loudspeaker installed one side of the tube, a material sample is positioned on the opposite side of the tube, and microphones to measure the reflected sound intensity are placed in between. By implementing the acoustic transfer function, two measurements are related, and the absorption and reflection coefficients can be calculated. Mainstream impedance tubes are limited to diameters ranging from 16 mm to 100 mm, requiring prototypes to be scaled due to the physical limitations. In contrast, a benefit of utilizing a virtual impedance tube is that geometries can be prototyped at 1 to 1 scale allowing for reliable simulation results. In this study a virtual impedance tube was created in COMSOL using a perfectly matched layer on the top of the tube, where an infinite tube was simulated. This yields more accurate results as compared to a finite tube length. A background pressure field is assigned to establish the incident plane wave, which interacts with the virtual hex samples that are positioned at the end of the tube. The reflected energy is virtually measured and the energy of the incident plane wave is known, then the reflection and absorption coefficient can be derived. The virtual impedance tube was set up to calculate the absorption and reflection coefficient as can be seen in Figure 3.

Digital Fabrication

New commercially available systems with varying levels of prefabrication have been created by the construction industry, which are targeting multi-story residential and commercial buildings (Kesik and Martin 2021). The prefabrication of mass timber assemblies within open floor building system that are flexible and customizable require the increased use of computer-controlled materials handling. The use of these fabrication equipment ensures the quality and precision of products for smooth assembly on site (Dangel 2017). The innovative use of mass timber in this context of prefabrication to a high degree of precision offers a stronger coordination between architects and fabricators. This high precision process and relative ease of multi-axis milling is providing new approaches to ornamentation and performance-based design.

RESULTS AND REFLECTION

Parametric Helmholtz Resonator Model

The Helmholtz resonator designs were based on results of select single band frequency simulations, where spatial parameters such as hex diameter, depth, and neck sizes were adjusted, see Figure 4. These results informed configuration layouts of hex clusters specialized for broadband absorptions. The frequencies of study ranged from 50 Hz to 500 Hz at 25 Hz intervals for single band frequencies and broadband frequencies ranged from 100 Hz to 750 Hz at 5 Hz and 25 Hz intervals.

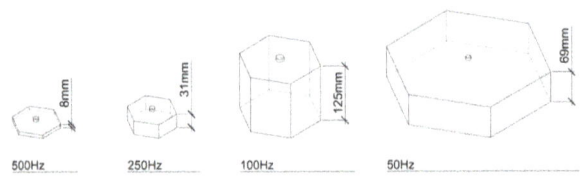

4 Hex resonator types, 1:1 Scale (100 mm)

Simulating Absorption at Single Frequencies

The single band frequency study achieved absolute absorption at 500 Hz as is seen from the absorption coefficient chart in Figure 5, Hex Resonator6. At 250 Hz and 100 Hz the frequency absorption coefficients of 0.95 and 0.92 were achieved respectively. In terms of interval rate, our prior experience conducting Helmholtz resonator studies for cubic and rectangular shaped volumes showed that 50 Hz intervals were sufficient. However,

3 Virtual Impedance Tube setup within COMSOL for pressure acoustics

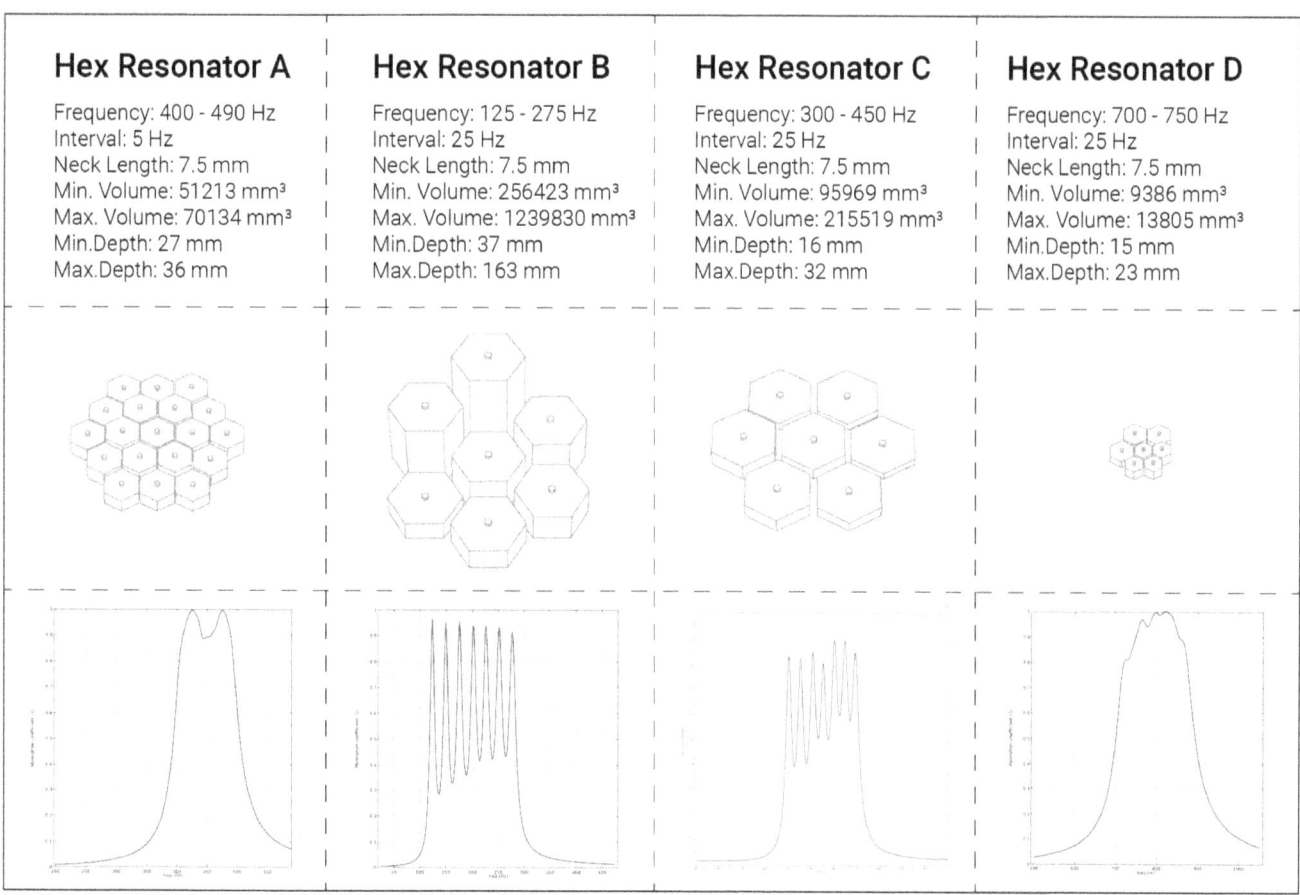

5 Single band frequency absorbers

6 Broadband frequency absorbers

ENVIRONMENTAL PERFORMANCE AND SIMULATION

this hexagonal geometry investigation revealed that an interval frequency of at least 25 Hz was required to achieve broadband absorption results as indicated in the charts in Figure 6.

Simulating Absorption at Multiple Frequencies

Using results extracted from the single band frequency absorbing geometries, hexagonal dimensions were modified to achieve broadband absorption as seen in Figure 6, Group A, where certain size hex resonators required lower than the average 25 Hz interval, at 5 Hz to achieve optimal performance. Neck lengths remained the same through the investigation due to the CLT layer thickness minumums after milling. The best performing group of broadband absorption occurred for the 700 to 750 Hz frequency as most of the absorption coefficient remained above 0.8, see Figure 6, Group D. The lower 125 Hz to 275 Hz and 300 Hz to 450hz frequencies did achieve suboptimal performance for broadband absorption as increased drops in absorption coefficient values are noted, as seen in Figure 6, Groups B and C.

Acoustic Absorption Performance of Coupled Hex

The coupling and sharing of hexagonal volumes were also tested in dual configuration to see if shallower volumes could be combined and leveraged to capture various frequencies that are not attainable of the individual volumes. In Figure 7, a comparison between a single and coupled hex can be seen where the Velocity Amplitude displays faster moving air at the neck and stale air inside the volume for the energy to dissipate. The Acoustical Pressure diagram shows us the pressure disparity between inside of the resonator volume and the outside, forming the air spring and oscillating actions to occur. The increased air velocity in the narrow region of the neck causes thermo-viscous damping to transpire at resonance at the boundary layer near the walls of the neck region.

Scattering Performance of Hexagon Geometries

An approach to study sound scattering performance is to visualize the procedure. A custom Finite Difference Time Domain program that was developed inside Processing to demonstrate the visualization approach. In Figure 8, stills from the animation were created to document how the sound energy is dispersed spatially and temporally as the reflections are

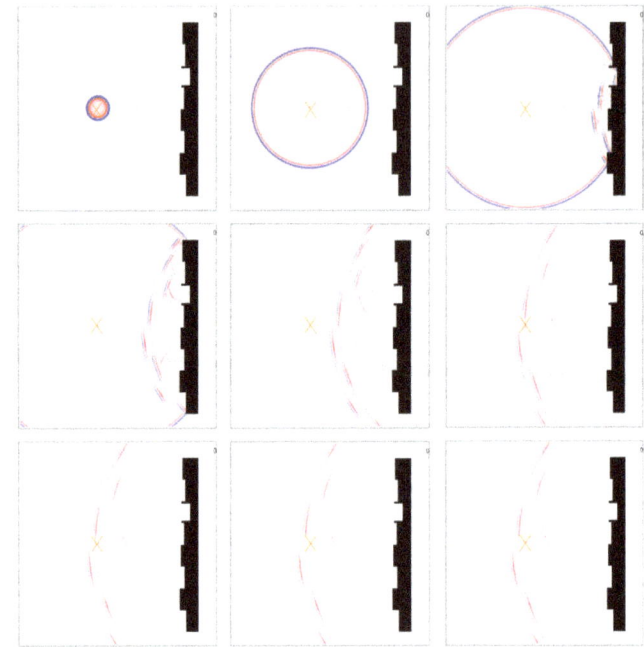

8 Finite Difference Time Domain scattering simulation against Hex section

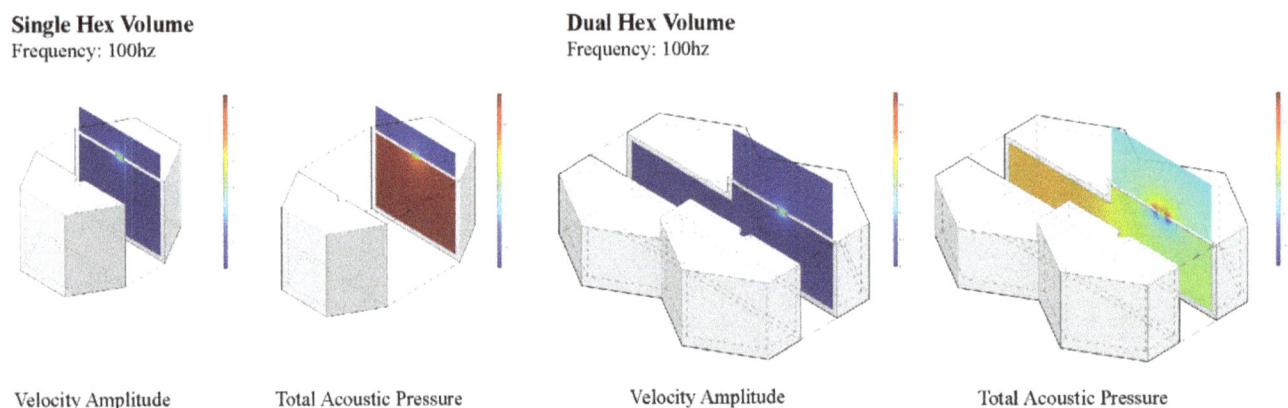

7 Single Hex and Shared Hex Volume studies

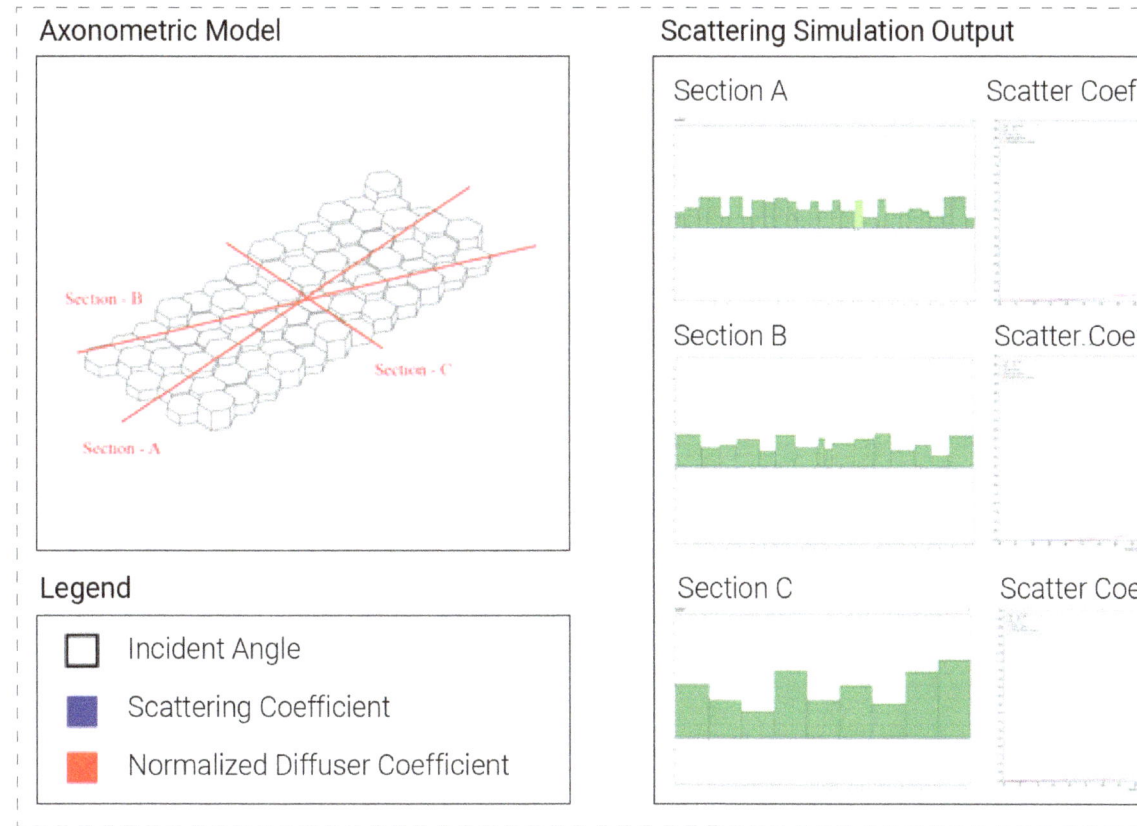

9 AFMG Reflex simulation output

spread throughout the space and come back to the listening positions numerous times as seen by the red and blue waves hitting the orange X.

Another approach to measure surface scattering is through the use of BEM simulation inside AFMG Reflex. The sound scattering simulation results showed that the hex geometries provided scattering in lower frequencies while maintaining strong scattering tendencies in the upper frequency ranges, refer to Figure 9. The shallow hexagon regions also showed good scattering from 200 kHz and above, and the deep hexagon from 630 Hz and above. The AFMG Reflex software also demonstrated that all the surfaces had good scattering performance at the incident angle. An issue encountered with the BEM simulation software Reflex is that it does not allow for geometries to be imported, forcing geometries to be recreated in a time-consuming manner within the software.

Fabrication Experiments
Using results from the Helmholtz resonators and surface scattering experiments, we proceeded to produce a 1 to 1 scale prototype out of CLT using a PushCorp 10hp spindle end effector connected to an industrial grade Kuka KR150 R2700 robotic arm situated on a KL4000 linear axis with a movement range of 4500 mm. Although the established design is supposed to be one unified CLT assembly, the fabrication procedure requires two separate CLT blocks to be milled with the scatter facing side requiring a flip mill process to maximize available hex well depths prior to gluing the pieces back together, see Figure 11. It is important to note that this specific project can also be manufactured using a computer numerically controlled (CNC) router with more optimized milling routines. The choice to utilize a robotic arm for this research was primarily due to accessibility and logistical reasons. The final result can be seen in Figure 10, where the image on the left side displays the finished product after being combined together into one CLT block. The middle and right image show how the scattering surface has neck openings drilled into the surface to create a direct connection to the honeycombed volumes of the accompanying CLT panel, allowing for air spring actions to occur. To increase broadband absorption for frequencies that dropped below an absorption coefficient of 0.5, it was decided to insert high density acoustic foam towards the base of all volumes on the honeycombed backplate.

Future Work
The next step in this research will be to use the data acquired from this investigation to develop new design patterns for acoustic testing, and to fabricate a full wall assembly to be installed within a space using CLT to record physical acoustic measurements for the validation of our simulation results.

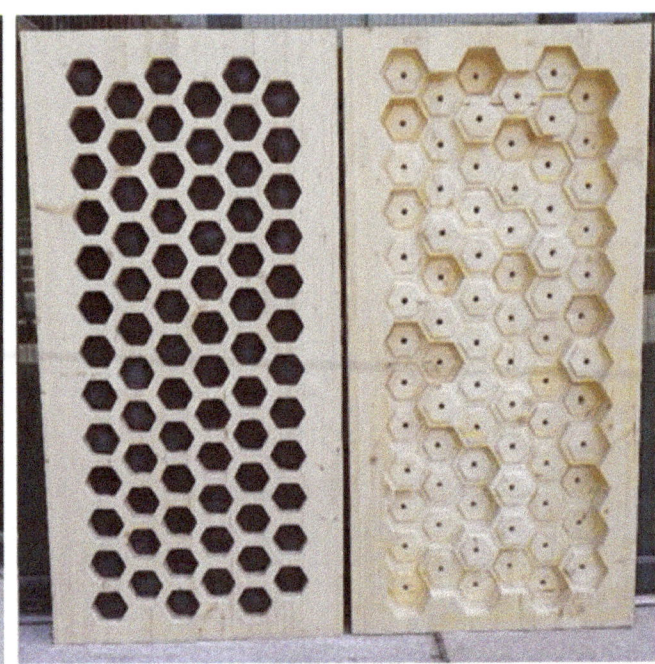

10 CLT Hex scatter panel flip milling using a Kuka robotic arm

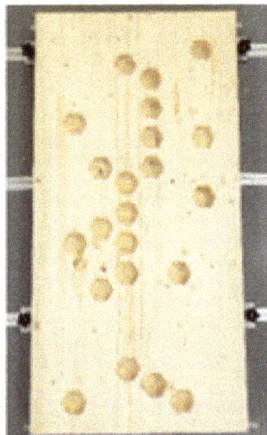
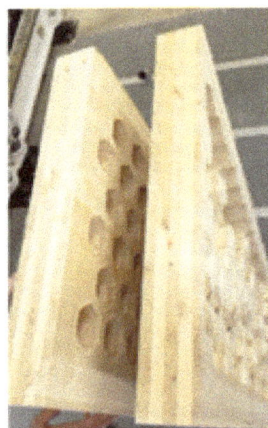

11 CLT Hex scatter panel flip milling using a Kuka robotic arm

Future scattering simulations will be tested in COMSOL Multiphysics using 3D Scattering to overcome AFMG Reflex's limitations of two-dimensionality to provide for improved acoustic performance accuracy. The outlook is to develop a system for mass timber assemblies that can combine Helmholtz resonators, surface scattering with acoustic metamaterials for optimization of tunable gradient acoustic surfaces as seen in Figure 12.

CONCLUSION

The key contribution from this paper is a novel approach where sound design strategies are combined to leverage well depth requirements of surface scatterers as Helmholtz resonant cavities to create prefabricated CLT assemblies that are sustainable and functional in improving room acoustics. Unlike systems with exclusive absorption or scattering behavior, our combined approach addresses both the reflected and transmitted sound fields in mass timber architecture to produce a more comfortable sound environment. The absorption behavior and the diffusive characteristics result from geometric proportions and are therefore directly addressable and tunable. Gradient acoustic performance across a surface ranging across frequency and sonic behavior is attainable. This research establishes a workflow where data from COMSOL, FDTD, and AFMG Reflex simulations inform CAD dimensions of 3D models adjusted through parametric modeling parameters inside Grasshopper to produce a milled CLT sonic surface. This paper displays the

12 Open office floorplan with a CLT wall consisting of hexagonal Helmholtz resonators coupled with Schröder diffusers and scattering surfaces

outcomes of our investigation of prefabricated mass timber assemblies to produce tuned gradient sound surfaces for sound absorption, diffusion and reflection. The relationship between digital forms to physical manifestation also revealed challenges and opportunities where two separate CLT panels could be strategically processed and combined as not just an acoustic performance solution, but also a space saving option for spaces with limited or restricted surface space such as airports or open floorplan offices (Figures 1 and 12).

ACKNOWLEDGMENTS

This research project was funded by Canada's National Sciences and Engineering Research Council's (NSERC) Discovery Grant. We would like to thank our industry partner ELEMENT 5 for their generous support in supply of CLT panels. A special thank you for our amazing summer research students, Edward Widjaja, Ross Cocks, and Nermine Hassanin for their help in assisting with the project. We also extend a big thank you to the many others from the University of Toronto's John H. Daniels Faculty of Architecture, Landscape, and Designs: Digital Fabrication Lab, Woodshop Team, and Facilities services that provided us with a helping hand throughout the research.

REFERENCES

Ahnert, Wolfgang and Rainer Feistel. *AFMC Reflex*. V. 10. AFMG Technologies GmbH. PC. 2011.

Arns, R. G., and B. E. Crawford. 1995. "Resonant Cavities in the History of Architectural Acoustics." *Technology and Culture* 36 (1): 104–35.

Brüel & Kjær. n.d. "Transmission Loss Tube / Impedance Tube Kits." Acoustic Material Testing Kits, Type 4206. Accessed May 15, 2022. https://www.bksv.com/en/transducers/acoustic/acoustic-material-testing-kits/transmission-loss-and-impedance-tube-kits-4206.

Brüel, Per V. 2002. "Models of ancient sound vases." *Journal of Acoustical Society America* 112 (2333).

Choi, Young-Ji. 2013. "Effects of Periodic Type Diffusers on Classroom Acoustics." *Applied Acoustics* 74 (5): 694–707.

Crunelle, Mark. 1993 "Is there an Acoustical Tradition in Western Architecture?" In *Proceedings of the 1st International Conference on Acoustic Ecology*. Banff, Canada.

Cop, Philipp, John Nguyen, and Brady Peters. 2021. "Modelling and Simulation of Helmholtz Resonators for Broadband Sound Absorption." In *SimAUD 2021; Proceedings of the 12th Annual Symposium on Simulation for Architecture and Urban Design*. San Diego, CA: Society for Computer Simulation International.

Cop, Philipp, Nguyen John and Brady Peters. 2022. "Modelling and Simulation of Acoustic Metamaterials for Architectural Application." In *Proceedings of DSMB22; Design Modelling Symposium Berlin: Toward Radical Regeneration*, edited by C. Gengnagel, O. Baverel, G. Betti, M. Popescu, M. R. Thomsen, and J. Wurm. Cham: Springer. https://doi.org/10.1007/978-3-031-13249-0_19.

Cox, T. J. and P. D'Antonio. 2009. *Acoustic Absorbers and Diffusers: Theory, Design and Application*. London: Taylor & Francis.

Dangel, Ulrich. 2017. *Turning Point in Timber Construction: A New Economy*. Basel: Birkhauser.

Everest, F. Alton. 2001. *Master Handbook of Acoustics*, 4th ed. New York: McGraw-Hill Education.

Fuchs, Helmut V. 2013. *Applied Acoustics: Concepts, Absorbers, and Silencers for Acoustical Comfort and Noise Control Alternative Solutions - Innovative Tools - Practical Examples*, 1st ed. Berlin, Heidelberg: Springer.

Frommhold, W., H. V. Fuchs, and S. Sheng. 1994. "Acoustic Performance of Membrane Absorbers." *Journal of Sound and Vibration* 170 (5): 621–36.

Godbold, Oliver. 2008. "Investigating broadband acoustic adsorption using rapid manufacturing." PhD thesis, Loughborough University, UK. https://hdl.handle.net/2134/8058.

Godman, Rob. 2007. "The Enigma of Vitruvian Resonating Vases and the Relevance of the Concept for Today." *The Journal of the Acoustical Society of America* 122 (5): 3054.

Haan, ChanHoon, and Fergus R. Fricke. 1997. "An Evaluation of the Importance of Surface Diffusivity in Concert Halls." *Applied Acoustics* 51 (1): 53–69.

International Organization of Standardization (ISO). 2012. *Acoustics—Sound-scattering properties of surfaces—Part 2: Measurement of the directional diffusion coefficient in the free field (ISO standard No. 17497-2:2012)*. Technical Committee: ISO/TC/SC 2 Building acoustics. https://www.iso.org/obp/ui/#iso:std:iso:17497:-2:ed-1:v1:en.

Kanev, Nikolay. 2020. "Resonant Vessels in Russian Churches and Their Study in a Concert Hall." *Acoustics* 2 (2): 399–415.

Kesik, Ted, and Rosemary Martin. 2021. *Mass Timber Building Science Primer 2021*. Toronto, Canada: Mass Timber Institute.

Lavrova, Marina, and Nikolay Kanev. 2020. "Sound Scattering Properties of Surfaces with Diffusers." *MATEC Web of Conference*, vol. 320, article 00024. https://doi.org/10.1051/matecconf/202032000024.

Lee, Taehwa, Tsuyoshi Nomura, and Hideo Iizuka. 2019. "Damped Resonance for Broadband Acoustic Absorption in One-Port and Two-Port Systems." *Scientific Reports* 9 (1): 13077–11. https://www.doi.org/10.1038/s41598-019-49222-w.

Littmarck, Svante and Farhad Saeidi. *COMSOL Multiphysics*. V. 6.1. COMSOL AB. PC. 2022.

Long, Marshall. 2006. *Architectural Acoustics*. Burlington, MA: Elsevier.

McNeel, Robert., et al. *Rhinoceros 3D*. V. 7.0. Robert McNeel & Associates. Windows. 2010.

Mijic, Miomir, and Dragana Sumarac-Pavlovic. 2004. "Analysis of Contribution of Acoustic Resonators Found in Serbian Orthodox Churches." *Building Acoustics* 11 (3): 197–212.

National Research Council Canada. 2021. "The State of Mass Timber in Canada 2021." Natural Resources Canada, Canadian Forest Service, Green Construction through Wood (GCWood) Program, Ottawa.

Orta, Belén, J. E. Martínez-Gaya, J. Cervera, and J. R. Aira. 2020. "Timber High Rise, State of the Art." *Informes de la Construcción* 72 (558): 346. https://doi.org/10.3989/ic.71578.

Peters, Brady, and Tobias Olesen. 2010. "Integrating Sound Scattering Measurements in the Design of Complex Architectural Surfaces." In *Proceedings of eCAADe 2010*. 481–491.

Peters, Brady, Nicholas Hoban, and Krystal Kraemer. 2020. "Sustainable Sonic Environments; The Robotic Fabrication of Mass Timber Acoustic Surfaces." In *CAADRIA 2020: Proceedings of the 25th CAADRIA Conference*. Bangkok, Thailand. 453–462. https://www.doi.org/10.52842/conf.caadria.2020.2.453.

Peters, Brady, Nicholas Hoban, Jay Yu, and Ziju Xian. 2019. "Improving Meeting Room Acoustic Performance through Customized Sound Scattering Surfaces." In *Proceedings of the International Symposium on Room Acoustics, ISRA 2019*. Amsterdam, Netherlands. 213–225.

Peters, Brady, John Nguyen, and Randa Omar. 2021. "Parametric Acoustics: Design Techniques That Integrate Modeling and Simulation." In *Proceedings of EuroNoise21; 21st Congress of the European Acoustics Association*. Madeira, Portugal.

Placid Instruments, n.d. "Impedance tube, Kundt's tube, sound absorption and transmission loss." Placid Instruments, Products. Accessed May 15, 2022. https://placidinstruments.com/product/impedance-tube.

Sabine, Wallace Clement. 1922. *Collected Papers on Acoustics*. Cambridge, Mass.: Harvard University Press.

Siltanen, Samuel, Tapio Lokki, and Lauri Savioja. 2010. "Rays or Waves? Understanding the Strengths and Weaknesses of Computational Room Acoustics Modeling Techniques." In *Proceedings of the International Symposium on Room Acoustics, ISRA 2010*. Melbourne, Australia.

Vitruvius. 1960. *The Ten Books on Architecture*, translated by M. H. Morgan. New York: Dover Publications.

Zakinthinos, Tilemachos and Dimitris Skarlatos. 2007. "The effect of ceramic vases on the acoustics of old Greek orthodox churches', *Applied Acoustics* 68 (11–12): 1307–1322.

John Nguyen is an academic researcher specializing in computational design and performance-based simulation. He is a PhD student at the University of Toronto, where his research investigates how computer simulations can elevate design processes rather than being a compromised coexistence of design and technology. Currently, he is exploring the topic of parametric acoustics in architecture with a focus on acoustic metamaterials and Helmholtz resonators using robotic fabrication strategies. Previously he worked as a computational designer at UofT's Platform for Resilient Urbanism working on collaborative projects with MIT's Urban Risk Lab. Prior to this, he was a Resident at Autodesk Technology Center in Toronto and a Computational Fluid Dynamics researcher at Seoul National University.

Philipp Cop is a computational designer and researcher at the University of Toronto. He was a member of the Daniels Acoustics Lab and Robotics Lab, where his work focused on acoustic metamaterials, digital fabrication using robotic methods, and acoustic simulation. His work was presented at international conferences and published in peer-reviewed papers. Philipp's academic projects include a design to fabrication workflow that was developed to produce area-minimized self-supporting surfaces for fabrication through robotic clay additive methods. In his current role as computational designer at Henningson, Durham, and Richardson (HDR, Inc.), he leverages digital strategies in architectural and transportation design projects.

Nicholas Hoban is a computational designer specializing in the field of digital fabrication, robotics, and computational workflows. He utilizes computer programming, simulation, and CAD/CAM programming to deliver data-driven design and prototypes. Nicholas is an instructor in the Technology stream of the John H. Daniels Faculty of Architecture at the University of Toronto, where he also oversees faculty and student digital fabrication projects as the Digital Fabrication Coordinator. Professionally, Nicholas is an Associate at Mesh Consultants Inc., a geometry studio and consultancy in Toronto specializing in solving critical problems in geometry through custom algorithms, software, and simulation.

Brady Peters, PhD is an architect and Associate Professor of architecture at the University of Toronto where he teaches design studio, computation, digital fabrication, and architectural acoustics. He is a director of Smartgeometry, an organization that promotes the use of computation in architecture. After graduating from Dalhousie (MArch 2001), Brady moved to London, England, and worked for Buro Happold and Foster + Partners, where he was an Associate Partner with their research and development team. He received his PhD in architecture from the Royal Danish Academy of Fine Arts in Copenhagen in 2015.

Ted Kesik, PhD is an Engineer and Professor of building science in the John H. Daniels Faculty of Architecture, Landscape and Design at the University of Toronto with a career focus on the integration of professional practice, research, and teaching. He has extensive experience in various aspects of building enclosure design, energy modeling, quality assurance, commissioning, performance verification, and building systems integration. Professor Kesik's research interests include resilience, sustainability, durability, high performance buildings, life cycle assessment, and building performance simulation. He continues to practice as a consulting engineer and is the author of books, studies, reports, and articles such as the "Mass Timber Building Science Primer" form the Mass Timber Institute at the University of Toronto.

The Sound of Kerfing

A New Approach to Integrating Geometry, Materials, and Acoustics to Build Invisibles

Alireza Borhani
Dr. Negar Kalantar
California College of the Arts

Erfan Rezaei Azari
Parametric House

Dr. Anastasia Muliana
Zaryab Shahid
Texas A&M University

Ed Green
Hottinger Bruel & Kjaer Inc.

1 Kerfonic Wall: Showcasing the application of "kerfD" in the acoustic field, Autodesk Gallery, San Francisco

ABSTRACT

Surfaces with kerf patterns are known for attaining various complex geometries to improve indoor acoustic performance. Currently, there is no available combined modeling and simulation tool to predict the acoustic effects of kerf surfaces. Due to the absence of such performance-based kerf design software, the authors explain how the outcomes of their acoustic simulation and experimental testing on kerf cells, panels, and a full-scale kerf wall have contributed to a script-based platform named "kerfD," which can serve as a decision-support setting for use during the early design stages. By integrating the parametric design environment, acoustic simulation tool, optimization process, and fabrication workflow into the design process, it is now possible to evaluate how kerf surfaces might manipulate sound waves to tune the acoustic characteristics of a space. The ultimate goal is to implement kerfD to serve as a generative design method for acoustic kerf form-finding. By examining the reciprocal relationship between kerf patterns and the soundscape of indoor spaces, this study presents a systematic approach to designing and fabricating kerf panels that address both aesthetic and acoustic qualities. Specifically, the authors demonstrate a potential application of kerfing in a permanent installation in a gallery space named "Kerfonic Wall" (Figure1). Furthermore, this research demonstrates how this wall modifies the acoustics characteristics of the gallery's open-plan space (e.g., reverberation time) through local tailoring of the geometric parameters of the kerf panels (e.g., pattern type, unit-cell size, cut density, cut thickness, etc.).

INTRODUCTION

Due to its economic benefits, ease of fabrication, and pleasing aesthetics, kerfing is now a ubiquitous technique in furniture making, boat construction, and architecture. Conventionally speaking, by using saw blades in multiple passes to create thin grooves close to one another, a piece of material can be bent into the desired radius, mostly in one direction. Since kerfing allows for the use of mass-produced materials of a standardized shape and size, acoustic panels have often been made this way for use in building interiors to serve as sound-dampening sheets (e.g., Linea & Dukta acoustic panels). Most of these acoustic panels contain multiple parallel groves produced by CNC routers to make flat or semi-cylindrical curved panels.

STATE OF THE ART

Adjustments to the kerf design features (such as orientation, length, width, frequency, depth, and density) allow for the production of various double-curvature assemblies, making delivery of their functional, structural, and aesthetic benefits possible. In the authors' design studios at Texas A&M University, a wide array of prototypes with different designs, dimensions, and patterns have been designed and fabricated, revealing the capacity for kerfing in architecture (Figure 2). All prototypes demonstrate kerfing's potential to address various functions, fabrications, aesthetic needs, and design specifications. Notwithstanding their great potential, these breeds of advanced kerfing have predominantly been limited to small- and medium-scale objects and temporary installations or notable research projects (Mamou-Mani and Jipa 2015; Konaković et al. 2016; Hoffer et al. 2012; Piker 2012; Lalvani 1996, 1998).

In kerfing, the material's properties, surface geometry, and treatment are all essential. Most available studies on kerfing have focused on the geometry of the cut pattern, kerf unit cell dimension, cut density, and amount of local stiffness and curvature (Hoffer et al. 2012; Ivanisević 2014; Kalantar and Borhani 2018; Chen et al. 2020, Shahid et al. 2022a, 2022b; Shahid et al. 2021). Despite the geometric development of these complex kerf patterns and their potential application in acoustics, their use in practice has been limited due to technical data appropriate for acoustic analysis. As a result of the complexity in modeling and simulating advanced kerfing, there is currently no software for acoustically evaluating kerf panels when integrating a parametric design environment with a fabrication workflow. The absence of such a performance-based platform causes difficulty in employing kerfing in the acoustic field.

A RECIPROCAL TRAJECTORY FOR ACOUSTIC DESIGN

The authors developed a script-based process named "kerfD," as an integrated tool to incorporate the outcomes of acoustic analyses with kerf design attributes (Figure 3). This

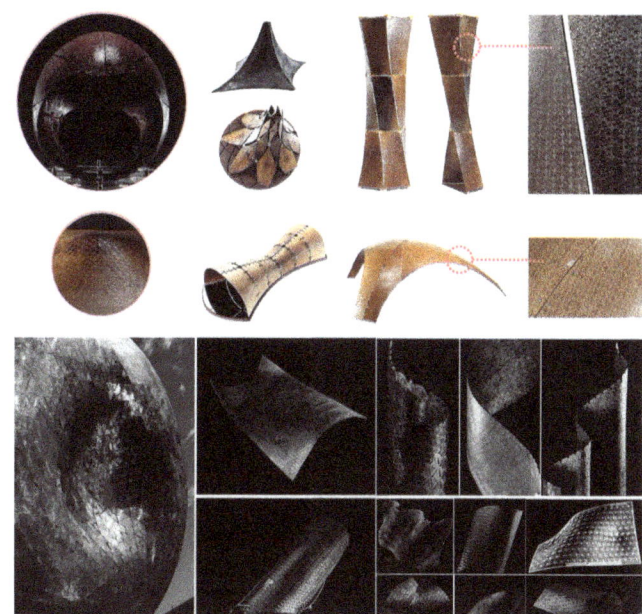

2 Fine-tuning the material's behavior via kerfing in the authors' design studios

performance-driven platform in Grasshopper is the product of four correlative processes, as follows:

- **Modeling:** Through a parametric design environment to control kerf design attributes
- **Simulation:** Via a ray-tracing acoustic simulation tool
- **Optimization:** Through an evolutionary solver to generate numerous design iterations
- **Fabrication:** Via an embedded fabrication workflow after finalizing the design

Here, the main objective is to establish a cyclical procedure, from forecasting sound distribution in an indoor room to suggesting kerf unit cells, their aggregated patterning, panel boundary conditions and topology, and the final form of clustered panels (Figures 4).

When creating different kerf panels with exclusive kerf patterns to form a full-scale installation with predicted acoustic properties, kerfD can be used in an iterative process to refine the proposed indoor soundscape. This iterative process involves a set of inputs and collection of outputs. To execute these approaches, a set of Grasshopper plugins and scripts are used, each complementing the others, including:

- **Relief-CUT (kerFIT):** An in-house plugin for Grasshopper that generates different kerf unit cell types, dimensions, and aggregated patterns within a defined boundary
- **kerFLUX:** A custom script that uses attractors to parametrize a heterogeneous kerf pattern's size, density, and regularity

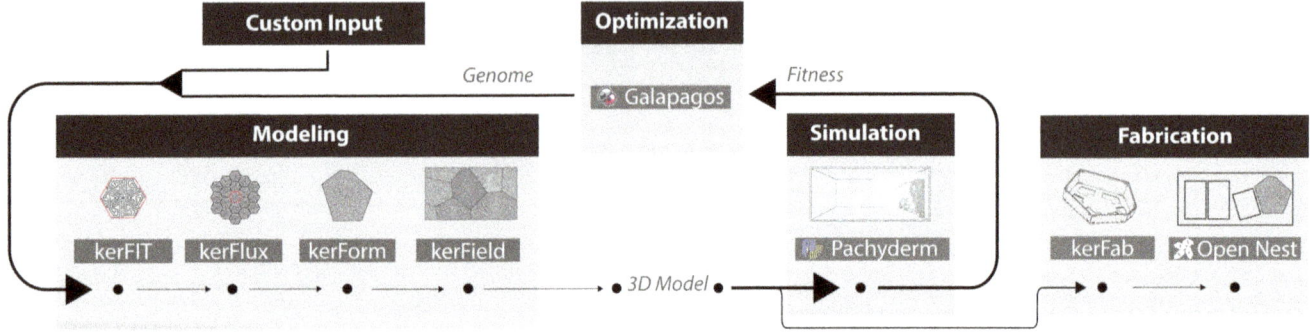

3   kerfD provides sufficient control on the modeling, simulation, optimization, and fabrication processes

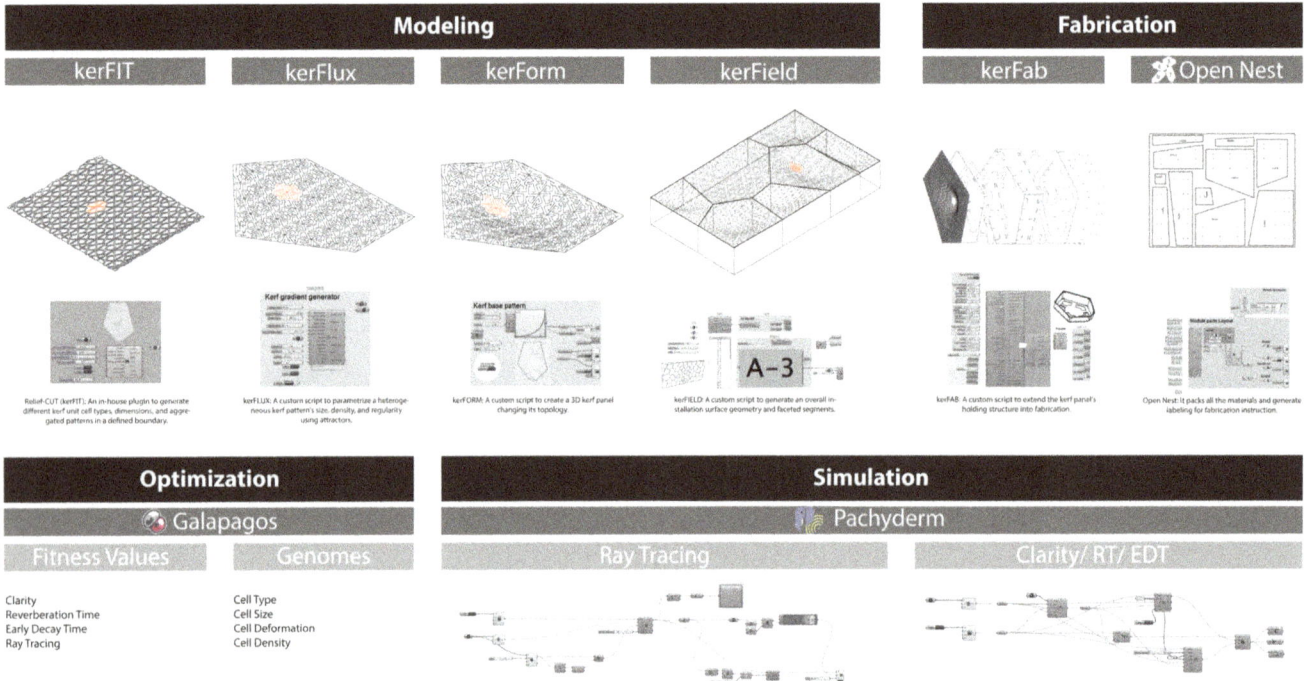

4   Informed by the material's logic and fabrication constraints, developing kerfD as a generative platform to make acoustically regulating kerf surfaces responds to sonic and spatial performance criteria in a correlative workflow

- **kerFORM:** A custom script that creates a 3D kerf panel by changing its topology
- **kerFIELD:** A custom script that generates an overall installation surface geometry and faceted segments
- **kerFAB:** A custom script that extends the kerf panel's holding structure into fabrication
- **OpenNest:** An open-source plugin for Grasshopper that fabricates 2D polyline packing via a laser or CNC cutter (Vestartas 2021)
- **Pachyderm:** An open-source geometrical acoustics laboratory for ray-tracing simulation (Harten 2015)
- **Galapagos:** An evolutionary solver used for acoustic design optimization (Rutten 2011)

Parametric Kerf Modeling

Depending on the desired acoustic output, several design attributes of a kerf panel can be simultaneously modified, including its unit cells, the cells' aggregated pattern, the kerfing plane's peripheral condition, the kerf panel topology, and the cluster of all kerf panels. These attributes are geometrically and acoustically dependent upon one another. The authors used an in-house plugin and custom scripts in Grasshopper to adjust these attributes and, ultimately, an interior space soundscape's characteristics. Here, the following design attributes will be reviewed.

UNIT CELLS AND THEIR AGGREGATION // As a series of cuts in close proximity, kerfing is able to turn a rigid planar material into various interconnected pliable elements, modifying the material's level of flexibility (Zarrinmehr et al. 2017a). Incisions made according to the kerfing method result in single or double-curved freeform shapes (Capone 2018). Relief-Cut (KerFIT), an in-house plugin, generates a parametric pattern of laser-cut lines (Figures 5). This plugin employs a remeshing process to produce a large subset of all possible 2D meander patterns that consist of interlocked

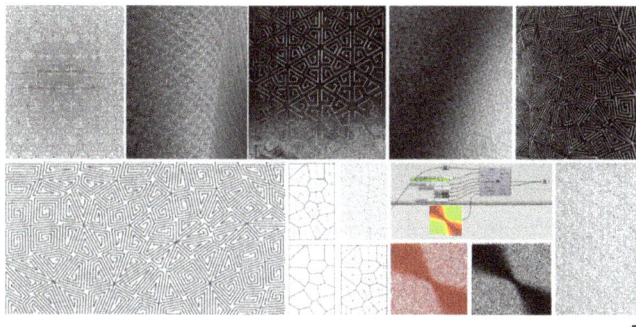

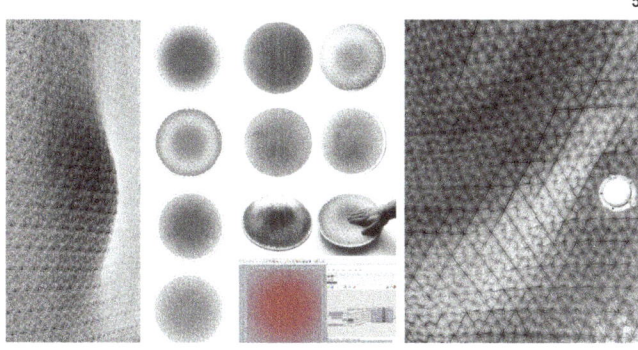

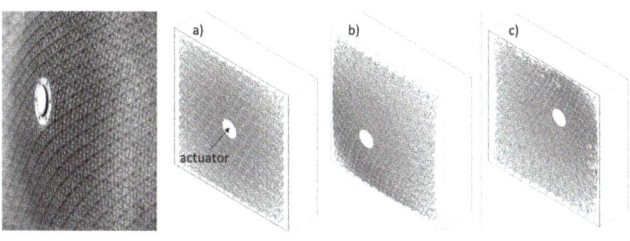

5 Relief-Cut (KerFIT): Using algorithms to obtain a large subset of meander patterns through a remeshing process

6 The size of each kerf density is parametrized to manipulate the overall flexibility and rigidity of the kerf surfaces

7 Proper attractors can be applied to change the kerf cut density and sheet bending stiffness locally

Archimedean spirals (Zarrinmehr et al. 2017a, 2017b; Kalantar and Borhani 2018). Although the use of uniform quad-based meander patterns is not a recent innovation (Ivanisević 2014), the proposed algorithm parametrically generalizes the method to any convex *n*-gon. Along with considering the material's properties, dividing a sheet into delicately linked spirals lets it be twisted along the axes of its torsional links. Adjustments to the cell type, length of cuts, and distance between each incision causes different torsional links and the attainment of various stiffness values for the material. The distribution of torsion stress along short links helps the surface bend in multiple directions (Porterfield 2016; Deffered Procrastination 2012). According to the material's resistance under a load, the denser the cuts are, the easier the cell will flex.

PERIPHERAL (BOUNDARY CONFINEMENT) // Although cell juxtaposition can impact soundscape characteristics, the kerfing plane boundary condition allows for different visual effects, structural performances, configurability, and acoustic outcomes. A kerf pattern can be extended on an infinite plane. However, limiting this pattern to a defined boundary creates a membrane. The boundary's geometry, size, and proportion change the membrane's flexibility and acoustic behaviors. When the patterns are restricted to defined boundaries, irregular shapes may appear where the cells are being trimmed.

ADVANCING FLEXIBILITY // Different cell shapes can potentially be combined when their scales and directions are modified. In the latter case, an attractor geometry can pull or push adjacent cells to change their scale and orientation. In the quest for finding the optimal geometry for flexibility, kerFLUX, a custom script, can be used to help morph a given tessellation. While maintaining the regularity of the kerf pattern, the cells are stretched and compressed in 2D to meet without gaps or overlaps. A series of attractors allows cells to jostle against one another to gradually shift in the desired direction. Varying the density of a homogeneous pattern impacts the local stiffness of any given region (Figure 6). The plugin's remeshing algorithm can be used to manipulate the local properties of the kerf pattern through the intensity of the grayscale gradient image as well. For example, darker areas can cause denser kerf patterns, making them locally more flexible at every surface zone.

PANELING (TOPOLOGICAL SURFACES) // A custom script called kerFORM can be used to support the out-of-plane topological transformation of flat kerf membranes and create nondevelopable panels with heterogeneous patterns. By using proper attractors at different heights, the entire kerf surface is subtly morphed in the direction perpendicular to the surface (Figure 7). In this way, the global properties of the membrane are changed, including its acoustic characteristics. In terms of both shear stiffness and stress-strain distribution, a piece of kerf s membrane is a structure that resists tension and compression. When constraining a flat kerf sheet by its box-holding perimeter and at least one circular ring on the sheet, keeping the sheet in tension forms it into a 3D form to minimize the waste of using a single-used mold (Figure 8). The circular ring depth (or height), located outside of the perimeter box plane, determines the amount of the sheet's tension. Although the material properties and applied kerf design affect the panel's internal forces, the ring radius and the kerf panel's concavity depth allow for a broad set of possible surface morphologies (Figures 9 & 10). When stretching the kerf pieces, the maximum stress is found on the circular ring edge (and exterior perimeter). Therefore, the flatter the kerf panel is, the less tension that appears in those areas (Figure 11). Also, the panel's shape and height difference between its perimeter and ring affect the stress concentration at the

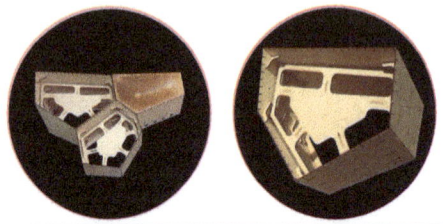

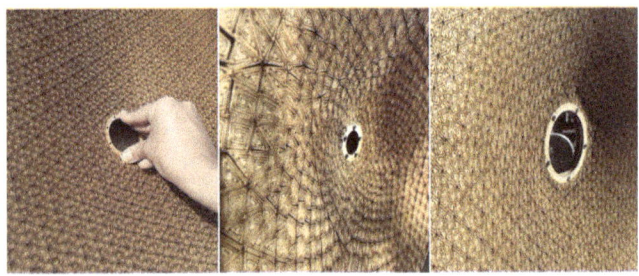
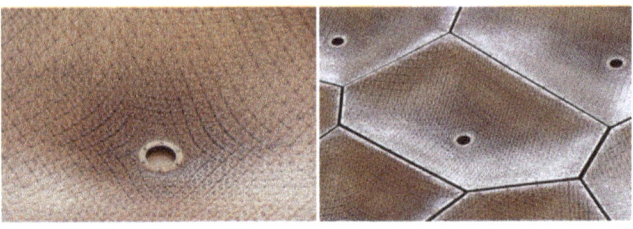

arrangements for the faceted kerf wall panels (Figure 12). The kerFIELD custom script is used to generate an overall installation surface geometry and faceted segments. Then, depending on the defined constraints and Gaussian curvature of the performative acoustic structure, kerFIELD parametrically adjusts the emergent panels' properties, including their perimeters, quantities, curvatures, strengths, and permeabilities (Figure 13). The regular and irregular arrangements of the double-curved panels meeting edge-to-edge results in an interplay between performance and appearance, together forming the indoor space.

Acoustic Kerf Simulation

kerfD is a platform to automate the generation process of performative kerf structures for the given acoustic requirements of the site. Various options should be explored to find the optimal kerf panels and patterns. The authors set up physical tests and simulations of different kerf unit cells and their aggregations to explore their mechanical and acoustical behaviors (Figure 14). The goal was to predict and manipulate how kerf patterns might help a sheet of material deform in certain ways to propagate sound waves.

Since the sound quality of a kerf-based design is based on the proper adjustment of the design attributes, a number of digital simulations and experimental measurements were conducted to support the authors' understanding of how sound is absorbed, reflected, or travels around and through a kerf piece. Several kerf factors such as cut thickness and bending curvature were found to influence the sound absorption coefficients and reverberation times of a kerf cell at frequencies of 125 to 4,000 Hz. The amount of this influence is frequency-dependent (Holterman 2018). Accordingly, numerous freeform kerf panels were simulated by altering these factors on kerf surfaces through the formulation of nonlinear beam element models (Darnal et al. 2021; Chen et al. 2020). These simulated models exhibited a wide range of wave manipulation associated with their periodicity, while capturing the modal responses of the kerf samples identified from experiments (Shahid et al. 2022c) (Figure 15).

corners. Therefore, the incremental manipulation of the cut lines' density creates a more gradual bending angle, avoiding a sharp kink transition and stress concentration.

CLUSTERING // Once the design constraints and acoustic requirements are identified, the kerf panels can be accumulated as a cluster. Various segmentation strategies can then be explored within the global wall geometry to find the best

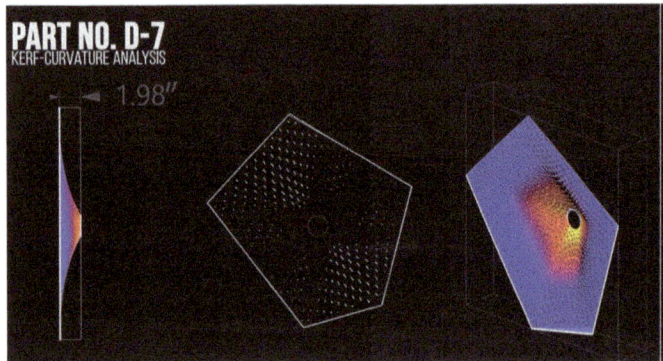
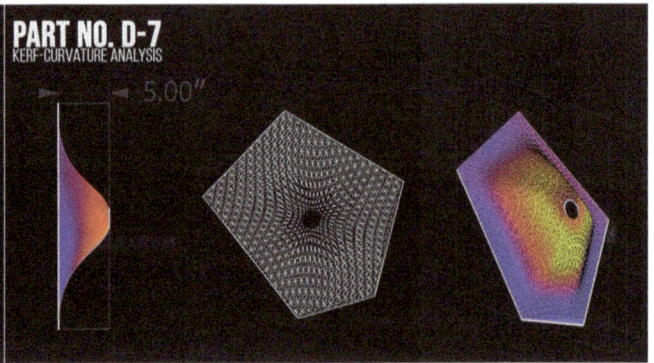

8 Using the holding box as a formwork to create a double-curve kerf panel

9 The ring location and diameter impact the panel's final form

10 Achieving different surface morphologies by manipulating sheet bending stiffness

11 The panel border and the ring height affect the stress concentration

12 Developing a parametric approach to generate a wall surface articulation, resulting in a differentiated acoustic subspace within an open-plan space

13 The global geometry of the kerf installation and panels' forms are fine-tuned when the kerf design attributes are calibrated to fulfill the acoustic performance criteria

In this way, any panel's corresponding sound wave propagation behavior could be forecast by adjusting the kerf factors (e.g., unit-cell size, cut density, orientation, pattern, etc.) and topology. Along with the simulated kerf models, all laboratory-scale tests and in-situ measurements helped to establish proper evaluation metrics for analyzing the acoustic performance of a kerf installation.

One main obstacle to developing an acoustically effective kerfing design is the lack of an integrated platform that combines kerf modeling, simulation, and fabrication workflow in a single environment. To simplify data exchange and interoperability across different software (Badino et al. 2020), Pachyderm, a ray-tracing plugin for Grasshopper (Harten 2015) can be employed to work within the kerfing modeling environment. In Pachyderm, kerf design attributes are utilized in two ways. To simulate energy reflection in the ray-tracing method, the outcome of the kerFIELD script is considered a 3D model within a room. Also, the results of both physical testing and simulation of various kerf cells with different densities are regarded as material specifications.

Acoustic Kerf Optimization

By defining the kerf design attributes as output relationships (or genomes) in Galapagos (an evolutionary solver in Grasshopper), multiple design iterations can be produced to meet optimized acoustic fitness goals during a preferred runtime (Wright et al. 2016). After connecting the pool of variable sliders of the kerf attributes' inputs, Galapagos compares their values to assign a score to the ray-tracing simulation (Jededia and Vlaun 2015). The design characteristic of the final performative kerf-based structure is the outcome of the interactive ray-tracing simulation process for acoustic optimization. The data derived from the acoustic analysis tool helps identify the most optimized kerf attribute geometries.

Fabrication Workflow

Capable of maintaining the geometric relations of the proposed kerf structure, kerFAB is another custom script that can be employed to create all box fabrication features to hold a kerf panel (e.g., spacers, holding arms, tilted plates, finger joints, nut holders, holes) when parametrically controlling the variables (e.g., material thickness, dimensions, corner conditions) (Figure 16). The kerf holding box is unfolded into a series of laser-cut pieces. OpenNest, an open-source plugin for Grasshopper, can be used to automate 2D layouts

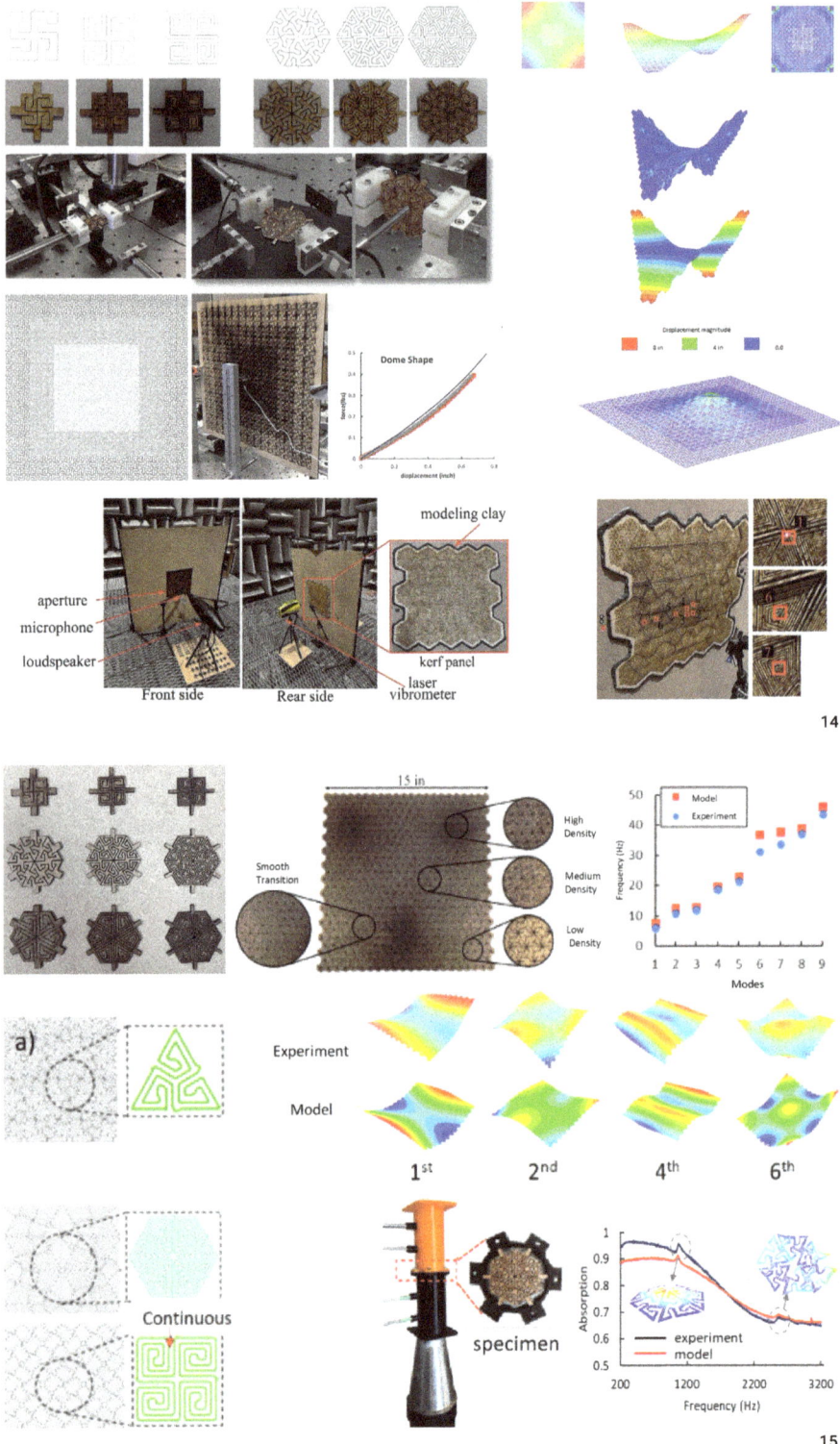

and packing for laser cutting. By combining the acoustic and fabrication optimization workflow, kerFAB delivers the kerf holding boxes' fabrication documents to subsequently be cut by a laser machine or CNC router (Figure 17).

THE BALANCE BETWEEN VISUAL AND AURAL

To showcase the application and benefit of "kerfD" in the acoustic field, the authors present one of their projects named Kerfonic Wall, a 20' x 8' permanent installation in a boardroom lounge of Autodesk Gallery in San Francisco (Figure 18).

Since the sound environment of the lounge area was not conducive to intimate and private conversations in the open-plan layout, the goal was to provide an aesthetically pleasing wall out of kerfing to improve the acoustic performance of the lounge, creating a diffuse soundscape within the gallery's echoey space. Besides simulating the acoustic effects of the wall at the early design stage, the sound quality of the lounge was physically tested and validated onsite to ensure the desired auditory experience.

In this project, by prioritizing architectural acoustics over a performance checklist, the sonic characteristics of the lounge area, its volume, wall surfaces, and materials (Long 2006), all became important. The overall wall is designed to create a subspace within the lounge, discerned subtly through the local acoustic environment. By building equal ground between visual and aural aesthetics, a multidimensional acoustical experience is made possible for those visiting the installation (Figure 19). One approach to efficiently diffusing sound is to have different wall panels reflect sound waves from their surfaces and concentrated near the surface. This minimizes any echo effect and reduces reverberation time.

14 Kerf cell and kerf panel testing and velocity spectra of different points on the kerf panel

15 Modal responses of a kerf surface with a varying cut density: an attempt to audibly recreate these behaviors digitally; impedance tube testing and absorption coefficient of an undeformed (flat) kerf unit cell

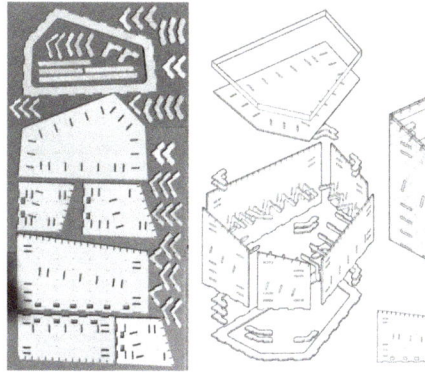
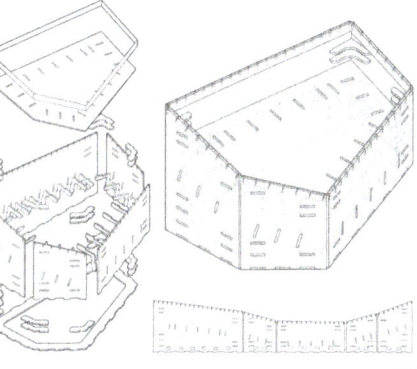

Besides modeling, simulation, and optimization of the Kerfonic Wall, KerfD was used to facilitate the fabrication process. For the Kerfonic Wall, due to the COVID-19 pandemic lockdown and lack of access to any fabrication equipment (except for a laser cutter), the authors had to develop a proper strategy for fabricating an entire three-dimensional wall installation with a 2.5D tool. In this project, the development of kerFAB was central to generating the fabrication files for producing the cell components, connectors, and spacers (Figure 20).

In the Kerfonic Wall, more than 54 boxes were custom-cut out of Plywood and distributed in an irregular pattern across the wall to serve as frames to hold kerf-cut surfaces. To incorporate all material and fabrication tolerances, a gap was left between each of the boxes (Figure 21). Since each kerf panel was unique in size, depth, and angle, a parametric model was employed to translate the surfaces into 2D planar sheets for laser cutting (Figure 22).

As explained, by establishing a parametric definition to create custom cuttings of meander patterns, unique volcano-shaped panels out of flat plywood sheets were made. In this way, the authors surpassed the underlying limitation of fabricating custom-made double-curve forms (Ohshima et al. 2013) without single-use molds when using the laser cutter (Figure 23). Using these structure boxes as formworks is a step towards sustainable fabrication.

In the Kerfonic Wall, the kerf panels have been locked into their shapes by applying an epoxy resin. The goal was to decrease the risk of damage caused by placing sudden stress on the kerf cells.

RESPONDING TO SONIC AND SPATIAL PERFORMANCE CRITERIA

The lounge space's sound quality was considered early on in the design process, both in terms of the wall's geometrical

16 Providing in-house parametric scripts to obtain fabrication layouts for cutting and assembling

17 kerFAB helped to control different phases of the fabrication process

18 Inventive in function and advanced in design, the wall utilizes as a design driver both formal and material approaches to satisfying sonic comfort

19 Diffusing the sound as much as possible and reducing its reverberation within the space: integration of material and geometry

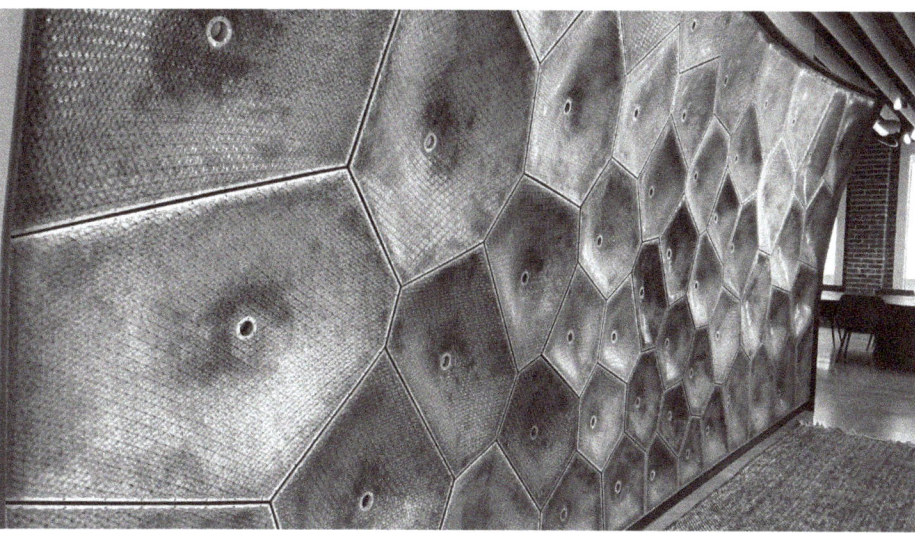

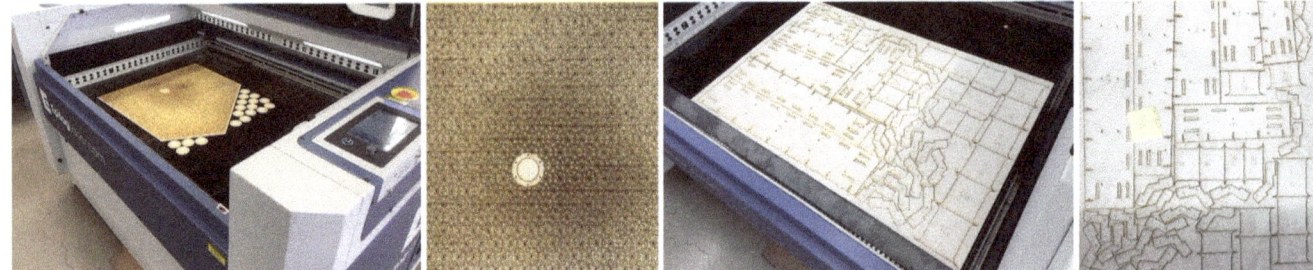

20 Limiting the fabrication process to the capacity of a laser cutter due to the pandemic lockdown

attributes and material behavior of the kerf surface. The final wall now plays an active role in improving the acoustic performance of the lounge space, as follows.

Geometric Attributes

In two ways, the geometric complexities of the wall and undulating surface of its boxes are imperative to the distribution of sound within the interior and attenuation of the reflected sound energy density.

WALL GEOMETRY // The control of sound reflection is a fundamental aspect of acoustic design. Therefore, the wall's overall curved geometry, combined with the irregularity and angles of its polygonal boxes, contributes to reflecting incident sound in as many directions as possible. A parametric strategy was developed to orchestrate several topological variations, make an acoustically heterogeneous space, and influence the characteristics of the acoustic field. Before the wall was installed, the lounge space had two flat parallel walls that acoustically produced an infinity mirror effect. Consequently, speech quality was degraded because the energy of middle- and high-frequency sounds was trapped between the ceiling and floor and moved back and forth between the parallel walls. In the proposed design, by inclining the parallel walls and turning one of them into a non-planar surface, sound energy is reduced when moving across the space. Beginning with the formation of four splines with different radii and tangent angles, the overall wall geometry in the gallery was swept between these curves, creating a Degree 3 surface with a smooth transition from bottom to top and left to right (Figure 24). To occupy the minimum amount of gallery space, the bottom edge of the wall was given a shallow curvature close to the existing onsite drywall. The wall's right edge had a slight curvature so as not to block the view or access. However, the left edge adapted itself to the geometry of the adjacent circular window. To maintain the tangential continuity of the wall's surface, the top wall edge was given an accentuated curve towards the window.

PANEL GEOMETRIES // Here, a family of discrete brick-like boxes was stacked with bolted connections to create a Voronoi tessellation when angled in different directions, to

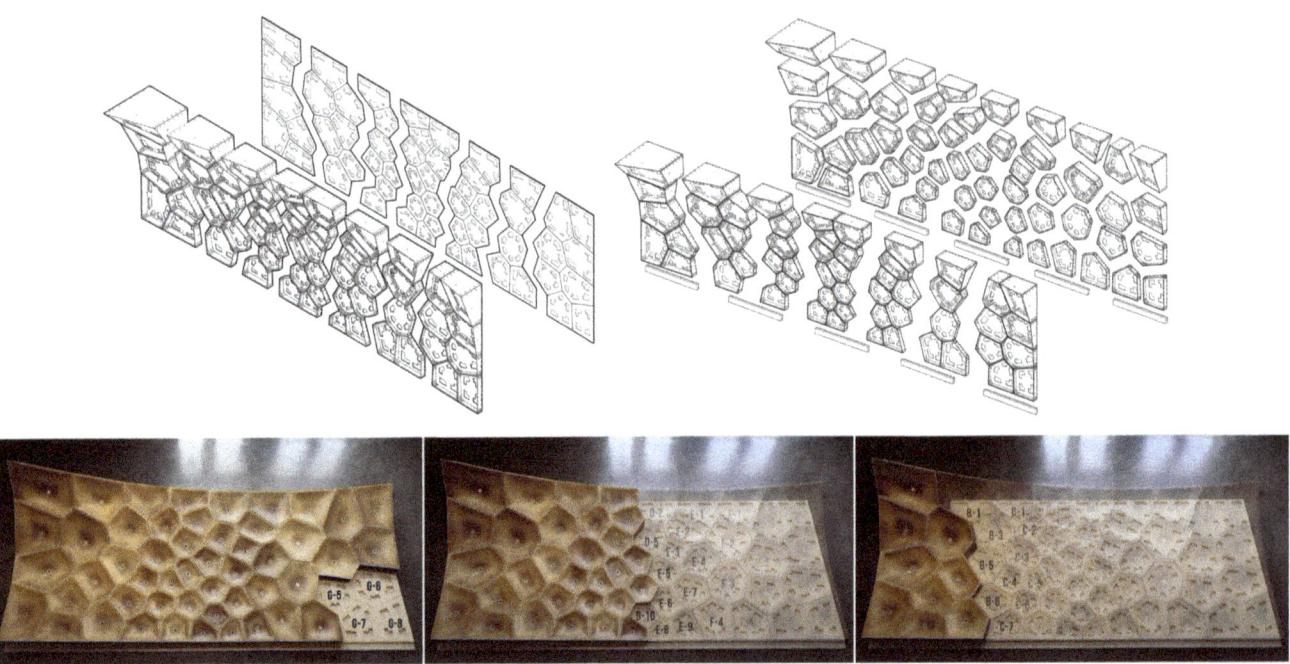

21 Assembling the aggregation of 54 irregular boxes

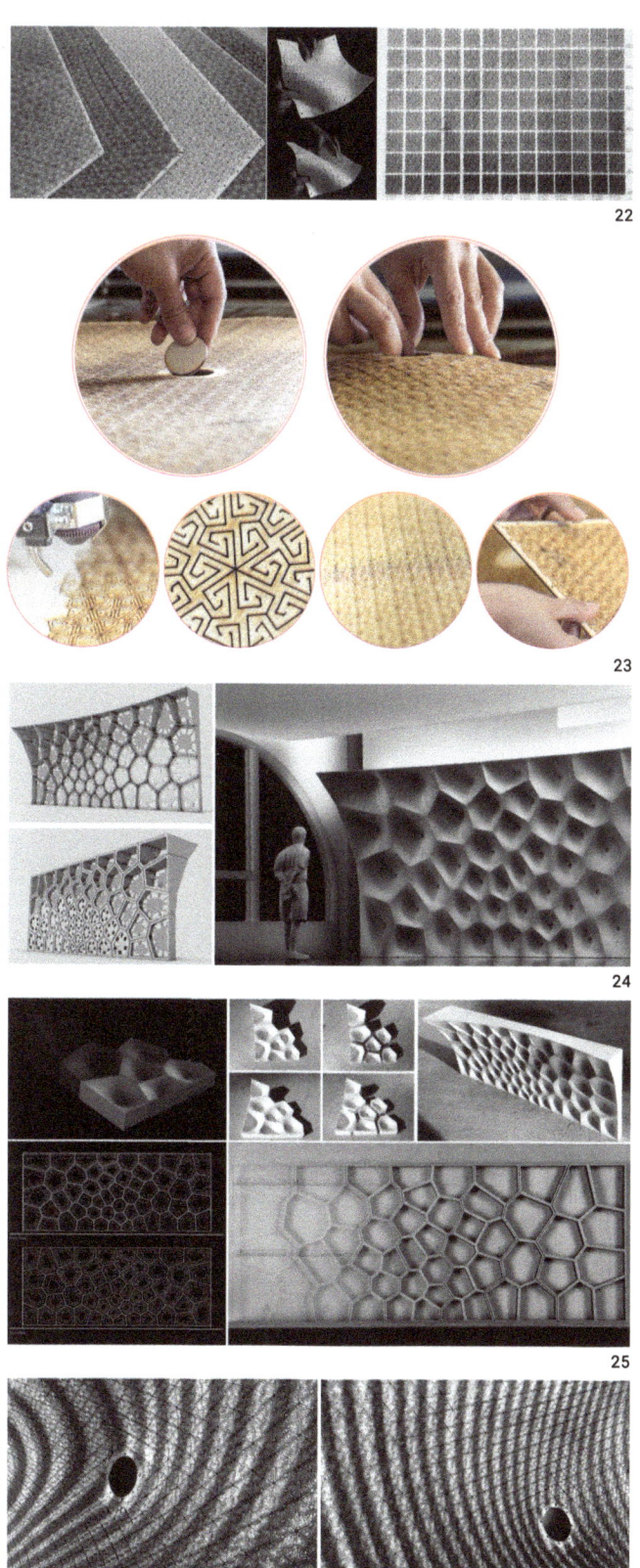

avoid sound-focusing effects (Figure 25). As mentioned above, to minimize the wall footprint and maximize the volumetric impressions, the wall was sculpted to shape a 3D structure projecting the overhung boxes in each section. Each box contains a double-curved kerf panel (or box cover). Varying the angle, curvature, and depth of each panel helps to deflect and scatter sound waves in multiple directions, due to local sound focusing. Since the wave distribution at various locations close to the surface has a critical impact on weakening the strength of the reflected sound and avoiding unwanted echoes (Belanger et al. 2018; Wulfrank et al. 2018), the geometry of each panel produces a better diffusion of sound and minimizes too strong a focus (Figure 26).

Material Behavior

Deploying kerf-cut surfaces as absorbing materials and employing pixilated cells to scatter sound (Holterman 2018) enhanced the auditory experience in the gallery. Holterman argued that acoustic performance depends on the kerf pattern, kerf width, curvature, depth of kerf surface, and depth of the air cavity behind the surface. A parametric model used in this project changes these factors to achieve the desired acoustic performance. Relying on the bending stiffness of the plywood, the kerf surfaces are kept in tension when pushed toward the bottom of the boxes. These inverted bubble soap-like surfaces not only look aesthetically pleasing but also improve acoustics by reducing the sound echoes in the space (Figure 27). Via a combination of sound absorption and diffusion, the kerfing technique improved the sound performance in two ways, while still leaving a live-sounding space (Bradley 2009; Williams et al. 2013):

PARTIAL ABSORPTION // A kerf panel resembles a porous surface that induces perforations. The corresponding solid portion of the kerf piece acts as multiple local resonators (Figure 28). When the kerf piece is deformed to a specific shape, its gap undergoes scale changes and rotations that alter the airflow, influencing sound wave propagation

22 Making and testing many kerfing patterns while utilizing a research-through-design methodology coupled with computational simulation and fabrication advancements.

23 Using a laser cutter to effortlessly transform a flat sheet of Plywood into a double-curved panel cost-effectively.

24 The overall wall geometry Responds to Sonic and Spatial Performance Criteria.

25 Comprised of a series of brick-like boxes with kerfed surfaces, the geometry of the final assembly and its components can be adjusted to impact acoustic performance.

26 The geometry of kerf panels facilitates sound scattering without producing loud and quiet spots.

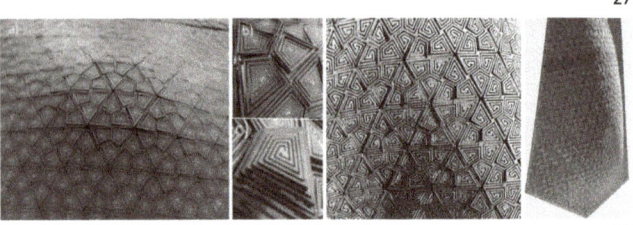

and absorption capacity. Modifying the flexibility of a stiff plywood sheet by kerfing it while also breaking it down into smaller parts makes it act as a resonator, contributing to the absorption of sound in ways much like Helmholtz absorbers in perforated panels (Komkin et al. 2017). The permeability of kerf surfaces allows for the dissipation of incident sound energy. As a perforated sheet, the kerf surface provides more absorption of the lower frequencies found in the human voice spectrum (Chile 2018). The balance between porosity level and sound absorption capacity of kerf panels allows the space to remain lively, without any deadening.

DIFFUSION // Kerfing helps to generate 3D patterns to guide sound diffusion. The slight unevenness of the kerf panels contributes to a better spreading of sounds across the lounge area, while also providing fewer opportunities for focusing acoustic energy. Also, by providing the possibility of effortless creation of double-curved surfaces, kerf patterns help to redirect sound waves and lessen distinct echoes and sound coloration (Long 2006).

RESULTS: ACOUSTIC PERFORMANCE MEASUREMENTS

In the lounge installation described before, in order to determine the effect of the kerf wall on the acoustic performance,

27 Altering material behavior and form to make performative surfaces, enhancing the interior soundscape and improving speech privacy

28 Kerf surfaces with a macro-micro scale topology

29 Conducting two acoustic tests before and after the wall installation

30 Reverberation time testing before and after kerf wall installation

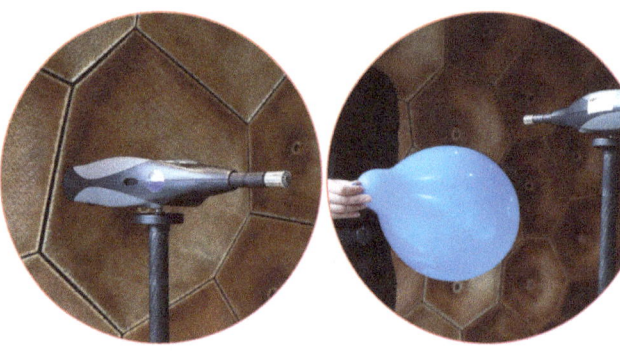

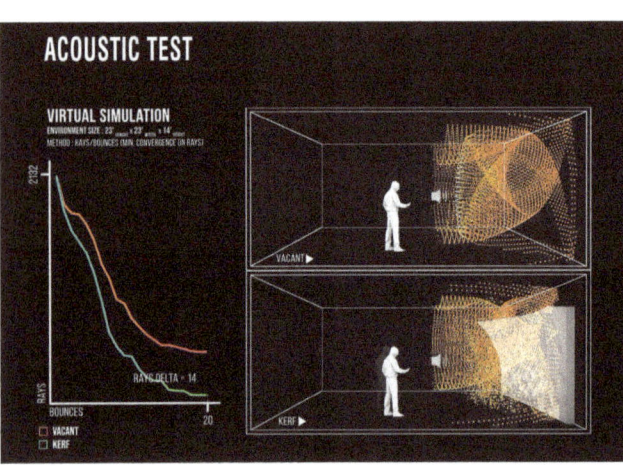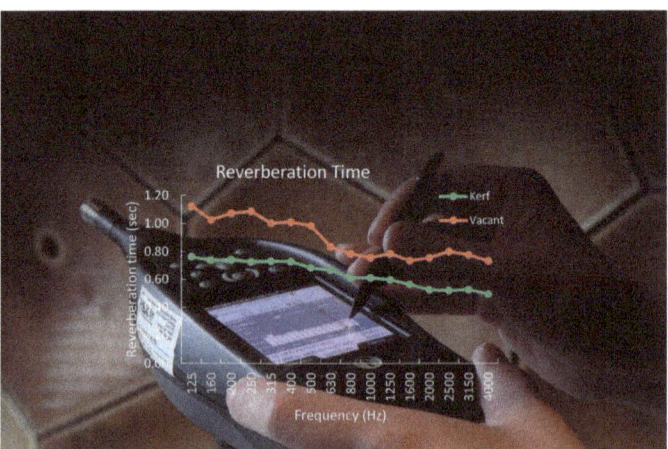

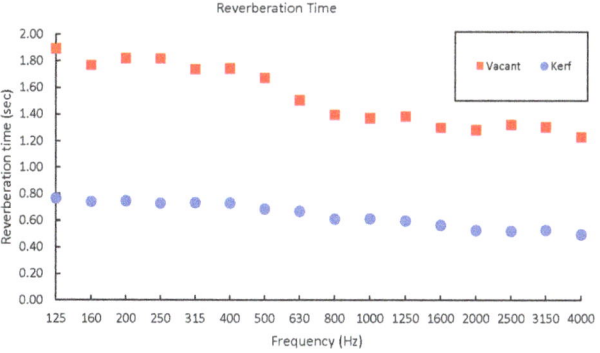

TABLE 1: Reverberation time results for vacant and kerfed rooms

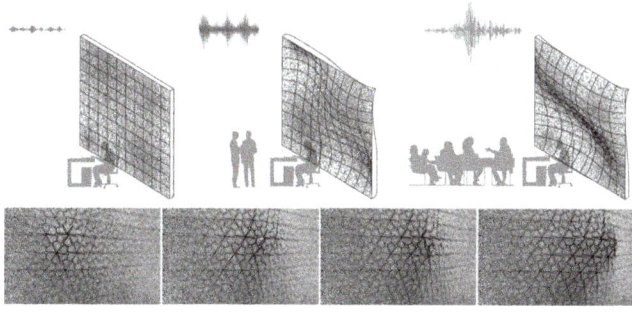

31 Creating adaptive sound space while using configurable kerf panels

reverberation time tests were performed in accordance with ISO 3382-2 (Figure 29).

The reverberation measurements were made using an Analyzer (Type 2270) from HBK (Hottinger Bruel and Kjaer, Naerum, Denmark). The analyzer used reverberation time software (BZ-7227) to measure the reverberation time at a 1/3 octave. As recommended in ISO 3382-2, the Impulsive Excitation Method (or Schroeder Method) with a balloon burst was used in the tests. The reverberation time was measured before and after the installation of the Kerfonic Wall in the space (Figure 30). All testing parameters (e.g., source and receiver locations) were kept the same throughout the test to avoid any bias in the measurement data.

Based on recommendations in ISO 3382-2, six Receiver (Analyzer) positions were chosen in the space. This ensured that there was at least half a wavelength distance between the receiver positions. Moreover, the receiver positions were chosen to be at least a quarter wavelength away from any reflecting surface. For each receiver position, two source positions (balloon bursts) were selected to obtain sufficient data. The frequency bandwidth chosen for the reverberation tests was 125 to 4,000 Hz. The average results for all positions at each frequency are shown in Table 1. The results show that the Kerfonic Wall significantly improved the acoustic performance of the space by decreasing the reverberation time. The deployment of the Kerfonic Wall increased the effective absorption of the space, thus decreasing the echo of the sound in the lounge.

DISCUSSION: FUTURE WORK

Fusing design complexity with construction simplicity, kerfing facilitates the possibility of designing for acoustic performance. In developing kerfD, the authors will share greater insight into the acoustic behavior of more advanced kerf installation walls and ceilings. In the future, through more in-depth physical testing and simulation of diverse kerf cells, patterns, and panel topologies, a reliable and complete materials library of macro- and micro-scale kerf samples can be obtained. The outcomes of this work can be used to predict and optimize the impact of complex designs on sound wave propagation across a wide range of audible frequencies.

As discussed, the kerfing approach proposed here indicates two possible types of sound treatment technique in one: diffusion and absorption. The improvement of kerfD will lead to the development of further iterations of acoustically reflective and absorptive surfaces, offering versatile and adaptable acoustic characteristics that focus and diffuse sound on demand. In this way, this research will support future adaptive acoustic designs that utilize responsive kerf surfaces. The authors believe that kerf patterns have the potential for active sound wave manipulation. Thus, an acoustically responsive surface is a legitimate possibility when reconfiguring surface topology with digitally operable components (Figure 31).

CONCLUSION

The interactions among auditory imperatives, geometric parameters, materiality, fabrication constraints, assembly rationale, and the space's programmatic considerations were included as design drivers. Besides outlining a digital workflow for the design, simulation, and fabrication of a wall in an open-plan gallery that minimizes speech distraction, this research presents the authors' investigation of the related acoustic potential of double-curved kerf panels. The faceted kerf wall installation in the gallery exemplifies how an aesthetically pleasing architectural element can guarantee a predictable auditory condition when kerf design attributes are incorporated into a global wall geometry and segmentation.

In conjunction with physical and simulation testing, this work outlines the necessity, process, and application of a script-based process named kerfD, for use when parametrically controlled kerf variations complement acoustic simulations in a correlative workflow. The utility of scripting helped establish a generative design tool for future acoustic kerf form-finding. By considering acoustic performance as a driver in the early design stages, the conventional distinction between architectural development and acoustic optimization is blurred.

ACKNOWLEDGMENTS

This research was partially supported by the National Science Foundation under EAGER: DREAM-B: Collaborative Research: Moldable and Wave Tunable Materials for Complex Freeform Structures (Award Abstract # 1912823 & 1913688). The views contained in this material are those of the authors and do not necessarily reflect the views of the National Science Foundation. The authors want to thank the Autodesk Technology Center in San Francisco for supporting the fabrication processes.

REFERENCES

Belanger, Zackery, Wes McGee, and Catie Newell. 2018. "Slumped Glass: Auxetics and Acoustics." In *ACADIA 18: Recalibration, On Imprecision and Infidelity; Proceedings of the 38th Annual Conference of the Association for Computer Aided Design in Architecture (ACADIA)*. Mexico City, Mexico. 244–249. https://doi.org/10.52842/conf.acadia.2018.244.

Badino, E. 2018, "Acoustic performance-based design: Exploration of the effects of different geometries and acoustical properties of an urban façade." Master thesis, Department of Energy, Politecnico di Torino, Italy. https://webthesis.biblio.polito.it/6746/1/tesi.pdf.

Badino, E., L. Shtrepi, and A. Astolfi. 2020, "Acoustic Performance-Based Design: A Brief Overview of the Opportunities and Limits in Current Practice" *Acoustics* 2 (2): 246-278. https://doi.org/10.3390/acoustics2020016.

Bradley, J. S. 2009. "A new look at acoustical criteria for classrooms." NRCC-51320, National Research Council Canada.

Capone, Mara, and Emanuela Lanzara. 2018. "Kerf Bending: Ruled Double Curved Surfaces Manufacturing." In *Blucher Design Proceedings*. São Carlos, BR: Editora Blucher. 653–660. https://doi.org/10.5151/sigradi2018-1389.

Chen, R., C. Turman, M. Jiang, N. Kalantar, M. Moreno, and A. Muliana. 2020. "Mechanics of kerf patterns for creating freeform structures." *Acta Mechanica* 231: 3499–3524.

Chile, Etex. 2018. "How To Improve Acoustic Comfort with Perforated Cardboard Plasterboards." *ArchDaily* (blog), translated by Z. Montano. Accessed May 27, 2022. https://www.archdaily.com/906544/how-to-improve-acoustic-comfort-with-perforated-cardboard-plasterboards.

Darnal, A., Z. Shahid, J. Han, M. Moreno, and A. Muliana. 2021. "Viscoelastic Responses of MDF Kerf Structures." In *Proceedings of the American Society for Composites; 36th Technical Conference on Composite Materials*. Texas A&M University. https://doi.org/10.12783/asc36/35749.

Deffered Procrastination. 2012. "Lattice Hinge Design — Minimum Bend Radius." Accessed May 25, 2022. https://www.defproc.co.uk/analysis/lattice-hinge-design-minimum-bend-radius/.

Dukta Acoustic Systems. n.d. Accessed May 25, 2022. https://dukta.com/en/products/acoustic-systems/.

Hoffer, B., G. Kahan, T. Crain, and D. Miranowski. 2012. "Kerf Pavilion". https://architecture.mit.edu/architecture-and-urbanism/project/kerf-pavilion.

Holterman, Andreas. 2018. "Pattern Kerfing for Responsive Wooden Surfaces: A Formal Approach to Produce Flexible Panels with Acoustic Performance." Master thesis, TUDelft University. https://repository.tudelft.nl/islandora/object/uuid%3Afecba343-a113-470b-8e41-1b1b02528bb7.

Ivanišević, D. 2014. "Super flexible laser cut Plywood." Kofaktor Lab (Experiments). Accessed May 25, 2022. http://lab.kofaktor.hr/en/portfolio/super-flexible-laser-cut-plywood.

Kalantar, Negar, and Alireza Borhani. 2018. "Informing Deformable Formworks; Parameterizing Deformation Behavior of a Non-Stretchable Membrane via Kerfing." In *Learning, Adapting and Prototyping, Proceedings of the 23rd International Conference of the Association for Computer-Aided Architectural Design Research in Asia (CAADRIA)*, vol. 2. Hong Kong: CAADRIA. 339–348. http://papers.cumincad.org/data/works/att/caadria2018_343.pdf.

Komkin, Alexander I., M. A. Mironov, A. I. Bykov. 2017. "Sound Absorption by a Helmholtz Resonator." *Acoustical Physics* 63 (4): 385–392. https://doi.org/10.1134/S1063771017030071.

Konaković, M., K. Crane, B. Deng, S. Bouaziz, D. Piker, and M. Pauly. 2016. "Beyond Developable: Computational Design and Fabrication with Auxetic Materials." *Proceedings of SIGGRAPH, ACMTransactions on Graphics* 35 (4): 1–11. https://doi.org/10.1145/2897824.2925944.

Lalvani, Haresh. 1996, Periodic and non-periodic tilings and building blocks from prismatic nodes, US Patent, https://patentimages.storage.googleapis.com/80/ed/e0/32956fcd5d0722/US5575125.pdf

Lalvani Studio, n.d. "XURF, 1998-present." Accessed May 24, 2022. http://lalvanistudio.com/architecture/xurf-purf/.

Linea Ceiling & Wall Systems. n.d. "LINEA Acoustic Kerf, T&G." Linea website specifications. Accessed May 20, 2022. https://www.lineaceilings.com/linea-acoustic-kerf-tg/.

Long, Marshall. 2006. *Architectural Acoustics*. London: Elsevier Academic Press.

Mamou-Mani, Arthur, and Andrei Jipa. 2015. "The Wooden Waves – Burohappold Engineering." *Mamou-Mani Ltd*. Accessed May 19, 2022. https://mamou-mani.com/project/wooden-waves/.

Ohshima, Taisuke, Takeo Igarashi, Jun Mitani, and Hiroya Tanaka. 2013. "WoodWeaver: Fabricating curved objects without moulds or glue." In *Proceedings of the 31st International Conference on Education and research in Computer Aided Architectural Design in Europe (eCAADe)*, vol. 1. 693–702

Piker, Daniel. 2012. Variation from Uniformity. *Space Symmetry Structure* (blog), October 15, 2012. Accessed May 23, 2022. https://spacesymmetrystructure.wordpress.com/2012/10/15/variation-from-uniformity/.

Porterfield, A. 2016. "Parametric-Kerf." *F=F* (blog). Accessed May 24, 2022. https://fequalsf.blogspot.com/p/parametric-kerf_5.html.

Vestartas, Petras. 2021. "OpenNest - 2D Polyline Packing for fabrication such as laser or CNC cutting." https://www.food4rhino.com/en/app/opennest.

Renzhe, C., C. Turman, M. Jiang, N. Kalantar, M. Moreno, and A. Muliana. 2020. "Mechanics of Kerf Patterns for Creating Freeform Structures." *Acta Mechanica* 231 (9): 3499–3524. https://doi.org/10.1007/s00707-020-02713-8.

Rutten, David. 2011. "Evolutionary Principles applied to Problem Solving." *I Eat Bugs for Breakfast* (blog), March 4, 2011. Accessed May 26, 2022. https://ieatbugsforbreakfast.wordpress.com/2011/03/04/epatps01/.

Shahid, Z., J. E. Hubbard, N. Kalantar, and A. Muliana. 2022a. "An investigation of the dynamic response of architectural kerf structures." *Acta Mechanica* (2022): 1–25.

Shahid, Z., C. G. Bond, M. S. Johnson, J. E. Hubbard, N. Kalantar, and A. Muliana. 2022b. "Dynamic Response of Flexible Viscoelastic Kerf Structures of Freeform Shapes." *International Journal of Solid and Structures* 254-255 (Nov): 111895.

Shahid, Z., E. Green, R. Hadjit, J. E. Hubbard, N. Kalantar, and A. Muliana. 2022c. "An Exploration of Acoustic Characteristics of Kerf Unit Cells." Under Review.

Shahid, Z., M. S. Johnson, C. G. Bond, J. Hubbard, N. Kalantar, and A. Muliana. 2021. "Dynamic Responses of Architectural Kerf Structures." In *Proceedings of the American Society for Composites; 36th Technical Conference on Composite Materials*.

van der Harten, Arthur. 2015. "Pachyderm acoustical simulation: an open-source geometrical acoustics laboratory." Github. https://github.com/PachydermAcoustic.

Vlaun, N. J. V. 2015. "Sound Working Environments; Optimizing the acoustic properties of open plan workspaces using parametric models." Master thesis, TUDelft. http://resolver.tudelft.nl/uuid:9ea83381-3a19-4486-ae16-f25a68811b6f.

Williams, N., D. Davis, B. Peters, A. Pena de Leon, J. Burry, and M. Burry. 2013. "FABPOD: An Open Design-To-Fabrication System." In Proceedings of the 18th International Conference on Computer-Aided Architectural Design Research in Asia (CAADRIA). Hong Kong: CAADRIA. 251–260.

Wright, O,. N. Perkins, M. Donn, and M. Halstead. 2016. "Parametric implementation of café acoustics." In *Proceedings of Acoustics 2016*. Brisbane, Australia. https://www.acoustics.asn.au/conference_proceedings/AASNZ2016/papers/p13.pdf.

Wulfrank, T., E. Green, E. Kahle, and J. O. García Gómez. 2017. "Creative possibilities and limitations of curved surfaces in the acoustic design of contemporary auditoria." In *24th International Congress on Sound and Vibration (ICSV24)*. https://kahle.be/articles/ICSV24-KahleAcoustics-CurvedSurfaces-final.pdf.

Zarrinmehr, S., M. Ettehad, N. Kalantar, A. Borhani, S. Sueda, and E. Akleman. 2017a. "Interlocked archimedean spirals for conversion of planar rigid panels into locally flexible panels with stiffness control." *Computers & Graphics* 66 (Aug): 93–102. https://doi.org10.1016/j.cag.2017.05.010.

Zarrinmehr, S., E. Akleman, M. Ettehad, N. Kalantar, and A. Borhani. 2017b. "Kerfing with Generalized 2D Meander-Patterns Conversion of Planar Rigid Panels into Locally-Flexible Panels with Stiffness Control." *CAADFutures 17*. 276–239.

IMAGE CREDITS

All drawings and images by the authors.

Alireza Borhani is an innovator, architect, educator, and co-principal of the transLAB. At the California College of the Arts, Texas A&M, and Virginia Tech, Alireza has taught architecture studios, concurrent with research and practice, for over a decade.

Dr. Negar Kalantar is an Associate Professor of Architecture and a Co-Director of the Digital Craft Lab at California College of the Arts (CCA) in San Francisco. Her cross-disciplinary research focuses on materials exploration, and robotic and additive manufacturing technologies to engage architecture, science, and engineering.

Erfan Rezaei Azari is an architecture enthusiast and curious about new methods of design and has been especially fascinated by parametricism. Erfan is also skilled in both weak and strong AI, which includes optimizing algorithms, machine learning, and deep learning.

Dr. Anastasia Muliana is working as Professor in the Department of Mechanical Engineering at Texas A&M University. In 2004, she completed her PhD in Structural Engineering and Mechanics at the Georgia Institute of Technology, Atlanta, Georgia.

Zaryab Shahid is currently a PhD candidate at Texas A&M University and works at Morpheus Lab. Zaryab joined Morpheus Lab in 2018 after finishing his Master of Science from Texas A&M University. Zaryab has experienced with numerical simulations, computer aided design modeling and programming.

Ed Green has a Doctorate in Mechanical Engineering from Herrick Laboratories, Purdue University.He has worked in the fields of noise, vibration, and ultrasonics for 35 years. His areas of expertise include sound material evaluation and modeling, and sound package optimization.

Session Introduction

Circularity, Carbon-Negative Design, and Fabrication

Franca Trubiano, Chair

The papers at the center of the session dedicated to "Circularity, Carbon-Negative Design and Fabrication" offer the field important contributions and innovations at the intersection of environment, material accountability, digital technology, and construction. Each of the research projects included articulate how building material choices and practices best contribute to reducing our ever-increasing global energy and carbon count. The papers presented do so by re-inventing the way in which building projects are conceived, measured, delivered, and valorized. Via the unmitigated adoption of robotics to disassemble existing light wood timber frames; the use of smaller scaled and possibly waste wood members for the efficient construction of structural spans; the coupling of Building Information Modeling (BIM) with robotic fabrication for empowering skilled crafts people at peril of losing their traditional building cultures; and by valorizing the most hyper localized building material with the highest carbon negative potential that is soil, all papers engage in a form of socio-ecological remediation and restoration that aimed to transcend the traditional limits of architecture and technology in their attention to material ethics.

Edvard Bruun et al. remind us of the embodied energy potential locked in existing timber frame buildings. Imagining a time when true material circularity is probable and not just possible, they identify a computational cooperative robotic sequence for disassembly and reuse of timber frame. Arash Adel demonstrates a new approach for the robotic construction of non-standard timber structures made from short timber elements, given what we know to be an ever-limited supply of old-growth wood members. Lien Kai Huang discusses the role of robot-based fabrication processes applied to the reproduction of traditional Chinese structures made of timber. Katie MacDonald and Kyle Schumann communicate their research on best ways to maximize our knowledge and use of non-orthogonal curved wood members. And lastly, Han Meng discusses their work in computationally organizing the design of large-scale landscapes using granular matter. In all cases, circularity, carbon-negative design and fabrication are the focus of important innovations.

Lithopic House

Computational Design Framework for Reshaping Ecologically-Attuned Architectural Systems Out of Granular Matter

Dana Cupkova
Carnegie Mellon University

Han Meng
Carnegie Mellon University

Jinmo Rhee
Carnegie Mellon University

1   Sand print of bio-integrated facade system (left); and digital 3D model with integrated biomass GAN-generated texture (right)

ABSTRACT

Lithopic House is the demonstration of a computational design workflow that intends to contribute to a biotechnological design approach enabled by disruptive in situ manufacturing technologies. The study expands on the first author's ongoing research into the additive manufacturing technology of binder-jetting—through the concomitant employment of construction waste and locally sourced granular soil materials—to propose a cradle-to-cradle design fabrication framework for a localized ecological construction of postindustrial sites. The focus of this research is on the procedural formation of architectural components relative to source material characterization, de-volumization strategies, and self-sustaining biomass integration, as primary inspiration for design speculation and prototyping. *Lithopic House* design studies serve as a series of workflow examples employing sand printing technology for early prototyping and a custom image synthesis interface based upon a generative deep learning model, to identify productive ecological features for component sculpting and surface figuration. Proposing an inter-scalar approach to developing an architectural project that is tightly coupled with landscape transformation and future potential soil remediation, this research interest aims to evaluate design limitations for architecture shaped by, and constructed from, local construction waste and soils. The material flow framework is rooted in a socio-ecological approach to collective living that intends to forge postindustrial urban resilience while proposing to remediate soil toxicity as a part of the architectural life cycle. This design study demonstrates a multi-platform computational design investigation aiming to address site specific socio-ecological issues while connecting cradle-to-cradle manufacturing of architectural form that facilitates biomass production.

DESIGN FRAMEWORK

"The major problems in the world are the result of the difference between how nature works and the way people think." — Gregory Bateson

The ambition of this design study, executed within an architectural design studio framework, is to examine architecture that probes embodied energy as a primary inspiration for matter formation. The goal is to resituate design within a hyperlocal framework of material resources / life cycle that positions architecture as a vehicle for ecological and communal restoration (Figure 2). Promoting a shift away from purely data-driven rationales, the desire is to engage in design that is framed by environmental ethics and sensory subjectivities as part of our collective aesthetic and ecological experience. Environmental aesthetics and aesthetics of nature are branches of philosophy that study appreciation of the 'world at large' as it is constituted—not simply by specific objects, but by holistic environments. Environmental empathy is rooted in the concepts of 'otherness' and 'difference.' Design grounded in environmental empathy leads to more diverse paradigms in the redistribution of resources, novel forms of co-shared domesticity, and social equity within collective urban space, while being closely entangled within its ecological functions. In this study, visualizing and understanding a larger set of multidimensional relationships within the design process intends to enable a projective design imagination that is tightly linked to the creation of biosynthetic and natural multispecies environments.

Faced with rapid urbanization and population growth, improving resource efficiency of the built environment has a significant impact on the global ecological outlook. Concrete remains environmentally attractive on a per-mass / unit-volume basis since the carbon dioxide (CO_2) emissions of concrete is lower than other commonly employed architectural materials (Reilly and Kinnane 2019). The negative environmental impact is not from a high embodied energy or CO_2 footprint per unit of concrete production, but is attributed to the large volume of concrete that is used worldwide. Consequently, to reduce this impact, we need to reduce the volume of virgin concrete together with offsetting the impact of current construction waste streams that contribute to industrial landfills. This project evaluates a design procedure to advance the implementation of the shape-factor in design relative to component strength and material composition,

2 Cradle-to-cradle manufacturing workflow diagram, employing in situ fabrication for component shaping and ecological integration

3 GAN-based image synthesis using custom interface

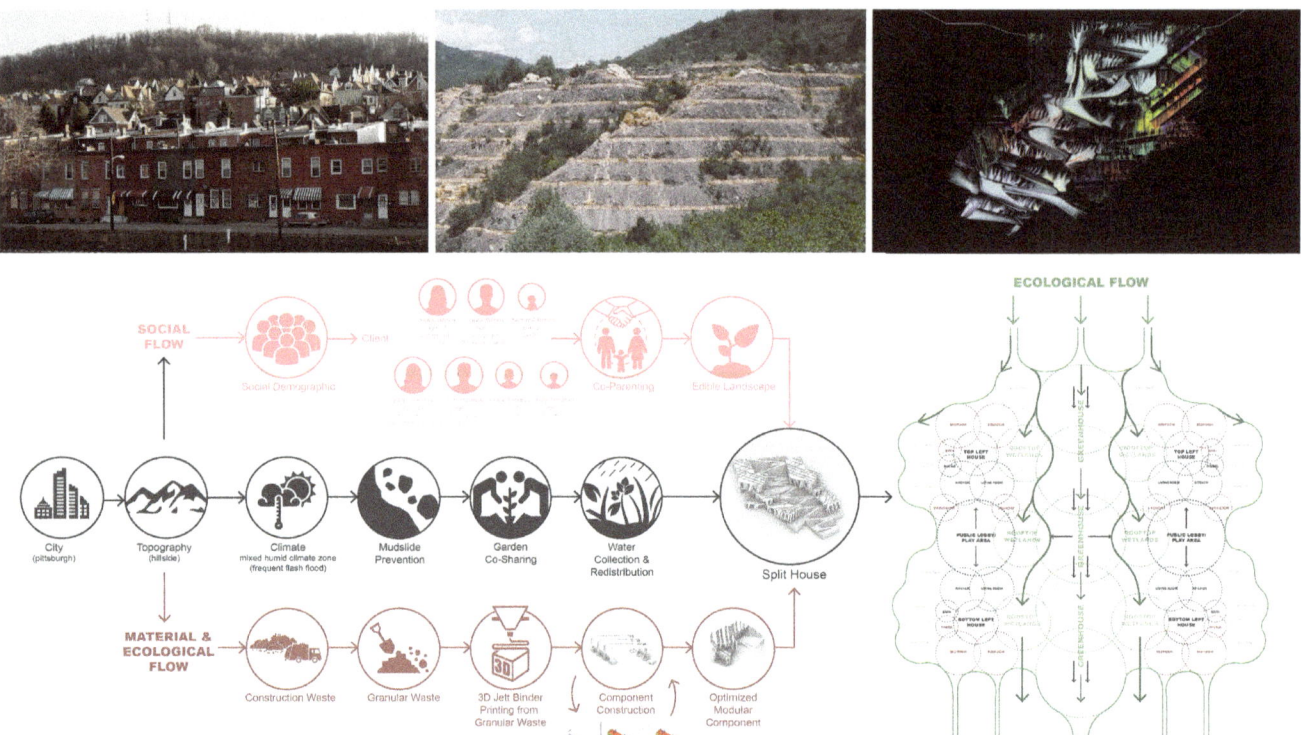

and to mitigate embodied carbon in design and construction by expanding on sand-printing technology and customized non-cementitious material mixtures.

Rather than relying on conventional simulation feedback, this study explores simulation combined with texture modeling techniques employing Generative Adversarial Networks (GAN) in a custom design interface (Figure 3). This approach utilizes voxel-based structural analysis feedback to reduce the volume of material employed, thus reducing the architecture's ecological footprint. Modeling techniques in this studio are founded in a rejection of reductionist modes of representation, incorporating drawing and data harvesting/mapping methodologies from cartography and landscape descriptions, and thus moving toward an evidence-based, data-rich design framework that embraces ecological restoration.

SITUATED SHAPING

The *Lithopic House* design framework operates across scales, acclimatized within the vertical topography of Pittsburgh's neighborhood (Hazelwood), which is prone to frequent landslides. Along with shifting and contaminated soils, Hazelwood's primary inhabitants belong to an ethnically diverse, underserved, and economically vulnerable demographic, living within food desert territory. This design study aims to develop inter-scalar connections between households and landscape stabilization, while concomitantly linking the design of architectural component systems to novel aesthetics, relative to de-massed structural logic, biomass retention, food production, component assembly and disassembly, and circular reuse. The process of uncovering the site's history of pollution patterns, which are deeply rooted in its current environmental conditions, intends to situate the health concerns of the community. Ecology acts across boundaries imposed by social and political systems. Historically, large-scale ecological patterns have been disregarded within the practice of architecture and urban planning. Modernist design thinking—as inherited from the era of industrialization—has been largely co-opted by ideologies of capital that organize social systems according to political and economic engineering rather than equitable access to resources. Mapping soil contamination relative to landslides, together with illustrating underground conditions of soil quality relative to slopes, runoff, and combined sewage outflows, helps to situate the connectedness of ecological design with a specificity of a place (Figure 4).

SITUATED CIRCULARITY

These design case studies rely on up-scaling sand 3D-printing technology centered on structural optimization, while also focusing on de-volumizing the assembly and disassembly sequence of architectural components. Aiming to embed architectural intervention in response to environmental dynamics of its larger ecological context (Figure 7), the initial investigation is focused on mapping landscape phenomena relative to the proliferation of soil toxicity, while also

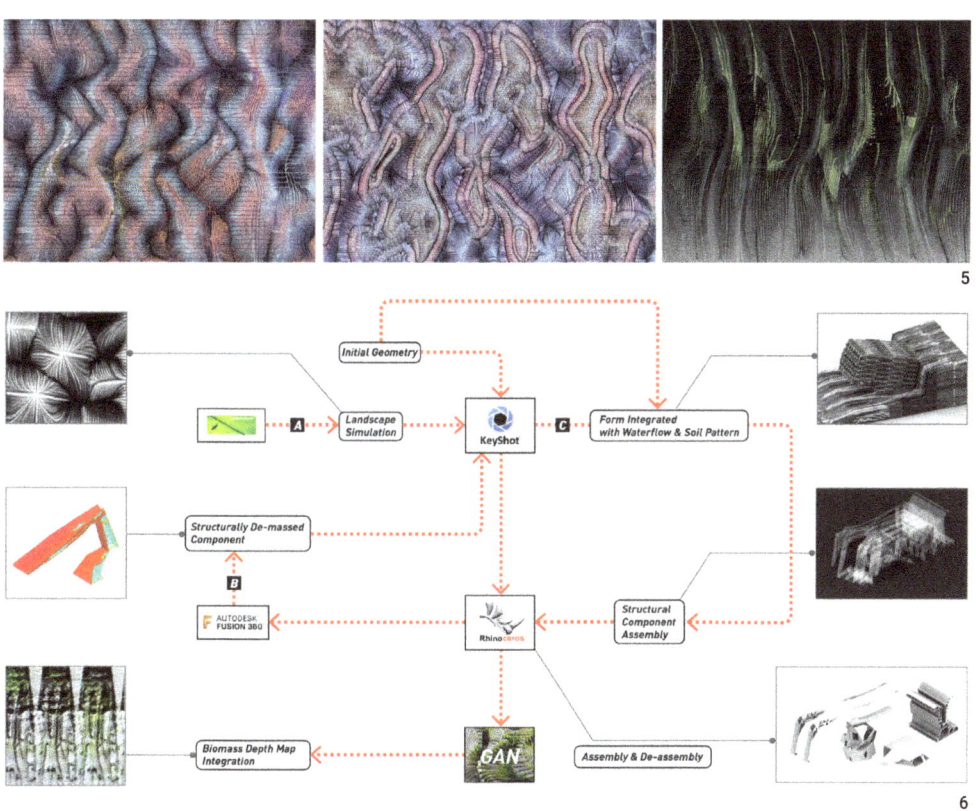

4 *Split House*: Site context in Hazelwood and dynamic terrain analysis

5 *House 25*: Predesign terrain simulation to determine patterns of water, toxicity, and topsoil accumulation, employed to situate architectural form within the terraformed landscape; this was performed by employing 3D mesh surface hydrology flow analysis EPIFLOW (Adaptive Analytical Design Framework for Resilient Urban Water Systems)

6 Computational workflow overview diagram

interacting with pervasive surface water flow. The proposed design framework is a method intended for the development of future sustainable communities within degraded land and soil areas, such as a former steel mill site that historically suffered from intensive air and soil pollution. The binder-jet-printed de-massed architectural components—intended to be produced as part of the remediation process—integrate surface figuration patterns that encourage self-sustaining biomass retention. This design-to-fabrication workflow opens novel possibilities for the direct use of toxic top-soil residue as a material resource for binder-jet printing within a phased soil remediation process, although more focused remediative studies will be required. Employing advanced computational modeling linked to emerging concepts of on-site robotic additive manufacturing, the *Lithopic House* studio speculates on localizing construction and material resource loops that offset the need for virgin materials, equipping future architects with novel approaches to address the complex issues of regenerative design strategies at the intersection of computational design, social sustainability, and building construction through additive manufacturing. Here, landscape restoration is a grounding framework for situating the architectural form. Architecture is conceived as a vehicle of urban 'acupuncture' that revitalizes its surrounding landscape. Embracing a climate-sensitive design approach, the project aims to examine a building's potential beyond human habitation and to attune the existing ecological context relative to the architecture's carbon footprint.

FROM SOCIO-HISTORICAL TO COMPUTATIONAL FRAMEWORKS

Lithopic House case studies are situated within a postindustrial site that suffered heavy exposure to air and soil pollution from 1869 to 1998, and consequently, a loss of population and functional businesses. Combined with steep hill topography, secondary pollution distribution has been exacerbated by surface water flow during extreme weather events, which intensifies poor soil quality and negatively affects the communities' living conditions and their ability to grow healthy local produce. This neighborhood suffers from a series of environmental disadvantages, with building sites sloped 25 degrees or greater, intensifying the likelihood of pesticide run-off and landslides. The presence of contaminated soils contributed to the area becoming a food desert, where it is more challenging to grow and harvest local produce and to establish access to local food supply (City-Data 2015).

Lithopic House tests a multiplatform design workflow (Figure 6) employing computational methods to: (a) investigate landscape behavior relative to pollution distribution patterns (Figure 5); (b) optimize topology or architectural components through material characterization and weight reduction (Figure 9); and (c) shape systems to enhance a building's ecological performance through biomass integration (Figures 8 & 10-12). This design process requires the integration of multiple software platforms and modeling techniques, namely: (a) computational simulation of surface

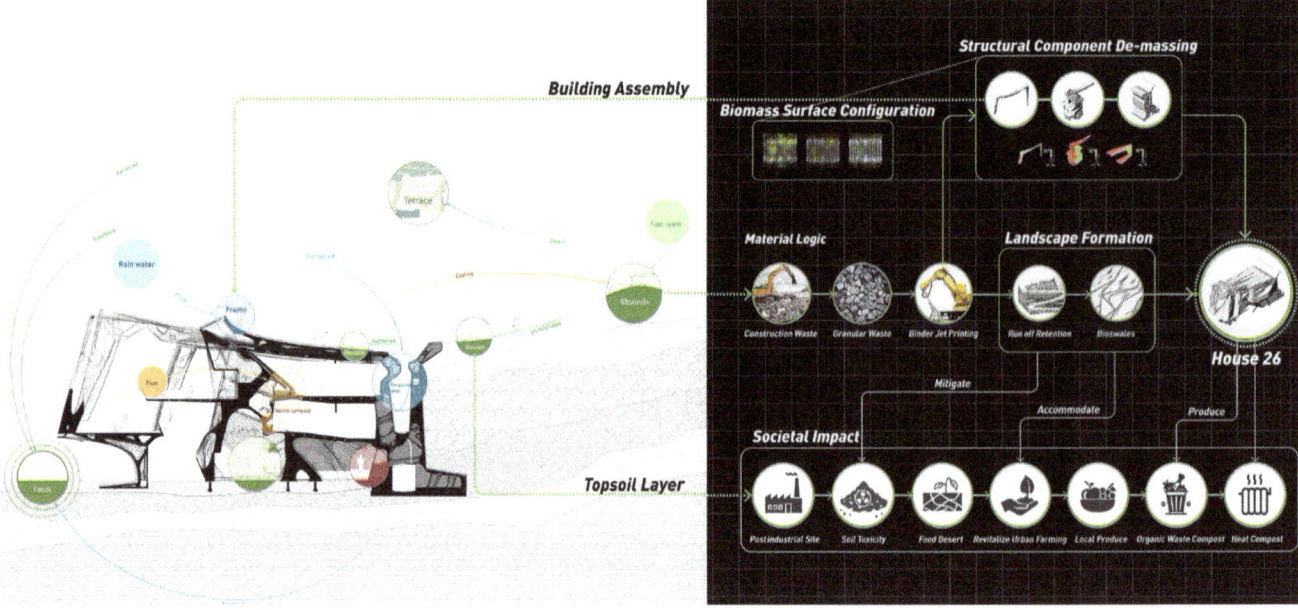

water flow to analyze pesticide distribution within large-scale landscape(s), and to visualize ecological behavior through geometric configuration (Cupkova, Azel, and Mondor 2015); (b) structural topological optimization (Craveiro et al. 2017) to de-volumize the overall building mass while leveraging additive manufacturing for bespoke shaping from granular material waste streams; (c) texture modeling, linked to the utilization of images using deep learning (DL), for probing the natural tendency of biomass to self-maintain within building surfaces (Cupkova and Clifford 2018). A neural-style transfer model in DL, CycleGAN (Zhu et al. 2017), was employed in combination with depth-map-transformation employing texture modeling. This model synthesized form and texture images to test emerging biomass growth patterns within architectural surfaces.

LITERATURE REVIEW

The foundation of the proposed design workflow lies within a contemporary biotechnological approach to ecological design thinking, supported by advancements in additive manufacturing. Mining possibilities of texture modeling (Gatys, Ecker, and Bethge 2015) linked to DL (Zhu et al. 2017; Zheng and Yuan 2021), we explored the shaping and prototyping of the complexity of architectural components that respond to robust ecological factors (Cupkova and Huber 2021). Expanding on the first author's previous research efforts that probed thermodynamically performative aspects of surface figuration (Cupkova and Promoppatum 2017), this workflow also harvests knowledge on design biodiversity, as previously examined in the construction of novel marine habitats (Dunn and Halpin 2009), together with aspects of ecological rugosity of surfaces (Marcus 2018) that promote more diverse habitat growth, via principles that manifest through geometric parameters. Furthermore, relying on topological optimization—a method that manipulates material distribution within a given design space for a given load path (Objectify 2022)—a topology optimization (TO) of architectural components (Beghini et al. 2014) is examined through the combination of granular material de-massing. Additive manufacturing not only enables effective materialization of the geometric complexity, but additionally offers the projection of lower carbon intensity within construction (Faludi et al. 2019). The first author's current material research combines the structural and ecological formation of matter through binder-jetting and material innovation (ExOne 2022), with a positive projection of lowering embodied carbon within manufacturing. The ecological surface characterization workflow proposes the use of DL, while also combining the topologically optimized component with ecological patterning (Li et al. 2017). This workflow relies on the analysis and synthesis of images in computer-vision technologies (Goodfellow et al. 2014) This is combined with examples of trained GAN machines in arts and design (Rhee and Veloso 2021) that demonstrate the ability to produce novel patterns of assimilation and characterization forms with a potential to retain performative abilities of the trained ecological data domain.

WORKFLOW OVERVIEW

This study probed an iterative design workflow (Figure 6) while bridging simulation, structural TO, and texture modeling, linked to a DL interface. Within the academic design studio, software complexity is limited to more accessible platforms, such as Rhinoceros, Fusion 360, Grasshopper, Keyshot depth map modeling, and a custom DL interface for texture synthesis.

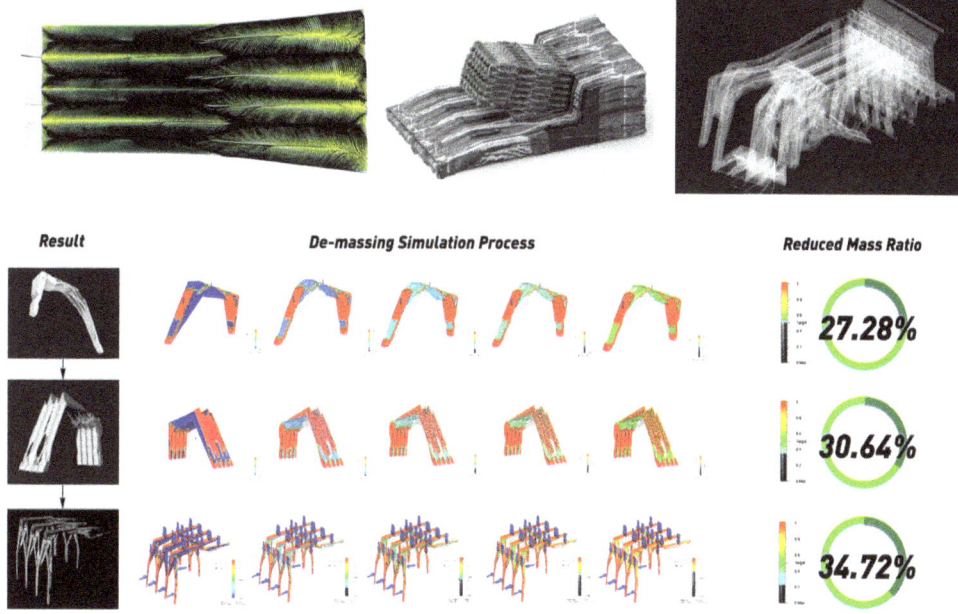

7   House 25: design, material, and manufacturing flow diagram

8   Two sequential 3D models employing depth map modeling with the intention to translate landscape water flow patterns into the architectural component system

9   House 26: structural optimization and component de-massing procedure with the calculation of the reduced mass ratio

The workflow was set up as follows:

(A) To identify nuanced phenomena of water accumulation or pesticide dispersion trends within specific terrain morphology, the Epiflow script was utilized (Cupkova, Azel, and Mondor 2015) in Grasshopper, which supports the simulation of water runoff, slope analysis, and water accumulation trends. This is a 3D mesh-based algorithm that was used to better understand ecological trends, expressions, and specific formal qualities in reshaping terrain morphology (Figure 5).

(B) The vector data of primitive geometries was combined with the rasterized format outcome of site analysis to experiment with textural modeling, using Keyshot—a render engine—that features an adaptable image 2D-to-3D depth-mapping displacement function (Figure 8). The intention was to use analytical site data to explore the direct formal modification, and to situate the architectural intervention within the site's formal language, while also enabling surface water flow, without erosive effects.

(C) The architectural component extracted from this design exercise was consequently manipulated within Autodesk Fusion 360, a 3D modeling software that is employed within industrial design and civil engineering for topology optimization (Autodesk 2022). Strategies of component reshaping relative to multiple load path scenarios for cementitious materials (Figure 9) were negotiated in comparison to the reduction in material mass, assembly/disassembly of components, and with the understanding that components will be fabricated using remediated granular soils within the additive manufacturing process.

The precision of the TO simulation is dependent on the mesh resolution that is evaluated within the voxel model. A reduced mesh subdivision produces more accurate results. The simulation generates novel versions of the component by eliminating unnecessary material volume. Calculation of the reduced embodied energy for material use will be a quantifiable measurement for energy saving.

(D) Employing depth map modeling, the mass-reduced component was combined with superimposed surface exuberance, for biomass retention (Figure 11). Since the formal logic of the component is a continuation of the landscape water-flow system, it was imperative to accentuate ecological processes within its overall expression. The concave structural frame is designed to function as a rainwater gutter, within which the ecological texture transformation is reintroduced.

Various water distribution patterns were tested to propose the component's surface figuration, for enhancing emergent biomass growth. A series of component figurations (Figures 10-12) were studied relative to component printability, while aiming to curate surface water behavior and integrate biomass features. In this case, the water channeling continuity, along with the strategic delay of water flow, were considered while also connected to landscape water patterns across scales.

(E) With de-volumized 'plug-and-play' components, a building envelope strategy relied on more intensive image synthesis investigation, employing a custom-designed interface and consequent depth map modeling (Figure 13). Additional textures were embedded into the aggregate component assembly surface, to mitigate ecological behavior at the envelope level. In

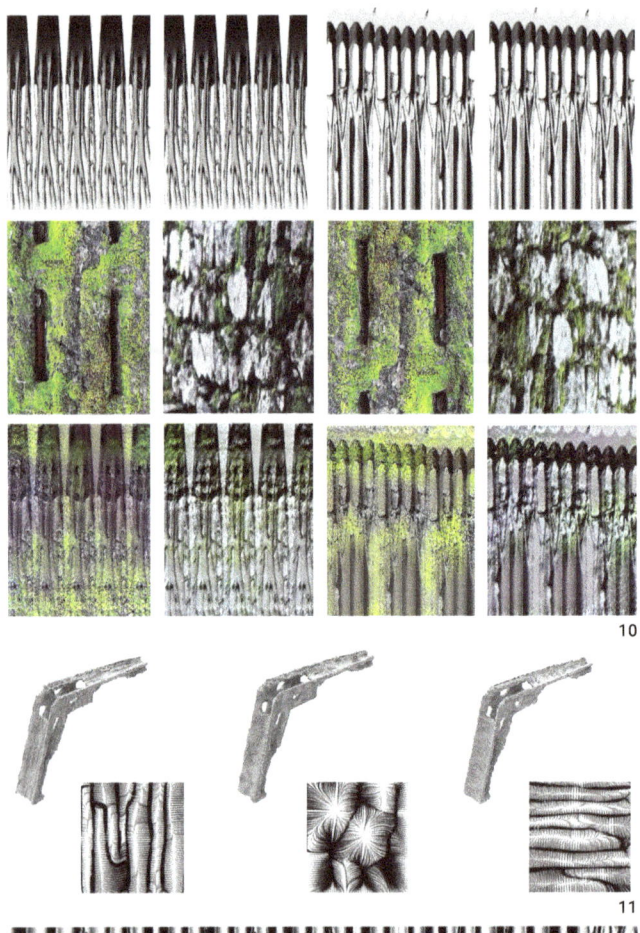

10

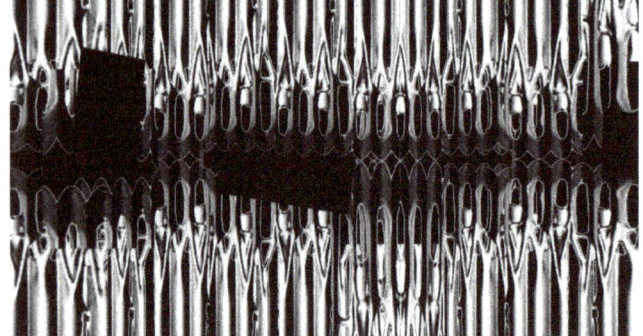

11

12

this synthesis, we established a library of ecological patterns collected from the site, as resources for image style transfer techniques.

The prototype of a custom machine learning (ML) interface has a GAN-based style transfer algorithm in the backend, enabling it to maintain the fundamental structure of the content image—in this case, the architectural assembly—while it transfers the ecological figuration within a novel style set.

The interface consists of three panes: control, viewer, and preview. Within the control panel, users can select a content image to be used as a shape, and a style image to be employed as a texture. In projects, the architectural components function as content images, while samples of biomass growth function as style images, thus allowing to visualize probable growth patterns of self-maintaining biomass on the proposed component set. By controlling the slider in the panel, users can set weights and other parameters that affect the transfer result. The weight values determine the intensity of the ecological features applied onto the architectural assembly; larger weights confer stronger applications for ecological features upon material assembly.

For this study, we developed a custom style transfer model, based on the Image Style Transfer Using Convolutional Neural Networks (Gatys, Ecker, and Bethge 2016). Once users run the algorithm with the selected images and parameters, the GAN model in the backend transfers the ecological texture into the material shape. Once the model finishes transferring, the resulting image is displayed in the viewer pane. The preview pane shows the variations of interpolated images with differing degrees of transformation. Users can select an image from these alternatives. Varied gradients of the ecological figuration onto the architectural assembly demonstrate a range of potential biomass growth.

(F) Physical prototyping process relied on sand 3D-printing as a manufacturing process that is parallel to large-scale binder-jet manufacturing using granular powders. The prototyping allowed for probing the viability of the proposed eco-structural shaping (Figures 14, 16), together with the 'plug-and-play' components' joinery, for easy assembly and disassembly (Figure 17). The prototyping was supported by industry collaboration with ExOne™ partner company. The assembled models were viable in communicating structural integrity and assembly sequence.

TOWARDS ARCHITECTURE OF ENVIRONMENTAL REMEDIATION AND STEWARDSHIP

The *Lithopic House* design workflow proposes to bridge a gap between global form-finding and microtexture characterization, rooted within the local ecology. Exploring the potential of architecture as an environmental filter and as a remediation device that supports emergent biomass integration, this project proposes an across-scale strategy for engaging landscape

10 Aggregate component biomass study simulation employing GAN synthesis

11 3D architectural component series modeled with simulated textures to evaluate water behavior, direction, and accumulation trends

12 3D study of textural character for de-massed structural system, employing depth map modeling of individual components for component assembly and biomass integration

13 Screen capture of the custom GAN interface for image synthesis with the trained neural style transfer model

transformation with soil remediation and on-site manufacturing for ecologically-conscious construction. By deploying cradle-to-cradle material flow relative to structural assembly, ecological performance, and potential for future remediation of local soils in use with binder-jet manufacturing, the embodied energy of construction would be reduced. The deep study of terrain morphology allows architecture to become an integral part of the urban-green infrastructure (Figure 18) that increases interaction with environments on multiple scales. A quantitative analysis of structural optimization allows for evaluating the reduction of material mass and structural weight, with a possible impact on CO_2 footprint. Due to the working mechanism of the binder-jet printing technique, the more volume the components have, the more energy it needs to apply the binder that holds the powder. Within the case study, the optimized framing components retained on average 46% of their original mass, while the optimized core components retained on average 37% of their original mass. This optimization was analyzed with cementitious material input for TO simulation. The texture treatment upon the facades has demonstrated its water channeling properties, ranging from directing, collecting, and retaining surface run-off. With such a microtextured feature, the building envelope can direct rainwater towards the processing system inside of the retaining wall, for further gardening uses. Similarly, the approach is being employed at differing scales for the design of communal garden landscapes. The terrain mounds of bioswales are situated based upon flow simulation in order to decelerate the toxic run-off and retain the erosion. Lastly, the textured facade surfaces provide growing habitats for emergent biomass. All the features mentioned above collectively establish a non-linear material loop that enables residents to grow produce by embracing ecological sustainable living.

CONCLUSION & DISCUSSION

The goal of this project was to propose and test a design framework for a circular waste stream construction scheme, utilizing the 3D binder-jet printing technique. The remediated topsoil, and/or up-cycled, excavated construction waste, can be processed into a granular size—in order to constitute printable building components. Structural optimization aims to reduce the embodied energy, de-massing the printed component, while also maintaining structural integrity. Since the reductive printing technique liberates the design from the restrictions of conventional geometry, the project affords a more nuanced consideration of ecological performance through formal configuration. For example, surface complexity was engaged to encourage biological growth and diversity (Figure 19). The textured micropattern surface channels surface run-off and, by increasing the surface area, promotes additional habitat growth.

The computational design framework introduced in this case study demonstrates the ability of design to iterate between quantitative analysis of material carbon impact and empirical evaluation of ecological systems. These tools enable the abstract representation of geometric properties, while also providing a substrate for an accurate analysis of structural and ecological behaviors. Relying on binder-jetting as a primary manufacturing technology makes the case for upscaling the technology by utilizing local soils for on-site fabrication. This would allow for phased environmental remediation and novel construction for productive landscapes, along with future eco-living. The authors believe that further Life Cycle Analysis (LCA) would reveal a significant decrease in embodied energy. Through developing a deeper understanding of computational design strategies, this workflow

14 *Chicken House*: Demonstration of component system developed from simulation to sand-printed prototype, utilizing a surface to mass increase with volume reduction calculation that intends to enhance thermal performance, while reducing embodied energy

15 *Filter House*: Workflow demonstration of integrated structure to facade system developed from TO simulation to depth map texture modeling coupled with GAN image

16 *Filter House*: Sand-printed prototype models

lies at the intersection of simulation, quantitative modeling, and empirical use of formal patterns. It allows greater agility within the design to move between qualitative analysis and intuitive gestures while remaining conscious of the ecological impact of site-specific construction.

ACKNOWLEDGMENTS

This study has been developed within a vertical design studio framework *Lithopic House*, led by Dana Cupkova at the Carnegie Mellon School of Architecture during the Spring semesters of 2020 and 2021. The design workflow has been conceived as part of the funded research project "CRuMBLE: Construction Rubble Manufacturing for Building Life-cycle and Environment," a research manufacturing framework that proposes a novel cradle-to-cradle design process for ecological architecture, manufactured directly out of construction waste and earthen materials by integrating recycled construction waste as a powder aggregate mixture to create novel pathways for direct non-toxic chemical activation, with water-based binders in binder-jet printing (PIs: Dana Cupkova, Josh Bard, CoPI: Robert Heard, Industry partners: ExOne, Michael Brothers Hauling, Research Associate: Mathew Huber). This work has been partially supported by The Manufacturing Futures Institute at Carnegie Mellon University (CMU), The Manufacturing PA Innovation Program, and ExOne Company.

The following SoA CMU students, some of whose work is featured in this paper greatly contributed to the development of design ideas and workflow testing: (Spring 2020) Harrison Branch-Shaw, Colleen Duong, Kirman Hanson, Tanvi Harkare, Gil Jang, Leah Kendrick, Ryu Kondrup, Lana Kozlovskaya, Longney Luk, Louis Suarez, Scarlet Tong, Alex Wang; (Spring 2021) Juhi Dhanesha, Lukas Hermann, Huang Yuxin, Jasmine Lee, Han Meng, Andy Qiu, Steve Wang, and Yingying Yan.

REFERENCES

Autodesk. 2022. "Lightweighting with Shape Optimization." *Autodesk*. https://knowledge.autodesk.com/support/fusion-360/learn-explore/caas/simplecontent/content/lightweighting-shape-optimization.html.

Beghini, L. L., A. Beghini, N. Katz, W. F. Baker, and G. H. Paulino. 2014. "Connecting architecture and engineering through structural topology optimization." *Engineering Structures* 59 (Feb): 716–726. https://doi.org/10.1016/j.engstruct.2013.10.032.

City-Data.com. n.d. "Hazelwood neighborhood in Pittsburgh, Pennsylvania (PA), 15207 subdivision profile; Real estate, apartments, condos, homes, community, population, jobs, income, streets." Accessed February 12, 2022. https://www.city-data.com/neighborhood/Hazelwood-Pittsburgh-PA.html.

Craveiro, F., H. Almeida, H. Bártolo, P. J. Bártolo, and J. P. Duarte. 2017. "Topology and Material Optimization of Architectural Components." In *Challenges for Technology Innovation: An Agenda for the Future*. Boca Raton: CRC Press.

Cupkova, D. and C. Clifford. 2018. "MOSS REGIMES: Embedded Biomass in Porous Ceramics." In *ACADIA '18: Recalibration, On Imprecision and Infidelity; Project Catalog of the 38th Annual Conference of the Association of Computer Aided Design in Architecture (ACADIA)*, edited by P. Anzalone, M. del Signore, and A. J. Wit.

Cupkova, D. and M. Huber. 2021. "Rocking Cradle: Interactive Urban Furniture in Pursuit of Environmental Attunement." Public Play Space Symposium (blog). https://www.publicplayspace.eu/symposium/.

Cupkova, Dana, Nicolas Azel, and Christine Mondor. 2015. "EPIFLOW: Adaptive Analytical Design Framework for Resilient Urban Water

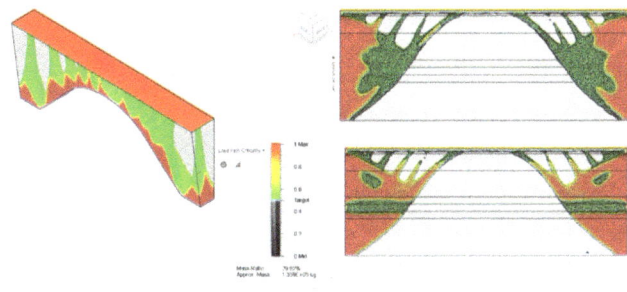

15

16

Systems." In *Modelling Behaviour*, edited by M. Thomsen, M. Tamke, C. Gengnagel, B. Faircloth, and F. Scheurer. Cham: Springer. 419–31. https://doi.org/10.1007/978-3-319-24208-8_35.

Cupkova, Dana, and Patcharapit Promoppatum. 2017. "Modulating Thermal Mass Behavior Through Surface Figuration." In *ACADIA 2017: Discipline & Disruption; Proceedings of the 37th Annual Conference of the Association for Computer Aided Design in Architecture*. Cambridge, November 2-4, 2017. 202–211.

Dunn, Daniel C., and Patrick N. Halpin. 2009. "Rugosity-Based Regional Modeling of Hard-Bottom Habitat." *Marine Ecology Progress Series* 377 (February): 1–11. https://doi.org/10.3354/meps07839.

ExOne. n.d. "Design Research Team Reinvents Eco-Friendly Architecture and Upcycled Materials." Accessed August 20, 2021. https://www.exone.com/Admin/ExOne/media/Case-Studies/2021_X1_CMU_CaseStudy_CRuMBLE.pdf.

Faludi, Jeremy, Corrie M. Van Sice, Yuan Shi, Justin Bower, and Owen M. K. Brooks. 2019. "Novel Materials Can Radically Improve Whole-System Environmental Impacts of Additive Manufacturing." *Journal of Cleaner Production* 212 (March): 1580–90. https://doi.org/10.1016/j.jclepro.2018.12.017.

Gatys, Leon A., Alexander S. Ecker, and Matthias Bethge. 2015. "Texture Synthesis Using Convolutional Neural Networks." *arXiv*. https://doi.org/10.48550/arXiv.1505.07376.

Gatys, Leon A., Alexander S. Ecker, and Matthias Bethge. 2016. "Image Style Transfer Using Convolutional Neural Networks." In *2016 IEEE Conference on Computer Vision and Pattern Recognition (CVPR)*. 2414–23. https://doi.org/10.1109/CVPR.2016.265.

Goodfellow, Ian, Jean Pouget-Abadie, Mehdi Mirza, Bing Xu, David Warde-Farley, Sherjil Ozair, Aaron Courville, and Yoshua Bengio. 2014. "Generative Adversarial Nets." In *Advances in Neural Information*

Processing Systems, vol. 27. https://papers.nips.cc/paper/2014/hash/5ca3e9b122f61f8f06494c97b1afccf3-Abstract.html.

Li, X., Y. Dong, P. Peers, and X. Tong. 2017. "Modeling Surface Appearance from a Single Photograph Using Self-Augmented Convolutional Neural Networks." *ACM Transactions on Graphics* 36 (4): 1–11. https://doi.org/10.1145/3072959.3073641.

Marcus, Adam, Margaret Ikeda, Evan Jones, Taylor Metcalf, John Oliver, Kamille Hammerstrom, and Daniel Gossard. 2018. "Buoyant Ecologies Float Lab. Optimized Upside-down Benthos for Sea Level Rise Adaptation." In *ACADIA '18: Recalibration, On Imprecision and Infidelity; Project Catalog of the 38th Annual Conference of the Association of Computer Aided Design in Architecture (ACADIA)*, edited by P. Anzalone, M. del Signore, and A. J. Wit. 414–423.

Objectify. n.d. "Topology Optimization and 3D Printing; Engineering Efficiency." September 30, 2021. Accessed November 1, 2021. http://objectify.co.in/topology-optimization-and-3d-printing-engineering-efficiency/.

Reilly, Aidan, and Oliver Kinnane. 2019. "Construction Is a Cause of Global Warming, but Is Concrete Really the Problem?" *Architects' Journal*, March 1, 2019. Accessed April 1, 2019. https://www.architectsjournal.co.uk/news/opinion/construction-is-a-cause-of-global-warming-but-is-concrete-really-the-problem.

Rhee, Jinmo, and Pedro Veloso. 2021. "Generative Design of Urban Fabrics Using Deep Learning." In *PROJECTIONS; Proceedings of the 26th CAADRIA Conference*, vol. 1, edited by A. Globa, J. van Ameijde, A. Fingrut, N. Kim, and T. T. S. Lo. The Chinese University of Hong Kong and Online, Hong Kong. 31–40. https://doi.org/10.52842/conf.caadria.2021.1.031.

Zheng, Hao, and Philip F. Yuan. 2021. "A generative architectural and urban design method through artificial neural networks." *Building and Environment* 205 (Nov): 108178. https://doi.org/10.1016/J.BUILDENV.2021.108178.

Zhu, Jun-Yan, Taesung Park, Phillip Isola, and Alexei A. Efros. 2017. "Unpaired Image-to-Image Translation Using Cycle-Consistent Adversarial Networks." In *2017 IEEE International Conference on Computer Vision (ICCV)*. 2223–2232. https://doi.org/10.1109/ICCV.2017.244.

IMAGE CREDITS

Figure 1: ©Lithopic House Studio 2021, (left) Juhi Dhanesha, (right) Han Meng

Figure 2: *Lichen House*, ©Lithopic House Studio 2020, Ryu Kondrup

Figure 3: ©Dana Cupkova

Figure 4: *Split House*, ©Lithopic House Studio 2020, Colleen Duong

Figure 5: *House 25*, ©Lithopic House Studio 2021, Yingying Yan and Han Meng

Figure 14: *Chicken House*, ©Lithopic House Studio 2020, Longney Luk and Louis Suarez

Figures 15 & 16: *Filter House*, ©Lithopic House Studio 2021, Juhi Dhanesha

Figure 17: *Radiant House*, ©Lithopic House Studio 2020, Gil Jang

Figure 18: *House 25,*: ©Lithopic House Studio 2021, Yingying Yan and Han Meng

Figure 19: *Lichen House*, ©Lithopic House Studio 2020, Ryu Kondrup

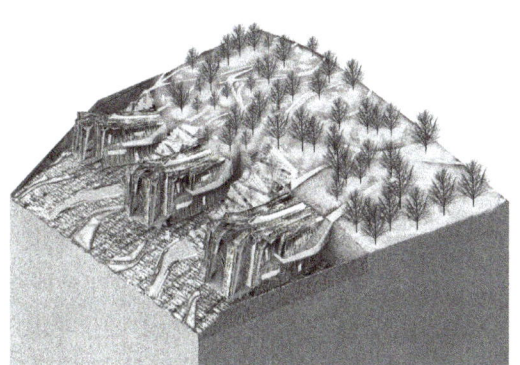
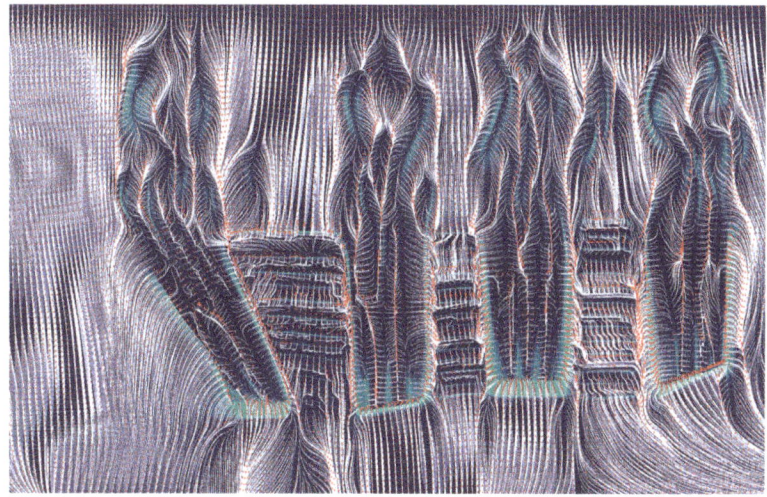

Dana Cupkova is an Associate Professor at Carnegie Mellon School of Architecture and directs Epiphyte Lab, Architectural Design & Research Collaborative. Engaging issues of environmental steward-ship in design, Dana's work is situated at the intersection of built environment and ecology, focused on computational methods, materiality, embodied energy, and advanced manufacturing frameworks, with a particular interest in thermodynamics, and construction waste streams.

Han Meng graduated with a Bachelor of Architecture from Carnegie Mellon University and is currently pursuing his Master of Science in Architecture and Urban Design at the UCLA IDEAS program. Han is interested in a multidisciplinary approach to architecture, exploring its agency at the intersection of emerging technologies and computational design.

Jinmo Rhee is a computational design researcher and architect, exploring the analytical and generative potential of artificial intelligence technologies in architectural design and research. Jinmo holds a Master of Science in Computational Design from Carnegie Mellon University and is currently pursuing his PhD within the same program, with a focus on machine learning and its use in the analysis of architectural form in an urban context.

17 *Radiant House*: Sand-printed prototypes with components optimized for a 'plug-and-play' assembly sequence and a resulting assembled 3D model of a structure in section

18 *House 25*: Situated shaping showing relationship between landscape patterns and architectural systems

19 *Lichen House*: Assembled section model and site view showing relationship between landscape retention and water collection system

ZeroWaste

Towards Computing Cooperative Robotic Sequences for the Disassembly and Reuse of Timber Frame Structures

Edvard P. G. Bruun
Princeton University

Erin Besler
Princeton University

Sigrid Adriaenssens
Princeton University

Stefana Parascho
École Polytechnique Féderale de Lausanne

1 Three industrial robotic arms (2xIRB4600 & 1xIRB7600) cooperatively sequenced to alternate in their function of supporting and removing elements during the scaffold-free disassembly of a conventional timber stick frame structure

ABSTRACT

ZeroWaste is a project about repositioning existing timber building stock within a circular economy framework. Rather than disposing of these buildings at the end of their life, the goal is to view them as stores of valuable resources that can be readily reused. By doing this, material life cycle becomes an integral design consideration alongside planning for the efficient disassembly and reuse of these structures. In this paper, the computational workflow is presented for the first phase of the project: planning a cooperative robotic disassembly sequence for the scaffold-free removal of members from existing timber structures. A pavilion-scale prototype is first constructed, in the Embodied Computation Lab at Princeton University, to represent an existing timber structure built according to conventional North American stick frame construction practices. A multi-directed graph data structure, representing structural member connectivity and support hierarchy, is then coupled with a breadth-first search algorithm to plan potential scaffold-free robotic disassembly sequences given a member removal target. In parallel, computer vision is integrated and implemented through the robotic setup to create an accurate as-built point cloud scan of the whole structure. This as-built information is then used to inform the evaluation of potential robotic sequences from the point of view of robotic reachability and structural performance. This work-in-progress paper first presents a high-level overview of the various components in this workflow, followed by its demonstration in planning the removal of a specific member in the prototype structure. Upcoming project developments will include the planning, and physical demonstration, of more complex disassembly sequences, coupled with reassembly and reuse of the removed members for various regions of the prototype structure.

INTRODUCTION

Construction and demolition processes continue to be among the largest contributors to our contemporary waste crisis (US EPA 2018). Among the many material systems used in North American building practices, conventional timber frame construction stands out as not only the most pervasive and ubiquitous, but also the most readily discarded (O'Brien et al. 2006; Diyamandoglu and Fortuna 2015). We address this problem by re-situating timber buildings as material depots, as a site of valuable material resources that can be utilized as part of a circular economy framework (Zimmann 2016). Identifying and cataloging existing building stock privileges the flow of material upstream on the construction site rather than downstream to the recycling and waste industry (Garcia et al. 2021).

Significant research activity has centered on creating models to better quantify the environmental benefits of material circularity and reuse potential of existing building stock (Cottafava and Ritzen 2021; Eberhardt et al. 2021). But there is need for further physical deconstruction and reassembly projects that combine existing buildings with modern digital fabrication tools (Brütting et al. 2019, 2021). With *ZeroWaste*, we aim to address this research gap by integrating 3D imaging technology with a cooperative industrial robotic fabrication setup, which can then be utilized for both information gathering and the physical disassembly and reuse of conventional timber frame structures.

Cooperative Robotic Assembly and Disassembly

Robots have been used in the design and construction industry for over forty years (Bock 2007). While initial developments focused on automating single human tasks (Bock and Linner 2016), in the last decades, researchers have begun using robots to enhance constructive work and expand the design space by making use of their specific capabilities, such as precise movement and accurate spatial placement of components (Gramazio and Kohler 2008). In a cooperative robotic setup, multiple robots are specifically sequenced to achieve outcomes that would not be possible with a single robot (Parascho et al. 2018). For example, geometrically complex structures can be assembled without temporary scaffolding when alternating the robotic placement of material and support of the structure during fabrication (Bruun et al. 2020, 2021; Han et al. 2020; Parascho et al. 2020, 2021). In *ZeroWaste*, we build on specific concepts from our most recent work on the scaffold-free cooperative robotic assembly and disassembly of a timber space frame structure (Bruun et al. 2022) and more conceptually on prior research about the robotic assembly of bespoke timber modules (Thoma et al., 2018). Thus, extending the capabilities of industrial robots beyond their traditional role in assembly.

ZeroWaste Project Description

The overall project is to develop a computational approach for determining how multiple industrial robotic arms should be sequenced for the scaffold-free disassembly and reconfiguration of an existing timber structure. Specifically, the goal is to plan a multi-robot sequence that leverages the robots as temporarily support to maintain stability of the structure during all stages of fabrication. While an assembly process can be completely pre-planned and simply executed by robots, for disassembly the robots take on the additional role of information gatherers. In *ZeroWaste*, the two robotic arms that are on tracks are first used to collect data and create a complete as-built point cloud of an unknown structure; this as-built information is then used in the computational workflow for planning a feasible robotic disassembly and reassembly sequence.

A timber structure prototype is used as a stand-in for an unknown existing structure built according to typical North American timber stick frame construction practices. While in this paper the robotic sequences are planned with this specific prototype structure in mind, the approaches developed are intended to be generic and thus transferable to similar discrete element structures. The complete project involves developing a workflow and implementing it in the planning of the robotic fabrication for the following four distinct phases:

- Phase 1: Removal of a simulated "damaged" member in the structure.
- Phase 2: Disassembly and partial reassembly of a region of the structure.
- Phase 3: Disassembly and one-to-one member reassembly of a single wall in the structure.
- Phase 4: Disassembly and partial reassembly of the whole structure.

This work-in-progress paper focuses on a high-level description of the generic computational workflow (Figure 2) and describe how it was implemented specifically for Phase 1. The approach is to use a support hierarchy topological representation of the structure to calculate possible stable disassembly sequences, after which a sequence is selected based on structural performance and robotic feasibility criteria evaluated using the as-built point cloud data. The subsequent phases, which will be presented in a future publication, increase in planning complexity but build on the methods described in this paper in the context of Phase 1.

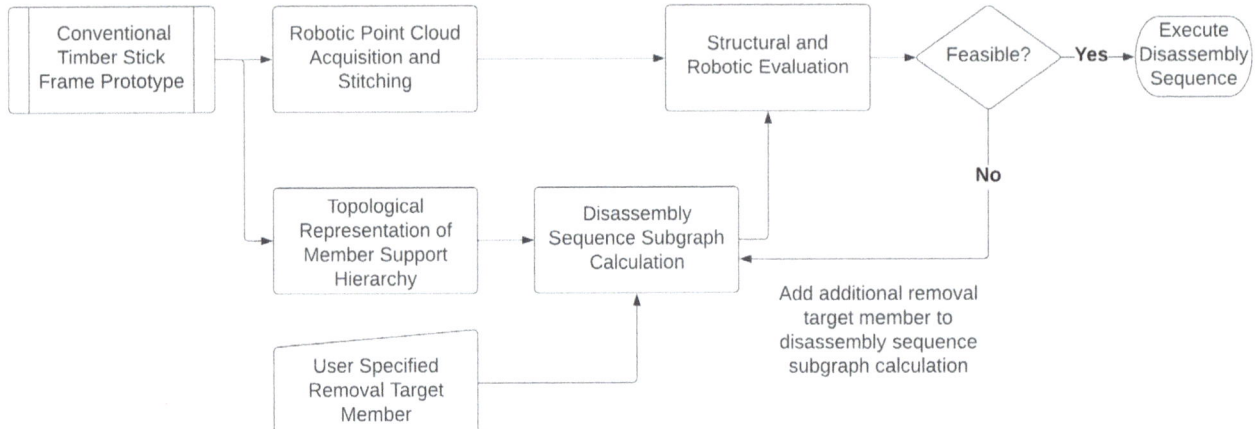

METHODS

Description of Timber Structure Prototype

A life-sized timber structure, built according to conventional American stick frame construction practices, serves as the experimental prototype for the computational methods presented in this paper. The as-built structure is shown in Figure 3 and 4, and measures 8 ft by 6 ft (~2.4 m by 1.8 m) in plan with a height of 9.6 ft (~2.9 m) built from SPF dimensional lumber. Members of the same type are grouped by color as shown in Figure 4. The following types of members are present in the structure: roof girder (x1, red), roof post (x2, brown), roof rafter (x14, blue), ceiling joint (x5, green), top plate (x4, pink), wall studs (x32, orange), header beams (x3, yellow), sheathing diagonals (x7, grey), bottom plate/floor (x4, black). These colors also be used to distinguish the different members in the support hierarchy graph representation (Figure 8). Wall sheathing is represented as diagonal members, which provide the necessary shear stiffness to the structure.

As-Built Structure Point Cloud

The first step in the disassembly sequence planning process is generating an accurate as-built digital representation of the structure. The prototype in the project is meant to represent an existing building for which detailed geometric information might not be available. In addition, as-built conditions might differ geometrically from what was planned even for known structures. Thus, design renders are not sufficiently accurate for planning robotic paths, support sequencing, and where to send the robots to grip members.

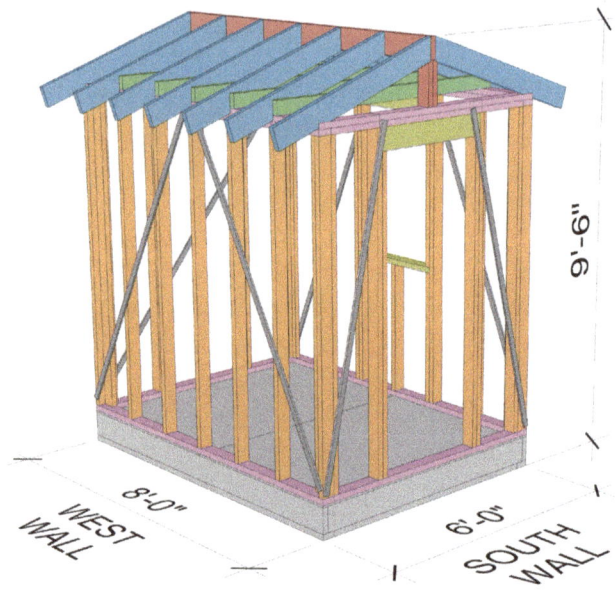

2 High-level workflow for Phase 1 of the *ZeroWaste* project corresponding to the different topics discussed in the Methods section

3 Photo of the conventional stick frame prototype structure in the Embodied Computation Lab at Princeton University

4 Rendering of timber structure where the various colors indicate the different types of members

The two IRB4600 robots (40 kg payload, 2.55 m reach) are each equipped with a Zivid 3D structured light camera with a spatial resolution of 0.39 mm at 700 mm distance (Zivid AS 2021). A point cloud of the full structure is created by moving the robots to various positions in space and capturing 3D images. These images are then transformed to the same coordinate system, stitched together using in-house developed scripts implementing the Zivid API, and downsampled to remove duplicate points. The transformation of a point cloud in the camera coordinate frame (P_{Object_Camera}) to a point cloud in the world0 (e.g., the 0 location in the CAD model) coordinate frame (P_{Object_World0}) is described as: $P_{Object\_World0} = H_3 \times H_2 \times H_1 \times P_{Object\_Camera}$

Where the 4x4 transformation matrices are the following:

- H_1: from robot tool center point (TCP) to camera location, calculated from the calibration routine.
- H_2: from robot base to the robot TCP, queried as a positional frame using the COMPAS RRC API (Fleischmann 2020).
- H_3: from World0 to the work object (WOBJ), user defined.

Figure 5 and 6 show the results of the point cloud scan of the structure, which is a combination of 100 separate image captures with Robot 1 (right), and 60 image captures with Robot 2 (left). A voxel size of 10 mm was used to downsample the stitched together point cloud, which resulted in a coarse model with approximately 0.5 million points. Additional point cloud fidelity can be preserved by reducing the voxel size, but this was not necessary at this stage when planning robotic paths.

Topological Representation of Member Support Hierarchy
Planning a stable and feasible disassembly sequence requires information about how the members in a structure are connected and supported by each other. A way to represent the connectivity and support hierarchy in a structure is through a multi directed graph (multidigraph) data structure (Valiente 2021). The vertices in the graph represent individual members, and the edges represent connections between members, with outgoing edges indicating the direction of support (e.g., S members are supported by the M member in Figure 7). When a member has multiple connection locations, it can be better represented by dividing it into its constitutive submembers. Vertices representing pieces of the same member are joined with two parallel but opposite edges to indicate the mutual support relationship between them (i.e., this is considered a fixed connection). An example of this is shown in Figure 7.

The graph of the full prototype structure is built manually and shown in Figure 8, where the colors of the vertices correspond to the member color scheme in Figure 4. For clarity, the structure is divided into five regions representing the roof and the four walls. The vertex naming convention is as follows AB#_$:

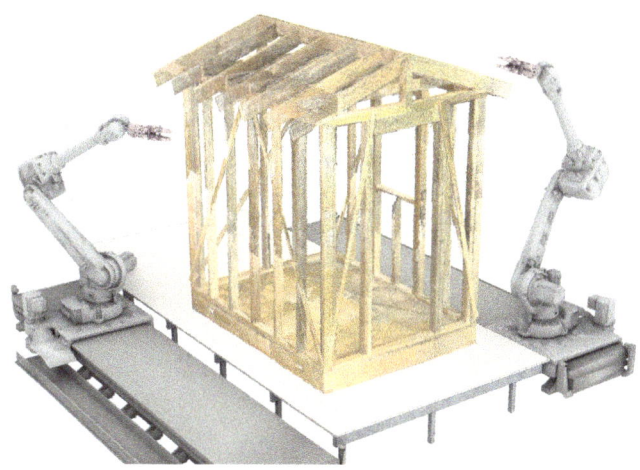

5

6

5   Point cloud of the as-built timber structure stitched together from multiple captures using two robots and transformed to the World0 coordinate frame

6   Zoomed in perspective of stitched together point cloud showing the low resolution downsampling based on a 10 mm voxel size

- A: The first digit of each vertex indicating if it is part of the roof (R) or one of the four walls (N, S, E, W). In Figure 5, the South wall is the short side closest to the camera.
- B: The second letter represents the type of member; roof girder (G), roof post (P), roof rafter (R), ceiling joint (J), top plate (P), wall studs (S), header beams (H), sheathing diagonals (D), foundation support (F).
- #: The third digit is used to number a unique member of a particular type.
- $: The fourth digit is used to indicate a unique subcomponent of a single member.

The graph terminates at the foundation supports, which are shown as hexagonal vertices and represent a bearing support between the bottom plate of each wall and the ground.

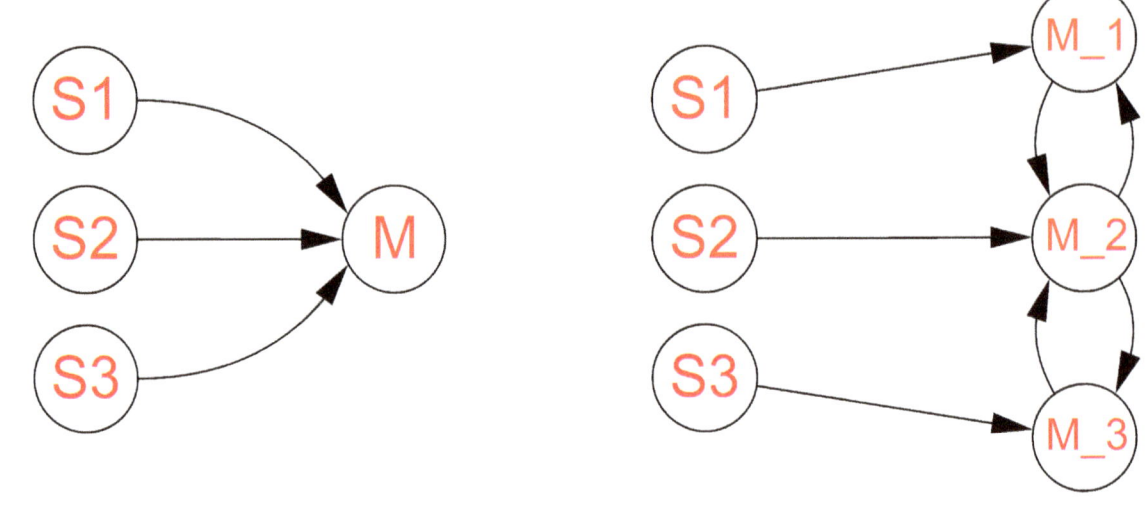

7   Directed edges show that members S1, S2, S3 are supported by member M. This support member can also be shown subdivided into its constituent submembers (M1, M2, M3) that are connected with parallel and opposite edges to represent a fixed connection.

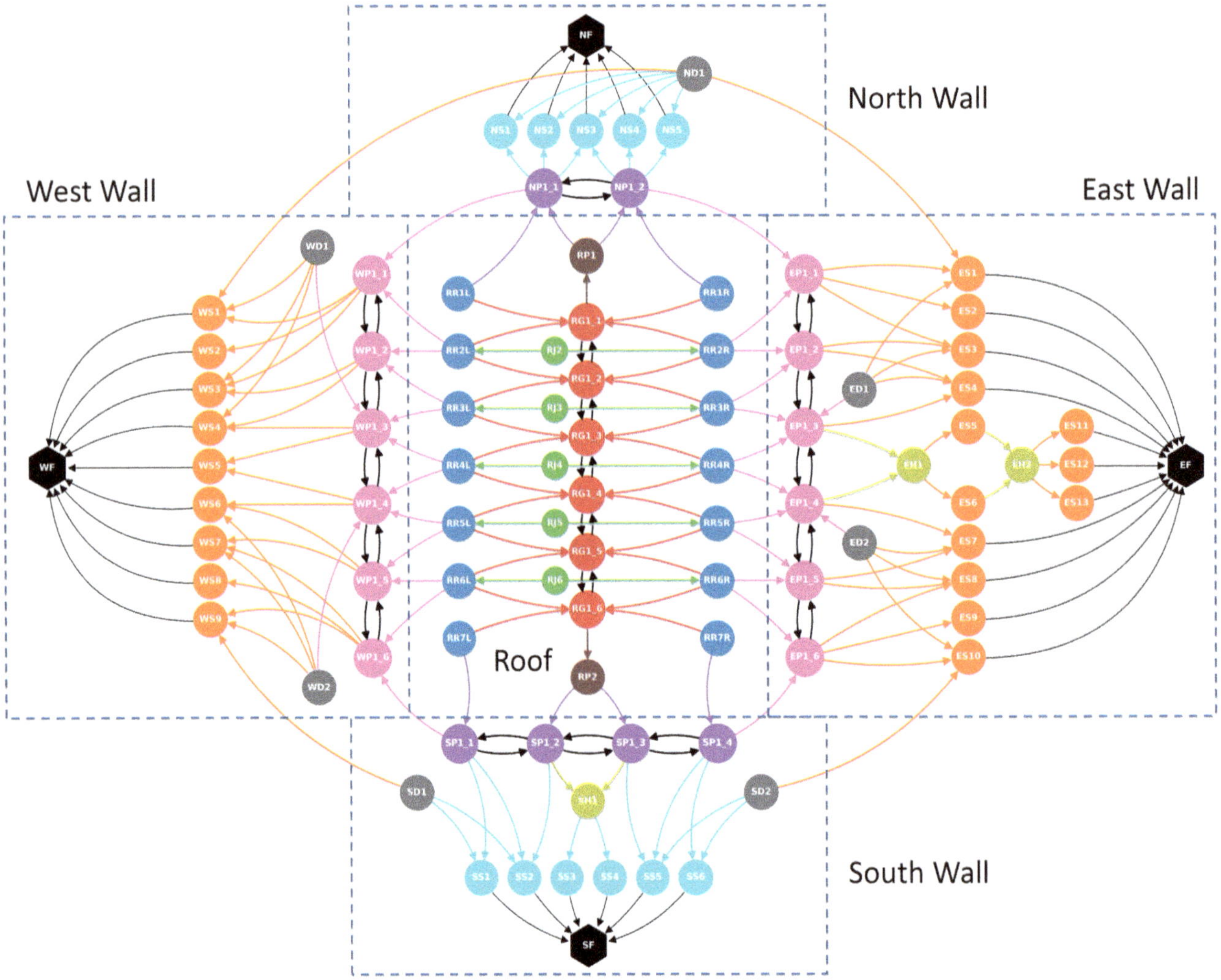

8   The connection hierarchy in the discrete element timber prototype structure is represented as a multidirected graph with outgoing edges indicating the direction of support. Vertices correspond to unique members or submembers in the structure and are organized into five regions (North, South West, East Walls, and the Roof).

Disassembly Sequence Subgraph

Given a target member for removal, a subgraph representing all the members in the structure that need to be removed and supported along the way is calculated through an algorithmic operation on the full support member hierarchy graph. The algorithm implemented is a modified breadth-first search, which explores regions in the graph adjacent to the member specified for removal (Valiente 2021). The logic is that if a member is supporting another member (i.e., has an incoming edge), then this supported member must first be removed or temporarily held in place by a robot before the supporting member can be safely removed.

The breadth-first search finishes when the queue of vertices to check by the algorithm is empty, which occurs before traversing the full graph since certain conditions result in an "end vertex" (i.e., its neighbors are not checked). For example, if a vertex only has outgoing edges, the member it represents can be removed without impacting any other parts of the structure. Conversely, if a vertex only has incoming edges, then it is a support vertex. There is a more complex end condition for a vertex representing a submember. When such a vertex is reached, it can trigger the end condition if the submember can be cut from its parent member while ensuring that the stability of the parent member is maintained (i.e., has at least two support points).

Structural and Robotic Evaluation of Disassembly Sequence

The disassembly subgraph calculated from the member support hierarchy graph can be thought of as a high-level plan for the removal of a member. However, additional checks related to structural stability and robotic reachability must be performed to verify that this sequence is feasible. This is done through two parallel processes: (1) a parametric finite element (FE) study of the structure in Rhino/Grasshopper using Karamba3D (Preisinger 2013), and a robotic path planning validation using the COMPAS and COMPAS FAB package with a ROS backend (Rust et al. 2018; Mele et al. 2017).

The point cloud of the as-built structure is used to build a finite element beam model of the structure, based on the centerline of the members identified in the point cloud. Working with this model in a parametric environment allows for a disassembly sequence to be fed directly into the analysis pipeline, by sequentially turning members off as they are removed from the structure. Temporary robotic support on the structure is represented as an additional pin support that can be assigned by the user.

The as-built structure point cloud is also used to test the robotic reachability and path planning related to a calculated disassembly sequence. Sets of three points are sampled from different locations along a member and are used to construct a plane in space where a robot would be sent to pick the member. Figures 9 to 11 show an example of this process. The X1 point represents the center of the plane, with the X-axis orientation defined as a vector between X1 and X2. Y1 is the third and final point anywhere on the surface of the member used to define the plane. These pick locations on the various members are then checked while simulating the disassembly sequence to see if a collision-free path is possible to reach them. If this path-planning check returns that no path is possible, either because of collisions with the rest of the structure or with the other robots, then the original disassembly sequence will require updating. These updates consist of removing additional members in the structure or moving the robots into less obstructive positions.

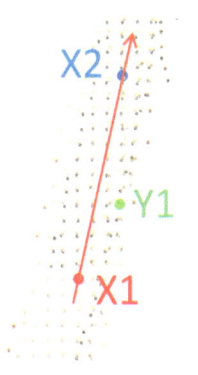

9   Front elevation of the structure point cloud with three points on the surface of a diagonal member highlighted

10   Sets of three points (X1, X2, Y1) are sampled from the point cloud of a member and are used to define the location and orientation of a robotic pick plane centered at X1

11   Each pick plane is checked to have a collision-free robotic path and final configuration kinematically possible to reach and grab the corresponding member

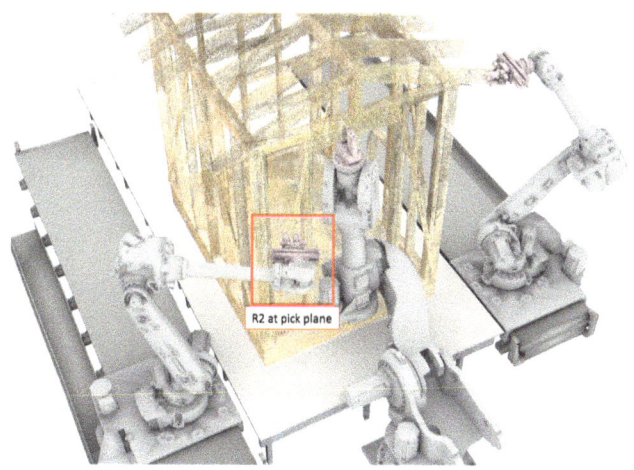

RESULTS AND DISCUSSION
Examples of Disassembly Subgraphs

Figure 12 shows several subgraphs generated from the overall member hierarchy graph with different member removal targets as inputs. The vertices here are highlighted to indicate different types of members in the structure:

- Red: specified member removal target
- Grey: regular member
- Black: support or a submember that is adequately supported in this sequence
- Green: start member (i.e., no member is supported on it)
- Yellow: submember that may require additional support in this sequence (i.e., not adequately supported)

In addition, parallel edges between submember are highlighted in red if the member must be cut at this location during the process of disassembly.

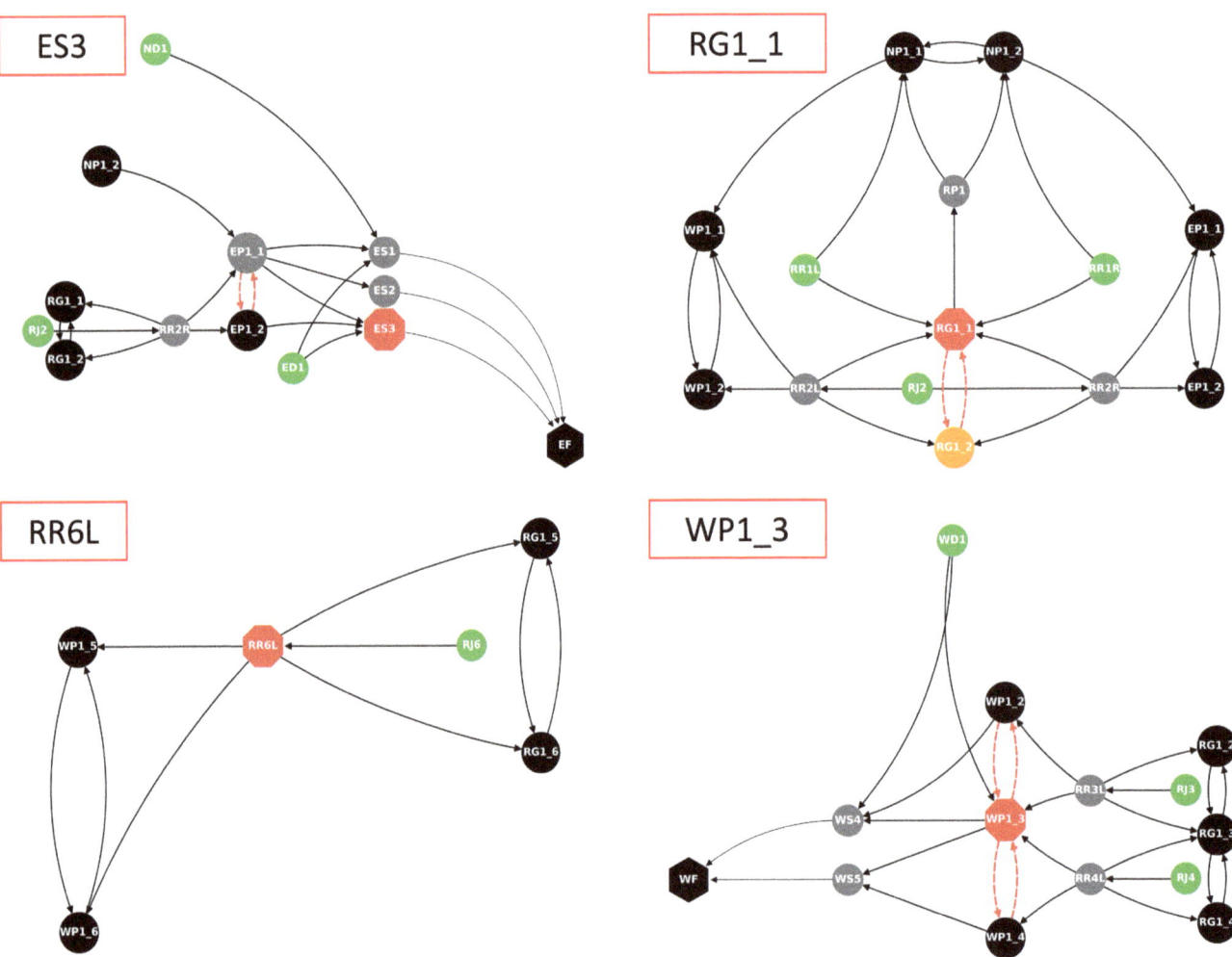

12 Example disassembly sequence subgraphs calculated with different member removal targets (red) specified as inputs: east face stud #3 (ES3), roof girder subcomponent #1 (RG1_1), roof rafter beam #6 (RR6L), and west top plate subcomponent #3 (WP1_3). These subgraphs represent all the members in the structure that are affected in the process of removing the target member.

Detailed Disassembly Planning for Member SS1

Member SS1 (South Wall, Stud #1) is chosen to be removed as part of Phase 1 of the *ZeroWaste* project. The process of calculating a disassembly sequence is meant to simulate the process of planning the removal of a potentially damaged member in a timber stick frame structure. Figure 13 show the disassembly subgraph calculated from the support hierarchy topological representation with the numerical approach described in the previous sections. The physical members that the graph represents are shown in Figure 14.

13 The calculated subgraph representing all the members part of a feasible disassembly sequence for member SS1 (South Wall Stud #1)

14 The members affected in the removal of member SS1 highlighted based on their connectivity and labelled in the rendering of the prototype structure

15 Aggregate disassembly subgraph when including member WS9 as a disassembly target

16 Aggregate disassembly subgraph when including member SS3 as a disassembly target

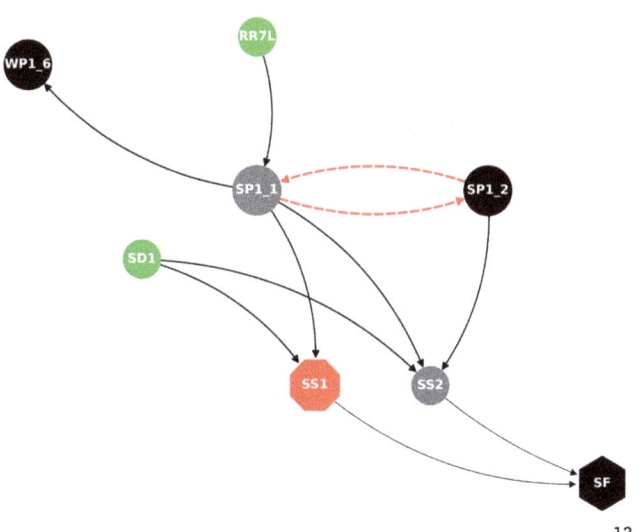

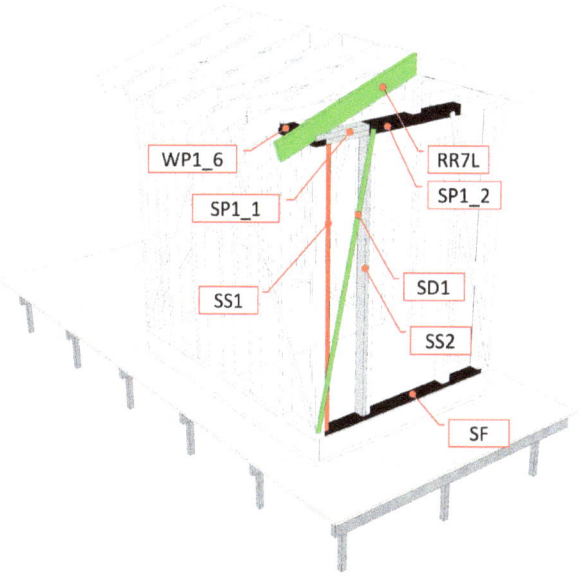

Structural and robotic kinematic evaluation of this sequence follow, which reveal that while the sequence is structurally feasible (i.e., can be executed without compromising stability) the robot is not able to reach member SS1 without colliding with either member WS9 or SS3. The subgraph for the removal of either of these members can be combined with the original SS1 subgraph to create two an aggregate disassembly sequences shown in Figure 15 and 16 respectively. While both sequences are feasible, the option with SS3 results in member SP1_2 not being adequately supported at the conclusion of this process (i.e., less than two supports remaining). Meanwhile, choosing the option with WS9 requires the removal of more members (12 vs. 9), but results in a structure that is self-stable at its termination. The structure at the end of the disassembly sequence is shown in Figure 17 and 18, with the eight individual steps in the cooperative robotic disassembly sequence shown in Figure 19.

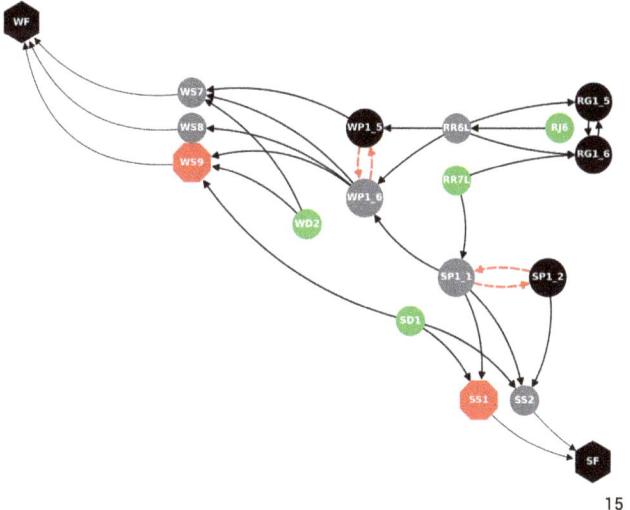

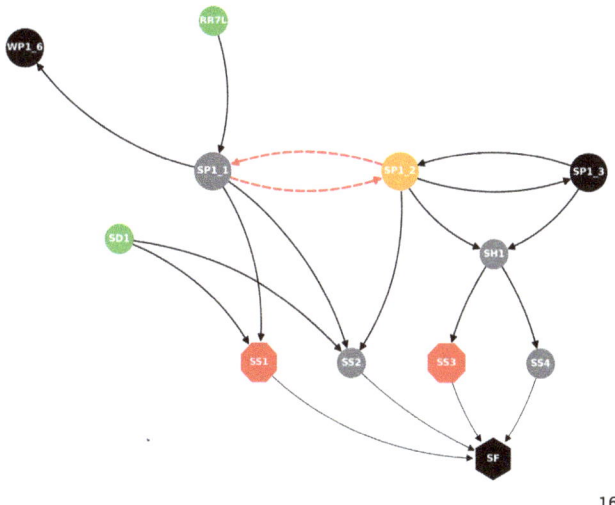

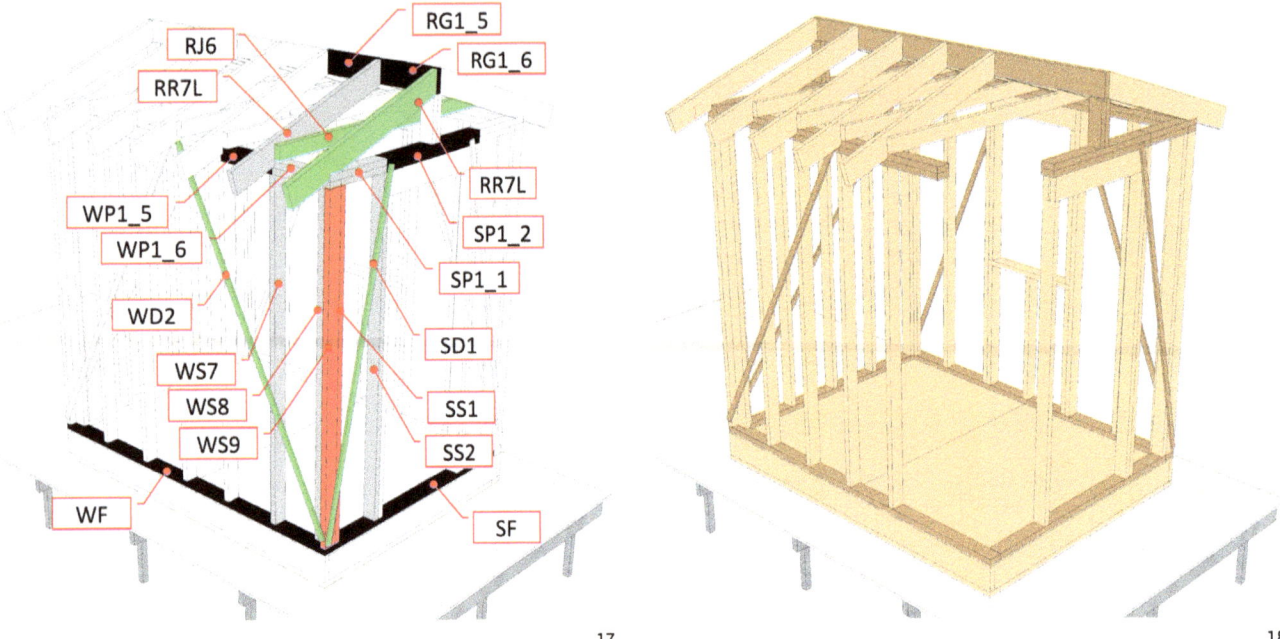

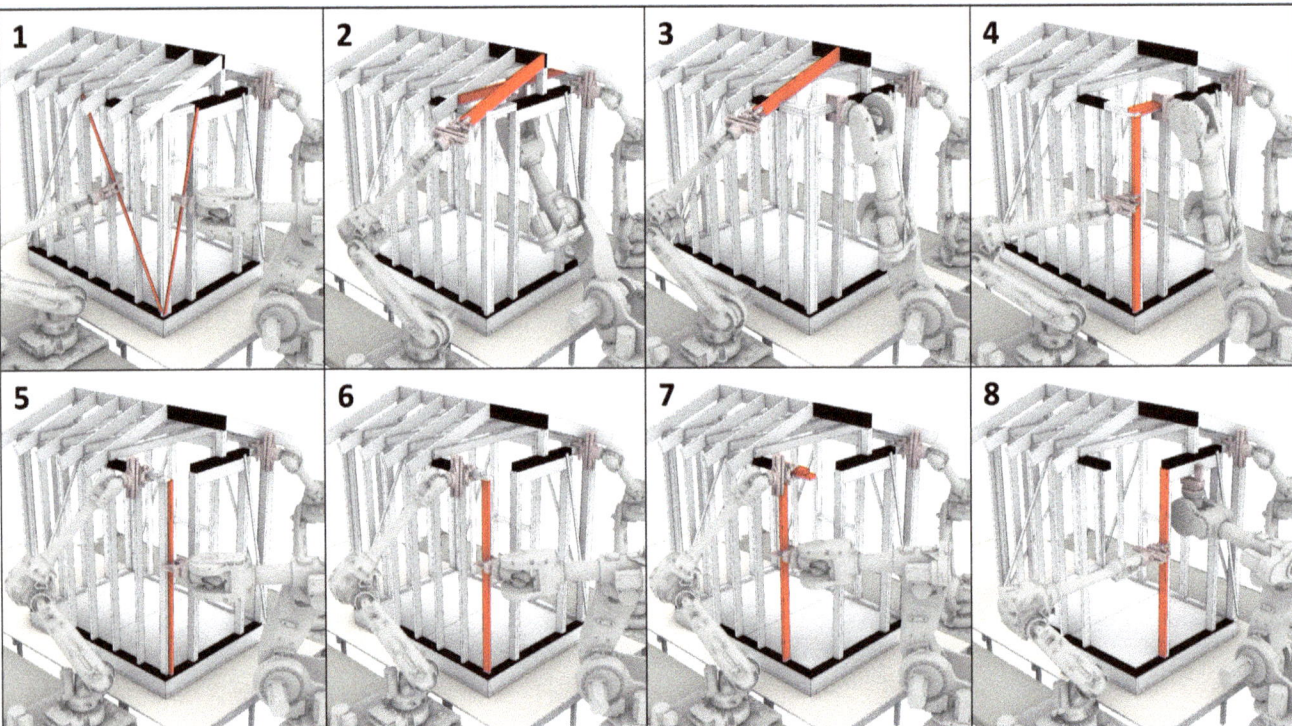

17 The physical members represented in the disassembly subgraph (Figure 15) calculated for members SS1 and WS9 as removal targets

18 The remaining part of the prototype timber structure after the disassembly sequence for members SS1 and WS9 is executed

19 The eight steps required in the cooperative robotic disassembly sequence for the removal of members SS1 and WS9. In each step the members to be removed are shown in red: (1) SD1 & WD2, (2) RR7L & RJ6, (3) RR6L, (4) SP1_1 & WS9, (5) SS1, (6) WS8, (7) WP1_6 & WS7, (8) SS2. Black members are end vertices or supports.

CONCLUSION

The *ZeroWaste* project is about creating a computational workflow for planning the scaffold-free cooperative robotic disassembly and reassembly of existing timber structures as a means for rethinking material circulation in the building industry. This work-in-progress paper focused on aspects of the disassembly planning, and presented the first steps in the development of this process, with a focus on explaining the information gathering methods for an unknown timber structure. We demonstrated how computer vision integrated with a robotic setup can be used to create an accurate as-built representation of an existing timber structure. This information, coupled with a topological representation of the member support hierarchy, was then used to compute various robotic disassembly sequences, and evaluate their robotic and structural feasibility. These methods were demonstrated in the planning of a feasible cooperative robotic disassembly sequence for the removal of a single member as part of Phase 1 of the project. The next project developments will involve the planning and physical demonstration of more complex disassembly sequences, coupled with reassembly of the removed members, for larger portions of the prototype structure (i.e., Phases 2-4).

ACKNOWLEDGMENTS

We would like to thank the following people for their help at various stages throughout the project: Chris Myefski, Chase Galis, Ian Ting, and John Mikesh. In addition, we would like to acknowledge the industry support from Zivid, in the form of reduced pricing for their high-end 3D color cameras. The project is funded in part by Princeton's Campus as a Lab fund and the National Science Foundation (Grant CMMI-ECI 2122271).

REFERENCES

Bock, Thomas. 2007. "Construction Robotics." *Autonomous Robots* 22 (3): 201–9. https://doi.org/10.1007/s10514-006-9008-5.

Bock, Thomas, and Thomas Linner. 2016. *Construction Robots: Elementary Technologies and Single-Task Construction Robots*, vol. 4. Cambridge, UK: Cambridge University Press.

Bruun, Edvard P. G., Sigrid Adriaenssens, and Stefana Parascho. 2022. "Structural Rigidity Theory Applied to the Scaffold-Free (Dis) Assembly of Space Frames Using Cooperative Robotics." *Automation in Construction* 141 (June): 104405. https://doi.org/10.1016/j.autcon.2022.104405.

Bruun, Edvard P. G., Rafael Pastrana, Vittorio Paris, Alessandro Beghini, Attilio Pizzigoni, Stefana Parascho, and Sigrid Adriaenssens. 2021. "Three Cooperative Robotic Fabrication Methods for the Scaffold-Free Construction of a Masonry Arch." *Automation in Construction* 129 (September): 103803. https://doi.org/10.1016/j.autcon.2021.103803.

Bruun, Edvard P. G., Ian Ting, Sigrid Adriaenssens, and Stefana Parascho. 2020. "Human–Robot Collaboration: A Fabrication Framework for the Sequential Design and Construction of Unplanned Spatial Structures." *Digital Creativity* 31 (4): 320–336. https://doi.org/10.1080/14626268.2020.1845214.

Brütting, Jan, Joseph Desruelle, Gennaro Senatore, and Corentin Fivet. 2019. "Design of Truss Structures through Reuse." *Structures* 18 (April): 128–37. https://doi.org/10.1016/j.istruc.2018.11.006.

Brütting, Jan, Gennaro Senatore, and Corentin Fivet. 2021. "Design and Fabrication of a Reusable Kit of Parts for Diverse Structures." *Automation in Construction* 125 (May): 103614. https://doi.org/10.1016/j.autcon.2021.103614.

Cottafava, Dario, and Michiel Ritzen. 2021. "Circularity Indicator for Residential Buildings: Addressing the Gap between Embodied Impacts and Design Aspects." *Resources, Conservation and Recycling* 164 (January): 105120. https://doi.org/10.1016/j.resconrec.2020.105120.

Diyamandoglu, Vasil, and Lorena M. Fortuna. 2015. "Deconstruction of Wood-Framed Houses: Material Recovery and Environmental Impact." *Resources, Conservation and Recycling* 100 (July): 21–30. https://doi.org/10.1016/j.resconrec.2015.04.006.

Eberhardt, Leonora, Julie Rønholt, Morten Birkved, and Harpa Birgisdottir. 2021. "Circular Economy Potential within the Building Stock; Mapping the Embodied Greenhouse Gas Emissions of Four Danish Examples." *Journal of Building Engineering* 33 (January): 101845. https://doi.org/10.1016/j.jobe.2020.101845.

Fleischmann, Philippe, Gonzalo Casas, and Michael Lyrenmann. 2020. "COMPAS RRC: Online Control for ABB Robots over a Simple-to-Use Python Interface." July 2020. https://doi.org/10.5281/zenodo.4639418.

Garcia, Anna Batalle, Irem Yagmur Cebeci, Roberto Vargas Calvo, and Matthew Gordon. 2021. "Material (Data) Intelligence - Towards a Circular Building Environment." In *Proceedings of the 26th International*

Conference of the Association for Computer-Aided Architectural Design Research in Asia, vol. 1. 361–70. http://papers.cumincad.org/cgi-bin/works/paper/caadria2021_088.

Gramazio, Fabio, and Matthias Kohler. 2008. *Digital Materiality in Architecture*, 1st ed. Baden: Lars Muller.

Han, Isla Xi, Edvard P. G. Bruun, Stuart Marsh, Sigrid Adriaenssens, and Stefana Parascho. 2020. "From Concept to Construction: A Transferable Design and Robotic Fabrication Method for a Building-Scale Vault." In *ACADIA '20: Distributed Proximities; Proceedings of the 40th Annual Conference of the Association for Computer Aided Design in Architecture*. Online. 614–623. http://papers.cumincad.org/cgi-bin/works/Show?acadia20_614.

Mele, Tom Van, Andrew Liew, Tomas Mendéz Echenagucia, and Matthias Rippmann. 2017. "COMPAS: A Framework for Computational Research in Architecture and Structures." COMPAS. 2017. https://compas-dev.github.io/.

O'Brien, Elizabeth, Bradley Guy, and Angela Stephenson Lindner. 2006. "Life Cycle Analysis of the Deconstruction of Military Barracks: Ft. McClellan, Anniston, AL." *Journal of Green Building* 1 (4): 166–83. https://doi.org/10.3992/jgb.1.4.166.

Parascho, Stefana, Isla Xi Han, Alessandro Beghini, Masaaki Miki, Samantha Walker, Edvard P. G. Bruun, and Sigrid Adriaenssens. 2021. "LightVault: A Design and Robotic Fabrication Method for Complex Masonry Structures." In *Advances in Architectural Geometry 2020*. Paris, France: Presses des Ponts. 350–75. https://thinkshell.fr/wp-content/uploads/2019/10/AAG2020_18_Parascho.pdf.

Parascho, Stefana, Isla Xi Han, Samantha Walker, Alessandro Beghini, Edvard P. G. Bruun, and Sigrid Adriaenssens. 2020. "Robotic Vault: A Cooperative Robotic Assembly Method for Brick Vault Construction." *Construction Robotics* 4 (3): 117–26. https://doi.org/10.1007/s41693-020-00041-w.

Parascho, Stefana, Thomas Kohlhammer, Stelian Coros, Fabio Gramazio, and Matthias Kohler. 2018. "Computational Design of Robotically Assembled Spatial Structures: A Sequence Based Method for the Generation and Evaluation of Structures Fabricated with Cooperating Robots." In *Advances in Architectural Geometry 2018*. Gothenburg, Sweden: Klein Publishing GmbH. 112–139. https://research.chalmers.se/en/publication/504188.

Preisinger, Clemens. 2013. "Linking Structure and Parametric Geometry." *Architectural Design* 83 (2): 110–13. https://doi.org/10.1002/ad.1564.

Rust, R., G. Casas, S. Parascho, D. Jenny, K. Dörfler, M. Helmreich, A. Gandia, Z. Ma, I. Ariza, and M. Pacher. 2018. "COMPAS FAB: Robotic Fabrication Package for the COMPAS Framework." COMPAS Fab. 2018. https://github.com/compas-dev/compas_fab/.

Thoma, Andreas, Arash Adel, Matthias Helmreich, Thomas Wehrle, Fabio Gramazio, and Matthias Kohler. 2018. "Robotic Fabrication of Bespoke Timber Frame Modules." In *Robotic Fabrication in Architecture, Art and Design 2018*. Zurich: Springer International Publishing. 447–458. https://doi.org/10.1007/978-3-319-92294-2_34.

US EPA (United States Environmental Protection Agency). 2018. "Construction and Demolition Debris Generation in the United States, 2015." Office of Resource Conservation and Recovery. Accessed September 2, 2022. https://www.epa.gov/facts-and-figures-about-materials-waste-and-recycling/studies-summary-tables-and-data-related.

Valiente, Gabriel. 2021. *Algorithms on Trees and Graphs: With Python Code*, 2nd ed. Cham: Springer.

Zimmann, Rainer, Harriet O'Brien, Josef Hargrave, and Marcus Morrell. 2016. "The Circular Economy in the Built Environment." Arup. Accessed September 2, 2022. https://www.arup.com/perspectives/publications/research/section/circular-economy-in-the-built-environment.

Zivid AS. 2021. "Zivid Two." Technical Specifications. Oslo, Norway. Accessed September 2, 2022. https://www.zivid.com/zivid-two?hsLang=en.

IMAGE CREDITS

All drawings and images by the authors.

Edvard P. G. Bruun is a PhD Candidate in the Form Finding Lab in the Civil and Environmental Engineering Department at Princeton University. Before starting his PhD, he worked as a structural engineer at Arup. His current research interests lie in exploring how multiple industrial robotic arms can be sequenced cooperatively when applied to the dis(assembly) of discrete-element structures. He develops approaches that leverage the multi-functionality of robotic setups to design aesthetic and efficient structures that can be (dis)assembled while preserving their stability without the need for any external scaffolding.

Erin Besler is an Assistant Professor of Architecture at Princeton University and cofounder of Besler & Sons, a design studio based in central New Jersey. They were named 2019 United States Artists Fellows in Architecture & Design. Their work has been exhibited at venues internationally, including the MAK Center for Art and Architecture, Chicago Architecture Biennial, and Shenzhen and Hong Kong Bi-City Biennale of Architecture/Urbanism. Recently, they were awarded a grant from the Graham Foundation for their debut book *Best Practices*. Erin's work is characterized by a particular interest in construction technologies, social media, and other platforms for producing and sharing content, where interactions rely less on expertise and more on ubiquity.

Sigrid Adriaenssens directs the Form Finding Lab at Princeton University, where she also teaches courses on (non-)linear mechanics and design of structures and the integration of engineering and arts. Her research interests lie in the mechanics of large-span structural surfaces under extreme loading and under construction. She works on advanced analytical formulations, numerical form finding and optimization approaches, fluid/structure interaction and machine learning models algorithms to open new avenues for accelerated discoveries and automated optimal designs. In 2021, she was named Fellow of the ASCE Structural Engineering Institute, elected IASS vice-president, received the DigitalFUTURES Matthias Rippmann and the Pioneers's Award.

Stefana Parascho is a researcher, architect, and educator whose work combines architecture, digital fabrication and computational design. She is currently Assistant Professor at EPFL where she founded the Lab for Creative Computation (CRCL). Through her research, she has explored multi-robotic fabrication methods and their relationship to architectural design. Her goal is to strengthen the connection between design, structure, and fabrication, and boost the interdisciplinary nature of architecture. Before joining EPFL, she was Assistant Professor at Princeton University. She completed her doctorate in 2019 at ETH Zurich and received her Diploma in Architectural Engineering from the University of Stuttgart.

The Reproduction of Chinese Traditional Timber Structure

The Development of Robot-based Fabrication Apply to the Reproduction of Chinese Traditional Timber Structure

Yang-Ting Shen
National Cheng Kung University

Mi-Chi Wang
National Cheng Kung University

Lien-Kai Huang
National Cheng Kung University

You-Min Gao
National Cheng Kung University

Chia-Chin Yen
National Cheng Kung University

1 Photos of the manufacturing process

ABSTRACT

In Chinese traditional timber building, "Dou-gong" stands as one of the most distinctive features to present the Chinese structure style. However, the preservation and reproduction of Dou-gong face difficulties due to the withering craftsman issue.

This paper proposes a method to digitize the structure into BIM (building information modeling) and reproduce it via robot-based fabrication. By modeling these Dou-gong components with BIM technologies, we can establish a geometrical and non-geometrical 3D database. Then we use Autodesk Fusion and Grasshopper to design the robotic fabrication information whose information is transferred from 3D database models. Based on the fabrication information, including work paths and tool parameters, the KUKA robotic arm with six axes can precisely mill the wood materials into Dou-gong components without any traditional craftsman's processing. The fabrication process we develop has produced the complete Dou-gong prototype which demonstrates the potential of robot-based craft fabrication in Chinese traditional timber structures.

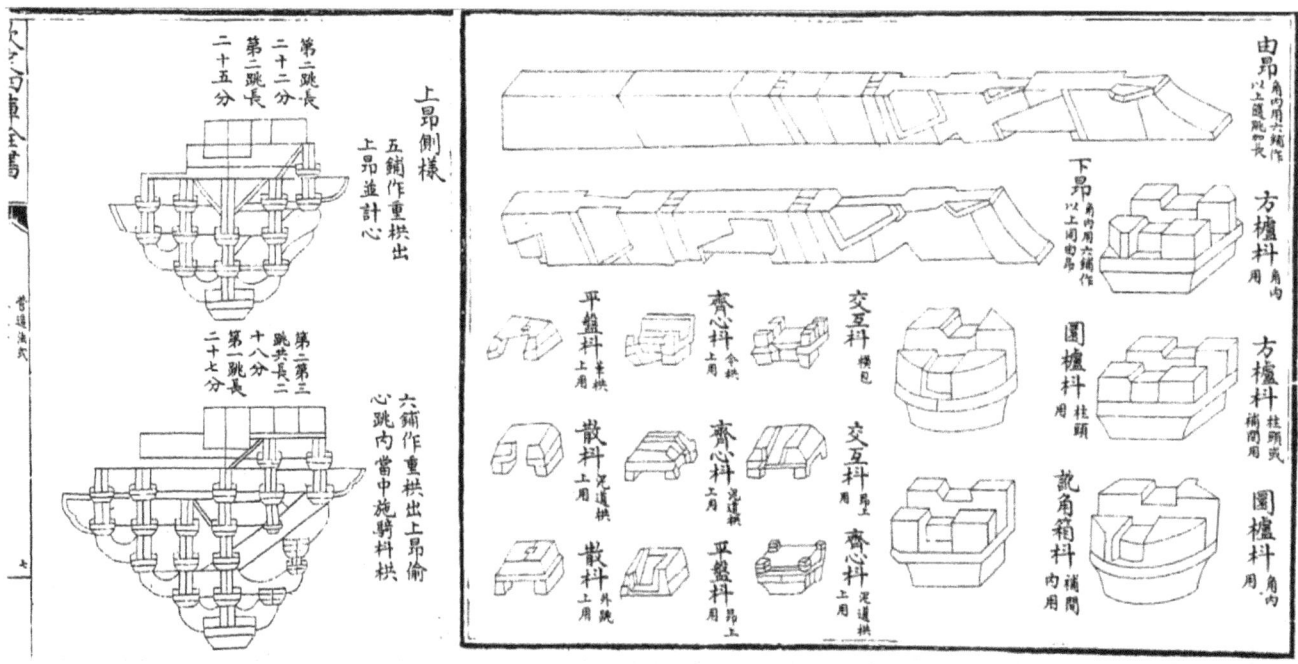

2   Chinese traditional timber structure recorded in Ying-Zao-Fa-Shi

INTRODUCTION

Chinese traditional timber structure plays an essential role in East Asian architecture. One of the most famous structural systems, called "Dou-gong," presents the beauty of wood and the skill of a craftsman, in the context of historical heritage. However, traditional temples that employ Dou-gong structures face a serious talent gap in restoration, where aging craftsmen and gradually declining apprenticeships present a critical challenge, further compounded by trending lower birth rates and the diminished pipeline for new craftsmen that presents less opportunity to reverse the phenomenon. Therefore, we might consider disruptive innovation that introduces collaboration between craftsman and technology. Based on this argument, this paper proposes a methodology to preserve and convert traditional knowledge and skill via digital fabrication technology.

In this paper, we focus on the preservation and reproduction of Dou-gong skill and structure. By studying the Dou-gong construction methods from literature, we digitize Dou-gong into BIM (Build Information Modeling), including geometry data and non-geometry data. Then we can build the auto-fabrication process to reproduce complex Dou-gong components via robot-based fabrication.

BACKGROUND

In Chinese traditional timber structure, "Dou-gong" is one of the most characteristic components. Dou-gong supports the roofs with complex members, leading to an elegant curve on the facade. Chinese traditional timber structure was recorded in the ancient Chinese language and shown as hand drawing in books like "Ying-Zao-Fa-Shi" (Figure 2). In the past, the production of Dou-gong relied on the craftsman's handicraft. It consumed lot of time to refine the components because the diversity parts need to be jointed together very precisely. As the development of digital parameter tools such as BIM (Building Information Modeling), the idea of applying BIM to digitize traditional architectures has become feasible. One of the first examples of using parametric design in ancient Chinese architecture indicates that Dou-gong is the crucial module of the whole structure (Chen 2002).

Dou-gong research is distributed in several countries, including Taiwan, Japan, and China, that adopted this structure in their ancient buildings or temples. Some research has demonstrated that digital fabrication with CNC can replicate the traditional timber joints. Takabayashi reproduces traditional Japanese joints with CNC equipped with various tools like circular saws, jigsaws, and reciprocating saws (Takabayashi et al. 2018). Zhao focus on using the parametric tool to optimize Dou-gong's structural behavior to be manufactured with a robotic system (Zhao, Jiangyang, et al. 2021). Most researchers report that one of the keys to the reproduction the Dou-gong structures must rely on the correct and precise digital modeling which can provide manufacturing information. Therefore, the digitization of Dou-gong for the intervention of digital fabrication tools will become the critical mission in our research.

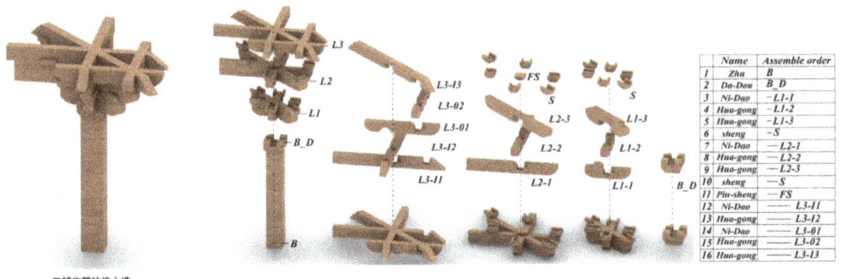
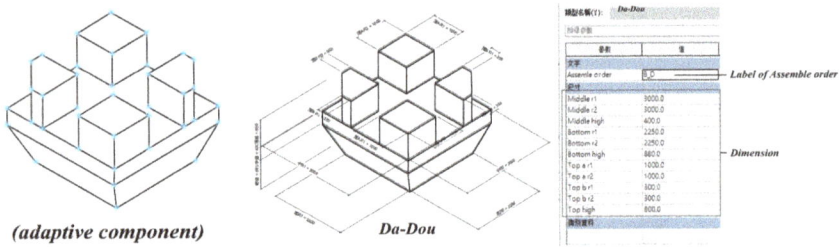

3 The process of the preservation of Dou-gong craftsmanship

4 Analysis process of fabrication method, and the workflow of the reproduction of Dou-gong components

METHOD

In our research, we take one of Dou-gong types from the Nan-Chang temple. The type we choose includes the roof structure and pillar, which is the typical combination of Chinese timber building. In this paper, we propose the integrated workflow of BIM and robot-based fabrication for the preservation and reproduction of Dou-gong craftsmanship. There are two goals we want to achieve: (1) the preservation of Dou-gong craftsmanship via digitalization, and (2) the reproduction of Dou-gong components via digital fabrication. We start the workflow by establishing the BIM Dou-gong model, which includes geometry and non-geometry data. The geometry part describes the shape of Dou-gong components. The non-geometry part describes the relationship of each component, such as the logic and order of assembly. Then, based on the BIM database, we can plan the robotic fabrication information and reproduce the real Dou-gong components via a robotic arm.

The Preservation of Dou-gong Craftsmanship via Digitalization

We use BIM technology to lead the preservation of Dou-gong craftsmanship. First, we document the geometrical and non-geometrical data of Dou-gong through related literature. Then, the components of Dou-gang are created by adaptive component modeling, which can link the model's node with dimension. After modeling, the components are sorted and labeled by their types and assembly order. Each component cab is built into the BIM model based on those data. Each component type in a traditional timber structure is organized into a BIM family database; thus, anyone can easily read the component data from the existing timber structure and change the parameters to meet different requirements. The BIM documents the component name, 3D shape, category, material, and assembly order which can support the further generation of fabrication information (Figure 3).

The Reproduction of Dou-gong Components via Digital Fabrication

In this step, we convert the BIM model into CAM data via AutoDesk Fusion 360. The critical action we need to do is geometry composition analysis of each Dou-gang. We analyze each Dou-gang component's geometry composition and separate it into different fabrication parts to match proper fabrication methods. In Figure 4, we take one of the Dou-gang components called Da-Dou as an example to demonstrate the workflow. In the beginning, we analyze the Da-Dou geometry shape and separate it into three parts, including the surface part, the sink part, and the slope part. Then we use three milling tools to mill the mass wood by order of face, sink, and slope. Firstly, a face milling cutter flattens the surface; secondly, the end milling cutter digs the sink with the rough milling; thirdly, the ball-nose cutter mills the curved slope and polishes all surfaces with the fine milling (Figure 4). Each part's fabrication may include a flipping process to account for the two-sided geometry.

In order to make this whole process easy to be promoted, the study used the common software in this field. The complete workflow of Dou-gong reproduction includes three major phases (Figure 4):

- **CAM Planning:** We import the BIM model to Autodesk Fusion 360, generating the tool path and exporting them as XML files.

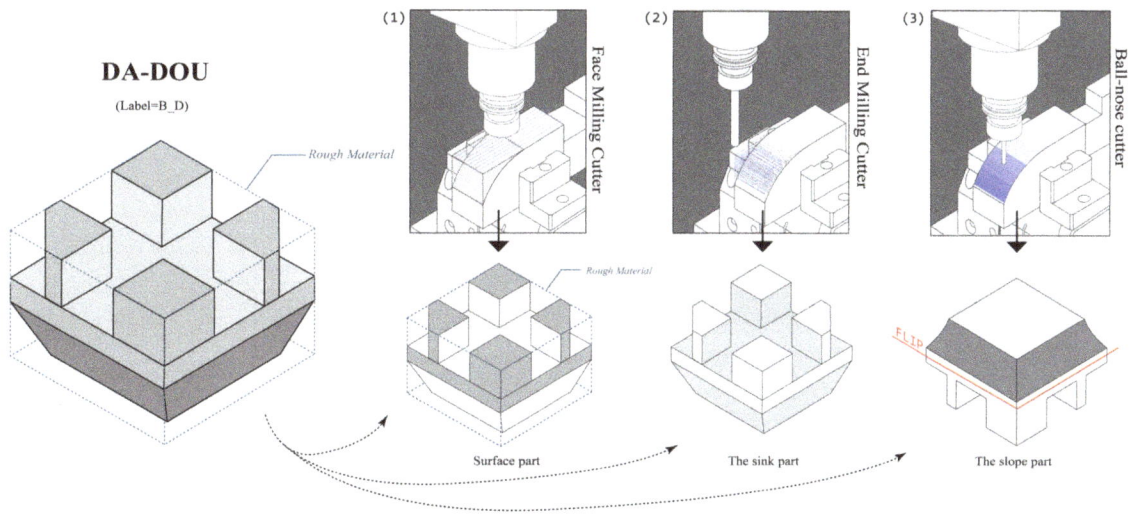

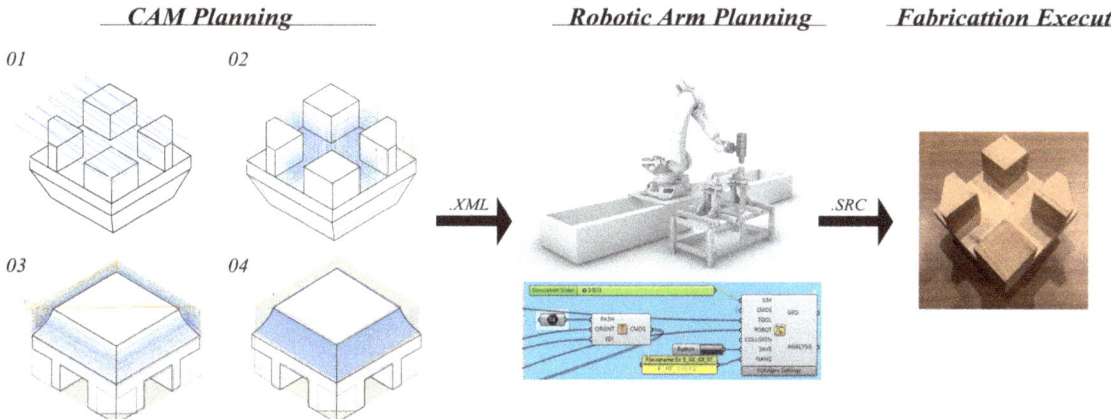

- **Robotic Arm Planning:** The XML files are then transferred to KUKA|prc in Grasshopper, setting up orientation and other robotic fabrication parameters, simulating the whole process of the robot movement to check for any collusion or singularity situation.
- **Fabrication Execution:** In each fabrication stage, material may need to flip during two-sided processing. After locating the material, the work path is converted to robot commands, executed with the robot teaching pendant to fabricate a replica of Dou-gang.

The Robotic Arm Fabrication Environment

The Robotic Arm Fabrication Environment consists of the following components:

- **Industrial Robot:** A KUKA KR300 with a spindle mounted on an external linear axis.
- **Fixture:** A pair of pneumatic grippers to prevent the timber from moving during the milling process.
- **Dust Shield:** Several transparent dust shields surrounding the fixture to keep the workspace clean while ensuring the milling process is visible to the operators.

RESULTS AND DISCUSSION

We apply the workflow to reproduce the Dou-gong of the Nan-Chang temple in our robot laboratory. In the end, we finish 27 components and assemble them to become a complete Dou-gong (Figure 6). Compared with the manual work by the craftsmen, there are several issues we can discuss:

Fabrication Time: The entire fabrication time is about 52 hours, about three times faster than the craftsman, according to the interview.

Accuracy: The dimensional tolerance of this prototype is fewer than 0.5 mm, which is relatively accurate, but further improvement is required. Through discussion with the monument restorer and cultural experts, doubts and concerns were encountered. The elaboration of the component may need improvement to reach the quality of finish flatness and glossiness.

Innovation and Current Limitation: The framework is still under development, and future studies will go on to address current limitations like the unique sense of handwork, finishing flatness, and glossiness different from the one made by artisans. In addition, it may face some problems when applied

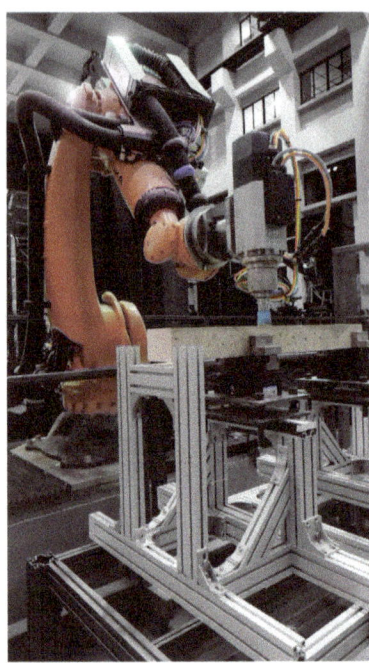

5 The fabrication environment

to heritage restoration. Such limitation needs further experiments and communication with Monuments Restorer and cultural experts about the requirement. Additionally, applying more traditional timber structure components is expected; in the future, we will develop more tools and fabrication methods to match more complex geometry.

Future Application: Collectively, our results will reduce the cost of heritage restoration. Furthermore, it allows the integration of traditional timber joinery with the modern timber structure.

- Eastern conventional timber frame structures in Taiwan are mainly driven by apprenticeship. Therefore, modern construction techniques have limited opportunity to be introduced into this field. However, optimistically, the future integration of intelligent construction into this field could save conventional timber frame craftsmanship from waning due to a declining artisan population.
- Modern timber frame structure heavily relies on metal connectors. By introducing traditional timber joinery into today's construction system, this research could bring beneficial effects to this industry in terms of sustainability and cost-efficiency.
- The reproduction process of Chinese traditional timber structures should work together with restoration experts to further develop the manufacturing technologies to be directly applied to the heritage restoration. It needs to meet the needs of restoration of ancient cultural architecture, such as the roughness of the finished surface of the material and the sense of handwork.

CONCLUSION

This paper presents a digitalization workflow and fabrication process for traditional Chinese timber structures to preserve the ancient knowledge of Chinese timber architecture. This ability to preserve traditional craftsmanship digitally and physically allows the historical architecture can be repaired and reproduced. In the fabrication aspect, we introduce a method to analyze the component's geometry and correspond it to the processing method, creating a standard operation process to reproduce Dou-Gong. The prototype examined in this paper demonstrates the potential of this research approach. It shows that robotic fabrication can be applied not only the innovation design or advanced technology but also to traditional Culture.

ACKNOWLEDGMENTS

This research is developed by Robot-Aided Creation and Construction (RAC-Coon) of National Cheng Kung University and the Department of Architecture. This paper is supported by the Ministry of Science and Technology (MOST) of Taiwan to National Cheng Kung University under MOST-110-2221-E-006-049, MOST-111-2221-E-006-050-MY3, and MOST-111-2420-H-006-007.

REFERENCES

Cokcan, B., J. Braumann, W. Winter, and M. Trautz. 2016. "Robotic production of individualised wood joints." In *Proceedings of the 21st International Conference on Computer-Aided Architectural Design Research in Asia (CAADRIA 2016)*. Melbourne. 559–568.

Chen, Yue. 2002. "Parametric Designing of Ancient Chinese Architecture." Chongqing University, Master's Thesis.

Chilton, John, and Gabriel Tang. 2016. *Timber Gridshells: Architecture, Structure and Craft*. London: Routledge.

6 The result of Dou-gong's reproduction

Heesterman, M., and K. Sweet. 2018. "Robotic Connections: Customisable Joints for Timber Construction." In *SIGraDi 2018: Proceedings of the 22nd Conference of the Iberoamerican Society of Digital Graphics*. São Carlos, Brazil. 644–652.

Hsu, J. S., Y. T. Shen, and F. C. Cheng. 2022. "The Development of the Intuitive Teaching-Based Design Method for Robot-Assisted Fabrication Applied to Bricklaying Design and Construction." In *International Conference on Human-Computer Interaction*. Cham: Springer. 51–57.

Jie, Li. 2018. *Yingzao Fashi (State Building Standards)*, translated by Mu Yu Fang. Chongqing, China: Chongqing University Press.

Svilans, T., P. Poinet, M. Tamke, and M. R. Thomsen. 2018. "A Multi-scalar Approach for the Modelling and Fabrication of Free-formGlue-laminated Timber Structures." In *Humanizing Digital Reality*. Singapore: Springer. 247–257.

Thoma, A., A. Adel, M. Helmreich, T. Wehrle, F. Gramazio, and M. Kohler. 2019. "Robotic Fabrication of Bespoke Timber Frame Modules." In *Robotic Fabrication in Architecture, Art and Design (ROBARCH) 2018*, edited by J.Willmann et al. Cham: Springer. 447–458.

Takabayashi, H., K. Kado, and G. Hirasawa. 2019. "Versatile Robotic Wood Processing Based on Analysis of Parts Processing of Japanese Traditional Wooden Buildings." In *Robotic Fabrication in Architecture, Art and Design (ROBARCH) 2018*. edited by J.Willmann et al. Cham: Springer. 221–230.

Zhao, J., D. Lombardi, H. Chen, and A. Agkathidis. 2021. "Reinterpretation of the Dougong Joint by the Use of Parametric Tools and Robotic Fabrication Techniques." In *Digital Heritage; Proceedings of the 39th eCAADe Conference*, vol. 2, edited by V. Stojakovic, and B. Tepavcevic. Novi Sad, Serbia: eCAADe. 223–242.

IMAGE CREDITS

Figure 1: ©Siao-Tung Chen, June 2022.
All other drawings and images by the authors.

Yang-Ting Shen is an Associate Professor and Vice CEO of RAC-Coon, at Department of Architecture, National Cheng Kung University, Taiwan.

Mi-Chi Wang is a PhD student and researcher at Department of Architecture, National Cheng Kung University, Taiwan, focusing on lightweight architecture systems, artificial intelligence, computer-aided design, and robotic fabrication.

Lien-Kai Huang is a graduate student and researcher at the Department of Architecture, National Cheng Kung University, Taiwan, focusing on wooden structures and robotic fabrication.

You-Min Gao is a graduate student and researcher at Department of Architecture, National Cheng Kung University, Taiwan.

Chia-Chin Yen is an architect and programmer with a passion for developing fabrication-informed design methods. He studied architecture at National Cheng Kung University in Taiwan. In 2020, he graduated with a degree of MS. in Architectural Design at University of Michigan. During his study in Michigan, he focuses on developing computational frameworks for robotic timber assembly. Now he works as the lead developer at the Working Space Raccoon at NCKU, where he leads the development of computational design and robotic fabrication.

Co-Robotic Assembly of Nonstandard Timber Structures

Arash Adel
University of Michigan

1 Co-robotic timber assembly

ABSTRACT

This paper presents a novel approach for the construction of nonstandard timber structures made from regionally sourced short dimensional lumber, which is enabled through human-robot collaborative assembly (HRCA). This approach is an attempt to address several challenges that exist in dominant timber frame construction practices, in particular: 1) Construction and manufacturing off-cuts that may not be used in the construction of full-height or full-span structural components, and 2) Short reclaimed lumber elements resulting from the deconstruction of buildings, which are limited (when not completely disposed) in their use for the construction of new structures. Therefore, to address these challenges, we ask the following research question: how can robotic assembly be integrated into a comprehensive design, planning, and construction process to facilitate the realization of building-scale structures made from short timber elements?

To address the research question, three main research objectives are identified and experimentally explored: 1) Characterization of a comprehensive construction process, which consists of off-site HRCA of bespoke timber sub-assemblies, 2) Development of a suitable constructive system for robotic assembly, making feasible the realization of articulated structures out of short timber elements, and 3) Incorporation of these techniques and their constraints into an integrative digital design and fabrication method and implementation of a continuous digital design-to-fabrication workflow. These objectives are developed through simulation and physical experimentation (e.g., prototyping) and validated in a real-world case study, Robotically Fabricated Structure (RFS).

INTRODUCTION

Addressing climate change requires significant innovations to reduce the carbon footprints of buildings and structures since the building industry contributes to one-third of global CO_2 emissions (Green 2012). Timber is a renewable material with a long history of use in creating structures and buildings and has the lowest embodied energy compared to other structural materials such as concrete and steel (Slavid 2006). However, several challenges exist in dominant timber frame construction practices, in particular: 1) Construction and manufacturing off-cuts that may not be used in the construction of full-height or full-span structural components, and 2) Short reclaimed lumber elements resulting from the deconstruction of buildings, which are limited (when not completely disposed) in their use for the construction of new structures. Therefore, to address these challenges, we ask the following research question: how can robotic assembly be integrated into a comprehensive design, planning, and construction process to facilitate the realization of building-scale structures made from short timber elements?

Robotic timber construction has been of interest in several research projects in recent years. For instance, Willmann et al. (2015), Helm et al. (2017), Adel et al. (2018), and Adel (2020) investigated the use of industrial robotic arms and other automated technologies (e.g., portal robots) for the fabrication and assembly of nonstandard timber structures. More specifically, the *Sequential Wall* (Oesterle 2009) and the *Sequential Roof* (Apolinarska et al. 2016) demonstrated the use of such technologies for assembling short timber slats into building-scale structural components. Their constructive system consists of layers of short timber slats connected by nails in a side-grain to side-grain configuration to form full-height wall modules in the case of the *Sequential Wall* and full-span beams in the case of the *Sequential Roof*.

The constructive system and the fabrication process of these two projects are highly interconnected. In the case of the *Sequential Wall*, a six-axis industrial robotic arm grips a raw timber element, moves it along the main axis of the element based on the predefined length of the structural element, and holds it until a human fabricator cuts it with a circular saw. The robotic arm then moves the processed element to its final position, where the human fabricator attaches it to the elements of the previous layer. This process repeats until the wall sub-assembly (or a portion of it) is fully fabricated. The fabrication process of the *Sequential Roof* is very similar, with some minor differences. A four-axis portal robot is attached to a telescopic base mounted on a two-axis gantry system. The robot grips a raw timber element and moves it along the main axis of the element based on the predefined length of the structural element, where a computer numerical controlled (CNC) saw cuts it. Subsequently, the portal robot moves the processed element into its final position and attaches it to the elements of the previous layer by shooting nails. This

2   The case-study project of this research, Robotically Fabricated Structure (RFS)

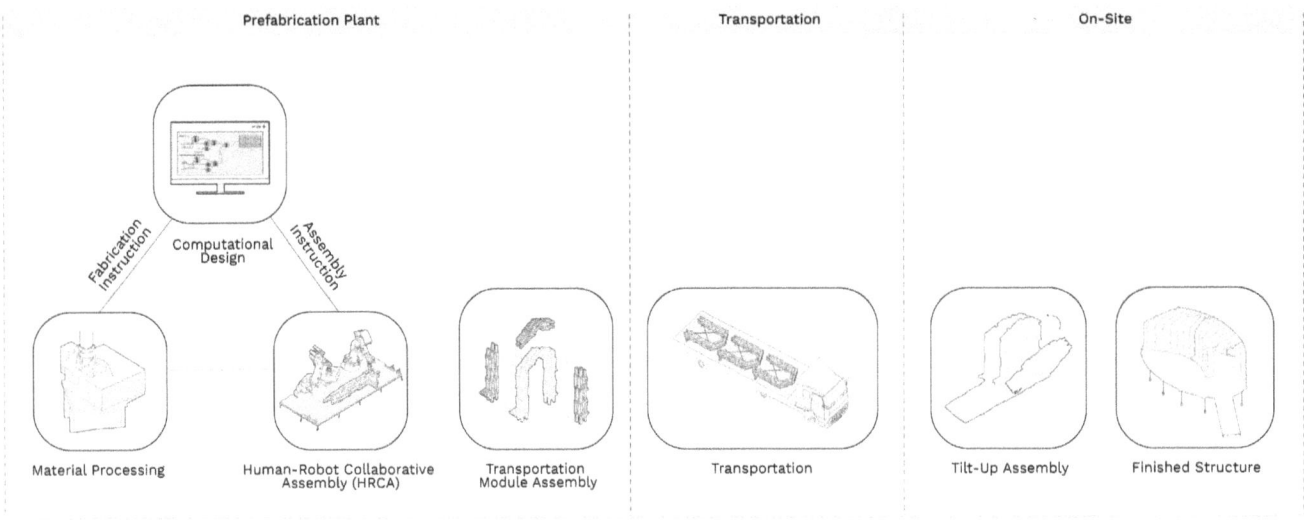

process repeats until the beam sub-assembly is fully fabricated. While these two projects are successful in developing a manufacturing process coupled with a constructive system for specific building components (i.e., wall sub-assemblies in the case of the *Sequential Wall* and ceiling beam sub-assemblies in the case of the *Sequential Roof*), further investigations are required to develop a complete building system consisting of floor, wall, and ceiling components.

More recently, Adel (2020) presented a comprehensive process to facilitate the design, planning, and construction of robotically assembled nonstandard modular timber frame buildings. The construction process includes cooperative robotic spatial assembly of bespoke timber frame modules, transportation of the prefabricated modules to the construction site, and mounting them on-site to realize nonstandard buildings. This process was tested and validated through a real-world case study, *DFAB HOUSE* (Adel et al. 2018; Adel 2020; Graser et al. 2021), which demonstrated the feasibility and potential of this process for fabricating nonstandard timber frame buildings.

Building on these projects, we present a novel approach for the construction of nonstandard timber structures made from regionally sourced short dimensional lumber, which is enabled through human-robot collaborative assembly (HRCA). More specifically, to address the main research question, three research objectives are identified and experimentally explored: 1) Characterization of a comprehensive construction process, which consists of off-site HRCA of bespoke timber sub-assemblies, 2) Development of a suitable constructive system for robotic assembly, making feasible the realization of articulated structures out of short timber elements, and 3) Incorporation of these techniques and their constraints into an integrative digital design and fabrication method and implementation of a continuous digital design-to-fabrication workflow.

These objectives are developed through simulation and physical experimentation (e.g., prototyping) and validated in a real-world case study, Robotically Fabricated Structure (RFS, Figure 2), which must satisfy strict building code requirements and constraints such as structural integrity. RFS is a timber pavilion located in the Matthaei Botanical Gardens in Ann Arbor, Michigan. It includes a raised platform that creates an opportunity for small public events and performances, an exterior seating area, and a semi-enclosed walkway that can be utilized for exhibitions and intimate conversations.

METHODS
Construction Process

We propose a comprehensive construction process, which facilitates the realization of building-scale structures made from short timber elements. Figure 3 illustrates the proposed construction process for the case-study structure of this research. This process includes the HRCA of bespoke timber sub-assemblies. We call these sub-assemblies fabrication modules. We will discuss the HRCA in detail in the following section. After the fabrication modules are completed, they are pre-assembled into an intermediate scale, which we call transportation modules. Subsequently, these transportation modules are braced for structural stability and transported to the construction site, where they are mounted together to form the whole structure. We designed this process to avoid crane usage on the construction site. The transportation modules can be carried by four to five people and put in place using a tilt-up action (Figure 3).

As illustrated in Figure 3, the construction process includes a prototypical just-in-time HRCA of bespoke timber sub-assemblies made from short timber elements.[1] The HRCA process (illustrated in Figure 4) consists of the following main steps. A robotic arm picks up a raw timber element positioned at the picking station, carries it to the CNC saw, and moves the timber element based on a computationally calculated path alongside the saw blade to cut it. When the robot reaches the end of the cutting path, the CNC saw turns off, and the robot carries the element and places it in its relative final position within the module on the assembly platform.[2] At this point, the robotic arm goes into HALT mode, and the human fabricator enters the workcell to shoot nails at the top face of the element to connect it with the previous timber layer and perform quality control. Elements are stacked to form a fabrication module, and therefore, the connection between two elements has a side-grain to side-grain configuration. We chose nails for connecting elements during this step due to the faster speed of using a nail gun for shooting nails compared to inserting screws with a compact driver. After connecting the timber element to the previous layer, the human fabricator exits the workcell, the robotic process resumes, and this process repeats until the module is fully assembled. A key consequence of this process is the elimination of error-prone and labor-intensive logistical steps such as labeling non-identical timber elements since the raw timber element is picked, cut, and placed in a single continuous process without any breaks (Adel et al. 2018; Adel 2020; Craney and Adel 2020).

Constructive System

We propose a constructive system made from short timber elements suitable for the discussed construction process. The proposed construction process and the HRCA procedure necessitate the development of strategies for dividing the structure into sub-assemblies at various scales (e.g., wall, floor, ceiling, and transportation modules). Each step of the construction process requires an in-depth analysis to define manufacturing constraints associated with that step and devise suitable joining techniques for that specific step. Our analysis reveals that this process includes several scales illustrated in Figure 5, which inform the development of the constructive system and characterization of its rules and constraints, as well as the implementation of the necessary data structure required for the digital design-to-fabrication workflow (discussed in the following sections).

In our proposed constructive system, short timber elements are stacked to form a fabrication module. These fabrication modules are then connected to each other to form a transportation module, and these transportation modules are connected to each other on the construction site to form the whole structure. We conducted simulation studies and performed physical prototyping experiments to identify the fabrication constraints of our prototypical fabrication setup, such as the minimum

3   The proposed construction process

4   Steps of the HRCA process (for clarity, the joining process by the human fabricator is not included in the diagram)

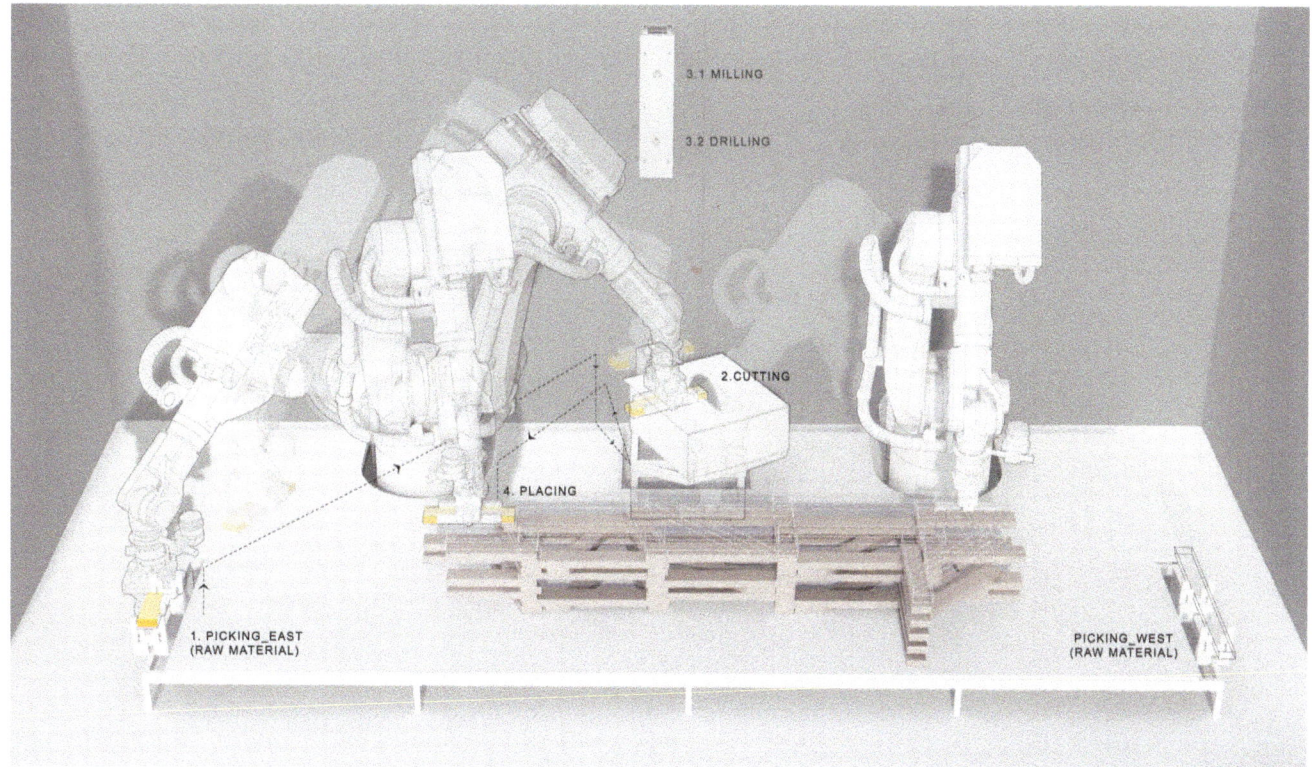

and maximum dimensions of the individual timber elements, directly impacting the manufacturability of sub-assemblies. These constraints must be integrated into the design-to-fabrication process and satisfied to guarantee manufacturability. We discuss a few key constraints here.

One of the main constraints is the minimum and maximum dimensions of the timber elements. Our constructive system consists of regionally sourced dimensional lumber, and for the case-study structure, we limit the elements to 2x4 lumber. The length of an element is measured as the distance along the element's central axis between each cut plane. Several key parameters were identified to precisely define the lower and upper length bounds of the 2x4s. The first is the length of the gripper, which, combined with the angle of the miter cut, defines the minimum length of the element. An element with a length smaller than the minimum length will result in the collision of the gripper with the saw blade during the cutting process. Figure 6 illustrates this constraint. A key observation is that the minimum length of an element can change depending on the angle of the miter cut; for instance, the minimum length of an element with perpendicular cuts is 330 mm, and the minimum length of an element with 45-degree miter cuts is 407 mm (including a safety margin). The maximum length of each element is defined such that the timber element does not collide with the body of the robot during the fabrication process. The moment of closest contact occurs during the cutting process due to the proximity of the saw to the robot, which limits the maximum length to 880 mm. The integrative computational design process performs this check and indicates the elements that do not satisfy the length constraint.[3] The design is then iteratively refined until all the elements satisfy this constraint.

Another key constraint is the maximum dimensions of a sub-assembly that can be fabricated in the workcell. Based on the results of simulation studies and physical experiments, we identified the overall working envelope of our workcell, which has maximum dimensions of 1.6 m wide by 4.6 m long by 2.0 m high. A key observation made during these studies indicates that this working envelope may shrink based on the desired angle of the sixth axis of the robot and the orientation of the tool center point (TCP) of the gripper for placing the elements. Since the modules of the case-study structure are assembled horizontally on the assembly stand, the orientation of the sixth axis needs to be parallel to the world z-axis, which reduces the maximum dimensions of sub-assemblies to be 1.0 m wide by 4.6 m long by 0.8 m high. These dimensions are much smaller than the transportation volume constraints; therefore, we devised an intermediate scale consisting of three to four fabrication modules joined together to form a transportation module (Figure 5). This approach reduces the time spent during the on-site assembly process since fewer modules need to be assembled on site.

For the pre-assembly of the transportation modules, we devised a suitable joining technique using alternating fingers coupled with screws, which enables the same side-grain to side-grain connection typology between timber elements of the two connecting modules (Figure 5). Here, we use screws since longer screws can easily penetrate several timber layers (fingers) to form a robust structural connection between two

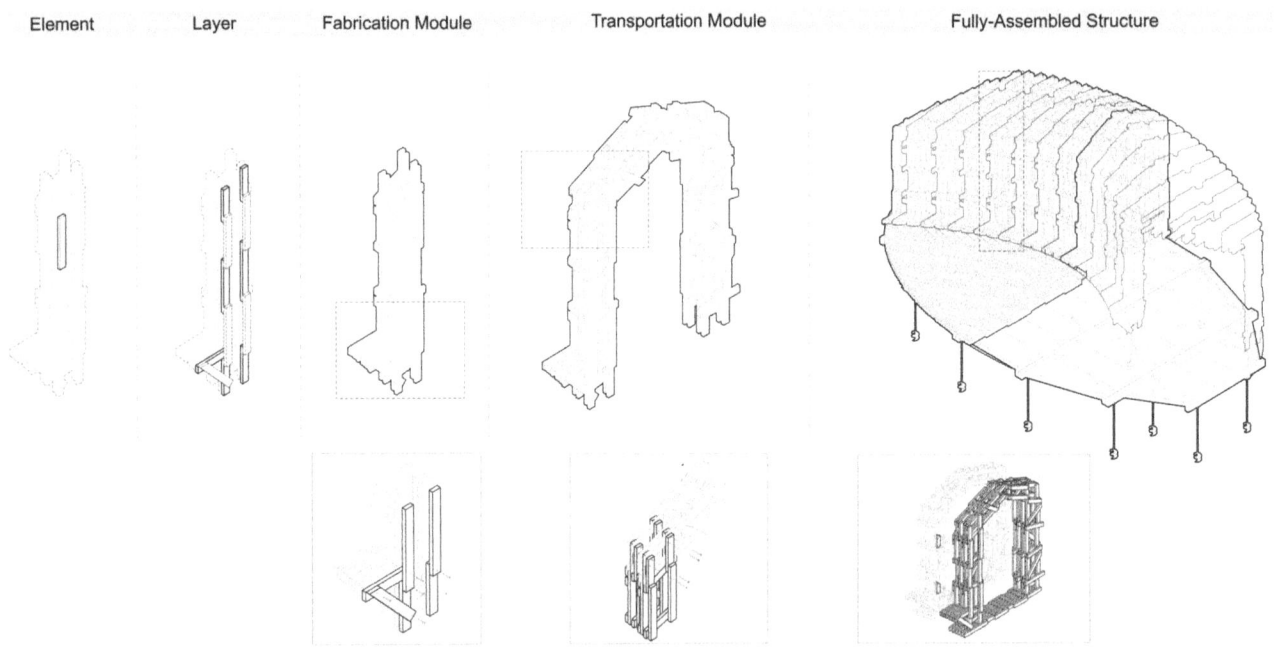

5 Constructive system and multiple scales of the structure

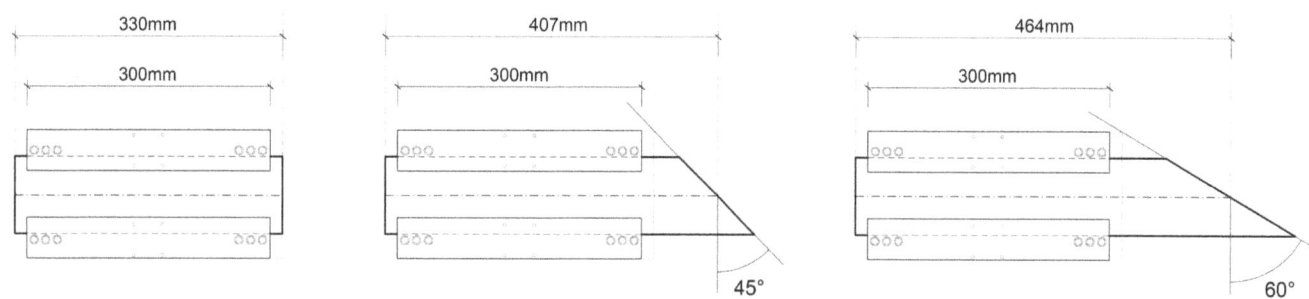

6 Calculation of the minimum length of the timber elements based on the length of the gripper fingers and the angle of the miter cut

modules. Furthermore, screws can be easily removed when disassembling the structure and relocating it or completely dismantling the structure and reclaiming its lumber for future use. On the construction site, transportation modules are connected to each other using module-connecting blocks coupled with screws (Figure 5).

Design-to-Fabrication Workflow and Process

The successful execution of this research requires the implementation of dynamic task and motion planning as well as a seamless digital design-to-fabrication workflow to automate the fabrication and assembly of non-identical timber elements. This workflow should enable the automatic generation of manufacturing instructions to control the robotic arms and the CNC saw (as well as the mill and the drill) based on the geometric attributes of each element. We implemented the necessary data structure and workflow to enable this process, which are discussed in detail in the rest of this section.

As previously discussed, the fabrication modules are assembled layer by layer horizontally (Figure 4). For each module, the human operator assigns a frame to that module (usually located at one of its corners) and defines its transformed counterpart on the assembly platform. Accordingly, a four-by-four transformation matrix is calculated and applied to the elements of that module to transform them onto the assembly platform. This approach simplifies the transformation of the selected module and the preparation of the manufacturing instructions for its timber elements. Each timber element is computationally represented as an instance of a custom object (Class), titled Element, which is implemented in Python (Python Software Foundation 2001-2018) and includes the necessary attributes (e.g., gripping frame, cut planes, etc.) to derive manufacturing instructions specific to that element. Based on these attributes, picking, cutting, and placing paths are calculated, and subsequently, manufacturing instructions are generated. For instance, the path for cutting each element is generated such that the gripping frame of the element has a specific distance to the frame of the saw blade, and the x-axis of the gripping plane forms a specific angle with the x-axis of the saw blade, both of which are derived from the desired length of the element and the angle of the miter cut.

RESULTS

We employed the discussed methods in the design, planning, and construction of RFS. The HRCA process proved effective in fabricating the bespoke sub-assemblies of this nonstandard structure. The primary structure of RFS is divided into 76 fabrication modules, which form 19 transportation modules (illustrated in Figure 7, not counting the floor sub-assemblies).[4] The transportation modules are connected to each other on site using 56 module-connecting blocks coupled with screws.

The primary structure consists of 4,045 timber elements connected by 17,336 nails. Out of these, 3,787 timber elements (93.62% of the total number of elements) are robotically cut and placed. The lengths of these elements vary from 345.20 mm to 861.80 mm, with a mean value of 615.70 mm and a median value of 600.20 mm. Besides these, there are 258 short elements (6.38% of the total number of elements), which are cut and placed by the human fabricator since these elements are too short to be cut and placed by our robotic setup. Most of these elements are fillers to fill the gap between the timber elements of the floor boundary and have a length of 110 mm.

Figure 8 includes a series of photos corresponding to the construction process illustrated in Figure 3. While the developed HRCA process proved effective, we observed several challenges during this process. One observation regards human intervention while nailing the elements. In our system, each connection requires two nails arranged on the diagonal of the overlap between the two timber elements. From one layer to the next, the arrangement of the nails alternates to the other diagonal of the overlap to avoid collisions between the nails of the two adjacent layers. While this simple system is effective for human interventions without any guides, bespoke nailing and screwing layouts will require additional guiding mechanisms and further research. Another observation regards the role of the human fabricator while handling exceptions, e.g., placing short timber elements that could not

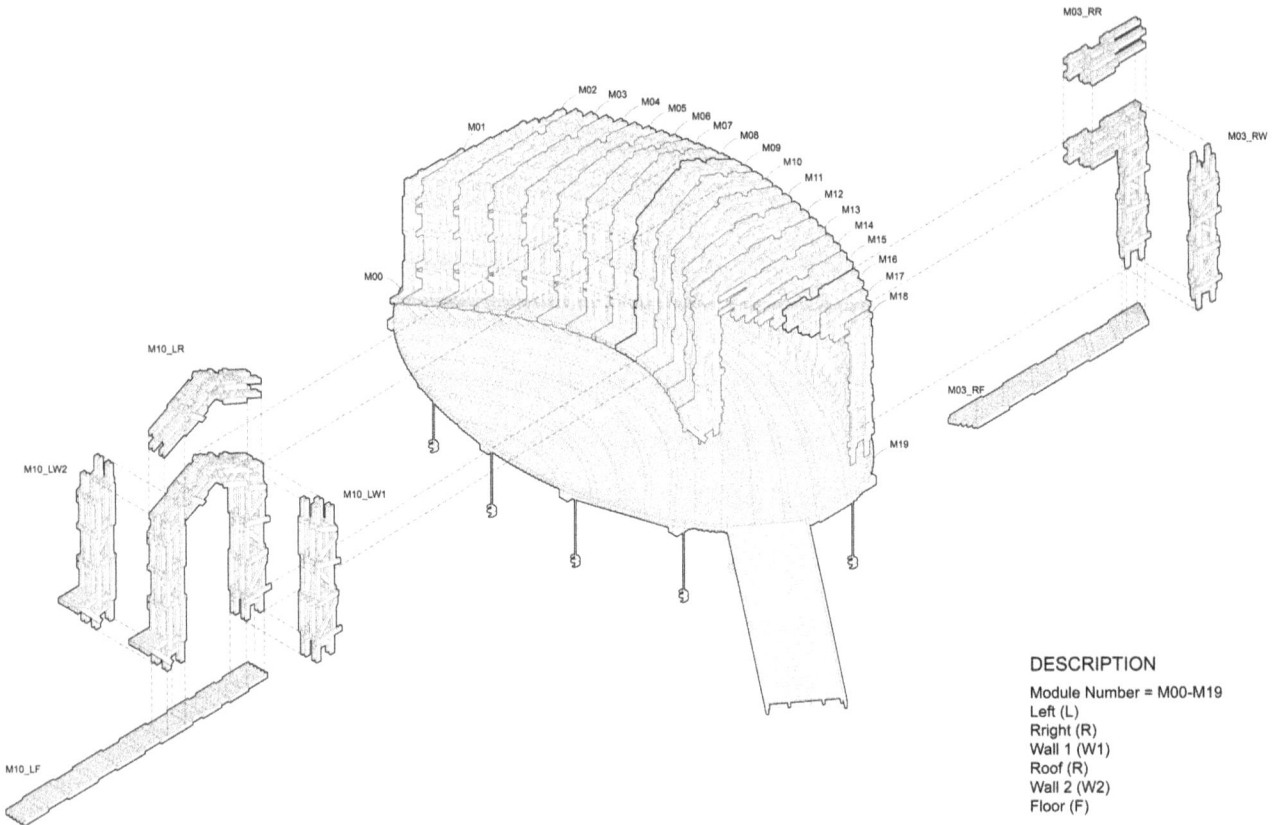

7 RFS and its modules

be cut and placed robotically. It can be argued that exceptions occur in most architectural projects beyond the constraints of the manufacturing process due to the specificities and requirements of each project. Fabricating RFS illustrated that human interventions could effectively complement robotic processes for handling exceptions. However, developing suitable digital communication channels for human interventions also requires further research.

A third observation regards the on-site tilt-up assembly of the transportation modules. Although this process is straightforward, the overall length of the RFS ended up being 212 mm longer than the digital model. This increase in length is due to the average thickness of the elements used (of which are not exactly consistent) being slightly larger than the modeled element thickness. Since the whole structure consists of 234 layers, the accumulative build-up of this tolerance resulted in a noticeable increase in the length of the structure. To avoid having this tolerance showing up only on one side of the finished structure, we assembled the transportation modules starting from the middle of the structure (module 10 in Figure 7) and built it outwards. This approach distributed the accumulative tolerance to both sides of the structure. Handling accumulative tolerance requires further research as well.

The proposed construction process, coupled with the developed constructive system, enables the construction of full-height wall and full-span floor and ceiling sub-assemblies using short timber elements, which respond to various programming requirements integrated into RFS such as a seating area, a raised platform, and an enclosed space. RFS also exemplifies the expressive qualities resulting from this approach (Figures 2 and 9) and its potential for creating novel architecture.

CONCLUSION

In this paper, we presented a comprehensive construction process coupled with a suitable constructive system and digital design-to-fabrication workflow to facilitate the design, planning, and construction of nonstandard timber structures made from short dimensional lumber. Furthermore, we demonstrated the application of developed methods in the realization of a case-study structure, RFS, which tested and validated the methods in a real-world setting beyond the laboratory environment. RFS illustrated the potential of the developed processes for the design and manufacture of highly articulated timber architecture. This research paves the way for further investigations utilizing short reclaimed lumber elements in the construction of new structures.

 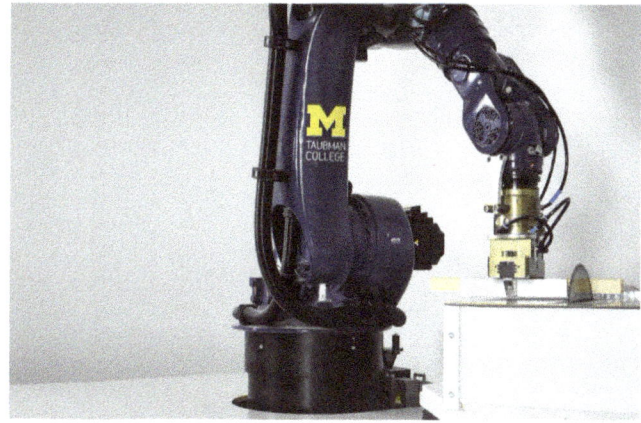 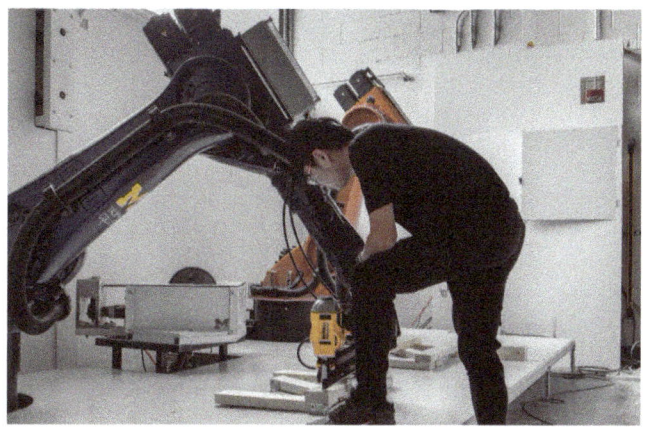 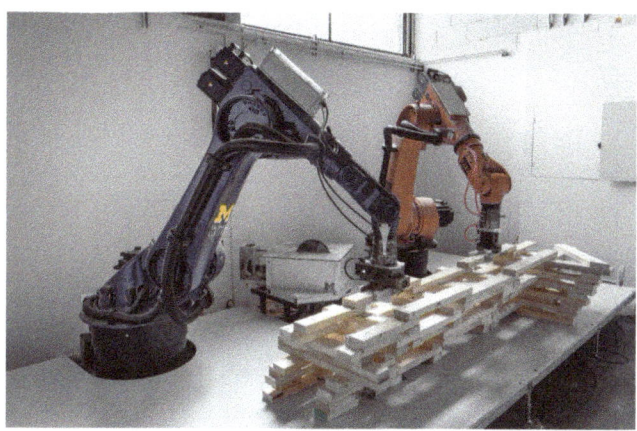    

8 The HRCA and construction process photos

9 RFS at night

Future Work

Several challenges were identified throughout the research, which could not be addressed within the scope of this project. We see a potential for integrating augmented reality (AR) to assist and guide human interventions for nailing and inserting screws, as well as providing a digital communication channel beyond conventional methods (e.g., two-dimensional drawings and measurement techniques) for handling exceptions. Moreover, we observed that the on-site assembly of the modules relied solely on drawings and manual measurements for placing the transportation modules in their correct position within the whole structure. We see an opportunity for integrating AR to assist with the on-site assembly process. Furthermore, the accumulative tolerance built-up (discussed in the previous section) could have been quantified during the prefabrication process using laser scanning or other digitization methods to develop a bidirectional digital twin of the built structure and could have been accounted for by employing adaptive fabrication methods. These approaches can be integrated into the future development of the research. Additionally, RFS demonstrated the application of the research in a single-story structure; future research could investigate the design and construction of multistory structures.

ACKNOWLEDGMENTS

The author would like to acknowledge and thank the people who directly and indirectly contributed to the research presented in this paper and the realization of RFS. The project credit for RFS is presented below:

Principal Investigator and Project Lead: Arash Adel, ADR Laboratory

Research, Design, and Fabrication Assistants: Ben Lawson, Ryan Craney, Sarah Nail, Gabrielle Clune, Andrew Hoover, Juliette Zidek

Construction Assistants: Abdallah Kamhawi, Tharanesh Varadharajan, Ali Fahmy, Elliot Smithberger, Qian Li, Nadim Hajj Ahmad, Joshua Powell, Ivan Gort-Cabeza de Vaca

Students (MS in Digital and Material Technologies, Taubman College of Architecture and Urban Planning): Ruxin Xie, Daniel Ruan, Xinran Li, Jingwen Song, Mehdi Shirvani, Mackenzie Bruce, Chris Humphrey

Structural Engineers: Robert Silman Associates Structural Engineers (Nat Oppenheimer, Omid Oliyan, Justin Den Herder, Paul Evans)

Diagrams: Yunyan Li

Photographers: Arash Adel, Bob Berg, Daniel Ruan, Matthew Weyhmiller, Jacob Cofer

Videographers: Jacob Cofer, Bob Berg

Video Editors: Yunyan Li, Jacob Cofer

Supported by the Herbert W. and Susan L. Johe Fund, Taubman College of Architecture and Urban Planning, University of Michigan

Special Thanks: Jonathan Massey, McLain Clutter, Catie Newell, Wes McGee, Cynthia Radecki, Earl Bell

NOTES

1. We designed and built an HRCA workcell, which includes two industrial robotic arms (Figure 4). However, the discussion of this workcell is beyond the scope of this paper.

2. After cutting the element, if necessary, milling and drilling procedure are performed on the element. However, milling and drilling procedures were not employed for the case-study structure.

3. The discussion of the computational design process is beyond the scope of this paper.

4. As illustrated in Figure 7, we used earth screws acting as the foundation of the structure. There is a conventional structure, that sits on these earth screws below the floor of the RFS.

REFERENCES

Adel, Arash. 2020. *Computational Design for Cooperative Robotic Assembly of Nonstandard Timber Frame Buildings*. Zurich: ETH.

Adel, Arash, Andreas Thoma, Matthias Helmreich, Fabio Gramazio, and Matthias Kohler. 2018. "Design of robotically fabricated timber frame structures." In *ACADIA '18: Recalibration: On Imprecision and Infidelity; Proceedings Catalog of the 38th Annual Conference of the Association for Computer Aided Design in Architecture (ACADIA)*. Mexico City: IngramSpark. 394–403.

Apolinarska, Aleksandra Anna, Ralph Bärtschi, Reto Furrer, Fabio Gramazio, and Matthias Kohler. 2016. "Mastering the "Sequential Roof."" In *Advances in Architectural Geometry*, edited by S. Adriaenssens, F. Gramazio, M. Kohler, A. Menges, and M. Pauly. Zurich: vdf. 240–258.

Craney, Ryan, and Arash Adel. 2020. "Engrained Performance: Performance-Driven Computational Design of a Robotically Assembled Shingle Facade System." In *ACADIA '20: Distributed Proximities; Proceedings Catalog of the 40th Annual Conference of the Association for Computer Aided Design in Architecture (ACADIA)*. 604–613.

Graser, Konrad, Arash Adel, Marco Baur, Andreas Thoma, and Daniel Sanz Pont. 2021. "Parallel Paths of Inquiry: Detailing for DFAB HOUSE." *Technology | Architecture + Design* 5 (1): 38–43.

Green, Michael. 2012. *The Case For Tall Wood Buildings*. Vancouver: Canadian Wood Council.

Helm, Volker, Michael Knauss, Thomas Kohlhammer, Fabio Gramazio, and Matthias Kohler. 2017. "Additive robotic fabrication of complex timber structures." In *Advancing Wood Architecture, A Computational Approach*, edited by A. Menges, T. Schwinn and O. D. Krieg. New York: Routledge. 29–43.

Oesterle, Silvan. 2009. "Performance as a design driver in robotic timber construction." In *CAADRIA 2009: Between Man and Machine*. Yunlin: Department of Digital Media Design. 663–671.

Python Software Foundation. 2001-2018. Python. Accessed September 18, 2019. https://www.python.org/.

Slavid, Ruth. 2006. *Wood Houses*. London: Laurence King.

Willmann, Jan, Michael Knauss, Tobias Bonwetsch, Anna Aleksandra Apolinarska, Fabio Gramazio, and Matthias Kohler. 2015. "Robotic timber construction — Expanding additive fabrication to new dimensions." *Automation in Construction* 61 (Jan): 16–23.

IMAGE CREDITS

Figures 1, 9: Daniel Ruan, 2022

Figure 2: Bob Berg, 2021

Figures 3-7: Yunyan Li/Arash Adel, 2022

Figure 8: Mehdi Shirvani, Jacob Cofer, Bob Berg, 2021

Arash Adel is an Assistant Professor of Architecture at the University of Michigan's Taubman College of Architecture and Urban Planning, where he directs the ADR Laboratory. His laboratory conducts interdisciplinary research at the intersection of design, computation, and robotics, contributing to resilient, sustainable, and low-carbon construction outlooks and achievements. At the core of his comprehensive research is investigating human-machine collaborative processes, which tackle fundamental questions related to the future of the design and construction industries and their potential to have a broader impact on inclusive and equitable building culture. Adel received his Master's in Architecture from Harvard University and his Doctorate in Architecture from the Swiss Federal Institute of Technology (ETH).

Tangential Timber

Nonlinear Wood Masonry

Kyle Schumann
University of Virginia /
After Architecture

Katie MacDonald
University of Virginia /
After Architecture

Abigail Hassell
University of Virginia

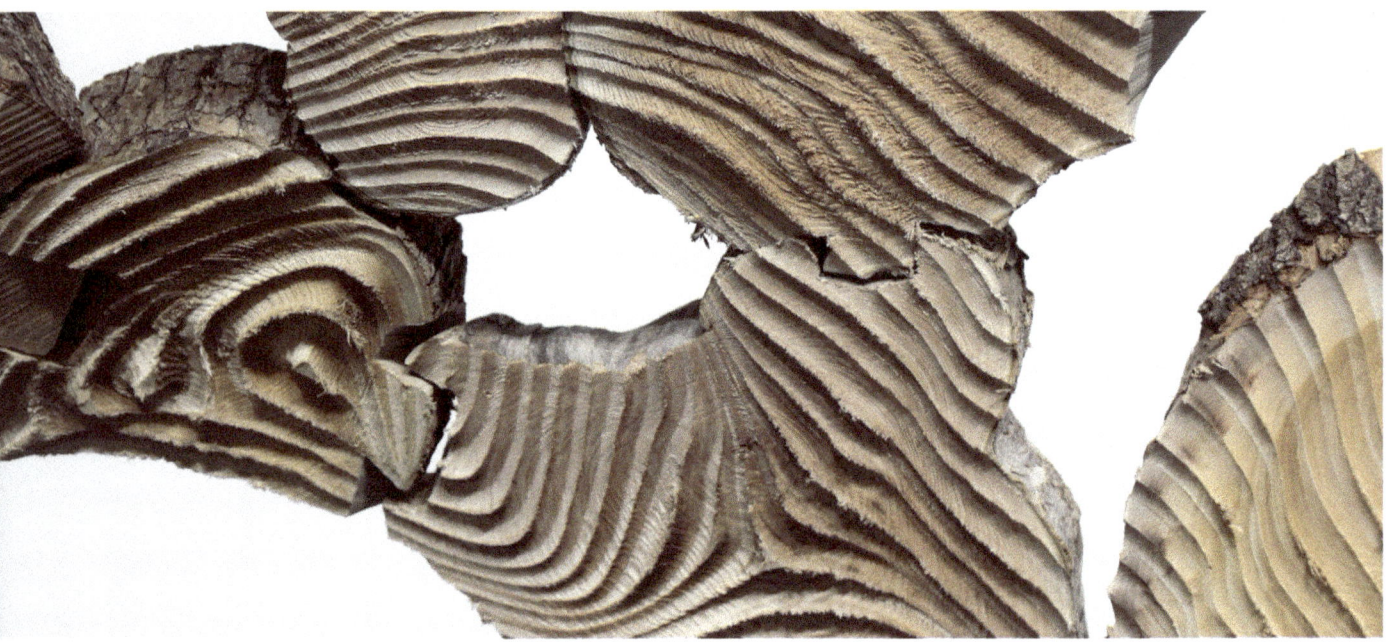

1   Detail of vault assembly.

ABSTRACT

This paper pilots a structural application for nonlinear wood through the development of a custom parametric workflow in which cross sections of logs are digitally imaged, analyzed, and manipulated, then physically manufactured into interlocking structural units. The project addresses resource scarcity and embodied carbon by defining a use for nonlinear wood, various species of which are found across the globe but are limited in use due to the constraints of conventional sawmilling.

The paper describes a methodology in which logs that are curved, branching, irregular in cross section, or otherwise unfit for milling into conventional lumber are cut into cross sections—'cookies' in woodworking terminology. Using a custom workflow, cookies are digitized through image tracing to create lean digital models, analyzed and sorted across a vaulted form, and inscribed with a set of joints primarily defined by the unique geometry of each cookie. The fabrication of this interlocking compression structure involves cutting irregular joints with a 5-axis waterjet and surfacing with a 3-axis CNC router. Together, the methods demonstrate how visual and spatial continuity can be implemented across a patchwork of irregular structural units. Specific innovations through the imaging, modeling, and fabrication processes are presented, along with challenges that suggest future improvements. The visual and aesthetic effects of the structure are discussed.

INTRODUCTION

The intensifying climate crisis has drawn attention to the building industry's contributions to global carbon emissions and the role industrialized building materials and construction processes play in emissions. Computation has upended the traditional order of operations of construction, opening new possibilities for reframing time and labor. Leaving standardized production behind creates opportunities not only for producing custom forms, but also inputting irregular materials.

Wood construction is a positive alternative to other energy intensive building techniques, such as concrete or steel production. In its growth phase, wood sequesters carbon and is locally available across the globe, decreasing transportation costs and environmental impacts. Some 48 percent of harvested timber is deemed unusable in construction due to irregularities and is instead shredded down for chips or pulp, composited into nonstructural panels with formaldehyde adhesives, or used for paper or energy production (Bowyer et al. 2012). This makes the lumber industry itself highly efficient, but energy intensive, in its material usage. Wood waste outside of the industrial timber production pipeline, including felled, disease-ridden, weather damaged, or otherwise irregular trees remain absent from building production. This paper addresses immediate pressures on global material supply chains, resource scarcity, and environmental impacts, focusing on nonlinear wood as a material stream produced in forests globally which is typically unsuitable for construction in its natural state due to the limitations of lumber processing and standards that shape it into industrialized products (Höglmeier et al. 2014). It follows that nonstandard and waste lumber provide an extensive, untapped, local material resource, which, if developed as an economical construction material, could relieve some pressure on lumber supply chain limitations and expand the global availability of this renewable, carbon-sequestering material for construction.

Tangential Timber demonstrates an application for this material by developing a low-cost, low-tech digital workflow to process cross sectional slices of logs (called 'cookies' in the timber industry) into structural blocks or voussoirs (wedge-shaped or tapered stone used to construct an arch) (Figure 1). A custom digital imaging system was developed to document the cookies and translate their images into a set of digital doppelganger models. In a custom parametric workflow, the digital cookies are analyzed and sorted across a form, then inscribed with a set of joints tailored to the nonstandard geometry of each cookie. Finally, cut files are sent to two computer numerical control (CNC) technologies to: (1) waterjet a precise joinery system, using a 5-axis OMAX CNC waterjet, and (2) route a continuous surface across the discrete parts using a 3-axis Onsrud CNC router. The resulting structural vault is designed for disassembly: each structural unit is inventoried and can only fit in one orientation, allowing it to be installed, dismantled, and repeatedly reassembled.

STATE OF THE ART
Digital Imaging

Increased access to 3D scanning and machine visioning technologies has resulted in a proliferation of construction projects in which architectural design responds to material stock. Such projects have used a range of imaging techniques, including professional or consumer-grade photogrammetry applications for analyzing branches and found material, strategies for distilling critical data from complex scans, and custom parametric workflows involving the collection and analysis of a small number of 2D photographs using edge detection (Saslawsky et al. 2021; MacDonald et al. 2019; MacDonald and Schumann 2021). *Tangential Timber* builds on the low-data, low-resolution approach of collecting 2D imagery and using edge detection to translate the face of each cookie into a digital doppelganger.

Compression Structures

Recent vaulted structural unit assemblies include compression-only structures which reduce standardized blocks of material into custom, interlocking parts such as the masonry assemblies of Block Research Group's *Armadillo Vault* and Höweler + Yoon's *Sean Collier Memorial,* as well as NADAAA's foam *Catenary Compression*, and Matter Design's plywood *La Voûte de LeFevre* (Block et al. 2018; Höweler and Yoon 2014; NADAAA 2015; Clifford and McGee 2014). Like the *Armadillo Vault* and *Catenary Compression*, *Tangential Timber* develops a compression-based vault geometry in order to avoid the need for adhesives and tensile forces within the joinery. Like *La Voûte de LeFevre, Tangential Timber* is composed of flat units that are joined and routed, resulting in an A-side and B-side of the assembly.

Heinz Isler developed compression structures by hanging fabric and inverting the forms (Chilton 2010; Isler 1980). *Tangential Timber* takes a similar approach, using Grasshopper's physics simulation plug-in Kangaroo to create a simulated hanging mesh, then inverting the form into a vaulted compression structure.

METHODS
Material Acquisition

The project team worked with a campus partner, UVA Sawmilling, to acquire material from a stockpile of landscaping waste including trees felled for construction, by weather, or due to disease. Selected logs were either too short, too curved, or too irregular to be milled into traditional

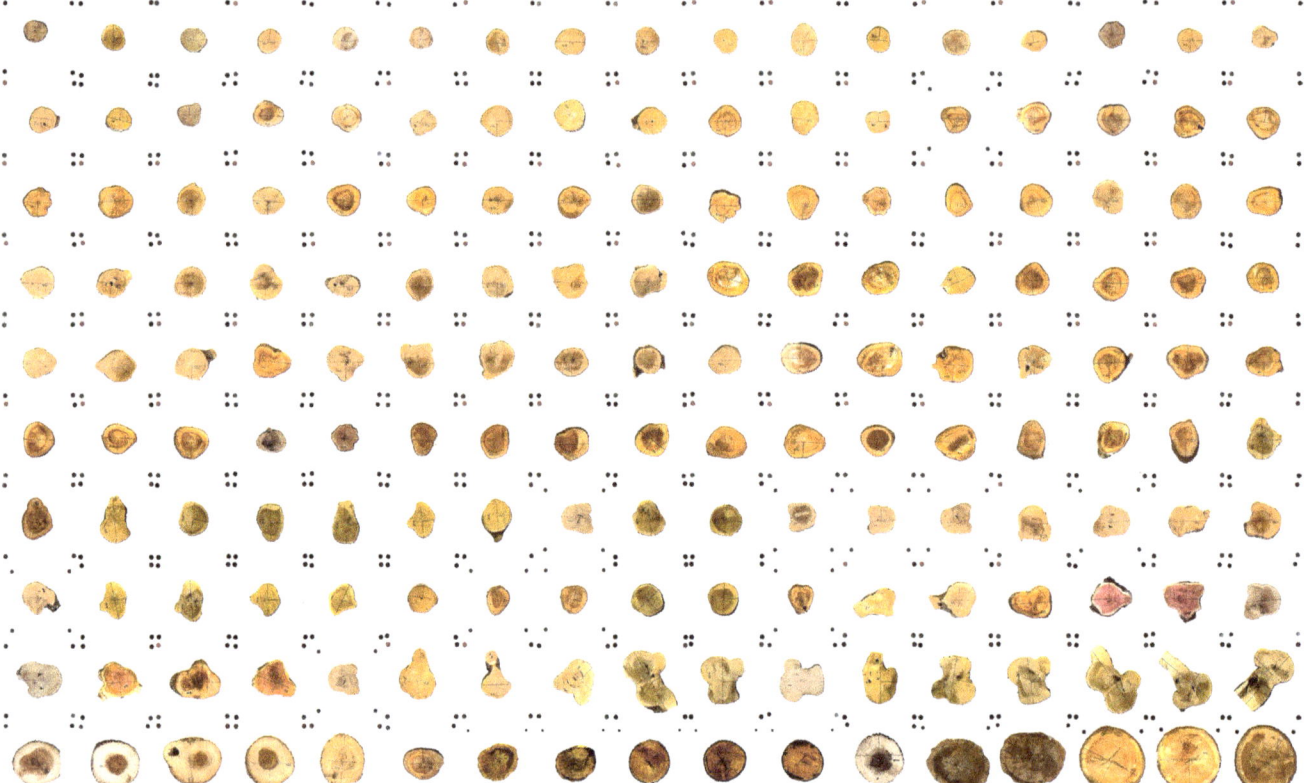

2   Inventory of cookies: timber rounds are inscribed with T-shaped registration marks and photographed against a white backdrop with stacks of 1" black registration dots

lumber, and would normally be chipped for mulch. A sawyer was hired to cut the selected logs into 3, 4, or 5" thick cookies using a mobile bandsaw sawmill. The resulting cookies were rough sawn, but the top and bottom faces were parallel, an important quality anticipating fabrication.

Material Treatment
The wood sourced for the vault came from an outdoor log pile of trees cut at various points in time and with varying moisture content. The team attempted to select primarily hardwoods when possible, but real-time species identification of the logs proved difficult. A range of species including a few softwood specimens of significant size were selected. The condition in which the material was acquired and the variety of species posed significant challenges for material treatment, specifically drying to avoid cracking.

Over time, wood distorts due to uneven rates of shrinkage and swelling. Of principal concern when working with log sections is radial cracking, which is caused by tangential shrinkage, with heartwood shrinking first because of its lower moisture content. The keys to avoiding cracking in cookies include selecting species with low shrinkage percentages, low tangential-to-radial shrinkage ratios, and low densities, and drying them slowly (Hoadley 2017). In the vaulted assembly, the joinery perforates the edge of each cookie, helping to relieve stress and discourage significant cracking.

The project team priced out and tested various methods for drying timber cookies without splitting or cracking. Attention was paid to testing methods that would not have negative environmental impacts or prevent degradation upon decommissioning and composting at end of life. This ruled out several options, including wood-plastic composite polymerization, a common method for preserving wood involving impregnation with chemicals that are then transformed into a rigid plastic (Hoadley 2017). The following methods were tested or investigated:

- Polyethylene Glycol 1000 (PEG-1000) is a stabilization treatment widely used by woodworkers. PEG prevents shrinkage and cracking and is applied to green wood by submerging it fully in liquid, causing its molecules to replace the water molecules in the wood. PEG was found to be cost, time, and space prohibitive, requiring 550 gallons of PEG solution to be heated at 140°F over 14 days of soaking (Hoadley 2017). This would require 2,619 pounds of solid PEG (over $17,000) and space for 550 gallons of liquid to be heated.
- Petracryl offers a similar method to wood-plastic composite polymerization and PEG, but at a higher and more prohibitive cost.
- Salt paste is a natural method of drying wood which reduces shrinkage and requires coating in lieu of soaking. Formal resources on this method are limited, so the team

3　Jig for CNC waterjet and routing

sourced techniques from a variety of online wood forums, which confirmed that this method may work for the thickness and size of the cookies. The team created paste by first incorporating a salt and water mixture which was thickened with cornstarch and egg whites. The mixture was applied to six test specimens, with three dried in a covered outdoor space and three in a conditioned indoor space. After one week, the moisture content of all cookies had decreased, but both batches of cookies had significant quantities of small surface cracks on both sides. The team mixed salt paste using salt, water, cornstarch, and eggs, starting first by combining water and salt and allowing the mixture to sit for a few hours. Next, cornstarch was slowly introduced into the mixture until a cake batter consistency was achieved. The cornstarch did not easily mix into the salt and water solution, and it failed to thicken the consistency even as more cornstarch was added. Finally, egg whites were mixed in. The mixture was applied to six test specimens, leaving three outside to dry, and moving three from outside to inside to dry. After one week, the moisture content of the cookies had decreased, both batches of cookies had surface cracks all over both sides. The failure of this test led the team to abandon this method.

- Kiln drying was evaluated. A test batch of nine cookies were kiln dried for one month. When they were pulled from the kiln, the cookies were at an ideal moisture content and had no cracks. However, over the next couple of weeks, left in an indoor space, the cookies started cracking severely. It is suspected that the speed of drying contributed to this failure.
- Air drying was selected as the best method for the project timeline and budget. After sawmilling, the cookies were stored outdoors underneath a loose semi-translucent tarp, which allowed the material to be heated by the sun while slowing the rate of drying, much like a passive solar kiln. Furthermore, the cookies were stacked without stickers (wood slats) separating them from one another, to decrease the surface area of end grain from which moisture could escape, further slowing the drying and shrinking process.

Digital Imaging

Each cookie was labeled with a unique identifier number and a T-shaped registration mark to track location and orientation. Each cookie was then photographed against a white backdrop with stacks of 1 inch black registration dots matching the height of the cookie to eliminate perspectival distortion through depth (Figure 2). The photographs were imported into Rhino (Robert McNeel & Associates 2018) via Grasshopper (Rutten 2018) and using a combination of plugins, namely Firefly (Payne and Johnson 2015), images were traced, scaled through a rectangular remapping operation that also removed perspectival distortion, and rotated based on the physical T-shaped registration mark. Once scanned, cookie models,

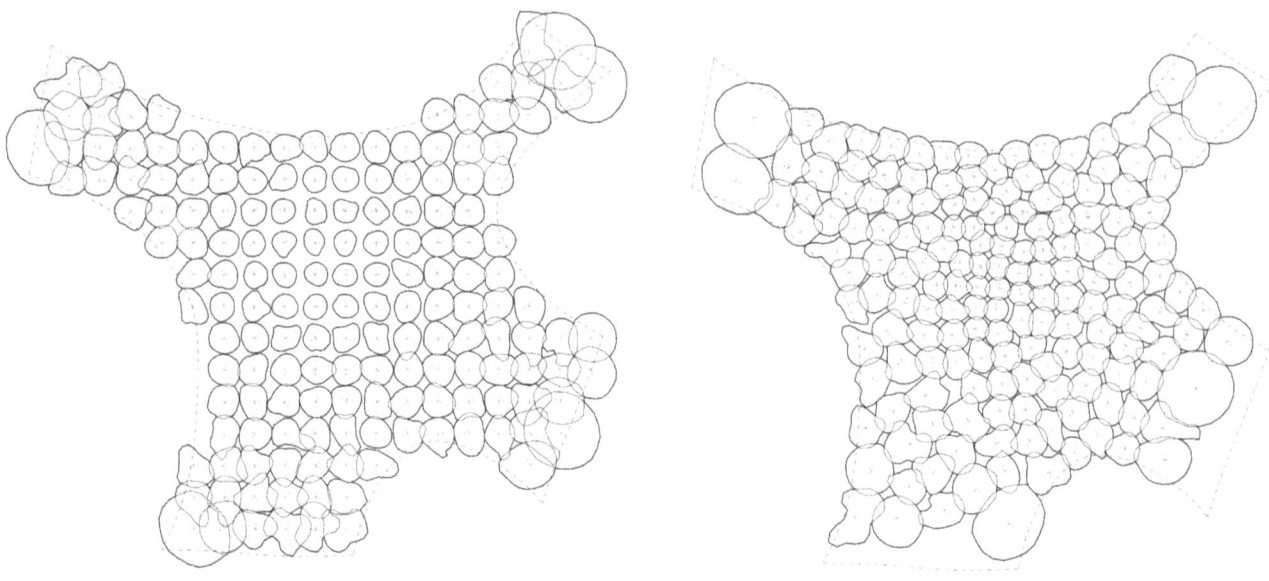

4   Cookies are sorted and fitted by size and geometry across flattened version of vaulted geometry

now located at the file origin, are exported as individual Rhino files containing a 2D outline and 3D extrusion. Additionally, model objects are assigned an object property name matching the cookie number, allowing Grasshopper to read this information in future steps.

Overall Form

The behavior of fabric informs both the structural schema and formal expression of the vault. Like the work of Heinz Isler, hanging fabric becomes a methodology for form-finding a compressive, structural form. Unlike Isler's work, this exploration is performed digitally through a Kangaroo gravity simulation, with the form inflated over a square base. On each side of the square, anchor points are removed to produce four arches of varying heights to allow passage into and out of the structure. These arches are later defined by a smooth twisting edge cut into the cookies.

Sorting and Fitting

Custom parametric workflows were developed for sorting and fitting each cookie precisely in the designed structure, adapting to the size and geometry of each cookie. This is accomplished in 2D with a flattened version of the 3D vault, made using the 'Squish' command in Rhino. In Grasshopper, a grid of points is populated with the digital cookie inventory, with the smallest cookies positioned in the center (later the top of the vault, allowing them to create a best fit to the tight radius of the vault's crest), and the largest cookies at the edge (later the bottom of the vault, along the more vertical sides and anchoring the assembly to the steel base) (Figure 4). Next, the grid of cookies was run through a Kangaroo (Piker 2017) spring simulation to nest and aggregate them evenly across the surface. Offset curves used in the simulation allowed for controlled overlaps with which to produce joinery. Simulation strength settings allowed for openings between cookies to create a more porous surface and allowed the natural edges of the cookies to be visible, producing a larger surface with less material. The resulting cookie distribution was translated back to 3D using the 'SquishBack' command in Rhino, while three points tracked the orientation of each cookie on the surface (Figure 5).

Joinery Adaptation

The joinery system necessitated connections that effectively secure the cookies to each other while allowing for a broad range in joint size and placement. The system draws from traditional wood joinery to create an interlocking joint without adhesive or fasteners (fasteners are used in the final assembly to provide a factor of safety for the prototype).

Two approaches were explored: (1) curving toolpaths that mimic conventional puzzle-piece connections, and (2) angled cuts that resemble dovetail joints. The decision to move forward with angled toolpaths was informed by the ease with which a ruled surface could be implemented and cut by

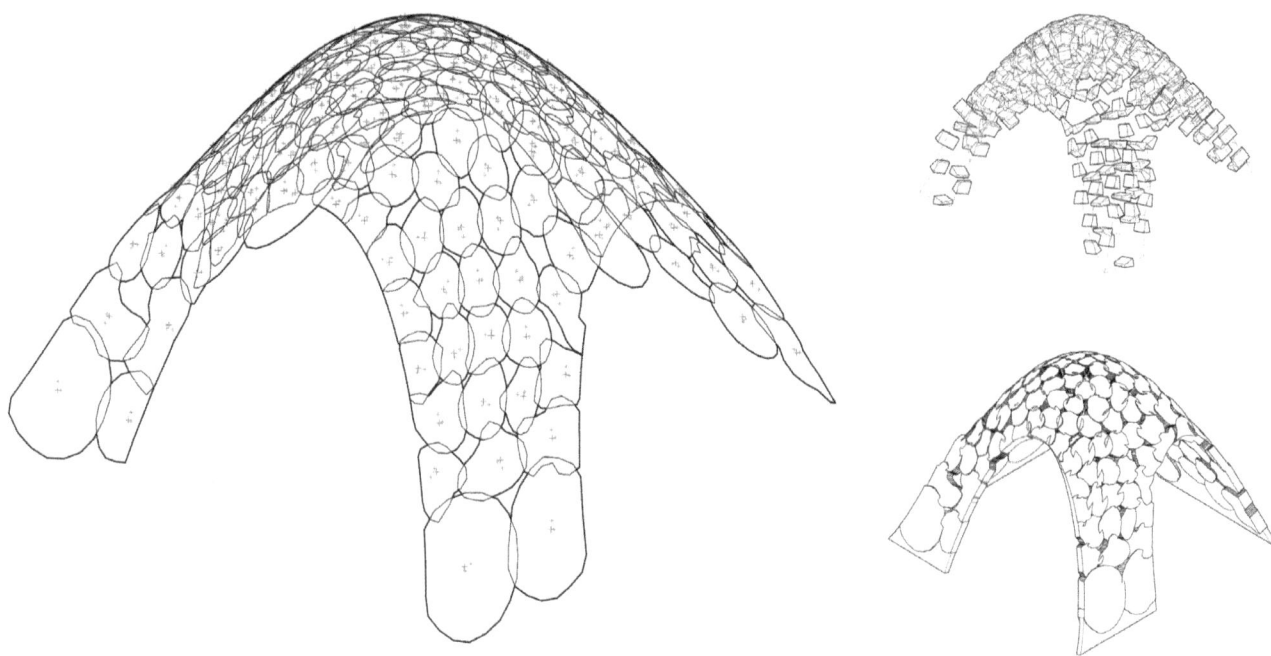

5   Once cookies have been sorted across 2D Vault, geometry is unflattened and populated with dovetail joinery

a waterjet. A uniform wedge-shaped dovetail geometry was combined with a lofted, inclined surface generated using each cookies' unique geometry, allowing for material expression and efficiency. The resulting geometry prevents cookies from falling through one another in the direction of the extruded wedge—which also enables assembly with minimal formwork or scaffolding—and produces a shingled appearance on the exterior of the structure (Figures 6 and 7).

Surfacing

On the interior of the vault, the stress lines of the compression structure are superimposed over each cookie's grain, taking shape as the pleated ridges of a CNC toolpath. A thickened surface is generated through a structural analysis in Karamba (Preisinger et al. 2022) to create a custom 2.5D topological optimization. Based on the density of stress lines generated, the resulting surface is thicker where greater structural stresses are present. The stress lines provide a fabric-like continuity of surface across the discrete timber units.

Digital Fabrication: Waterjet

A 5-axis CNC waterjet was selected for cutting the three-dimensional joinery into each cookie due to the speed with which it can cut through dense material and the joinery's design as ruled surface cuts running through the material stock. This machine is typically used for precision fabrication in metals, making its use for the application of wood joinery atypical. The waterjet is limited to a maximum tilt angle of 60 degrees from vertical. The cut behavior of the waterjet demands a straight line that pierces all the way through the material stock.

A custom jig was built with an aluminum extrusion frame and sliding mounts for a series of adjustable clamps that can be calibrated to each cookie. One jig was used for workholding on the CNC router, and another on the waterjet during fabrication (Figure 3).

The waterjet ran on OMAX IntelliCAM, Layout, and Make software (OMAX 2022). The joinery cut files were designed with the limitations of the waterjet in mind: all cuts are designed as either straight extrusions or ruled surfaces, so that the cut path will always be a straight line. The software itself, which was designed for repetitive machining of conventional industrial parts, posed significant challenges during fabrication, with only basic options available when generating and editing toolpaths. Files of all types transferred from Rhino to OMAX experienced a proliferation of issues upon import, which the software struggled to process. The software also struggled to interpret complex part cut files with multiple joints, often producing unanticipated angles and curves upon cutting that reduced part accuracy.

Several strategies were employed as a workaround to these issues: (1) Digital cookie models were reduced to

6   Inside and outisde of a joinery prototype for vaulted assembly, with interior angular dovetail side (left) and exterior shingle-like side (right)

straight-segment polylines with 100 control points during the scanning process, which helped to reduce complexity for both readability by the CAM software and data intensity while modeling the assembly. This is a variation on similar approaches to distill digital scans for easier workability (MacDonald and Schumann 2021); (2) Joint surfaces were extensively remodeled in Rhino, producing straight lofted surfaces composed only of quadrangles with a leading edge on both the top and bottom of the cut, readable by the CAM software; (3) Cookies with multiple joints were divided into several discrete cut files; and (4) Joints with cut angles greater than 60 degrees were modified to reduce the overlap distance.

Throughout fabrication, it was critical to ensure the proper location and rotation of the material. The physical T-shaped registration mark on the cookie was used to position it on the waterjet and served at the cut origin. This same data, incorporated into the digital model during scanning, informed the placement and rotation of the model at the file origin point. The OMAX software does not have an origin input function; instead, it generates an origin based on the corner of an object bounding box. *X*- and *Y*- offsets for the object bounding boxes were read and made into a spreadsheet in Grasshopper, and used to manually move the origin for each cut file. Care was taken to use the same bounding box geometry for cookies with multiple cut files.

Digital Fabrication: CNC Router

Initial trials for surfacing cookies on the Onsrud 3-axis CNC Router produced several issues, including workholding issues and vibration, milling time, and edge tearout and surface quality. Milling tests used traditional roughing passes and a spiral finishing toolpath with 1/2 inch end mills, and milling time was approximately 40 minutes for a medium-large cookie. This proved unreasonable for milling the entire inventory, so an alternative was developed. First, a custom toolpath was created in Grasshopper for each part. Consistent coverage was ensured by interpolating the stress lines with a Pufferfish (Pryor 2021) 'tween through curves' component. The curves were joined into a single curve that was used as the toolpath for an engraving operation in RhinoCAM.

A 1-1/2 inch ball nose bit was used for the milling, both to accentuate the varying texture and density of the lines, and to reduce machine time. With 1 inch deep roughing passes and a 0.05 inch finishing pass, most cookies were able to be milled in five to ten minutes. The cuts ran with an average spindle speed of 6,000 rpm and an actual feed rate of around 300 ipm. The biggest challenge with this method was maintaining a sharp bit, as constant milling of end grain resulted in rapid dulling. Once dull, the bit produced burn marks on harder wood and furry edges on softer wood. Professional sharpening services were tested and proved effective but too time intensive (two-week turnaround). In the end, bits were manually sharpened on a sharpening stone after every few cookies (Figures 8 and 9).

In lieu of waterjet access, alternative methods for cutting the joinery include: (1) 3-axis CNC milling, which would require flip milling, and (2) a router end effector on a robotic arm. Both options would require substantially more machine time, and joint geometries may have to be modified to preclude sharp

7   Test wall assembly with 18 cookies

corners. Alternatively, mixed reality methods with a bandsaw, or a bandsaw end effector on a robotic arm could be explored.

Assembly

After parts are prepared through the waterjet and CNC mill stages, the structural units are ready for assembly. Assembly can be accomplished by a team of two people working with minimal scaffolding. Cookies are assembled from the ground up upon a custom waterjet steel base scribed to the bottom cookie geometries. This base serves as a tension ring at the bottom of the structure, preventing the legs of the vault from sliding outward. A few wood braces help support each leg of the structure and prevent leaning as it is built. Once a single row of cookies is complete in each of the four arches, that part of the structure becomes self-supporting.

The first assembly of the structure took additional time, requiring some manual adjustments where waterjet inaccuracies occurred. Trim screws were used to provide an additional factor of safety at the connections. If the waterjet issues are resolved in the future, it is anticipated that both of these steps will become unnecessary.

RESULTS AND DISCUSSION

The joining and surfacing of the cookies give the vault assembly a dramatic expression that, while informed by structure, builds on an architectural lineage of vaults. Unlike the stucco coffered ceilings characteristic of classical domes that obscure brick or concrete vaults, the timber units double as both structure and ornament. The language of wood is emphasized in the design and articulation of the vault: the structural stress lines inscribed into the interior surface of the vault register as a second wood grain layered over the natural texture of the cookies. Rather than be defined by age (tree rings), the grain is defined by structural path (stress lines). The overlay of grains results in an intensity of visual linework on each cookie.

The greatest challenges were presented by the proprietary OMAX waterjet software, which demanded extensive testing, troubleshooting, and workarounds that added a substantial amount of time during fabrication. Even when all went according to plan, some cuts still produced unanticipated variations since the software lacked a way to visualize the operation. For future work, it would likely be worth investing significant time to develop a custom CAM workflow compatible with the OMAX waterjet. Any kind of editable G-code or a Grasshopper workflow similar to KUKA|prc or Taco ABB would be a significant improvement and allow for a greater degree of accuracy and control over cut parameters.

Cookie tearout at the edges during CNC milling also proved a constant issue, which was somewhat reduced by lowering the feed rate when the toolpath exits the cookie perimeter. The

8   Detail of two cookies after waterjetting and CNC routing

team considered milling before cutting joints on the waterjet in the future, but the uniform thickness of the raw cookies meant that the distance to the waterjet head was more consistent, producing cleaner cuts. Ideally, a future custom CAM workflow for the waterjet would also allow for adaptation to the milled cookie height using the z-axis—a physical motion the machine is capable of and meaning cookies could be milled before cutting joinery. Finally, the irregularity of the material presented some challenges. Since several months passed between cookie scanning and fabrication, the material shrunk and moved from its original geometry. This did not prevent accurate fabrication with the CNC or waterjet but produced some looser or tighter fits than intended during assembly. Ideally, cookies would be completely dried over a long period of time before any scanning or fabrication work began.

CONCLUSION

This paper demonstrates a construction process that can be adapted to a global supply of unused material—nonlinear timber specimens—adding to a lineage of projects which have explored tree forks, branches, and live edge boards (Self and Vercruysse 2017; Larsen and Aagaard 2019; Johns and Foley 2014). While the prototype vault assembly is a small-scale investigation closer to a pavilion than an enclosed building, it demonstrates structural performance and human inhabitability, and suggests possibilities for larger architectural applications including structural walls, structural vaults, and nonstructural expressive, aesthetic, or spatial applications (Figures 10 and 11).

The ability to create a digital doppelganger of a series of irregular log sections from a 2D photograph expands the digital imaging space to include low-tech, low-cost solutions, building on the authors' prior work aimed at democratizing technology in architecture (MacDonald and Schumann 2021). At the same time, the assemblies designed using these democratized doppelgangers require access to CNC waterjet and routing machines, which limit the accessibility of this fabrication methodology for widespread implementation in construction. Instead, these assemblies belong to a group of expanding projects focused on the possibilities enabled by mass customization. Within this area of scholarship, they demonstrate how customization technologies allow designers and builders to work more closely with and responsively to irregular natural materials. This is a notable development relative to an industry-wide push to identify the potentials of renewable building materials.

Future work will explore which nonlinear tree species have the desirable characteristics identified in material preparation (low density and low shrinkage) as well as how this methodology might be applied to an expanded catalog of irregular inputs including renewable and reused materials.

ACKNOWLEDGMENTS

The work presented in this paper was produced with funding from the University of Virginia's Jefferson Trust and Center for Global Inquiry and Innovation. The authors would like to thank students Audrey Lewis, Jacob McLaughlin, Rohan Singh, and Abbie Weissman, who, along with Hassell, developed an initial version of the project in the 'Material Cybernetics' studio taught by Schumann, as well as research assistants Sonja Bergquist, Sophie Depret-Guillaume, Cecily Farrell, Alex Hall,

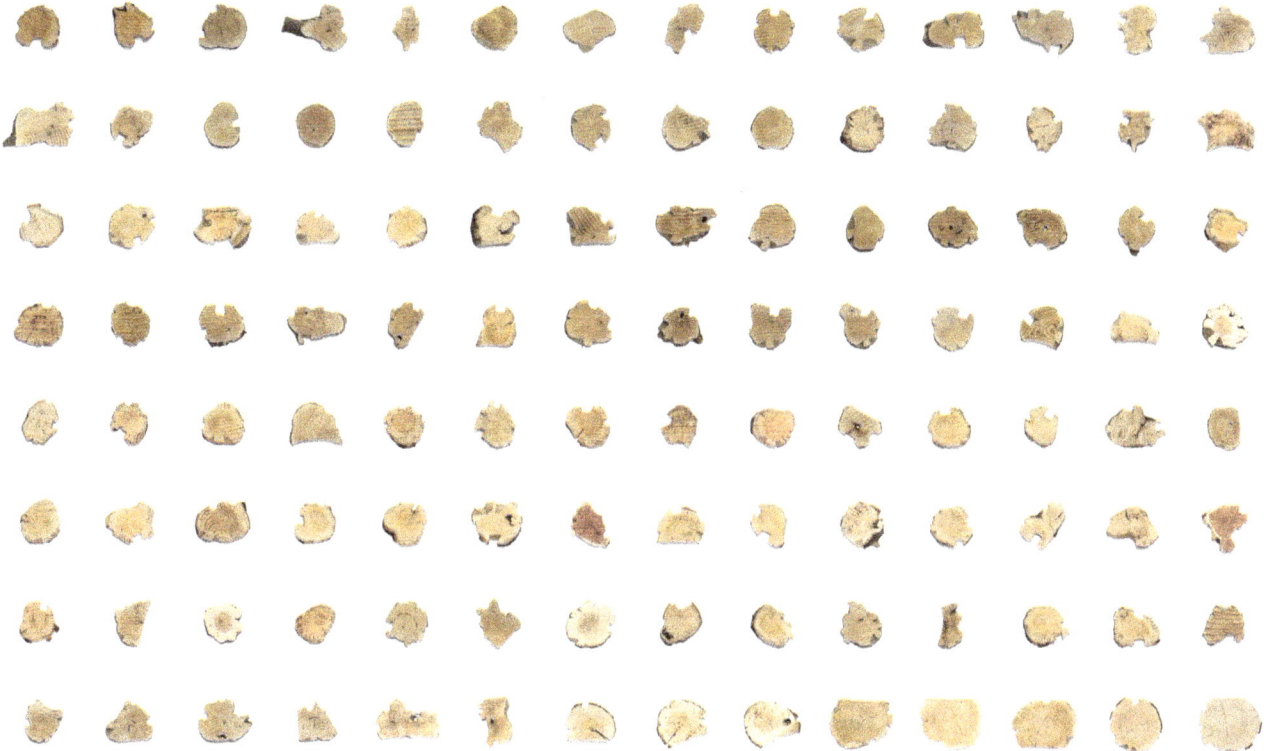

9 Inventory of waterjet and CNC milled cookies, each surfaced with expressive pleats that describe stress lines through the compression structure

Caleb Hassell, Dillon Mcdowell, Russell Petro, Emily Ploppert, Yianni Spears, Jolie Talha, and Annabelle Woodcock for their work on the final vault. The authors would like to thank School of Architecture Dean Malo Hutson and UVA fabrication staff members Melissa Goldman, Trevor Kemp, and Sebring Smith for their support.

REFERENCES

Block, Philippe, Tom Van Mele, Andrew Liew, Matthew DeJong, David Escobedo, and John Ochsendorf. 2018. "Structural Design: Fabrication and Construction of the Armadillo Vault." *The Structural Engineer* 96 (5): 10–20.

Bowyer, Jim, Steve Bratkovich, and Kathryn Fernholz. 2012. "Utilization of Harvested Wood by the North American Forests Products Industry: Understanding and Supporting the Benefits of Zero-Waste." Minneapolis: Dovetail Partners Inc.

Chilton, John. 2010. "Heinz Isler's Infinite Spectrum: Form-Finding in Design." *Architectural Design* 80 (4): 64–71.

Clifford, Brandon, and Wes McGee. 2014. "La Voûte de LeFevre: a Variable-Volume Compression-Only Vault." In *FABRICATE: Negotiating Design & Making*, edited by F. Gramazio, M. Kohler, and S. Langenberg, Verlag, Zurich. 146–153.

Hoadley, Bruce R. 2017. "Chapter 6: Water and Wood." In *Understanding Wood: a Craftsman's Guide to Wood Technology*. Newtown, CT: Taunton Press. 111–131.

Höglmeier, Karin, Gabriele Weber-Blaschke, and Klaus Richter. 2014. "Utilization of Recovered Wood in Cascades versus Utilization of Primary Wood—A Comparison with Life Cycle Assessment Using System Expansion." *The International Journal of Life Cycle Assessment* 19, 1755–1766.

Höweler, Eric, and J. Meejin Yoon. 2014. "Material Computation: The Collier Memorial Design Using Analogue and Digital Tools." In *Paradigms in Computing: Making Machines, and Models for Design Agency in Architecture*, edited by D. J. Gerber and M. Ibanez. Los Angeles, CA: eVolo. 106–111.

Isler, Heinz. 1980. *Heinz Isler as Structural Artist: Catalogue of an Exhibition*. Princeton: Princeton University Art Museum.

Johns, Ryan Luke, and Nicholas Foley. 2014. "Bandsawn Bands: Feature-Based Design and Fabrication of Nested Freeform Surfaces in Wood." In *Robotic Fabrication in Architecture, Art and Design 2014*, edited by W. McGee and M. Ponce de Leon. Cham: Springer. 17–32.

Larsen, Niels Martin, and Anders Kruse Aagaard. 2019. "Exploring Natural Wood: A workflow for using non-uniform sawlogs in digital design and fabrication." In *ACADIA '19: Ubiquity and Autonomy; Proceedings of the 39th Annual Conference of the Association for Computer Aided Design in Architecture (ACADIA)*. Austin, Texas. 500–509.

MacDonald, Katie, Kyle Schumann, and Jonas Hauptman. 2019. "Digital Fabrication of Standardless Materials." In *ACADIA '19: Ubiquity and Autonomy; Proceedings of the 39th Annual Conference of the Association for Computer Aided Design in Architecture (ACADIA)*. Austin, Texas. 266–275.

10 Detail of final vault assembly

MacDonald, Katie, and Kyle Schumann. 2021. "Twinned Assemblage: Curating and Distilling Digital Doppelgangers." In *Projections: Proceedings of the 26th International Conference of the Association for Computer-Aided Architectural Design Research in Asia (CAADRIA)*. Hong Kong, China (virtual). 693–702.

NADAAA. 2015. "Catenary Compression." Accessed May 1, 2022. https://www.nadaaa.com/portfolio/catenary-compression/.

OMAX. *IntelliCAM, Layout, and Make*. OMAX. Windows 10. 2022.

Payne, Andy, and Jason Kelly Johnson. *Firefly*. V. 1.0.0.70. Firefly Experiments. Windows 10. 2015.

Piker, Daniel. *Kangaroo*. V. 2.42. Windows 10. 2017.

Preisinger, Clemens et al. *Karamba3D*. V. 2.2.0. Windows 10. 2022.

Pryor, Michael. *Pufferfish*. V. 3.0. Windows 10. 2021.

Robert McNeel & Associates. *Rhinoceros 3D*. V. 6. Robert McNeel & Associates. Windows 10. 2018.

Rutten, David. *Grasshopper 3D*. Robert McNeel & Associates. Windows 10. 2018.

Saslawsky, Kevin, Tyler Sanford, Katie MacDonald, and Kyle Schumann. 2021. "Branching Inventory: Democratized Fabrication of Available Stock." In *Projections: Proceedings of the 26th International Conference of the Association for Computer-Aided Architectural Design Research in Asia (CAADRIA)*. Hong Kong, China (virtual). 513–522.

Self, Martin, and Emmanuel Vercruysse. 2017. "Infinite Variations, Radical Strategies." In *Fabricate: Rethinking Design and Construction*, edited by A. Menges, B. Sheil, R. Glynn, and M. Skavara. London: UCL Press. 30–35.

Kyle Schumann is cofounder of After Architecture and Assistant Professor of Architecture at the University of Virginia, where he co-directs the Before Building Laboratory. Schumann seeks to advance the accessibility of digital fabrication techniques, leveraging democratized technologies in his teaching and research.

Katie MacDonald is cofounder of After Architecture and Assistant Professor of Architecture at the University of Virginia, where she co-directs the Before Building Laboratory. MacDonald's work reconsiders construction standards and materials in light of new lifecycle questions and overextended supply chains. MacDonald pioneers new biomaterial assemblies, with the aim of creating building material systems that sequester carbon.

Abigail Hassell is a recent graduate of the University of Virginia School of Architecture. She holds a Bachelor of Science in Architecture with a concentration in Design Thinking. She is a Project Manager in the Before Building Laboratory.

11 Final vault assembly

Session Introduction

Machine Learning and Artificial Intelligence

Matias del Campo, Chair

The Machine Learning and Artificial Intelligence session of the *Hybrid & Haecceities* conference interrogates the emergent field of Artificial Intelligence in architectural design. The rise of learning systems provides a large territory for exploration on aspects such as convolutional neural networks for the optimization of fabrication methods, to the use of Bayesian networks for site optimization to more theoretical considerations such as the rise of adversarial networks.

As this entire novel field is currently a moving target, the papers presented in this session can be considered pioneering work—more akin to the deciphering of the Rosetta stone, than solid evidence of completely explored ideas, methods, and concepts. There is still plenty of space in this new terrain for novel explorations, whether they be practical, theoretical, or aesthetical. This being said, the papers in this section allow for other branches to sprout out of this novel research in architecture, providing the discipline with the basic information necessary to pick up the pace, and contribute further with novel insights on the theoretical and practical use of artificial intelligence in architecture design.

Design Contextualism by AI

Encoding Architectural Site Layouts Using a Bayesian Network

Woongki Sung
Massachusetts Institute of Technology

Takehiko Nagakura
Massachusetts Institute of Technology

Daniel Tsai
Massachusetts Institute of Technology

1 Site layouts in different locations (Kyoto, Vicenza, New York, and Barcelona)

ABSTRACT

Architectural site layouts in a city are seldom exactly the same, but they often share similarities in size, proportion, and inner placement of buildings, garages, gardens and other design components. At the same time, these properties often appear very different from one city to another. Architects can explain these differences in their own sensible architectural terms, but it is not easy to articulate the differences in measurable manners. The goal of this project is to quantify the architectural context including the properties of a site that are similar within a region but distinct between different geographical locations. This paper presents a data-driven method for encoding and representing the statistical information of an architectural site layout in the form of a Bayesian network. Given a set of simplified satellite photos and maps, the site layout model is formulated that consists of variables of interest. Structured learning is performed to find an optimal Bayesian network structure that best fits the dataset and is then trained to calculate its parameters. The resulting network is then examined to verify if it effectively summarizes the statistical information of the given site layouts in terms of the variables of interest. In addition, the network is presented to be of further use for making site predictions where only a few variables are specified. The results show that the trained Bayesian network can successfully complete the partially defined site context, and propose a plausible building site layout based on the regional data.

INTRODUCTION

"No house should ever be on a hill or on anything. It should be of the hill. Belonging to it. Hill and house should live together each the happier for the other."
— Frank Lloyd Wright

Architectural site layouts in the same location have some common characteristics such as size, dimensional proportion, and placement of the building, garden, garage, and other components. At the same time, as a collection, the characteristics of site layouts vary from one place to another. For example, the site layouts in a residential area in New York City look similar. Still, such properties are very different from those found, for example, in suburban areas in Italy. The leading causes of this difference are the distinct local building codes, typical site conditions, and lifestyles unique to cultures (Figure 1). As Frank Lloyd Wright's statement suggests (Wright 1932), architects, in the spirit of architectural contextualism, are aware of their project contexts, and buildings are designed in response to them. And while we often hear architects eloquently talk about such contextual rationale behind their design decisions in their own terms, it is not easy to establish a quantifiable model on which to perform computational procedures.

The motivation of this project is twofold. First, from an architectural standpoint, many generative computational design projects to propose a new design based on existing data often overlook the site's contextual conditions. Because the consideration of the building site is indispensable to the creation of the building, it is also essential to create a computational model that encodes the site context. Second, in making such a computational model of architectural sites, Bayesian networks (BNs) are examined as a knowledge representation of an architectural site. As a knowledge representation, BNs have a benefit in that they can be interpreted as a probabilistic dependency map between variables. Moreover, the concise graphical representation of their structure is relatively understandable compared to other machine learning paradigms, such as neural networks or numeric regression systems. It is straightforward to incorporate prior architectural knowledge about the subject in the form of a constraint graph.

In this project, the architectural site layout is modeled as a BN and then trained using a dataset that comprises residential site layouts in different cities. Then, it is investigated whether the learned BN could successfully capture the statistical properties of the site layout models. The project has found that BNs can summarize and encode numeric dependencies hidden in the dataset. Furthermore, their structure can be easily interpreted as interactions between the site layout model variables.

RELATED WORK

In the early work in computational research, design knowledge was scripted by a domain expert and encoded in various forms. Shape grammar provides a well-formed rule-based framework where a domain expert creates a set of rules and uses them to procedurally create a new design or analyze existing designs (Stiny and Mitchell 1978). Expert systems serve as a knowledge base where an expert encodes design knowledge into a database, responding to new queries in solving future problems (Harber 1986). While these traditional approaches summarize design knowledge by human experts, recent advancements in AI and machine learning facilitate new processes where existing data form the design knowledge itself.

In these new approaches, statistical properties hidden in data samples play a significant role. At the same time, the representation and the interpretation of the formed knowledge widely vary depending on the computational framework they use. In recent computational research projects, researchers have investigated various topics in design and architecture using data-driven techniques. Shape grammars were revisited to parse building facade images with a reinforced learning technique (Teboul 2011) and formulate the generative design rules from a dataset (Talton 2012). Meanwhile, several researchers have focused on generating new building layouts based on existing data without specifying any rule sets. Neural networks and generative neural networks (GANs) became very popular in this area due to their strength in image parsing and prediction. ArchiGAN used *pix2pix* software to create a plausible indoor image based on the user's inputs (Chaillou 2020). Based on the building outline and a few architectural components, such as windows and doors, it predicts the indoor space image that best fits the specification. HouseGAN produces possible building layouts constrained by a connectivity graph (Nauata et al. 2021).

Outside of the neural network territory, Bayesian graph learning and sampling optimization was used to produce a new cabin-type building (Merrell et al. 2010).

Generating indoor scenes was also investigated using machine learning techniques. For example, sampling-based optimization was employed to find an optimal indoor furniture arrangement (Yu et al. 2011), and a graph convolutional neural network (GCN) was used to create a plausible room instantiation given a few architectural specifications, such as walls, floor, and room types (Wang 2019).

While a long line of interesting work integrates data-driven techniques and architectural subjects as above, little effort has been made to have a structured representation of site

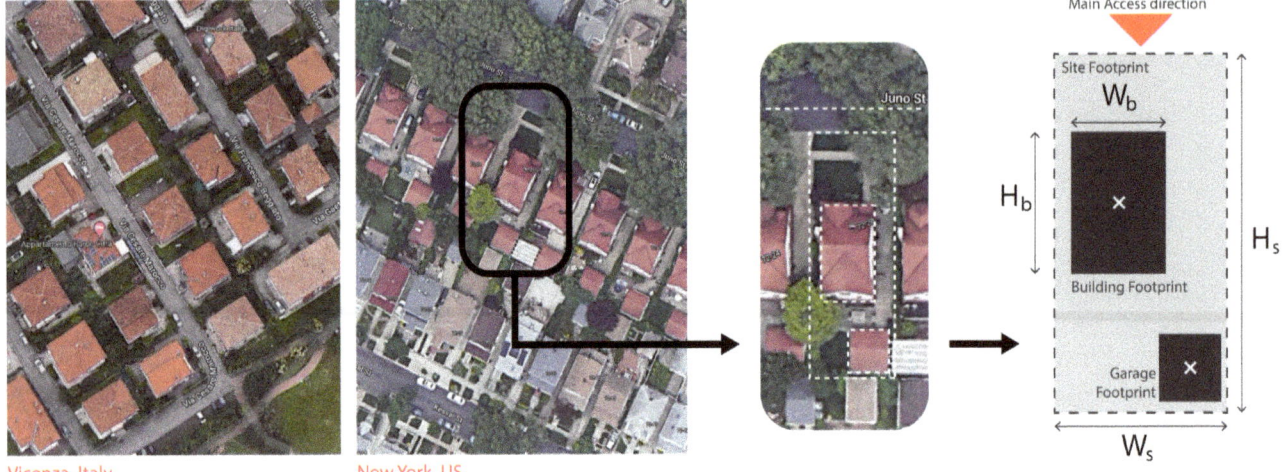

2   Site layout model scheme

layouts using machine learning techniques. A missing perspective is the way most architects design by considering (i) the design context, (ii) the conditions of the project imposed by physical, cultural, and social settings, and then (iii) generating the design from the outside in. This project applies machine learning of geometric components to architectural design, which engages with the regional heuristics of a house layout on a site and the site's physical conditions such as streets, views, and landmarks. ActiveContext proposed a rule-based expert system that can generate and score architectural plans based on site conditions (Fred 2001). While this project is not purely data-driven, it offers an interesting way of symbolizing contextual elements of architecture, such as views, landmarks, and access, which interact with the building plans. The work proposed by Wu (Wu 2020) used pix2pix to create a building layout image given the site image. While this approach is very similar to ArchiGAN, its generation is based on a site image rather than a building footprint outline. This line of image-based learning is successful in classification and prediction. However, it has limitations in formulating highly structured architectural knowledge. For example, projects based on pix2pix work on rasterized data constrained by representing spatial components as color codings. The ambiguous image output is hard to process further to architectural compositions of primitives.

In this project, a Bayesian network is used as a tool to encode a site condition in architecture. Several projects have shown that a Bayesian network can encapsulate different aspects of architectural designs, such as room configurations and building performance (Merrell et al. 2010; Naticchia 2007). A Bayesian network has its strength in modeling causal dependencies and creating a structured representation. Because its graphical structure represents probabilistic dependencies, its interpretation is straightforward, making it easy to assess the learned model. As a comparative illustration, a neural network can also work on a vectorized dataset and in an architectural setting to make a good prediction (Zheng 2020). However, its structure has to be predefined before training, and the interpretation of its structure with fitted parameters gets difficult as the number of hidden layers increases. Meanwhile, the structure and parameters of a Bayesian network can be learned directly from data, and the connection complexity of the resulting graph can be controlled. Moreover, prior knowledge about the structure can be embedded as a constraint in the learning process, which gives an additional layer of control in finding the graph structure that fits both the data and the prior knowledge about the subject.

METHODOLOGY
Bayesian Network
A Bayesian network (BN) is a probabilistic graphical model that represents a set of variables and their probabilistic dependencies via a directed acyclic graph (DAG). The edge in BNs represents the dependency between the connected nodes, and the node represents the conditional probability of the variable given its connected parent variables. A BN is widely used in artificial intelligence due to its ability to make inferences and predictions given complete/incomplete data.

Structure Learning and Parameter Estimation
Identifying the highest-scoring DAG models from large datasets using consistent scoring criteria is NP-hard (Chickering et al. 2004). However, several techniques to find an optimal structure given the dataset utilize dynamic programming and heuristic search (Yuan 2011). Because the experiment's model of an architectural site is relatively simple, an exact algorithm that performs an exhaustive search over possible graph structures given data is implemented.

TABLE 1. Variables of interest: the site model includes eleven parameters, which represent several dimensional aspects of a site layout

Variable	Value
Site Label	'NY' or 'Vicenza'
Site Area	Ws * Hs
Site Ratio	Ws Hs
Building Area	Wb * Hb
Building Ratio	Wb / Hb
Building Placement X	X coordination of the building in 9 by 9 grid within the site shape
Building Placement Y	X coordination of the building in 9 by 9 grid within the site shape
Garage?	Binary value indicating the existence of a garage building
Garage Placement X	X coordination of the building in 9 by 9 grid within the site shape
Garage Placement Y	Y coordination of the building in 9 by 9 grid within the site shape
Access	Main access direction to the site (N, S, E, W)

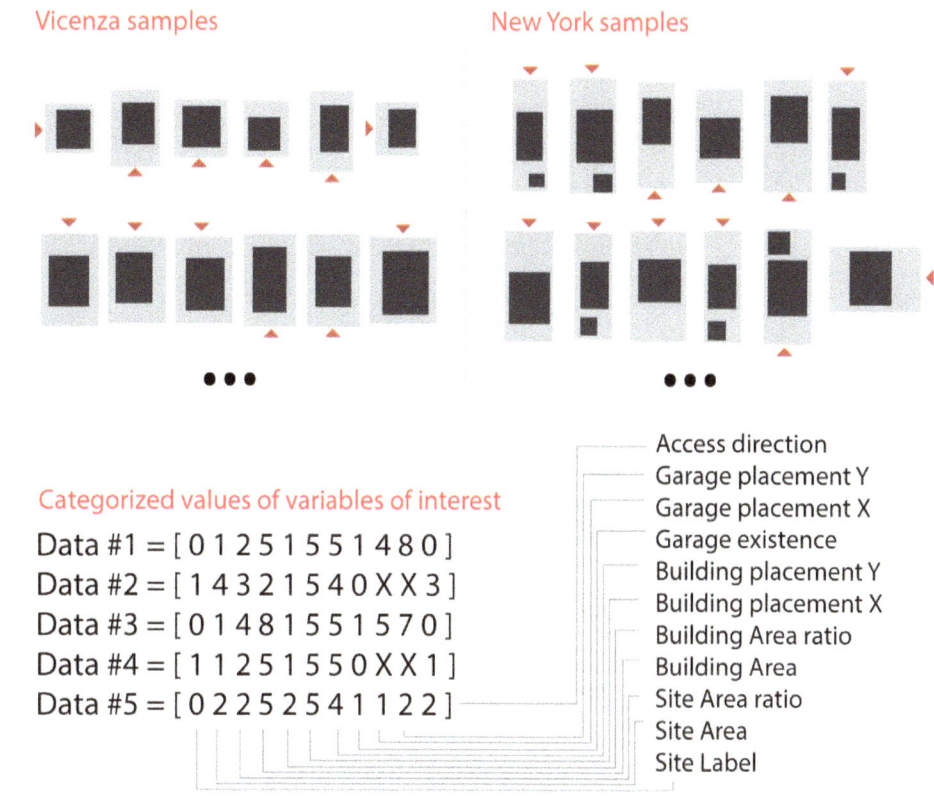

3   Categorized data samples: the parameters of data samples were mapped to a categorical index to form a multinomial distribution

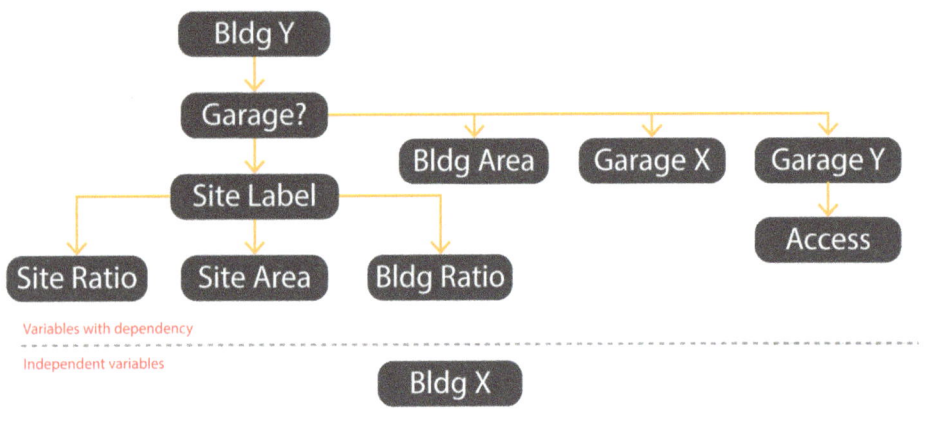

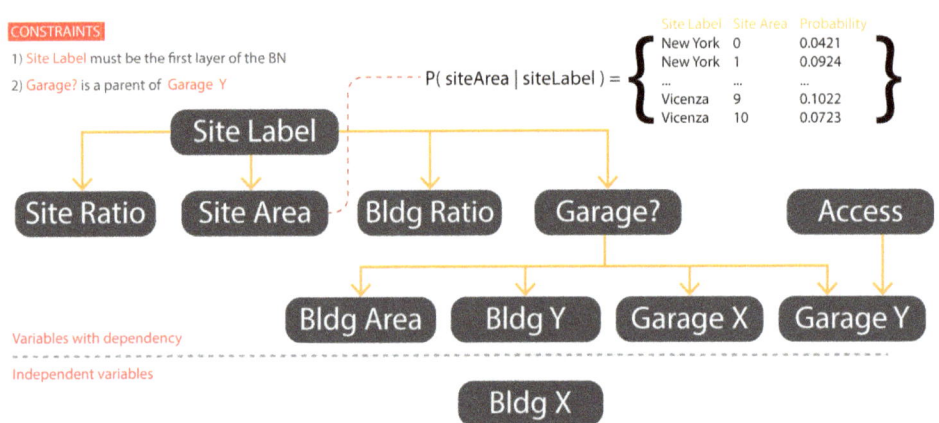

4 Graph structure learned without constraints

5 Graph structure learned with constraints

Inference and Prediction

After finding the optimal structure from the dataset, the parameters of the nodes are estimated. Since the variables of the site model are modeled as a multinomial distribution and use maximum likelihood estimates for their associated parameters, the parameters can be calculated by tabulating the corresponding counts in the dataset. In addition, having the learned structure and parameters of the BN makes predicting straightforward. In this case, prediction queries take the form of a data instance with unknown variable values, and the learned BN predicts these missing values. A loopy belief propagation algorithm is implemented to calculate the marginal probabilities needed to predict unseen data (Murphy 2013).

EXPERIMENT

Dataset Encoding

In the experimental scenario, the goal was to find the statistical relationship between the location of the site and the variables of interest. As the first step, the variables of interest are formulated and a site layout model is created, as shown in Figure 2. This model is simplified for two reasons. First, the experimental dataset consists of only 300 data samples. Thus, a minimum number of variables is chosen to avoid data fragmentation (Koller 2009). Second, this approach enables clearer interpretation and understanding of the working of BNs in terms of the variables of interest. In doing so, maintaining a small number of variables helps track how the variables interact with each other. Two places are selected, New York in the United States and Vicenza in Italy, as the locations where data is sampled. Because site layouts in these two cities exhibit very different characteristics, it is easy to verify the result of the experiment. Figure 3 illustrates a few representative site layouts found in each city.

After processing each site layout data instance into a simplified representation, the eleven variables of each data instance are manually encoded. The site layout consists of eleven variables of interest in our site model, as listed in Table 1.

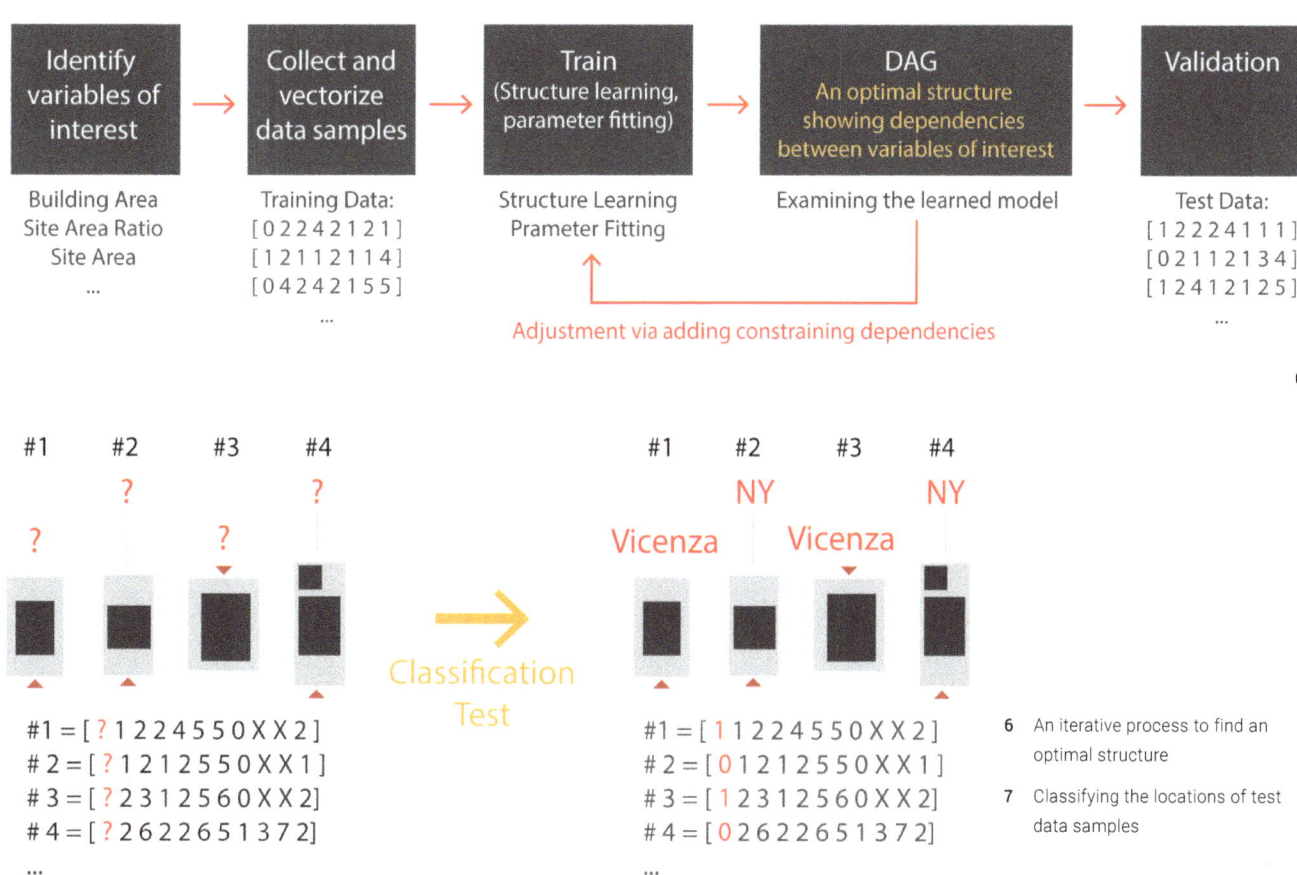

6 An iterative process to find an optimal structure

7 Classifying the locations of test data samples

Formulating the model's variables is equivalent to selecting the vocabulary to express the subject. In this sense, they are the variables of interest to analyze the site layout. The variables enclose three aspects of the site: (1) location of the site, (2) dimensional properties, and (3) contextual information. Since the experimental model uses multinomial distribution for each variable, all of the continuous parameters were discretized to be represented as categorical data. For example, the range of the site area's values is divided into ten intervals. Then, for the site area variable of each data sample, the bin that included the value of the corresponding site area was allocated. All the other numeric measurements of variables were encoded in the same way. The site's location, the existence of a garage, and the access direction to the site are modeled as categorical variables. In this project, a garage was the only extra building, but it is also possible to add more buildings to the site representation by adding corresponding variables at the cost of expanding the sample space.

Structure Learning and Parameter Estimation
A DAG-structure learning algorithm with the encoded dataset was run to find an optimal structure. From a dataset of 300 instances, 240 instances were used as a training set and the remaining 60 instances were used as a test set. The multinomial parameters of each variable were then estimated using maximum likelihood estimation. In BN, the parameters of a variable are conditional probabilities given its parent.

Experiment Result
After finding the structure and fitting the network's multinomial parameters, the learned structure and the dependency of the graph are examined. Figure 4 shows the found structure of the graph. The found structure reveals how the variables are interrelated. The main observation of the graph is that the dimensional properties of the site depend on the site's location, and the existence of a garage affects, obviously, the location of a garage and the size of the building. Meanwhile, the horizontal placement of the building is not related to any

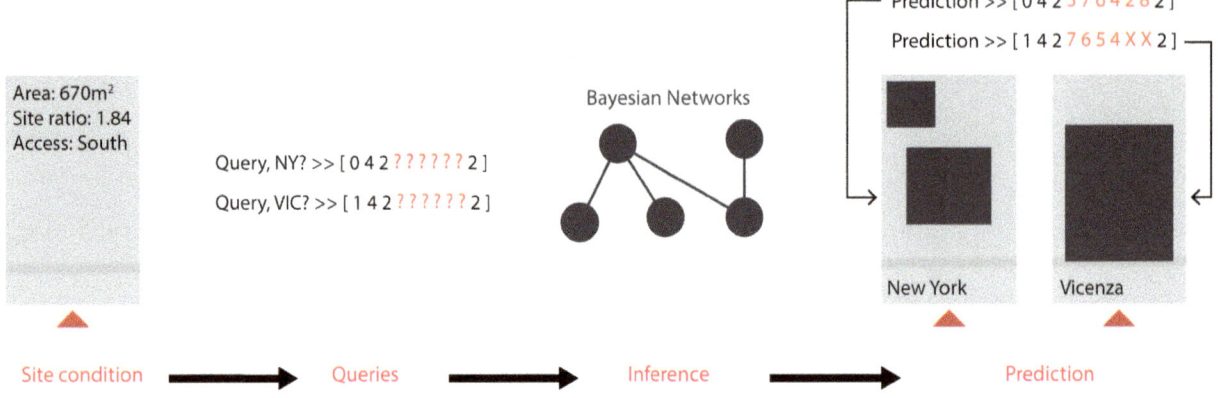

8 Predicting site layouts in different locations from the same site configuration

other variables, which makes sense because most buildings are placed near the center of the site on the X-axis.

The initial found graph structure reveals major dependencies between variables in the model. However, the organization of the network does not fit what is known about the site layout formulation. For example, the graph presents the causality that the access direction depends on the Y location of a garage and that the location of a site depends on the existence of a garage. While this graph structure resulted in the best score in the structure-finding algorithm, and the connected variables have statistical interdependency, it is known that these causality directions must be reversed. To cope with this issue, the prior domain knowledge is added into the learning process, which is a strength of Bayesian structure learning. Specifically, two constraints were added to the learning process: (1) the site location must be on the first layer of the graph, and (2) the access direction to the site must be the parent of the garage Y location. Figure 5 shows the learned graph structure that satisfies the constraints given. This process of revising the learning steps by adding constraints forms a cycle where the learned system can be enhanced incrementally in iterations, and it is illustrated in Figure 6.

Although the individual dependencies found in the two graphs are similar, the graph in Figure 5 better captures the site layout structure because it satisfies what is already known about the site layout formulation. By observing the dependencies of the final network, it can be seen how the variables interact with each other in the site model. First, depending on the location, the site layout exhibits different properties in the size of a site area, the dimensional ratio of a site area, the size of a building, and the existence of a garage. Second, the presence of a garage affects the size and the Y position of a building. In addition, the main access direction to the site also affects the Y position of a garage. Lastly, the X position of a building is not related to any other aspects of a site layout.

The learned BN was tested on whether it could successfully classify the site location of an unseen site layout using the test set as a quantitative assessment. For 60 unseen site layouts, the accuracy of the classification of where the site is located was 98% (Figure 7). In addition, how the learned BN generated a complete site layout given partially specified site variables was tested as a qualitative assessment. As shown in Figure 8, the dimensions of a site and the access direction to the site were specified. The learned BN generated two different site layouts that were most plausible in the two locations, New York and Vicenza. Upon visual inspection, it can be seen that the two resulting layouts exhibit similar configurational properties shared within the samples with the same regional label.

In conclusion, despite the small amount of data, a BN could successfully capture the major statistical relationships hidden in the dataset. By embedding prior knowledge about the site layout, the structure learning process could derive the BN structure that revealed the interactions between variables and satisfied domain knowledge. The observations on quantitative and qualitative tests confirm that the learned BN could function as a knowledge representation used in classification and generation tasks.

DISCUSSION
Limitations
As described in the experiment section, the experiments show that the BN constructed from the dataset successfully

revealed latent parametric dependencies in the dataset, classified the location of the unseen site layout, and generated plausible site layouts. However, there are limitations to this method.

First, because maximum likelihood inference is used to make predictions, the model fails to parse the site data whose encoded data values are not in the training set. For example, a prediction query with an unseen data instance that has a Site Area value larger than 500 m² fails when no sample instances in the training set have such value. In that case, the prediction for this new data instance equals zero. This is a well known limitation of MLE estimates. One immediate solution to this issue could be to use a Bayesian estimator to estimate parameters and to make a prediction. Since our model was based on multinomial distributions, the inference model can be easily transformed into a Bayesian estimator with Dirichlet prior.

Second, the prediction of a variable in BNs uses the information contained in the Markov blanket of that variable. That means the amount of the information used in the prediction depends on how many connections the variable has. For example, in the test BN, the site label variable has connections to site area ratio, building area ratio, site area size, and garage existence, which means that in predicting the site label of unseen data, there are five variables involved. Thus, the prediction accuracy was high. However, suppose one predicts the independent variables such as building locations X. In that case, the prediction will be just the value with the largest population in that variable without referencing any other information. In this case, the prediction does not involve considering the interactions between variables but rather a simple mean calculation of the variable itself. Thus, when using the BNs for prediction, one should be aware of this aspect of the MLE estimate of BNs.

Third, the site layout model used in the paper is an experimental simplification for proof of concept, and it simplifies on-site geometries into rectilinear shapes and includes only a few contextual variables. While this scheme worked in the experimental setting where we only focused on residential sites, a more flexible site layout model needs to be constructed to deal with intricate site layout conditions.

Future Work

Future work involves developing a flexible site layout model and further extending the application of the site encoding method. The test site model is simplified to rectilinear shapes and has only a few contextual variables. It is anticipated that using convex-hull-based shape estimation can replace the rectilinear base shape assumption, so that the model can cope with irregular shapes of a site and a building. In this scheme, the shapes of a site and a building are converted to convex-hull polygons, and then, the dimensional parameters, such as width and height, are extracted by orienting the shapes to the cardinal directions. The work-in-progress model aims to include the views, site landmarks, and adjacency conditions of a place to introduce more contextual information to the method.

In extending the application of encoding site conditions, the method presented in this paper can be incorporated into many generative design applications that miss the contextual consideration of an architectural design. For example, the site layout encoding method presented in this paper can be immediately integrated into the work of Merrell et al. (2010) to create BNs that include the statistical information of both the contextual knowledge of the site and the building in it. In addition, because BNs represent the full probability distribution of a subject, they will well fit computational applications where sampling is involved, such as generating a new site layout based on MCMC sampling.

In addition, the number of samples that we used in this pilot project was very small in comparison to the overall sample space. We only used 300 examples where the sample space includes more than 109 sample points. In general, more data render more accuracy in ML. Although the process of adding prior knowledge in the form of a constraint graph helped learn the structure of a generalized graph from a very limited number of samples, it is important to have more data for a better structure finding accuracy and a larger prediction space. A strategy for future work is to include an automated pipeline that can use a large number of samples from widely available geospatial data, such as a vectorized map and GIS dataset. Another strategy is to develop a method to take advantage of a small number of significant sources, such as the standard typologies described in Architects' Data and Architectural Graphic Standards or built projects of prestigious architects (Nagakura and Sung 2017). Such a method would require an additional crowd sourcing tool to generate enough samples of parametric variations.

Lastly, it is worth exploring different underlying distributions of a graph and graph structure learning techniques to reduce the dimension of the sample space and learning accuracy. The multinomial BN used in this project has a limitation in scaling. The number of parameters to learn increases exponentially as more variables are introduced and their discretization gets finer. It is expected that incorporating continuous distributions into the model and exploring alternative learning mechanisms, such as GAN, can potentially aid this issue (Wang 2017; Wang 2019).

CONCLUSION

This paper described a method to encode an architectural site layout using BNs. The contextual condition of the site is indispensable to the design of the building. However, this aspect is often overlooked, especially in projects that aim to generate building layout designs. The experiment showed that BNs could serve as a computable knowledge representation of site conditions. Using this method, one can construct the variables of interest and apply a learning technique to a dataset to derive the graph structure expressed in terms of the variables. It was shown that one could embed prior knowledge of the site layout to find the graph structure that satisfies what is already known. Despite the small sample size and simplified model, the learned BN could successfully reveal the statistical dependencies hidden in the dataset, which can be further used in classification and generation tasks. This project identified a step towards bringing the intelligence of contextualism into architectural design using AI.

REFERENCES

Chaillou, Stanislas. "Archigan: Artificial intelligence x architecture." In *Architectural Intelligence: Selected Papers from the 1st International Conference on Computational Design and Robotic Fabrication (CDRF 2019)*. Singapore: Springer Nature Singapore, 2020. 117–127.

Cheng, Jie, and Russell Greiner. 2013. "Comparing Bayesian network classifiers." *arXiv*:1301.6684. https://doi.org/10.48550/arXiv.1301.6684.

Chickering, Max, David Heckerman, and Chris Meek. 2004. "Large-sample learning of Bayesian networks is NP-hard." *Journal of Machine Learning Research* 5: 1287–1330.

Haber, David, and Saeed Karshenas. 1987. "An expert system for conceptual design of buildings." In *1987 Proceedings of the 4th International Symposium on Automation and Robotics in Construction (ISARC)*. Haifa, Israel. 799–810.

Koller, Daphne, and Nir Friedman. 2009. *Probabilistic Graphical Models: Principles and Techniques*. Cambridge, MA: The MIT Press.

Mbruru, Fred Andrew. 2001. "Context Modeling: Extending the Parametric Object Model with Design Context." MSArchS diss., Massachusetts Institute of Technology.

Merrell, Paul, Eric Schkufza, and Vladlen Koltun. 2010. "Computer-generated residential building layouts." *ACM Transactions on Graphics* 29 (6): Article 181.

Murphy, Kevin, Yair Weiss, and Michael I. Jordan. 2013. "Loopy belief propagation for approximate inference: An empirical study." *arXiv* preprint arXiv:1301.6725.

Naticchia, B., A. Fernandez-Gonzalez, and A. Carbonari. 2007. "Bayesian Network model for the design of roofpond equipped buildings." *Energy and Buildings* 39 (3): 258–272.

Nauata, Nelson, Sepidehsadat Hosseini, Kai-Hung Chang, Hang Chu, Chin-Yi Cheng, and Yasutaka Furukawa. 2021. "House-GAN++: Generative Adversarial Layout Refinement Networks." *arXiv* preprint arXiv:2103.02574.

Stiny, George, and William J. Mitchell. 1978. "The Palladian Grammar." *Environment and Planning B: Planning and Design* 5 (1): 5–18.

Teboul, Olivier, Iasonas Kokkinos, Loic Simon, Panagiotis Koutsourakis, and Nikos Paragios. 2011. "Shape grammar parsing via reinforcement learning." In *Conference on Computer Vision and Pattern Recognition (CVPR) 2011*. 2273–2280.

Wang, Hongwei, Jia Wang, Jialin Wang, Miao Zhao, Weinan Zhang, Fuzheng Zhang, Xing Xie, and Minyi Guo. 2017. "Graphgan: graph representation learning with generative adversarial nets." *arXiv* preprint arXiv:1711.08267.

Wang, Kai, Yu-An Lin, Ben Weissmann, Manolis Savva, Angel X. Chang, and Daniel Ritchie. 2019. "Planit: Planning and instantiating indoor scenes with relation graph and spatial prior networks." *ACM Transactions on Graphics (TOG)* 38 (4): 1–15.

Wright, Frank Lloyd, 2005. *Frank Lloyd Wright, An Autobiography*. Petaluma, CA: Pomegranate.

Wu, Chaoyun. 2020. "Machine learning in housing design: exploration of generative adversarial network in site plan/floorplan generation." MArch thesis, Massachusetts Institute of Technology.

Yu, Lap Fai, Sai Kit Yeung, Chi Keung Tang, Demetri Terzopoulos, Tony F. Chan, and Stanley J. Osher. 2011. "Make it home: automatic optimization of furniture arrangement." *ACM Transactions on Graphics (TOG), Proceedings of ACM SIGGRAPH 2011* 30 (4), Article 86.

Yuan, Changhe, Brandon Malone, and Xiaojian Wu. 2011. "Learning optimal Bayesian networks using A* search." In *IJCAI'11: Proceedings of the Twenty-Second International Joint Conference on Artificial Intelligence*. 2186–2191.

IMAGE CREDITS
All drawings and images were created by the authors.

Woongki Sung is a PhD student in the Design and Computation group at the Massachusetts Institute of Technology, School of Architecture and Planning. His research interest lies in applying A.I. and data-driven techniques to architectural design. His current research focuses on developing a data-driven framework for designers in creating architectural layout designs, which can help designers better understand architectural variables and their interactions in computational design practice.

Takehiko Nagakura is Associate Professor in the Design and Computation group at the Massachusetts Institute of Technology, School of Architecture and Planning. He teaches courses related computer-aided design, and his research focuses on the representation and computation of architectural space and formal design knowledge.

Dr. Daniel Tsai is a research fellow at the Massachusetts Institute of Technology, School of Architecture and Planning, where he teaches and conducts research on design and computation. His cultural heritage research interest focuses on the architecture of Andrea Palladio. He also mentors student startup ventures in the MIT DesignX innovation accelerator.

Integrated Reconfigurable Autonomous Architecture System

Distributed Robotic Material System Integrated with an Intelligent Spatial Planning Algorithm and Interactive Platform for Continuous Architecture Adaptation

Tyson Hosmer
Bartlett School of Architecture, UCL

Jiaqi Wang
School of Architecture, South China University of Technology

Wanzhu Jiang
School of Architecture, South China University of Technology

Ziming He
College of Design and Innovation, Tongji University

1 The rendering of project TESSERACT developed based on the IRAAS

ABSTRACT

Advances in state-of-the-art architectural robotics and artificially intelligent design algorithms have the potential not only to transform how we design and build architecture, but to fundamentally change our relationship to the built environment. This system is situated within a larger body of research related to embedding autonomous agency directly into the built environment through the linkage of AI, computation, and robotics. It challenges the traditional separation between digital design and physical construction through the development of an autonomous architecture with an adaptive lifecycle. Integrated Reconfigurable Autonomous Architecture System (IRAAS) is composed of three components: 1) an interactive platform for user and environmental data input, 2) an agent-based generative space planning algorithm with deep reinforcement learning for continuous spatial adaptation, 3) a distributed robotic material system with bidirectional cyber-physical control protocols for simultaneous state alignment. The generative algorithm is a multi-agent system trained using deep reinforcement learning to learn adaptive policies for adjusting the scales, shapes, and relational organization of spatial volumes by processing changes in the environment and user requirements. The robotic material system was designed with a symbiotic relationship between active and passive modular components. Distributed robots slide their bodies on tracks built into passive blocks that enable their locomotion while utilizing a locking and unlocking system to reconfigure the assemblages they move across. The three subsystems have been developed in relation to each other to consider both the constraints of the AI-driven design algorithm and the robotic material system, enabling intelligent spatial adaptation with a continuous feedback chain.

INTRODUCTION

The system we present is situated within a larger body of work related to embedding agency and autonomy directly into the built environment through the linkage of AI, computation, and robotics. The underpinnings of this research are rooted in the transdisciplinary field of cybernetics, largely pioneered in the 1960s and 1970s through the work of Gordon Pask, Cedric Price, Nicholas Negroponte, Christopher Alexander, John Frazer, and Archigram (Steenson 2010). Pask stated, "The role of the architect here, I think, is not so much to design a building or city as to catalyze them; to act that they may evolve" (Frazer 1995). Pask challenged the paradigm that architecture is a static material artifact produced through a linear series of processes from design to fabrication to construction and reconsidered it as a composition of interrelated active systems regulated and controlled through feedback within a constantly shifting environment (Pask 1969).

Today, the climate crisis, housing crisis, and COVID-19 pandemic are radically transforming the way we live and work, while automation, consumer platforms, and AI are changing how we interact with and personalize our world. As Sean Hanna states, "for the first time our environment is no longer seen as fixed, or shaped by forces beyond our control, but as in constant and noticeable change, and that our relationship with it is one of mutual interaction" (Hanna 2020). Computational design and construction robotics research focus primarily on automation, customization, and optimization within linear, separated processes of design and construction rather than fundamentally changing the interrelationship between them to enable open-ended and physically adaptive architecture. Robotic buildings must consider not only design to production, but design-to-production-to-operation chains from a lifecycle perspective relating to the socio-economic and ecological impacts (Bier and Mostafavi 2018; Bier et al. 2018).

Our research challenges the separation between digital design and physical construction processes through the development of an integrated cyber-physical architectural system with a feedback-based adaptive lifecycle. Integrated Reconfigurable Autonomous Architecture System (IRAAS) is a semi-autonomous reconfigurable architecture integrating three main components: 1) an interactive platform for user and environmental data input, 2) an agent-based space generation algorithm with deep reinforcement learning for continuous spatial adaptation, and 3) a distributed robotic material system with a bidirectional control protocol for simultaneous state alignment. The three subsystems have been developed in relation to each other to consider both the constraints of the AI-driven design algorithm and the robotic material system, enabling intelligent spatial adaptation with a continuous feedback chain.

BACKGROUND

Historically cybernetics, defined as the study of communication and control within systems (Wiener 1948), and artificial intelligence (McCarthy 1955), defined as the study of machines that exhibit and simulate intelligent behavior (Oxford English Dictionary 1989), are closely interrelated. They have led to two primary threads of architecture research: 1) intelligent computational processes for design, and 2) physically adaptive and responsive environments. Nicholas Negroponte was interested in developing a symbiotic relationship between designers and "architectural machines," introducing the notion of an intelligent agent in the design process (Negroponte 1970), while Cedric Price developed systems where "the act of engaging and interacting with the architecture would change the user" (Steenson 2010). Both Pask and Price were interested in managing "indeterminacy," considering architecture's ability to adapt to, be adapted by, and impact its inhabitants (Pask 1965; Landau 1968; Steenson 2010). Development of embodied adaptive architecture requires reappraisal of linear building lifecycles. Rather than automate known, predefined patterns of construction, our overriding aim is to develop an autonomous architecture that continuously adjusts itself to actively maintain a symbiotic relationship between people, the natural environment, and the architectural environment.

State-of-the-art research in autonomous robotic systems for architecture has investigated two primary strategies: 1) small, distributed robots manipulating building assemblies, and 2) large, monolithic embodied robotic spaces and buildings. Monolithic systems enlarge the robot to the size of spaces or whole buildings (Oosterhuis 2012; Kilian 2018; Maierhofer 2019; Hosmer 2019). Adaptive spatial behaviors are triggered through the sequencing of actuators integrated into larger bodies. Alternatively, strategies have been developed using distributed robots to collaboratively manipulate building assemblies. Principles of swarm intelligence have been applied through embodied swarm construction (Rubenstein et al. 2012; Petersen and Nagpal 2017; Petersen et al. 2011). Full scale collaborative robotic assembly of timber structures is demonstrated using industrial robotic arms hung in a mobile gantry system (Adel et al. 2018). BILL-E robotic platform demonstrates a type of "relative robot in a structured environment," enabling robots to climb on the same structure they assemble and reconfigure (Jenett and Cheung 2017). A similarly symbiotic dependency between distributed robots and static elements demonstrates self-assembly of timber structures (Leder et al. 2019). Monolithic embodiments offer opportunities for controlled adaptation through constraints with less radical shifts in topology while distributed robotic strategies offer more flexibility enabling radical changes in topology at the cost of organizational complexity requiring additional spatial design inputs to be effective.

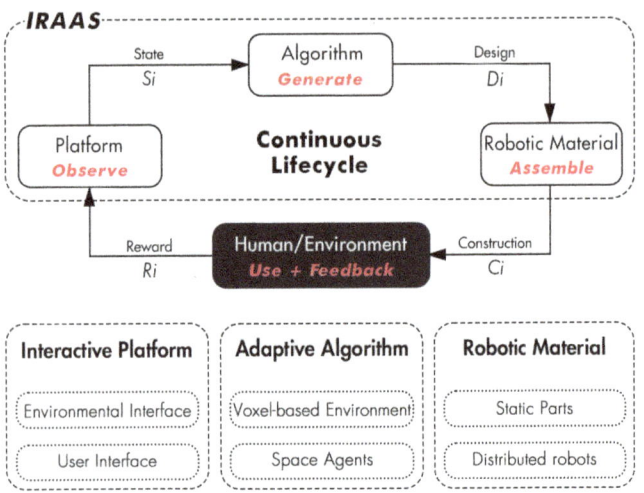

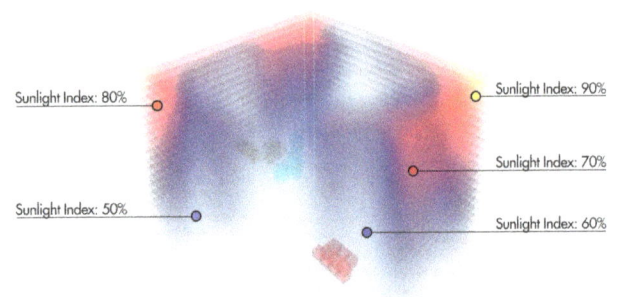

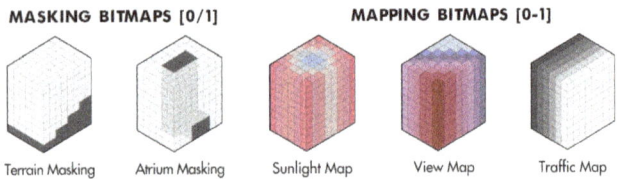

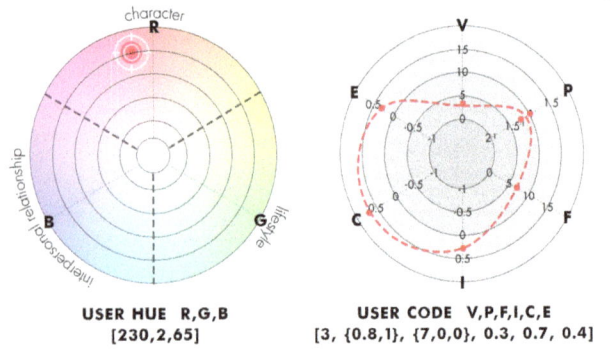

2　Continuous lifecycle and system composition

3　The environmental data, types and examples

4　The user interface, data structure and example

Generative space planning research focuses on a range of strategies for semi-autonomously computing spatial organization. We divide relevant strategies into procedural geometry, rule-based, and data-based algorithms. Liggett gives a historical overview of automated space planning methods (Liggett 2000). Procedural geometry is used for automating desk layouts (Anderson 2018) while procedural geometry with a multi-objective genetic algorithm was developed for 2D office space planning (Nagy 2017). Rule-driven models have been explored such as shape grammars (Hua 2017), cellular automata (Dinçer et al. 2014), semi-autonomous constraint satisfaction (Hosmer et al. 2020), and various multi-agent systems achieving 2D or 3D space planning by coupling design goals with geometric or topological constraints (Veloso 2019; Meyboom and Reeves 2013). Multi-agent approaches can be divided into three main groups: (1) agents as moving spatial units, (2) agents that occupy a space, and (3) agents that partition a space (Veloso 2019). One multi-agent method learns space planning behaviors in 2D with reinforcement learning (Veloso and Ramesh 2020). Data-based models have been developed using generative adversarial networks (GANs) (Goodfellow et al. 2014). To learn significant features and their relationships across image-based datasets (Isola et al. 2017) and have demonstrated 2D space planning strategies (Zheng and Huang 2018; Chaillou 2019; Chaillou 2020).

Most research in semi-autonomous robotics places emphasis on methods of construction or adaptability through the lens of the robotic constraints without apt consideration for how spaces would be designed, organized, and adjusted. Results of purely bottom-up robotic methods tend to be limited to behaviors for assembling abstract structures. Generative space planning strategies tend to overemphasize methods for spatial organization, resulting in digital models with neither ample alignment with the constraints of the robotic material system nor an appropriate communication protocol between the digital design processes and physical assembly processes, leaving gaps that prevent a continuous chain.

Autonomy requires a system with an ability to self-manage through a degree of self-awareness. It can be defined as having an effective interdependency between properties of facilitated variation, situated and embodied agency, and intelligence (Hosmer 2019). This research embeds principles of autonomous architecture by developing an interdependency between the intelligent agency of the space planning algorithm and situated and embodied agency in the design of the robotic material system as a structured environment. Facilitated variation is achieved through the design of effective constraints in the robotic material system to reconfigure elements that simultaneously enable locomotion patterns. The development of the control system protocols enables

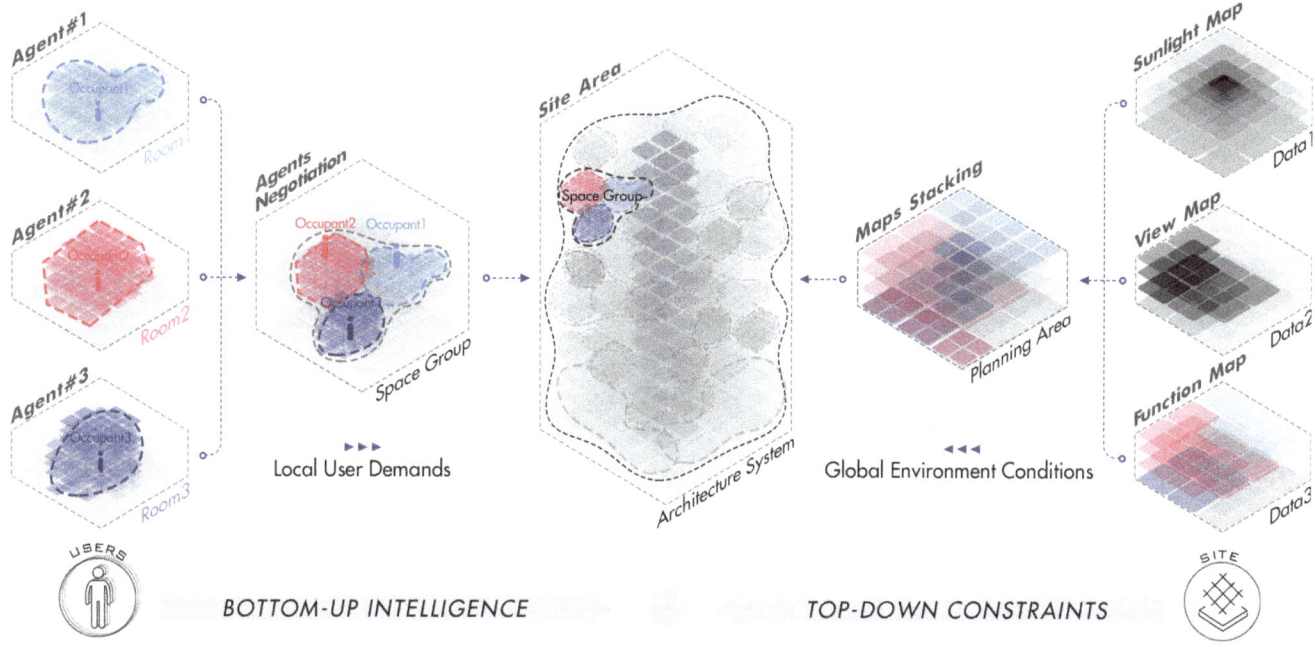

5 The basic logic of the Agent-Based Spatial Planner

simultaneous state alignment between the virtual simulator and physical world in a continuous feedback cycle.

METHODS
System Composition and Lifecycle

IRAAS is a semi-autonomous architecture that operates in a continuously adaptive lifecycle through an "Observe, Generate, Assemble" feedback loop. IRAAS is implemented through a combination of three closely related components: 1) interactive platform, 2) agent-based space generation algorithm, and 3) robotic material system. At a discrete moment (T_i), environment data and user goals are collected by the interactive platform and sent to the space generation algorithm. It observes its current state (S_i) in relation to processed inputs to generate an adaptation as a virtual space design (D_i), guiding the adjustment or reconfiguration of the actual construction (C_i) through the robotic control system. Changes in the environment and user goals provide constant feedback (R_i) in a closed loop (Figure 2).

Interactive Platform

The interactive platform serves as a data collection port that triggers adjustments in the system from 1) environmental data and 2) multi-user space planning goals collected through the interface. Environmental data such as sunlight exposure is collected through digital simulations and processed as 3D bitmaps. Fixed elements in the environment such as exterior facades are also processed through bitmaps. Bitmaps are translated into 3D data matrices establishing global constraints and environmental factors that influence the space planning algorithm (Figure 3).

The user interface is responsible for collecting user characteristics and space planning goals, processed as inputs that drive the behaviors of the space generation algorithm. Each user is translated into a "user code" and "user hue." User hue reflects the user's characteristics in relation to other users. The user code captures their spatial goals and willingness for negotiating their spaces with others (Figure 4).

Agent-Based Spatial Planner with Self-Play Reinforcement Learning

The spatial planner component is responsible for generating 3D volumetric space boundaries within a virtual environment. The computational model is trained using reinforcement learning to learn adaptive policies for adjusting the scales, shapes, and relational organization of spatial volumes by processing changes in the environment and user requirements in near real-time (Figure 5). It is designed as a multi-agent system with programmable agents, each representing an independent space within the environment. We extend principles of the Stigmergic Space Adjacency Software for multi-agent space planning with stigmergic communication in a 3D environment (Meyboom and Reeves 2013). Our algorithm generates spatial organization through intelligent negotiation behaviors of agents driven by user and environment data, outputting structured environments readable by the robotic material control system.

The environment is a 3D grid of voxel arrays. Each node contains metadata, including its position, size, geometry type (cubic, tetrahedral, etc.), color, and layers of site/environmental information imported from the platform. The size and

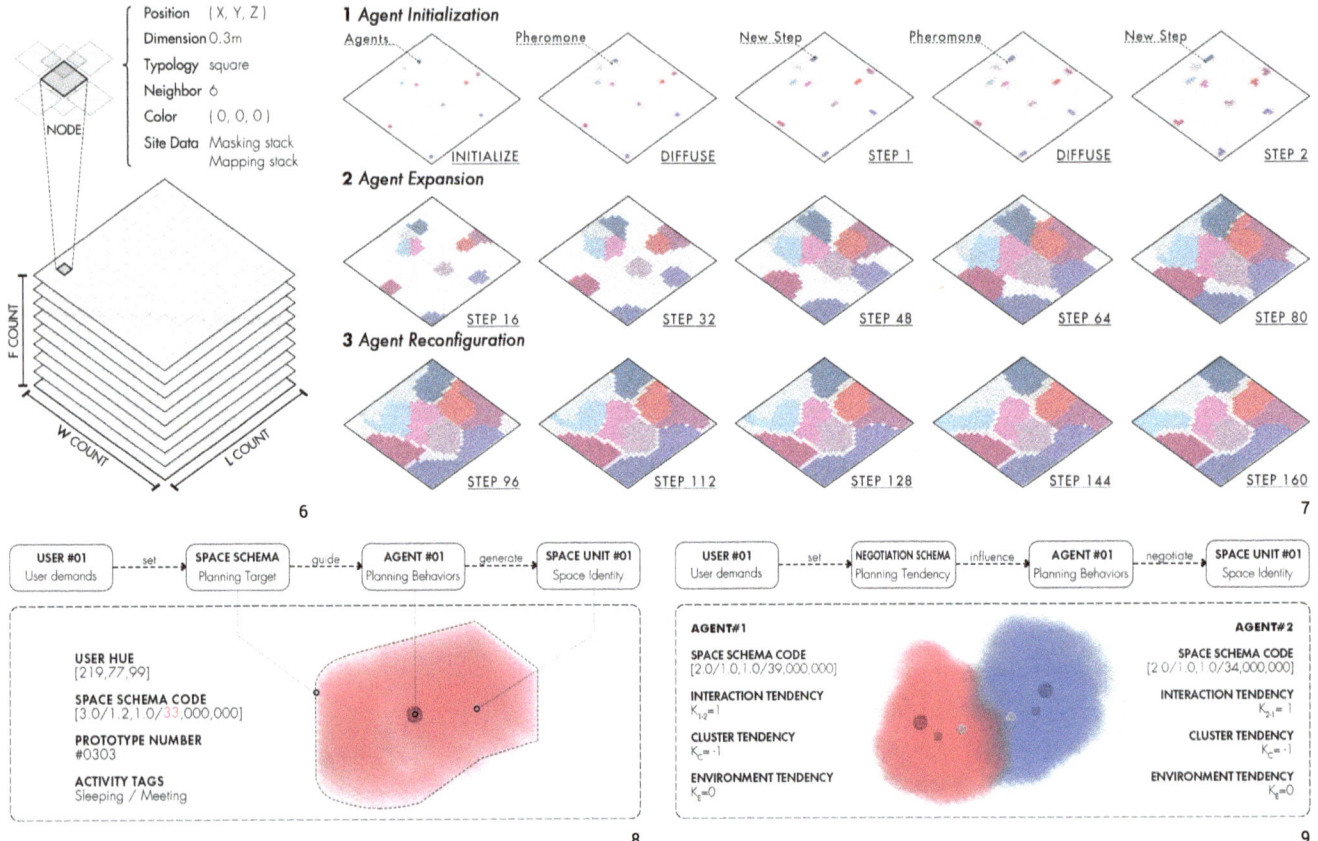

geometry type have a consistent relationship with the parts and degrees of freedom in the robotic material system. Color represents a "pheromone," distinguishing an association of agents occupying the node with an initial value of [0,0,0] for unoccupied nodes (Figure 6).

Each spatial agent contains a collection of nodes which form a volume. Its basic parameters include position, color, capacity, diffusion rate, and cohesion rate. Position is the agent's center of mass. Color represents agent type as a degree of association to other agents. Capacity is the maximum territory the agent will contain. Diffusion rate is the speed the agent passes its pheromone through the environment. Cohesion rate is the speed of balancing force of the agent boundary toward its center of mass. The agent expands the three-dimensional territory it occupies from its starting location by adding or releasing adjacent nodes until reaching its capacity in a state of dynamic equilibrium with its neighbors, thereby gradually forming an internally closed volume (Figure 7).

Behaviors defining how the agent moves and adapts its territory are dictated by adjustments to internal parameters in relation to neighboring agents and its environment. To achieve a mapping from user space planning goals to the agents' space planning behavior, we introduce three "Schema" as collections of properties of the agent that influence changes in its behavior in response to design objectives: Relational Schema, Space Schema, and Negotiation Schema.

Relational schema defines an agent's degree of relationality or association to other agents through an RGB value. The degree of similarity of each agent's color to other agents forms a closer or more distant association.

Space schema contains parameters that influence the agent's behavior related to the local space it generates and adjusts. These include "Volume," "Proportion," and "Form," which respectively determine the size of the space volume, the aspect ratio of the space bounding box, and the space morphology type (Figure 8).

Negotiation schema parameters influence its behavior related to negotiating space with its neighbors and environment, including three parameters described as "tendencies." Interaction Tendency represents the degree of resistance or attraction along the boundary with a neighbor. For each adjacent agent, it is expressed as a weight K_{I-A} in the range of [-2,2], applied as a repulsive or attractive force according to the value of K_{I-A}. Cluster Tendency defines the relationship between the agent and spatial clusters together with the Relational Schema. The weight K_c in [-1,1] dominates the agent's tendency to groups with similar colors. Environment

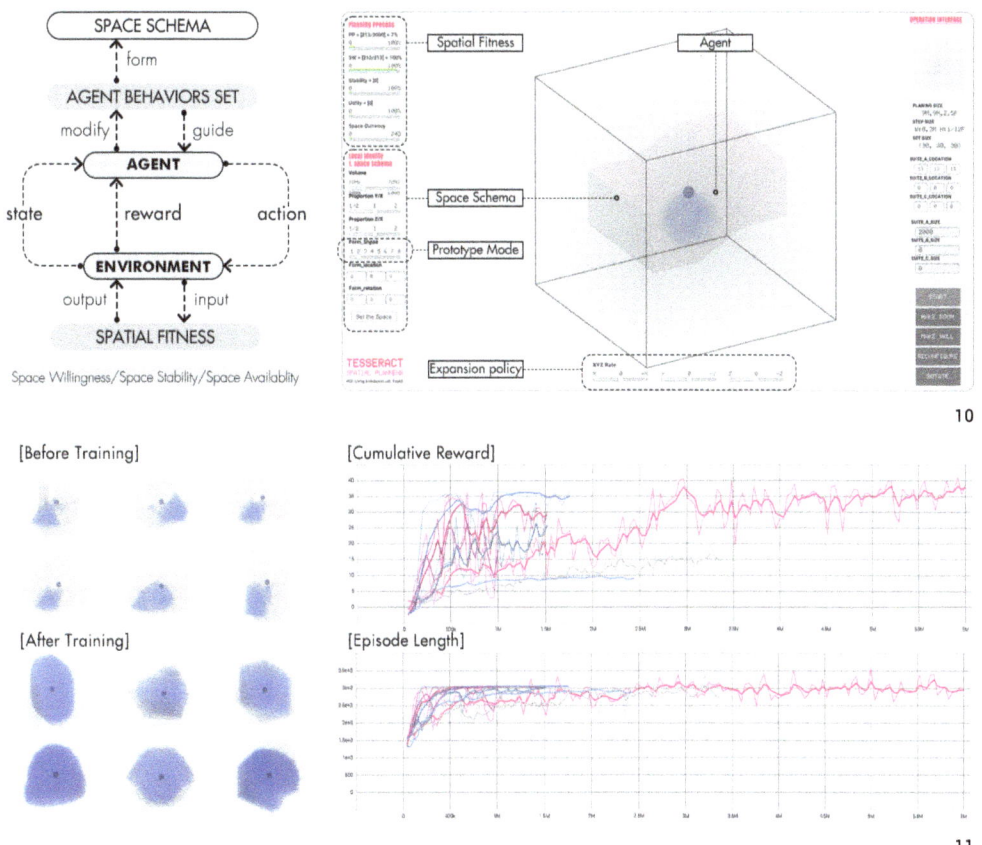

6 The discrete planning environment and the metadata in each node
7 The space planning process
8 Space Schema, principle, and example
9 Negotiation Schema, principle, and example
10 The principle of reinforcement learning and the training model
11 Reinforcement learning results: (left) comparison of behaviors; (right) graphs (pink one success) with the cumulative reward and the episode length

Tendency quantifies the agent's demand in relation to environmental resources such as natural light. The set K_E aggregates the agent's reaction to site information, making agents exhibit specific behaviors such as phototaxis or light avoidance (Figure 9).

The spatial planner is designed to autonomously negotiate multiple users' goals and environmental factors. To manage the high dimensionality of this problem, we leverage Reinforcement Learning with Self-Play to train a neural network to learn adaptive decision-making strategies for the agents (Sutton and Barto 2018). Each agent learns a behavioral policy to continuously maintain a close mapping between its space schema goal parameters and the generated volume while negotiating its territory with other agents through the negotiation schema. Training begins with 600 groups of {V, P, F} random combinations within a limited range as the initial dataset. Adjustments in stigmergic parameters control the agent's three-dimensional growth direction and intensity based on the centroid position and a policy {x, y, z} (Figure 10). Observations include the space schema parameters, the agent's current position, the last occupied node, and its stigmergic parameters. A real-time reward is given when each captured point is inside the target shape, and a penalty is added when it is outside. A staged reward is set to double the real-time reward when the proportion of qualified points (defined as "space fitness") increases beyond a threshold (30%, 60%). When the ratio reaches 90%, the task is considered complete, and a set of global rewards from additional analysis such as "structural stability" and "space availability" are given. After $6*10^6$ episodes, the training curve typically reaches a state of relative stability (Figure 11).

Semi-Autonomous Robotic Material System

To enable semi-autonomous reconfiguration of building parts in direct relation, a novel voxel-based robotic material system was developed with a structured environment operating in the same voxel grid as the spatial planner. Rather than using a gantry system for locomotion, the robots are designed symbiotically with passive blocks to slide over the dynamically changing host structure while reconfiguring it through a system of tracks, dynamic knobs acting as gates and switches, and locking and unlocking mechanisms. This enables simple collaborative robots with low numbers of degrees of freedom to efficiently adapt to the spatial assembly.

Modular passive parts are cubic with a side length of 0.3 m. Each passive part is divided into passive actuation, structural, and panel components. Passive actuation components enable relative sliding and locking, equipped with female cross slots on one side of each axis and male flat knobs on the opposite. Structural components form a frame with tracks for gears in

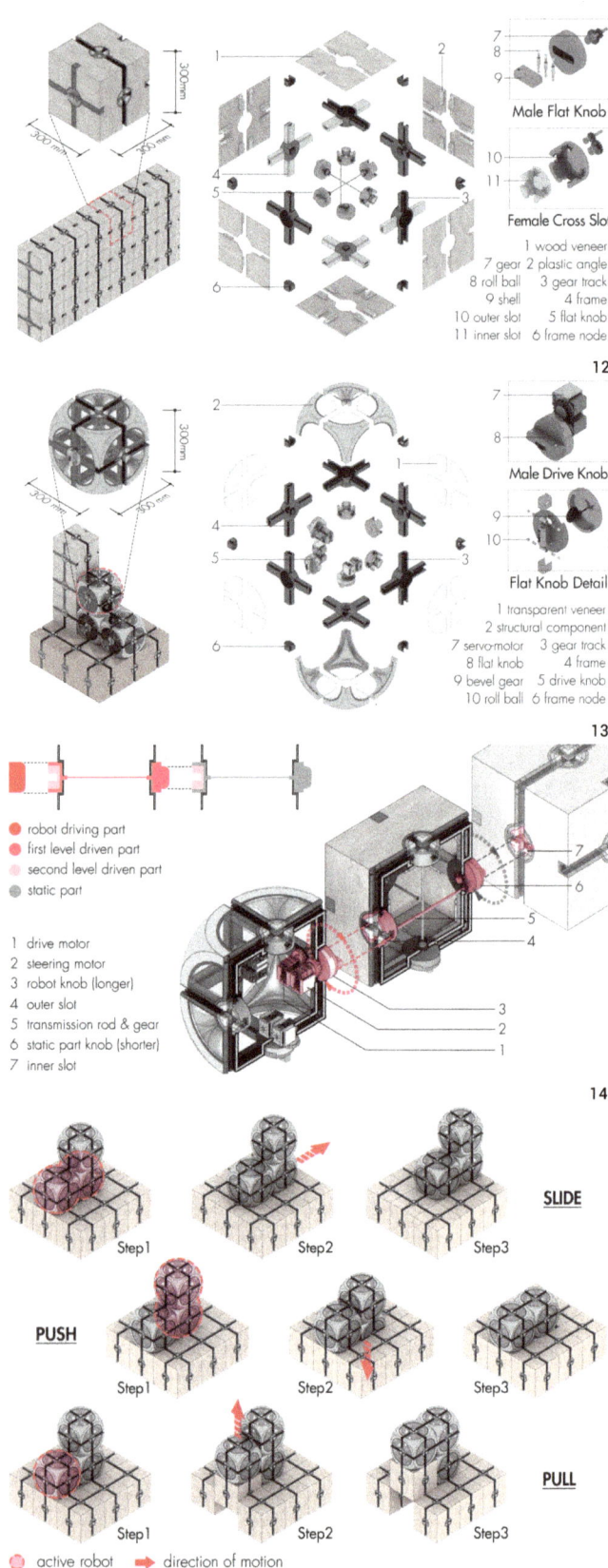

12 Static parts and the construction details
13 Distributed robots and the construction details
14 Joint details for lock and unlock
15 Robot actions

the knobs. Panel components provide the architectural infill for the assemblies (Figure 12).

Modular robots are the same size as passive parts with similar fabrication details, but robotic structural components are filleted to avoid collisions during movement (Figure 13). Physical prototypes are manufactured with 3D printing at a scale of 1:1 (Figure 16). Two servomotors are installed behind each knob for driving and steering. The body of one robot consists of three linked cubic parts forming an L-shaped combination with an action mode for sliding its cubic parts along tracks in its body or sliding itself along tracks in static parts. Robots can slide, change direction, push and pull, and lock and unlock through mutual collaboration for various reconfiguration tasks. In Figure 15, for example, the robot on the top can connect to a passive part, drive the motor and slide it down along the tracks provided by the adjacent passive parts, pushing the object one side length into place.

The robotic control system is setup with the principle of simultaneous state alignment. Simple "relative" robots cooperate with a relative localization strategy synchronized through bidirectional communication between the simulation environment and physical environment. The simulation environment is built with Unity3D (Unity 2020), a game engine with a comprehensive multi-physics module and high extendibility. Assemblies of robots and passive blocks are designed as a digital twin with encoded actions computed through multi-physics constraints and inverse kinematics directly related to the actions of the physical system. The actions developed in the simulator include: 1) sliding on static parts, 2) changing direction, 3) sliding segments of their body, 4) locking/unlocking passive blocks, 5) pushing/pulling, and 6) carrying passive blocks (Figures 16, 17).

Communication is established wirelessly with a local network through a UDP protocol between our robotic simulator loaded on a PC and Raspberry Pi microcomputers mounted to the physical robots. Each Raspberry Pi is loaded with custom-built control software developed in Python. Custom commands are streamed back and forth as packets through a wireless port, converted to python functions on the Raspberry Pi and C# functions in the PC simulator. Dynamixel AX-12A servomotors were installed on the robotic knobs with high precision angle control with data feedback including angle position, load, and speed, enabling semi-realtime cyber-physical state alignment. The simulator hosts the server and connects the physical modules as clients. Robotic actions in the simulator are calculated with inverse kinematic functions, mapped to precise motor speeds and rotations, and sent wirelessly as instruction sequences to the Raspberry Pi to drive the servos on the physical robots. Each motor takes 0.196 seconds to rotate 60° in 10V voltage. When the robot slides one unit length

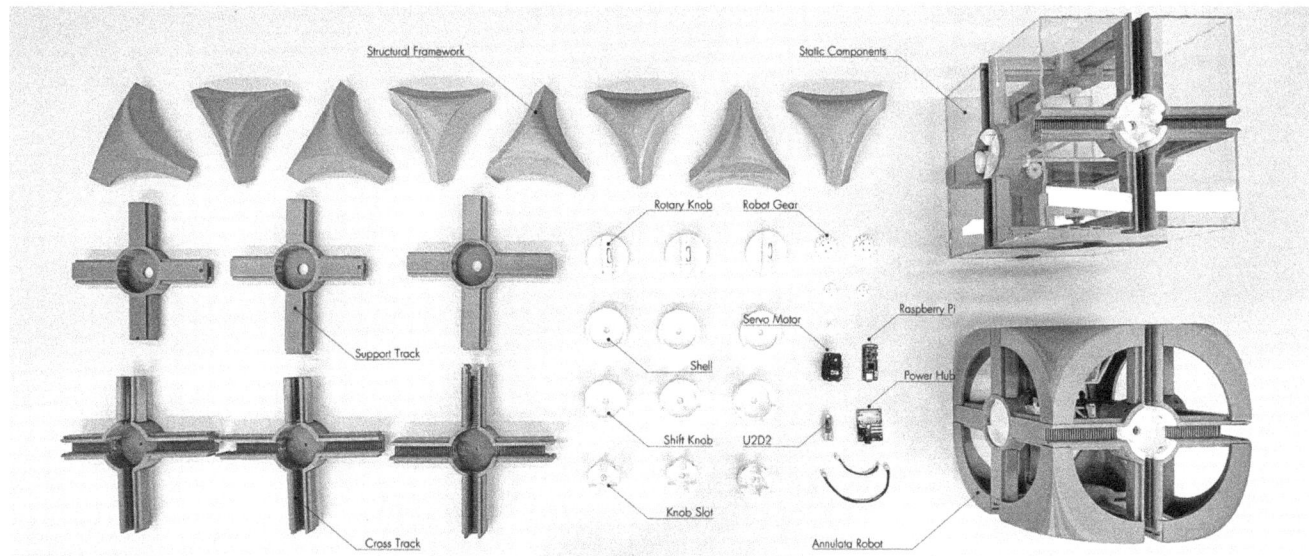

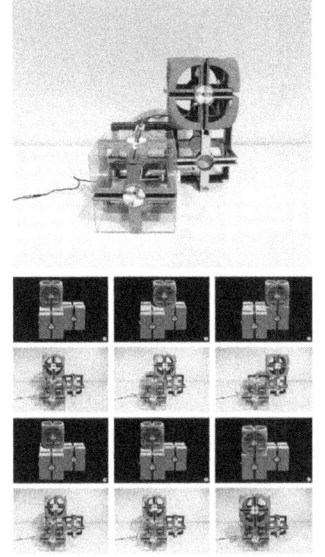
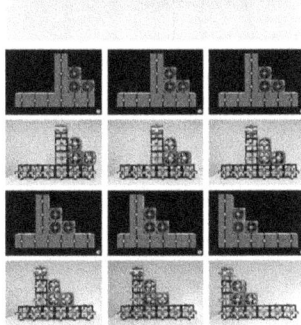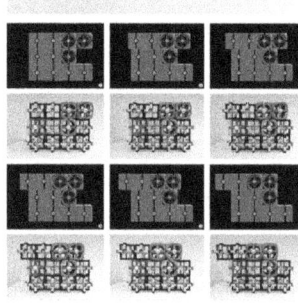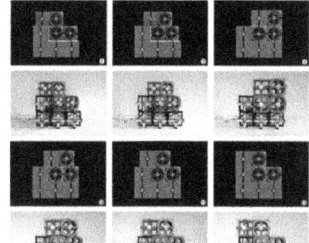

16 The component list and physical prototypes.

17 The robotic actions: 1) sliding on static parts and changing direction 2) sliding segments of their three-part body 3) pushing/pulling passive blocks, and 4) carrying passive blocks reinforcement learning with self-play for collaborative robotic behavior

(300 mm), the bearing set rotates 3.18 circles in 3.75 seconds. Sensors in the servos then collect physical state data (torque, load, speed, and temperature) and send it back to the simulator as observations from the physical environment.

Assembly constraints are introduced through the rotating male-female connections in the passive parts while constraints of locomotion and actuation are introduced through the robots three-part body and male-female knob and groove system. While pathfinding algorithms such as A* (Hart 1968) and Dijkstra's algorithm (Dijkstra 1959) have proven to be efficient for solving simple path planning problems through weighted graph traversal, we introduced deep reinforcement learning in our simulator to formulate adaptive strategies involving collaboration between multiple robots coordinating the biased and constrained behaviors in dynamically changing structured environments. The simulator was setup with L-shaped robots composed of three independent embodied agents taking observations of the current state of each of the six faces of each part. A curriculum learning strategy was developed as three stages of increasing difficulty.

Stage 1 is carried out in an environment over a 2D plane with a range of 8m*8m. The goal is to transport the static parts from

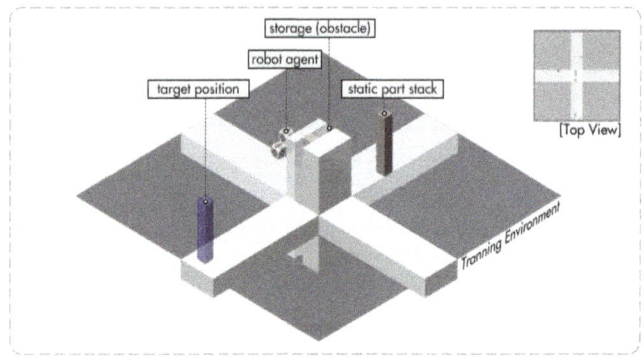
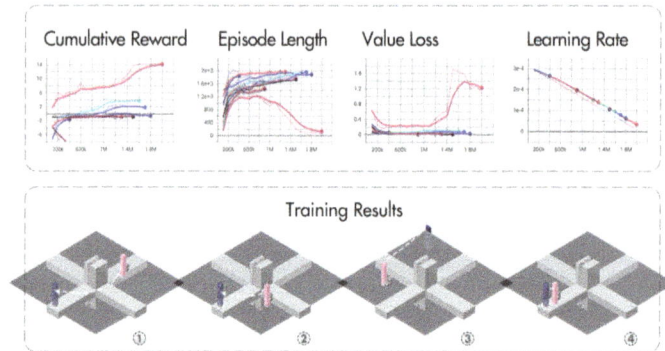
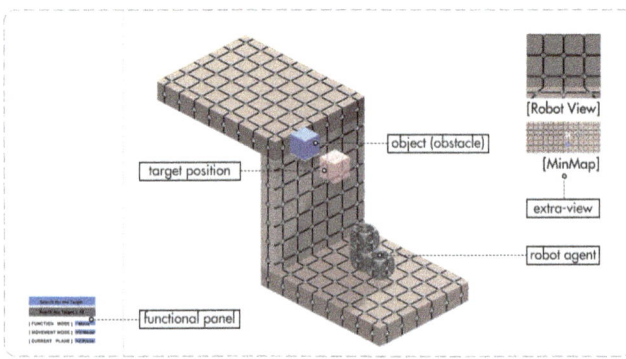
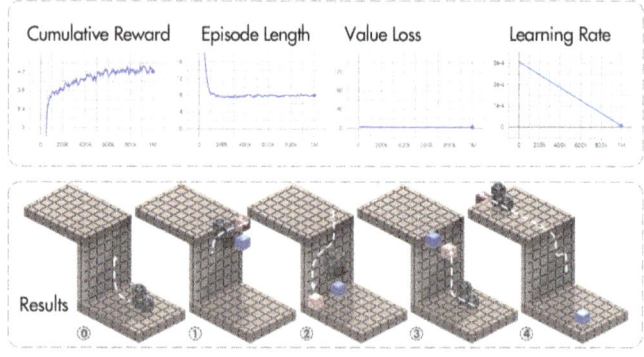
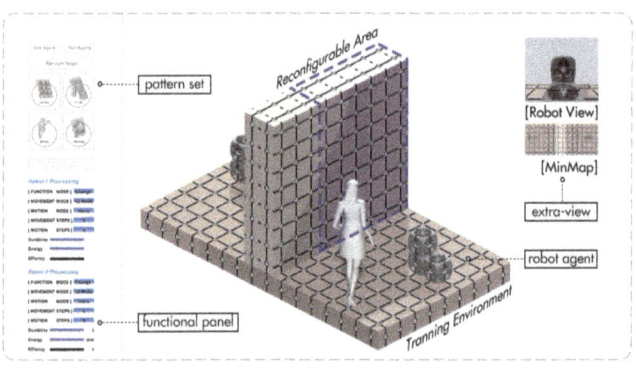
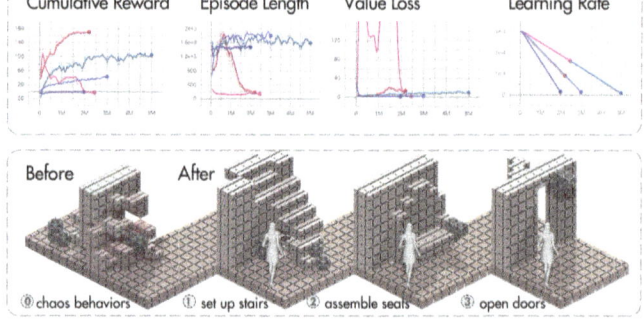

an initial position to a target position. Observations include the robot sensors and positions of the agent, starting point, and target point. If the robots or the static parts fall out of the boundaries of the environment, the agents receive a penalty of -1. The reward is set to be inversely proportional to the distance from the target, and 10 extra points are added when the target position is reached (Figure 18).

Stage 2 is performed in a Z-shaped 3D environment composed of 22*6 static parts to train the agents to reach a designated position while avoiding obstacles. The position of the obstacle is added to the observations. When agents collide with an obstacle, they receive a penalty of -1, and the other rewards remain the same (Figure 18).

Stage 3 aims to train two robotic agents to collaborate in assembling passive parts into 3D design goals. The training uses four spatial forms as target goals. Observations are the robot positions, sensor activation, and all static parts' current and target positions. Rewards are added for energy consumption and accuracy of the result in relation to the target goal (Figure 18).

RESULTS AND REFLECTION

This methodology was developed in a research through teaching context in the Living Architecture Lab (RC3) at the Bartlett School of Architecture led by Tyson Hosmer, Octavian Gheorghiu, and Philipp Siedler. Its implementation was tested through a case study project for a community called TESSERACT, which proposes a new paradigm of distributed living and working by providing spaces that continuously adapt and reconfigure in near real-time according to the changing requirements of the shared community of users. Within the context of the global COVID-19 pandemic, this enables people to enjoy a variety of activities in a tiny scope of life.

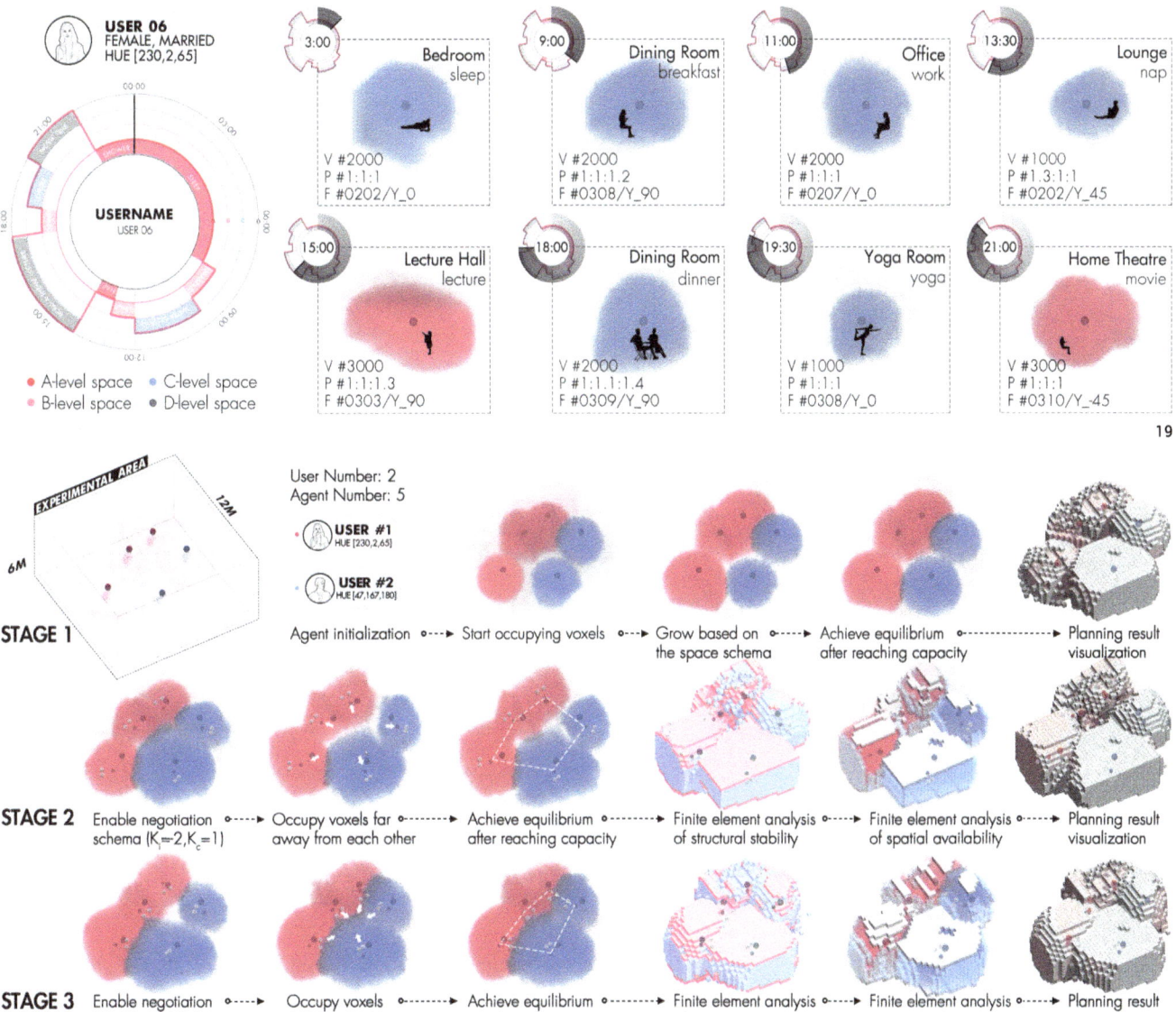

18 The training environment and training results of Stage 1, Stage 2, and Stage 3

19 Single-user continuous space planning experiments

20 Multi-user continuous space negotiation experiments

The interactive platform was developed with interfaces for the user to input their preferences. A "user hue" is generated by personal characteristics, lifestyle, and interpersonal relationships. A "user code" is generated by personal schedules, interaction needs, and spatial preferences. These are processed as inputs for each agent's relational, spatial, and negotiation schema.

The algorithm was tested through a series of experiments. Individual agents were successfully trained through curriculum learning to achieve policies first for space schema goals for volume, proportion, and shape only (Figures 11, 19), and next to achieve space schema goals while adapting through relational and negotiation parameters with multiple agents. The proportion of overlap between the resulting territory and the target territory reached more than 85% typically after $6*10^6$ episodes. In further experiments, we input dynamically changing user requirements and the algorithm proved successful in autonomously adjusting its behavioural patterns accurately (Figure 20).

The trained model was tested in a 30m*18m*18m site volume. Environmental data and constraints including fixed structure, daylight intensity, and view analysis were loaded as maps that agents responded to. User codes were auto-generated as virtual user demand parameters to initialize various types of agents, forming a community social network. We modeled a 50% occupancy rate resulting in the agents easily achieving

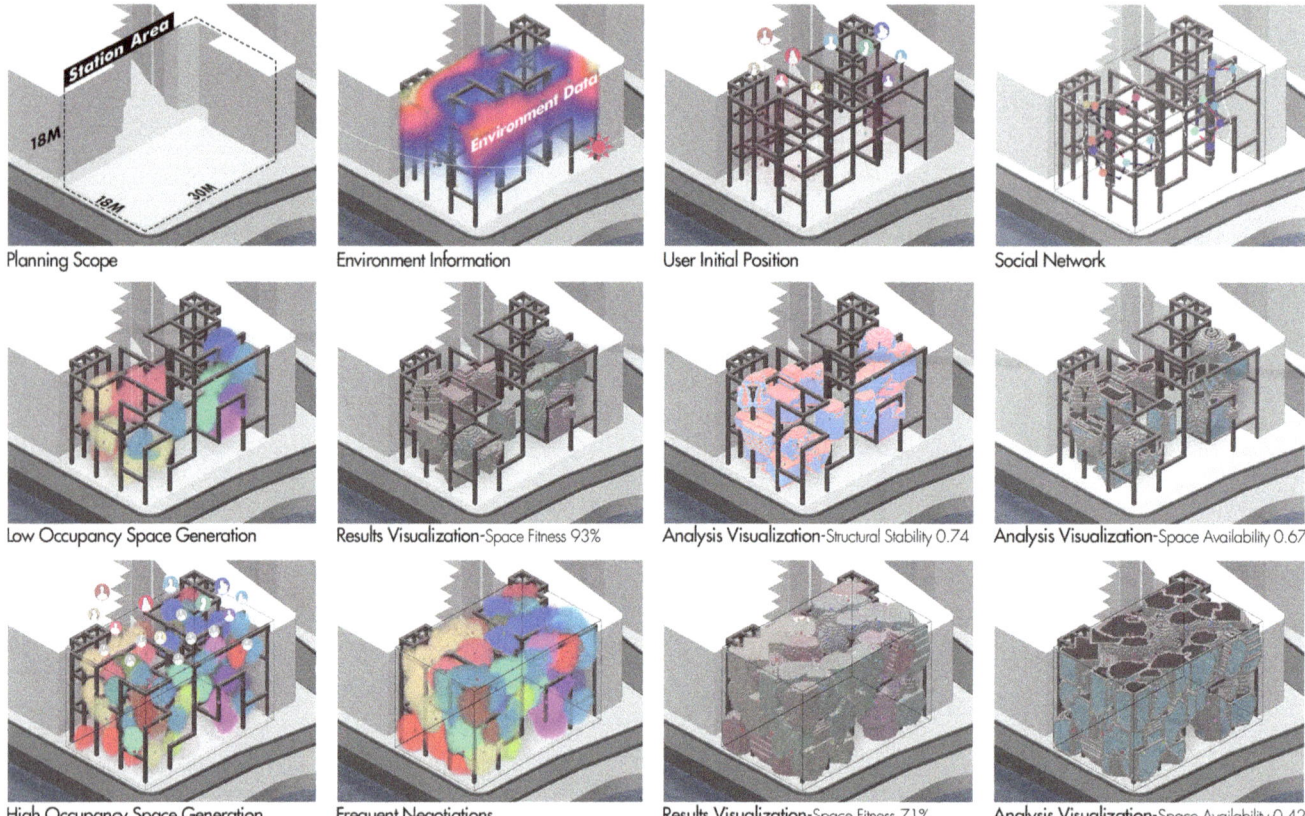

21 The space generation of project TESSERACT in different occupancy rate

spatial goals at over 90% accuracy with lots of open space and little negotiation behaviours. Next, we tested a 100% occupancy rate with much more negotiation behaviours required which achieved approximately 70% spatial accuracy (Figure 21).

Finally, the robotic material system was iteratively developed through a series of prototype configurations directly controlled by the virtual simulator which successfully demonstrated simple reconfiguration sequences combining sliding, locking/unlocking, and pushing/pulling behaviours (Figures 16, 17). The system has been successfully tested to convert agent boundaries as voxels directly linked with the agent simulator sending instruction sequences to multiple robots while receiving sensor data back wirelessly. Self-play reinforcement learning was tested with two simulated robots to coordinate reconfiguration of a wall into a series of goals, successfully improving from random outcomes to efficient sequences closely matching the goals (Figure 18).

CONCLUSION

IRAAS shows enormous potential for negotiating the changing spatial requirements of multiple users in a dynamic environment by learning to adapt itself in near real time. Rather than simply automating known construction patterns, this system begins to leverage the potential for intelligent robotic architecture. Through the agile development of the three components of our system and interrelational communication protocols between them, we have successfully demonstrated a semi-autonomous adaptive system.

While our physical prototypes successfully demonstrate simple reconfiguration behaviours, scaling up the system for building construction poses the next challenge requiring a more robust material and locking system along with stronger mechatronics. In future work, we intend to add sensing technology to the robotic system to increase its degree of self-awareness. Additionally, we are developing more sophisticated spatial adaptation behaviours with reinforcement learning with a focus on multi-agent collaboration, user feedback as a fitness criteria, and integration of structural stability prediction analysis. This model has the potential to disrupt and reorganize the connection between architecture, humans, and the environment, defining a new paradigm for a self-regulating living environment that continuously adapts itself in a dialogue with its users.

22 The section of project TESSERACT at a certain stage

ACKNOWLEDGMENTS

Living Architecture Lab, The Bartlett AD Research Cluster 3 (RC3), BPro, UCL

Tesseract Team: Jiaqi Wang, Wanzhu Jiang, Ying Lin, and Zongliang Yu

Studio Masters: Tyson Hosmer, Octavian Gheorghiu, Philipp Siedler

Machine Learning Tutor: Panagiotis Tigas

Technical Tutors: Ziming He and Baris Erdincer

Theory Tutor: Jordi Vivaldi Piera

REFERENCES

Adel, Arash, Andreas Thoma, Matthias Helmreich, Fabio Gramazio, and Matthias Kohler. 2018. "Design of Robotically Fabricated Timber Frame Structures." In *ACADIA 2018: Recalibration: On Imprecision and Infidelity; Proceedings of the 38th Annual Conference of the Association for Computer Aided Design in Architecture (ACADIA)*, edited by P. Anzalone, M. del Signore, and A. J. Wit. Mexico City, Mexico. 394–403.

Anderson, Carl, Carlo Bailey, Andrew Heumann, and Daniel Davis. 2018. "Augmented Space Planning: Using Procedural Generation to Automate Desk Layouts." *International Journal of Architectural Computing* 16 (2): 164–177.

Bier, Henriette, Alexander Liu Cheng, Sina Mostafavi, Ana Anton, and Serban Bodea. 2018. "Robotic Building as Integration of Design-to-Robotic-Production and -Operation." In *Robotic Building*, edited H. Bier. Cham, Switzerland: Springer. 97–119.

Bier, Henriette, and Sina Mostafavi. 2018. "Robotic Building as Physically Built Robotic Environments and Robotically Supported Building Process." In *Robotic Building*, edited H. Bier. Cham, Switzerland: Springer. 253–271.

Chaillou, Stanislas. 2019. "AI + Architecture: Towards a New Approach." Master's thesis, Harvard School of Design.

Chaillou, Stanislas. 2020. "ArchiGAN: Artificial Intelligence x Architecture." *Architectural Intelligence*. Singapore: Springer. 117–127.

Chaillou, Stanislas. 2019. "Archigan: a generative stack for apartment building design." *NVIDIA Developer* (blog), Jul 17, 2019. Accessed June 1, 2022. https://developer.nvidia.com/blog/archigan-generative-stack-apartment-building-design/.

Dijkstra, E.W. 1959. "A Note on Two Problems in Connexion with Graphs." *Numerische Mathematik* 1 (1): 269–271.

Dinçer, Ahmet Emre, Hakan Tong, and G. Çagdas. 2014. "A computational model for mass customized housing design bu using cellular automata." *A/Z ITU Journal of the Faculty of Architecture* 11 (2): 351–368.

23 A rendering of project TESSERACT

Goodfellow, Ian, Jean Pouget-Abadie, Mehdi Mirza, Bing Xu, David Warde-Farley, Sherjil Ozair, Aaron Courville, Yoshua Bengio. 2014. "Generative Adversarial Networks." *arXiv*:1406.2661.

Frazer, John. 1995. *An Evolutionary Architecture*. London: Architectural Association.

Hanna, Sean. 2020. "Architecture as Agent." In Home Architecture Agency, edited by Z. Brown, M. Denig, H. S. Kim, M. McClelland, and J. Peponis. Atlanta: Georgia Institute of Technology School of Architecture.

Hart, Peter E., Nils J. Nilsson, and Bertram Raphael. 1968. "A Formal Basis for the Heuristic Determination of Minimum Cost Paths." *IEEE Transactions on Systems Science and Cybernetics* 4 (2): 100–107.

Hosmer, Tyson, and Panagiotis Tigas. 2019. "Deep Reinforcement Learning for Autonomous Robotic Tensegrity (ART)." In *ACADIA '19: Ubiquity and Autonomy; Proceedings of the 39th Annual Conference of the Association for Computer Aided Design in Architecture (ACADIA)*, edited by K. Bieg, D. Briscoe, and C. Odom. Austin, TX: ACADIA. 16–29.

Hosmer, Tyson, Panagiotis Tigas, David Reeves, and Ziming He. 2020. "Spatial Assembly with Self-Play Reinforcement Learning." In A*CADIA '20: Distributed Proximities; Proceedings of the 40th Annual Conference of the Association for Computer Aided Design in Architecture (ACADIA)*, edited by B. Slocum, V. Ago, S. Doyle, A. Marcus, M. Yablonina, and M. del Campo. Online + Global: ACADIA. 382–393.

Hua, Hao. 2017. "A Bi-Directional Procedural Model for Architectural Design: A Bi-Directional Procedural Model." *Computer Graphics Forum* 36 (8): 219–231.

Huang, Weixin, and Zheng Hao. 2018. "Architectural drawings recognition and generation through machine learning." In *ACADIA '18: Recalibration: On Imprecision and Infidelity; Proceedings of the 38th Annual Conference of the Association for Computer Aided Design in Architecture (ACADIA)*, edited by P. Anzalone, M. del Signore, and A. J. Wit. Mexico City: ACADIA. 156–165.

Isola, Phillip, Jun-Yan Zhu, Tinghui Zhou, and Alexei A. Efros. 2017. "Image-to-Image Translation with Conditional Adversarial Networks." *arXiv*:1611.07004.

Jenett, Ben, and Kenneth Cheung. 2017. "Bill-e: Robotic platform for locomotion and manipulation of lightweight space structures." In *25th AIAA/AHS Adaptive Structures Conference*. 1876.

Kilian, Axel. 2018. "The Flexing Room Architectural Robot." In *ACADIA '18: Recalibration: On Imprecision and Infidelity; Proceedings of the 38th Annual Conference of the Association for Computer Aided Design in Architecture (ACADIA)*, edited by P. Anzalone, M. del Signore, and A. J. Wit. Mexico City: ACADIA. 232–241.

Landau, Royston. 1968. *New Directions in British Architecture*. New York: G. Braziller. 11.

Leder, Samuel, Ramon Weber, Dylan Wood, Oliver Bucklin, and Achim Menges. 2019. "Distributed robotic timber construction." In *ACADIA '19: Ubiquity and Autonomy; Proceedings of the 39th Annual Conference of the Association for Computer Aided Design in Architecture (ACADIA)*, edited by K. Bieg, D. Briscoe, and C. Odom. Austin, TX: ACADIA. 510–519.

Liggett, Robin S. 2000. "Automated facilities layout: past, present and future." *Automation in Construction* 9 (2): 197–215.

Maierhofer, Mathias, Valentina Soana, Maria Yablonina, Seiichi Suzuki, Axel Koerner, Jan Knippers, and Achim Menges. 2019. "Self-choreographing network: towards cyberphysical design and operation processes of adaptive and interactive bending-active systems." In *ACADIA '19: Ubiquity and Autonomy; Proceedings of the 39th Annual Conference of the Association for Computer Aided Design in Architecture (ACADIA)*, edited by K. Bieg, D. Briscoe, and C. Odom. Austin, TX: ACADIA. 654–663.

McCarthy, John, Marvin L. Minsky, Nathaniel Rochester, and Claude E. Shannon. 2006. "A Proposal for the Dartmouth Summer Research Project on Artificial Intelligence: August 31, 1955." *AI Magazine* 27 (4): 12–14.

Meyboom, AnnaLisa, and Dave Reeves. 2013. "Stigmergic Space." In *ACADIA 13: Adaptive Architecture; Proceedings of the 33rd Annual Conference of the Association for Computer Aided Design in Architecture (ACADIA)*, edited by P. Beesley, M. Stacey, and O. Khan. Toronto: ACADIA. 200–206.

Nagy, Danil, Damon Lau, John Locke, James Stoddart, Lorenzo Villaggi,

Ray Wang, Dale Zhao, and David Benjamin. 2017. "Project Discover: An Application of Generative Design for Architectural Space Planning." In *SIMAUD '17: Proceedings of the Symposium on Simulation for Architecture and Urban Design*. https://doi.org/10.22360/simaud.2017.simaud.007.

Negroponte, Nicholas. 1970. *The Architecture Machine*. Cambridge, MA: MIT Press.

Oosterhuis, Kas, et al. 2012. Hyperbody: First Decade of Interactive Architecture. Heijningen, the Netherlands: Jap Sam.Pask, Gordon. 1969. "The Architectural Relevance of Cybernetics." *Architectural Design* 39: 494–496.

Oxford English Dictionary. 1989. Oxford: Oxford University Press.

Pask, Gordon. 1965. "Comments on an indeterminacy that characterizes a self-organising system." *Cybernetics of Neural Processes, Consiglio Nazionale delle Richerche* (1965): 1–30.

Pask, Gordon. 1969. "The Architectural Relevance of Cybernetics." *Architectural Design* 7 (6): 494–496.

Petersen, Kirstin Hagelskjaer, Radhika Nagpal, and Justin K. Werfel. 2011. "Termes: An autonomous robotic system for three-dimensional collective construction." In *Robotics: Science and Systems VII*, edited by D. Hugh, N. Roy, P. Abbeel. Cambridge, MA: The MIT Press. 257–264.

Petersen, Kirstin, and Radhika Nagpal. 2017. "Complex Design by Simple Robots: A Collective Embodied Intelligence Approach to Construction." *Architectural Design* 87 (4): 44–49.

Rubenstein, Michael, Christian Ahler, and Radhika Nagpal. 2012. "Kilobot: A Low Cost Scalable Robot System for Collective Behaviors." In *2012 IEEE International Conference on Robotics and Automation*. 3293–3298.

Sutton, Richard S., and Andrew G. Barto. 2018. *Reinforcement Learning: An Introduction*. Cambridge, MA: The MIT Press.

Unity. Unity3D. V. 2020.3.16f1. Unity. Windows. 2020.

Veloso, Pedro, Rhee Jinmo, and Krishnamurti Ramesh. 2019. "Multiagent Space Planning; A Literature Review (2008-2017)." In *Hello, Culture! Proceedings of the 18th International Conference on Computer Aided Architectural Design Futures (CAAD Futures)*, edited by Ji-Hyun Lee. Daejeon, South Korea: CAAD. 130–152.

Veloso, Pedro, and Ramesh Krishnamurti. 2020. "An Academy of Spatial Agents-Generating spatial configurations with deep reinforcement learning." In *Proceedings of the 38nd eCAADe Conference*. Berlin: eCAADe. 191–200.

Wiener, Norbert. 1948. *Cybernetics: Or Control and Communication in the Animal and the Machine*. Cambridge, MA: The MIT Press.

Wright Steenson, Molly. 2010. "Artificial Intelligence, Architectural Intelligence: The Computer in Architecture, 1960-80," PhD. diss. Proposal, Princeton University.

IMAGE CREDITS

All images by the authors as well as Living Architecture Lab, The Bartlett BPro Research Cluster 3, Jiaqi Wang, Wanzhu Jiang, Ying Lin, and Zongliang Yu.

Tyson Hosmer is an architect, researcher, and software developer working at the intersection of design, computation, AI, and robotics. He is the Director of the Architectural Design masters program and a Lecturer at the Bartlett School of Architecture in London, where he directs the Living Architecture Lab (RC3). He is a Senior Associate Researcher with Zaha Hadid Architects, leading grant-funded research development of cognitive agent-based technologies and machine learning for design. His fifteen years of experience in practice include working for Asymptote Architecture, Kokkugia, AXI:OME, and serving as Research Director with Cecil Balmond Studio for over six years. Tyson was previously a Course Tutor with the AADRL for seven years and has been a visiting professor in several institutions internationally.

Jiaqi Wang is a PhD student and researcher at the South China University of Technology, focusing on artificial intelligence aided architectural design, generative design, and robotics. He received his master's degree from UCL in Bartlett BPro RC3, guided by Tyson Hosmer.

Wanzhu Jiang is a PhD student and researcher at the South China University of Technology, whose research direction is artificial intelligence and autonomous architecture. She received her master's degree from UCL in Bartlett BPro RC3 in 2021, supervised by Tyson Hosmer.

Ziming He is a designer and a software developer focusing on computational design, generative design, architectural visualization, and robotics. He is currently working as a lead designer in Zaha Hadid Architects and is a technical tutor with RC3, Bartlett School of Architecture, UCL.

Deep Relief

Integrating Convolutional Neural Networks & Industrial Robotic Hot-Wire Fabrication of Ruled Surfaces

Andrew Saunders
Weitzman School of Design,
University of Pennsylvania

Riley Studebaker
Weitzman School of Design,
University of Pennsylvania

Claire Eileen Moriarty
Weitzman School of Design,
University of Pennsylvania

1 Ruled surfaces distributed with a convolutional neural network

ABSTRACT

Deep Relief combines generative artificial neural network computational methods with industrial robotic manufacturing procedures to explore the potential of designing, fabricating, and expressing novel part to whole relationships at an architectural scale. More specifically, the relief tests the capacity of convolutional neural networks (CNNs) to generate a language of three-dimensional geometry to drive tool specific robotic fabrication at the intersection of painterly representation and non-authorial composition. While artificial neural networks are not as quick or agile as biological ones, heavy brute-force cloud computing allows them to continually role over information and slowly "learn" through non-parametric algorithms. By referencing main features of ruled surface geometry, rails and rulings, as image features, a VGG-19 CNN can redistribute pre-embedded construction logic indicative of robotic hot-wire fabrication routines in new and unanticipated hybridized feature sets. Unimagined, self-similar ruled surface feature clusters can then be cut from blocks of EPS foam by an ABB IRB4600 6-axis robot carrying a large-scale custom hot-wire cutter end effector. *Deep Relief* navigates real-world constraints while simultaneously operating within both the rigorous geometric construction logic of ruled surfaces and a unique twenty-first century meronomy enabled by artificial neural networks.

INTRODUCTION

To generate a three-dimensional relief from two-dimensional images generated by artificial neural network methods, *Deep Relief* develops through a transpositional methodology informed by an understanding of the illusive Baroque techniques of the 'painterly.' Historically, Heinrich Wölfflin situates the problem of creating illusion of three-dimensions and movement in two-dimensional painting as a unique characteristic of the Baroque. He describes the concept of painterly: "The painterly style, with its chiaroscuro, gives an illusion of physical relief, and different objects seem to project or recede in space" (Wölfflin 1864). Three-dimensional figures are mined and depicted from two-dimensional images by artificial neural networks in much the same way. Convolutional neural networks (CNNs) generate new images by "learning" and redistributing feature hierarchies. Through pretrained routines (including VGG-19), CNNs can identify recognizable three-dimensional features and redistribute them according to principles of light and shadow (Gatys, Ecker, and Bethge 2015).

The fabrication method for *Deep Relief* employs a hot-wire cutter attached to a 6-axis robotic arm. The process inherently produces ruled-surface geometry by sweeping a single line through space.

Due to the innate tectonic signature of the tool and how hyperbolic surfaces can produce gradients of light and shadow, the initial stage of design begins with analysis and explicit digital modeling of ruled hyperbolic precedent. The discipline of architecture provides ample examples of architects deeply invested in developing ruled surfaces including Antoni Gaudi and twentieth century reinforced concrete pioneers such as Miguel Fisac, Eduardo Torroja, Uegine Freyssinet, Le Corbusier and Iannis Xenakis, Robert Maillart (Saunders 2004). However, none are as extreme or intricate as the sculptures of Constructivist brothers Naum Gabo and Antoine Pevsner. For this reason, linear constructions by Naum Gabo were chosen to serve as primary feature hierarchies for CNN output offering the most potential to produce depth through light and shadow.

While there are overlapping qualities of Constructivism and the Baroque, including effects of dynamism and movement, the project borrows from different sources but is not aligned strictly to either historic paradigm as it quickly moves away from direct derivation. For example, in the design process, rather than an additive construction of the surfaces (wires and acrylic for Naum Gabo), the hot-wire process is used to produce a subtractive carving of the Gabo ruled surfaces from solid EPS foam blocks, which contradicts the Realistic Manifesto mandate: "We disown volume as a plastic form of space" (Gabo and Pevsner 1920). Regardless, the rendered two-dimensional image of the resultant carved foam figure

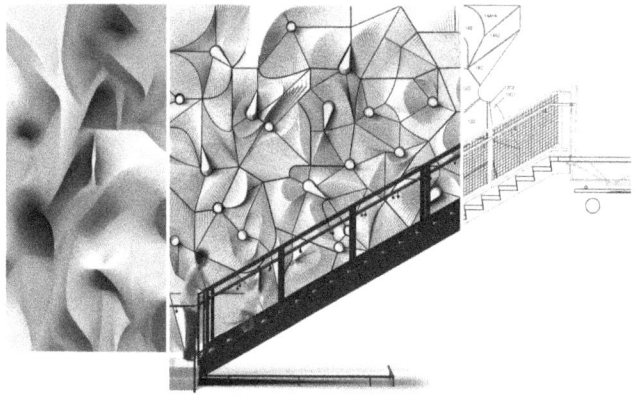

2 *Deep Relief* in site

3 *Deep Relief* process triptych

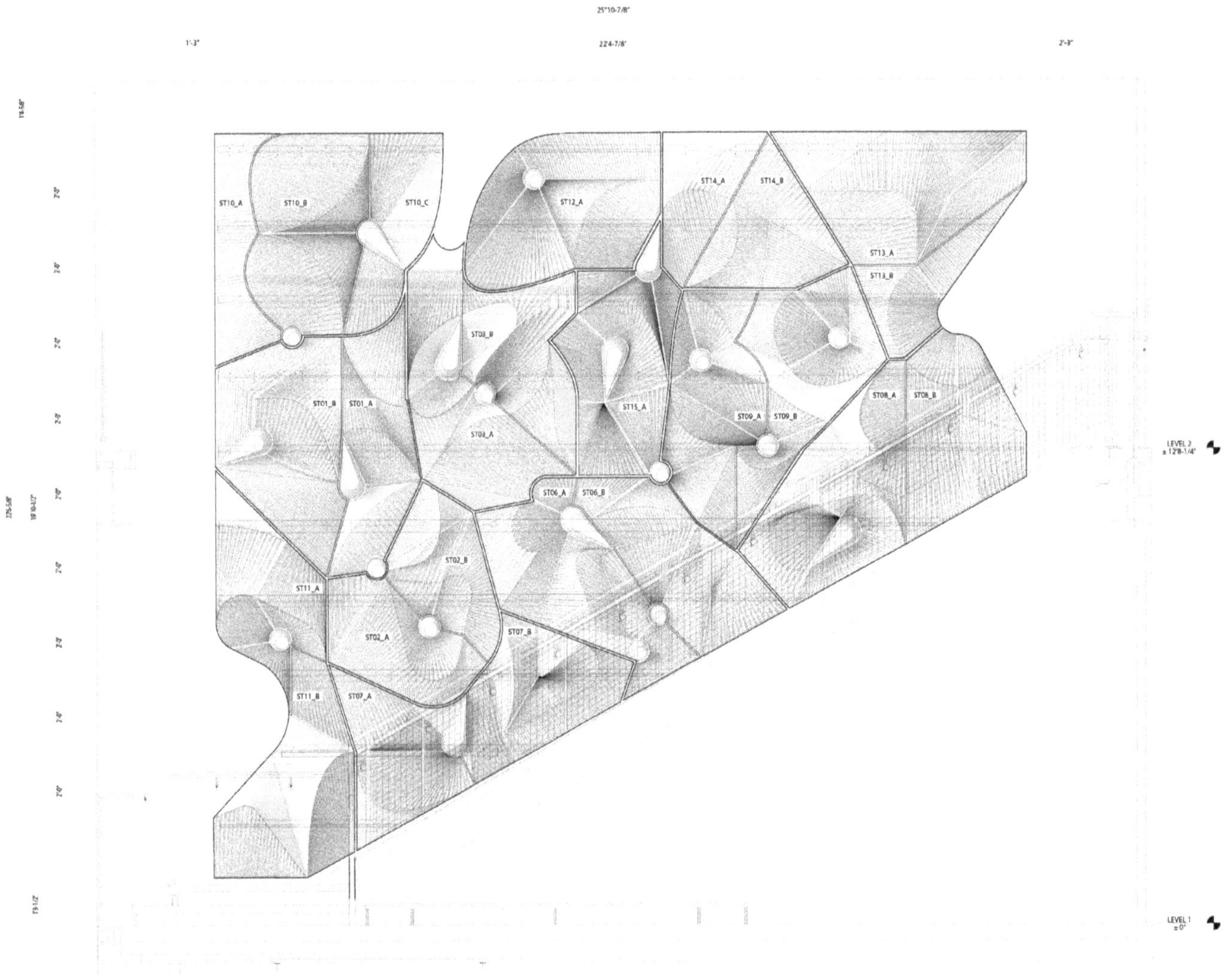

4   Deep Relief elevation

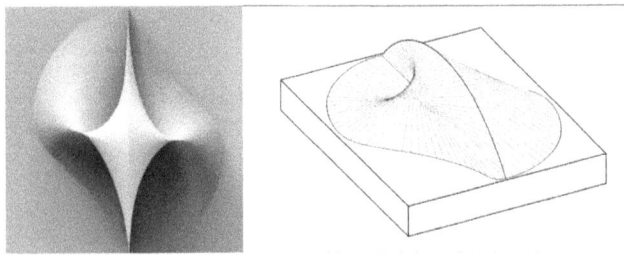

5   Two-dimensional light and shadow to three-dimensional geometric relationship

embeds three-dimensional traits of construction: the ruled geometry of the surface, the sweeping directionality, and the choreography of the robotic arm's motion. These specific preloaded fabrication traits play an important role in the later stage of design process. Due to their recognizability, they can be recovered three-dimensionally once the feature hierarchies have been redistributed two-dimensionally via CNN.

Through the synthesis of design and fabrication processes *Deep Relief* (Figures 1, 2, and 3) marks a new possibility for composition of formal and spatial aesthetics that embraces and champions contradictions and hybridizations (see theme of *ACADIA 2022, Hybrids & Haecceities*). This paper describes the process being used to imagine, fabricate, and realize a 900 square foot *Deep Relief* in Philadelphia (Figure 4).

STATE OF THE ART

This paper focuses on the intersection of two processes: images made by artificial neural networks, and robotic hot-wire fabrication. Following recent pioneering work with artificial neural networks by Ian Goodfellow in 2014 (Goodfellow et al. 2014), this paper uses a VGG-19 artificial neural network with TensorFlow libraries (Abadi et al. 2016) commonly used for CNN image generation and style transfers. This paper also built upon the existing field of robotic hot-wire fabrication. Using a hot-wire cutter as an end of arm tool, or EOATs is an established method of robotic fabrication,

6 Three-dimensional geometric signatures in a field produces with a convolutional neural network

and is being expanded into more easily accessible and designer-friendly applications (Pigram and McGee 2011). One of the most common uses for this tool and system is to quickly and efficiently remove large volumes of material (Feringa, Jelle, and Søndergaard 2014). Often this is the first part of a two-step process of subtractive fabrication, with a hot-wire cutter removing very large amounts of volume quickly, before a milling tool is used to make slower, more precise finishing passes (Brooks, Hadley, and Aitchison 2010). In applications like this paper's project where the finished surface is made by the hot-wire cutter, a larger amount of calibration and precision is needed in the tool and robot setup. Tuning the heat of the wire, linear and relative speed of the wire, and accuracy of the robot are all considerations taken in this paper and laid out in previous work addressing fine-tuning the robotic hot-wire cutting control for more precise surface finishes (Park et al.). Often in architecture, the hot-wire will be used to create surfaces as formwork for secondary fabrication methods. These secondary processes can use the formwork as molds (Feringa, Jelle, and Søndergaard 2014), or use the formwork as a volumetric substrate for additive processes such as 3D printing (Zaha Hadid Architects 2018). These examples include projects that have pushed the physical fabrication method of hot-wire cutting, such as projects that control the curvature of the wire to create more complex forms (Rust et al. 2016); this paper discusses taut hot-wires which is intrinsic to the geometric language. Other projects have explored the geometric possibilities of hot-wire cutting; this paper uses the natural and intrinsic results of using a hot-wire cutter (Bidgoli 2015). Because of the fragility of EPS foam there are fewer examples of the foam being the architectural building material, often in acoustic applications (Walker, James, and Foged 2018), leading to some of our considerations for material coating to allow the foam to withstand architectural wall applications.

METHODS

Isolating Specific Topological Traits

Beginning with the context that the project would be fabricated with a large hot-wire cutter, our geometric language originated from an understanding that ruled surfaces can be created by the line of the hot-wire sweeping through space. All ruled surfaces can be described as a lofting between two rails with the shape of the rails directly creating the geometry of the lofted surface, and the geometry of the lofted surface directly creating conditions of light and shadow. These conditions of light and shadow, when viewed in two-dimensions, directly show the visual signatures of the three-dimensional shape of the rails (Figure 5), allowing for a three-dimensional understanding of a two-dimensional image of light and shadow.

Feeding Specific Topological Traits into the Artificial Neural Network as Feature Hierarchies

Using this understanding of the two-dimensional to three-dimensional relationship of ruled surfaces, images of ruled surfaces were fed into a VGG-19 CNN. By supplying the neural network with these understood topological traits as feature hierarchies (Gatys, Ecker, and Bethge 2015), the images produced were wholly defamiliarized, but granularly composed of identifiable and familiar signatures embedded with three-dimensional geometric information (Figure 6).

Rebuilding Ruled Surface Geometry from Embedded Instructions

From the autonomously generated field, researchers reconstructed the embedded three-dimensional geometry using the same rulesets as when making the original input geometry. By having made the geometry that was fed into the neural network as latent space, the researchers were able to recognize the original ruleset and reconstruct a new geometry based on the embedded instructions in the images produced by the neural network. Though the input image consisted of ruled surfaces, the algorithm cannot completely contextualize that restraint, and will generate new imagery that seems similar given its understanding of light and shadow. Therefore, considerations for fabrication had to be taken into account during the three-dimensional reconstruction process. The field of geometric traits depicted in the two-dimensional

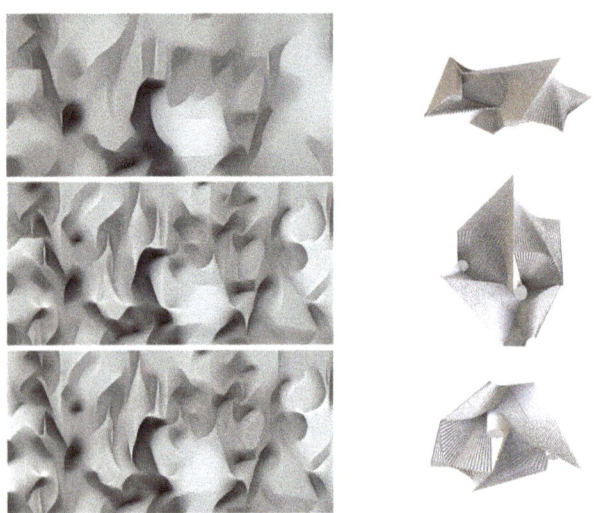

image is produced using software unaware of any real-life constraints; the images produced are without scale, materiality, or physical restrictions. The researchers, as discerning designers, utilize these considerations to create a more constrained and detailed final proposal with greater clarity of the initial topological features. In the process of reconstruction, the researchers also made detailed modifications to the surface condition, creating micro-reliefs on the large figural surfaces, writing the signature of the construction method in highlights and shadow lines in the surface along the hot-wire's path. Through these conscious decisions the researchers reclaim a degree of authorship over the non-authorial methodology.

Critical Explicit Evaluation of Part To Whole Geometry for Fabrication

After the global surface geometry has been generated, individual parts are evaluated and discretized based on constraints endemic to the hot-wire cutting process, namely geometric conditions, and stock material size. The two geometric conditions are intersections between surfaces at valley conditions, and intersections from the extension of isocurves between rails. The material size constraint is based on the size of EPS foam stock logistically available. Based on these constraints, the global geometry was rationalized into local regions with self visual clarity, or 'states' (Figure 7). Initial borders between all surfaces are placed along all valley conditions. Next, surfaces are divided into groups smaller than the stock material size. Finally, the groups are evaluated for self-intersection, and further fragmented if necessary. These final fragments are composed of one or more surfaces, and are colloquially referred to as 'counties.'

Custom Fabrication Environment for Bespoke Part Production

For the fabrication of a set of highly variable and bespoke elements, a robotic fabrication environment was designed to allow for maximum flexibility within one 'cut,' as well as among a sequence of cuts. Each face of each 'county' was programmed as a distinct and continuous 'cut'—with the nature of the stereometric process not allowing for any change of the environment during the cut. Therefore, each face's geometry was limited by the range that the robot could reach within one static robot environment. Increasing the range was considered a design challenge, with success criteria being measured in maximizing the range of the robot. Constraints include the large size of the block and size of the tool relative to the robot, extreme parabolic cuts with undercuts causing large range of joint movement, the need to access all sides of the block, and collisions between the tool, the environment, the stock foam block, and the robot. The range of cuttable geometry was maximized through a series of physical considerations in the setup of the robotic

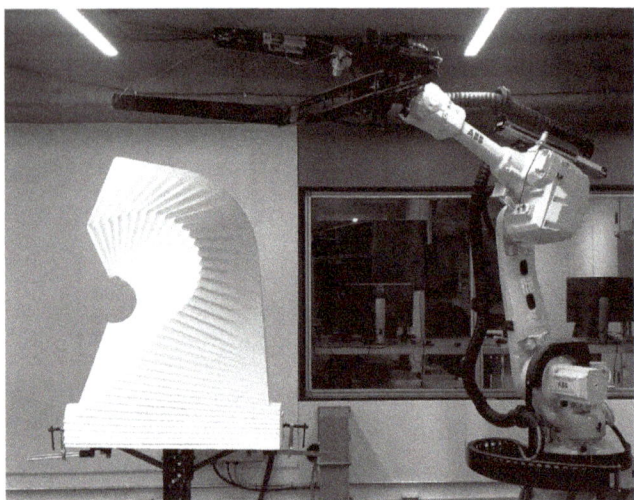

7   Self-similar surfaces discretized into 'states' and 'counties' for production
8   Completed county on custom 7th axis jig

9   Partial arrangement of wall relief

environment. The flexibility of the hot-wire cutter's position in relation to the stock material is critical to the project's bespoke geometric nature, and therefore, a controllable 7th axis facilitated the most substantial increase in geometric complexity (Figure 8). This 7th axis was introduced to this specific system as a custom jig to minimize environmental collisions and maximize block orientation options. The jig consists of an elevated platform that can be rotated around its vertical axis. The foam block is fastened with the majority of the block's mass cantilevered away from the jig to maximize the area in which the hot-wire cutter can pass below the bottom face. This 7th axis allowed for a single controllable variable (rotation around the vertical axis) to greatly increase the robotic 'range.' This was specifically useful for single extreme hyperbolic cuts, as well as sequencing individual cuts on multiple sides of the block (Figures 9 and 10).

Custom Script for Bespoke Part Analysis and Fabrication

A highly adaptable but generally standardized Grasshopper script was crucial to efficient production of code for robotic fabrication. By deconstructing the process and geometric language we were able to streamline the simulation and production process while allowing for customization in response to bespoke fabrication conditions.

The volume and construction rails of each element or 'county' are input into the custom Grasshopper script and automatically positioned in the build space. A visualization of the hot-wire's cut path is automatically generated to help the researcher

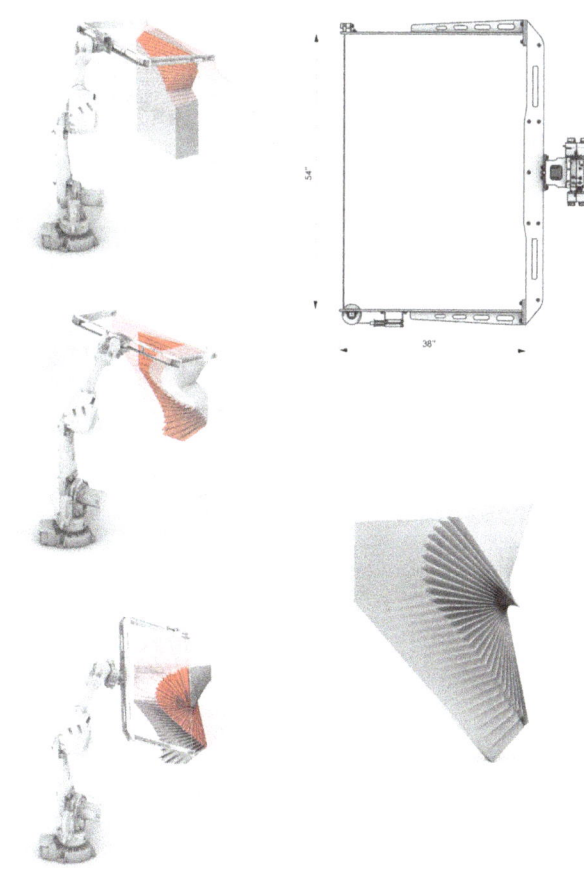

10   Single 'county' from sequence of three articulation cuts with associated robotic movement

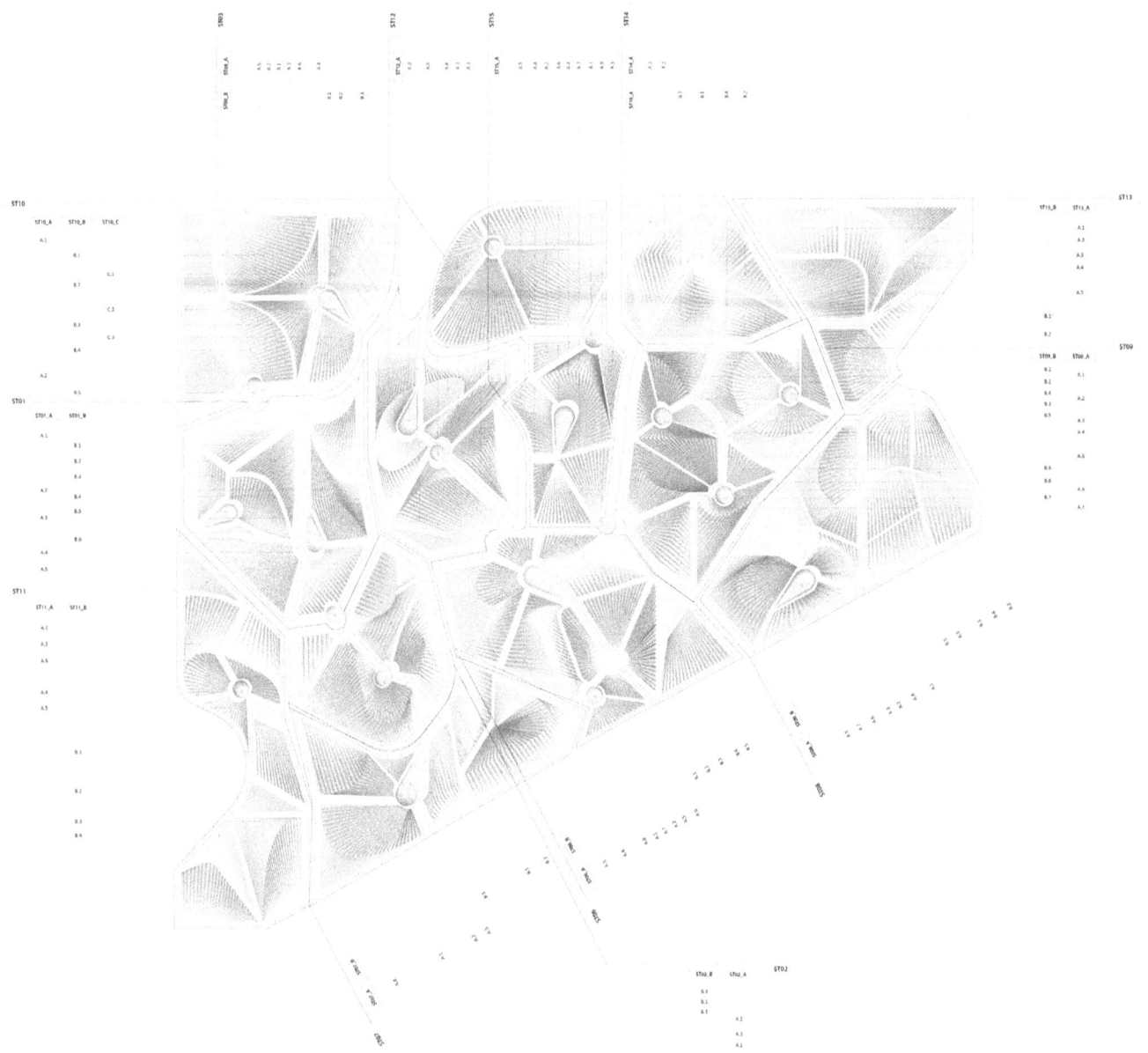

11 120 'counties' composing the wall relief system

identify possible environmental collisions and strategically modify the geometry's location and orientation inside of the stock foam block.

While the bespoke surfaces of the counties are highly variable, the script allows for the general categorization of each surface as one of four standardized types of stereometric action, allowing for efficient production of robotic code. The two-dimensional boarder of each 'county' is used to automatically generate several standardized cuts: a first 'rough cut' to remove extra material around the intended geometry, a 'back cut' to control the depth of the piece, and a 'profile cut' of the exact shape of the county. The 4th type of cut generally creating most of the figure and depth is the 'articulation cut,' using the three-dimensional construction rails originally developed during the modeling of the ruled surface geometry. Each 'county' has one or more articulation surfaces, with some articulation cuts have completely smooth surface sweeps while others display a tooled pattern generated through modifications on one or both of the construction rails (Figures 11 and 12). All of these conditions are processed by the same custom Grasshopper script for an efficient and largely automated production of robot code. Because of the highly variable nature of the surfaces within the relief wall, the script was specifically designed to facilitate bespoke adaptations of the general script in response to unique conditions.

From the boundary and construction lines the path of the hot-wire cutter and robotic program can be extrapolated. One of the harder things at the beginning of our process, and

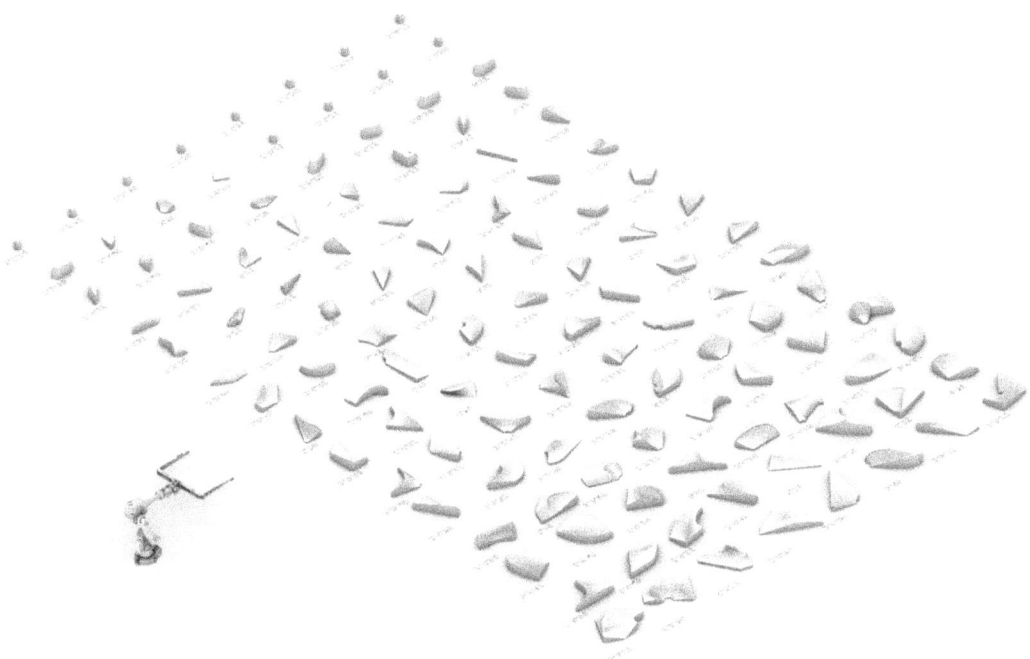

12 120 'counties' composing the wall relief system arranged by scale

commonly in robotic hot-wire fabrication is robotic collisions and out of reach errors. At the beginning of our process development, one of the most time-consuming aspects of the robotic program generation was modifying the target planes to avoid robotic arm collisions and out-of-reach errors during fabrication. Using the Visose Grasshopper plug-in, the robot's movement can be simulated from target planes and predict positioning errors. The highly varied and large sweeps required to cut the relief surfaces lead to a narrow range of successful positioning options. We were able to greatly reduce simulation time by generating planes based on the orientation of a vector between the cut point and robot base, with parametric control built in for unique and more extreme conditions. The precise rotation of the custom jig's 7th axis can also be controlled during the simulation process to avoid out of reach errors. From this custom script, an efficiency and automation in unique part fabrication was achieved.

RESULTS AND DISCUSSION

Currently, there are many architects exploring the role of artificial neural networks in design. Few at this stage are attempting to integrate robotic fabrication at the architectural scale. Since many of the convolutional neural network applications reside in the realm of two- dimensional pixel reading and redistribution, architects are faced with the classic problem of transposing two-dimensional images into three-dimensions. Direct computational automation of this transposition using image height translations, extrusions, and voxels to name a few common techniques result in a non-discriminate and reductive 2.5D figuration and cannot reconstruct content geometry of initial feature hierarchies embedded in source images. One possible trajectory of this research would be the development of a more automated transposition of raster-based two-dimensional images that can recognize and depict geometric feature hierarchies to translate.

Within the defined scope of *Deep Relief* as it exists, more serious consideration would be given to the waste produced by a subtractive fabrication process. Currently, offcut pieces are being used for packaging and transportation as well as recycled for additional academic projects in the lab. In addition, labor can and should be drastically reduced having established and streamlined a more efficient workflow through the learning process of completing an architecture scale project based on the original hypothesis of merging artificial neural network processes of design with robotic fabrication. Further integration and pairing of unique robotic tool capabilities would be ideal. The feedback loop between all three methods, brute-force computational feature distribution, three-dimensional geometric transposition, and real-world constraints of fabrication at an architectural scale make *Deep Relief* a cohesive and novel step in the field of artificial neural networks and architectural application.

CONCLUSION

The success of *Deep Relief* lies in expressing the potential of novel part to whole relationships in an architecture that are motivated by the hybridization and organizational capacity of artificial neural networks with disciplined geometric, material and fabrication constraints (Figures 13 and 14). In this

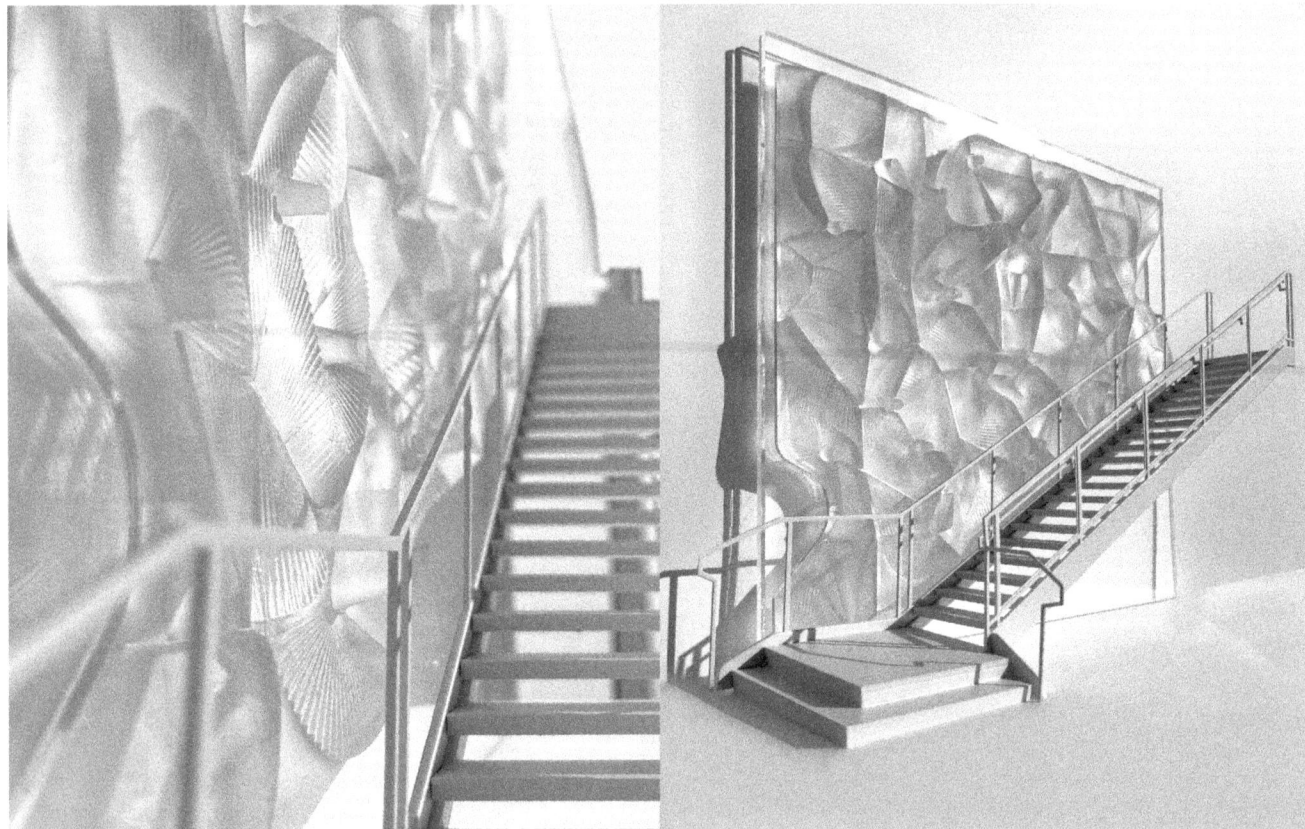

13 Resin assembly model of *Deep Relief* showing states with tooling

context, the part to whole "problem" goes beyond assembly and marks new possibilities for composition of formal and spatial aesthetics moving away from previous paradigms including mass standardization or the more recent structural lattice of parametricism (Thenhaus 2015). The project is a real-world application that serves as a vehicle for further exploration in scale and scope. Such procedures can be expanded to inform approaches to building facades, landscapes and even urban design. As a result, *Deep Relief* embodies a twenty-first century meronomy that is not (yet) recognizable as an episode in the long catalog of part-to-whole relationships expressing paradigmatic turns in the industry of architecture and the construction of its discourse.

ACKNOWLEDGMENTS

Project Director: Associate Professor Andrew Saunders
Robotic Research Assistants: Penn Praxis Fellow Riley Studebaker, Penn Praxis Fellow Mathew White, and Claire Eileen Moriarty
Research Assistants: Penn Praxis Fellow Caleb Ehly, Penn Praxis Fellow Benjamin Hergert, Yujie Li, Jesse Allen, Macarena De La Piedra, and Cecily Nishimura

Support by:
The Stuart Weitzman School of Design
Dean Frederick Steiner and Architecture Chair Winka Dubbeldam
The ARI Lab and Weitzman MSD-RAS Program

Robert Stuart-Smith
Penn Praxis
Erdy McHenry Architecture
Scott Erdy, Drew Kmetz, Nathan Barlett, and Elena Mangigian
Universal Foam
L. J. Paollela Construction
Middletown Library & Community Center

REFERENCES

Abadi, Martín, Paul Barham, Jianmin Chen, Zhifeng Chen, Andy Davis, Jeffrey Dean, Matthieu Devin et al. 2016. "{TensorFlow}: A System for {Large-Scale} Machine Learning." In *12th USENIX Symposium on Operating Systems Design and Implementation (OSDI 16)*. Savannah, GA. 265–283.

Brooks, Hadley, and David Aitchison. 2010. "A review of state-of-the-art large-sized foam cutting rapid prototyping and manufacturing technologies." *Rapid Prototyping Journal* 16 (5): 318–327.

Feringa, Jelle, and Asbjørn Søndergaard. 2014. "Fabricating architectural volume: Stereotomic investigations in robotic craft." In *Fabricate: Negotiating Design & Making 2*, edited by F. Gramazio, M. Kohler, and S. Langenberg. London: UCL Press. 76–83.

Gabo, Naum, and Antoine Pevsner. 1920. "Realistic Manifesto (Realisticheskii Manifest)." Accessed August 22, 2022. https://www.moma.org/collection/works/173291.

Gatys, Leon A., Alexander S. Ecker, and Matthias Bethge. 2015. "A neural algorithm of artistic style." *arXiv* preprint arXiv:1508.06576.

Goodfellow, Ian, Jean Pouget-Abadie, Mehdi Mirza, Bing Xu, David Warde-Farley, Sherjil Ozair, Aaron Courville, and Yoshua Bengio. 2014. "Generative adversarial nets." *Advances in Neural Information Processing Systems 27 (NIPS 2014)*, edited by Z. Ghahramani, M. Welling, C. Cortes, N. Lawrence, and K.Q. Weinberger.

Pigram, David, and Wes McGee. 2011. "Formation Embedded Design: A methodology for the integration of fabrication constraints into architectural design." In *ACADIA 2011: Integration through Computation; Proceedings of the 31st Annual Conference of the Association for Computer Aided Design in Architecture (ACADIA)*. Banff (Alberta): ACADIA. 122–131.

Rust, Romana, David Jenny, Fabio Gramazio, and Matthias Kohler. 2016. "Spatial Wire Cutting: Cooperative robotic cutting of non-ruled surface geometries for bespoke building components." In *CAADRIA 2016; Living Systems and Micro-Utopias*. Melbourne, Australia. 529–538.

Saunders, Andrew. 2004. "Early Twentieth Century Pioneers in Reinforced Concrete." SOM Traveling Fellowship. Accessed August 22, 2022. https://somfoundation.com/fellow/andrew-saunders/

Thenhaus, Clark, 2015. "Part Problems / Problem Parts." In *103rd ACSA Annual Meeting Proceedings*. Toronto, Canada. 614–621.

Walker, James, and Isak Worre Foged. 2018 "Robotic Methods in Acoustics-Analysis and Fabrication Processes of Sound Scattering Acoustic Panels." In *Proceedings of the 36th eCAADe Conference 2018*. Lodz, Poland. 835–840.

Wölfflin, Heinrich. 1964. *Renaissance and Baroque [Renaissance und Barock]; Fontana Library*, vol. 873L. London: Collins. 31.

Zaha Hadid Architects. 2018. "Thallus for White in the City." *Zaha Hadid Architects* (web site), April 5, 2017. Accessed August 27, 2022. https://www.zaha-hadid.com/2017/04/05/thallus-for-white-in-the-city/.

IMAGE CREDITS

All drawings and images by the authors.

Andrew Saunders is an Associate Professor and Director of the Master of Architecture Program of Architecture at the University of Pennsylvania Stuart Weitzman School of Design and founding principal of Andrew Saunders Architecture + Design, an internationally published, award winning architecture, design, and research practice committed to the tailoring of innovative digital methodologies to provoke novel exchange and reassessment of the broader cultural context. The practice innovates at a number of scales ranging from product design, exhibition design, and residential and large-scale civic and cultural institutional design.

14 *Deep Relief* 'State 1' at site for install

Riley Studebaker is the Manager of the Advanced Research and Innovation Lab at the Weitzman School of Design, as well as a Research Associate with the AML, a Guest Lecturer with the Weitzman School of Design, and a Praxis Fellow of the University. Riley graduated with the inaugural class of the Weitzman School of Design's Master of Science in Design in Robotics and Autonomous Systems, and holds a BArch from the College of Architecture and Urban Planning at Virginia Tech.

Claire Eileen Moriarty holds a Master of Science in Design in Robotics & Autonomous Systems from the Weitzman School of Design at the University of Pennsylvania, as well as a BFA in Architectural Design from the Maryland Institute College of Art. Claire has previously been a designer at Jenny Sabin Studio, and currently works with Treeswift, a drone-mapping startup at Pennovation Works, as well as being a Guest Lecturer with the Weitzman's MSD-RAS program and a Research Fellow of the University.

Towards an Adversarial Architecture

Antonio Furgiuele
Wentworth Institute
of Technology

Dr. Mehmet Ergezer
Wentworth Institute
of Technology

Dr. Cagri Hakan Zaman
Massachusetts Institute
of Technology

1A A series of everyday objects, served as the elements of research, that were transformed to become adversarial

ABSTRACT

A key technological weakness of artificial intelligence (AI) is adversarial images, a constructed form of image-noise added to an image that can manipulate machine learning algorithms but is imperceptible to humans. Over the past years, we developed *Adversarial Architecture*, a scalable systems approach to design adversarial surfaces, for physical objects, to manipulate machine-learning algorithms. *Adversarial Architecture* explores the application of adversarial images to the built environment and develops a new method of design agency to directly engage artificial intelligence. Embedding a layer of information to physical surfaces that is only perceptible to machines has many potential applications, such as uniquely identifying and tracking objects, embedding accessibility features directly to surfaces, and counter-surveillance systems in different scales. To construct an adversarial architecture, a series of objects were selected: a cup, a banana, and a table. These everyday objects were chosen because their geometric, material, and spectral differences offer complex characteristics repeatedly found in the built environment. To transform the built environment to become adversarial, a workflow was developed: scanning existing objects, unrolling objects, generating adversarial noise, printing noise onto surfaces, wrapping objects with noise, testing objects, and reception of feedback from neural networks. The research offers a new framework of design possibilities to construct and apply adversarial surfaces to control how the built environment is captured by computer vision, understood by machine learning, and acted upon by neural networks. While AI has and will continue to profoundly affect the built environment, the future of design can critically inform AI.

INTRODUCTION

Becoming Adversarial

So many images and so many types: images that are originals, images that are meant to fool you, images designed to authenticate, images that train algorithms and help artificial intelligence (AI) compute faster and with more certainty.[1] We are entering a condition where the technologies of artificial intelligence have become ubiquitous and with that has come a growing necessity to use images to train machine learning algorithms.[2] This has rendered new forms of power through algorithmic decision-making and control by various actors and their networks, such as locating objects and people in urban settings (Figure 1B).[3]

A key technological weakness of neural network models is adversarial images, which are constructed forms of image-noise added to an image that is only seen by computer vision. They are constructed from layers of information, typically placed on images, and can manipulate machine learning algorithms and AI systems. Because they can add and manipulate information in a predictable way, adversarial images have the potential to be integrated into the built environment as patterns on everyday objects. This would allow the design of the built environment to uniquely manage how it is captured as data, understood as information, and acted upon by AI.

Over the past year, we developed an *Adversarial Architecture*, a scalable systems approach on how the design of surfaces, for physical objects, can affect artificial intelligence. Our research introduces a complete workflow of making physical objects adversarial: (1) digitizing 3D objects and structures, (2) unraveling objects, (3) generating adversarial information specific to the scanned form, and (4) fabricating and applying adversarial surfaces. In our current study, we treat adversarial information as a surface feature that can be applied to existing objects and structures to manage and control machine learning algorithms. However, our approach can also be used in digital fabrication processes where the adversarial information is embedded directly to the fabricated object. *Adversarial Architecture* empowers designers by allowing them to embed visually imperceptible information in their designs—leading to seamless integration of digital infrastructures to the built environment, to add a new program, to manage and control (non-human) AI systems.[4]

To study and illustrate adversarial architecture, a series of objects were selected: a cup, a banana, and a table. These everyday objects were chosen because their geometric, material, and spectral differences offer up deceptively complex characteristics that are found in the built environment. The project developed a feedback loop to transform these select objects into adversarial objects, and with this construct the elements of an adversarial architecture.

1B AI processing an urban environment in Hamburg, Germany, using the You Only Look Once (YOLO) model (Redmon et al. 2016). YOLOv3 is a favored object detection algorithm due to its accuracy and speed (Redmon and Farhadi 2018).

A HISTORICAL REVIEW

Artificial intelligence, and its subfield computer vision, have been finding increasing success in their performance and applications owing to advances in deep learning. Adversarial machine learning and its application to image misclassification were proposed in 2013, where a perturbation that was imperceptible to humans was shown to change the predicted object category of an image (Szegedy et al. 2013). An adversarial misclassification is shown in Figure 2. It evidences the limits and weaknesses of a series of technologies. First, it shows the limits of machine vision by only targeting the image frequencies embedded within image noise. Second, it shows how classification algorithms use images to train machine learning that can also be used to mislead machine learning models. Adversarial images make legible the way confidence, and machine certainty are constructed through recognition, classification, and neural network exchanges.

Deep neural networks (DNNs) are arguably the most popular algorithms used in artificial intelligence, especially in the contemporary field of computer vision. The ease with which they can be fooled by adversarial images evidences a series of compound vulnerabilities and opens new forms of a design agency to manage and control these technologies, all of which are essential to digital environments.[5] Since its development, there have been various types of adversarial imaging methods and attack types.[6] From these early precedents, there have also been physical world attacks on deep learning models (Kurakin et al. 2016).

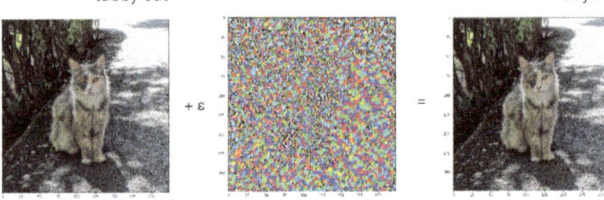

2   Missy becomes a coyote: leftmost is the original image of a stray cat from Istanbul, Turkey, accurately classified as a "tabby cat" by MobileNetV2 (Sandler et al. 2018). Rightmost is the adversarial image where a perturbation that is imperceptible to humans is added that causes the AI to misclassify the cat as a "coyote." The middle image is the perturbation noise, scaled up for visibility, that is calculated to break the prediction algorithm. "ε" indicates a small positive number employed to scale this noise. For instance, using the same noise with a slightly larger "ε" causes the AI to change its prediction yet again to a "timber wolf."

The increased access to artificial intelligence has become a key force multiplier, increasing both the scale and types of processes possible.[7] Recent years have been a testament to the various ways actors, both humans and non-humans, have tried to find forms of power in images through authentication, manipulation, and weaponization. The moment an image is captured and recognized, processed, and classified by artificial intelligence systems, is also a potential moment of power and control by various actors and their networks. Within this discourse of the power of images, a key sequence circumvents this power structure, which is the creation of new objects that are informatic and use adversarial logic. These objects have control over their future ability to be converted into data and information through the design of their surface(s). These adversarial surfaces control auto-recognition (computer vision) and allow objects to have a different relationship with how they are captured and classified, and how network relationships are configured. This is an important restructuring of relationships in that adversarial surfaces allow for objects to be active agents in controlling how computer vision, machine learning, and neural networks operate on them.

Architecture composes an essential set of boundaries within the built environment, a key construct to help measure, frame, and influence human and environmental actions and interactions. The project actively modifies and expands upon this defining architectural characteristic to engage digital actions and interactions. It allows adversarial architectural surfaces to add a new program, to manage and control (non-human) AI systems.

The design of adversarial images on objects allows for an emergent design agenda. What would you like to communicate to machines and how does that differ from humans? Adversarial images allow the built environment to be an active agent engaging the way AI systems capture, process, and act upon it. The adversarial prefigures control and can be both a form of attack on existing systems, a method to manage future ones, and for the identification and tracking of existing objects.[8]

METHODS

To best understand and develop a technical control of an adversarial architecture, a series of objects were selected: a cup, a banana, and a table (Figure 3). They were chosen because they hold complex individual and interrelationships, cloaked within their everyday ubiquity. The cup has manifold curves, a reflective and even surface, and is dimensionally similar to other objects because of their relationship to the human hand. A banana has small surface irregularities. A wooden table's size and form make it difficult to see in its entirety with a top and underside that create pronounced shadows.

To sequence the project, a workflow was designed to move between digital and analog forms of production. It was necessary to synthesize a variety of disciplinary knowledge and skills from architecture and computer science, and it necessitated the following phases: scanning existing objects (Figure 4),[9] elegantly unrolling/developing the objects (Figure 5),[10] generation of adversarial noise (Figure 6), printing the noise onto surfaces (Figure 7), wrapping objects with noise (Figure 8),[11] testing objects, and reception of feedback from neural networks (Figure 9).[12]

Amongst the various phases of the project workflow, the generation and testing of the adversarial layer presented the highest degree of computational complexity and energy. The adversarial objects surfaces, while seemingly simple to humans, are encoded with noise information which allows it to perform and render recognizable other objects, other than what it is (Figure 10). To begin this stage of the project, a survey of the most popular image classification algorithm models was conducted (eg. VGG, Google's Inception, Xception, and MobileNetV2). To allow for increased openness, accessibility, and transparency, we selected models that were trained on the ImageNet dataset.

3   Images of the select objects: cup, banana, table; each object presents a challenge for digitization, surface development, and visual recognition

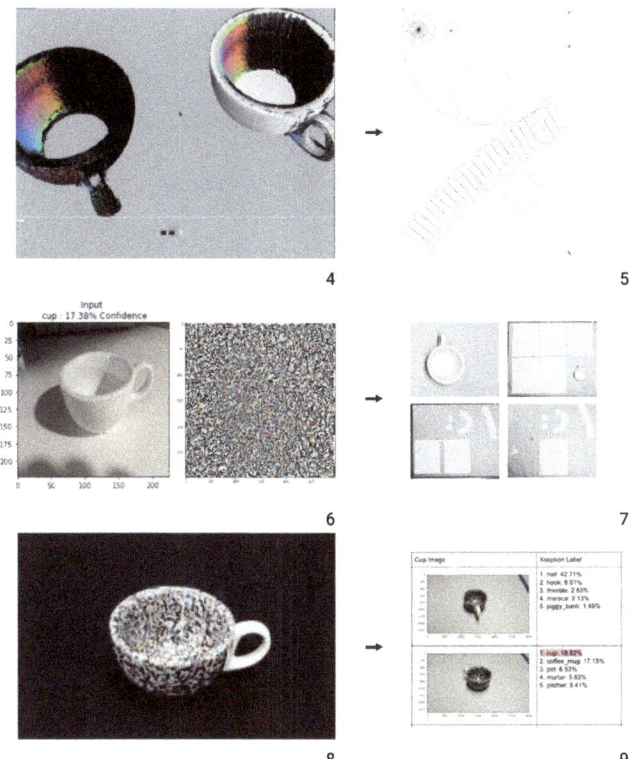
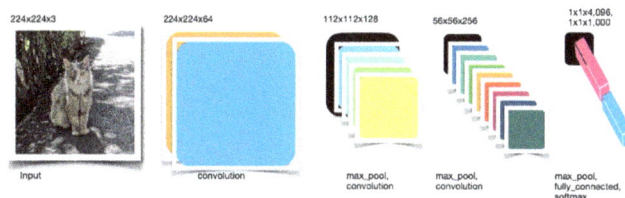

FIGURES 4 - 9 WORK FLOW:

(4) Scanning: Both laser scanning and photogrammetry used to capture the objects; (5) Unraveling: Scanned objects were re-meshed and unraveled into templates; (6) Generation of adversarial image: A neural network trained to produce noise to manipulate the original image; (7) Application of adversarial images to the template; (8) Adversarial Object; (9) Classification results by various neural networks

These visual classification algorithms are commonly used to recognize objects within images, powered by convolutional neural networks (CNN). Each network is composed of convolutional layers with various weights, and various coefficients. Each image is multiplied, cyclically, and edited many times, through each layer, until the network creates a prediction (Figure 11). The classification results of our unaltered three objects by the deep learning models Inception (GoogLeNet 2014), Xception (Google 2017), and MobileNetV2 (Google 2018) are presented in Figure 12a.

10 Experimental setup: objects with and without adversarial noise were placed in the same environment

11 Diagram of a convolutional network: The input shape is affected while going through a network. At each layer, while the shape of the image is getting deeper, its width and height are halved. The image starts with a shape of 224 by 224 by 3 as width, height, and depth, respectively, and the output of the network is 1 by 1 in width and height, with a depth in thousands.

The adversarial noise breaks this network, through its ability to work on this sequence in reverse. We selected the fast gradient sign method (FGSM), largely recognized in the computer science discipline as the most direct and functional method to construct adversarial images (Figure 11) (Goodfellow et al. 2014). The FGSM method was created for generating adversarial images and operates on the gradient of the classification model's loss function. As it moves through each of the layers, it monitors the change in gradient weights and steps in the opposite direction as it creates perturbations that maximize a loss for each pixel value. The collective set of perturbations creates a layer of noise that is an extraction of the image frequencies recognized by machine vision.

RESULTS
Adversarial Objects

The three objects, the cup, banana, and table went through a similar process of becoming adversarial. They yielded different results, which helped the research team to understand which parts of the process were affecting the strength of the adversarial attacks.[13]

We first employed the MobileNetV2 classifier on the photo of the original cup and confirmed that it was classified as a cup as shown in Figure 12a. We then generated a small-scale adversarial noise via FGSM, ε of 0.01, added this perturbation to the image, and (mis)classified this adversarial image as shown in Figure 12b. By increasing the noise scale, ε to 0.05, we transformed the prediction to be a potter's wheel; we ran the test one additional time and increased ε to 0.1, and rotated the image, and it classified the image as a toilet. Thus, we exhibited how a correctly classified object could be misclassified and the effects of the adversarial noise scaling parameter, ε.

Slight changes in the image angle produced slightly varying degrees of confidence. We photographed the cup at 35-degree intervals and then produced adversarial layers from 10 vantage points to maximize the ability of it being recognized and (mis)classified in the round (Figure 13).

12A Classification results of our objects pre-adversarial attacks by the three neural networks, Xception, Incepton V3, and MobileNet

12B Classification results of the adversarial cup, banana, and table by three neural networks

We then took images of the cup wrapped with the adversarial image and then processed it again through MobileNet, from varying angles, and were able to fool the neural network into believing it was a necklace (27.05% confidence), letter opener (18.80% confidence), bottle cap (62.95% confidence), and bolo-tie (10.89 % confidence). To compare the results to other neural networks we then processed the same images in Google's Xception, and Inception, and compared them to our results via MobileNet (Figure 12b).

The adversarial banana went through a similar process. In this example, the adversarial layer generated was a targeted attack.[14] The adversarial image was encoded with a noise image of a wall clock. The results of the adversarial attack on the banana were largely unsuccessful, in contradistinction to the other objects. Inception and Xception both classified the object, from various distances; from three feet away as a slug (99.5%, 89.36%), and six inches away as an eel (6.79%) (Figure 12b). This proves that the adversarial layer did not perform, as the encoded adversarial wall clock was not recognized. This failed example of an adversarial attack helped the research team locate essential variables for all future adversarial objects: first, the noise was printed on transparent film, and the yellow of the banana prevented the image contrast to register and the noise from being detected by computer vision. Secondly, and perhaps more importantly, because of the complexity of the multiple dimensional curvatures of the surface of the banana's form, it was developed (unrolled) into complex series of slices, approximately one-quarter inch in width and nine inches in length. The number of slices distributed the necessary continuity of the adversarial noise, preventing it from recognizing the embedded noise pattern of the wall clock.

The adversarial table went through a similar process as that of the cup and banana. In this example, the adversarial layer generated was a targeted attack. The adversarial image was encoded with a noise image of a toilet tissue roll. A toilet tissue roll was recognized with a high degree of confidence by both Xception (12.59%) and MobileNet (56%) (Figure 12b). The success of the table, helped us understand that the scale and continuity of the adversarial noise, was essential to successful recognition.

To facilitate the project, the computation team created a website to review the classification results from all other neural networks (VGG, Google's Inception, Xception, and MobileNet). The computational team's specialized knowledge, skills, and forms of production required the development of select tools, the website, for interdisciplinary engagement. The website was a key for the team's interaction and project development.

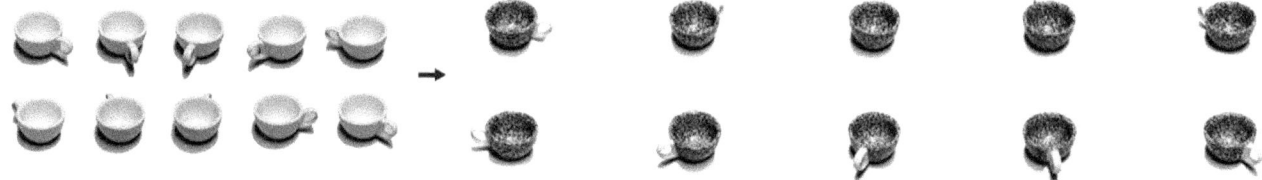

13 An array of adversarial images: multiple vantages are processed to allow adversarial strength from multiple angles

To compound the strength of adversarial architecture, we incorporated unique spatiotemporal characteristics. Two key spatial factors, depth, and angle are elements that two-dimensional images do not have as "live" contingent variables.[15] One study to allow adversarial images to operate on the depth of field used scale invariance (Figure 14):[16] it aggregates a series of nested interrelated scales to allow it to be understood from a variety of focal distances and angles, and to operate on depth of field. This ability to have multiple scaled adversarial imagery on an object, (mis)classifiable from various distances, allows the adversarial object to have a series of strategic attack advantages (Figure 15).[17] The two most operative advantages include, "identity differences" and "information overload." The first attack advantage, identity differences, allows for many scaled and nested adversarial images on a single object to present itself with shifting differences. As computer vision changes its proximity and position to the adversarial object it focuses on different scales of adversarial noise, this in turn (mis)classifies it based on changes in spatial position. This strategy allows for the design of multiple programmable identities that would fluctuate based on the depth of field and point of view. The second attack advantage, information overload, is the ability to present information excess, a system overload, which allows objects to produce incomprehensible results, initiate computational glitches, and crashes, and can become weaponized.

The development of an adversarial architecture is now taking on its own reproducibility. The project is currently continuing its successes by utilizing the workflow and methods designed to transform various objects to become adversarial while continuing to intensify the adversarial attack strength through developing specific unique spatiotemporal properties. As we scan existing objects, elegantly unroll objects, generate more adversarial noise, print noise onto surfaces, wrap objects with noise, test objects, and receive feedback from neural networks.

CONCLUSION
A Temporary Ending in the Form of a Future Agenda

The research highlighted an emergent agenda on the process, organizational strategies, and project scenarios.

ON PROCESS

In the development of a process to transform the built environment to become adversarial, a workflow was developed that privileges how to transform existing objects: scanning existing objects, unrolling/developing objects, generating adversarial noise, printing noise onto surfaces, wrapping objects with noise, testing objects, and reception of feedback from neural networks. The production of new adversarial objects could make use of other production methods and technologies, namely, color, and multi-material 3D printing. It would allow for a less convoluted process of integration of analog and digital forms of production, by allowing the process to be controlled digitally. The objects initially selected to become adversarial, a cup, a banana, and a table helped to build a process and fundamental knowledge about how to transform existing objects with a diverse array of formal and material characteristics. A host of discursive everyday materials and objects could help to push the process on how to become adversarial; materials that create various scaled boundaries for environments and cloak objects: fabrics (i.e., drapes, tablecloths, etc.), wall & floor materials (wallpaper, panels, etc.). The shift in process of modifying existing objects to the production of materials and elements that create boundaries for larger environments allows the technology to move from being tactical, and small scale, to be strategic, which allows for large-scale possibilities and long-term effects. The shift in focus from the production of singular objects to scalable materials, that cover surfaces, would allow for the construction of both adversarial architectures and environments.

ON ORGANIZATIONAL STRATEGIES

The organization of adversarial surfaces on objects directly affects their performativity. Currently, the adversarial surface is applied to a series of objects. Further development of principles of organization of how adversarial surfaces can be composed will directly open strategic possibilities. Studies are underway to review how to nest various scaled adversarial patterns, which will help it operate with changes in depth of field; how to position many adversarial surfaces from various vantage points onto a single object, which will help it operate with shifting angles; how to use multi-spectral materials to allow changes in light angle and intensity to make recognizable varying amounts of adversarial information, which will allow it

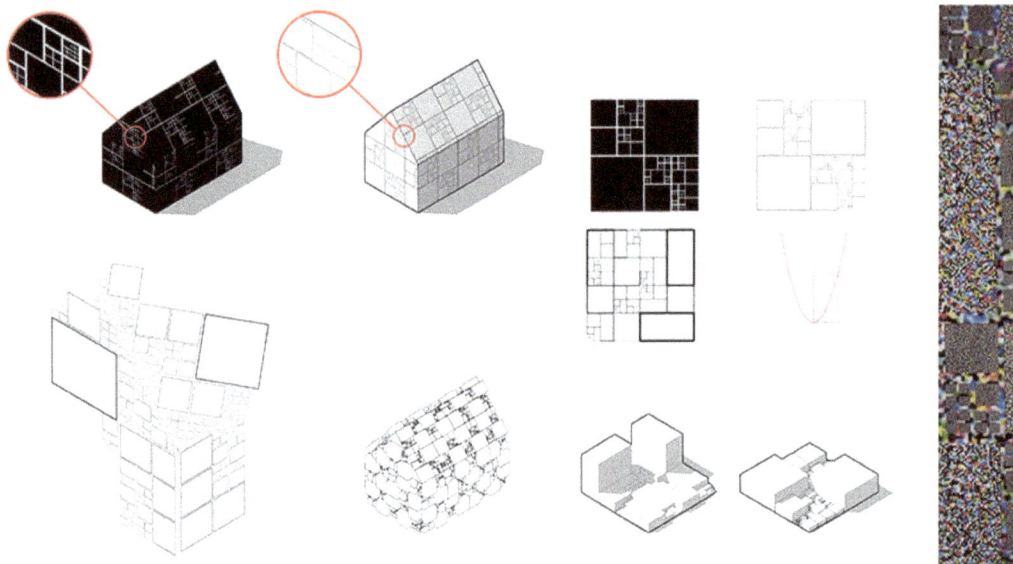
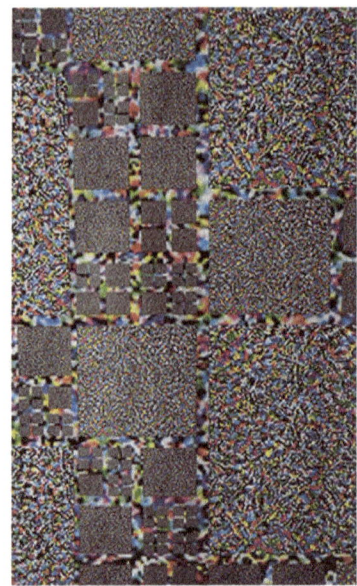

14 Diagrams of scale invariance applied to the three-dimensional object.

15 Adversarial scale invariance.

to change based on shifts in light-levels. The organizational logic of the adversarial information on the surface compounds its strength by allowing the complexities of spatio-temporal and material relationships to be part of the attack strength. By further developing the organizational possibilities it allows for unique adversarial surfaces to be organized for specific sites and scenarios.

ON PROJECT SCENARIOS

The use and value of adversarial architecture vary for communities, disciplines, and users. It presents varying values and possibilities for system designers, and computer scientists, strengthening the architecture of neural networks and computer vision by locating system boundaries. For designers, architects & urban planners, it helps to locate a new site for the design of informatic surfaces to structure the meaning of the built environment. It allows for the encoding of volumes of information on to surfaces, for others to make machines privy to material histories or ornamental messages. For marketeers, it locates a new site to insert communication for goods and services to AI systems. For politics and institutions, it helps locate the surface for propagandistic possibilities, influence, and how meaning can be plastically constructed. For the public, it helps to satiate a growing desire for agency to redefine how privacy from systems that auto-recognize and process their environments can be integrated into the surfaces of everyday life. The project is in the development of a series of scenarios of how various communities, select disciplines, institutions, and users, may use an adversarial architecture to find new forms of agency and build new relationships with AI. The value of being adversarial and using adversarial architecture in the built environment frames a series of scenarios that evidence the project's own boundaries and emergent situations.

The development of an adversarial architecture established a workflow and method for how existing objects can become adversarial. It offers an outline of how objects from their inception can gain new forms of control over their digital presence and manage their relationship with AI. The assembly of an interdisciplinary team propelled a synthesis of knowledge and helped to frame a more critical relationship between the built environment and AI.

The industrialization of artificial intelligence may someday also be understood as a new age of the world image.[18] The widespread and pervasive use of these AI technologies, embedded within the environment within a short amount of time has allowed its system vulnerabilities to have compound value and become a matter of concern. An adversarial architecture allows for objects and environment surfaces to be informatic, a source of technical control with the systems that are engaging it, from computer vision, machine learning algorithms, and neural networks. Adversarial logic presents direct access to AI systems, an emergent discourse, and a method to allow the built environment to critically engage digital interactions.

ACKNOWLEDGMENTS

Adversarial Architecture research has been fortunate to have many contributors, and support from colleagues and from institutions.

Many thanks to graduate architecture students Jack Foisey, Angelica Samson, and undergraduate computer science student Tino Cheung, and mathematics student Matt Maresca. A special thanks to Wentworth Bistline Grant Committee for the financial support.

NOTES

1. While the average human can process images in as little as 13 milliseconds, convolutional neural network recognition on a cloud-based system aims to be the fastest method to process and learn from images. Many systems have propelled this turn to the use of images for neural networks and deep learning, which include the establishment of large-scale data storage (2008-2011) and networks (1995, 2008).

2. Neural networks are powerful tools for visual processing and have enabled many exciting applications from self-driving cars to urban analysis and design tools. However, these powerful models too have weaknesses, which attracted researchers in recent years to develop particular attacks to manipulate them.

3. This "condition" referenced is the industrialization of artificial intelligence, a defining moment in the Information Age. It is signaled by the widespread use, access to artificial intelligence, as well as the ability to embed it into objects in ways that escapes collective consciousness. The term "actors" in the text is specific to include both human and non-humans; for an expanded discourse on what constitutes an "actor," refer to Actor-Network Theory (M. Callon, M. Akrich, B. Latour, J. Law 2005). The image presented is an example of the use of YOLO - You Only Look Once. It is a favored object detection algorithm due to its accuracy and speed. This image illustrates the bounding boxes and associated class probabilities generated by Joseph Redmon in YOLOv3 trained on the COCO dataset with 80 labels, including cars, bicycles, people, traffic lights, and umbrellas. Note, subsequent versions of YOLO have been constructed: YOLOv4 (Alexey Bochkovsky, with C. Yan, Hon-Yuan), YOLOv5 (Glenn Jocher). The project team aims to continue the research with new forms of speed and accuracy provided by these subsequent YOLO models.

4. The research is an interdisciplinary collaboration between Associate Professor Antonio Furgiuele at Wentworth, School of Architecture and Design (WIT), Dr. Mehmet Ergezer at Wentworth, School of Computing and Data Science, Dr. Cagri Hakan Zaman (MIT, Department of Architecture), and a team of undergraduate and graduate students: Jack Foisey (WIT, MArch '22), Angelica Samson (WIT, MArch '21), Ryan Maresca (WIT, Mathematics Department '21), and Tino Cheung (WIT, Computing Program '22).

5. The phrase "essential to the digital-environment" is a reference to the ways that autorecognition and AI play a large role, often in cities, and are projected to play an increasing role with emergent smart city technologies of driverless cars.

6. There are four categories of adversarial attacks, which highlight intended goals and targeted weaknesses: 1) Confidentially attack, attacks on training data that helps authenticate confidential data; 2) Attack timing, which attacks the decision versus training time; 3) Attacker info, which attacks the full versus blackbox info about the user; and 4) Attacker goals, which attacks the targeted versus reliability of goals.

7. "Force multipliers" are often social, technical, or socio-technical system(s) or technology(s) that allows for a large increase in effectiveness, or power over subjects(s), objects(s), or situations. The increased accessibility to AI technologies allows individuals and various communities access to multiply their force.

8. The discourse of adversarial imaging is understood for its ability to attack machine learning algorithms and artificial intelligence; the configuration into the built environment allows for it to be positioned both offensively (attack) and defensively (management of information). The categories of possible use-values is still emergent, but the discourse of potential is focused in three areas: identification, information, and countersurveillance.

9. The scanning of each of the objects necessitated a capture of its true dimensions and geometry. Because the objects selected had a range of material, textural, and spectral differences, this seemingly simple task presented instant challenges. To produce the scans required multiple scanning technologies, the use of laser scanning and photogrammetry, to be within approximately 10 microns (.01 mm). The difference in object scale and spectral differences in materials became a key factor in using both scanning types. This spectral uniqueness in objects can be a future source to add adversarial effectiveness and uniqueness (signature) to objects.

10. The term "elegant" is used to problematize a notion of efficiency. After each scan was created, the surfaces were unraveled or "developed" with a goal to unroll the object to create a surface for the adversarial image. While the developed surface patterns were constructed computationally, they also had to work with elements of hand production, cutting, and wrapping an object. The developed surface had to present not only the least amount of material, but also the most achievable realization given the limits of hand production. The software Blender (Roosendal 2002) was used to develop the computer model surface. As this surface was being prepared, the adversarial noise was being generated in parallel.

11. The adversarial noise information was laser-printed onto the unrolled surface and manually wrapped onto the cup. We see the current process of unraveling objects as a future function of computation and automation. We are currently exploring scripting methods to automate this process from scanning, to unrolling the surface, to cutting this surface out of sheet materials. Additional studies with full color multi-material 3D printing are currently

underway, which will help build two discrete methods, one that privileges wrapping existing objects and 3D printing which can privilege the production of new objects.

12. The team tasks were organized around the types of task and technologies they could engage with: architecture team, Professor Antonio Furgiuele, and graduate student Jack Foisey largely took on the analog and visual communication tasks, unrolling, object wrapping, documentation, communication; the computation team, with Professor Cagri Zaman, championed scanning and unrolling, and technical development; Professor Mehmet Ergezer developed the adversarial image surfaces.

13. The "strength" of attack, is a term that refers to the confidence in-which is it fooled, that is, the ability of the adversarial network to fool the network. This is measurable in the percentage of confidence in the network's classification.

14. The notion of "targeted" adversarial attack refers to the specific coded information of an adversarial noise image. In this example, a toilet paper roll was encoded in the noise. This is in contradistinction to a "non-targeted" adversarial attack, where the noise itself simply disrupts the recognizability of the base image or object. Note, the adversarial cup, presented in this paper, is an example of a non-targeted adversarial attack. An unsuccessful adversarial attack differs for targeted and non-targeted types. An unsuccessful non-targeted attack would allow the source image/object (largely) unaffected by the adversarial image. A targeted attack is more difficult to produce and easier to understand its success, since the adversarial image encoded would be recognized with a certain percentage of confidence by the network selected.

15. "Live" here is being used to evoke contingent factors that are spatial (depth), temporal (changes), and spectral (light). These three factors allow objects to be contingent to one another and to actors that affect these conditions. While images have a fixed focal distance, elements in the image are stable.

16. Scale invariance, a key concept within modern mathematics and computing is applied for its ability to assert self-similar computational patterns.

17. Another current study underway foregrounds the unique temporal qualities of objects. If one set of adversarial architecture examples is developed to focus on the spatial strengths, the other dimension of exploration is the unique temporal characteristics of objects. The adversarial noise generated by the change in gradient weight based on a collective set of perturbations to create a layer of image-noise; it is printed as a two-dimensional image this gradient remains fixed and constant. Gradients in lived space, on both objects, are constantly shifting because of changes in light and transmission. A method to control shifting light gradients is explored in "counter-shading." The concept is associated with the coloration of animals, which allows mammals the ability to counter their shadows. It effectively erases the gradient produced through shifting light patterns and allows animals to perceptually appear flat and therefore more difficult to position in actual space.

18. A reference to Martin Heidegger's "Age of the World of the Picture," often translated as "Age of the World Image" (1938). The philosophical text highlights the foundational questions within modern scientific methods by inquiring about what constitutes fact and objectivity and their tie to image-based evidence.

REFERENCES

Bochkovsky, Alexey, C. Hon-Yuan Yan. 2021. "Traffic Sign Detection using YOLOv4". International Journal of Creative Research Thoughts 9 (5): 891–897.

Callon, M., M. Akrich, B. Latour, J. Law. 2001. "Actor-Network Theory." In *International Encyclopedia of the Social & Behavioral Sciences*, edited by. N. J. Smelser and P. B. Balte. Amsterdam: Elsevier. 62–66.

Goodfellow, Ian J., Jonathon Shlens, and Christian Szegedy. 2014. "Explaining and harnessing adversarial examples." *arXiv* preprint arXiv:1412.6572.

Jocher, Glenn, K. Nishimura, T. Mineeva, R. Vilariño et al. *YOLOv5 by Ultralytics*. V. 7.0. Github. 2020. https://github.com/ultralytics/yolov5.

Kurakin, Alexey, Ian Goodfellow, and Samy Bengio. 2016. *Adversarial examples in the physical world*. Technical Report, Google, Inc. arXiv:1607.02533v1 [cs.CV] 8 Jul 2016. 1–15.

Redmon, Joseph, Santosh Divvala, Ross Girshick, and Ali Farhadi. 2016. "You Only Look Once: Unified, Real-Time Object Detection." In *2016 IEEE Conference on Computer Vision and Pattern Recognition (CVPR)*. 779–788.

Redmon, Joseph, and Ali Farhadi. 2018. "Yolov3: An Incremental Improvement." *arXiv* preprint arXiv:1804.02767.

Roosendal, Ton. *Blender*. v. 2.25. Windows. 2002.

Szegedy, Christian, Wojciech Zaremba, Ilya Sutskever, Joan Bruna, Dumitru Erhan, Ian Goodfellow, and Rob Fergus. 2013. "Intriguing properties of neural networks." *arXiv* preprint arXiv:1312.6199.

Sandler, Mark, Andrew Howard, Menglong Zhu, Andrey Zhmoginov, and Liang-Chieh Chen. 2018. "Mobilenetv2: Inverted Residuals and Linear Bottlenecks." In *2018 IEEE/CVF Conference on Computer Vision and Pattern Recognition*. 4510–4520.

Simonyan, Karen, and Andrew Zisserman. "Very deep convolutional networks for large-scale image recognition." *arXiv* preprint arXiv:1409.1556 (2014).

Antonio Furgiuele is an Associate Professor at Wentworth Institute of Technology. He is an architect, educator, and scholar, whose work investigates the histories and theories of the architecture of data and information and the disciplinary changes they propel in the Information Age. Antonio has lectured, exhibited, and taught internationally at MIT, Pratt Institute, The City College of New York, Columbia University, and Parsons. He was a Research Fellow at the University of Wisconsin, Milwaukee (2014-16) and the MacDowell Colony (2017). He has a Master's degree in History and Criticism from MIT.

Mehmet Ergezer holds a Doctor of Engineering degree from Cleveland State University in Cleveland, Ohio. He is an Associate Professor of Computing and Data Science at Wentworth Institute of Technology in Boston, Massachusetts, with interest in embedded systems and computational intelligence. He has co-authored publications on artificial intelligence and computer science education and holds consumer electronics patents in the USA and Europe. Dr. Ergezer is also focused on increasing undergraduate students' inter-cultural competence and led a semester-long study abroad in Berlin, Germany, and held the leadership role in broadening participation in AI as the co-founder for AAAI Undergraduate Consortium.

Cagri Zaman is the Director of MIT Virtual Experience Design Lab and a Lecturer of Design and Computation at MIT. His interdisciplinary research focuses on the study of human spatial experiences in physical and virtual spaces. His dissertation "Spatial Experience in Humans and Machines" offers a novel approach to spatial experience from a story-understanding perspective. Dr. Zaman has extensive research experience in artificial intelligence, immersive media, and computational design. A recipient of the MIT DesignX challenge grant in 2017, Dr. Zaman founded Mediate, a research and innovation laboratory, which develops technologies that empower people in physical spaces.

Latent Isovist

Discovery of Machine-Human Interpretable Spatial Properties Using Inverted GANs

Mikhael Johanes
École Polytechnique Fédérale de Lausanne

Jeffrey Huang
École Polytechnique Fédérale de Lausanne

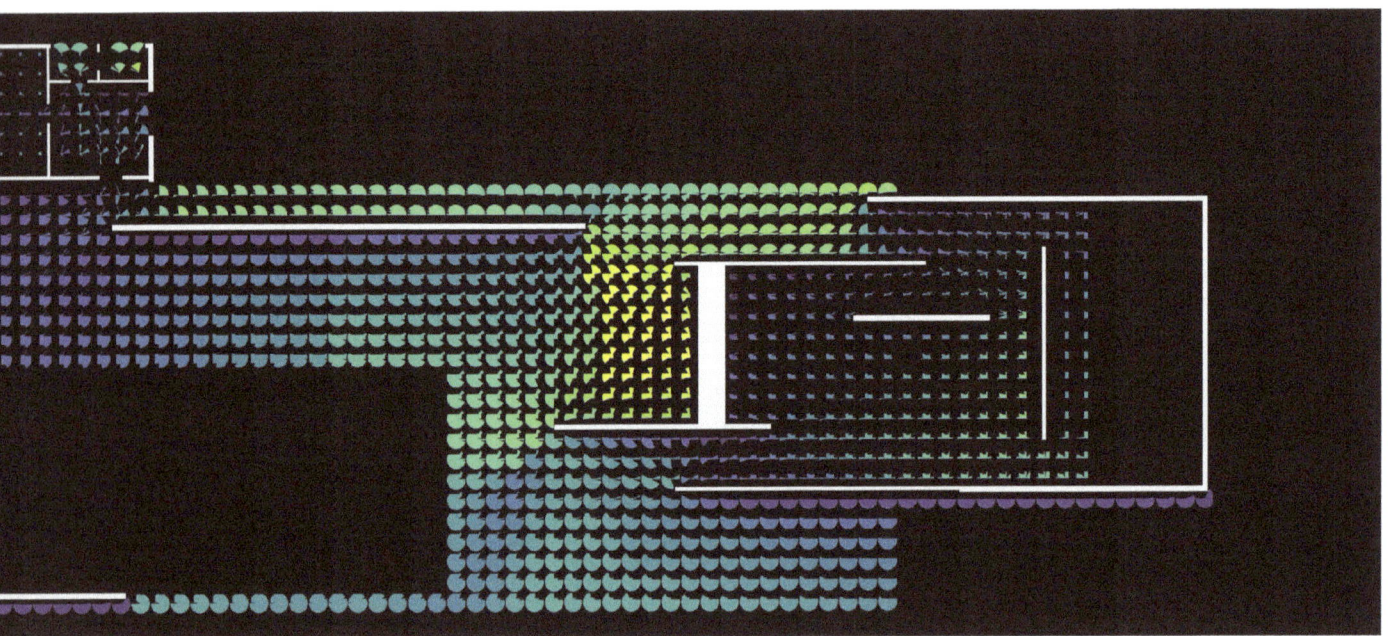

1 Latent isovist mapping of Barcelona Pavilion

ABSTRACT

Isovist provides a situated visuospatial representation of space commonly used to quantify the spatial properties of an environment. While the development of isovist-based spatial analysis has advanced over the last decades, most isovist measures are still abstract and distanced from the human experience of architectural space. This research leverages the development in deep learning research to develop an experimental framework for discovering machine-human interpretable spatial properties from latent isovist, a reduced dimensionality isovist representation obtained from generative adversarial networks (GANs). GAN latent space contains a wide range of semantically interpretable directions, potentially being used to quantify the spatial properties encoded in isovist representation. The experiment aims to facilitate the bridging of machine computation and human representation of space by discovering directions on the latent space from an isovist-trained GAN that are human interpretable. The framework expands and combines the techniques for GAN training, inversion, and interpretation of latent space with isovist representation to enable the mapping of spatial properties from a floorplan. Our result suggests that latent representation of isovists could provide a novel way for measuring spatial properties that are both machine computable and human interpretable useful for spatial pattern indexing and classification to support a human-centric AI-assisted architectural design.

INTRODUCTION

Isovist is a visuospatial representation that provides a computational model for examining spatial experience in an architectural floorplan, defined by visible spaces from a particular vantage point in an environment (Benedikt 1979). Isovist makes it possible to quantitatively represent the visibility pattern from a local point in an environment (Figure 2a). Some measures can be derived from the shape of the isovist to give a numerical representation of the spatial apprehension of the observer to the environment, such as degree of openness, perceptual uncertainty, sense of exposure, and safety or power (Ostwald and Dawes 2019). Measurements from uniformly distributed isovists generate a scalar field that gives a global reading for a specific spatial property to create a visuoperceptual map of architectural space. While proven to correlate with how we perceive or behave in space, the developed measurements transform the rich spatial information from isovist into a set of abstract mathematical calculations that are sometimes distanced by how humans identify architectural space.

It would be valuable if isovists could be used to provide a more nuanced architectural reading that is closer to the human interpretation to bridge machine-human understanding of architectural space. A potential solution is leveraging machine learning techniques that can learn the rich, high-dimensional information in an unsupervised manner, such as GANs. One of GANs' advantages is the ability to fit the data distribution from a given training sample into a much lower-dimension latent space that contains a wide range of interpretable directions (Cherepkov, Voynov, and Babenko 2021). Currently, most GANs applications for architectural space focus on the pixel representation of the floorplan, which allows the machine learning of symbolic or abstract aspect of the space such as forms (Asmar and Sareen 2020), architectural elements (Cho et al. 2020), and programs (Huang and Zheng 2018), but less emphasis on the experiential aspect that the isovist representation is trying to capture.

This research combines the isovist representation and GAN in learning human interpretable spatial properties from architectural floorplans. We propose the notion of latent isovist, a reduced dimensionality isovist representation obtained from GANs that facilitates the bridging of machine computation and human interpretation of space. The contribution of this research is twofold. The first contribution is the extension of GANs human interpretable direction discovery from 2D image representation to isovist periodic function (Figure 2b) to allow the learning of situated visuospatial experiences from architectural floorplans. The second contribution is the development of machine-human interpretable isovist measures that utilize the discovered directions of the GAN latent space to provide novel spatial properties quantification. By extending the current isovist computing techniques closer to human spatial interpretation, the developed technique is potentially useful for mapping and spatial indexing that could lead to an intuitive AI assistive tool for architectural design.

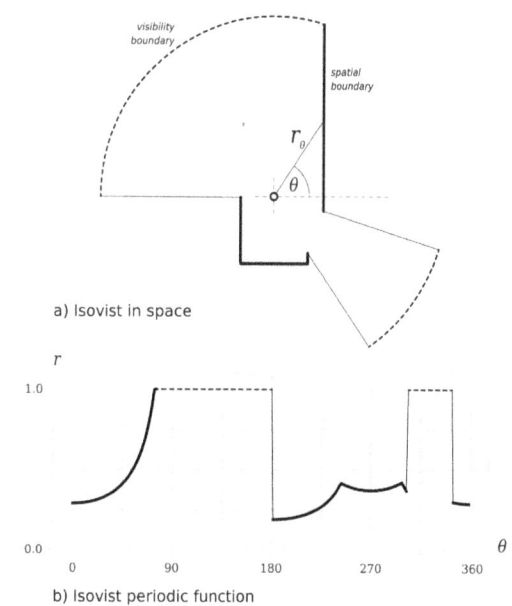

2   Isovist representation of space

BACKGROUND

Machine Learning Isovist

A few machine learning approaches have been explored for isovist representation, of which the general objective is to extract specific spatial information from the environment for classification and recognition purposes. A supervised convolutional neural network (CNN) classification model has been implemented for 3D isovists depth map image representation to recognize specific predefined spatial prototypes out of a complex architectural 3D environment (Peng 2018). While the capture of 3D information into an isovist gives comprehensive spatial information, the research focuses on predefined spatial prototypes, potentially excluding some other essential spatial patterns.

Recognition of recurring spaces using unsupervised learning of 2D isovist representation has been demonstrated using a set of isovist shape quantitative descriptors described in (Benedikt 1979) as feature vectors for machine learning (Sedlmeier and Feld 2018). By using two basic clustering algorithms, the approach can extract meaningful isovist clusters corresponding to various spaces such as small rooms, big rooms, corridors, and doorways. The research shows the potential use of the coupling of 2D isovist with machine learning as an effective and cheap spatial recognition technique. Our approach differs by dismissing the feature engineering stage and directly learning the discrete, ray-based isovist periodic function, as illustrated in Figure 2b. As such, we expect to be able to learn more nuanced information in comparison to isovist conventional quantitative descriptors.

Inverted GAN for Learning Architectural Floorplan

GAN, in the beginning, is designed to generate images from the given samples by training two competing neural networks, generator *G* and discriminator *D*, that are simultaneously trying to generate and identify fake samples from the real ones in an unsupervised manner (Goodfellow et al. 2014). Once trained, the generator is used to synthesize data from an input noise *z* drawn from a latent space with a prior distribution. This generative capacity provides a new experimental ground to exploit machine creative potential in generating images such as building facades and floorplans (Newton 2019; Huang et al. 2021). A type of conditional GAN (cGAN) is designed to enable a controlled, conditional generation based on an input image (Isola et al. 2017). This particular class of GAN has gained popularity for machine learning architectural floorplans, as exemplified by the work of (Huang and Zheng 2018; Chaillou 2019; Cho et al. 2020) to generate different types of floorplans conditioned by particular input images.

In contrast to the common interest in the generative capacity of GANs, we focus our attention on the understanding of GAN latent space, which can have a wide range of human interpretable directions that correspond to the generated outputs (Meng 2022). Access to the latent space enables us to tap into GAN's discriminative potential to support architectural spatial classification and recognition. Several inversion methods have been developed to access GAN latent space, either by using an optimization or learning approach (Xia et al. 2021). We suggest that the discriminative potential of GAN latent space could be combined with the isovist's ability to encode the rich spatial information as an effective way to enable machine-human recognition of architectural space. Compared to the previous research on using GANs to learn architectural floorplans from images, we focused on the isovist representation as a proxy for visuospatial experiential information derived from the floorplans.

Latent Space Structure

Latent space is a reduced-dimensional vector space embedding that has been shown to capture high-level information within a defined domain, enabling the reading of meaningful relationships between data points (Liu et al. 2019). Recent efforts have been invested in discovering the GAN latent space's human interpretable directions for semantical image generation control (Cherepkov, Voynov, and Babenko 2021; Voynov and Babenko 2020; Jahanian, Chai, and Isola 2020). The research's common focus is finding directions in the latent space that correlate with human interpretable generated image transformation such as translation, zooming, and adding or removing elements. The discovered direction and latent space arithmetic are then used to enable high-level image editing and transformation.

The discovery of semantically meaningful latent space direction can be done in both supervised and unsupervised manner. In a supervised manner, a linear SVMs classifier is trained with different high-level predefined attributes to identify hyperplanes in the latent space that are perpendicular to the corresponding attribute transformation (Shen et al. 2020). The unsupervised manner of direction discovery works without predefined attributes by seeking direction in the latent space in which the generated images' transformations are distinguishable from each other by using a CNN classifier (Voynov and Babenko 2020). While the technique in principle does not have any semantic input from the labeling in the supervised method, the research shows that some of the discovered transformations are humanly interpretable and potentially useful in data exploration.

We extend the research of unsupervised latent space discovery for images (Voynov and Babenko 2020) to isovist representation in reading architectural space and propose the notion of latent isovist as the embedding representation of the isovists in the GAN latent space. As the information in isovist is not directly interpretable as in images, we propose a mapping framework to better evaluate the discovered latent directions by visualizing the measures from latent isovist to a floorplan, thus facilitating their architectural interpretation.

MAPPING OF ISOVIST LATENT REPRESENTATION

Framework Development

To enable machine-human bridged interpretation of isovists representation, we developed a latent isovist experimental machine learning framework that consists of (a) GAN training and inversion, (b) unsupervised direction discovery, and (c) latent space projection and architectural mapping (Figure 3). The experiment utilizes the current GAN research techniques adopted to isovist periodic function by flattening its 2D convolutional layers to 1D. The dataset is developed by randomly sampling isovists from a floorplan dataset comprised of approximately 4000 architectural housing floorplans collected from Finland regions (Kalervo et al. 2019). The isovist sample is represented as 256 discrete features, measuring the distance from the center to the physical boundary in a radial manner. The shape of the isovist can be recovered from the function by plotting it in a polar coordinate (Figure 2). The resulting dataset is comprised of a 300k training set and a 34k evaluation set of isovist samples.

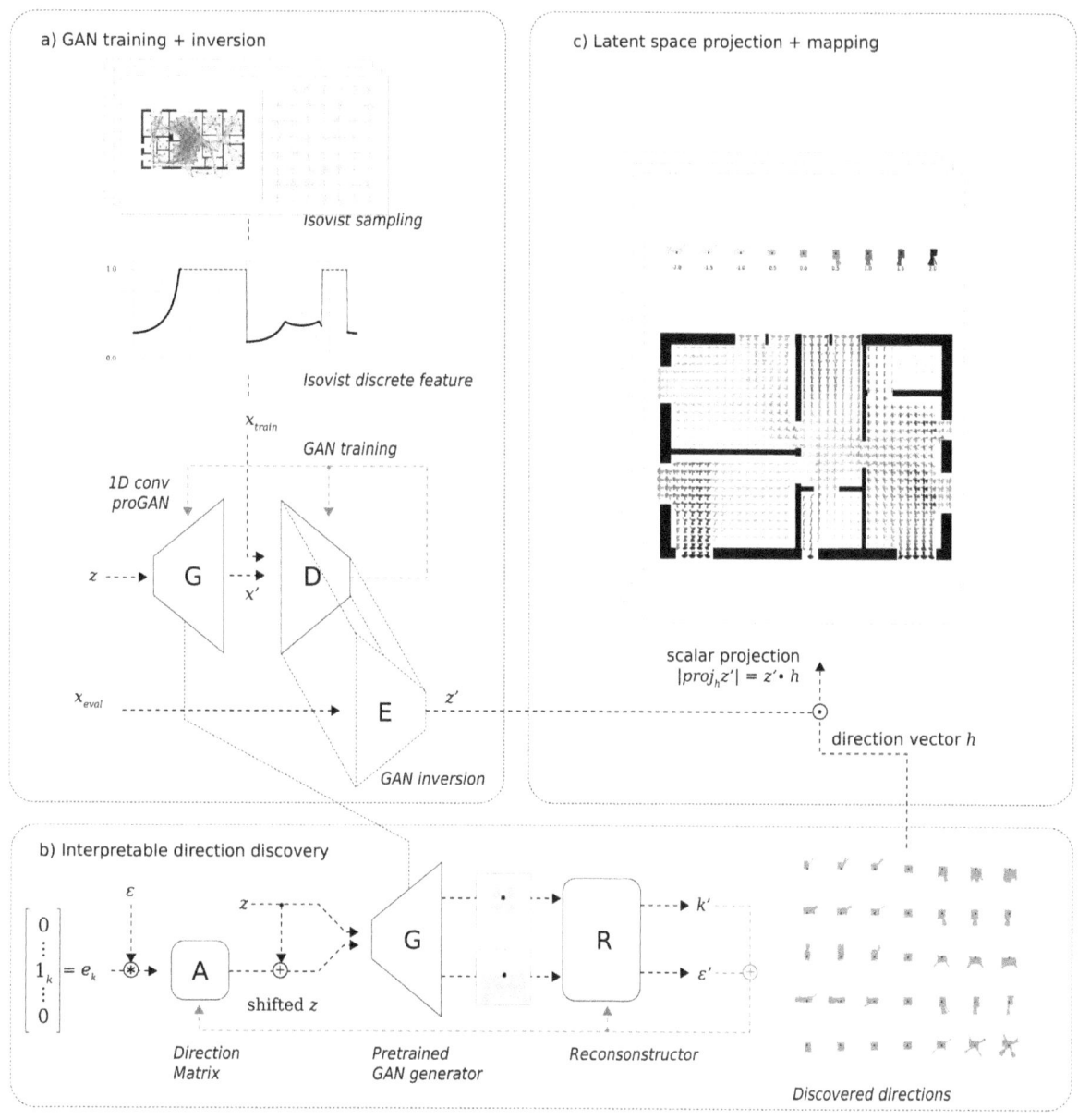

3 Framework latent space isovists mapping experiment

We employ a 1D convolutional implementation of progressive growing GAN (ProGAN) to learn the isovist samples (Figure 3a). ProGAN learns the samples from the low to full resolution of isovist in a hierarchical manner, stabilizing the training and increasing the network's capacity to learn the entire distribution of the dataset (Karras et al. 2018). At the end of the training, the GAN provides a mapping of 16-dimensional latent space with normal prior distribution to 256-feature isovist. As the inverse of the mapping is not directly available, we transform the discriminator network D into encoder E by changing its last layer to match the dimensionality of GAN latent space and re-train it in a self-supervised manner using generated samples from G (Yu and Wang 2022). We use the encoder to project the isovist samples into the latent space for evaluation and mapping purposes. The trained generator is further explored in an unsupervised manner to extract the meaningful direction from the latent space (Figure 3b). Based on the scalar projection of the discovered directions with the projected isovists within the latent space, quantitative measures that correspond to particular spatial property is calculated and visualized (Figure 3c).

Unsupervised Latent Space Direction Discovery

The trained GAN latent spaces contain potentially human interpretable directions relevant to spatial properties represented in the isovist data. Building on the work of Voynov and

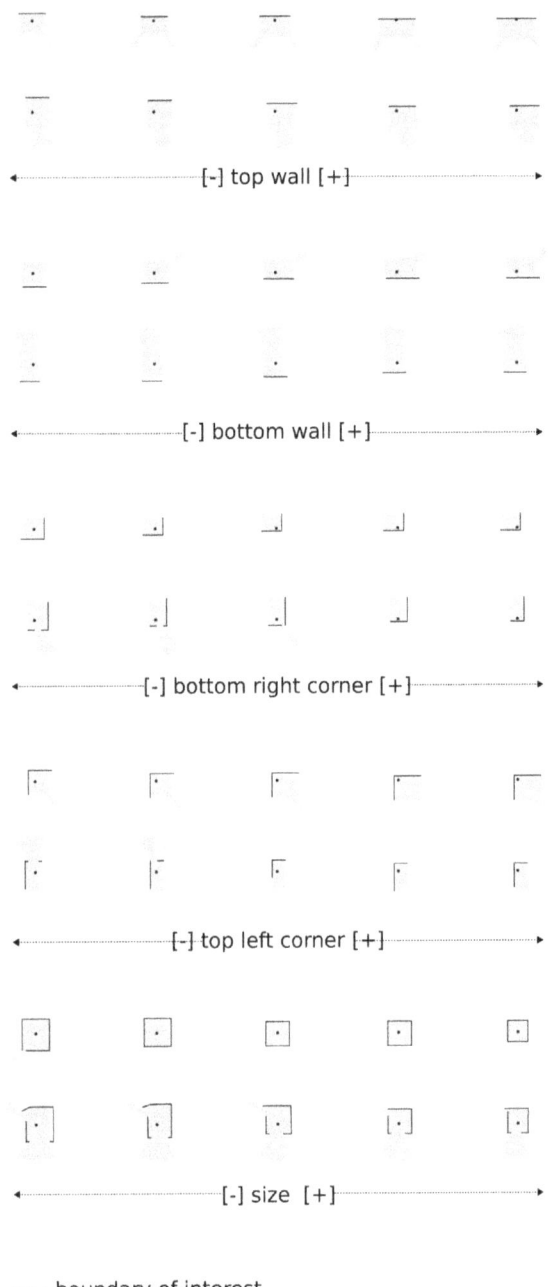

4 Discovered directions for the isovist transformation

Babenko (2020) on the unsupervised discovery of interpretable direction of images-trained GANs latent space, we used the method to recover architecturally meaningful directions from isovist-trained GAN latent space. The goal is to develop a way of measuring the presence of specific spatial properties by using the discovered latent space directions. While the semantic content in images is arguably different from what isovists contain, the algorithm is essentially domain agnostic, which only seeks to extract as many transformations in the latent space that are different from each other and happened to be human interpretable, thus potentially applicable for isovist representation.

The intuitive idea behind the learning algorithm is to obtain latent space direction from columns of A so that the transformations in the generated output are easily distinguished by reconstructor R (Figure 3b). A Matrix A with a size of $d \times K$ whose all columns have unit length contains the latent space directions, where d denotes the dimensionality of the latent space and K determines the number of directions we would like to discover. The generator G generates two isovists from the latent noise z and the shifted $z + A(\varepsilon e_k)$, where the shift is the k-th column in matrix A with magnitude ε. The column index k and magnitude ε have uniform prior distributions. A reconstructor R is a flattened, 1D convolutional version of the ResNet-18 model (He et al. 2016), which obtains the pair of the generated isovists and predicts both column index k' and magnitude ε'. The optimization is performed by minimizing the cross-entropy loss $L_{CE}(k, k')$ and mean absolute loss $L_{MAE}(\varepsilon, \varepsilon')$ together. A more detailed explanation of the algorithm could be found in the original implementation (Voynov and Babenko 2020).

As the learning algorithm recovers distinguishable directions from the latent space, we pick up the most significant directions by human interpretation. Some of those directions is shown in Figure 4. From the observation, the 'disentangled' directions from the latent space often show spatially meaningful changes in isovist representation, such as the degree of proximity to walls, corners, openings, and space size. The discovered directions are admittedly exploratory, but the interpretation of the directions could evolve in the architectural mapping stage as the floorplan give a more contextual reading of its spatial perception. These directions are assumed to be isotropic to some degree. A scalar projection in the latent space translates those directions into quantitative measures to provide an architectural reading of the discovered latent isovist directions.

Isovist Latent Representation Projective Mapping
Figure 5 illustrates the projection of the latent isovist into a discovered direction for mapping the spatial qualities in a floorplan. We use transformed encoder E to project the isovists into GAN latent space. The scalar projection of the latent vector z' to an interpretable latent direction h denotes a correlation between the isovist samples with the measured spatial interpretation (Figure 5a). This technique is developed based on the intuition that the orthogonality in the latent space is analog to isovists representation, which means that if a direction in the latent space is perpendicular to an isovist latent vector, the corresponding spatial property is not

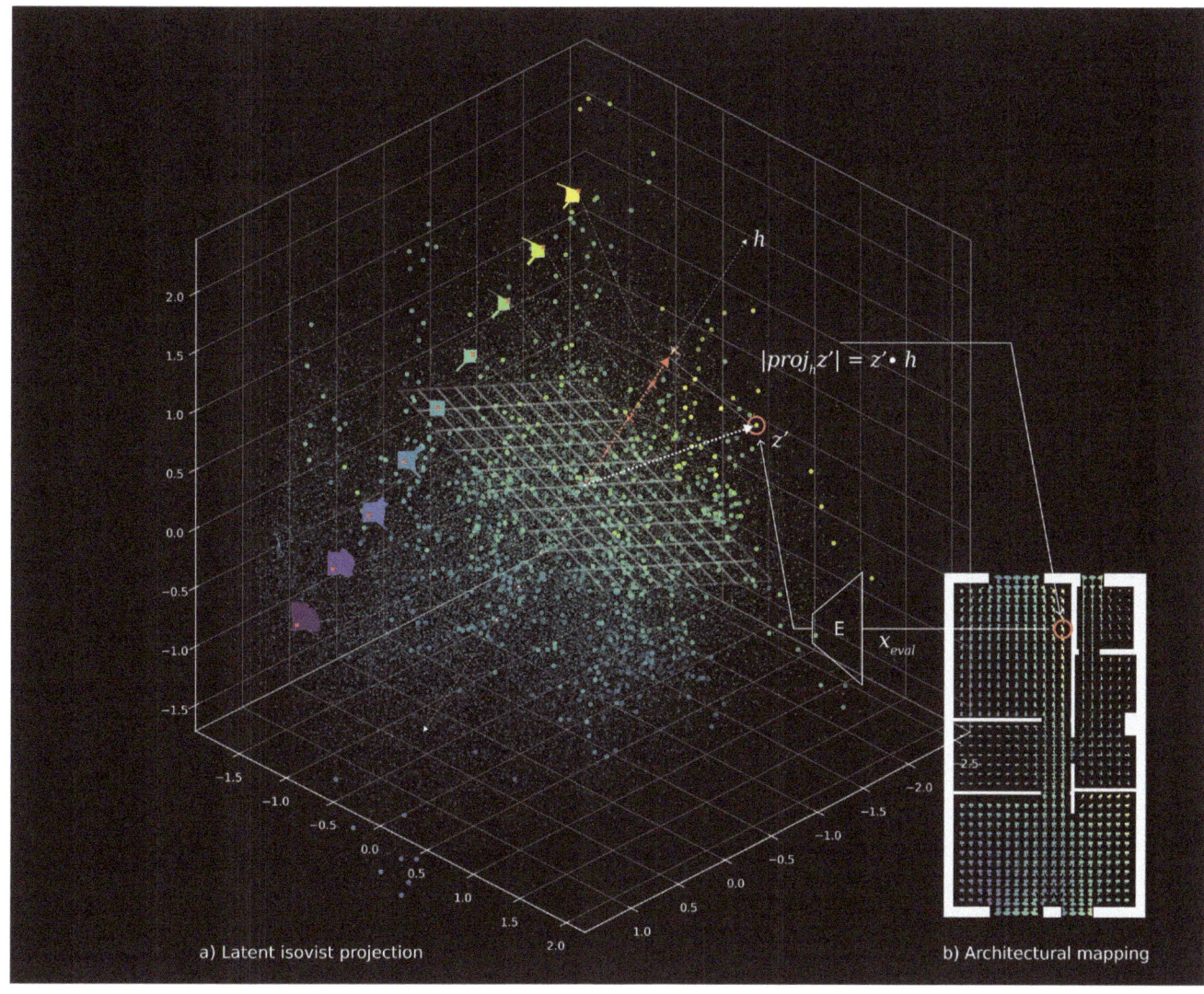

a) Latent isovist projection   b) Architectural mapping

5   Projection in the latent space (a) and architectural plan (b)

present in the measured isovist. While the generality of this technique is subject to future validation and improvements, it indicates the possibility of spatial property quantification that is machine computable and human interpretable. We plot the quantitative representation of the latent isovists to indicate the presence of observed spatial property in the architectural context (Figure 5b).

RESULTS AND DISCUSSION: MACHINE-HUMAN UNDERSTANDING OF SPACE
Latent Isovists as Measures of Spatial Properties

Latent representation of isovists obtained from GAN allows the discovery of particular spatial patterns. As shown in Figure 6, some basic architectural moments such as proximity to the (a) walls, (b) openings, and (c) corners with four orientations can be identified from the latent isovists. The technique opens up the possibility of bridging everyday human understanding of space such as boundary, opening, orientation, and scale with machine processes, thus enabling a more humanistic approach to architectural spatial analysis. While the obtained spatial interpretations are arguably generic and straightforward, this technique could pave a more complex and nuanced reading architectural space if the latent space feature disentanglement and the GAN inversion precision are improved.

The fixed prior distribution of the latent sampling restricts the features in the dataset from being fully disentangled, as the variation of those features from the dataset will be 'warped' in the latent space (Tzelepis, Tzimiropoulos, and Patras 2021). Intermediate latent space *w* introduced by StyleGAN can undo the warping and provide better disentanglement properties (Karras, Laine, and Aila 2021). The disentangled features in this intermediate latent space also allow a more precise latent embedding (Abdal, Qin, and Wonka 2019), thus giving a more accurate spatial interpretation. Our results suggest that the entangled latent space and the imprecise GAN inversion hinder the more nuanced reading of spatial properties encoded in isovists.

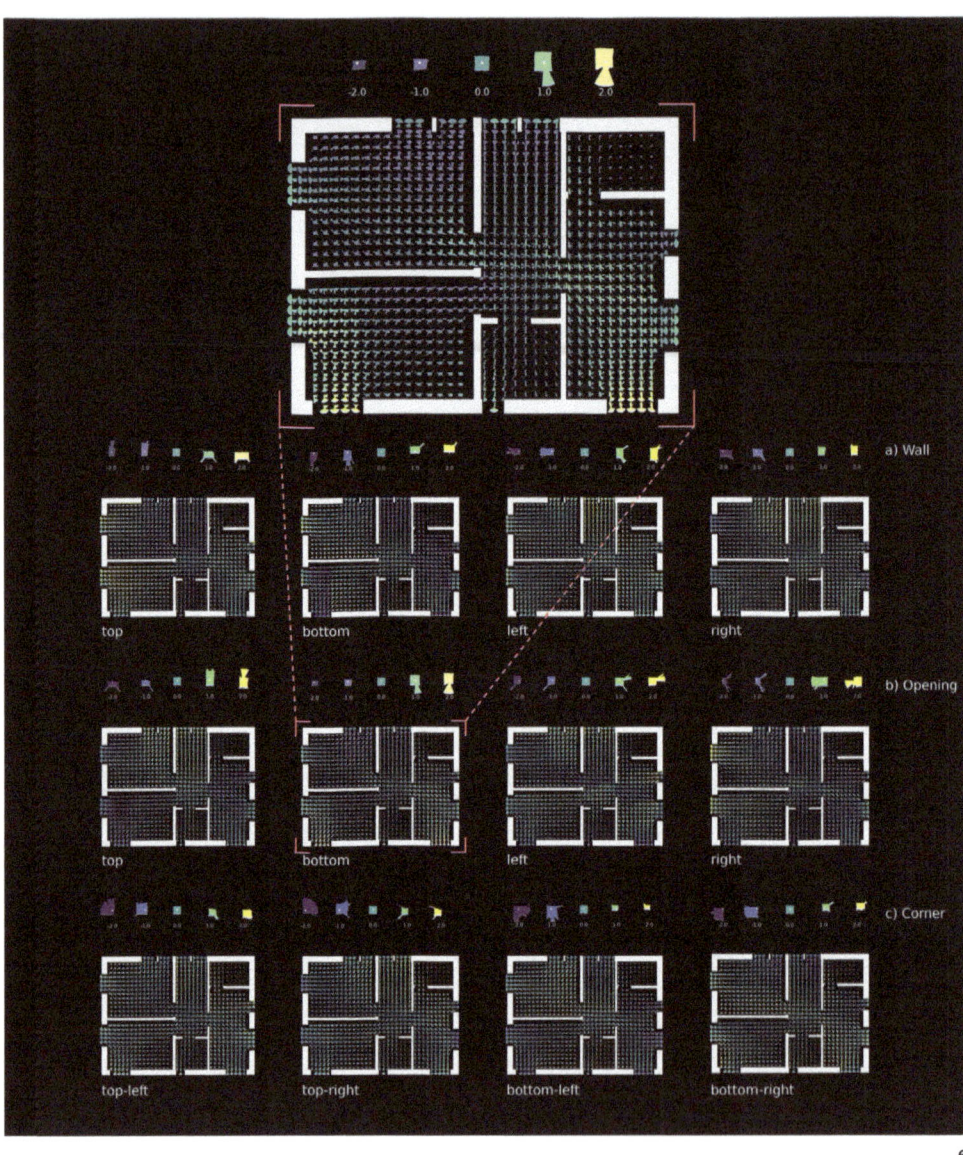

6   Latent isovist mapping of discovered directions

Artificial Critique

One critical aspect of deep learning techniques is the ability to generalize beyond the training samples. The experiment uses isovist samples from apartment floorplans collected from the Finland regions. Consequently, the GAN mainly learned the spatial properties of Western housing typology, which are composed of enclosed spaces and corridors. To understand the ability of the algorithm to generalize, we extend the evaluation beyond the training dataset by mapping a modernist villa, Mies van der Rohe's Barcelona Pavilion (Figure 7). Compared to traditional apartments, Barcelona Pavilion is characterized by its free-flowing spaces, which give an ever-changing spatial experience.

The mapping of spatial properties inferred from latent isovist suggests the ability of the algorithm to appreciate the spatial configuration of Barcelona Pavillion's architecture. The discovered latent isovist directions could be used to grasp (a) the opposite directionality that is defined by the long walls, (b) notice a corner space in the middle of the building, and (c) highlight the contrast of scale of spaces between interior and exterior. It underlines some of the most defining architectural features of the pavilion by learning from a relatively limited architectural typology. The capacity to generalize suggests a possibility of 'artificial critique' whose task automatically measures the desired spatial properties. Bridging this method with the development of language models for spatial concepts could extend this method to bring the computational spatial analysis of architecture closer to human interpretation.

CONCLUSION AND FUTURE WORK

This research extends and combines the research on GAN latent space interpretation from images into isovist architectural spatial analysis. It is already known that GANs can encode many meaningful directions in their latent space corresponding to human interpretation. By assuming that

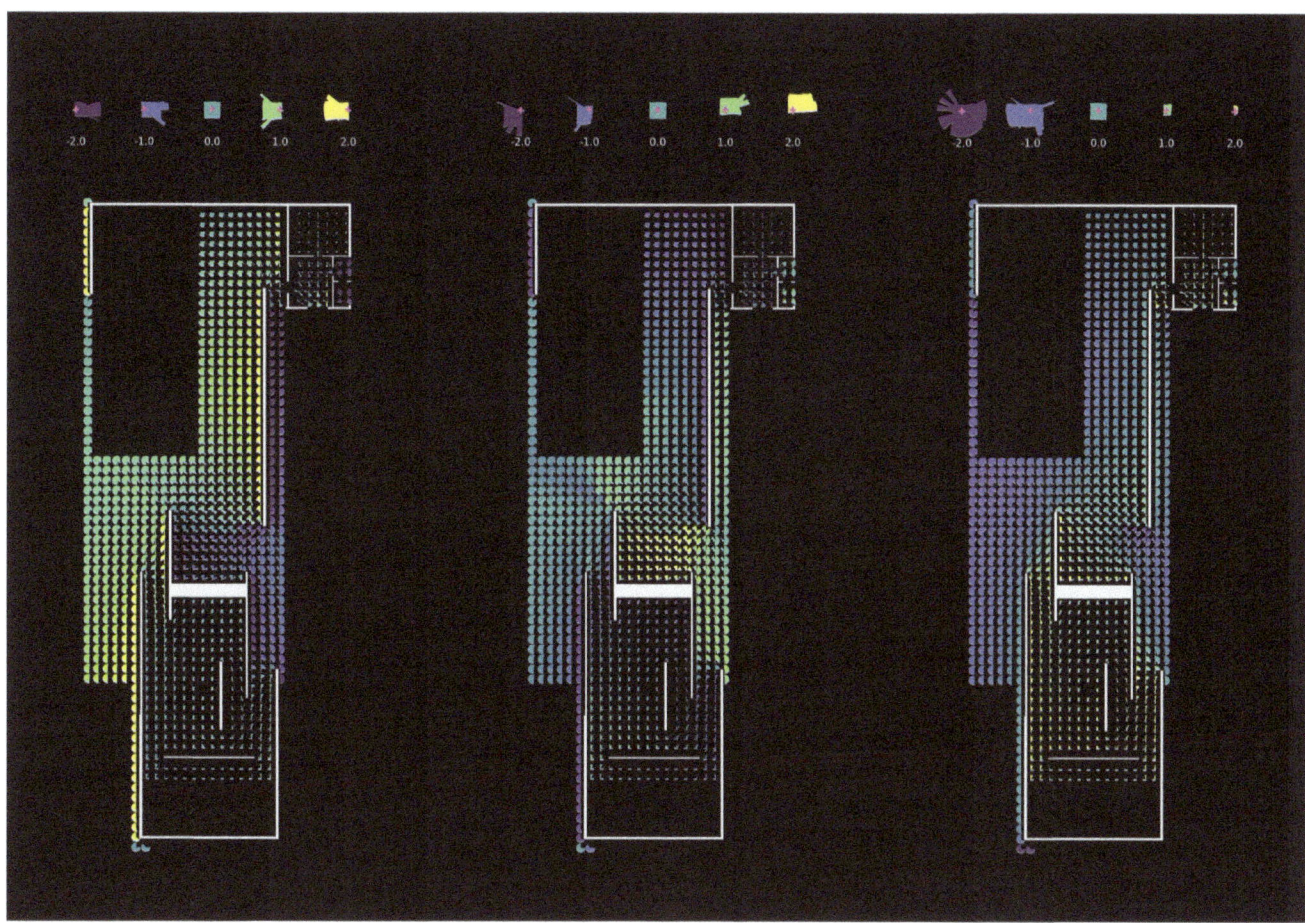

7   Latent isovists as artificial critique

the discovered interpretable direction in the latent space is isotropic, a scalar projection in the latent space could be used to measure the presence of specific spatial property in an isovist representation. We aim to find a way to measure spatial properties that are both machine computable and human interpretable by revealing and utilizing the structure of GANs latent space for isovist representation. The developed technique could provide a foundation for discriminative tasks that are both human interpretable and machine computable, thus supporting the development of a human-centric AI-assisted architectural design framework.

The research is currently exploring an experimental framework for GAN training, latent space direction discovery, and latent space projection strategy customized for isovist representation. The proposed framework has demonstrated that the interpretable direction in GAN latent space can be disentangled and used for recognizing specific architectural spatial properties in a floorplan. Some of the discovered interpretable directions correspond to the proximity to architectural elements such as walls, corners, and openings in different orientations. The experiments also indicate that our technique can be extended to evaluate the spatial properties of floorplans based on learning from a limited set of architectural typologies.

The difficulty of GANs in learning the entire distribution of the isovist samples and the imprecision in the inversion of input isovists to the latent space restrict the developed technique to measure more nuanced spatial properties. The GAN latent space's entangled representation also hinders the 'isolation' of specific, interpretable directions. This research could benefit from the research of latent space disentanglement, which would be the focus of our future work. Furthermore, the integration with the current design software ecosystem will expand our technique further in supporting architectural design by providing spatial indexing capability.

REFERENCES

Abdal, Rameen, Yipeng Qin, and Peter Wonka. 2019. "Image2StyleGAN: How to Embed Images Into the StyleGAN Latent Space?" In *2019 IEEE/CVF International Conference on Computer Vision (ICCV)*. 4431–40.

Asmar, Karen El, and Harpreet Sareen. 2020. "Machinic Interpolations: A GAN Pipeline for Integrating Lateral Thinking in Computational Tools of Architecture." In *Proceedings of the 24th Conference of the Iberoamerican Society of Digital Graphics*. Medellín, Colombia. 60–66.

Benedikt, Michael L. 1979. "To Take Hold of Space: Isovists and Isovist Fields." *Environment and Planning B: Planning and Design* 6 (1): 47–65.

Chaillou, Stanislas. 2019. "AI + Architecture: Towards a New Approach." Master thesis, Harvard University, Graduate School of Design.

Cherepkov, Anton, Andrey Voynov, and Artem Babenko. 2021. "Navigating the GAN Parameter Space for Semantic Image Editing." In *2021 IEEE/CVF Conference on Computer Vision and Pattern Recognition (CVPR)*. Nashville, TN, USA: IEEE. 3670–79.

Cho, Dahngyu, Jinsung Kim, Eunseo Shin, Jungsik Choi, and Jin-Kook Lee. 2020. "Recognizing Architectural Objects in Floor-Plan Drawings Using Deep-Learning Style-Transfer Algorithms." In *Proceedings of the 25th CAADRIA Conference*. 717–725.

Goodfellow, Ian, Jean Pouget-Abadie, Mehdi Mirza, Bing Xu, David Warde-Farley, Sherjil Ozair, Aaron Courville, and Yoshua Bengio. 2014. "Generative Adversarial Nets." In *Advances in Neural Information Processing Systems*, 2672–80.

He, Kaiming, Xiangyu Zhang, Shaoqing Ren, and Jian Sun. 2016. "Deep Residual Learning for Image Recognition." In *2016 IEEE Conference on Computer Vision and Pattern Recognition (CVPR)*. Las Vegas, NV, USA: IEEE. 770–778.

Huang, Jeffrey, Mikhael Johanes, Frederick Chando Kim, Christina Doumpioti, and Georg-Christoph Holz. 2021. "On GANs, NLP and Architecture: Combining Human and Machine Intelligences for the Generation and Evaluation of Meaningful Designs." *Technology|Architecture + Design* 5 (2): 207–24.

Huang, Weixin, and Hao Zheng. 2018. "Architectural Drawings Recognition and Generation through Machine Learning." In *ACADIA '18; Proceedings of the 38th Annual Conference of the Association for Computer Aided Design in Architecture*. Mexico City, Mexico. 156–165.

Isola, Phillip, Jun-Yan Zhu, Tinghui Zhou, and Alexei A. Efros. 2017. "Image-to-Image Translation with Conditional Adversarial Networks." In *2017 IEEE Conference on Computer Vision and Pattern Recognition (CVPR)*. 5967–5976.

Jahanian, Ali, Lucy Chai, and Phillip Isola. 2020. "On the 'Steerability' of Generative Adversarial Networks." In *International Conference on Learning Representations*.

Kalervo, Ahti, Juha Ylioinas, Markus Häikiö, Antti Karhu, and Juho Kannala. 2019. "CubiCasa5K: A Dataset and an Improved Multi-Task Model for Floorplan Image Analysis." In *Image Analysis: 21st Scandinavian Conference, SCIA 2019*. Norrköping, Sweden. 28–40.

Karras, Tero, Timo Aila, Samuli Laine, and Jaakko Lehtinen. 2018. "Progressive Growing of GANs for Improved Quality, Stability, and Variation." In *International Conference on Learning Representations*.

Karras, Tero, Samuli Laine, and Timo Aila. 2021. "A Style-Based Generator Architecture for Generative Adversarial Networks." *IEEE Transactions on Pattern Analysis and Machine Intelligence* 43 (12): 4217–28.

Liu, Yang, Eunice Jun, Qisheng Li, and Jeffrey Heer. 2019. "Latent Space Cartography: Visual Analysis of Vector Space Embeddings." *Computer Graphics Forum* 38 (3): 67–78.

Meng, Shengyu. 2022. "Exploring in the Latent Space of Design: A Method of Plausible Building Facades Images Generation, Properties Control and Model Explanation Base on StyleGAN2." In *Proceedings of the 2021 DigitalFUTURES*, edited by P. F. Yuan, H. Chai, C. Yan, and N. Leach. Singapore: Springer. 55–68.

Newton, David. 2019. "Generative Deep Learning in Architectural Design." *Technology|Architecture + Design* 3 (2): 176–189.

Ostwald, Michael J., and Michael J. Dawes. 2019. "Isovists: Spatio-Visual Mathematics in Architecture." In *Handbook of the Mathematics of the Arts and Sciences*, edited by B. Sriraman. Cham: Springer International Publishing. 1–13.

Peng, Wenzhe. 2018. "Machines' Perception of Space." PhD thesis, Massachusetts Institute of Technology.

Sedlmeier, Andreas, and Sebastian Feld. 2018. "Learning Indoor Space Perception." *Journal of Location Based Services* 12 (3–4): 179–214.

Shen, Yujun, Jinjin Gu, Xiaoou Tang, and Bolei Zhou. 2020. "Interpreting the Latent Space of GANs for Semantic Face Editing." In *2020 IEEE/CVF Conference on Computer Vision and Pattern Recognition (CVPR)*. Seattle, WA, USA: IEEE. 9240–49.

Tzelepis, Christos, Georgios Tzimiropoulos, and Ioannis Patras. 2021. "WarpedGANSpace: Finding Non-Linear RBF Paths in GAN Latent Space." In *2021 IEEE/CVF International Conference on Computer Vision (ICCV)*. 6373–82.

Voynov, Andrey, and Artem Babenko. 2020. "Unsupervised Discovery of Interpretable Directions in the GAN Latent Space." In *Proceedings of the 37th International Conference on Machine Learning*. PMLR. 9786–96.

Xia, Weihao, Yulun Zhang, Yujiu Yang, Jing-Hao Xue, Bolei Zhou, and Ming-Hsuan Yang. 2021. "GAN Inversion: A Survey." *ArXiv*:2101.05278 [Cs], August.

Yu, Cheng, and Wenmin Wang. 2022. "Fast Transformation of Discriminators into Encoders Using Pre-Trained GANs." *Pattern Recognition Letters* 153 (January): 92–99.

IMAGE CREDITS

All other drawings and images by the authors.

Mikhael Johanes is a Doctoral Assistant at Media x Design Lab, EPFL. His works examine the computational medium with its generative and analytical possibilities in architectural design. Since he finished his Master of Architecture at Universitas Indonesia in 2013, he has continued his career as a lecturer and research staff in the same University. He is part of the editorial team of *Interiority* and *ARSNET* journals, and he has also published in several international journals.

Jeffrey Huang is the Director of the Institute of Architecture at EPFL, Head of the Media x Design Lab, and a Full Professor in Architecture and Computer Science at EPFL. His research examines the convergence of physical and digital architecture. His recent work on artificial design (Design Brain) is featured at the Seoul Biennale of Architecture and Urbanism 2021.

Session Introduction

Big Data and Augmented Environments

Biayna Bogosian, Chair

In the past several decades, advancements in data collection, classification, correlation, and processing have converted the world into a live information platform. In this context, the future trajectory of disciplines and professions shaping the building industry depends on fostering critical and applied data-driven approaches. One of the most important branches of this pursuit is research in spatial computing systems to enhance our understanding of the built environment. The "Big Data and Augmented Environments" paper session foregrounds research in sensory and vision-based data acquisition, analysis, and representation that synthesize and contextualize the relationship of data to its referents in various architectural and urban scales.

This session highlights: 1) how to approach affordances and limitations of Extended Reality (XR) technologies for rethinking architectural representation and documentation conventions; 2) how to leverage interaction and immersion in XR technologies to provide guided feedback for enhancing fabrication and structural analysis; 3) how to empower expert and non-expert users to engage with high-dimensional data and analytics; and 4) how to incorporate machine learning and other statistical models for improving computational workflows.

The roundtable conversation of this paper session further focused on the reliance on interdisciplinary research and industry partnerships for closing software and hardware knowledge gaps as well as resource accessibility, especially for young researchers and marginalized institutes. The closing discussion of this paper session also paid homage to ACADIA conferences in the early 1990s which focused on understanding the role of Virtual Reality and Augmented Reality media in the field of architecture and design. Now, three decades later, it is important to revisit these early conversations in order to minimize research blindspots while planning research continuity.

Depth Camera Feedback for Guided Fabrication in Augmented Reality

Gwyllim Jahn
Fologram

Cameron Newnham
Fologram

Nick van den Berg
Fologram

1   A proof of concept prototype assembled from free-formed paper strips using depth camera feedback in augmented reality

ABSTRACT

Augmented reality environments have been demonstrated to assist with architectural fabrication tasks by displaying construction information at full scale and in context. However, this information typically needs to be sparse in order to prevent virtual models occluding a fabricators view of the physical environment, and this limits the application of augmented reality to tasks such as surface forming. To address this issue, we propose a method for guided fabrication in augmented reality using real time comparisons between depth scans of as built conditions and target conditions defined by design models. Through the design and fabrication of a small proof of concept prototype from paper strips, we demonstrate that guided fabrication is adequate for high speed, approximate and ad-hoc fabrication of complex surface geometries without the need for extensive rationalization for fabrication constraints or explicit documentation of parts. We further show how this method generalizes to other processes such as additive fabrication or part placement and speculate on the implications of accessible real time depth data from the HoloLens within Grasshopper.

INTRODUCTION

Fabrication in augmented reality has proven to be an effective and expedient method for forming and assembling complex structures from steel (Jahn et al. 2018), bricks (Mitterberger et al. 2020; Jahn et al. 2019) and steam bent timber (Jahn, Wit, and Pazzi 2019). Recent work has also demonstrated augmented reality applications working with flexible materials such as bent bamboo (Goepel and Crolla 2020) or nonuniform and unprocessed timber (Lok 2022). However, there remain few examples of augmented reality fabrication approaches to forming complex surface topologies due to the challenge of correctly judging surface depth with current generation augmented reality displays. This is a limitation on the design space afforded by augmented reality fabrication, as surface forming is an efficient way to describe geometry typically generated by CAD modeling software. While there exists several computer aided manufacturing approaches to materializing surface geometry, each of these approaches is not without limitations.

Subtractive forming complex surface models introduces the challenge of avoiding undercuts, while additive manufacturing is notoriously slow and introduces geometric constraints such as draft angles and requirements for infill and support material. Surfaces can be rationalized into smaller parts such as panels, shingles, or strips, but this rationalization introduces design, logistics, and assembly complexity. By comparison, ad hoc fabrication of complex surface geometries has been demonstrated to be possible by artists like Henrique Oliviera and Lucy Irvine, though this negates the advantages of digital design to production workflows. This work is motivated by the desire to capitalize on the efficiency and formal possibilities of ad hoc fabrication of complex surface geometries while also facilitating the approximate adherence to digital design models. We are also motivated by the desire to overcome the current limitations of augmented reality approaches to surface fabrication by demonstrating a generalizable approach to guiding fabricators using real time comparisons between point cloud scans of as-built conditions and target conditions defined by digital design models.

This research demonstrates a method for fast and approximate forming of surfaces with arbitrarily complex topology and geometry by hand in augmented reality. Within our approach, the only documentation required is the design model of the target surface, thereby simplifying or eliminating requirements for rationalizing surfaces for fabrication or discretizing design models to parts and joints (Figure 3). We aim to improve the feasibility of fabricating complex surface geometry in augmented reality compared to existing approaches, measured by improvements in fabrication time and formal complexity, and provide evidence of this method

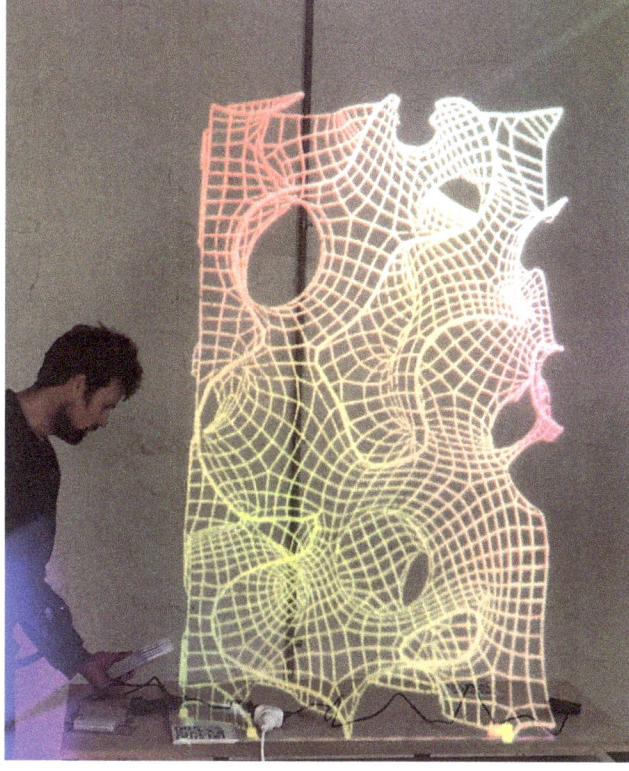

2 A detail view of the paper prototype

3 Hologram of the design model at 1:1; mesh triangles are rendered as occlusion geometry with lines representing the mesh quads on top

in a proof-of-concept prototype Figure 1) completed over the course of a single day by a team of three inexperienced fabricators. The proof-of-concept project is used to explore how real time feedback can assist fabricators in forming complex surface structures from flexible, lightweight materials by improving legibility of digital models viewed in augmented reality and enabling the improvisation of ad hoc strategies for their materialization.

BACKGROUND
Formwork-free Approaches to Fabricating Surfaces from Strips

Fabrication of curved surfaces typically involves generating a supporting substructure that is then clad to form the surface. These tectonic systems constrain the design space of fabricable geometry by limiting design topology to that which can be derived from substructures (such as waffle grids or other contoured supports) and cladding geometries (such as panels, strips, or shingles). More complex double curved topologies can be fabricated by developing approaches that do not rely on permanent formwork or substructure. For example, work by Mark Fornes shows that complex, branching mesh topologies can be assembled from thin aluminum strips by using the inherent stiffness of the surface to support the form during assembly (Fornes 2014). Ayers, Martin and Zwierzycki work with the hand craft of Kagami weaving illustrates how lattice structures can be fabricated by working from a modification of the edge topology of arbitrary mesh geometries (Ayres, Martin, and Zwierzycki 2018). Extensive examples can be found in the literature of fabricating simpler surface topologies from elastically bent timber lamellas, such as work by Djordje Stanojevic (Quartara and Stanojevic 2019) or Oliver David Krieg (Meyboom, Correa Zuluaga, and Krieg 2019). Each of these examples requires exactly determining the geometry of parts to fabricate a desired surface.

Simple Approaches to Fabricating Curved Surfaces in Augmented Reality

Despite the recent proliferation of examples of augmented reality fabrication in academic literature (Song, Koeck, and Luo 2021), there are relatively few demonstrated examples of methods for fabricating curved surfaces. Artillion Studio utilized augmented reality to form a complex double curved sculpture from hand-bent pieces of stainless steel flat bar (*Inferno Redux/Wind Stone* 2018). Kristof Crolla and Garvin Goepel have demonstrated the use of augmented reality to form double curved surfaces from interwoven strips of bent bamboo (Goepel and Crolla 2020). Yang Song completed a project called *BloomShell* as part of Soomeen Hahm's research lab at the Bartlett that explored the use of thermoplastic sheets to form and assemble double curved surfaces from augmented reality guides (Song 2020). In each of

4

these examples, surfaces are first discretized into explicitly modeled parts which are used to guide simple forming processes. Parts are then assembled together to form larger curving surfaces.

Guided Fabrication in Augmented Reality

Many technical approaches exist to taking dimensional measurements from physical objects (Peggs et al. 2009). The comparison of metrology data from physical objects to target dimensions defined by digital models can be used to guide fabrication processes in augmented reality. By mapping the distance between current and target vertex positions of a design to a color gradient and projecting these colors directly on the workpiece, Skeels and Rehg demonstrated that guided augmented reality could enable fabricators to sculpt accurate recreations of digital models by hand (Skeels and Rehg 2007). Yoshida et al. demonstrated that similar real-time projection-based methods could be used to assist in additive fabrication of building scale structures from chopsticks (Yoshida et al. 2015). However, projection based methods rely on the work surface being in the line of sight of the projector, and limit the capacity to work with surfaces with large numbers of undercuts or other complex topologies. Chun et al. demonstrated a method for carving arbitrary designs from clay in augmented reality that utilized photogrammetry to create 3D models of as-built objects and indicating differences to target geometry by colorizing mesh vertices. However, the authors note that this process is time consuming and could be improved with real-time methods for scanning the as-built object (Chun et al. 2020). Object detection and tracking algorithms have also been shown to assist

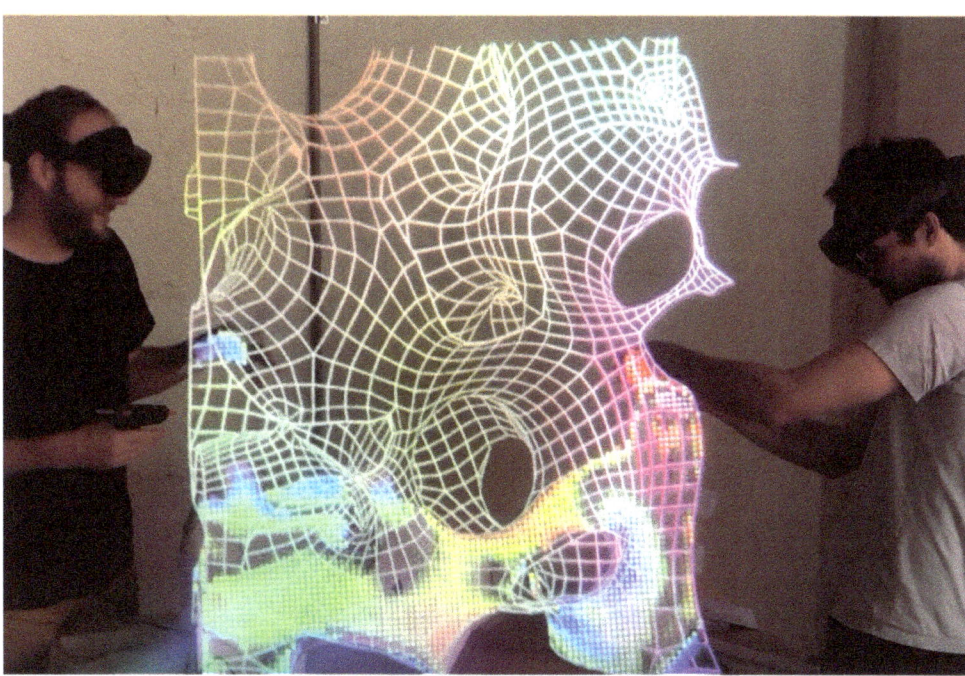

4   Aruco markers stenciled on the surface of an acrylic sheet to approximately track changes in the geometry of the surface during fabrication

5   Simultaneous fabrication by two fabricators; color indicates deviation of physical material from the target surface

fabricators with positioning parts in arbitrary orientations (Sandy and Buchli 2018), eliminating the need to calibrate a 3D scan with the workpiece. The limitation of object tracking methods is that the position of parts in the digital design must be defined prior to fabrication, preventing the possibility of working with subtractive or deposition-based manufacturing processes, or assembly from ad-hoc distribution of parts.

Simple Approaches to Fabricating Curved Surfaces in Augmented Reality

We have previously attempted several approaches to fabricating complex double curved surfaces without explicitly designing parts and instead using only the surface model viewed in augmented reality on the HoloLens. The HoloLens is an augmented reality display produced by Microsoft that renders virtual objects called "holograms" by projecting a stereoscopic image of the object onto a transparent lens located a few centimeters in front of the wearer's eyes. Because the light representing a hologram originates from this 2D display and not from the hologram's true position in physical space, physical objects that are between the hologram and the display are occluded by the additive light of the hologram and disappear behind them. Thus, trying to use a hologram for fabrication of surface models is challenging because it is very difficult to correctly judge the precise depth of the surface of a hologram without the ordinary reference points of knowing if something (like a part, construction material or your own hand) is in front or behind it.

We have previously attempted to work around this limitation by constructing curved structures in layers. The surface model is first contoured or sliced into layers that are then displayed to the user one at a time to reliably view the correct shape of the form at any given height without creating large areas of holographic surface that would occlude the physical structure being built. However, this results in design detail being directly correlated to the number of layers (as in additive manufacturing) and introduces additional labor if small details or accurate curvature are required. Furthermore, the representation of double curved surfaces as planar layers prevents fabricators from observing and understanding changes in the geometry and curvature of the surface that would be useful to efficiently plan fabrication tasks and improve workmanship.

We have also explored simple approaches to digitizing existing parts and structures during fabrication using fiducial markers directly embedded in construction materials (Figure 4). Fiducial markers such as printed Aruco or QR codes provide a low resolution and approximate indication of the geometry described as planes within Grasshopper. The main limitation of this approach is that to increase the resolution of the digitized geometry, markers must decrease in size thereby reducing precision. Digitizing geometry with printed fiducial markers also does not generalize very well to double-curved surfaces, as markers deform out of plane when fixed to twisting strips, and surface curvature results in most markers appearing at oblique angles to the tracking camera on the HoloLens, all of which result in suboptimal tracking precision. The methods outlined in this paper make use, instead, of the long-throw depth camera on the HoloLens to provide fabricators with real time visual feedback on the current position of construction material relative to the desired geometry of the surface.

METHODS

Mixed Reality Application

An augmented reality application was developed by the authors in Unity (Unity Technologies 2021) for viewing models on the HoloLens 2. The Unity application includes: (1) a model loader for downloading GLTF models hosted remotely on the cloud; (2) utilities for locating models in physical space, and thereby enabling shared experience of virtual content by viewing the same model in the same location on multiple HoloLens 2 headsets; (3) rendering models as surfaces or outlines; and (4) displaying depth stream data as a colored point cloud. The HoloLens 2 provides access to the long throw and short throw depth stream data behind a C# API available when Research Mode is enabled. Our method utilizes long-throw depth data at a maximum resolution of 1024 x 1024 depth points updated at five frames per second (Bamji et al. 2018).

Rendering Point Clouds on the HoloLens

The long throw point cloud was clipped to 350 x 410 to approximately match the 52-degree field of view provided by the display. These points were then further filtered to remove background features by culling points further than 150 mm from the target geometry. Due to the performance cost of exact mesh proximity calculations, we developed a method of approximate mesh proximity testing by raycasting from the depth camera location through the geometry, followed by a second raycast from the detected point position in the inverse direction of the mesh normal (Figure 6).

This reduced the inaccuracy of distance measurements on curved surfaces or those at oblique angles to the camera position. The point cloud is rendered as a gradient through red, green and blue, signifying in front of, equal to, and behind the surface of the digital mesh model. The rendered opacity of each point decreases with proximity to the surface, and a completely transparent point indicates that the point is within 3 mm of the digital mesh model.

Proof of Concept Surface Design

A proof-of-concept prototype was designed by filling a volume of 900 x 600 x 1250 mm with 38 randomly distributed points that were then isosurfaced to generate a uniformly curving mesh with an arbitrarily complex topology. The design of the surface aimed to balance estimated structural stiffness due to double curvature with the size of strips required to match this curvature; however, there were otherwise very few intended design constraints. The mesh of the surface was then offset to create geometry for ray casting point clouds against, and the edges of a low-resolution quad mesh approximation of the design model were extracted to provide fabricators with an approximate guide to the design intent during fabrication.

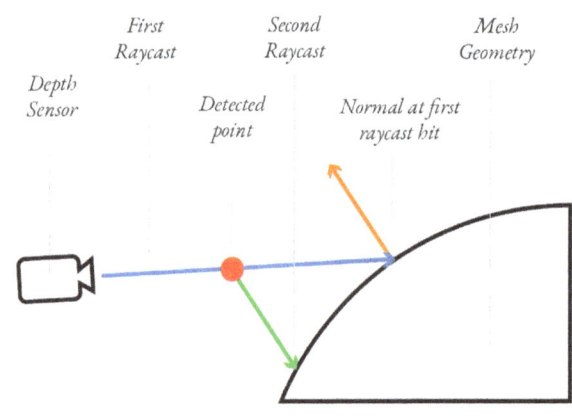

6   Diagram illustrating raycasting algorithm; note the distance along the second raycast to the mesh (green) is less than the distance along the first raycast (blue)

Fabrication

The prototype was fabricated by three people working simultaneously from the same digital model viewed in augmented reality on the HoloLens 2. Each fabricator began to construct a part of the model by locating 50 x 300 mm strips of 300 gsm paper such that they approximated the geometry of the surface indicated in augmented reality. Each strip would typically need to bend and twist into arbitrary curves to match the geometry of the surface. While the strip of paper was visible to the depth camera on the HoloLens and within a distance threshold of 150 mm to the mesh, any points that intersected the strip would be displayed on the HoloLens. As the fabricator moved the strip closer to the surface the color of these points would change to indicate a change in the proximity of the physical strip to the surface (Figure 4).

In practice no strip perfectly matched the geometry of the design model due to limitations of forming strips by hand, the flexibility of the material and the work time of the adhesive. Instead fabricators would optimize for fabrication speed rather than fabrication precision. Once a strip was within acceptable tolerances of the surface it would be fixed in place with minor imprecision occasionally overcome by inserting subsequent strips. The visual feedback on the precision of the strip was therefore a means to understand the desired geometry of the design model rather than a means to improve precision over unguided processes. Acceptable tolerances varied over the course of fabricating the prototype as it became necessary to work with large variations due to deformation from self-weight.

7   Image of the point cloud scan and capture locations; colors indicate deviation of points from the target digital model

Fabricators would typically test multiple strip orientations and shapes using this visual feedback to balance several fabrication goals. These included maximizing covered surface area, strip and surface curvature, connections to existing strips for stiffness and accessibility for the purposes of applying adhesive. Because of the inherent flexibility of the material, fabricators would attempt to avoid adding strips that could not be supported by existing structure or that would add cantilevers causing the structure to sag. The geometry of strips was not defined a priori by the design model, and therefore, the patterns produced by the strips were determined on the fly by each fabricator. Adding strips to the model following the heuristics previously described required skill, and each fabricator developed a unique approach to inserting strips into the structure with a corresponding and recognizable aesthetic (Figure 5).

Once a satisfactory geometry for a strip was found, one end of the strip was pinned in place using hot glue. The remainder of the strip was then formed into the planned geometry and tacked in place with additional adhesive. This allowed some workability of the strip and prevented accidentally fixing material in unplanned locations due to the very short work time of the adhesive. Because the visual feedback ran as a background process and was "always on," any part of the as-built structure could easily be compared to the design model for precision and small changes could be made such as adding more strips to reinforce slumping areas, or to work on a completely different part of the model (Figure 5).

Because no parts were defined by the design model, there was also no predetermined and explicit sequence of assembly, and fabrication could occur at any part of the design model at any time. A sparse distribution of strips was produced that approximated the form of the prototype while holding its form as much as possible under self-weight. This sparse distribution was then reinforced with additional strips that could be added without using an augmented reality guide.

RESULTS AND DISCUSSION

The prototype was completed in 6 hours by the team of three fabricators using approximately 1000 strips of 50 x 300 mm 300 gsm paper (Figure 2). The average deviation of the fabricated surface was 27 mm. This deviation was calculated by recording 24 point clouds from 8 vantage points and performing a closest point analysis for the approximately 200,000 points (Figure 7). Because this deviation is distributed over large surfaces rather than local to the geometry of individual strips, we expect deviation to be due to deformation of low-curvature areas of the surface under self-weight rather than any technical limitations to the method described for guided fabrication. This deviation could be reduced by increasing the stiffness of construction material, increasing the curvature of the design or avoiding cantilevering and unsupported geometry. However, we expect that deviation and construction time could be reduced further simply by increasing the skill of the fabrication team.

Our fabrication team had no prior experience constructing such complex surfaces from paper, and different ad hoc

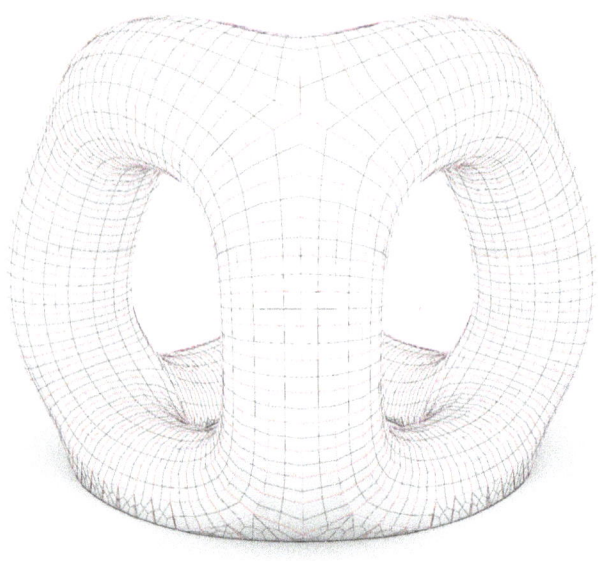

8 A simple double-curved surface model for fabrication from timber veneer

9 A prototype fabricated with our method from long strips of timber veneer

methods for locating strips following the surface of the design model were invented and refined while fabricating the prototype. These included attempting to follow the edges of the quad mesh, following estimations of minimum and maximum curvature, following and reinforcing topological loops or simply optimizing for simplicity of installation. Providing fabricators with some additional guidance on optimal strip orientation—for instance, by displaying principal curvature directions of the mesh or estimating geodesic curves for a given starting vector of a strip—would accelerate the acquisition of skills that would assist in fabricating a maximally stiff representation of arbitrary curving surfaces. However, this would also introduce additional design complexity and necessitate an understanding of these computational tools and approaches.

We contribute a novel method for approximately fabricating arbitrarily curving surfaces in augmented reality that does not require an underlying substructure or rationalization to explicit parts. Because the method does not require the assembly of sub structures, 2D documentation or any form of digital design rationalization we further contribute a fast, low cost, accessible and unconstrained design to production workflow compared to additive manufacturing or other computer aided manufacturing processes such as assembly from CNC cut strips. Our method affords an intuitive approach to fabricating surfaces from flexible and lightweight materials in the same fashion as artists such as Henrique Oliviera, while also affording the possibility of relative precision, analysis and iteration provided by digital design workflows. Mixed reality environments provide an additional benefit and contribution to these types of projects, whereby bottlenecks of design expertise are eliminated by simultaneous and unambiguous access to design intent. This enables complex structures to be fabricated efficiently in parallel and suggests that this would enable even larger and more complex structures to be completed simply by increasing the size of the fabrication team.

To evidence that the method described in this paper is generalizable to other fabrication approaches we have explored two other small proof of concepts including additive fabrication from river pebbles (Figure 10) and arbitrary part placement with six degrees of freedom (Figure 11). In both cases, providing fabricators with visual feedback on the delta between the location of physical material and the target geometry improves the ability of fabricators to complete tasks. We speculate that reducing the complexity of these fabrication tasks in augmented reality will impact the discourse on digital fabrication by broadening the design space of fabricable structures in augmented reality without introducing the need to rationalize surfaces to parts or substructures. This in turn will have follow on implications by improving the feasibility of constructing geometry from arbitrary, heterogenous, and non-standard materials such as those in the *Stik Pavilion* (Yoshida et al. 2015), as the digital model does not need to explicitly model material behaviors and constraints for this material to be guided into a target form.

Our guidance system enabled the fabrication team to approximately complete the prototype in a very short span of fabrication time compared to alternative methods. The speed of the method is achieved by enabling a theoretically unlimited

10 Guidance system for assembly of curved surfaces from pebbles

11 Guidance system for positioning a part with 6 degrees of freedom

number of fabricators to work on construction simultaneously. As evidence of the efficacy of the guidance system, the fabrication team was not able to construct any part of the prototype by working with the hologram of the design model alone. This was attributed to the difficulty of matching strip geometry to surface geometry in augmented reality. Fabricating the prototype would, therefore, ordinarily require explicitly defining parts making up the surface and fabricating each part from rigid material to within tolerances of the digital model as has been demonstrated by other approaches to realizing complex surface models in augmented reality.

While completing the prototype we learned that forming double curved surfaces from flexible strips of paper requires craft skill that needed to be acquired from practice. This resulted in initially poor workmanship and precision that improved gradually over time. Poor decisions in forming parts of the structure introduced areas of low curvature that then deformed under self-weight and could not be reinforced with additional material. The primary failure in our method, as demonstrated, is in the appropriateness of the design language of the proof-of-concept model. Large areas of the model lacked sufficient double curvature or topological support to hold their form when constructed from paper, introducing deviation from the digital model that had little to do with the precision of the guidance system or the accuracy of individual strips. The marching cubes algorithm used to create the mesh of the proof-of-concept model also contained many details too small to form with the 50 mm wide strips, leading to further deviations that could easily have been avoided.

CONCLUSION

This proof-of-concept project represents the very first experiment with real-time guided fabrication of surface models in augmented reality without discretization to explicitly defined parts. A fundamental cause of error and limitation of our prototype was deformation of the structure under self-weight, and in future work we aim to address this limitation by working with more rigid materials and design models. However, we also recognize the opportunities afforded by light weight and readily available materials such as paper, and the possibility of exploring ad hoc design processes by coloring or patterning the paper strips as an additional expressive device during fabrication.

We have conducted additional experiments forming curved surfaces from strips of timber veneer (Figures 8, 9). Our method should generalize to pliable construction scale materials such as bent timber or steel, as well as to sculpting materials such as clay and plaster. Guided feedback in augmented reality may also address limitations observed by other fabrication approaches within the discourse. For instance, bricklaying with simple augmented reality guides describing courses of bricks in a design requires maintaining a top-down view to accurately position each brick. This in turn necessitates frequently moving scaffolding to construct larger walls, and this is unfamiliar to bricklayers used to working up to and beyond eye height. Guided feedback should enable the accurate placement of bricks in arbitrary orientation and eliminate the need for top-down views. Future work intends to explore these opportunities.

While we have aimed to implement and demonstrate a generalizable and accessible method for guided fabrication in augmented reality, future work will refine this approach to address other observed limitations. By performing analysis on design models, it may be possible to reduce the risk of poor structural performance by displaying structural analysis, geodesic curves or principal stress directions in real time based on the proposed position of a strip of material by a fabricator. Similarly, providing designers with access to the HoloLens 2 depth stream within parametric modeling software such as Grasshopper would enable the design of customizable guidance algorithms. One could imagine fitting parts to point clouds, displaying correctional information with arrows, or colorizing mesh surfaces with various proximity information, and these will be explored in future work and collaborations.

REFERENCES

Ayres, Phil, Alison Martin, and Mateusz Zwierzycki. 2018. "Beyond the Basket Case: A Principled Approach to the Modelling of Kagome Weave Patterns for the Fabrication of Interlaced Lattice Structures Using Straight Strips." In *Advances in Architectural Geometry 2018*. Vienna: Chalmers University of Technology. 72–91.

Bamji, Cyrus S., Swati Mehta, Barry Thompson, Tamer Elkhatib, Stefan Wurster, Onur Akkaya, Andrew Payne, et al. 2018. "IMpixel 65nm BSI 320MHz Demodulated TOF Image Sensor with 3μm Global Shutter Pixels and Analog Binning." In *2018 IEEE International Solid - State Circuits Conference - (ISSCC)*. 94–96. https://doi.org/10.1109/ISSCC.2018.8310200.

Chun, Jacky, Adabelle Long, Wing Sze, Chung Hei, Garvin Goepel, and Kristof Crolla. 2020. "Augmenting Craft with Mixed Reality: A Case Study Project of AR-Driven Analogue Clay Modelling." In *ACADIA 20: Distributed Proximities; Proceedings of the 40th Annual Conference of the Association of Computer Aided Design in Architecture (ACADIA)*, edited by B. Slocum, V. Ago, S. Doyle, A. Marcus, M. Yablonina, and M. del Campo. 436–444.

Farquhar-Still, Geoff. 2018. "Inferno Redux/Wind Stone." Geoff Farquhar-Still (website). Accessed July 1, 2022. https://geofffarquharstill.com/exhibitions/inferno-redux-wind-stone.

Fornes, Marc. 2014. "Double Agent White." In *ACADIA 14: Design Agency; Projects of the 34th Annual Conference of the Association for Computer Aided Design in Architecture (ACADIA)*. Los Angeles, USA. 157–160.

Goepel, Garvin, and Kristof Crolla. 2020. "AUGMENTED REALITY-BASED COLLABORATION - ARgan, a Bamboo Art Installation Case Study." In *RE: Anthropocene, Design in the Age of Humans; Proceedings of the 25th CAADRIA Conference*. Bangkok, Thailand. 313–322.

Jahn, Gwyllim, Cameron Newnham, Nick Berg, and Matthew Beanland. 2018. "Making in Mixed Reality." In *ACADIA 18: On Imprecission and Infidelity; Proceedings of the 38th Annual Conference of the Association for Computer Aided Design in Architecture (ACADIA)*, edited by P. Anzalone, M. del Signore, A. J. Wit. Mexico City, Mexico. 88–97.

Jahn, Gwyllim, Cameron Newnham, Nick Berg, Melissa Iraheta, and Jackson Wells. 2019. "Holographic Construction." In *Impact: Design With All the Senses*, edited by C. Gengnagel, O. Baverel, J. Burry, M. Ramsgaard Thomsen, and S. Weinzierl. Berlin: Springer. 314–324.

Jahn, Gwyllim, Cameron Newnham, and Nick van den Berg. 2022. "Augmented Reality For Construction From Steam Bent Timber". In *Proceedings of the 27th International Conference of the Association for Computer-Aided Architectural Design Research in Asia*. Sydney, Australia.

Jahn, Gwyllim, Andrew Wit, and James Pazzi. 2019. "[BENT] Holographic Handcraft in Large-Scale Steam-Bent Timber Structures." In *ACADIA 19: Ubiquity and Autonomy; Proceedings of the 39th Annual Conference of the Association for Computer Aided Design in Architecture (ACADIA)*, edited by K. Bieg, D. Briscoe, and C. Odom. Austin, Texas. 438–447.

Lok, Leslie. 2022. "Timber De-Standardized 2.0 : Mixed Reality Visualizations And User Interface For Processing Irregular Timber." In *Proceedings of the 27th International Conference of the Association for Computer-Aided Architectural Design Research in Asia*. Sydney, Australia.

Meyboom, Annalisa, David Correa Zuluaga, and Oliver Krieg. 2019. "Stressed Skin Wood Surface Structures: Potential Applications in Architecture." In *ACADIA 19: Ubiquity and Autonomy; Proceedings of the 39th Annual Conference of the Association for Computer Aided Design in Architecture (ACADIA)*, edited by K. Bieg, D. Briscoe, and C. Odom. Austin, Texas. 470-477.

Mitterberger, Daniela, Kathrin Dörfler, Timothy Sandy, Foteini Salveridou, Marco Hutter, Fabio Gramazio, and Matthias Kohler. 2020. "Augmented Bricklaying: Human–Machine Interaction for in Situ Assembly of Complex Brickwork Using Object-Aware Augmented Reality." *Construction Robotics* 4 (December): 151–161. https://doi.org/10.1007/s41693-020-00035-8.

Peggs, G. N., Paul Maropoulos, Ben Hughes, Alistair Forbes, Stuart Robson, Marek Ziebart, and Bala Muralikrishnan. 2009. "Recent Developments in Large-Scale Dimensional Metrology." *Proceedings of the Institution of Mechanical Engineers, Part B: Journal of Engineering Manufacture* 223 (6): 571–595. https://doi.org/10.1243/09544054JEM1284.

Quartara, Andrea, and Djordje Stanojevic. 2019. *Computational and Manufacturing Strategies; Experimental Expressions of Wood Capabilities*. Singapore: Springer. https://link.springer.com/book/10.1007/978-981-10-8830-8.

Sandy, Timothy, and Jonas Buchli. 2018. "Object-Based Visual-Inertial Tracking for Additive Fabrication." *IEEE Robotics and Automation Letters* PP (99): 1–1. https://doi.org/10.1109/LRA.2018.2798700.

Skeels, Christopher, and James M. Rehg. 2007. "ShapeShift: A Projector-Guided Sculpture System." *Proc. UIST* 7: 2.

Song, Yang. 2020. "BloomShell: Augmented Reality for the Assembly and Real-Time Modification of Complex Curved Structure." In *Anthropologic: Architecture and Fabrication in the Cognitive Age; Proceedings of the 38th ECAADe Conference.* Berlin. 345–354.

Song, Yang, Richard Koeck, and Shan Luo. 2021. "Review and Analysis of Augmented Reality (AR) Literature for Digital Fabrication in Architecture." *Automation in Construction* 128 (August): 103762. https://doi.org/10.1016/j.autcon.2021.103762.

Unity Technologies. *Unity Engine*. V. 2021. Unity Technologies. PC. 2021.

Yoshida, Hironori, Takeo Igarashi, Yusuke Obuchi, Yosuke Takami, Jun Sato, Mika Araki, Masaaki Miki, Kosuke Nagata, Kazuhide Sakai, and Syunsuke Igarashi. 2015. "Architecture-Scale Human-Assisted Additive Manufacturing." *ACM Transactions on Graphics* 34 (4): 88:1–88:8. https://doi.org/10.1145/2766951.

IMAGE CREDITS
All drawings and images by the authors.

Gwyllim Jahn is a co-founder of Fologram and a Lecturer in Architecture at RMIT University in Melbourne. His work focuses on designing for mixed reality fabrication, most notably in the design and construction of the 2019 Tallinn Architecture Biennale Pavilion. Gwyllim's research has been published in leading computational design conferences and journals including IJAC, ACADIA, and RobArch, and he has given talks, presentations, and workshops at international institutions including MIT, Stuttgart ICD, UCL, SciArc, Tongji and Tsinghua University.

Cameron Newnham is the co-founder and CTO of Fologram where he leads the technical development of mixed reality software for the design and construction industry. His experience lies in the creation of novel tools for designing and fabricating complex geometric systems, ranging from code libraries to mixed reality interfaces and extending to machine design and robotic fabrication. Cameron has experience as a computational designer in internationally renowned architectural practices, and academic experience as an Associate Lecturer–Industry Fellow at RMIT University. Cameron has led numerous international design and build workshops in Shanghai, New York, Paris, Boston, Sydney, and Melbourne.

Nick van den Berg is the co-founder and CEO of Fologram, a design research practice and technology startup building a platform for designing and making in mixed reality. Fologram's platform is being used by world leading architecture practices, product design houses, manufacturers and design schools internationally. Nick is especially interested in building solutions that are utilized by a large user base around the world, and he has assisted with workshops focusing on utilizing augmented and mixed reality as a design and fabrication tool at DMS, CAADRIA, Cooper Union, McNeel Europe, UDK & TU Berlin.

BIMxAR: Building Information Modeling-Powered Augmented Reality

Ziad Ashour
King Fahd University of Petroleum and Minerals

Wei Yan
Texas A&M University

Accurate Building-Scale AR Registration and Novel Mixed-Reality Visualization for Understanding Buildings and BIM

1 A BIMxAR user in the physical environment (left) and a view in AR scene where the physical building is superimposed by its virtual Building Information Model with cross-sections visualized (right)

ABSTRACT

Literature review shows limited research investigating the utilization of Augmented Reality (AR) to improve learning and understanding architectural representations, specifically section views. In this study, we present an AR system prototype (BIMxAR), its new and accurate building-scale registration method (DL-3S-BIM) for aligning BIM and physical buildings, and its novel visualization features that facilitate the comprehension of building construction systems, materials configuration, and 3D section views of complex structures through the integration of AR, Building Information Modeling (BIM), and physical buildings. We present our approach to enable the user to understand the orientation and the coordinate system of the virtual model with respect to its location in the physical environment through body movement in the environment, facilitating embodied learning. Moreover, the study presents an innovative method to create sections within an AR setting. The method enables users to further inspect the building from different views through fully controlling the location and orientation of the sectional planes. Additionally, it allows the user to retrieve the building component's related information through section poches. The study developed and showcased a novel mixed-mode of real and virtual worlds (mixed reality) to provide a better understanding of the spatial relationships in a building.

INTRODUCTION

The ability to translate virtual information and relate it to the physical world is a crucial skill in the domain of architecture. Cognitive mental loads on the students are anticipated during the process of translating and relating components of a 2D or 3D drawing to their locations in the physical world due to the differences in views, perspective angles, and scales (Diao and Shih 2019). The mental effort required to process multiple sources of information that are distant from each other can increase the extraneous cognitive load (Schroeder and Cenkci 2018). One key feature of Augmented Reality (AR) is superimposing virtual content relative to its correct location in the physical world.

There is scant research investigating the utilization of AR in facilitating learning and creation of building sections, which are important in building design, construction, and modeling. Additionally, the limited examples of BIM-enabled AR in the literature lack the level of interaction needed for building components inspection. Thus, further investigation in these particular areas is required. The research asserts the necessity to explore new methods that improve spatial abilities in the domain of architecture education.

Further, this research seeks to explore the AR effects on assisting students to comprehend and reproduce architectural sections by utilizing AR: augmenting physical buildings by virtual building models. The developed prototype (BIMxAR) utilizes the physical-virtual overlay feature to facilitate spatial learning using existing physical buildings and their Building Information Modeling (BIM) analogues (Figure 1). It enables the user to virtually cut and see through the building being inspected, in a way similar to magnetic resonance imaging (MRI), to provide better visualization that enables embodied learning for improved understanding of the internal elements behind finishes and how they integrate with other systems.

In this paper, we present: (1) the workflows, (2) the model registration methods that we have explored and developed in BIMxAR, (3) extraction of the BIM metadata and its utilization in AR, (4) the user interface and the graphical representations inside an AR environment, (5) user interaction with the AR environment, and (6) the section creation function.

STATE OF THE ART
Augmented Reality (AR)

AR can be defined as an interactive display system that enhances reality in real-time by contextually aligning virtual objects with the physical world (Azuma et al. 2001). Virtuality continuum (VC) is a continuous scale spanning from a real-world to a virtual environment, and anything in between is a potential combination of real and virtual objects (mixed realities); one such combination is AR (Milgram and Kishino 1994). Unlike virtual reality (VR), where the user is completely immersed in a synthesized environment that is disconnected from the real world around the user, AR enhances real-world perception by complementing it with virtual objects (Azuma 1997). Physical environment tracking and virtual information registration (alignment between virtual and physical objects) in the real world are the key functions of an AR system (Li, Nee, and Ong 2017). The tracking unit in an AR system must understand the environment and track the camera relative to the real world in order to correctly align virtual information with a real-world environment (Van Krevelen and Poelman 2010). AR utilizes one or multiple tracking technologies, e.g., digital cameras, optical sensors, Global Positioning System (GPS), electronic compasses, and accelerometer-gyroscopes (Chatzopoulos et al. 2017). The selection for a tracking or registration method depends on the application for which it will be used and the environment in which it will be used (Meža, Turk, and Dolenc 2014). Registration (model alignment) methods can be categorized into three categories: vision-based, sensor-based, and hybrid methods (Chatzopoulos et al. 2017).

Augmented Reality in Architectural Education

AR has the potential to reform the architecture, construction, and engineering education (Shanbari, Blinn, and Issa 2016). It has been already explored in several areas in architecture and construction education. For example, AR has been employed in project presentation (Fonseca et al. 2014; Sánchez Riera, Redondo, and Fonseca 2015), design (Chen and Wang 2008), teaching CAD (Wen et al. 2021; Devedzic and Bari 2016; Chandrasekera and Yoon 2015; Kim and Irizarry 2020), geometric transformations (Shaghaghian et al. 2021; Shaghaghian, Yan, and Song 2021), architectural history (Chu et al. 2019), structural analysis (Turkan et al. 2017), and architectural lighting (Birt, Manyuru, and Nelson 2017). In spite of that, our review of the literature indicates a little emphasis on the utilization of AR in teaching students about building construction system integration, material assemblies, and section view creation. Moreover, many studies, such as (Vasilevski and Birt 2020; Vassigh et al. 2018; Shanbari, Blinn, and Issa 2016), lack the alignment of the virtual and the physical building objects, which is a core feature of a true AR experience. Few examples in the literature utilize this core feature. Additionally, the amount of interaction that allows students to inspect the virtual content (building components) is very limited, in examples such as Vasilevski and Birt (2020) and Sánchez et al. (2015). Furthermore, other examples, such as Vassigh et al. (2014), provide limited visualizations through axonometric views from one single angle.

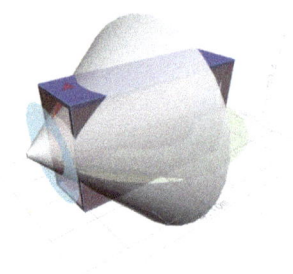

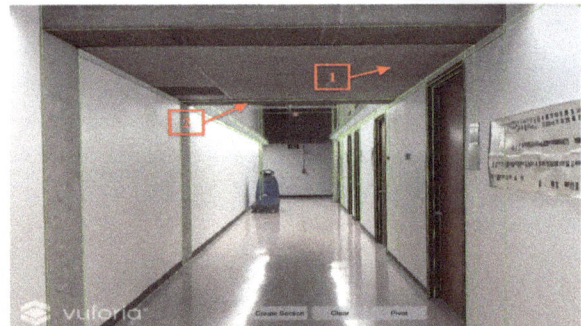
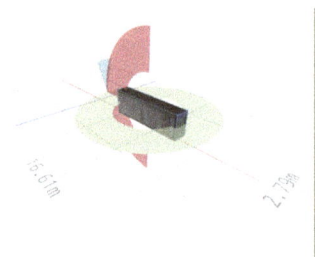
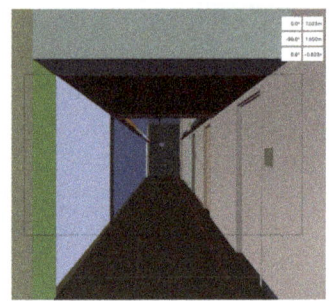

METHODS

The current research showcases the performance and the technical aspects of BIMxAR in terms of the workflows, registration methods, BIM metadata extraction and retrieval in AR, user interface and graphical representation, and section creation. To develop BIMxAR, we used Unity®, which is a common AR platform and gaming engine. Unity houses AR Foundation, which contains core features of ARCore and ARKit. Programming in Unity was done using the C# language, and the developed prototype is an iOS application deployed to devices including iPhone and iPad.

AR Model Registration

We considered different solutions to register the virtual model in the physical space. One solution is based on our previous prototype, which utilizes GPS, inertial measurement unit (IMU) sensors, and manual transformation controls (Ashour and Yan 2020). The solution can support registration in outdoor and part of indoor environments, but accurate alignment cannot be maintained in the indoor environment when relying only on the device's motion sensors. This necessitates the integration of other types of tracking, such as computer vision and artificial intelligence, specifically deep learning methods.

Our new solution utilizes computer vision and 3D model-based AI/Deep Learning (DL), e.g., Vuforia Model Target. We employed Vuforia Model Target, which is normally used for registering small-scale 3D objects, e.g., artifacts and cars, but is not designed for registering large environments, such as a spaces or buildings, in the physical environment. The adopted method requires an accurate reconstructed 3D model of the physical building in order to generate a model target database in Vuforia Model Target Generator (MTG) that will later be utilized by BIMxAR to recognize and track the physical building, as shown in Figure 2 (right). However, such method normally does not perform well in larger environments.

Through extensive experiments, we developed an integrated 3D model-based Deep Learning with 3D Scanning-corrected BIM (DL-3S-BIM) as our registration method for the scale of buildings, and this method has been proven to provide the best solution in terms of accuracy and robustness, as shown in Figure 3, with only minor errors throughout the virtual model (explained in detail below).

The reconstructed 3D model in Figure 2 (center) was created based on measurements taken manually. We noticed that the generated model target database from the 3D model based on manual measurements could only work with small spaces. However, in larger spaces, the 3D model of the physical space must be based on more accurate measurements using a professional 3D scanner, e.g. Matterport's Structured Light Scanner. Therefore, we used the Matterport Pro2 scanner to scan the entire space and reconstruct its 3D model.

During the training, a cone view (virtual camera view) in Vuforia MTG is placed inside the 3D model. The location is defined to cover most of the physical space by setting the cone view at the midpoint of the space height. For smaller spaces, the azimuth range was set at 360 degrees, and the

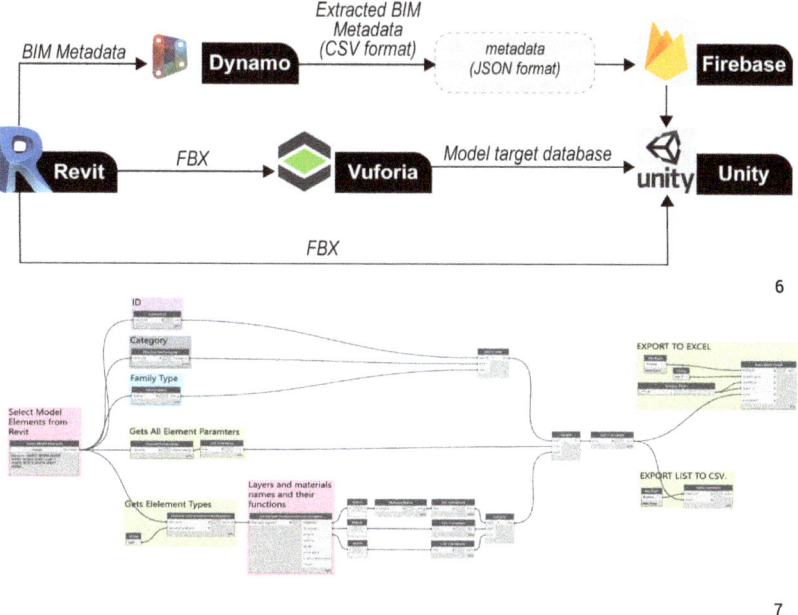

2   (left) Actual physical space; (center) 3D BIM model, shaded, of the physical space based on manual measurements, built for Vuforia DL; (right) the BIM model, wireframe, registered in the physical space

3   3D BIM model (green wireframe) registered in the physical space with high accuracy using the DL-3S-BIM method; the physical suspended false ceiling (1) occludes the front face of the virtual concrete beam (2)

4   The imported virtual model, based on manual measurement of the physical space (in FBX format) in Vuforia MGT; the Azimuth Range is set to 360 degrees, and the Elevation Range is set from -40 to +50 degrees

5   The imported virtual model, corrected with Matterport Pro2 3D scan of the physical space (in FBX format) in Vuforia MGT; the Azimuth Range is set to 360 degrees, and the Elevation Range is set from -90 to +90 degrees

6   Workflow BIM (Revit) to AR development in Unity

7   Approach for extracting BIM metadata from Revit via Dynamo

elevation range from -40 to +50 degrees, as shown in Figure 4. For larger spaces (DL-3S-BIM), the azimuth range was set at 360 degrees, and the elevation range from -90 to +90 degrees, as shown in Figure 5. The last step is to align the 3D model with the generated database (target model) in Unity in order to enable BIMxAR to spatially register the 3D model in its correct location and orientation in the physical world.

Workflow

The workflow utilizes BIM (Revit) files, in which the geometric and non-geometric information can be both accessed in Unity, as seen Figure 6. The geometric information (3D model) is exported as an FBX file format, while preserving the building components' IDs, to be used in Vuforia MTG and Unity.

The extraction of BIM metadata is accomplished through Dynamo, a visual programming tool for Revit®, as seen in Figure 7. The proposed approach collects the building model metadata, including the building components' IDs, categories, families, and all related parameters, and exports them into a CSV file format. The CSV file is then converted to the JSON format in order to be stored in a real-time database, Firebase®. A script was developed to enable Unity to retrieve building objects' metadata through their IDs directly from the real-time database.

Shaders

By default, the assignment of opaque shaders for the virtual model will always occlude the physical building on the AR screen, no matter what spatial relations (front or back) exist between the virtual and physical objects. To handle this AR occlusion problem, we decided to use a transparent yet occlusive shader highlighted with a wireframe and assigned it to the virtual model as used in (Yan 2022). As a result, the user can simultaneously view the physical and virtual objects with correct occlusions between them—objects in front occlude those on the back, no matter whether the objects are physical or virtual. Using Figure 3 as an example, the physical suspended false ceiling occludes the front face of the virtual concrete beam. As for building object selection, if the user touches an object of interest, it will be highlighted with a red wireframe shader, and a table of relevant information will be displayed, as shown in Figure 8.

Section Mode

The section creation function allows the user to spatially slice the building to create architectural section views. When creating a section, BIMxAR does not change the geometry. Instead, a Unity Asset shader named crossSection (Tomekkiez 2020) is adopted to create a rendering effect that can be designed to show the section views. The shader allows BIMxAR to create sections by only rendering the part behind the sectional plane and the rest of the model in front of the plane is hidden. It also provides hatch patterns for the section poche. The previous examples in the literature review enable a user to examine a building from specific section views, but preventing the user from examining other parts of the building or revealing internal building elements at specific locations, and thus the user cannot thoroughly inspect the internal parts. In contrast, BIMxAR enables the user to freely control

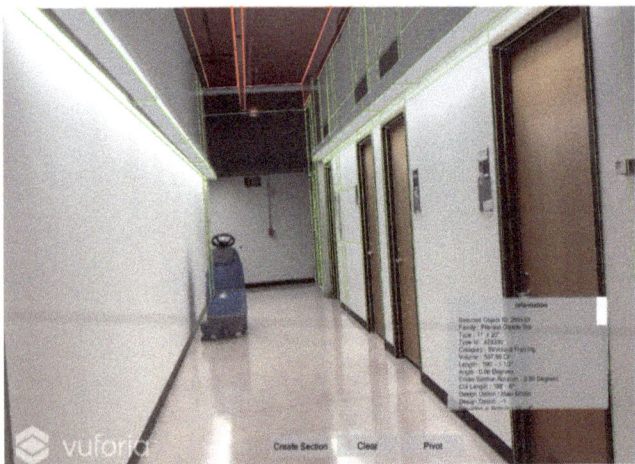

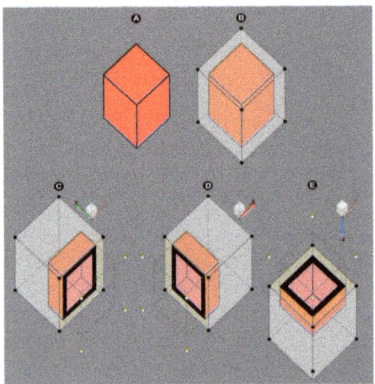

8   BIM model rendered with a transparent-occlusive shader (highlighted with a green wireframe), selected BIM object (T-Beam) rendered with a red transparent-occlusive shader (highlighted with a red wireframe), and BIM metadata window of the selected object

9   (A) Virtual model box rendered in red; (B) Six sectional planes surrounding the virtual model (box); (C) Section box is moved along the y-axis in the positive direction; (D) Section box is moved along the x-axis in the positive direction; (E) Section box is moved along the z-axis in the negative direction

10 A section view along the x-axis (parallel to the left and right walls); sliders on the bottom left corner control the location and the orientation of the sectional plane

the sectional plane location and orientation, allowing the user to inspect the building from different architectural section views, supported by other advanced visualization features described in User Interaction. BIMxAR contains six sectional planes to create a bounding box that surrounds the virtual model or a part of it. This configuration enables the user to create sections at all three axes (x, y, and z) with two orientations (left-right and front-back). To elaborate more, Figure 9A shows a virtual model (box) rendered in red, and when the section mode is enabled, six sectional planes (section box) will surround the virtual model, as shown in Figure 9B. Two sectional planes along each axis, where each sectional plane is responsible for one 3D section view. For example, if a section box is moved along the y-axis in the positive direction, the sectional plane will be facing the negative direction of the y-axis, as shown in Figure 9C. Similarly, if the section box is moved along the x-axis in the positive direction, the sectional plane will be facing the negative direction of the x-axis, as shown in Figure 9D. If a section box is moved along the z-axis in the negative direction, the sectional plane will be facing the positive direction of the z-axis, as shown in Figure 9E.

To control the location of the sectional planes, the interface has three pairs of transformation sliders (x, y, and z). Also, multiple (up to three) sectional views can be simultaneously viewed to inspect the model from different sides. Multiple tests have been conducted to examine the visualization performance of BIMxAR in an AR environment. During the section creation mode, we noticed that if a large portion of the model is discarded or more than one section view is created, the user cannot know if the virtual model is still correctly registered in the physical environment. Therefore, we decided to include the discarded part of the model in the rendering pipeline during the section creation mode. The discarded part is rendered with a transparent shader highlighted with a wireframe, as shown in (Figure 10).

We also needed to support the touch feature and metadata retrieval during the section creation mode, through the section poche. Moreover, this feature becomes valuable when a building component consists of multiple elements, e.g., a wall with multiple layers between the two wall surfaces. Since the virtual model is not modified in terms of geometry when a section is created (a section poche is added onto the wireframe virtual model), highlighting a building component or one of its elements becomes problematic because of how Unity handles ray casting. For example, if a user wants to touch a building object through its poche, the casted ray will hit the first object it will collide with and return its ID or name. Depending on the location of the user in the environment, the ray might hit first the object (rendered invisibly) in front of the poche and eventually highlights the wrong building component

 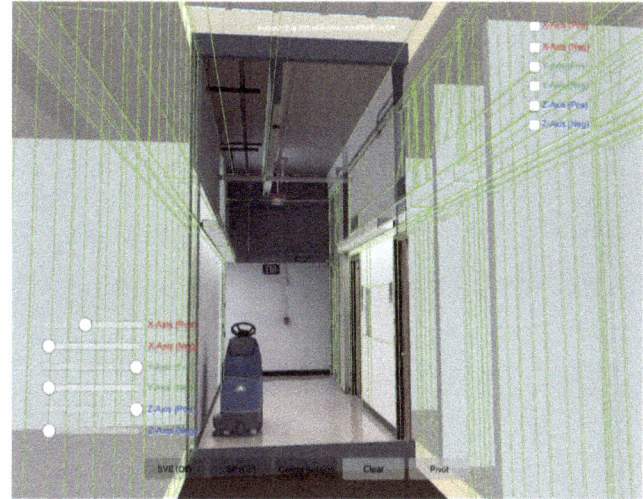

11 Selection of building objects or elements (layers) is enabled through the section poche; the selected element is highlighted with a red shader and its metadata are retrieved and displayed in the right bottom corner of the screen: notice the pivot's orientation (located on the right side of the screen) is aligned with the virtual model's orientation

12 Section view revealing the spaces behind the physical objects being virtually sliced; the walls, floor, and soil in front of the section poche are rendered virtual models, instead of physical building objects

or element. To overcome this problem, we adopted a solution (Tomekkiez 2021) that sorts all the objects that were hit after a ray is cast from the AR camera towards the objects. The solution sorts the hit objects by their distance from the AR camera and checks which hit objects are behind the sectional plane and removes them from the list and finally uses the closest hit object to the sectional plane.

At the section creation mode, the user interface displays six toggles, each of which represents a sectional plane and its orientation. The user must choose one of these toggles to enable the section poche touch feature according to the section view the user is working on. If a building object is highlighted from the section poche, only the part behind the sectional plane will be highlighted, and it will be rendered with a solid red shader. The poche can accommodate multiple patterns (shaders) where each one represents an element (layer), as shown in Figure 11. The UI design allows every single BIM component to be selected and highlighted for examination, even if the AR device screen (iPad) has a very limited area for user interaction. Figure 11 also demonstrates another example of occlusion: the light gray wall on the left, which is a physical wall (and virtually cut), occludes the virtual building elements, such as the framing studs.

Spatial and Context Awareness

The first consideration is to enable the user to understand the orientation and the coordinate system of the virtual model with respect to its location in the physical environment. To achieve this, a three-axis pivot (x, y, and z) has been added to the side of the UI, and its orientation is frequently updated with respect to the AR camera. The pivot becomes handy when the user switches to the Section Mode, as it allows the user to understand the location and orientation of the sectional planes (Figure 11).

The second consideration is to render the context space behind the physical objects being sliced, so that the virtual context space (e.g., a room behind the wall) becomes visible through the "cut openings" on the physical building, while the uncut portion of the physical building component (e.g., the wall) occludes parts of the virtual context space, as shown in Figure 12. This effect produces a new mixed mode of real and virtual worlds that has not been exhibited in the literature before.

DISCUSSION

Although the Vuforia Area Target method provided fairly accurate and robust registration, the scan of the physical space is not sufficiently accurate at corners and edges (rounded instead of sharp), making it difficult to accurately align the virtual model with the scanned space in Unity as shown in Figures 13 and 14. Moreover, the rounded corners of the walls in the scanned space made BIMxAR suffer from misalignment issues whenever the user approaches a corner. Our experiments suggested that the misalignment was due to the reason that Vuforia Area Target was constantly trying to match the corners and edges seen by the AR camera with the scanned space.

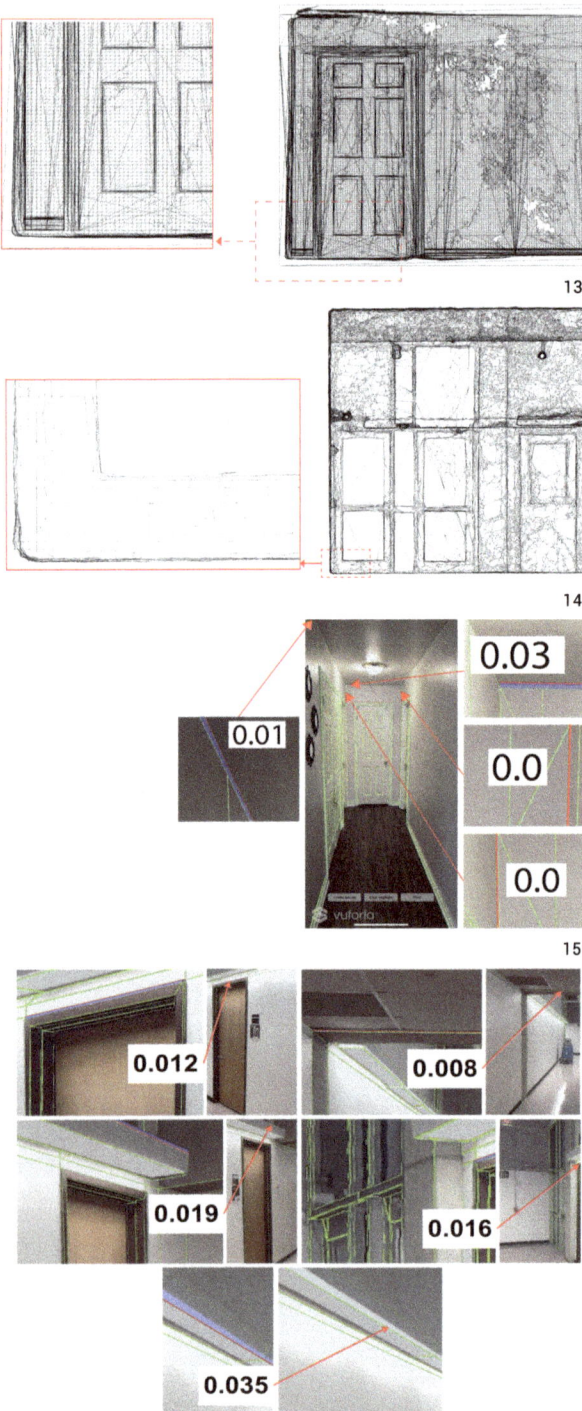

The Vuforia Model Target method for smaller spaces (using a virtual model based on manual measurement) and the DL-3S-BIM method developed in the BIMxAR project provided great results. The method was quantitatively evaluated to measure the performance in registering the virtual model in the physical environment. The error of the registration is defined as the distance measured in the 2D projection of the 3D edges of the physical building and its virtual model. The error is not defined and measured as the 3D distances of the edges, because the measurements are 2D view-dependent and not truly measurable in 3D. The analysis was conducted by visually inspecting multiple screenshots (2D perspective images) and manually measuring the distances at the edges between the virtual model and the physical building. The manual measurement was done in Adobe Illustrator® by first creating a vertical line representing the real height of the physical building, then scaling the screenshot image to match the corner-edges of the physical building with the vertical line, and finally measuring the difference between the virtual model and the physical space, as shown in Figures 15 and 16. The average errors throughout the virtual model in smaller spaces (manual measurement) and larger spaces (DL-3S-BIM) are 0.016 meters and 0.015 meters, respectively.

The highly accurate registration of BIMxAR facilitates the new mixed mode; otherwise, misaligned virtual and physical rooms/walls will not help understand the spatial relationship. In Figure 12, the walls, floors, and soil are rendered virtual models, instead of physical building objects. The virtual models are rendered to reveal the spaces behind the physical building as if the physical building is physically sliced. This is an innovative and improved visualization compared with Figure 10, in which the relationship between the virtual sections and the physical building does not appear to be natural. For example in Figure 10, the portions of the physical door, walls, and T-beam in front of the section poche is still visible, but in reality, if these physical building objects are cut to show the poche, the front portions of these physical building objects should not be visible; instead, the spaces behind them should be partially visible as achieved in Figure 12. We expect that the utilization of the new mode enables a better understanding of the physical context or BIM components being explored and enhances spatial awareness.

13 Scan quality using the iPhone 12 Pro Max built-in LiDAR sensor; corners of the walls are rounded instead of sharp

14 Scan quality using the Matterport Pro2 scanner; corners of the walls are still rounded instead of sharp

15 Measuring the performance of the registration method (Vuforia Model Target) in Adobe Illustrator; the figure shows the alignment differences (in meters) at multiple locations in one of the screenshots

16. Measuring the performance of the registration method DL-3S-BIM in Adobe Illustrator; the figure shows the alignment differences (in meters) at multiple locations in one of the screenshots

CONCLUSIONS AND FUTURE WORK

The purpose of the study is to investigate new methods to improve spatial abilities in the domain of architecture education. Specifically, this research seeks to explore the AR effects on assisting students to comprehend and reproduce architectural section views. We presented our AR system prototype (BIMxAR), its highly accurate registration method (DL-3S-BIM), and its novel visualization features that facilitate the comprehension of 3D section views of complex structures through the integration of AR, BIM, and the physical building. The study developed and showcased multiple novel AR technologies, visualization functions, and applications, as major contributions of this research:

- A highly accurate building-scale AR registration method (DL-3S-BIM) integrating 3D model-based Deep Learning (e.g., Vuforia Model Target), 3D scanning (e.g., Matterport Structured-Light), and BIM
- A complete solution for utilizing a BIM project and its metadata in AR
- Virtual building section views created and registered (aligned) with a physical building in AR through a full control of sectional planes' location and orientation
- A mixed-mode of real and virtual worlds to show the correct spatial relationship among rooms or BIM components related to the section views.

Promising use of BIMxAR has been observed, for example, in the automated, highly accurate alignment between the virtual model (BIM), and the physical building was conveniently utilized by invited students when exploring BIMxAR in the building. For future work, a comprehensive test case and user studies will be conducted to measure more detailed performance of BIMxAR and the effect of our approach on students' learning gains in architectural representations. Moreover, we will explore other AR display systems, specifically, hands-free devices, such as HoloLens, to enhance the users experience and enable additional types of interactions, e.g., eye-gaze and hand gestures.

ACKNOWLEDGMENTS

The authors thank Dr. Mark Clayton's class at Texas A&M University for the BIM (Revit) model of the Langford A building used in the research. This material is based upon work supported partially by the National Science Foundation under Grant No. 2119549 and the Mattia Flabiano III AIA/Page Southerland Page Design Professorship at Texas A&M University.

REFERENCES

Ashour, Z. and W. Yan. 2020. "BIM-Powered Augmented Reality for Advancing Human-Building Interaction." In *Proceedings of the Conference of Education and Research in Computer Aided Architectural Design in Europe (eCAADe)*, vol. 1. Berlin, Germany. 169–178.

Azuma, R., Y. Baillot, R. Behringer, S. Feiner, S. Julier, and B. MacIntyre. 2001. "Recent Advances in Augmented Reality." *IEEE Computer Graphics and Applications* 21 (6): 34–47. https://doi.org/10.1109/38.963459.

Azuma, Ronald T. 1997. "A Survey of Augmented Reality." *Presence: Teleoperators and Virtual Environments* 6 (4): 355–85. https://doi.org/10.1162/pres.1997.6.4.355.

Birt, James, Patricia Manyuru, and Jonathan Nelson. 2017. "Using Virtual and Augmented Reality to Study Architectural Lighting." In *ASCILITE 2017; Proceedings of the 34th International Conference of Innovation, Practice and Research in the Use of Educational Technologies in Tertiary Education*. 17–21.

Chandrasekera, Tilanka, and So Yeon Yoon. 2015. "Adopting Augmented Reality in Design Communication: Focusing on Improving Spatial Abilities." *International Journal of Architectonic, Spatial, and Environmental Design* 9 (1): 1–14. https://doi.org/10.18848/2325-1662/CGP/v09i01/38384.

Chatzopoulos, Dimitris, Carlos Bermejo, Zhanpeng Huang, and Pan Hui. 2017. "Mobile Augmented Reality Survey: From Where We Are to Where We Go." *IEEE Access* 5: 6917–50. https://doi.org/10.1109/ACCESS.2017.2698164.

Chen, Rui, and Xiangyu Wang. 2008. "An Empirical Study on Tangible Augmented Reality Learning Space for Design Skill Transfer." *Tsinghua Science and Technology* 13 (SUPPL. 1): 13–18. https://doi.org/10.1016/S1007-0214(08)70120-2.

Chu, Hui Chun, Jun Ming Chen, Gwo Jen Hwang, and Tsung Wen Chen. 2019. "Effects of Formative Assessment in an Augmented Reality Approach to Conducting Ubiquitous Learning Activities for Architecture Courses." *Universal Access in the Information Society* 18 (2): 221–30. https://doi.org/10.1007/s10209-017-0588-y.

Devedzic, Goran, and Politecnico Bari. 2016. "Engineering Design Education for Industry 4.0: Implementation of Augmented Reality Concept in Teaching CAD Courses." In *Proceedings of 2016 International Conference on Augmented Reality for Technical Entrepreneurs (ARTE'16)*. 11–16.

Diao, Pei Huang, and Naai Jung Shih. 2019. "Trends and Research Issues of Augmented Reality Studies in Architectural and Civil Engineering Education; A Review of Academic Journal Publications." *Applied Sciences* 9 (9): 1–19. https://doi.org/10.3390/app9091840.

Fonseca, David, Nuria Martí, Ernesto Redondo, Isidro Navarro, and Albert Sánchez. 2014. "Relationship between Student Profile, Tool Use, Participation, and Academic Performance with the Use of Augmented Reality Technology for Visualized Architecture Models." *Computers in Human Behavior* 31 (1): 434–45. https://doi.org/10.1016/j.chb.2013.03.006.

Kim, Jeffrey, and Javier Irizarry. 2020. "Evaluating the Use of Augmented Reality Technology to Improve Construction Management Student's Spatial Skills." *International Journal of Construction Education and Research* 00 (00): 1–18. https://doi.org/10.1080/15578771.2020.1717680.

Li, Wenkai, A. Nee, and S. Ong. 2017. "A State-of-the-Art Review of Augmented Reality in Engineering Analysis and Simulation." *Multimodal Technologies and Interaction* 1 (3): 17. https://doi.org/10.3390/mti1030017.

Meža, Sebastjan, Žiga Turk, and Matevž Dolenc. 2014. "Component Based Engineering of a Mobile BIM-Based Augmented Reality System." *Automation in Construction* 42 (1): 1–12. https://doi.org/10.1016/j.autcon.2014.02.011.

Milgram, Paul, and Fumio Kishino. 1994. "A Taxonomy of Mixed Reality." *IEICE Transactions on Information and Systems* 77 (12): 1321–29.

Sánchez, Albert, Ernest Redondo, David Fonseca, and Isidro Navarro. 2014. "Academic Performance Assessment Using Augmented Reality in Engineering Degree Course." In *2014 IEEE Frontiers in Education Conference (FIE) Proceedings*. 1–7. https://doi.org/10.1109/FIE.2014.7044238.

Sánchez Riera, Albert, Ernest Redondo, and David Fonseca. 2015. "Geo-Located Teaching Using Handheld Augmented Reality: Good Practices to Improve the Motivation and Qualifications of Architecture Students." *Universal Access in the Information Society* 14 (3): 363–74. https://doi.org/10.1007/s10209-014-0362-3.

Schroeder, Noah L., and Ada T. Cenkci. 2018. "Spatial Contiguity and Spatial Split-Attention Effects in Multimedia Learning Environments: A Meta-Analysis." *Educational Psychology Review* 30 (3): 679–701. https://doi.org/10.1007/s10648-018-9435-9.

Shaghaghian, Zohreh, Heather Burte, Wei Yan, and Dezhen Song. 2022. "Learning Geometric Transformations for Parametric Design: An Augmented Reality (AR)-Powered Approach." In *19th International Conference, CAAD Futures 2021; Computer-Aided Architectural Design; Design Imperatives: The Future is Now*, vol. 1465. Singapore: Springer. https://doi.org/10.1007/978-981-19-1280-1_31.

Shaghaghian, Zohreh, Wei Yan, and Dezhen Song. 2021. "Towards Learning Geometric Transformations through Play: An AR-Powered Approach." In *The 5th International Conference on Virtual and Augmented Reality Simulations*. Melbourne, Australia, March 20-22, 2021.

Shanbari, Hamzah, Nathan Blinn, and Raja R.A. Issa. 2016. "Using Augmented Reality Video in Enhancing Masonry and Roof Component Comprehension for Construction Management Students." *Engineering, Construction and Architectural Management* 23 (6): 765–81. https://doi.org/10.1108/ECAM-01-2016-0028.

Tomekkiez. 2020. "CrossSection: VirtualPlayground." 2020. https://assetstore.unity.com/packages/vfx/shaders/crosssection-93478.

Tomekkiez. 2021. "CrossSection: VirtualPlayground." 2021. https://forum.unity.com/threads/released-crosssection-tool.223790/page-7#post-6434111.

Turkan, Yelda, Rafael Radkowski, Aliye Karabulut-Ilgu, Amir H. Behzadan, and An Chen. 2017. "Mobile Augmented Reality for Teaching Structural Analysis." *Advanced Engineering Informatics* 34 (July): 90–100. https://doi.org/10.1016/j.aei.2017.09.005.

van Krevelen, D. W. F., and R. Poelman. 2010. "A Survey of Augmented Reality Technologies, Applications and Limitations." *International Journal of Virtual Reality* 9 (2): 1–20. https://doi.org/10.20870/ijvr.2010.9.2.2767.

Vasilevski, Nikolche, and James Birt. 2020. "Analysing Construction Student Experiences of Mobile Mixed Reality Enhanced Learning in Virtual and Augmented Reality Environments." *Research in Learning Technology* 28 (1063519): 1–23. https://doi.org/10.25304/rlt.v28.2329.

Vassigh, Shahin, Debra Davis, Amir H. Behzadan, Ali Mostafavi, Khandakar Rashid, Hadi Alhaffar, Albert Elias, and Giovanna Gallardo. 2018. "Teaching Building Sciences in Immersive Environments: A Prototype Design, Implementation, and Assessment." *International Journal of Construction Education and Research* 16 (3): 180–196. https://doi.org/10.1080/15578771.2018.1525445.

Vassigh, Shahin, Winifred E. Newman, Amir Behzadan, Yimin Zhu, Shu-Ching Chen, and Scott Graham. 2014. "Collaborative Learning in Building Sciences Enabled by Augmented Reality." *American Journal of Civil Engineering and Architecture* 2 (2): 83–88. https://doi.org/10.12691/ajcea-2-2-5.

Wen, Jing, Masoud Gheisari, Sambhav Jain, Yuanxin Zhang, and R. Edward Minchin. 2021. "Using Cloud-Based Augmented Reality to 3D-Enable the 2D Drawings of AISC Steel Sculpture: A Plan-Reading Educational Experiment." *Journal of Civil Engineering Education* 147 (3): 04021006. https://doi.org/10.1061/(asce)ei.2643-9115.0000046.

Yan, W. 2022. "Augmented reality instructions for construction toys enabled by accurate model registration and realistic object/hand occlusions." *Virtual Reality* 26: 465–478. https://doi.org/10.1007/s10055-021-00582-7.

IMAGE CREDITS

All drawings and images by the authors.

Ziad Ashour, PhD is an Assistant Professor in the Architecture Department at King Fahd University of Petroleum and Minerals. His research interests include building information modeling, augmented reality, spatial learning, and computational design.

Wei Yan, PhD is the Mattia Flabiano/Page Southerland Design Professor at Texas A&M University. He teaches computational methods in design and conducts research in BIM, simulation, optimization, AR, and AI, with projects funded by the NSF, NEH, DOE, Autodesk, etc.

Visualization Methods for Big and High-Dimensional Acoustic Data

Achilleas Xydis
ETH Zurich

Chaoyu Du
ETH Zurich

Romana Rust
ETH Zurich

Fabio Gramazio
ETH Zurich

Matthias Kohler
ETH Zurich

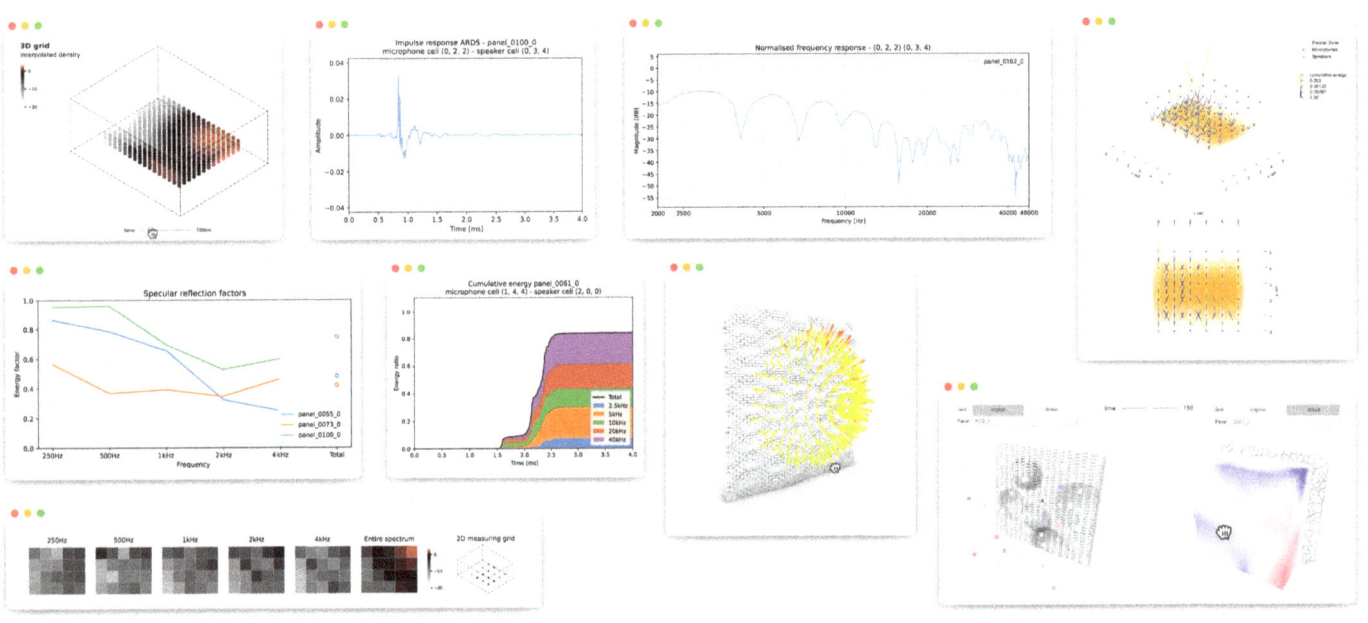

1 A selection of low and high-dimensional visualizations for acoustic data available in the presented computational workflow

ABSTRACT

Acoustics are rarely included in architectural design because available acoustic analysis tools are cumbersome and require expert knowledge in acoustics. This exclusion from the design phase could lead to late-stage design modifications, potential delays, and increased building costs. On the contrary, their inclusion can improve the acoustic properties of spaces and ensure a seamless design integration. This can be achieved by providing architects with easy-to-use visualization tools to study the relationship between geometry and sound without expert knowledge in acoustics. Available acoustic datasets can enable the development of such visualizations, but recent technological advances have increased their complexity and size. Although existing data science methods can process and analyze them, it remains challenging to develop easy-to-use and informative visualizations for architects and non-acoustic experts.

This research proposes a novel approach for interactive visualizations of acoustic datasets for architects and non-acoustic experts. It introduces a series of simple acoustic properties for users with basic knowledge of acoustics and describes methods for low- and high-dimensional data visualizations. It describes the computational workflow and uses a design scenario to demonstrate the proposed visualizations. Finally, it discusses the challenges of developing such methods, their advantages, limitations, and future work.

INTRODUCTION

Sound visualizations are an integral part of the study of room acoustics. Since the introduction of Ultrasonic-Schlieren photography by Wallace C. Sabine in 1912 (Sabine 1922), acousticians have developed multiple methods to visualize the invisible-to-the-naked-eye propagation of sound and its interaction with surfaces. Advancements in measurement techniques and computer simulations make it easier today to measure and simulate sound propagation. However, our display mediums are still two-dimensional, and visualization tools face the challenge of projecting high-dimensional data on flat screens or paper. This challenge becomes apparent when we try to display even the most essential acoustical features (frequencies, amplitude, phase, direction) simultaneously, making the visualization unreadable or overwhelming. Looking at recent developments in acoustic measurement technology and advances in computing power and storage capabilities, acoustic data collection has the tendency to increase in complexity and size rather than simplify. Large and often heterogeneous data sets make it challenging to develop informative visualizations (Genender-Feltheimer 2018) because such datasets have surpassed human cognitive capabilities when explored trough simple data analysis tools (Van Long and Linsen 2011). Therefore, the challenge of visualizing large acoustic datasets to make them accessible and readable to humans becomes more and more pressing.

Data visualization can facilitate meaningful analysis, accessibility, and interpretation of large datasets because it relies on human cognitive capabilities to process visual information (Gisbrecht 2015). Furthermore, it can support unanticipated discoveries by visually exploring and analyzing the data (Lowe and Matthee 2020). These discoveries could be a valuable resource in steering the creative process within architectural design. One reason for the necessity of data visualization in architecture is the potential to include acoustics in early design stages. Although architects already include performance as an early-on design driver for building components such as structure or facades, room acoustics are rarely included in the early design process (Badino, Shtrepi, and Astolfi 2020; Peters 2010). The reason for this is twofold. On the one hand, there is a lack of acoustic visualizations for architects and users with basic knowledge in acoustics. On the other hand, available visualizations are decoupled from the geometry that influences the sound. Nevertheless, geometry is essential for architects to understand the relationship between sound and geometry. This understanding would enable them to develop design workflows where they can manipulate the design and intuitively understand how this manipulation affects the room's acoustics.

Therefore, this research focuses on developing acoustic visualizations for big acoustic datasets for architects and non-acoustic experts that can be integrated into the early design processes to enable acoustically informed design explorations.

BACKGROUND

As big data becomes more prevalent in acoustic research, data visualizations have become increasingly important to interpret it. Lowe and Matthee describe that dimensionality reduction, interactivity, readability, and user assistance are key requirements of data visualization tools to interpret big data (Lowe and Matthee 2020). Building upon the key requirements introduced by them, we focus on four points to develop novel acoustic visualization for architectural applications: low- and high-dimensional visualizations, interactive visualizations, and usability of acoustic visualizations in architectural applications.

Low-dimensional Visualizations

From a data science perspective, an impulse response (IR) is a time series of float numbers that describe the sound's energy and phase over time. All acoustic parameters (called descriptors) that derive from it can be grouped into: (a) single value descriptors and (b) series-of-numbers descriptors. Single value descriptors include reverberation time (RT), clarity (C), strength (G), definition (D), center time (TS), and more. Series-of-numbers descriptors include frequency response and energy over time. Although acoustic analysis software such as Odeon and CATT-Acoustic (Odeon A/S 2022; CATT 2022) provides methods to calculate these acoustic descriptors, it requires expert knowledge in acoustics to decode their meaning.

High-dimensional Visualizations

High-dimensional data pose another challenge for accurate and readable visualization methods. Humans can only visually perceive three dimensions. Traditional data science visualization techniques, such as scatter plots and heat maps, can represent small or intermediate datasets in two or three dimensions. Although these visualizations are intuitive and may be used to identify bivariate correlations between variables, they require dimensionality reduction to arrange the data points in a lower-dimensional space. Dimensionality reduction methods such as Principal Component Analysis (Wold, Esbensen, and Geladi 1987) and *t*-distributed stochastic neighbor embedding (Van der Maaten and Hinton 2008) can compress attributes and reduce complexity. Although this compression is necessary to lower the dimensions down to two or three dimensions, it could lead to projection losses (Lowe and Matthee 2020). Projection loss describes a scenario where well-spread points in high-dimensional space appear falsely close in the low-dimensional projection (Lowe and Matthee 2020). Other limitations of dimensionality

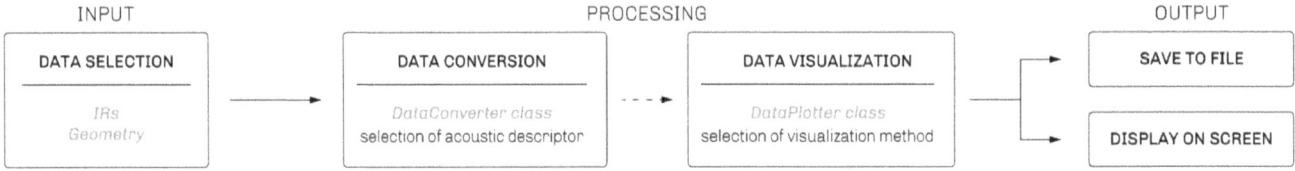

2 Computational pipeline

reduction methods are that they treat the values as pure numbers, completely decoupled by the geometry that influenced them. This makes it especially difficult to include these methods in architectural applications as geometry is one of the key design components.

Interactive Visualizations
Interactive visualizations combine human and machine intelligence (Chen et al. 2019) to explore and uncover unexpected patterns in datasets (Cho et al. 2014). This visual analysis benefits "visual perception, interactive exploration, improved understanding, informed steering and intuitive interpretation" (Liu 2019). Furthermore, this type of analysis can have two approaches: a bottom-up approach that can potentially uncover patterns in the data (Ruan and Zhang 2017) and a top-down approach to test theories and search for evidence in the data (Genender-Feltheimer 2018; Mwangi, Soares, and Hasan 2014).

Usability of Acoustic Visualizations in Architectural Applications
During the design phase, architects explore various alternative design ideas. Early-stage design decisions have a significant impact on the final design's quality and performance (Badino, Shtrepi, and Astolfi 2020). In contrast, late-stage design modifications can rarely compensate for poor early-stage choices. Therefore, it is essential to consider all factors early on to avoid potentially delaying the project, increasing the building cost, or impairing the overall design. To employ acoustic performance as a design driver, we must be able to quantify and interpret the acoustic effects of our geometric design choices. Currently, most of visual analytics are performed by acousticians equipped with the necessary knowledge and specialized acoustic analysis software. The reason for this is that this type of software is cumbersome to use and requires expert knowledge in acoustics. As a result, architects are discouraged from using them to evaluate their design, especially early on.

Commercially available acoustic analysis and visualization software are black boxes, providing insufficient feedback to the user and often no description of how the results were calculated (Kim et al. 2015). Furthermore, most commercial software do not provide APIs[1] or ways for external software to interface with them. This limitation, for example, hinders form-finding studies using computational design because each design must be exported from the design software and imported into the acoustic analysis software for analysis. Therefore, acoustic visualization tools should be flexible for customization and easily expandable to accommodate study-specific requirements. The increased popularity of web-based python programming allows the development of visualization tools that are easily accessible and easy to use. Furthermore, their open-source nature allows users to customize them and extend their capabilities according to their needs.

This research proposes a novel approach for interactive visualizations of acoustic datasets for architects and non-acoustic experts. It introduces a series of simpler descriptors for users with basic knowledge in acoustics and describes methods for low- and high-dimensional data visualizations. It introduces visualization methods that incorporate the geometry that influenced the sound under study. The inclusion of the influential geometry provides a more detailed insight into the relationship between geometrical characteristics and the sound properties they influence. Moreover, this research proposes methods for interactive visualizations that allow users to explore the data from different angles and zoom closer to reveal more details. Interactivity also allows for animated content to display how sound properties change over time. Furthermore, it describes methods that allow the simultaneous display of multiple data points, enabling the user to compare data within the same visualization. Lastly, all visualizations are customizable and extendable, and the entire code is open-sourced and available at https://github.com/gramaziokohler/sdsc_data_driven_acoustic_design.

METHODS
Based on the topics described in the background, this section describes the computational workflow of the proposed visualization pipeline and demonstrates its use through a design scenario. For the design scenario, we used the open-source GIR Dataset (Xydis, Perraudin, Rust, Heutschi, et al. 2021). The dataset contains 920,712 real impulse responses (IRs) from 312 surfaces (2951 per surface). More details about how the dataset was collected can be found in (Rust et al. 2021).

Computational Workflow
The visualization pipeline is written in python and contains two main classes the DataConverter and the DataPlotter. The

TABLE 1 Available acoustic descriptors inside DataConverter class

Descriptor	Value type
Reverberation time	float
Clarity	float
Definition	float
Energy over time	list of floats
Cumulative energy over time	list of floats
Total cumulative energy	float
Frequency response	list of floats
Absorption coefficient	list of floats
Scattering coefficient	list of floats

TABLE 2 Available visualizations in DataPlotter class

Visualization method	Interactive	Comparable
2D IR	no	yes
Energy over time	no	yes
Cumulative energy over time	no	yes
Frequency response	no	yes
Absorption coefficient	no	yes
Scattering coefficient	no	yes
2D grid	yes	yes
3D grid	yes	yes
3D polar	yes	yes

DataConverter handles the data retrieval from the dataset and can convert the IR data to the desired acoustic descriptors, and the DataPlotter handles all the visualization computing. Open-source libraries are used to extend the core code. The *numpy* library handles the mathematical operation and the *scipy* library manages the audio-related computation such as Fast Fourier Transform (FFT) analysis and resampling. The visualization part of the code uses *seaborn* for computing heat maps, *matplotlib* for constructing all the static visualizations, and *pythreejs* for interactive visualizations. *Ipywidgets* is used to generate graphic widgets such as number sliders, button, and text inputs. The computational pipeline comprises of three main steps (see Figure 2). First, the input step where the data selection takes place, second, the processing step with data conversion and data visualization, and finally, the output step where the visualization is displayed or saved to a file.

Visualization Types
There are two types of visualizations possible with the computational workflow, explanatory visualizations, and comparative visualizations.

Explanatory Visualizations
The acoustic data visualization process begins with the selection of an input. For low-dimensional data the input refers to an IR measurement, in high dimensional visualization the input refers to several IRs. Afterwards the user selects the desired acoustic descriptor, and the DataConverter class returns the converted data. This data then can be passed to the DataPlotter for visualization. Table 1 shows all the available acoustic descriptors that DataConverter class can compute. Besides the standard output, each of these descriptors can also be normalized (from 0.0-1.0 or 0%-100% depending on the descriptor) or scaled using a scale factor. Table 2 shows all the available visualizations the DataPlotter class can generate. All visualizations have the option to display the output for the entire audio spectrum or per user-defined frequency bands (see Figure 5 and 6). This option, for example, allows the user to analyze the relationship between a geometrical design and groups of frequencies. This enables them to adjust their design to target specific frequencies. Finally, the output of the visualization can be either saved directly to a file, displayed on the screen, or both.

All visualizations follow the same principle. A sound is emitted at the source position, then the sound wave hits the room's surfaces and finally arrives at the receiver position. The visualizations show how these surfaces influenced the sound wave when the wave came in contact with them. Except for IR, Absorption coefficient, Scattering coefficient, and Frequency response, all visualizations can display absolute and relative values. For relative values, the values of a flat and smooth surface are taken as a baseline, and all values of the chosen surface relate to them. This way, the visualizations enable the evaluation of acoustic properties without professional assistance.

Comparative Visualizations
Several low- and high-dimensional descriptors can be used in comparative visualizations. The process is similar to the one described in the computational workflow section, except that for low-dimensional data (see Figure 4 top), users select

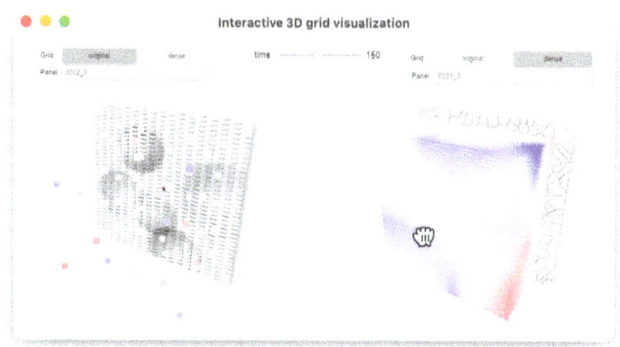

3 Interactive comperative visualization of the energy over time: the left surface has the original density of the measuring grid and the right surface the denser interpolated grid; the time slider can be changed to show the energy for a specific time

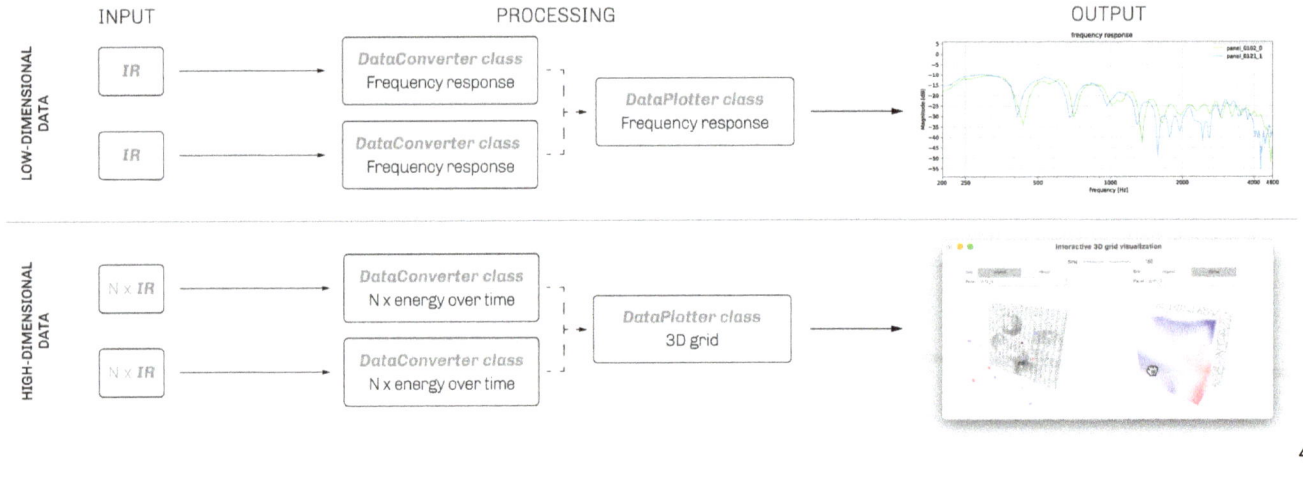

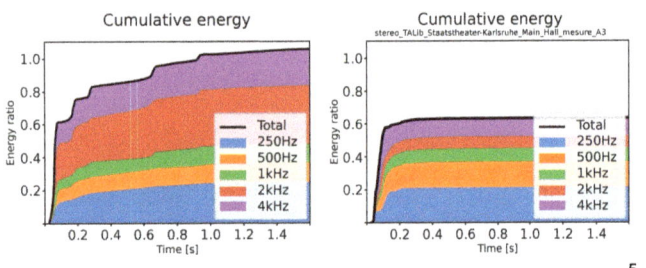

4 Computational workflow for comparative visualizations: to illustrate both output options, the low-dimensional visualization outputs a file and the high-dimensional visualization opens a window where users can interact with the data

5 Cumulative energy over time visualization: (left) The current acoustic conditions of the scenario's room; the room is highly reverberant and has an unbalanced frequency response, and the 2kHz and 4kHz bands contain more energy than the other bands; (right) The Staatstheater in Karlsruhe as a reference of an acoustically treated space; black line represents the cumulative energy of the entire audio spectrum in the IR, and the colored sections for each filter band

two or more IRs and for high-dimensional data, two sets of multiple IRs. Then, they select the desired acoustic descriptor and generate the comparative visualization.

Interactivity
All high-dimensional data visualizations, explanatory, and comparative are displayed in a three-dimensional interactive window and include the surface geometry (Figure 3). They allow the user to pan, zoom, and rotate around the displayed data, breaking the barrier of a static data representation enabling users to study the relationship between geometry and sound intuitively. Furthermore, these visualizations support animated content. If a specific acoustic descriptor is selected, they can display the descriptor's value over time, adding an extra dimension to the visualization. When the comparative mode is active, the user can apply the same transformation to both visualizations and see the animated data simultaneously for both surfaces.

DESIGN SCENARIO
To show the entire workflow of the presented research, we are converting an existing room into an open-plan office space as an example. This section goes through several architectural design phases and demonstrates how the proposed visualizations enable a more acoustically informed design outcome. The design scenario has the architect as a user, and it consists of two main steps: (a) analysis and evaluation of the existing acoustic conditions and (b) exploration of acoustic design interventions.

Analysis and Evaluation
The process starts by analyzing the existing acoustic conditions of the room to understand the design actions necessary to improve the acoustic properties of the room for our open-plan office. After acquiring the impulse response measurements of the existing room, the user is ready to start the analysis. The computational workflow supports single IR measurements or multiple IR measurements recorded at several positions inside a room. Multiple IR measurements can be arranged in orthogonal two-dimensional or three-dimensional grids. Bellow we describe the three different arrangement options.

Single IR Measurements
Figure 5 shows explanatory two-dimensional visualizations of the Cumulative energy over time of single IRs. In this case, the user defines five filter bands, with a center frequency of 250 Hz, 500 Hz, 1 kHz, 2 kHz, and 4 kHz. The 250 Hz band is a low-pass filter (LPF), and the 4 kHz is a high-pass filter (HPF). The values are normalized from 0.0 to 1.0, with 0.0 representing no energy at the receiver position, and 1.0 that all the energy emitted by the source arrived at the receiver position. The Cumulative energy inside the room (see Figure 5 left) goes up quickly when the direct sound arrives, but

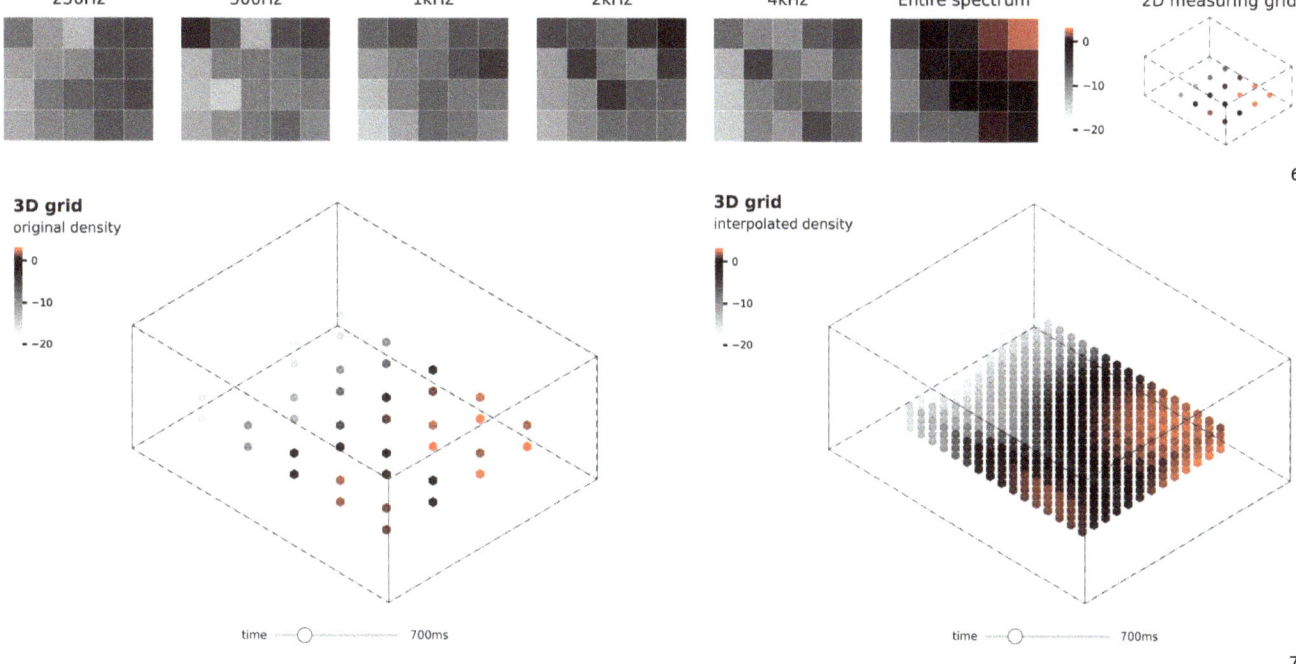

6   2D grid: total energy per filter band and for the entire audio spectrum; black represents energy values similar to the transmitted energy, and white 20dB less energy (equivalent to 100 times less energy); red squares are positions where the received energy is higher, indicating a local focusing effect

7   Interactive 3D grid visualization: (left) the original density grid with 40 measuring positions; (right) higher-density linear interpolated grid with 884 positions; in the interpolated grid the energy pattern is more visible

because the room is very reverberant due to its big volume and the lack of absorptive surfaces, the energy continues to rise for another 1.5 seconds. The small "steps" in the plot indicate when strong reflections arrive at the receiver position. These reflections can cause coloration, image shift, and flutter echoes, all negative characteristics for meeting rooms, lecture halls, and offices where speech clarity is important. Comparing it to the Cumulative energy of the Staatstheater in Karlsruhe,[2] we see that the energy rises quickly and then remains stable. This shows that there are no more reflections arriving at the receiver position. Furthermore, the total energy is lower because a portion of it (around 40%) was absorbed by the rooms surfaces. Rooms with Cumulative energies like Figure 5 (right) will have a clearer sound and be more relaxing working environments.

Multiple IR Measurements: 2D Measurement Grids

Multiple IR measurements of the same room enable a higher resolution analysis of the existing acoustic conditions. The visualization in Figure 6 can be used when the measurement positions are arranged in a two-dimensional grid. For this visualization we chose the total energy as a descriptor and each square represents one total energy value. This value defines the total amount of sound energy that arrived at that location for the duration of the measurement. Looking at the entire spectrum grid we can clearly see the high sound energy concentrated at the corner of the room.

Multiple IR Measurements: 3D Measurement Grids

The visualization in Figure 7 can display values that were measured using a three-dimensional grid. The values are color-coded and displayed inside the room's geometry. Users can pan and rotate the geometry to look at the values from a different angle or zoom in to take a closer look when the data is too dense. The grid's density can be increased to make patterns more visible or decreased to reduce the visual complexity. Because this visualization also supports animated content, the energy values over time can be displayed by moving a time slider.

The analysis and evaluation show us clearly that the existing acoustic conditions are not suitable for a comfortable working environment. More precisely, some of the energy needs to be removed by means of absorptive materials. The ceiling and the floor are ideal locations for installing such materials. To balance the sound energy inside the room, limit disturbing flutter echoes, and prevent the room from becoming very unnatural from excess absorption, the walls can be designed in such a way that will diffuse sound. Bellow, we introduce different design explorations that the user can use to address the acoustic problems.

Design Explorations

This part of the scenario focuses on design interventions that

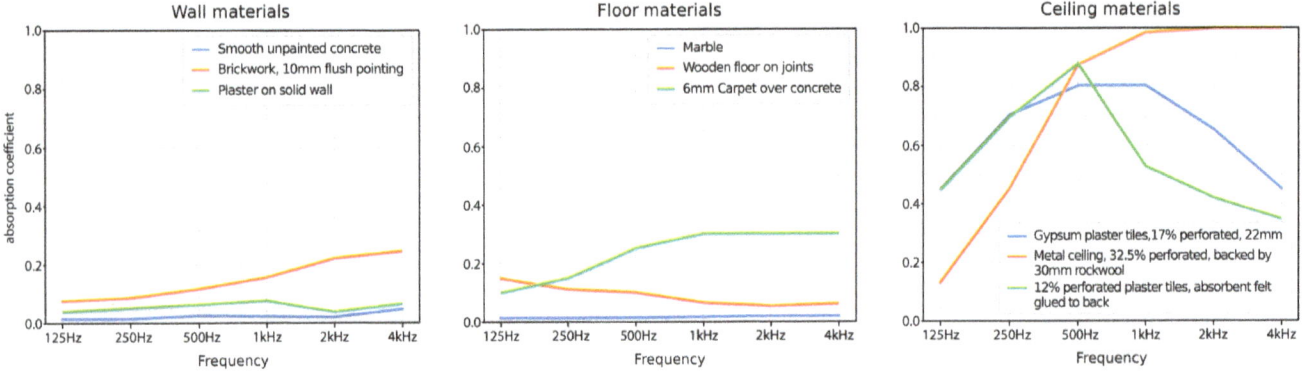

8 Comparative visualizations of absorption coefficients of different materials for the wall surfaces, the floor, and the ceiling of the room; values close to 0.0 represent no sound absorption, therefore, all energy is reflected, and values close to 1.0 complete absorption

can improve the acoustic conditions of the room. This part consists of two main steps, a) general acoustic treatment through proper surface material selection and b) design and adjustment of the wall's geometry to maximize the acoustic performance.

General Acoustic Treatment

The sound we hear is a combination of the direct sound coming straight from the source and indirect reflections from the surrounding surfaces. Sound hitting a surface is either transmitted, absorbed, or reflected; the ratio depends on the surface's acoustic properties. Based on the analysis of Figure 5, the user needs to lower the sound energy inside the room, emphasizing on the mid and high frequencies. This issue can be addressed by choosing appropriate materials that absorb the sound energy of mid- and high frequencies. Figure 8 shows three comparative two-dimensional visualization of Absorption coefficients for common architectural materials that can be used on the walls, floor, and ceiling of the room. The algorithm either receives the coefficients as a list of floats or computes them from IRs measured according the ISO 354:2003 standard (ISO 2003). In this case, the coefficient values for these visualizations are taken from a dataset that was created using data from www.acoustic.ua (Acoustic Traffic LLC 2022). Analyzing the three visualization reveals that a Brickwork would be preferable for the walls of the room (Figure 8 left, in orange), a thin carpet for the floor (Figure 8 middle, in green), and 32.5% perforated thin metal sheets, backed by 30 mm rock wool for the ceiling (Figure 8 right, in orange).

Design and Adjustment of Architectural Geometries

After deciding for brickwork as a wall material, the exact design of its geometry must be defined. Brickwork has a structured surface texture. Parts of the texture diffuse sound and other parts reflect sound in a specular way (Figure 9). The brickwork geometry can be optimized to redirect the reflected energy towards the ceiling (Figure 10). This way, more sound is reflected towards the ceiling where it gets absorbed.

A recent research introduced a computational workflow to generate various acoustically informed diffusive surfaces, including brickworks (Xydis, Perraudin, Rust, Lytle, et al. 2021) The research uses self-organizing maps (SOM) to arrange sound diffusive architectural surfaces based on chosen acoustic properties. Using its design workflow, the SOM cell with the best matching unit contains two surfaces, a flat and a textured brickwork. To decide which of the two surfaces addresses better our acoustic requirements, the user can evaluate the direction of the reflected energy. Figures 11, 12, and 13 show interactive visualizations that display the direction and intensity of the reflected sound energy. The direction is represented as a ray leaving the surface and the intensity by the length and color of the line. Users can pan, zoom, or rotate to explore the sound directivity from different angles. Furthermore, the descriptors can be displayed for the entire audio spectrum or per user-defined frequency bands (see Figure 13). By comparing the two visualizations in Figure 11 and Figure 12 we can clearly see that the surface on the right redirects more energy upwards making it an ideal option for the design scenario.

RESULTS AND DISCUSSION

We have presented a novel approach to visualizing low and high-dimensional acoustic data. We described the computational workflow to produce these visualizations, its components, and how each of them contributes to the entire workflow. We demonstrated a series of visualizations and described how they could be used for acoustic studies. These visualizations show that thanks to their intuitive visual implementation, they can be used both by expert and non-expert users in acoustics. The proposed workflow addresses both experts and non-experts in acoustics users. Expert users can use standard and familiar acoustic descriptors and visualizations, while non-expert users are presented with a range of newly proposed and simplified descriptors and visualizations. Furthermore, both users benefit from the intuitive layout of the visualizations, especially from the interactivity of the high-dimensional visualizations. Finally, including the geometry that influenced the sound in the visualizations allows for a deeper analysis of the mutual relationship between geometry and sound. We believe that these visualizations will help bring acoustics closer to the early phases of architectural design and enable a more integrative acoustic and architectural design exploration.

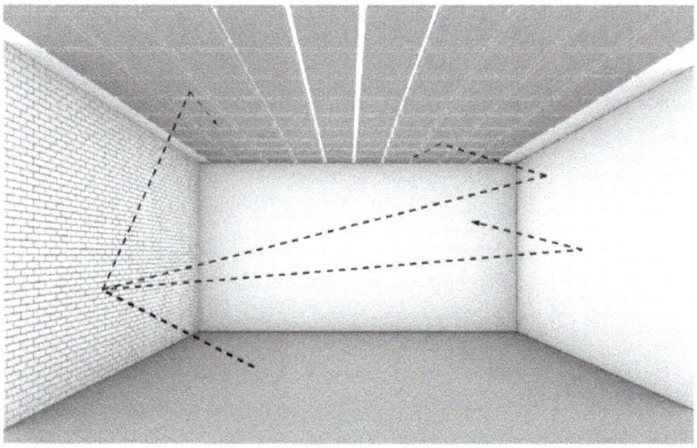

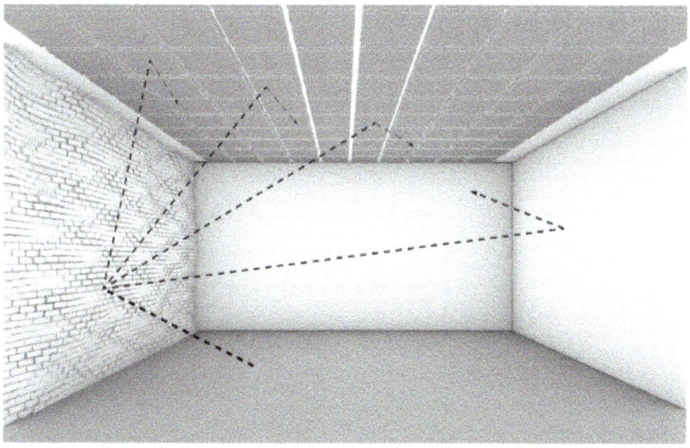

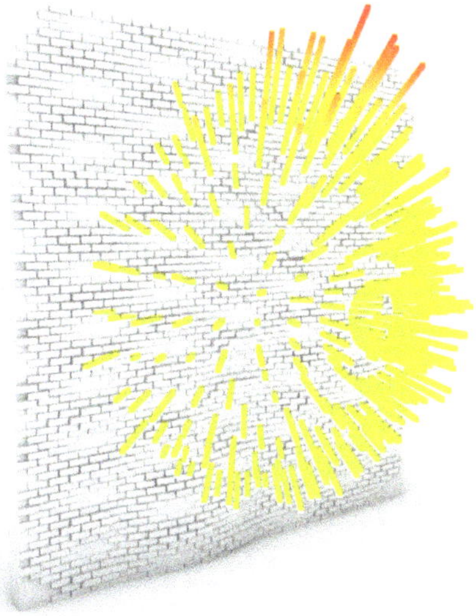

Limitations and Future Work
Despite the intuitive workflow, users still required basic knowledge of python to run the scripts. That is also true for extending and further customizing the visualizations. However, this visualization pipeline could be turned into a plugin for CAD software, eliminating the need for programming knowledge. In future steps, we are committed to continuing extending the visualization pipeline by adding more acoustic descriptors and visualization methods to it. Finally, we are confident that the proposed visualization methods will encourage researchers to create more open-source high-dimensional acoustic datasets.

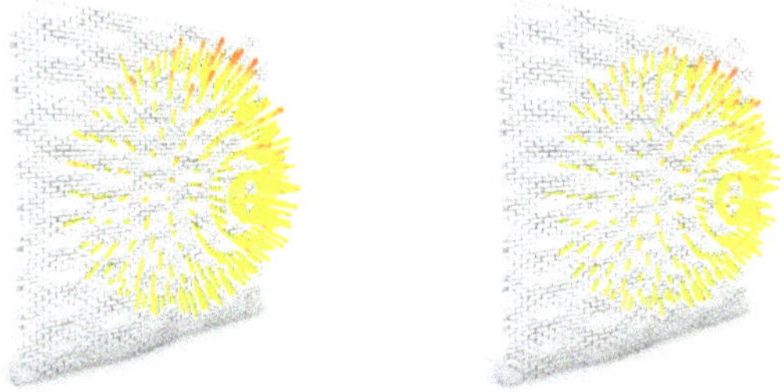

9   The room from the design scenario with the flat brickwork. The black dashed lines illustrate the direction and intensity of the reflected sound according to the data from Figure 11. Most of the sound energy bounces back into the working area.

10 The room from the design scenario with the optimized brickwork. The black dashed lines illustrate the direction and intensity of the reflected sound according to the data from Figure 12. Here a significant portion of the sound energy gets reflected towards the absorbent ceiling.

11 3D polar sound directivity visualization of the total energy of the entire audio spectrum for the flat brickwork.

12 3D polar sound directivity visualization of the total energy of the entire audio spectrum for the textured brickwork

13 3D polar sound directivity visualization of the total energy of two frequency bands for the textured brickwork, left: 4kHz, right: 1kHz.

ACKNOWLEDGMENTS

This research steamed out of a collaborative and multidisciplinary project between Gramazio Kohler Research at ETH Zurich, the Swiss Data Science Center, the Laboratory for Acoustics/Noise Control at EMPA, and STRAUSS ELEKTROAKUSTIK GMBH. Therefore, the authors would like to thank Dr. Fernando Perez-Cruz, Dr. Nathanaël Perraudin, Dr. Kurt Heutschi, Kurt Eggenschwiler, and Jurgen Strauss for their inputs. Furthermore, we would like to thank Daniela Mitterberger for her valuable help and support preparing this paper.

NOTES

1. API stands for application programming interface, which is a set of definitions and protocols for building and integrating application software.
2. The IR is taken from the "Théâtre Acoustique Room Impulse Response Library," https://www.lieuxperdus.com/convolver/download/.

REFERENCES

Acoustic Traffic LLC. "Absorption Coefficients." Accessed June 2, 2022. https://www.acoustic.ua/st/web_absorption_data_eng.pdf.

Badino, Elena, Louena Shtrepi, and Arianna Astolfi. 2020. "Acoustic Performance-Based Design: A Brief Overview of the Opportunities and Limits in Current Practice." *Acoustics* 2 (2): 246–78.

CATT. CATT-Acoustic. V.9.1. CATT. PC. 2022.

Chen, Yi, Zeli Guan, Rong Zhang, Xiaomin Du, and Yunhai Wang. 2019. "A Survey on Visualization Approaches for Exploring Association Relationships in Graph Data." *Journal of Visualization* 22 (3): 625–39.

Cho, Wonhee, Yoojin Lim, Hwangro Lee, Mohan Krishna Varma, Moonsoo Lee, and Eunmi Choi. 2014. "Big Data Analysis with Interactive Visualization Using R Packages." In *BigDataScience '14; Proceedings of the 2014 International Conference on Big Data Science and Computing*. New York, NY, USA: Association for Computing Machinery. 1–6.

Genender-Feltheimer, Amy. 2018. "Visualizing High Dimensional and Big Data." *Procedia Computer Science, Cyber Physical Systems and Deep Learning Chicago* 140 (January): 112–21.

Gisbrecht, Andrej. 2015. "Advances in Dissimilarity-Based Data Visualisation," PhD diss., Bielefeld University, Germany.

ISO, International Organization for Standardization. 2003. *ISO 354:2003 Acoustics — Measurement of Sound Absorption in a Reverberation Room*. Geneva, Switzerland: ISO.

Kim, Seokyeon, Seongmin Jeong, Sung Uk An, Jae Seok Yoo, Sang Min Han, Hanbyul Yeon, Sangbong Yoo, and Yun Jang. 2015. "Big Data Visual Analytics System for Disease Pattern Analysis." In *BigDAS '15; Proceedings of the 2015 International Conference on Big Data Applications and Services*. New York: Association for Computing Machinery. 175–79.

Liu, Zhanping. 2019. "A Prototype Framework for Parallel Visualization of Large Flow Data." *Advances in Engineering Software* 130 (April): 14–23.

Lowe, Joy, and Machdel Matthee. 2020. "Requirements of Data Visualisation Tools to Analyse Big Data: A Structured Literature Review." In *Responsible Design, Implementation and Use of Information and Communication Technology; Lecture Notes in Computer Science*, edited by M. Hattingh, M. Matthee, H. Smuts, I. Pappas, Y. K. Dwivedi, and M. Mäntymäki. Cham: Springer International Publishing. 469–80

Mwangi, Benson, Jair C. Soares, and Khader M. Hasan. 2014. "Visualization and Unsupervised Predictive Clustering of High-Dimensional Multimodal Neuroimaging Data." *Journal of Neuroscience Methods* 236 (October): 19–25.

Odeon A/S. ODEON. V.17. Odeon A/S. PC. 2021.

Peters, Brady. 2010. "Acoustic Performance as a Design Driver: Sound Simulation and Parametric Modeling Using SmartGeometry." *International Journal of Architectural Computing* 8 (3): 337–58.

Ruan, Guangchen, and Hui Zhang. 2017. "Closed-Loop Big Data Analysis with Visualization and Scalable Computing." *Big Data Research, Tutorials on Tools and Methods using High Performance Computing resources for Big Data* 8 (July): 12–26.

Rust, Romana, Achilleas Xydis, Kurt Heutschi, Nathanael Perraudin, Gonzalo Casas, Chaoyu Du, Jürgen Strauss, et al. 2021. "A Data Acquisition Setup for Data Driven Acoustic Design." *Building Acoustics* 28 (4). https://doi.org/10.1177/1351010X20986901.

Sabine, Wallace Clement. 1922. *Collected Papers on Acoustics*. Cambridge: Harvard University Press.

Van der Maaten, Laurens, and Geoffrey Hinton. 2008. "Visualizing Data Using T-SNE." *Journal of Machine Learning Research* 9 (11): 2579–2605.

Van Long, Tran, and Lars Linsen. 2011. "Visualizing High Density Clusters in Multidimensional Data Using Optimized Star Coordinates." *Computational Statistics* 26 (4): 655.

Wold, Svante, Kim Esbensen, and Paul Geladi. 1987. "Principal Component Analysis." *Chemometrics and Intelligent Laboratory Systems, Proceedings of the Multivariate Statistical Workshop for Geologists and Geochemists* 2 (1): 37–52.

Xydis, Achilleas, Nathanaël Perraudin, Romana Rust, Kurt Heutschi, Gonzalo Casas, Oksana Riba Grognuz, Kurt Eggenschwiler, Matthias Kohler, and Fernando Perez-Cruz. 2021. *GIR Dataset: A Geometry and Real Impulse Response Dataset*. V2. August 27, 2021. Distributed by Zenodo. https://www.zenodo.org/record/5500519.

Xydis, Achilleas, Nathanaël Perraudin, Romana Rust, Beverly Ann Lytle, Fabio Gramazio, and Matthias Kohler. 2021. "Data-Driven Acoustic Design of Diffuse Surfaces Using Self-Organizing Maps." In ACADIA 2021 Realignments: Toward Critical Computation; Proceedings of the 41st Annual Conference of the Association for Computer Aided Design in Architecture (ACADIA), edited by K. Dorfler, S. Parascho, and J. Scott.

Achilleas Xydis is an architect and currently a doctoral researcher at the Chair of Architecture and Digital Fabrication (Gramazio Kohler Research) at ETH Zurich. He received his Diploma in Architecture from the University of Patras in 2010. In 2013, he completed the post-graduate Master of Advanced Studies in Architecture and Information at ETH Zurich. He focuses on combining machine learning techniques with architectural acoustics.

Chaoyu Du is a computational architect and doctoral researcher at the Block Research Group, ETH Zurich. She obtained her Master's degrees in architecture from Tongji University and urban design from the Technical University of Berlin, a MAS in Architecture and Digital Fabrication, and a CAS in Computer Science from ETH Zurich. Her research interests include computational structural design, complex fabrication, and machine intelligence in architecture.

Romana Rust is a computational architect and senior researcher at Gramazio Kohler Research, ETH Zurich within the Design++ initiative Centre for Augmented Computational Design in AEC. She is the co-coordinator of the Immersive Design Lab, a lab for collaborative research and teaching in the field of extended reality and machine learning in architecture and construction. Her particular interest is the development of innovative computational methods that integrate multiple design objectives such as geometry, acoustics, materiality, and robotic fabrication.

Fabio Gramazio and Matthias Kohler are professors of Architecture and Digital Fabrication at ETH Zurich. In 2000, they founded the architecture practice Gramazio & Kohler, which realized numerous award-winning projects. Opening the world's first architectural robotic laboratory at ETH Zurich, Gramazio & Kohler's research has been formative in the field of digital architecture, setting precedence and de facto creating a new research field merging advanced architectural design and additive fabrication processes through the customized use of industrial robots.

Measuring Street Vitality Based on Video-image Using Deep Learning

A Case Study of a Commercial Complex in Osaka, Japan

Yunqin Li
Nanchang University

Jiaxin Zhang
Nanchang University

Xueqiang Wang
Nanchang University

Kai Ma
Nanchang University

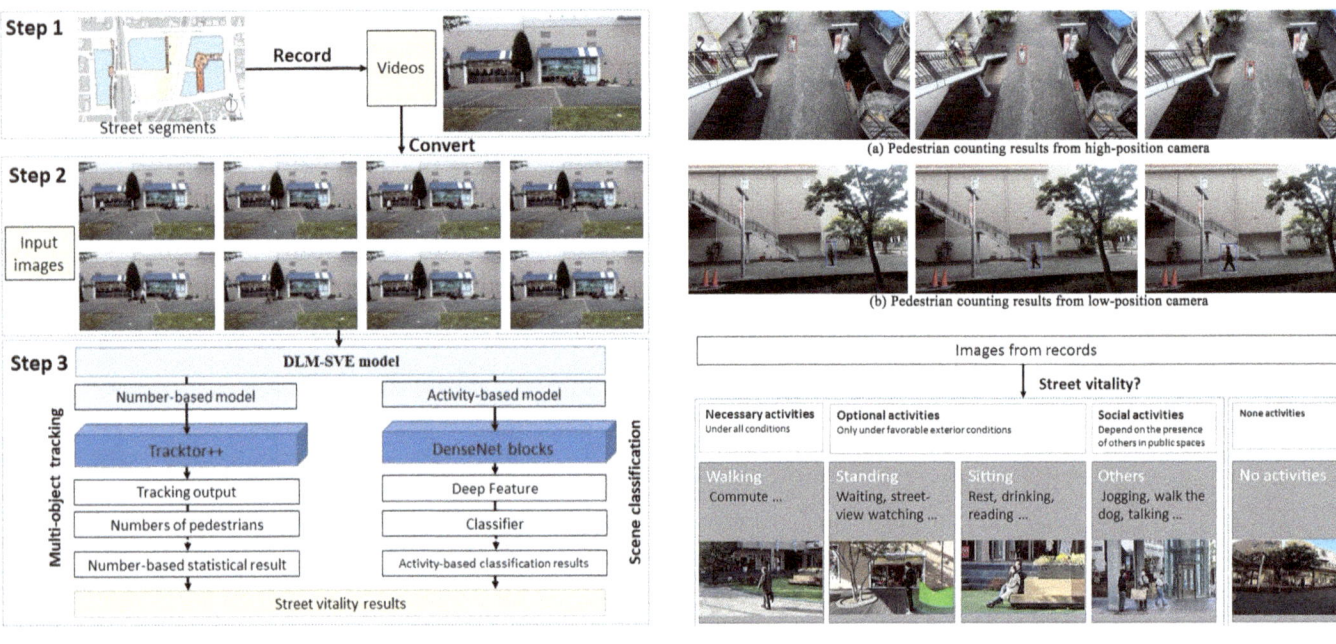

1A Key steps in DLM-SVE model: (left) workflow of DLM-SVE model; (upper right) results samples of number-based model; (lower right) classification hierarchy of activity-based model

ABSTRACT

Street vitality plays an important role in advancing urban sustainable development and improving residents' quality of life, and it can be reflected by the type and the frequency of pedestrian activities in street level. Previous studies have often relied on subjective perceptions and observational assessments that are difficult to quantify results or quantitative measurements at the city scale.

This paper proposes a deep convolutional neural network-based framework for fine-scale studies on automatic evaluation of street-level vitality using multiple object tracking and image segmentation with video data. A deep learning model for street vitality evaluation was proposed based on the intensity and complexity of pedestrian activities. To verify the model's accuracy and to illustrate the applicability of our method, a commercial complex in Osaka was implanted as an example. The visualized the street vitality results showed that the proposed method is high-efficiency, reliable, and feasible with acceptable accuracy compared to conventional manual on-site surveys. Compared to previous city-scale quantitative evaluations of street vitality, our method is suitable for fine-scale studies on street-level vitality that help to improve street built environment design from a pedestrian perspective and that can be transferred to other urban streets with the supported video data of pedestrian behavior. Moreover, our approach combines definitions for street vitality perception in traditional architecture, embraces emerging computer vision technologies, and can be further extended for application in urban vitality perception with the Internet of Things.

INTRODUCTION

Streets are not only major pedestrian corridors, but also valuable urban public spaces for the social, cultural, and economic life of the residents. Street vitality has attracted much academic attention and has become an important sustainable indicator for evaluating the appeal and potential of urban neighborhoods for livable urban neighborhood development from a human perspective (Li, Yabuki, and Fukuda 2022). To create a lively urban street space of attractiveness, diversity, and vitality, many design theories have been proposed, such as Jane Jacobs's discussion on the diversity of urban public space (Jacobs 1961) and Jan Gehl's suggestion on improving the vitality of street space (Gehl 1987). However, while there is an increasing corpus of scientific and theoretical research on street vitality assessment, urban street vitality is still difficult to define and relies on planners' and designers' experience and intuition.

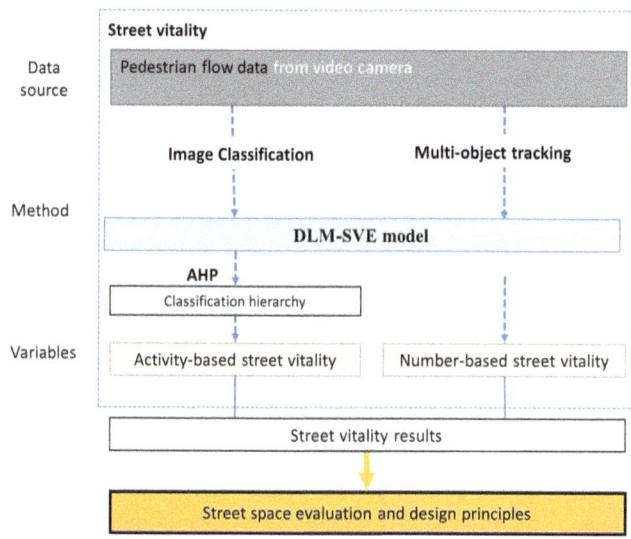

1B The framework of the study

Urban sociologists and architects have long held unique opinions on how to interpret the concept of urban vitality. Generally, urban sociologists consider urban vitality as an indicator to measure whether a space can activate economic, social, and cultural activities, which depends heavily on people's gatherings and activities (Jacobs 1961). Architects, on the other hand, believe that urban vitality can be constructed by physical entities of urban form and can be improved through design (Ye, Li, and Liu 2018). It can be defined in terms of both the intensity and complexity of residents' activities, especially the intensity of optional activities (Gehl 1987). Street vitality in this paper focuses on a pedestrian perspective and is measured by the intensity and complexity of the activity, i.e., in the type and frequency of pedestrians performing on the street.

Previous research on street vitality has mostly been qualitative and descriptive from a human perspective, lacking strong data support, or has been quantitative from the urban perspective that may differ from residents' real needs. Their qualitative methods are mostly sociological and statistical methods—such as field surveys, questionnaire surveys, and cognitive maps to collect data on the characteristics of the pedestrian activity (including activity type, activity route, and activity frequency)—that are both time-consuming and labor-intensive. The "Public Space, Public Life" (PSPL) survey (Gehl and Svarre 2013) is a typical subjective perception and observation assessment approach based on field studies. Many researches has been carried out using this method to measure street vitality based on observers and demonstrate the link between street vitality and street built environment features, such as street commercial density, greenery ratio, interface continuity, and street transparency.

Quantitative methods of street vitality evaluation are using computer-assisted technologies based on urban big data such as traffic smart card data, Street View Images (SVIs), Point of Interests (POIs), and social media check-in data to conduct a large-scale measurement of street vitality features (Zhang, Fukuda, and Yabuki 2021). For example, Liu et al. (2019) used the georeferenced time series data clustering combined with POI-based land use analysis to measure urban vitality. Ye et al. (2018) used small catering data as an indicator of urban vitality, representing the urban attractiveness of a place, especially in densely populated cities. The data sources for these quantitative analyses are not uniform, making it difficult to show complex human behavioral activities in street-level studies. However, fine-scale studies on street-level vitality are important in street-built environment design, especially for those low-vibrancy street spaces hidden in large-scale cities.

Mobile signal data, GPS data, Location-Based Service data, wearable device data, and embedded sensors data, make it possible to observe and evaluate people's activities, movements, and preferences. For example, some studies analyzed pedestrian activities using temporal-spatial GPS data and Wi-Fi access points to evaluate pedestrians' space perceptions and assess the street vitality (Kim 2018; Zhang, Zhang, and Yin 2020). Nevertheless, cell phone data and GPS data can only be used for large-scale studies due to the location accuracy, and sensor-based human behavior studies with high data accuracy suffer from limited experimenter sample size with high equipment costs.

The recent development of deep learning in computer vision has piqued interest in video-image content analysis by allowing low-cost but high-accuracy analysis of human activities using video or photo data. There have been some attempts to use surveillance to record people's activities to count and classify

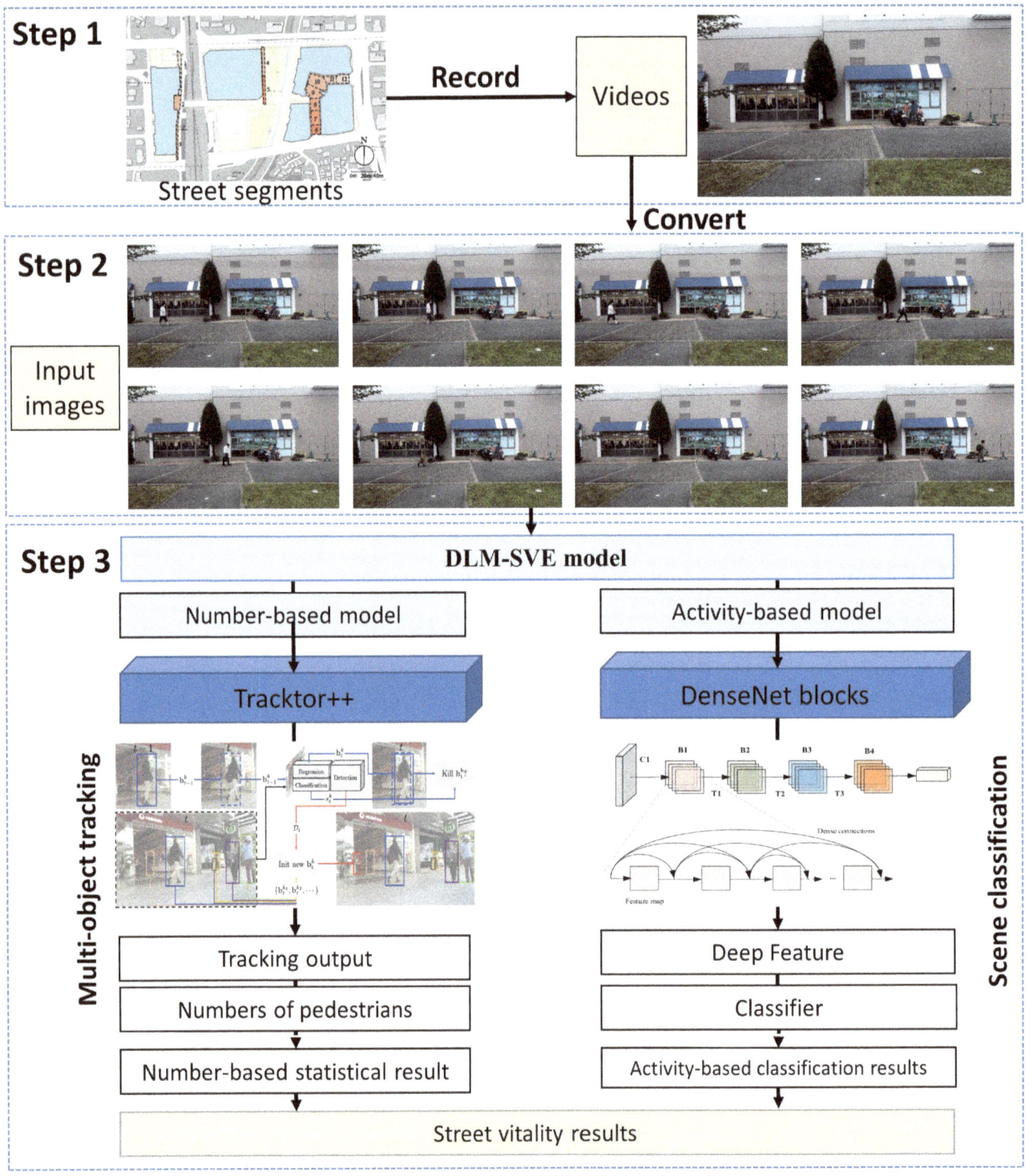

2 Workflow of the DLM-SVE model for the street vitality evaluation, including number-based model using MOT and activity-based model using image classification

(a) Pedestrian counting results from high-position camera

(b) Pedestrian counting results from low-position camera

3 Pedestrian counting result examples from cameras of multiple positions using multiple-object tracking

pedestrian routes in public areas, providing a new approach to studying the use of small public areas (Hou et al. 2020). However, these early investigations were rarely used to evaluate street vitality based on pedestrian activity and mostly used cameras with fixed angles and heights, making it difficult to address the obscuration of some pedestrian behaviors.

In this paper, we propose a novel approach based on video-image data from multiple position cameras for quantifying street vitality that combines image classification and multiple object tracking. Using image classification for video-image processing based on on-site auditing from a human perspective could help classify human activity intensity and investigate pedestrian preferences in the physical environment (Barchiesi et al. 2015). Multiple object tracking for processing video images makes the pedestrian volume measurement possible and saves labor with acceptable accuracy. The purpose of this study is to develop a deep learning-based urban street vitality evaluation framework using video-image data from multiple position cameras. A street vitality classification hierarchy was proposed based on the intensity of pedestrian behavior and an automatic system for pedestrian counting was applied to measure the density of pedestrians. A deep learning model for street vitality evaluation (DLM-SVE) is built on this foundation. In our experiments, video data of street activities from a commercial complex in Osaka was collected to validate the proposed framework and the suggested DLM-SVE model.

METHODS

The evaluation of street vitality in this paper in terms of both the intensity and complexity of pedestrians' activities. In this section, we present the research framework (Figure 1B) for measuring street vitality based on video-image using deep learning. First, pedestrian flow data is collected from a video camera. Then, the street vitality evaluation is achieved automatically by the proposed DLM-SVE model. In this model, both activity-based and number-based street vitality are included. The pedestrian number is calculated in the activity-based model using multi-object tracking (MOT) and pedestrians' activity is classified into five categories in the number-based model using image classification according to classification hierarchy with analytic hierarchy process (AHP).

Data Collection

The data processing of the proposed street vitality evaluation framework is divided into two steps. We first collected raw data in the form of video recordings for multiple streets throughout time. The video data provided precise and sufficient video-image information with spatial and temporal data about the pedestrian activity. The videos are then pre-processed to prepare for a MOT model of number-based vitality and an image classification model of activity-based vitality. The recorded videos are converted and filtered into photos by selecting specific time intervals that ensure the continuity of people's actions in the photos according to the model characteristics. In this way, the number of images required for MOT and image classification data preparation is reduced.

For different video-image analysis tasks, the video data are recorded by low-position and high-position cameras, respectively. For example, the high position camera can effectively avoid the problem of pedestrians in the foreground blocking the pedestrians behind during high pedestrian flow, while the low position camera can avoid the problem of foliage occlusion that tends to occur in the high position camera. In particular, in the image classification task of activity-based street vitality evaluation, the camera equipment should be mounted parallel to the street façade to identify the motion and behavior of each pedestrian. Furthermore, for the MOT task in number-based street vitality evaluation, video scenes with simple backgrounds are selected when placing cameras to avoid interference with object tracking, such as the occlusion from very large obstructions and reflection from glass windows.

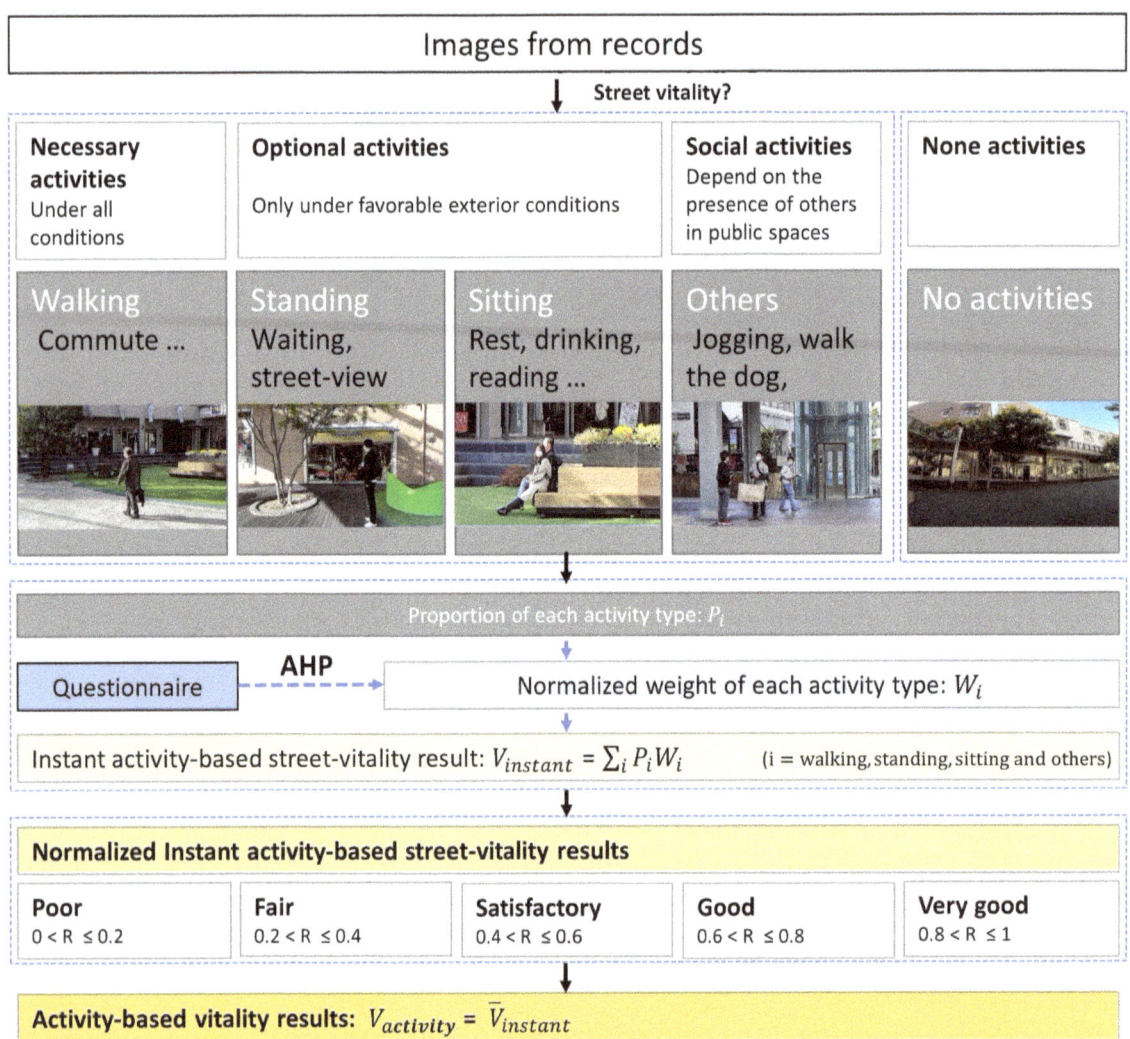

4 A classification hierarchy for pedestrian activities

DLM-SVE Model

Emerging deep convolutional neural networks have been successfully applied to several disciplines, including image classification, image attribute prediction, image segmentation, and object recognition and tracking. We present a DLM-SVE model based on this research that may infer street vitality from two different perspectives: counting pedestrians (number-based model) and categorizing pedestrian activity (activity-based model). As Figure 2 shows, videos are first needed to be recorded during survey time and to be converted into images as raw data. Then, to evaluate street vitality from both pedestrian numbers and activity types, the converted images are incorporated into the number-based model and activity-based model.

Multiple Object Tracking for Number-based Street Vitality

In the number-based model, we use a Faster R-CNN-based Multiple object tracking (MOT) model, Tracktor++, to detect, track, and record ID of multiple interest items in a video at the same time (Bergmann, Meinhardt, and Leal-Taixe 2019). The multi-object tracking model we used for counting pedestrians was proposed by Bergmann et al. (2019). In our case, we trained a Faster R-CNN with ResNet-101 and Feature Pyramid Networks on MOT17 datasets (Milan et al. 2016) for pedestrian detection and tracking. Figure 3 shows a pedestrian counting example of the proposed model. According to previous studies (Gehl 1987), the maximum allowable pedestrian street density is 10 to 15 people per minute per meter of street width. As a result, we scaled the results to a 0 to 5 scale and quantified the video's number-based vitality outcomes as:

$$V_n = 5 \cdot \left(1 - \left|N_m - \frac{N}{T}\right|/N_m\right) \qquad (1)$$

where V_n is the number-based street vitality, N_{max} is the maximum acceptable number of pedestrians per minute (here N_{max} is 15); N is the total number of people passing through the video, and T is the number of video minutes.

TABLE 1 Weightings of each activity type in the questionnaire results

Activity type	Walking	Sitting	Standing	Others
Weight	13.620%	28.238%	22.257%	35.886%

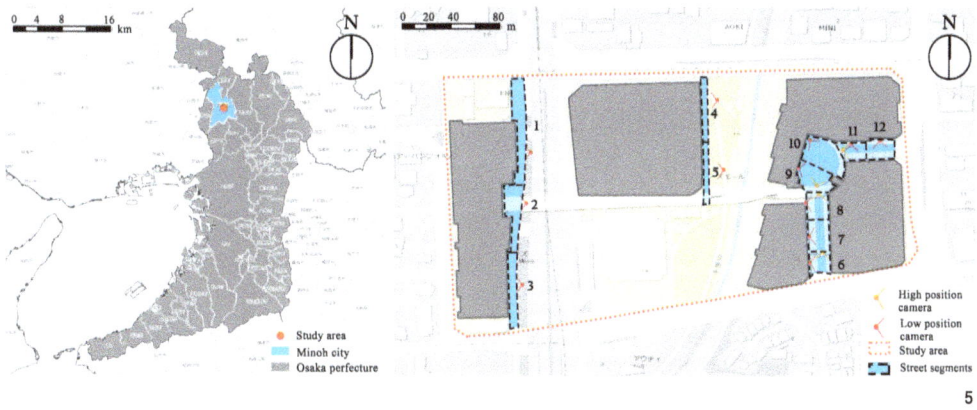

5 (a) The region map of the study area, and (b) 12 street segments and locations of camera installation in the study area

6 The characteristics of 12 street segments

7 Installed viewing angle and height of the high position and low position cameras

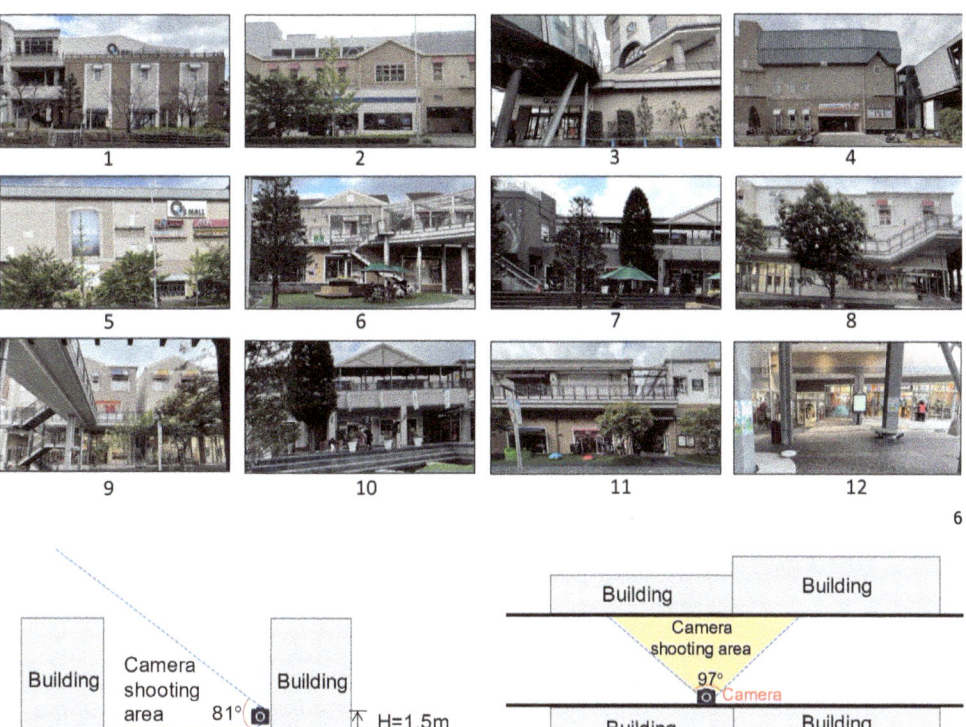

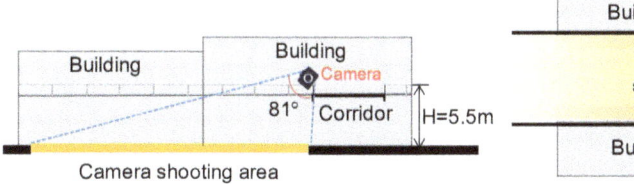

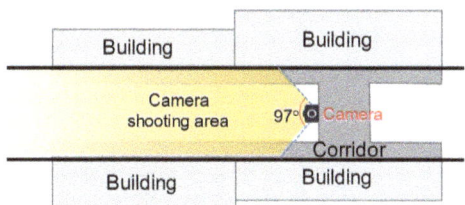

(a) Vertical view angle of low position camera

(b) Horizontal view angle of low position camera

(c) Vertical view angle of high position camera

(d) Horizontal view angle of high position camera

Classification Hierarchy and Image Classification for Activity-based Street Vitality

The activity-based model is an automatic street vitality classification method based on pedestrians' activity using video images according to the classification hierarchy. Using pedestrians' activity for vitality evaluation can trace back to the 1960s when Gehl classified pedestrian activities into social activities, optional activities, and necessary activities (Gehl 1987). Following Gehl's work, we classified walking, standing, and sitting, as well as other activities, into these three categories. Figure 4 presents the classification hierarchy for the activity-based model. To better understand the activity-based street vitality data for deep neural networks, a five-level classification hierarchy for street vitality based on the proportion of each activity was created, namely P, F, S, G, VG, i.e., Poor, Fair, Satisfactory, Good, and Very Good, with scores ranging from 1 to 5.

The classification criteria are based on the proportions of each activity category in each video image. We first invited one hundred raters who are all designers or students with a background in urban design education to rate the weight of each pedestrian activity type contributing to the street vitality. AHP calculates weightings for each activity category by comparing pairs of responses to a questionnaire. Participated evaluators are required to use pairwise comparisons to assess the relative importance of indicators, limiting the possibility of selection errors and providing more trustworthy results. The questionnaire survey included 60 students with an architecture or urban planning education background, and the AHP results are provided in Table 1.

With these weighted information and the average percentage of each activity type in each video image, we can obtain the vitality level of each video image, that is, the activity-based instant street vitality by Equation 2. Then, the activity-based street vitality results from recorded videos of the site during the survey time can be obtained by the mean value of each activity-based instant street vitality.

$$V_{instant} = \sum_i P_i W_i \qquad (2)$$

where $V_{instant}$ is the instant activity-based street vitality of a video image, i is the activity type, P_i is the proportion of each activity type in a video image, and W_i is the weight of each activity type in Table 1.

In the activity-based model, we use image classification with DenseNet based on the deep convolutional neural network (DCNN) model (Zhu and Newsam 2017). Following the classification hierarchy, a tailored video image-based benchmark dataset was created to train the image classifier in an activity-based model. Statistical pedestrian behavior data is not required in this method. Instead, the developed DenseNet will automatically learn the deep features of input images to support the classification task according to the corresponding benchmark dataset based on the classification hierarchy. Finally, number-based and activity-based street vitality results are standardized and averaged in a range of 0 to 5 as street vitality res.ults

RESULTS

The Study Area of the Experiment

The proposed method and the DLM-SVE model were implemented on the streets of an anonymous commercial complex in Osaka, with twelve street fragments chosen from a total field of 0.53 km² as Figure 5a shows. The length of the street segments is 12 meters, and Figure 5b illustrates the street view of twelve street segments. Figure 6 shows the characteristics of twelve street segments. Compared to other neighborhoods, commercial complexes are more focused on enhancing the vitality of the street-built environment to drive economic and other development. The pedestrian flow here is complex, with a wide variety of street-built environment elements and pedestrian behavior. The streets are sidewalks, with road widths ranging from 5 to 15 meters. At this distance, the recorded video at an angle parallel to the street facade can better reflect all pedestrians' movements.

Experimental Setup

Video data were collected from 10:00 to 18:00 for a total of eight days. We deployed camera equipment to record videos covering the entire streets of both the study area. Four cameras were used to take eight consecutive shots at sixteen shooting locations in twelve street segments. Each shot lasted 15 minutes per location at one hour intervals. Low-position cameras with a height of 1.5 m were mounted at each street segment for the DLM-SVE model. High-position cameras with a height of 5.5 m were mounted in second-floor pedestrian corridors to collect supplemental video data for the MOT model. The viewing angle and height of multiple position cameras are depicted in Figure 7.

These videos of the sixteen shooting locations were then processed into images. These images were then labeled based on the classification hierarchy. Video images from other commercial complexes are used as a training subset (75%), and those from the study area are used as a testing set (25%). The trained model was validated on the testing subset after the training procedure. Finally, the well-trained model correctly predicted all 34,560 photos.

The DLM-SVE Model Accuracy

To validate the accuracy of the DLM-SVE model, a random sample of 20 videos and 200 images were selected and

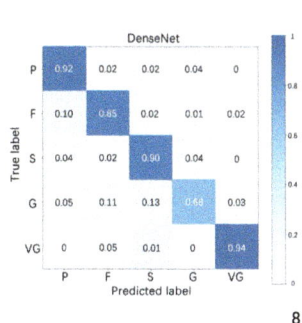
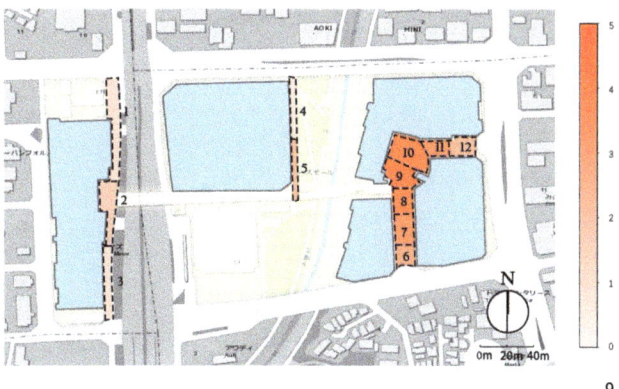

8 Confusion matrices of the activity-based model in the proposed DLM-SVE model

9 Street vitality visualized result in the study area activity-based model in the proposed DLM-SVE model

TABLE 2 Street vitality evaluation result of 12 street segments

Street segment number	1	2	3	4	5	6	7	8	9	10	11	12
street vitality	0.9	1.7	0.8	0.7	2.2	2.8	3.6	4.2	4.5	4.3	3.7	2.4

labeled with the number of pedestrians and the percentage of pedestrian activity types, respectively. Then, the statistical results were compared with the prediction results. For the number-based model, the multiple object tracking accuracy (MOTA) is 70.7% based on accuracy in the number of tracking targets. For the activity-based model, in the classification at the five categories, the model had an overall accuracy of 78.3%. The activity-based model's confusion matrices are shown in Figure 8. The values in the matrix represent the percentage of samples from one category that the model properly classified into another.

Street Vitality Results

The vitality results of 12 street segments predicted by the well-trained DLM-SVE model are shown in Figure 9 and Table 2. In Segments 9 and 10, the variety of pedestrian activities is greater, with a higher share of rest and viewing. The street space is made up of a small circular square that is orientated, open, and comfortable. At the same time, the location is at the intersection of three buildings, where it is easy to gather people, allowing people to see and be seen by each other. Street segments 4 and 5 have a high-quality interior supermarket with mostly solid walls and low transparency. Even though they are near to a beautiful park with plenty of greenery, overall activity is modest. It suggests that transparency could be a key component in attracting activities to commercial complex streets. Small units of visible commercial storefronts provide product information so that people can approach goods and businesses, while solid walls visually prevent people from engaging with goods.

DISCUSSION

This research investigated a deep learning-based method for quantitatively evaluating street vitality using video-image data. The proposed DLM-SVE model could measure street vitality based on the intensity and complexity of pedestrian activities. Video data from cameras of multiple positions were a dependable, efficient, and cost-effective data source. All the images were taken under local laws, and experiments did not obstruct pedestrians.

The major contributions are described below. First, we built an autonomous assessment approach for street vitality evaluation using deep convolutional neural network-based models and video data to reflect pedestrian activities and preferences. Compared to previous city-scale quantitative evaluations of street vitality, our method is suitable for fine-scale studies on street-level vitality which helps to improve street built environment design from a pedestrian perspective. Second, the experiment of a commercial complex and verification results of the DLM-SVE model showed that both the number-based and activity-based sub-models in our proposed DLM-SVE model are effective. Our method is high-efficiency, reliable, and feasible with acceptable accuracy compared to conventional manual on-site surveys. The proposed DLM-SVE model can be transferred to other urban streets with the supported video data of pedestrian behaviors. Third, a custom video-image dataset for activity-based model training in the DLM-SVE model was created.

Some limitations in this study needed to be noticed. First, the activity-based model in the DLM-SVE model ignores the temporal order of the images in the video. Using action recognition rather than image classification can avoid the missing of temporal information. Second, the experiment results may be different when applied to other streets, necessitating further debate.

In future practical applications, we can further collect street-built environment data based on current street vitality analysis, investigate the association between vitality and built environment at the street level, and extract positive street built environment characteristics, to conduct targeted optimization research on improving street vitality and to give some practical human-centric solutions for architects and urban planners to revitalize the space. In addition, although the current research results in the experimental area can only be extended to urban streets with similar characteristics, the construction of smart cities provides favorable conditions for the application of the method in a large scale.

CONCLUSION

We present a deep learning-based framework for automatically quantifying street vitality using video-image data from a human perspective, and our methodology was validated with a commercial complex experiment. Specifically, a DLM-SVE model was built including a number-based model and activity-based model using multi-object tracking and image segmentation. Our approach is centered around data analysis, and is both highly efficient and reliable, delivering acceptable accuracy. By evaluating the pedestrian traffic and activities on the streets, we can assess the vitality levels of different neighborhoods and offer recommendations for street-level space design that prioritizes human needs. This data-driven method provides an effective tool for identifying and understanding the vitality of urban areas. With the construction of smart cities and the development of urban big data, the proposed approach has the prospect of being applied at intercity and larger research scales.

REFERENCES

Barchiesi, Daniele, Dimitrios Giannoulis, Dan Stowell, and Mark D. Plumbley. 2015. "Acoustic Scene Classification: Classifying Environments from the Sounds They Produce." *IEEE Signal Processing Magazine* 32 (3): 16–34.

Bergmann, Philipp, Tim Meinhardt, and Laura Leal-Taixe. 2019. "Tracking without Bells and Whistles." In *Proceedings of the IEEE International Conference on Computer Vision*. 941–51.

Gehl, Jan. 1987. *Life between Buildings: Using Public Space*. New York: Van Nostrand Reinhold.

Gehl, Jan, and Birgitte Svarre. 2013. "Public Space, Public Life: An Interaction." In *How To Study Public Life*. Washington, DC: Island Press. 1–8.

Hou, Jingxuan, Long Chen, Enjia Zhang, Haifeng Jia, and Ying Long. 2020. "Quantifying the Usage of Small Public Spaces Using Deep Convolutional Neural Network." *PLoS ONE* 15 (10): e0239390.

Jacobs, Jane. 1961. *The Death and Life of Great American Cities*. New York: Vintage.

Kim, Young-Long. 2018. "Seoul's Wi-Fi Hotspots: Wi-Fi Access Points as an Indicator of Urban Vitality." *Computers, Environment and Urban Systems* 72: 13–24.

Li, Yunqin, Nobuyoshi Yabuki, and Tomohiro Fukuda. 2022. "Exploring the Association between Street Built Environment and Street Vitality Using Deep Learning Methods." *Sustainable Cities and Society* 79 (April): 103656.

Liu, Shaojun, Ling Zhang, and Yi Long. 2019. "Urban Vitality Area Identification and Pattern Analysis from the Perspective of Time and Space Fusion." *Sustainability* 11 (15): 4032. https://doi.org/10/gk9ms2.

Milan, Anton, Laura Leal-Taixé, Ian Reid, Stefan Roth, and Konrad Schindler. 2016. "MOT16: A Benchmark for Multi-Object Tracking." *ArXiv* (preprint):1603.00831.

Ye, Yu, Dong Li, and Xingjian Liu. 2018. "How Block Density and Typology Affect Urban Vitality: An Exploratory Analysis in Shenzhen, China." *Urban Geography* 39 (4): 631–52.

Zhang, Jiaxin, Tomohiro Fukuda, and Nobuyoshi Yabuki. 2021. "Development of a City-Scale Approach for Façade Color Measurement with Building Functional Classification Using Deep Learning and Street View Images." *ISPRS International Journal of Geo-Information* 10 (8): 551.

Zhang, Lemin, Ruoxi Zhang, and Biao Yin. 2020. "The Impact of the Built-up Environment of Streets on Pedestrian Activities in the Historical Area." *Alexandria Engineering Journal* 60 (1): 285–300.

Zhu, Yi, and Shawn Newsam. 2017. "DenseNet for Dense Flow." In *2017 IEEE International Conference on Image Processing (ICIP)*. Beijing, China. 790–794.

IMAGE CREDITS
All the drawings and images by the authors.

Yunqin Li is a lecturer in Architecture & Design College, Nanchang University. She received the PhD degree from Osaka University. She specializes in spatial auditing, measurement, perception, understanding, and interaction supported by new data, new technologies and new methods. She holds the BArch degree from Nanchang University, China, and the MArch degree from Southeast University, China.

Jiaxin Zhang holds the BArch degree from Nanchang University, China, and the MArch degree from Southeast University, China. He received the PhD degree from Osaka University, Japan. He is also aiming at cross-boundary researches related to the urban environment and computer science. His research interests include architecture design, urban environment visualization, smart cities, and machine learning in urban data analytics.

Xueqiang Wang is Associate Professor at the Architecture and Design College, Nanchang University, China.

Kai Ma is Associate Professor at the Architecture and Design College, Nanchang University, China.

Session Introduction

Critical Analysis of Architectural Design and Production

Kathy Velikov, Chair

Computational design is hardly new. As Theodora Vardouli's paper in this session reminds us, it now sixty years ago, in 1962, that Christopher Alexander and Marvin Manheim developed a computer system for design decision-making. Yet computational design practices still foreground an ethos of novelty, with this conference being no exception. While critical positions and reflections are present in a handful of papers throughout the conference, this is the only session specifically dedicated to history, theory, and critical practices. The four papers that follow are heterogenous, ranging in topic, methodology, and approach. Vardouli's paper, "Setting Historic Computer Systems in Motion" is remarkable for being the only paper in the past several years to undertake original archival scholarship in computation. Focusing on the HIDECS2 computer program by Alexander and Manheim, the paper makes contributions to not only the contextual understanding of this program but also advances contemporary historiographic methods for the study of computer systems drawn from the field of digital humanities: *data visualization* and *software reconstruction*. The former is used to situate the software in its intellectual actor-network, while the latter opens myriad questions regarding acquiring knowledge of the history of computation through software enactment.

Neil Leach's paper, "What is Creativity," draws on literature in cognitive science, aesthetic philosophy, and artificial intelligence to challenge assumptions and definitions of creativity. Taking the form of a position paper, Leach contributes to the discourse and debate of AI and the impact of large generative models; topics that been prominent at this conference. The papers by Paul Howard Harrison and Behnaz Farahi differ in approach in that both use their own creative work as a device through which to address critical questions. Harrison's "Parsed Precedent" directs a GAN process toward vernacular construction techniques, mobilizing machine learning toward forms of knowledge-making that can engage with cultural heritage, repair, and rehabilitation. Farahi's paper, "Returning the Gaze," describes a recent robotic installation piece and situates it within cybernetic theory and feminist discourse of the male gaze. The paper bridges theory and practice by articulating the work's own agency in relation socio-political questions, and thus contributes to a growing body of critical computation in the field.

Setting Historic Computer Systems in Motion

Theodora Vardouli
McGill University

Mobilities and Reenactments of the Hierarchical Decomposition System 2
(HIDECS 2, 1962)

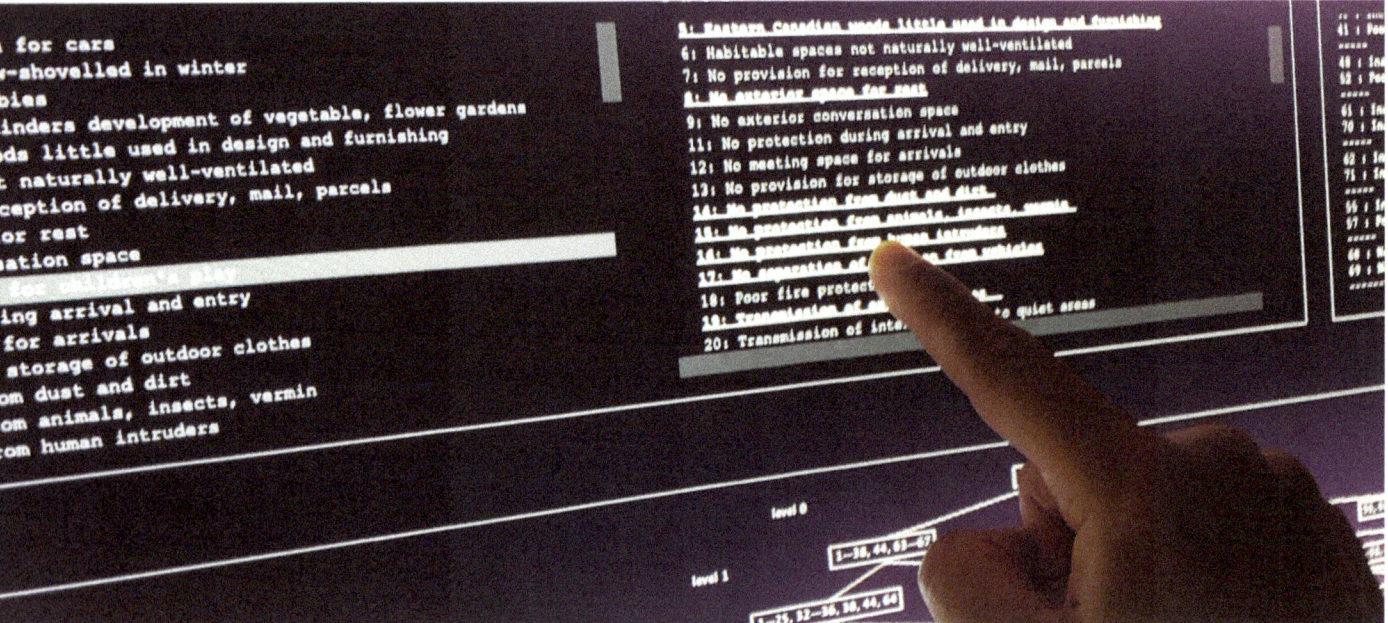

1 Interactive reconstruction of the historic HIDECS program using a list of 72 "misfits" (situations to avoid) in designing a Canadian house, listed in a 1966 paper by Allen Bernholtz and Edward Bierstone

ABSTRACT

This paper engages digital humanities methods, namely data visualization and software reconstruction, to shed new light on a landmark computer system for design decision-making presented in 1962 by Christopher Alexander and Marvin Manheim. The paper puts the historic computer system in motion in two interconnected ways. First, it moves the system outside the context of its initial development and tracks its trajectories and adaptations within a larger network of researchers in North American postwar research institutions. It does this through a newly built database on activity in "rational design methods" from 1966–1971, which the paper also details. Second, it reenacts a hybrid of the system's multiple versions unearthed within the Design Methods Network in a contemporary technical context and makes it available for manipulation by contemporary audiences. The paper begins with a brief overview of the HIDECS 2 system and its significance. Then, it discusses the development of the Design Methods Network database and its use to identify implementations, applications, and versions of the HIDECS 2 system developed outside the context of its origin. Finally, the paper presents the interactive reconstruction of a hybrid version of HIDECS that synthesizes features discovered through querying the database, and its presentation in a public exhibition on histories and contemporary practices of computer-aided design. Ultimately, the paper contributes new insights on the history of this impactful computer system and offers productive historiographic methods for the study of other computer systems and programs from the early years of design and architectural computing.

INTRODUCTION

This paper focuses on a historical computer system for design decision-making developed by architect and mathematician Christopher Alexander and civil engineer Marvin Manheim in 1962 and consecutively reworked by various researchers in North American research institutions operating under the broad umbrella of "rational design methods" (Moore 1966). The development and uses of the first version of the system, known as Hierarchical Decomposition System 2 or HIDECS 2, have been well documented (Upitis 2013; Steenson 2017; Vardouli 2020; Cristobal Olave 2021). The current paper adds new insights to the study of this important artifact in the history of architectural computing by deploying methods from the interdisciplinary field of digital humanities, namely data visualization and software reconstruction (Cardoso Llach and Donaldson 2019). These insights include a renewed understanding of the program as a mobile and malleable technical artifact whose trajectory extended beyond its initial setting of conception and development. New insights also emerge from the reconstruction of a hybrid form of the program—one that crossbreeds different contexts of implementation and application as well as distinct technical conditions. The reconstruction takes the form of an interactive artifact available for manipulation by contemporary audiences.

Using data visualization, the paper situates HIDECS 2 within an intellectual ecology consisting of multiple researchers and institutions who adopted and adapted it. An interactive software reconstruction of HIDECS 2 using the programming language Java reenacts a hybrid version of the system that synthesizes versions and adaptations as it moved across different institutions and research settings, while also reinterpreting it in a new programming language and computer hardware. Presented in the public exhibition *Vers un imaginaire numérique* (Centre de Design de l'UQAM, Montreal, 2021) and displayed together with historical documents around the program's inception and development, the interactive reconstruction of HIDECS 2 activates critical debates around software as a product of specific historical settings. On one level, the interactive nature of the reconstruction serves a instructive function, allowing audiences to acquire an intuitive understanding of its algorithmic logic and the design process it modeled. The reconstruction also operates on a critical register: it renders visible tensions and collusions between messy, at times even arbitrary, data and the ostensibly objective algorithms that structured them for design and dramatizes the notion of "failure" as a constitutive category of the program's computational logic. That is in juxtaposition with the positive categories of "goals" and "optima" that drive much of contemporary data-driven design.

BACKGROUND: A BRIEF HISTORY OF HIDECS 2

Although not developed specifically for architectural or urban design, HIDECS 2 was one of the first computer systems applied to architectural decision-making. The system became known to architectural audiences through Christopher Alexander's influential book *Notes on the Synthesis of Form*, published in 1964 and based on Alexander's doctoral dissertation at Harvard University. In the book Alexander famously presented an implementation of the computer system for the determination of so-called "components" of an agricultural village of six hundred people in Bavra, India. Components were groups of design considerations (or "requirements") that the designer addressed through simple schematic drawings (or "diagrams"). The computer system was developed to identify these components and indicate the structure by which the "diagrams" should be combined to produce a complete design that responded to the requirements.

HIDECS 2 achieved this by analyzing relationships between design requirements in terms of the data they shared and the ways they influenced each other. This analysis allowed grouping together requirements that were strongly connected and separating ones that were relatively independent to each other. Following this process, HIDECS 2 broke down ("decomposed") an unstructured set of requirements into

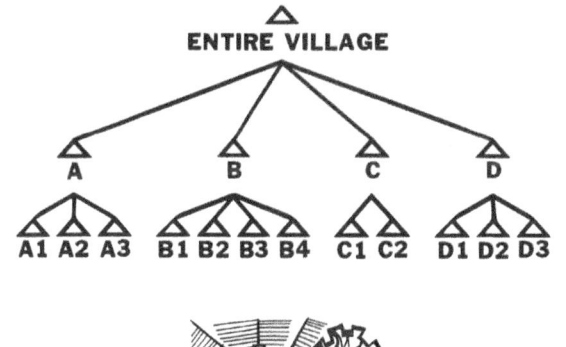

2

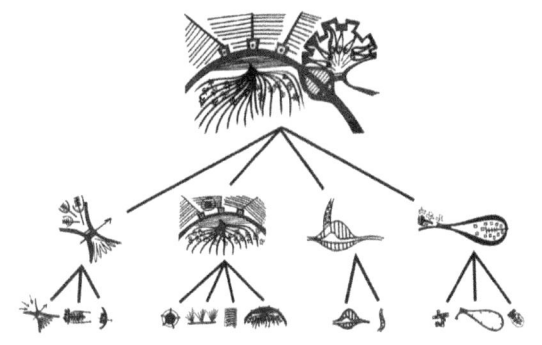

3

2   Tree decomposition of 141 design requirements for an agricultural village in India into 12 subgroups using HIDECS 2

3   Composition of diagrams, defined by the designer, addressing the subgroups defined by HIDECS 2 to produce the design of the village

a hierarchical tree: a graph in which every two vertices are connected by exactly one path. The tree indicated the order by which a designer ought to combine the component diagrams they developed for each grouping of requirements.

For instance, in the "Worked Example" of the village in India presented in Notes, HIDECS 2 broke down 141 requirements including wildly diverse statements such as "67. Drinking water to be good, sweet" and "117. Spread of information about birth control, disease, etc." into 12 independent components (Alexander 1964, 140-41) that Alexander recomposed to produce a design for the village.

Although in Notes HIDECS 2 was presented as the implementation of a theory of design presented in the book, the computer system preceded and, as scholars of Alexander have convincingly argued, influenced the theory (Upitis 2013). HIDECS 2 was developed during a consultancy at the Civil Engineering Systems Laboratory from 1960–1962 in collaboration with civil engineer Marvin L. Manheim. It was implemented in the IBM 709 of the MIT Computation Center, under the control of the Fortran Monitor System in use at the Center during the second half of 1961. Since the beginning of his doctoral studies, Alexander had been concerned with the organization of empirical data related to design and how this organization could indicate a well-ordered sequence of decision-making steps for the designer (Vardouli 2020). Alexander called this decision-making sequence a design "program" (1964, 69). HIDECS 2 computed the design "program" through first, making trial cuts of an initial unordered graph (the set of requirements) into subgraphs, then calculating an "INFO" parameter based on the number of links that the partition cut and the number of vertices at each side of the partition, and finally performing a heuristic optimization method to minimize the INFO parameter. The smaller INFO, the more independent the groups partitioned, achieving the principle of breaking down the design problem in independent components.

Aside from the village in India, HIDECS 2 was also used to locate a section of the I-91 Interstate Highway System in Western Massachusetts (Alexander and Manheim 1962b). This project brought forward tensions arising between the analysis to identify design subproblems for which to develop diagrams and the synthesis of the diagrams, which Alexander and Manheim envisioned as transcending the mere combination of discrete entities (Alexander and Manheim 1962b, 91–92). The history of HIDECS 2 continues with the development of a new version, HIDECS 3, by Alexander in June 1963. HIDECS 3 accounted for anomalies stemming from hierarchical decomposition and instead approached decomposition by searching for maximal simplices. Simplices were shape-like topological constructs like triangles and tetrahedra defined by a set of requirements (represented as vertices of a graph). HIDECS 3 searched for triangles or tetrahedra whose number of vertices was not smaller than other triangles or tetrahedra in the graph (Alexander 1963; Vardouli 2017). The crucial attribute of this method was that a vertex (representing a design requirement) could be part of multiple subsystems. This method produced as its output not a tree, but instead a "lattice." Alexander discussed the theoretical implication of this move from trees to lattices in the awarded 1965 article "A City is Not a Tree," which ushered a new image of the city as a complex system not as a single hierarchy but as multiple overlapping ones.

METHODS: DIGITAL HISTORY TACTICS FOR THE STUDY OF COMPUTER-AIDED DESIGN PROGRAMS

Historians and technology studies scholars have long urged to move beyond the "inventor" narratives and to study technological artifacts, including computer programs, in their multiple and diverse contexts of use (for example, Pinch and Bijker 1980; Oudshoorn and Pinch 2003). This becomes challenging with influential artifacts in the history of design computing such as HIDECS 2, which resulted in publications and demonstration

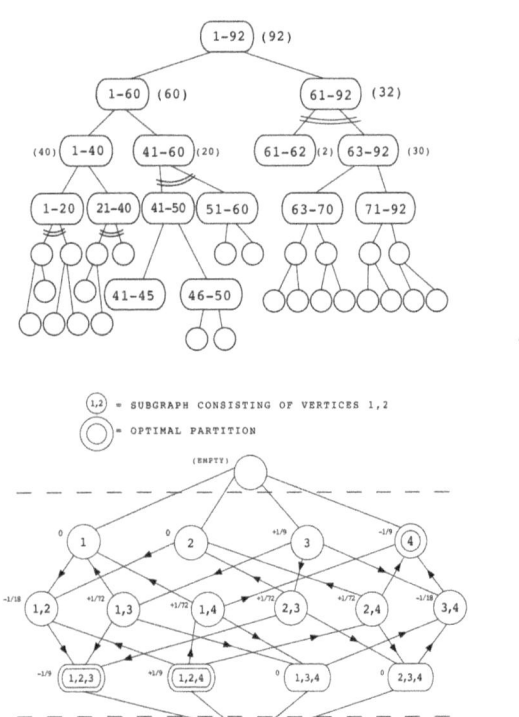

4 Tree drawn to represent the output of HIDECS 2 (printed as verbal statements) after analyzing relationships between 92 requirements

5 HIDECS 3 replaced the hierarchical tree with a semi-lattice

6 Front matter of the *Design Methods Group Newsletter*, Volume 2, Number 1-2, 1968

projects but were not taken up by professional architects and designers, thus mostly remaining academic experiments. How to mobilize such historic computer programs beyond the confines of their invention? This paper presents two interconnected methods for historically mobilizing HIDECS 2 drawing from the interdisciplinary field of digital humanities, specifically digital history and software reconstructions. First, it presents a method for tracking mobilities of HIDECS 2 within a cross-institutional and cross-disciplinary network of North American academics who developed little known subroutines and implementations of the computer system. Then, it discusses an interactive software reconstruction developed to reenact HIDECS 2 in an alternate technical and cultural setting opening it up for new uses and critical readings. "Mobilizing" here refers both to a historical revealing of HIDECS 2 as mobile but also to rendering of the system capable for action in the present through a reenactment of a hybrid version that recalls its multiple versions and implementations.

Mobilities Within the Design Methods Network

Alexander's work was received with avid interest from a short-lived but influential movement that proliferated across sites of architectural education and research in the second half of the 1960s under the umbrella of "rational design methods" (Montgomery 1970). Design methods filtered interwar mandates of "rational architecture"—buildings designed under the tenets of material economy or functional purpose—through the lens of a goal-oriented, rule-based rationality characteristic of postwar intellectual life (Erickson et al. 2015; Vardouli 2020). Design methods research took place mainly in the setting of British and North American universities. Working with mathematicians and engineers, architects developed methods that viewed architectural design as a stepwise process, amenable to mathematical analysis, and directed toward articulable goals. This produced a new disciplinary focus for architecture, from the physical form of the final artifact (object, building, city) to the steps and decisions leading to it, ultimately paving the path for algorithmic and computational approaches to design. Ardent engagement with mathematical analysis and rigorous theory produced not only new styles of studying and talking about design and architecture, but also a proliferation of societies and associations during the 1960s, founded with the mission to advance communication among design methods activity performed in disparate settings.

Acknowledging design methods as an intellectual ecology whose study requires attention to mobilities and exchanges across institutions and geographies as opposed to their study in isolation, the CoDEx research team at McGill University led by the author, produced an interactive digital database mapping the design methods network in North America. We built the database using the entries of a monthly periodical headquartered at the University of California Berkeley and founded in 1966 with the aim to establish a network of researchers working on design methods in North America and around the world. The *Design Methods Group Newsletter*, as the periodical was called, is a rich resource for identifying cross-disciplinary and cross-institutional transactions, tentative computer experiments, technical languages, controversies, and trends, giving a lively view of research-in-the-making throughout the six published volumes of the *Newsletter* from 1966-1971.

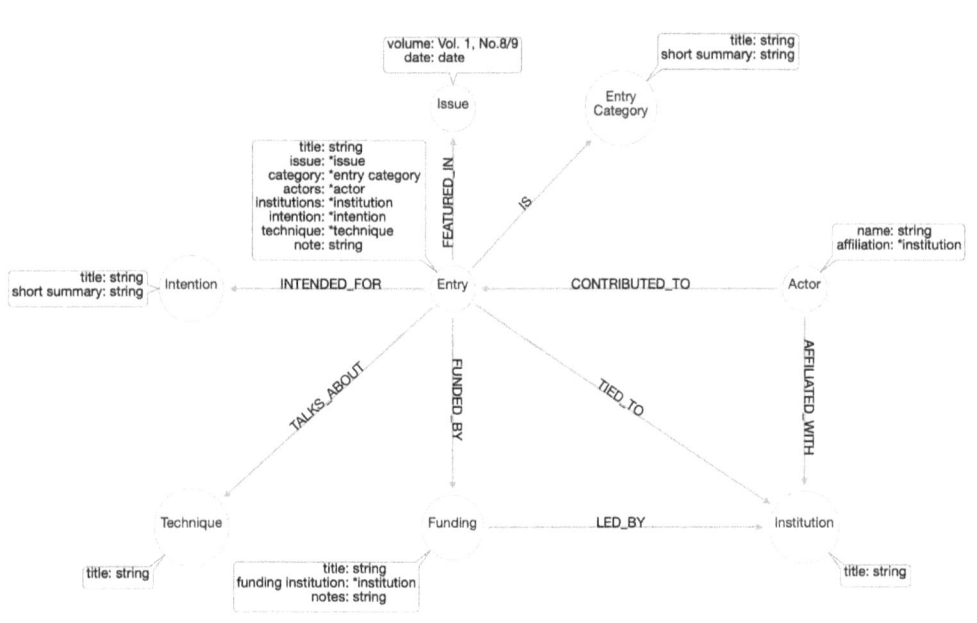

7 Database architecture of the Design Methods Network in Neo4j

8 Mapping mobilities of HIDECS 2 and its successors within the Design Methods Network

We used the *DMG Newsletter* as the foundation of the Design Methods Network database, which we began building using the HEURIST data management system (https://heurist-network.org). From each *Newsletter* entry we extracted the following data: 1. "title" (entry title), 2. "entry category" (the newsletter accepted news, abstracts, work in progress outlines, reviews and criticisms, bibliographies of design methods, and computer program abstracts), 3. "actor" (entry author), 4. "institution" (location that each entry refers to), 5. "funding" (funding organization associated to the entry). We also added interpretative information such as "technique" and "intention," following the study of each entry's content. We use "intention" to account for the conceptual drivers of each entry and statements for what the author was trying to accomplish. "Technique," in turn, reflects the algorithmic, mathematical, and calculative tools enlisted to achieve these goals and how concepts were operationalized. The assignments of "intention" and "technique" were performed based on the researcher's interpretation of each entry's content and awareness of the recurrence of certain keywords and themes. We are currently exploring ways to combine the *close reading* of the entries that we performed with *distant reading* (Moretti 2013) techniques that would allow us to trace the prominence of certain themes or keywords in a more systematic way. To improve usability and functionality of the database, we transferred the entries to the graph database system Neo4j (https://neo4j.com). The use of directed graphs allowed us to also characterize relationships and dependencies between the different data types and led us to refine the database architecture as indicated in Figure 7.

The motivation behind building the database was to generate visualizations that revealed connections between actors, sites, or techniques. We have explored both static images made by hand that condense findings from the database within graphic conventions of maps, timelines, and diagrams of relationships, as well as dynamic visualizations such as digital stories using the markdown-based app Obsidian (https://obsidian.md). The Design Methods Network database can be mined for historical projects concerned with mobilities of technical practices within academic research networks. In the case of HIDECS 2, the database revealed 15 related entries by 21 authors who engaged the development, implementation, or application of the system in 14 distinct institutions.

Figure 8 shows a static visualization, made by hand and for printing, of manifestations of the HIDECS 2 system and its successors within the Design Methods Network. The figure focuses on one specific technique ("hierarchical decomposition") and visualizes its associated entries, actors, institutions, and funding organizations. As elaborated on in the "Results and Discussion" section of the paper, the visualization establishes HIDECS as a mobile entity that can be followed across distinct geographic, institutional, and material contexts. A closer look at the associated data shows a multiple and diverse engagement with the system: some projects entailed the development of specific subroutines for decomposition and recomposition such as DECOMP, RECOMP, and VTCON 2 as well as adjusting these routines to languages and computer systems other than FORTRAN and the IBM 709. Other entries focused on the practical advantages and limitations of the program in specific application contexts such as the Coventry Community Nursing home. Finally, other entries discussed versions of the system, such as CLSTR, adapted for use in educational settings. Contexts identified in this study, such as the University of Toronto in which a version of HIDECS 3

CRITICAL ANALYSIS OF ARCHITECTURAL DESIGN AND PRODUCTION

Hybrids & Haecceities

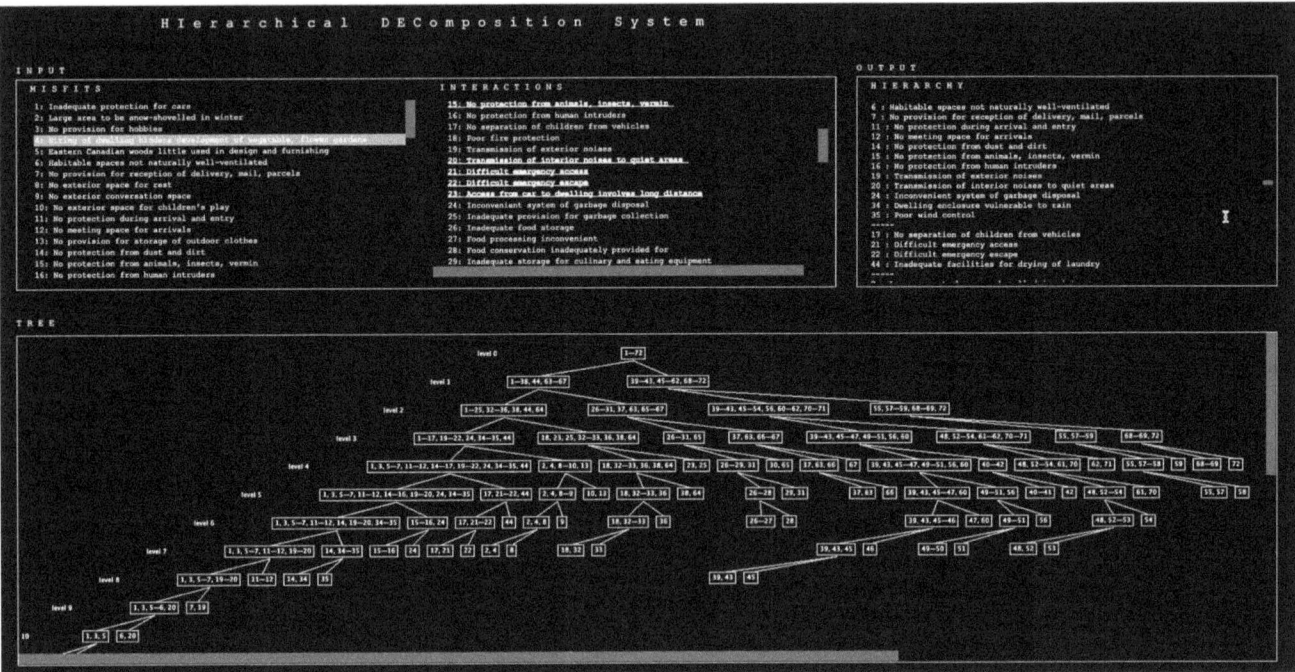

9 A screenshot from the HIDECS interactive reconstruction

was adapted for the development of Canadian housing under a National Research Council Grant, became the launching pad for thinking about the interactive reconstruction of the computer system.

Reconstructing the HIDECS Hybrid

Design and computing scholar Daniel Cardoso Llach has put forward software reconstruction as a key tactic in a media archaeological approach to computer-aided design. He and Scott Donaldson have framed software reconstruction "as a method to shed new light into the material, gestural, and sensual dimensions of computer-aided design technologies" (Cardoso Llach and Donaldson 2019). Some projects developed under the related project "Experimental Archaeology of CAD" have paid attention to the hardware of early computer aided design systems to highlight the embodied, as opposed to just cognitive, interaction between users and the systems, as well as to their spatial, aesthetic, material and design sensibilities. Other projects, such as the Coons Patch reconstruction, have focused on making visible and palpable abstract mathematical concepts (Cardoso Llach and Donaldson 2019). Our reconstructive work of HIDECS 2 aligns more with the latter approach, which we characterize as a software *reenactment* to convey liberties we take in materializing an abstract algorithm into an interactive artifact. HIDECS 2 was not an interactive system: as Upitis (2013) has detailed, the system's input was a deck of punch cards and its output was a sheet of printed verbal statements. Users of the system then had to translate these statements manually into the iconic tree or semi-lattice. Although these material conditions were key for the development and function of the system, we decided to temporarily set them aside and focus on HIDECS 2 as the computer implementation of a theory and method for designing.

A similar approach was taken by Pablo Miranda Carranza, who in 2020 published a Python implementation of HIDECS 2. His intention was, as he wrote, to "deemphasise the material and technical conditions behind the code and foreground the abstractions and concepts implemented" (Miranda Carranza 2020). Exploring different algorithms for finding a minimum cut in a graph, as HIDECS 2 was programmed to do, Miranda Carranza concluded that Alexander and Manheim did not look for an efficient solution to the graph partition problem but sought to demonstrate the intractability of a design problem without the heuristic algorithms that they put forward. Our reconstruction of HIDECS carves a middle ground between seeing the system solely through the algorithmic abstractions it performed or exclusively through the material settings and conditions in which it was embedded. We aimed to produce a reconstruction that was, on the one hand, *instructive*, communicating the fundamental tenets of HIDECS as design methodological and theoretical proposition, and on the other hand, *critical*, rendering visible biases and assumptions embedded into the system.

To make the reconstruction, we used the 1962 report of HIDECS 2 by Alexander and Manheim in conjunction with a 1966 paper by Allen Bernholtz and Edward Bierstone from the University of Toronto that applied a version of HIDECS in problems of Canadian housing. The paper was included in the Proceedings of the 1966 Design and Planning Conference at the University of Waterloo, which was the founding event of the Design Methods Group. As such, our reconstruction of the HIDECS computer system holds together multiple sites and events of historical significance. The blending of these disparate but related contexts renders the reenactment a *hybrid*. This is both in the more direct sense of cross-polinating two distinct things for the production of a third, but also in a broader and perhaps theoretically more evocative sense of creating an assemblage that is capable of holding together, but also producing, difference in its unpredictable uses by audiences today.

We coded the graph partition algorithm in Java using hill climbing and implemented interactive features using touch events. The "user" of our reconstruction is presented with a touch screen organized in an INPUT and OUTPUT frame, which include a "Misfits" and "Interactions" frame and a "Hierarchy" and "Tree" frame respectively. As I discuss in the following section, the notion of "misfits" was central in Alexander's theory and in the logic of HIDECS 2. Unlike traditional architectural briefs that typically list desired situations, goals, or qualities to achieve, Alexander proposed, and Bernholtz and Bierstone took on, the idea that the design brief ought to list all situations that an architect ought to *avoid*.

The "Misfits" frame is prepopulated with 72 requirements for Canadian housing listed in the 1966 paper by Bernholtz and Bierstone, such as "16. No protection from human intruders," "22. Difficult emergency escape," and "54. Poor love-making facilities for parents." The "Interactions" frame default state includes interactions between these requirements as they were identified in the same paper. Based on these default conditions, the reconstruction shows the calculated "Hierarchy" decomposing these requirements into subgroups and automatically draws the corresponding "Tree." Users can use the interactive touch screen to change the links between requirements by clicking on one of the 72 requirements in the "Misfits" window and then clicking on requirements in "Interactions" window to select or deselect them. Every touch event triggers the program to compute a new hierarchy and a new tree. Users can also type in new requirements using a keyboard, which are automatically added in the "Misfits" and in the "Interactions" frames. The interactive reconstruction (missing the keyboard implementation) was shown in the 4,000 sq ft exhibition *Vers un imaginaire numérique* in Montreal, alongside historical material showing the origins and trajectories of HIDECS 2 and its subsequent versions.

RESULTS AND DISCUSSION: HISTORIOGRAPHIC AND THEORETICAL IMPLICATIONS

Considering HIDECS 2 within the Design Methods Network destabilizes it as a fixed or singular system and reveals it to be a mobile and malleable entity that can be followed across multiple contexts. This is in keeping with approaches in adjacent fields in the humanities that foreground specific artifacts as objects to be followed as they move between diverse settings. Anthropologist Arjun Appadurai, for instance, has compellingly proposed that focusing on things and their trajectories rather than on social actors and settings can shed light on the human operations that enliven them (1986). In the history of science, Lorraine Daston has spoken of "biographies of scientific objects" as shedding light on the formation of knowledge through attentiveness to the fluctuating cultural, material, and theoretical meanings of objects of scientific inquiry (2000). Hans-Jörg Rheinberger (1997), in turn, proposed tracking "things embodying concepts"—what he termed "epistemic things"—as a way to navigate sites of knowledge production.

In a similar spirit, mobilities and agencies of techniques and technological artifacts within the Design Methods Network enable threading together disparate contexts while preserving their social, intellectual, and institutional specificities. In other words, the mobilities of HIDECS, its versions, implementations, failures and applications, do not only shed new lights on the computer system itself and provide new sites and cases for historical scrutiny, but they also weave together disparate and heterogeneous settings as parts of the same story— a story about tensions between human reason and algorithmic rationality (Vardouli 2017) and negotiations between empirical data collection and the algorithmic orders that were argued to undergird that data.

The interactive reconstruction of HIDECS hybridizes key contexts of its development and adaptation, while also highlighting some of its key theoretical categories. One of these categories, which puts HIDECS in stark contrast with

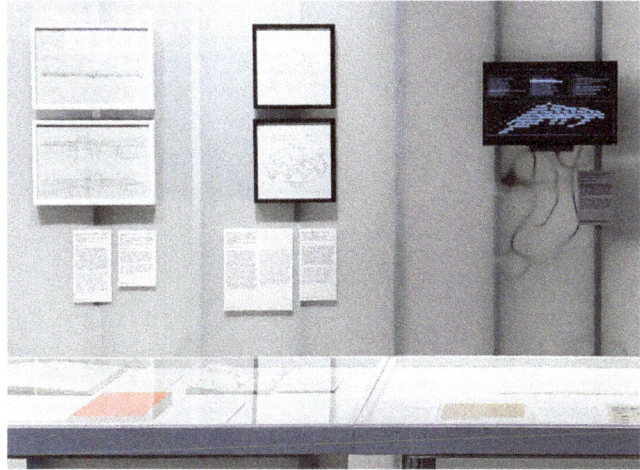

10 The hybrid HIDECS interactive reconstruction and accompanying historical material in the gallery space

contemporary data-driven design projects that are predicated on goals and optima, is the notion of "failure." Through its list of 72 "misfits" reconstruction activates questions about what it means to consider design through a list of situations to avoid and to use algorithmic rationality not to maximize efficiency or productivity of various kinds but instead to avoid a failed architecture. The historical documents accompanying the reconstruction, such as for example a sketch a so-called "failure card" from 1961 also position failure as a foundational technical category pertaining to the organization of building-related information (Alexander 1960a; 1960b). "Failure" denoted a kind of physical event that prevented a need from being satisfied (for example, sleep prevented by bioclimatic discomfort). Sometimes failures shared data, other times they were corrected by the same operations, and other times the correction of one failure aggravated the other. Similar relations of overlap, reinforcement, or conflict were then established among the failures' corresponding requirement. Depending on the data they shared, failures could be established as "interacting" in certain ways. Grouping together failures that interacted more strongly and setting apart those that were independent from each other, would reveal what Alexander would refer to as the non-arbitrary "structure" of the design problem, that HIDECS 2 was designed to compute (Vardouli 2017).

The interactive reconstruction dramatizes the contingent relationship between data and their organization. The list of 72 misfits loaded in the reconstruction reads like a cornucopia of aphorisms, intriguing in its unapologetic arbitrariness—why 72 misfits, and why these 72? It soon becomes evident that the semantic content of these requirements matters little for the algorithm and that the key consideration are their stated relationships. One could type in non-design related, even nonsensical, statements and the program would unproblematically incorporate them in its tree calculations. At the same time, the reconstruction renders visible the multiple layers of bias and human intervention in an algorithmic process: humans collect the data, describe them as "misfits", and establish their relationships.

CONCLUSION

In this paper, I have presented two interconnected methods that complement text-based scholarship on historic computer systems with digital humanities approaches such as database visualization and with software reconstructions. These methods have mobilized the historic computer system of HIDECS 2 in two cross-fertilizing ways: first, by tracking its trajectories and adaptations outside its initial contexts of inception and second, by reconstructing a hybrid of the system's multiple versions and making it available for manipulation and critical contemplation.

Responding to the call behind the *Hybrids & Haecceities* conference theme to dissolve binary conditions and immanent hierarchies and to juxtapose specificities to totalizing abstractions, the paper casts the historic computer program as a *hybrid* absorbing but also revealing the specificities of the heterogeneous sites it traversed. Setting HIDECS 2 *in motion* brings forward critical arguments about cultural meanings and social exchange of computer systems and algorithmic techniques, entanglements of institutional and intellectual formations, and stories about theoretical commitments cloaked under algorithmic objectivity and software pragmatism.

Moving forward, we are planning further work on the software reconstruction to implement the recomposition process: the use of the tree to combine user-defined schematic diagrams. We anticipate this to be challenging and exciting in terms of the interface design, and to present opportunities for further hybridizing the system with the contemporary digital landscape (for instance, one possibility is using web databases for identifying diagrams instead of the users developing them). We are also eager to delve into some of the contexts we have identified by mapping the HIDECS trajectory within the Design Methods Network and to consider digital storytelling as a means for narrating the history of this landmark computer program in its many reconfigurations in the second half of the 1960s.

ACKNOWLEDGMENTS

I would like to gratefully acknowledge Jiaqi (Arlene) Chen for her work on the HIDECS reconstruction, Max Leblanc and George-Étienne Adam for their work on the Design Methods Network database and visualizations, and Eliza Pertigkiozoglou for her feedback on the database structure and querying and for reconstructing drawings from the HIDECS 2 and 3 reports. I would also like to warmly thank the anonymous reviewers for their detailed and constructive comments. This work has been made possible by a grant by the Social Sciences and Humanities Research Council of Canada.

REFERENCES

Alexander, Christopher. 1960a. "Letter to Chermayeff Re: Failure Cards." Box 4, Folder "Alexander, Christopher, 1958-1966." Serge Ivan Chermayeff Architectural Records and Papers, 1909-1980. Department of Drawings & Archives, Avery Architectural and Fine Arts Library, Columbia University.

Alexander, Christopher. 1960b. "Letter to Chermayeff Re: Failures Interlock, IBM Group Extraction." Box 4, Folder "Alexander, Christopher, 1958-1966." Serge Ivan Chermayeff Architectural Records and Papers, 1909-1980. Department of Drawings & Archives, Avery Architectural and Fine Arts Library, Columbia University.

Alexander, Christopher, and Marvin L. Manheim. 1962a. *HIDECS 2: A Computer Program for the Hierarchical Decomposition of a Set Which Has an Associated Linear Graph*. Cambridge, Mass.: Massachusetts

Institute of Technology, Civil Engineering Systems Laboratory Publication 160.

Alexander, Christopher, and Marvin L. Manheim. 1962b. *The Use of Diagrams in Highway Route Location: An Experiment*. Cambridge, Mass.: Massachusetts Institute of Technology, Civil Engineering Systems Laboratory Publication 161.

Alexander, Christopher. 1963. *HIDECS 3: Four Computer Programs for the Hierarchical Decomposition of Systems Which Have an Associated Linear Graph*. Cambridge, Mass.: Massachusetts Institute of Technology, Civil Engineering Systems Laboratory Publication Report R63-27.

Alexander, Christopher. 1964. *Notes on the Synthesis of Form*. Cambridge, Mass.: Harvard University Press.

Alexander, Christopher. 1965. "A City Is Not a Tree." *Architectural Forum* 122 (1): 58–62 (Part I) and 122 (2): 58–62 (Part II).

Appadurai, Arjun, ed. 1986. *The Social Life of Things: Commodities in Cultural Perspective*. Cambridge, U.K.: Cambridge University Press.

Bernholtz, Allen, and Edward Bierstone. 1966. "Computer-Augmented Design." *Design Quarterly*, Special Issue: Design and the Computer 66/67: 40–51.

Cardoso Llach, Daniel, and Scott Donaldson. 2019. "An Experimental Archaeology of CAD: Using Software Reconstruction to Explore the Past and Future of Computer-Aided Design." In *Computer-Aided Architectural Design*, edited by J.-H. Lee. Singapore: Springer. 105–119.

Cristobal Olave, Diana. 2020. "The Computer-Aided Rough Patterns of Christopher Alexander." *Nexus Network Journal* 23: 113–128.

Daston, Lorraine. 2000. *Biographies of Scientific Objects*. Chicago: University of Chicago Press.

Erickson, Paul, Judy L. Klein, Lorraine Daston, Rebecca Lemov, Thomas Sturm, and Michael D. Gordin. 2015. *How Reason Almost Lost Its Mind: The Strange Career of Cold War Rationality*. Chicago: University of Chicago Press.

Miranda Carranza, Pablo. 2020. *HIDECS2 Python*. GitLab. Accessed May 28, 2022. http://www.diva-portal.org/smash/record.jsf?pid=diva2%3A1565218&dswid=-7330.

Moretti, Franco. 2013. *Distant Reading*. London: Verso.

Montgomery, Roger. 1970. "Pattern Language - The Contribution of Christopher Alexander's Centre for Environmental Structure to the Science of Design." *Architectural Forum* 132 (1): 52–59.

Moore, Gary, ed. 1966. "What Is the Design Methods Group?" *DMG Newsletter* 1 (1): 1.

Oudshoorn, Nelly, and Trevor Pinch, eds. 2003. *How Users Matter: The Co-construction of Users and Technologies*. Cambridge, Mass.: The MIT Press.

Pinch, Trevor J., and Wiebe E. Bijker. 1984. "The Social Construction of Facts and Artefacts: Or How the Sociology of Science and the Sociology of Technology Might Benefit Each Other." *Social Studies of Science* 14 (3): 399–441.

Steenson, Molly Wright. 2017. Architectural Intelligence: How Architects and Designers Created the Digital Landscape. Cambridge, Mass.: The MIT Press.

Rheinberger Hans-Jörg. 1997. *Toward a History of Epistemic Things: Synthesizing Proteins in the Test Tube*. Writing Science Series. Stanford, Calif.: Stanford University Press.

Vardouli, Theodora. 2017. Graphing Theory: New Mathematics, Design, and the Participatory Turn. Doctoral Dissertation, Massachusetts Institute of Technology.

Vardouli, Theodora. 2020. 'Bewildered the Form-Maker Stands Alone': Computer Architectures and the Quest for Design Rationality. In Vardouli T and Olga Touloumi (eds.) Computer Architectures: Constructing the Common Ground. Routledge Research in Design, Technology and Society Series.

Upitis, Alise. 2013. Alexander's choice: How Architecture Avoided Computer-aided Design c. 1962. In Dutta, A. (ed.) A Second Modernism: MIT, Architecture, and the "Techno-Social" Moment. Cambridge, Mass: The MIT Press.

IMAGE CREDITS

Figures 2, 3: *Notes on the Systhesis of Form* by Christopher Alexander, Cambridge, Mass.: Harvard University Press, Copyright © 1964 by the President and Fellows of Harvard College. Copyright © renewed 1992 by Christopher Alexander. Used by permission. All rights reserved.

Figures 4, 5: Reconstructed drawings by Eliza Pertigkiozoglou from Alexander and Manheim (1962a).

Figure 6: *Design Methods Group Newsletter*, Volume 2, Numbers 1-2, January-February 1968.

All other drawings and images by the author and the CoDEx research group, McGill University.

Theodora Vardouli is an Assistant Professor at the Peter Guo-hua Fu School of Architecture, McGill University. She is co-editor of *Computer Architectures: Constructing the Common Ground* (Routledge 2020) and *Designing the Computational Image, Imagining Computational Design* (forthcoming AR+D Publishing), and co-curator of the exhibition "Vers un imaginaire numérique" (Centre de Design de l'UQAM, Montreal 2021). Her in-progress book (under contract with the MIT Press) examines postwar architecture's relationship with mathematics with focus on the physical and symbolic prevalence of graphs. Vardouli is an External Examiner at the Bartlett (UCL, UK) and an Editorial Board member of the journal *Technology, Architecture and Design* (TAD).

Parsed Precedent

Paul Howard Harrison
HDR / University of Toronto

Parameterization, Vernacular, and Machine Intelligibility

1 Simulation-validated genetic optimization within the latent space of a neural network trained on precedent images; this figure shows six generations of the genetic optimization, with GAN-generated inputs (bottom) and rigid-body simulated outputs (top)

ABSTRACT

The built environment is our best repository of architectural knowledge, but the lessons embedded within it are often overlooked—at least through a computational lens. Recent developments in artificial intelligence allow neural networks to engage directly with precedent, generating new images that reflect and mutate a dataset of existing conditions; while this is useful in and of itself, an unsung byproduct is the smooth parameterization of the neural network's latent space. This enables designers to deploy genetic algorithms directly within the latent space itself, bypassing the need to create bespoke parametric models for optimization.

Within the neural network, this optimization is often critical; the frequently marginal quality of AI outputs means that a degree of validation is required to find acceptable outputs. This paper describes a method of using rigid-body physics simulation to evaluate the structural stability of AI-generated archways and feed stability data back to a genetic algorithm as fitness criteria. The optimization can consistently find stable versions of most inputs, leveraging the small mutations found in neural-network outputs to find solutions that are novel and performative but also contextual in nature. The paper also describes a secondary technique that redeploys the outputs of the latent-space genetic algorithm to train a predictive AI model that estimates the structural stability of existing masonry structures. Both methods combine to form a kind of hybrid toolkit for making precedent machine-intelligible, allowing computation—and ourselves—to learn from the knowledge embedded within our built environment.

INTRODUCTION

The architectural style known as 'Parametricism' started in aspirational fashion: designers would harness 'objective' data describing the world around them, we were told, and create radical new forms by feeding this data into a set of rational parametric models (Schumacher 2008). Architectural reality was somewhat divorced from this sentiment; the radical forms were often sculpted in Maya, not generated procedurally (Schumacher 2020), and the connection between input data and resultant form was sometimes tenuous. This is not to diminish the power of the work, or its considerable technical achievement—rather, we can say that the architecture and theory were mismatched, if equally bombastic.

The central thesis of 'Parametricism' was furthered by the next generation of practitioners, who worked to build data-driven parametric models with clear connections between input and output. While the parametric model itself is useful for quickly generating design options, it is perhaps best used to define the design space of an optimization algorithm; in the case of a genetic optimization, the model's parameters are used as an input genotype while the output phenotype is evaluated for a given fitness criteria.

This method is elegant, but the geometric basis of most parametric architectural models can sometimes limit their applicability to real-world problems—many typologies can be too complex to model parametrically, making genetic optimization of these conditions difficult. Consider the simple corbeled masonry arch: building a parametric model that successfully emulates the typology's diverse real-world examples would require many hours of expert labor and would likely miss the outliers and edge conditions that often make parametric design difficult.

As an alternative, we can look towards a neural-network approach to encode and parameterize existing formal phenomena. Artificial intelligence's ability to find patterns and organize disparate precedents into a smoothly parameterized 'latent space' effectively automates the production of parametric models when enough precedent examples are used in the initial model training. The latent space of the trained network contains outputs that both mimic and recombine the initial training dataset in ways that are frequently novel; the process is roughly analogous to the biological mutation that drives real-life evolutionary processes.

Like all mutations, these novel outputs are often imperfect. A layer of validation is required to separate successful outputs from those that do not meet design criteria; in this case, neural-network outputs are physically simulated to test for validity, and the results are fed back into a genetic algorithm as fitness criteria. We can think of this neural network-generated parametric model (or neuro-parametric model) as a kind of tool for hybridity, taking the most successful elements from various precedents and recombining them into novel, performative solutions.

2 Interpolating between two vectors in the StyleGAN2-Churches neural network by Karras, Laine and Aila (2019)

3 GAN-generated corbeled archways

This is a contrasting approach to the traditional user-generated parametric model. Those models are often broadly validated across their design space due to the consistency of their mathematical inputs, which allows human users to infer and predict the relationship between inputs and outputs and ultimately provides a rich, interactive user experience. That experience is lost almost entirely with neuro-parametric models, as their organizing latent parameters are not easily comprehended and the overall quality of the outputs tends to be hit-or-miss. As a result, neuro-parametric models are better suited for machine-learning approaches that automate the process of controlling the model itself—perhaps a worthy trade-off given the speed with which they are trained. If the traditional parametric model makes complex design systems human-intelligible, we can think of the neuro-parametric model as making those systems machine-intelligible.

While this method is an effective way of producing new form via optimization, a second new approach refocuses on the precedent itself. Training an image-to-image translation AI model on the inputs and outputs of the latent-space genetic algorithm results in a neural network that can effectively predict the results of compute-heavy physical simulation in a small fraction of the time. Importantly, this method works on any image—not just the output of the precedent-trained neural network. The result is an AI model that can quickly validate real-world conditions (in this case, structural stability) without expert knowledge or specialized modeling.

4   On the left, the GAN-generated input archway; on the right, various steps in the rigid-body simulation; note that falling bricks are automatically removed from the simulation once a velocity threshold is reached

5   Input/output image pairs used in training the pix2pix model

These retroactively parametric approaches are a kind of inversion of the promise of capital-P 'Parametricism'. Whereas that movement sought to use 'objective' contextual data as a kind of justification for formal exuberance, enabled by human-authored parametric models, these new methods work in reverse: the predictive model, trained on AI-generated precedent, can repurpose this data to aid in review or repair; the parametric model, this time authored by AI, produces novel form by reshaping context itself.

SIMULATION-VALIDATED LATENT-SPACE GENETIC OPTIMIZATION

An image-generation Generative Adversarial Network (GAN) (Karras, Laine and Aila 2019) allows users to train a neural network on a disparate set of visual precedent images. These AI models recombine formal elements of the training dataset into a set of novel, reinterpreted images. While this raw image output has merit as a means of design ideation—or as a kind of AI collage technique—what is perhaps most interesting about the trained neural network is that the resulting outputs are arranged into a smooth 'latent space' organized by 512-parameter vectors. These float values are unnamed, making the resulting parametric models somewhat opaque to human users, but visually similar images do maintain mathematically similar input vectors. As a result, the neural network's structure maps formal similarities between architectural precedents and generates novel forms to fill in the gaps. The popular 'latent walk' visualization linearly interpolates between two input vectors to produce an animation that 'morphs' between two output images (Figure 2).

This smooth parameterization has significant benefits when viewed through the lens of algorithmic optimization. Since the GAN can find visual patterns in a disparate set of precedent images, its latent space can be used directly by a genetic algorithm as a design space of possible outcomes. This computational approach—similar to that employed by Wen et al. (2021)—points to an exciting new methodology of general applicability to architectural design computation.

What sets this method apart from other genetic algorithm/neural network hybrids (Fuhrimann et al. 2018; Danhaive 2020) is its precedent-based approach. In bypassing the expert knowledge required to create parametric models, this new GAN-based method requires only the selection of precedent images to develop a coherent, navigable design space. In conjunction with a genetic solver, this latent-space optimization allows for the discovery of novel solutions within a design space populated by precedent-informed options.

The current implementation has an obvious limitation: most Generative Adversarial Networks are limited to two-dimensional inputs and outputs. As a result of this technical limitation, the paper focuses on the freestanding corbelled arch, which can be accurately represented by only a single elevational image. That said, the technique outlined in this paper could be easily adapted to a future three-dimensional GAN workflow given sufficient 3D training data, allowing the optimization algorithm to be applied to a broader and more applicable set of masonry loading conditions.

In the example described here, a two-dimensional StyleGAN2 model (Karras, Laine and Aila 2019) was trained using a dataset of 449 images of corbeled arches. The network for this study was trained locally in Ubuntu using Tensorflow (Abadi 2016), Python, and an NVIDIA RTX 2060 GPU. The network's trained latent space is populated by images of new arches that visually resemble the training set, but recombine the elements in ways that are often novel and surprising. Many of the new arches produced by the neural network appear structurally viable, but a significant portion are clearly unstable (Figure 3); as a result, validation and optimization is required to find viable GAN-generated archways. A purpose-built image interpreter, written in Processing, takes the GAN's photorealistic raster output and makes it three-dimensional, adding bonded bricks to pixel locations that meet color criteria. Once modeled in their pre-collapse state, the bricks are exposed to gravitational force and allowed to 'settle' (Figure 4); non-viable arches collapse to the ground, as expected, but successful iterations maintain the form depicted in their GAN-generated input images.

Per Heyman (1995), masonry structures can be idealized as a collection of rigid bodies; although this approach ignores the stress within a given masonry structure, it can quickly evaluate its stability, which is often an acceptable measure of structural performance in masonry applications. The interaction of multiple rigid bodies can be simulated using a multi-agent method known as Contact Dynamics, a technique popularized by computational physics engines like Bullet (Coumanns 2003) and validated for provisional applications in structural analysis (Whiting, Ochsendorf and Durand 2009) and civil engineering (Pytlos, Gilbert and Smith 2015). Most dynamic simulation for structure uses the Discrete Element Method (Cundall 1971), which adopts a more accurate soft-contact model at the expense of additional compute time (Izadi and Bezuijen 2018), but the faster Contact Dynamics approach allows for a workable approximation of the real-world structural performance with considerably less overhead. This implementation of Contact Dynamics uses bRigid (Kohler 2012), a Processing port of jBullet (Jezek 2008).

The computational efficiency found in rigid-body analysis allows for rapid, iterative deployment—a prerequisite for evolutionary optimization. Contact Dynamics has seen use in rigid-body structural optimization via genetic algorithm (Harrison 2016); like many architectural applications of genetic algorithms (Rutten 2013), this approach used a genotype that describes multiple parameters with phenotype outputs produced using a parametric model.

The method described here uses a similar rigid-body simulation approach to phenotype evaluation, but using a GAN to establish the design space results in a faster, more intuitive approach than traditional parametric modeling. The genetic algorithm here is a relatively simple random-walk implementation; in each generation, random genotypes within a certain parameter range of the initial iteration are tested for their overall height and the number of stable bricks remaining after

6   Outputs of the predictive stability model benchmarked against Contact Dynamics simulation. The left column shows a selection of GAN-generated input images, the middle column shows the actual results of dynamically simulating those forms, and the right column shows the results predicted by the image-to-image translation model.

simulation. Genotypes with fitness criteria higher than the previous generation are adopted as the baseline for the next generation. The result is a structurally-sound version of any output found within the neural network.

In the example shown here (Figure 1), six generations are required to progress from unstable initial input to structurally viable final product. Overall, the random-walk approach shown in this example required 34 simulation iterations, or 5.6 iterations per generation. This result appears to outperform classically parametric genetic algorithms, but further testing is required to benchmark the performance increase.

PREDICTIVE MODELING WITH IMAGE-TO-IMAGE TRANSLATION

The iterative method described above is a relatively fast way to determine a structure's stability, but an even more efficient method may be the use of a paired image-to-image translation neural network like pix2pix (Isola et al. 2017). These models can interpolate between two types of images by training the neural network on a set of paired input and output images;

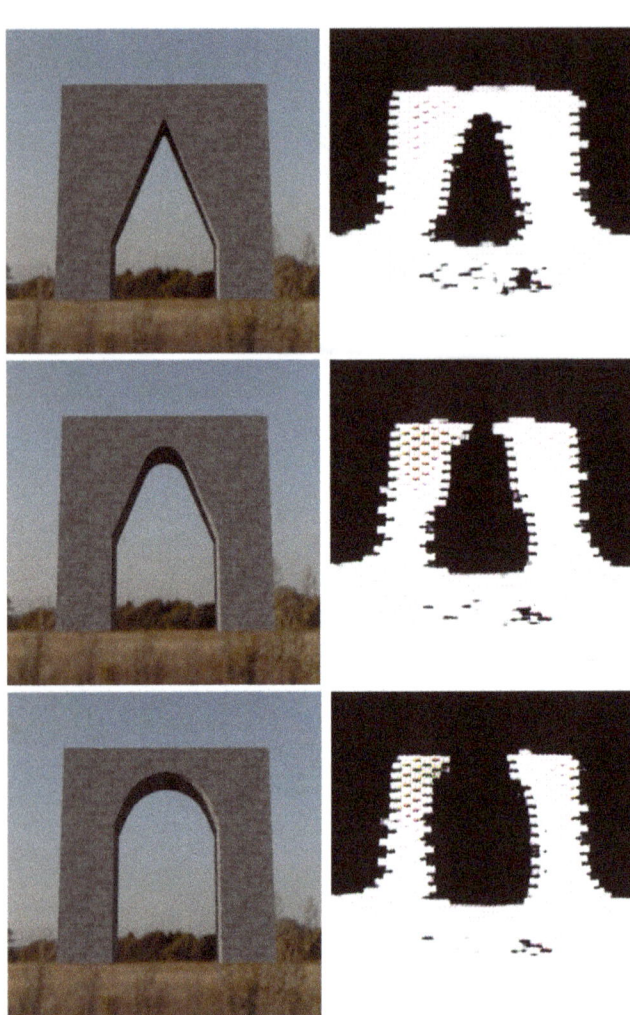

7 Benchmarking the error found in the predictive model using a parametric photorealistic model as input; only corbeled archways are interpreted correctly; vaulted archways are interpreted as structurally unsuccessful

once trained, the network can infer an output based on only an input image—or vice versa.

The latent-space genetic optimization described above takes a raster image as its input (i.e., a GAN-generated corbeled arch) and can express its output as an image (i.e., a screen capture of the final result of the rigid-body simulation). Structurally-sound outputs of the solver closely resemble their input images, but unstable configurations produce outputs that appear partially or fully collapsed. When aligned, these input/output image pairs (Figure 5) are an ideal candidate for image-to-image translation; the datasets can be populated by simply repeating the genetic optimization, which provides the required cross-section of both successful and unsuccessful results.

Once a model is trained using these paired datasets, it can infer an appropriate output for any given input; when presented with an image of an archway, the trained network described here will output a prediction of what that archway might look like after undergoing the dynamic simulation technique described above—a process that takes only a few milliseconds, making the tool effective for real-time feedback.

In practice, this means that the network can skip the computationally intensive process of dynamic simulation. Perhaps most importantly, the trained network accepts any photorealistic image of a masonry structure as its input—not simply those created by the corbeled archway GAN. As a result, the tool can be used to validate the stability of two-dimensional masonry structures through a simple photographic input.

Predicting the results of complex simulation techniques using image-to-image translation shows considerable promise. The results of this initial study are similarly compelling: the image-to-image translation model can approximate the outputs of Contact Dynamics simulation in a fraction of the time. This initial model was trained with a dataset of 1163 image pairs produced using the genetic algorithm described above; the results are a close match to the expected results of dynamic simulation (Figure 6).

One considerable source of error is the AI solver's inability to understand vaulted archways. Since the initial GAN was trained only on images of corbeled archways, the new predictive neural network interprets a photograph of a vaulted arch as structurally impossible and outputs an image of that archway in partial ruin (Figure 7). This is a fascinating result, but reflects the bias and inaccuracy that can beset all AI tools—our results can only be as good as our input datasets. This error would likely be mitigated by training the dataset on a more diverse set of masonry loading conditions.

Even in its initial state, the tool provides a compelling example of the broad use cases for image-to-image translation. Using AI to effectively simulate other forms of compute-intensive simulation reduces dependencies on specialized hardware or cloud computing; while the Contact Dynamics solver described earlier requires a Linux workstation and high-end GPU to run, the predictive image-to-image model could be easily ported to run on an iPhone or iPad. This ease of access means that a simulation tool trained to identify structurally unstable conditions could be deployed broadly with minimal compute cost. In time, this kind of real-time analysis could act as a key tool for the diagnosis, repair and rehabilitation of our built environment.

CONCLUSION

The approaches outlined in this paper are contextual and incremental in nature—by design. Our intersecting climate and inequality emergencies have rightfully reduced the demand for formal complexity in architecture, but the underlying tools of computational design are a crucial component in building an equitable and sustainable future. Another key element is our existing built environment, and the many millennia of vernacular and Indigenous knowledge it describes. By turning our computational tools towards this environment—and learning from and within it—we can hone and repurpose these lessons for the next epoch.

ACKNOWLEDGEMENTS

The initial version of the genetic optimization tool demonstrated here—originally presented at ACADIA 2016—was developed as a master's thesis under the guidance of Dr. Benjamin Dillenburger in 2014.

REFERENCES

Abadi, Martín. 2016. "TensorFlow: Learning Functions at Scale." In *Proceedings of the 21st ACM SIGPLAN International Conference on Functional Programming*. New York: Association for Computing Machinery. 1.

Bernhard, Mathias, Reza Kakooee, Patrick Bedarf, and Benjamin Dillenburger. 2020. "Topology Optimization with Generative Adversarial Networks." In *Proceedings of Advances in Architectural Geometry 2020*. Paris: Ecole des Ponts ParisTech. 208–229.

Box, George E. P. 1979. "Robustness in the strategy of scientific model building." In *Robustness in Statistics*. Madison: Academic Press. 201–236.

Coumanns, Erwin. 2003. "Bullet." *Bullet Physics Library*. Accessed May 10, 2022. http://www.bulletphysics.org.

Cundall, Peter A. 1971. "A computer model for simulating progressive, large-scale movement in blocky rock systems." In *Proceedings of the International Symposium on Rock Mechanics*. Paris: Society for Rock Mechanics (ISRM). 2–8.

Danhaive, Renaud Aleis Pierre Emile. 2020. "Structural Design Synthesis Using Machine Learning," PhD diss., Massachusetts Institute of Technology. https://hdl.handle.net/1721.1/138590.

Fuhrimann, Lukas, Vahid Moosavi, Patrick Ole Ohlbrock, and Pierluigi D'acunto. 2018. "Data-driven design: Exploring new structural forms using machine learning and graphic statics." In *Proceedings of International Association for Shell and Spatial Structures (IASS)* 2018 (2). City: IASS. 1–8.

Harrison, Paul Howard. 2016. "What Bricks Want: Machine Learning and Iterative Ruin." In *ACADIA 2016: Posthuman Frontiers: Data, Designers, and Cognitive Machines; Proceedings of the 36th Annual Conference of the Association for Computer Aided Design in Architecture*. 72–77.

Heyman, Jacques. 1995. *The Stone Skeleton: Structural Engineering of Masonry Architecture*. Cambridge, UK: Cambridge University Press.

Isola, Phillip, Jun-Yan Zhu, Tinghui Zhou, and Alexei A. Efros. 2017. "Image-to-image translation with conditional adversarial networks." In *Proceedings of the IEEE Conference on Computer Vision and Pattern Recognition*. New York: IEEE. 1125–1134.

Izadi, Ehsan, and Adam Bezuijen. 2018. "Simulating direct shear tests with the Bullet physics library: A validation study." *PLoS ONE* 13 (4): e0195073.

Jezek. 2008. *jBullet*. http://jbullet.advel.cz.

Karras, Tero, Samuli Laine, and Timo Aila. 2019. "A style-based generator architecture for generative adversarial networks." In *Proceedings of the IEEE/CVF Conference on Computer Vision and Pattern Recognition*. New York: IEEE. 4401–4410.

Kohler, Daniel. 2012. *bRigid, a rigid-body physics java-library for the Processing environment*. http://github.com/djrkohler/bRigid.

Pytlos, M., M. Gilbert, and C. C. Smith. 2015. "Modelling granular soil behaviour using a physics engine." *Géotechnique Letters* 5 (4): 243–249.

Rutten, D. 2013. "Galapagos: On the logic and limitations of generic solvers." *Architectural Design* 83 (2): 132–135.

Schumacher, P. 2008. "Parametricism as Style - Parametricist Manifesto." Talk presented and discussed at the Dark Side Club, 11th Architecture Biennale, Venice 2008. Accessed May 10, 2022. http://www.patrikschumacher.com/Texts/Parametricism%20as%20Style.htm.

Schumacher, P. 2020. "A Soaring Success: Patrik Schumacher on Zaha Hadid Architects' Beijing Daxing International Airport." Interview by Paul Keskeys. Architizer, October 16 2020. Accessed May 10, 2022. https://architizer.com/blog/inspiration/stories/patrik-schumacher-beijing-daxing-zaha-hadid-architects/.

Wen, Jeffrey, Fabian Benitez-Quiroz, Qianli Feng, and Aleix Martinez. 2021. "Diamond in the rough: Improving image realism by traversing the GAN latent space." *arXiv* preprint, arXiv:2104.05518.

Whiting, Emily, John Ochsendorf, and Frédo Durand. 2009. "Procedural modeling of structurally-sound masonry buildings." In *Papers of the ACM SIGGRAPH Asia 2009*. New York: Association for Computing Machinery. 1–9.

IMAGE CREDITS

All drawings and images by the author.

Paul Howard Harrison is a designer, researcher, and practitioner. Trained as both architect and engineer, his research aims to build equitable and sustainable futures through the integration of simulation, labour, and automation. Paul directs HDR's Design Computation Group, a team of researchers and designers headquartered in Toronto and Vancouver, and lectures at the University of Toronto's John H. Daniels Faculty of Architecture, Landscape, and Design.

What is Creativity?

Learning from AlphaGo

Neil Leach
FIU/ Tongji/ EGS

1   Neil Leach, Architectural Studies generated by MidJourney (2022). Prompt engineering has become the new skill to be mastered by anyone using MidJourney. This particular prompt consisted of several references, including 'in the style of Zaha Hadid Architects'.

ABSTRACT

Many discussions about AI and architecture end up addressing the question of creativity. Can computers be considered creative? Or is it impossible for any entity to be considered creative if it does not possess consciousness? Indeed, what exactly is creativity? Might AI even offer us some insights into the true nature of creativity? This paper explores what we can perhaps begin to understand about the nature of creativity in the mirror of AI, with reference to the now famous Go match between AlphaGo and Lee Sedol. It argues that one particular famous move in that match sheds light on some of the crucial questions regarding creativity. It compares this move to the 'smart' architectural designs generated by AI, and asks whether computers can be creative, or whether they are simply conducting a 'search and synthesis' operation. Finally, the paper asks the provocative question, as to whether creativity even exists, or whether it is a myth that can now be debunked, thanks to our insights from the world of AI.

INTRODUCTION

There are many definitions of creativity. For the sake of brevity, however, let us simply adopt the definition given by Margaret Boden, one of the first to address the question of creativity in the context of computation: 'Creativity is the ability to come up with ideas or artifacts that are new, surprising and valuable' (Boden 2003).

Can computers be considered creative?

Discussions about computer creativity are becoming increasingly topical, as computers are beginning to generate ever more novel outcomes. There has been a series of recent books and articles addressing computer creativity in the context of AI (Boden 2009; Miller 2019; du Sautoy 2020). Within architecture and art circles, too, there has been considerable interest in the topic (Bolojan 2022; Manovich 2022). Is the reason why we often call computers 'creative' that we have a tendency to anthropomorphize technology, and project human attributes on to it? Or is it simply that we do not understand what they are doing? What about creativity itself? Indeed the whole notion of creativity and the so-called 'creative economy' have been called into question (Mould 2018; Manovich 2022). Others, such as Rochelle King, Elizabeth Churchill, and Caitlin Tan (King, Churchill, and Tan 2017) have even claimed that creativity is 'something of a myth'.

Initial assumptions, for sure, were that computers could never be truly creative. For example, Ada Lovelace, daughter of the poet Lord Byron, and herself no stranger to creativity, predicted that the Analytical Machine, an early proto computer on which she collaborated with Charles Babbage, would never be able to produce anything beyond what it was programmed to do: "The Analytical Machine has no pretensions whatsoever to originate anything. It can do whatever we know how to order it to perform. It can follow analysis, but it has no power of anticipating any analytical revelation of truths. Its province is to assist us in making available what we are already acquainted with" (Fuegi and Francis 2003). More recently, Japanese computation expert, Makoto Sei Watanabe, reinforces this view: "Machines are better than people at solving complex problems with many intertwined conditions. In that realm, people are no match for machines. But people are the only ones who can create an image that does not yet exist. Machines do not have dreams" (Watanabe 2017).

It has now become increasingly apparent, however, that computers can indeed 'synthesize' or generate novel images using neural networks. The assumption that they can even be creative, moreover, is written into the title of the TED talk by Blaise Agüera y Arcas, "How Computers are Learning to be Creative" (Agüera y Arcas 2016). Here Agüera y Arcas outlines

2, 3 Neil Leach, Architectural Studies generated by MidJourney (2022). With MidJourney, the process of generating images often involves several iterations that allow the author to coax the platform in a broad general direction, but without full control, by choosing one variation over another.

4   Neil Leach, Furniture Studies generated by MidJourney (2022). MidJourney can often produce high quality images that give a real sense of materiality and three dimensional form.

early explorations into inverting the network and using neural networks to generate artworks through DeepDream and other technique (Leach 2022, 21). Since then Generative Adversarial Networks (GANs) have become a popular mechanism for synthesizing images in a more controlled fashion (Leach 2022, 26). More recently still, DALL.E 2 has begun to generate some astonishingly convincing images that show how the domain of AI continues to develop at a rapid pace.

This article, then, seeks to interrogate the nature of creativity through the lens of AI. The premise is that AI can potentially offer us a mirror in which to understand what it is to be human. As Hiroshi Ishiguro notes in the context of robots: "The robot is a kind of mirror that reflects humanity and by creating intelligent robots we can open up new opportunities to contemplate what it means to be human" (Ishiguro 2019, 179). Could the same be said of AI? And might AI even help us to understand the nature of human creativity?

AlphaGo versus Lee Sedol
We could argue, however, that some of the most fascinating insights into the question of creativity are raised by a high profile match of Go that has proved to be one of the most iconic moments in the history of AI. In March 2016 a match of Go was staged in Seoul, Korea, between AlphaGo, a deep learning computer program developed by DeepMind of London, and Korean 9-Dan professional Go player, Lee Sedol, one of the greatest Go players of all time. The match consisted of five games, with most Go experts—including Lee himself—predicting that Lee would win easily. However,

much to their surprise, AlphaGo won the first game and went on to win the match by four games to one. The match was watched by millions of viewers in Asia. It effectively alerted humankind—especially Go playing nations—to the extraordinary potential of AI, and led to the famous 'Sputnik' moment, as Kai-Fu Lee has termed it, when China woke up to the extraordinary potential of AI: "Overnight, China plunged into an artificial intelligence fever" (Lee 2018, 3).

Yet what was most remarkable about the match was not the fact that AlphaGo beat Lee Sedol, but the manner in which it beat him. Game 2 proved to be the turning point. After Game 1, Lee was surprised, but after Game 2 he was lost for words: "Yesterday, I was surprised. But today I am speechless. If you look at the way the game was played, I admit, it was a very clear loss on my part. From the very beginning of the game, there was not a moment in time when I felt that I was leading' (Metz 2011). The biggest talking point of the whole match was a remarkable move played by AlphaGo in this game: "In Game 2, Lee exhibits a different style, attempting to play more cautiously. He waits for any opening that he can exploit, but AlphaGo continues to surprise. At move 37, AlphaGo plays an unexpected move, what's called a 'shoulder hit' on the upper right side of the board. This move in this position is unseen in professional games, but its cleverness is immediately apparent" (Moyer 2016). Hassabis goes even further: "Anyone can play an original move on a Go board by simply playing randomly. Yet a move can only be considered truly creative if it's also effective. In that sense, Move 37's decisive role in game two represents a move of exquisite computational

5   Neil Leach, Office of the Future, generated by MidJourney, 2022. The office of the future is likely to be dominated by AI, as AI techniques become increasingly capable and pervasive

ingenuity that not only changed the game of Go forever, but also came to symbolize the enormous creative potential of AI" (Hassabis and Hui 2019, 84).

European Go champion, Fan Hui, also recognised the 'creativity' of the move—Move 37, as it has become known: "When AlphaGo chose that move, I assumed that it had made a mistake. I immediately looked to see Lee's reaction. At first, he seemed to smile—as though he too thought it had made a mistake—but as the minutes rolled by it was clear that he was starting to realise its brilliance. In fact, after the match, he said that when he saw this move he finally realised that AlphaGo was creative" (Hassabis and Hui 2019, 89). Interestingly, Hui describes the move as also being beautiful: "I've never seen a human play this move. So beautiful" (Moyer 2016). Meanwhile, Lee himself also describes the move as being both creative and beautiful: "I thought that AlphaGo was based on probability calculation and that it was merely a machine. But when I saw this move, I changed my mind. Surely AlphaGo is creative. This move was really creative and beautiful" (Hassabis 2018).

Leading Go players acting as commentators for the match initially thought that Move 37 was a mistake (Hassabis 2018). Move 37, however, was no mistake, and it paved the way for victory, as 100 moves later that stone connected with a series of other stones on the other side of the board and won the match. It was as though AlphaGo could operate not just one move at a time, but at a level of strategic far sightedness that no human being could ever hope to match. In fact it was only one of a series of moves that have been referred to as 'slack moves,' because their strategic brilliance did not become apparent until much later. Not only did these 'slack moves' effectively change the game of Go forever, challenging accepted approaches to playing the game, but they also showed that AI could demonstrate a level of 'creativity' well beyond human creativity, assuming, of course, that it is permissible to describe these moves as 'creative.' As Lee comments: "AlphaGo showed us that moves humans may have thought are creative, were actually conventional" (Kohs 2017). Indeed, we could even say that it demonstrated the limits of the human imagination, as no one was even capable of even recognizing the strategic brilliance of the move until many moves later. We could also say, however, that the AlphaGo match illustrates the limits of what we call 'creative.' Clearly 'creativity' exists only in so far as we can recognize it. This, surely, is the crucial point.

AI, Architecture, and Creativity

What, then, are we to make of all this, from an architectural perspective? First of all, and most obviously, we can see that parallels present themselves with the domain of architecture. Similar incidents have been noticed when Spacemaker AI software has been used in architectural design (Leach 2022, 122). It would seem that AI has the capacity to outperform architects in certain design challenges—especially those that require a strong strategic input, not unlike AlphaGo. Just as AlphaGo was able to come up with moves that no humans would have thought about, so too, according to Håvard Haukeland, Spacemaker AI software used within an architectural context has been able to make 'smart' design proposals that no architects would have thought about: "The places

6, 7 Neil Leach, Office of the Future, generated by MidJourney (2022). The chairs and tables in these images might have no legs, but nonetheless the wistful, suggestive shapes generated seem to invite the viewer to make sense of them.

where the architects thought that it would be smart to build tall buildings, and the places where they thought it would be smart to build a dense wall, all the things that they intuitively thought would be smart—because they had hundreds of projects of experience—were flipped around. Because when you get the complexity of thinking of a multi-objective organizational problem...you are really not able to see the patterns that a computer can find. So what happened was that the computer was able to find a pattern as to how to solve that site that you would never come up with yourself" (Haukeland 2020).

Secondly, however, we have to question whether a computer can produce 'creative' designs or merely 'smart' solutions. Interestingly, Haukeland refers to these design proposals as 'smart,' which is perhaps a better term than 'intelligent,' in that 'smart' is associated perhaps more with materials ('smart materials' etc.), whereas the term 'intelligence' is associated with human beings. Indeed, according to neuroscientist and computational expert, Jeff Hawkins, the term 'intelligence' can only refer to human beings and not to computers (Hawkins 2021). More importantly, however, Haukeland does not refer to these proposals as being 'creative.' What, then, might we say about AlphaGo? Was AlphaGo actually being creative, or was it simply conducting a search for the smartest possible solution? In other words, computers might be able to generate smart or novel outcomes, but can we consider them to be genuinely creative? Or is the question of computer creativity an impossible one to answer, as Boden argues: "Whether a computer could ever be 'really' creative is not a scientific question but a philosophical one. And it's currently unanswerable, because it involves several highly contentious—and highly unclear—philosophical questions" (Boden 2009). Memo Akten, at any rate, seems to think that computers can be creative: "By saying that a machine can be creative, you are not anthropomorphising the machine, but liberating it by expanding the term 'creativity' to go beyond humans. Creativity is not limited to people. I'm a biological machine. Humans can create art. Why not machines?" (Akten 2022).

Others are not so sure. Indeed, if we are to follow the thinking of Melanie Mitchell, AI cannot be said to be creative, in that it does not possess consciousness: "I also believe that being creative entails being able to understand and judge what one has created. In this sense of creativity, no existing computer can be said to be creative" (Mitchell 2019, 272). Of course, we could argue that we human beings might not be fully conscious of an idea, when it suddenly pops up in our mind in a flash of inspiration. Indeed, Max Tegmark notes: "Neuroscience experiments suggest that many behaviors and brain regions are unconscious, with much of our conscious experience representing an after-the-fact summary of vastly larger amounts of unconscious information" (Tegmark 2017,

8   Hotel Room, images generated by MidJourney AI (2022). MidJourney seems to be particularly good at adding lighting effects, reflections and shadows. Even though these effects are not exactly accurate, they often look convincing.

315). However, even if this might be the case initially, very soon afterwards—the 'aha' moment—the brilliance of the idea dawns on us, as we use our consciousness to appraise and appreciate the idea. From this perspective, AI cannot be said to be creative, since AI is not capable of appreciating its 'creativity.'

Perhaps the most famous argument about the importance of consciousness has been made by philosopher John Searle, using his famous Chinese Room thought experiment (Searle 1980). Imagine that a computer program has been designed that seems as though it is capable of understanding Chinese. The program is able to take English text and translate it into Chinese text so convincingly that it passes the Turing test, in that no one realizes that it is just a computer program. Now imagine if Searle himself were to find himself in a closed room equipped with an instruction manual in English that described the translation process. He would be able, surely, to process the material in a similar way. The only problem is that he does not understand Chinese. This is analogous to the question of consciousness. A computer might appear to be conscious, but it does not understand what it is doing. Searle might appear to be able to translate Chinese, but he does not understand Chinese. The Chinese Room experiment can therefore be understood as a critique of the Turing test. For Searle, the Turing test is inadequate, because it only tests whether a computer appears to be human. It does not test whether the computer is actually human. There is, in other words, a significant difference between an entity appearing to be something, and actually being it.

AlphaGo, according to this argument, might have 'won' the match of Go, but it cannot have been aware of the fact that it was even playing Go, because it does not possess consciousness. As such, AlphaGo might be good at processing 'symbols,' but it has no idea what they mean. Indeed it has no more capacity to 'think' than a pocket calculator.

What if we were to judge AI, however, not in terms of human intelligence, but on its own terms? As such, the question of consciousness becomes irrelevant. The issue becomes what tasks AI can perform, not whether AI can display human-level consciousness. As Yuval Noah Harari notes, "There might be several alternative ways leading to superintelligence, only some of which pass through the straits of consciousness" (Harari 2017, 314). This leads Harari to pose the important question, "Which of the two is really important, intelligence or consciousness?" (Harari 2017, 314). As far as corporations and armies are concerned, the answer is clear: "intelligence is mandatory, but consciousness is optional" (Harari 2017, 314). Harari goes on to give the example of a taxi driver: "The conscious experiences of a flesh-and-blood taxi driver are infinitely richer than a self-driving car, which feels absolutely nothing" (Harari 2017, 314). But as far as the passenger is concerned, the consciousness of the taxi-driver is irrelevant. Soon, Harrari notes, we will have autonomous taxis that are far more reliable than taxis driven by conscious drivers. He comments that "Taxis are highly likely to go the way of horses" (Harari 2017, 315). In fact Harrari notes that a Google self-driving car was once involved in an accident, when hit from behind by a sedan 'whose careless human driver was perhaps

9   Neil Leach, Furniture Studies generated by MidJourney (2022). For this particular series, the word 'shark' was included in the prompt, and might account for the interesting black textures generated

contemplating the mysteries of the universe, instead of concentrating on the road" (Harari 2017, 314). Consciousness, in other words, is not only optional; it might also become a liability.

Surely, the question of whether computers can mimic human intelligence and consciousness is largely irrelevant. The important question is whether or not AI can produce a good design. "Is our goal," asks Daniel Bolojan, "to create machines that mimic human intelligence and creativity, or are we aiming to create machines that are capable of being intelligent and creative in their own right?" (Bolojan 2022). Yet the important issue of creativity still remains.

The Myth of Creativity

As Margaret Boden has noted, "Thanks in part to AI, we have already begun to understand what sort of phenomenon creativity is" (Boden 2009). The key question raised by the match between AlphaGo and Lee Sedol; however, is whether human beings would even be able to recognize the 'creativity' of AI.

Alan Turing seemed to acknowledge this, when he commented on the potential creativity of computers: "We have to have some experience with the machine before we really know its capabilities. It may take years before we settle down to the new possibilities, but I do not see why it should not enter any one of the fields normally covered by the human intellect, and eventually compete on equal terms. I do not think that you can even draw the line about sonnets, though the comparison is perhaps a little bit unfair, because a sonnet written by a machine will be better appreciated by another machine" (Turing 1950). His final comment—that a machine could best appreciate the output of another machine—is an interesting observation. Of course, we could question whether a machine could ever actually 'appreciate' a sonnet, since it does not possess consciousness, but the point still stands.

It would seem that machines can reach a level of 'creativity' that far exceeds 'human creativity,' much as a dog can sense a range of sounds and smells far beyond the range that a human can. This evokes the interesting principle—often referred to as the 'Dunning-Kruger Effect'—that the dumb do not know how dumb they are. We could even extend this principle to infer that the intelligent do not know how unintelligent they are, or —perhaps more appropriate in this context—the creative do not know how uncreative they are. Put simply, there are levels of 'creativity' that we simply cannot grasp. 'Creativity', in other words, is nothing, if we do not recognize it.

One way to think about this important issue is to reconsider Lee's comment about AlphaGo cited above: "AlphaGo showed us that moves humans may have thought are creative, were actually conventional" (Kohs 2017). The important word here is 'thought.' In other words, Lee is implying that creativity is simply a question of perceived creativity.

What becomes clear is that our understanding of creativity is somewhat subjective. Margaret Boden has attempted to define creativity in terms of three types of outcomes list, 'combinatorial,' 'exploratory,' and 'transformational,' as

10 Neil Leach, Hotel Room, images generated by MidJourney AI, 2022. MidJourney consists of little more than a sketching tool, but it gives us an interesting glimpse of the potential of AI to generate convincing architectural designs in the future.

though creativity could be grasped in such objective terms (Boden 2009). Meanwhile, Demis Hassabis has attempted to outline three 'levels' of creativity: 'intensive,' 'extensive,' and true 'invention' (Hassabis 2018). But can creativity be judged in such as objective way? Who is in a position to even judge creativity?

François Pachet, believes that it makes no sense for creativity to be defined in objective terms, in that it can only be understood subjectively (Miller 2019, 26). And what if a creative act is not recognized immediately? And how are we to appraise the creativity of Move 37, if the creativity of that move was only appreciated later? Certainly, as Arthur Miller notes, it is difficult to come up with a clear, objective description of creativity, when the creativity of some artists, such as Vincent Van Gogh, was not recognized in their own lifetime (Miller 2019, 26). Indeed, the art world is dependent on curators who validate artworks, so that art itself could be understood as a form of curatorial discourse in which artworks are inscribed. As such, artistic creativity needs to be recognized by that discourse if it is to be recognized at all. Presumably, what is true for the art world should also apply to the domain of architecture.

Could we even see parallels between creativity and beauty, since both concepts were evoked by commentators of the AlphaGo match? Is creativity a little like beauty? Indeed, David Hume believed that beauty is highly subjective, "Beauty is no quality in things themselves: It exists merely in the mind which contemplates them; and each mind perceives a different beauty" (Hume 1742). Could not creativity—like beauty—lie in the eye of the beholder? Or might it also exist in the mind of the creative individual?

As Margaret Boden has commented, there is nothing magical about creativity (Boden 2009). Could not 'creativity,' however, be viewed in a similar way to magic? We could argue that there is no such thing as magic. Magicians do not perform magic. They simply conceal the processes that actual happen, so that the audience comes to attribute them to magic. "Like the conjurer's trick, where the magician conceals the true devices at work, so as to fool the audience into attributing them to magic, so technology, in effacing itself, invites us to believe in its magical potential" (Leach 1999). Nor indeed is technology magical. But is there not a risk, when dealing with advanced technology (whose operations we might not fully comprehend), of mistaking it for magic? Here I refer to the famous quote of Arthur C. Clarke, "Any sufficiently advanced technology is indistinguishable from magic" (Clarke 1973, 14, 21, 36). This is related to a further common quote, often attributed to Clarke although there is no evidence, "Magic's just science that we don't understand yet."

Here we have three propositions:
1. There is nothing magical about creativity.
2. There is nothing magical about magic.
3. There is nothing magical about technology.

But could we not push the connection even further? If there is nothing 'magical' about magic, is there anything 'creative' about creativity? Might we even use 'creativity' to describe

11 Neil Leach, Hotel Room, images generated by MidJourney AI (2022)). The question that the quality of these AI generated images raises is what impact this will have on the way that architect works in both academia and the profession.

processes that we don't understand yet? Could we even say that the term 'creativity' is a form of mystification?

Let us return to the example of Searle's Chinese Room experiment. What Searle was arguing was that there is a difference between perceived consciousness and actual consciousness. Could we not use the same experiment to argue that there is a difference between perceived creativity and actual creativity? Indeed what would happen if all 'creativity' were simply a question of perceived creativity? And what if we were to discover, for example, that in being 'creative' we are just following a straightforward search and synthesis process—much like AI?

Is creativity, then, no more than the appearance of creativity, just as magic is merely the appearance of magic? Is 'creativity,' then, simply a convenient term to describe a 'search and synthesis' process that we do not fully understand, but that is actually quite straightforward?

Could we even say that there is no such thing as creativity?

REFERENCES

Adorno, Theodor W. 1978. *Minima Moralia*, translated by E. F. N. Jephcott. London: Verso.

Agüera y Arcas, Blaise. 2016. "How Computers are Learning to be Creative." TED Talk. Accessed May 10, 2022. https://www.ted.com/talks/blaise_aguera_y_arcas_how_computers_are_learning_to_be_creative/transcript?language=en.

Akten, Memo. 2022. "Distributed Consciousness." Lecture at Univeristy of California Santa Barbara, May 2022.

Benjamin, Walter. 1978. Reflections, trans Edmund Jephcott. New York: Schocken.

Blackmore, Susan. 1999. *The Meme Machine*. Oxford: Oxford University Press.

Boden, Margaret. 2003. *The Creative Mind: Myths, Mechanisms*. London: Routledge.

Boden, Margaret. 2009. "Computer Models of Creativity." *Association for the Advancement of Artificial Intelligence* 103 (I 998): 347–356.

Bolojan, Daniel. 2022. "Creative AI: Augmenting Design Potency." In *Architectural Design: Machine Hallucinations: Architecture and AI* 92 (3): 22–27, edited by Matias del Campo and Neil Leach.

Clarke, Arthur C. 1973. "Hazards of Prophecy: The Failure of the Imagination." In *Profiles of the Future: An Enquiry into the Limits of the Possible*. London: Pan. 14, 21, 36.

Cuarón, Alfonso, dir. *Gravity*. 2013. Burbank, CA: Warner Bros. Pictures, 2014. Blu-ray Disc, 1080p HD

Dawkins, Richard. 1989. *The Selfish Gene*. Oxford: Oxford University Press.

du Sautoy, Marcus. 2020. *The Creativity Code: How AI is Learning to Write, Paint and Think*. London: Fourth Estate.

Elgammal, Ahmed. "Art and Artificial Intelligence Laboratory at Rutgers: Advancing AI Technology in the Digital Humanities." DigiHumAn Lab at CBIM - Rutgers. Accessed May 10, 2022. https://sites.google.com/site/digihumanlab/home.

Fuegi, John, and Jo Francis. 2003. "Lovelace & Babbage and the creation of the 1843 'notes'." *Annals of the History of Computing* 25 (4): 16–26.

Harari, Yuval Noah. 2017. *Homo Deus: A Brief History of Tomorrow*. New York: HarperCollins.

Hassabis, Demis. 2018. "Creativity and AI." The Rothschild Foundation Lecture, Royal Academy of Arts, London, September 17, 2018. Accessed May 10, 2022. https://www.youtube.com/watch?v=d-bvsJWmqlc.

Hassabis, Demis, and Fan Hui. 2019. "AlphaGo: Moving Beyond the Rules." In *AI: More Than Human*, edited by Chloe Wood, Chloe, Suzanne Livingston, and Maholo Uchida. London: Barbican International Enterprises. 84.

Haukeland, Håvard. 2020. Zoom conversation, May 6, 2020.

Hawkins, Jeff. 2021. *A Thousand Brains: A New Theory of Intelligence*. New York: Basic Books.

Hume, David. 1742. *Essays, Moral and Political*. Edinburgh: R. Fleming and A. Alison.

Ishiguro, Hiroshi. 2019. "A Reflection of Ourselves: Robots and the Journey Towards Understanding Human Intelligence." Conversation with Maholo Uchida. In *AI: More Than Human*, edited by Chloe Wood, Chloe, Suzanne Livingston, and Maholo Uchida. London: Barbican International Enterprises. 179.

Kohs, Greg, dir. 2017. *AlphaGo*. https://www.alphagomovie.com/.

Leach, Neil. 2022. *Architecture in the Age of Artificial Intelligence: An Introduction to AI for Architects*. London: Boomsbury.

Leach, Neil. 1999. *Millennium Culture*. London: Ellipsis.

Lee, Kai-Fu. 2018. *AI Superpowers: China, Silicon Valley and the New World Order*. New York: Houghton Mifflin Harcourt.

Manovich, Lev. 2022. "AI and Myths of Creativity." In *Architectural Design: Machine Hallucinations: Architecture and AI* 92 (3): 60–65, edited by Matias del Campo and Neil Leach.

Metz, Cade. 2016. "The Sadness and Beauty of Watching Google's AI Play Go." *Wired*, March 11, 2016.

Miller, Arthur I. 2019. *The Artist in the Machine: The World of AI-Powered Creativity*. Cambridge, Mass.: The MIT Press.

Mitchell, Melanie. 2019. *Artificial Intelligence: A Guide for Thinking Humans*. New York: Farrar, Straus and Giroux.

Mould, Oli. 2018. *Against Creativity*. London: Verso.

Moyer, Christopher. 2016. "How Google's AlphaGo Beat a Go World Champion." *The Atlantic*, March 18, 2016. Accessed May 10, 2022. https://www.theatlantic.com/technology/archive/2016/03/the-invisible-opponent/475611/.

King, Rochelle, Elizabeth F. Churchill, and Caitlin Tan. 2017. *Designing with Data: Improving the User Experience with A/B Testing*. Sebastopol, CA: O'Reilly Media.

Schwartz, Hillel. 1996. *The Culture of the Copy*. New York: Zone Books.

Searle, John. 1980. "Minds, Brains and Programs." *Behavioral and Brain Sciences* 3 (3): 417–457.

Seth, Anil. 2021. *Being You: A New Science of Consciousness*. London: Dutton.

Tegmark, Max. 2017. *Life 3.0: Being Human in the Age of Artificial Intelligence*. New York: Vintage.

Terzidis, Kostas. 2006. *Algorithmic Architecture*. London: Routledge.

Terzidis, Kostas. 2014. *Permutation Design*. London: Routledge.

Turing, Alan. 1950. "Computing Machinery and Intelligence." *Mind* 59 (236): 433–60.

Walton, Kendell. 1990. *Mimesis as Make-Believe*. Cambridge, Mass.: Harvard University Press.

Watanabe, Makoto Sei. 2017. "AI Tect: Can AI Make Designs?" In *Computational Design*, edited by Neil Leach and Philip Yuan. Shanghai: Tongji University Press. 68–75.

IMAGE CREDITS

All images are generated by the author

Neil Leach is a British architect and theorist. He is a Professor at FIU, Tongji and EGS, and has alsotaught at Harvard, AA, SCI-Arc, Columbia, USC, and Cornell. He is a co-founder of DigitalFUTURES and an academician within the Academy of Europe. He has also the recipient of two NASA research grants exploring 3D printing technologies for the Moon and Mars. Neil Leach has published over forty books, including *On the Art of Building in Ten Books* (The MIT Press 1988), *Rethinking Architecture* (Routledge 1997), *The Anaesthetics of Architecture* (The MIT Press, 1999), *Camouflage* (The MIT Press 2006), *Digital Cities* (Wiley 2009), *Computational Design* (Tongji 2017), and *Architecture in the Age of Artificial Intelligence* (Bloomsbury 2022).

Returning the Gaze

Behnaz Farahi
CSULB Department of Design
Human-Experience Design
Interaction (HXDI)

Robotic Installation for Milan Fashion Week

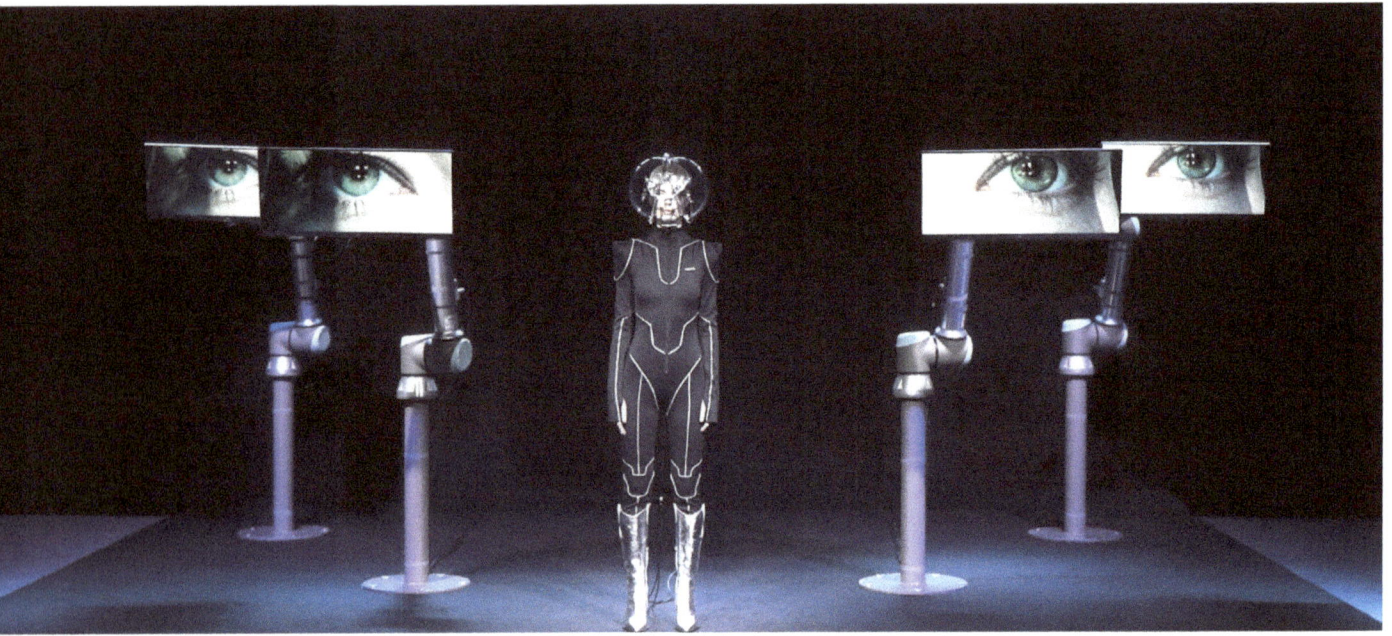

1 'Returning the Gaze' is an cyber-physical robotic installation by Behnaz Farahi supported by Universal Robots for ANNAKIKI's Milan Fashion Week. 2022.

ABSTRACT

Industrial robots have long been used for manufacturing and research purposes around the world. More recently they have also been used in the fields of art, performance, dance and interaction design, where they have served as a collaborator, autonomous agent and, in some cases, as an extension of the artist. Seeing machines as an extension of the self is nothing new. In fact, back in 1985, Donna Haraway had already claimed that we are all cyborgs.

This paper provides a brief overview of how technology might be considered an extension of the self from early cybernetics research onwards. It then explores how a robotic extension could engage with the critical feminist issue of "the male gaze." In doing so, it provides a brief theoretical context to the notion of the gaze and visual behavior from a cognitive science and neuroscience point of view, and then explores it in the context of a discourse on visual culture and feminism. Finally, it illustrates the application of such a critical concept through the example of a robotic installation developed by the author for Milan Fashion Week in collaboration with Universal Robots.

This is a hybrid paper where theory and practice are intertwined. It engages with ways in which robots might extend the gaze through to discourses of resistance whereby the gaze could be used to empower those who typically face discrimination.

CYBORG AND EXTENDED SELF

"A cyborg is a cybernetic organism, a hybrid of machine and organism, a creature of social reality as well as a creature of fiction." — Donna Haraway

In 1960 during the height of the Cold War Space Race, laboratory mice at the Rockland State psychiatric hospital were injected with chemicals at a controllable rate using an osmotic pump. On the basis of this experiment, Manfred Clynes and Nathan S. Kline, two research scientists at Rockland, introduced the term cyborg in their paper "Cyborg and Space" (1960). They use the term cyborg to refer to the notion of a human being who had been technologically enhanced in order to survive in extra-terrestrial environments. They argue that, in order to prepare for space travel, we would need human-machine systems where the self-regulatory function of human beings is enhanced by drugs or regulatory devices.

In May 1963, a NASA report identified the benefits of cyborg technology for space exploration in terms of "reducing metabolic demands and the attendant life support systems" (Driscoll 1963, 7). As O'Mahony notes, one outcome was the development of the spacesuit, which serves as a second skin and enhances the wearer's bodily functions such as walking, breathing and vision (O'Mahony 2007, 44). To perform activities outside the aircraft, the astronauts have to wear an Extravehicular Mobility Unit (EMU), which is a spacesuit providing life-support systems, mobility, and communications. The development of EMU took over fifty years. It was a complicated design incorporating many different electrical components and functioning almost as a one-person aircraft (O'Mahony 2007, 46).

Augmenting human bodily functions through technology is not limited to the space explorations. In the medical industry, surgical-assistant robots allow surgeons to have minimally invasive surgery with higher precision and speed while performing complex tasks. In this type of procedure, a surgeon operates from a robotic console with his/her hand movements translated into micro-movements that guide the robotic arms (Lanfranco 2004, 15).

While the aforementioned example demonstrates an extension to a surgeon's hands, a research team from MIT's Computer Science and Artificial Intelligence Laboratory (CSAIL) and Boston University have pushed this idea even further. They created a system that works using an electroencephalography (EEG) cap that records brain activity and relays it to a robot in real time, causing the robot to move. In this situation the robot literally becomes an extension of one's being (Conner-Simons 2017).

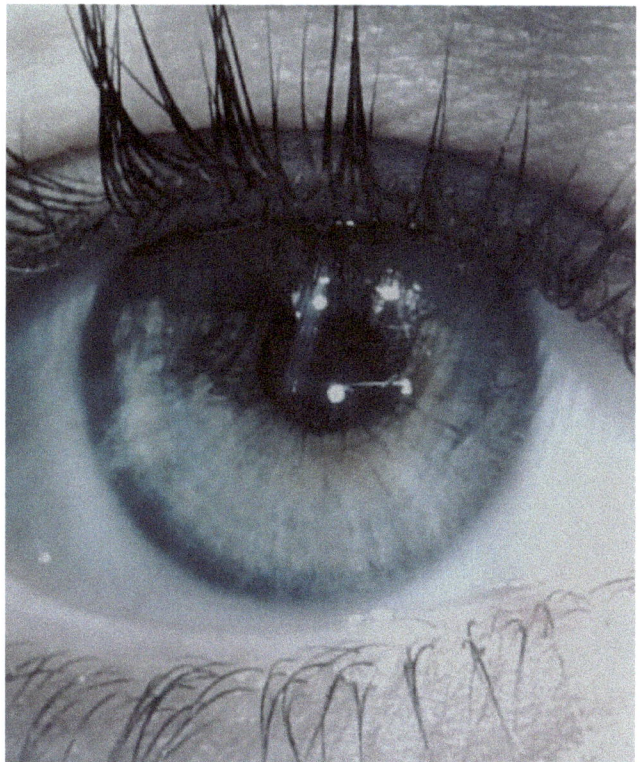

2

3

2 Two microscopic cameras are mounted on the headpiece; they track the performer's eye movements and stream video to the four monitors attached to robotic arms

3 The gaze of the model thereby extends to the robotic arms

4 The audience is looking at models and the performer with four extended robotic eyes is staring back at onlookers. Photographer: Nick Soland.

Australian performance artist, Stelarc, uses robotic technology to extend the operations of his body. Leading philosopher and cognitive scientist, Andy Clark, explains how he does this: "Stelarc routinely deploys a 'third hand,' a mechanical actuator controlled by Stelarc's brain through commands to muscle sites on his legs and abdomen… Stelarc reports that, after some years of practice and performance, he no longer feels as if he has to actively control the third hand to achieve his goals" (Clark 2007, 33).

Beyond these highly technological examples, another instance in which artifacts can become a physical extension of our bodies can be found in our relationship with tools. Extended physiological proprioception (EPP) is a concept proposed by D.C. Simpson to describe the ability to perceive the external world through the tip of a tool, in this case a prosthetic limb (Simpson 1974, 148). When we are using a tool such as a stick, perceptual experience can be extended to the end of the tool. Similarly, Clark gives an example of a blind person navigating their way through the world. He argues that the interface with the cane eventually fades away, so that the subject has an extended sense of self: "In so far as they succeed at this task, the new agent-tool interface fades from view, and the proper picture is one of an extended or enhanced agent confronting the wider world" (Clark 2007, 31).

All examples discussed show that the brain constructs a body image and body schema, which do not always match the biological physical body. The body image and body schema can be extended to incorporate external objects such as prostheses, implants and wearables.

In fact, new technologies including robotic technologies are changing the notion of our bodies by allowing them to become augmented, enhanced and expanded in terms of their functionality. This will change fundamentally what it means to be human and make us closer to being post-human. In this context we should ask how our bodies should be extended and for what purposes? How could different modalities of our sensory experiences such as vision be extended through technology?

Visual Behavior in Space, Bias, and the Male Gaze

"When I turn towards perception, and pass from direct perception to thinking about that perception, I reenact it, and find at work in my organs of perception a thinking older than myself of which those organs are merely the trace. In the same way I understand the existence of other people. Here again I have only the trace of a consciousness which evades me in its actuality and when my gaze meets another gaze, I re-enact the alien existence in a sort of reflection." — M. Merleau-Ponty (1945/2008, 410)

As we gaze out into the world, we actively perceive and make sense of that world. Certain elements draw our attention and hence our gaze, while some remains unnoticeable and stay in our blind spot. As we move in the space we often move where we are looking and as we speak, we prioritize our gaze to a specific face or point in space. Our gaze changes the value of the world before our eyes by bringing certain elements forward

5 The performer stares back at the audience with two more pairs of eyes. Photographer: Nick Soland.

and leaving the rest in the background. Handing the relationship between attention, direction and gaze is intuitive. In many cases our gaze shows our preference. We look at the world through our gaze, make sense of our surrounding environment and select where we would like to direct our attentions. Our gaze shows an object of preference and displays how we give attention to the outside world.

In social settings, we move and are moved by the gaze of others. Our gaze creates signals for our social interactions. It displays something about what the onlooker is looking at. In combination with our facial expression and bodily gestures, our gaze contributes to our emotional expressions. In *Brain and the Gaze: On the Active Boundaries of Vision*, neuroscientist Jan Lauwereyns states, "The gaze actively produces effects outside the body of others, who gaze at the gaze—an act that becomes reciprocal when the two gazes fixate on one another. The effect of the gaze on others, or of the gaze of others on us, can be thought of as a form of emission. The effects involve the emission of signals, the gaze of others contains information that influences us…" (Lauwereyns 2012, 187).

In his lecture "Architectures of the Entangled Mind," Andy Clark discusses visual behavior in relation to the environment (Clark 2022). He shares an interesting finding where different pottery styles capture different patterns of eye-movement. Interestingly there is a direct connection between social organization and different patterns of visual eye movement. He explains the "correlation between social complexity and organization at different periods and the different patterns of visual examination that the different styles encouraged. Societies with more hierarchical nesting their civic organization had pottery styles that encouraged more upwards and downwards visual exploration—this was quantified by the eye-trackers as the 'vertical index' of different types of pottery…." According to Clark radical entanglement theory explains "the core idea is that the built environment—from decorated objects to monuments, buildings, and city-plans—alter not just scan-paths but more fundamentally patterns of attention at every level of neural processing."

This means that the organization of power structure in our societies affects our gaze, attention direction, and basically how we see the world. In simple words, it is not only 'visual sensitivity' which affects our 'attention' but our internal and external biases that are informing it. Lauwereyns explains "the actual underlying neural mechanism looks like a prioritization through bias, yet it is labeled 'attention,' at the risk of confusion with the conventional notion, with us since William James, that attention improves visual sensitivity" (Lauwereyns 2012, 198). To put it simply, Lauwereyns explains that it is only accepted to include both bias and sensitivity effects in our concept of 'attention.'

Let's look into how these biases are manifested in our visual culture.

In his book *Ways of Seeing*, art critic John Berger (1990), the ways in which men and women are represented in visual culture through different gazes; different ways in which they

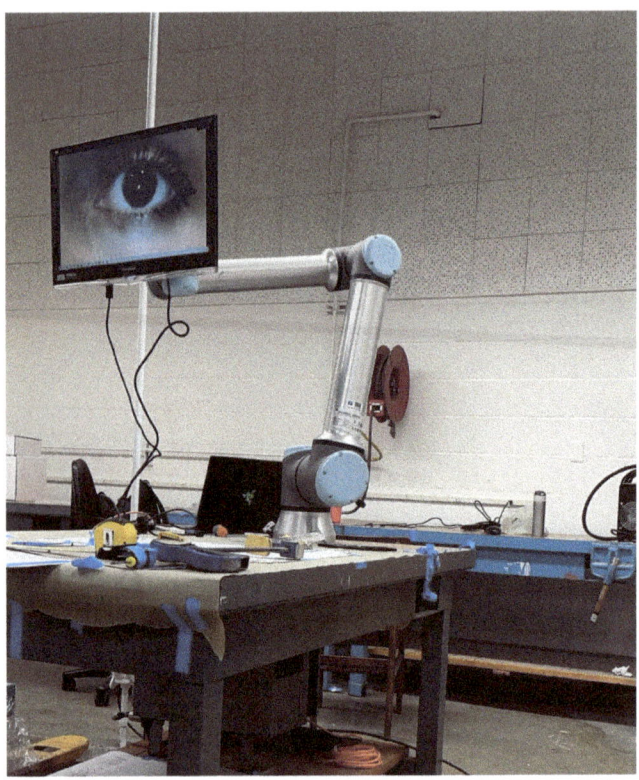

6 Prototyping process: attaching a monitor to a UR10 robotic arm and streaming real-time video data.

7 Close up shot of the helmet. Photographer: Nick Soland.

8 Using C# in Unity we controlled robots using TCP/IP socket protocols. This interface was also developed
   to easily recognize which robot is running which animation.

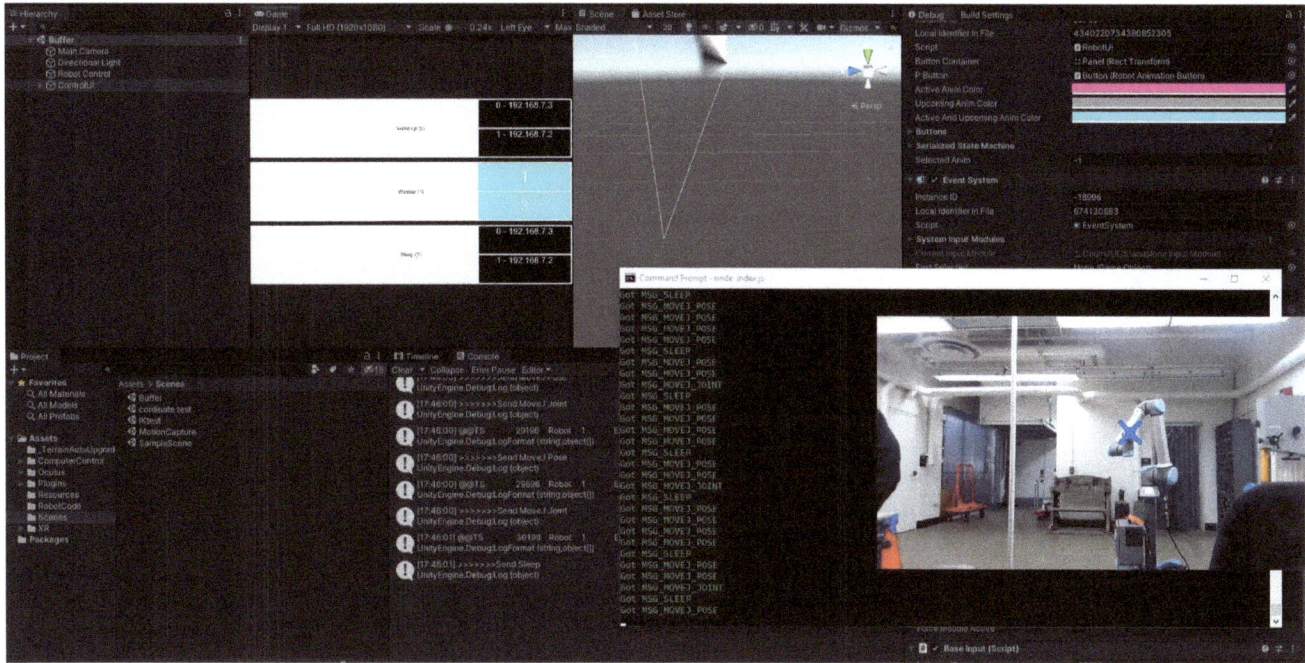

8

are looked at. He argues that historically men have been portrayed as though they are allowed to examine women, while women must continually watch themselves. As he puts it "Men act and women appear. Men look at women. Women watch themselves being looked at. This determines not only most relations between men and women but also the relation of women to themselves. The surveyor of woman in herself is male: the surveyed female. Thus, she turns herself into an object—and most particularly an object of vision: a sight" (Berger 1972, 20).

Three years later, feminist film theorist, Laura Mulvey, published her seminal article "Visual Pleasure and Narrative Cinema" (1975). As she observes, the male gaze serves to depict women as the object of pleasure for the heterosexual male viewer. It is as though women are under perpetual surveillance from the male gaze. She demonstrates the asymmetry of power relations between the observer and the observed where the unwanted gaze may be experienced as a form of violation. She explains, "Thus the woman as icon, displayed for the gaze and enjoyment of men, the active controllers of the look, always threatens to evoke the anxiety it originally signified" (Mulvey 1975, 12).

The gaze could be intrusive. It may lead to anxiety, fear, and harassment. In this context, we could ask whether it might be possible for women to be liberated from the objectification of the male gaze. Might women be able to use technology to 'return the gaze'?

'Returning the Gaze' Installation

Fashion is an important medium for the production of culture. Yet the fashion industry has long been complicit with a tradition of female objectification and sexual harassment. Women's bodies are regularly objectified within the fashion industry; moreover, women have come to absorb this condition unconsciously as a form of the internalized male gaze.

But what if we women were to subvert this through the power of our gaze? What if we could use technology to extend and amplify our response? And what if we were to do so on the catwalk itself, where the female model is often subject to sexual harassment through various forms of 'looked-at-ness'? What uncanny feelings would this evoke? And what strategies of resistance would it promote?

'Returning the Gaze' is an exploration of this scenario. It is a robotic performance installed during Milan Fashion Week in collaboration with the Universal Robots and commissioned by ANNAKIKI. In the center, a female model wears a spacesuit-like outfit and a headpiece fitted with two tiny cameras. The cameras track and capture the movements of the model's eyes, enlarging and displaying them on four monitors mounted on moving robotic arms glaring back at the observer. The gaze of the model is thereby directed back at the viewer, extended and enhanced through cyborgian technologies.

From a technical perspective, this project consists of four industrial robotic arms (UR10e) which are controlled using online programing. Online programming (versus offline) uses

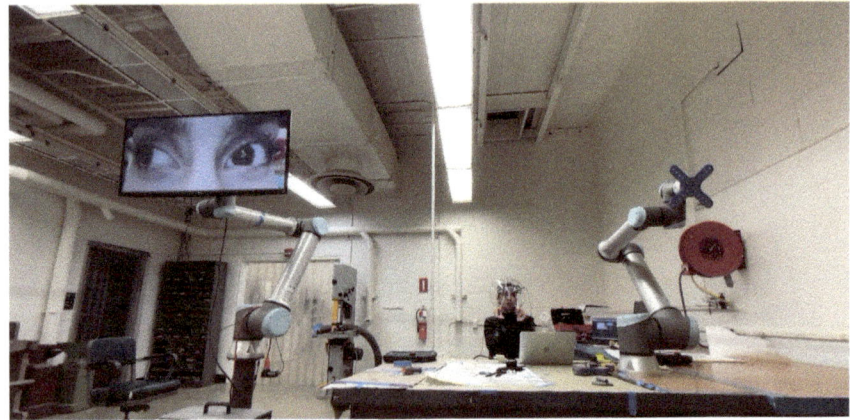

9 Testing the real-time streaming of eyes tracking video to the monitor mounted on the robotic arm

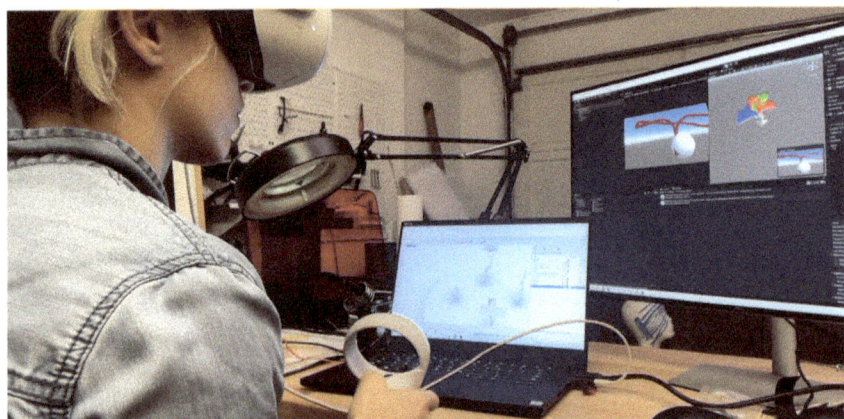

10 Using VR Oculus Quest2, we tracked hand movement and define robotic movement behaviors

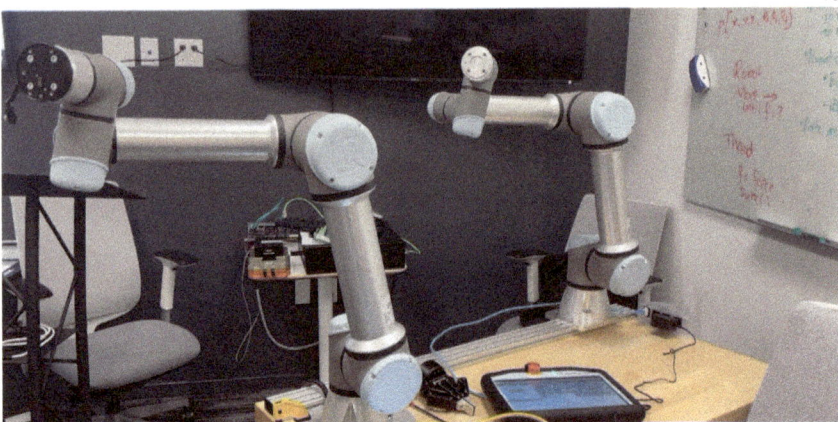

11 In collaboration with the Universal Robots team in California and Italy, we synchronized the movements of multiple robots to make sure they start and finish specific animations at the same time

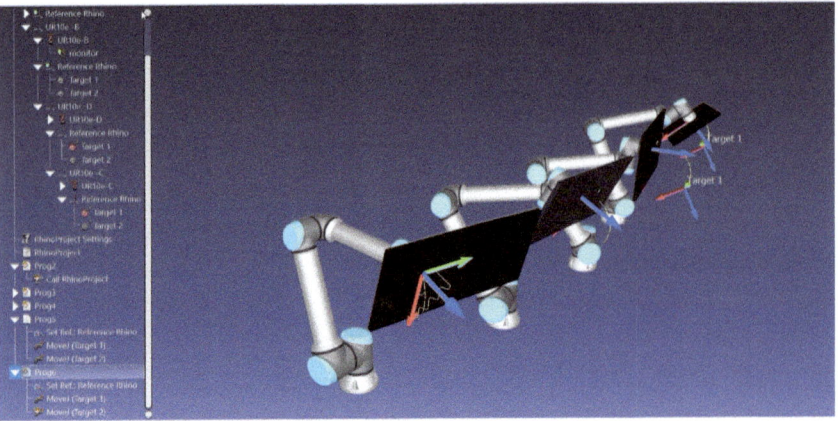

12 Synchronization of the robots: sometimes they perform the same movement, sometimes they mirror each other

13 The audience is looking at models, and the performer with four extended robotic eyes is staring back at onlookers

13

robot drivers for programming while it is being simulated. In this, robot drivers can control the robots that are connected to a central computer using ethernet cables via an ethernet switch hub. Robot drivers use a TCP/IP socket communication protocol. C# in Unity (Farahi 2022) was used to control and monitor a specific robot controller, enabling a computer to directly control robotic arms. In this case UR10e provides socket communication by default, thereby allowing the remote control on the robot.

In order to produce a life-like, organic and anthropomorphic robotic behavior, the movement of a performer's hands was tracked using a Oculus Quest2 VR headset. In this case a performer embodies the robot's movement. Specifically, the performer's hand location is used as a tool path, dividing the path into a series of points and using "MoveJ" (Move Joint Instruction) to control the joint movement of each robot.

A program of robot simulation is accomplished by adding a sequence of instructions. Four different animation behaviors ('wake up,' 'curiosity,' 'staring down,' and 'sleeping') are produced and sent to each individual robots using their specific IP address. Each animation includes specific IP address for a specific controller written in C# programming language. A Graphical User Interface (GUI) was also developed to control the robotic programs, and check which robot is running which animation.

In order to make sure that the robots are synchronized with one another, check points were defined between the animations to make sure that all the robots reach the same position after finishing a given instruction. The first movement of each program was kept as a Joint Move using a Joint target. This helped to set up the desired configuration from the first movement onwards and to ensure that the robot was moving the same as the simulation.

One of the challenges of this project was to detect any collisions between the mounted monitors (34-inch ultra-wide monitor) and the robot's body. which was achieved through simulation using RobotDK software as well as Robot plugin for Rhino's Grasshopper (Farahi 2022), along with a lot of physical prototyping. In this process, carefully defining the TCP (Tool Center Point), orientation, center of gravity, payload, and inertia matrix were crucial to make sure that the robot was running smoothly.

Data captured from the digital microscopic cameras mounted on the headpiece, was streamed to the monitors attached to the robot using HDMI cables. The total of two cameras (to capture right and left eye), four HDMI cables, a HDMI hub transfer the data to a central computer. Using OBS Studio (Farahi 2022), each video is modified, adjusted and streamed to the appropriate monitor.

While the eye movements are streamed to the monitors in real time, the future direction of this project benefitted from direct control over the behavior of the robot using eye tracking, such as controlling the robot's behavior through the performer's eye movements. This would make the project a more natural extension of the human. This goal was not achieved due a number of challenges:

- Problems with fast track commission
- The high risk of real time control and collision of a large tool (34-inch monitor) with the robot
- Safety concerns during the 15-minute runway show

Nonetheless, this project brings together robotics, fashion, design, feminism and critical thinking in order to critique the asymmetry of social and political power relations between men and women.

CONCLUSION

"The gaze is an instrument of social cognition, a communicative device with which we express signals and from which we acquire information _there is sending as well as receiving. From a utilitarian perspective we can ask the practical question of how we use this device most effectively." — Jan Lauwereyns (2012, 215)

Decades after Haraway's "A Cyborg Manifesto," Elon Musk said "We are already cyborgs." Some might not realize it, but devices and computational systems are already extending our biological capabilities. Using our devices, we memorize names/phone numbers, navigate in our cities, and remember our loved one's birthday.

In further developing this cyborgian future, we should raise critical questions as to whether technology could be used to empower those who typically have faced injustice and discrimination. This could engage with culturally and historically sensitive topics such as the male gaze.

By describing the robotic installation developed by the author for Milan Fashion Show, this paper attempts to build upon the field of critical computation where computational tools could be used to address larger socio-political issues.

Contextualized in the feminist discourse of the male gaze, the aim of this project is to reflect on ways we could learn to improve our implicit gaze biases in a manner which might be more inclusive and fair. Through a robotic installation where the gaze of a female model is returned back to the onlookers using four industrial robotic arms as an extension of the performer, the intention is to show that those who are historically victimized could be empowered.

Furthermore, through their augmented gaze they could force us to reflect on our own very real biases. In the future we might hopefully be able to use our gaze in a manner that might resist the disturbing biases rooted deep within our subconscious.

ACKNOWLEDGEMENTS

I would like to thank Julian Ceipek for their valuable contributions to this project. The video could be watched at: www.behnazfarahi.com. Also special thanks to Universal Robots team Gloria Sormani, Rafael Mancilla, Courtney Fernandez and Cliff Tsugawa.

REFERENCES

Berger, John. 1990. *Ways of Seeing: Based on the BBC Television Series*, 1st ed. London: Penguin Books.

Clark, Andy. 2022. "Architectures of the Entangled Mind," YouTube video, 2:06:40, March 27, 2022, https://www.youtube.com/watch?v=O0ggXvICpv8.

Clark, Andy. 2015. "Out of our brains." In The Stone Reader: Modern Philosophy in 133 Arguments, 1st ed., edited by P. Catapano and S. Critchley. New York: Liveright.

Clark, Andy. 2007. "Re-Inventing Ourselves: The Plasticity of Embodiment, Sensing, and Mind." Journal of Medicine and Philosophy 32 (3): 263–282. https://doi.org/10.1080/03605310701397024.

Clark, Andy. 2004. *Natural-Born Cyborgs: Minds, Technologies, and the Future of Human Intelligence*, 1st ed. Oxford, New York: Oxford University Press.

Clark, Andy, and David Chalmers. 1998. "The Extended Mind." Analysis 58 (1): 7–19. http://www.jstor.org/stable/3328150.

Clarke, Sarah E. Braddock, and Marie O'Mahony. 2008. *Techno Textiles 2: Revolutionary Fabrics for Fashion and Design*, 2nd ed. London: Thames & Hudson.

Clynes, Manfred E. and Nathan S. Kline. 1960. "Cyborgs and Space." *Astronautics* (Sept): 26–27, 74–76.

Conner-Simons, Adam. 2017. "Brain-controlled robots." MIT Computer Science & Artificial Intelligence Laboratory blog. Accessed May 10, 2022. https://www.csail.mit.edu/news/brain-controlled-robots.

Driscoll, Robert W. 1963. "Engineering Man For Space: The Cyborg Study." NASA (OART) Biotechnology and Human Research, Washington, D.C. Accessed May 10, 2022. http://cyberneticzoo.com/wp-content/uploads/2012/01/cyborg-nasa-driscoll-1963.pdf.

Harraway, Donna. 1987. "A manifesto for Cyborgs: Science, Technology, and Socialist Feminism in the 1980s." *Australian Feminist Studies* 2 (4): 1–42. https://doi.org/10.1080/08164649.1987.9961538.

Lanfranco, A. R., A. E. Castellanos, J. P. Desai, and W. C. Meyers. 2004. "Robotic Surgery: A Current Perspective." *Annals of Surgery* 239 (1): 14–21. https://doi.org/10.1097/01.sla.0000103020.19595.7d.

Lauwereyns, Jan. 2012. *Brain and the Gaze: On the Active Boundaries of Vision*, 1st ed. Cambridge, MA: The MIT Press.

Merleau-Ponty, M. (1945/2008). *Phenomenology of Perception*, translated by C. Smith. New York: Routledge Classics.

Mulvey, Laura. 1975. "Visual Pleasure and Narrative Cinema." *Screen* 16 (4): 6–18. https://doi.org/10.1093/screen/16.3.6.

Simpson, David C. 1974. "The choice of control system for the multimovement prosthesis: extended physiological proprioception." In *The control of upper-extremity prostheses and orthoses*, edited by P. Herberts, R. Kadefors, R. Magnusson, and I. Petersen. Springfield, IL: Charles C. Thomas. 146–150.

Behnaz Farahi is trained as an architect and an award-winning designer and critical maker based in Los Angeles. She holds a PhD in Interdisciplinary Media Arts and Practice from USC School of Cinematic Arts and is currently Assistant Professor of design at California State University, Long Beach. She is a coeditor of an issue of *Architectural Design: 3D Printed Body Architecture* (2017) and *Interactive Futures* (forthcoming). www.behnazfarahi.com.

Keynotes

KEYNOTE EVENT

Hybrids & Haecceities Prologue Panel

Marcos Cruz
Bartlett

Winka Dubbledam
University of Pennsylvania

Rashida Ng
University of Pennsylvania

Mette Ramsgaard Thomsen
Royal Danish Academy

Robert Stuart-Smith
University of Pennsylvania
Moderator

PROLOGUE PRESENTATIONS & PANEL DISCUSSION
Robert Stuart-Smith

The architectural profession has embraced computational approaches that have impacted how we conceive, develop, document, fabricate, inhabit or interact with architecture. As a profession, we attempt to identify issues, challenges, and opportunities where design can make a difference. Similar to the scientific fields, we tend to generalize problems and sometimes solutions. Researchers Joy Buolamwini and Cathy O'Neill have shown in different ways that generalizations made in computer software have had a significant impact on equity in our civil liberties. It begs the question: is the Fordist mass production of building systems or repetitively used design algorithms equally biased or discriminatory? Where computational design has often been championed for its ability to produce variability, in this conference, we questioned whether variability is as important as specificity.

While the bespoke is historically an elitist domain, the nascent Fourth Industrial Revolution is perceived to be ushering in smart factories and autonomous technologies that will enable manufacturing of the bespoke to be more economically and practically scalable to industrial levels of production. Although efforts must be taken to avert Industry 4.0 exacerbating consumption, waste, environmental impact, and equity, this paradigm shift will offer opportunities for a more considered, personalized, site-specific, and less wasteful built environment.

The ACADIA 2022 theme of *Hybrids & Haecceities* embodies a diversity of untapped possibilities within design, manufacturing, and use. Hybrids have characteristics that are enhanced by combining two or more elements with different properties. Functional hybrids exist in varying ratios and can often perform better than their constituent parts. Hybrids dispel binary thinking and offer opportunities for combining materials, technologies, objects, methods, disciplines, and more. Haecceities describe a non-qualitative property that defines something as unique and indivisible. Thus, the theme seeks to motivate a move towards more inclusive and specific forms of computational design, that rejects binary thinking in any form, and offers a critical optimism towards technology, with an open and diverse engagement with the world at large, while seeking to make each engagement specific and expansive.

For the Prologue session, we were honored to have: **Rashida Ng**, Penn's Undergraduate Chair and Presidential Associate Professor of Architecture; **Winka Dubbeldam**, Chair of Graduate Architecture at Penn and Director of Architectonics; **Mette Ramsgard Thomsen**, Professor of Digital Technologies and Director of CITA in Copenhagen's Royal Danish Academy; and **Marcos Cruz**, Professor of Architecture at the Bartlett School of Architecture and Co-Director of the MSC degree in Bio-Integrated Design. These distinguished guest speakers spoke briefly about the conference theme through their own work and manifestos, covering topics ranging from architectural education and practice to material circularity and bio-integrated design.

Rashida Ng discussed how Philadelphia is one of the poorest and most racially segregated large US cities, with a history of redlining and infrastructural neglect that has caused communities of color to be disproportionately impacted by floods, urban heat, and pollution. Rashida questioned architecture's agency, and its social and political relevance within this context, and in relation to the climate crisis, social inequities, health disparities, conflicts across the globe, the Covid-19 global pandemic, and protests sparked by the murder of George Floyd. In response to this, Rashida seeks to evolve architectural pedagogy to cultivate a more creative social agency of architecture, enlisting faculty to operate more as facilitators of learning.

Rashida advocates for *"a professional responsibility, if not a moral and ethical one, to educate global citizens who contend with the consequences of their practice. In my studios, students have a say in the format of critique and engage in a more public debate about the relative merits of design, reinforcing a civic notion of the social impact of architecture. We aim to cultivate individuality, leverage collective creativity, and provoke innovation in an evolving educational model, that is as humbling as it is rewarding"*.

Winka Dubbeldam presented a recently completed 116-acre park project for the 2023 Asian Games in China. The project sought to explore architecture as an object that gains intelligence, identity, and character over time, while operating as a social space for both humans and non-humans. All the buildings in the park were designed as hybrids, each developed as a mutant shape produced by the intersection of several geometries. The overall project does not delineate between building and landscape but instead offers a more ambiguous and productive hybrid. Consisting of 85% greenspace together with a river and aqueduct, seven of its buildings lie beneath the expansive park.

Winka describes that *"the park's hills are also formed from the site's own excavated soil, while the project's 65,000m2 of greenspace operates as a sponge city that absorbs and filters water in its reclaimed wetlands, producing 84 kilograms of oxygen and absorbing 150kg of CO_2 each year. In addition to its ecological benefits, our recreation of native vegetation has also restored the local biome, bringing back plants and insects"*.

This ecological design agency aligned with presentations from Mette Ramsgard Thomsen and Marcos Cruz. Mette Ramsgard Thomsen discussed the need to build a more sustainable architecture through Biodesign. In response to our co-existence with finite resources, Mette seeks to challenge the foundations of industrial production by exploring circular design, considering entropy in building materials and developing material composites. Having already developed digital processes that link advanced modelling, integrated analysis, and adaptive fabrication to support material-based craft, Mette now seeks to develop new resource-aware strategies that address material resources, their origin, transferal, pollution, and end-of-life.

In contrast to manufactured materials, biological materials are fundamentally differentiated with haecceities. Mette suggests we can use computation to capture, steer, and deploy bio-based material heterogeneity in response to design performance as a means of material grading while incorporating temporalities lasting from days to years. Her work with bioluminescent bacteria explores such durational qualities, leveraging 3D printing as a medium in which shape and surface-to-volume ratios steer the design of colonies, their access to oxygen, and the brightness and duration of their light emission. Through this work Mette questions whether we can: *"expand our anthropocentric gaze and open up to other kinds of interspecies co-construction? When we design for other species, how do we shape, predict, and formalize the interactions of living organisms, not just to functionalize them but to truly co-create and co-exist? If design control is replaced with new interacting, listening, or co-creating methods, we still need to find ways to characterize and represent these correlations and engage with heterogeneity, and threshold adaptive, remedial, or restorative design action"*. Mette hopes this work will foster greater environmental sensitivity by enabling designers to become aware of their impact on ecology and landscape.

Marcos Cruz together with biochemical engineering colleague – Brenda Parker, conducts interdisciplinary research and teaching focused on the 3-dimensional growth of small species such as cyanobacteria in designs that have demonstrated an ability to absorb carbon, produce oxygen, and other capabilities. Marcos presented research exploring a wide range of bio-enhancements for architecture including more variable and controllable forms of translucency, an up-cycled calcium carbonate waste project, the robotic additive manufacturing of large-scale hydrogel-algae multi-material prints, and the use of biomineralization to manufacture tiles that control the runoff of water and the intake of heavy metals. Marcos is also exploring an "architectural bark for buildings," that serves as a host for an evolving ecology that controls material porosity, water retention and growth without the need for irrigation or maintenance. The research into callus tissue by one of his PhD students was a potent example of a haecceity. Related to the genotype of the carrot, Marcos explained that callus tissue is "responsive to the input of external environmental factors that cause the phenotypological variation of the plant to change", suggesting bio-designs can be radically tailored to site and environmental conditions. Marcos also advocated that we need to start designing for aging buildings.

Marcos conjectures that *"in bio-integrated design, we are starting to think about the age of the Biocene, and how we can fuse the built and natural environments in a new way. Our own microbiome, for example, is essential to our survival, and we now know that we don't live in sterile environments. In fact, more porous systems can actually be better for the development of biodiverse microbiomes in urban environments. There's a lot we can learn from this, but it's not just about mimicking these natural systems. We need to bring that knowledge into a new light with the technologies and capabilities we have now."*

A recently built poikilohydric wall in central London explores this, currently being monitored to see how its mossy surfaces weather and age.

Marcos and Mette's research is suggestive of a Posthuman architecture that is co-developed in partnership with biological colonies, requiring different levels of scalar and temporal perception than architects are accustomed to. Similarly, Winka's Asian Games park is not dominated by an architectural geometry or tectonic, instead, it is a multifarious and evolving urban-natural hybrid ecology that is a scaffold for biological growth. Architecture is both more hybrid and yet more distinctive, actively creating its own microclimatic conditions. The unconventional approaches discussed during the Prologue session appear to be quite distinct from the typical US educational model for training architects. Fortunately, Rashida proposes an alternative approach that empowers students with more agency over their education and careers, yet also provides agency to the communities her studios engage with. Collectively, these presentations spoke of greater degrees of multi-disciplinary collaboration, consideration and inclusivity, engaging with other agencies that proffered new hybrid approaches and effects to produce more highly specific and sensitive responses within architectural education, research and practice. In this sense, the thought-provoking and inspiring work of the presenters provided an exciting initial foray into the conference theme *Hybrids & Haecceities* that stimulated further discussion throughout the conference.

KEYNOTE EVENT

Origins and Destinations Beyond Midjourney

Chigozie Nri
Stability AI

Joel Simon
Morphogen

Kyle Steinfeld
UC Berkeley

Masoud Akbarzadeh
University of Pennsylvania
Moderator

**KEYNOTE PRESENTATIONS
& PANEL DISCUSSION**
Masoud Akbarzadeh

Most recently, there has yet to be a single day when each one of us, in our daily interactions with news or social media, encounters images generated by AI. In his preface at the *Architecture and Computer Conference* in 1964, Sanford Greenfield pronounced that computers and their upcoming impact on architecture are changing the profession into an irresistible force. Whether we plan for it or not, this force will radically alter architectural practice. In the special topics keynote panel of the *Hybrids & Haecceities*, as panelists and scientists, we discussed this impact of the advent of AI tools in architecture how how they are about to change the practice yet again.

Kyle Steinfeld, a computational designer near Silicon Valley and Associate Professor of Architecture at UC Berkeley, refers to architects as a group who often find themselves adapting technology not originally designed for their discipline. The summer of 2022 witnessed a surge in synthetic architectural imagery generated by text-image tools like DALL-E, MidJourney, Stable Diffusion, and ImageGen. These tools, while impactful, emerged from a specific technical and socio-technical history.

Let's explore three critical timeframes: the 1950s-60s, the 2000s-2010s, and the rapid developments of the past 18 to 20 months. These scales encompass technological evolution, social context, and infrastructure.

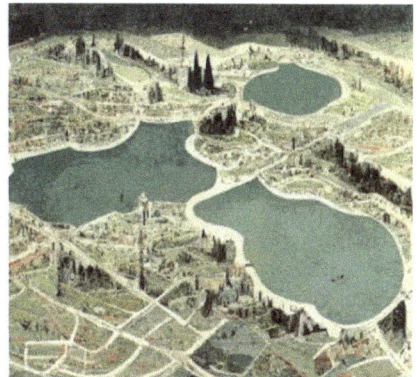

In the 1940s-50s, AI's foundation was laid with a "systems theory" approach to human cognition, including early experiments like the Perceptron. AI research initially seemed distant from practical applications but thrived on military funding, notably ARPA's contributions to MIT. The social context of this era involved academic labs and government grants, with limited private funding. Access to resources was restricted due to expensive computing technology.

In the early 2000s and around 2010, faster computers and the rise of "big data" spurred technical innovations. Foundational datasets and cloud infrastructure became key, with companies like AWS and Google playing crucial roles. AI applications like image search and natural language processing emerged, impacting practical domains and public perception of technology through events like Deep Blue and AlphaGo. This period also saw an expansion of academic and private research labs, fostering innovations like DeepDream and GANs, which fueled generative aesthetics from 2015 to 2020.

Today, we're in a period marked by accessibility to AI technologies. Non-affiliated research labs, casual technologists, and collectives like ELutherAI contribute significantly. These independent actors have introduced diffusion models, such as "CLIP-guided diffusion."

CLIP, developed by OpenAI, plays a pivotal role. Small experiments by independent actors, such as Ryan Murdock's work connecting CLIP to image generation functions, drove progress. Katherine Crowson's VQGAN-CLIP project became a widely adopted image generation system, leading to Disco Diffusion.

Collaborative notebooks like CoLab were critical in enabling more involvement in these projects. Contributions from individuals like Daniel Russell, Max Ingham, Adam Letts, and Chigozie Nri have had a substantial impact. Companies like Midjourney and Stability AI have emerged by hiring talent from this creative community.

In summary, creative machine learning is currently more accessible than ever. Both established institutions and creative communities contribute significantly to this space. We all have the potential to make meaningful contributions to this exciting field.

Panelists Joel Simon, Chigozie Nri, and Kyle Steinfeld participated in a residency program at Stochastic Lab, focusing on ML-based creativity support tools. Joel's GAN Breeder was created in just three days at a Manhattan coffee shop, while Chigozie contributed to Disco Diffusion as a creative community member.

Joel Simon, the founder of Morphogen, is a creative visionary bridging the realms of generative biological art and simulations. In a recent discussion, he delved into the world of creative tools based on generative methods and questioned the prevailing text-image paradigm. This exploration sheds light on where this technology and these tools might be headed and how they could impact the creative process.

The conventional text-image paradigm involves translating text into images but grapples with inherent limitations concerning composition and color. Joel invokes Wittgenstein's idea that uttering a word is like playing a note on the keyboard of the imagination, emphasizing that language operates in communication and interpretation through art.

He introduces the concept of variability and unpredictability in creativity. Imagine a single prompt that generates a different rendering each time it's used. Switching to another model opens up entirely new creative possibilities but also restricts control and intentionality in the creative process.

Joel presents three tools that exemplify the intersection of generative methods and creativity. The first is Art Breeder, akin to a breeding ground for images. It uses BigGAN to collaboratively and interactively create images from many sources. Everything is open and remixable, akin to a biological metaphor for the technical processes underneath. This approach gamifies image creation, appealing to both casual and serious creators.

The second tool, Prose Painter, transforms text into spatial elements in an image, offering precise control. It resembles composing ink from prompts and allows individuals to discover unique textures and paint with them.

The third tool, Art Breeder Collage, harnesses the diffusion methodology. Users cut and arrange images like a collage, providing the guiding input for the generative process. This blend of generation and rendering falls between two creative processes. These tools showcase the role of machine learning as a serendipity engine, enabling a different form of generative creativity and fostering collaboration, inspiration, and abstraction in the creative process.

A recurring question is whether these tools automate creativity or stifle human ingenuity. Joel's perspective centers on human creativity, where these tools act as collaborators. They're akin to creative directors who communicate with a team, providing guidance and direction within the creative process's abstraction levels. These tools don't diminish creativity; they operate at a different level of abstraction, enhancing the creative experience.

Joel emphasizes the importance of designing intuitive interfaces that enable more people to engage with these systems while maintaining control and guidance. It's a fusion of interface design and metaphorical representations to shape raw technologies into user-friendly tools.

He presents two intriguing ideas for future creative interfaces. The first involves text-based collaboration between an architect and a client. A visual interface reminiscent of Figma allows clients to select preferred images and combine text, images, and references in a collaborative and iterative process. The aim is to facilitate consensus-building, especially in situations with language barriers or divergent perspectives.

The second idea leans towards a chatbot-style interaction. Imagine a conversation where a chatbot presents options or references based on preferences. Users can provide feedback and specific references, creating a dialogue with the system. This approach combines iterative language, guidance, and reference imagery, adding an element of serendipity without surrendering authorship.

In conclusion, Joel's exploration of generative tools in the creative process challenges the traditional text-image paradigm and offers a glimpse into a more abstract and collaborative future of creativity. These tools, rooted in machine learning, act as partners rather than replacements for human creativity, opening up exciting possibilities for the creative process and its interfaces.

Chigozie Nri, currently associated with Stability AI, an organization known for its open-source text-image model called Stable Diffusion, delivered a presentation shedding light on the future of text-image technology. While Nri comes from a bioengineering background, she considers herself a creative outsider immersed in the open-source community. Her talk focused on emerging trends and projects, particularly in the realm of "inference tricks" within the field of machine learning.

Nri acknowledges that predicting the future in the rapidly evolving field of machine learning is challenging. The sheer volume of research papers in this domain is doubling approximately every twenty-three months, with new models emerging almost daily or weekly, consistently surpassing expectations. As such, she approached the topic by surveying projects in the prototype stage, with a particular focus on what she referred to as "inference tricks."

She emphasizes a distinction between innovation in the "foundation model space" and these "inference tricks." Foundation models, which are trained on extensive data and adapted for various applications, typically require significant resources and are often the domain of large companies or well-funded academic institutions. In contrast, inference, where image generation takes place, is more computationally affordable. Innovations in this space can originate from unexpected sources and may not necessitate entirely new model training. This makes the field accessible to individuals who can contribute to innovation.

Nri stresses the role of open-source models in accelerating innovation by facilitating collaboration and experimentation. These models provide resources for individuals to innovate, opening up possibilities for new ideas to flourish. She cites her own contribution as an example—an inference-side trick that generates animations from a network designed for two-dimensional static images. These tools are created to empower artists, complementing their vision rather than replacing it.

Further, she discusses alternative approaches, such as using rendered 3D environments as initial imagery. This approach, demonstrated by Ukilodeon Plane using Disco, enables textures to be reprojected onto 3D scenes and later used as decals. This technique aligns with neural radiance fields (NERF), a machine learning-powered photogrammetry method that allows the rendering of novel views without the need for traditional 3D mesh models.

Nri then delves into "Dream Fusion," a paper by Ben Poole from Google and Ajay Jain at UC Berkeley, which exemplifies the collaboration between researchers, open-source initiatives, and hackers. This project utilizes Google's text-to-2D image network (Imagen) to render 3D NERF environments without requiring additional training data.

She highlights that foundation models might encode hidden information beyond their apparent training objectives, allowing consistent results to be achieved by switching between NERF and 2D models. Open-source resources, such as Stable Diffusion, have accelerated such achievements.

Nri also discusses "Dream Booth," a project offering greater control over subject matter using text prompts. Similar projects—"Textural Inversion" by Garlatale from NVIDIA and Tel Aviv University and "ResetAR" by Google—were released, allowing users to incorporate images into text prompts. This approach differs from traditional fine-tuning and opens up possibilities for personalized concepts and interpretations.

The availability of open-source weights and code, similar to Google Colab's resources, accelerates open-source projects. Hugging Face maintains a library of published concepts based on Stable Diffusion, further facilitating experimentation and creativity.

Nri touches upon Google's recent releases, "Unitune" and "Imagic," enabling image editing with just a single image and minimal training. However, she noted that some techniques require full retraining on new datasets, like full-text-to-video generation, which remains a relatively unexplored field.

In conclusion, Nri emphasizes that the next big breakthroughs in text-image technology can come from creative outsiders, artists, and architects who push the boundaries of existing networks. Providing open-source models and code accelerates artistic expression and innovation within the field, with the potential to shape the future of text-image technology.

KEYNOTE EVENT

Artificial Intelligence and the Future of Architectural Design

Antoine Picon
Harvard GDS

ACADIA NOTE
During the conference, the keynote lecture by Antoine Picon was followed by the Epilogue Panel that follows this section.

KEYNOTE INTRODUCTION

Hina Jamelle: Very early in the conference planning, as we were structuring the conference, Masoud, Robert, Dorit, and I, as the Co-Chairs, wanted the theme, Hybrids and Haecceities, to permeate throughout the conference. We decided that the best way to carry that theme through the conference was through an epilogue and a prologue, as well as specialized panels as a probe into special topics.

The first Hybrids and Haecceities Prologue Panel, led by conference chair Robert Stuart-Smith, set up the larger themes of the conference and acknowledged the current state of affairs in the new normal of technology in every aspect of our life and work, our immediate climate concerns, and the need for a larger focus on new ecologies and biomaterials. The panel also set up the importance of new educational and collaborative models that are needed both on building sites and in academia and practice. We then carried the theme onward to a special round table session led by Masoud Akbarzadeh on AI, which took us inside the mindset of the engineers behind one of the most hotly discussed topics in architectural design today. Through that probe, we had a unique opportunity to understand what goes on so that we can stay ahead as we begin to think through an architect's approach and response.

Today, we will take a look at future technologies and their implications for both architectural design and culture. As AI has emerged as a topic, moving from analytic to generative, we are cognizant that this became our focus. However, we also recognize that no new technology emerges from a vacuum; there is always a context and a trajectory, and there are key patterns and cycles that occur as new technologies are adopted, adapted, and then implemented in design and practice. In my mind, there is no better person to help us think about this topic than our keynote speaker for today, Antoine Picon. I am delighted that he accepted our invitation to speak about these ideas, and Antoine, I have personally enjoyed our conversations and appreciate your time in thinking through artificial intelligence and its broader impact on both architectural history, our position in history, and practice.

So let me take a moment to introduce Antoine Picon to you. Antoine Picon is currently the Jiwara Travel State Professor of the History of Architecture and Technology, as well as the Director of Research at the Harvard Graduate School of Design. His background is quite diverse, as he has been trained as an engineer, architect, and historian. His main focus of research is on the history of architectural and urban technologies from the 18th century to the present day. He has published extensively on this subject, with a particular interest in the changes brought to cities and architecture by the development of digital tools and digital culture. Antoine has written several books dealing extensively with this question. Three of these books are "Digital Culture in Architecture: An Introduction for the Design Profession" (2010), "Ornament: The Politics of Architecture and Subjectivity" (2013), and "Smart Cities: A Spatialized Intelligence" (2015). Antoine's latest book, "The Materiality of Architecture" (2021), is an ambitious attempt to place the digital in architecture within a more global theoretical and historical frame. He has received a number of awards for his writing, including being elected as a member of the French Academy of Technologies in 2010 and a member of the French Academy of Architecture in 2015. In addition, Antoine is the chairman of the Foundation Le Corbusier. I invite you all to join me this evening in giving a very warm welcome to Antoine Picon.

KEYNOTE ADDRESS: ANTOINE PICON
A New Phase or a Return to the Origins of the Digital Revolution?

Artificial Intelligence presents itself from a double perspective. On the one hand, it is the new hot topic. It appears as the third phase in the rise of the digital in architecture. The first had been marked by the exploration of new geometries. The second had seen the development of digital fabrication. The third that we have begun to enter is about the possibility to have a true participation of the computer to the core of the design process. What this participation exactly entails remains however to be determined.

On the other hand, the current investigations regarding the role of AI give an impression of déjà vu. Indeed, when the possibility to use computers in design initially developed in the early 1960s, the goal was to use the machine as a true partner rather than a mere tool enabling to complete differently but above all more easily the tasks involved in the conception and realization of buildings. As Molly Wright Steenson has shown, such was in particular the ambition of Negroponte's Architecture Machine Group at MIT. The recentering of the work of the group on questions of interface, that would eventually lead to the creation of the Media Lab, was emblematic of the then receding seduction of a conception of computer aided design that would give to the machine the role of a true partner. Instead of this partnership, because of the successive so-called 'winters' that hampered the development of Artificial Intelligence, we simply got machines that could draw and at best manage the variation of key parameters of the project. [1]

It looks more and more today as if the situation was finally going to change. Are we about to recapture the original ambition of the digital in architecture or will we have again to reconsider our objectives? To what extent is this a rediscovery of long neglected issues or the beginning of a new era? These questions are worth asking, if only to better identify what is unique about the present circumstances.

Like many participants at this conference, I believe that we are indeed entering a new and exciting phase in the development of the digital in architecture, a new phase that is not reducible to a rediscovery of forgotten ambitions, although one should not underestimate what is at stake in the impression of déjà vu that I mentioned earlier.

There are in my opinion two factors that go in the direction of a new phase that will not end by a withdrawal to more modest ambition. The first is evidently the technology itself. With the dramatic rise of computing power, the possibility to manipulate extremely large quantities of data without equivalent in the past and approaches such as neural networks and deep learning, it looks like we have finally overcome limitations that had plagued former episodes in the history of AI. Of course, we are still far from artificial general intelligence and Raymond Kurzweill's singularity is not for tomorrow, but AI has become a pervasive reality.

Equally important are the new challenges that await architecture in the context created by the environmental crisis. With the need to consider embodied and operational carbon and the full cycle leading from the extraction of materials to the construction site, and from the realized building to its disassembly and recycling, design is about to become an extremely difficult if not impossible task. We are reaching a level of complexity that needs more than the existing parametric tools to be manageable.

For architecture, past the exciting character of experiments conducted with Dall-E or Midjourney, this second aspect may prove the more essential. One of the tasks ahead of us is by the way to better organize the convergence between the use of advanced digital tools such as AI and the need for architecture to meet with environmental requirements and even to find in those a source of creative renewal.

An Archetypal Impulse

Now, I said that the impression of déjà vu shouldn't be underestimated. It harks back to the fact that the use of AI is connected to something broader, the quest for automation, that possesses deep psychological roots. As Pierre Naville, a French sociologist who had been a Surrealist in his youth (the detail has its importance), once noted, the quest for automation is actually rooted in some deep structures of the human psyche. In a 1963 visionary text entitled, *Vers l'Automatisme Social, (Towards Social Automation)*, he declared in particular:

> "I am not far to believe, despite the violent rejections that it often provokes, that automation represents an archetype as ancient and radical as our sense of symmetry or cycle, that it touches something intimate in us, a vibrant chord of our creative power, that it has to do with enchantment, with our will to power, and many other impulses that move us, starting from our unconscious." [2]

There is something archetypal in the quest for automation. To return to the Surrealists, this is what their attempts at automatic writing and drawing was trying to reveal. From that perspective, AI is only the latest and most sophisticated expression of an archetypal impulse that has led in the past to the invention of various machines, including computers and robots.

Archetypes are recurring; they come back periodically to haunt us. As Naville noted aptly, automation has actually to do with the depth of human psyche and its unconscious desires. This might account for the dreamlike quality of some of the manifestations of AI, from Google Deep Dream to the strange combinations that Dall-E or Midjourney enable us to create. The expression 'machine hallucination' that is often employed today might not capture what is really at stake; for it is us, humans, that ultimately hallucinate with the help of machines.

One thing is sure, the quest for automation is as much about us as about what machines can achieve. This conclusion had been already reached by philosopher Denis Diderot, the main editor of the *Encyclopédie* in the second half of the 18th century. Studying machines was for Diderot inseparable from nagging questions regarding what might be mechanical in the way the brain functions. Likewise, AI raises all sorts of interrogations regarding human creativity, its similarities and differences with the generating power of machines. This is especially true in architecture today. In this lecture, I would like to discuss these questions by following three main threads: the way artificial intelligence leads to reopen the intricate question of authorship, the problem of intention in architecture and what it becomes when the project is designed partially or even totally by means of machines, and finally the aesthetic dimension of the current AI experiments and where they may lead.

Authorship, intention, and aesthetics: these questions are of course related one to another. They nevertheless enable to discuss different aspects of the current craze for AI in architecture. Whereas authorship raises questions related to the figure of the architect, intention is about both the objectives that architecture tries to accomplish and its effect on those who inhabit it. Finally, aesthetics seems almost unavoidable, given the multiplication of strange, sometimes provocative, always a bit disturbing images that have invaded the pages of magazines and books on architecture.

My point of view is obviously not that of a specialist of artificial intelligence, but rather that of an architectural theorist and historian wondering about what is at stake through its rapid development. This is not an entirely new problem. Since its initial stage, the digital has triggered inquiries regarding the direction that the architectural discipline was taking. The difficulty is to avoid techno-determinism when doing that. As I have argued in a recent article, we should not approach the digital as an external factor impacting architecture, but rather consider the digital and architecture as reflecting both, in connected ways, a massive ongoing social and cultural change. In other words, architecture in the age of rising AI is as much AI in the age of a rapidly evolving architectural discipline. Both capture something of the current social and cultural change, a change that affects primarily the way we understand ourselves as well as our relation to the planet. It is not a coincidence if interrogations regarding a possible posthuman condition as well as questions regarding the relationship between humans and non-humans have taken such an importance in the past decade or two.

Expanding Authorship
Authorship, creative authorship that is, not the unimaginative reinterpretation of well-known models, was always a complex question. It involves the capacity to let something or someone unknown and alien develop inside oneself as an irrepressible force of innovation. Indeed, how could the truly new be based only on what we believe that we know about us? The poet Arthur Rimbaud accurately captured this strange splitting of the creative mind with his famous formula 'I Is Another.'

Could AI play the role of this catalytic other? Would it entail a radical departure from our received conception of the creation process? An intermediary between this received conception and its most radical transformation with AI could be to train the machine from a database of past drawings, projects, and realizations by the designer. This is what Coop Himmelb(l)au has attempted with DeepHimmelblau that mobilizes deep-learning tools in order to train the machine to make proposals from the past production of the office. [3]

Now, with the collective character of many of its processes, authorship in architecture was always especially complex. Because of the presence of design collaborators, because of the important role that engineers, craftsmen and more generally labor, not to mention the input of the client, otherness laid not only in the mind of the architect. Authorship was always more distributed than generally acknowledged by architectural history and criticism, which focused on the principal designer as if she/he operated in a vacuum. Until the dawn of modernist architecture and its demise in the name of a new rigor, ornament often bore the mark of this distributed authorship by owning something both to the designer and to the craftsman. Ornament was a 'trading zone,' to use historian of science Peter Galison's concept. [4] Design is more than ever full of trading zones enabling actors with different cultures and goals to cooperate.

From such a perspective, one of the immediate consequences of AI is to dramatically expand the field of architectural authorship and the number of 'trading zones' to the point that the traditional discourse centered on an allegedly sole author can no longer hold. Indeed, in addition to the 'intelligent' computer, the mobilization of AI corresponds to the arrival of a series of other contributors, from the programmers who wrote the algorithms to the curators of the datasets used for machine learning. Actually, the fully trained machine is only part of a story that involves various cyborg assemblages of humans, computing and informational resources. With the development of the recourse to AI in digital fabrication, one should also add to the list of authors the 'intelligent' robots as well as the various people involved in their building and training.

I said earlier that AI and architecture had to be treated on equal footage as stakeholders in a more global social and cultural change instead of trying to ascertain the impact of the former on the latter. It is striking to observe how the still recent rise of AI in architecture has been preceded by a growing interest taken by younger generations of architects in collective, distributed, and networked forms of authorship. On a more academic ground, this trend accounts for the success met by Bruno Latour's Actor-Network Theory to account for these forms of authorship, a success epitomized by the recent publication of a 'Latour for Architects.' By giving a large place to non-human actors, Actor-Network Theory proposes a convenient frame to consider the increased role of machines in the design process.

Another way to account for this dramatically expanding authorship is to go back to a principle mobilized by one of the founders of the architectural discipline, at least in its Western definition, Leon Battista Alberti, namely the dialogic character of architecture.[5] Dialogue or conversation: instead of approaching design as the creation of an isolated mind, this principle leads to the exploration of its roots in a series of dialogues or conversations. Architecture emerges from conversations between designers, experts, working force and of course clients. The 'I Is Another' may be also interpreted as a dialogue or conversation. From this perspective, AI is expanding the range of conversations, which are constitutive of design. It raises by the same token the question of how to truly dialogue with the machine instead of simply giving it orders. Beyond the seduction of the uncanny hybrids that they generate, this might be one of the major interests to the text-to-image AI, namely to force us to start the conversation.

So far, the type of conversation that we have with these text-to-image AI remains constrained by various factors. In addition to the limitation of current machine intelligence, one finds its difficulty with three-dimensional space, as well as our lack of knowledge as average users about the precise modality of its training. AI still tends to produce 2D images that appear even more intriguing that we don't fully understand their generation process. Confronted with these images and constantly oscillating between fascination and disillusion, we are in a position reminiscent of the public watching a magic trick in a state of mind which blurs the distinction between belief and disbelief. But the desire to have a true conversation might be by the end stronger than our flickering state of minds and the somewhat narcissistic pleasure that we experience at times facing what we ultimately consider as our creations to the point that we may be tempted to post them on Facebook or Instagram.

Returning to the Latourian approach or the dialogic one, both tend to neglect the importance of power structures. The design process bears the mark of various economic power relationships and professional hierarchies. Ultimately, an individual or a small group of people make decision. The design process is both made of thousands of these mutual adjustments between actors that Actor-Network Theory documents well, and of decisions, sometimes consensual, sometimes brutal, that interrupt the continuous fabric of mutual adjustments.

One of the consequences of the diffusion of digital tools was to reinforce the decision-making dimension. Indeed, computers generate continuous flows of geometric and technological conditions, like a moving image in which every moment represent a possible solution to the design problem, leaving to the architect the task to choose where to stop, which instance or still to select in order to finalize the project. AI will probably reinforce this dimension at various moments in the design process. At the start of this process, it will for instance propose multiple suggestions, something like prompts the diversity of which may go far beyond what traditional sketching allowed, thus calling for a selection to be made. Later in the elaboration of the project, it will enrich the content of possible solutions in ways that we have barely begun to explore.

One thing seems certain in any case: the role of humans in the design process will look more and more like that of a curator managing not only the variations theorized a decade ago by the proponents of the so-called 'non-standard' but arrays of possibilities that extend far beyond the range of references mobilized traditionally by architects. AI will also contribute to make text to image or music to image transitions look almost natural, there again calling for humans to set limits to its multimedia and synesthetic powers by making selections.

This evolution will probably present a series of very concrete professional consequences. We may not be facing the prospect of a 'death of the architect,' as Neil Leach sometimes suggests, but a drastic reduction of the numbers employed in the profession, at least in its classic definition. It is by the way striking how the professional dimension of the changes brought by the digital are rarely evoked in the literature. The age of AI-induced expanded authorship may mean a radical distribution of architectural competences and functions.

In architecture, choices impact by the end the way humans experience space and live their lives both individually and collectively. The reinforcement of their role goes with a shift in what matters in design, a move from the how to the why. Whereas the machine will play a much greater role in how the design process unfolds, the why, which conditions the solution ultimately chosen, should remain a human prerogative, at least for the foreseeable future, if only because human intelligence is embodied, unlike intelligence based on bits of information and silicon. This suggests than more than form making, design is increasingly akin to an action, a political action that is, since it involves decision regarding the way humans live their lives. Curation and politics, this polarity might define better than the traditional Vitruvian triad what authorship may mean in the new phase of the digital revolution that we have entered.

Intentionality in Architecture

Evoking authorship and the increasingly strategic character of choice leads to the difficult question of what is the goal to achieve, or what is the intention. Using the word intention is of course not neutral. It suggests a connection with the very definition of intelligence. Revealingly, the rise of artificial intelligence has been accompanied by a more and more frequent use of the expression 'architectural intelligence', in close relationship with the issue of intention in architecture. It was used as a title by Molly Wright Steenson for her 2017 magistral overview of early developments of the digital in architecture.

More recently, it has been used again by specialist of artificial intelligence Immanuel Koh for his stimulating book entitled *Artificial and Architectural Intelligence in Design*.

There are actually two ways to approach the question of the objectives that architecture assigns to itself.

One is to consider that the purpose of architecture is to address and solve problems. After having referred to Le Corbusier's famous analogy between the house and the machine, Immanuel Koh defines it as follows, I quote:

> "Architectural intelligence is that activity devoted to designing intelligent environments, and architectural intelligence is also the quality that enables an environment to function appropriately and with foresight towards any entity or entities."

From that perspective, the promises of AI are evident, from design to construction, and from the evaluation of the embodied energy of a project to the streamlining of its digital fabrication.

Now, when referring to Le Corbusier, Koh tends to forget that the architect quickly corrected himself by stating that architecture was primarily about emotion. His arch-rival for the leadership of the modernist movement, Walter Gropius, was following a very similar track when he wrote in his 1936 book *The New Architecture and the Bauhaus* that, I quote, architecture "should use the products of science as the materials of expression", and that, I quote again, "the aesthetic satisfaction of the human soul, is just as important as the material."

Now, there might be an even broader way to define what architectural intention is beyond enabling a built environment to function appropriately, even beyond emotion and aesthetic.

Architectural intention may have to do with the ambition to design environments which suggest that what humans think and do matters. For the Italian humanist Daniele Barbaro, the patron of Andrea Palladio, the task of architecture was to institute a human realm distinct from the purely natural one as well as from the domain where the gods ruled. Both nature and the gods were infinite powers. According to him, the task of architecture was to create a finite environment where humans were the measure of what surrounded them. Emotion and aesthetic were part of this project, but the latter went much further. It had an existential ambition.

Another way to grasp this approach of intention is to consider that architecture is comparable to a theater. Even when it is empty, the theater is organized in such a way that it suggests that the words pronounced on the stage have a special relevance. Architecture stages human thoughts and above all actions. It does not necessarily possess in itself a meaning. It rather suggests that human actions have a meaning and nudges them in certain directions. This nudging is what the political character of architecture means most of the time.

Understood from such a perspective, architectural intention does not imply a return to individual creativity. Collectively shaped spaces or spaces shaped by tradition rather than design in the modern sense, take Italian piazzas or vernacular architecture like the buildings of the M'zab region in Algeria, show intention.

This has of course a relation with the symbolic, but the symbolic understood in a much broader sense than the use of ornaments and allegorical devices. The symbolic should also be approached in connection to our embodied form of intelligence. The best way to understand could it very well be in reference to Adolf Loos' famous reflection on the evocative power of forms and circumstances:

> "If we find a mound six feet long and three feet wide in the forests, formed into a pyramid, shaped by a shovel, we become serious and something says, "someone lies buried here"... Now that is architecture."

No one might actually be buried here, just like no one lies under Eisenman memorial in Berlin, but there again something like an intention is conveyed.

Dimension Human Body Scale

Now, what kind of intention lies behind our intensifying interest in the use of AI in architecture? Again, it cannot be only about improving the quality of the response to the various constraints that a built environment has to respect. In that regard, it is worth remembering after Pierre Naville that automation "touches something intimate in us, a vibrant chord of our creative power, that it has to do with enchantment, with our will to power, and many other impulses that move us, starting from our unconscious."

The answers to such a question are of course multiple. What is however common to most of them is the enchantment, to use Naville term, the enchantment of moving on the both exciting and dangerous ridge that allegedly separates the human from the machine, a ridge that could very well be in reality a surface of contact and possibly a zone of blurring and even fusion. Whether we like it or not, the posthuman perspective is never very far in our excitement regarding the character both familiar and uncanny of what machines can already do.

Following Diderot, are we ultimately trying to understand better ourselves, the inhuman part of what it means to be human (I prefer the term inhuman to non-human because of its disturbing connotation), when we try so hard to substitute machines to women and men for certain tasks?

What could be the intention conveyed by the architecture produced in such conditions? Could it have to do with a new way to interrogate what matters, or even more radically, what makes sense to the mix of human and inhuman that we are?

Tell me what makes sense for you, and I will tell you who you are. In my book on the *Materiality of Architecture* I have insisted on the role played by architecture in helping us to define who we are. This may look more modest at first than Beatriz Colomina and Mark Wigley's suggestion that "Design always presents itself as serving the human but its real ambition is to redesign the human." But to help defining the human instead of aiming to forcefully redesign it may have actually to do with the most formidable power of architecture: its capacity to make us inhabit. For what is inhabiting but this opportunity to define ourselves in relation to place, built masses and the experience of space, materials, and light? In certain circumstances, architecture may aim to reform or redesign the human. Most of the time it enables it to inhabit, thus suggesting that what it thinks and does matters.

From such a perspective, the current explorations of what AI does to architecture, the fascination exerted by what Neil Leach and Matias del Campo calls its capacity for 'defamiliarization' in a recent article for *Architectural Design* may have to do ultimately with the intention to explore what it means to inhabit in an age of intelligent machines and possibly posthuman subjects.

To continue discussing what intention in architecture may mean in relation with the diffusion of AI, let's make a thought experiment, let's fast forward and imagine a time when machines are intelligent enough to produce entirely by themselves convincing plans, façades, and cross-sections, following or not received stylistic guidelines. Whether these machines have acquired some form of consciousness or not does not really matter. Let's simply assume that humans are no longer needed to make crucial choices. The machines would be sufficiently expert to be able to dispense with their input.

After Bernard Rudofsky's architecture without architects, we would be observing the development of an architecture without humans involved in its design. Would such an architectural production still carry intentions in the human sense of the term or would it be the result of a purely mechanical process of optimization?

A pessimistic take on the question would be to answer no. Humans would be actually living in settings that could be extremely well suited to their needs in the way artificial environments designed by humans and imposed on animals in zoos are adapted to the latter requirements. But this would hark back to the definition of intention as environmental optimization that I discussed earlier.

A more optimistic answer would be to observe that human intentions would still be deposited in the training datasets that these machines would have been exposed to as well as in some of the rules that they would have been obliged to follow. The human would then become in a literal way the ghost in the machine, a haunting presence behind algorithms, whereas until now it is rather the algorithms which have been haunting us. One thing is sure, with this kind of speculation, we are back to the issue of authorship that I discussed earlier.

But how, even if that was the case, would architecture evolve? Architecture is one of the ways humans have found to make sense of their own history, as testified by the role played by monuments. With due respect to Victor Hugo, the book has never really killed the monument. Architectural intention is never entirely separable from the necessity for the designer to insert her or himself in the history of mankind as it is reflected by the history of the architectural discipline, whether she/he accepts or rejects key episodes of these intertwined histories. Even the modernist dream of the tabula rasa owes something to historical considerations. Can we delegate to machines the task to curate our own history through the unavoidable evolution of the architectural discipline?

Finally, allow me an additional thought experiment, perhaps a tad more provocative. In an age of advanced, perfectly autonomous machines capable of producing good architecture, a least if we consider that good means following received architectural precedents, respecting established principles of composition, and fulfilling constraints in an arresting way, the entire built environment could finally become ruled by the architectural discipline. How not to think in this regard of the totally built Coruscant, the siege of republican and imperial regimes in Star Wars, where architecture, despite its grand character, is not much more by the end than a series of protuberances like those that texture an orange skin. The earth could finally become this total work of planning and architecture that designers have been dreaming of at least since the Renaissance, a dream the modernist version of which was castigated by Manfredo Tafuri in *Architecture and Utopia*. But wouldn't this perfect planetary environment actually sign the death of architecture as a practice that makes a difference in the world? In a totally well-designed world, no significant difference could ever be made.

Contrary to the somewhat ritualistic complaint of architects about the limitations of the recourse to their competence, a complaint heard from Vitruvius to the present, architecture feeds upon the fact that its global project is always left unfinished, unrealized. We do not only live in historical times. We live in the accumulated ruins of all the perfect worlds that planners and architects from the past failed to realize. Our territories and cities are probably better as superimposition of such failed worlds, as palimpsests, than as the result of the writing of a single period and group of designers. True intention in architecture appears always as an unfulfilled dream.

Labyrinths and Hybrids: Toward a New Aesthetics?

Let me turn to the last theme that I want to evoke, aesthetics. This is a difficult question for at least two reasons. First, there is a striking contrast between the strong visual presence of AI, between its evident aesthetic dimension and the incertitude regarding the direction that architecture might be eventually taking under its influence. Will the future development of the discipline confirm or quash the tendencies that we see at work today? The answer is far from evident.

The second reason has to do with the domination exerted presently by 2D. AI tends to produce mostly images, a tendency exacerbated by the current craze for generative AI, for text to image AI in particular. In other words, an aesthetic universe is unfolding under our eyes that is only partially architectural.

Observing the numerous AI aided productions of architects, one can observe a number of recurring features. Of course, there are not systematically present in the astoundingly varied design landscape that is taking shape under our eyes, but their recurrence gives food for thought.

Repetition and variation, sometimes oblivious of usual spatial constraints, seem to lead to a somewhat labyrinthic condition that evoke at times hallucinatory and even nightmarish ambiances. The blurring of contours and the effects produced by partial superimpositions and not entirely compatible perspective points of view add to this hallucinatory character. One is frequently reminded of German expressionists film decors or Maurits Cornelis Escher's impossible objects.

Another striking feature is hybridity. Text-to image AI have reinforced the capacity to generate hybrids of pretty much everything with everything, stone façade and feathers, balloons and gothic vaults, human figure, birds, and modernist interior spaces. Interestingly, this power of hybridization is different from the capacity to morph a thing into another that excited Greg Lynn in his early writings, beginning with the 1993 *Architectural Design* famous issue on "Folding in Architecture." Contrary to earlier morphing effects inspired by film, AI generated hybridity preserves a palpable tension between the objects and above all textures that are hybridized. The effect is often close to collage, but a collage the edges of which have begun to melt.

An ornamental impulse seems often at work, but as suggested by the term impulse, one is far from the controlled ornament of old. There again, proliferation and hybridity seem more frequently the case than the careful disposition of ornaments that was supposed to reinforce the main line of the composition. Ornament is more

1   House Made of Feathers, Midjourney V1, example of hybridity, ©SPAN (Matias del Campo, Sandra Manninger), 2022; used by permission of the authors

2   Sublime Giant Machines, contemplations on CERN, ©SPAN (Matias del Campo, Sandra Manninger), 2022; used by permission of the authors

1

textural than compositional, thus reinforcing the dreamlike overall impression conveyed by the image.

What is sometimes suggested is that we see portals leading to unknown worlds, which waiting to be discovered by the viewer. A distinctly sci-fi dimension seems at work there, a dimension that evokes novels such as Dan Simmons' Hyperion series, or TV shows like Stargate. This is related to another characteristic, the tendency to blur the limits between genres. The video games universe seems often close to some of the images produced with Midjourney.

I started my career in research by working on the late Enlightenment. So please forgive me if I make use of a comparison between then and now. History doesn't repeat itself, nor does it really stutter, but it is brims with strange resonances that may sometimes provide clues to understand what we have under our eyes.

It is tempting to evoke a number of productions of the late Enlightenment. I am thinking in particular of Piranesi's labyrinthic prisons, and of his strange and proliferating assemblages of ornaments that seem to dissolve the very possibility of received architectural compositions. There is a hallucinatory and even nightmarish character in some of his caprices that seemed echoed by the definitely strange ambiances of Jean-Jacques Lequeu projects. Lequeu is probably the ultimate virtuoso of hybridity.

Speaking of hallucination and nightmare, another well-know reference comes to the mind, that of the Swiss-born British painter Johann Heinrich Füssli and his oniric work that seems to announce some features of Surrealism. Füssli's universe is evidently linked to the rising genre of the Gothic novel with its specific blend of the horrific and the grotesque which announces some features of soon to come early Romanticism.

To end this rapid evocation by a painter may be justified by the fact that at the turn of the 18th and 19th centuries, the relation between architecture and painting intensifies to the point that the question of the image and its status is at the forefront of architectural reflections. The relation between architecture, painting and literature is also a subject of speculation at the time.

It is tempting to draw analogies between this past period and the present, and this all the more that it is also a period when computation, the machine, and its relation to subjectivity are repeatedly probed by philosophers as well as artists and writers. It is worth

1

remembering that this relation looms in the background of the marquess of Sade's novels.

Without overstating the case, let's simply remark that these late 18th-century and early 19th century productions are inseparable from a crisis of the compositional and aesthetic codes that had ruled design since the Baroque era, a crisis epitomized by Piranesi's fantastic reconstitution of the Campo Marzio, which was brilliantly analyzed by Tafuri.

Dwelling on such precedent, the question could be what is dissolving today under our eyes through the experimental use of AI generative power? Modernism is certainly not the answer, for it has been gone for quite a long time. Given the acceleration of the pace of change of society, culture, and architecture, could it be rather the dissolution of the aesthetic codes that ruled for a while digital architecture? Is this the end of the parametric expression that Patrik Schumacher tried so hard to transform into a style?

More radically, could it be the very possibility of style that is dissolving with the tsunami of AI-generated images, ambiances and atmospheres that rely on radical hybridization? Going even further, are we on the verge of a crisis of composition as profound if not more than what took place at the end of the Enlightenment with the demise of the Baroque guidelines for design.

It is probably too early to answer these questions, but they are definitely worth raising. Interestingly, speaking of the end of composition as we knew it, some of the AI generated images echoes the complexity of contemporary informal settlements with their aggregation of volumes that are profoundly non compositional. Are we on the eve of a convergence between the architecture without and with architects?

Now, it is worth remembering that the late 18th-century architectural and artistic turmoil was followed by the radical return to order of neo-classicism. Will such a return to the normative take place eventually? The question has been with us since the beginning of the digital age with its capacity to periodically disrupt established codes.

Meanwhile, all sorts of exciting perspectives are beginning to emerge from the mist of unbridled novelty. A new albeit twisted connection between imagery, pleasure, and knowledge ranks among these perspectives. I mentioned earlier the ornamental dimension at work in so many of AI inspired experiments. The relation between pleasure and knowledge was a characteristic of traditional ornament that we seem to have recaptured.

There is also the possibility of new sublime that goes well with the somewhat scarry character of many contemporary visual and spatial experiments with AI. Like the sublime of old, this new sublime is based on the blurring between the human and the non-human, or the human and the inhuman. As I said earlier, I prefer this term to the tepid non human expression. Contrary to the 18th century sublime, which dealt mostly with the relation between subjectivity and nature, we are now dealing with machines and wondering whether we need them to hallucinate as if their effect was comparable to that of some new and powerful substance. Whereas the sublimity of old was about the inhuman pureness of the natural, hence the fascination for pristine snowy mountains or unsullied oceans, the sublimity of labyrinthic hybridity tells us something about the impossibility to return to a state of pureness, an impossibility linked to the environmental crisis, but not only. We humans are not even sure to be truly humans.

Do we now need machines to dream? How not to think of Philip K. Dick famous novel *Do Androids Dream of Electric Sheep?* Are we actually androids? The vertigo that we experience evokes there again past episodes, as if this question had been with us for far longer than what we usually imagine. It is worth remembering that before Amazon, the Mechanical Turk was an 18th century fake automaton supposed to play chess better than most humans. The Turk fascinated crowds and raised the question of the definition and limits of the human. Confronted with what looked like a machine which could think, spectators were torn between the desire to be looking at a true mechanical wonder and the impulse to look for a hidden hoax. Multiple explanations were brought forward before the illusion was finally revealed.

Have we ever abandoned this inner dilemma? For sure, we are entering a new age in the complex history of ourselves as both alike and different from machines that we both design and find mysterious. How will we live in such context? For architects, the question should ultimately remain how to inhabit in an age of renewed wonder and puzzlement.

NOTES

1. Sébastien Chaillou, Artificial Intelligence and Architecture. From Research to Practice (Basel, Berlin, Boston: Birkhäuser, 2022).

2. Pierre Naville, Vers l'Automatisme Social (Paris: Gallimard, 1963).

3. Wolf dPrix, Karolin Schmidbaur, Daniel Bolojan, Efilena Baseta, "The Legacy Sketch Machine: From Artificial to Architectural Intelligence," Architectural Design, "Machine Hallucination: Architecture and Artificial Intelligence," n° 3, vol. 92, 2022, p. 14-21.

4. Antoine Picon, Ornament: The Politics of Architecture and Technology (Chichester: Wiley, 2013); Peter Galison, Image & logic: A material culture of microphysics (Chicago: The University of Chicago Press).

5. See Françoise Choay, introduction to the French translation by François Choay and Pierre Caye of Leon Battista Alberti, *L'Art d'Édifier* (Paris: Le Seuil, 2004), pages 9-39, and page 23 in particular.

KEYNOTE EVENT

Hybrids & Haecceities Epilogue Panel

Ferda Kolatan
University of Pennsylvania

Andrew Kudless
University of Houston

Antoine Picon
Harvard GSD

Jenny Sabin
Cornell University

Hina Jamelle
University of Pennsylvania
Moderator

NEW TECHNOLOGIES AND THEIR EFFECT ON ARCHITECTURAL DESIGN AND CULTURE

EPILOGUE PRESENTATIONS & PANEL DISCUSSION
Hina Jamelle

Following Antoine Picon's keynote lecture of ACADIA's *Hybrids and Haecetties* symposium at the University of Pennsylvania titled "Artificial Intelligence and the Future of Architectural Design," the epilogue panel participants convened to discuss the provocative themes raised by the keynote. Participants included: **Antoine Picon**, G. Ware Travelstead Professor of the History of Architecture and Technology and Director of Research at the Harvard Graduate School of Design; **Jenny Sabin**, Arthur L. and Isabel B. Wiesenberger Professor in Architecture and Associate Dean for Design at Cornell College of Architecture, Art, and Planning; **Ferda Kolatan**, Associate Professor at the University of Pennsylvania Weitzman School of Design and Founding Director of su11; **Andrew Kudless,** Professor of Architecture at the University of Houston where he holds the Bill Kendall Memorial Endowed Professorship and is the Director of the Construction Robotics and Fabrication Technologies (CRAFT) Lab, and **Hina Jamelle,** Associate Professor of Practice at the University of Pennsylvania Weitzman School of Design and Director of Contemporary Architecture Practice. New York and Shanghai.

Generative AI was top of mind during discussion, as it has been predicted to change every industry that requires humans to create original work, from gaming to advertising to law and more. The panel conversation covered several topics within the sphere of AI including intentionality, authorship, bias, and new approaches to architectural design and practice.

The discussion began by querying intentionality in the use of generative AI. Within architectural design, recent experiments with AI-powered text-to-image diffusion models like DALL-E, Craiyon, Stable Diffusion, and Midjourney indicate a fascination with dramatic juxtapositions that elicit viewer responses. Some of these responses to the technology emerge from a disjunction between form and aesthetic predispositions in two-dimensional outputs. Whether or not they lead to an aesthetics of radical hybridization was investigated by the panel. Contextual insights included the long history of collage and juxtaposition as creative tools for architects and the fact that AI simply makes techniques that were already in place faster and easier to access. What was considered novel is the estranged hybrid condition of images that lie in a position between the past, present, and a fictitious future. Another important point was the need for a clear imperative within the artistic project itself as a prerequisite to producing meaningful work with generative AI. A comparison was made between the technological black box of diffusion models, where text is transformed into images, and that of the automated processes undergirding traditional architecture-specific software like Revit. In both cases, intentionality serves as the mediating ground for a technological dialectic where biases of the designer and the technology are at play. The entry barriers to that dialectic—in terms of technical knowledge and skills—suggest a politics of access.

The question of authorship was also at the forefront of the discussion. Generative AI tools are powered by countless texts and images scraped from the internet. AI is using images of works made by specific architects and images taken by photographers without acknowledging them in any way. Questions posed to the panel included how this should be considered and what the visible marks of authorship in generative AI are. How should architecture approach this conundrum of authorship, and are we moving towards new forms of collective authorship that defy the current design culture? The panel discussed these questions at length, including the nature of source material and data sets themselves. Generative AI tools have been made so generic and accessible that linkages are being made to source material across fields and disciplines from art to engineering to design and beyond. People globally have been putting data out into the world for decades now, and large swaths of historic human creative production have been digitized. On one optimistic level, generative AI tools allow people to reclaim and use that data for their own new and novel purposes. Crafting, steering, and helping to curate data sets in order to train AI models and processes in sync with architectural aspirations and principles is a meaningful role the discipline can play. The panel discussed the need to go beyond a pure fascination with the seductive nature of AI-generated images to clearly analyze what is possible with these tools and take an authored stance. This then puts the designer in the role of a curator learning to talk to the machine and understand its inner workings.

An important conversation ensued regarding bias embedded in training data in intentional, unintentional, and unpredictable ways. At a global level, there are increasing concerns regarding misinformation, misuse, and unknown negative ramifications for publicly releasing these tools. Data sets have an implicit bias towards those crafting them, and that has resulted in a first-world slant that glosses over and/or excludes many cultures and histories. The worry is that there will be a deeper compounding and acceleration of existing inequities across design and visual culture. The notion of a two-way interrogation of bias between the designer and the algorithm was discussed as a critical working method. When language and image results are used to distill biases of the data set and oneself, a creative third space opens up to negotiate and move between and beyond the two.

The call of the conference, *Hybrids and Haecceities*, posed new approaches to theory and practice, to design and construction that transgress disciplinary boundaries. While the construction industry is notoriously slow to innovate, the panel inquired about what new modes of architectural practice technologies- including AI can help discern and develop. The potential of AI is exciting to architects as well as the general public, opening up new possibilities for public dialogue for architecture. At present, the public dialogue must deal with and negotiate through a tension between images and materiality, which challenges architects and designers to actually deliver on the promise of what is being imagined. There is a question of whether this leads to more involvement in the role of the architect, which would be considered positive, or a significant change in building practices and methods. The panelists also emphasized the rich nature of working with data in an AI-powered world in ways that defy image-based approaches or create subsurface connections to images and image processing. Projects working with environmental phenomena like airflow or thermal comfort for not just optimization or simulation but to create new effects and experiences can create new relationships to materiality and pose a shift in questions around which data sets are relevant and why. AI and machine learning research with a focus on health and wellness such as that led by Daniel McDuff for Microsoft, track relationships between the body, facial patterns, and environmental factors. Keying this type of research into architecture's aesthetic and material design approaches is a potential launch point for new architecture and architectural feedback.

Finally, an important point at the conclusion of the panel was that above and beyond specific techniques and methodologies, the architectural discipline must reckon with the implications of generative AI for its own modality, its own way of being. There is great optimism and excitement about collaborative and collective modes of work that call more for a shift in attitudes rather than a shift in technologies or techniques. We still live in an economy of superstars with roots in Modernism and the notion of the genius artist. It will require reprogramming ourselves and changing our behaviors to fully grapple with the productive modality implicit in the collective nature of data and the generative methods that have become the hallmarks of AI.

Awards

AWARDS EVENT

Design Excellence Award

Verify in Field

J. Meejin Yoon
and Eric Höweler
Höweler + Yoon Architecture

We use the term verify in field as a lens through which to view the agency of architecture at different scales; to examine the techniques and processes that both constrain and enable us to create, make, and build today. VIF not only acknowledges the gap between the idealized realm of conception and the messy and material realm of construction but allows the gap to be a productive space of inquiry. Verification involves double-checking—measure twice, cut once. Verification is about curiosity and expanding knowledge. It requires an openness to inputs. It is investigative. It acknowledges contingencies and externalities, as well as expertise that may be located elsewhere, in another discipline, or outside the realm of the experience of the designer. Verification entails doubt. Verification is essential to the translation of design intent into built reality.

Verify in Field (VIF), as a notational convention indicates that some information on an architectural drawing is incomplete, contingent, and subject to unknown conditions in the field. VIF highlights the gap between design intent and built reality. The term acknowledges both the disciplinary investment in the instruments of design and the specific processes and protocols that inflect, alter, and modify design intent. Verification may refer to a field dimension, an alignment, or survey point. It may also refer to a process of interaction and feedback where information and expertise are incorporated into the design process in ways that adapt, modify, or recast the original design intent. While VIF can be understood as architecture's fine print, the notation speaks volumes about the discipline and the profession—and, our agency to act on the physical world.

Likewise, we see the field as necessarily both territorial and operational. The field of construction and craft, of discourse and debate, or of political and environmental forces that act on the built environment. While VIF technically addresses the physical conditions of construction and shifts responsibility for executing intent from the

1   Collier Memorial, Courtesy of Höweler + Yoon, Photographer Iwan Baan

architect to the contractor, we employ the term more expansively to frame our process and projects. Here we examine the term as an ethos—one that develops means for more agency, responsibility, and engagement.

If our early work argued for media as material, the last decade of our work examines material after media. While the construction industry is historically slow to embrace innovation, there have been significant transformations on many digital design and fabrication fronts. Design workflows are now entirely mediated by digital technologies from early concept through prototyping, coordination, and construction. The contemporary design workflow begins as a digital model, it is then engineered via computational software, coordinated on a cloud-based building information model, transmitted to the fabricator via file transfer platforms, translated from a 3D model to a scripted tool path, milled with robotic equipment, installed with digital surveying equipment, documented and post-produced, and shared on social media. Indeed, given our experience with the impact of the digital, and in our contemplation of the post-digital, we ask ourselves: what about architecture is not media, or mediated?

At the same time, material—and the way we conceptualize material—is also rapidly evolving. Thinking materially today is less about understanding material as having essential qualities that produce inevitable correspondences between, say, bricks and arches. Rather, it is more about understanding material qualities and potential—strength and durability, embodied energy, their upstream and downstream impacts, as well as new composites and hybrids. Material thinking in design now is a matter of expanding and recasting, using digital tools to work with existing materials in new ways, and finding productive uses for new materials. Material thinking today is thinking material after media.

The world today is more complex, and the contexts in which we work are more saturated with information and disinformation, challenging us to speculate on how to navigate contemporary fields and rethink how we act on the world. Verification has also taken on new relevance. Not only because there are new digital tools that enable simulation, analysis, and verification, but because these digital tools enable us to doctor, modify, fake, and disseminate misinformation. The cold war quip "trust, but verify" in the context of nuclear detente assumed that truth could be verified. There either were or were not missiles in the silos. Fifty years later, a common understanding of our shared reality, a shared understanding of truth, and the articulation of common values seem remote. The age of Post-truth coincides with Big Data and the Information Age. In theory, we have never known more. And yet in practice, the need for verification, systems of feedback, and techniques for translation, has never seemed more urgent.

Höweler + Yoon Architecture is a creative studio committed to transformational impact in the built environment and design excellence across multiple scales. Led by Eric Höweler and J. Meejin Yoon, the studio has built a reputation for work that is research-driven, socially engaged, and formally innovative. Our work asks how design can produce a sense of place, create sustainable solutions, bring communities together, and support institutional missions. Their projects ask how design fits within contemporary culture, how it can affect behavioral and social norms, and how it can produce a sense of place or create environmental awareness. Projects range from cultural and institutional buildings, mixed-use residential and commercial buildings, to public spaces, interactive environments, and research projects. Specializing in special projects, Höweler + Yoon is a dedicated group of 25+ architects, designers, and researchers.

NOTES

The ACADIA Design Excellence Award is given by ACADIA's Board of Directors to exceptional architects, designers, and researchers who have made significant, innovative, and impactful contributions to the fields of architecture and computational design.

Essay adapted from Höweler and Yoon's book, *Verify in Field: Projects and Conversations, Höweler and Yoon*. Versions of this text have appeared in other publications.

2   212 Stuart Street, Courtesy of Höweler + Yoon, Photographer Chuck Choi

3   Moongate Bridge, Courtesy of Höweler + Yoon, Photographer Shrimp Studio

4   University of Virginia Memorial to Enslaved Laborers, Courtesy of Höweler + Yoon

AWARDS EVENT

Society Award for Leadership

Community

Jason Kelly Johnson
California College of the Arts
and Futureforms

For all of us, the last two years have been filled with immense stress and uncertainty. After a period dominated by the pandemic, it is incredible to be back together in person. There have been many challenges and setbacks in terms of family, health, career, and projects, but we have managed to persevere, reconnect and rebuild our community.

First, I want to acknowledge and thank the conference Chairs and hosts of this year's conference. After having to postpone the conference due to the pandemic two years ago, your perseverance has paid off. The ACADIA community is incredibly thankful for your team's work to produce an amazing 2022 conference.

Second, I would like to thank the ACADIA Board of Directors. The board does an immense amount of behind-the-scenes work to keep this community thriving even during the shutdown. It is an honor to be recognized by a group of people I respect and value so deeply. I have gained so much from this community over the years.

The title of this talk is simply "Community." I want first to share a few thoughts about ACADIA, explore how we got here, and think a little about this community's future. I will share a few anecdotes about what has compelled me to contribute over the years and how it, in turn, has inspired my creative practice. It has been a give-and-take. I have given a lot of time to the ACADIA community over the years, but I have gained much more than I have given. I believe it has been more than worth it. This conference is certainly a testament to that.

It was 40 years ago that my father brought home an Apple II+ for my brother and me to use, then in 1988, when my parents moved from Canada to the US, they appeased me with with an Apple 2GS (the GS was an acronym for "Graphics and Sound") for my birthday. It had a mouse and 1MB of memory, and a cool drawing program called Paintworks Plus. This was a transformative moment for me.

It was fifteen years ago that I attended my first ACADIA conference and presented a project called Vivisys, and ten years ago that I served as a conference chair (with Kyle Steinfeld and Mark Cabrinha) for the San Francisco conference hosted by CCA (where I still teach today), and five years ago that I was serving as ACADIA's President.

During this period, I have witnessed this community grow and evolve; it has survived various crises, including hosting virtual conferences during the global pandemic. Now ACADIA is over 40 years old, and I believe the community is stronger than ever.

How did we get here?

Founded in 1981, ACADIA was originally conceived as a research organization focused on emerging "Computer-Aided Design," aka CAD software tools. Over time it mutated to include a broader range of topics - virtual reality, digital fabrication, and parametric design. Today the community continues to be at the forefront of engaging emerging topics - AI, robotics, cutting-edge material science, and much more.

Layered within these interests, our community is also now at the forefront of exploring the historical and theoretical implications of computation design and AI. As the world is being remade before our eyes, it is taking a leadership role in using computation to engage ecology, sustainability, and climate change. ACADIA's board and its members are also leading conversations regarding diversity, equity, and inclusion within the field. ACADIA's Cultural History Project is a testament to that. In this regard - this community is not only at the forefront of computational design research, but it is pioneering conversations that are now central to society.

Each year the ACADIA community distills these activities into publications and presentations of peer-reviewed papers and projects. In fact, ACADIA remains the most selective and prestigious peer-reviewed conference of its kind in the world. It hosts workshops, sponsors an exhibition, and cultivates an online community through its website, job listings, video, and social media channels. It uses the bulk of its sponsorship funds to support and celebrate emerging research; it awards student scholarships and discounts to attend the conference. It continues to promote the conference through media partnerships. These are many of the things that I have helped contribute to building while I was President and a part of the ACADIA Board, and I am excited to see many of them continue to thrive.

My own involvement with ACADIA began around 2004 or so. I was splitting my time teaching between UVA and UPenn. One of our faculty was an early member of ACADIA. He walked into my office one day as I was using 3d modeling software to rotate a little hobby servo motor. He said something like, "Oh wow - that is really cool - you need to share that at ACADIA!" I had no idea what he was talking about. Before social media - that was how it happened. It was very much a word-of-mouth type of thing.

The following year I was fortunate to have a project accepted for presentation at the conference. The biggest takeaway or memory of that conference was something I recorded in big BOLD text in my sketchbook: "Do you want to be a user or a creator?" "To be a creator, you need to get under the hood, you need to peel back the layers to understand the origin of things, but also understand whose shoulders you are standing on ...". I do not remember if this was a direct quote or a personal takeaway, but it was clear that the conference profoundly affected the way I was thinking about my creative and technical practice.

The quote runs parallel with something I discovered Steve Jobs had written in an email, ironically to himself, in 2010. I'll paraphrase what he said:

I speak a language I did not invent or refine.

I did not discover the mathematics I use.

I did not invent the transistor, the microprocessor, object-oriented programming, or most the technology I work with.

I love and admire my species, living and dead, and am totally dependent on them for my life and well-being.

So, as I now reflect on my own work and contributions to this community - I feel the need to add humbly:

I did not invent C#, Arduino, Visual Studio or Python.

I did not develop CATIA, Revit, Rhino, or Grasshopper, but I rely on them for my research, teaching, and professional practice.

I honestly have a limited understanding of how AI, Machine Vision, or Dali and Midjourney really work. Yet I am inspired, thrilled - and also a little terrified - by their creative potential.

When I needed help programming our Kuka, I posted questions to a forum and built upon examples I found or settings provided to me by others.

I have produced some original creative work and scholarship, but in reality, much of it would not have been possible without the pioneering work this community has shared and continues to share today.

I am supported and inspired by the work of this community, and without it, I would have accomplished very little.

Thank you, and I look forward to being a part of this community for many years to come.

Orbital is a contemporary garden folly, exploring geometric and material exuberance. It evokes organic forms found in nature, but also giant robots and futuristic space vehicles. The structure is composed of three coiled legs that spiral towards the sky. The exterior surface is defined by stainless steel origami skins, while the interior space is wrapped by a vortex of colorful tactile shingles. The project's dynamic form evokes an era of rapid change and uncertainty, while also inspiring curiosity and playful interaction.

ACKNOWLEDGEMENTS

Project Name: Orbital
Date: 2021
Location: Mission Bay, San Francisco, CA
Dimensions: 24' x 12' x 34'
Materials: Stainless steel, aluminum
Lead Artists: Jason Kelly Johnson & Nataly Gattegno (Futureforms)
Artist Team: Jason Kelly Johnson, Nataly Gattegno with Carlos Sabogal, Brian McKinney, Clayton Williams, Natalie Abbott, Chris Leo, Valerie Tse; Assisted by Lee Marom, Ki Schmidt, Sam Higgwe; Structural Engineer: Arup (Nick Sherrow-Groves, Lead); Fabrication Team: Futureforms (San Francisco, CA), Olson Steel (San Leandro, CA), Seaport Stainless (Richmond, CA), Standard Sheet Metal; Skin Installers: Pacific Erectors, Sheedy Crane Co.; Aluminum Shingles: Neal Feay (Santa Barbara); Art Consultant: Dorka Keehn; Select Photography: Matthew Millman

Jason Kelly Johnson is a California based artist, designer, and professor of architecture. Jason served as the chair of the 2012 ACADIA conference "Synthetic Digital Ecologies." He then served as a member of the ACADIA Board of Directors from 2013-20, serving as President in 2016 and 2017. Jason has produced a range of award-winning projects exploring the intersections of art and design with public space, advanced fabrication technologies, robotics, and computation. Recent projects have included fine art objects, furniture and lighting fixtures, art pavilions and sculptural shade canopies, as well as large scale urban art installations and art master plan consulting. Jason was born and raised in Canada. He received his Master of Architecture degree from Princeton University, and his Bachelor of Science in Architecture degree from the University of Virginia. Johnson is currently a tenured Professor at the California College of the Arts (CCA) in San Francisco, California.

NOTES

This award recognizes extraordinary contributions and service to the ACADIA community. The abbreviated text included here was transcribed from Prof. Johnson's acceptance speech at the ACADIA 2022 conference.

1 *Orbital* from above (Image courtesy of Futureforms)

2 *Orbital* streetview (Image courtesy of Futureforms)

3 View from below (Image courtesy of Futureforms)

4 *Orbital* detail photographs (Image courtesy of Futureforms)

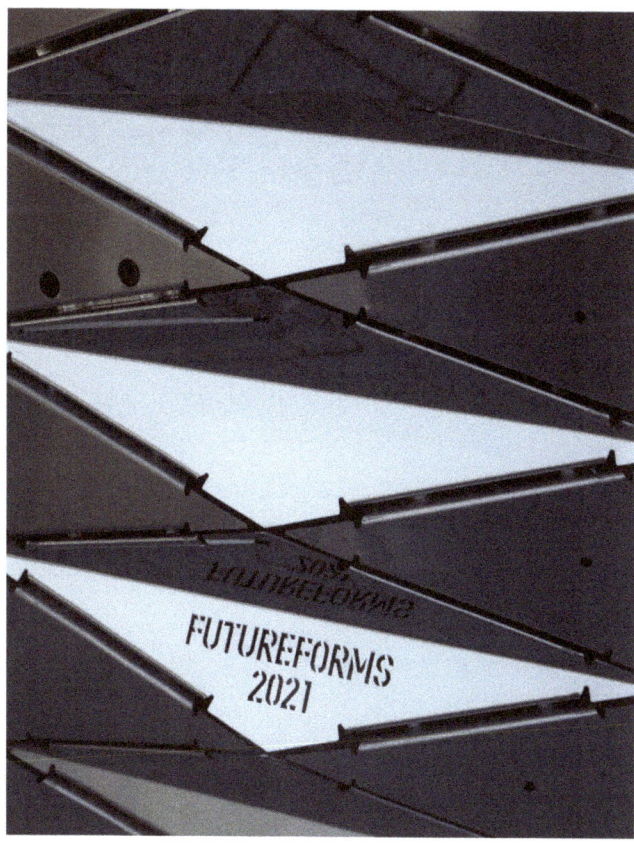

5 *Orbital* detail photographs (Image courtesy of Futureforms)

AWARDS EVENT

Teaching Award of Excellence

Computational Design in Architecture

Rajaa Issa
Robert McNeel & Associates

Computational Design in architecture is a multidisciplinary field that is deeply rooted in science and technology. Building its tools and workflows requires specialized training, but also an intimate understanding of the creative design process. Promoting wider adoption of computational design methods needs a community of users who are well-versed in digital technologies and are able to utilize, criticize and influence the next generation of development.

Early in my career, I retrained from an architect to a software engineer, and worked to build digital design tools and workflows. I witnessed a generation of talented designers struggling with the analytical skills needed to bridge the gap between intuitive and algorithmic design. My teaching philosophy acknowledges that a solid foundation in mathematics and algorithmic thinking is necessary for students to build their computational design skills and develop the confidence to instantiate their design intent using the digital medium.

The first challenge to teaching analytical subjects came with the scarcity of resources that were suitable for creative designers. In the course of my work developing professional architectural tools, and having formally studied both architecture and computer science, I was well-situated to curate the essential material needed, and to present it in a format that was intuitive to designers. To this end, I developed a series of essential references that covered mathematics, algorithms, and programming basics. These materials were made openly accessible online and became widely used by students and professionals alike. They later became part of the standard material referenced by educators in schools of architecture around the world.

Teaching the science and technology behind computational design became my subsequent passion. In many programs, the students are exposed to a fragmented set of tools and technologies without learning the underlying principles that motivate their integration into the design process. Although these resources are still important,

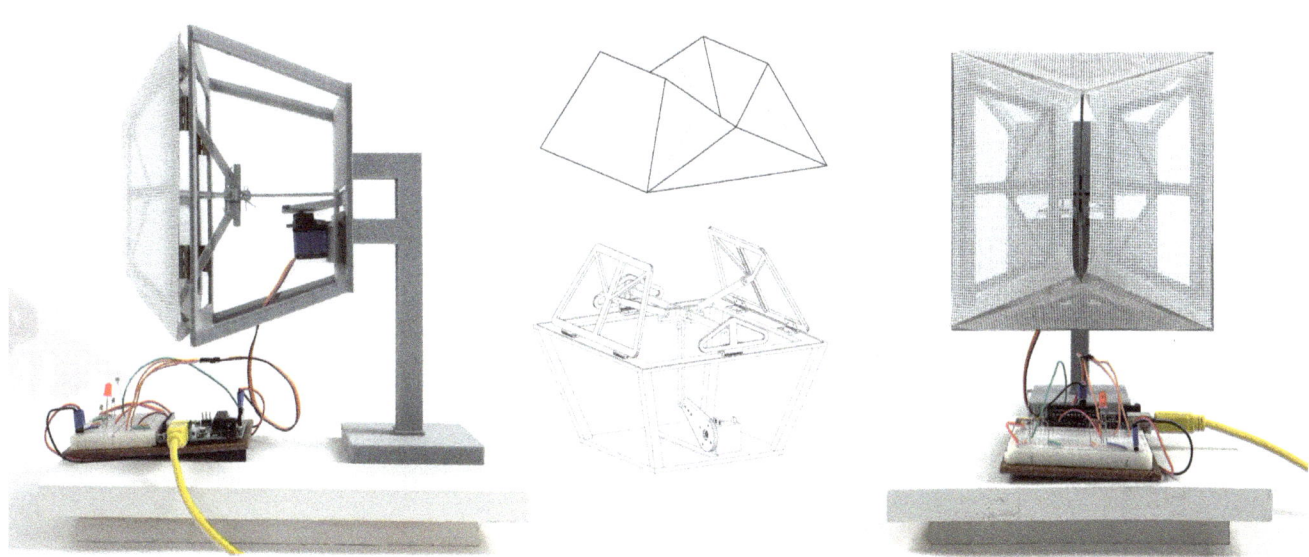

1   CDM III, NewSchool of Architecture and Design: Prototype for kinetic facade system (Student: Barrak Darweesh, Fifth Year Undergraduate, 2016).

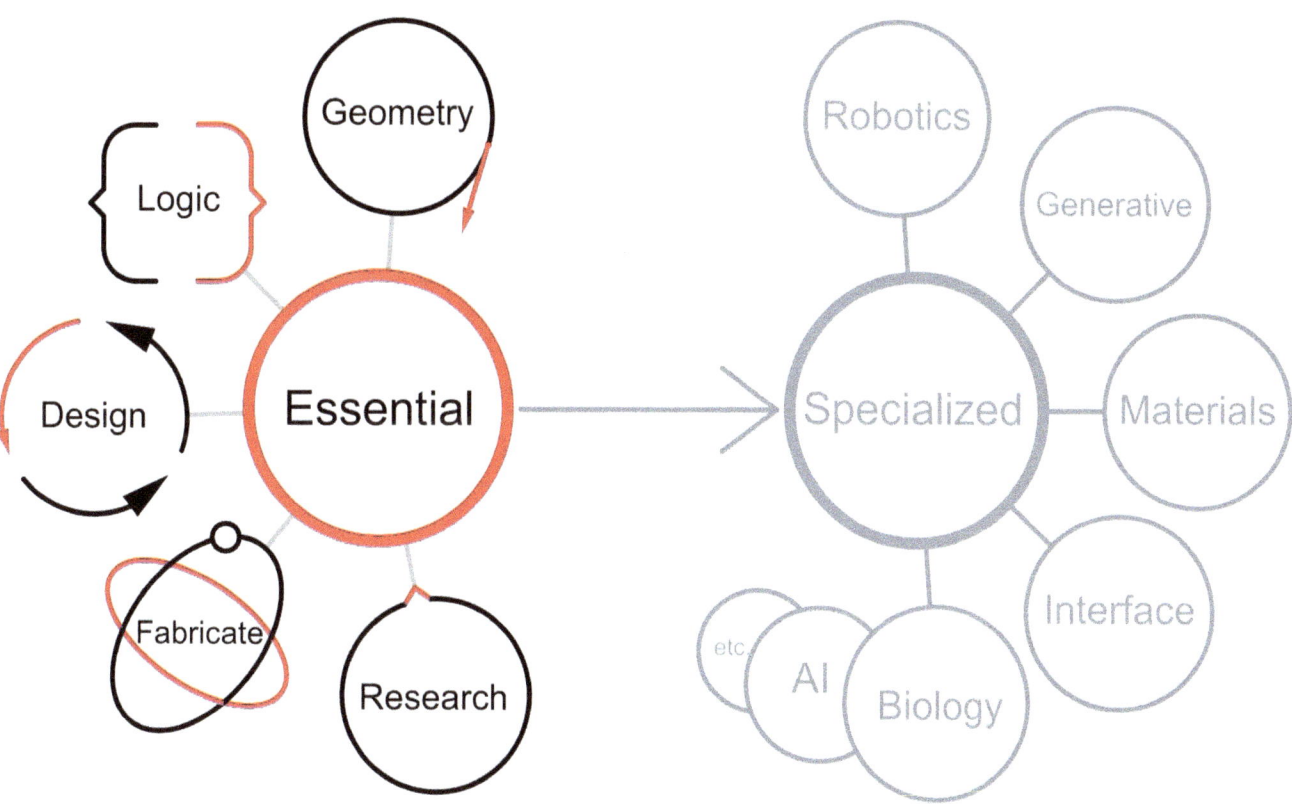

2   Core subjects of Essential Education in Computational Design (EECD): 1) Geometry and vector mathematics, 2) Algorithms & data structures, 3) Design cycles from concept to analysis to optimization, 4) Digital to physical modeling cycles, and 5) Theory and scientific research.

students with strong analytical training learn new tools faster and are better able to keep up with the fast pace of industry. While tool-centric curricula often become obsolete with each new generation of development, foundational analytical curricula remain relevant because they aim to teach the universal intuitions of computational design. To this day, many design programs do not incorporate a rigorous instructional framework to teach core principles of math and algorithmic design. Students are left to learn the bulk of the foundational material independently, which raises the barrier to entry and limits the talent pool in the field.

During my work at NewSchool of Architecture and Design, supported by their willingness to experiment with new teaching methodologies, I established a program that identified five core subjects to provide the *Essential Education for Computational Design (EECD)*. The program was organized in the form of a vertical series of classes extending over one academic year. It separated science studies from creative design activities, and covered a wide range of tools, technologies, and workflows. It aspired to achieve the difficult balance between acquiring fluency in the digital medium while developing the vision to

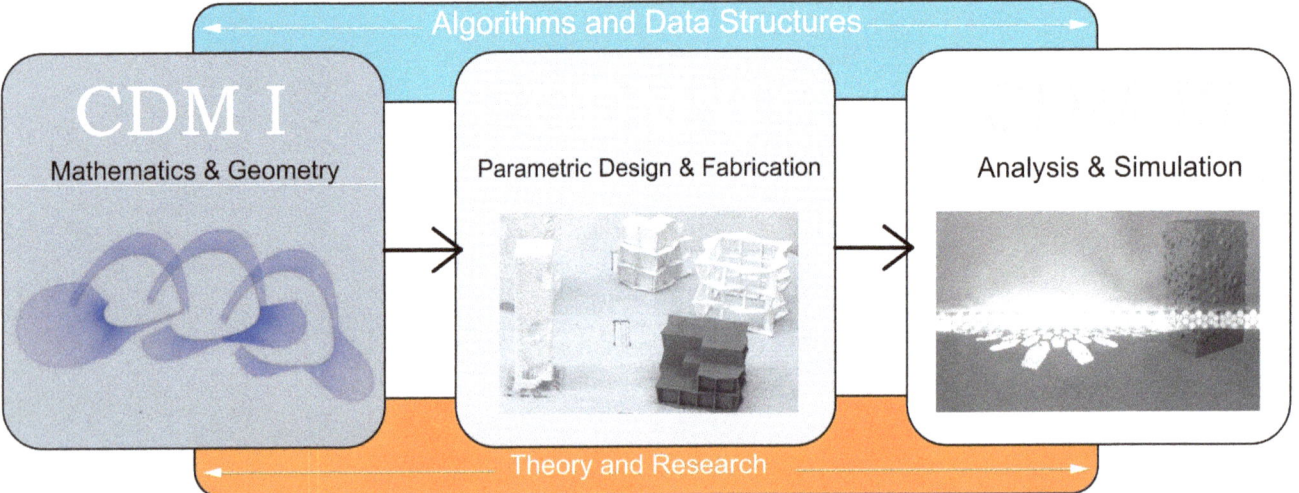

3   Program structure and flow as implemented in NewSchool of Architectural Design: **Computational Design Methodology (CDM)** classes were offered in a sequence of three classes over one academic year. **CDM I:** Knowledge base of math and geometry; **CDM II:** Parametric design and digital fabrication; **CDM III:** Design analysis and optimization. Algorithms and data structures start with the basics, then increase in complexity with time. The study of literature and research were integrated across all classes (Images from graduate students work: Steven Hansen and Ryan Stangl).

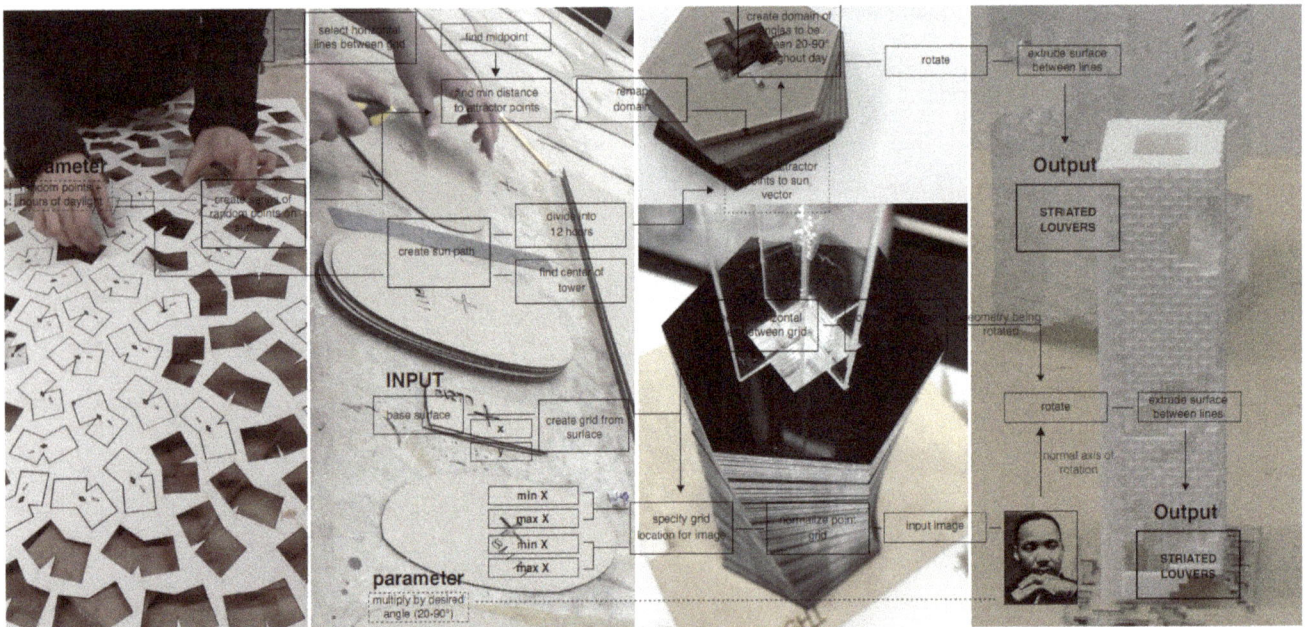

4   Computational Design Methodology II at NewSchool of Architecture and Design: Multiple cycles using different fabrication techniques to develop design concepts from simple mass to a fully rationalized adaptive parametric solution (Images from students work at Newschool: Ryan Conner, Cortney Fronberg, Moises Robles, Sanbir Sidhu, Heiarii Li Cheng, Steven Hansen, Tony Salamone, 2014-2016).

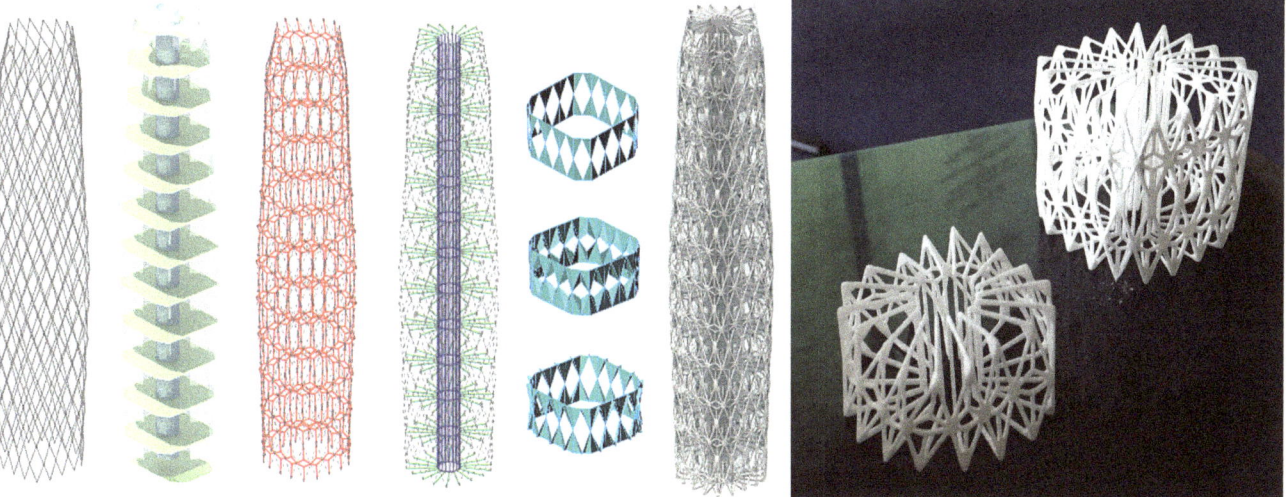

5   CDM III, NewSchool of Architecture and Design: After mastering all knowledge-based topics and parametric design and fabrication, students were able to engage in exploratory studies to examine additional constraints such as structure, performance, and others. This is an example of a comparative structural analysis using different module designs and parametric structural analysis tools (Student: Yangi Situ, Third Year Undergraduate, 2015).

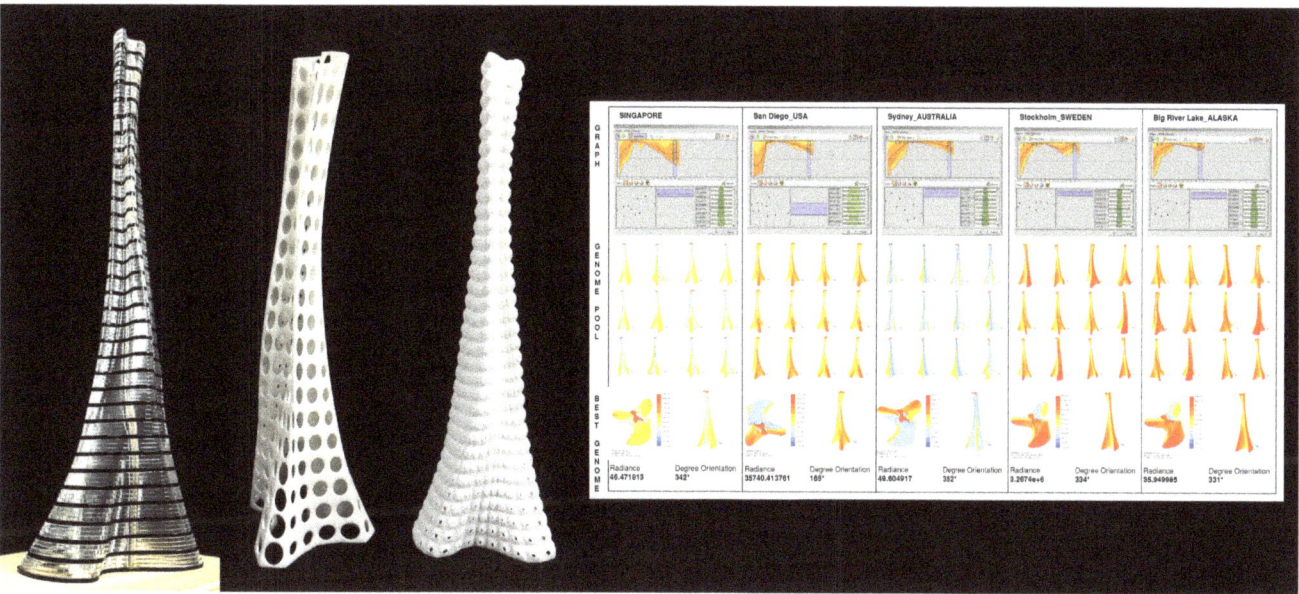

6   CDM II, NewSchool of Architecture and Design: Initial building mass (left) evolved into fully rationalized solutions (center) as an example on how students' digital ideas matured when transitioning between digital and physical environments and the use of advanced tools to analyze and rationalize form. Once a form was finalized, students could learn optimization workflows through the selection and investigation of site-specific performance criteria (right) (Student: Heiarii Li Cheng, Third Year Undergraduate, 2016).

see beyond the rigid frameworks and assumptions of any one tool or technology. Student works evolved with time, reflecting their level of maturity and mastery of concepts. Toward the end of the program, the students were able to conduct complex multi-constrained analysis and optimization from the ground up using a variety of tools and technologies. Developing analytical skills was grounded in the study of theory and history of computational design, where the students researched its workflows and application in practice. Despite variations in their backgrounds, all students in the program were able to make the transition to building complex algorithms and developing critical understanding of the tools and workflows.

While it is useful to teach how to use computational methods, it is more important to actively engage students in questioning *why* and *when* they should be used, to help develop critical views in regards to their applications, limitations and potentials. My experience has shown that without strong analytical skills, computational designers struggle to develop fluency in algorithmic thinking affecting their ability to take advantage of digital technologies. My pedagogical approach to teaching computational design systematically imparts an analytical mindset while respecting and leveraging students' contextual background in design. It strongly advocates for the inclusion of science-based programs and research in architectural education to ensure that the next generation can take full advantage of emerging digital design technologies, and drive its future development.

Rajaa Issa is a developer, researcher, and educator in the field of computational design in architecture. As a Software Developer at Robert McNeel & Associates, she builds solid modeling and surface editing tools to support the architectural design and fabrication processes. She is the developer of PanelingTools, one of the earliest tools for concept paneling and complex surface rationalization. She is Adjunct Professor at the NewSchool of Architecture and Design, where she established a new program in computational design methodologies. Her pedagogical method is centered around building strong foundational knowledge in mathematics, algorithms, and fabrication to bridge the gap between intuitive and algorithmic thinking among creative designers. She has authored a series of publications on parametric and computational design essentials, which are popular among both students and professionals. As a firm believer in inclusion and the accessibility of knowledge, she has made all of her work openly accessible.

NOTES

This award recognizes innovative teaching in the field of digital design in architecture. Teaching approaches that can be adopted by other educators are recognized in particular.

AWARDS EVENT

Innovative Research Award of Excellence

Soft Systems: Crafting an Architecture

Felecia Ann Davis
Penn State University
SOFTLAB@PSU

In my lab SOFTLAB@PSU in the Stuckeman Center for Design Computing or the SCDC at Penn State University, we develop computational textiles, which are textiles that respond to their environment via programming, embedded sensors, and electronics, in addition to using the responses from the natural properties of fibers where the response is used by the designer to communicate some type of information to people. Communication, for our purposes, means the sending of an intentional message or information through an aesthetic expression (Frijda 1982, 112). My work is to relook at the role of textiles in building, not just as a way to achieve lightweight buildings but understand how we can use the fibers of the textile itself to communicate and how that impacts architecture.

We believe material itself shapes a way of thinking and working that leads to considering architecture and its practices in a different way. As Tim Ingold and J.J. Gibson and many others show and argue, working with materials changes our brains and change us (Ingold and Gibson 1983). In other words, the affordances of any one material shapes a way of thinking about the world and can shape a world. Materials change us as we change them.

The work we have produced in SOFTLAB@PSU shows some of the entanglements, so to speak of working with textiles, computation and soft materials. The lab has been working towards an understanding of a soft system, which thus far to us, is a system that accounts for and accommodates a technological system that can register design in relation to specific bodies in specific places engaging specific social, cultural and political constructions.

In the three projects mentioned below I pull out three characteristics or qualities that we have understood to be indicative of a soft system and are issues that we think about in SOFTLAB@PSU.

THE FABRICATING NETWORKS QUILT AND THE BLACK FLOWER ANTENNA

These two projects, *Fabricating Networks Quilt* and the *Black Flower Antenna* were commissioned by MoMA for the 2021 show *Reconstructions: Blackness and Architecture in America* exhibition curated by Sean Anderson and Mabel Wilson with assistance from Ariel Dionne Krosnick and Anna Burkhardt at MoMA.

Open System & Reaching Out_The Fabricating Networks Quilt, is about electronic transmissions. The quilt was a way to understand the research we were finding on the Hill District, a Black community in Pittsburgh. It was not intended to be the architecture of the project. Rather it was used as a method of questioning, operating as a metaphor or heuristic device for making architecture in a context that acknowledges an overlaid invisible urbanity, a hybrid cyber world that we inhabit but do not often recognize. The interchangeable panels of the quilt starts a conversation with people about making and is a repository of history. When touched, the copper in each panel activates a speaker that tells a story of what happened in the panel. The quilt was and is meant to travel outside the gallery and connect with people who were not traditionally part of the museum's spaces. Because we could not do this during the height of the pandemic now we would like to take this quilt on the road to the Hill District and continue to hear and make new photographs and stories with the residents on the Hill. To us storytelling and conversation is an important part of making architecture, as important as bricks and mortar, especially in communities of color. The softness in this material system is how we defined the concept of construction by communication which is collaborative and incrementally built. The Lilypad designed by Leah Buechley a designer and computer scientist at University of New Mexico, permits the fashioning of do it yourself work. This is supported by a network of on online tutorials and programs or an open-source system. This network is also part of what we understand to be the material. For many of the works we do, the ability to access and use these in expensive, do it yourself components is the difference between connecting or not (Figures 1, 2).

Acknowledging Invisible Cyber Architecture That Shapes Space_ The Black Flower Antenna is about electronic receptions to the museum and is a receiving antenna that picks up invisible electromagnetic waves in the gallery. Thirty-four industrially knitted cones embedded with pink copper yarns transmit the presence of electromagnetic waves via high pitched, live sounds in this gallery. The 'Black Flower' serves to make visible and present the workings of an invisible cyber

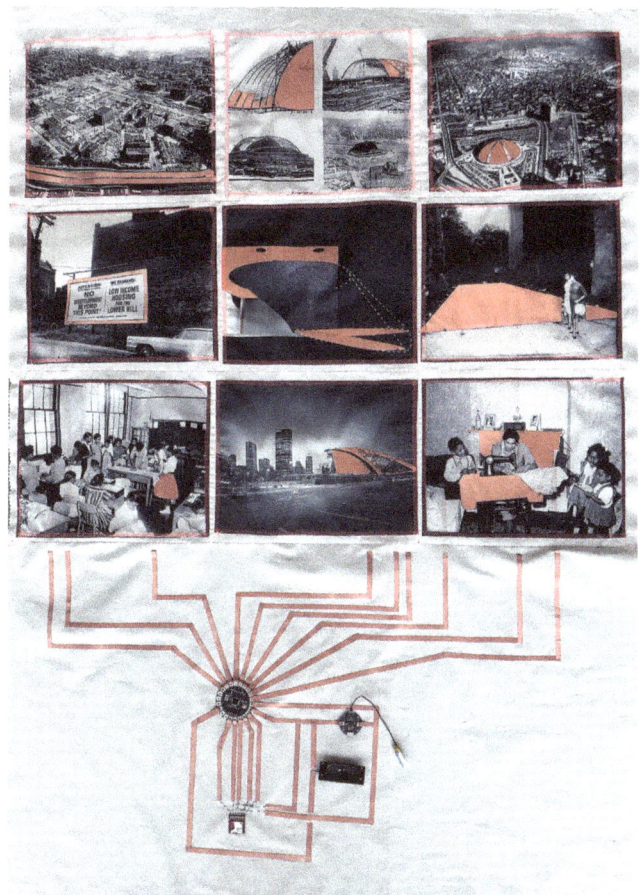

1   Fabricating Networks Quilt

2   Freedom Corner

3 Black Flower Antenna

urbanism that floats above the physical city supported and constructed by electromagnetic waves. Its tangible knit material is a meshing structure that points to how invisible and intangible spaces are being constructed and interlinked with our physical spaces and need to be considered if we do not want the perpetuation of technological redlining, electronic extraction of information from people for use in profit and classifications of people for purposes which are not revealed to them or wanted by those people (Figure 3).

FELT_FEELING EMOTION LINKED TO TOUCH PANEL

What we Touch Shapes our Brains_The FELT Panel demonstrates the communicative power of textures using vision and touch as well as motility in projecting emotions to people through a textile. Analogue and memory play a large part of how people understand what is communicated by a texture. These are personal and specific experiences; therefore it is not possible to tag one emotion or a set of emotions to a texture. As a designer one can get sense of what some people will understand and take away as a communicated emotion from a texture. In addition, people's sense of touch provides a choice on understanding a situation. A person can toggle between understanding a touch as subjective, being touched or objective,

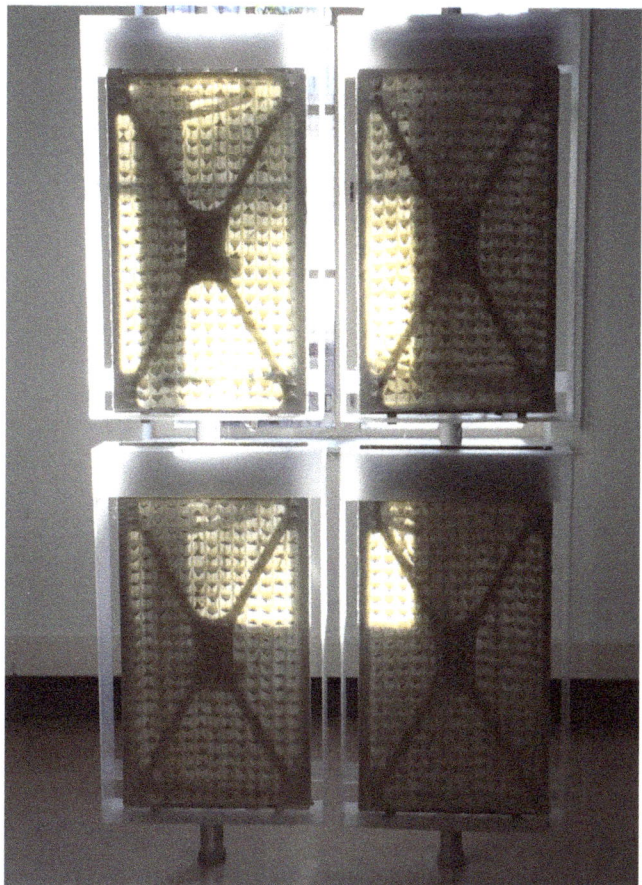

4 FELT_Feeling Emotion Linked to Touch Panel

5 FELT_Ways of Touching

touching something. Returning to one of our original ideas, materials shape what fires off between our neurons and our neurons are shaped by that firing of sense. What we understand is a conversation between our environment and our bodies that are individually shaped and tuned vessels (Figures 4, 5).

The contribution of this work is an evolving soft system, and offers ways of making and thinking about architecture and design computing.

REFERENCES

Frijda, N. H. 1982. *The Meanings of Emotional Expression*, editedy by Mary Ritchie. In *Nonverbal Communication Today: Current Research. Berlin*: Mouton Publishers. 103-119, Xiv.

Gibson, J.J. 1983. *The Senses Considered as Perceptual Systems*, 1st edition, 1966. Westport, C.T.: Reprinted by Greenwood Press.

Ingold, T. 2010. "Bringing Things to Life. Creative Entanglements in a World of Materials." *Working Paper* 15, University of Aberdeen, Scotland. https://eprints.ncrm.ac.uk/id/eprint/1306/1/0510_creative_entanglements.pdf.

Felecia Davis's work in computational textiles questions how we live and she re-imagines how we might use textiles in our daily lives and in architecture. Davis is interested in developing computational methods and design in relation to specific bodies in specific places engaging specific social, cultural and political constructions. Davis is an Associate Professor at the Stuckeman Center for Design Computing in the School of Architecture at Pennsylvania State University and is the director of SOFTLAB@PSU. She completed her PhD in Design Computation at MIT. Davis' work in architecture connects art, science, engineering and design and was featured by PBS in the Women in Science Profiles series. Davis' work was recognized in 2022 by the New York Architectural League's Emerging Voices Awards. Davis' work was part of the Museum of Modern Art's exhibition Reconstruction: Blackness and Architecture in America. She is a founding member of the Black Reconstruction Collective a not-for-profit group of Black architects, scholars and artists supporting design work about the Black diaspora.

NOTES

This award recognizes innovative research that contributes to the field of digital design in architecture. The award distinguishes research with the potential to transform contemporary architecture.

Workshops

Workshops

Andrew Saunders, Chair

The 2022 ACADIA Conference hosted fourteen total simultaneous three-day workshops: seven in person and seven remote. The workshops engaged with the conference theme Hybrids & Haecceities: Haecceities describe the qualities or properties of objects that define them as unique; concurrently, Hybrids are entities with characteristics enhanced by the process of combining two or more elements with different properties.

Workshops content represented a range of current interest in computational design including advanced robotic fabrication techniques, data-driven procedural workflows at multiple scales and more speculative stances toward recent advances in artificial intelligence via computational neural networks as well as advanced representational workflows designed to operate with photogrammetry and Lidar (Light Detection and Ranging) high density mesh model output. The workshop format allowed for the showcasing of the newly formed Advanced Research & Innovation (ARI) Lab at the Weitzman School of Design. Its full capacity allowed the facilitation of multiple intense robotic research workflows simultaneous over the short three-day period.

The hybrid in-person and online format allowed an inclusive and wide audience both nationally and internationally to participate in the ACADIA conference via the workshop format. Sponsorship support fostered a unique dialogue and exposure to contemporary workshops that were led by internationally established practices, industry experts and academics at the forefront of the computational design in architecture. At the conclusion of the workshops a public colloquium was held on Wednesday, October 26 to celebrate and discuss the outcomes of the workshops. In addition, the workshop outcomes were exhibited during the ACADIA conference as part of the overall exhibition.

Colloquium Presentations / Group I

Robotic Fabrication for Building Components // *ZHACODE - Cesar Fragachan and Tim Fu*

Augmented Vision: Guided Fabrication // *FOLOGRAM - Cameron Newnham and Nick van den Berg*

Bio-Polymer 3D Printing // *Mette Ramsgaard Thomsen, Paul Nicholas, Gabriella Rossi, and Carl Eppinger*

A Molten Gesture // *Claire Moriarty and Riley Studebaker*

Robotic Mark Making // *Sara Codarin and Karl Daubmann*

Colloquium Presentations / Group II

Form-finding Explorations in CLT // *Amin Adelzadeh and Hamed Karimian A., and Christopher Robeller*

Data-driven Urban Design // *Grimshaw - Jorge Sainz de Aja and Esther Rubio Madronal*

Marginalized Craft Traditions // *Duane McLemore*

Power Automate Data Workflows in AEC // *Cesar Escalante, Philippe Videau, Geng Wang, and Sagar Baver*

Colloquium Presentations / Group III

Diffusion: Architecture, AI // *Dr. Matias del Campo and Sandra Manninger*

AI, Literature, and the Mind's Eye // *Karel Klein*

Generating Spatial Hybrids: 3DGANS // *Benjamin Ennemoser and Ingrid Mayrhofer-Hufnagl*

Digital Sculpting // *Patrick Danahy and Caleb Ehly*

Vibrant Artefacts // *Barry Wark*

Robotic Fabrication for Building Components

Zaha Hadid Architects Computation and Design Group (ZHACODE) /
Cesar Fragachan and Tim Fu

Architectural Geometry (AG) focuses on the synthesis of shapes that guarantee structural and fabrication optimality. It is also closely aligned with and complementary to the development of robotic and digital fabrication (RDF) technologies and design methods. The workshop explores the relevance of this state-of-the-art design and construction paradigm in the realm of complex, multi-objective, precision manufactured, computational geometry projects. Specifically, the workshop is a collaborative design exercise exploring the design space of topologies and seeks synergies of tacit mesh modeling with computational form-finding, digital timber, and RDF. The workshop provides a hands-on, introductory experience to prepare for both the imminent future of architectural design and construction and architectural practice. The content and tool-chains of the workshop are representative of the state-of-the-art in the AEC industry as it shifts from Building Information Modeling (BIM) for documentation to Design for Manufacturing and Assembly (DfMA)/ industrialized construction (IC) paradigm.

Instructors

Shajay Bhooshan is co-founder of CODE, Zaha Hadid Architects in the Computation and Design Group (ZHACODE).

Henry David Louth is an Associate at Zaha Hadid Architects in the Computation and Design Group (ZHACODE).

Cesar Fragachan is a registered Architect in Venezuela. He holds a Master in Architecture and Urbanism degree with distinction from the Architectural Association Design Research Laboratory (AADRL).

Tim Fu is a part of Zaha Hadid Architects in London, UK. He is a designer at ZH CODE (Computational Research Group), where he specializes in algorithmic design and parametric facade research.

Participants Ali Fahmy, Abdallah Kamhawi, Zaid Marji, Mohammad Karkoutly, Tuli Bhattacharjee, Sarah Kusuma Rubritz, Chunze Li, Seyed Hossein Zargar

Sponsors ZHACODE

Image Credits ACADIA 2022 *Robotic Fabrication for Building Components* workshop participants

Augmented Vision
Realtime Feedback for Guided Fabrication

FOLOGRAM / Cameron Newnham and Nick van den Berg

This workshop provides participants with skills and experience in mixed reality applications that utilize depth camera data for guiding fabrication processes in real-time. Depth camera data from HoloLens 2 can be used to inform parametric models with the precise geometry of physical materials and provide information to fabricators on how to manipulate material to better match digital models. Participants are introduced to Fologram for creating interactive mixed reality applications in Grasshopper, collaborate on the design and fabrication of a large prototype using real-time feedback for complex surface forming, and finally explore extensions of this approach to suit their own research interests and agendas, including clash and part detection, quality assurance, subtractive or additive material forming, engagement with highly non-uniform or unpredictable materials, training applications or other ideas that arise through discussions.

Instructors

Cameron Newnham and **Nick van den Berg** are co-founders of Fologram, a design research practice and technology startup building a platform for designing and making in mixed reality. Fologram's clients include leading universities, multinational architectural firms, industrial designers, engineers and artists who are building mixed reality applications for full scale construction, public art, architectural fabrication, sculpture, automotive design, and cyber-physical visualization systems.

Fabrication Assistant Andrew Wit

Participants Matt Ward, Lee-Su Huang, Michael Riggin, John Doria, Melissa Goldman, Troy Malmstrom, Ted Madden, Ertunc Hunkar, Dave Lee, Arielle Spencer, Sina Mostafavi, Carolina Myers, Sihan Li

---
Image Credits ACADIA 2022 *Augmented Vision* workshop participants

Bio-polymer 3D Printing
Strategies for Material Grading of Bio-Based Material

Mette Ramsgaard Thomsen, Paul Nicholas, Gabriella Rossi, and Carl Eppinger

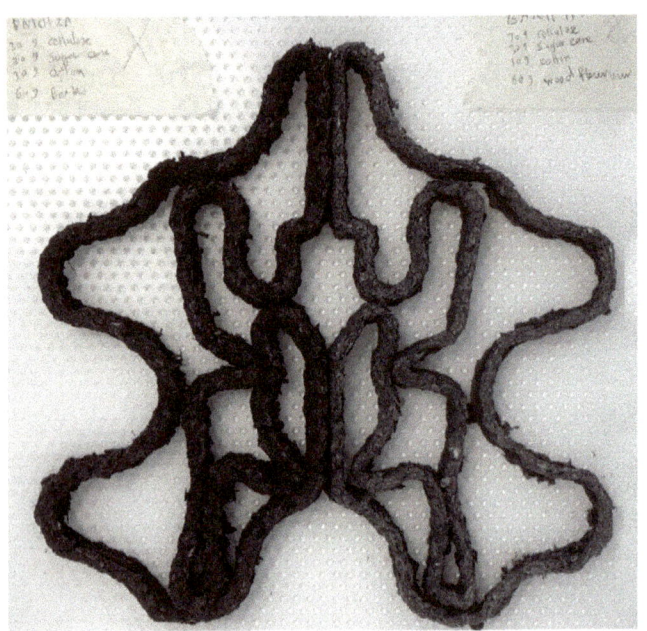

Bio-polymer 3D Printing asks what new concepts, designs and workflows a bio-based material paradigm may provoke. With focus on the robotic 3D extrusion of cellulose reinforced bio-polymers—materials made from material waste streams—we question how architectural materials are sourced, composed and formed, and how rethinking these axioms can lead to the imagination of new ways of understanding the ecological contexts, performance and expression of buildings. The workshop examines the composition, printing, sensing and grading of bio-polymers. With an outlook to local materials, participants develop customized recipes and examine how these can be characterized for strategic deployment in graded material composites. We define ways to robotically steer 3D extrusion and work with machine vision to characterize how geometry and drying behavior interact, all in order to explore how fabrication parameters can drive new material languages challenging what architecture can be.

Instructors

Mette Ramsgaard Thomsen examines the intersections between architecture and advanced computational design processes, with a lens toward the profound changes that digital technologies instigate in the way architecture is thought, designed, and built.

Paul Nicholas is Associate Professor at the Centre for Information Technology and Architecture (CITA) at the Royal Danish Academy.

Gabriella Rossi is a PhD Fellow at CITA, Royal Danish Academy.

Carl Eppinger is a Research Assistant at CITA, Royal Danish Academy.

Participants Chavi Gupta, Maho Kobayashi, Xinlin Lu, Zach Keller, Wei Wu, Shan Sutherland, Amber Klinger, Sinead Nicholson

Image Credits ACADIA 2022 *Bio-polymer 3D Printing* workshop participants

A Molten Gesture
Expanding 'Hand-Craft'

Claire Moriarty and Riley Studebaker

 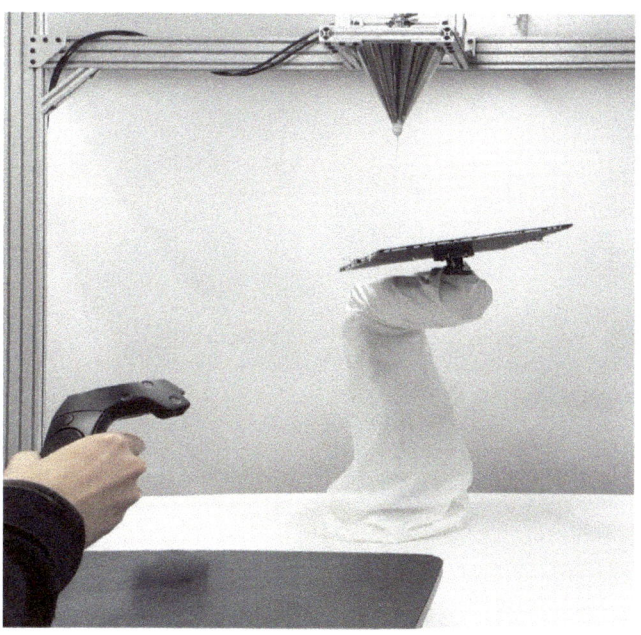

Objects made by the human hand are inherently seductive. This workshop explores an augmentation of bodily agency in fabrication using computation and robotics to bring hazardous processes into the realm of hand-craft. Participants use body-tracking software, real-time robotic control, and methods of neural network-based gestural redistribution to create a bespoke poured metal lighting element. Each lighting element is intended to be specific to a site chosen by the participant, creating directly responsive light and shadow. Participants made a series of gestural studies in dripped metal using real-time VR control of a robotic arm. A selection of these physical results are mapped around the surface of the final pendant light using artificial neural networks for final robotic fabrication, redistributing the arrangement of dripped metal, porosity, and intrinsic and familiar qualities of the maker's hand, for a responsive and bespoke lighting effect and an object that challenges the conception of hand-craft.

Instructors

Riley Studebaker is the Manager of the Advanced Research and Innovation Lab at the University of Pennsylvania Weitzman School of Design, as well as a Research Associate with the AML, a Guest Lecturer with the Weitzman School of Design, and a Praxis Fellow of the University of Pennsylvania.

Claire Moriarty holds a Master of Science in Design in Robotics & Autonomous Systems from the Weitzman School of Design at the University of Pennsylvania, as well as a BFA in Architectural Design from the Maryland Institute College of Art.

Participants Chavi Gupta, Maho Kobayashi, Xinlin Lu, Zach Keller, Wei Wu, Shan Sutherland, Amber Klinger, Sinead Nicholson

---
Image Credits ACADIA 2022 *A Molten Gesture* workshop participants

Robotic Mark Making

Sara Codarin and Karl Daubmann

When we learned to paint, we had to understand how to hold the brush, test the pressure against the resulting brush strokes, and build confidence in the dexterity of our gestures. The workshop called "Robotic Mark Making" traces our process of re-learning to paint but this time using a retired automotive welding robot as an extension of our hands. With this proposal, we are seeking the productive confluence between an old form of analog expression and the opportunities afforded by new digital technologies. Parametrically designed drawings will be implemented robotically using a soft brush as a means to share different attempts and reflections on the successes and failures of this type of work. The soft brush will navigate the 2.5D space and dissolve the rigorous digital inputs into unique ink marks. The participants' outcomes will document the various attempts, iterations, parameters, and variables against the circumstances of the pen, paper, humidity, and whim of the designers.

Instructors

Sara Codarin is an Assistant Professor in Architecture at the Lawrence Technological University, College of Architecture and Design. She earned her PhD from the University of Ferrara, Italy, with a dissertation investigating robotic manufacturing processes for the conservation of Cultural Heritage.

Karl Daubmann is an architect at the forefront of digital design. He is Dean of the College of Architecture and Design at Lawrence Technological University.

Participants Namjoo Kim, Amir Mohammad Azizi, Seyedeh Gelareh Sanei, Alec Naktin

Image Credits ACADIA 2022 *Robotic Mark Making* workshop participants

Form-finding Explorations in CLT Construction

Amin Adelzadeh, Hamed Karimian A., and Christopher Robeller

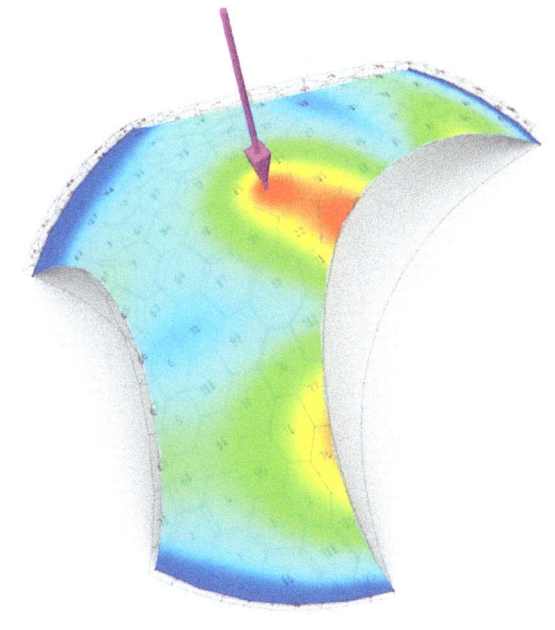

The application of Cross-Laminated Timber (CLT) in the building industry has a major role in shaping the recent renaissance in timber construction; however, despite the benefits, CLT fabrication produces a huge amount of offcuts that despite the high quality are too small for the timber building applications. For the first time, these CLT offcuts were turned into a segmented shell demonstrator *Recycleshell* by using algorithmically-generated form-fit connectors and screws. Inspired by such a research philosophy, this workshop aims to provide fundamental knowledge and essential computational skills for the design-to-assembly of lightweight segmented wood-only shell structures made of CLT plates. Participants gain knowledge of innovative CLT construction techniques, material systems, joinery, rapid and precise assembly, and improve computational skills in developing modeling workflows consisting of form-finding, discretization, rationalization, joint generation, and fabrication data production. The acquired knowledge and skills would benefit research and practices in the CLT construction industry.

Instructors

Amin Adelzadeh is an educator, researcher, critic, and currently scientific staff at the Augsburg University of Applied Sciences.

Hamed Karimian A. is a member of the "Timber Structure Interface" project led by Prof. Dr. Christopher Robeller and hosted by the Technical University of Kaiserslautern, Bremen University of Applied Science, and Augsburg University of Applied Science.

Christopher Robeller is a Professor for Digital Design and Production at the University of Applied Sciences Augsburg, teaching and researching as part of the new and innovative program Digital Building Masters program.

Participants Christo van der Hoven, Andres Obregon, Christos Baknis, Luis Antonio Lopez Montiel, Ronan Bolaños, Bahar Moradi, Sepehr Alipour, Divya Rastogi, Maedeh (Kania) Kazemi, Alex Woodhouse, Habeeb Muhammad, Jessica Elizalde, Xiaodong Li, Luis León-Escoda, Bentian Wang, Mark Cooper, C. Beniamin Herring, Zeno Zoppi

Image Credits ACADIA 2022 *Form-finding Explorations in CLT Construction* workshop participants

Data-Driven Urban Design and Master Planning Development

Grimshaw / Jorge Sainz de Aja Curbelo and Esther Rubio Madronal

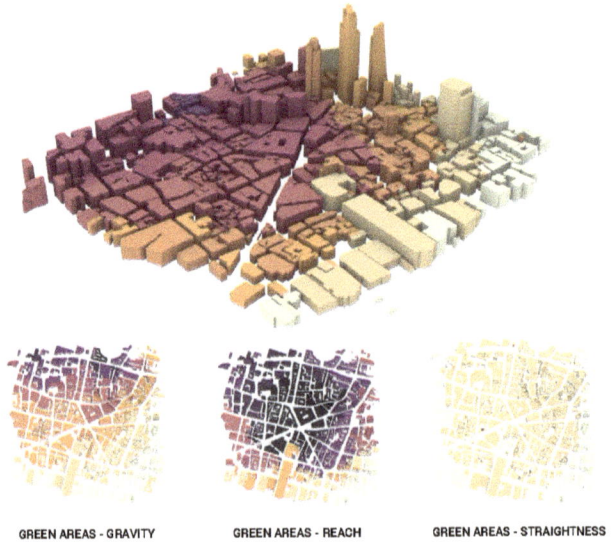

GREEN AREAS - GRAVITY   GREEN AREAS - REACH   GREEN AREAS - STRAIGHTNESS

The workshop proposes an overview of different techniques and methodologies around spatial analysis for urban design and master-planning projects, through the implementation and processing of different spatial open data sets that will ultimately inform a theoretical urban development proposal. Using a scripting language (R) for ETL processes, and its interoperability with Rhinoceros and Grasshopper, the workshop explores different geocomputation techniques and methodologies to create a data-driven approach to urban design and master planning modeling that is capable of generating multiple scenarios—based on dedicated KPIs and more specifically the use of Space Syntax theory to allocate different land use and its quantum distribution on a parametric urban design model.

The final part of the workshop explores other possible development environments and outcomes, such as web apps or Rhino compute integration to trigger a Q&A session about next steps and integration of other relevant workflows.

Instructors

Jorge Sainz de Aja Curbelo is Grimshaw's urban computation manager, specializing in parametric and generative methodologies as well as geospatial data analysis.

Esther Rubio Madronal is a Computational Design Specialist and a trained architect. She completed her MArch degree with honors at the University of Seville in 2016.

Participants Sahar Gohari Moghadam, Andrew Smith, John Chun, Aswin Indraprastha, Daniel Luegering, Vivek Rathore, Reza Fouladi, Diva Rastogi, Alexa Resendiz, Amber Chen, Justin Wang, C. Beniamin Herring, Manfred Saberbein

Image Credits ACADIA 2022 *Data-Driven Urban Design and Master Planning Development* workshop participants

Marginalized Craft Traditions and Advanced Fabrication

Duane McLemore

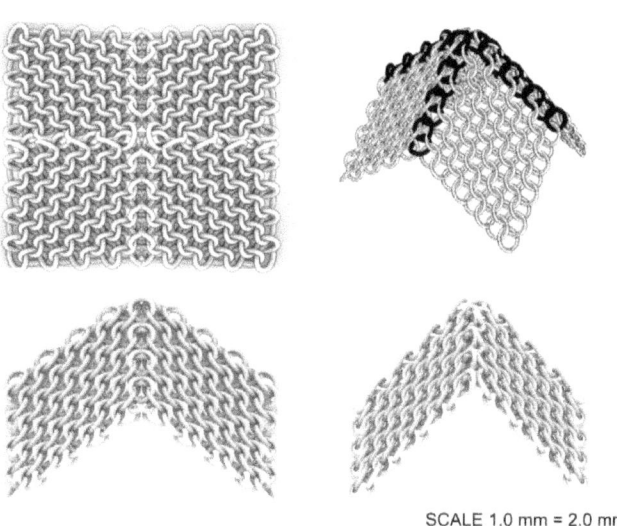

SCALE 1.0 mm = 2.0 mm

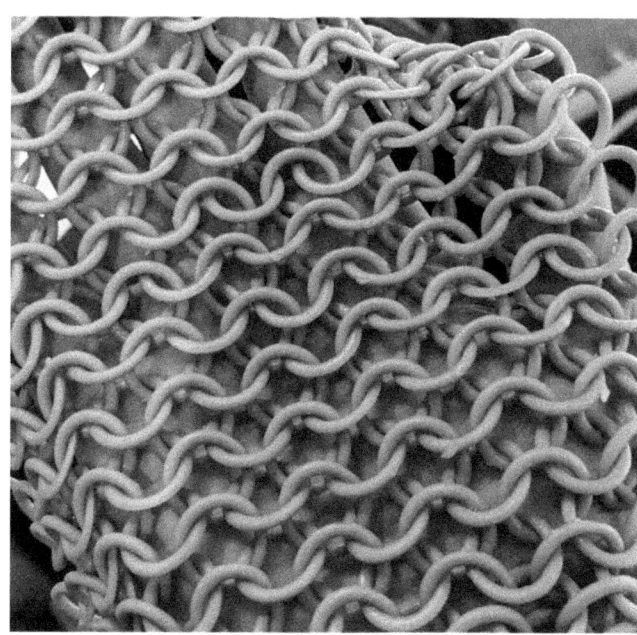

This software-agnostic workshop will help participants build a workflow to research and transpose knowledge and fabrication practices in novel ways. Within Western modernity a deliberate rift was created between 'Art' and 'Crafts' with the latter term used to demote work to secondary status due to abundance, commonness, and useful function. But can these instead be seen as virtues which give Craft value otherwise overlooked by the canon? Is there a way of looking at Art and the history of fabrication that re-situates both?

This course proposes that Craft traditions outside Western modernity can be reexamined on equal footing with Western Art. Similarly, what if Advanced Fabrication is viewed as part of these Craft traditions? The inherent intelligence of materials and techniques within Craft traditions can be a new departure point for the lineage of Advanced Fabrication. The most important requirement is that participants have access to high-quality information resources such as a University or City library and / or online resources.

The proposed workshop outcomes for each student is a small body of individual findings and potential experiments designed / prototyped prior to the end of the course which participants present to the group.

Instructor

Duane McLemore, AIA is an Assistant Professor of Architecture at Mississippi State University. He is a California Registered Architect and co-founder of X Over Zero, an architecture and design firm best known for their 3D printed jewelry. Duane's academic research uses mathematics and geometry to unite design computation and contemporary fabrication methods.

Participants Alston Brown, Jacob Bryson, Spence Colmer, Ethan Harvey, JD Jaggers, Elle Mason, Gabbi Morelli, Luke Murray, Jessica Ninnis, Tyler Pipkins, Evelyn Ramirez, Jeb Thomas

---

Image Credits ACADIA 2022 *Marginalized Craft Traditions and Advanced Fabrication* workshop participants

Power Automate Data Workflows in AEC

Cesar R. Escalante, Philippe Videau, GengWang, and Sagar Baver

The tectonic shift towards real-time rich data-driven generative design solutions requires embracing the power of computing as a partner in the design process and discovering new ways to integrate our diverse contributions from transdisciplinary teams using a myriad of apps and analytical tools. The end game of this trend towards convergence is marked by the idea of superintegrated apps. What are the implications of this convergence in practice? How can design professionals leverage this convergence when designing buildings and cities and how we fabricate and construct? In this workshop, participants think about these implications as they learn how to design and build streaming connections to transfer granular segments of information from a cloud-hosted generative solution to a system of interconnected downstream apps designed to do it all in real-time. Participants learn how to unlock the data from a conceptual exercise in Revit and pack the information as data exchanges. Using Microsoft Power Automate, participants learn to bridge model information to downstream applications, build triggers on MS Teams, refresh data on Excel, and display live data insights in Power BI reports. Attendees bring their own projects from academia or practice or are given the opportunity to explore an iterative design solution and present at the end the various methods used for automating the streamed content. This network of operating systems and cloud services—traditionally made with sophisticated API investments—are now more accessible to architects and computational designers with entry-level coding skills and poised to challenge digital delivery as we know it.

Instructors

Cesar Escalante works as Global Technical Marketing Manager at Autodesk, where he builds compelling technical content on new and emerging technologies in Architecture. Prior to Autodesk, he worked as Design Technology leader at GENSLER and HOK, where he provided technical leadership in the design technology strategy, execution, and delivery of the large complex multi-million dollar projects including the new Santa Clara City Center, Salt Lake City Airport Project, and the Apple Central & Wolfe Campus in Sunnyvale.

Philippe Videau is a Product Manager on Autodesk's cloud data platform, working alongside the team charged with developing cloud services that integrate data across a variety of design, engineering, construction, and manufacturing software tools. Since joining Autodesk he has made strides on investigating and communicating the impact of future technologies on the industries Autodesk serves.

Participants Andrew Heilman, Andrew Cocke, Zach Kron

Sponsors Autodesk

Image Credits ACADIA 2022 *Power Automate Data Workflows in AEC* workshop participants

Diffusion
Architecture, Artificial Intelligence and Synthetic Imaginations

SPAN / Dr. Matias del Campo and Dr. Sandra Manninger

"The limits of my language mean the limits of my world." The famous quote by Ludwig Wittgenstein has gained an entire new meaning in the emerging context of natural language text-to-image applications driven by artificial intelligence algorithms. Applications such as Midjourney, Disco Diffusion, and Stable Diffusion are spreading like wildfire in the architecture community. Two specific results can be observed in the Cambrian explosion of a novel design tool. On the one side, the generation of thousands of astonishing images in a very short period of time, on the other the emergence of a critical interrogation of architecture theory in the face of a posthuman design method. The workshop takes on the challenge of exploring this new design method, using diffusion models as a starting point. The premise for the project is the design of a series of interconnected pixel images elaborating aspects of elevation, plan, perspective, and atmosphere of a house—components of a mood board. The ability of diffusion models to take advantage of cultural, aesthetic, practical, technical, and even political resources is used to co-imagine an architectural solution. The focus of the discursive component of the workshop is on the immense contribution of the method to the architecture discipline in terms of its transformative power in practice and theory. Questions such as the ontology of the architecture emerging from diffusion models: "Can a neural network based on existing datasets create something genuinely new?" The epistemology deduced from the results are part of the conversation, including the ethical implications of the method. Aspects of authorship, agency, and sensibility in this new context of design are explored in particular towards its inherent abilities to tap into the estrangement and defamiliarization of architectural objects. Participants learn how to use diffusion models, how to strategize methods to convert the resulting images into 3D models, how to perform a critical forensic examination of the results, and how to formulate a theory around their designs.

Instructors

Dr. Matias del Campo is a registered architect, designer, and educator. He is an Associate Professor at Taubman College of Architecture and Urban Planning, University of Michigan, director of the AR2IL – The Architecture and Artificial Intelligence Laboratory at UoM, and affiliate faculty member of Michigan Robotics, Computer Science and Data Science.

Dr. Sandra Manninger is a registered architect, teacher, and researcher. She is co-principal of SPAN. The focus of the practice lies the integration of advanced design and building techniques that fold nature, culture, and technology into one design ecology.

Participants Luis Sanabria, Maria Paula Aranzales Villamil, Elton Gjata, Jinsil Seo, Umar Mahmood

Image Credits ACADIA 2022 *Diffusion* workshop participants

AI, Literature, and the Mind's Eye

Karel Klein

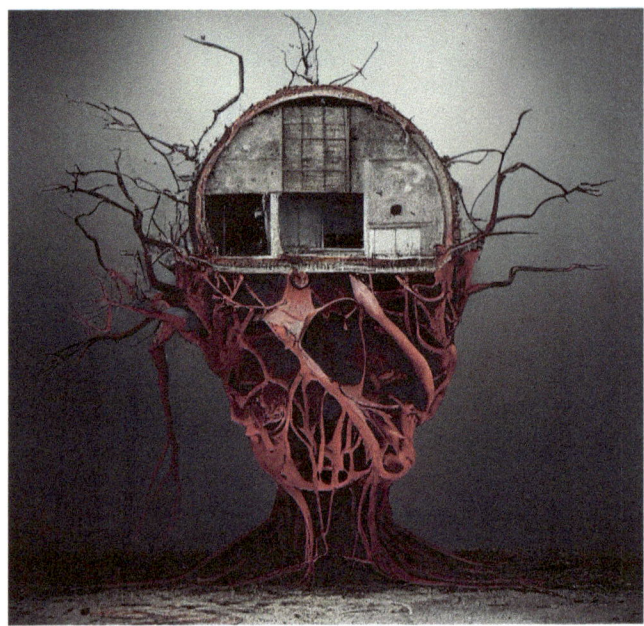

"She had pewter-colored hair set in a ruthless permanent, a hard beak, and large moist eyes with the sympathetic expression of wet stones." —Raymond Chandler

As our eyes scan the book's page, the words we read evoke an array of sensory perceptions in the brain. But what the brain does best in this case is see. What we are doing when we read is engaging the mind's "eye." And when we enter language prompts into text-to-image generators such as Midjourney or Dall-E 2, the strange images that are thus produced appear to be from the imagination of AI. Does AI, too, have a mind's "eye"?

In this workshop, we look at several short stories that elicit vivid imagery by way of richly descriptive language and use fragments of that language in Midjourney prompts, attempting to evince mind's "eye" imagery. We discuss memory, aesthetics, and AI as an externalized proxy for the human faculty of imagination.

Instructor

Karel Klein is an architect and co-director of Ruy Klein. Investigating craft, precision, and the evolution of design expertise in the digital age, Karel continues to foreground the persistence of the designer in contemporary culture. She received her MArch degree from Columbia University and also holds a BS degree in civil engineering from the University of Illinois Urbana-Champaign. Karel currently teaches design studios at SCI-Arc, Pratt Institute, and University of Pennsylvania.

Participants Leila Delafrooz, Omid Dorrani, Jinkui Gao, Sutanuka Jashu, Yuanyuan Lin, Elena Mai, Shun Sasaki, Matthew Scholtz, Reginald Wilson

Image Credits ACADIA 2022 *AI, Literature, and the Mind's Eye* workshop participants

Generating Spatial Hybrids with 3D Generative Adversarial Networks

Benjamin Ennemoser and Ingrid Mayrhofer-Hufnagl

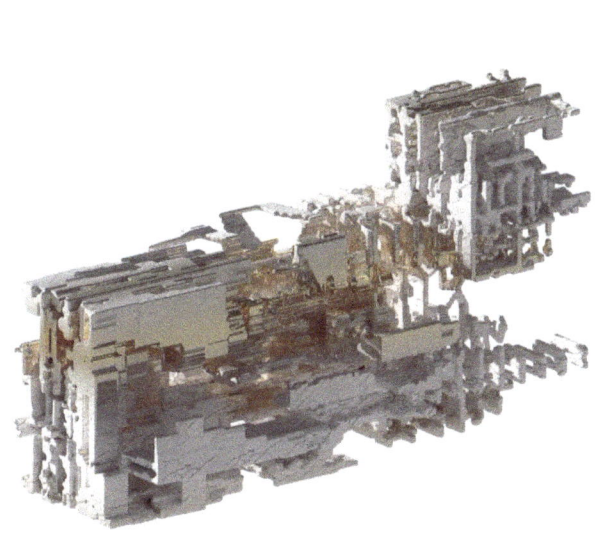

The invention of Generative Adversarial Networks (GANs) seems to have a profound influence on the field of architecture. GANs have been used to generate 2D drawings and images, but their deployment to 3D generation has rarely been examined. Therefore, this workshop tackles the fairly uncharted territory of 3DGANs and the application of architectural 3D models as datasets. We explore a novel method of 3DGAN that we developed over the last couple of years and has the ability to train and generate spatial configurations as an output that goes beyond building mass and 3D shapes. We harness the power of the machine by using unsupervised learning to avoid pre-classification of the different buildings that make up the dataset. This approach allows us to eliminate biases and conventions to generate unprecedented hybrid spatial configurations.

Instructors

Ingrid Mayrhofer-Hufnagl is a multidisciplinary architect, researcher, and educator whose work focuses on the intersection of architecture, technology, and science. She received her PhD from the University of Innsbruck, where she currently also holds a postdoc position leading grant-funded research projects on artificial intelligence for architectural and urban design.

Benjamin Ennemoser is an Assistant Professor at Texas A&M University and has previously taught and lectured at UCLA, the University of Innsbruck, the University of Applied Arts in Vienna, and the Yun-Tech University in Taiwan.

Teaching Assistant Quinn McCormack

Participants Ahmed Hesham, Aseel Alamoudi, Carlos Rios Toto, Carmen Hull, Christianna Bennett, Christina Doumpioti, Dana Cupkova, Daniele Paulino, Diego Macias, Franklin Wu, Katerina Touska, Matthew Mansour, Oliver Hamedinaer, Jozef Sedlacek, Raffael Stegfellner, Runjia Tian, Samuele Agrimi, Shunta Moriuchi, Tarique Ali, Tilman Fabini, Zhijuan Liu

Image Credits ACADIA 2022 *Generating Spatial Hybrids with 3D Generative Adversarial Networks* workshop participants

Digital Sculpting:
Advanced Modeling High Polygon Workflows

Patrick Danahy and Caleb Ehly

With the prevalence of contemporary surveying tools, including photogrammetry, lidar scanning, and neural radiance mapping, comes a set of workflow obstacles including irregular mesh topologies, and the high polygon mesh models. These obstacles limit the potential use of these tools for surveying and design workflows, where geometric control is often sacrificed for high resolution artifact retrieval. In this workshop, we will leverage explicit and procedural modelling techniques to develop pseudo architectural elements from a high-resolution 3D scanned mesh collection. Base geometric primitives are used to build out formal elements, while custom brushes and alphas will be used for explicit texturing, and finally the operative design models will be retopologized, cleaned and UV mapped using conformal mapping techniques to be sent into rendering software for final image production. This workshop offers a full design workflow from simulated discovery, surveying, compositing, advanced modeling, and image production with procedural material shaders based on geometric properties.

Instructors

Patrick Danahy is an Assistant Research Professor and the Design Innovation Fellow at Ball State University, and a Lecturer in the Landscape Architecture Department at the University of Pennsylvania.

Caleb Ehly is a designer at Young Projects in NYC and part-time lecturer at UPENN. He has previously worked at MAD, ODA, Uptic Studios, and held teaching positions in the topics of studio, media, and research at UPENN, and at the University of Idaho.

Participants Al Ahmed Hesham, Christian Belanger, Tinghui Mo, Vidan Wang, Jeffrey Liao, Joseph Anaya, Shuoxuan Su, Courtney Ward, Ghassan Alserayhi, Moritz Riedl, Xaver Roos, Lorenz Andexer, Justin Wang, Mehr Dandiwal, Shruti Hippalgaonkar, Tanisha Lahane, Erick Vernon-Galindo, Jingyu Luo

Sponsors Chaos Group

Image Credits ACADIA 2022 *Digital Sculpting* workshop participants

Vibrant Artefacts

Barry Wark / Bartlett, Weitzman

The workshop explores the creation of architectural artefacts designed to weather and display their interconnectedness with their environments—encompassing notions of erosion, non-determinate plant growth, and aging and patina. Students engage in a series of tasks working with a combination of 3D modeling and procedural design using Houdini FX. The outcomes are not fixed, singular designs, but a series of objects produced from the procedural systems within Houdini. The proposals are then represented on a site of students' choice, forming a collection of follies.

Instructor

Barry Wark is an architect and designer who combines practice with research and teaching activities at the Bartlett School of Architecture, UCL. Following over a decade working for reputable practices in London, he established his eponymous studio in 2018. His work explores the development of ecological aesthetics, moving past the sweetness of nature led design; it strives towards ecocentrism, rejecting that humans and their artifacts are impervious to or separate from the natural world, instead exploring permissible forms of coexistence between human and nonhumans within the built environment.

His work has been published internationally in both physical and digital media. He has given lectures and workshops about his research to architects and to other design disciplines, such as film and game development. Barry is also the cofounder of the academic atelier Biophile, a space for students and educators to contribute to the evolving nature-architecture dialog.

Participants Ahme A. Hanyan Chen, Lixue Cheng, Irene Dole, Darlene Farris-LaBar, Zihua Mo, Casey Stamm, Milad Saboori, Jingwen Wu, Jingxiao Zhou

---

Sponsors Nvidia, Dell, Intel

Image Credits ACADIA 2022 *Vibrant Artefacts* workshop participants

Credits

Conference Chairs

DR. MASOUD AKBARZADEH
Assistant Professor of Architecture, Weitzman School of Design, University of Pennsylvania

Masoud Akbarzadeh is a designer with academic background and experience in architectural design, computation, and structural engineering. He is Assistant Professor of Architecture focusing on Structures and Advanced Technologies and the Polyhedral Structures Laboratory (PSL) director. He holds a DSc from the Institute of Technology in Architecture, ETH Zurich where was a research assistant in the Block Research Group. In addition, he has two degrees from MIT: a Master of Science in Architecture Studies (Computation) and a MArch, the thesis for which earned him the renowned SOM award. He also has a degree in Earthquake Engineering and Dynamics of Structures from the Iran University of Science and Technology and a BS in Civil and Environmental Engineering. His main research topic is Three-Dimensional Graphical Statics, a novel geometric method of structural design in three dimensions. In 2020, he received the National Science Foundation CAREER Award to extend the methods of 3D/Polyhedral Graphic Statics for Education, Design, and Optimization of High-Performance Structures. He is also a Co-PI in a $4.6 million grant funded by National Science Foundation to investigate high-performance, self-morphing building blocks across scales toward a sustainable future. Has also received a $2.4 Million ARPA-E Grant to Research the Design of Carbon-Negative Buildings starting September 2022.

HINA JAMELLE
Associate Professor of Practice, Weitzman School of Design, University of Pennsylvania
Director, Contemporary Architecture Practice, NY SH

Hina Jamelle teaches final year Graduate Option Studios and directs the Graduate Program's Urban Housing Studios at the University of Pennsylvania Weitzman School of Design. She has held the Visiting Schaffer Practice Professorship at the University of Michigan. Jamelle is the co-director of the New York and Shanghai-based architectural firm Contemporary Architecture Practice and has co-edited issues of *Architectural Design AD* titled IMPACT (2020) as well as *Elegance* (2007). Hina Jamelle's book *UNDER PRESSURE* on urban housing was published in 2021. Founded in 1999, Contemporary Architecture Practice [CAP] is known for futuristic designs using digital techniques and the latest technologies for the design and manufacturing of architecture. Projects include commissions by The Museum of Modern Art [New York]; Reebok Shanghai, Lijia Smart Park, Chongqing, Wenjin Hotels, Beijing, NJCTTQ Pharmaceuticals, Nanjing, AMEC Technologies, Nanchang [China]; Samsung, Seoul [South Korea]; and IWI Orthodontics Clinic, Tokyo [Japan]. Contemporary Architecture Practice's projects have been exhibited extensively at the Museum of Modern Art, New York; the London, Beijing, and Shanghai Biennales; and the Tel Aviv Museum of Art, among others. They also have been featured in more than 250 major publications around the world. Co-Directors Rahim and Jamelle have won the Architectural Record Design Vanguard Award and were featured in Phaidon's *10x10x2* as one of the world's top 100 emerging architects. Their project, IWI Orthodontics in Tokyo, Japan was featured in Phaidon's *ROOM 100* as one of the most creative interior design projects of the century. In 2015 she was recognized as 50 Under 50 Innovators of the 21st Century by a distinguished jury.

DR. DORIT AVIV
Assistant Professor of Architecture, Weitzman School of Design, University of Pennsylvania

Dorit Aviv, PhD, AIA, is Assistant Professor of Architecture at the University of Pennsylvania's Weitzman School of Design, where she directs the Thermal Architecture Lab, an interdisciplinary laboratory focused on the intersection of thermodynamics, architectural design, and material science. Her work examines how architectural materials and forms can impact airflows, energy interactions, and human health. A recipient of a 2020 Holcim Award for Sustainable Design and Construction, Aviv's recent projects include a combined evaporative and radiative cooling prototype for desert climate, development of radiant cooling for hot-humid climates, a blockchain-enabled distributed environmental sensing network, and indoor environmental quality control and assessment technologies. She is currently working on a Department of Energy grant for designing carbon-negative buildings. Aviv holds a PhD in architectural technology (energy and computation track) from Princeton University, an MArch degree with a certificate in urban policy from Princeton University, and a BArch from The Cooper Union. She is a licensed architect and has practiced in design roles at Tod Williams Billie Tsien Architects, KPF, and Atelier Raimund Abraham. Her work was exhibited in the 2021 Venice Architecture Biennale, and she was the co-curator of the energy pavilion in the 2017 Seoul Biennale for Architecture and Urbanism. Her research papers have been published in leading scientific journals such as Applied Energy, Indoor Air, and Energy and Buildings.

ROBERT STUART-SMITH
Assistant Professor of Architecture, Weitzman School of Design, University of Pennsylvania

Robert Stuart-Smith is Program Director for the Masters of Science in Design: Robotics and Autonomous Systems degree (MSD-RAS), Assistant Professor of Architecture in the Weitzman School of Design, and an Affiliate Faculty member of Penn Engineering's GRASP Lab. He directs the Autonomous Manufacturing Lab in Penn's Department of Architecture (AML-PENN), and co-directs its sister lab in University College London's Department of Computer Science (AML-UCL). Stuart-Smith's research operates at the intersection of algorithmic design, robotic fabrication, and collective robotic construction—developing an integrated approach to design, manufacturing, and robot behavior through varying degrees of programmed autonomy. Stuart-Smith is Principal Investigator (PI) for the £1.2mil EPSRC project "Applied Off-site and On-site Collective Multi-Robot Autonomous Building Manufacturing" and Co-PI for a £2.9mil EPSRC research project into "Aerial Additive Building Manufacturing," involving collaborations with industry partners including Cemex, Skanska, Mace, Burohappold, Arup, MTC, Ultimaker, Kuka, and others. Stuart-Smith is a co-founder of the experimental research collaborative Kokkugia and architectural practice Robert Stuart-Smith Design. Prior to joining the faculty at Penn in 2017, Stuart-Smith was a Studio Course Master in the AA School's Design Research Laboratory (2009-17), and held visiting professorships at Washington University, RMIT, University of Innsbruck, among others. Stuart-Smith's work has been published in journals including *Nature*, *Science Robotics*, *AD Architectural Design*, and *Architecture D'Aujourd'hui*. He has lectured and presented in symposia at institutions including AA, Sci-Arc, CCA, MIT, RMIT, Angewandte, Strelka Institute, Tsinghua University, Texas A+M, and others.

Workshop Chair

ANDREW SAUNDERS
Director of Master of Architecture Program
Associate Professor of Architecture, University of Pennsylvania

Andrew Saunders is Associate Professor of Architecture at the University of Pennsylvania Weitzman School of Design and founding principal of Andrew Saunders Architecture + Design, an internationally published, award winning architecture, design, and research practice committed to the tailoring of innovative digital methodologies to provoke novel exchange and reassessment of the broader cultural context. The practice innovates at a number of scales ranging from product design, exhibition design, and residential and large-scale civic and cultural institutional design.

He received his Bachelor of Architecture from the University of Arkansas and a Masters in Architecture with Distinction from the Harvard Graduate School of Design. His current practice and research interests lie in computational geometry as it relates to aesthetics, emerging technology, fabrication, and performance. He has significant professional experience as project designer for Eisenman Architects, Leeser Architecture, and Preston Scott Cohen, Inc.

He has taught and guest-lectured at a variety of institutions, including Cooper Union and the Cranbrook Academy of Art, and, most recently, he was Assistant Professor of Architecture & Head of Graduate Studies at Rensselaer Polytechnic Institute in New York.

In 2004 he was awarded the SOM Research and Traveling Fellowship for Masters of Architecture to pursue his research on the relationship of equation-based geometries to early 20th century pioneers in reinforced concrete. His current practice and research interests lie in computational geometry as it relates to emerging technology, fabrication, and performance. He is currently working on a book using parametric modeling as an analysis tool of 17th century Italian Baroque architecture. Most recently Andrew won the ACADIA international fabrication competition for the production of the Luminescent Limacon. The design for this lighting fixture was inspired by Flemish baroque portraits of the Dutch ruff and builds on computational and material research from his seminar Equation-based Morphologies.

Exhibition & Media Chairs

FERDA KOLATAN
Associate Professor of Architecture, University of Pennsylvania

Ferda Kolatan is Associate Professor at the University of Pennsylvania Weitzman School of Design and the founding director of SU11 Architecture + Design. He received his Architectural Diploma from the RWTH Aachen and his M.S.AAD from Columbia University. SU11 is an internationally acclaimed practice based in Brooklyn, New York and is dedicated to the conceptual and material exploration at the intersection of contemporary culture, technology, and design. SU11's projects have been exhibited at renowned venues such as MoMA, FRAC Center, Walker Art Center, Vitra Design Museum, Art Basel, Artists Space NY, and SU11 has participated in the SIGGRAPH and ACADIA conferences, and the Venice, Beijing, and Istanbul Biennales. Ferda Kolatan has taught, lectured, and written extensively about architecture. In 2010 he co-authored the book *Meander: Variegating Architecture* with Jenny Sabin, and his new book *Misfits and Hybrids* is forthcoming in 2023. In 2016, his Penn Research-Studio on Cairo received the inaugural 2017 *ARCHITECT Magazine* Studio Prize. In 2011 Ferda was selected as a Young Society Leader by the American-Turkish Society in New York for his achievements as an educator and designer.

NATE HUME
Senior Lecturer, University of Pennsylvania

Nathan Hume is a licensed architect and principal of Hume Architecture. His design work and writings have been published in journals and periodicals including *Project*, *Log*, *Posit*, *Tarp*, *Paprika*, and *The New York Times*. Nathan has exhibited work in shows at The Druker Gallery, the A+D Museum, Yale Architecture Gallery, CAED Gallery, Land of Tomorrow, One Night Stand, and the New York Center for Architecture. He is a senior lecturer at The University of Pennsylvania and has previously taught at Yale University and Pratt Institute. He received a Bachelor of Architecture from The Ohio State University and a Master of Architecture from Yale University.

Departmental Chair

WINKA DUBBELDAM
Chair of Architecture and Miller Professor, University of Pennsylvania

Winka Dubbeldam, MArch MS-AAD, is a seasoned academic and design leader, serving as Chair and Miller Professor of Architecture at the University of Pennsylvania Stuart Weitzman School of Design, where she has gathered an international network of innovative research and design professionals. She also taught advanced architectural design studios at Columbia University and Harvard University, among other prestigious institutions. Dubbeldam was the External Examiner at the Architectural Association London (2006-2009) and is currently the External Examiner at the Bartlett UCL in London (2019-present). Professor Dubbeldam was named one of the *DesignIntelligence* 30 Most Admired Educators 2015. She has been a juror and chaired many international and national award juries, and was keynote speaker at international conferences. Professor Dubbeldam is one of the creative directors for CityX for the Virtual Italian Pavilion at the Venice Architecture Biennale (2021). Winka Dubbeldam is also the founder and partner of the WBE certified firm Archi-Tectonics NYC LLC, widely known for their award-winning work, recognized as much for its design excellence as for its use of smart building systems, sustainable materials, and innovative structures. A recent book *Strange Objects, New Solids, and Massive Things* was published by ACTAR Publishers, Spain, in Fall 2021.

About ACADIA

The Association for Computer Aided Design in Architecture (ACADIA) is an international network of digital design researchers and professionals that facilitates critical investigations into the role of computation in architecture, planning, and building science, encouraging innovation in design creativity, sustainability, and education.

ACADIA was founded in 1981 by some of the pioneers of the field of design computation including Bill Mitchell, Chuck Eastman, and Chris Yessios. Since then, ACADIA has hosted over 40 conferences across North America and has grown into a strong network of academics and professionals in the design computation field.

Incorporated in the state of Delaware as a not-for-profit corporation, ACADIA is an all-volunteer organization governed by elected officers, an elected Board of Directors, and appointed ex-officio officers.

PRESIDENT
Jenny E. Sabin

VICE-PRESIDENT
Kathy Velikov

SECRETARY
Tsz Yan Ng

TREASURER
Phillip Anzalone

MEMBERSHIP OFFICER
Vernelle A. Noel

DEVELOPMENT OFFICER
Matias del Campo

COMMUNICATION OFFICER
Melissa Goldman

TECHNOLOGY OFFICER
Jose Luis Garcia del Castillo López

IJAC ACADIA OFFICER
Dana Cupkova

2021 ELECTION BOARD OF DIRECTORS
Term: January 1st, 2022 - December 31st, 2023

Shelby Doyle, *Iowa State University*
Behnaz Farahi, *California State University, Long Beach*
Maria Yablonina, *University of Toronto*
Leslie Lok, *Cornell University*
Kathrin Dorfler, *Technical University of Munich*
Sina Mostafavi, *TU Delft (alternate)*
Daniel Bolojan, *Florida Atlantic University (alternate)*
Leighton Beaman, *Cornell University (alternate)*

2020 ELECTION BOARD OF DIRECTORS
Term: January 1st, 2021 - December 31st, 2022

Matias del Campo, *University of Michigan*
Tsz Yan Ng, *University of Michigan*
Jose Luis Garcia del Castillo López, *Harvard University*
June A. Grant, *blink!LAB Architecture*
Stefana Parascho, *EPFL*
Biayna Bogosian, *Florida International University (alternate)*
Melissa Goldman, *University of Virginia (alternate)*
Vernelle A. A. Noel, *University of Florida (alternate)*

Conference Management

**WEITZMAN SCHOOL OF DESIGN
AT THE UNIVERSITY OF PENNSYLVANIA**

Fritz Steiner, *Dean and Professor*
Winka Dubbeldam, *Chair of Architecture and Professor*

CONFERENCE CHAIRS
Dr. Masoud Akbarzadeh, *Assistant Professor of Architecture*
Dr. Dorit Aviv, *Assistant Professor of Architecture*
Hina Jamelle, *Associate Professor of Practice*
Robert Stuart-Smith, *Assistant Professor of Architecture*

WORKSHOP CHAIR
Andrew Saunders, *Associate Professor*

EXHIBITION & MEDIA CHAIRS
Ferda Kolatan, *Associate Professor*
Nate Hume, *Senior Lecturer*

WEBSITE & TECHNICAL
Christine Khouri Sader
Kevin He
Leon Yi-Liang Ko

EXHIBITION
Lauren Hanson
Reem Abi Samra

MEDIA
Jorge Couso

MERCHANDISE DESIGNERS
Peik Shelton
Yasmin Goulding

GRAPHIC IDENTITY DESIGNERS
Madison Green
Paul Germaine McCoy
Peik Shelton

EDITORIAL ASSISTANTS
Anna Ji-Eun Lim
Mingyang Yuan

**WEITZMAN SCHOOL OF DESIGN
AT THE UNIVERSITY OF PENNSYLVANIA
ADMINISTRATION**

Scott Loeffler, *Director of Administration*
Michael Grant, *Director of Communications*
Kait Ellis, *Executive Secretary to the Dean*
Hanna Finchler, *Associate Director of Communications*
Christopher Cataldo, *Director of Finance*
Nadine Beauharnois, *Coordinator of Finance & Budget*

ACADIA STEERING COMMITTEE
Biayna Bogosian
Matias Del Campo
Shelby Doyle
Behnaz Farahi
Melissa Goldman
Sina Mostafavi
Cameron Nelson
Vernelle A. A. Noel
Jenny E. Sabin
Kathy Velikov

COPYEDITOR
Gabi Sarhos

STUDENT VOLUNTEERS

Reen Abi Samra
Cem Akgun
Aisha Alshehri
Adhityan Anbumozhi
Jorge Couso
Joseph Depre
Yasmin Goulding
Madison Green
Lauren Hanson
Kevin He
Nicholas Houser
Christine Khouri Sader
Bohan Lang
Chunze Li
Sihan Li

Jeffrey Liao
Anna Lim
Mahsa Masalegoo
Paul McCoy
Cheuk Ming
June Mingyang Yuan
Clayton Monarch
Shunta Moriuchi
Sophia O'Neill
Hrishi Rajasekar
Peik Shelton
Nicholas Sideropoulos
Pouria Vakhshouri
He Wai Valerie Tse
Franklin Wu

Peer Review Committee

Arash Adel
Princeton University

Sigrid Adriaenssens
Princeton University

Mania Aghaei Meibodi
Taubman College of Architecture + Urban Planning, University of Michigan

Viola Ago
UCLA

Chandler Ahrens
Washington University in St. Louis

Mostafa Akbari
University of Pennsylvania Weitzman School of Design

Aysegul Akcay Kavakoglu
Istanbul Technical University

Suleiman Alhadid
MIT Media Lab

Ali AlYousefi
University of Pennsylvania Weitzman School of Design

Jeffrey Anderson
University of Pennsylvania Weitzman School of Design

Iman Ansari
Knowlton School at The Ohio State University

Phillip Anzalone
New York City College of Technology

German Aparicio
Trimble

Inés Ariza
ETH Zurich

Imdat As
Istanbul Technical University

Ehsan Baharlou
University of Virginia School of Architecture

Kristy Balliet
Southern California Institute of Architecture

Efilena Baseta
TU Wien

Mathias Bernhard
University of Pennsylvania Weitzman School of Design

Kory Bieg
University of Texas at Austin School of Architecture

Ezio Blasetti
University of Pennsylvania Weitzman School of Design

Biayna Bogosian
University of Southern California

Ronan Bolaños
National Autonomous University of Mexico

Mohammad Bolhassani
City College of New York

Daniel Bolojan
Florida Atlantic University School of Architecture

William Braham
University of Pennsylvania Weitzman School of Design

Johannes Braumann
Robots in Architecture

Danelle Briscoe
University of Texas at Austin School of Architecture

Nicholas Bruscia
University at Buffalo SUNY

Edvard Bruun
Princeton University

Timo Carl
Frankfurt University of Applied Sciences

Jason Carlow
American University of Sharjah

Gonçalo Castro Henriques
Federal University of Rio de Janeiro

Hua Chai
University of Pennsylvania Weitzman School of Design

Kian Wee Chen
ETH Zurich

Mike Christenson
University of Minnesota

Brandon Clifford
MIT

Christopher Connock
KieranTimberlake

Greg Corso
Syracuse University

David Costanza
Cornell University

Kristof Crolla
Hong Kong University

Brandon Cuffy
KieranTimberlake

Dana Cupkova
Carnegie Mellon University

Pierre Cutellic
ETH Zürich

Mahesh Daas
Boston Architectural College

Pierluigi D'Acunto
Technical University of Munich

Patrick Danahy
Ball State University

Matias del Campo
Taubman College of Architecture + Urban Planning, University of Michigan

Marcella Del Signore
New York Institute of Technology

Antonino Di Raimo
University of Portsmouth

Liyang Ding
Marywood University

Nancy Diniz
University of the Arts London

Mark Donohue
California College of the Arts

Kathrin Dörfler
Technical University of Munich

Shelby Doyle
Iowa State University

Stylianos Dritsas
Singapore University of Technology and Design

Emre Erkal
Erkal Architects

Alberto T. Estévez
iBAG-UIC Barcelona

Behnaz Farahi
California State University Long Beach

Wendy W Fok
USC School of Architecture

Pia Fricker
Aalto University, School of Arts, Design and Architecture

Richard Garber
University of Pennsylvania Weitzman School of Design

Jose Luis García del Castillo y Lopez
Harvard University Graduate School of Design

Guy Gardner
University of Calgary

David Gerber
University of Southern California

Andrei Gheorghe
Harvard University Graduate School of Design

Melissa Goldman
University of Virginia School of Architecture

Rhys Goldstein
Autodesk Research

Marcelyn Gow
Southern California Institute of Architecture

Derya Gulec Ozer
Istanbul Technical University

Isla Xi Han

Sean Hanna
The Bartlett School of Architecture

Erik Herrmann
Knowlton School, The Ohio State University

Tyson Hosmer
The Bartlett School of Architecture

Miaomiao Hou
University of Pennsylvania Weitzman School of Design

Yasushi Ikeda
The University of Tokyo

Gwyllim Jahn
Fologram

Ryan Luke Johns
Institute for Advanced Architecture of Catalonia

Nathaniel Jones
Arup

Damjan Jovanovic
SCI-Arc

Negar Kalantar
California College of the Arts

Lydia Kallipoliti
Cooper Union

Jyoti Kapur

Neil Katz
Skidmore, Owings & Merrill LLP

Ted Kesik
University of Toronto

Joachim Kieferle

Axel Kilian
Massachusetts Institute of Technology

Simon Kim
University of Pennsylvania Weitzman School of Design

Jihun Kim
University of Pennsylvania Weitzman School of Design

Nathan King
Harvard Graduate School of Design

Christoph Klemmt
University of Cincinnati

Daniel Koehler
University of Texas at Austin School of Architecture

Andreas Körner
University of Innsbruck

Anna Kostreva
Plural Studio Books & Builds

Evangelos Kotsioris
Museum of Modern Art

Sotirios Kotsopoulos
MIT

Sarah Aipra Kott-Tannenbaum
Quezada Architecture

Oliver David Krieg
Intelligent City

Riccardo La Magna
Karlsruhe Institute of Technology

Christian Lange
University of Hong Kong

Julie Larsen
Syracuse University

Carla Leitao
Rensselaer Polytechnic Institute

Brian Lonsway
Syracuse University

Yao Lu
University of Pennsylvania

Gregory Luhan
Texas A&M University

Nam Ma
University of Pennsylvania

Katie MacDonald
University of Virginia

Arthur Mamou-Mani
University of Westminster

Ryan Vincent Manning
Quirkd33

Sandra Manninger
Taubman College, University of Michigan

Mara Marcu
University of Cincinnati

Adam Marcus
Tulane University

Nikola Marincic
ETH Zurich

Bob Martens
TU Wien

Matan Mayer
IE University

Wes Mcgee
Taubman College of Architecture
& Urban Planning

Forrest Meggers
Princeton University

Frank Melendez
City College of New York

AnnaLisa Meyboom
University of British Columbia

Saurabh Mhatre
Harvard University Graduate School of Design

Clayton Miller
National University of Singapore

R. Scott Mitchell
University of Southern California

Laia Mogas-Soldevila
University of Pennsylvania Weitzman
School of Design

Philippe Morel
University College London

Sina Mostafavi
University of Huddersfield

Stephen Mueller
Texas Tech University

Zoltan Nagy
UT Austin

Alicia Nahmad Vazquez
University of Calgary School of Architecture,
Planning and Landscape

H. Burçin Nalinci
Studio Bits2Atoms - Zahner

Taro Narahara
New Jersey Institute of Technology

Sabrina Naumovski
KieranTimberlake

Andrei Nejur
University of Montréal

Catie Newell
University of Michigan

Tsz Yan Ng
Taubman College, University of Michigan

Ted Ngai
Pratt Institute

Vernelle Noel
Georgia Tech University

Betul Orbey
Dogus University

Stefana Parascho
CREATE Laboratory at Princeton University

Ju Hong Park
Pohang University of Science and Technology

Vera Parlac
New Jersey Institute of Technology

Andrew Payne
Robert McNeel and Associates

Santiago Perez
University of Western Australia

Brady Peters
University of Toronto

Mariana Popescu
TU Delft

Ebrahim Poustinchi
Kent State University

Marshall Prado
University of Tennessee

Eleanor Pries
San Jose State University

Nick Puckett
OCAD University

Carolina Ramirez-Figueroa
Royal College of Art

Mette Ramsgaard Thomsen
Royal Academy of Fine Arts

Casey Rehm
SCI-Arc

Mariana Righi
University of Pennsylvania

Christopher Romano
University at Buffalo

Rhett Russo
Rensselaer Polytechnic Institute

Jenny Sabin
Cornell University

Jose Sanchez
University of Michigan

Anton Savov
ETH Zurich

Marc Aurel Schnabel
Victoria University of Wellington,
New Zealand

Alexander Schofield
California College of the Arts

Mathew Schwartz
New Jersey Institute of Technology

Tobias Schwinn
University of Stuttgart Institute
for Computational Design

Jane Scott
Newcastle University

Jason Scroggin
University of Kentucky College of Design

Zhan Shi
Carnegie Mellon University

Brian Slocum
Universidad Iberoamericana

Valentina Soana
University College London

Rajat Sodhi
Orproject

Aldo Sollazzo
Institute for Advanced Architecture
of Catalonia

Kyle Steinfeld
University of California, Berkeley

Satoru Sugihara
Architectural Technology Laboratory Venture

Martin Tamke
The Royal Danish Academy of Fine Arts,
Schools of Architecture, Design
and Conservation

Eric Teitelbaum
AIL Research

Skylar Tibbits
MIT

Daniel Tish
Harvard University Graduate School of Design

Kenneth Tracy
Singapore University of Technology
and Design

Franca Trubiano
University of Pennsylvania Weitzman
School of Design

Hans Tursack
Washington University in St. Louis

Richard Tursky
Ball State University

Donna Vakalis
University of British Columbia

Theodora Vardouli
McGill University

Lauren Vasey
ETH Zurich

Shota Vashakmadze
UCLA

Kathy Velikov
Taubman College of Architecture
& Urban Planning

Tom Verebes
New York Institute of Technology

Joshua Vermillion
University of Nevada

Hans Jakob Wagner
ICD University of Stuttgart

Gabriel Wainer
Carleton University

Zherui Wang
University of Pennsylvania
Weitzman School of Design

Nick Williams
Aurecon

Andrew John Wit
Tyler School of Art and Architecture

Jun Xiao
University of Pennsylvania Weitzman
School of Design

Shai Yeshayahu
Ryerson University

Lei Yu
Raytheon Technologies

Machi Zawidzki
Polish Academy of Sciences

Catty Dan Zhang
UNC Charlotte

Qi Zhang
Tongji University

Hang Zhang
Digital Building Technologies, Institute of
Technology in Architecture, ETH Zurich

Hao Zheng
University of Pennsylvania Weitzman
School of Design

Sasa Zivkovic
Cornell University

Sponsorship

PLATINUM

Zaha Hadid Architects

SILVER

GRIMSHAW

BRONZE

EVENTSCAPE

SPONSOR

 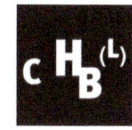 

MEDIA

The Architect's Newspaper

The Graphic Identity of Hybrids & Haecceities

Madison Green, Paul Germaine McCoy, Peik Shelton

The amalgamation of a hybrid and a haecceity is to recognize the existence of polarity and uniqueness while enabling relationships between such entities to simulate and blend qualities from one another which overtime appear to be "new" and "only" while exisiting in "multiplicities" and "varieties." The graphics proposal for the University of Pennsylvania Weitzman School of Design's hosting of the ACADIA 2022 conference titled *Hybrids & Haecceities* illustrates that things can exist as a thing in a state of transition or in between. In this way, lines that describe contemporary tools and discrete methods also outline familiar figures and forms. A graphics library extracts euclids from splines and hatches from colors to generate tweens of curvatures that—when sampled with AI learning—open up a relevant conversation of the computational, material, objective, aesthetic, genetic, biological, environmental, robotic, and operational to a new reading. A reading where the graphic aesthetic of the digital is informed by multiple layers of graphics and computer-aided design in architectural practice. This reading develops through constant curation, editing, tracing, adding, removing, fading, and overlaping of graphic elements. The compositions create dense overlay assemblies of rasterized and vector images. A library of cropped editions is created from this image to focus on different regions of information. Overall, *Hybrids & Haecceities* is an evolution of the familiar graphic reading of ACADIA into the Ivy League and the second decade of the 20th century.

www.ingramcontent.com/pod-product-compliance
Ingram Content Group UK Ltd.
Pitfield, Milton Keynes, MK11 3LW, UK
UKHW061623240426
12048UKWH00051B/1730